AF512946

Ananthanarayan and Paniker's Textbook of Microbiology

THIRTEENTH EDITION

About the Authors

R ANANTHANARAYAN (1913–1998)

Dr R Ananthanarayan graduated from Madras Medical College in 1940, and began his illustrious medical career in the Indian Army, serving as Quarter Master and Medical Officer in charge of wards in Sudan and Egypt. He also served in Iraq, Iran, Palestine, Syria, Malaya and Java. He was appointed Assistant Surgeon at the Government hospital in Calicut in 1946, moving to Stanley Medical College the following year. He obtained a Diploma in Bacteriology (DB) from London in 1951 and his PhD in 1953. For the next eight years, he worked as Professor of Bacteriology in the newly established Trivandrum Medical College, after which he assumed the post of Principal and Professor at Calicut Medical College. In May 1967, he was transferred to Kottayam Medical College from where he retired in April 1969. Retirement did not slow him down, however, and he served as Professor of Microbiology at Gulbarga Medical College till 1972. From 1972–1975, he served as Professor Emeritus at Trivandrum Medical College and Adviser to the Kerala Government on Laboratory Services.

During his career, Dr Ananthanarayan served as Examiner for Microbiology in different universities, was a member of several commissions and was awarded the Arun Banerjee Memorial Award in 1979. A Founding Member of the Indian Association of Pathologists and Microbiologists, he was elected President in 1976 and also served as President of the Association of Microbiologists in India in 1980. A greatly respected and revered teacher, he published more than 64 papers in reputed journals.

C K JAYARAM PANIKER (1932–2014)

Dr C K Jayaram Paniker graduated from Madras Medical College in 1955 and completed his postgraduate degree from All India Institute of Medical Sciences in 1962. During the initial phases of his career, he worked at the Pasteur Institute of India, Coonoor, Trivandrum Medical College and as Founder–Professor of Department of Microbiology at Kottayam Medical College. He was Principal, Director and Head of the Department of Microbiology at Calicut Medical College and Dean, Faculty of Medicine, Calicut University.

Well known for his command over the English language and for his sparkling sense of humour, Dr Paniker performed yeoman service for Microbiology in India. He was the recipient of the BC Roy Award, and the Kerala Government honoured him with the post of Professor Emeritus in Microbiology. He headed many advisory committees and published several research papers, especially in the fields of typhoid and influenza.

Ananthanarayan and Paniker's Textbook of Microbiology

THIRTEENTH EDITION

As per the CBME Curriculum (NMC)

Editor

Sonal Saxena MD, FAIMER

Director Professor and Head, Department of Microbiology
Maulana Azad Medical College, New Delhi
WHO Fellow on Antimicrobial Resistance
Secretary, Indian Association of Medical Microbiologists (IAMM)

Universities Press

ANANTHANARAYAN AND PANIKER'S TEXTBOOK OF MICROBIOLOGY, THIRTEENTH EDITION

UNIVERSITIES PRESS (INDIA) PRIVATE LIMITED

Registered Office
3-6-747/1/A & 3-6-754/1, Himayatnagar, Hyderabad 500 029, Telangana, India
info@universitiespress.com; www.universitiespress.com

Distributed by
Orient Blackswan Private Limited

Registered Office
3-6-752, Himayatnagar, Hyderabad 500 029, Telangana, India

Other Offices
Bengaluru, Chennai, Guwahati, Hyderabad, Kolkata, Mumbai, New Delhi, Noida, Patna, Visakhapatnam

First published 1978: Reprinted 1979, 1980; Second edition 1981: Reprinted 1982, 1983, 1984, 1985
Third edition 1986: Reprinted 1987, 1988, 1989; Fourth edition 1990: Reprinted 1991, 1992 (twice), 1994, 1995 (twice)
Fifth edition 1996: Reprinted 1996, 1997 (thrice); Reissued with revision 1998: Reprinted 1999
Sixth edition 2000: Reprinted 2001 (twice), 2002 (twice), 2003 (twice), 2004 (thrice)
Seventh edition 2005: Reprinted 2005 (five times), 2006 (twice), 2007, 2008 (thrice), 2009
Eighth edition 2009: Reprinted 2009, 2010, 2011, 2012
Ninth edition 2013: Reprinted 2013 (twice), 2014, 2015, 2016
Tenth edition 2017: Reprinted 2017, 2018 (twice), 2019
Eleventh edition 2020
Twelfth edition 2022: Reprinted 2023
Thirteenth edition 2025: Reprinted 2025

ISBN 978-93-93330-65-9

Cover and book design

Cover image source: FREEPIK.com

502123

Typeset in Life BT 10.5/12.5 pt *by*
e-Leaf Technologies, Chennai

Printed in India at
Rasi Graphics Pvt Ltd., Chennai 600 014

Published by
Universities Press (India) Private Limited
3-6-747/1/A & 3-6-754/1, Himayatnagar, Hyderabad 500 029, Telangana, India

Contents

Note on AETCOM and Pandemic Management

- **AETCOM elements:** From the 2024 batch, only the topics listed in competencies 2.2A–E, 2.3, and 2.5 are to be covered under microbiology. These are detailed in Chapter 36. The rest are to be taught under pharmacology (2.1 and 2.8) and pathology (2.4A, 2.4B, 2.6, and 2.7).
- **Pandemic management:** In Chapter 9, there is a section on '**Management of Pandemics**' (page 140). The topic is also covered in Chapters 20, 23, and 35 (pages 309, 455, and 755). The competencies in the pandemic module have been covered contextually as presented in the table below.

Competency	Coverage in book
Phase 1 (covered in foundation course): Infection control practices—Handwashing, decontamination, use of PPEs	Pages 184–190 (Chapter 12)
Phase 2.1: Air-borne precautions—Contact precautions, infection control committee	Pages 184–190 (Chapter 12)
Phase 2.3: Sample collection, microbial diagnosis, serologic tests and their performance parameters	Pages 72–76 (Chapter 4) (Serology and performance parameters covered in Chapter 16)
Skills suggested under this module are covered in multiple chapters	

Chapter Location of Major Organisms Associated with Various Infective Syndromes

Organism	Chapter	Page
Bacterial agents		
Staphylococci	4, 21, 25, 29	44, 317, 325, 328, 564, 665
Streptococci	4, 21, 23, 24, 29	46, 328, 401, 434, 495, 663
Pneumococci	4, 23, 24	47, 433, 495
Enterococci	4, 21, 28	47, 320, 442, 495, 653
Neisseria meningitidis	4, 24	53, 492
Neisseria gonorrhoeae	4, 27	53, 624
Corynebacterium	4, 23, 29	49, 405, 666
Bacillus anthracis	4, 29	51, 666
Actinomycetes	4, 29	51, 676
Clostridia	4, 24, 25, 29	52, 497, 500, 564, 670
Bacteroides	4, 24	66, 491
Mycobacterium tuberculosis	4, 23	67, 468
Non-tuberculous mycobacteria	4, 29	66, 682
Mycobacterium leprae	4, 29	67, 684
Coliforms (*E. coli*, *Klebsiella*, *Proteus*)	4, 23, 27	55, 59, 442, 549
Shigella	4, 25	58, 558
Salmonella	4, 21, 25	58, 330, 545
Yersinia	4, 21, 30	58, 336, 567
Vibrio	4. 25	60, 551
Pseudomonas, *Acinetobacter* and other non-fermenters	4, 23	63, 320, 444
Haemophilus	4, 21, 23, 24, 27	64, 319, 437, 640
Brucella	4, 21	65, 342
Bordetella	4, 23	65, 439
Spirochetes	4, 21, 23, 27, 29	68, 339, 410, 630, 681
Mycoplasma and *Ureaplasma*	4, 23, 27	68, 339, 410, 630, 681
Rickettsiae	4, 21, 31	70, 347, 729
Helicobacter, *Campylobacter*	4, 25	560, 561
Viral agents		
Bacteriophages	5	95
Herpes virus	5, 27, 29	84, 427, 641
Papova and Parvo virus	5, 27, 29, 30, 34	84, 641, 693, 724, 746
Adeno virus	5, 23, 25, 34	84, 411, 569, 749
Picorna virus	5, 23, 425, 26	84, 425, 502, 503,
Reo and Rota virus	5, 25	84, 568
Orthomyxo virus	5, 23	84, 413
Paramyxo and Rubella virus and other respiratory viruses	5, 23	84, 411, 413, 425
Arthropod- and rodent-borne virus infections	5, 21, 24	84, 351, 518
Rhabdo virus (Rabies virus)	5, 24	84, 510
Slow virus	5, 24	84, 516
Prion agents	5, 24	84, 517
Hepatitis virus	5, 26	84, 598
Human immunodeficiency virus	5, 22	84, 383
Other retroviruses	5, 22, 34	84, 383, 749
Oncogenic viruses	5, 34	84, 749
Fungal agents		
Agents of deep mycosis: *Aspergillus*, *Penicillium*, *Pneumocystis*, *Histoplasma*, *Mucor*, *Coccidiodes*, *Paracoccidiodes*, *Blastomycetes*	6, 23, 24, 29	101, 322, 456, 459, 465, 537, 712
Opportunistic mycoses: *Candida*, *Cryptococcus*	6, 22, 23, 25	101, 358, 537, 712
Mycotoxins and mycetism	25	567
Agents of superficial and subcutaneous mycoses: *Dermatophytes*, *Sporothrix*, *Madurella*, etc.	6, 29	101, 702, 707
Parasitic agents		
Protozoa: *Plasmodium*, *Leishmania*, trypanosomes, etc.	7, 21, 23, 25, 27, 29	110, 361, 368, 371, 535, 696
Nematodes (intestinal): *Ascaris*, hookworm, *Strongyloides*, *Trichuris*, etc.	7, 25	110, 579, 580, 581. 583
Nematodes (tissue): *Trichinella*, *Onchocerca*, *Loa loa*, *Mansonella*, *Dracunculus*	7, 21, 27	110, 579, 580, 581, 583
Cestodes: *Taenia*, *Hymenolepis*, *Echinococcus*, etc.	7, 24, 25, 26	113, 585, 587, 613
Trematodes: Schistosomes, *Fasciola*, *Clonorchis*	7, 21, 23, 25, 26, 29	114, 377, 466, 589, 599, 616, 617

Acknowledgements

In addition to all those whose assistance was invaluable in the preparation of the last edition, the editor and the publishers sincerely thank the following individuals whose contributions, inputs, and support have helped prepare this edition of the book.

Dr Himanshu Bhim Khatri, Associate Prof, Department of Microbiology, Surabhi Institute of Medical Sciences, Siddipet, Telangana
Dr G. Sumathi, MBBS, MD, PhD, Head of the Department of Microbiology, Sri Muthu Kumaran Medical College & Hospital (SMMCH&RI), Chikkarayapuram, Kundrathur Road, Near Mangadu, Chennai
Dr Neetu Shree, MBBS, MD, DHM, Associate Professor, Department of Microbiology, HIMSR & HAHC Hospital, New Delhi
Dr Meena Mishra, MD, Professor and Head, Department of Microbiology, AIIMS, Nagpur
Dr Nandkishore Bankar, Professor & Vice Dean – Research & Development, Department of Microbiology, Datta Meghe Medical College, Nagpur
Dr Haripriya Kar, Associate Professor, Department of Microbiology, MGM Medical College, Navi Mumbai
Dr Kalpana T. Suryavansi, Associate Prof, HBTMC and Dr RN Cooper Hospital, Mumbai
Dr Kanchan V. Joshi, Additional Professor, Department of Microbiology, Lokmanya Tilak Municipal Medical College and General Hospital, Sion, Mumbai
Dr Neelam Pandiya, Assistant Prof, Baroda Medical College, Gujarat
Dr Shripad Murlidhar Taklikar, Add. Associate Professor, Lokmanya Tilak Municipal General Hospital and Lokmanya Tilak Municipal Medical College, Sion, Mumbai
Dr Sunita Raj Gajbhiye, Professor, Govt Medical College, Nagpur
Dr Arani Debnandi, Assistant Professor, Rampurhat Government Medical College & Hospital, Kolkata
Dr Manas Sarkar, Professor, Department of Microbiology, Murshidabad Medical College & Hospital, Murshidabad
Dr Subhenjit Ray, Tutor, Diamond Harbour Govt. Medical College & Hospital, South 24 Parganas, West Bengal
Dr Mallika Sengupta, Assistant Professor Microbiology, AIIMS Kalyani, Basantapur, Saguna, West Bengal
Dr Kiran Gopal, Assistant Professor, Department of Microbiology, Government Medical College, Kollam
Dr Indrani Mohanty, Professor and HOD, Department of Microbiology, PRM Medical College, Baripada

Preface To The Thirteenth Edition

Welcome to the 13th edition of *Ananthanarayan and Paniker's Textbook of Microbiology*, which has been meticulously developed to align with the principles of the competency-based medical education (CBME) curriculum prescribed by the National Medical Council (NMC).

As educators and practitioners in the field of microbiology, we understand the evolving landscape of medical education and the need to equip future healthcare professionals with the knowledge and skills necessary for clinical practice. My primary goal, while preparing this edition, was to equip learners with the knowledge, skills, and attitudes they need to excel in their medical careers.

I recognise the immense responsibility entrusted to me and my fellow educators to shape the next generation of healthcare professionals and am committed to providing a robust educational resource that aligns with the principles of CBME. In this edition, I have meticulously curated the content not only to cover the essential aspects of microbiology but also to foster the development of core competencies that are vital for medical practice. My approach integrates foundational knowledge with clinical applications, encouraging learners to bridge the gap between theoretical understanding and practical patient care.

The introductory chapter on bacteriology has been split into two—Basic Concepts of Bacteriology, which deals with concepts such as staining, growth characteristics, and identification methods, and Systematic Bacteriology, which has been expanded to include the virulence factors and diseases caused by various organism groups. I request teachers to allocate 4–6 hours to cover the basics and features of different bacteria before proceeding to the syndromes. Content has been revised to include updated information on plasma sterilisation, CSSD, and the care bundle approach. New techniques such as VITEK, BioFilm Array, and AST have been added. In the section on applied microbiology, various infective syndromes have been discussed, along with their causative organisms, epidemiology, relevant investigations, treatment, and prevention, in keeping with the syndromic approach prescribed by the NMC. Parasitology has been integrated into all the syndromes. Topics such as HIV, direct-acting antivirals, National Hepatitis Control Programme guidelines, and zoonotic diseases have also been updated. A new chapter has been added on the AETCOM elements applicable to microbiology.

I am ever grateful to Dr Reba Kanungo for her exemplary work on the 12th edition. I would also like to acknowledge Dr Masoom Nathani, my Senior Resident who worked on the MCQs, and Dr Amala Andrews, Assistant Professor, GMC, Trivandrum, and Dr Arpita Saxena, 1st year PG resident at SMS Medical College, Jaipur, for their work on the PowerPoint presentations for this edition.

I am grateful to all my teachers, particularly my guide, Dr S. Krishna Prakash, for the encouragement, constructive criticism, and motivation he provided throughout my academic pursuits. I am forever indebted to my parents for inculcating the habit of reading in me and for imparting values that have shaped me so profoundly. I am very grateful to my family, especially my husband, Manish, for his patience, understanding, and support during the preparation of this edition. A special word of appreciation for my daughter, Arpita, for being my greatest critic and a tremendous source of strength.

I would like to acknowledge the team at Universities Press, who pulled out all the stops to revise this classic textbook. I am indebted to Dr Sudha Ganesan, Managing Consultant, Medical Publishing, whose vast technical expertise and inputs were invaluable to me. Her efforts in collating feedback from faculty across the country helped in fine-tuning several portions of the book. Ms Malini Gopalakrishnan's eye for detail contributed significantly to the finer nuances of technical editing and proofreading. I extend my gratitude to Mr Madhu Reddy, Director, Universities Press, whose steadfast support guided the entire process.

My heartfelt thanks are extended to the educators, students, and healthcare professionals whose contributions have been instrumental in shaping this edition. Their insights, feedback, and dedication to excellence have been invaluable in this endeavour.

I hope that this textbook will serve as a valuable companion to readers on their journey toward mastering microbiology and ultimately contributes to the delivery of safe, effective, and compassionate patient care.

Sonal Saxena, MD, FAIMER
Director Professor and Head, Department of Microbiology
Maulana Azad Medical College, New Delhi
Secretary, Indian Association of Medical Microbiologists
sonalsaxena3@gmail.com

Preface to the First Edition

Many of the health problems in developing countries like India are different from those of developed countries. Bacterial diseases still play a considerable role in our country. Topics such as cholera and enteric diseases are important to us though only of less or academic interest to the developed countries. The increasing importance of the newer knowledge in immunology to health and disease is not adequately stressed in most of the extant textbooks. Virus diseases which are responsible for nearly 60 per cent of human illness require wider coverage. The general approach to the teaching of microbiology in our country has also been rather static. All these factors called for a textbook of microbiology more suited to countries like India.

We therefore undertook this endeavour based on our experience of teaching undergraduates and postgraduates for over two decades. We omitted the discipline of parasitology from our book since we already have an excellent textbook on the subject published in India.

This book has taken us over three years to write and over a year in publication. Naturally, we would be out of date to a certain and inevitable extent. We do not claim any perfection. On the contrary, we have requested medical students and teachers all over the country to write to us about any shortcomings and give us suggestions as to how to improve the book. We shall spare no pains in seeing that their valuable suggestions are given effect to in our second edition.

R Ananthanarayan
C K Jayaram Paniker

Note to Teachers and Students

On using the 13th edition of *Ananthanarayan and Paniker's Textbook of Microbiology* in keeping with the competency-based medical education (CBME) curriculum

Great care has been taken to ensure a logical flow in the arrangement of chapters in this book to establish a sound understanding of the basic concepts of microbiology and the broad features of microorganisms before proceeding to a syndromic approach to systemic infections. Faculty may choose on a sequence of teaching the subject based on their curricular design.

The first section of the book, '**General Microbiology**' (MI1.1–1.6), lays the foundation for the study of microbiology; it deals with the fundamentals and presents a brief history of the subject. Microorganisms are introduced through overviews of bacteriology, virology, mycology, and parasitology. These introductory chapters carry information on the classification, general morphology, habitats, and growth requirements and the infections caused by the significant members of these groups. In the chapter on parasitology, the life cycles of the broad groups of parasites and the systems affected by them are also discussed.

In light of the giant strides being taken in the fields of basic and applied research in genetics and the rapidly increasing role of molecular methods in understanding disease processes and making rapid diagnoses, a brief chapter on **microbial genetics** has been included in the first section of the book.

The epidemiology of infectious diseases has been discussed next to help the student understand the roles of the host, pathogen, and environment in the development of disease and methods to break these chains. Subsequent chapters deal with sterilisation and disinfection, as applied to **healthcare settings** (MI8.5–8.7), followed by the general concepts of infection control practices in such settings. Antimicrobial agents have been introduced at this stage for students to understand the principles and mechanisms of action of such agents and the development of drug resistance among microorganisms. We recommend that this chapter be taught in integration with pharmacology.

The second section of the book—'**Immunology**' (MI1.7–1.11)—introduces students to the tenets of immunology as applied to human health and disease, including the broad immune mechanisms of the human body and their detection, immunological disorders (hypersensitivity, immunodeficiency, autoimmune disorders, etc.), immunisation, immunohematology, transplantation, and graft reactions and tumour immunity. This section has scope for integration with pathology.

The third section, '**Microbiology as Applied to Infectious Diseases**' (MI2–MI8), which is aligned with the competencies prescribed by the National Medical Council of India, elaborately and comprehensively deals with infections in a system-wise manner. The pattern of discussion is uniform across syndromes, detailing the significant etiological agents of each system, with the focus on important infectious syndromes and the corresponding causative agents, pathogenesis, clinical features, epidemiology, and the approach to the laboratory diagnosis and treatment of each of the clinical entities described. Each of the chapters in this section carries one or more tabular columns summarising the causative agents of the various syndromes of the respective organ system for easy reference and recall. The organisms listed in these tables are **colour-coded** in keeping with their relative etiological significance with respect to the organ system—**red denotes the most significant, orange indicates a slightly less etiological value and blue, lesser so**. This is a unique feature of the book that we believe will be very helpful for readers.

This section of the book deals with the following systems:

- **CVS and blood (MI2):** In line with this competency, the book deals with bloodstream infections that lead to fever (e.g., typhoid, brucellosis, dengue, scrub typhus) and the causes of anemia (e.g., common parasites causing anemia). Organisms commonly associated with cardiovascular and bloodstream infections have been discussed individually in Chapter 21. Chapter 22 deals with HIV and AIDS and includes information on the current national guidelines and policies.
- **Respiratory system (MI6):** Chapter 23, which deals with infections of the respiratory system, has been split into upper and lower respiratory tract infections. The organisms commonly associated with these infections have been elaborated on. Tuberculosis as a respiratory tract infection has been dealt with in detail.

- **Central nervous system infections (MI5):** In Chapter 24, the common causes, pathogenesis, clinical features, diagnosis and treatment of acute and chronic meningitis, viral encephalitis, and other CNS infections are discussed.
- **Gastrointestinal and hepatobiliary system (MI3):** In Chapter 25, the causes and pathogenesis of diarrhea/ dysentery have been discussed in detail (e.g., viral and bacterial causes of diarrhea, including cholera and shigellosis). In-depth information on various parasitic infections of the gut, their modes of infection and life cycles, the methods of detecting ova and cysts, and the treatment of parasitic infections has been presented. Food poisoning and the methods employed to diagnose it have also been detailed in this chapter. **Hepatobiliary infections** have been discussed in Chapter 26, which highlights the viral and bacterial causes of hepatitis.
- **Genitourinary and sexually transmitted infections (MI7):** The common causes of sexually transmitted infections, including syphilis, gonorrhea, non-gonococcal urethritis and vaginosis have been elucidated in Chapter 27. Urinary tract infections, divided into acute and chronic urinary tract infection, and the predisposing conditions for these infections have been discussed in Chapter 28.
- **Skin, soft tissue and musculoskeletal system infections (MI4):** Microbial agents causing superficial and deep infections as well as the pathogenesis, clinical features and diagnosis of these infections have been dealt with in Chapter 29. The major anaerobic organisms causing skin and soft tissue infections have been described in detail in this chapter. Bone and joint space infections have also been emphasised.
- **Infections of the eyes and ears and congenital infections (MI8.15):** Chapter 30 discusses the most significant clinical entities affecting the eyes and ears. The mechanisms of congenital infections and their prevention have also been discussed in this chapter.
- **Zoonotic diseases (MI8.1):** These have been discussed in Chapter 31 in relation to the factors responsible for infection, the common diseases affecting humans, as well as the associated epidemiology, pathogenesis, clinical features, diagnosis, treatment, and prevention.

The subsequent chapters of the third section address miscellaneous infections: **opportunistic infections** (MI8.2)—the bacterial, viral, fungal, and parasitic agents responsible for opportunistic infections and the predisposing conditions facilitating them—**oncogenic viruses** (MI8.3), **emerging infections** (MI8.4), and **national programmes** for disease detection and reporting (MI8.16).

Laboratory-specific **AETCOM components**, based on the ethical issues of confidentiality and patient consent and other aspects, have been elaborated in Chapter 36 and are also integrated into the laboratory aspects of patient investigations within individual chapters.

With all these developments and modifications, we hope that the 13th edition of *Ananthanarayan and Paniker's Textbook of Microbiology* will meet the requirements of medical faculty and the medical graduate.

Part I: General Microbiology

Part I: General Microbiology

CHAPTER 1

History and Scope of Microbiology

(MI 1.1)

Chapter Navigation

- Historical perspective
- Scope of microbiology

Learning objective

- Describe the brief history and scope of microbiology

INTRODUCTION

Microorganisms are found abundantly almost everywhere in the environment, from mountaintops to the ocean bed. In the early 19th century, the association between disease and microbes was established, thus laying the foundation for the **science of microbiology**. Later, microorganisms infectious to humans were discovered, and their role in causing infections was identified. Today, microorganisms play a vital role in the agriculture, dairy, pharmaceutical, and brewing industries.

Microbiology deals with the diagnosis and prevention of infectious diseases. Microorganisms are classified into the following broad groups to identify and study them:

- Bacteria
- Viruses
- Fungi
- Protozoa
- Helminths
- Prions

HISTORICAL PERSPECTIVE AND NOTABLE MILESTONES IN MICROBIOLOGY

As most microbes are invisible to the unaided eye, definitive knowledge about microorganisms had to await the development of the microscope. The credit for having first observed and reported bacteria goes to **Antonie van Leeuwenhoek** of Delft, Holland—a draper whose hobby was grinding lenses and observing diverse materials through them.

Some microbiologists who were responsible for discoveries that shaped microbiology are discussed here (Fig. 1.1a–d).

- **Louis Pasteur** (1822–95), who is known as the **Father of Microbiology**, proved that microorganisms cause diseases and established the principles of **immunisation**. His important contributions include the following:
  - Development of sterilisation techniques and pasteurisation of milk
  - Development of the steam steriliser, hot air oven, and autoclave
  - Knowledge of the growth needs of different bacteria
  - Attenuation of chicken cholera bacilli
  - Development of fermentation techniques
  - Development of vaccines for hydrophobia (rabies) and anthrax
  - Postulated the germ theory of disease and disproved the idea of spontaneous generation of life
- **Joseph Lister** (1867) furthered Pasteur's work by introducing **aseptic techniques** that resulted in a drop in mortality and morbidity due to surgical sepsis. He used carbolic acid as an antiseptic agent.
- **Robert Koch** (1843–1910) is known as the **Father of Bacteriology**. His contributions include:
  - Discovery of the anthrax disease cycle
  - Development of staining techniques using dyes to visualise bacteria
  - Development of methods of obtaining bacteria in pure culture using agar as solid media
  - The discovery of bacteria such as *Bacillus anthracis*, tubercle bacilli, and *Vibrio cholerae*
  - The invention of the hanging drop technique to observe bacterial movement
  - Development of Koch's postulates (see box)

(a) Antonie van Leeuwenhoek (b) Louis Pasteur (c) Robert Koch (d) Paul Ehrlich

Fig. 1.1 Some eminent microbiologists whose contributions laid the foundations of microbiology, the concepts of infectious disease, and antisepsis

Koch's postulates

According to Koch, a microorganism can be accepted as the causative agent of an infectious disease only if certain conditions are satisfied.

- The microorganism should be consistently associated with the lesions of the disease.
- It should be possible to isolate the microorganism in a pure culture obtained from the lesions.
- The inoculation of such a pure culture into suitable laboratory animals should reproduce lesions of the disease.
- It should be possible to re-isolate the microorganism in a pure culture from lesions produced in experimental animals.
- An additional criterion introduced subsequently requires that specific antibodies to the microorganism be demonstrable in the serum of patients suffering from the disease.

- **Paul Ehrlich** (1898) studied **toxins and antitoxins** and laid the foundation for their biological standardisation. He also discovered Salvarsan, a chemotherapeutic agent that was found to cure syphilis.
- **Emile Roux** and **Alexandre Yersin** (1988) discovered the diphtheria toxin.
- **Ernst Ruska's** invention of the **electron microscope** (1934) paved the way for the discovery of many viruses that were not visualised by light microscopy.
- **Niels Jerne** (1955) proposed the **natural selection theory of antibodies**, which could explain the specificity and biological basis of antibody synthesis.
- **Frank Burnet** (1957) modified Jerne's theory into the **clonal selection theory**.

Fig. 1.2 presents a brief timeline of discoveries that shaped microbiology.

NOBEL LAUREATES IN MICROBIOLOGY

For the discoveries and innovations that became milestones in the field of microbiology, a number of scientists have been awarded the Nobel Prize in Medicine and Physiology. Some of these Nobel laureates and their contributions to the field of microbiology are listed in Table 1.1.

THE ANTIBIOTIC ERA

- In 1928, **Alexander Fleming** accidentally discovered an antibacterial substance. He found that the fungus *Penicillium* produces a substance that destroys staphylococci.
- A decade later, **Howard Florey** and **Ernst Chain** successfully isolated the active substance 'penicillin'. Its subsequent mass production during the Second World War saved the lives of many wounded soldiers.
- Other similar antibiotics were then discovered in rapid succession. With the development of a wide variety of antibiotics against the whole spectrum of pathogenic bacteria and the advent of vaccines against some bacterial and viral diseases, the scientific community began to foster hope for the elimination of all infectious diseases.
- The global eradication of smallpox inspired visions of similar campaigns against other major infections. But soon, the development of newer pathogens and antibiotic resistance in bacteria presented serious difficulties.

SCOPE OF MEDICAL MICROBIOLOGY

The clinical implications of microbiology include the following:

- Diagnosis of infectious diseases
- Prescription of antibiotics as per sensitivity testing
- Development of vaccines and monoclonal antibodies

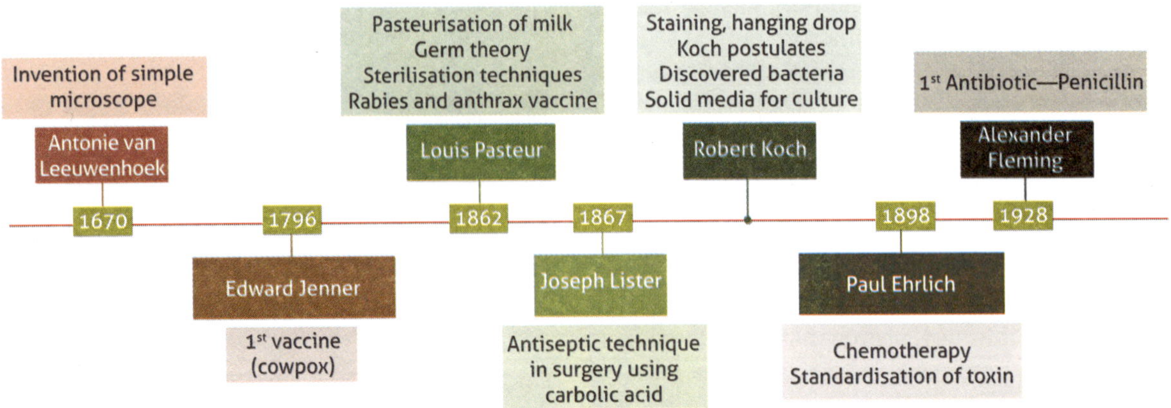

Fig. 1.2 A timeline of the major discoveries that shaped microbiology

Table 1.1 Some notable Nobel laureates and their contributions to the science of microbiology

Year	Nobel laureate	Work
1902	Ronald Ross	Malaria
1905	Robert Koch	Tuberculosis
1908	P. Ehrlich and E. Metchnikoff	Antibodies and phagocytosis
1913	Charles Robert Richet	Anaphylaxis
1919	Jules Bordet	Immunity
1928	Charles Nicolle	Typhus
1930	Karl Landsteiner	Discovery of human blood groups
1939	Gerhard Domagk	Discovery of the antibacterial effects of prontosil
1945	A. Fleming, E. Boris Chain, and Howard Walter Florey	Discovery of penicillin
1951	Max Theiler	Yellow fever
1952	Selman A. Waksman	Discovery of streptomycin
1954	Franklin Enders, T. H. Weller, and F. C. Robbins	Poliomyelitis growth in tissue
1958	Joshua Lederberg	Genetic recombination and the organisation of the genetic material of bacteria
1960	F. M. Burnet and Peter B. Medawar	Acquired immunological tolerance
1966	Peyton Rous	Tumour-inducing viruses
1969	M. Delbrück, A. D. Hershey, and S. E. Luria	Replication mechanism and the genetic structure of viruses
1975	D. Baltimore, R. Dulbecco, and H. Martin Temin	Interaction between tumour viruses and the genetic material of the cell
1976	Baruch S. Blumberg and D. Carleton Gajdusek	New mechanisms of infectious disease dissemination (kuru and hepatitis)
1983	Barbara McClintock	Mobile genetic elements—'jumping genes'
1984	Niels K. Jerne, Georges J. F. Köhler, and César Milstein	Specificity and control of the immune system and principle of the production of monoclonal antibodies
1987	Susumu Tonegawa	Generation of antibody diversity
1989	J. Michael Bishop and Harold E. Varmus	Cellular origin of retroviral oncogenes
1993	Karry B. Mullis (Chemistry)	Polymerase chain reaction
1996	Peter C. Doherty and Rolf M. Zinkernagel	Specificity in cell-mediated immune defence
1997	Stanley B. Prusiner	Discovery of prions
2005	Barry J. Marshall and J. Robin Warren	Discovery of *Helicobacter pylori* and its role in peptic ulcer disease and gastritis
2008	Harald zur Hausen	Discovery of human papillomaviruses
	Françoise Barré-S and L. Montagnier	Discovery of human immunodeficiency virus
2011	Bruce A. Beutler and Jules A. Hoffmann	Activation of innate immunity
	Ralph M. Steinman	Discovery of the dendritic cell and its role in adaptive immunity
2015	William C. Campbell and Satoshi Ōmura	Discovery of a novel therapy against roundworm infection
	Youyou Tu	Discovery of a novel therapy against malaria
2018	James P. Allison and Tasuko Honjo	Cancer therapy by inhibition of negative immune regulation
2020	Harvey J. Alter, Michael Houghton, and Charles M. Rice	Discovery of hepatitis C virus

- Elucidation of the immunological characteristics of a disease
- Increased scope of understanding of cancer and measures to treat certain malignancies
- Study of infectious diseases in the community and outbreaks in the form of epidemics and pandemics and public health programmes to control these diseases
- Infection control and prevention in healthcare facilities
- Development of recombinant DNA technology, genetic manipulation, and molecular engineering

CLINICAL MICROBIOLOGY FOR DOCTORS

- Microbiology deals with microorganisms and their relationships with humans in health and disease. This book deals with microbiology as applied to clinical medicine.
- The importance of microbial flora (**microbiome**) is being increasingly recognised for its role in infectious diseases.
- Antimicrobial resistance is a growing problem and is likely to result in treatment failures in the future. Microbiology laboratories assist clinicians in decision making regarding the judicious use of antimicrobials.
- Infection control measures and their knowledge can assist clinicians in controlling infections in both the community as well as in hospitals.

Key Points

✓ Over centuries, individuals from many different countries have, with their pioneering work, laid the foundation for the science that is currently known as microbiology.

✓ Louis Pasteur (1822–1895) discovered methods of sterilisation and developed methods for the culturing of microbes. He demonstrated that microorganisms cause disease and established the principles of immunisation.

✓ Joseph Lister (1827–1912) introduced the concept of 'antisepsis', wherein he sprayed the patient and operating field with carbolic acid. Robert Koch (1843–1910) defined the criteria used to attribute a disease to a microorganism. These criteria—known as Koch's postulates are as follows:
  - The microorganism should be found in all cases of the disease.
  - It should be possible to isolate the microorganism from the host and grow it in a pure culture for several generations.
  - The microorganism should produce the same disease in other susceptible animals.
  - It should be possible to isolate the microorganism in a pure culture from lesions in infected animals.
  - An additional postulate, not formulated by Koch, was added later—a specific antibody to the organism should develop during the course of the disease.

✓ Ruska (1934) developed the electron microscope, enabling the visualisation of the microbes we now call viruses. The development of tissue culture techniques has permitted the cultivation of viruses.

✓ Paul Ehrlich (1854–1915) pioneered the study of antitoxin and toxin neutralisation.

✓ Sir Alexander Fleming (1881–1955) discovered that the fungus *Penicillium* produces a substance—penicillin—which destroys staphylococci; this discovery led to the formulation of other antimicrobials.

✓ Microbiology laboratories assist clinicians in decision making regarding judicious use of antimicrobials.

✓ Infection control measures and their knowledge can assist clinicians in controlling the infections in both community as well as in hospitals.

Questions

Short answer questions

Write short notes on the following:

1. Robert Koch
2. Louis Pasteur
3. Joseph Lister
4. Koch's postulates
5. Scope of medical microbiology

MCQs

1. **Which of the following is NOT one of Koch's postulates?**
   a) The microorganism should be constantly associated with the lesions of the disease caused by it.
   b) It should be possible to isolate the microorganism in pure culture from the lesions.
   c) Inoculation of such pure cultures into suitable laboratory animals should reproduce the disease.
   d) The microorganism need not be re-isolated from lesions produced in experimental animals.
2. **Who is the father of antiseptic surgery?**
   a) Robert Koch b) Paul Ehrlich
   c) Joseph Lister d) Alexander Fleming
3. **Who discovered the active component of *Penicillium*—penicillin?**
   a) Alexander Fleming b) Flory and Chain
   c) Domagk d) Frank Burnet
4. **Which of the following is NOT a contribution of Louis Pasteur?**
   a) The introduction of sterilisation techniques
   b) The discovery of a vaccine for rabies
   c) The introduction of antiseptic techniques in surgery
   d) The discovery of the principles of fermentation
5. **Who invented the cowpox vaccine?**
   a) Edward Jenner b) Joseph Lister
   c) Niels Jerne d) Paul Ehrlich
6. **Which of the following organism does not follow Koch's postulates?**
   a) *Mycobacterium leprae* b) *Escherichia coli*
   c) *Klebsiella* species d) *Staphylococcus aureus*
7. **Stanley B. Prusiner was awarded the Nobel prize for his discovery of:**
   a) Prions b) Cell-mediated immunity
   c) Smallpox vaccine d) Penicillin
8. **The first scientist to have observed bacteria using a single lens microscope is:**
   a) Edward Jenner b) Antonie van Leeuwenhoek
   c) Alexander Fleming d) Louis Pasteur
9. **Who is credited with the discovery of electron microscope?**
   a) Antonie van Leeuwenhoek
   b) Ernst Ruska
   c) Niels Jerne
   d) Frank Burnet
10. **Which of the following bacteria was discovered by Robert Koch?**
   a) *Mycobacterium leprae* b) *Vibrio cholerae*
   c) *Treponemes* d) *Clostridium tetani*

Answers: 1:d, 2:c, 3:b, 4:c, 5:a, 6:a, 7:a, 8:b, 9:b, 10:b

CHAPTER 2

Part I: General Microbiology

Microscopy

(MI 1.1)

Chapter Navigation

- Optical or light microscopy
- Phase contrast microscopy
- Dark-field microscopy
- Fluorescence microscopy
- Differential interference contrast microscopy
- Confocal scanning laser microscopy
- Electron microscopy
- Atomic force microscopy

Learning objective

- Describe the functions and uses of various types of microscopes in healthcare settings

MICROSCOPY

- The size of microorganisms is expressed in **microns** (or micrometres, i.e., µm)—the unit of measurement of bacteria, parasites, and fungi—and **nanometres**, which is the unit of measurement of viruses, which are ultramicroscopic structures.
- Viruses range in size from 20–300 nanometres, whereas some worms can reach a length of up to a metre.
- Yeast cells are usually 4–8 µm long, while fungal hyphae can reach up to a few centimetres in length.
- Bacteria can be visualised only under magnification; their detection in clinical material or laboratory cultures and the study of their morphology and physiology is made possible by microscopy.
- **The bacteria of medical importance are generally 0.2–1.5 µm in diameter and about 3–5 µm in length.**
- In 1590, a Dutch father–son duo, **Hans and Zacharias Janssen**, developed the first compound microscope. Over the years, several modifications have resulted in different types of microscopes, ranging from simple to compound, phase-contrast, dark-field, fluorescence, and electron and scanning microscopes.

Terms Used in Microscopy

Resolution power

The resolving power of a microscope is the **minimum distance between two objects placed close to each other that can be distinguished as separate images**.

- The limit of resolution of the unaided eye is about 200 microns and that of a light microscope is about 200 nm.
- It is inversely proportional to the wavelength of the light used, i.e., the shorter the wavelength employed, as in the case of UV rays, the more is the resolving power (Fig. 2.1).

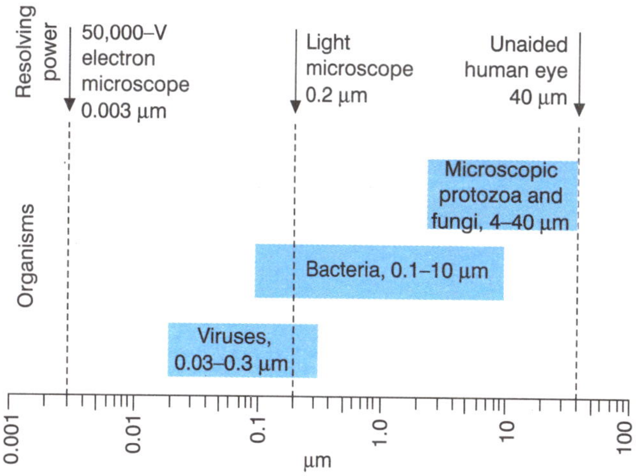

Fig. 2.1 Resolution power of light and electron microscopes and the microorganisms visualised through them

Magnification

This refers to the degree of visual enlargement of an observed object.

- **Total magnification** = Magnification by the objective lens × magnification by the eyepiece (in an oil-immersion lens, magnification is 100 × 10 = 1,000 times).
- **Empty magnification** refers to the degree of magnification of an object that can be achieved without any additional details being visible. This occurs when higher magnification lenses are used to enlarge objects.

Contrast

- This is the difference between the light intensities of the image and the adjacent background.
- This difference in intensity and/or colour allows individual features and details of the specimen to become visible. Contrast can be improved by staining the object or the background.

Optical or Light Microscopy

Principle

- A light microscope uses visible light to illuminate an object.
- It consists of two series of lenses—the **objective** and the **ocular** lenses (Fig. 2.2). The objective lens has a magnification of 10X, while the ocular lens can have a magnification of 4X, 10X, 40X, or 100X.
- Light rays pass through a **condenser** to illuminate the object (**specimen**) mounted on the stage. These rays then pass through the **lens** of the objective to the eyepiece, and then to the observer's eye. The diaphragm of the iris controls the amount of light that passes through the condenser.
- Together, the two sets of lenses of a light microscope may magnify the specimen up to 1,000 times.

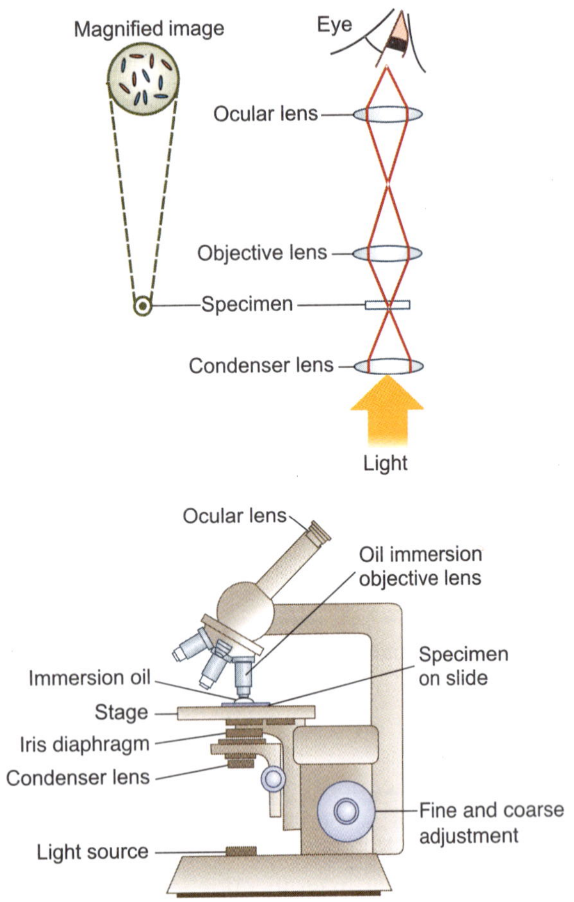

Fig. 2.2 Light microscopy

Uses of a light microscope

- It is a **rapid and presumptive diagnostic method**, which can be used in the outpatient wing, in the field, or in the laboratory as the first step towards identifying a microorganism.
- The **detection of bacteria** and inflammatory cells in a gram-stained preparation reflects on the type of infection—whether acute inflammatory response (neutrophils) or chronic (lymphocytes).
- **Stool microscopy** is used to detect ova and cysts and other additional structures like RBCs, pus cells, etc.
- **Fungal elements** are isolated from clinical material either by potassium hydroxide (KOH) or lactophenol cotton blue stain.
- **Peripheral blood smears** are used to detect parasites like *Plasmodium*, *Leishmania*, and microfilariae.
- The identification of an organism from a culture medium is the first step towards further characterisation—to choose appropriate media and biochemical tests as well as antibiotics for susceptibility testing.

Types of light microscopy

Dark-field microscopy and **phase-contrast microscopy** are variations of using visible light to enhance contrast and improve visibility.

Phase-contrast Microscopy

This method improves the contrast in the field and makes evident the structures within cells, which differ in thickness or refractive index.

Principle

- The phase-contrast technique uses an optical mechanism to **translate small variations in the phases of light waves into corresponding changes in amplitude**, which translate into differences in image contrast. These differences in refractive index (RI) between bacterial cells and the surrounding medium make them clearly visible.
- The retardation (by a fraction of wavelength) of rays of light that pass through the object compared to those that pass through the surrounding medium can be brought about with a special ring known as the **annular ring** (Fig. 2.3a and b). This produces 'phase' differences between the two types of rays.
- The phase plate (**diffraction plate**) placed between the objective lens and the eyepiece enhances the difference in the diffracted rays. This produces a three-dimensional effect, highlighting the depth and thickness of the structure and parts of the bacteria.

Uses of a phase-contrast microscope

- Living microorganisms, cells, and tissues can be visualised as **high-contrast images**.
- This method is generally used for **research purposes** and rarely in diagnostic laboratories.

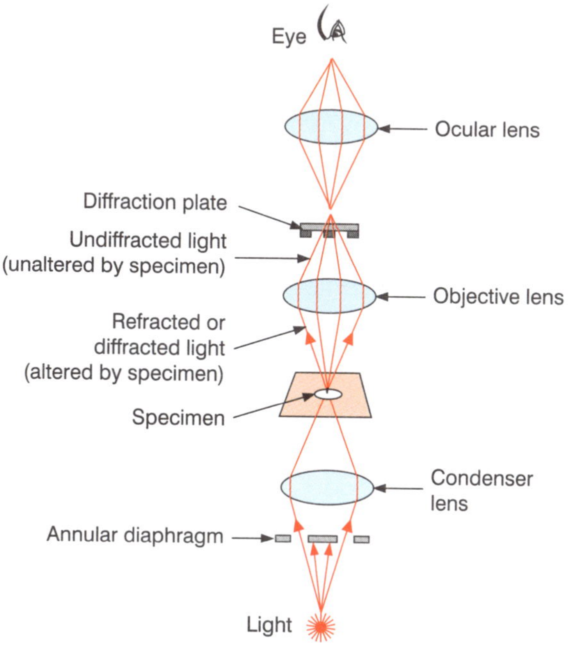

(a)

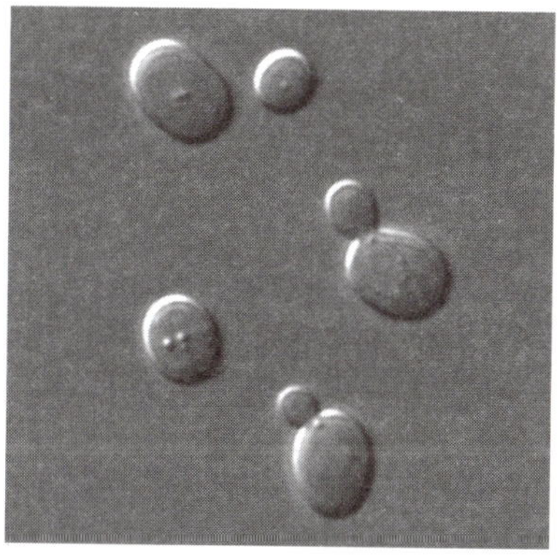

(b)

Fig. 2.3 (a) Pathway of light waves in a phase-contrast microscope and (b) image of *Saccharomyces cerevisiae* with a 3D effect (*Source*: Masur, Public domain, via Wikimedia Commons)

Dark-field/Dark-ground Microscopy

One method of improving contrast in microscopy is by **using reflected light instead of transmitted light** (as in the light microscope).

Principle

- A dark-field condenser, which has a **central, circular stop**, directs a cone of light towards the stage. The object on the stage is illuminated by this cone of light. Only light reflected or scattered by the object reaches the objective lens. As a result, the **object appears luminous against a dark background**.
- This contrast gives an illusion of increased resolution and allows very slender organisms such as spirochetes, which are not visible in ordinary illumination, to be visualised (Fig. 2.4a and b).

Uses of a dark-field microscope

- Useful for examining clinical material or cultures of slender bacteria including **spirochetes**.
- To examine **discharge from syphilitic ulcers** for the presence of treponemes.
- To detect ***Leptospira*** in suspected cases of leptospirosis from urine samples.

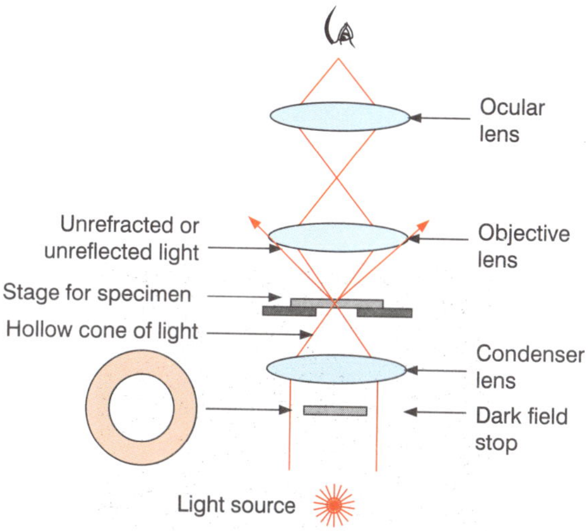

(a)

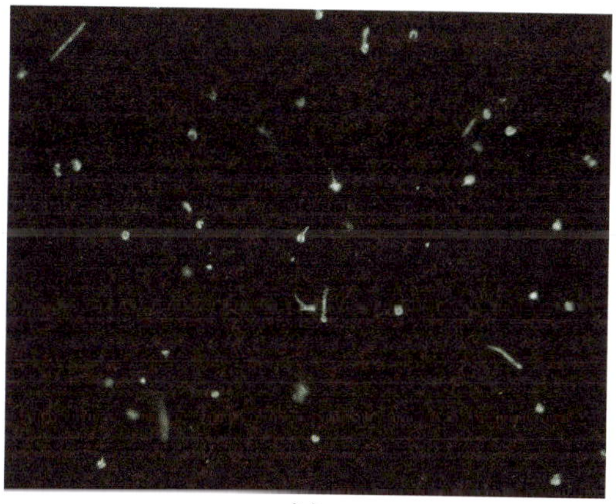

(b)

Fig. 2.4 (a) Pathway of light rays passing through a dark-field microscope and (b) *Treponema pallidum* spirochetes visualised by dark-field microscopy (*Source*: CDC, PHIL, Image ID 21116)

Fluorescence Microscopy

Principle

- A fluorescence microscope uses **light from a high-intensity source to excite a fluorescent agent**, which in turn, emits a low-energy light of a longer wavelength to produce an image.
- Filters designed to the specific wavelength produced by the fluorescing dye are used to block the surrounding radiation, allowing the viewer to visualise only the organisms/structures that have taken up the dye (Fig. 2.5a and b).

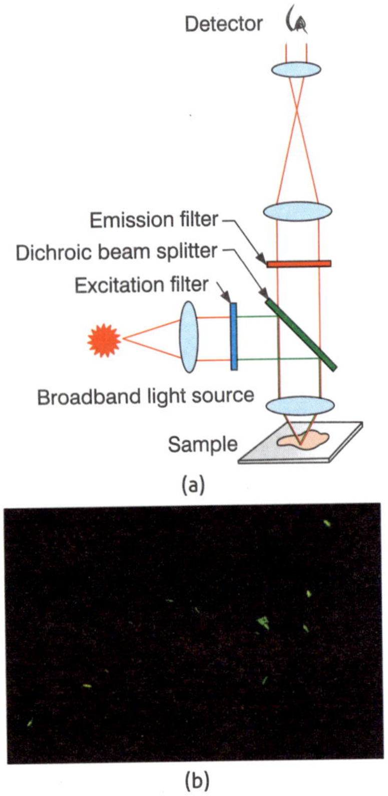

Fig. 2.5 (a) Pathway of light in a fluorescence microscope and (b) fluorochrome-stained *Mycobacterium tuberculosis* (*Source*: [b] Centres for Disease Control and Prevention [CDC], Public Health Image Library [PHIL], Image ID 22206)

- **Fluorochromes**, which are fluorescent dyes such as auramine O and rhodamine, are used to stain microorganisms.
- **Immunofluorescence** is a technique that uses antibodies labelled with a fluorescent dye to aid in the visualisation of microorganisms (Fig. 2.6).
- **Fluorescein isothiocyanate (FITC)** is a conjugate used to specifically stain a particular bacterial species or infected cells.

Uses of a fluorescence microscope

- Direct fluorescence (light-emitting diode [LED] microscope) is used for the examination of sputum smears for ***Mycobacterium tuberculosis*** (auramine–rhodamine staining) and fungal filaments (calcofluor white and acridine orange staining).
- This method is used for the **rapid diagnosis** of infections like diphtheria, viral respiratory infections, and *M. tuberculosis* (MTB) in sputum or a CSF smear.
- It is used for the detection of the **malaria parasite** in peripheral blood smears.
- It is used for the **early detection of bacteria** in blood cultures.
- Fluorescent antibodies provide a wide variety of immunologically specific, rapid diagnostic tests for infectious diseases.

Differential Interference Contrast Microscopy

- This is a specialised type of microscopy that not only reveals cell organelles, but also enables the **quantitative measurement of the chemical constituents of cells** such as lipids, proteins, and nucleic acids.
- It is used in specialised areas of cell biology.

Confocal Scanning Laser Microscopy

- The term 'confocal' is used when **two lenses are arranged to focus on the same point**.
- The presence of confocal pinholes allows only light from the plane of focus to reach the detector.

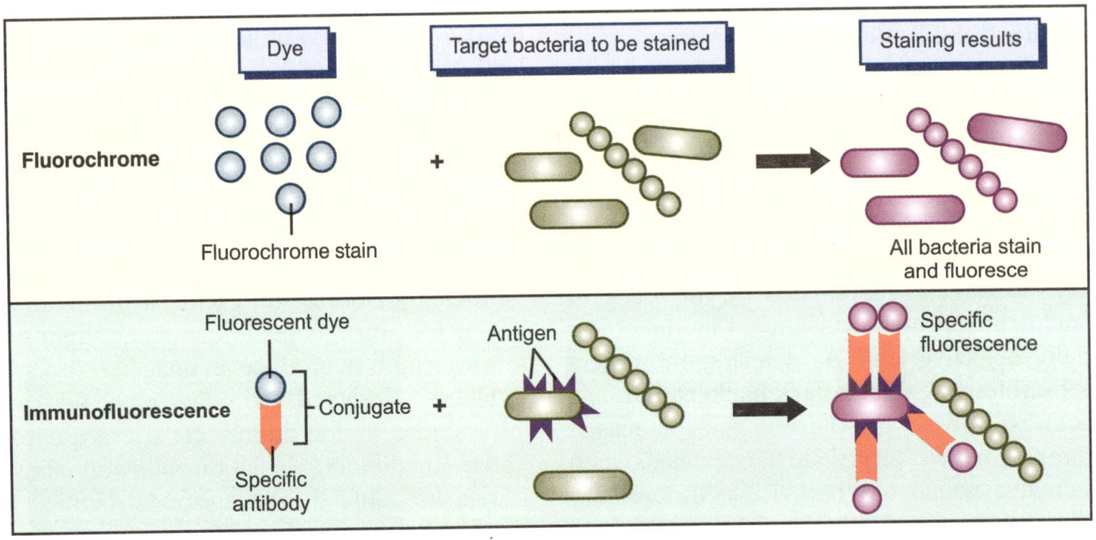

Fig. 2.6 Fluorochrome staining and immunofluorescence

- It is equipped with a laser beam that bounces off a mirror and is directed through a scanning device.
- The beam is then directed through a pinhole, which keeps out most of the light that is not from the microscope's focal plane. In a thick specimen, various layers can be seen by adjusting the plane of focus of the laser beam.
- Cells are stained with fluorescent dye to make them more visible.
- The microscope is equipped with computer software to digitally process the images.
- It is used to examine tissue sections, in stem cell research, and in fluorescence in-situ hybridisation.

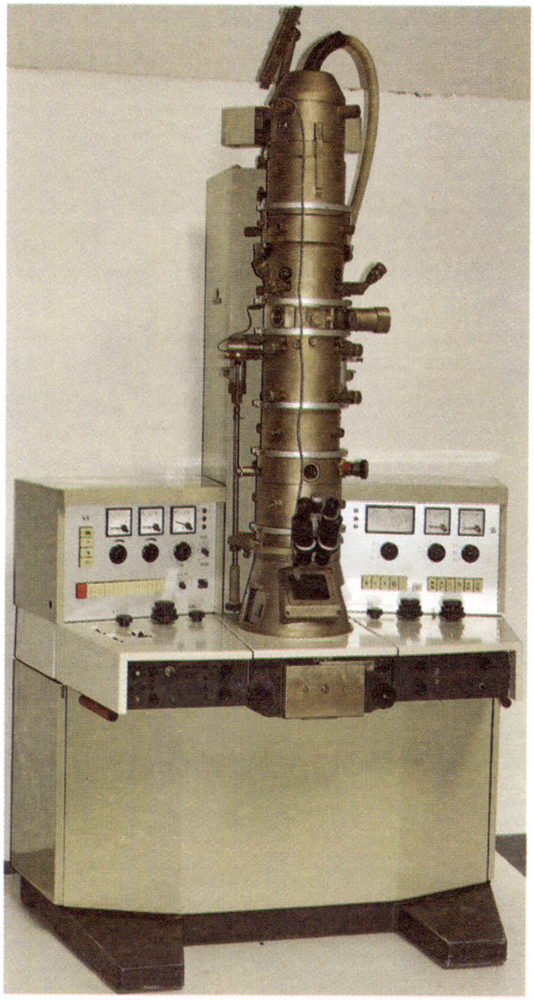

Fig. 2.7 Electron microscope (*Source*: Miloš Jurišić, CC BY-SA 3.0 <https://creativecommons.org/licenses/by-sa/3.0>, via Wikimedia Commons)

Electron Microscopy

Principle

- It uses a beam of **electrons instead of light rays** to visualise microscopic objects.
- The electron beam is focused by **electromagnets** (analogous to the lenses in the light microscope). The object held in the path of the electromagnetic field scatters the electrons and produces an image, which is focused on a fluorescent viewing screen (Fig. 2.7 and 2.8).
- As the wavelength of electrons used is approximately 0.005 nm (visible light has a wavelength of 500 nm), the resolving power of the electron microscope is **about 0.1 nm**.
- Since gas molecules scatter electrons, electron microscopes create a vacuum environment to visualise specimens. This requires specimens to be dried and fixed, which can lead to the distortion of cell morphology.

Uses of an electron microscope

Electron microscopes are highly specialised devices used mainly for research purposes to study the following:

- Intracellular structures
- Pathogen–host structural interactions
- Virus structures

Modifications of electron microscopy

- **Transmission electron microscope**—a beam of electrons passes through the object and creates an image.
- **Shadow-casting** with vapourised heavy metals has made taking pictures with good contrast and three-dimensional effect possible.
- **Negative staining** is useful in studying the fine structures of microorganisms. It involves staining the specimen with phosphotungstic acid.
- **Freeze-etching**, a process involving the deep-freezing of specimens in liquid gas and subsequently forming a carbon-platinum replica of the material, avoids distortion of the image. As frozen cells may remain viable, freeze-etching enables the study of the cellular ultrastructure as it appears in the living state.
- **Scanning electron microscopy** produces images of a sample by scanning the surface with a focused beam of electrons. The electrons interact with atoms in the sample, producing various signals about the surface topography and composition of the sample. The electron beam is scanned to study the cell surface with greater contrast and higher resolution (Fig. 2.9).

Atomic Force Microscopy (AFM)

- It is also known as **scanning force microscopy (SFM)**.
- It is a very-high-resolution type of scanning probe microscopy (SPM), with a resolution power amounting to **fractions of a nanometre**.

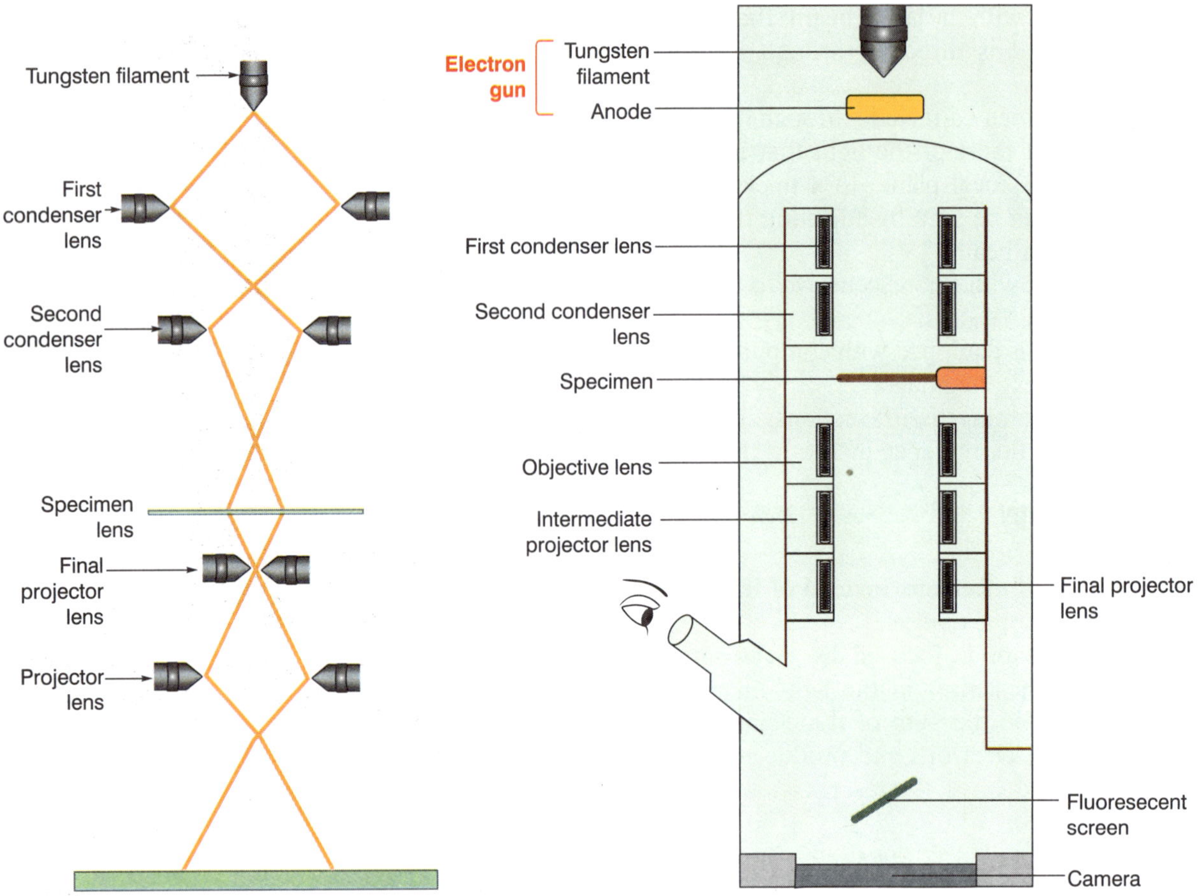

Fig. 2.8 Line diagrams showing the path of electron beams and the electromagnets, which function as a series of condenser lenses in an electron microscope

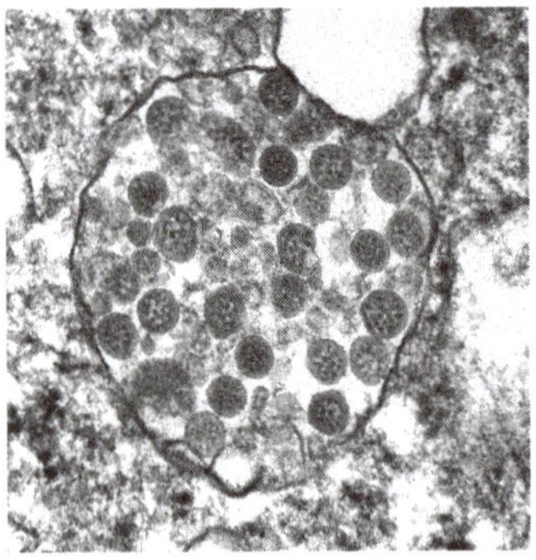

Fig. 2.9 Scanning electron micrograph of the COVID-19 virus in the cytoplasm of an infected cell (*Source*: CDC, PHIL, Image ID 23591)

- A small cantilever probe moves across the surface of a sample; topographic differences in the sample are translated into differences in the amount of laser light reflected into the photodiode.
- This method is used in molecular biology, cell biology, and medicine.

Key Points

✓ Light or optical microscopy is most commonly used in diagnostic microbiology. It consists of two series of lenses, the objective and the ocular lenses.

✓ Phase-contrast improves the contrast and makes evident the structures within cells, which differ in thickness or refractive index.

✓ In fluorescence microscopy, the organism to be visualised is stained with a fluorescing agent. On exposure to excitation light, organisms emit fluorescent light and can be visualised.

✓ By using reflected light instead of transmitted light in the dark-field microscope, the visibility of the small microorganism is enhanced. The most important part of the dark-field microscope is the dark-field condenser.

✓ In an electron microscope, an electron beam is focused by electromagnets. The resolving power of an electron microscope is about 0.1 nm.

Questions

Short answer questions

1. Describe the working principle of a light microscope.
2. Trace the pathway of light in a fluorescence microscope.
3. How does a dark-field microscope enhance the visualisation of microorganisms?
4. How does an electron microscope magnify the image of an ultramicroscopic organism?

MCQs

1. **Which microscope is based on the principle that differences in the refractive indices of bacterial cells and the surrounding medium make them clearly visible?**
   a) Dark-ground microscope
   b) Phase-contrast microscope
   c) Electron microscope
   d) Confocal microscope
2. **The resolving power of a microscope depends upon the:**
   a) Focal length and aperture of the eye lens
   b) Focal length and objective of the eye lens
   c) Apertures of the objective and the eye lens
   d) Wavelength of light illuminating the object
3. **The magnification power of a compound microscope does not depend on the:**
   a) Focal length and aperture of the eye lens
   b) Apertures of the objective lens
   c) Tube length of the microscope
   d) None of the above
4. **The resolving power of a light microscope is:**
   a) 2 mm b) 0.2 mm
   c) 0.1 mm d) 1 mm
5. **Which of the following microscopes is most useful for viewing the internal structures of an unstained specimen?**
   a) Phase contrast
   b) Confocal
   c) Electron
   d) Any of the above
6. **The resolving power of a microscope is the:**
   a) Degree of visual enlargement of an observed object
   b) Difference between the light intensities of the image and the adjacent background
   c) Minimum distance between two objects placed close to each other that can be distinguished as separate images
   d) Degree of magnification of an object that can be achieved without any additional details being visible
7. **Which of the following is correct regarding fluorescence microscopy?**
   a) It uses light from a low-intensity source to excite a fluorescent agent.
   b) It uses light from a high-intensity source to excite a fluorescent agent.
   c) It can use both low-as well as high-intensity light.
   d) It is not of much use in microbiology.
8. **Which of the following is not true regarding electron microscopy?**
   a) It uses a beam of electrons.
   b) It uses a beam of laser.
   c) The beam is focused by electromagnets.
   d) Only dried and fixed specimens can be visualised.
9. **The resolving power of the electron microscope is about:**
   a) 1 nm b) 0.1 nm
   c) 1 μm d) 0.1 μm

Answers: 1:b, 2:d, 3:b, 4:b, 5:a, 6:c, 7:b, 8:a, 9:b

Part I: General Microbiology

CHAPTER 3

Basic Concepts of Bacteriology

(MI 1.2-1.3)

Chapter Navigation

- Staining
- Morphology and physiology of bacteria
- Growth and nutrition, isolation, and identification for the laboratory diagnosis of bacterial infections

Learning objectives

- Describe the bacterial growth curve and enumerate the factors influencing bacterial growth
- Describe various techniques to detect bacteria and their structural components, including staining and molecular techniques performed on clinical material

Bacteria are a diverse group of unicellular organisms that constitute one among several distinct classes of microorganisms. They belong neither to the plant nor the animal kingdom, but to a third kingdom, **Protista**. The kingdom Protista is further divided into two groups:

- **Prokaryotes**—unicellular organisms without true branching (except the actinomycetales); they have a cell wall, cell membrane, and nucleoid with intracellular organelles (Fig. 3.1). Bacteria are prokaryotic cells, which differ from the eukaryotic cells seen in humans and animals.
- **Eukaryotes**—multicellular and distinct from prokaryotes (Fig. 3.1). They do not have a rigid cell wall; instead, they possess a cell membrane and a well-defined nucleus with a nucleolus and mitochondria.

The salient differences between prokaryotes and eukaryotes are presented in Table 3.1.

Table 3.1 Important differences between prokaryotic and eukaryotic cells

Characteristic	Prokaryotes	Eukaryotes
Nucleus	Not well-defined	Well-defined
Nuclear membrane	Absent	Present
Nucleolus	Absent	Present
Deoxyribonucleoprotein	Absent	Present
Chromosome	One (circular) DNA molecule	More than one DNA molecule
Mitotic division	Absent	Present
Cytoplasm		
Cytoplasmic streaming	Absent	Present
Pinocytosis	Absent	Present
Mitochondria	Absent	Present
Lysosomes	Absent	Present
Golgi apparatus	Absent	Present
Endoplasmic reticulum	Absent	Present
Chemical composition		
Sterols	Absent	Present
Muramic acid	Present	Absent
Examples	Bacteria and blue-green algae	Fungi, other algae, slime moulds, and protozoa

MORPHOLOGY OF BACTERIA

- Morphology depends on the strain studied, the nature of the culture medium, the temperature and time of incubation, the age of the culture, and the number of subcultures it has undergone.
- Bacterial morphology is best studied by microscopy after appropriately staining the specimen.

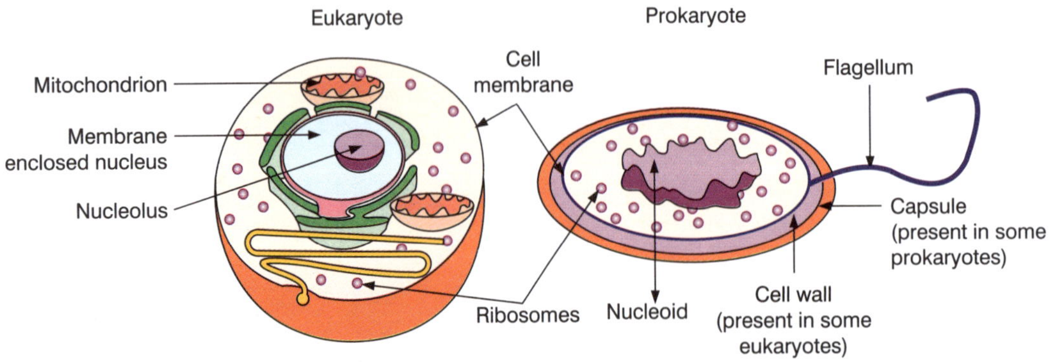

Fig. 3.1 Schematic diagram displaying the differences between a eukaryote and a prokaryote

Characteristic Features of Bacteria

Shape

- Bacteria may be spherical (cocci), filamentous, rod-shaped (bacilli), comma-shaped, or spiral.
- The axis of the organism may be straight or curved.
- Depending on their shape, bacteria are classified into several types as presented in Fig. 3.2 and Table 3.2.

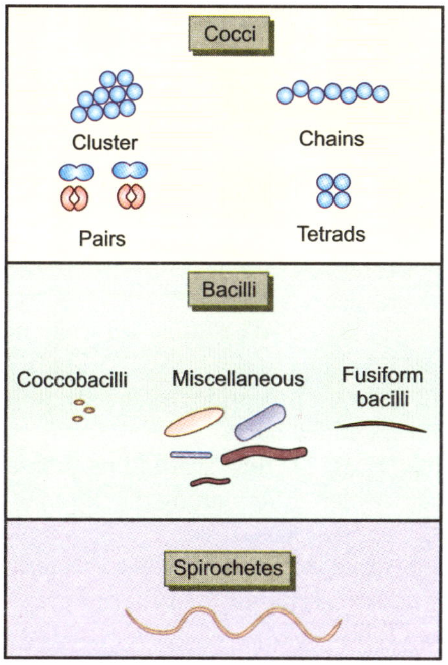

Fig. 3.2 Shapes of bacteria

Table 3.2 Classification of bacteria based on their shapes

Group	Examples
Cocci (from *kokkos*, meaning berry)	*Staphylococcus, Streptococcus*
Bacilli (rods)—from *baculus*, meaning rod	*Escherichia coli, Proteus, Salmonella*
Vibrio (comma-shaped)	*Vibrio cholerae*
Spirilla	*Spirillum minus*
Spirochetes (from *speira*, meaning coil and *chaite*, meaning hair)	*Borrelia, Leptospira, Treponema*
Filamentous bacteria	Actinomycetes
Amorphous morphology—round or oval bodies and interlacing filaments	*Mycoplasma*

Size

The length and breadth of an organism may vary. Some species of bacteria exhibit great variation in the shapes and sizes of individual cells. This is known as **pleomorphism.** Pleomorphic and involuted forms are often caused by defective cell wall synthesis.

Sides and ends

- The sides of the organism may be parallel, convex, concave, or irregular.
- The ends may be truncated or blunt, rounded, or tapering.
- Considerable variation in shape and size may lead to clubbing or swollen or giant forms.

Arrangement

- The arrangement of groups of bacteria is determined by the plane through which binary fission takes place and the tendency of the daughter cells to remain attached even after division.
- Bacteria may be arranged singly, in pairs, in tetrads, or in packets of eight. Alternatively, they may be arranged in chains, short or long, as in the case of cocci.
- Bacilli may be arranged at random, in short or long chains, in Chinese letter patterns, as palisades, or in bundles.
- Vibrios may be visualised singly or in S or spiral forms.

Flagella

- Some bacteria have no flagella, i.e., they are **atrichate**.
- Flagellated bacteria are called **monotrichate, lophotrichate, amphitrichate, or peritrichate** depending on the number and arrangement of their flagella.
- Flagella cannot be visualised by light microscopy unless they are stained by special methods.

Spores

- These, when present, may be oval, spherical, or ellipsoidal and may be as wide as or wider than the bacillary body.
- Spores may be equatorial, subterminal, or terminal.

Capsules

- These may or may not be present.
- A capsule can be demonstrated by negative staining, dark-ground illumination, phase-contrast, or electron microscopy.

Motility

Bacteria may be non-motile, sluggishly motile, actively motile or they may exhibit darting motility.

Bacterial Cell Structure

A bacterial cell (Fig. 3.3) is composed of an outer layer or cell envelope, internal structures, and some additional appendages.

Cell envelope

The cell envelope consists of two components:

- A rigid cell wall
- A cytoplasmic or plasma membrane (beneath the cell wall)

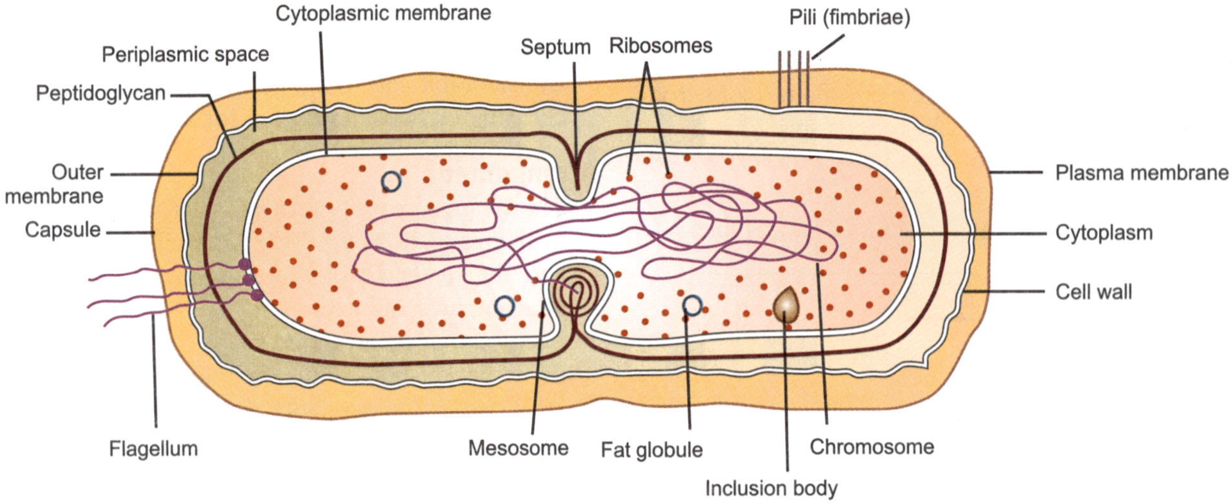

Fig. 3.3 Parts of a bacterial cell

Cell wall

- It is responsible for the shape, rigidity, and plasticity of a bacterial cell.
- It does not stain with simple stains and hence cannot be visualised by light microscopy. It can be demonstrated by:
  - Plasmolysis (cytoplasm shrinks in a hypertonic solution, whereas the cell wall remains intact)
  - Microdissection
  - Reaction with specific antibodies
  - Mechanical rupture of the cell
  - Differential staining procedures
  - Electron microscopy

Structure

- Bacterial cell walls are about 10–25 nm thick and account for about 20–30% of the dry weight of the cell.
- Chemically, the cell wall is composed of **mucopeptide** (peptidoglycan or murein) scaffolding formed by chains of alternating N-acetyl glucosamine (NAG) and N-acetyl muramic acid (NAMA) molecules cross-linked by peptide chains (Fig. 3.4).

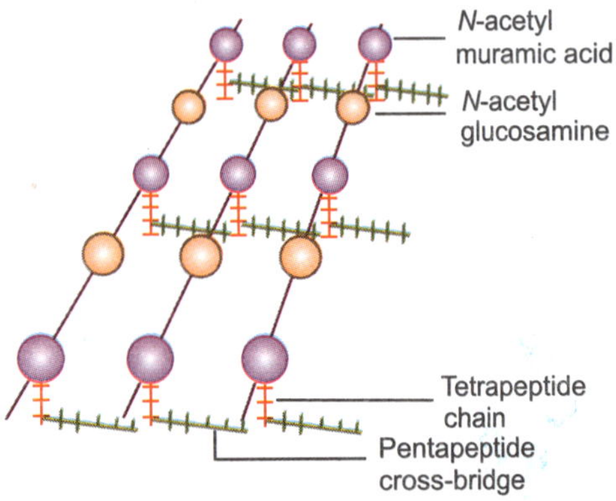

Fig. 3.4 Chemical structure of the bacterial cell wall

- Each molecule of NAG is attached to a tetrapeptide structure. Pentapeptide cross bridges interlink adjacent layers of peptidoglycans by attaching to these tetrapeptide chains.
- The interstices of this scaffolding contain other chemicals, which differ among the different bacterial species.
- The cell walls of gram-positive bacteria have a simpler chemical nature than that of gram-negative bacteria (Table 3.3).
- The cell wall carries bacterial antigens that are important for virulence and immunity.

Table 3.3 Comparison of the cell walls of gram-positive and gram-negative bacteria

Characteristic	Gram-positive bacteria	Gram-negative bacteria
Thickness (peptidoglycan)	Thick (15–80 nm)	Thin (2–3 nm)
Variety of amino acids	Few	Several
Aromatic and sulphur-containing amino acids	Absent	Present
LPS and outer membrane	Absent	Present
Teichoic acid	Present	Absent

The cell wall of gram-positive bacteria (Fig. 3.5a)

- **Peptidoglycan layer:** The peptidoglycan layer of gram-positive bacteria is significantly thicker (15–80 nm) than that of gram-negative bacteria (2–3 nm).
- **Teichoic acid:** The cell wall contains a significant amount of teichoic acid, which is a major surface antigen of gram-positive bacteria, containing ribitol or glycerol polymers. Two types of teichoic acid are found in gram-positive bacteria:
  - Cell wall teichoic acid—linked to peptidoglycans
  - Membrane teichoic acid (lipoteichoic acid)—linked to membrane glycolipid

Cell wall of gram-negative bacteria (Fig. 3.5b)

- **Outer membrane:** The outermost layer of the gram-negative bacterial cell wall is called the outer membrane. It contains various proteins known as **outer membrane proteins** (OMP). Among these, are **porins**, which form transmembrane pores that serve as diffusion channels for small molecules. They also serve as specific receptors for some bacteriophages.
- **Lipopolysaccharides (LPS):** These are integral to the cell walls of gram-negative bacteria and account for their endotoxic activity and O antigen specificity. This LPS layer was formerly known as the **Boivin antigen**. It consists of three regions:
  - Region I is the O polysaccharide portion, which determines O antigen specificity.
  - Region II is the core polysaccharide.
  - Region III is the glycolipid portion (**lipid A**) and is responsible for the **endotoxic** activities (pyrogenicity, lethal effect, tissue necrosis, anticomplementary activity, B cell mitogenicity, immunoadjuvant property, and anti-tumour activity).
- **Lipoprotein:** This component attaches the protein of the peptidoglycan layer to the lipid of the outer membrane.
- **Peptidoglycan layer:** This is a thin (2–3 nm) layer bound by the lipoprotein and plasma membrane.

Functions of the cell wall

- Provides shape to bacteria
- Act as a virulence factor
- Antigenic in nature
- Protects bacteria

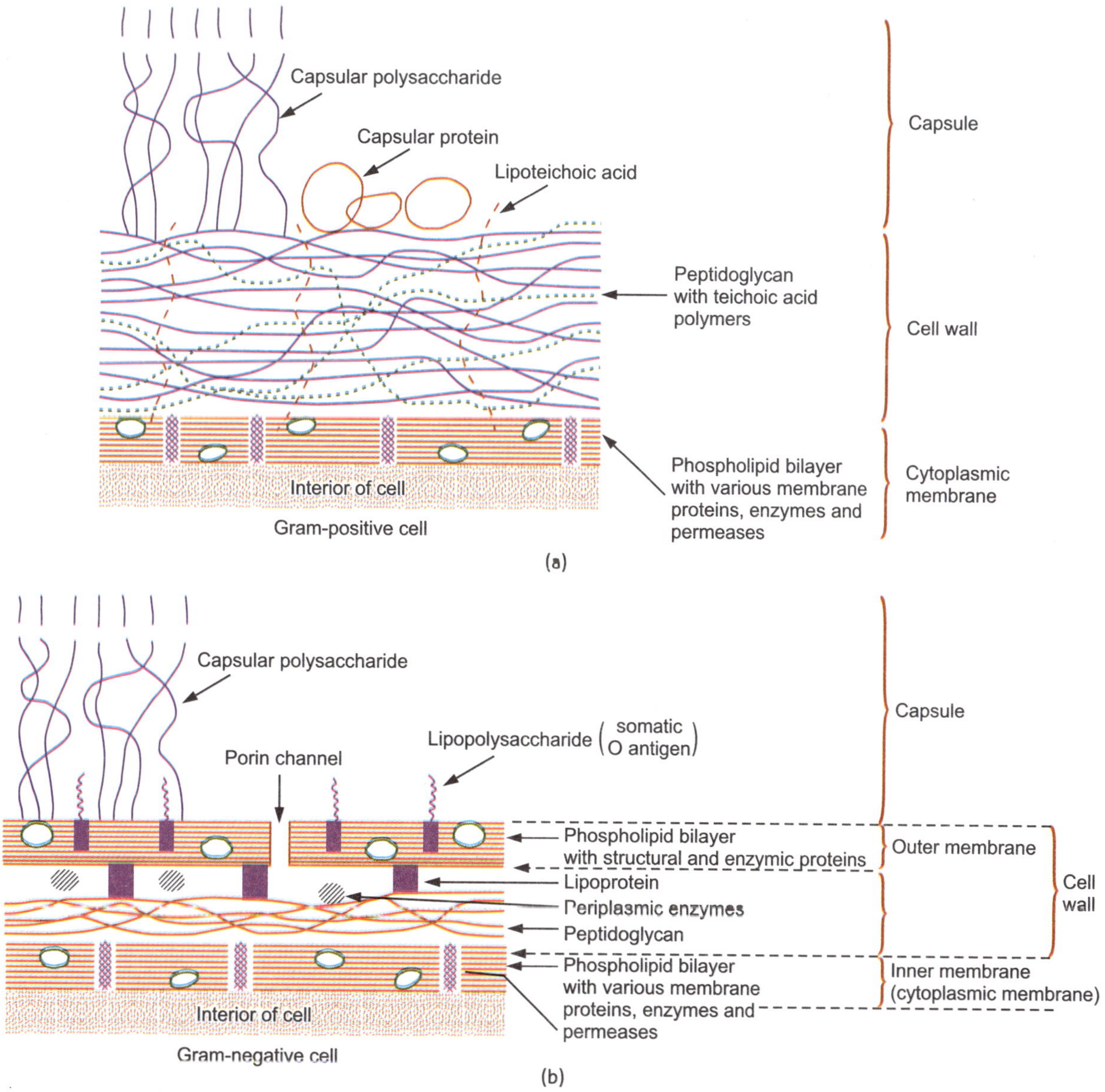

Fig. 3.5 (a) Gram-positive and (b) gram-negative cell wall

Inhibition of cell wall synthesis

- An enzyme, **lysozyme**, which is normally present in many human tissue fluids, lyses susceptible bacteria by splitting the cell wall mucopeptide links. This can result in the following:
  - **Protoplasts:** When lysozyme acts on a gram-positive bacterium in a hypertonic solution, a protoplast is formed, consisting of the cytoplasmic membrane and its contents.
  - **Spheroplast:** When lysozyme acts on gram-negative bacteria, the result is a spheroplast which differs from the protoplast in that some cell wall material is retained. Protoplasts and spheroplasts are spherical, regardless of the original shape of the bacterium.

L-forms

- In 1935, while studying cultures of *Streptobacillus moniliformis*, **Emmy Kleineberger–Nobel** observed certain swollen and morphologically aberrant bacterial cells which were devoid of cell walls. She named these L-forms after the Lister Institute, London, where she made the observation.
- L-forms are seen in several species of bacteria, developing either spontaneously or in the presence of penicillin or other agents that interfere with cell wall synthesis.

Types

- **Unstable L-form:** The morphological abnormality is maintained only in the presence of penicillin or other inducing agents.
- **Stable L-form:** The aberrant form becomes a permanent feature of the strain and is retained in serial subcultures.

L-forms resemble *Mycoplasma* in several ways, including morphology, type of growth on agar, and filterability. It is possible that *Mycoplasma* represents stable L-forms of parent bacteria that have not yet been identified. Cell-wall deficient forms may have a role in chronic persistent infections.

Cytoplasmic membrane

- It is a thin (5–10 nm) layer that lines the inner surface of the cell wall and separates it from the cytoplasm.
- It acts as a semipermeable membrane, actively controlling the flow of metabolites to and from the protoplasm. Passage through the membrane is not solely a function of the molecular size of the particles but also depends on the presence of specific enzymes (permeases) in the membrane.
- Chemically, it consists of lipoproteins with small amounts of carbohydrates. Sterols are absent, except in *Mycoplasma*.

Cell interior

This is constituted by the **protoplasm**—comprising the cytoplasm, cytoplasmic inclusions such as ribosomes and mesosomes, granules, and vacuoles—and **the nuclear body**.

Cytoplasm

- It is a colloidal system comprising a variety of organic and inorganic solutes in a viscous, watery solution.
- It contains ribosomes, mesosomes, vacuoles, and other inclusions such as volutin granules.
- Unlike eukaryotic cytoplasm, it does not exhibit internal mobility (protoplasmic streaming) and does not possess an endoplasmic reticulum or mitochondria.
- The cytoplasm stains uniformly with basic dyes in young cultures but becomes increasingly granular with age.

Intracytoplasmic inclusions

Intracytoplasmic materials present inside the bacterial cells are as follows:

- **Ribosomes:** These act as the **centres of protein synthesis**. They are slightly smaller than the ribosomes of eukaryotic cells (sedimentation constant 70S) and are seen integrated into linear strands of mRNA to form polysomes. Bacterial ribosomes exist as 30S and 50S subunits (Fig. 3.6).
- **Mesosomes (chondroids):** These are vesicular, convoluted, or multi-laminated structures formed as invaginations of the plasma membrane into the cytoplasm. They are more prominent in gram-positive bacteria. Mesosomes are the principal sites of respiratory enzymes in bacteria and are analogous to the mitochondria of eukaryotes. They are often seen in relation to the nuclear body and the site of synthesis of cross-wall septa, suggesting that they coordinate nuclear and cytoplasmic division during binary fission.

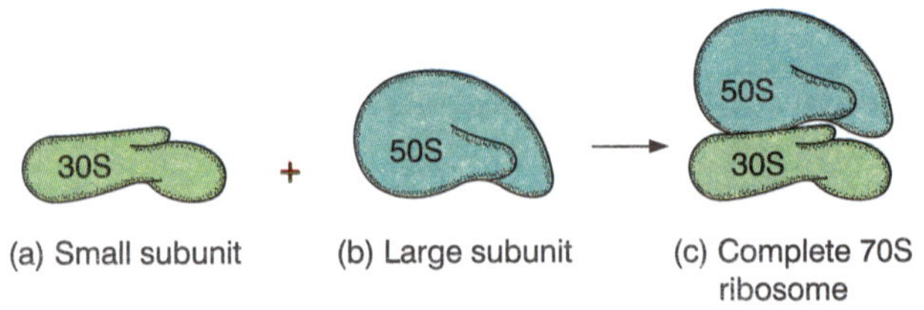

Fig. 3.6 Ribosomal subunits

- **Metachromatic (volutin) granules (seen in some bacterial cells):** These are storage products for metabolism and energy. They exist either as volutin (polymetaphosphate in diphtheria bacilli) granules or as polysaccharides or lipids. These inclusions differ in different species and depend on the age and condition of the culture. These can be demonstrated by special stains like Albert's and Neisser's stains.

Nuclear material

- Bacteria, being prokaryotic cells, do not have a membrane-bound nucleus. Instead, their DNA is found in the nucleoid.
- It can be demonstrated by acid or ribonuclease hydrolysis and subsequent staining for nuclear material and by electron microscopy.
- The nucleus appears as an oval or elongated body. Generally, there is one nucleus per cell.
- Bacterial DNA is not associated with basic protein.
- The genome consists of a **single molecule** of double-stranded DNA arranged in the form of a circle, which may open under certain conditions to form a long chain, about 1 mm in length.
- The bacterial chromosome is haploid and replicates by simple fission and not by mitosis, which is the mode of division in eukaryotic cells.

Plasmid

- Plasmids are extranuclear genetic elements, which are also called episomes (Fig. 3.7). Episomes can integrate into the chromosome while plasmids remain free in the cytoplasm.
- Besides being transmitted to daughter cells during binary fission, they may be transferred from one bacterium to another either through conjugation or by bacteriophages.
- They may carry genes responsible for toxigenicity and drug resistance.

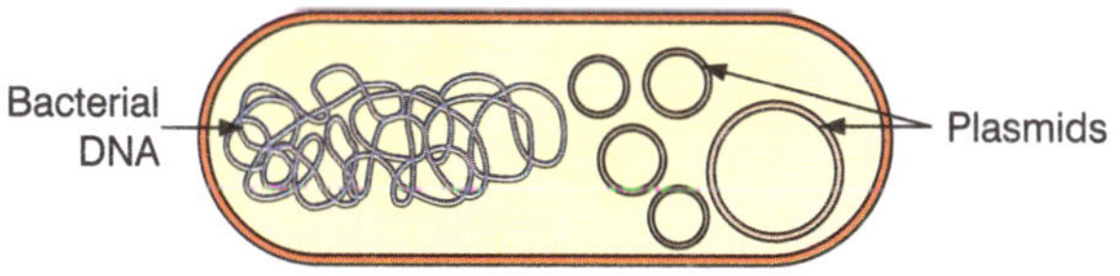

Fig. 3.7 Plasmids, which are extrachromosomal genetic material

Additional structures

- A bacterial cell may be enclosed in a viscid layer, which may either be a loose slime layer (**glycocalyx**) or tightly organised in the form of a **capsule**.
- Some bacteria carry filamentous appendages protruding from the cell surface—**flagella** (organs of locomotion) or **fimbriae** (organs for adhesion).

Slime layer and capsule

- Many bacteria secrete viscid material around the cell surface. When this material is organised into a sharply defined structure, as in *Streptococcus pneumoniae*, it is known as a **capsule**.
- When it is a loose, undemarcated secretion, as in *Leuconostoc*, it is called a **slime layer.** Capsules too thin to be seen under the light microscope are called **microcapsules**.
- The slime may be polysaccharide (*S. pneumoniae*) or polypeptide (anthrax bacillus) in nature.
- Some bacteria may possess a capsule as well as a slime layer (e.g., *Streptococcus salivarius*). Bacteria that secrete large amounts of slime produce mucoid growth on agar (e.g., *Klebsiella pneumoniae*).

Demonstration of the capsule

- **Staining:** Slime has little affinity for basic dyes and is not visible in Gram-stained smears.
- **Negative staining in wet films with India ink:** Capsules are seen as clear halos around bacteria against a black background (Fig. 3.8a). India ink is often used to demonstrate capsulated microorganisms from clinical material. For example, this method is used to demonstrate *S. pnuemoniae* or a fungus known as *Cryptococcus* in the CSF of patients who have meningitis.
- **Quellung reaction:** When a suspension of a capsulated bacterium is mixed with its specific anticapsular serum, the capsule appears very prominent and 'swollen' due to an increase in its refractivity. This capsule swelling is also called the **Quellung reaction** (Fig. 3.8b).

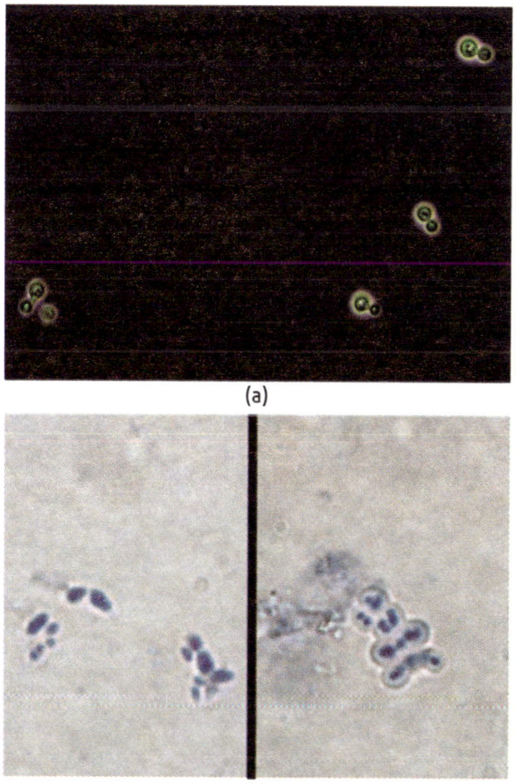

(a)

(b)

Fig. 3.8 (a) *Cryptococcus* capsule (seen as a halo) visualised by India ink staining (*Source*: Dept. of Microbiology, PIMS) and (b) capsular swelling reaction (Quellung phenomenon)

Function of the capsule

- Capsules protect bacteria from lytic enzymes and drying.
- They contribute to the virulence of pathogenic bacteria by inhibiting phagocytosis.
- Repeated subcultures in vitro lead to the loss of the capsule and, consequently, virulence.

Flagella

Flagella are filamentous appendages 3–20 μm long. They are of uniform diameter (0.01–0.013 μm) and terminate in a square tip.

Structure

- Flagella consists of a basal body and a hook that is attached to a filament.
- The **basal body** is embedded in the cell envelope.
- The hook and basal body are antigenically different (Fig. 3.9).
- The length and thickness of the filament are characteristic of each species. However, some bacteria exhibit **biplicity**, i.e., they have flagella of two different lengths.

Composition

- Flagella are made up of a protein (**flagellin**), which is similar to keratin or myosin.
- Though flagella of different genera of bacteria have the same chemical composition, they are antigenically different.
- Flagellar antigens (known as **H antigens**) induce specific antibodies in high titres. Though flagellar antibodies are not protective, they are useful in serodiagnosis.

Arrangement (Table 3.4; Fig. 3.10)

- The presence or absence of flagella and their number and arrangement differ in different bacteria.
- Flagella may be arranged singly (**monotrichous**), in tufts (**lophotrichous**), at both poles (**amphitrichous**), all around the cell (**peritrichous**), or at one or both ends of the cell (**polar**).

Table 3.4 Different types of flagellar arrangements

Arrangement	Example
Monotrichous	*Vibrio cholerae*
Lophotrichous	*Pseudomonas fluorescens*
Amphitrichous	*Alcaligenes faecalis*
Amphilophotrichous	*Spirillum minus*
Peritrichous	*Salmonella* Typhi

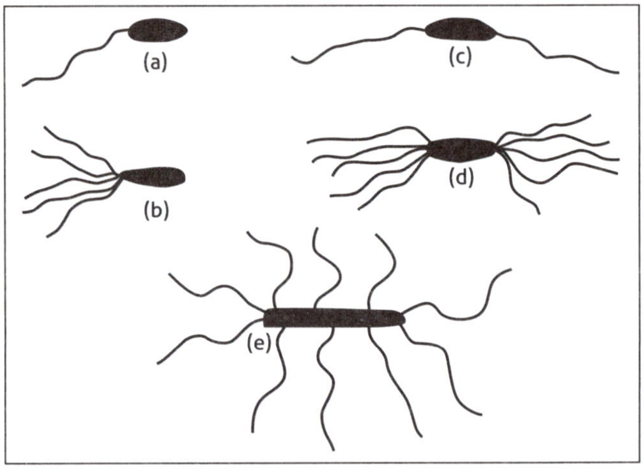

Fig. 3.10 Types of flagellar arrangements (a) monotrichous, (b) lophotrichous, (c) amphitrichous, (d) amphilophotrichous, and (e) peritrichous

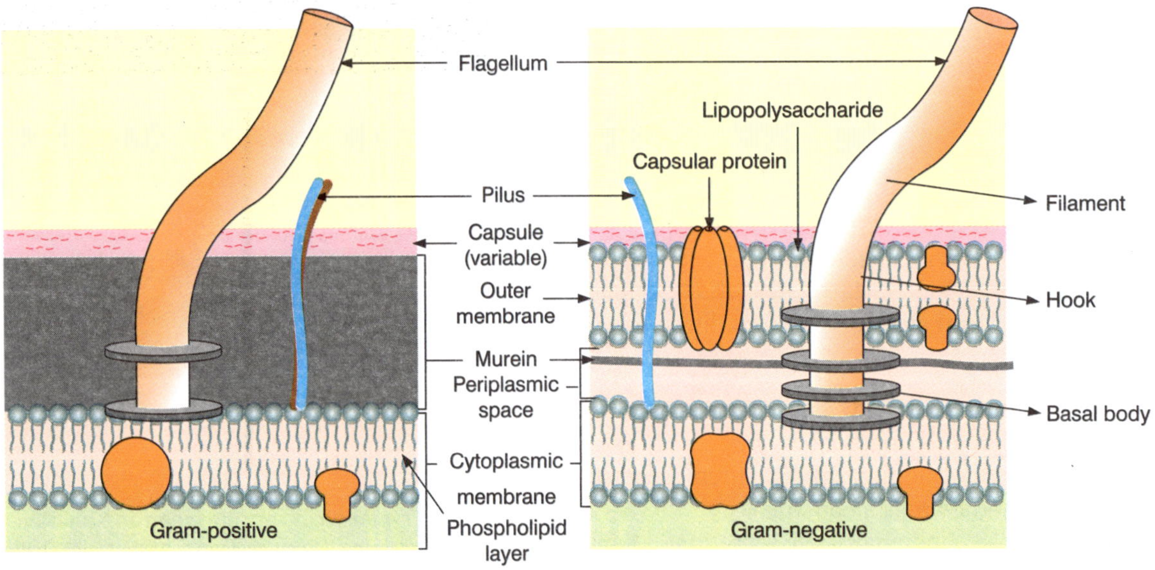

Fig. 3.9 General structure of the flagellum of gram-positive and gram-negative bacteria

Demonstration

- **Microscopy:** Flagella are less than 0.02 µm in thickness and hence beyond the limit of the resolution of the light microscope. They may, in some instances, be seen against dark-ground illumination.
  - **Staining:** Flagella can be demonstrated by special staining techniques in which their thickness is increased by mordanting (e.g., silver impregnation) or by electron microscopy.
- **Motility testing:** Due to the difficulty of demonstrating flagella directly, their presence is usually inferred from the motility of bacteria. The **hanging drop method** is used to detect motility:
  - A loopful of bacterial suspension is placed in the center of a cover slip.
  - A glass side with a plasticine ring at its centre is placed over the coverslip so that coverslip sticks to the ring on the slide.
  - Immediately, the glass slide is lifted and turned around.
  - The drop of bacterial suspension now 'hangs' on the lower surface of the coverslip. The drop is then observed under low power (10X) to focus on the edge of the drop, where bacteria tend to accumulate.
  - Once the edge is located, it is observed under 40X high power objective.
- Under the microscope, active motility must be differentiated from the passive movements of the cells, either due to air currents or due to Brownian movement.
- Bacterial motility may range from the slow, 'stately' motion of peritrichate bacteria (e.g., *Bacillus*) to the darting movement of polar flagellated vibrios (200 mm per second).
- Motility can also be inferred by the spreading type of growth of bacteria on a semisolid agar medium.

Fimbriae

- Some bacteria have very fine, hair-like surface appendages called **fimbriae** or **pili.**
- They are present on some gram-negative bacilli.
- Fimbriae are shorter and thinner than flagella (about 0.5 µm long and less than 10 nm thick) and project from the cell surface as straight filaments.
- At least **eight morphological types** of pili are known, classifiable as either **common** or **sex pili** based on their function.
- Fimbriae are made up of self-aggregating monomers of pilin.
- They originate in the cell membrane.
- They function as organelles of adhesion, helping cells to adhere firmly to particles of various kinds. This property may serve to anchor the bacteria in nutritionally favourable microenvironments.
- Fimbriated bacteria form surface pellicles in liquid media.
- Fimbriae are antigenic.
- A special type of fimbriae is the **sex pilus** (Fig. 3.11). Sex pili are longer and fewer in number than other fimbriae. They are found on 'male' (donor F^+) bacteria and help them to attach to 'female' (recipient F^-) bacteria. This attachment results in the formation of **hollow conjugation tubes** through which, it is assumed, genetic material is transferred from the donor to the recipient cell.

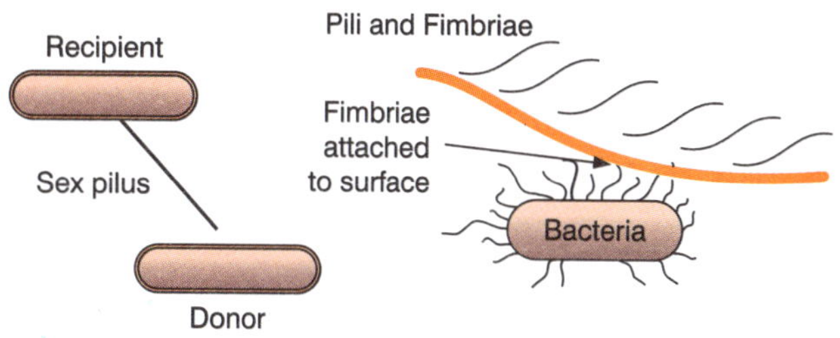

Fig. 3.11 Bacterial fimbriae and sex pili

Spores

- Some bacteria possess the ability to form highly resistant resting stages called spores (e.g., *Bacillus* and *Clostridium*).
- Each bacterium forms one spore, which, on germination, forms a single vegetative cell. Sporulation in bacteria, therefore, is not a method of reproduction.
- As bacterial spores are formed inside the parent cell, they are called **endospores.**
- While the exact stimulus for sporulation is not known, it is known to occur after a period of vegetative growth and is presumed to be related to the depletion of exogenous nutrients.

Process of sporulation and structure of the spore (Fig. 3.12a and b)

- **Forespore:** Sporulation is initiated by the appearance of a clear area, usually near one end of the cell, which gradually becomes more opaque to form the 'forespore'.
- **Spore wall:** The fully developed spore has, at its core, the nuclear body, surrounded by the spore wall, a delicate membrane from which the cell wall of the future vegetative bacterium will develop.
- **Spore cortex:** Outside the spore wall, is the thick spore cortex.
- **Spore coat:** This is a multilayered, tough coat that encloses the spore cortex.
- **Exosporium:** Some spores have an additional outer covering called an exosporium which may have distinctive ridges and grooves.

Shape and position

- Young spores are seen attached to the parent cell.
- The shape and position of the spore and its size relative to the parent cell are species-specific characteristics.
- Spores may be **central (equatorial), terminal, or subterminal**. They may be oval or spherical and may or may not distend the bacillary body (Fig. 3.13).

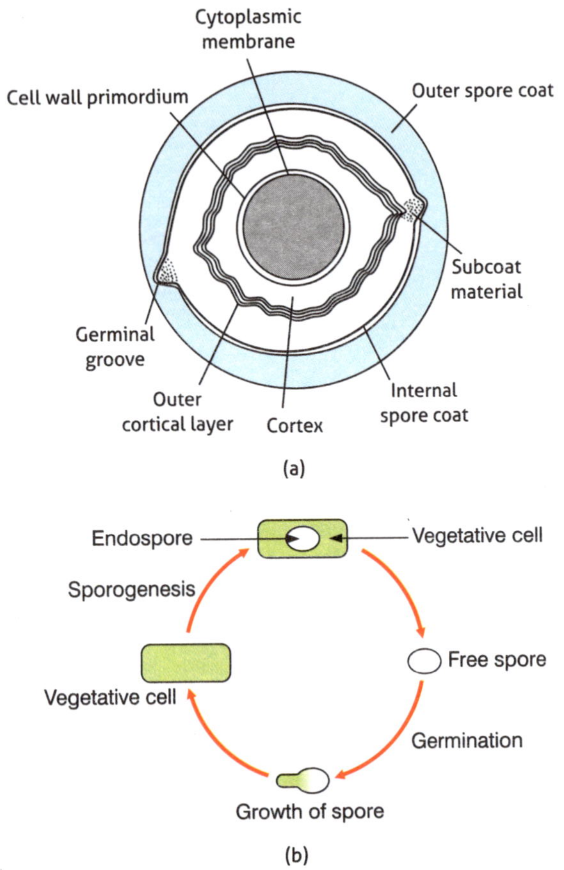

Fig. 3.12 (a) Diagrammatic representation of a bacterial spore and (b) steps of sporulation

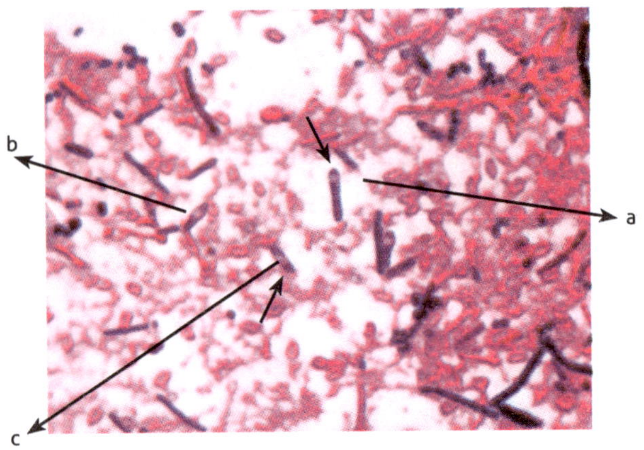

Fig. 3.13 Spores seen in stained mixed clostridial smear: (a) Terminal spore, (b) subterminal spore, and (c) central spore (*Source*: Department of Microbiology, PIMS)

Resistance

- Bacterial spores constitute some of the most resistant forms of life. They may remain viable for centuries under unfavourable conditions. They are extremely resistant to desiccation and relatively so to chemicals and heat.
- Though some spores may resist boiling for prolonged periods, spores of all medically important species are destroyed by **autoclaving at 120°C for 15 minutes**.
- Methods of sterilisation and disinfection should ensure that spores are also destroyed.
- When conditions become conducive for growth, spores germinate. First, the spore loses its refractility and swells. Then, the spore wall is shed, and the germ cell emerges after rupturing the spore coat. The germ cell then elongates to form the vegetative bacterium.

Demonstration

- Spores may be seen in unstained preparations as refractile bodies.
- The forespore stains intensely, but once the spore envelope is formed, the spore does not stain readily.
- Spores appear as unstained areas in Gram-stained preparations. However, being acid-fast, they can be stained by a **modification of the Ziehl–Neelsen technique**.

STAINING METHODS

In order to visualise bacteria either in a clinical specimen or a culture, they need to be stained and viewed under a light microscope. This visualisation is facilitated by the staining of the bacteria or the background.

- The study of morphology and staining characteristics helps in the preliminary identification of the isolate. Bacteria are stained by different dyes to produce colour contrast.
- Routine methods of staining bacteria involve drying and heat-fixing smears.
- Heat-fixing makes the smear adhere firmly to the slide and allows the sample to readily take up stains. It also protects the specimen from getting washed away by fluid stains.
- Effective staining depends on the freshness of the specimen. When smears are made from old cultures, staining characteristics either vary or are not brought out well.
- Bacteria have an affinity for basic dyes due to the acidic nature of their protoplasm.
- Some staining techniques commonly used in the bacteriology laboratory are detailed below.

1. Simple Stains

These provide colour contrast but impart the same colour to all bacteria. Simple stains are the best way to bring out the morphological features. Two simple stains are:

- Methylene blue
- Basic fuchsin

2. Negative Staining

This method creates a **dark background against which the unstained bacteria stand out in contrast**. It is useful in the demonstration of bacterial capsules, which do not take up simple stains and very slender bacteria such as spirochetes, which are not demonstrable by simple staining. The commonly used negative stains are:

- India ink
- Nigrosin

3. Impregnation Method

Cells and structures that are too thin to be seen under the ordinary microscope may be rendered visible by impregnating their surface with silver to increase the thickness. This method is used to demonstrate spirochetes and bacterial flagella. For example, the **Levaditi stain is used to visualise spirochetes** in tissues.

4. Differential Stains

These stains impart different colours to different bacteria or bacterial structures. Differential and special stains are necessary to study characteristics like flagella, capsules, spores, and metachromatic granules. The most widely used differential stains are as follows:

- **Gram stain:** It characterises bacteria as gram-positive (Fig. 3.14a and b) or gram-negative (Fig. 3.15).
- **Ziehl–Neelsen stain/acid-fast stain:** It helps to identify bacteria based on their acid-fastness.
- **Albert stain:** It imparts different colours to bacteria and granules inside them.
- **Fluorescent antibody technique:** This enables the identification of surface antigens on microorganisms with specific antibodies tagged to a fluorescent dye.

Gram stain

- It is an important preliminary step in the initial characterisation and classification of bacteria.
- It was originally devised by a Danish bacteriologist, **Hans Christian Joachim Gram**, in 1884. The procedure has undergone several modifications subsequently (Table 3.5).

Principle

Gram stain differentiates bacteria into two types:

- Gram-positive bacteria, **which resist decolourisation and retain the primary stain**, appearing violet (Fig. 3.16a)
- Gram-negative bacteria, **which are decolourised by organic solvents and take up the counter stain**, appearing red (Fig. 3.16b)

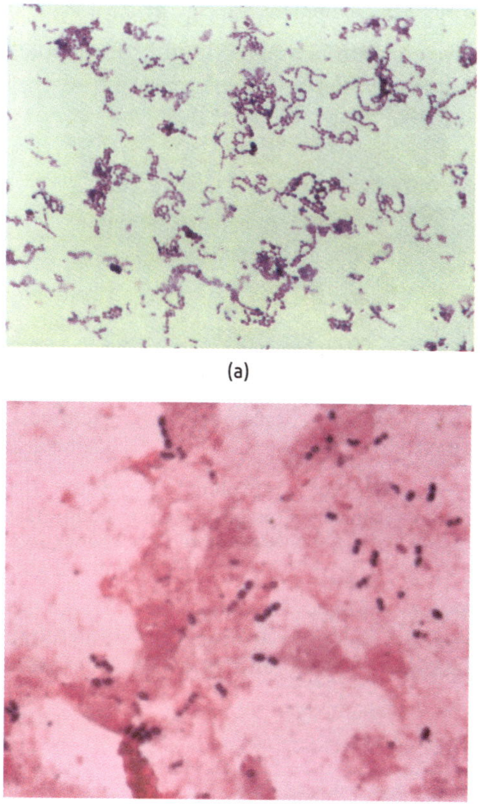

Fig. 3.14 Gram stain of a pus smear: Gram-positive, violet-coloured cocci in (a) short chains and (b) pairs with pus cells which are stained pink (*Source*: Department of Microbiology, PIMS)

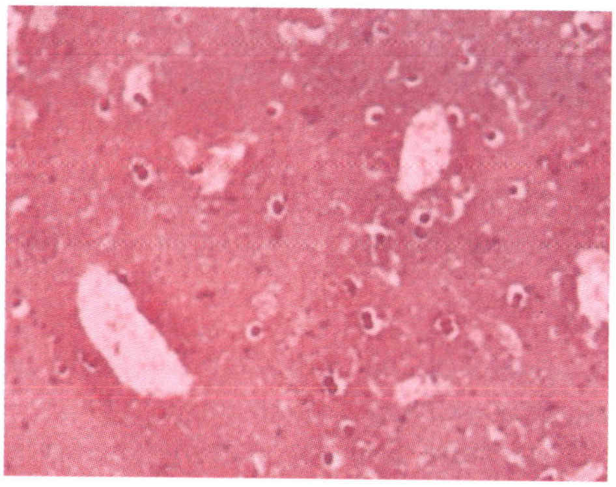

Fig. 3.15 Gram-negative diplococci (*Source*: Department of Microbiology, PIMS)

The exact mechanism of the Gram reaction is poorly understood. Various theories have been suggested:

- **pH theory:** Gram-positive cells have a more acidic protoplasm, which may account for their retaining the primary basic dye more strongly than gram-negative bacteria.
- **Cell wall theory:** The peptidoglycan layer of gram-positive bacteria is thick, and thus is able to retain the dye–iodine complex.

- **Lipid content:** The high lipid content of gram-negative bacteria makes them permeable to the secondary dye after decolourisation with organic solvents like acetone.

Table 3.5 Modifications of the Gram stain

Modification	Primary stain	Decolouriser/ Counterstain	Use
Weigert's modification	Carbol gentian violet	Aniline xylol	Useful for tissue sections
Preston and Morrell's modification	Crystal violet	Iodine acetone	Strengthens gram-positive staining
Jensen's modification	Methyl violet	Absolute alcohol/ Neutral red	Better counterstain for gonococci and meningococci
Kopeloff and Beerman's modification	Methyl violet	Acetone/ Acetone alcohol	Useful for the identification of anaerobes and bacterial vaginosis

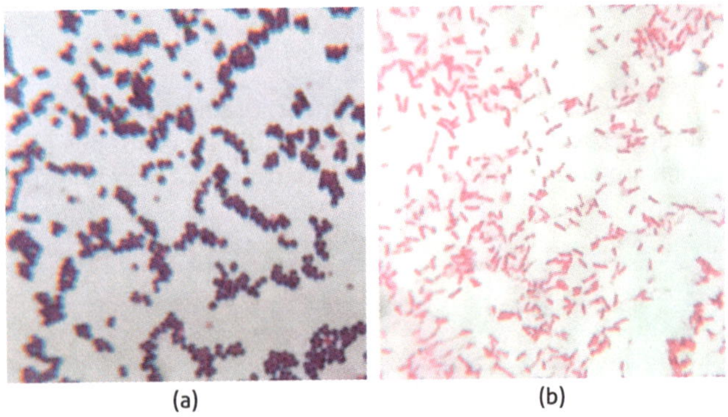

Fig. 3.16 Colour images of bacteria stained using Gram stain: (a) Gram-positive cocci and (b) gram-negative bacilli

Procedure

1. Thin smears of the specimen are made on clean, grease-free slides and air-dried.
2. The smears are **fixed** by passing the slide over a flame or by pouring cold methanol over the smear and allowing it to evaporate.
3. **Primary staining** is done with a pararosaniline dye such as crystal violet, methyl violet, or gentian violet.
4. Iodine solution (Gram's iodine) is used as a **mordant**.
5. **Decolourisation** is brought about with an organic solvent such as ethanol, acetone, or aniline.
6. **Counterstaining** is done with a dye of a contrasting colour such as carbolfuchsin, safranine, or neutral red.

Decolourisation is not an all-or-none phenomenon. Gram-positive cells may be decolourised by prolonged treatment with the organic solvent. Conversely, inadequate decolourisation may cause all cells to appear gram-positive. It should also be noted that gram-positive bacteria become gram-negative when their cell wall is damaged.

Applications of Gram staining

- Identification of bacteria and yeasts from clinical material
- Provisional differentiation of the organism as gram-positive or -negative
- Based on this empiric evidence, the appropriate empiric antibiotics may be initiated
- Rapid diagnosis of pyogenic meningitis and other acute infections by the detection of the organism from sterile body fluids and aspirates
- Other information, e.g., the presence of inflammatory cells (which may need to be quantified to see the appropriateness of the specimen)

Ziehl–Neelsen stain (acid-fast stain)

- **Paul Ehrlich** found that once they were stained with aniline dyes, tubercle bacilli resisted decolourisation with acids.
- Ehrlich's method was modified by **Franz Ziehl and Friedrich Neelsen** into the method commonly used today.

Principle

- Acid-fastness in tubercle bacteria is attributed to the presence of a variety of lipids, fatty acids, and higher alcohols in the cell envelope (Fig. 3.17a and b).
- A high-molecular-weight hydroxy acid (mycolic acid), which has wax-containing carboxyl groups, renders this acid-fast property to the cell wall.
- However, acid-fastness is not a property of lipids alone; it also depends on the integrity of the cell wall.

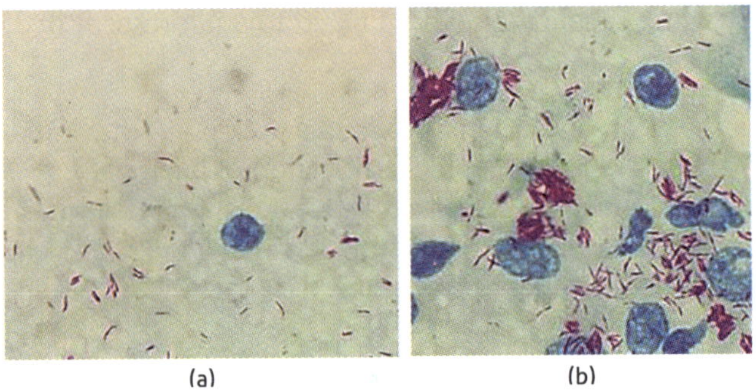

Fig. 3.17 (a) Ziehl–Neelsen acid-fast *M. tuberculosis* and (b) acid-fast *M. leprae* (*Source*: Department of Microbiology, PIMS)

Procedure

1. The smear is first stained with the **primary stain**—a solution of strong carbolfuchsin—and heat-fixed.

2. It is then **decolourised** with 25% sulphuric acid (as per the guidelines of the National Tuberculosis Elimination Programme [NTEP]).
3. The smear is further **counterstained** with a contrasting dye such as methylene blue.
4. Acid-fast bacteria retain the fuchsin (red) colour, while the others take up the counterstain.

Modified acid-fast staining

- **Gabbet's methylene blue stain:** The smear is flooded with basic fuchsin–phenol stain and allowed to stand at room temperature for ten minutes. It is then washed in running tap water and counterstained with Gabbet's methylene blue for two minutes. Finally, the smear is washed in tap water and air-dried.
- **Acid-fast stain using 5% sulphuric acid:** *Mycobacterium leprae* resist decolourisation with 5% sulphuric acid.
- **Modified Kinyoun's method:** This method is used for organisms like *Nocardia* and *Legionella*, which resist decolourisation with 1% cold sulphuric acid. The procedure is the same as that for Z–N staining, except that cold 1% sulphuric acid is used as the decolourising agent instead of a 5% or 20% solution. The bacteria appear pink against a blue background.

Table 3.6 lists the percentages of sulphuric acid used for the decolourisation of various acid-fast structures.

Table 3.6 Percentages of sulphuric acid used for the decolourisation of various acid-fast structures

Acid-fast organism/structures	Percentage of sulphuric acid
Mycobacterium tuberculosis	25
Mycobacterium leprae	5
Nocardia species, *Cryptosporidium parvum*, *Cyclospora*, *Cystoisospora*, hooklet of hydatid cyst, *Taenia saginata* eggs and segments, egg of *Schistosoma mansoni*, *Sarcocystis hominis*	1
Bacterial spore	0.25–0.5
Head of sperm, *Legionella micdadei*	0.5–1

5. Stains for Demonstrating Intracellular Granules

Some special stains are required to demonstrate the intracellular granules present in some bacteria.

Albert's stain

Principle

The granules of *Corynebacterium diphtheriae* (Fig. 3.18) take up a bluish-purple colour. These granules are called **metachromatic granules** (on account of this colour).

- Metachromasia is a **characteristic colour change that certain aniline dyes exhibit when bound to particular substances**.
- In Albert's stain, toluidine blue changes colour to bluish-purple after binding to polymetaphosphate granules.

Procedure

1. The smear is stained with Albert's I stain (which contains toluidine blue, malachite green, glacial acetic acid, and ethyl alcohol in distilled water); the smear is then drained and washed.
2. Then, Albert's II stain (which contains iodine crystals in potassium iodide dissolved in distilled water) is poured over the smear and drained.

Application

- To demonstrate *Corynebacterium diphtheriae* in throat swabs or cultures
- Neisser's and Ponder's stains are also used to demonstrate granules

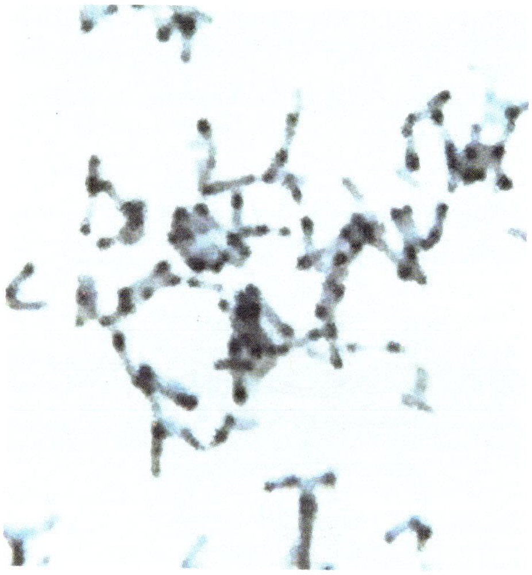

Fig. 3.18 Metachromatic granules of *C. diphtheriae* in Albert's stain (*Source*: Department of Microbiology, PIMS)

BACTERIOCINS

In 1925, **Andre Gratia** observed the production of a highly specific bacterial substance by one strain of *E. coli* which was active against another strain of the same species. These substances were later named bacteriocins.

- Chemically, they are **proteins**, but some have been associated with a lipopolysaccharide-derived substance (from cell walls).
- The synthesis of bacteriocins by bacteria is determined by the presence of colicinogenic factors (**Col factors**).
- Col factors are **episomes**, which can be transmitted from cell to cell by conjugation or transduction.

- Certain physical and chemical agents (UV rays, nitrogen, and mustard) induce colicin production by cells harbouring Col factors.
- A cell producing a bacteriocin is immune to it but may be sensitive to bacteriocins produced by other bacteria.
- Bacteriocins have specific activity, with the capacity to kill some but not all strains of a species.
- Bacteria can be identified based on the ability of bacteriocins produced by the test strain to kill standard indicator strains of bacteria.

PHYSIOLOGY OF BACTERIA

In order to understand the pathophysiology, treatment, and prevention of infectious diseases, it is necessary to understand the physiology, growth, and nutrition of bacteria.

Growth of Bacteria

Bacteria divide by **binary fission** (Fig. 3.19). When a bacterial cell reaches a certain size, it divides to form two daughter cells. Nuclear division precedes cell division and, therefore, in a growing population, many cells carrying two nuclear bodies can be seen.

The cell divides by a constrictive or pinching process, or by the in-growth of a transverse septum across the cell. In some species, the daughter cells may remain partially attached after division.

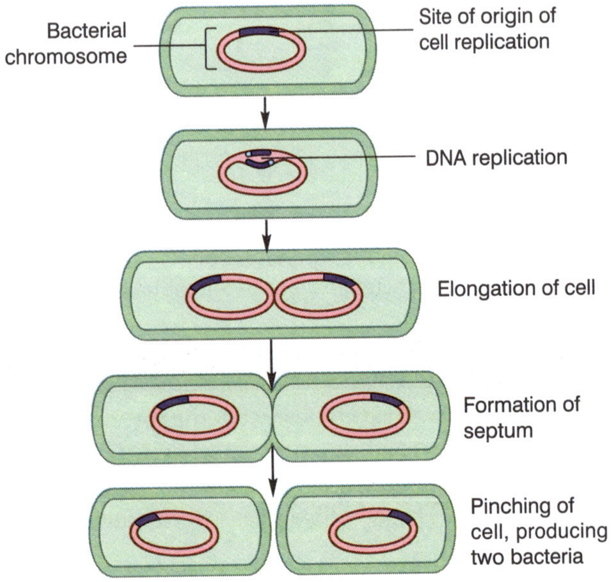

Fig. 3.19 Bacterial cell division by binary fission

Generation time

- It is the time required for a bacterium to give rise to two daughter cells under optimum conditions. It is also known as the **population doubling time**.
- In coliform bacilli and many other medically important bacteria, the generation time is about **20 minutes**.
- Some bacteria are slow-growing—the generation time in tubercle bacilli is about 20 hours and in lepra bacilli, it is as long as 20 days.
- As bacteria reproduce rapidly and by geometric progression, a single bacterial cell can theoretically produce 1,021 progeny in 24 hours!

Bacterial growth curve

- When grown on a solid media, bacteria form colonies. **Each colony represents a clone of cells derived from a single parent cell.** In liquid media, growth is diffuse.
- When a bacterium is seeded into a suitable liquid medium and incubated, its growth follows a definite course.
- If bacterial counts are made at regular intervals after inoculation and plotted in relation to time, a growth curve is obtained. The various stages of the growth curve are associated with **morphological and physiological alterations** of the cells (Fig. 3.20).

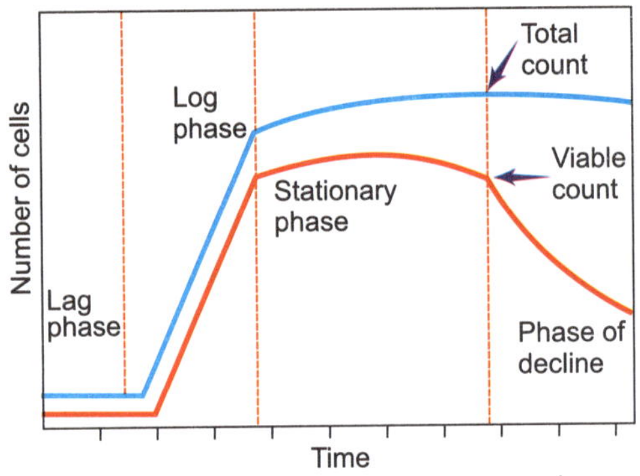

Fig. 3.20 Bacterial growth curve showing the lag, log, stationary, and decline phases of bacterial growth; in the total count, the phase of decline is not evident

The bacterial growth curve is characterised by the following phases:

Lag phase

- During the time immediately following the seeding of a culture medium, there is no increase in the number of bacteria, though there may be an increase in the size of the cells.
- Bacteria adapt to their new environment and produce the necessary enzymes and metabolic intermediates required for multiplication.
- The duration of this phase varies with the species of the bacteria, size of the inoculum, nature of the culture medium, and environmental factors such as temperature.
- **Bacterial cells attain their maximum size at the end of the lag phase.**

Log (logarithmic) or exponential phase

- The bacterial cells start dividing, and their numbers increase with time, either exponentially or by geometric progression.

- When the viable count is plotted against time, a straight line is obtained.
- In this phase, cells are smaller and stain uniformly.
- **Bacteria are more susceptible to the action of antibiotics during the log phase.**

Stationary phase

- After a varying period of exponential growth, cell division stops due to the depletion of nutrients and the accumulation of toxic products.
- The number of progeny cells is just enough to replace the number of dying cells in this phase.
- The viable count remains stationary as an equilibrium exists between the dying cells and new cells.
- Cells are frequently gram-variable and have intracellular storage granules.
- Some bacteria undergo **sporulation** at this stage while some bacteria produce secondary metabolic products such as exotoxins and antibiotics during the stationary phase; others undergo sporulation at this time.

Phase of decline

- The bacterial population decreases due to cell death.
- Besides nutritional depletion and toxin accumulation, cell death is also brought about by the presence of autolytic enzymes.
- For most bacteria. the total count parallels the viable count until the stationary phase and continues steadily without any phase of decline.
- However, autolytic bacteria show a phase of decline in total count also.
- **Involution forms** are common.

Bacterial counts

The **total count** gives the total number of cells in the sample, irrespective of whether they are living or not. This number is obtained by:

- Direct counting under the microscope using counting chambers
- Counting in an electronic device such as the **Coulter counter**

The **viable count** is a measure of the number of living cells. Viable counts are obtained by the dilution and plating methods.

- **Dilution method:** The bacterial suspension is diluted to a point beyond which unit quantities do not yield growth when inoculated into suitable liquid media (extinction). This method does not give accurate values but is used widely in water bacteriology for the estimation of the '**presumptive coliform count**' in drinking water.
- **Plating method:** Appropriate dilutions of the bacterial suspension are inoculated on solid media, either on the surface of plates or as pour plates. The number of colonies that develop after incubation is used to estimate the viable count.

Bacterial Nutrition

- Nutritional requirements for the growth and multiplication of bacteria are **water** (80% of the total bacterial weight), a source of **carbon**, a source of **nitrogen,** and some **inorganic salts**.
- All bacteria require small amounts of carbon dioxide for growth. This requirement is usually met by the carbon dioxide present in the atmosphere or that which is produced endogenously by cellular metabolism.
- **Capnophilic bacteria** like *Campylobacter* and *Brucella abortus* require much higher levels of carbon dioxide.

Energy Requirement

- **Phototrophs:** Derive energy from sunlight
- **Chemotrophs:** Derive energy from chemical reactions
- **Autotrophs:** Synthesise their own organic material
- **Heterotrophs:** Unable to synthesise their own metabolites, and depend on preformed ones; they also cannot utilise atmospheric carbon dioxide as the sole source of carbon

Oxygen Requirement

Based on the requirement of oxygen for growth and viability, bacteria can be categorised as follows:

- **Obligate aerobes:** These bacteria grow only in the presence of oxygen and obtain their energy and intermediates only through oxidation, with oxygen as the ultimate hydrogen acceptor. The carbon and energy source may be completely oxidised to carbon dioxide and water. Examples are *Vibrio cholerae* and *Pseudomonas*.
- **Facultative anaerobes:** These organisms are ordinarily aerobic but can also grow in the absence of oxygen, though less abundantly. Most bacteria of medical importance are facultative anaerobes.
- **Anaerobic bacteria:** These do not grow in the presence of oxygen. Anaerobes use hydrogen acceptors, nitrates, or sulphates instead of oxygen. Anaerobic bacteria produce energy by producing energy-rich phosphate bonds and converting adenosine diphosphate (ADP) to adenosine triphosphate (ATP). This process is known as **oxidative phosphorylation**. Facultative anaerobes do the same in the absence of oxygen. The reason for the apparent toxicity of oxygen for anaerobic bacteria is not well understood. The following are some theories that have been proposed to explain this phenomenon:
  - In the presence of oxygen, hydrogen peroxide and other toxic peroxides accumulate in bacterial cells. In most aerobic bacteria, the enzyme catalase splits this hydrogen peroxide. Anaerobic bacteria

do not produce this enzyme and hence are not equipped to break down the toxic peroxides.
  - Another possible explanation is that obligate anaerobes possess essential enzymes that are active only in the reduced state.
- **Microaerophilic bacteria:** These grow best in the presence of low oxygen tension. *Helicobacter* belongs to this group of bacteria.

Fermentation

This is a process of anaerobic metabolism leading to the formation of organic end products such as acids, alcohols, and gases (carbon dioxide and hydrogen), producing energy-rich phosphate bonds as a result of organic phosphates getting incorporated into intermediate metabolites. This process is known as **substrate-level phosphorylation**. These phosphates are used to convert ADP to ATP.

Temperature Requirement

Bacteria vary in their requirement of temperature for growth and have a distinct 'temperature range'. **Optimum temperature** is the temperature at which growth occurs best. Pathogenic bacteria have an optimum temperature of 37°C.

- **Mesophilic bacteria:** These grow best at temperatures of 25– 40°C. All bacteria are mesophilic.
- **Psychrophilic bacteria:** These grow best at temperatures below 20°C, sometimes even as low as –7°C. These bacteria are soil and water saprophytes and, though not of direct medical importance, may cause the spoilage of refrigerated food, e.g., *Chryseobacterium* and *Pseudomonas*.
- **Thermophiles:** These grow best at high temperatures of 55–80°C. They may cause spoilage of under-processed canned food. Some thermophiles (like *Bacillus stearothermophilus*) form spores that are exceptionally thermoresistant. Extremely thermophilic bacteria can grow at temperatures as high as 250°C.

Thermal death point

This is the **lowest temperature that kills a bacterium under standard conditions within a given time.** Under moist conditions, most vegetative, mesophilic bacteria have a thermal death point between 50 and 65°C; most spores have a thermal death point between 100 and 120°C.

At low temperatures, some bacterial species die rapidly, whereas others survive well. Storage in the refrigerator (3–5°C) or the deep freezer (–30 to –70°C) preserves cultures for a variable length of time.

Rapid freezing with solid carbon dioxide or a stabiliser such as glycerol minimises the death of cells on freezing.

Other Requirements for Growth

- **Moisture:** Water is an essential component of the bacterial protoplasm and hence, drying is lethal to cells. However, the effect of drying varies in different species. Some delicate bacteria like *Treponema pallidum* are highly sensitive, while others like staphylococci withstand drying for months. Spores are particularly resistant to desiccation and may survive in the dry state for several decades.
- **Hydrogen ion concentration:** Most pathogenic bacteria grow best at neutral or slightly alkaline conditions (pH 7.2–7.6). Some acidophilic bacteria such as lactobacilli grow in acidic conditions. Strong solutions of acid or alkali readily kill most bacteria, though mycobacteria are exceptionally resistant to them.
- **Light:** Bacteria (except the phototrophic species) grow well in the dark. They are sensitive to ultraviolet light and other radiations.
- **Osmotic effect:** Bacteria are more tolerant to osmotic variation than most other cells due to the mechanical strength of their cell walls.
- **Mechanical and sonic stress:** Though bacteria have tough cell walls, they may be ruptured by mechanical stress and exposure to ultrasonic vibration.

CULTURE AND IDENTIFICATION OF BACTERIA

Microorganisms need to be cultivated in pure form for several reasons. This is done by inoculating samples on growth media.

Culture Media

Essential ingredients of media

- **Peptone**—a common ingredient of culture media; it is a complex mixture of partially digested proteins and is made up of proteoses, polypeptides, and amino acids
- A variety of **inorganic salts** including phosphates, potassium, and magnesium
- Certain accessory growth factors such as riboflavin
- Blood, serum, and yeast extract are other common ingredients of culture media
- **Solidifying agents**—usually agar (or agar-agar); other solidifying agents include egg, serum, or gelatin

Agar forms the base of solid media. It is obtained from a type of seaweed. It has virtually no nutritive value and is not affected by the growth of bacteria. It melts at 98°C and usually solidifies at 42°C. Approximately 2% agar is used in most solid media.

Classification

Bacterial culture media are classified based on the functional and growth requirements they fulfil (Table 3.7).

Table 3.7 Classification of culture media

Type of culture media	Examples
Based on consistency	
Liquid	Brain–heart infusion broth, peptone water, nutrient broth
Solid	Nutrient agar, blood agar, chocolate agar
Semi-solid (sloppy 0.5-1% agar)	Motility agar, bacterial stock cultures
Based on constituents	
Simple	Non-nutrient agar, nutrient agar
Complex	Thiosulphate citrate bile salt sucrose agar (TCBS)
Synthetic or defined	Hank's balanced salt solution (used for virus transport)
Based on the functional requirements they meet	
Enriched	Blood agar, chocolate agar, brain–heart infusion agar and broth, Todd–Hewitt broth
Enrichment	Tetrathionate broth, Selenite F medium (for fecal samples—to suppress gut commensals)
Selective	*Salmonella—Shigella* agar (to isolate *Salmonella* and *Shigella* from fecal samples), deoxycholate citrate agar (DCA), thiosulphate citrate bile sucrose agar (TCBS; Fig. 3.21a and b), Lowenstein–Jensen medium (Fig. 3.22)
Indicator	MacConkey agar (differentiates lactose fermenters from non-lactose fermenters), potassium tellurite blood agar (PTBA; Fig. 3.26).
Differential	Mannitol salt agar (differentiates *Staphylococcus aureus* from *Staphylococcus epidermidis*) MacConkey medium differentiates lactose fermenters, which form pink colonies (Fig. 3.27a and b) from non-lactose fermenters (Fig. 3.28), which produce colourless or pale colonies Cysteine lactose electrolyte deficient (CLED) medium (Fig. 3.29)
Transport	Stuart's transport medium (for transporting throat and genital tract swabs collected to isolate pathogens from), buffered glycerol saline, Cary–Blair medium
Anaerobic media	Robertson's cooked meat medium, thioglycolate medium (for the cultivation of *Bacteroides*, *Clostridium*, etc.)

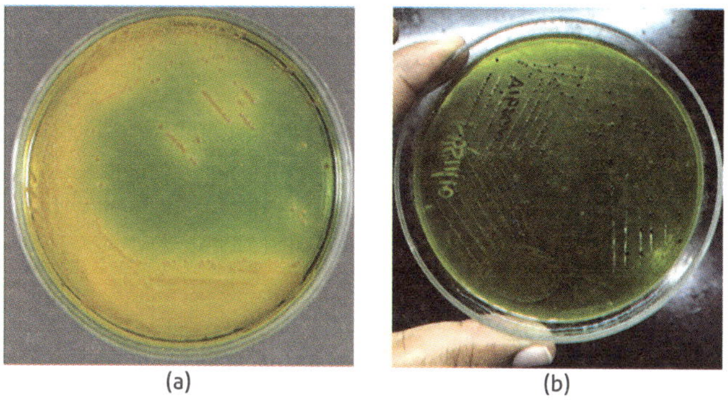

Fig. 3.21 TCBS agar with (a) yellow colonies of *V. cholerae* (due to fermentation of sucrose in the medium) and (b) green colonies of *V. parahaemolyticus*—sucrose is not fermented (*Source*: Department of Microbiology, PIMS)

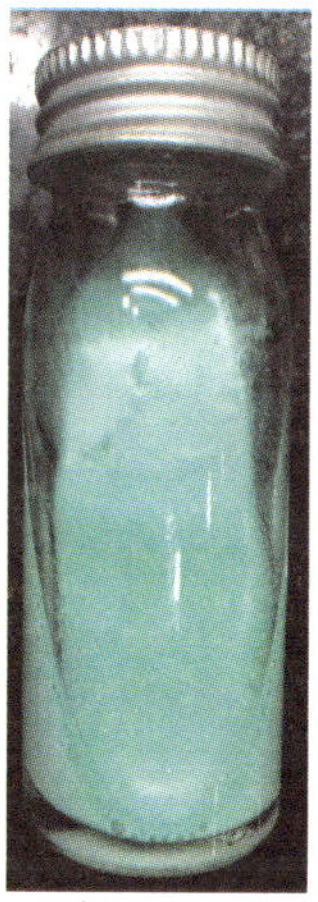

Fig. 3.22 Lowenstein–Jensen medium (*Source*: Department of Microbiology, PIMS)

Grouping based on constituents

- **Simple media (basal media):** Nutrient broth consists of peptone, meat extract, sodium chloride, and water. It is used to grow bacteria to demonstrate motility. **Nutrient agar** is made by adding 2% agar to nutrient broth.
- **Complex media:** These have added ingredients for special purposes such as bringing out certain characteristics or for providing special nutrients required for the growth of the bacterium under study.
- **Synthetic or defined media:** The constituents are clearly defined in terms of each ingredient. These chemically defined media are prepared from pure chemical substances, e.g., minimum essential medium (MEM), which is used in tissue culture.

Grouping based on functional requirement

1. Enriched media

Nutritional requirements, such as blood, serum, or eggs are added to a basal medium to cultivate exacting or fastidious bacteria. Examples are blood agar, chocolate agar, and brain–heart infusion broth (BHIB).

- **Blood agar:**
  - This is a solid culture medium consisting of agar, peptone, and sheep blood.
  - It supports the growth of most aerobic and anaerobic bacteria (vitamin K, cysteine, and hemin supplementation enhances the growth of anaerobic bacteria) and fungi.
  - It can indicate the degree of hemolysis caused by hemolysin. It is used to differentiate gram-positive cocci. For this reason, it is also known as a **differential medium**.
    - **Beta-hemolysis:** This refers to the complete lysis of red blood cells and hemoglobin and appears as the complete clearing of the blood agar medium surrounding the colonies (Fig. 3.23).
    - **Alpha-hemolysis:** This is the partial lysis of red blood cells and hemoglobin, which produces a greenish discolouration of the blood agar around the colonies (Fig. 3.24).

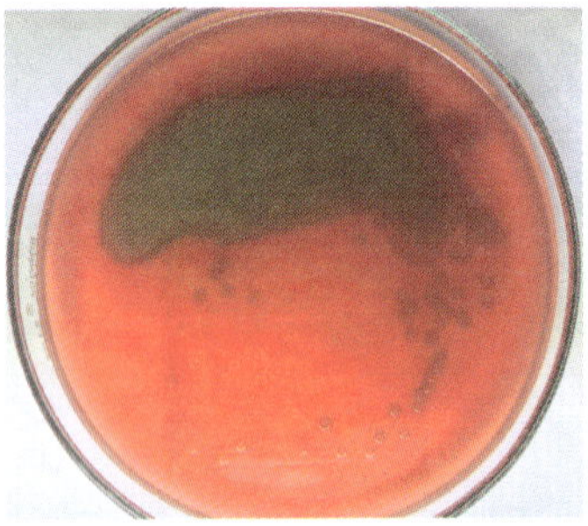

Fig. 3.23 Alpha-hemolytic *Streptococcus pneumoniae* cultured on blood agar (*Source*: Department of Microbiology, PIMS)

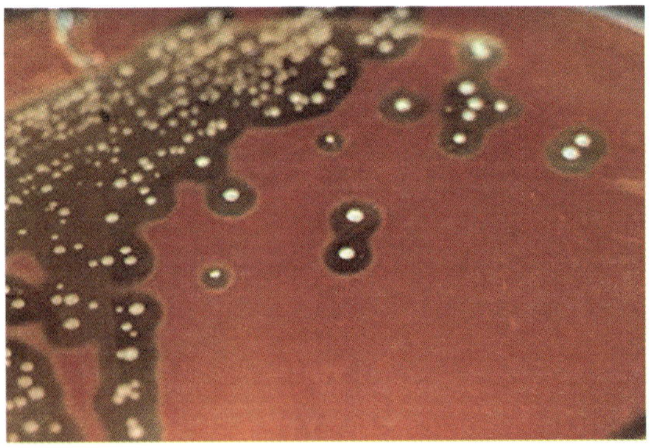

Fig. 3.24 *Streptococcus pyogenes* colonies exhibiting beta-hemolysis on blood agar (*Source*: Department of Microbiology, PIMS)

- **Chocolate agar:** This is prepared by heating a mixture of sheep blood and nutrient agar (Fig. 3.25). This results in the release of hemoglobin, a related substance, hemin (also called X factor), and nicotinamide adenine dinucleotide (**NAD; called the V factor**) during the process of heating. Chocolate agar is used to grow fastidious organisms.

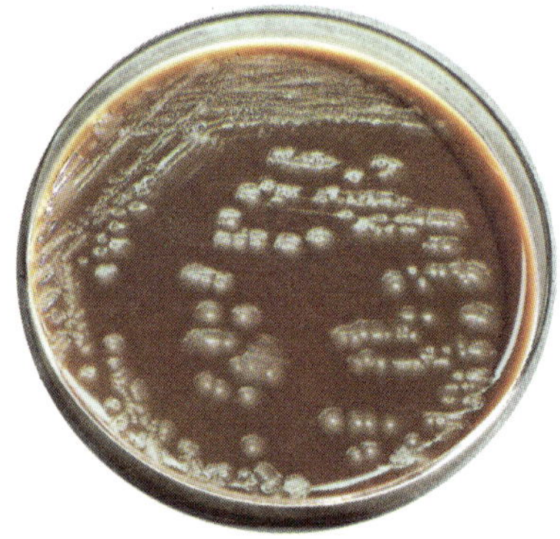

Fig. 3.25 Chocolate agar with luxuriant growth (*Source*: CDC, PHIL, Image ID 12472)

- **Brain–heart infusion broth:** This is a highly nutritious, buffered fluid culture medium that is suitable for the cultivation of fastidious organisms.

2. *Enrichment media*

- Commensal bacteria are suppressed while allowing the pathogen to remain viable and grow.
- These media are used for specimens with mixed flora, e.g., for the isolation of diarrheagenic bacteria from a fecal sample.
- Substances that have an inhibitory effect on the commensals are incorporated into the medium.

3. *Selective media*

An inhibiting substance added to a solid medium enables the required bacterium to form colonies while inhibiting other bacteria.

4. *Indicator media*

- These media contain an indicator that changes colour when a bacterium grows on them, e.g., sulphite is the indicator in the Wilson–Blair medium; *Salmonella typhi* reduces this sulphite to produce colonies that have a black, metallic sheen; PTBA is used for the identification of *C. diptheriae* (Fig. 3.26).

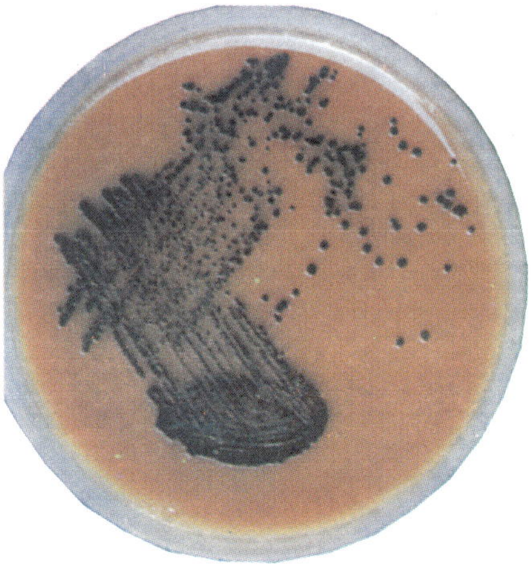

Fig. 3.26 Potassium tellurite agar (McLeod's medium) with black colonies of *C. diphtheriae*

5. Differential media

These contain indicator dyes that help to differentiate different bacteria (Fig. 3.27a and b, Fig. 3.28, and Fig. 3.29).

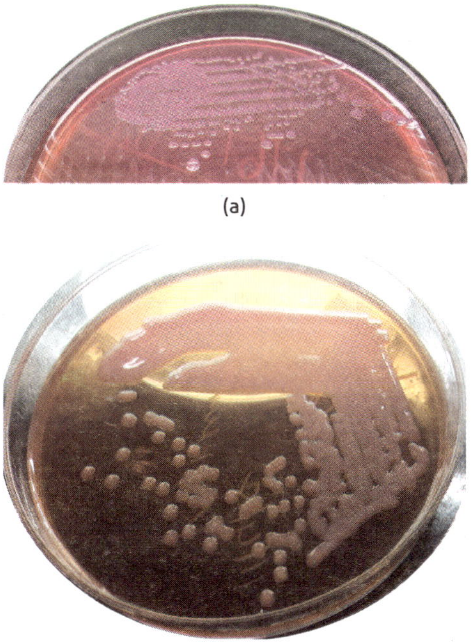

(a)

(b)

Fig. 3.27 (a) MacConkey agar with flat, smooth, pink colonies of *Escherichia coli* and (b) large, mucoid colonies of *K. pneumoniae* (*Source*: Department of Microbiology, PIMS)

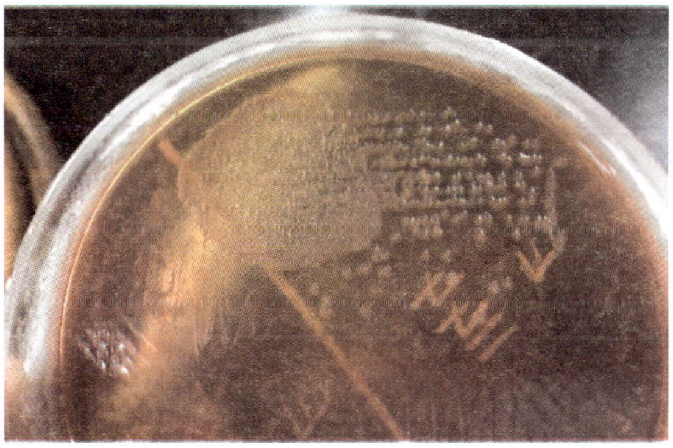

Fig. 3.28 MacConkey agar with flat, smooth, colourless colonies of *Salmonella* (*Source*: Department of Microbiology, PIMS)

6. Transport media

- These are devised to transport specimens suspected to contain fastidious organisms.
- They may contain a reducing agent such as sodium thioglycolate to prevent oxidation and charcoal to neutralise certain bacterial inhibitors.

7. Anaerobic media

- These are used to cultivate anaerobes like *Clostridium*, *Bacteroides*, etc.
- These media contain reducing agents.

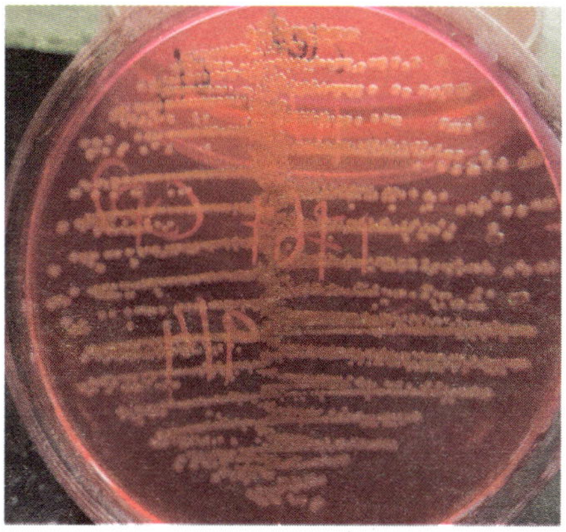

Fig. 3.29 CLED medium showing pink colour colonies of *Escherichia coli* cultured from urine (*Source*: Department of Microbiology, PIMS)

8. Defined or synthetic media

- These are media whose constituents are clearly defined in terms of each ingredient.
- These chemically defined media are prepared from pure chemical substances, e.g., minimum essential medium (MEM—used in tissue culture).
- Some media commonly used in a clinical microbiology laboratory are presented in Table 3.8.

Table 3.8 Common media used in the clinical microbiology laboratory

Media	Specimen	Organisms
Nutrient agar/broth Peptone water	Not used for direct specimen inoculation	All non-fastidious bacteria Nutrient agar: To demonstrate pigment Broth: Used for culture to demonstrate motility
Blood agar	All aspirates, body fluids, blood, exudates, urine (except feces)	All common pathogenic bacteria and yeast (except *Haemophilus* and nutritionally-deficient bacteria)
Chocolate agar	All samples (except feces and urine)	All common pathogenic bacteria and yeast
MacConkey agar	All samples, including feces	All common gram-negative pathogens, *Enterococcus*
DCA	Feces	Enteric pathogens—*Salmonella, Shigella*
CLED medium	Urine	All urine pathogens
TCBS agar	Feces (suspected cholera)	*V. cholerae*
PTBA	Throat swab (in suspected diphtheria)	*Corynebacterium diphtheriae*
RCM/ Thioglycolate	All samples for anaerobic pathogens (except feces)	For most anaerobic pathogens—*Clostridium, Bacteroides*

Culture Methods

Bacteria need to be isolated in the laboratory for the following reasons:

- To establish etiology based on clinical samples
- For antibiotic susceptibility testing to start appropriate treatment
- To study virulence factors
- To study antigenic properties and antigen preparation
- For pharmaceutical, industrial, and agricultural purposes

Inoculation methods on culture media

- **Loops or straight wires** made of platinum or nichrome (24 SWG) are used to transfer material onto culture plates or tubes.
- The loops are sterilised by heating them over a flame until they are red hot and then cooled before picking up the material. One loopful of the specimen is transferred onto the surface of a well-dried plate. The inoculum is then distributed thinly over the plate by streaking it with the loop in a series of parallel lines on different segments of the plate (Fig. 3.30a and b).

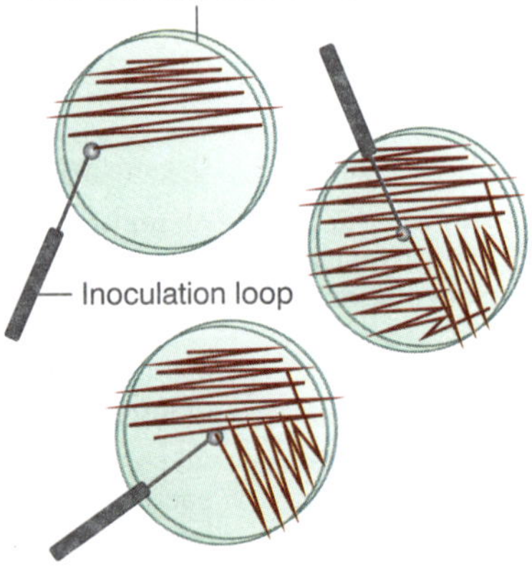

(a)

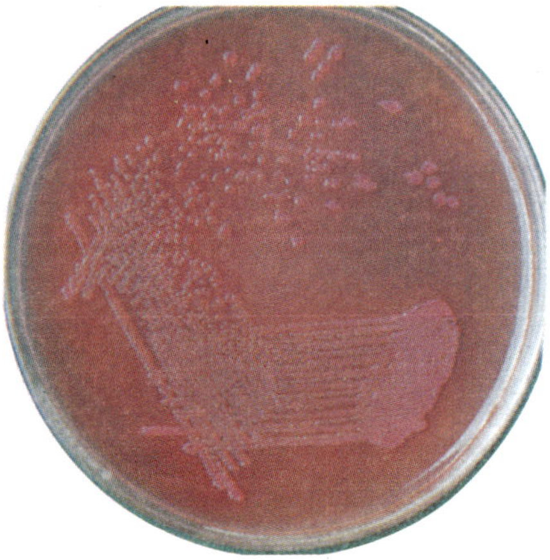

(b)

Fig. 3.30 (a) and (b) Streak cultures

Incubation

- Culture plates and tubes with liquid culture are incubated at 37°C in a **bacteriological incubator** (Fig. 3.31) overnight or longer before reading the plates/tubes to look for growth in the form of colonies (on solid media) or turbidity.
- **Candle jar:** It is used to provide an atmosphere containing 5–7% CO_2. This jar is used for the culture of capnophilic bacteria like *Haemophilus influenzae, Streptococcus pneumoniae*, and *Neisseria meningitidis*. It is a container into which a burning candle is introduced before sealing the container's airtight lid. The candle consumes oxygen and produces carbon dioxide. Sophisticated CO_2 incubators can also be used to provide excess carbon dioxide.

Fig. 3.31 Bacteriological incubator set to a temperature of 37°C for optimal growth of most pathogenic bacteria (*Source*: Department of Microbiology, PIMS)

Anaerobic culture methods

Anaerobic bacteria differ in their requirements of and sensitivity to oxygen. Growth conditions may be different for strict anaerobes (*Clostridium tetani*) and aerotolerant bacteria (*C. histolyticum*). Some of the commonly used methods of anaerobiosis are listed below.

McIntosh and Filde's anaerobic jar

- Inoculated culture plates are placed in the jar, and the lid is clamped tight.

- After suctioning the air out of the jar, it is filled with a mixture of gases (consisting of 5% CO_2, 10% H_2, and 85% N_2).
- Alumina pellets coated with palladium, kept dry in a sachet within the jar, act as catalysts at room temperature. They catalyse the conversion of any remaining oxygen to water.
- Reduced methylene blue is used as an indicator for verifying the anaerobic condition in the jar (Fig. 3.32). It remains colourless anaerobically but turns blue on exposure to oxygen.

Fig. 3.32 McIntosh and Filde's anaerobic jar

Anoxomat

It is an automated microprocessor-controlled system for the cultivation of anaerobic, microaerophilic, and capnophilic bacteria.

Gaspak

The Gaspak (Fig. 3.33a and b) is commercially available as a disposable envelope containing chemicals that generate hydrogen and carbon dioxide upon the addition of water. A catalyst permits the combination of hydrogen and oxygen to produce an anaerobic environment. The Gaspak is simple and effective.

Pre-reduced anaerobic system (PRAS)

Pre-reduced media are commercially available for the cultivation of fastidious anaerobes.

Anaerobic chamber ('glove box')

- An anaerobic chamber is an airtight, glass-fronted cabinet filled with an inert gas.
- It has a locked entry for the introduction and removal of materials, and glove fittings (Fig. 3.34) to allow the operator to manipulate objects inside without disturbing the environment.

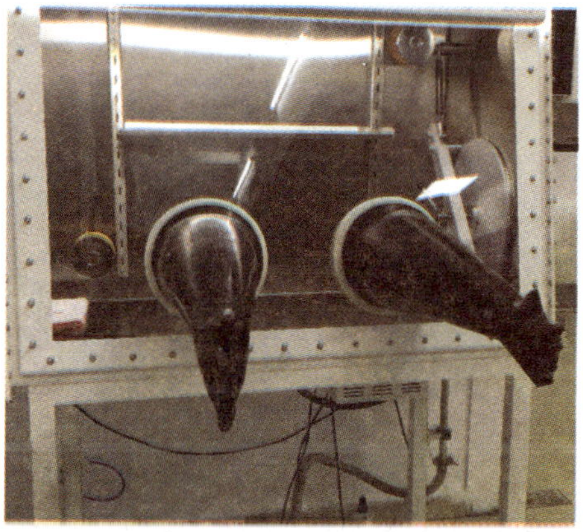

Fig. 3.34 A glove box (*Source*: Wikimedia Commons)

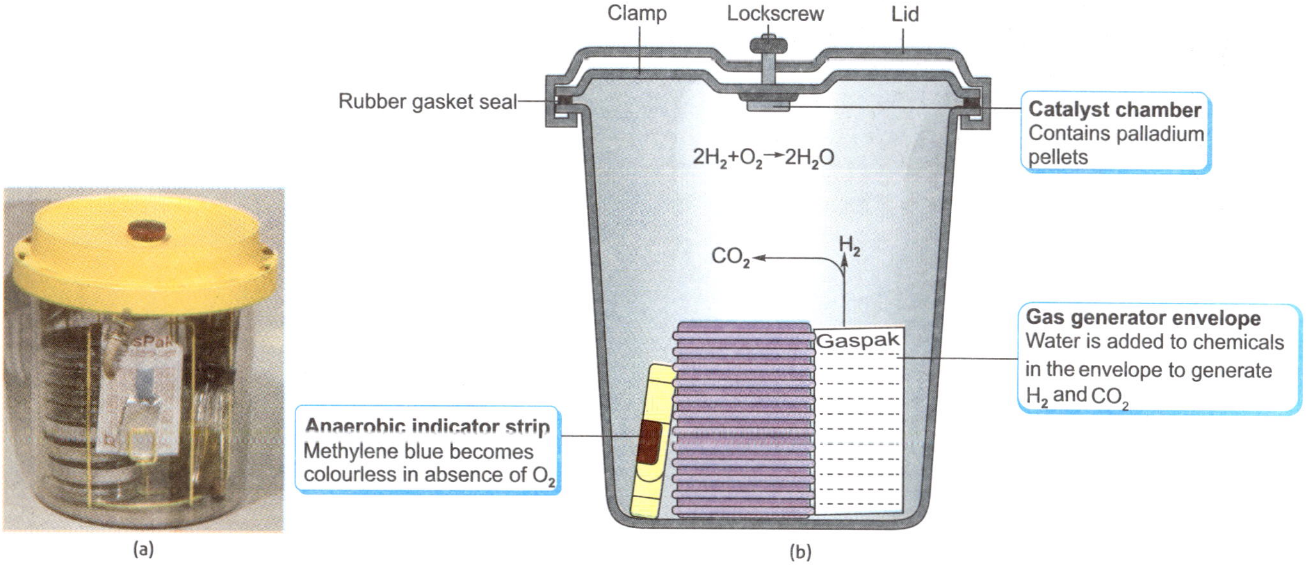

Fig. 3.33 (a) and (b) Gaspak airtight anaerobic jar (*Source*: [a] Department of Microbiology Laboratory, PIMS)

Other methods of anaerobiosis

- **Robertson's cooked meat (RCM) medium:** This is a liquid medium used for the culture and transport of anaerobes. It consists of fat-free minced cooked meat in a broth (Fig. 3.35). It permits the growth of even strict anaerobes. Additionally, saccharolytic or proteolytic activities of such organisms are indicated by the meat turning red or black, respectively.
- **Thioglycolate broth with hemin and vitamin K:** This is an enriched liquid medium used for culturing and transporting anaerobic and microaerophilic bacteria (Fig. 3.36a, b, and c).

Fig. 3.35 Robertson's cooked meat medium—widely used medium for transport and culture of anaerobic organisms (*Source*: Dept of Microbiology, PIMS)

Fig. 3.36 Growth in thioglycolate broth: (a) Uniform turbidity, (b) surface growth, and (c) granular turbidity

Preservation of Microorganisms

Bacteria and viruses can be preserved by the following methods:

- **Freezing:** Microorganisms can be frozen at –80°C or preserved in liquid nitrogen at a temperature of –195.79°C (–320°F).
- **Lyophilisation/cryodessication:** Rapid freezing and thawing using a lyophiliser is another method of preserving microorganisms. In cryodessication, bacteria are dried in a vacuum at very low temperatures.
- **Sub-culturing:** Bacteria may be sub-cultured on solid/ liquid media to preserve them for a short duration.

Identification of Bacteria (Laboratory Diagnosis)

Once a pure culture of a bacterium is obtained, it must be identified.

- Identification comprises determining the morphology of the organism by staining, noting cultural characteristics, performing biochemical tests, confirming the organism by immunological methods, and testing for pathogenicity (virulence) in some instances (Fig. 3.37).
- Molecular detection and confirmation by nucleic acid-based methods are recent advances in diagnosis.
- Several automated systems have become useful in speeding up the process of isolation and confirmation of clinical isolates.
- The following methods based on bacterial characteristics are used for identification.

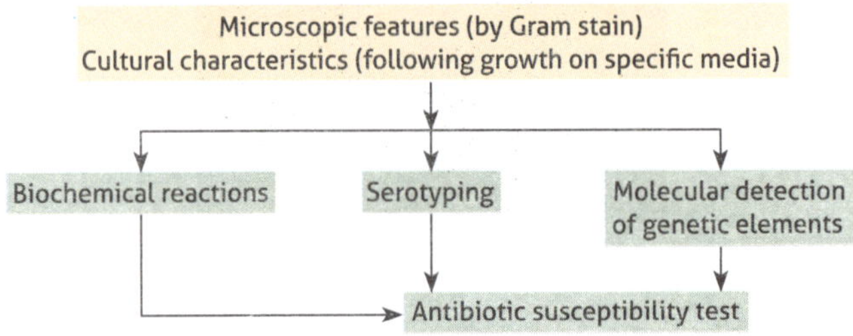

Fig. 3.37 Flowchart representing the steps involved in the identification of bacteria

Cultural characteristics

These provide additional information for the identification of the bacterium. The characteristics revealed in different types of media are noted.

Solid media

The following colony characteristics are noted on solid media:

- Shape—circular, irregular, or rhizoid
- Size—in millimetres
- Elevation—effuse, elevated, convex, concave, umbonate, or umbilicate
- Margins—bevelled or otherwise
- Surface—smooth, wavy, rough, granular, papillate, or glistening
- Edges—entire, undulate, crenated, fimbriate, or curled
- Colour—golden-yellow, white, buff, pink, etc.
- Structure—opaque, translucent, or transparent
- Consistency—membranous, friable, butyrous, or viscid
- Emulsifiability

Liquid media

In liquid culture, the quantum of growth, turbidity, presence of deposit and its character, and the nature of surface growth such as pellicle formation is noted.

Biochemical tests

Tests based on the fermentation of carbohydrates and other biochemical properties are used for the characterisation of bacteria. Some important and commonly used tests for bacterial identification and reference and described below.

Carbohydrate fermentation

- Media containing a single carbohydrate (sugar) are used for these tests. **The term 'sugar' in microbiology denotes any fermentable substance.**
- The usual sugar media consist of 1% sugar in peptone water along with an appropriate indicator.
- A small tube (**Durham's tube**) is kept inverted in the sugar tube to detect the production of gas, which is indicated by the presence of a bubble inside the tube.
- Acid production is indicated by a change in the colour of the medium to pink or red depending on the indicator used.
- For organisms that are exacting in their growth requirements (for example, *N. gonorrhoeae*), Hiss' serum sugars (containing 3% serum along with the carbohydrate) are used.

Indole production (Fig. 3.38)

This test demonstrates the production of indole from tryptophan.

- **Indole-positive:** A red ring indicates a positive reaction when Kovak's reagent is added (e.g., *E. coli*).
- **Indole-negative:** A yellow ring is formed on the surface of the broth (e.g., *Klebsiella*).

Fig. 3.38 Indole production test—the red ring indicates a positive reaction, whereas the yellow ring indicates a negative result (*Source*: Department of Microbiology, PIMS)

Methyl red test (MR)

This test is used to detect the production of acid during the fermentation of glucose and the maintenance of pH below 4.5.

- **MR-positive:** A red colour denotes positivity (e.g., *E. coli*).
- **MR-negative:** Yellow signifies a negative test (e.g., *Klebsiella* species).

Voges–Proskauer test (VP)

This test demonstrates the production of acetyl methyl carbinol from pyruvic acid as an intermediate stage, which reacts with the peptone of the broth to produce a red colour.

- **VP-positive:** In a positive reaction, a pink colour appears in 2–5 minutes, deepening to magenta or crimson in half an hour (e.g., *Klebsiella*).
- **VP-negative:** In a negative reaction, it remains colourless for half an hour. Traces of pink colouration should be ignored.

Citrate utilisation (Fig. 3.39)

The utilisation of citrate as a sole source of carbon is tested in Simmon's citrate medium (solid medium; colour changes to blue to indicate growth).

Fig. 3.39 Simmon's citrate medium—blue indicates positive and green is negative for citrate utilisation (*Source*: Department of Microbiology, PIMS)

Indole, MR, VP, and citrate (IMViC) tests are useful in the identification and classification of enteric gram-negative bacteria. These tests are often used in clinical laboratories.

Nitrate reduction (NR)

This is performed to test for the presence of the enzyme nitrate reductase which reduces nitrate to nitrite (e.g., Enterobacteriaceae).

Urease test (Fig. 3.40)

This is performed to test the ability of bacteria to reduce urea to ammonia. This test is performed in Christensen's urease medium.

- **Urease-positive:** Cultures (e.g., genus *Proteus*) produce a pink colour. Urease-producing bacteria reduce urea to ammonia, which is responsible for the change in colour of the medium.
- **Urease-negative:** Cultures are yellow in colour.

Fig. 3.40 Urease test (*Source*: Department of Microbiology, PIMS)

Hydrogen sulphide production (Fig. 3.41)

- Some organisms (e.g., *Proteus mirabilis*) break down amino acids like cysteine to release H_2S during protein degradation. The H_2S reacts with iron in the medium (in which the bacteria are grown) to produce black precipitates of ferric sulphide.
- This is demonstrated in composite media like triple sugar iron (TSI) agar or Kligler's iron agar (KIA).

Fig. 3.41 Blackening of the medium indicating H_2S production in TSI medium (*Source*: Department of Microbiology, PIMS)

Catalase production

- Some bacteria (e.g., staphylococci) produce the catalase enzyme, which can be used as an identifying characteristic.
- Colonies from nutrient agar are picked up using a glass rod and placed in a tube containing 0.5 mL of 3% H_2O_2. Prompt effervescence indicates catalase production.

Oxidase reaction

- Cytochrome oxidase enzyme produced by aerobic organisms (e.g., *Pseudomonas aeruginosa*) catalyses the oxidation of reduced cytochrome by oxygen.
- A freshly prepared solution of 1.0–1.5% tetramethyl p-phenylenediamine hydrochloride (oxidase reagent) is poured over the colonies. Alternatively, a filter paper soaked in the oxidase reagent is smeared with the colony.
- **Oxidase-positive** colonies turn purple or black in 30 seconds to a minute.

Lecithinase production

- Organisms that produce lecithinase (e.g., *C. perfringens*), when grown on a solid egg yolk agar, form colonies surrounded by a zone of clearing.
- The lecithin of the egg gets digested by lecithinase produced by the organism.

Litmus milk

- When litmus milk is used, there may either be no change, or the organism may produce acid or alkali. Clotting of milk, peptonisation, or saponification may occur.
- The clot may be disrupted by the gas produced (stormy clot fermentation, e.g., *Clostridium perfringens*).

Composite media for biochemicals

- **Triple sugar iron (TSI) medium** (Fig. 3.42a–d)—a bacterium may ferment only glucose, or glucose, lactose, and sucrose, with or without gas formation; this is indicated by a change in the colour of the slant or butt. A composite medium may also indicate H_2S production.

Fig. 3.42 Triple sugar iron (TSI) agar showing (a) alkaline slant and acid butt without any gas or H_2S (K/A^{--}), (b) acid slant and acid butt with gas but no H_2S (A/A^{+-}), and (c) alkaline slant and acid butt with no gas with abundant H_2S and (d) alkaline slant

- **Bile esculin medium** is used to differentiate *Enterococcus* (group D) from other streptococci (for example *Streptococcus mitis*). Enterococci demonstrate growth in the presence of 40% bile and have the ability to hydrolyse esculin to esculetin and glucose. Esculin combines with ferric ions to produce a black complex (phenolic iron complex).

Other tests

Tests such as the fermentation of organic acids, oxidation of gluconate, amino acid decarboxylation, and hydrolysis of sodium hippurate are some of the other tests employed to characterise some bacteria. Apart from these, the following tests are also employed in the differentiation and identification of bacteria:

- Resistance to heat and disinfectants, e.g., *Enterococcus faecalis* is resistant to heat at 60°C for half an hour; clostridial spores are resistant to boiling for long periods
- Resistance to antibiotic and chemotherapeutic agents and bacteriocins

Immunological methods

These tests are based on the surface antigenic structure of bacteria.

- Serotypes or serovars are detected by using specific antisera.
- Immunological tests are based on agglutination or other suitable serological reactions. Immunofluorescence tests are also useful in some cases.

Pathogenicity tests in laboratory animals

Often, it becomes necessary to perform in vivo tests to demonstrate the production of toxins or other virulence characteristics.

- Pathogenicity/virulence tests are conducted by inoculating the organism in question into laboratory animals like guinea pigs, rabbits, rats, and mice.
- The routes of inoculation are as follows:
  - Intradermal
  - Subcutaneous
  - Intramuscular
  - Intraperitoneal
  - Intracerebral
  - Intravenous
  - Oral or nasal spray

These tests were performed commonly in the past. However, they are rarely used these days because of the availability of simpler in vitro tests in tissue cultures and due to stringent animal ethics guidelines.

Automated systems for isolation and identification of bacteria

With the advances in technology and the increasing use of computers in microbiology, the use of automation is slowly replacing the conventional methods of identification. The advantages of automated systems are as follows:

- Rapid turnaround time
- Uniformity in methods
- However, since they depend on a database, there is a need to keep updating the database as more information gets generated.

Automated culture systems

- These are used for detecting bacteria in cultures of blood and body fluids.
- They detect the presence of bacteria based on the production of CO_2 as an end product of the metabolism of bacteria. CO_2 production is detected by sensors, which transmit information to the software for analysis.
- With the first evidence of growth, bottles containing the bacteria are flagged (by a beep). They are then removed from the system and used to perform smears and cultured on a solid medium to identify the type of organism.
- Examples of available commercial systems are BacT Alert (Fig. 3.43a and b) and BACTEC.

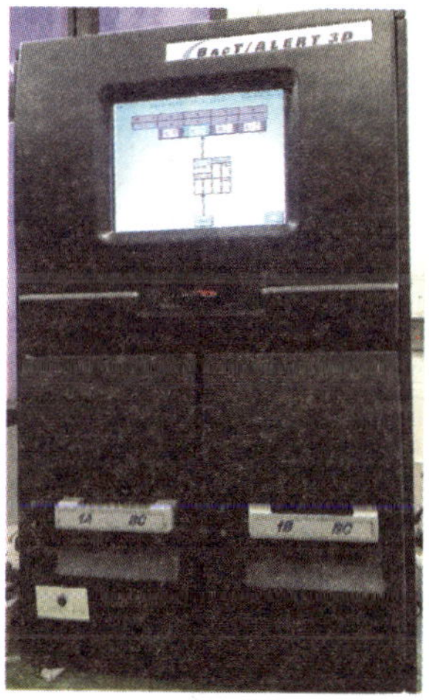

(a)

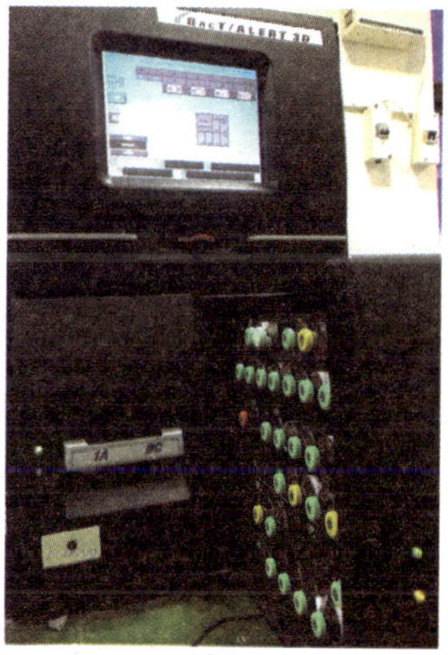

(b)

Fig. 3.43 (a) Automated blood culture system and (b) slots with bottles (*Source*: Department of Microbiology, PIMS)

Automated identification systems

- These identify bacteria based on the characteristics or metabolic properties of bacteria.
- Multiple wells in the identification card contain a test substrate that measures metabolic activity.
- The pattern produced is analysed based on the database in the computer software.
- These systems are also equipped to perform antimicrobial susceptibility tests. Examples of automated identification systems are VITEK™ (Fig. 3.44), Phoenix™ auto-microbic system™, and Alfred 60™ system.

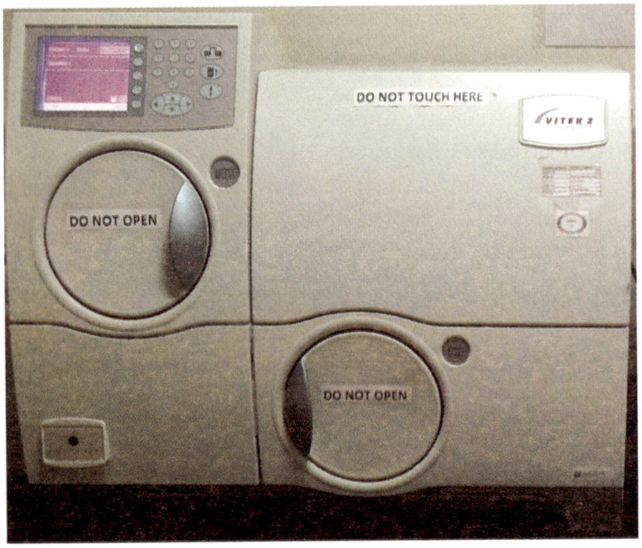

Fig. 3.44 VITEK™ rapid bacterial identification system (*Source*: Apollo Specialty Hospitals, Vanagaram)

MALDI-TOF

Matrix-assisted laser desorption/ionisation-time of flight mass spectrometry (MALDI-TOF-MS) identifies organisms based on the unique protein composition of the cells (Fig. 3.45).

Fig. 3.45 MALDI-TOF-MS (*Source*: Apollo Hospitals, Greams Road)

Advantages of MALDI-TOF

- It is a rapid, simple, and inexpensive method of identifying bacteria and yeasts.
- It has a short turnaround time and the ability to screen large quantities of samples (high throughput) by the proteomic technique.
- It can be implemented in a conventional laboratory setting in its current application.
- It can identify new, previously unknown bacteria from a clinical specimen, which would otherwise have to be identified at the generic level by a conventional method.
- It can directly identify microorganisms from body fluids like blood and urine.
- It has a high accuracy for microbial identification in general and performs as well as, or better than, conventional techniques.
- It is also used in research laboratories, veterinary, and defence laboratories, and in the pharmaceutical industry.

Nucleic acid-based identification

- Bacteria can be identified based on specific DNA or RNA sequences extracted from them.
- Nucleic acid-based methods are used to identify the species or subspecies of organisms by identifying the nucleic acid sequences in a sample. These methods use nucleic acid probes to detect target genetic materials such as RNA or DNA.
- **Real-time polymerase chain reaction (RT-PCR)** or sequence-based gene amplification methods (Fig. 3.46) are highly specific and sensitive and can be used for all categories of microbes.
- They also produce rapid results.

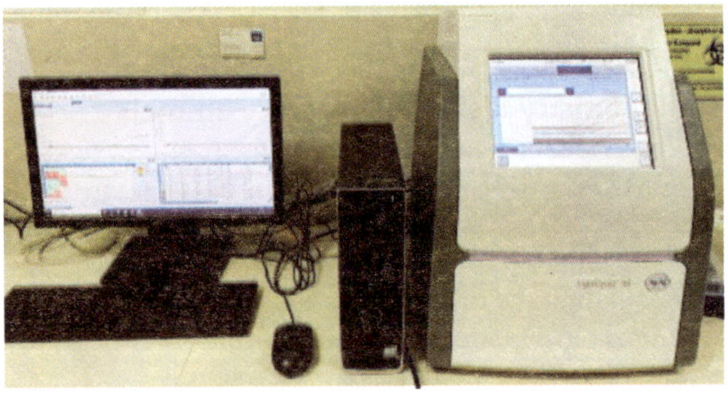

Fig. 3.46 Real-time PCR (*Source*: Molecular Laboratory, Pondicherry Institute of Medical Sciences)

- Nucleic acid-based methods may be of the following types:
  - **Qualitative**—to identify the target gene
  - **Quantitative**—to quantify RNA, which reflects the severity of the disease or the viral load, e.g., HIV or HCV; sequential testing following anti-viral therapy also reflects on the response to therapy; further, it can pick up drug resistance in patients who do not respond to therapy

Key Points

✓ Bacteria and blue-green algae are prokaryotes. Fungi and protozoa are eukaryotes.

✓ Staining techniques used for the study of bacteria are as follows:
- Simple stains (methylene blue, basic fuchsin)—all bacteria are stained the same colour
- Negative stains
- Impregnation methods
- Differential stains—stains impart different colours to different bacteria or bacterial structures (Gram stain, acid-fast stain)

✓ Based on the shape bacteria are classified as follows:
- Cocci (spherical or oval)
- Vibrios (comma-shaped, curved rods)
- Spirochetes (flexuous spiral forms)
- Bacilli (rod-shaped)
- Spirilla (rigid spiral forms)
- Actinomycetes (branching filamentous bacteria)

✓ The following are components of a bacterial cell:
- A rigid cell wall
- Cytoplasm containing nucleoids, ribosomes, mesosomes
- Fimbriae and pili
- A phospholipid cytoplasmic membrane
- Flagella

✓ Bacteria divide by binary fission.

✓ Doubling time may vary from 20 minutes (coliform bacilli) to 20 hours (tubercle bacilli) to 20 days (lepra bacilli).

✓ The bacterial growth curve consists of the following phases:
- Lag phase (no appreciable increase in number of bacteria)
- Log phase (an exponential increase in number of bacteria)
- Stationary phase (no increase or decrease in number)
- Decline phase (decrease in the bacterial population due to cell death)

✓ Bacteria vary in their requirements of temperature for growth.

✓ Bacteria are identified based on the morphology of the bacterial colony on culture and the detection of bacterial products released during metabolism.

✓ Culture techniques are required to isolate bacteria so they can be precisely identified and antibacterial susceptibility tests can be run.

✓ Automated systems identify bacteria using computer-based software.

✓ Molecular methods employing PCR are used for the detection of microorganisms, which are then identified based on sequencing.

Questions

Short answer questions

1. Describe the principles of Gram staining and its relevance in clinical microbiology.
2. Describe the bacterial growth curve and its clinical relevance.
3. Draw a labelled diagram of the structure of a bacterial cell.
4. Describe the structure and function of flagella.
5. Describe the relevance of bacterial spores.
6. Why is there a need to isolate bacteria in the laboratory?

MCQs

1. All the following statements are TRUE for the outer membrane layer of the gram-negative cell wall EXCEPT:
   a) It forms the outermost layer.
   b) It has abundant teichoic acid.
   c) It has receptors for some bacteriophages.
   d) Porins in this layer serve as diffusion channels for small molecules.

2. Which of the following is the organ of adhesion in bacteria?
   a) Flagella b) Capsule
   c) Fimbria d) Mesosome

3. Which of the following are used to demonstrate bacterial capsules?
   a) Albert's stain b) India ink
   c) Ziehl-Neelsen stain d) All of these

4. During which of the following phases are bacteria most susceptible to antibiotics?
   a) Lag phase b) Log phase
   c) Stationary phase d) Phase of decline

5. Which of the following is a characteristic feature of prokaryotes?
   a) There is no nuclear membrane.
   b) Cytoplasmic streaming can be seen unidirectionally.
   c) Intracellular organelles such as Golgi apparatus, mitochondria, and endoplasmic reticulum are present.
   d) There are sterols in the cell membrane.

6. Which of the following bacteria resist decolourisation with 5% sulphuric acid in acid-fast staining?
   a) *Nocardia*
   b) *Legionella*
   c) *Mycobacterium leprae*
   d) *Mycobacterium tuberculosis*

7. All the following are components of the cell wall of gram-negative bacteria except:
   a) Peptidoglycan b) Lipopolysaccharide
   c) Outer membrane d) Teichoic acid

8. All of the following statements are true for the outer membrane layer of gram-negative cell wall except:
   a) It forms the outermost layer.
   b) It is responsible for endotoxic activities.
   c) It has receptors for some bacteriophages.
   d) Porins in this layer serve as diffusion channels for small molecules.

9. Peritrichous arrangement of flagella refers to:
   a) A single flagellum at one pole
   b) Flagella all around the cell
   c) Flagella at both poles
   d) Tufts of flagella at one pole

10. Bacteria that grow at temperatures below 20°C are called:
   a) Mesophiles b) Thermophiles
   c) Psychrophiles d) Capnophiles

11. All the following statements are true about L-forms of bacteria except that they:
   a) Are aberrant morphological forms
   b) Develop in the presence of penicillin
   c) Are always stable
   d) Are named after Lister Institute, London

12. All the following statements are true about bacterial spores except that:
   a) They are a method of reproduction.
   b) They are resistant to desiccation.

c) They can be destroyed by autoclaving at 120°C for 15 minutes.
d) They are produced by the genus *Bacillus*.

13. In which phase of bacterial growth does sporulation occur?
a) Lag phase b) Log phase
c) Stationary phase d) Phase of decline

14. Bacteria that are unable to synthesise their own metabolites and depend on preformed organic compounds are called:
a) Phototrophs b) Chemotrophs
c) Autotrophs d) Heterotrophs

15. Mesosomes function as:
a) Site for storage of products for metabolism and energy
b) Site for bacterial respiration
c) Site for protein synthesis
d) Diffusion channels for small molecules

16. Which of the following is an enrichment medium?
a) Nutrient broth b) Tetrathionate broth
c) Stuart's medium d) Thayer-Martin medium

17. Which one of the following statements is true?
a) Agar has nutrient properties.
b) Chocolate agar is a selective medium.
c) The addition of selective substances in a solid medium forms an enrichment medium.
d) Nutrient broth is a basal medium.

18. MacConkey agar is an example of:
a) Differential medium b) Transport medium
c) Enriched medium d) Anaerobic medium

19. All the following are enriched media except:
a) Loeffler's serum slope b) Blood agar
c) Chocolate agar d) Nutrient agar

20. A solid medium containing substances which allow the required bacteria to grow and inhibit the growth of unwanted bacteria is called:
a) Differential medium b) Enriched medium
c) Enrichment medium d) Selective medium

21. Stuart's transport medium is used for the transport of specimens containing:
a) *Salmonella* b) Gonococci
c) *Vibrio cholerae* d) *Shigella*

22. Robertson's cooked meat medium is used to grow:
a) Anaerobes b) Viruses
c) Yeasts d) Aerobes

23. Agar-agar is used in microbiological media for:
a) Nutritive value b) Solidifying media
c) Carbon source d) All of the these

24. Which of the following selective medium is used to isolate *M. tuberculosis*?
a) Deoxycholate citrate medium
b) Thayer-Martin medium
c) Lowenstein-Jensen medium
d) Potassium tellurite blood agar

25. Which of the following require chocolate agar to grow:
a) *Neisseria meningitidis* b) *E. coli*
c) *Salmonella* Typhi d) *Mycoplasma pneumoniae*

26. McIntosh and Filde's anaerobic jar is used for the isolation of:
a) *Clostridium tetani* b) *Pseudomonas aeruginosa*
c) *Nocardia asteroides* d) None of these

27. Triple sugar iron medium contains all the following sugars EXCEPT:
a) Glucose b) Sucrose
c) Lactose d) Maltose

28. Lecithinase production on egg yolk agar is a characteristic feature of:
a) *Clostridium botulinum* b) *Clostridium tetani*
c) *Clostridium perfringens* d) *Clostridium difficile*

29. An example of an anaerobic transport medium is:
a) Cary-Blair medium
b) Crystal violet medium
c) Pike's medium
d) Robertson cooked meat medium

30. Routine blood agar for anaerobic culture contains:
a) Penicillin b) Vitamin K
c) Less quantity of blood d) X and V factors

31. The principle of bacterial identification in MALDI-TOF (matrix-assisted laser desorption /ionization-time of flight) mass spectrometry is:
a) Analysis of unique bacterial protein composition
b) Formation of a complex between the probe and target DNA
c) Amplification of target nucleic acid
d) Nucleic acid sequence-based amplification

32. What is TRUE regarding MALDI-TOF technique?
a) Rapid technique to identify from sample
b) Requires culture of the organism
c) Targets the cell wall component
d) Less specific

33. All the following are true regarding conventional microbiological techniques except:
a) Labour intensive b) Less expensive
c) Greater turn-around time d) Do not require expertise

34. The following substances can be added to produce an anaerobic environment EXCEPT:
a) Ascorbic acid b) Glutathione
c) Glycerol d) Cysteine

35. Phenyl deaminase test is used to identify which of the following?
a) *Shigella* b) *Salmonella*
c) *Proteus* d) *Pseudomonas*

Answers: 1:b, 2:c, 3:b, 4:b, 5:a, 6:c, 7:d, 8:b, 9:b, 10:c, 11:c, 12:a, 13:c, 14:d, 15:b, 16:b, 17:d, 18:a, 19:d, 20:d, 21:b, 22:a, 23:b, 24:c, 25:a, 26:a, 27:d, 28:c, 29:d, 30:b, 31:a, 32:b, 33:d, 34:c, 35:c

CHAPTER 4

Part I: General Microbiology

Systematic Bacteriology

(MI1.1–1.3, 8.9–8.13)

Chapter Navigation

- Classification, nomenclature, and taxonomy of bacteria
- General concepts of pathogenesis and disease spectrum
- Gram-positive cocci and bacilli
- Gram-negative cocci and bacilli
- Mycobacteria
- Miscellaneous bacteria
- Normal flora of the human body
- Principles of laboratory diagnosis

Learning objectives

- Describe various methods of classifying bacteria
- Explain the pathogenesis of bacterial infections and associate different bacteria with infectious syndromes
- Describe the diagnosis of bacterial infections and various methods of sample collection, processing, and reporting
- Recognise the role of quality control in the diagnostic laboratory

INTRODUCTION

Bacterial diseases are among the most common causes of morbidity and mortality world over despite significant advances in medical sciences. They affect all ages and sects of the human race.

These diseases are a global concern because they can be transmitted across countries and continents.

This chapter presents an overview of the classification and taxonomy of bacteria associated with human infections; their natural habitats, morphology, and pathogenesis; the organ- or system-specific infections caused by these bacteria and the laboratory diagnosis of such infections.

Details of the systemic infections and the clinical features and management thereof will be discussed in subsequent, disease-specific chapters (see the organism mapping table at the beginning of the book).

CLASSIFICATION, NOMENCLATURE, AND TAXONOMY OF BACTERIA

- The primary purpose of classifying organisms is to enable **easy identification**.
- Classification systems aim to group organisms with similar properties and to distinguish them based on their differences.
- The basic taxonomic unit of bacteria is the **species**. Species differ from one another in several features, which are determined by genes. **Bergey's Manual of Determinative Bacteriology** is widely accepted for its systematic classification of microorganisms.

Linnaeus Classification

In the 1700s, **Carolus Linnaeus** introduced the hierarchical system of classification of bacteria—**domain, kingdom, phylum, class, order, family, genus, and species** as shown in Fig. 4.1.

Naming of Organisms

The scientific name of an organism usually consists of two words, the first being the name of the genus and the second, that of the species (e.g., *Bacillus subtilis*).

- A **clone** of a bacterium is the population derived by the successive binary fission of a single cell.
- A single bacterial **colony** represents a clone. Though all the cells in a clone are expected to be identical in all respects, a few may exhibit differences due to mutation.
- A **strain** is the population of bacteria derived from a particular source, such as a patient.

Microorganisms are classified based on several characteristics which help in their identification. The different approaches to classification are as follows:

1. **Phylogenetic classification:**
    - The hierarchical classification is represented by a branching, tree-like arrangement, one characteristic being employed for division at each branch or level.
    - It implies an **evolutionary arrangement** of species.
    - Based on the **ribosomal RNA, 16S rRNA, and 23S rRNA gene sequences, phylogenetic**

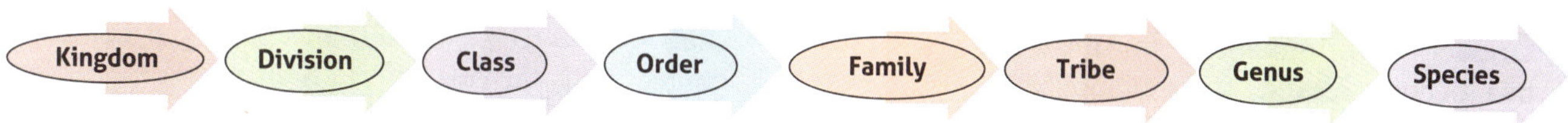

Fig. 4.1 Linnaeus classification

relatedness among different bacteria can be revealed.

2. **Adansonian classification (numerical taxonomy):**
   - This system takes into account **all the characteristics expressed at the time of study**.
   - The availability of computers has widened the scope of this system by permitting the comparison of very large numbers of properties of several organisms at the same time.
3. **Molecular or genetic classification:**
   - This is based on the **degree of genetic relatedness** of different organisms.
   - Since all the properties of an organism are ultimately based on its genes, this classification is said to be the most natural or fundamental method. DNA relatedness can be tested by studying the nucleotide sequences of DNA by DNA hybridisation or recombination methods.
   - The nucleotide base composition and base ratio, i.e., adenine–thymine (A–T):guanine–cytosine ratio (G–C), remains constant for members of the same species but varies widely among different groups of microorganisms.
   - Molecular classification has been employed more with viruses than with bacteria.
4. **Intraspecies classification:** For diagnostic or epidemiological purposes, it is often necessary to subclassify bacterial species. This classification may be based on the following:
   - **Biochemical properties (biotypes)**
   - **Antigenic features (serotypes)**
   - **Bacteriophage susceptibility (phage types)**
   - **Production of bacteriocins (colicin types)**

A species may be divided first into groups and then into types based on certain components (such as proteins). For example, group A streptococci are subclassified into M, T, and R types.

- Bacteria are further classified into **subspecies and sero-variants, i.e., serotypes**, within the species based on the organism's antigenic properties, e.g., genus name *Salmonella*; specific name enterica; subspecies enterica; serotype Typhi—i.e., *Salmonella enterica* enterica Typhi.
- The properties of a population of bacteria are studied in toto and not with respect to an individual bacterium or microorganism as is done for other life forms.

Molecular Epidemiology

- Greater discrimination in intraspecies typing has been achieved by the application of newer techniques of immunology, biochemistry, and genetics.
- **Investigations of epidemiology and pathogenesis** using these techniques are collectively referred to as molecular epidemiology.
- These methods can largely be classified as follows:
  1. **Phenotypic (study of expressed characteristics) methods:** Include electrophoretic typing of bacterial proteins and immunoblotting
  2. **Genotypic (direct analysis of genes, chromosomal and extrachromosomal DNA) methods:** Includes plasmid profile analysis, restriction endonuclease analysis of chromosomal DNA with Southern blotting, polymerase chain reaction, and nucleotide sequence analysis

Type cultures

- As a point of reference, type cultures of bacteria are maintained in international reference laboratories.
- These type cultures contain **representatives of all established species and the variants of these organisms that have been characterised.**
- The original cultures of any new species described are deposited in type collections.
- They are made available by the reference laboratories to other workers for study and comparison.
- The important type collection centres in the world currently are:
  - **ATCC** (American Type Culture Collection)
  - **NCTC** (National Collection of Type Cultures, Public Health England)

GENERAL CONCEPTS OF PATHOGENESIS AND DISEASE SPECTRUM

Bacterial infections may be of the following types:

- **Acute**, life-threatening infections, e.g., bacterial meningitis, anthrax, and plague
- **Chronic**, indolent infection for prolonged periods of time, e.g., tuberculosis
- **Mild** infection, e.g., streptococcal sore throat
- **Primary** (occurring for the first time) or secondary infections (following some predisposing factor)

Details of epidemiology of infections can be found in Chapter 9.

Pathogenesis

- The first event in the chain of pathogenesis is the interaction between the source, host, and pathogen. This leads to the establishment of disease in the host (Fig. 4.2).
- To establish an infection, bacteria must enter the host through an appropriate route such as—via direct contact, ingestion, inhalation, or penetration of the skin/mucosa (Fig. 4.3). If the pathogen comes in contact with the host through an uncommon route (e.g., intact skin), it may not be able to establish the infection.

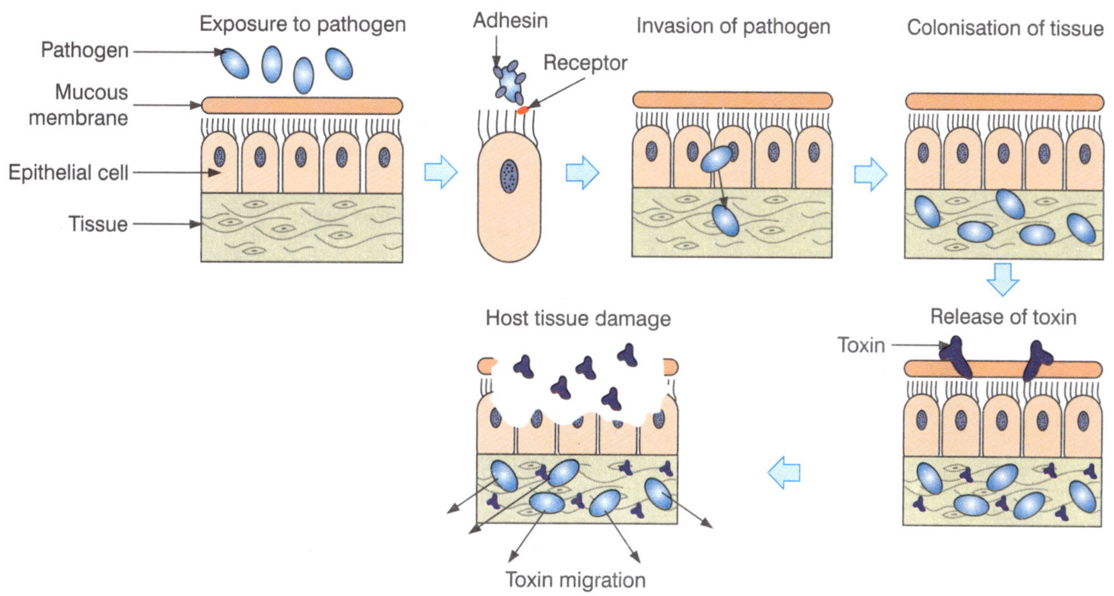

Fig. 4.2 Diagram showing the entry, invasion, and damage to host tissue by pathogenic bacteria

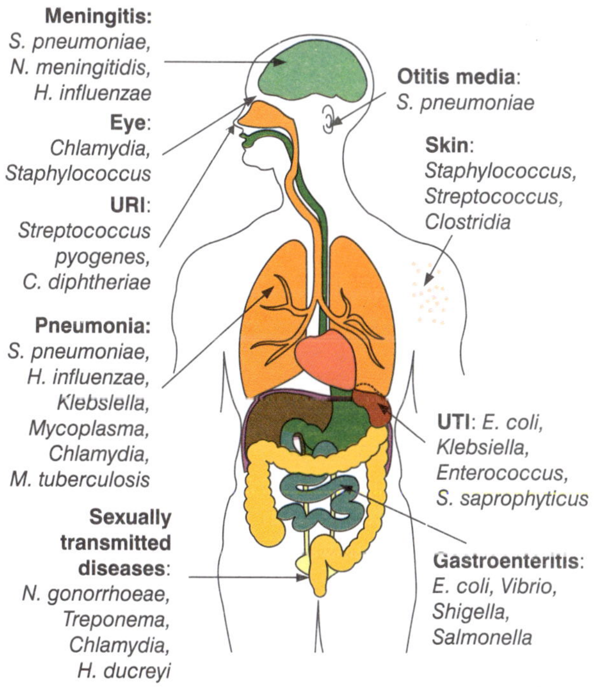

Fig. 4.3 Disease—pathogen association

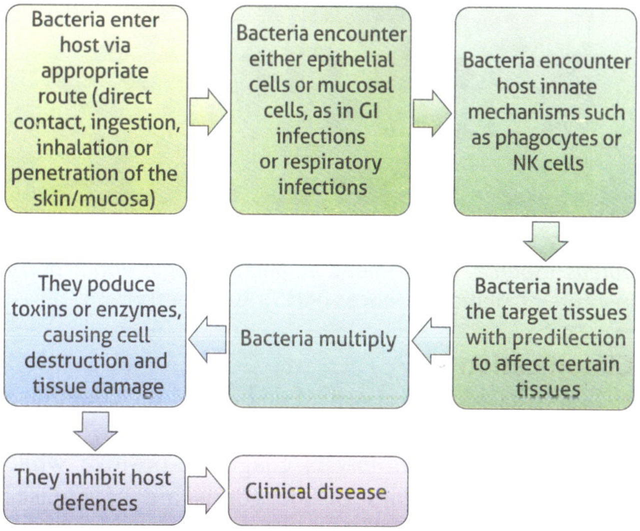

Fig. 4.4 Pathway of bacterial infections

- Following entry, bacteria encounter either epithelial cells (as in the case of infections with staphylococci and streptococci, which result in abscesses or pustules) or mucosal cells, as in GI infections (e.g., *Shigella* and *Vibrio*) or respiratory infections (e.g., *Pneumococcus*).
- In brief, the pathway of bacterial infections is shown in Fig. 4.4.

Incubation period: This is the period from the time of exposure (entry of the pathogen) to the first manifestation of symptoms (and clinical signs).

OVERVIEW OF IMPORTANT BACTERIA ASSOCIATED WITH HUMAN HEALTH AND DISEASE

The grouping of medically important bacteria is presented in Table 4.1.

GRAM-POSITIVE COCCI

I. Staphylococci

Morphology

- Staphylococci are **spherical bacteria**, approximately 1 μm in diameter and characteristically arranged in **grape-like clusters** (Fig. 4.5).
- This cluster formation is a result of cell division occurring in three planes, with daughter cells tending to remain in close proximity.

Table 4.1 Classification based on Gram staining

Gram-positive bacteria		Gram-negative bacteria		Miscellaneous
Cocci	Bacilli	Cocci	Bacilli	
Staphylococcus *Streptococcus* *Pneumococcus* *Enterococcus*	*Bacillus* *Corynebacterium* *Actinomycetes* *Clostridium*	*Neisseria* *Moraxella*	Enterobacteriales Vibrios Non-fermenters	Mycobacteria Spirochetes *Mycoplasma* *Chlamydia* Rickettsiae

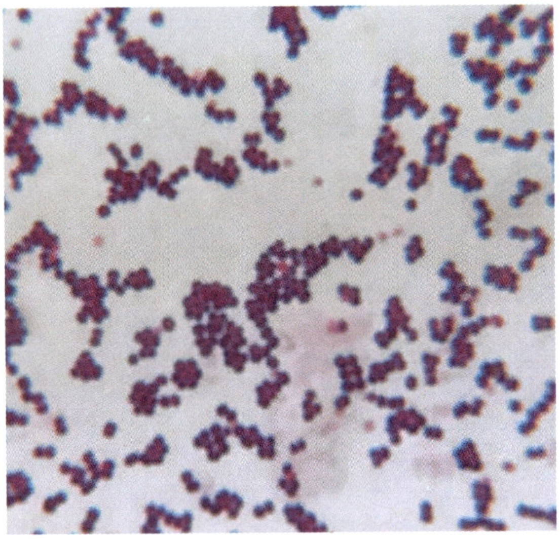

Fig. 4.5 Gram-positive cocci in clusters: *Staphylococcus aureus* (*Source*: Department of Microbiology, Pondicherry Institute of Medical Sciences [PIMS], Puducherry)

Classification

- Staphylococci are classified on basis of the **coagulase test** as coagulase-positive or -negative.
- *Staphylococcus aureus* is coagulase-positive while many other species of the genus are coagulase-negative.

Culture characteristics

Staphylococci grow readily on ordinary media within a temperature range of 10–42°C, the optimum being **37°C**, and a pH of **7.4–7.6.**

Virulence factors

Streptococci produce potent toxins, enzymes, and other metabolites, which play a major role in the pathogenesis of the infections they cause (Table 4.2).

Habitat

S. aureus colonise the skin of healthy individuals, becoming a source of infection for such an individual as well as others.

Common infections

Staphylococcus aureus is one of the most important human pathogens. The most common staphylococcal infections caused by this agent are presented in Table 4.3.

Identification and confirmation

S. aureus tests positive for the tube and slide coagulase test.

Antimicrobial resistance

The ability of staphylococci to develop resistance to multiple antibiotics makes staphylococcal infections difficult to treat.

Coagulase-negative staphylococci (CoNS)

These include:

- *S. epidermidis*
- *S. lugdunensis*
- *S. haemolyticus*
- *S. hominis*

Habitat

These agents are found as commensals in the throat, anterior nares, the skin of the perineum, and the web-spaces of the hands.

Virulence of CoNS

- **Biofilm formation** is an important factor in the pathogenesis of infections by *S. epidermidis*.
- A biofilm is an **extracellular polysaccharide matrix**, which protects bacteria from antibacterial agents and aids colonisation and resistance of infections.

Common pathogenic species

- *S. saprophyticus* may be present on healthy human skin and the periurethral area and can cause urinary tract infection, particularly in sexually active young women.
- *S. epidermidis* is the most common cause of CoNS infections in humans.
- These bacteria cause disease when the host's defences are breached.
- They can also cause healthcare-associated infections in immunocompromised hosts

II. Micrococci

- These are gram-positive cocci which can be mistaken for *S. aureus* in culture but are larger and mostly arranged in pairs, tetrads, or irregular clusters.
- They rarely cause infections in a healthy host.

Table 4.2 Virulence factors and characteristic features of *S. aureus*

Virulence factor	Constituents	Features
Cell-associated factors	Polysaccharide peptidoglycan	Activates the complement and induces the release of inflammatory cytokines
	Teichoic acid	Facilitates adhesion to the host cell surface and protects the bacteria from complement-mediated opsonisation
Cell-surface proteins	Protein A	Chemotactic Anti-phagocytic Anti-complementary effects Induces platelet damage and hypersensitivity (It is often used in staphylococcal co-agglutination tests)
	Clumping factor	Facilitates staphylococcal adhesion to host cells and tissues (Responsible for the slide coagulase test)
Extracellular enzymes	Coagulase	Acts with a coagulase-reacting factor (CRF) present in plasma to bind to prothrombin and convert fibrinogen to fibrin (This forms the basis for the tube coagulase test)
	Lipid hydrolases	Helps infect skin and subcutaneous tissues
	Hyaluronidase	Breaks down the connective tissue Staphylokinase (fibrinolysin) and proteases help in the spread of infection
	Nuclease	Breaks down nucleic acid (Used in the identification of the organism in the DNAse test)
Toxins	Cytotoxin	Membrane-active substances
	Alpha-hemolysin	Penetrates host cell membranes causing osmotic swelling, rupture, lysis and subsequently, cell death
	Beta-hemolysin	Hemolytic for sheep cells but not for human or rabbit cells
	Gamma-hemolysin	Composed of two separate proteins, both of which are necessary for hemolytic activity
	Delta-hemolysin	Has a detergent-like effect on the cell membranes of erythrocytes, leukocytes, macrophages, and platelets
	Panton–Valentine leukocidin (PVL)	Increased association with community-acquired MRSA (methicillin-resistant *S. aureus*) It causes leukocyte lysis and tissue necrosis
	Enterotoxin	Exotoxin that causes food poisoning Superantigen
	Toxic shock syndrome toxin (formerly, enterotoxin F) or TSST-1	Superantigen Potent activator of T lymphocytes; being a Vb-restricted T cell mitogen, it stimulates large numbers of T cells causing excessive and dysregulated immune response with the release of Il-1, IL-2, TNF, IFN-γ
	Exfoliative (epidermolytic) toxin or ET or exfoliatin	Dermonecrotic and exfoliation action

Table 4.3 Staphylococcal diseases

System affected	Disease caused
Cardiovascular system and bloodstream	Bacteremia, septicemia, pyemia, and endocarditis
Skin and soft tissue	Folliculitis, furuncle (boil), abscess (particularly breast abscess), wound infection, carbuncle, impetigo, paronychia, and less often, cellulitis
Musculoskeletal system	Osteomyelitis, arthritis, bursitis, and pyomyositis
Respiratory system	Tonsillitis, pharyngitis, sinusitis, otitis, bronchopneumonia, lung abscess, empyema and rarely, pneumonia
Central nervous system	Abscess, meningitis, and intracranial thrombophlebitis
Urinary system	Infection in association with local instrumentation, implants or diabetes
Common toxin-mediated staphylococcal diseases	Food poisoning Toxic shock syndrome Scalded skin syndrome

III. Streptococci

- **Morphology:** Streptococci are gram-positive cocci arranged in chains or pairs.
- **Culture characteristics:** They require blood or serum for growth and are nutritionally more exacting. Growth is better under 5–10% CO_2. The optimum temperature is 37°C (range 22–42°C).
- **Habitat:** They are part of the normal flora of human and animal bodies. Some of them are human pathogens (Table 4.4).

Classification

Several systems of classification have been used for streptococci (Fig. 4.6).

1. Based on **oxygen requirement**, they are classified as follows:
   - Facultative anaerobes
   - Obligate anaerobes
2. Based on **hemolysis produced on sheep or horse blood agar**, streptococci are classified as hemolytic and non-hemolytic (Fig. 4.7):

i. Hemolytic streptococci

a. *Alpha (α)-hemolytic streptococci*
- They produce a greenish discolouration with partial hemolysis around the colonies.
- The zone of lysis is small (1 or 2 mm wide).
- Most alpha-hemolytic streptococci are commensals in the throat and may cause opportunistic infections in rare cases.
- **Examples**: Members of the viridans group of streptococci and pneumococci.

b. *Beta (β)-hemolytic streptococci*
- They produce a sharply defined, clear, colourless zone of complete hemolysis, 2–4 mm wide, around their colonies.
- The term 'hemolytic streptococci' applies strictly to beta-hemolytic strains.
- Most pathogenic streptococci belong to this group.

ii. Non-hemolytic streptococci:
- These do not produce hemolysis on blood agar.

Table 4.4 Medically important streptococci and their characteristics

Species or common name	Lancefield group	Hemolysis	Habitat in human hosts	Common diseases caused	Laboratory tests for identification
S. pyogenes	Group A Griffith M-types (80) Nephritogenic pyoderma: M types 2, 49, 57, 59, 60, and 61	Beta	Throat and skin	Upper respiratory tract infections, impetigo, pyoderma, neonatal meningitis, septicemia Streptococcal **toxic shock syndrome** **Sequelae:** Rheumatic fever, glomerulonephritis	Bacitracin-sensitive; PYR test-positive; ribose not fermented
S. agalactiae	Group B	Beta	Female genital tract and rectum	**Neonatal infections:** Meningitis or septicemia **Adults:** Sepsis, wound infections	CAMP test, hippurate hydrolysis
S. equisimilis (*S. dysgalactiae*)	Group C	Beta	Throat	Severe invasive infection	Ribose- and trehalose-fermenting
S. anginosus has two subspecies: *S. anginosus* subsp anginosus and *S. anginosus* subsp whileyi	Non-typable	Beta (or may be alpha- or non-hemolytic)	Throat, colon and female genital tract	Abscesses and systemic infection Must be considered a pathogen when isolated from human infections	Bacitracin-resistant; PYR-negative; minute colony variants of other groups
Enterococcus spp. (*S. faecalis* and other enterococci)	Group D	Non-hemolytic	Colon, female genital tract and throat	Urinary tract infections, endocarditis, suppurative infections	Grow in 6.5% NaCl; PYR-positive
Non-enterococcal group D species (*S. bovis*)	Group D	Non-hemolytic	Colon	Endocarditis (*S. sanguinis*); dental caries (*S. mutans*)	No growth in 6.5% NaCl
Viridans streptococci (many species)	Non-typable	Alpha	Mouth, colon and female genital tract	Infective endocarditis	Optochin-resistant; species classification on biochemical properties

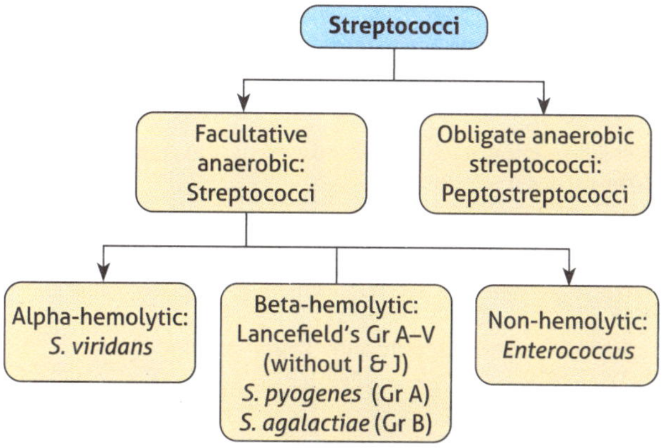

Fig. 4.6 Classification of streptococci

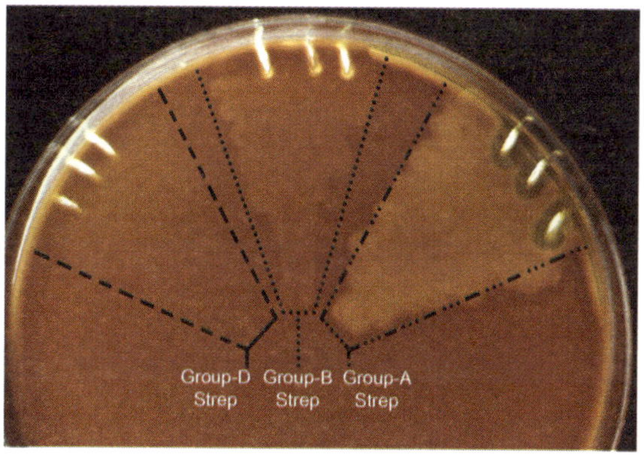

Fig. 4.7 Hemolysis on blood agar (*Source*: Centers for Disease Prevention and Control [CDC], Public Health Image Library [PHIL], Image ID 10861)

- This group largely includes fecal streptococci (*Enterococcus*), which are currently classified as a separate genus.

3. **Lancefield's grouping**
    - Based on the **carbohydrate antigen** (C) on the cell wall, **Rebecca Lancefield** (1933) serologically grouped beta (β)-hemolytic streptococci.
    - These groups are known as Lancefield's groups. So far, 20 of these groups have been identified, named A–H and K–V (without I and J).
    - *Streptococcus pyogenes* (Group A) is the most important member of group A.
4. **Griffith typing:**
    - It is used for the further classification of *S. pyogenes*.
    - Based on the M proteins on the cell surface, they are subdivided into **eighty M types** (types M1, M2, M3, and so on).

Infections

Streptococcus pyogenes or Group A Streptococcus

- These are **human pathogens**, which produce a diverse array of diseases, ranging from self-limited conditions (pharyngitis, impetigo, and erysipelas) to severe, invasive infections (bacteremia, necrotising fasciitis, myositis, and streptococcal toxic shock syndrome; Table 4.4).
- **Streptococcal pharyngitis or skin infections** may have lasting sequelae such as post-streptococcal rheumatic disease or glomerulonephritis respectively.

Group B streptococci (GBS)—Streptococcus agalactiae

- It is normal inhabitant of the vaginal tract and is associated with **neonatal meningitis and septicemia**.
- It is also an important pathogen of cattle, producing **bovine mastitis**.
- Nine capsular serotypes have been identified, antibodies to which confer type-specific protection.

Virulence factors

Virulence factors produced by streptococci are listed in Table 4.5.

IV. Pneumococci

Morphology

- *S. pneumoniae* (pneumococci) are typically small (1 μm), slightly elongated diplococci, with one end broad or rounded and the other pointed, bearing a **flame-shaped or lanceolate appearance**.
- Each pair is surrounded by a **capsule** that may be lost on repeated cultivation

Culture characteristics

- *S. pneumoniae* have complex growth requirements. They grow only in enriched media such as sheep blood agar and chocolate agar with 5–10% CO_2.
- Growth appcars as 0.5–1 mm, **dome-shaped, glistening colonies showing alpha-hemolysis**, which subsequently become flat with raised edges, central umbonation, and concentric rings (**Draughtsman or carom coin** appearance).

Habitat

S. pneumoniae is found in the throats of approximately half the population sampled at any time.

Infections

Pneumococci are associated with meningitis, lobar pneumonia, and bacteremia.

Virulence factors

The virulence factors of pneumococci are presented in Table 4.6.

V. Enterococci

Morphology

- Enterococci typically appear as pairs of **oval cocci** with paired cells arranged at an angle to each other (Fig. 4.8).

Table 4.5 Virulence factors of streptococci and their features

Virulence factor	Constituents	Features
Cell wall-associated factors	Capsule (hyaluronic acid)	Inhibits phagocytosis
	Carbohydrate antigen	Shows cross-reactivity with some human tissues, leading to post-streptococcal sequelae
	M protein Lipotechoic acid	Helps in attachment
Toxins	Hemolysin Streptolysin is predominantly produced by group A, though groups C and G may also produce them	Streptolysin O—antigenic and oxygen-labile Streptolysin S—non-antigenic and oxygen-stable
	Streptococcal pyrogenic exotoxin (SPE)	Act as a superantigen
	Streptokinase (fibrinolysin) Streptodornase (DNAse B)	Promotes the lysis of human fibrin clots by activating a plasma precursor (plasminogen), thereby causing the spread of infection
	Pyrogenic (erythrogenic) toxin	Causes the rash of scarlet fever
Enzymes	Deoxyribonucleases (streptodornase, DNase)	Cause depolymerisation of DNA and help to liquefy the thick pus in exudate
	Hyaluronidase	Breaks down the hyaluronic acid of the tissues, favouring the spread of infection along the intercellular spaces
	Proteinase, phosphatase, esterase, amylase, N-acetyl glucosaminidase, neuraminidase, serum opacity factor	Pyogenic

Table 4.6 Virulence factors of pneumococci

Virulence factor	Function
Capsular polysaccharide	Protects the cocci from phagocytosis because of its acidic and hydrophilic properties Inhibits complement and recognition by immunoglobulins Allows adherence to and colonisation of the nasopharynx
Pneumolysin	Membrane-damaging toxin—forms pores which cause cell lysis Induces inflammation Drives host-to-host transmission Can activate complement and modulate chemokine and cytokine production
Autolysin	Cell lysis Break down peptidoglycan Exposes hosts cell to pneumolysin and teichoic acid Aids bacterial colonisation
Pneumococcal surface protein A	Protects against the host's complement system Promotes adherence to epithelial cell membrane
Pili	Promotes adherence and colonisation of the epithelial cells within the nasopharynx Inhibits phagocytosis
Neuraminidase	Degrades mucus Promotes growth and survival Aids with cell adherence

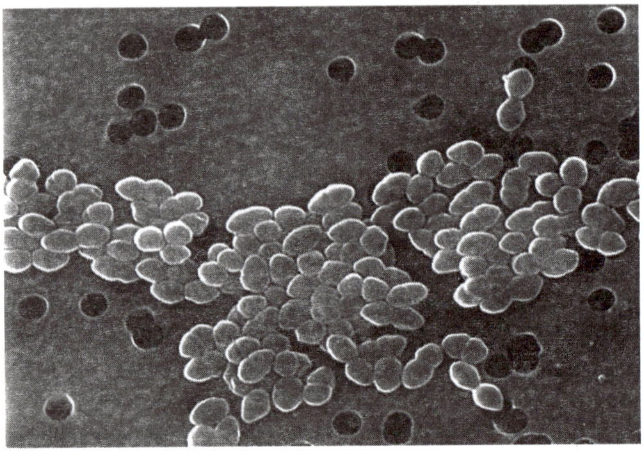

Fig. 4.8 Enterococci—electron micrograph showing oval cells arranged in pairs (*Source*: CDC, PHIL, Image ID 266)

- They are usually non-hemolytic, though some strains may show alpha- or beta-hemolysis.
- The *Enterococcus* group includes *E. faecalis*, *E. faecium*, and *E. durans*.

Habitat

E. faecalis is most often isolated from human sources; it is a part of the microbiota of the human gut.

Identification

- Enterococci possess several features that are distinctive from those of streptococci.
- They are identified by their **ability to grow on bile** and cause **esculin hydrolysis**.

- They do not grow in 6.5% NaCl (helps in differentiating them from enterococci).
- They can grow on tellurite blood agar, on which they produce black colonies.

Infections

Enterococci are frequently isolated from cases of the following infections:

- Urinary tract infections
- Wound infections
- Endocarditis
- Infections of the biliary tract, septicemia. and intra-abdominal abscesses complicating diverticulitis and peritonitis

The virulence factors of enterococci are presented in Table 4.7.

Anerobic Gram-Positive Cocci

Peptostreptococci

- **Morphology**: These are relatively small cocci (0.2–2.5 μm).
- **Culture:** Many of them are aerotolerant and grow well under 10% CO_2 in an aerobic atmosphere.

Table 4.7 Virulence factors of enterococci

Virulence factor	Function
Cell associated factors	
Aggregation substances	Adhesion
	VanA co-transference in *E. faecalis*
	Biofilm formation
Microbial surface components recognising adhesive matrix molecules (MSCRAMMs)	Bind to extracellular matrix proteins
Pili	Biofilm formation
	Bind to extracellular matrix proteins
	Colonisation of both the kidneys and the bladder
Enterococcal surface protein (ESP)	Helps in the development of infectious endocarditis and urinary tract infections
	Biofilm formation
	Associated to *E. faecium*'s susceptibility to ß-lactams
Secreted factors	
Cytolysin	Bactericidal and cytolytic activity
Secreted Antigen A	Interferes with cell-wall metabolism
	Binds to extracellular matrix proteins
	Activation of the host immune system
Gelatinase	Interferes with complement-mediated immunity
	Participates in the development of infectious endocarditis

- **Habitat**: They are normal inhabitants of the vagina, intestines, and mouth.
- **Infections:** They may cause several clinical infections such as **puerperal sepsis and other genital infections, wound infections, gangrenous appendicitis, urinary tract infections, osteomyelitis, and abscesses in the brain, lungs, and other internal organs**. Peptostreptococci are often seen in large numbers in **pus from suppurative lesions** and hence a Gram-stained smear may be helpful in diagnosis. Infections are usually polymicrobial.
- **Species:** *Peptostreptococcus anaerobius* is most often responsible for puerperal sepsis and *P. magnus* for abscesses. *P. asaccharolyticus*, *P. tetradius*, and *P. prevotii* are some other species commonly present in clinical specimens.

GRAM-POSITIVE RODS

I. Corynebacteria

Morphology

- Corynebacteria are gram-positive, **non-sporing, non-acid-fast, and non-motile rods**.
- On staining, they exhibit irregularly stained segments due to the presence of granules in some species.
- They frequently show **club-shaped swellings**—hence, the name *Corynebacterium* (from *coryne*, meaning 'club').
- They are pleomorphic, measuring approximately 3–6 × 0.6–0.8 μm.
- Although gram-positive, they tend to be decolourised easily.
- They possess polymetaphosphate granules called **Babes–Ernst granules or metachromatic granules**, which serve as energy stores. These give the bacilli a **beaded appearance** when stained with aniline dyes.
- The bacilli are arranged in characteristic pairs, palisades (resembling the stakes of a fence), or small groups. They often appear at angles to each other, resembling the letters V or L. This arrangement, called the **Chinese letter or cuneiform arrangement,** occurs due to the incomplete separation of the daughter cells after binary fission (Fig. 4.9).

Infection

Corynebacterium diphtheriae causes **diphtheria**, a life-threatening childhood respiratory infection.

Virulence factors

- **Diphtheria toxin**—virulent strains of diphtheria bacilli produce a very powerful exotoxin known as the diphtheria toxin. It inhibits protein synthesis by inactivating elongation factor-2.
- **Adhesion factors** such as pili and non-fimbiral proteins also play a key role in virulence.

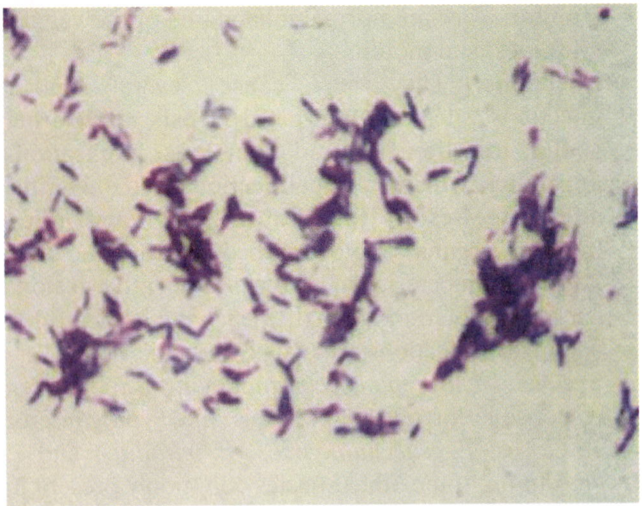

Fig. 4.9 *C. diphtheriae* showing cuneiform arrangement (*Source*: Dept of Microbiology, PIMS)

- **Phospholipase D** is the major virulence factor in *C. pseudotuberculosis* and *C. ulcerans*. It is a secreted exotoxin that possesses sphingomyelinase activity.

Non-lipophilic corynebacteria

- These bacteria resemble the gravis type of the diphtheria bacillus.
- They produce two types of toxins, one probably identical to the diphtheria toxin and the other resembling the toxin of *C. pseudotuberculosis*.
- Examples include *C. pseudotuberculosis* and *C. ulcerans*, which can cause diphtheria-like lesions.

Lipophilic corynebacteria

- They often infect immunocompromised hosts.
- They are usually multidrug-resistant, responding only to vancomycin.
- These bacteria are increasingly being associated with infective endocarditis.
- Examples include *C. jeikeium* and *C. urealyticum*.

Diphtheroids

- Some corynebacteria exist as commensals in the throat, skin, conjunctiva, and other areas.
- These may sometimes be mistaken for diphtheria bacilli and are called diphtheroids.
- In general, diphtheroids stain more uniformly than *C. diphtheriae*, possess few or no metachromatic granules, and tend to be arranged in parallel rows (palisades) rather than in a cuneiform pattern.
- They **cause infection in immunocompromised patients**.

II. Bacillus

Morphology

- These are **gram-positive bacilli that produce spores**.
- The spores are central or subterminal and elliptical or oval and do not bulge beyond the diameter of the vegetative cell.
- Oxygen is required for the formation of spores but not for germination. The optimum temperature for sporulation is 25–30°C.
- Spores are formed in culture or in the soil but not in a living animal body.

Culture characteristics

- The members of this group are psychrophilic, mesophilic, and thermophilic species, with maximum temperatures for vegetative growth ranging from about 25°C to above 75°C and the minimum from about 5°C to 45°C.
- The salt tolerance of these bacteria ranges from less than 2% to 25% NaCl.
- They constitute the commonest contaminants in bacteriological culture media.

Susceptibility

- The **spores are resistant to physical and chemical agents. They resist dry heat at 140°C for 1–3 hours and boiling for 10 minutes.**
- Vegetative bacilli, on the other hand, are found in tissues and are destroyed at 60°C in 30 minutes.
- Normal heat fixation of smears may not kill the bacilli in blood films.
- Spores have been isolated from naturally infected soil after as long as 60 years.
- Spores are destroyed by glutaraldehyde, formaldehyde, iodine, chlorine compounds, hydrogen peroxide, and beta-propiolactone at higher concentrations.

Bacillus anthracis

- It is the major pathogen of this group.
- **Infection**—*B. anthracis* causes **anthrax**, which is primarily a disease of cattle and sheep and may affect other animals also.
- Humans get infected when they come in contact with or inhale spores shed by animals or their products.
- *B. anthracis* is an **aerobic, gram-positive, non-motile, sporulating bacillus** which measures 1–1.5 × 3–5 µm.
- In clinical material and culture, *B. anthracis* **appears in chains** (end to end) with a characteristic **bamboo stick or boxcar appearance** (Fig. 4.10).
- It grows readily on conventional microbiology media, including blood agar.
- Spores are not generally seen on direct Gram stain of specimens, but they appear in culture when nutrients are depleted. They grow in 6–24 hours.
- The organism is easily recovered if samples are obtained before the administration of antibiotics.

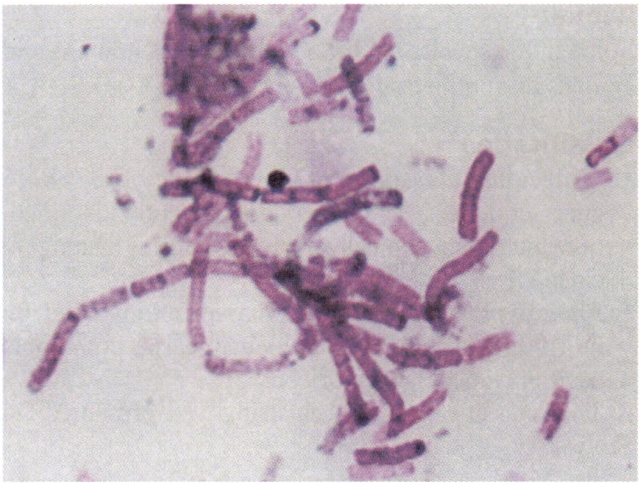

Fig. 4.10 Bamboo stick appearance of *B. anthracis* in Gram stain (*Source*: Dept of Microbiology, PIMS)

- **Virulence factors**
  - *Polypeptide capsule*: The capsule inhibits phagocytosis.
  - *Anthrax toxin*: It has three fractions: the **protective antigen (PA), the lethal factor (LF), and the edema factor (EF)**. PA binds toxin to cells while EF activates adenyl cyclase inside the target cells, leading to intracellular accumulation of cyclic AMP and edema. LF causes cell death.

Bacillus cereus

- *B. cereus* is widely distributed in nature and may be readily isolated from soil, vegetables, and a wide variety of foods including milk, cereals, spices, meat, and poultry.
- It is commonly associated with food poisoning.
- *B. cereus* is generally motile but non-motile strains may occur.
- It is not capsulated and not susceptible to the gamma phage.
- **Infection**—*B. cereus* is an important cause of food poisoning. It is associated with contaminated cooked food (e.g., Chinese fried rice) that is warmed and eaten next day.
- **Virulence factors**
  - **Enterotoxins** such as **hemolysin BL (Hbl), non-hemolytic enterotoxin (Nhe), and cytotoxin K (CytK)** are responsible for the diarrheal syndrome. These enterotoxins kill cells by forming pores in their cell membranes. Hemolysin BL can destroy immune cells such as macrophages and dendritic cells, which accelerates the spread of *B. cereus* and tissue damage. Widespread destruction of epithelial tissue in the intestine leads to ineffective absorption of water, causing diarrhea.
  - The **emetic toxin (cereulide)** is produced in food products and ingested as a preformed toxin. It interferes with the mitochondrial membrane potential, halting energy production in the form of ATP. The dysfunctional mitochondria in intestinal muscles causes cell death. The progressive degeneration of the muscles leads to gastrointestinal dysmotility and symptoms like vomiting and nausea.

Bacillus thuringiensis

- *B. thuringiensis* is found in the soil.
- It produces a toxin that is fatal to mosquitoes and other herbivorous insect larvae. For this reason, it has been exploited as a **biological pesticide and larvicidal agency**.
- *B. thuringiensis* is also a source of genes used to genetically modify food crops to make them resistant to pests.

Bacillus species as sterilisation indicators (controls)

Several *Bacillus* species are used as biological indicators in sterilisation controls, e.g., the spores of *B. atrophaeus* (subtilis var niger) and *Geobacillus stearothermophilus*.

III. Actinomycetes

Morphology

- Actinomycetes are slender, gram-positive bacteria, 0.4–1 μm in size, that possess a cell wall containing muramic acid.
- They bear a superficial resemblance to fungi as they form branching filaments.
- Actinomycetes are related to mycobacteria and corynebacteria.
- They are non-motile, non-sporing, non-capsulated filaments that break up into bacillary and coccoid forms.

Habitat

Most members of this group are free-living, particularly in the soil. Actinomyces are normally present in the human mouth, intestine, and vagina as commensals.

Classification

- **Anaerobic/Microaerophilic**: *Actinomyces, Bifidobacterium, Arachnia,* and *Rothia*
- **Aerobic**: *Nocardia, Actinomadura, Dermatophilus,* and *Streptomyces*

Infections

- Actinomycetes produce a **chronic, granulomatous infection** in human beings and animals.
- The disease is characterised by the development of indurated swellings, chiefly in the connective tissue and by suppuration and the discharge of pale granules called sulphur granules.
- The lesions often have multiple sinuses.

- Trauma, the presence of foreign bodies, or poor oral hygiene may favour tissue invasion.
- *A. israelii* is the most common infective agent.
- **Actinomycosis** in human beings is an endogenous infection. It is usually a mixed infection, which may involve *Bifidobacterium dentium*, *Actinobacillus actinomycetemcomitans*, *Eikenella corrodens*, *Haemophilus aphrophilus*, *Bacteroides*, fusobacteria, staphylococci, and anaerobic streptococci.

Other species

A. naeslundii, *A. viscosus*, *A. odontolyticus*, *A. meyeri*, *A. gerencsonei*, and *Propionibacterium propionicum* may sometimes be responsible for infections.

Nocardia

- **Morphology:** They are aerobic actinomycetes, catalase-positive, gram-positive bacteria with branching, filamentous forms that are generally acid-fast and resist decolourisation with 1% sulphuric acid (**Kinyoun's method**).
- **Infections:** *Nocardia* produce infections in the lungs, which may be confused with tubercular lesions. They often cause **opportunistic infections in immunocompromised individuals** including those who have HIV. The infection often resembles tuberculosis.

IV. Clostridia

Morphology

- Clostridia are gram-positive anaerobic bacteria that form heat-resistant spores.
- They are highly pleomorphic, usually 3–8 × 0.4–1.2 µm in size, and are rod-shaped.
- Clostridia are **motile with peritrichate flagella**, the exceptions being *C. perfringens* and *C. tetani* type VI, which are non-motile. The motility is described as **'stately' or slow**.
- *C. perfringens* and *C. butyricum* are capsulated, while others are not.

Habitat

Some of the species are found in the gut of humans and animals as commensal flora.

Classification

- **Molecular classification**: Based on 16S rRNA gene sequences, 19 clusters are currently identified, of which, the clinically relevant species belong to cluster 1.
- **Phenotypic classification**: Classification based on morphology, culture, and biochemicals is commonly used in clinical laboratories.

The classification of clostridia is presented in Table 4.8.

Spores

- The spores of clostridia are wider than the bacillary bodies, giving the bacillus a swollen appearance resembling a **spindle**—hence, the name *Clostridium* (from *kloster*, meaning 'spindle').
- Some (such as *C. sporogenes*) sporulate readily, while others (such as *C. perfringens*) do so inconsistently. The shape and position of spores vary in different species and are used in the identification and classification of clostridia (Table 4.8; Fig. 4.11).
- The arrangement of spores can be:
  - Central or equatorial, giving the bacillus a **spindle shape** (*C. bifermentans*)
  - Subterminal, the bacillus appearing **club-shaped** (*C. perfringens*)
  - Oval and terminal, resembling a **tennis racket** (*C. tertium*)
  - Spherical and terminal, giving the bacillus a **drumstick appearance** (*C. tetani*)
- Viability of spores in the environment:
  - Spores of clostridia are relatively more resistant to external influence than the vegetative forms.
  - The viability of the spores of various clostridia are detailed in Table 4.9.

Table 4.8 Morphological and biochemical classification of clostridia

Position of spores	Both proteolytic and saccharolytic	Saccharolytic	Neither proteolytic nor saccharolytic
Central or subterminal	**Proteolytic predominance** *C. bifermentans* *C. botulinum* A, B, F *C. histolyticum* *C. sordellii* *C. sporogenes*	**Asaccharolytic predominance** *C. perfringens* *C. septicum* *C. chauvoei* *C. novyi*	*C. fallax* *C. botulinum* C, D, E *C. tertium* *C. tetanomorphum* *C. sphenoides*
Oval and terminal	*C. difficile*	—	*C. cochlearum*
Spherical and terminal	—	*C. tetani*	—

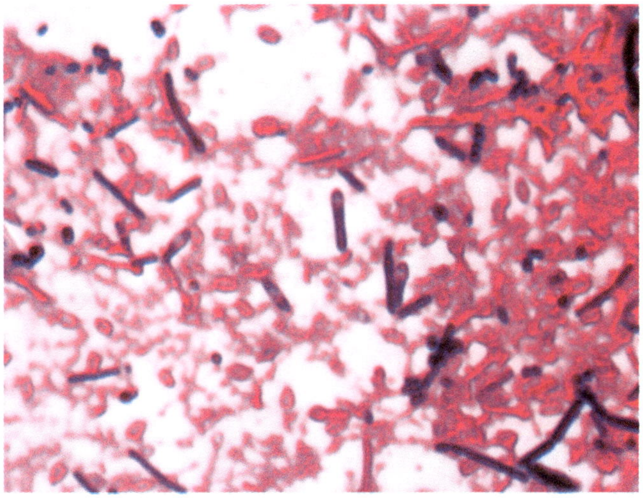

Fig. 4.11 Sporulating clostridia (*Source*: Dept of Microbiology, PIMS)

Virulence factors

Clostridia produce disease chiefly by toxin production.

- **Exotoxin:** Seen in tetanus, botulism, and gas gangrene.
- **Invasive toxin**: *C. perfringens*, besides being toxigenic, is also invasive and can spread along the tissues and even cause septicemia.

GRAM-NEGATIVE COCCI

- They are often arranged in pairs (diplococci).
- They are aerobic, non-sporulating, non-motile, and oxidase-positive cocci.
- The gram-negative cocci of medical importance are:
  - *Neisseria*, some species of which cause infection in humans, while others may be commensals
  - *Moraxella*, which are generally colonisers and may occasionally cause pneumonia

I. *Neisseria*

The genus *Neisseria* consists of gram-ncgativc, acrobic, non-sporulating, non-motile, oxidase-positive cocci arranged typically in pairs (diplococci).

Morphology

- These are gram-negative, oval or spherical cocci, 0.6–0.8 μm in size, arranged typically in pairs, with the adjacent sides flattened.
- The long axis of the coccus is at right angles to a line joining the two cocci in a pair.
- The **intracellular presence of these bacteria inside polymorphs** in patient samples is a characteristic finding.
- A polysaccharide capsule is present in *Neisseria meningitidis*.

Classification

The genus *Neisseria* includes two important human pathogens:

i. *Neisseria meningitidis*

Table 4.9 Viability of spores of clostridia

Clostridium species	Viability
C. botulinum spores	Survive 100°C up to 3–4 hours At 105°C—not killed completely in less than 100 minutes
C. perfringens spores Spores of type A strains cause food poisoning	Destroyed by boiling for less than five minutes Survive for several hours
C. tetani spores	Persist for years in dried earth Resist boiling for 15–90 minutes

Spores of all species are killed by autoclaving at 121°C within 20 minutes.

- Spores are resistant to phenolic disinfectants.
- Halogens are effective in killing spores.
- 1% aqueous iodine solution kills spores within three hours.
- Glutaraldehyde (2% at pH 7.5–8.5) is very effective in killing spores.

ii. *Neisseria gonorrhoeae*

- The other members of this genus are normal commensals rarely causing infection. They include *N. lactamica*, *N. flavescens*, *N. subflava*, and *N. sicca*. These are commensals in the mouth or upper respiratory tract. Several species of neisseriae inhabit the normal respiratory tract.
- Their pathogenic significance is uncertain though some of them (e.g., *N. flavescens*, *N. catarrhalis*) have been reported occasionally as having caused meningitis.

Culture characteristics

- *Neisseria* have **exacting growth requirements** and do not grow on ordinary media.
- Growth occurs on media enriched with blood or serum which promote growth by neutralising certain inhibiting substances in culture media rather than by providing additional nutrition.
- They are aerobes, but growth is facilitated by 5–10% CO_2 and high humidity. The optimum temperature for growth is 35–36°C and optimum pH 7.4–7.6.
- Blood agar, chocolate agar, Thayer–Martin medium, and Mueller–Hinton starch casein hydrolysate agar are used to culture meningococci.
- Selective medium—the **modified Thayer–Martin medium (with vancomycin, colistin and nystatin)** is used for the culture of *Neisseria*.
- The characteristic culture features of some of the common species are listed in Table 4.10.

Habitat

- The human nasopharynx is the only reservoir of the meningococcus.
- Several species of neisseriae such as *N. lactamica* and *N.flavescens* inhabit the normal respiratory tract.

Table 4.10 Differential characteristics of neisseriae

Species	Colonies	Growth on nutrient agar and at 22°C	Serological classification
N. meningitidis	Round, smooth, shiny, creamy consistency	Negative	Thirteen antigenic groups
N. gonorrhoeae	Same as above, but smaller and more opalescent	Negative	Antigenically heterogeneous
N. catarrhalis (*Branhamella catarrhalis*) Resembles meningococcus but is pigmented yellow	Variable, smooth, and translucent or adherent and opaque, not easily emulsifiable	Grows on nutrient agar and also at 22°C	Autoagglutinates

Infections

- **Meningitis and meningococcal septicemia** are caused by *N. meningitidis*.
- *N. gonorrhoeae* causes **gonorrhea**, a sexually transmitted infection.

Virulence factors

- In meningococci, the **endotoxin lipopolysaccharide** released by autolysis acts on the vascular endothelium. All major inflammatory cascade systems as well as cytokines and nitric oxide are triggered and upregulated, causing shock.
- In gonococci, pili promote adherence, and the trilaminar outer membrane contains many different proteins.
- Both gonococci and meningococci produce IgA protease, which splits and inactivates IgA.

II. *Moraxella*

- The genus *Moraxella* belongs to the family Moraxellaceae.
- Members of this genus are gram-negative diplococci or coccobacilli.
- Species of medical importance include *Moraxella catarrhalis*, *M. lactamica*, *M. bovis* and *M. oslonensis*. *M. catarrhalis* resembles commensal *Neisseria* species in culture.

Morphology

These are gram-negative coccobacilli or diplococci, 1–1.5 μm in size, which are catalase- and oxidase-positive.

Culture

- These organisms are inert biochemically (asaccharolytic).
- *Moraxella* grows on ordinary media, unlike the pathogenic *Neisseria*.

Habitat

Moraxella catarrhalis is found in the nasopharynx of human beings.

Infections

- *Moraxella catarrhalis* is an exclusively human opportunistic pathogen and is a common cause of **otitis media in infants and children**. The tendency of this organism to colonise the upper respiratory tract is high in infants and children but decreases in adulthood.
- It can cause sinusitis, laryngitis, bronchopneumonia, and meningitis.
- It can also cause **respiratory tract infections in immunocompromised hosts**.
- In a significant proportion of elderly patients with chronic obstructive pulmonary disease and chronic bronchitis, it causes lower respiratory tract infections.

Anaerobic Gram-Negative Cocci

Veillonella

- **Morphology:** *Veillonella* are gram-negative cocci of varying sizes occurring as diplococci, short, or groups.
- **Habitat:** They are normal inhabitants of the mouth and the intestinal and genital tracts.
- **Infections:** *Veillonella parvula* has been reported from clinical specimens, but its pathogenic role is uncertain.
- **Antibiotic sensitivity**: All anaerobic cocci are generally sensitive to penicillin, chloramphenicol, and metronidazole, and resistant to streptomycin and gentamicin.
- **Classification:** Medically important anaerobic gram-negative bacilli belong to the family Bacteroidaceae and are classified into the genera *Bacteroides*, *Porphyromonas*, *Prevotella*, *Fusobacterium*, and *Leptotrichia*.

GRAM-NEGATIVE BACILLI

They can be non-fermenters or fermenters. A broad classification is presented in Fig. 4.12.

Enterobacterales

General features

- These are a large group of **non-sporing, non-acid-fast, gram-negative rods that are heterogeneous**

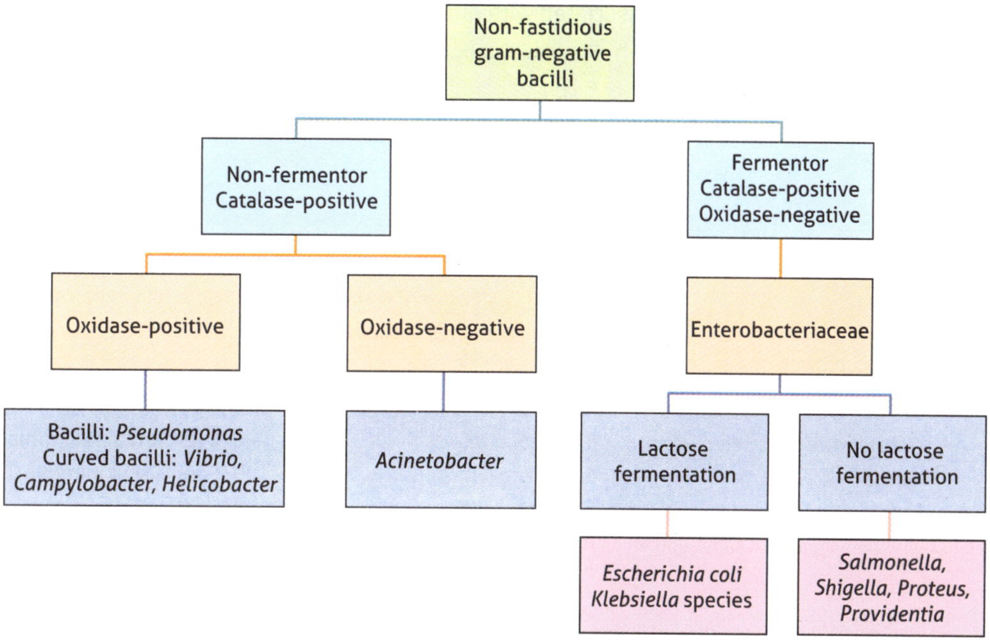

Fig. 4.12 Broad classification of non fastidious gram-negative bacilli

and may be **motile** with peritrichous flagella. Capsules and fimbriae are found in some virulent strains.

- They are found in the **guts of humans and animals.**
- They are **aerobic, facultatively anaerobic, and grow readily in ordinary media.**
- They are known to **break down glucose fermentatively**, producing acid and gas or acid only and to reduce nitrates to nitrites; members of this group **form catalase but not oxidase.**
- Within the family, they vary widely in their biochemical and antigenic properties.
- Most members of the genera are commensals in the gut and cause infections only in certain circumstances, except *Salmonella* and *Shigella*, which are always pathogenic.

Classification

Enterobacterales is divided into seven families:

- Enterobacteriaceae
- Erwiniaceae
- Pectobacteriaceae
- Yersiniaceae
- Hafniaceae
- Morganellaceae
- Budviciaceae

The species in each family are further classified into biotypes, serotypes, and bacteriophage types based on biological and serological characters and susceptibility to specific bacteriophages. The commonly used classification for clinically significant infections is as presented in Table 4.11.

I. *Escherichia coli*

Morphology

- *E. coli* are **gram-negative, straight rods** measuring 1–3 × 0.4–0.7 μm without any specific arrangement.
- They are motile by peritrichate flagella, though some strains may be non-motile.
- Capsules and fimbriae are found in some virulent strains.

Habitat

Unlike other coliforms, *E. coli* are parasites living only in the human or animal intestine. They are generally gut commensals.

Culture characteristics

- *E. coli* grows aerobically and is a facultative anaerobe.
- The temperature range required for growth is 10–40°C (optimum 37°C).
- It grows well in ordinary media, forming large, greyish-white, smooth, moist, opaque or partially translucent, and discoid colonies.
- On blood agar, many strains are **hemolytic.**
- On MacConkey medium, colonies are bright pink due to **lactose fermentation**.
- *E. coli* are **indole-** and **MR-positive**, **VP** and **citrate-negative** (IMViC ++--).
- They do not liquefy gelatin; nor do they form H_2S or split urea.

Infections

- Urinary tract infections
- Diarrhea (with several virulence and toxigenic factors)
- Life-threatening bloodstream infections

Table 4.11 Enterobacteriaceae—major genera associated with human infections

Tribe	Genus	Species examples	Common infections
Escherichiae	*Escherichia* *Shigella*	*E. coli* *S. dysenteriae* *S. boydii* *S. flexneri* *S. sonnei*	Diarrhea, UTI, bloodstream infections (BSI), dysentery
Edwardsiellae	*Edwardsiella*	*E. tarda*	BSI, wound infections
Salmonellae	*Salmonella*	*S. typhi*, *S. paratyphi* A and BR *S. typhimurium* and *S. enteritidis*	Enteric fever, diarrhea, food poisoning
Citrobacteriaceae	*Citrobacter*	*C. freundii* *C. diversus*	UTI, BSI
Klebsielleae	*Klebsiella Enterobacter Serratia* *Hafnia* *Pantoea*	*K. pneumoniae* *E. cloacae* *S. marcescens* *H. alvei* *P. agglomerans*	UTI, BSI, and pneumonia UTI and BSI in immunocompromised persons HAI
Proteeae	*Proteus* *Providencia* *Morganella*	*P. vulgaris* *P. mirabilis* *P. stuartii* *M. morganii*	UTI, HAI
Yersinieae	*Yersinia*	*Y. enterocolitica*	Diarrhea

- Voided in feces, it remains viable in the environment only for a few days
- The detection of thermotolerant *E. coli* in drinking water indicates fecal contamination of the water

Virulence factors

The virulence factors of *E. coli* are presented in Table 4.12.

II. *Klebsiella*

Morphology

- They are gram-negative, non-motile bacilli, which are capsulated short, plump, and straight, measuring about 1–2 × 0.5–1.8 μm in size.
- Their mucopolysaccharide capsules are often prominent, producing **mucoid colonies on solid media**.
- They can be made out even in Gram-stained smears as halos around the rods.

Habitat

Klebsiellae are widely distributed in nature, occurring both as **commensals in the intestines and as saprophytes in soil and water.**

Classification

They have been classified into three species—*K. pneumoniae*, *K. ozaenae*, and *K. rhinoscleromatis*—based on biochemical reactions and into over 80 serotypes based on the K antigens.

Table 4.12 Virulence factors of *E. coli*

Virulence factor	Function
Surface antigens	
Somatic lipopolysaccharide O antigen	Protection from phagocytosis and the bactericidal effects of complements. Endotoxic activity
Envelope or K antigens	Protection against phagocytosis and antibacterial factors in normal serum
Fimbriae	Initial attachment and colonisation
Colonisation factor antigens (enterotoxigenic *E. coli* causing human diarrhea)	Adherence
P fimbria (uropathogenic strains)	Binding to uroepithelial cells
Toxins	
Cytotoxic necrotising factor-1 and **siderophores** (in uropathogenic *E. coli*)	Biofilm production Adhesion
Enterotoxins	
Heat labile toxin (LT)	Causes diarrhea similar to cholera by activation of adenyl cyclase
Heat-stable toxin (ST)	Activation of cyclic guanosine monophosphate (cGMP) in the intestine to induce fluid accumulation
Verotoxin (or Shiga-like toxin)	Cytotoxic by inhibition of protein synthesis

Culture characteristics

- They grow well on ordinary media, forming **large, dome-shaped, mucoid colonies** of varying degrees of stickiness.
- On MacConkey agar, they grow as **lactose-forming** colonies.
- Fresh clinical isolates are often very mucoid, almost flowing down the surface of the medium.

Infections (Table 4.13)

Table 4.13 Infections caused by various species of *Klebsiella*

Species	Infection
K. pneumoniae	Pneumonia (community- or healthcare- or ventilator-associated), urinary tract infections including catheter-associated UTI, septicemia, and several pyogenic infections
K. ozaenae	Ozena, a disease characterised by foul-smelling nasal discharge
K. rhinoscleromatis	Rhinoscleroma, a chronic granulomatous hypertrophy of the nose

III. *Enterobacter*

Morphology

Enterobacter are **gram-negative, motile, capsulated, lactose-fermenting rods** which are **indole- and MR-negative** and **VP and citrate-positive**.

Habitat

These are normally found in feces, sewage, soil, and water and rarely in urine, pus, and other pathological materials.

Classification

Two clinically relevant species are *E. cloacae* and *E. aerogenes*.

Infections

Enterobacter may be responsible for healthcare-associated infections such as:

- Urinary tract infections (UTI)
- Healthcare-associated infections

Enterobacter has recently been shown to be resistant to several classes of antibiotics, which makes the treatment of infections caused by these organisms difficult.

IV. *Shigella*

Morphology

- Shigellae are short, gram-negative rods, about 0.5 × 1–3 µm in size. They are **non-motile, non-sporing, and non-capsulated.**
- Fimbriae may be present.

Classification

Shigella are classified into four species or subgroups based on a combination of biochemical and serological characteristics. Serotypes are distinguished within the species.

- ***S. dysenteriae* (subgroup A):** This species of mannitol non-fermenting bacilli comprises 12 serotypes.
  - ***S. dysenteriae* type 1** causes fulminant bacillary dysentery and hemolytic uremic syndrome (HUS). Type 1 was originally described by Kiyoshi Shiga (*S. shigae*). It is indole-negative and is the only member of the family that is always catalase-negative.
  - ***S. dysenteriae* type 2** (*S. schmitzi*) forms indole and ferments sorbitol and rhamnose. Serotypes 3–7 are responsible for dysentery.
- ***S. flexneri* (subgroup B):** This group is antigenically the most complex among shigellae. Based on type- and group-specific antigens, the members of this group have been classified into six serotypes (1–6) and several subtypes (1a; 1b; 2a, 2b; 3a, 3b, 3c, 4a, 4b, 5a, 5b). In addition, two antigenic 'variants' called X and Y are recognised, which lack type-specific antigens.
- ***S. boydii* (subgroup C):** This group consists of dysentery bacilli, which resemble *S. flexneri* biochemically but not antigenically. Eighteen serotypes have been identified so far.
- ***S. sonnei* (subgroup D):** It ferments lactose and sucrose late and is indole-negative. It is antigenically homogeneous but may occur in two forms—phase I and phase II. Organisms in phase II may be isolated from patients but are more common in convalescents and carriers.

Habitat

Human beings are the only natural hosts for shigellae.

Infections

- *Shigella* species cause bacillary dysentery.
- *Shigella dysenteriae* is associated with acute infection and may sometimes be associated with hemolytic uremic syndrome.
- *S. sonnei* causes the mild form of bacillary dysentery. In many cases, the disease may only be a mild diarrhea. However, *S. sonnei* infection continues to be the most common shigellosis in developed countries.

Virulence factors

- *S. dysenteriae* type 1 forms a exotoxin (**Shiga toxin**) that has exhibits **neurotoxicity, enterotoxicity, and cytotoxicity.**
- **Endotoxin** due to the LPS of gram-negative cell wall is also produced.

- Other cell surface virulence factors similar to *Esherichia coli* may be present.

V. *Salmonella*

The genus *Salmonella* consists of bacilli that parasitise the intestines of a large number of vertebrate species and infect human beings causing enteric fever, gastroenteritis, septicemia with or without focal suppuration, and the carrier state.

Morphology

- Salmonellae are gram-negative rods, about 1–3 × 0.5 μm in size.
- They are **motile with peritrichate flagella** (except *S. gallinarum* and *S. pullorum*).
- They **do not form capsules or spores** but may possess fimbriae.

Classification and nomenclature

DNA hybridisation studies have shown that there are two species in the genus *Salmonella*:

1. **Species *enterica***—which is further divided into six subspecies. Most human infections are caused by subspecies enterica which includes *Salmonella* Typhi, the causative agent of typhoid fever. As per the nomenclature system recommended by the WHO Collaborating Centre, the organism is named *Salmonella*, subspecies *enterica*, serovar Typhi or in short, *Salmonella* Typhi or *S.* Typhi (the serovar is not written in italics and also starts with a capital letter).
2. **Species *bongori***: Regarded as the *Salmonella* of lizards, it is known to cause diarrhea with fever and acute enteritis in children.

Habitat

- Salmonellae are strict parasites of animals or humans. *S.* Typhi, *S.* Paratyphi A and usually, but not invariably, *S.* Paratyphi B are confined to human beings.
- Other salmonellae are parasitic to various animals domestic animals, rodents, reptiles, and birds.
- Some species are host adapted, *S.* Abortus-equi found only in horses, *S.* Abortus-ovis in sheep, and *S.* Gallinarum in poultry. Others such as *S.* Typhimurium have a wide host range, affecting animals, birds, and humans.
- Infection in animals may vary from an asymptomatic condition to fatal and sometimes epizootic disease. *S.* Typhimurium and *S.* Enteritidis cause fatal septicemia in rats and mice.

Infections

Salmonellae currently comprise above 2,000 serotypes or species, all of which are potentially pathogenic. Members of the genus *Salmonella* cause the following diseases:

- **Typhoidal:** Members of the enteric fever group, which comprises the typhoid and paratyphoid bacilli, are exclusively or primarily human parasites. An infection may lead to a carrier state.
- **Non-typhoidal:** The food poisoning group essentially comprises animal parasites that can also infect human beings, producing gastroenteritis, septicemia, or localised infections.

Salmonellae cause the following clinical syndromes in humans:

- **Enteric fever**
- **Gastroenteritis or food poisoning**
- **Septicemia**, with or without local suppurative lesions

VI. *Yersinia*

Species that are incriminated in human infection:

i. *Y. enterocolitica*
ii. *Y. pseudotuberculosis*
iii. *Y. pestis*

Morphology

- *Y. pestis* is a short, plump, ovoid, gram-negative bacillus, measuring about 1.5 × 0.7 μm in size, arranged singly, in short chains or in small groups.
- In Giemsa or methylene blue stain, it shows **bipolar staining (safety pin appearance) with the two ends densely stained and a clear central area** (Fig. 4.13a and b).
- Pleomorphism is very common and is characteristically enhanced in media containing 3% NaCl.
- The bacillus is surrounded by a **slime layer** (envelope or capsule).
- It is **non-motile, non-sporing, and non-acid-fast**.
- *Y. pseudotuberculosis* and *Y. enterocolitica* closely resemble *Y. pestis* but can be distinguished based on their motility at 22°C (but not at 37°C).

Culture characteristics

- *Y. pestis*, the plague bacillus is **aerobic and facultatively anaerobic**. Growth occurs over a wide

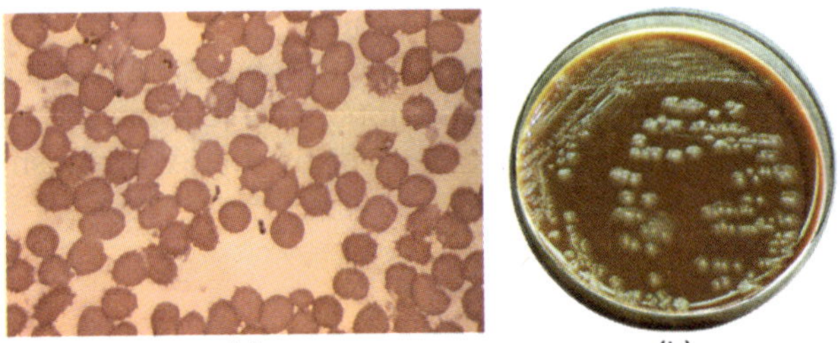

(a) (b)

Fig. 4.13 (a) *Y. pestis* in a blood smear exhibiting a safety pin appearance (bipolar staining) and (b) *Y. pestis* grown on chocolate agar (*Source*: CDC, PHIL, Image IDs 14555 and 12472)

range of pH (5–9.6; the optimum pH being 7.2) and temperature (range 2–45°C).
- The optimum temperature for growth (unlike most pathogens) is **27°C** but the envelope develops best at 37°C.
- They are not nutritionally exacting and grow on basal media.
- *Y. pseudotuberculosis* and *Y. enterocolitica* show relatively poor growth on MacConkey agar.

Infections
- *Yersinia pestis* causes **plague**, which has the potential to become an epidemic or a pandemic.
- *Y. enterocolitica* causes an acute bacterial enteric disease characterised by febrile diarrhea and enterocolitis that is indistinguishable from *Salmonella* and *Shigella* gastroenteritis.
- *Y. enterocolitica* and *Y. pseudotuberculosis* cause **acute mesenteric lymphadenitis** that mimics appendicitis.

Virulence factors
The virulence of *Yersinia pestis* is attributed to the following factors:
- **Fraction I or F-1:** A heat-labile protein envelope antigen that inhibits phagocytosis
- **V and W antigens:** Inhibit phagocytosis and intracellular killing of the bacillus
- **Bacteriocin (Pesticin I):** Inhibits strains of *Y. pseudotuberculosis*, *Y. enterocolitica* and *E. coli*
- **Endotoxin:** A lipopolysaccharide similar to the endotoxins of enteric bacilli

Yersinia enterocolitica produces heat-stable chromosomally encoded enterotoxin.

VII. *Serratia*
Serratia form a characteristic pink, red, or magenta, non-diffusible pigment called **prodigiosin** (Fig. 4.14), which is formed optimally at room temperature. Only one species of this genus is of medical importance—*S. marcescens*.

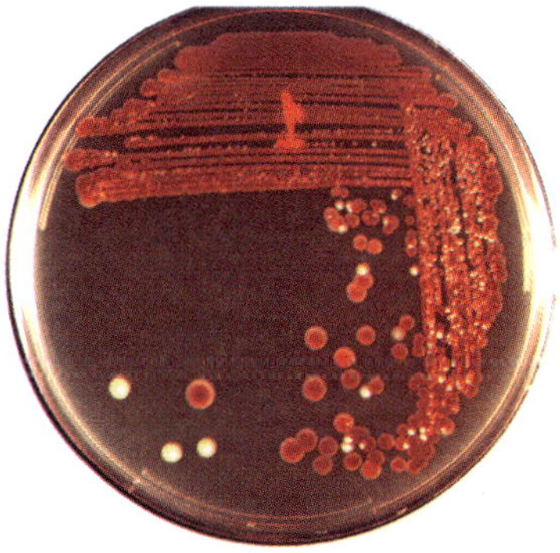

Fig. 4.14 Non-diffusible red pigment-producing *Serratia* spp. on blood agar with white colonies of *S. aureus* (*Source*: CDC, PHIL, Image ID 10544)

Morphology
It is pleomorphic, with minute gram-negative coccobacillary and rod forms.

Habitat
- They are saprophytes found in water, soil, and food.
- They may grow in sputum after collection. The prodigiosin pigment may be mistaken for hemoptysis ('**pseudohemoptysis**').
- They have been reported to contaminate IV fluids, surgical instruments, and antiseptic solutions.

Infections
- The bacillus has been associated with healthcare-associated infections such as meningitis, endocarditis, septicemia, peritonitis, respiratory infection, and many other conditions.
- Multiple drug resistance is common in hospital strains.

VIII. Proteeae
This group consists of three species:
i. *Proteus*
ii. *Morganella*
iii. *Providencia*
- *Proteus* is a group of bacilli that constitute the tribe Proteeae, which are non-lactose fermenters and are not considered as coliform bacilli.
- Most of these organisms, except for some Providencia strains, produce a powerful enzyme, **urease**, which rapidly hydrolyses urea to produce ammonia and carbon dioxide.
- A characteristic feature that distinguishes Proteeae from other coliforms is the presence of the enzyme phenylalanine deaminase which converts phenylalanine to phenyl pyruvic acid (PPA reaction); it is produced by all members of the tribe.
- Proteeae comprises intestinal commensals and opportunistic pathogens like the other coliforms. The name *Proteus* is inspired by the Greek god by the same name who could assume any shape; it alludes to the pleomorphism of these organisms.

i. *Proteus*
The genus *Proteus* comprises two medically important species and a third recently identified one.
- *P. mirabilis*—an important urinary and nosocomial pathogen
- *P. vulgaris*—found much less commonly in human infections
- *P. hauseri* (a biotype of *P. vulgaris* genomospecies 3)

Morphology
- All members of this species are gram-negative, non-capsulated, pleomorphic, and motile rods.

- They ferment several carbohydrates.
- *P. vulgaris* and *P. mirabilis* produce abundant hydrogen sulphide.

Habitat

- *Proteus* species are widely distributed in nature as saprophytes.
- They are found in decomposing animal matter, sewage, manured soil, and human and animal feces.
- They are frequently present on the moist areas of the body (perineal and axillary regions, and anterior urethra).
- They are opportunistic pathogens, often nosocomial, commonly responsible for urinary and septic infections.

Culture characteristics

- Smooth, colourless colonies are formed on MacConkey medium.
- A characteristic putrid odour described as a **'fishy' or 'seminal' odour** is produced by *Proteus* species in cultures.
- Some species (*P. vulgaris*) swarm on culture media.

Swarming (Fig. 4.15)

- *P. mirabilis* and *P. vulgaris* exhibit swarming on solid culture media due to the presence of flagella.
- Methods to inhibit swarming are:
  - Increasing the concentration of agar to 6%
  - Incorporating chloral hydrate (1:500), or sodium azide (1 :500), or alcohol (5-6%), or sulphonamide, or surface active agents or boric acid (1:1000) in the medium
  - Swarming does not occur on MacConkey medium due to the presence of bile

Infections

- *Proteus* species are commonly associated with **urinary tract and hospital-acquired infections**.

Fig. 4.15 Swarming of *Proteus* on blood agar (*Source*: Wikimedia Commons)

- Its ability to split urea to ammonia by urease causes the pH to increase in urine making treatment difficult.

ii. Morganella

- The genus *Morganella* has only one species, *M. morganii*.
- It is commonly found in human and animal feces and causes urinary infection or healthcare-associated wound infections.
- It does not swarm in culture.

iii. Providencia

The genus *Providencia* comprises three species that are associated with infections:

- *P. alcalifaciens* is sometimes seen in normal human feces but far more frequently in diarrheal stools, though its role in the disease is uncertain.
- *P. stuartii* is a common cause of urinary infection and infection in burns.
- *P. rettgerii* is part of the normal fecal flora of reptiles and amphibians, and sometimes causes nosocomial infections of the urinary tract, wounds, burns and blood.

Vibrio

The name '*Vibrio*' alludes to the characteristic **vibratory motility** of these organisms (from *vibrare*, meaning 'to vibrate').

Morphology

- Vibrios are gram-negative, curved bacilli that are actively motile by means of a polar flagellum.
- They are non–acid-fast and non capsulated.

Habitat

- Vibrios are present in marine environments and surface waters worldwide.
- The most important member of the genus is *Vibrio cholerae*, the causative agent of **cholera**.

Vibrio cholerae

Morphology

- The cholera vibrio is a short, curved comma-shaped rod, about 1.5 × 0.2–0.4 μm in size (Fig. 4.16).
- Pleomorphism is common in old cultures.
- In stained films of mucus flakes taken from acute cholera cases, the vibrios are seen arranged in parallel rows—this is described as a **'fish in a stream'** appearance.
- *V. cholerae* are **actively motile (darting motility)**, with a single, sheathed polar flagellum.
- This darting type of motility—resembling a **'swarm of gnats'**—can be seen microscopically in stool collected in cases of acute cholera or in a young culture.

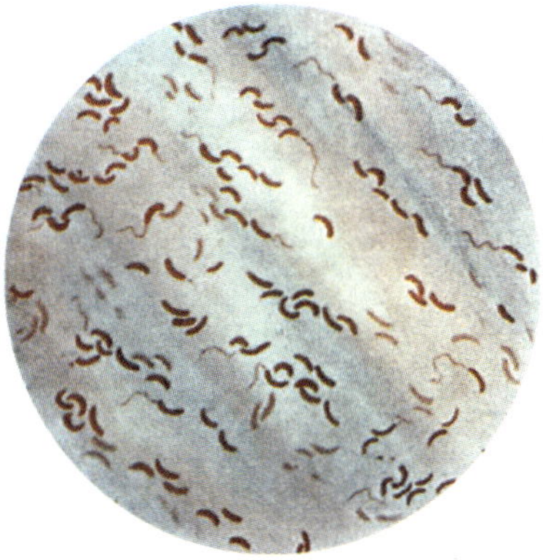

Fig. 4.16 *Vibrio cholerae*—comma-shaped gram-negative rods (*Source*: Wikimedia commons)

Culture characteristics

- *Vibrio cholerae* are aerobic; growth is slow and scanty in anaerobic conditions.
- They grow within a temperature range of 16–40°C (optimum 37°C).
- Growth is better in an alkaline medium, the range of pH being 6.4–9.6 (optimum 8.2) .
- NaCl (0.5–1%) is required for optimal growth, though high concentrations (6% and above) are inhibitory.
- They grow well on ordinary media.

Susceptibility

- Cholera vibrios are susceptible to heat, drying, and acids, but resist high alkalinity.
- They are destroyed at 55°C in 15 minutes.
- Survival in water is influenced by pH, temperature, salinity, the presence of organic pollution and other factors.
- Vibrios are killed in a few minutes in gastric juices of normal acidity but may survive for 24 hours in achlorhydric gastric juice. In clean tap water, they survive for thirty days.
- In untreated night soil, they may survive for several days.
- Vibrios are susceptible to common disinfectants.

Infections

V. cholerae are responsible for causing cholera, both as isolated cases as well as epidemics.

Halophilic vibrios

- Vibrios that have a **high requirement of sodium chloride** are known as halophilic vibrios.
- Their natural habitats are marine environments—rivers, deltas, shorelines, and ocean surfaces—and marine life.
- Some halophilic vibrios have been known to cause human disease—*V. parahaemolyticus*, *V. alginolyticus*, and *V. vulnificus*.

Vibrio parahaemolyticus

Morphology

- It is an enteropathogenic halophilic vibrio.
- It resembles the cholera vibrio except that it is capsulated and shows bipolar staining.
- Unlike other vibrios, it produces peritrichous flagella when grown on solid media and polar flagella when it is grown in liquid cultures.

Culture

- *V. parahaemolyticus* grows only in media containing NaCl.
- It tolerates salt concentrations up to 8% but not 10% with optimum salt concentration of 2–4%.
- **TCBS agar**—*V. parahaemolyticus* produce green colonies (as they do not ferment sucrose) with an opaque, raised centre and flat, translucent periphery (note that *V. cholerae* colonies are yellow in colour).
- The **string test is positive** for these organisms.
- **Kanagawa phenomenon**—strains isolated from environmental sources (such as water, fish, crabs, or oysters) are nearly always non-hemolytic when grown on a special high-salt blood agar (**Wagatsuma agar**). Strains isolated from human patients are almost always hemolytic due to a heat-stable hemolysin produced by these organisms. The significance of hemolysis is not known but it is used as a laboratory test of pathogenicity; only the Kanagawa-positive strains are pathogenic. No enterotoxin has been identified.

Susceptibility

- *V. parahaemolyticus* are killed at 60°C in 15 minutes.
- They do not grow at 4°C but can survive refrigeration.

Habitat

V. parahaemolyticus are common in the coastal seas, where they are found in fish and arthropods such as shrimps and crabs, and molluscs such as oysters.

Infections

They cause food poisoning and gastroenteritis.

Vibrio vulnificus

- There are three distinctive syndromic presentations of *V. vulnificus* infections—primary septicemia, wound infection, and gastroenteritis to a limited extent.
- *V. vulnificus* enters the stomachs of shellfish (primarily oysters) as well as the intestines of fish.
- Humans acquire the infection either by the consumption of raw or partially cooked contaminated seafood or via direct wound contact with contaminated water.

- *V. vulnificus* has a capsular polysaccharide, a lipopolysaccharide, and a flagellin homologous protein as **virulence factors**.
- Patients with chronic hepatic disease, diabetes, malignancies, renal disease, or HIV present with primary sepsis.
- Patients with primary wound infections present with swelling, blister formation, ulceration, hemorrhagic bullae, and abscesses; they may develop necrotising fasciitis and myonecrosis.

Vibrio alginolyticus

- It causes **otitis, cellulitis, and wound infection** and less often, gastroenteritis.
- It is most often seen in swimmers.

Vibrio vulnificus and *V. alginolyticus* can be isolated in alkaline pH and high salt content (10% NaCl). They differ in their carbohydrate fermentation reactions. Halophilic vibrio infections are treated with **third-generation cephalosporins, fluoroquinolones, or minocycline**.

Non-fermenting Gram-Negative Bacilli

Gram-negative non-fermenters are a group of unrelated bacteria that have gained notoriety in the recent past for acquiring antibiotic resistance across a wide spectrum of drugs and due to their involvement in healthcare-associated infections, which are associated with high morbidity and mortality.

Habitat

- They are ubiquitous in nature, often found in the environment, soil, and water.
- They can survive adverse conditions and are known to contaminate and survive in antiseptics and disinfectant solutions used in the hospitals that are diluted and prepared in-house. They are generally resistant to commonly used antibiotics in the hospital.
- *Pseudomonas* and *Burkholderia* are large groups of aerobic, non-sporing, gram-negative bacilli widely distributed in soil and water, while *Acinetobacter* is a gram-negative coccobacillus found in moist environments, both in the hospital and the community.

I. Pseudomonads

Classification

Based on molecular analysis, pseudomonads have been re-classified, and many former *Pseudomonas* species have been re-allocated to new genera such as *Burkholderia* and *Stenotrophomonas*.

Pseudomonas

Morphology

- *Pseudomonas* is a slender, gram-negative bacillus, 1.5–3 × 0.5 µm in size, actively motile by a polar flagellum.
- Mucoid strains, particularly isolates from cystic fibrosis patients, have an abundance of extracellular polysaccharides composed of alginate polymers.
- These form a loose capsule (glycocalyx), in which microcolonies of the bacillus are enmeshed and protected from host defences.

Culture characteristics

- It is an obligate aerobe and grows at a wide range of temperatures, 6–42°C, the optimum being 37°C.
- It grows well in ordinary media, producing **large, opaque, irregular colonies, with a distinctive, musty, mawkish or earthy smell.**

Habitat

- Pseudomonads are hardy organisms; they are mostly saprophytic and frequently colonise hospital environments, equipment, and moist areas.
- Some members of this group are pathogenic to plants, insects, and reptiles.

Virulence factors

- **Loose capsule (glycocalyx)**—microcolonies of the bacillus are enmeshed in this and protected from host defences
- **Exotoxin A**—inhibits protein synthesis
- **Other extracellular enzymes and toxins** such as proteases, elastases, hemolysins, and enterotoxin
- **Slime layer** and **alginate glycocalyx** and the ability to form **biofilms**
- **Pili**—enhances adhesion
- **Pigment production (pyocyanin, fluorescein, pyorubin, and pyomelanin)** causes tissue injury

Susceptibility

- *Pseudomonas* is killed at 55°C in one hour but exhibits a high degree of resistance to chemical agents.
- It is **resistant to common antiseptics and disinfectants** such as quaternary ammonium compounds, chloroxylenol, and hexachlorophene and **may even grow profusely in bottles of such antiseptic solutions** kept for use in hospitals.
- *P. aeruginosa* can grow in Dettol or cetrimide selective media.
- It is sensitive to acids, beta-glutaraldehyde, silver salts, and strong phenolic disinfectants.
- Its **susceptibility to silver** has been applied clinically in the use of silver sulphonamide compounds in topical creams for burns.

Infections

- A few pseudomonads cause human infections, typically, healthcare-associated infections and infections in immunocompromised individuals.

- This biofilm-producing bacterium is one of the **commonest causes of healthcare-associated infections**, including **catheter-associated infections and ventilator-associated pneumonia**.

II. *Burkholderia*

Burkholderia cepacia

- This organism is increasingly being recognised as an **opportunistic pathogen**, particularly among individuals who have cystic fibrosis or chronic granulomatous disease. In such individuals, *B. cepacia* causes fatal necrotising pneumonia.
- It is nutritionally very versatile.
- It can grow in many common disinfectants and can even use penicillin G as its sole source of carbon.
- It is inherently resistant to most antibiotics.

Infections

B. cepacia is commonly associated with the following:

- Cystic fibrosis
- Hospital-acquired urinary, respiratory, and wound infections
- Peritonitis, endocarditis, and septicemia

Burkholderia mallei

- **Morphology:** This is a slender, non-motile, gram-negative bacillus, 2–5 × 0.5 µm in size. It stains irregularly and often has a beaded appearance.
- **Culture:** It is an aerobe and facultative anaerobe, growing on ordinary media within a wide temperature range. Its colonies, which are small and translucent initially, become yellowish and opaque on ageing. It is quite inactive biochemically, breaking down only glucose.
- **Infection**: The human infection, **glanders**, is usually occupational—it affects ostlers, animal cleaners (grooms), and veterinarians.

Burkholderia pseudomallei

- This is a slender, motile, gram-negative bacillus, 2–5 × 0.5 µm in size. It resembles *B. mallei* but differs in its biochemical characters and motility.
- In fact, the name **melioidosis** is derived from *melis*, a disease of asses (glanders) and the word *eidos*, meaning resemblance.
- *B. pseudomallei* stains irregularly and often has a **beaded or a safety pin appearance**.

Infections

- *B. pseudomallei* is the causative agent of **melioidosis**, a glanders-like disease, epizootic in rodents in Southeast Asia, India and North Australia.
- The disease came into prominence after the Vietnam war, and hence, is also called **Vietnam time-bomb disease.**
- In India, cases are reported from the western and southern belts. Infections usually spike after the rainy season.

III. *Acinetobacter*

Morphology

- *Acinetobacter* are strictly aerobic, gram-negative coccobacillary organisms (Fig. 4.17), measuring 11.5 × 1.5–2.5 µm in size. They often appear in pairs, mimicking neisseriae in appearance.
- These organisms are non-motile, non-sporing, and non-acid-fast.

Culture characteristics

- They are aerobic and grow on routine laboratory culture media.
- *A. baumanii* can grow best at 44°C.
- They are oxidase-negative, nitrate-negative, and do not ferment sugars.

Classification

The classification of *Acinetobacter* has undergone many changes. Within the genus, the strain most commonly isolated in clinical laboratories is *Acinetobacter baumannii*.

Habitat

- Acinetobacters are opportunistic pathogens and are frequently present on normal skin.
- They are an important cause of healthcare-associated infections like **ventilator-associated pneumonia, meningitis, and bacteremia.**
- They can survive for long in the hospital environment and colonise almost all patients on prolonged hospitalisation.

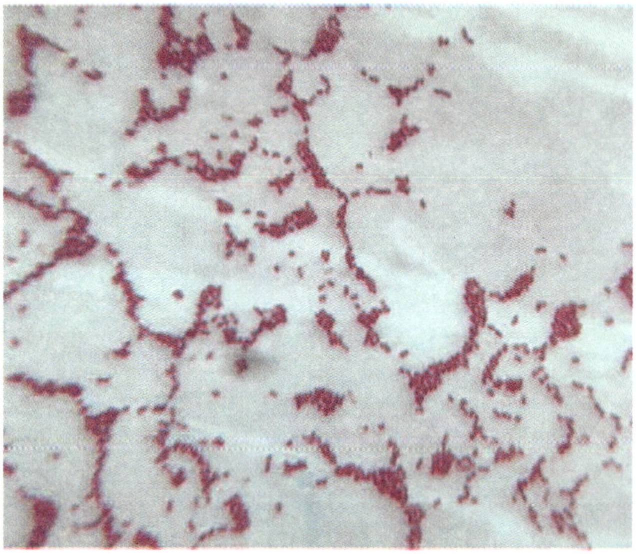

Fig. 4.17 *A. baumannii*—gram-negative coccobacillary forms (*Source*: Dept of Microbiology, PIMS)

Infections

Acinetobacter are usually multidrug resistant, causing **healthcare-associated infections, biofilm-based catheter-associated infections, bloodstream infections and ventilator-associated pneumonia**.

Fastidious Gram-Negative Bacteria

I. *Haemophilus*

The genus *Haemophilus* includes a number of species of small, pleomorphic, oxidase-positive, fastidious organisms which can cause infection in humans and animals.

Morphology

Haemophilus are gram-negative coccobacilli that are non-motile and non-sporing.

Culture

- *Haemophilus* have fastidious growth requirements. They require two accessory growth factors (**X or hemin and V or NAD**) that are present in blood. The V factor is a co-enzyme of nicotinamide adenine dinucleotide (NAD) or NAD phosphate (NADP), which acts as a hydrogen acceptor in the metabolism of the cell.
- They are aerobic but grow anaerobically also. The optimum temperature is 37°C; they do not grow below 20°C.
- Some strains require 10% CO_2, especially for primary isolation from the clinical specimen.
- They grow on blood agar if a source of the V factor is also provided.

H. influenzae

It is the first **free-living organism whose complete genome has been sequenced**.

Morphology

- *H. influenzae* are small (1.0 × 0.3 μm), gram-negative bacilli that exhibit considerable pleomorphism. In sputum, they usually occur as clusters of coccobacillary forms, while in CSF from meningitis cases, long, bacillary and filamentous forms predominate.
- Cells from young cultures (18–24 hours) are usually coccobacillary, while older cultures are distinctly pleomorphic. Strains isolated from acute infections are often capsulated.
- **Capsular typing** is of medical importance as about 95 per cent of *H. influenzae* isolated from acute invasive infections like meningitis and pneumonia belong to type B.

Culture characteristics

- The optimum temperature for their growth is 37°C. Some strains require 10% CO_2, especially for primary isolation from the clinical specimen. They grow on blood agar if a source of the V factor is also provided.
- The X factor is hemin or other porphyrins, while V factor is NAD. These are required for the synthesis of cytochrome along with other heme enzymes such as catalase and peroxidase involved in aerobic respiration.
- *H. influenzae* are is heat-stable.
- **Satellitism**—*H. influenzae* colonies exhibit satellitism when grown near *S. aureus* (Fig. 4.18a). This demonstrates the dependence of *H. influenzae* on the V factor, which is available in high concentrations near staphylococcal growth. Another differentiating test is growth around discs incorporated with X, V, and XV factors (Fig. 4.18b). This helps in differentiating the species of *Haemophilus*.

Virulence factors

- Capsular polysaccharide
- Outer membrane protein
- Lipooligosaccharides

Infections

They commonly cause **pneumonia and meningitis in children**.

H. ducreyi

Morphology

- It is a short, ovoid, gram-negative bacillus (1–1.5 × 0.6 μm) with a tendency to occur in end-to-end pairs or short chains.

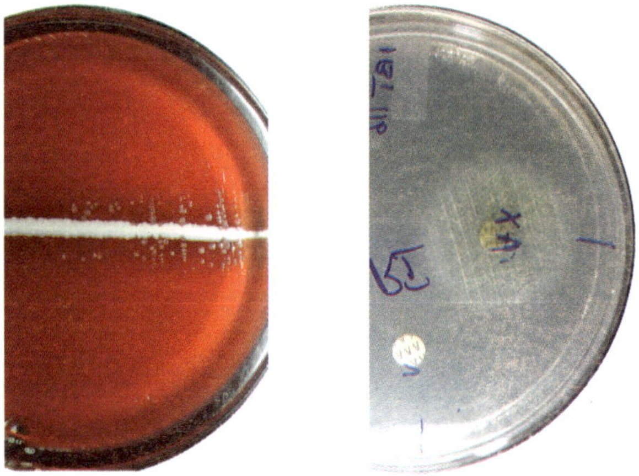

(a) (b)

Fig. 4.18 *H. influenzae*: (a) on blood agar showing satellitism around *S. aureus* streaks (*Source*: Dept of Microbiology, PIMS) and (b) demonstration of the requirement of XV factor

- It appears in groups or whorls or in parallel chains, giving a **'school of fish' or 'railroad track' appearance**.
- Smears made after 24–48 hours of incubation show tangled chains of bacilli.

Culture

- It can be grown on fresh clotted rabbit blood, on the chorioallantoic membrane of the chick embryo or on chocolate agar.
- *H. ducreyi* forms small, grey, translucent colonies after incubation at 35°C under 10% CO_2 and high humidity in 2–8 days.

Infections

It causes **chancroid or soft sore**—a sexually transmitted infection

II. *Brucella*

The species of importance are *Brucella melitensis*, *B. abortus*, and *B. suis.*

Morphology

- The genus consists of very **small (0.5–0.7 × 0.5–1.5 μm in size), non-motile, aerobic, gram-negative coccobacilli** (Fig. 4.19).
- In older cultures, irregular forms appear.
- They are non-capsulated, non-acid fast, and non-sporing.
- They are gram-negative, and bipolar staining is not uncommon.

Culture

- Brucellae are strict aerobes and do not grow anaerobically.
- *B. abortus* is capnophilic, many strains requiring 5–10% CO_2 for growth.

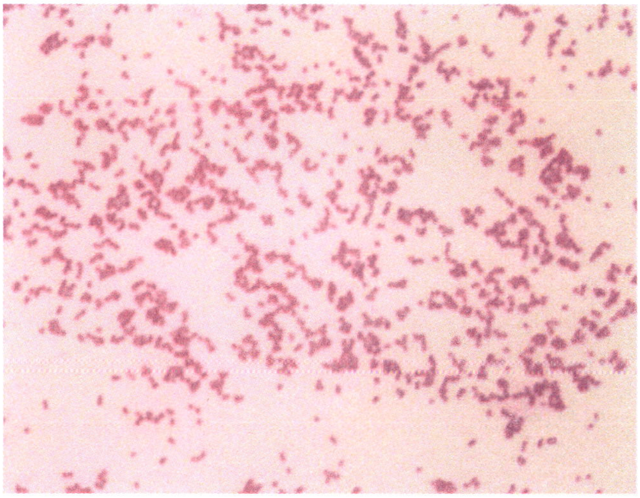

Fig. 4.19 *Brucella*—gram-negative coccobacilli (*Source*: Dept of Microbiology, PIMS)

- The optimum temperature is 37°C (range 20–40°C) and pH 6.6–7.4.
- They grow poorly on ordinary media and have little or no fermentative powers.
- **Erythritol** has an especially stimulating effect on the growth of brucellae.

Habitat

The organism primarily affects goats, sheep, cattle, buffaloes, pigs, and other animals and is transmitted to humans by **contact with infected animals** or their products.

Infections

The three species *Brucella melitensis*, *B. abortus*, and *B. suis* cause a zoonotic infection in humans called **brucellosis**, otherwise known as **undulant fever, Malta fever, or Mediterranean fever**.

III. *Bordetella*

- The genus *Bordetella* is named after the Belgian immunologist and microbiologist Jules Bordet.
- The species responsible for human infection is *B. pertussis*.
- *B. parapertussis* has been isolated from mild cases of whooping cough; *B. bronchiseptica* may occasionally infect human beings, producing a condition resembling pertussis.

Bordetella pertussis

Morphology

- It is a small, ovoid, gram-negative coccobacillus, 0.8 × 0.4 μm in size, of uniform size and shape.
- It is non-motile, non-sporing, and capsulated but tends to lose its capsule upon repeated cultivation.
- Bipolar metachromatic granules may be demonstrated on staining with toluidine blue.

Susceptibility

- *B. pertussis* is a delicate organism, readily killed by heat (55°C for 30 minutes), drying, and disinfectants. It survives at 0–4°C.
- Outside the body, *B. pertussis* survives for five days in dried droplets, three days on cloth and a few hours on paper.

Infections

Bordetella pertussis causes **pertussis or whooping cough** (meaning intense cough) in human beings.

Bordetella parapertussis

- This is an infrequent cause of whooping cough. The disease is mild.
- The pertussis vaccine does not protect against *B. parapertussis* infection.

Anaerobic Gram-Negative Bacilli

I. *Bacteroides*

- These are non-motile, strict anaerobes, usually pleomorphic, which appear as slender rods, branching forms or coccobacilli 0.5–1.5 × 1–6 µm in size.
- They are seen singly, in pairs, or in short chains.
- They grow well on media such as **brain–heart infusion agar** in an anaerobic atmosphere containing 10% CO_2.
- They possess **capsular polysaccharides**, which are the virulence factors. They evoke an antibody response that can be detected in patients.
- **Bacteroides** species cause deep-seated infections and brain abscesses.
- *Bacteroides* are normal inhabitants of the intestinal, respiratory, and female genital tracts.
- They are susceptible to metronidazole, clindamycin, and chloramphenicol.
- *B. melaninogenicus* is susceptible to penicillin, while *B. fragilis* is not.

II. *Fusobacterium*

- *Fusobacterium* is a long, thin, or spindle-shaped bacillus with pointed ends.
- *F. nucleatum* is a normal inhabitant of the mouth and is found in **oral infections and pleuropulmonary sepsis**.
- *F. necrophorum* produces a wide range of exotoxins and has been responsible for **liver abscesses** and other abdominal infections in animals and, less often, in humans.

III. *Leptotrichia*

- It contains a single species, *L. buccalis*, which was formerly known as Vincent's fusiform bacillus or *Fusobacterium fusiforme*.
- It is a long, straight, or slightly curved rod, 0.5–1.5 × 5-20 µm in size, often with pointed ends.
- It is part of the normal oral flora and is seen in acute necrotising lesions in the mouth.
- *Fusobacterium* cause **Vincent's angina**, which may resemble diphtheria with inflamed pharyngeal mucosa showing a greyish membrane that peels easily.
- Stained smears show large, fusiform and spiral *Borrelia bacilli*.

MISCELLANEOUS BACTERIA

I. Mycobacteria

Morphology

- Mycobacteria are slender rods that do not stain readily, but once stained, resist decolourisation with dilute mineral acids due to the presence of mycolic acid in their cell wall. For this reason, they are called **acid-fast bacilli (AFB)**.
- Mycobacteria are slow-growing, aerobic, non-motile, non-capsulated, and non-sporing bacteria.

Virulence factors

- The mycobacterial cell wall is rich in long-chain fatty acids called **mycolic acids**. These mycolic acids, when complexed with peptidoglycan, are responsible for **granuloma formation**. The pattern of these lipids on gas chromatography has been used to classify different species.
- Another factor called the **cord factor** is also responsible for the virulence of bacteria.
- **Lipoarabinomannan** is a cell wall-associated glycolipid and a virulence factor that can inhibit macrophage activation, leading to the persistence of mycobacteria within mononuclear phagocytes.

Culture characteristics

In liquid cultures, they form a **mould-like pellicle** (hence the name mycobacteria, meaning **fungus-like bacteria**).

Susceptibility

- Mycobacteria are killed at 60°C in 15–20 minutes.
- Bacilli in sputum may be viable for 20–30 hours, and in droplet nuclei, up to 8–10 days under suitable conditions.
- Cultures remain viable at room temperature for 6–8 months and may be stored for up to two years at –20°C.

Classification

The genus *Mycobacterium* comprises three groups—obligate parasites, opportunistic pathogens, and saprophytes.

Table 4.14 presents some common mycobacteria and their habitats.

- The microscopic demonstration of acid-fast bacilli provides only presumptive evidence of tuberculosis as even saprophytic mycobacteria may bear a similar appearance.
- Saprophytic mycobacteria stain uniformly without a barred or beaded appearance and are usually only acid-fast.
- These mycobacteria may be present in the environment and tap water and can contaminate clinical materials. Saprophytes may pose problems with gastric aspirates, feces, and urogenital specimens.
- Mycobacteria are also classified based on their ability to produce pigments when grown in the presence or absence of light.

Non-tuberculous mycobacteria

Some of these produce infections in immunocompromised individuals (Table 4.15).

Table 4.14 Some common mycobacteria and their habitats

Type	Members	Features
Obligate parasites	*M. tuberculosis, M. bovis, M. leprae*	Always considered pathogenic Transmitted from person to person There is no other known source of infection
Opportunistic pathogens	*M. scrofulaceum, M. kansasii, M. marinum, M. ulcerans, M. avium*-intracellulare complex (MAC)	Uncommon causes of human disease Infect immunocompromised individuals
Saprophytes	*M. smegmatis, M. gordonae, M. flavescens*	Found in soil and water Produce environmental contamination

Table 4.15 Runyon classification of non-tuberculous mycobacteria

Group	Features	Species	Human infection
Group I—photochromogens	Form colonies that do not produce a pigment in the dark, but when the young culture is exposed to light for one hour in the presence of air and re-incubated for 24–28 hours, a yellow-orange pigment appears	*M. kansasii* *M. marinum* *M. simiae*	*M. marinum*: Swimming pool granuloma *M. kansasi*: Chronic human pulmonary disease resembling tuberculosis
Group II—scotochromogens	Form pigmented colonies (yellow-orange-red) even in the dark	*M. scrofulaceum* *M. xenopi* *M. gordonae* (called 'the tap water scotochromogen')	*M. scrofulaceum* may cause scrofula (cervical adenitis) in children
Group III—non-photochromogens	Do not form pigment even on exposure to light	*M. avium* *M. intracellulare* *M. xenopi* *M. ulcerans*	Opportunistic infections in immunocompromised individuals Skin infections
Group IV—rapid growers	Colonies appear within seven days of incubation at 25°C or 37°C	*M. fortuitum* *M. abscessus* *M. chelonae* *M. smegmatis*	Opportunistic infections in immunocompromised individuals

Mycobacterium tuberculosis and *M. bovis*

Morphology

- *M. tuberculosis* is a straight or slightly curved rod, about 3 × 0.3 µm in size. Its size depends on the conditions in which it grows.
- Long, filamentous, club-shaped, and branching forms may sometimes be seen.
- *M. bovis* is usually straighter, shorter, and stouter.

Culture characteristics

- It is a obligate aerobe and a slow grower with generation time of 14–15 hours.
- Colonies appear from 2–8 weeks.
- The optimum temperature for growth is 37°C. Temperatures below or above 40°C do not favour growth.
- The optimum pH is 6.4–7.
- These are highly susceptible to traces of toxic substances like fatty acids in culture media.
- Special media for growth include **Lowenstein-Jensen (LJ) medium**.

Mycobacterium leprae

Morphology

- *M. leprae* are straight or slightly curved rods, 1–8 × 0.2–0.5 µm in size, showing considerable morphological variation.
- These mycobacteria may exhibit polar bodies and other intracellular elements. Clubbed forms, lateral buds, or branching may be observed.
- They are acid-fast, but less so than the tubercle bacillus. Hence, 5% sulphuric acid instead of 20% is used for decolourisation after staining with carbol fuchsin.
- The bacilli are seen singly and in groups, intracellularly or lying free outside the cells.
- The bacilli are bound together by a lipid-like substance, the glia. These masses are known as **'globi'**.

- The parallel rows of bacilli in the globi present a **'cigar bundle'** appearance.
- In tissue sections, the clumps of bacilli resemble cigarette ends. The globi appear in undifferentiated histiocytes called **Virchow's lepra cells or foamy cells.**

Culture characteristics

- Till date, it has not been possible to cultivate lepra bacilli either in bacteriological media or in tissue culture.
- **The generation time** of the leprabacillus is 12–13 days on an average but may range from 8–42 days. On the other hand, *Mycobacterium tuberculosis* has a generation time of about 14 hours and *E. coli* has a generation time of about 20 minutes.

Infections

- *M. tuberculosis* causes pulmonary and extrapulmonary tuberculosis, while atypical mycobacteria cause opportunistic pulmonary infections and skin infections.
- *M. leprae* causes leprosy.
- *M. bovis* produces tuberculosis in cattle, humans, primates, and carnivores including dogs and cats, badgers, swine, parrots, and some birds of prey.
- Saprophytic mycobacteria such as *M. phlei* are incapable of infecting human beings or animals and are distinct from non-tuberculous mycobacteria (NTM).

II. Spirochetes

Morphology

- Spirochetes are elongated, motile, flexible bacteria, twisted spirally along the long axis (the name is derived from *speira*, meaning 'coil' and *chaite*, meaning 'hair'; Fig. 4.20).
- They are structurally more complex than other bacteria.
- They have **endoflagella** (axial filament), which are polar and wound along the helical protoplasmic cylinder and are situated between the outer membrane and cell wall.
- Spirochetes vary widely in size, some being as long as 500 μm and others as short as 5 μm.
- Some of them can be stained by Gram stain and appear gram-negative (e.g., *Borrelia*, although they are NOT classified as gram-negative bacteria).
- Reproduction is by transverse fission.

Habitat

Many are free-living saprophytes, while some are obligate parasites.

Culture characteristics

Pathogenic treponemes do not grow in artificial culture media.

Classification

Among spirochetes, human pathogens are found in the genera *Treponema*, *Borrelia*, and *Leptospira* (Table 4.16).

III. *Mycoplasma*, *Ureaplasma*, *Chlamydia*, and *Chlamydophila*

- *Mycoplasma* and *Ureaplasma* are the smallest free-living bacteria that can be grown in a cell-free medium.
- They cause two categories of infections in humans—**pneumonia** caused by *M. pneumoniae* and **urogenital** infections by *Ureaplasma urealyticum*, *M. hominis*, and *M. gentalium*.
- Mycoplasmas may be saprophytic, parasitic, or pathogenic.
- More than 100 species of *Mycoplasma* are known to cause disease in a variety of mammalian, insect, and plant hosts.
- About 16 species belonging to three families are associated with human infection or colonisation.

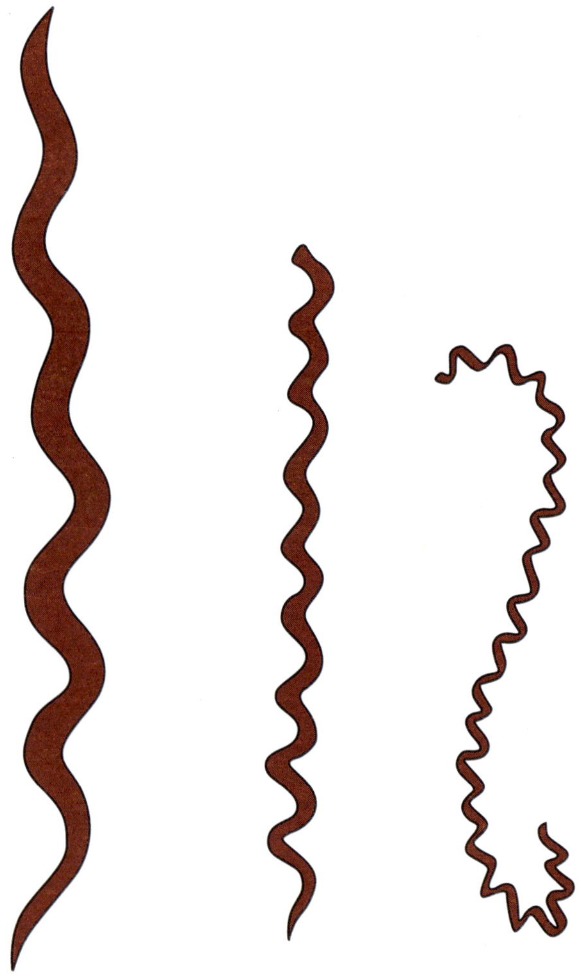

Fig. 4.20 Schematic representation of the comparative morphology of different spirochetes

Table 4.16 Characteristics of important human spirochetes

Properties	*Treponema*	*Borrelia*	*Leptospira*
Structure	• Short, fine, tight spirals at intervals of 1 μm with tapering ends • Movement is forward and backward	• Loosely coiled with spiral morphology and presence of periplasmic flagellum	• Tightly coiled spirals with hooked ends • Have a corkscrew movement
Staining	• Silver impregnation, Giemsa staining, or immunofluorescence • In unstained clinical material, they can be visualised by dark-field microscopy	• Giemsa, Gram stain • Appear gram-negative in isolates from humans • May be gram-positive when isolated from tissue culture	• Do not take up any stain • Visualised by dark-field microscopy
Culture	• Tissue culture • They can be propagated in rabbit testes for use in research	• Media containing serum or tissue extracts	• Ellinghausen, McCullough, Johnson, Harris (EMJH) medium
Habitat	• Non-pathogenic species can be found in oral, intestinal, and genital tracts	• Non-pathogenic species are found in the oral cavity	• Soil and water • Rats and other rodents
Pathogenic strains and disease spectrum	• Venereal: Syphilis • Non-venereal (endemic syphilis): Yaws, pinta, and bejel	• *B. burgdoferi*: Lyme disease (fever with rash and joint pain, headache, and swollen lymph nodes) • *B. recurrentis*: Relapsing fever	• *Leptospira interrogans*: Several subspecies cause leptospirosis, Weil's disease

Mycoplasmas

Morphology

- Mycoplasmas are bacteria that are devoid of cell walls and hence are highly pleomorphic, with no fixed shape or size.
- They **lack cell wall precursors** like muramic acid and diaminopimelic acid.
- The cells are bound by a soft, trilaminar membrane containing sterols.
- *Mycoplasma* occur as granules and filaments of various sizes. They may be of coccoid, balloon, disc, ring or star forms; the filaments are slender and of varying length and show branching.
- **Multiplication** is by binary fission.
- A distinctive feature seen in some species is a **bulbous enlargement with a differentiated structure of the tip**, by means of which the organisms attach to suitable host cells carrying neuraminic acid receptors. This enlargement may be responsible for the hemadsorption shown by some species.
- Mycoplasmas are gram-negative but are better stained by Giemsa stain.

Culture characteristics

- They have a specific requirement of **sterols** for growth and membrane synthesis.
- Mycoplasmas are chemoorganotrophs, their metabolism being mainly fermentative.
- They can pass through filters.

Classification

Mycoplasma comprise two genera of human importance (Table 4.17).

1. Genus *Mycoplasma*, which utilises glucose or arginine but does not split urea
2. Genus *Ureaplasma*, which hydrolyses urea

Susceptibility

- Mycoplasmas are destroyed at 45°C in 15 minutes.
- They can be lysed by surface-active agents and **lipolytic agents** such as taurocholate and digitonin.
- They are **resistant to penicillin and cephalosporins and to lysozymes** that act on bacterial cell walls.
- They are **sensitive to tetracycline, erythromycin, and other macrolides** as well as other antibiotics acting on intracellular organelles.

Table 4.17 Mycoplasmas responsible for human infections or colonisation: diseases, species, and site of colonisation

Established pathogen	Infections	Non-pathogenic	Site of colonisation
M. pneumoniae	Pneumonia	*M. orale, M. buccale, M. salivarium, M. faucium*	Oropharynx
M. hominis, U. urealyticum, and *M. genitalium*	Genital infections	*M. fermentans, M. genitalium, M. penetrans*	Genital tract

Infections

Mycoplasma cause a variety of infections including **atypical pneumonia and non-gonococcal urethritis along** with *Ureaplasma*.

Chlamydiae

- They have a **predilection for squamous epithelial cells and macrophages of the respiratory tract**.
- They were considered to be viruses on account of their filterability and failure to grow in cell-free media. However, they are now recognised as bacteria as they have many of their attributes, including:
  - Presence of both DNA and RNA
  - Presence of cell wall and ribosomes
  - Replication by binary fission without an 'eclipse phase'
  - Susceptibility to antibiotics

Habitat

- Chlamydiae are obligate intracellular bacteria that infect humans, animals, and birds.
- *C. pneumoniae* is an exclusively human pathogen with no animal or avian host.

Morphology

- The chlamydial cell wall resembles that of gram-negative bacteria, consisting of an outer layer, outer membrane complex (which possibly confers rigidity and stability), and an inner layer.
- A lipopolysaccharide is present in the outer membrane complex, with a small amount of peptidoglycan.
- Unlike other bacterial cells, chlamydiae do not have peptidoglycan in the cell walls.
- They lack enzymes for electron transport, can survive in an intracellular environment and cannot be cultured on artificial media.
- Chlamydiae occur in the following two forms:
  i. **Elementary body**: It is an extracellular, infective form. It is a spherical particle, 200–300 nm in diameter, with a rigid trilaminar cell wall similar to the cell walls of gram-negative bacteria, and an electron-dense nucleoid.
  ii. **Reticulate body (formerly called the 'initial body'):** It is the intracellular, growing and replicative form, measuring 500–1,000 nm in size. Its cell wall is fragile and pliable, leading to pleomorphism.

Classification

- Taxonomically, chlamydiae have been divided into two distinct genera based on their molecular nature:
  - Genus *Chlamydia* includes a single species, *Chlamydia trachomatis*
  - Genus *Chlamydophila* includes two species—*Chlamydophila pneumoniae* and *Chlamydophila;* however, from the medical perspective, these organisms are still known as *Chlamydia*
- Species differentiation is based on growth characteristics, nucleic acid profile, antigens, plasmids, and the nature of the inclusion body.
- *C. trachomatis* strains form compact inclusions with the glycogen matrix. They are sensitive to sulphonamides.
- *C. psittaci* strains form diffuse, vacuolated inclusions without the glycogen matrix, and are resistant to sulphonamides.

Infections

- *C. trachomatis* strains are natural parasites of humans, often causing localised infections of the eyes and genitals (trachoma and non-gonococcal urethritis).
- *C. psittaci* strains are natural parasites of birds and animals, capable of causing pneumonia and generalised infection in humans.
- *C. pneumoniae* is a common cause of acute respiratory disease worldwide and atypical pneumonia in humans.

IV. Rickettsiae

- The family Rickettsiaceae (Table 4.18) includes a diverse group of organisms that share the common features of intracellular growth and transmission by bloodsucking arthropod vectors (lice, fleas, ticks, and mites).
- Rickettsiae are small, gram-negative bacilli.
- They cannot be seen by the ordinary light microscope and are obligate intracellular parasites; in this aspect, they bear some similarity with viruses.
- However, they are categorised as bacteria based on the following features:
  - Cell wall made of peptidoglycan
  - Metabolic enzymes
  - Both DNA and RNA
  - Reproduction by binary fission
  - Susceptibility to antibacterial agents

Classification

- **Rickettsiaceae**—*Rickettsia* and *Orientia* (comprise several human pathogens)
- **Anaplasmatacea**—*Ehrlichia*, *Wolbachia*, *Anaplasma*, and *Neorickettsia*
- Of these genera, few are pathogenic to humans and have been liseted in Table 4.18.

Other allied arthropod-borne bacterial agents which have similar epidemiological and clinical features are dealt with in this chapter and include the following:

- *Bartonella* (*Rochelimia*), *Grahamella*
- *Coxiella*, *Rickettsiella*, *Piscirickettsia*
- *Eperythrozoon*

Table 4.18 Some common rickettsial infections in humans

Group	Species	Disease	Vector	Vertebrate reservoir	Distribution
Typhus	*R. prowazekii*	Epidemic typhus	Louse	Human beings	Worldwide
	R. prowazekii	Brill–Zinsser disease	Louse	Human beings	America, Europe, Australia
	R. typhi	Endemic typhus	Rat flea	Rat	Worldwide
	R. felis	Endemic typhus	Cat flea	Opossum	USA
Spotted fever	*R. rickettsii*	Rocky Mountain spotted fever	Tick	Rabbit, dog, small rodents	North America
group	*R. conori*	Indian tick typhus	Tick	Rodents	India
	R. akari	Ricketsial pox	*Gamasid* mite	Mouse	USA, Russia
Scrub typhus	*O. tsutsugamushi*	Scrub typhus	*Trombiculid* mite	Small rodents, birds	Asia, Pacific Islands, Australia

Other disease-causing bacteria of significance in humans include *Pasteurella*, *Francisella*, *Alcaligenes*, *Flavobacterium*, and *Klebsiella* (*Calymmatobacterium*) *granulomatis.*

NORMAL BACTERIAL FLORA OF THE HUMAN BODY

- The healthy human body possesses a natural flora of microorganisms (chiefly bacteria, and some fungi, protozoa, and viruses) that continuously inhabit it and may even be beneficial.
- These organisms are also called **commensals**, i.e., organisms that dine together, or microflora.
- The human body, which contains about 10^{13}cells, routinely harbours about 10^{14} bacteria. This bacterial population constitutes the normal microbial flora.
- The normal microbial flora is relatively stable, with specific genera populating various body regions during particular periods in an individual's life.
- The commensals from the normal flora of the mouth, nasopharynx, and intestinal tract may gain entry into the blood and tissues. Normally, they are quickly eliminated by the normal defence mechanisms of the body.

Types of Normal Flora

1. Resident flora

- It includes microbes that are always present at a body site.
- It promptly re-establishes itself even when removed.
- It consists of relatively fixed types of microorganisms regularly found in a given area at a particular age.

2. Transient flora

- It includes organisms that are present for very short time.
- It consists of non-pathogenic or potentially pathogenic microorganisms.
- These organisms may be temporarily established due to exogenous factors such as food.
- Example—transient bacteremia occurs after the brushing of teeth.

Factors Determining the Composition of the Normal Flora

- **Nature of the local environment** (age, pH, temperature, redox potential, and oxygen, water, and nutrient levels)
- **Local factors** such as site of body, peristalsis, saliva, lysozyme secretion, and secretion of immunoglobulins
- **Exogenous factors** such as clothing worn, occupation, antibiotic consumption, nutritional status, immunosuppressive, and metabolic disorders

Advantages

- It provides the host with **some essential nutrients** (e.g., microbial flora of the intestinal tract synthesise vitamin K and several B group vitamins).
- It aids in the **metabolism** of indigestible compounds.
- It provides **defence against colonisation** by opportunistic pathogens by providing a local barrier against them.
- It **aids the host's immunity**—the immune system is dependent on interactions with the human microbiome, as suggested by recent studies.
- It helps in the **prevention of inflammatory disease** during colonisation (as shown by some symbiotic bacterial species); the production of metabolic waste products such as acetic, butyric, and lactic acids by strict anaerobes inhibits the growth of pathogenic bacteria in the large bowel.
- In newborns, it stimulates the development **of the immune system**.

Disadvantages

- Some of the normal microbial flora **can act as opportunistic pathogens** and cause disease in immunocompromised individuals.

- Normal microbial flora often **contaminate clinical samples** collected from sterile sites (due to improper collection), causing confusion in laboratory diagnosis.

Site-Specific Normal Flora (Table 4.19)

The bacteria that make up the natural flora of various sites of the human body are presented in Table 4.19.

The longitudinal distribution, frequency of occurrence and densities of the bacteria making up the normal flora of the human gastrointestinal tract is presented in Fig. 4.21.

Changes in Normal Flora with Age

- A fetus is sterile when born (no normal flora). After birth, the newborn starts acquiring the normal flora from its mother, and the air, and the environment.
- The nasopharynx of the infant is sterile at birth but within 2–3 days, it acquires the common commensal flora and the pathogenic flora carried by the mother and her attendants.
- The mouth of the infant is not sterile at birth. A mixture of micrococci, streptococci, coliform bacilli, and Doderlein's bacilli—in about the same relative numbers as are present in the mother's vagina. These organisms diminish in number during the first 2–5 days after birth and are replaced by the types of bacteria present in the mouths of the mother and the nurse/attendant.
- Within 12 hours of birth, alpha-hemolytic streptococci are found in the upper respiratory tract and become the dominant organisms of the oropharynx and remain so for life.
- The nature of the flora in the vagina depends on the pH of its secretions and its enzyme content.
  - At puberty, the pH becomes acidic due to the metabolic activity of Doderlein's bacilli, *E. coli*, and yeasts. This change probably helps prevent colonisation by possibly harmful microorganisms.
  - During pregnancy, there is an increase in *Staphylococcus epidermidis*, Doderlein's bacilli, and yeasts.
  - After menopause, the flora resembles that before puberty.

Sterile Sites in the Body

Few bacteria are found in the normal bronchi. The smaller bronchi and the alveoli are normally sterile. The other sterile sites of the body are as follows:

- Spleen
- Liver
- Pancreas
- Bladder
- CSF
- Blood

Table 4.19 Microorganisms that are a part of normal flora

Site	Microorganisms constituting natural flora
Skin, outer ear, auditory canal	Staphylococci Streptococci Corynebacteria Propionibacteria Yeasts
Conjunctiva (sparse flora)	*Staphylococcus epidermidis* *Staphylococcus aureus* *Propionibacterium acnes* *Corynebacterium* species *Streptococcus* species *Neisseria* species *Haemophilus influenzae*
Nasopharynx	*S. pneumoniae* Coagulase-negative staphylococci *Haemophilus influenzae* *Moraxella* Commensal *Neisseria* species Diphtheroids Mixed anaerobes *Candida* *Actinomyces*
Mouth and oral cavity	Viridans streptococci Lactobacilli Staphylococci (*S. aureus* and *S. epidermidis*) *Corynebacterium* spp. *Bacteroides* spp. *Borrelia* and anaerobes (*Fusobacterium* and *Bacteroides*) *Streptococcus sanguis* (dental plaque) *Streptococcus mutans* (dental plaque) *Actinomyces* species *Candida* species
Nares	Staphylococci *Corynebacterium* spp.
Stomach	*Helicobacter pylori*
Small intestine	Enterobacteriaceae Enterococci *Bifidobacterium* *Candida*
Colon	*Bacteroides* *Bifidobacteria* *Clostridium* Peptostreptococci
Anterior urethra (sparse flora)	Staphylococci *Corynebacteria* Enterobacteriacae
Vagina	Lactobacilli Streptococci *Corynebacteria* *Candida* *Actinomyces* *Mycoplasma hominis*

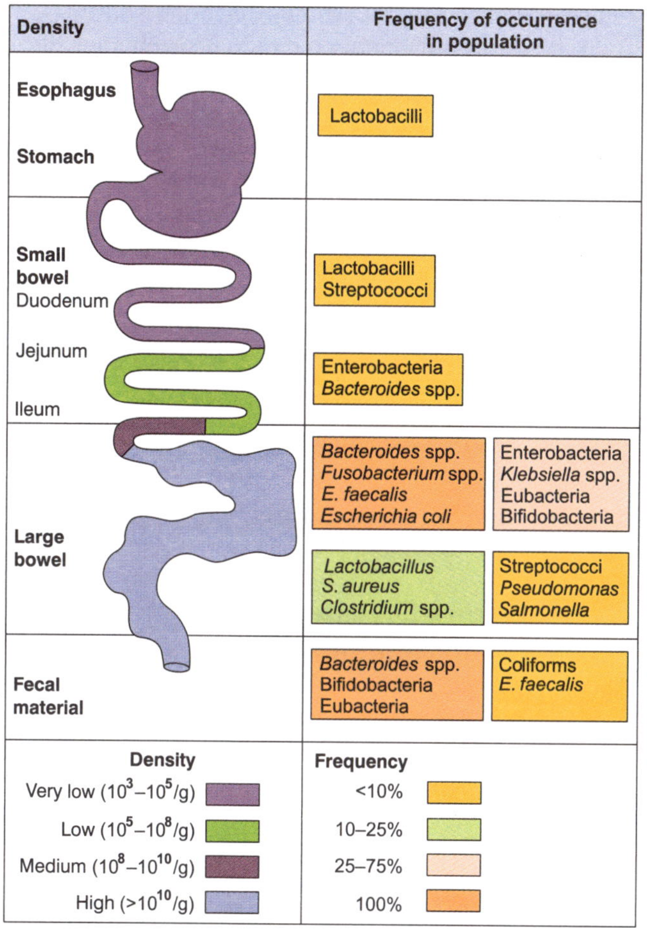

Fig. 4.21 Longitudinal distribution, frequency of occurrence, and densities of the bacteria making up the normal flora of the human gastrointestinal tract

- Upper genitourinary tract (male and female genitals, kidneys, ureters, and urinary bladder)
- Middle and inner ear

Probiotics

- Probiotics are live microorganisms found in the normal human body.
- These organisms can have multiple interactions with the host, including competitive inhibition of pathogenic bacteria.
- When administered in adequate amounts, they confer health benefits to the host. Some of their functions are as follows:
  - Inhibition of other pathogenic microbes in the gut by rapid colonisation
  - Effects on the mucosal barrier function and interaction with antigen-presenting dendritic cells
  - They have been successfully used in antibiotic-associated diarrhea
- The two most common probiotics are:
  - *Bifidobacterium*
  - *Lactobacillus*

Normal Flora and the Immune System

- The normal bacterial flora of the human body induces an immune response at lower levels than pathogenic bacteria.
- It provides the minimum stimulation for the immune system of the human body to remain attuned.
- However, the exact mechanism and the role of the immune system regulation brought about by the normal flora is not known.

Implications of Alteration of Normal Bacterial Flora

- There exists a delicate balance between the human body and the microbial flora that live symbiotically or as commensals in the host.
- Any factor affecting this balance results in adverse effects.
- Some of the major factors responsible for alterations in the balance, which result in infection, are as follows:
- **Prolonged antibiotic therapy, immunosuppressive agents, and immunodeficiency diseases**: Indiscriminate administration of antibacterial agents may completely eliminate the resident microorganisms, permitting exogenous and endogenous pathogens to gain the upper hand and cause disease.
- **Alteration of pH (as in the genitourinary tract**: This occurs in the elderly and during pregnancy.
- **Increased virulence of the commensals, production of toxins, and change in microbial flora** also occur.

Pseudomembranous colitis (PMC)

- When patients are **treated with broad-spectrum antibiotics** administered orally, diarrhea may be a side effect due to the overgrowth of yeast or staphylococci.
- When certain drugs, especially clindamycin, penicillins, fluoroquinolones, and cephalosporins are administered, the anaerobic gram-positive bacillus *Clostridium difficile* multiplies due to the suppression of other members of the flora; this may result in a serious, life-threatening condition known as pseudomembranous colitis.

Prevention

Rational use of antimicrobial agents should be prioritised and, wherever possible, narrow-spectrum antibiotics should be used at the correct dosage and for the correct duration of time to prevent the suppression of gut flora.

Treatment

- **First episode**: Oral vancomycin (500 mg/day) is recommended.
- **Initial recurrence**: Oral vancomycin—pulse or tapered therapy. In case of three or more recurrences

despite treatment with a tapered/pulsed dose oral vancomycin regimen, **fecal microbiota transplant (FMT)** should be considered.

Fecal microbiota transplantation

- Gut bacteria from a healthy donor are introduced into a patient by infusing a fecal sample via nasogastric tube, nasoduodenal tube, rectal enema, or the biopsy channel of a colonoscope. These bacteria colonise the gut.
- Fecal microbiota transplantation is now used in the treatment of conditions like irritable bowel syndrome and recuurent non responding PMC.

DIAGNOSIS OF BACTERIAL INFECTIONS

CLINICAL DIAGNOSIS

- Knowledge of the natural history of the infection, its geographic preponderance and clinical presentation is essential to the diagnosis of bacterial infections.
- In addition to thorough physical examination, a **detailed history** of the geographic location of the person (to rule out endemic infection, e.g., malaria), travel (to consider an infection from other areas, e.g., arbovirus), occupation (zoonotic infections, e.g., anthrax) and behaviour (addiction, e.g., HIV in IV drug abuse) is key to the diagnosis of infectious diseases.

LABORATORY INVESTIGATIONS

- These are aimed at establishing an etiological diagnosis. Hence, they must be carefully considered to get the best result in the shortest possible time and must be cost-effective.
- Laboratory investigations use scientific methods for the rapid and accurate diagnosis of infectious diseases.
- Once the causative agent of a particular infection has been identified, appropriate and rational therapy can be instituted.
- The details of laboratory diagnosis have been discussed in individual chapters. This chapter gives an overview of the laboratory process (Fig. 4.22), which consists of the following steps.

The factors that determine the correct identification of the bacterial agent and sound patient care are as follows:

- **Collection of appropriate clinical specimens** at the right stage (time) of the disease and from sites most likely to yield the microorganism
- **Performing tests** using standard operating procedures
- **Meticulous documentation** of the test results and effective communication of these results to the treating clinician
- **Application of test results** while deciding on appropriate treatment

Pre-test Process (Pre-analytical Phase)

Specimen collection and transport

The outcome of a laboratory test depends, to a large extent, on the appropriateness of the specimen in terms of selection, quality, quantity, time (prior to antibiotics or other treatment), and method of collection.

Precautions for sample collection

- **Asepsis** during the collection of the sample is essential to avoid contamination by skin flora or environmental organisms.
- The container in which the sample is to be collected must be sterile and should have sufficient capacity to contain the material.
- Samples for anaerobic culture require an anaerobic container or transport media.
- The clinical staff collecting the sample must ensure adequate **personal protection** to avoid getting exposed to the infection, e.g., wearing personal protective equipment (PPE) while collecting samples from patients with aerosol-generating respiratory infections (e.g., viral respiratory infections like H1N1 or SARS-CoV).

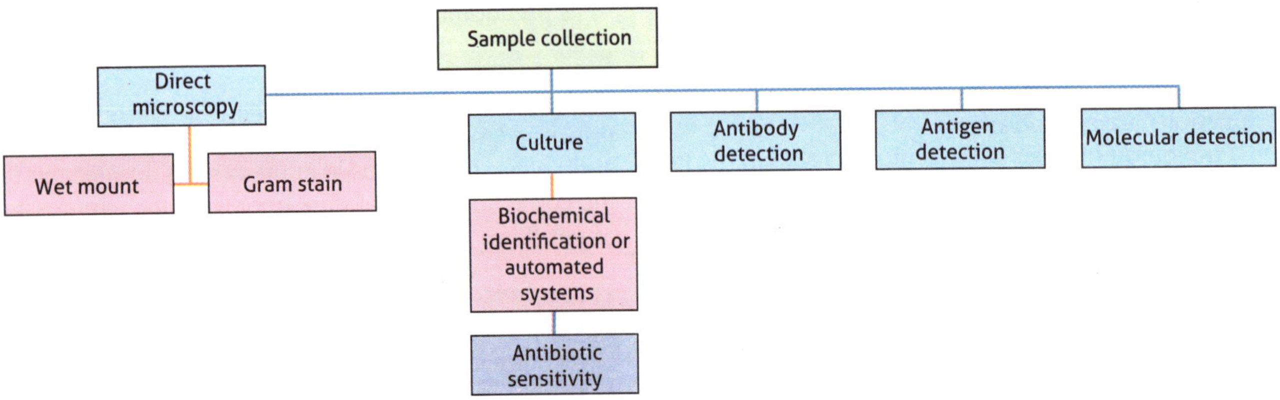

Fig. 4.22 Outline of the laboratory diagnosis of bacterial infections

- The specimen should be transported without undue delay to the laboratory in tightly closed containers to avoid spillage.
- If this is not possible, it should be stored at an appropriate temperature for a specified time until it can be transported (CSF and blood culture at room temperature while other samples may be refrigerated).

Sites of sample collection

- The sample should be representative of the infection. For example, sputum and not saliva should be collected when pulmonary tuberculosis is suspected.
- The type of specimen to be examined is determined by the clinical presentation of the infection (Table 4.20).

Test Procedures and Processing of the Specimen in the Laboratory (Analytical Phase)

- Once the appropriate specimens have been collected, they are **processed as speedily as possible** to ensure that the organisms do not die before being transferred to the culture media.
- Doing so also ensures that the reports are available as early as possible.
- The specimens processed in the microbiology laboratory are potentially hazardous since the etiological agents are unknown at the time of processing. For this reason, it is advisable to **process all biological samples in a biosafety cabinet.**
- A sample could be of moderate or high individual risk but of low risk to the community (biosafety level 1 and 2). On the other hand, organisms that have high individual risk and also pose high risk to the community need to be processed in containment laboratories (level 3) with high levels of biosafety.
- Based on the category of the infectious agent to be processed, biosafety cabinets are designed to control air circulation by HEPA (high-efficiency particulate air) filters within the cabinet. In a diagnostic microbiology laboratory, class I and class II biosafety cabinets are used.

Table 4.20 Sites of infection and corresponding samples for microbiological tests

Site	Sample
Skin	Scrapings, swabs, and biopsies
Pus	Aspirates and drained fluid
Nail and hair	Clippings
Conjunctiva	Swabs
Ear	Swabs
Respiratory tract • Upper tract • Lower tract	Throat swabs, nasopharyngeal/ nasal swabs/nasal wash/sinus aspirates Sputum (not saliva), bronchoalveolar lavage
Gastrointestinal tract	Stool samples, vomitus, endoscopic biopsies
Genitourinary tract	Vaginal and cervical swabs, urethral discharge
Urinary tract	Urine, suprapubic aspirate
Central nervous system	Cerebrospinal fluid (CSF)
Bloodstream infections	Blood, bone marrow
Intraoperative samples from internal organs	Clotted blood (serum) or plasma for immunological tests to demonstrate antigens or antibodies

The steps for the isolation and identification of the causative agent of an infection are detailed below.

Gross examination of the specimen

This may help to correlate the specimen with the clinical condition, e.g., bloodstained CSF may suggest traumatic tap or subarachnoid bleed due to anthrax meningoencephalitis; larvae or segments of worms may be seen; cobweb formation in CSF may be seen; high protein may be seen in tubercular meningitis.

Microscopy

- This often helps in giving a rapid diagnosis.
- If bacteria and fungi are swiftly detected, specific therapy can be started at once.

Unstained preparations

- A portion of the specimen is placed on a clean slide. Immediately, a coverslip is applied over the sample for examination and then the slide is examined under the microscope.
- This provides the visualisation and observation of microorganisms and cells in real-time. It can be used to identify the presence of bacteria, fungi, parasites, and other microscopic organisms in biological samples.
- KOH can be added to wet mounts to visualise fungal elements.

Stained preparations

- **Gram stain**
  - It is the most common stain used in diagnostic microbiology.
  - It helps in provisionally identifying the organism as gram-positive or -negative. Based on this empiric evidence, antibiotics may be started.
  - The Gram stain is found most useful for the rapid diagnosis of pyogenic meningitis.
  - Sterile body fluids and aspirates can also be stained for rapid diagnosis.

- This can also be used to detect inflammatory cells in the specimen, thus providing additional information.

- **Acid-fast staining (Ziehl–Neelsen and Kinyoun stain):** This is used for the rapid identification of acid-fast bacilli (mycobacteria) and oocysts of coccidian parasites.
- **Fluorochrome staining (calcofluor white or acridine orange):** Helps in the visualisation of fungal elements.
- **Special stains**: Albert stain or India ink can be used.

Culture and identification

- Based on the outcome of the preliminary microscopy result, a specific medium and method of incubation (at 37°C, ambient temperature or in CO_2) can be selected for the culture of the specimen.
- Generally, after overnight incubation, the colonies on culture plates are processed for the identification of the organisms.

Identification

Some of the common tests used to identify bacteria are as follows:

- Hanging drop test for motility
- Breakdown of carbohydrates, glucose, sucrose, and lactose using composite media like triple sugar iron agar (TSI)
- Tests for oxidase, the production of gas and hydrogen sulphide in gram-negative baccili (GNB)
- Catalase, coagulase test, hippurate hydrolysis, bacitracin susceptibility for gram-positive cocci (GPC)
- Utilisation of citrate as the sole source of carbon and tests for the production of enzymes like oxidase and urease
- Serotyping with specific antisera (*Salmonella* and *Shigella*)

These tests identify most of the common gram-negative pathogenic bacteria.

Antibiotic susceptibility test (AST)

- In this test, organisms are simultaneously subjected to antibiotic susceptibility tests by the Kirby–Bauer disc diffusion method to check their susceptibility.
- Some laboratories identify microorganisms by automated VITEK/MicroScan systems.

Rapid culture and identification methods

- To decrease the turnaround time or TAT (the time from the collection of the sample to the reporting of the results to the physician), several automated and semi-automated rapid diagnostic methods such as automated culture and identification systems (MALDI-ToF, VITEK, MicroScan, etc.) are used.
- These tests are especially useful in the diagnosis of life-threatening infections like acute pyogenic meningitis, peritonitis, sepsis, and some viral infections, which need to be diagnosed rapidly to start treatment immediately.

Immunological tests for rapid antigen detection

Latex agglutination test can be performed for the detection of *Haemophilus*, *Pneumococcus*, *Meningococcus*, group B *Streptococcus*, or *E. coli* antigen from CSF in acute pyogenic meningitis.

Molecular methods

- These methods detect the DNA or RNA of the infectious agents.
- Among the various molecular methods available today are the conventional PCR, real-time PCR, multiplex PCR and a recent innovation, the BioFire FilmArray, which is a multiplex PCR system that integrates sample preparation, amplification, detection, and analysis (see Chapter 8 for details).

Serological tests for the detection of IgM and IgG

- Etiological agents cannot always be isolated or detected in the laboratory by conventional or automated culture systems.
- This challenge is overcome by detecting specific antibodies in the patient's serum.
- The presence of IgG denotes past exposure to the infection (this is the principle of the *Brucella* agglutination test), while IgM indicates recent infection.

Post-test Process (Post-analytical Phase)

- This is a crucial step in the laboratory process.
- The results of tests must be conveyed to the treating physician who requested the investigation.
- They must be conveyed in standard reporting formats in such a way that the physician or patient is able to get an accurate and reliable picture in a manner that is clear and easy to understand.
- An inaccurate or incomplete report might put the patient in danger of receiving the wrong treatment or inadequate management.
- To make the most of laboratory investigations, there needs to be effective communication between the clinician and laboratory personnel.
- It is, therefore, prudent to consult the laboratory on the tests to be performed, the types of specimens to be collected for the tests, and the mode of transport of these specimens.

Ethical Issues Involved in Processing and Reporting Test Results

- Laboratory investigations are a part of the management of an individual seeking healthcare. This process is bound by the same ethical principles as the rest of the doctor–patient relationship.
- As the laboratory procedures involve patient's biological material in the form of samples, they should be treated with respect under the ambit of ethical principles.
- Several ethical principles are applicable in the laboratory with respect to the following:

1. **Collection of samples:** Respect for a person's autonomy and ensuring the patient's privacy are important. Before collecting a sample, informed consent must be taken from the patient.
2. **Processing and storage of samples**: All samples collected from patients must be handled with care and processed as required. Generally, additional unwarranted testing must not be performed without the prior consent of the patient/their legally authorised representative. Samples may be stored only up to the period specified.
3. **Reporting of results:** Confidentiality of the test results must be maintained. The results of investigations must be conveyed only to the treating doctor/requisitioner and the patient. Any breach of confidentiality can have legal implications and also impact the doctor–patient relationship. In the case of HIV and genetic tests, post-test counselling is required.
4. **Use of biological material for a purpose other than what was intended**: This is applicable to areas of human research and clinical trials. Material that is collected as part of the study with due consent must be used only for what it is intended for, and not for any other purpose. The patient's autonomy is of paramount importance.
5. **Sharing of results, data, and patient's details**: An important component of confidentiality in the process is to ensure that the patient's details and test results are not disseminated to others. If the results are intended to be published in scientific literature, then prior permission is sought from the patient. This is a component of ethical human research.
6. **Access to laboratory and data**: It must be ensured that the laboratory is accessed only by authorised personnel. This is essential to safeguard patient data stored electronically or as hard copies.

These guidelines are bound by the medical code of ethics and must be adhered to by all investigative laboratories. All accrediting bodies (for laboratories) make it mandatory for laboratories to strictly adhere to these principles.

Quality Control in a Clinical Microbiology Laboratory

- Clinical microbiology laboratories play an important role in ensuring the quality of care.
- This is ensured by adhering to quality control in all laboratory procedures. In clinical microbiology laboratories, a number of procedures are used to generate test results.
- These tests include direct smears and the use of stains such as the Gram stain, and the use of different aerobic and anaerobic media, including fungal and mycobacterial media for the culture of the suspected pathogens.
- Numerous tests are performed for the identification of these pathogens—biochemical tests, enzymatic tests, and antibiotic sensitivity tests are routinely performed. Immunological tests are performed to detect antigens or antibodies from the blood/serum or other specimens from patients.
- Clinicians and infectious disease specialists need to be aware of the quality control systems used in microbiology laboratories.
- Such awareness helps in understanding the laboratory process, right from specimen collection to the receipt of test results.

Quality check

Procedure

- Each test and procedure, each type of medium used, and even the reporting of results must adhere to prescribed standards.
- The quality of the procedures must be regularly checked by quality control (QC) measures, wherein procedures are tested against standard controls whose range of activity is known.
- When tested in laboratory conditions, they must give predetermined results.
- These checks should be performed periodically to ensure the quality of the laboratory procedure.
- The frequency and type of QC varies according to the test, medium, or reagent. For some tests like coagulase or catalase, QC must be performed for every test. For other tests, QC need only be performed when a new lot number is to be used (e.g., when a new batch of antibiotic discs is used).
- Automated systems in the laboratory like the ELISA or blood culture systems require the calibration of the instruments to give reliable results.

Results

Quality checks for reliable results, which ultimately benefit the patient, need to be mandatorily performed and documented for verification.

Laboratory Accreditation

- It is a process by which an external agency (a regulatory body) assesses a laboratory's practices by checking the processes and procedures and certifying specific tests to ensure that they meet the standards of the **International Organization for Standardization (ISO).**
- In India, the **National Accreditation Board for Testing and Calibration Laboratories (NABL)** provides certification for tests after having them assessed by a team of assessors and ensuring that they fulfil the standard quality criteria set down by the ISO.
- This certification is valid for a specified period of time and pertains only to the tests assessed. NABL accreditation of laboratories is gradually becoming mandatory for laboratories to meet the quality criteria for patient care.

Key Points

- ✓ Taxonomy is the science of classification. The classification of microorganisms is based on their arrangements into related groups or taxa. The hierarchical levels of classification are as follows: Kingdom → Division → Class → Order → Family → Tribe → Genus → Species.
- ✓ Since all organisms are classified primarily to enable easy identification, all classification systems aim to group organisms with similar properties and to distinguish them based on their differences. The basic taxonomic unit of bacteria is species; two species differ from one another in several features determined by genes.
- ✓ The diagnosis of infectious diseases relies upon the isolation and identification of the causative agent or demonstration of the specific antigen in infected tissues.
- ✓ Gram-positive cocci—*Staphylococcus, Streptococcus, Pneumococcus,* and *Enterococcus.*
- ✓ Gram-positive bacilli—*Bacillus, Corynebacterium, Actinomycetes,* and the anaerobic *Clostridium.*
- ✓ Gram-negative bacilli—enterobacteria, vibrios, and non-fermenters.
- ✓ A large group of bacteria with different growth characteristics and habitats are grouped as miscellaneous bacteria, as they do not have any common properties. They include mycobacteria, spirochetes, *Mycoplasma, Chlamydia,* and rickettsiae.
- ✓ The normal microbial flora is more or less constant for each species and is broadly divided into residents and transients.
- ✓ This flora prevents or interferes with colonisation/invasion of the body by pathogens and raises the overall immune status of the host against pathogens. However, these organisms can become pathogenic when host defences falter.
- ✓ Normal flora cause confusion in diagnosis due to their ubiquitous presence in the body and their resemblance to some pathogens.
- ✓ General procedures followed in the laboratory include:
  - Microscopic examination of the sample smeared and stained on a slide by Gram, acid-fast, or other specific stains
  - Culture on a battery of solid (basal/differential, enriched) or liquid media
  - Processing of culture isolates for further characterisation and identification
  - Antibiotic susceptibility test
  - Generation of final report for the clinician or patient
  - Quality control in all test procedures must be adhered to for reliable reports

Questions

Short answer questions

1. Mention the important reasons for classifying bacteria.
2. What are the major morphological groups under which gram-positive bacteria are classified with examples of their clinical implication?
3. Outline the procedures adopted in the laboratory to isolate and identify pathogens.
4. Give an account of normal flora in health and disease.

MCQs

1. ***S. aureus*** **infections are mediated by the following except _______.**
   a) Exotoxins b) Enzymes
   c) Endotoxins d) Metabolic products
2. **Which of the following is the Lancefield grouping of streptococci based on?**
   a) Protein antigen b) Carbohydrate antigen
   c) Capsular antigen d) Teichoic acid antigen
3. **Which of the following describes meningococci?**
   a) Non-capsulated b) Oxidase-negative
   c) Gram-positive diplococci d) Gram-negative diplococci
4. **The best agent for destroying clostridial spores is ______________.**
   a) 1% phenol b) 0.5% iodine
   c) 75% alcohol d) 2% glutaraldehyde
5. **Which of the bacteria belonging to the family Enterobacteriaceae are always pathogenic to humans?**
   a) *E. coli* b) *Shigella*
   c) *Klebsiella* d) *Yersinia*
6. **Phenotypic classification refers to classification based on:**
   a) Direct analysis of genes in microbes
   b) Chromosomal and extrachromosomal DNA analysis of microbes
   c) Expressed characteristics of microorganisms
   d) Nucleotide sequencing of microbes
7. **Incubation period refers to time duration:**
   a) Between entry of pathogen and exit of pathogen in body
   b) Between entry of pathogen in vector and entry in the host
   c) Between entry of pathogen and manifestation of symptoms or signs in the host
   d) Between entry of pathogen and its detection in the host
8. **Which of the following is not a gram-positive cocci?**
   a) *Streptococci pneumoniae*
   b) *Neisseria meningitidis*

c) *Enterococcus* species
d) *Staphylococcus* species

9. Normal microbial flora are also called:
a) Resident flora b) Commensal flora
c) Transient flora d) All of these

10. All the following are constituents of normal flora except:
a) Gram-positive bacteria b) Gram-negative bacteria
c) Yeast d) Viruses

11. The major components of the intestinal flora constitute:
a) Coliform bacteria b) *Clostridium* species
c) Yeast d) *Enterococcus* species

12. Sterile body sites include all the following EXCEPT:
a) Central nervous system b) Nasopharynx
c) Lungs d) Liver

13. Normal flora of the skin includes all the following EXCEPT:
a) α-Streptococci b) Staphylococci
c) *Candida* d) Lactobacilli

14. The relationship between a host and the normal flora wherein both co-exist is called:
a) Symbiosis b) Commensalism
c) Parasitism d) Opportunistic infection

15. Normal flora in the human body, when administered in adequate amounts, confer a health benefit to the host. These organisms are known as:
a) Probiotics b) Gnotobiotic
c) Symbiotics d) Prebiotics

16. Most common probiotics include:
a) *Fusobacterium* and *Lactobacillus*
b) *Bacteroides* and *Fusobacterium*
c) *Lactobacillus* and *Bifidobacterium*
d) *Bifidobacterium* and *Propionibacterium*

17. Disturbance to the normal flora can be caused by:
a) Prolonged antibiotic therapy
b) Alternation of pH in the mucosa
c) Immunosuppressive therapy
d) All of these

18. Which of the following statements is not true?
a) Normal flora cause confusion in the laboratory diagnosis.
b) Clindamycin administration helps in the multiplication of normal flora.
c) Probiotics are used in treating antibiotic-associated colitis.
d) Fecal transplantation is attempted for irritable bowel syndrome.

19. Microorganisms that fail to grow in the presence of as low as 0.03% oxygen are called:
a) Aerotolerant b) Obligate anaerobes
c) Facultative anaerobes d) Facultative aerobes

20. The false statement about anaerobic cocci is:
a) They are always pathogenic.
b) They are generally sensitive to penicillin.
c) They are normal inhabitants of the vagina, intestine, and mouth.
d) They occur as cocci in pairs, groups, and chains.

21. The most commonly isolated anaerobe from clinical specimens is:
a) *Fusobacterium* b) *Lactobacillus*
c) *Bacteroides* d) *Prevotella*

22. Usually, pus from anaerobic infections is putrid, with a nauseating odour. An exception to this is infection with:
a) *Bacteroids fragilis* b) *Prevotella*
c) *Porphyromonas* d) *Fusobacterium*

23. The predominant flora in the human GIT is anaerobic. The number of anaerobic bacteria in the colon is:
a) 10,000/g b) 10^5/g
c) 10^7/g d) 10^{11}/g

24. Most members of enteric bacilli are motile, except:
a) *Salmonella* b) *E. coli*
c) *Klebsiella* d) *Proteus*

25. Serotyping or antigenic typing of *E. coli* is based on:
a) Somatic antigen O b) Flagellar antigen H
c) Capsular antigen K d) All of these

26. The following are true about the somatic antigen (antigen O) of *E. coli* except:
a) These are lipopolysaccharides
b) Heat-stable
c) Associated with virulence
d) 'Early' groups are pathogenic

27. Commonest culture media used for *Leptospira* is:
a) Korthof's c) Stuart's
b) Fletcher's d) EMJH

28. *Clostridium perfringens* has the shape of a:
a) Spindle b) Club
c) Tennis racket d) Drumstick

29. Epidemic relapsing fever is caused by:
a) *Borrelia recurrentis* b) *B. burgdorferi*
c) *B. duttonii* d) *B. vincentii*

30. The vector for endemic relapsing fever is:
a) *Pediculus humanus corporis*
b) *Pediculus humanus capitus*
c) *Ornithodoros* ticks
d) *Ixodes* ticks

31. Migrating skin lesions in Lyme disease are called:
a) Erythema marginatum b) Erythema migrans
c) Erythema nodosum d) Erythema toxicum

32. Motility of borreliae is described as:
a) Darting b) Tumbling
c) Lashing d) Twitching

33. *Bifidobacterium* is a:
a) Prebiotic b) Probiotic
c) Parabiotic d) Postbiotic

34. Which of the following sample for bacterial culture must be collected in a sterile container?
a) Stool b) Rectal swab
c) Pleural fluid d) All of the above

35. A lumbar puncture was performed in night on a patient suspected of acute pyogenic meningitis. The CSF sample for culture will have to wait till morning when the lab opens. Where should the sample be stored overnight for best results.
a) In the freezer b) In the fridge
c) At –20°C d) In an incubator at 37°C

36. A blood sample was collected from a patient admitted in the fever ward for NS1 and IgM dengue test. However, when the specimen arrived in the microbiology lab, the patient's complete residential address and contact number were not mentioned on the requisition slip. This error pertains to an error in the:
a) Pre-analytical phase b) Analytical phase
c) Post-analytical phase d) None of the above

37. HEPA refers to:
a) Highly effective particulate air
b) High efficiency particulate air
c) Highly effective partition air
d) High efficiency part air cycle

38. Which of the following is false regarding sample collection from a patient?
a) Informed consent must be taken before sample collection.
b) Material with due consent must be used only for what it is intended, and not for any other purpose.
c) Results can be shared with family members.
d) Patient privacy must be ensured while collecting samples.

39. The vector for epidemic typhus is:
a) Rat flea b) Mite
c) Louse d) Tick

40. Which of the following is not true regarding Rickettsiae:
a) Do not possess peptidoglycan cell wall
b) Have metabolic enzymes
c) Reproduce by binary fission
d) Possess both RNA and DNA

41. Disease that does not occur in the typhus fever group is:
a) Epidemic typhus b) Endemic typhus
c) Q fever d) Brill-Zinser disease

42. Griffith typing of streptococci is based on:
a) Carbohydrate C antigen
b) M proteins
c) Hemolysis on blood agar
d) Oxygen requirement in culture

43. Which of the following is not gram-positive?
a) *Staphylococcus*
b) *Corynebacterium*
c) *Bacillus*
d) *Clostridium*

Answers: 1:c, 2:b, 3:d, 4:d, 5:b, 6:c, 7:c, 8:b, 9:d, 10:c, 11:b, 12:b, 13:d, 14:a, 15:a, 16:c, 17:d, 18:b, 19:b, 20:a, 21:c, 22:a, 23:d, 24:c, 25:d, 26:d, 27:d, 28:b, 29:a, 30:c, 31:b, 32:c, 33:b, 34:c, 35:d, 36:a, 37:b, 38:b, 39:c, 40:a, 41:c, 42:b, 43:a

Part I: General Microbiology

CHAPTER 5

Basic Concepts of Virology

(MI1.1–1.3, 8.3, 8.9–8.13)

Chapter Navigation

Learning objectives

- Describe the general properties of viruses
- Enumerate the methods of classification of viruses
- Describe host responses to viral infections
- Enumerate laboratory methods of detecting viruses from clinical material
- Describe bacteriophages and enumerate their clinical applications

- The infectious nature of viruses was first demonstrated by **Adolf Mayer** in the tobacco mosaic virus, a plant virus.
- Viruses are responsible for numerous human diseases, ranging from the common cold to the highly fatal rabies and AIDS.
- They have been associated with cancers (HPV) and other metabolic syndromes (e.g., diabetes with coxsackievirus B).
- Viral infections may be sporadic like mumps, endemic like infectious hepatitis, epidemic like dengue fever, or pandemics like influenza and the recent SARS-CoV-2.
- They may be geographically localised (such as dengue fever) or distributed worldwide (like influenza).

Some of the defining characteristics of viruses are:

- Viruses do not fall strictly into the category of unicellular microorganisms as they lack cellular organisation.
- They are the **smallest living organisms**, ranging from 20–300 nm in size.
- They contain **only one type of nucleic acid**, either DNA or RNA, but never both.
- They are obligate intracellular parasites.
- They lack the enzymes necessary for protein and nucleic acid synthesis and are **dependent for replication on the synthetic machinery of host cells.**
- They multiply by a complex process and not by binary fission.
- They are unaffected by antibacterial antibiotics.
- The chief differences between viruses and other microorganisms are presented in Table 5.1.

MORPHOLOGY OF VIRUSES

Size

- The extracellular infectious virus particle is called a **virion**.
- These particles are too small to be seen under the light microscope (hence, they are known as ultramicroscopic agents) and can only be seen under the **electron microscope**.

Table 5.1 Important differentiating characteristics of viruses and bacteria

Microorganism	Organised cell	Growth in artificial media	Binary fission	DNA and RNA	Ribosomes	Sensitivity to antibiotics	Sensitivity to interferons
Viruses	–	–	–	–	–	–	+
Bacteria	+	+	+	+	+	+	–
Mycoplasmas*	+	+	+	+	+	+	–
Rickettsiae**	+	+	+	+	+	+	–
Chlamydiae**	+	–	–	+	+	+	+

*Cell-wall deficit bacteria
**Obligate intracellular bacteria

- Some of the larger viruses, such as poxviruses, can be seen under the light microscope when suitably stained (Fig. 5.1).
- Viruses vary widely in size. The largest (poxvirus) measures about 300 nm and is as large as the smallest bacteria (*Mycoplasma*). The smallest viruses (e.g., parvovirus) measure about 20 nm.

Structure

- The virion consists of a nucleic acid surrounded by a protein coat called the **capsid**.
- The capsid and the enclosed nucleic acid together are known as the **nucleocapsid**.
- Virions may be **enveloped or non-enveloped (naked)**.

Capsid

- The morphological unit of the capsid is the **capsomere**.
- Capsids are polypeptide molecules arranged symmetrically to form an impenetrable shell around the nucleic acid core.
- One of the main functions of the capsid is to introduce the viral genome into host cells by adsorbing readily onto host cell surfaces.
- Three kinds of symmetry are seen in the capsid (Fig. 5.2):
  - **Icosahedral (cubical):** An icosahedron is a polygon with **12 vertices or corners and 20 facets or sides**. Each facet is an equilateral triangle. Two

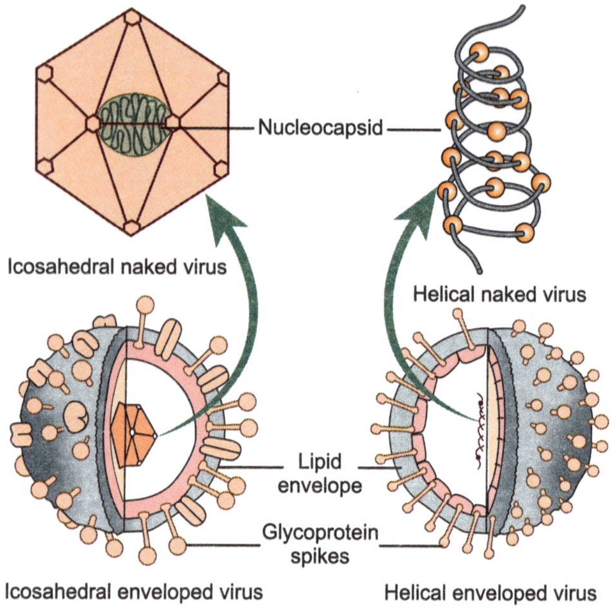

Fig. 5.2 The structure of a viral particle

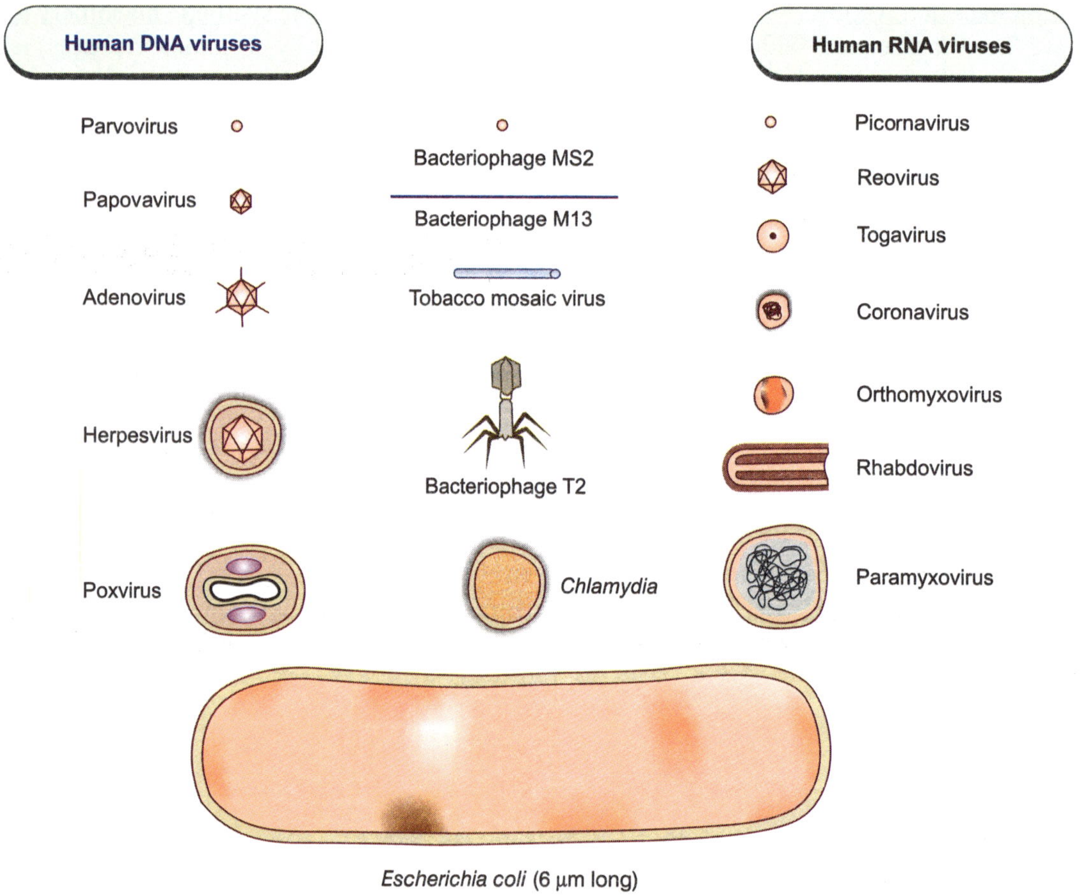

Fig. 5.1 Relative sizes of representative viruses, bacteriophages (bacterial viruses), and bacteria, including chlamydia

types of capsomers constitute the icosahedral capsid—the pentagonal capsomers at the **vertices (pentons)** and the hexagonal capsomers making up the **facets (hexons)**. Icosahedral capsids always have 12 pentons; the number of hexons, however, varies with the virus group.
  - **Helical:** In **nucleocapsids with helical symmetry**, the capsomers and nucleic acid are wound together to form a helical or spiral structure.
  - **Complex:** Some capsids, like those of the poxviruses, exhibit complex symmetry.

Envelope

The envelope or outer covering is derived from the host cell membrane when the progeny virus is released by budding. It is made up of lipoproteins (Fig. 5.2).

- Protein subunits may be seen as projecting spikes on the surface of the envelope. These structures are called **peplomers** (from *peplos*, meaning 'envelope'). A virus may have more than one type of peplomer.
- Envelopes possess chemical, antigenic, and biological properties.
- Enveloped viruses are susceptible to the action of lipid solvents like ether, chloroform, and bile salts.
- Specific neutralisation of virus infectivity depends on antibodies to the surface antigens.
- Biological properties such as attachment to the host cell surface or hemagglutination depend on the envelope.

Shape

- Viruses are roughly spherical.
- Some are **irregular and pleomorphic**.
- The rabies virus is bullet-shaped; Ebola virus and hepatitis virus are filamentous; poxviruses are brick-shaped; the tobacco mosaic virus is rod-shaped.

Viral Components

- **Nucleic acid:** Viruses contain only one type of nucleic acid, either single or double-stranded DNA or RNA. The viral RNA may itself act as mRNA and is known as **positive-sense**. In negative-sense viruses, the RNA is 'antisense', and the complementary strand synthesised by a viral RNA transcriptase serves as a mRNA.
- **Proteins:** Viruses also contain proteins that make up the capsid. Viral proteins, besides protecting the nucleic acid, also determine the antigenic specificity of the virus.
- **Non-structural proteins:**
  - Enzymes that initiate and maintain the synthesis of viral components
  - May induce the shutdown of host protein and nucleic acid synthesis
- **Other components:** Some viruses also contain small amounts of carbohydrates. Most viruses do not possess any enzymes for the synthesis of viral components or for energy production, but some have other enzymes such as the following:
  - **Neuraminidase** (produced by the influenza virus), which destroys the sialic acid on the host cell receptors
  - **RNA-dependent DNA polymerase** or 'transcriptase', which can transcribe RNA into DNA in retroviruses

SUSCEPTIBILITY OF VIRUSES TO PHYSICAL AND CHEMICAL AGENTS

- **Temperature:**
  - Viruses are very **heat-labile**. In general, they are inactivated within seconds at 56°C, within minutes at 37°C, and within days at 4°C. They are stable at low temperatures.
  - For long-term storage, they are frozen at –70°C.
  - **Lyophilisation or freeze-drying** (drying the frozen virus in a vacuum) is best for prolonged storage for years. Some viruses (such as the poliovirus) do not stand freeze-drying.
- **pH:** Some viruses like enteroviruses are very resistant to acidic pH, whereas rhinoviruses are very susceptible. All viruses are disrupted by alkaline conditions.
- **Radiations:** Viruses are inactivated by sunlight, UV rays, and ionising radiation.
- **Disinfectants:** In general, viruses are more resistant than bacteria to chemical disinfectants, probably because they lack enzymes (see box). Non-enveloped viruses are more resistant to disinfectants. Chlorination of drinking water kills most viruses, but its efficacy is greatly influenced by the presence of organic matter. Some viruses such as the hepatitis virus and polioviruses are relatively resistant to chlorination.

Disinfectants active against viruses

- Weakly virucidal—phenolic disinfectants
- Actively virucidal—oxidising agents such as hydrogen peroxide, potassium permanganate and hypochlorites, organic iodine, formaldehyde, and beta-propiolactone

- **Lipid solvents:** The action of lipid solvents such as ether, chloroform, and bile salts is selective, the enveloped viruses being sensitive and the naked viruses resistant to them.
- **Antibiotics:** These are completely ineffective against viruses, a property used to eliminate bacteria from specimens before attempting virus isolation.

CLASSIFICATION OF VIRUSES

Viruses are classified primarily into DNA viruses and RNA viruses (Table 5.2) on the basis of the nucleic acid present. These may be further divided into single-stranded (ss) and double-stranded (ds).

Table 5.2 Viruses that commonly cause human infections

Family	Features	DNA structure	Envelope	Genus (only includes genera with human pathogens)	Members causing human disease
DNA viruses					
Poxviridae	Large, brick-shaped/ ovoid viruses (300 × 240 × 100 nm)	ds, linear	+	Orthopox	Variola virus, vaccinia virus, cowpox, orf
			+	Molluscipox	Molluscum contagiosum, monkeypox
Herpesviridae	Medium-sized (70–90 nm), icosahedral with 252 capsomeres	ds, linear	+	Herpesvirus	Herpes simplex 1 and 2, EBV, CMV, varicella zoster, HSV-6, 7, 8
Adenoviridae	Medium size (90-100 nm), icosahedral	ds, linear	+	Adenovirus	Human adenoviruses A–G
Papovaviridae	Small (40–55 nm), with 72 capsomers	ds, circular	–	Papillomas	Human papillomavirus
		ds, circular	–	Polyoma	JC virus, BK virus, SV40
Hepadnaviridae	Spherical, 42 nm with a 27 nm core; replicate with reverse transcriptase	ds, incomplete, circular	+	Orthohepadnavirus	Hepatitis B virus
Parvoviridae	Very small (18–26 nm), with 31 capsomers	ss, linear, single-stranded	–	Parvovirus	B19 virus
RNA viruses					
Picornaviridae	Small (20–30 nm), icosahedral	ss, linear	–	Enterovirus	Enteroviruses—poliovirus, coxsackie, echovirus, EV-68-71
				Rhinovirus	Human, bovine, and equine rhinoviruses
				Hepatovirus	Hepatitis A virus
Orthomyxoviridae	Medium-sized (80–120 nm), spherical/elongated with hemagglutinin and neuraminidase peplomers and segmented genome	ss, linear	+	Alphainfluenzavirus	Influenza virus A–C
Paramyxoviridae	Pleomorphic virions (150 nm)	ss, linear	+	Paramyxovirus	Newcastle disease virus, mumps, parainfluenza viruses
				Morbillivirus	Measles virus, pneumovirus
				Canine distemper, rinderpest	Respiratory syncytial virus of humans and related viruses
Togaviridae	Spherical, 40–70 nm	ss, linear	+	Alpha virus (formerly classified as group A arboviruses)	Respiratory syncytial virus of humans and related viruses
				Chikungunya virus, Eastern equine encephalitis virus, Western equine encephalitis virus, Sindbis virus	Rubella virus
Flaviviridae	Spherical (50 nm), icosahedral symmetry	ss, linear	+	Flaviviruses (group B arboviruses)	Yellow fever, dengue viruses, Japanese encephalitis virus, West Nile virus
				Hepacivirus	Hepatitis C virus

(Table 5.2 Contd. . .)

Table 5.2 Contd.

Family	Features	DNA structure	Envelope	Genus (only includes genera with human pathogens)	Members causing human disease
Bunyaviridae	Spherical (90–100 nm)	ss, circular	+	Bunyavirus Hantavirus, Nairovirus, Phlebovirus, Ukuvirus	Sandfly fever, hantaviruses, CCHF virus, Hantaan virus
Arenaviridae	Spherical or pleomorphic viruses (50–300 nm) containing a number of electron-dense ribosome-like particles, which give them a sandy appearance	ss, circular	+	Arenaviruses	LCM, Lassa fever
Rhabdoviridae	Bullet-shaped, (130–300 nm long and 70 nm wide) with peplomers on the surface	ss, linear	+	Vesiculovirus	Vesicular stomatitis virus, Chandipura virus
				Lyssavirus	Rabies virus and related viruses such as Lagos bat, Mokola, Duvenhage, and other viruses
Reoviridae	Icosahedral (60–80 nm), double- layered capsids with segmented genome in 10–12 pieces	ds, linear	–	Reovirus	Reovirus
				Orbivirus	Species of arboviruses such as blue tongue virus, African horse sickness virus, and Colorado tick fever virus
				Rotavirus	Human rotavirus
Coronaviridae	Pleomorphic, enveloped (100 nm), with unique club-shaped peplomers projecting like a fringe from the surface, resembling the solar corona	ss, linear	+	Coronavirus	Human coronaviruses, SARS virus, MERS-CoV, SAS-CoV2
Retroviridae	Icosahedral (100 nm) with RNA-dependent DNA polymerase (reverse transcriptase)	ss, linear	+	Deltaretrovirus	HTLV-1,2
				Spumivirinae, the foamy virus group (*spuma* = foam)	Human foamy virus
				Lentivirinae (*lenti* = slow)	HIV 1–2, visna and maedi viruses of sheep
Calciviridae	Naked spherical particles (35–39 nm) with 32 cup-shaped depressions arranged symmetrically on the surface	ss, linear	–	Calcivirus	Norwalk virus, hepatitis E virus
Hepeviridae	Small (32–34 nm), icosahedral and spherical	ss, linear	–	Orthohepevirus	Hepatitis E virus
Filoviridae	Long, filamentous, enveloped viruses (80 nm diameter and up to 14,000 nm long) with a helical nucleocapsid	ss, linear	+		Marburg, Ebola viruses
Astroviridae	Spherical (28–30 nm) with T-3 icosahedral symmetry	ss, linear	–	Avastrovirus	Avastrovirus 1-3
				Mamastrovirus	Mamastroviruses 1-19
Deltaviridae	Smallest animal virus; spherical (36 nm); uses HBsAg as an envelope to package HDAg	ss, circular	+	Deltavirus	Hepatitis deltavirus

REPLICATION OF VIRUSES

- The genetic information necessary for viral replication is contained in the viral nucleic acid. However, viruses lack biosynthetic enzymes, and hence, depend on the synthetic machinery of the host cell for replication.
- The viral replication cycle can be divided into six sequential phases (Fig. 5.3), though some of these phases may overlap.

Attachment and Adsorption

- Adsorption takes place only if there is affinity between the virion and the host cell.
- The host cell surface should bear specific receptor sites to which the virus can gain attachment. This determines the **host range (cell specificity)**. For example, in the case of HIV, the viral surface glycoprotein gp 120 gets attached to the CD4 receptor on host cells.
- Differences in susceptibility to viral infection are, to a large extent, based on the presence or absence of receptors on cells.
- If the phase of adsorption can be bypassed, cells normally insusceptible to viruses may be rendered susceptible to them.

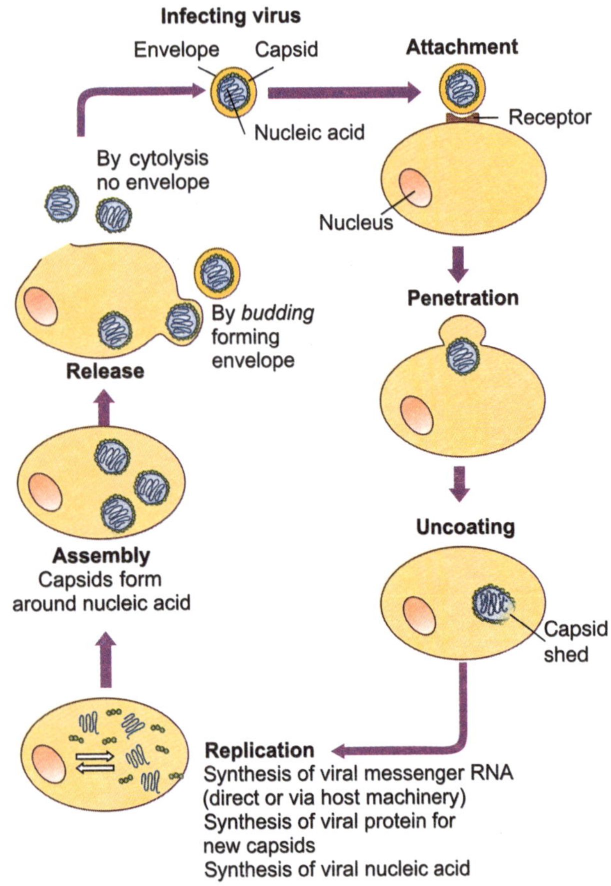

Fig. 5.3 Stages in the infection of a host cell and replication of a virus

Penetration

- Animal cells do not have rigid cell walls; this allows the whole virus to enter such cells.
- Virus particles may be engulfed by a process known as **endocytosis**.
- Alternatively, in enveloped viruses, the viral envelope may fuse with the plasma membrane of the host cell and release the nucleocapsid into the cytoplasm.

Uncoating

- The virus sheds its outer layers and capsid so that the nucleic acid is released into the cell.
- With most viruses, uncoating is affected by the action of the lysosomal enzymes of the host cell.

Replication

- The viral nucleic acid, capsid protein, and enzymes required for the various stages of viral synthesis, assembly, and release are synthesised during this stage.
- Some '**regulator proteins**' are also synthesised, which serve to shut down normal cellular metabolism and direct the sequential production of viral components.
- **DNA viruses** synthesise their nucleic acid in the host cell nucleus (Fig. 5.4) except poxviruses, which synthesise all their components in the host cell cytoplasm.
- In **RNA viruses**, all the components are synthesised in the cytoplasm. The exception to this are orthomyxoviruses, some paramyxoviruses, and retroviruses, which are synthesised partly in the nucleus (Fig. 5.5).

Steps involved in viral genome biosynthesis

- There is early **transcription of messenger RNA** (mRNA) from the viral nucleic acid—once this is achieved, the virus can utilise host cell resources for translating mRNA into viral components.

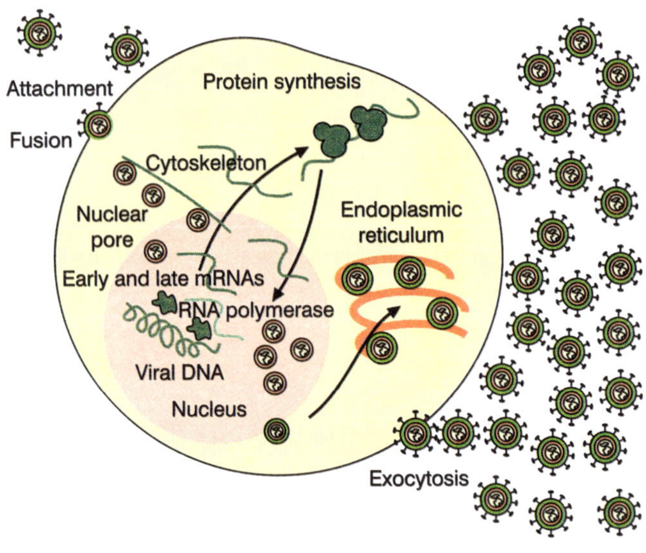

Fig. 5.4 Replication of a DNA (herpes) virus

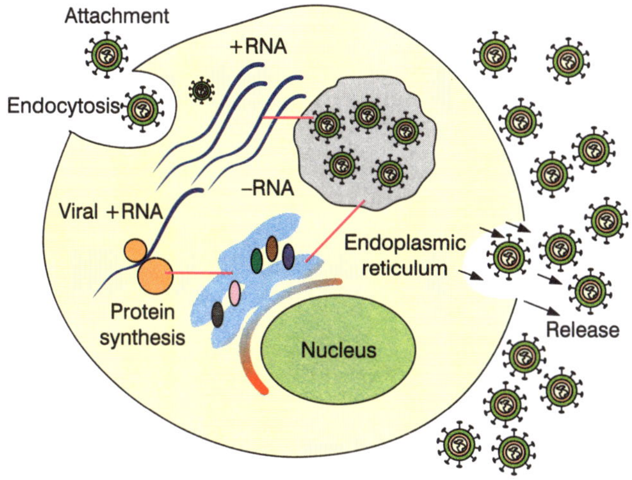

Fig. 5.5 Replication of RNA viruses

- This is followed by the early **translation** of the mRNA into 'early' or **'non-structural' proteins**; these proteins:
  - Are enzymes that initiate and maintain the synthesis of virus components
  - May also induce the shutdown of host protein and nucleic acid synthesis
- The double-stranded RNA is transcribed to viral mRNA by **RNA-dependent RNA polymerase (RdRp)** viral polymerases and then translated into proteins.
- Depending on the structure of their genome, viruses may use any of the following six strategies for the transcription of mRNA (Table 5.3).

Assembly and Maturation

- Once the viral nucleic acid and proteins have been synthesised, the assembly of progeny virions begins.
- Virion assembly may take place in the host cell nucleus (herpes and adenoviruses) or in the cytoplasm (picorna and poxviruses).
- At this stage, non-enveloped viruses are present intracellularly as fully developed virions, but in the case of enveloped viruses, only the nucleocapsid is complete.
- **Viral envelopes are derived from the host cell membrane during the process of budding.**

Release

- Though virus release usually occurs without cell lysis, many non-enveloped human viruses are released following the lysis of the host.
- Non-enveloped viruses exit the cell via exocytosis.
- Enveloped viruses assemble in the cytoplasm and then 'bud out', utlising the host plasma membrane to form the envelope.
- The time taken for a single cycle of replication is about 15–30 minutes for bacteriophages and about 15–30 hours for animal viruses. A single infected cell may release a large number of progeny virions.

Eclipse phase

From the stage of penetration until the appearance of mature progeny virions, the virus cannot be demonstrated inside the host cell. This period, during

Table 5.3 Different types of mRNA replication

Class of mRNA transcription	Type of virus	Example	Method of transcription
Class 1	dsDNA viruses	Adeno, Herpes, Papovaviruses	The mRNA is directly transcribed from the viral DNA.
Class 2	ssDNA viruses	Parvovirus	The ssDNA is converted to dsDNA by DNA polymerase. RNA polymerase II transcribes viral genes, which are then translated into viral proteins, and DNA polymerase replicates the genome so that the assembly of nascent virions can occur.
Class 3	dsRNA viruses	Reoviruses	The double-stranded RNA is transcribed to viral mRNA by RNA-dependent RNA polymerase (RdRp). This mRNA is then translated into proteins.
Class 4	ssRNA viruses	Picorna and Togaviruses	The viral RNA itself acts as the mRNA. Viral RNA is infectious by itself and is translated directly into viral proteins in the host cell cytoplasm.
Class 5	ssRNA viruses	Rhabdo, Prthomyxo, and Paramyxoviridae	The RNA is 'antisense', with polarity opposite to mRNA. Negative-sense RNA must be copied into positive-sense RNA by a viral RdRp before it can be translated by ribosomes. Extracted nucleic acids from these viruses are not infectious.
Class 6	ssRNA viruses	Retroviridae	The single-stranded RNA genome is converted into an RNA:DNA hybrid by the viral reverse transcriptase (RNA-directed DNA polymerase) enzyme. Double-stranded DNA is then synthesised from the RNA:DNA hybrid. The double-stranded DNA form of the virus (provirus) is integrated into the host cell chromosome. This integration may also lead to the transformation of the cell and development of neoplasia.

which the virus seems to disappear, is known as the 'eclipse phase'.

Abnormal replication cycles

- **Von Magnus phenomenon:** A proportion of daughter virions produced is not infective due to defective assembly. Such 'incomplete viruses' are seen in large numbers when cells are infected with a high dose of the influenza virus. In such cases, the virus yield will have a high hemagglutinin titre but low infectivity.
- **Defective viruses:** Some viruses are genetically deficient and therefore incapable of producing infectious virions without the help of another virus.
- **Abortive infection:** In some cases, infection by a virus does not lead to the production of infectious progeny. The viral components may be synthesised, but there may be defects in maturation or assembly. As a result, there is either no release of progeny, or the progeny released is non-infectious. Here, the defect is in the type of host cell and not the parental viruses.
- **Co-infection:** Some viruses are genetically defective in that when they infect cells, they are unable to produce fully formed progeny. The yield of progeny virions occurs only if the host cells are simultaneously infected with a helper virus, which can supplement the genetic deficiency. For example, hepatitis D virus replicate only in the presence of the helper virus, hepatitis B.

VIRAL GENETICS

The chief mechanisms for genetic modification in viruses are discussed here.

Mutation

The frequency of mutation in viruses is about 10^{-4} to 10^{-8}, approximately the same as that in bacteria. Mutations, therefore, occur during every viral infection. They may occur spontaneously or may be induced by mutagens, physical agents such as irradiation, or chemical agents such as 5-fluorouracil.

Conditional lethal mutant: These mutants can grow under certain conditions (called **permissive conditions**) but not in certain other specified conditions, which are lethal (called **non-permissive or restrictive conditions**). These mutants can be classified as follows:

- **Ts or temperature-sensitive mutant:** These mutants can grow at a low (permissive) temperature (28–31°C) but not at a higher (restrictive) temperature (37°C).
- **Host-dependent mutants:** These can grow in permissive cells but cause abortive infection in non-permissive cells.

Genetic Interactions

1. Recombination

- Genetic recombination may occur when two different but related viruses infect a cell simultaneously.
- The two viruses exchange segments of their nucleic acids so that a hybrid results, possessing genes from both parents.
- This may occur naturally, or it may be produced by recombining pieces of DNA using recombinant DNA technology.

2. Reassortment

- Reassorment may occur when multiple viruses with segmented genomes co-infect the same cell.
- The progeny virions are assembled from a mix of segments from the parental virions, creating a novel assortment of genome segments.

3. Reactivation

- When a latent virus that has infected a host cell switches to the lytic stage, undergoing productive viral replication and allowing the virus to spread, it is called reactivation.
- It can occur naturally or artificially (UV irradiation).

As a rule, virus capsids enclose viral nucleic acids. Sometimes, segments of host nucleic acid become encapsidated instead. For example, in a papovavirus capsid, a linear piece of host DNA roughly the same size as the papovavirus genome may be found. Such a capsid is known as a **pseudovirion**.

PATHOGENESIS OF VIRAL INFECTIONS

Entry

Viruses enter the body through various routes to establish organ-specific infections (Fig. 5.6).

- Inhalation (droplet infection—influenza, SARS-CoV)
- Ingestion (hepatitis A, rotavirus)
- Direct transfer from an infected host
- Sexual contact (HIV, hepatitis C)
- Transplacental (rubella, HIV)
- Direct inoculation (needlestick injuries)
- Bites of vectors (arbovirus), animals
- Blood transfusion (hepatitis B and C, HIV)

Most viruses show **tissue tropism**, i.e., they infect or replicate only in specific cells (rabies in neurons; hepatitis virus in hepatocytes).

Host–Virus Interactions

- The initial phase of viremia results in the entry of the virus into the host cell.
- Then, the virus settles down in the target cells to replicate, causing cell death and tissue damage.

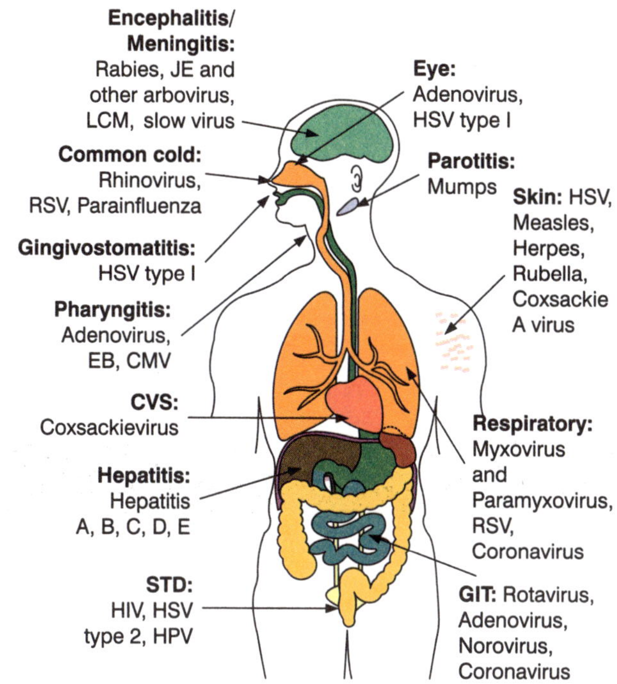

Fig. 5.6 Organ-specific infections by some common viral agents

- The pathogenesis of viral infections is largely attributed to the mounting immune response of the host and the cytokines released from activated T cells.
- These interactions are influenced by various defence mechanisms of the body and can result in:
  - **Cell death:** Cytocidal effect or lysis cytolysis, as seen with poliovirus
  - **Cellular proliferation:** As seen with molluscum contagiosum
  - **Malignant transformation:** As seen with oncogenic viruses
  - **Steady-state infection:** The virus and host cells co-exist peacefully, both replicating independently; in such cases, there is no cellular injury or chronic diffuse infection

Outcomes of Viral Infections

Three types of infections may be seen (Fig. 5.7):

- **Full-blown acute infection:** This occurs due to cell lysis and tissue damage with resultant pathology, clinical features, and full-blown disease.

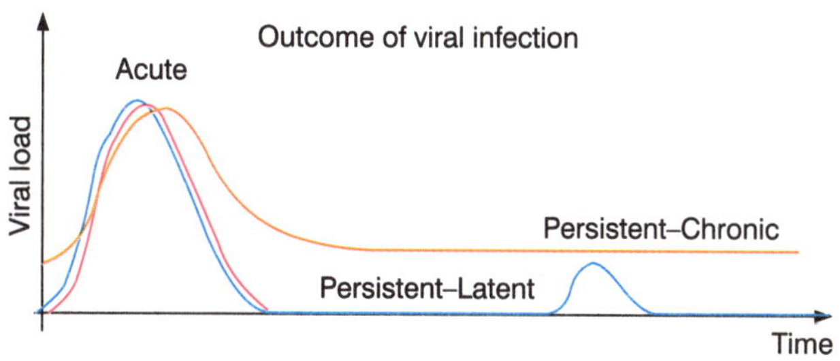

Fig. 5.7 Outcomes of a viral infection

- **Persistent (chronic) infection:** After the initial manifestation of the full-blown disease, some viruses may persist in the cell cytoplasm (e.g., HIV).
- **Latent infection:** Some viruses exist in the host in a dormant state and get activated in the presence of some trigger to release them from latency. Such infections are called latent infections (e.g., herpes).

Action of Viruses on Cells

- Shutdown of host protein and DNA synthesis during virus replication
- Accumulation of large amounts of viral macromolecules in the infected cell, which distorts the cellular architecture and exerts a toxic effect
- Alteration in the permeability of the plasma membranes, resulting in the release of lysosomal enzymes and autolysis
- Fusion of adjacent cell membranes, leading to polykaryocytosis or syncytium formation (e.g., respiratory syncytial virus)
- Appearance of virus-coded antigens on the surface of the infected cells that may confer new properties on the cells (e.g., viral hemagglutinins appear on the surface of cells infected with the influenza virus and cause the adsorption of erythrocytes to the cell surface—hemadsorption)
- Damage to the chromosomes of host cells (measles and mumps, adenoviruses, cytomegalovirus, and varicella)
- Formation of inclusion bodies in infected cells

Inclusion bodies

- The most characteristic histological feature of virus-infected cells is the presence of inclusion bodies (Fig. 5.8).
- These are structures with a distinct size, shape, location, and staining properties, which can be histologically demonstrated in virus-infected cells with the help of light microscopy.
- They may be crystalline aggregates of virions or made up of virus antigens present at the site of virus synthesis.
- Some inclusions represent degenerative changes produced by viral infection which confer altered staining properties on the cell.
- They may be situated in the cytoplasm (as with poxviruses), nucleus (herpes viruses), or both (measles virus).
- They are generally acidophilic (seen as pink structures upon Giemsa staining).
- Some viruses (e.g., adenovirus) form basophilic inclusions (Table 5.4).
- The demonstration of inclusion bodies helps in the diagnosis of some viral infections.

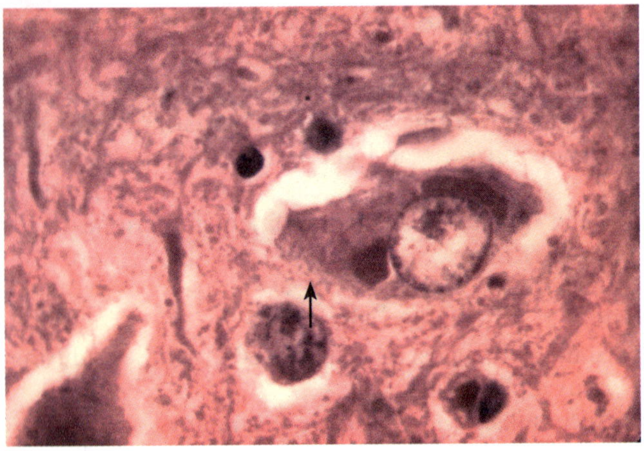

Fig. 5.8 Negri bodies (intracytoplasmic inclusion bodies) in a case of rabies virus infection seen in the pyramidal cells of Ammon's horn (*Source*: CDC, PHIL, Image ID 3372)

Table 5.4 Some viral inclusion bodies, their staining characteristics, and sites in the cell

Inclusion body	Site	Nature	Virus/ Disease
Negri bodies	Intracytoplasmic	Eosinophilic	Rabies
Guarnieri bodies	Intracytoplasmic	Acidophilic	Vaccinia virus
Henderson–Paterson bodies (molluscum bodies)	Intracytoplasmic	Eosinophilic	Molluscum contagiosum
Cowdry type A inclusions	Intranuclear	Acidophilic	Herpes virus, yellow fever virus
Type B inclusions	Intranuclear	Basophilic	Adenovirus, poliovirus
Warthin–Finkeldey cell	Intracytoplasmic and intranuclear	Eosinophilic	Measles

VIROIDS

This term was introduced by **Theodor Diener** (1971) to describe a new class of subviral agents. The characteristics of viroids are as follows:

- Apparent **absence of an extracellular dormant phase** (virion)
- Very small genome
- **Protein-free, low-molecular-weight RNA**, which is resistant to heat and organic solvents but sensitive to nucleases
- Cause diseases in plants
- Disease-causing potential in humans is doubtful

PRIONS

Stanley B. Prusiner received the Nobel Prize for his discovery of prions in 1982. Some of the characteristics of prions are as follows:

- They are **small proteinaceous infectious particles** that have a MW 50,000 and diameter of approximately 4–6 nm.
- They lack any detectable nucleic acid.
- They are resistant to heat (90°C for three minutes), UV rays, and nucleases but are sensitive to proteases, hypochlorites, and autoclaving.
- They may be associated with some chronic neurological degenerative diseases of humans, resulting in fatal dementia.
- They also cause slow degenerative disorders such as scrapie, Kuru, and Creutzfeldt–Jakob disease.

LABORATORY DIAGNOSIS OF VIRAL DISEASES (Fig. 5.9)

The diagnosis of viral infections was limited in the past due to the following challenges:

- The technical difficulties involved in the isolation of viruses and their identification
- The methods to diagnose viral infections were so time-consuming that the patient would either have recovered from the infection or succumbed by the time results became available
- The lack of availability of a specific diagnosis of a viral infection did not impact management

With the advent of molecular technologies, diagnostic virology has rapidly become a part of the routine scope of clinical laboratories now. Viral diagnostics is helpful in:

- Early institution of specific antiviral therapy whenever indicated
- Initiation of suitable prophylactic therapy in cases of viral infections that cause immunosuppression
- Follow-up and monitoring of chronic viral infections like HIV and HCV
- Screening of populations at risk for viruses such as HIV, HBV, and HCV
- Detection and prediction of epidemics and pandemics
- Control of outbreaks through vaccination of vulnerable populations and estimation of protective antibodies in these populations

Sample Collection and Transport (Table 5.5)

- The successful diagnosis of viral infections depends as much on the awareness of the physician as on the capability of the diagnostic laboratory.
- It is crucial that appropriate specimens are collected from patients, preserved suitably, and transported to the laboratory in the proper manner along with pertinent clinical and epidemiological information.

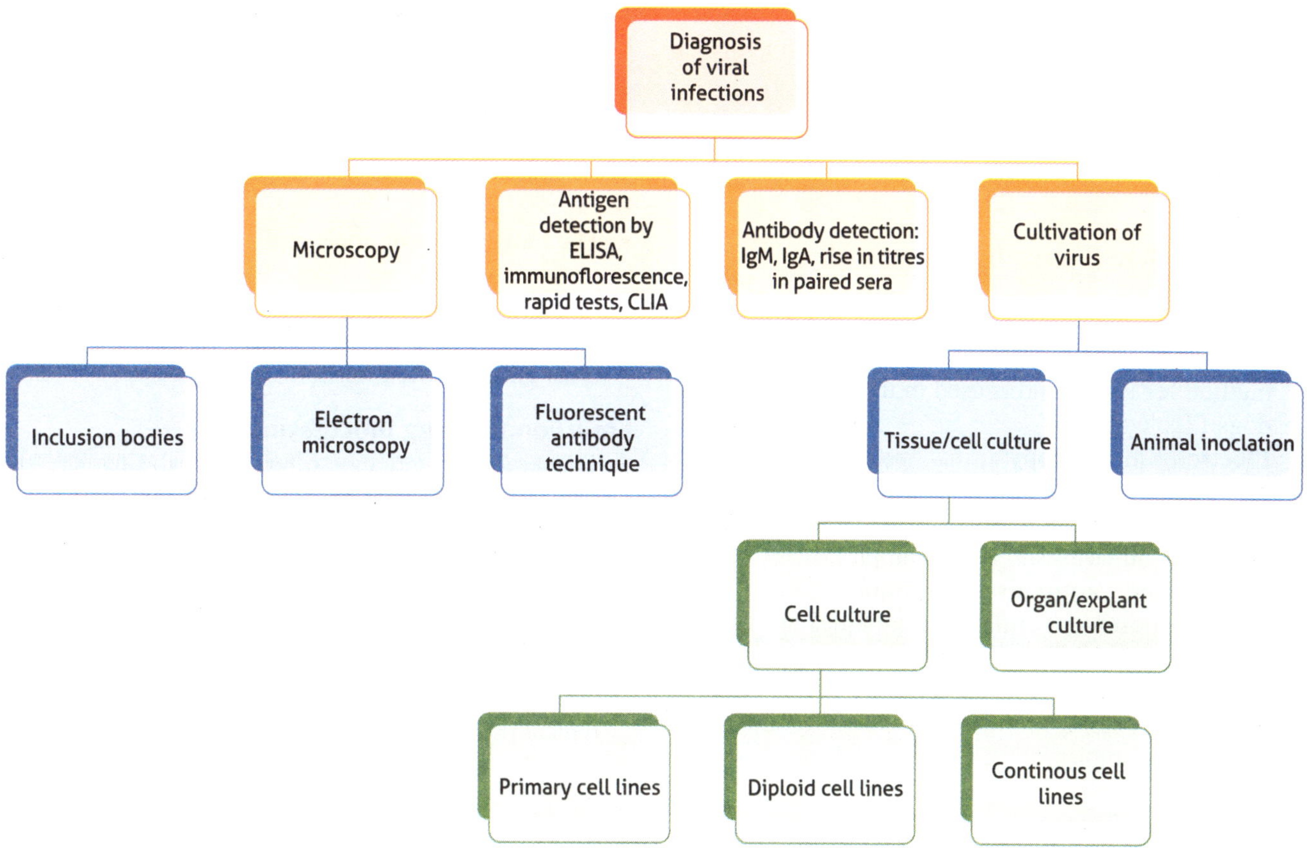

Fig. 5.9 A brief outline of laboratory diagnosis of viral infections

Table 5.5 Types of specimens to be sent to the laboratory for the diagnosis of viral infections (adapted from the WHO's guidelines)

System affected	For isolation	For direct examination and detection of viral genome	For serology
Respiratory system	Throat swab	Nasopharyngeal aspirate (IF[2])[1] molecular diagnosis	Paired sera
Nasopharyngeal	Throat washings, aspirates	IF and molecular diagnosis	Paired sera
Central nervous system	CSF, brain biopsy, throat swab, rectal swab, feces	Brain biopsy (IF and EM[3]) and molecular diagnosis	Paired sera
		CSF (EM and IF) and molecular diagnosis	
Cardiovascular system	Blood (for arbovirus isolation)	Molecular diagnosis	Paired sera
Skin	Macular/papular scrapings, vesicular/pustular fluid, ulcer scrapings, crust, feces, throat swabs	Vesicular/pustular fluid (EM and ID), ulcer scrapings (EM), crusts (EM and ID[4]) and molecular diagnosis	Paired sera
Eye	Conjunctival scrapings or swabs	Conjunctival scrapings as smears on microscope slides (LM[5] and IF) Molecular diagnosis	Paired sera
Liver	Blood (for yellow fever)	Serum (feces)	Serum
General; congenital infections	Throat swab (in some cases, products of conception)	Molecular diagnosis	Single sera (mother and baby)
General; PUO[6]	Heparinised blood (arbovirus and arenavirus infections), throat swabs, feces (sometimes, fresh urine)	Molecular diagnosis	Paired sera

1. Specimens within brackets are not appropriate for routine diagnosis but may be indicated in particular circumstances; 2. *IF*, immunofluorescence; 3. *EM*, electron microscopy; 4. *ID*, immunodiffusion; 5. *LM*, light microscopy; 6. *PUO*, pyrexia of unknown origin

- The specimen for virus isolation should be collected aseptically and transported without delay for further processing.
- As most viruses are heat-labile, refrigeration is essential during transport (for most specimens).

Microscopy

- **Light microscopy:** Inclusion bodies can be demonstrated by light microscopy, e.g., Cowdry type A bodies in herpes and Negri bodies in rabies (Fig. 5.8). Seller's staining is a routine diagnostic method for rabies and is used in the post-mortem of dogs (Table 5.4).
- **Electron microscopy:** This method allows for the detection of virus particles and aggregates. Immune-electron microscopy has been used to detect Nipah viruses in tissue samples.
- **Fluorescent antibody technique:** Direct and indirect fluorescent antibody techniques are used to examine material from lesions as well as for the early demonstration of viral antigens in tissue cultures inoculated with specimens. It is used for the rapid diagnosis of rabies.

Cultivation of Viruses

- The methods used for isolation depend on the virus sought.
- Inoculation is performed after the specimen is processed to remove bacterial contaminants.
- After cultivation, the isolates are identified by neutralisation or other suitable serological procedures.
- The mere recovery of a virus from a patient cannot confirm it as the causative agent of the illness given that many viruses (e.g., adenoviruses and enteroviruses) are known to exist in healthy individuals.
- Since viruses are obligate intracellular parasites, they cannot be grown on any artificial culture medium.
- Three methods are employed for the cultivation of viruses—animal inoculation, embryonated egg inoculation, and tissue culture.

Animal inoculation

- Live inoculation was first used on human volunteers for the study of the yellow fever virus. Later, animals were utilised for studying viruses.
- The choice of animals and the route of inoculation (intracerebral, intraperitoneal, subcutaneous, intradermal, or intraocular) depends largely on the type of virus to be isolated.
- The laboratory animals used include monkeys, rabbits, guinea pigs, rats, hamsters, and infant (suckling) mice (coxsackie and arboviruses).
- Apart from diagnosis, animal inoculation is used for the study of pathogenesis, immune response, epidemiology, and oncogenesis.
- The growth of the virus in inoculated animals may be indicated by death, disease, or visible lesions.

Disadvantages

- The handling of animals and inoculation by various routes requires special experience and training.
- The host animal's immunity may interfere with viral growth.
- Host animals often harbour latent viruses.

Embryonated egg inoculation

- The embryonated egg offers several sites for the cultivation of viruses (Fig. 5.10 and Table 5.6).
- Specimens are inoculated into pathogen-free fertilised eggs that are 10–11 days old.
- The eggs are incubated for 2–9 days before harvesting the viruses.
- Growth and multiplication of the viruses are indicated by the death of the embryo or by the formation of typical pocks or lesions on the egg membranes.

> Embryonated hen's egg was first used for the cultivation of viruses by **Ernest William Goodpasture** (1931); the method was further developed by **F. M. Burnet**.

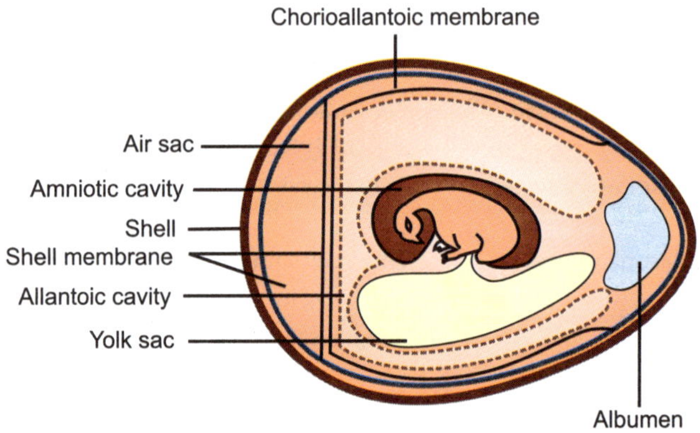

Fig. 5.10 Anatomy of a 10-day-old embryonated hen's egg

Table 5.6 Sites and organisms cultured in hen's eggs

Site	Viruses cultured
Chorioallantoic membrane	Poxvirus, herpes zoster virus, and roux sarcoma virus
Yolk sac inoculation	Herpes simplex virus, arbovirus (e.g., Japanese encephalitis virus), bacteria such as chlamydiae and rickettsiae
Allantoic cavity	Vaccine production for the influenza virus, some paramyxoviruses, yellow fever (17D strain), and rabies (Flury strain) virus
Amniotic sac	Primary isolation of the influenza virus, mumps virus

Tissue culture

Human viruses can be grown in tissue culture (Fig. 5.11).

Types of tissue culture

- **Organ culture:** Small bits of organs can be maintained in vitro for days and weeks, preserving their original architecture and function. Organ cultures are useful for the isolation of some organotrophic viruses. For example, the tracheal ring organ culture is employed for the isolation of coronavirus, a respiratory pathogen.

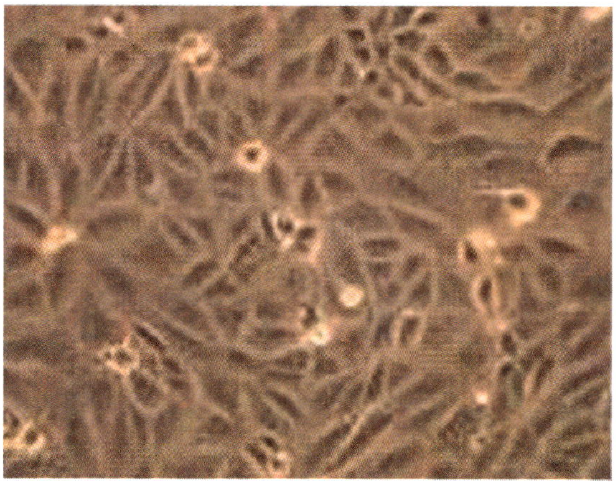

Fig. 5.11 Vero cell monolayer

- **Explant culture:** Fragments of minced tissue can be grown as explants embedded in plasma clots. They may also be cultivated in suspension. For example, adenoid tissue explant cultures were used in the past for the isolation of adenoviruses.
- **Cell culture:** This type of culture is routinely employed for growing viruses. The procedure is as follows:
  - Tissues are dissociated into component cells by the action of proteolytic enzymes such as trypsin and mechanical shaking.
  - The cells are washed, counted, and suspended in a growth medium with all the required nutrients such as essential amino acids, vitamins, salts, glucose, and a buffering system. They are grown in an atmosphere containing about 5% of carbon dioxide. The medium is supplemented with up to 5% calf or fetal calf serum and antibiotics. Such a medium will enable most cell types to multiply with a division time of 24–48 hours.
  - The cells adhere to the glass surface, and on incubation, divide to form a confluent monolayer sheet of cells covering the surface within about a week.
  - Cell cultures are classified into three types (Table 5.7).

Detection of virus growth in cell cultures

Viruses grown on cell lines may be detected by any one or more of the methods:

Cytopathic effect

- The morphological changes in the cultured cells used to cultivate the virus can be readily observed by examining the cultures microscopically.
- These changes are known as cytopathic effects (CPE) and the viruses that produce CPE are called **cytopathogenic viruses**.

Table 5.7 Classification and characteristics of cell cultures

Type of cell culture	Growth characteristics	Uses	Examples
Primary cell lines	Normal cells freshly taken from the body Capable of limited growth in culture and cannot be maintained in serial culture	Isolation of viruses and vaccine production	Rhesus monkey kidney cell culture Human amnion cell culture, human embryonic kidney, chick embryo cell cultures
Diploid cell lines	Cells of a single type that retain their original diploid chromosome number and karyotype over a limited number, usually 50, of serial subcultivations	Production of viral vaccines	WI-38 (Wistar Institute-38): Human embryonic lung cell strain MRC-5 (Medical Research Council cell strain-5): Fibroblast from lung tissue
Continuous cell lines	Single type, usually derived from cancer cells Capable of continuous, serial, and indefinite proliferation May be maintained by serial sub-cultivation or stored in the cold (–70°C) for use when necessary	Diagnostic and research Some cell lines are now permitted to be used for vaccine manufacture, e.g., Vero cell lines for rabies vaccine or Japanese encephalitis vaccine	HeLa: Human carcinoma of cervix cell line HEp-2: Human epithelioma of larynx cell line McCoy: Human synovial carcinoma cell line Vero: Vervet monkey kidney cell line BHK-21: Baby hamster kidney cell line

CPE produced by viruses are characteristic and help in the presumptive identification of virus isolates (Fig. 5.12). The following are some examples:

- **Enteroviruses:** Rapid CPE with crenation of cells and degeneration of the entire cell sheet
- **Measles virus:** Syncytium formation
- **Herpes virus:** Discrete focal degeneration
- **Adenovirus:** Large granular clumps resembling bunches of grapes
- **SV40:** Prominent cytoplasmic vacuolation

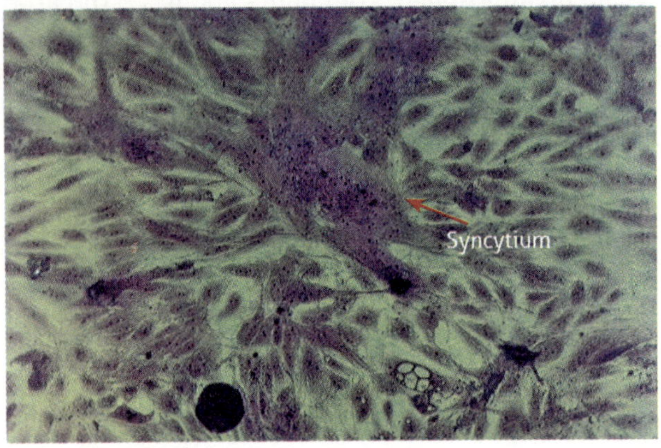

Fig. 5.12 Vervet monkey kidney cells (Vero cell line) infected with measles virus stained with crystal violet and viewed under 100 × magnification (note syncytium formation)

Metabolic inhibition

- In normal cell cultures, the medium turns acidic due to cellular metabolism.
- When viruses grow in cell cultures, cell metabolism is inhibited, and hence there is no acid production.
- The absence of acid production is detected by an indicator (phenol red) incorporated in the medium.

Hemadsorption

- It is seen during the growth of hemagglutinating viruses (such as influenza and parainfluenza viruses).
- When guinea pig erythrocytes are added to the cultures, they are adsorbed onto the surface of the cells.

Interference

- The growth of a non-cytopathogenic virus in a cell culture can be tested with a subsequent challenge with a known cytopathogenic virus.
- The growth of the first will inhibit infection by the second virus by interference.

Transformation

Tumour-forming (oncogenic) viruses induce cell transformation and a loss of contact inhibition so that growth appears in a piled-up fashion and produces **microtumours.**

Immunofluorescence

Cells from virus-infected cultures can be stained with a fluorescent-conjugated antiserum and examined under the UV microscope to look for the presence of the virus antigen. This method produces positive results earlier than other methods and, therefore, finds wide application in diagnostic virology.

Immunological Tests

Demonstration of virus antigens

In cases where the virus antigen is abundant in lesions, a rapid diagnosis can be made by demonstrating the viral antigen by the following immunological methods:

- Immunofluorescence
- Enzyme immune assays, e.g., NS1 in serum to diagnose dengue
- Chemiluminescence immune assays (CLIA)
- Lateral flow assays, e.g., COVID-19 antigen in nasopharyngeal swabs

Antibody detection (serological tests)

By itself, the detection of antibodies in the serum is of limited diagnostic significance as an individual may have been exposed to a virus at any time in his life and would, therefore, have antibodies. However, serological diagnosis is done with following conditions:

- To demonstrate a **rise in the titre of antibodies** to a virus during the course of a disease based on the examination of paired sera. Depending on the stage of the infection, the following types of samples are collected:
  - The **acute phase** sample is collected early in the course of the disease.
  - The **convalescent phase** sample is collected 10–14 days later.
- To **detect IgM** by specific tests—the presence of IgM denotes a recent or ongoing infection as it is the first antibody to be produced in an infection. The serological techniques (these methods are discussed in detail in Chapter 13) in general use are the following:
  - **Enzyme-linked immunosorbent assay (ELISA)** and other immunoassays, e.g., IgM in dengue
  - **Neutralisation test** (often performed to demonstrate protective neutralising antibodies)
  - **Complement fixation test** (rarely performed these days due to the labourious and cumbersome nature of the process)
  - **Hemagglutination inhibition test**—red cells are added to serial dilutions of a viral suspension; the highest dilution that produces hemagglutination is taken as the **hemagglutination titre** (Fig. 5.13); this test is performed for the detection and assay of the influenza virus

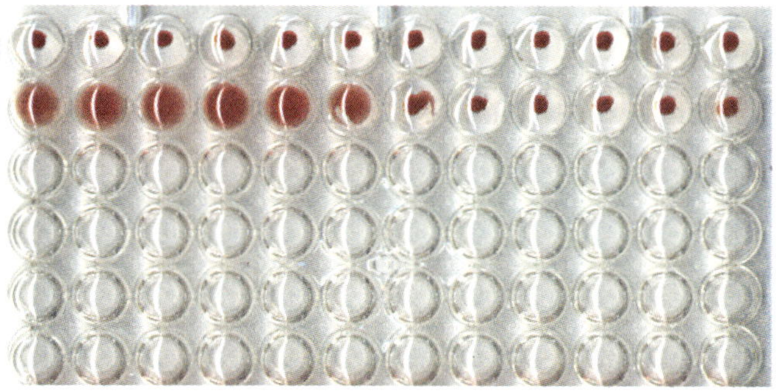

Fig. 5.13 Viral hemagglutination test

Molecular Diagnosis

The availability of molecular methods has transformed the diagnosis of viral diseases, enlarging the scope, sensitivity, and specificity of such tests.

- **In situ hybridisation** using nucleic acid probes has been used to diagnose human papillomaviruses.
- Viral nucleic acid amplification techniques like PCR, **reverse transcriptase PCR (RT-PCR), and real-time PCR** have increased the sensitivity of diagnostic assays.
- **Quantitative RT-PCR (qRT-PCR)** is used to determine the viral load in a patient's blood for monitoring his response to treatment (e.g., HIV treatment and follow-up).
- **Multiplex PCR** is a type of polymerase chain reaction that is used to amplify several DNA sequences simultaneously. This process amplifies DNA in samples using multiple primers and a temperature-mediated DNA polymerase in a thermal cycler. It is useful in detecting multiple causative agents.
- The **automated integrated multiplex PCR platform**, which multiples the PCR reaction, has been automated into a single tube to detect multiple pathogens directly from samples. The detection of amplified targets is done by microarray-based assays (Biofire™ [Fig. 5.14a], QIAstat-Dx™). The sample preparation, amplification, detection, and analysis have been integrated in one tube or reaction.
- In the **rapid or one-tube molecular diagnostic platform**, the pathogen can be detected directly from a sample, and the result is available in a few minutes to one hour, for example, GeneXpert™ (Fig. 5.14b).
- Following PCR, DNA or RNA sequencing of the product can be performed to obtain the full sequence of a viral genome. This is useful in detecting mutants and antiviral drug resistance and also for epidemiological studies and vaccine development.

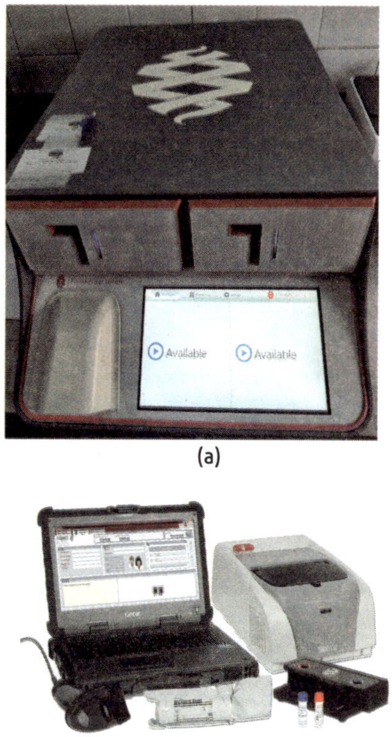

(a)

(b)

Fig. 5.14 (a) Biofire™: Microarray-based multiplex PCR platform and (b) GenXpert™: Rapid/single tube molecular diagnostic platform (*Source*: [a] Dr Poonam Loomba, Govind Ballabh Pant Institute of Postgraduate Medical Education and Research, New Delhi, and [b] photo by Joint Project Manager for CBRN Medical Graphics, from Flickr)

BACTERIOPHAGES

- A bacteriophage is a **virus that infects bacteria**. Such viruses occur widely in nature in close association with bacteria.
- They can be readily isolated from feces, sewage, and other natural sources of mixed bacterial growth.

Morphology

- Bacteriophages come in varying shapes and sizes, but they are typically tadpole-shaped (Fig. 5.15a and b). They can be spherical or filamentous and possess either single-stranded DNA or RNA.
- A bacteriophage consists of a head and tail attached to a base plate and prongs.
  - **Head:** It consists of a tightly packed core of nucleic acid (double-stranded DNA) surrounded by a protein coat or capsid with variable sizes of 28–100 nm.
  - **Tail:** It has a hollow core, a contractile sheath surrounding the core, and a terminal base plate which has prongs, tail fibres, or both attached to it.

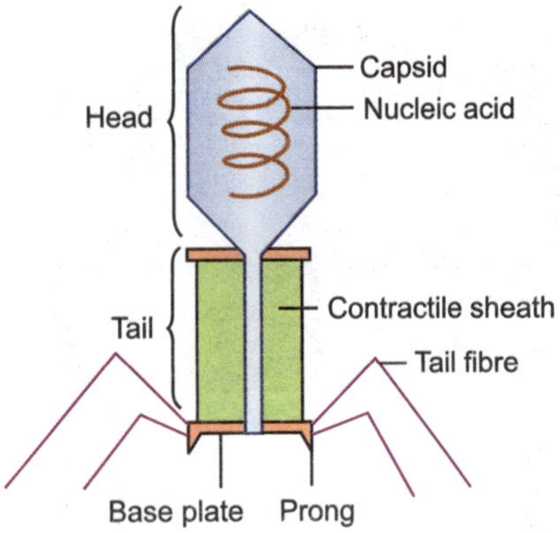

(a)

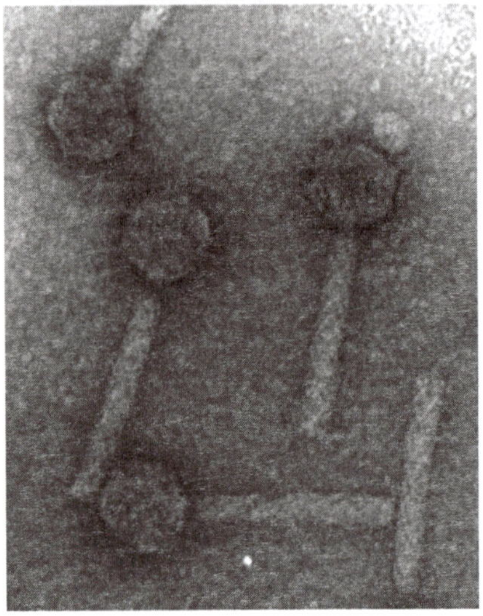

(b)

Fig. 5.15 (a) Morphology of bacteriophage and (b) electron micrograph of a bacteriophage

Life Cycle

Phages exhibit two types of life cycles (Fig. 5.16).

- **Lytic or virulent cycle:** The intracellular multiplication of the phage culminates in the lysis of the host bacterium and the release of progeny virions. The replication of a virulent phage occurs in the following stages—**adsorption, penetration, synthesis of phage components, assembly, maturation, and release of progeny phage particles,** like most other viruses.
- **Lysogenic or the temperate cycle:** The phage DNA becomes integrated with the bacterial genome, replicating synchronously with it, causing no harm to the host cell. The integrated phage nucleic acid is known as the **prophage.** The prophage replicates synchronously with the host chromosome. This phenomenon is called **lysogeny** and a bacterium that carries a prophage within its genome is called a **lysogenic bacterium**. The prophage confers certain new properties on the lysogenic bacterium. This is known as lysogenic conversion or **phage conversion**. This occurs due to the synthesis of new proteins coded for by prophage DNA. For example, toxin production by *C. diphtheriae* occurs due to the presence of a **beta prophage.** Elimination of the prophage abolishes the toxigenicity of the bacillus.
- **Superinfection immunity:** UV rays, hydrogen peroxide, and nitrogen mustard convert a lysogenic cycle to a lytic cycle. A lysogenic bacterium is resistant to re-infection by the same or related phages.

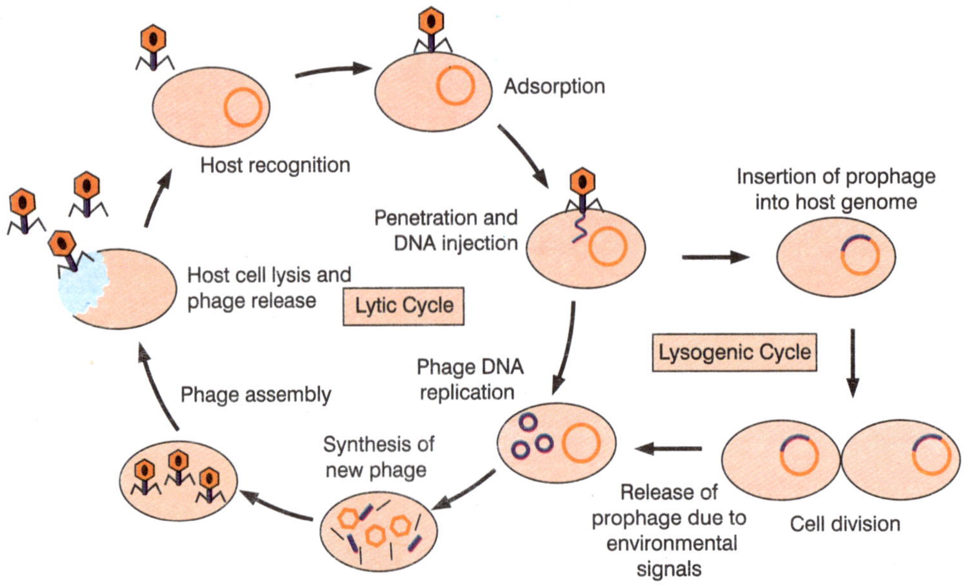

Fig. 5.16 Lytic and lysogenic cycles of bacteriophages

Functions of Bacteriophages

The important functions of bacteriophages are as follows:

- **Transmission of genetic information**—bacteriophages may carry genes from one bacterium to another by transduction (elaborated in Chapter 8).
- **Transduction** has been demonstrated in many genera of bacteria and is one of the most important mechanisms of genetic exchange among bacteria in nature. For example**, plasmid-mediated drug resistance** in many bacteria is a medically important property that is transmitted by a bacteriophage. There are two types of transduction:
  - **Restricted transduction:** Only bacterial genes contiguous to the prophage are transmitted. For example, transduction by the prophage lambda in *E. coli* K12 transfers only the *gal*+ gene (determining the fermentation of galactose).
  - **Generalised transduction:** Any bacterial gene may be transferred to the prophage.
- **Phage conversion**—the integration of specific genes of the phage genome with bacterial chromosomes confers certain properties on the bacteria.

Significance in Medical Microbiology and Clinical Practice

Phage assay

- When phages are applied on the lawn culture of a susceptible bacterium, areas of clearing appear after incubation. These zones of lysis are called **plaques**.
- Under optimum conditions, a single phage particle is capable of producing one plaque.
- The plaque assay can be employed for titrating the number of viable phages in a preparation.

Phage typing

- The specificity of the phage–bacterium interaction is applied to the identification and typing of bacteria. Phages exhibit different degrees of host specificity. Some phages possess wide host ranges, spanning many bacterial genera, while others have a narrow range limited to certain strains of bacteria.
- Sometimes, serial passage in a strain of bacterium makes a phage specific for that strain and related strains (adaptation of host range).
- Phages may lyse members of a bacterial genus (e.g., genus-specific bacteriophage for *Salmonella*) or a species (e.g., specific bacteriophage for *B. anthracis*), a biotype or a subspecies (e.g., Mukherjee's phage IV which lyses all strains of classical *V. cholerae* but not *V. cholerae* biotype El Tor [Fig. 5.17]).

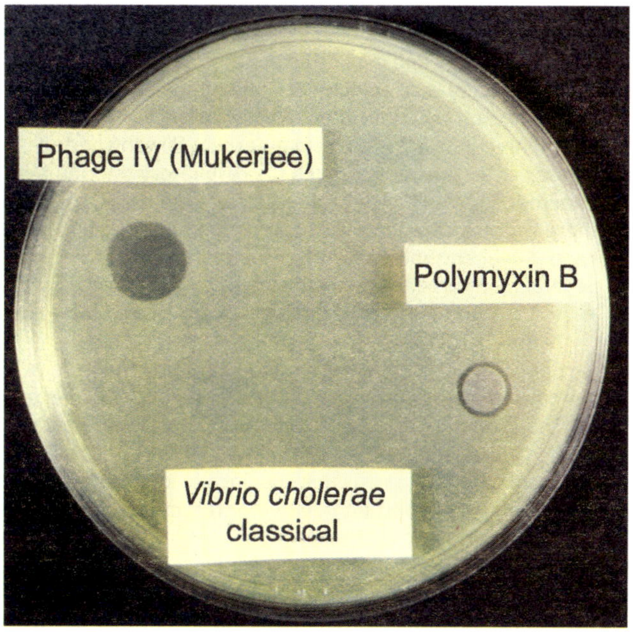

Fig. 5.17 Mukerjee's phage IV classical lysis of *V. cholera* culture (*Source*: CDC, PHIL, Image ID 3915)

Cloning vector

Phages have been used as cloning vectors in genetic manipulations.

Phage therapy

They are increasingly being researched for treating antibiotic-resistant organisms as an alternate to antibiotic therapy in multidrug-resistant infections.

- Phage therapy using bacteriophages was practiced for almost a century to treat bacterial infections.
- However, this was not pursued further. The rise of multidrug-resistant organisms and the failure of antibiotics in recent years has led to renewed interest in using them for treating infections.
- Phage therapy relies on the use of naturally occurring bacteriophages that infect and lyse bacteria at the site of infection.
- The use of bioengineered phages resulting from biotechnological advances has expanded the repertoire of treatment modalities with novel strategies such as using whole phages or purified phage lytic proteins.

Key Points

- ✓ Viruses are obligate intracellular parasites—they are dependent on the synthetic machinery of the host cell for replication.
- ✓ They are considered to be the smallest 'living units not affected by antibacterial antibiotics'.
- ✓ Viruses are much smaller than bacteria and can be visualised directly by electron microscopy.
- ✓ The extracellular infectious virus particle is the virion.
- ✓ Viruses either have DNA or RNA and never both.
- ✓ Virions may be enveloped or non-enveloped. The envelope helps in attachment to the host cell surface or erythrocytes.

- ✓ Protein subunits (peplomers) may appear as projecting spikes on the envelope surface (hemagglutinin and neuraminidase of the influenza virus).
- ✓ The viral multiplication (replication) cycle consists of six sequential phases (overlaps may occur):
  - Adsorption (attachment) • Penetration into the cell • Uncoating • Biosynthesis • Maturation • Release of progeny
- ✓ The virus cannot be demonstrated within the host cell from the stage of penetration to the stage of release—this is called the eclipse phase.
- ✓ The most characteristic histological feature of virus-infected cells is the appearance of inclusion bodies; these can be seen by light microscopy and may be acidophilic or basophilic and intracytoplasmic or intranuclear.
- ✓ The Negri body is an acidophilic, intracytoplasmic inclusion body in cells infected by the rabies virus.
- ✓ Depending on the clinical outcome, viral infections can be classified as inapparent (subclinical), apparent (clinical or overt), or latent infections, which remain quiescent for long periods.
- ✓ Viruses enter the body through the respiratory tract, alimentary tract, skin, conjunctiva, or the genital tract.
- ✓ The infecting virus usually spreads from the point of entry and multiplies in the sites of selection. In systemic virus infections, there is entry, spread to the local lymph nodes, multiplication, entry into the bloodstream (primary viremia), and spread to the central liver and spleen, which results in secondary viremia, and consequently, spread to the target organs.
- ✓ Several methods are used in the laboratory for the diagnosis of viral infections:
  - Viruses in infected specimens may be detected by electron microscopy. Light microscopy is used to detect viral inclusion bodies in Giemsa-stained smears.
  - Specific viral antigens in the blood or body fluids of an infected individual can sometimes be detected.
  - Viruses can be cultivated in animals, embryonated eggs, or tissue cultures.
  - The growth of a virus in cell cultures can be detected by tests that demonstrate cytopathic effect, metabolic inhibition, hemadsorption, and transformation, or the viruses can be detected by immunofluorescence.
- ✓ A bacteriophage is a virus that specifically infects bacteria and fungi.
- ✓ Phages play an important role in the transmission of genetic information between bacteria by the process of transduction. They can carry antibiotic resistance genes or virulence genes. Bacteria can be typed by phage-typing methods for epidemiological purposes.
- ✓ Bacteriophages exhibit two different types of life cycles:
  - In the lytic (virulent) cycle, intracellular multiplication of the phage ends in the lysis of the host bacterium and release of progeny virions.
  - In the lysogenic (temperate) cycle, phage DNA becomes incorporated within the bacterial genome and then replicates synchronously with it, causing no harm to the host cell.

Questions

Short answer questions

1. List the general characteristics of viruses.
2. Describe viral multiplication.
3. What are the various routes of viral entry into the host? Give some examples.
4. What are the methods of cultivating viruses?
5. Describe hemagglutination in viruses.
6. List five RNA viruses and the diseases caused by them.
7. List DNA viruses and the diseases caused by them.
8. Briefly describe prions.
9. What are the uses of bacteriophages in medical science?

MCQs

1. **Which of the following is a receptor-destroying enzyme?**
   a) Neuraminidase b) Reverse transcriptase
   c) Transcriptase d) Hemagglutinin
2. **Which of the following is an early step in the viral replication cycle?**
   a) Adsorption b) Syncytium formation
   c) Uncoating d) Biosynthesis
3. **Which of the following is not an intracytoplasmic inclusion body?**
   a) Guarnieri body b) Negri body
   c) Henderson–Paterson body d) Type A Cowdry body
4. **Phages transmit genetic information between bacteria by the process of ________.**
   a) Transduction b) Transformation
   c) Conjugation d) Transfection
5. **Which is of the following cell lines is not used for virus cultivation for vaccine production?**
   a) Primary monkey kidney
   b) Human embryonic kidney
   c) Human carcinoma cervix
   d) Human fibroblast of lung tissue
6. **Prolonged storage of viruses can be achieved by:**
   a) Freezing at -70°C b) Storing at -20°C
   c) Lyophilisation d) Any of the above
7. **The following virus cannot withstand freeze-drying:**
   a) Varicella b) Rabies
   c) Polio d) CMV
8. **Viruses are inactivated by:**
   a) Antibiotics b) Acidic pH
   c) Ionising radiation d) Organic iodine compounds
9. **Elution is the characteristic feature of which of the following virus groups?**
   a) Polio b) Arbo
   c) Myxo d) Pox
10. **Von Magnus phenomenon is:**
    a) An incomplete virus assembly
    b) Eclipse phase
    c) Elution
    d) Abortive infection
11. **Pock-forming viruses belong to which of the following groups?**
    a) Influenza b) Vaccinia
    c) Yellow fever d) Paramyxovirus
12. **Cytoplasmic vacuolation is the characteristic feature of:**
    a) Herpes virus b) SV40 virus
    c) Influenza virus d) Parainfluenza virus
13. **Hemadsorption test uses RBCs of:**
    a) Sheep b) Mouse
    c) Guinea pig d) Hamster
14. **RBCs used for the estimation of viral hemagglutination titre are from:**
    a) Chicken b) Sheep
    c) Rabbit d) Male mouse

15. When the genome of a virus is covered by the entire capsid of the other virus it is called:
a) Stable variation b) Transcapsidation
c) Complementation d) Interference

16. Which of the following is a single-stranded DNA virus?
a) Pox b) Papova
c) Paramyxo d) Parvo

17. Which one of the following groups of viruses belongs to Picornaviridae family?
a) HAV, Polio, Echo b) Denso, Polio, Rhino
c) Rhino, Nairo, Lassa d) Rubi, Coxsackie, Echo

18. The process of viral uncoating is favoured by:
a) RDE enzyme b) Neuraminidase
c) Transcriptase d) Lysosomes

19. Which one of the following is false regarding inclusion bodies?
a) They are crystalline aggregates of virions
b) They represent degenerative changes produced by viral infections
c) They can be demonstrated under an electron microscope
d) They can be acidophilic or basophilic

20. Guarnieri bodies are seen in which viral infection?
a) Rabies b) Polio
c) Vaccinia d) Herpes

21. Enveloped viruses are inactivated by:
a) Gastric pH b) Bile
c) Saliva d) Tears

22. The common route of entry for all the following viruses is through the alimentary tract except:
a) Adenoviruses b) Reoviruses
c) Hepatitis C d) Poliovirus

23. The important initial response of a human body to a viral infection is:
a) B cell stimulation b) Antibody production
c) Interferon production d) T cell stimulation

24. Seller's stain is used for the demonstration of which of the following viruses?
a) *CMV* b) Rabies
c) Measles d) Herpes

25. Which of the following agents was developed as an antiparasitic but is presently used as an antiviral?
a) Amantadine b) Acyclovir
c) Suramine d) Virazole

26. All of the following viruses produce lesions at the site of entry except:
a) Orf b) Vaccinia
c) Papilloma d) Adeno

27. Which one of the following is false regarding bacteriophages?
a) They have no role in genetic manipulation.
b) They confer certain properties to bacteria.
c) They are used for bacterial typing.
d) They control the total number of bacteria in natural environment.

28. Phage typing is not useful for:
a) Tracing the carrier during an epidemic
b) Identification of bacteria
c) Differentiating classical *V. cholera* from El tor
d) Pathogenicity testing

29. The process of transmission of genes between bacteria through bacteriophages is called:
a) Transduction b) Eclipse
c) Conjugation d) Transformation

30. Which of the following event/s occurs in the temperate cycle of a bacteriophage life cycle?
a) Intracellular multiplication
b) Intracellular multiplication and host cell death
c) Multiplication, incorporation into host DNA and cell death
d) Incorporation in host DNA and no harm to host cell

31. The interval between the entry of phage into the cell and the appearance of the first infectious intracellular phage particle is called:
a) Latent period b) Eclipse phase
c) Rise period d) Burst phase

32. Beta-prophage is associated with:
a) *E. coli* b) Classical vibrios
c) El tor Vibrios d) *C. diphtheriae*

33. Super infection immunity is seen in the case of:
a) MRSA b) El tor Vibrios
c) Bacteriophage d) *E. coli*

34. The highest dilution of the phage preparation that produces confluent lysis is called:
a) Routine test dose b) Prophage beta
c) Eclipse phase d) Latent phase

35. Which of the following is true about lysogenisation?
a) Inhibits bacterial multiplication
b) Abolishes toxigenicity of bacteria
c) Does not interfere with bacterial metabolism
d) Interferes with bacterial motility

36. Bacteriophages cannot be used for:
a) Tracing the disease carrier
b) Identification of bacteria
c) Prophylaxis of bacterial infections
d) Differentiating classical *V. cholera* from El tor

37. Yolk sac inoculation technique is used for cultivation of:
a) Influenza virus b) Smallpox virus
c) *Chlamydia* d) *Mycoplasma*

38. Which of the following is an example of primary cell lines?
a) HeLa b) HEp-2
c) KB d) Monkey kidney

39. Which continuous cell line is now permitted to be used for vaccine production?
a) Vero cell lines b) HeLa
c) HEp-2 d) KB

40. Which virus results in syncytium formation in cell lines?
a) Adenovirus b) Herpesvirus
c) Enterovirus d) Measles

41. Hemagglutination inhibition test is used to detect and assay which virus?
a) Influenza virus b) Herpesvirus
c) Enterovirus d) Rubella virus

42. Warthin–Finkeldey cell is an:
a) Acidophilic intracytoplasmic and intranuclear inclusion body seen in measles
b) Eosinophilic intracytoplasmic and intranuclear inclusion body seen in rabies
c) Eosinophilic intracytoplasmic and intranuclear inclusion body seen in measles
d) Acidophilic intracytoplasmic and intranuclear inclusion body seen in vaccinia

43. Which of the following is not true regarding viral replication?
a) Adsorption takes place only if there is affinity between the virion and the host cell.
b) Viral envelopes are derived from the host cell membrane during the process of budding.
c) There is early transcription of messenger RNA (mRNA) from the viral nucleic acid.
d) In most viruses uncoating is due to action of specific viral protein.

Answers: 1:a, 2:a, 3:d, 4:a, 5:c, 6:c, 7:c, 8:c, 9:c, 10:a, 11:b, 12:b, 13:c, 14:a, 15:b, 16:d, 17:a, 18:d, 19:c, 20:c, 21:b, 22:c, 23:c, 24:b, 25:c, 26:d, 27:a, 28:d, 29:a, 30:d, 31:b, 32:d, 33:c, 34:a, 35:c, 36:c, 37:c, 38:d, 39:a, 40:d, 41:a, 42:c, 43:d

CHAPTER 6

Basic Concepts of Mycology

(MI1.1–1.3, 8.9–8.13)

Chapter Navigation

- General characteristics of fungi
- Classification of fungi
- Pathogenesis of fungal infections
- Laboratory diagnosis of fungal infections

Learning objectives

- Describe the general properties of fungi
- Enumerate the various methods to classify medically important fungi
- Discuss the pathogenesis of fungal infections
- Outline the modalities for the laboratory diagnosis of fungal infections

Mycology (from the Greek word *mykos*, meaning fungus) is the branch of microbiology that deals with the study of fungi (yeasts and moulds).

- **Medical mycology** deals with the etiology, pathogenesis, detection, prevention, and treatment of fungal infection in humans.
- Of the 3,000 odd fungal species on earth known to humans, **only about 300 are pathogenic to human beings**. The majority of fungi are chiefly of concern as plant pathogens, which cause huge losses in the agricultural economy.
- Fungi are gaining importance due to their omnipresence in nature and their association with human infections, some of which lead to life-threatening conditions with high mortality. The **chief causes** of the increasing incidence of fungal infections include:
  - The increasing use of chemotherapeutic agents, steroids, immunosuppressive agents, and antibiotics to increase life expectancy in moribund patients
  - Medical advances such as organ transplant, robotic surgery, etc.
  - Increasing incidences of immunocompromised conditions and chronic debilitating diseases like tuberculosis and diabetes
- Fungal infections cause more fatalities today than whooping cough, diphtheria, typhoid, dysentery, and malaria put together.
- Considering that most fungi are saprophytes in soil, human mycotic infections are usually **opportunistic**, barring a few infections that are caused by true pathogens.

CHARACTERISTICS OF FUNGI

Fungi are eukaryotic **Protista** that differ from bacteria and other prokaryotes.

The following are some of the defining characteristics of fungi:

- Fungi possess **rigid cell walls** containing chitin, mannan, and other polysaccharides.
- Their cytoplasmic membrane contains **sterols**.
- They possess **true nuclei** with a nuclear membrane and paired chromosomes.
- They divide asexually, sexually, or by both processes.
- They may be unicellular or multicellular.
- Fungal cells exhibit various degrees of specialisation.

CLASSIFICATION OF FUNGI

Like most microorganisms, different types of fungi possess different characteristics, based on which they are classified. Some of these characteristics are as follows:

- Morphological forms
- Type of infection
- Site of infection
- Type of sporulation produced by the fungus

Classification Based on Morphological Forms

From a diagnostic point of view, fungi may be classified into four groups based on their cell morphology (Fig. 6.1 and 6.2).

1. **Yeasts (e.g., *Cryptococcus neoformans*)**
   - **Unicellular fungi**; do not produce germ tubes or pseudohyphae
   - Reproduce by **budding** (known as **blastospore** or **blastoconidium**)
   - In culture, the macroscopic appearance of the colony is **pasty**, resembling bacterial colonies
   - Microscopically, the yeast cell—in tissues and culture—is **oval or spherical** (3–15 mm in diameter)
   - Filamentous (hyphae-like) structures are **not seen** in tissues or culture (Fig. 6.3a and b)

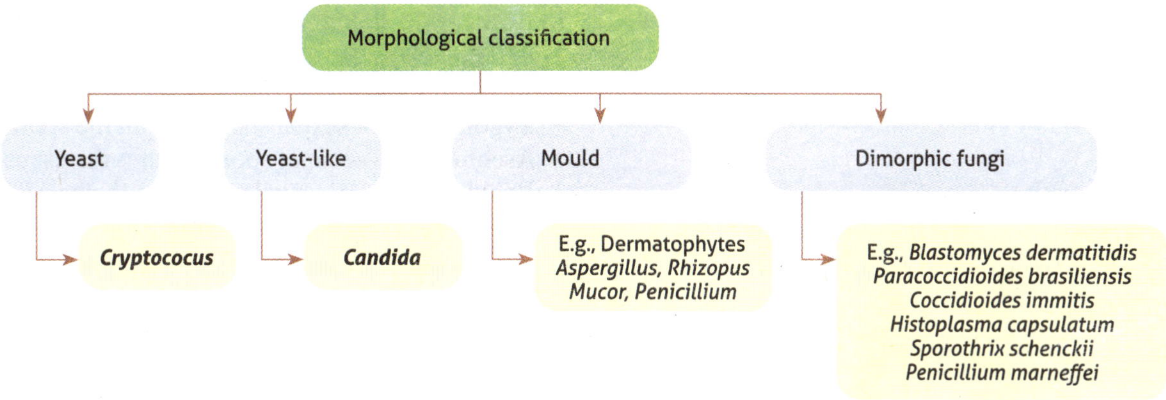

Fig. 6.1 Morphological classification of medically important fungi

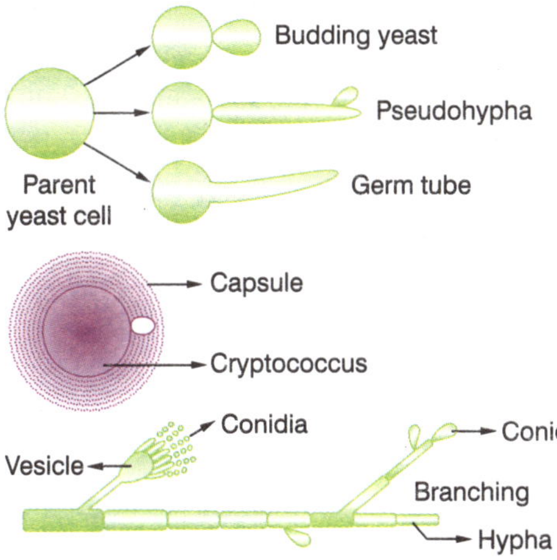

Fig. 6.2 Morphological forms of fungi—budding yeast and branching mould, *Aspergillus fumigatus*

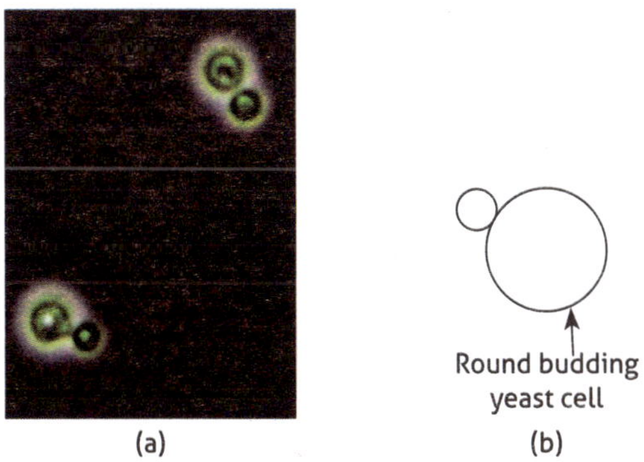

Fig. 6.3 (a) Budding yeast cell: *Cryptococcus neoformans* in India ink preparation and (b) round, budding yeast cell

2. **Yeast-like fungi (e.g., *Candida albicans*)**
   - **Unicellular** fungi
   - Reproduce by **budding or fission**
   - **Macroscopically**, colonies in culture are similar to those of other yeasts—pasty (resembling bacterial colonies)
   - **Microscopically**, cells are spherical or oval (in tissues as well as in culture); filamentous structures (pseudohyphae) may be seen in clinical material (Fig. 6.4a, b, and c)

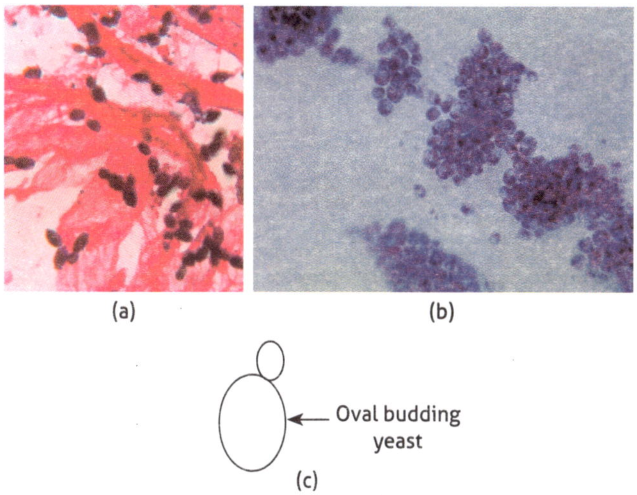

Fig. 6.4 (a) Oval budding cells of *Candida albicans* in Gram smear (*Source*: Dept of Microbiology, PIMS), (b) pseudohyphae (*Source*: Centers for Disease Control and Prevention [CDC], Public Health Image Library [PHIL], Image ID 2918), and (c) oval, budding yeast cells

3. **Filamentous fungi or moulds** (e.g., *Aspergillus*, *Penicillium*, and *Dermatophytes*)
   - Made up of hyphae which may have cross-walls or septa (**multicellular**) or may be devoid of septa (**coenocytic**) with conidiophores bearing fruiting bodies and asexual spores
   - **Reproduce** by asexual means (spore formation); some exhibit sexual reproduction
   - **Macroscopically**, colonies may have a surface texture that is cottony/woolly/velvety/granular; **pigmentation** may be observed on the reverse side of the plate
   - **Microscopic appearance:**
     - **Thread-like, filamentous hyphae** (2–10 mm) seen in tissues and culture
     - These may be **aseptate** (without cross-walls, e.g., *Zygomycetes*) or **septate** (with cross-walls, e.g., *Aspergillus fumigatus*; Fig. 6.5a)
     - Mycelium may be **vegetative** or **aerial**

- Hyphae may be racquet-shaped, nodular, pectinate, spiral, root-like (rhizoid), or chandelier-like (a repeatedly branched cluster of hyphal apices that resemble a chandelier; Fig. 6.5b)

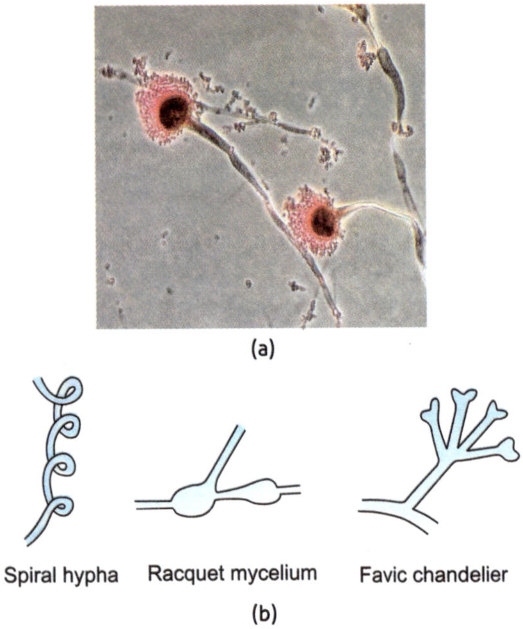

Fig. 6.5 (a) *Aspergillus* (*Source*: CDC, PHIL, Image ID 15145) and (b) different forms of hyphae

4. **Dimorphic fungi (e.g., *Histoplasma capsulatum*)**
   - They grow as **filamentous forms** in cultures incubated at 22–25°C (Fig. 6.6a) and in the natural environment and as yeasts in cultures incubated at 37°C and in tissues (Fig. 6.6b).

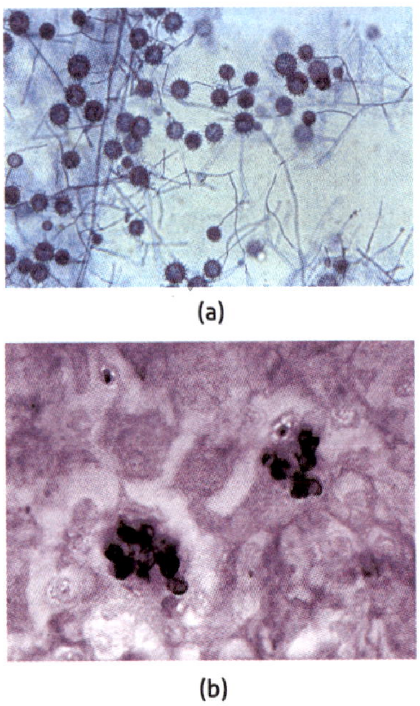

Fig. 6.6 *Histoplasma capsulatum*: (a) Mycelial (filamentous) form (*Source*: CDC, PHIL, Image ID 19420) and (b) yeast form in a tissue section

Classification Based on Sporulation

Fungi may also be classified based on the type of sporulation they exhibit, whether sexual or asexual.

- **Zygomycetes**—sexual spores within the sporangium
- **Ascomycetes**—sexual spores within the ascus
- **Basidomycetes**—sexual spores on the basidium
- **Deutermycetes** or **fungi imperfecti**—fungi with no known sexual spores (most clinically important spores belong to this group)

Classification Based on the Type of Infection

Based on the pathogenic potential of a fungus and the immune status of the host, **the infection caused by it is classified as primarily pathogenic or opportunistic.**

Primarily pathogenic

- These infections are generally caused by dimorphic fungi that produce infections in **healthy individuals**.
- These infections may be superficial, cutaneous, subcutaneous, or deep or systemic.
- Examples—the infection produced by primary pathogen *Coccidioides immitis* (Fig. 6.7).

Opportunistic

- These infections are caused by fungi that can, by various mechanisms, cause infections in **immunocompromised individuals** (e.g., individuals receiving immunosuppressive drugs, those infected with HIV, and those who have diabetes mellitus or other debilitating conditions).
- Examples—aspergillosis, zygomycoses, and cryptococcosis.

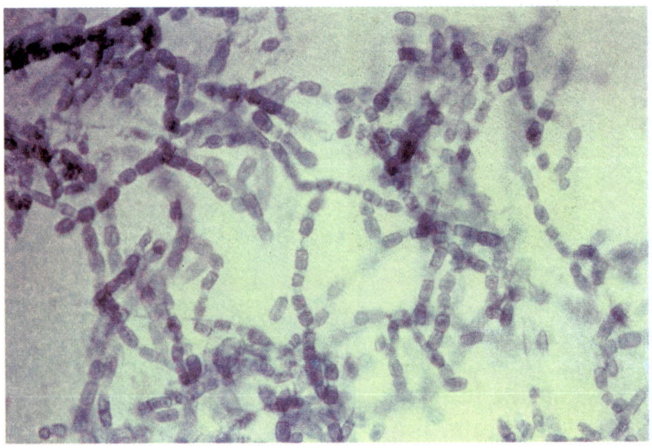

Fig. 6.7 Chains of arthrospores of *Coccidioides immitis* arising from the mycelia (*Source*: CDC, PHIL, Image ID 16303)

PATHOGENESIS OF FUNGAL INFECTIONS

The pathogenesis of fungal infections differs according to the site of the infection. The various types of fungal infections are presented in Table 6.1 and Fig. 6.8.

Table 6.1 Common sites of major fungal infections

Type of infection	Site of infection	Disease
Superficial	Skin, hair shaft	Pityriasis, piedra
Cutaneous	Skin, hair, and nails	Dermatophytoses
Subcutaneous	Subcutis, dermis	Mycetoma, sporotrichosis, chromomycoses
Deep	Internal organs	Histoplasmosis, coccidiosis
Opportunistic	Disseminate in internal organs	Candidiasis, cryptococcosis, aspergillosis

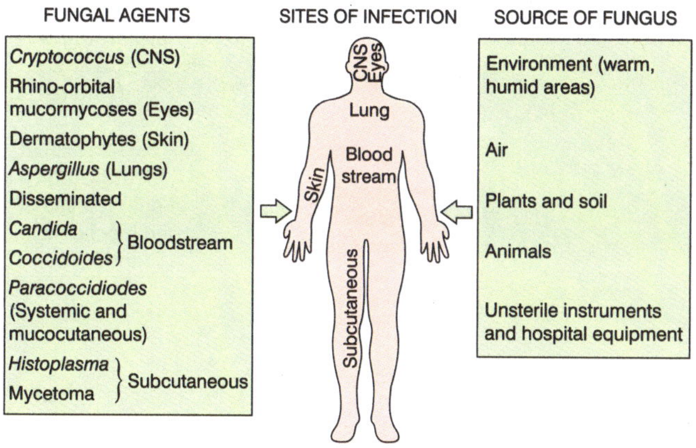

Fig. 6.8 Fungal agents and sources and major sites of fungal infections

Superficial Mycoses

- These involve the superficial surfaces of the body (**dermis**).
- They are acquired by **direct contact**, either person-to-person or animal-to-person (dermatophytes).
- These infections rarely disseminate to the deeper structures.
- They produce **local inflammatory reactions** and spread contiguously.
- These fungi affect the skin, hair shaft, and nails.

Subcutaneous Mycoses

- These involve the subcutaneous tissues (**dermis and subcutis**).
- They are acquired by a **penetrating injury**, whereby the fungal spores enter the subcutaneous tissues, where they establish an infection with inflammatory cell infiltration and host tissue reaction.
- They may **spread via the draining lymphatics** to the regional lymph nodes (e.g., infections caused by *Sporothrix* and mycetoma).

Deep or Opportunistic Fungal Infections

- These are **generally disseminated**.
- They are usually acquired by **inhalation**.
- In immunocompromised or metabolically impaired individuals with impaired cell-mediated immunity, the normally saprophytic fungi gain entry to the body and establish infection (e.g., pulmonary aspergillosis and mucormycosis).

HOST RESPONSE TO FUNGAL INFECTIONS

- Most fungal infections are **chronic,** with a long incubation period (except disseminated opportunistic infections); they present with systemic manifestations and florid clinical features (e.g., disseminated candidiasis).
- The host responds by launching an immune response with chronic inflammatory cells (lymphocytes and plasma cells) and **granuloma formation.**
- The components of the innate immune system (e.g., dendritic cells and macrophages) bind to the components of fungal cell walls using **pattern recognition receptors (PRRs)** on their surface. After binding, they signal using their intracellular tails or associated molecules (FcRγ), resulting in phagocytosis and the triggering of killing mechanisms (e.g., production of reactive oxygen). This also helps in the development of adaptive immunity.
- The mechanisms involved in the development of adaptive immunity against fungi are not fully understood, although it seems they are brought about by $CD4^+$ T cells which produce IFNγ (Th_1) or IL-17 (Th_{17}) and cause effective killing by innate effector cells such as neutrophils and macrophages.

LABORATORY DIAGNOSIS OF FUNGAL INFECTIONS

- A detailed clinical examination, history of occupation, and geographic location are important in diagnosing a fungal infection.
- The demonstration of fungal elements in the clinical specimen is one of the most effective methods of correlating the agent with the infection, as many fungi are commonly found in the environment. Fig. 6.9 outlines the steps involved in the laboratory diagnosis of fungal infections.
- **All laboratory procedures must be carried out in biosafety cabinets (class IIB with negative pressure and HEPA filter).**

Specimen

The specimen to be collected for the laboratory depends on the site of involvement (Table 6.2).

Microscopy

Fungal structures can be detected by the direct microscopic examination of material collected from

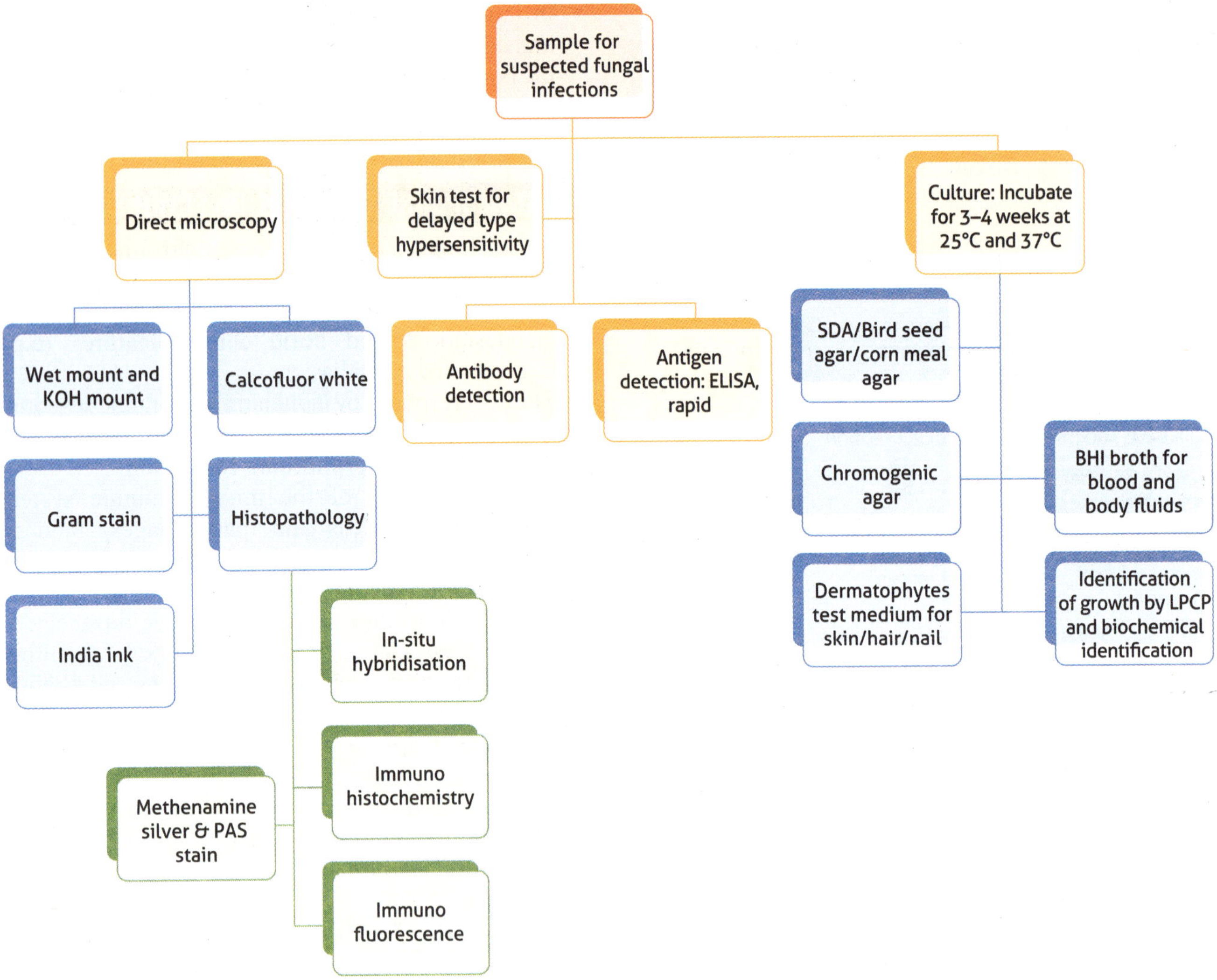

Fig. 6.9 Laboratory diagnosis of fungal infections

Table 6.2 Specimens for different types of fungal infections

Site of infection	Clinical material
Superficial	Skin scrapings and hair and nail clippings
Subcutaneous	Tissue biopsy
Mycetoma	Granules from the sinus or tissues from the sinus tract
Deep and opportunistic infections	Blood, CSF, pus discharge, sputum, pleural and peritoneal fluid, other aspirates, or any other material from the site of infection

lesions and with the help of morphological studies of fungal isolates. The following methods are used to detect these fungal elements:

- **Wet mount and potassium hydroxide (KOH) preparation**—tissue specimens such as skin scrapings are generally examined as wet mounts in saline and after treatment with 10% potassium hydroxide (20% for nail clippings). KOH digests keratin, cells, and other tissue materials, enabling the fungal elements to be seen clearly.
- **Calcofluor white (CFW)** is a fluorescent dye that binds to glucan and chitins, which are abundant in the fungal cell wall.
- **Gram stain** is useful in the identification of yeast and yeast-like fungi.
- **India ink** is used for the negative staining of capsulated yeast, e.g., *Cryptococcus*.
- **Tissue sections**:
  - Hematoxylin–eosin (H&E), Gomori methenamine silver stain, and periodic acid–Schiff (PAS) are routinely used to demonstrate fungi in tissues
  - Immunofluorescent staining
  - In-situ hybridisation with specific probes is used to visualise fungi in paraffin-fixed tissues
  - Immunohistochemistry for the detection of invasive fungi in tissue sections

Culture

Fungi are slow-growing, and therefore, need to be cultivated in media that can support their growth for two to three weeks without drying. Fungal growth requirements are as follows:

- Most fungi are grown at a **low pH (5.4–6)** in media similar to those used for bacteria.
- All fungi require **oxygen** for growth. They are usually incubated in BOD (biochemical oxygen demand) incubators.
- Fungi grow better at temperatures ranging from **25–30°C** (except fungi that cause deep mycoses or dimorphic fungi that grow well at 37°C and *Aspergillus fumigatus*, which can grow even at 50°C).

Media used for fungal culture

Sabouraud dextrose agar

- **Most commonly used** culture medium for fungi.
- It permits the growth of yeasts and most filamentous fungi.
- It has a high concentration of either glucose or maltose and also contains mycological peptone.
- Its low pH (5.4) is inhibitory to the growth of most bacteria.
- Antibacterial agents (chloramphenicol 40 mg or gentamicin 50 mg per litre of the medium) can also be added to inhibit bacteria; cycloheximide (actidione) incorporated in the medium inhibits contaminant moulds.

Some other media used for the cultivation of fungi are as follows :

- **Czapek–Dox medium**
- **Dermatophyte test medium**
- **Bird seed agar**—brown colonies are seen
- **Corn meal agar**—for detecting sporulation, and chlamydospore formation
- **Chromogenic agar (CHROM agar)**—different species of *Candida* produce different coloured colonies (Fig. 6.10)
- **Blood agar, brain–heart infusion agar, or brain–heart infusion broth**—for culturing dimorphic and other fastidious fungi from blood and body fluids

Cultures are incubated in parallel at room temperature (25°C) and 37°C for 3–4 weeks.

Fig. 6.10 CHROM agar showing different coloured colonies produced by different species of *Candida* (*Source*: Department of Microbiology, PIMS)

Identification of Fungi from Culture

Microscopic identification of fungi

1. **Staining:** Lactophenol cotton blue (LPCB) is used for the microscopic study of fungal colonies that are teased onto a slide and mounted.
2. **Preparation of slides**
   - **Teased mounts:** A small portion of the fungal colony is placed in a drop of lactophenol cotton blue on a clean, grease-free slide. The mount is then 'teased out' using a bent wire, covered with a coverslip, and viewed under the microscope.
   - **Disadvantage:** This technique breaks the hyphae and scatters the conidia, making them difficult to identify.
3. **Slide culture:** It preserves the 'in situ' appearance of the fungal structures without any breakage of hyphae or attachment of conidia to the conidiogenous cell.
   - **Method:** A sterile slide is placed on a bent glass rod placed inside a petridish. A square block of SDA measuring 1 cm^2 is placed on the slide. The fungal strain to be identified is inoculated at the four sides of the agar block. The inoculated block is covered with a sterile coverslip and incubated at 25°C. After 48 hours, growth is visible.
   - A drop of LPCB is then placed on a fresh slide and the coverslip is transferred from the block of SDA onto this slide. The preparation is then viewed under the microscope (Fig. 6.11).

Growth characteristics of fungi

The following characteristics of fungal culture are noted:

- The **rate of growth** at 37°C or room temperature (25°C)—yeasts and saprophytic moulds grow within 48 to 72 hours, whereas pathogenic fungi such as *Paracoccidioides*, which are obligate pathogens, grow very slowly.
- **Colour and morphology**—the appearance of the colony (pale, creamy, or pigmented) and its texture (velvety, granular, or cottony) are noted.
- The obverse (front) and reverse (back) sides of the culture plate should be **examined for pigmentation**; if present, it should be determined whether the pigmentation is diffusible or non-diffusible.

Fig. 6.11 Slide culture showing agar blocks with fungal culture (*Source*: Dept. of Microbiology, PIMS)

- **Morphology of various structures**—hyphae (diameter, type, septa), conidiogenous cells, fruiting bodies, conidia (spores), the manner of their attachment to the conidiogenous cell, and other special structures should be noted.

Other Tests for the Identification of Fungi

Yeasts can be identified based on carbohydrate fermentation and assimilation tests.

- **Fermentation test**
  - A medium containing 1% yeast extract and 2% of the test carbohydrate in distilled water is used.
  - The ability of yeasts to ferment carbohydrates is determined strictly on the basis of gas production from the substrate. Carbohydrate fermentation occurs within seven days.
- **Assimilation test**
  - The ability of a yeast isolate to use a particular carbohydrate substrate as its sole carbon source for its metabolic processes can be used to identify it.
  - The medium used is a yeast nitrogen base onto which the yeast is inoculated as a carpet. Filter paper discs containing 2% of the test carbohydrates are placed on the plate.
  - Colonies grow around the disc of the carbohydrate it assimilates.
- **Automated systems**—VITEK™ and Microscan™ can be used to rapidly identify *Candida* species.
- **Antifungal susceptibility tests—disc diffusion tests**
  - Inexpensive and easy to set up
  - Used as a screening test.
  - Tests against azoles (e.g., fluconazole) and echinocandins (e.g., caspofungin) for *Candida* spp. and some dermatophytes
  - **Mueller–Hinton agar supplemented with 2% glucose** is used as a medium
- **Microbroth dilution methods**
  - Used for yeasts and some filamentous fungi, against antifungal agents like amphotericin, fluconazole, and echinocandins.
  - Automated methods (e.g., VITEK™) are used to test the antifungal susceptibility of *Candida*.

Immunological tests

Antibody detection

- **Immunodiffusion tests**—antibodies to *Coccidioides* can be detected by immunodiffusion.
- **Complement fixation test**—antibodies to some invasive fungal infections become detectable in 2–3 weeks, e.g., antibodies to the histoplasmin antigen.
- **Immunoassays** are highly sensitive in detecting antibodies to the soluble antigens of some fungi, e.g., *Histoplasma capsulatum*, and *Cryptococcus*.
- The **mannan antibody** is used for the detection of *Candida* infection.

Antigen detection

Indirect fluorescent antibody tests are performed on tissue sections.

- **Latex agglutination test** is used to detect the cryptococcal capsular antigen and the mannan antigen of *Candida* and *Aspergillus*.
- **Galactomannan antigen** is detected in aspergillosis.
- **Lateral flow assays** are used for rapid detection.
- **(1,3)-β-D-glucan** is a polysaccharide in the cell wall of fungi such as *Aspergillus*, *Fusarium*, and *Candida*. Calorimetric assays are available for its detection.
- **Skin test** is used for the detection of delayed-type hypersensitivity (DTH) for pathogens like *Histoplasma* or *Coccidioides*.
- **Newer rapid diagnostic tests** are available for the identification of newer fungal species causing human infections:
  - Nucleic acid hybridisation
  - Polymerase chain reaction

Key Points

✓ Fungi are eukaryotic Protista that possess a rigid cell wall containing chitin, mannan, and other polysaccharides. They have true nuclei with a nuclear membrane and paired chromosomes.

✓ Medically important fungi are classified based on their morphology as yeasts, yeast-like fungi, moulds (filamentous fungi), and thermally dimorphic fungi.

✓ From the standpoint of disease-causing potential in humans, fungi are classified as primary pathogens and opportunistic pathogens.

✓ Fungi are also classified based on the method of sexual reproduction into zygomycetes (asexual spores in the sporangium), ascomycetes (sexual spores within the ascus), basidiomycetes (sexual spores borne on the basidium), and deuteromycetes ('fungi imperfecti' with no known sexual form).
✓ Pathogenic fungi cause true infections or mycoses, mycotoxicosis, and allergic diseases (type I and type III hypersensitivity reactions).
✓ Superficial mycoses include pityriasis versicolor, tinea nigra, and black and white piedra.
✓ Cutaneous infections include dermatophytosis (caused by three genera of dermatophytes—*Trichophyton*, *Microsporum*, and *Epidermophyton*). Dermatophytosis manifests as ringworm (tinea). Dermatophytes grow relatively slowly on laboratory culture media at room temperature.
✓ Subcutaneous mycoses are fungal infections of the subcutaneous tissue. They remain localised or spread by contiguity. The main types are mycotic mycetoma (eumycetoma), chromoblastomycosis, pheohyphomycosis, sporotrichosis, rhinosporidiosis, and entomophthoromycosis.
✓ Specimens for diagnosis are usually collected from the affected site (and from blood in disseminated infections). They are either examined by direct microscopy or cultured (on Sabouraud dextrose agar).
✓ New diagnostic methods like VITEK™ and Microscan™ systems are increasingly being used to identify yeasts and perform antimicrobial sensitivity testing.

Questions

Short answer questions

1. How are fungi classified? Describe the various groups with examples.
2. Describe the general principles of the laboratory diagnosis of fungal infections.

MCQs

1. **Which of the following is a characteristic of filamentous (moulds) fungi?**
   a) They possess multiple conidia arising from conidiophores.
   b) They have long, septate pseudohyphae.
   c) They produce abundantly mucoid colonies.
   d) They are always associated with infections in healthy individuals.
2. **Which of the following stains is used to stain filamentous fungi in a clinical specimen?**
   a) India ink b) Leishman's stain
   c) Lactophenol cotton blue d) Acid-fast stain
3. **Why are the antibiotics gentamicin and chloramphenicol added to Sabouraud dextrose agar?**
   a) To test the susceptibility of the fungus.
   b) To inhibit contaminating saprophytic fungi.
   c) To inhibit contaminating bacteria.
   d) To inhibit contaminating yeasts.
4. **Which of the following statements is not true for dermatophytes?**
   a) These fungi affect the keratinised layer of the skin.
   b) They affect the hair and nails.
   c) They may be geophilic.
   d) They do not grow at room temperature.
5. **The following are saprophytic fungi associated with infections in immunocompromised individuals except:**
   a) *Candida albicans* b) *Aspergillus fumigatus*
   c) *Coccidioides immitis* d) *Rhizopus arrhizus*
6. **Which of the following statement is not correct regarding yeasts?**
   a) These are unicellular fungi that do not produce germ tubes or pseudo-hyphae.
   b) The reproduce by budding.
   c) Filamentous (hyphae-like) structures are seen in tissues or in culture.
   d) In culture, the colonies are pasty, resembling bacterial colonies.
7. **The cytoplasmic membrane of fungi contains:**
   a) Carbohydrate b) Protein
   c) Sterol d) Mannan
8. **The rigidity of fungal cell walls is due to presence of:**
   a) Protein b) Chitin
   c) Teichoic acid d) Lipopolysaccharide
9. **Which of the following is a subcutaneous mycosis?**
   a) Favus b) Pityriasis versicolor
   c) Tinea capitis d) Chromomycosis
10. **Which of the following is a superficial fungal infection?**
    a) Favus b) Pityriasis versicolor
    c) Tinea capitis d) Chromomycosis
11. **Which one of the following fungi is not a mould?**
    a) *Rhizopus* b) *Cryptococcus*
    c) *Aspergillus* d) *Penicillium*
12. **Which of the following is the morphological characteristic that refers to fungi imperfecti?**
    a) Lack of cell wall
    b) Lack of septate hyphae
    c) Unidentified sexual phases
    d) Unidentified microconidia
13. **Which of the following is not a fungal stain?**
    a) Leishman b) Calcofluor white
    c) Periodic acid Schiff d) Gomori's methenamine silver
14. **All the following media are used for identifying fungi except:**
    a) Cornmeal b) Czapek–Dox medium
    c) Bird seed d) Horse serum agar
15. **The function of cycloheximide in Sabouraud dextrose agar is to prevent the growth of the contaminant:**
    a) Gram positive bacteria b) Gram negative bacteria
    c) Moulds d) Yeasts
16. **Which of the following is correct regarding the Sabouraud dextrose agar?**
    a) Yeast do not grow on this medium.
    b) It has a high concentration of either glucose or maltose.
    c) The pH of the medium is high (8.4).
    d) It is prepared in petridishes.
17. **The most common direct microscopy method to visualise fungal elements from clinical samples is:**
    a) KOH preparation b) LPCB staining
    c) Iodine mount d) Albert stain
18. **Yeasts can be identified based on carbohydrate fermentation tests. Which of the following is correct regarding the fermentation test?**
    a) Fermentation test is performed in a medium containing 10% yeast extract and 2% test carbohydrate in distilled water.
    b) Fermentation test is performed in a medium containing 10% yeast extract and 2% test carbohydrate in thioglycolate broth.

c) Fermentation test is performed in a medium containing 1% yeast extract and 20% test carbohydrate in distilled water.
d) Fermentation test is performed in a medium containing 1% yeast extract and 2% test carbohydrate in distilled water.

19. Which of the following is not a medium to cultivate fungi from clinical samples?
a) Czapek–Dox medium b) Bird seed agar
c) MacConkey's agar d) Brain–heart infusion agar

20. Which of the following is not a prerequisite for the growth of fungi in the laboratory?
a) Most fungi are grown at a low pH (5.4–6).
b) All fungi require oxygen for growth.
c) Fungi grow best at 50°C.
d) They are usually incubated in BOD (biochemical oxygen demand) incubators.

21. Which stain can be used to visualise fungi from culture?
a) Lactophenol cotton blue b) KOH
c) ZN stain d) Methenamine silver stain

22. Which stain can be used to visualise fungi from tissues?
a) Lactophenol cotton blue
b) KOH
c) ZN stain
d) Methenamine silver stain

23. Reynolds-Braude phenomenon refers to the formation of:
a) Chlamydospores b) Germ tubes
c) Rhizoids d) Chains of conidia

24. Which of the following fungus is capsulated?
a) *Candida* b) *Rhodotorula*
c) *Pneumocystis* d) *Cryptococcus*

25. Which statement is not true regarding *Cryptococcus neoformans*?
a) Grows at 37°C
b) Grows on SDA
c) Possesses polysaccharide capsule
d) Negative urease test

Answers: 1:a, 2:c, 3:c, 4:d, 5:c, 6:c, 7:c, 8:b, 9:d, 10:b, 11:b, 12:c, 13:a, 14:d, 15:c, 16:b, 17:a, 18:d, 19:c, 20:c, 21:a, 22:d, 23:b, 24:d, 25:d

Part I: General Microbiology

CHAPTER 7

Basic Concepts of Medical Parasitology

(MI1.1–1.3, 8.9–8.13)

Chapter Navigation

- General parasitology: nomenclatures and definitions
- Classification of parasites
- Protozoa
- Helminths
- Characteristics, life cycles, and modes of transmission of parasites
- Concepts of the pathogenesis of parasitic infections
- Laboratory diagnosis of parasitic diseases

Learning objectives

- Discuss parasite–host relationships
- Classify parasites based on their morphological structures and habitats
- Describe the pathogenesis of various types of parasites
- Enumerate various laboratory techniques to detect parasites or their different stages from clinical material

GENERAL CONCEPTS OF PARASITOLOGY

- Parasites are probably the most complex of all microbes infecting humankind. The literal meaning of the word parasite is **'eating from another's plate'**.
- The famous biologist **Edward Osborne Wilson** defined parasites as 'predators that eat prey in units of less than one'.

Characteristic Features

Parasites are eukaryotic organisms with a true nucleus; they are considered to be higher organisms than prokaryotic bacteria, which are primitive. Parasites may be of the following types:

- **Unicellular**—these can be visualised with a microscope, e.g., protozoa such as *Entamoeba histolytica* and the *Plasmodium* species, which causes malaria
- **Multicellular**—these can be seen with the naked eye with sizes extending up to 10 meters, e.g., helminths such as roundworms and the tapeworm

Life Cycle

While a number of parasites have a simple or direct life cycle involving a single host, some require two or more hosts to complete their life cycles. Unlike other unicellular microorganisms, parasites can enter the host and establish infection only at specific stages in their life cycle, e.g., as larvae, embryonated eggs, etc. These stages are known as the **infective forms of a parasite**.

Host

A living entity that provides nutrition and harbours a parasite is known as its host. In the process, the host may suffer injury and lose nutrients.

Depending on their relationship with the parasite, hosts can be categorised as follows:

- **Definitive host:** The host in which the adult stage of the parasite lives (as in the case of helminths) or the organism in which the parasite undergoes sexual reproduction (as in the case of protozoa) is known as the definitive host.
- **Intermediate host:** A host that harbours the larval stage of the parasite (as in the case of helminths) or in which asexual reproduction takes place (as in the case of protozoa) is known as its intermediate host. For example, humans are intermediate hosts in malaria and hydatid disease.
- **Reservoir host:** The host harbours the parasite and also acts as a source of infection is its reservoir host.
- **Paratenic host:** The parasite is unable to develop further in this host. A paratenic host can act as a transport or carrier host.

Habitat

A place where the parasite naturally lives and grows is known as its habitat, e.g., the malarial parasite resides in the liver and RBCs of human beings.

Types of Parasites

- **Ectoparasites:** Parasites that live on the external surfaces of the host's body are known as ectoparasites. Head and body lice and ticks are examples of ectoparasites.
- **Endoparasites:** These live inside the host's body. Roundworms and hookworms are endoparasites that live in the intestine.
- **Obligate parasites:** These organisms cannot exist without a host under any circumstance. Examples include *Toxoplasma gondii* and *Entamoeba histolytica.*

- **Facultative parasites:** These organisms can exist in the free-living state or as parasites. For example, *Naegleria* and *Acanthamoeba* exist in the free-living state in soil and water but cause debilitating diseases in humans.

Host–Parasite Relationship

The relationship between a host and parasite can be one of the following types:

- **Symbiosis:** In this type of relationship, the parasite and its host are mutually dependent on each other. Both entities benefit from such an association, with neither sustaining any damage or injury.
- **Commensalism:** This refers to an association from which the parasite benefits, while the host is neither benefitted nor harmed.
- **Parasitism:** This is a situation in which a living organism is fully dependent on the host for survival and cannot live independently. Parasites derive the benefits of nutrition and shelter from the host they live in. In the process, the host is harmed either by the loss of nutrients or by pathologic changes caused by the parasite.

Routes of Entry of Parasites

- As with those produced by other microbial agents, diseases caused by parasites are site- and organ-specific.
- The routes of entry and modes of infection vary with the nature of the parasites. Some of the common routes of entry of parasites are presented in Fig. 7.1.

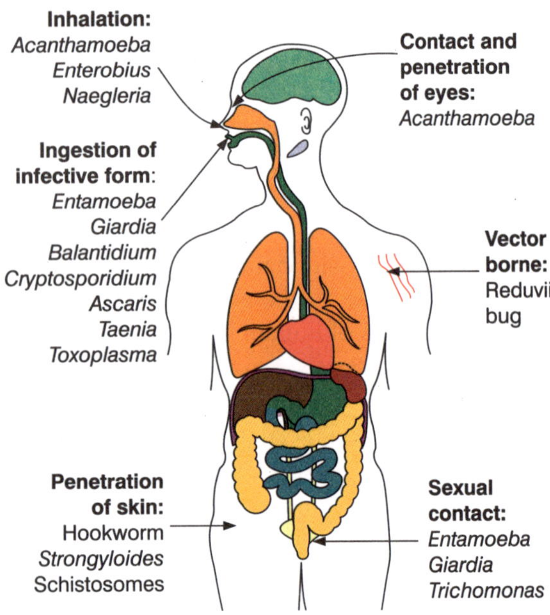

Fig. 7.1 Routes of entry of disease-causing parasites

Classification

Parasites that cause disease in humans are classified into three kingdoms—**(i) Protozoa, (ii) Stramenopila, and (iii) Animalia**.

> **Stramenopila**
> These are unicellular, plant-like animals. This term refers to a diverse group of organisms including diatoms and algae. The pathogenic parasite in this group is *Blastocystis*, which causes gastrointestinal infections.

PROTOZOOLOGY

Protozoa are unicellular, eukaryotic, simple, microscopic parasites (e.g., *Plasmodium*), which obtain nutrition by ingesting nutrients present in the host.

- Intestinal protozoa are unicellular organisms with very little structural differentiation. Each contains at least one or even several nuclei. They may be motile—on account of the presence of pseudopodia, cilia, or flagella—or non-motile.
- The single protozoal cell performs all the life functions like respiration, digestion, reproduction, etc. Its generation time is short, and the reproduction rate is high. Most protozoa are harmless and exist in nature in soil and water.

The important human disease-causing parasites are presented in Table 7.1.

HELMINTHOLOGY

Helminths are **multicellular worms** belonging to the kingdom **Animalia**. They are elongated, flat, or round worms that can range in size from a few millimetres to metres. They affect several systems in the body and cause mild asymptomatic to severe, life-threatening infections.

Classification

Helminths are classified into two phyla (Table 7.2), which are as follows:

- **Phylum Nemathelminths**—includes a single group of pathogens, the nematodes.
- **Phylum Platyhelminths**—includes two classes of human pathogens, namely cestodes and trematodes.

Clinical diseases associated with helminths

Most patients infected with helminths are asymptomatic or mildly symptomatic. The clinical presentation depends on the habitat of the parasite and the damage caused by it (Table 7.3).

GENERAL CHARACTERISTICS, LIFE CYCLES, AND MODES OF TRANSMISSION OF PARASITES

Nematodes

General characteristics

- Adult worms range in size from **3 mm to 1 m**.
- They are long, cylindrical, bilaterally symmetrical, and unsegmented, with both ends pointed and without any appendages (Fig. 7.2).

Table 7.1 Common protozoa associated with human infections and their characteristics

Phylum	Characteristic	Pathogenic species	Clinical disease caused
Amoebozoa	Amoeboid organisms with pseudopodia as organs of locomotion and feeding	*Entamoeba histolytica*	Diarrhea, dysentery, amoebic liver abscess, satellite lesions in the CNS, lungs, and skin (Chapters 24 and 25)
		Acanthamoeba spp.	Granulomatous amoebic encephalitis (GAE), ulcerative keratitis, cutaneous infection, sinusitis (Chapter 24)
		Naegleria fowleri	Primary amoebic meningoencephalitis (PAM) (Chapter 24)
Metamoneda (intestinal and genital flagellates)	May be parasitic (*Entamoeba*) or free-living	*Giardia duodenalis*	Diarrhea (Chapter 25)
		Trichomonas vaginalis	Vaginitis (sexually transmitted disease) (Chapter 27)
Euglenozoa (blood and tissue flagellates)	Flagellated organisms that lack mitochondria	*Leishmania* spp.	Visceral leishmaniasis (kala-azar), cutaneous leishmaniasis, post-kala-azar leishmaniasis (Chapters 25 and 29)
		Trypanosoma brucei	Sleeping sickness (Chapter 21)
		Trypanosoma cruzi	Chagas' disease (Chapter 21)
Apicomplexa (Sprorozoan)	Possess a whip-like flagellum with a subcellular structure called kinetoplast at the base of the flagellum, as well as mitochondria and Golgi apparatus	Cryptosporidium spp., *Cyclospora cayetanensis, Cystoisospora belli*	Opportunistic GI infections (Chapter 25)
		Toxoplasma gondii	Encephalitis, brain calcifications, blindness (toxoplasmosis) (Chapter 24)
		Plasmodium spp.	Malaria (Chapter 21)
		Babesia spp.	Babesiosis, anemia (Chapter 21)

Table 7.2 Phylogeny of helminths

Phylum	Class	Pathogenic species with habitats
Phylum Nemathelminths	Nematodes (roundworms) Chapter 25	Small intestine: *Ascaris lumbricoides, Necator americanus, Ancylostoma duodenale, Strongyloides* spp.
		Large intestine: *Trichuris trichiura, Enterobius vermicularis*
		Tissue nematodes: *Wuchereria bancrofti, Brugia malayi, Loa loa, Onchocerca, Mansonella, Trichinella, Dracunculus medinensis*
Phylum Platyhelminths	Cestodes (tapeworms) Chapters 25 and 26	Intestine: *Taenia saginata, Taenia solium, Diphyllobothrium, Hymenolepis nana,* and *H.diminuta*
		Liver: *Echinococcus granulosus*—causes hydatid disease (liver, lung, CNS), *E. multilocularis*
	Trematodes (flukes) Chapters 21, 23, 25, and 26	Intestine: *Fasciolopsis buski, Heterophyes, Gastrodiscoides*
		Liver: *Fasciola hepatica, Clonorchis sinensis, Opisthorchis* spp.
		Blood: *Schistosoma* spp.
		Lung: *Paragonimus*

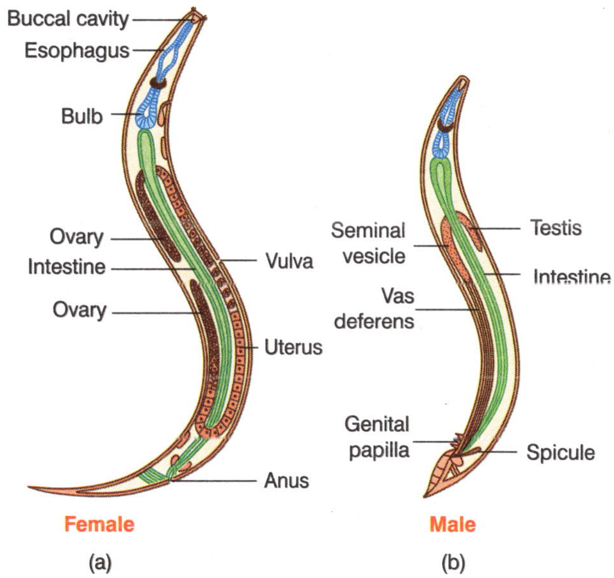

Fig. 7.2 Internal structure of nematodes (*Source*: Dr Sumeeta Khurana and Dr Abhishek Mewara, *Textbook of Medical Parasitology*, Universities Press, 2021)

- A **pseudocele (body cavity)** is present between the body wall and intestine.
- The **sexes are separate**, and females are generally longer than males.
- The body wall has a tough cuticle layer on the outside and a longitudinal muscle which is responsible for the movement of the worm.
- The alimentary canal is well-developed, with a mouth at the anterior end and a subterminal anus. Hookworms have **teeth or cutting plates**. Internally,

Table 7.3 Common helminths and the diseases caused by them

Helminth	Clinical disease/syndrome
Nematodes	
Ascaris lumbricoides *Trichuris trichiura* *Enterobius vermicularis*	Gastrointestinal infection with vague abdominal symptoms, Löffler's syndrome, malnutrition (Chapter 25)
Necator americanus *Ancylostoma duodenale*	Anemia, malnutrition (Chapter 25)
Strongyloides stercoralis	Gastrointestinal infection with vague abdominal symptoms, larva currens In immunocompromised individuals; it can cause hyperinfection with complications (Chapter 25)
Trichinella spiralis	Subcutaneous nodules, brain cyst (Chapter 24)
Wuchereria bancrofti *Brugia malayi* *Brugia timori*	Lymphatic filariasis (Chapter 21)
Loa loa, *Onchocerca* *Mansonella streptocerca*	Subcutaneous filariasis (Chapter 29)
Mansonella ozzardi *M. perstans*	Serous cavity filariasis (Chapter 29)
Dracunculus medinensis	Guinea worm disease (Chapter 29)
Cestodes	
Taenia saginata	Intestinal taeniasis (tapeworm infection) (Chapter 25)
Taenia solium	Intestinal taeniasis, cysticercosis (Chapter 25)
Diphyllobothrium latum	Megaloblastic anemia (Chapter 25)
Hymenolepis nana *H. diminuta*	Gastrointestinal infection with vague abdominal symptoms (Chapter 25)
Echinococcus granulosus	Hydatid disease (liver, lung, CNS) (Chapter 26)
E. multilocularis	Alveolar echinococcosis (Chapter 23)
Trematodes	
Fasciolopsis buski	Gastrointestinal infection with vague abdominal symptoms, malabsorption syndrome (Chapter 25)
Fasciola hepatica *Clonorchis sinensis* *Opisthorchis* spp.	Hepatitis, cholangitis, cholecystitis, cholelithiasis, liver rot, cirrhosis, pancreatitis (Chapter 26)
Schistosoma haematobium	Urogenital disease (bilharziasis), bladder carcinoma (Chapter 21)
S. japonicum, S. mansoni	Intestinal and hepatosplenic disease (Chapter 21)
Paragonimus	Lung cyst, extrapulmonary lesions in the CNS and skin (Chapter 23)

nematodes have a well-developed glandular esophagus, intestine, and rectum.

- The reproductive system in males comprises a long, convoluted tube, which can be differentiated into the testes, vas deferens, seminal vesicle, and ejaculatory duct. The ejaculatory duct opens at the posterior end along with the rectum.
- The female reproductive system consists of an ovary, oviduct, seminal receptacle, and uterus; both tubes open into the vagina, which opens through a genital pore near the mouth or in the middle of the body.
- The nervous system is rudimentary, consisting of a brain which is a circular nerve ring with six nerve trunks running along it longitudinally.
- The excretory system too is rudimentary, sometimes ending in an anus and other times with waste being excreted through the body wall. In some members, the excretory canal is located near the esophagus.
- Intestinal nematodes complete their life cycle in a single host (Fig. 7.3). Details of the life cycles of nematodes are discussed in chapters dealing with the diseases caused by individual members of the group.
- Nematodes either lay eggs (oviparous, e.g., *Ascaris*) or larvae (viviparous, e.g., *Wuchereria*), or lay eggs that immediately hatch into larvae (ovoviviparous, e.g., *Strongyloides*).

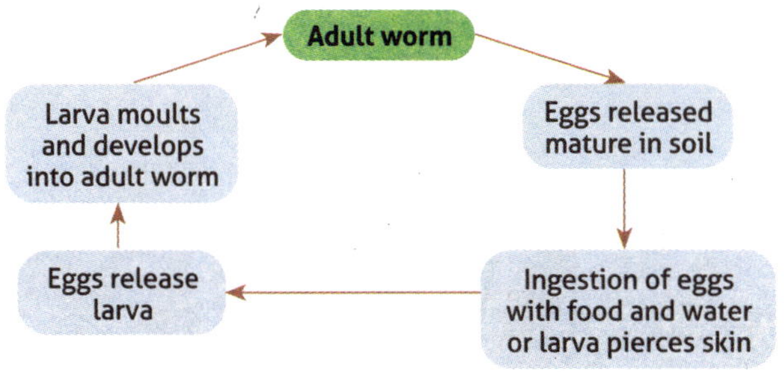

Fig. 7.3 Schematic representation of the life cycle of nematodes

Life cycle

- For their life cycles, nematodes require a definitive host and, in some cases, also an intermediate host or a certain environment where certain phases of their sexual or asexual cycles take place respectively. (Details of the life cycles and the hosts of various nematodes will be discussed in chapters pertaining to the diseases produced by them.)
- The general life cycle of nematodes is illustrated in Fig. 7.3.

Clinical implications of nematode infections

- Most patients infected with intestinal helminths remain asymptomatic.
- The clinical presentation may be limited to vague abdominal pain and occasional diarrhea.
- In children, the **habit of eating mud (pica)**, growth retardation, and malnutrition may be the only complaints.
- The bolus of adult worms can block the intestine, causing **abdominal pain, intestinal obstruction, perforation, intussusception, or volvulus**.
- The migration of adult worms into the biliary tract can cause biliary colic, cholecystitis, cholangitis, and pancreatitis.
- Allergic manifestations such as fever, urticaria, angioneurotic edema, and conjunctivitis can occur in all nematode infections.
- Patients infected with hookworms may present with anemia and malnutrition.
- The migration of larvae through the skin may cause cutaneous presentations.
- Filariasis is characterised by the enlargement of the feet, hydrocele, or nodule formations.
- Details of the common nematodes and the clinical manifestations of the diseases they cause are presented in Table 7.3.

Cestodes

General characteristics

- The adult worm, which usually resides in the intestine of its host is segmented, long, and flattened dorsoventrally. Due to their tape-like form, cestodes are called **tapeworms**.
- The body is covered with a tough cuticle layer; there is no body cavity.
- Broadly, the morphological features of the adult worm are—a **head (scolex), a neck, and a body (strobila)** (Fig. 7.4).
- The scolex or head is the organ of attachment. It can bear suckers (as in *Taenia*) or a pair of longitudinal grooves called **bothria**. The head continues into a neck.
- The strobila or body is made up of multiple segments called **proglottids**. Small tapeworms like *Echinococcus granulosus* possess 3–4 segments, while longer ones like *Taenia saginata* may have up to 1,000 segments.
- Each segment—called **proglottid**—has its own reproductive organs and excretory systems.
- These worms lack an alimentary canal; they absorb nutrients through their outer coat.
- Excretory tubules are present in each segment. Waste products are eliminated through the terminal flame cells.
- The sexes are not separate. Both male and female reproductive systems are present in each proglottid, i.e., these worms are hermaphrodites.

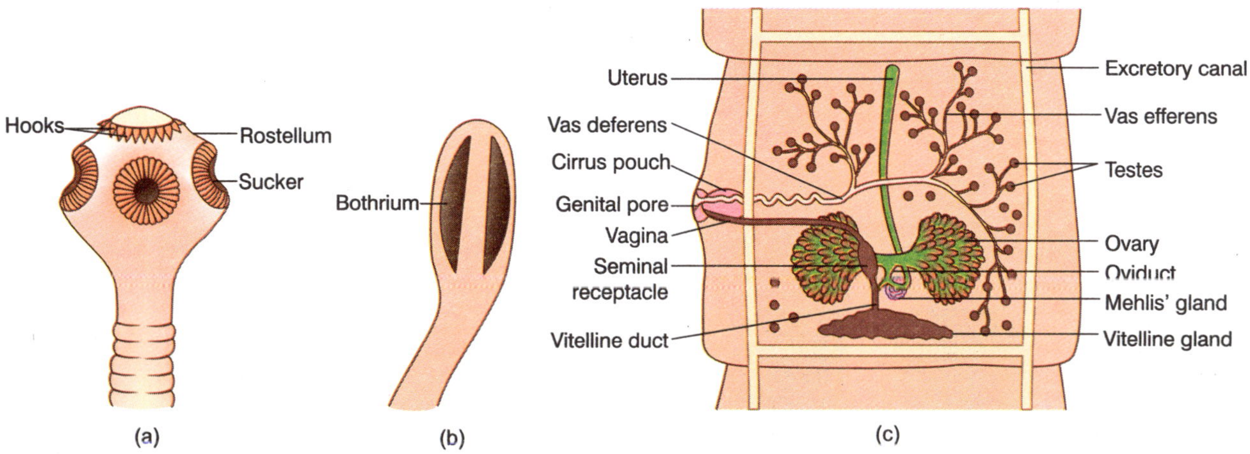

Fig. 7.4 Morphology of cestodes (*Source:* Dr Sumeeta Khurana and Dr Abhishek Mewara, *Textbook of Medical Parasitology*, Universities Press, 2021)

- Cestodes have highly developed reproductive systems. Both male and female genital openings are located in a common genital pore. Fertilisation occurs between segments of the same worm. Gravid proglottids containing hundreds of eggs are released into the environment by the host.
- The nervous system is primitive and consists of the cephalic ganglion in the scolex, which is connected with sensory and motor nerves across proglottids.
- In *Diphyllobothrium*, the **eggs are operculated** (bearing a lid), whereas in *Taenia*, the eggs may contain a **hexacanth embryo**.
- The life cycle of cestodes is complex (Fig. 7.5), often requiring more than one host. Human beings are the definitive hosts, except in the case of *Echinococcus granulosus* (which causes hydatid disease).

Clinical implications of *Taenia* infection

- *Taenia* infections of the intestines are usually asymptomatic.
- Cysticerci may be seen in the brain, skin, eye, muscles, and subcutaneous tissues.
- Cysts of *Echinococcus granulosus* (hydatid cysts) can also be found in the brain and other organs and closed spaces (such as the eye); they may produce space-occupying lesions with symptoms related to the organ.
- The diseases caused by cestodes are presented in Table 7.3.

Trematodes

General characteristics

- Adults are **leaf-like, flat, unsegmented** worms ranging in size from 1 mm to several cm (Fig. 7.6). Females are longer and thinner than males.
- Trematodes reside in the intestine, liver, lungs, or blood vessels.
- Their body is covered with an integument; they lack a body cavity.
- Eggs are oval to round, operculated, and embryonated when laid.
- Several larval forms are seen—**metacercaria**, which are the infective forms, **cercaria, redia, sporocysts, and miracidia**.

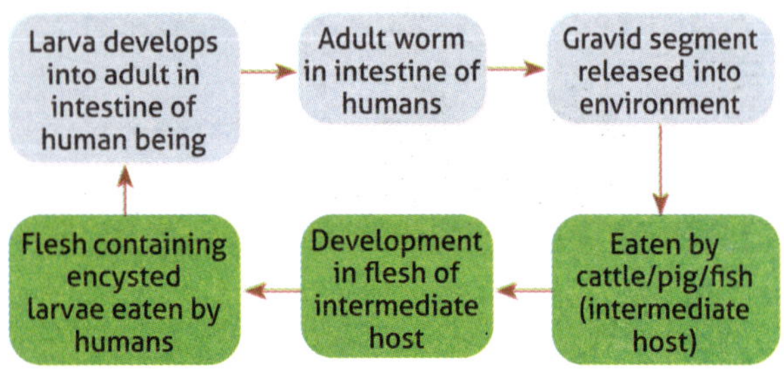

Fig. 7.5 Schematic representation of the life cycle of cestodes

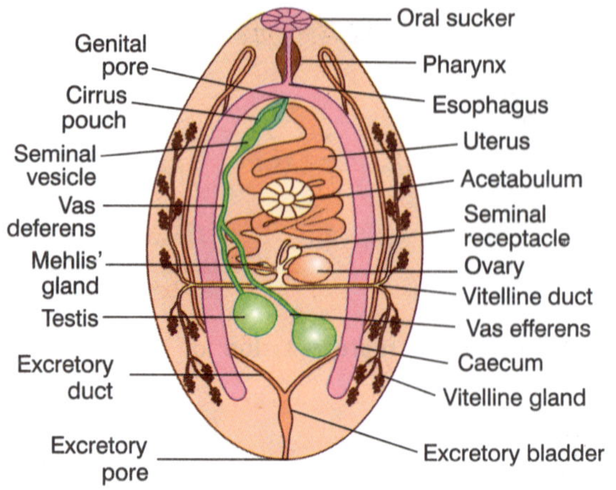

Fig. 7.6 Morphology of trematodes (*Source*: Dr Sumeeta Khurana and Dr Abhishek Mewara, *Textbook of Medical Parasitology*, Universities Press, 2021)

- The sexes are not separate except in schistosomes.
- The alimentary tract consists of a blind, bifurcated intestinal tract without an anus.
- Trematodes have a primitive nervous system and well-developed excretory and reproductive systems.
- Trematodes are **oviparous**. Eggs are operculated (lidded), containing a small ovum with yolk granules. *Schistosoma* eggs are not operculated and have a spine.
- Trematodes are present in many animals besides humans and are often associated with zoonoses.
- The life cycles of trematodes are complex, often requiring two intermediate hosts. The primary host is a vertebrate, while the intermediate host, in which asexual reproduction occurs, is usually a mollusc (Fig. 7.7).

Clinical implications of trematodes

- Trematodes comprise liver flukes, blood flukes, lung flukes, and intestinal flukes.
- In the acute stages, liver flukes cause vague abdominal symptoms like abdominal pain, anorexia, nausea, diarrhea, and flatulence. Later, the worm causes inflammation, granuloma, and abscesses,

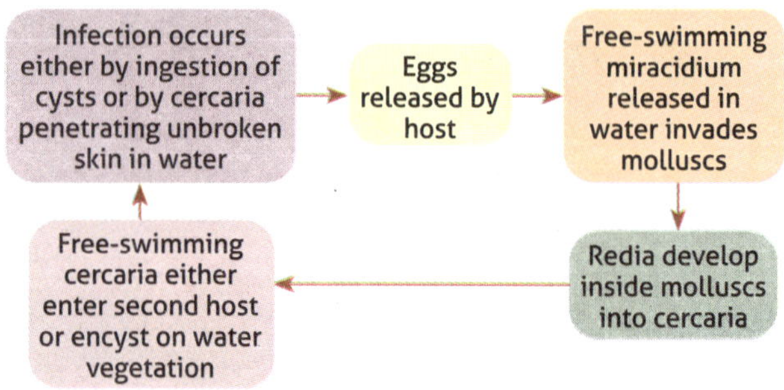

Fig. 7.7 Schematic diagram of the life cycle of trematodes

followed by fibrosis and bile duct hyperplasia, causing cholangitis, cholecystitis, cholelithiasis, liver rot, cirrhosis, and pancreatitis.
- Blood flukes can cause urogenital disease and predispose to urinary bladder carcinoma.
- Table 7.3 presents the common trematodes and the diseases caused by them.

Modes of transmission

The modes of transmission of the common disease-causing parasites are presented in Table 7.4.

PATHOGENESIS OF PARASITIC DISEASES

Physical Obstruction, Irritation, and Trauma

- A high load of worms in the intestine can cause intestinal obstruction (*Ascaris*).
- Attachment to the intestinal mucosa causes irritation and bleeding from the villi (hookworm).

Table 7.4 Modes of transmission common parasites infecting humans

Mode	Examples
Ingestion	
Food or water contaminated with parasitic cysts or eggs	*Entamoeba histolytica, Giardia lamblia, Ascaris lumbricoides, Trichuris trichiura, Enterobius vermicularis*
Undercooked meat containing larval forms of the parasite	*Taenia solium, Taenia saginata, Trichinella spiralis*
Water containing the intermediate host	*Dracunculus medinensis*
Aquatic plants as food containing metacercarial forms	*Fasciola hepatica*
Penetration of the skin	
The filariform larval forms present in damp soil enter the body by piercing intact skin	*Necator americanus, Ancylostoma duodenale, Strongyloides* species
Contact with water containing cercarial forms	*Schistosoma haematobium, S. mansoni, S. japonicum*
Vector-borne	
Bite of an arthropod (mosquito or sandfly)	*Plasmodium* species, *Leishmania* species, *Wuchereria bancrofti, Brugia malayi*
Direct transmission	
Sexual transmission	*Trichomonas vaginalis*
Vertical transmission	*Toxoplasma gondii, Trypanosoma cruzi*

- Migration of worms through the lungs as part of the natural life cycle causes mechanical irritation and damage to the pulmonary capillaries (e.g., *Strongyloides*).

Vascular or lymphatic obstruction

- **Elephantiasis**—edematous swelling of the limbs—is caused by the blockage of the lymphatics by the adult worms in filariasis.
- **Cerebral malaria** can occur due to the blockage of the blood vessels of the brain.

Enzymatic damage

- Several parasites secrete enzymes that damage tissues and allow penetration, e.g., flask-shaped ulcers are seen in *Entamoeba histolytica* infections of the gut.
- Intracellular parasites cause the necrosis of cells.

Allergic reactions

- The presence of larval forms and adult worms in the host elicits an increase in IgE production, which results in allergic manifestations.
- The rupture of hydatid cysts to release hydatid fluid can trigger an anaphylactic reaction.

Inflammation

- Many parasites provoke cellular infiltration and proliferation, resulting in an inflammatory reaction.
- Granuloma formation can be seen in infections caused by some parasites such as *Entamoeba histolytica*.

Neoplastic reactions

Some parasites such as *Schistosoma* are associated with carcinoma.

Anemia

- Intestinal parasites are associated with anemia as they produce blood loss or the deficiency of essential nutrients.
- Malaria may result in anemia due to the parasitisation and subsequent destruction of RBCs.

IMMUNITY IN PARASITIC DISEASES

- The immune response against parasitic infections consists of cytotoxic T cell response, natural killer cells, activated macrophages, and antibodies.
- IgM is the first antibody to appear in the circulation, followed by IgG. Helminths elicit the production of high titers of IgE antibodies. It should be noted that immunological responses against parasites are lower than those against bacteria and viruses. This is attributed to multiple factors:
  – Parasites have large and complex antigenic structures, against which the immune system

is not able to mount an effective and consistent immune response.

- The intracellular location of parasites (as in malaria) protects them from the host's defenses.
- Reticuloendothelial system damage can also result in immunological tolerance as seen with *Leishmania donovani*.
- Parasites may exhibit multiple antigenic variations, which enables their prolonged persistence in the body.
- The cuticles covering the bodies of nematodes are antigenically inert and evoke little immune response.
- Further, many nematodes reside only in the intestine, where the immune system is not stimulated.
- In most parasitic infections, premunition immunity is observed. This refers to immunity which lasts only while the original infection is active.

LABORATORY DIAGNOSIS OF PARASITIC DISEASES

The diagnosis of parasitic disease depends on the collection of appropriate samples (Table 7.5) at the right time.

Examination of Fecal Samples

Collection of samples

i. The **NIH swab** is used to collect samples when there is a suspicion of *Enterobius infection*. An NIH swab consists of a glass rod to which a cellophane tape is fastened using a rubber band such that the sticky side of the tape is exposed (Fig. 7.8). The upper end of the rod is fixed to a rubber stopper fixed to the mouth of the test tube. To collect the sample, the cellophane end is rubbed against the patient's perianal region (to allow the eggs to stick to the tape) and placed back in the tube. The tape is directly examined under a microscope for eggs.

ii. Alternately, cellophane tape is applied to the perianal area of the patient early in the morning. The tape is then placed on a slide and examined for eggs.

iii. The **string test** for duodenal samples consists of a gelatin capsule attached to the thread, which is swallowed by the patient. The other end of the thread is taped to the patient's cheek. The thread is pulled out in 4–5 hours, by which time the gelatin dissolves in stomach acid. After immersing the thread in saline, it is examined for trophozoites.

Table 7.5 Recommended sample to demonstrate commonly encountered parasites

Recommended sample	Parasites commonly detected
Blood and bone marrow	*Plasmodia, Leishmania,* microfilaria
Feces or rectal swabs	Trophozoites, cysts, ova, larvae or, at times, adult worms
CSF	Free-living amoebae
Genitourinary samples	*Schistosoma haematobium, Trichomonas*
Biopsy tissues, including endoscopic biopsy	Parasites harboured in the viscera or gut
Aspirated pus from liver abscess	Trophozoites of *Entamoeba histolytica*
Sputum	Migrating larva of *Strongyloides,* roundworm, and hookworm, hooklets of cestodes, and rarely, *Entamoeba histolytica*
Aspirates from lymph nodes, spleen, and bone marrow	Malaria parasites and Leishman–Donovan (LD) bodies
Skin snips or biopsy	*Onchocerca volvulus* (on keeping the snips in saline, the microfilariae can be seen emerging)

Stool concentration techniques

The purpose of concentration techniques is to separate parasites from fecal matter. This increases the chances of detecting parasitic organisms, especially when they are present in small numbers. Floatation and sedimentation are used to concentrate fecal samples.

i. **Floatation method:** The stool sample is mixed with solutions that have a higher specific gravity, so that the ova rise to the top, and the debris sinks to the bottom.
   - ***Saturated salt technique*:** About 1 g of feces is mixed with a saturated salt solution in a test tube. The tube is filled to the brim, and a coverslip is placed on top. The tube is allowed to stand for 30–35 minutes. The coverslip is then carefully removed and placed face down on a slide. The slide is examined microscopically. Since eggs of hookworms and roundworms float in a saturated

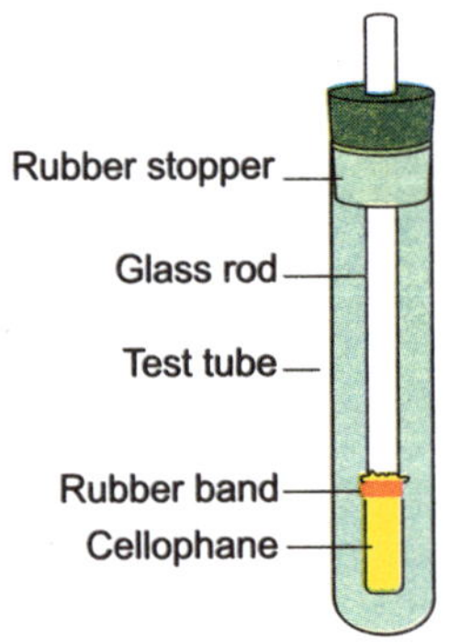

Fig. 7.8 NIH swab to collect a sample for demonstrating the eggs of *E. vermicularis* (*Source:* Dr Sumeeta Khurana and Dr Abhishek Mewara, *Textbook of Medical Parasitology*, Universities Press, 2021)

salt solution, they adhere to the coverslip and can be visualised.

- ***Zinc sulfate floatation***: The procedure is the same as that of the saturated salt technique except that a 33% zinc sulfate solution is used in place of the saturated salt solution. Cysts of *Giardia* and *Entamoeba histolytica* float in a zinc sulfate solution.

ii. **Sedimentation techniques** use solutions having a low specific gravity so that larvae, ova, and cysts settle down in the sediment.

- **Formol ether sedimentation method:** A 10% solution of formalin is emulsified with 1 g of feces in a test tube and left for 10 minutes. After straining this mixture through a wire mesh, 3 mL of ether is added to the filtrate. The filtrate is vigorously shaken and centrifuged at 2,000 rpm for two minutes. The supernatant is decanted, and the sediment is used for the detection of parasites. Sedimentation techniques are easier to perform and less prone to technical errors. They can be used for egg counting and have a better sensitivity for the detection of eggs and cysts.

Wet Mount

The wet mount identification technique can be used to identify protozoan trophozoites, cysts, oocysts, helminth eggs, and larval forms.

- A small amount of the fresh stool sample or its sediments (after concentration) is placed on a slide. A drop of saline is added to it and mixed. To make an iodine mount, a drop of iodine is added to the sample instead of saline. Both smears can be made on a single slide and covered with coverslips and scanned systemically—initially under low power, and then under high power to look for ova and cysts.
- Saline mounts are useful for visualising eggs, motile trophozoites, and cysts.
- Iodine mounts provide information on internal ultrastructure as iodine stains the glycogen mass of trophozoites and cysts.
- Cellular elements such as pus cells, red blood cells, and Charcot–Leyden crystals may be seen in these preparations. These findings help in differentiating bacillary from amoebic dysentery. Charcot–Leyden crystals are seen in amoebic dysentery as colourless, needle- or diamond-shaped refractile forms.

The common findings of parasitic agents in stool samples are presented in Table 7.6. Other staining techniques used to demonstrate ova and cysts are discussed below.

Modified Acid-Fast Staining

This technique is used for the identification of the oocysts of the following:

- *Cryptosporidium*
- *Cystoisospora*
- *Cyclospora*

A smear is made with a fresh or concentrated stool sample on a slide and fixed with absolute methanol for 30 seconds. The smear is stained with Kinyoun's carbol fuchsin and decolourised with a weak (3%) acid–alcohol solution. Malachite green or methylene blue is used for counterstaining.

Chromotrope Staining

This technique is used to differentiate microsporidial spores from background fecal elements.

Permanent Staining

- This method is used when a record of the specimen needs to be maintained and can be used for training purposes also.
- Slides are prepared with fresh or polyvinyl alcohol-(PVA) or sodium acetate-acetic acid formalin (SAF)-fixed stool specimens. The smears are air-dried and fixed with methanol or ethanol. This method can be used to visualise coccidian parasites, trophozoites, and ova of protozoa and eggs and larvae of the helminth.

Trichrome Staining

- Used to stain smears made from PVA- or SAF-preserved stool samples.
- It is a simple and rapid method used to visualise intestinal protozoa, human cells, yeasts, and artefact material in stool.
- The smears are fixed with ethanol and stained with trichrome.

Table 7.6 Stool findings commonly seen in infected patients

Trophozoites	Cyst	Larva	Adult worm	Eggs
Entamoeba histolytica *Giardia lamblia* *Balantidium coli*	*Entamoeba histolytica* *Giardia lamblia* *Balantidium coli*	*Trichinella spiralis*	*Ascaris lumbricoides* *Enterobius vermicularis*	*Ascaris lumbricoides* *Enterobius vermicularis* Hookworms *Trichuris trichiura* Various trematodes and cestodes

Iron–Hematoxylin Staining

- The fixed smear is stained with a working solution of iron–hematoxylin.
- It is then washed with 1% acid–alcohol; any excess of the stain is dehydrated and cleared by placing it in xylene.

Preservation of Stool Samples

When samples cannot be examined within a few hours, they need to be preserved. This is also important in cases where the specimen for molecular diagnosis must be shipped to a referral laboratory.

The following agents are used for the preservation of stool samples:

- **Formalin 10% aqueous solution**—this is a good preservative for visualising the morphology of helminth eggs, larvae, protozoan cysts, and coccidian parasites. Formalin-fixed samples are suitable for concentration procedures, immunoassay kits, UV fluorescence microscopy and acid-fast, safranin, and chromotrope staining. One part of the stool sample is added to three parts of formalin solution to preserve the sample. However, this method is not suitable for PCR and provides inadequate preservation of the morphology of protozoan trophozoites.
- **PVA (polyvinyl alcohol)**—this is a resin prepared in a modified Schaudinn fixative. It preserves the morphology of protozoan trophozoites and cysts well. Samples remain stable in it for years. However, it is expensive and not suitable for concentration procedures, immunoassay kits, and acid-fast, safranin, and chromotrope stains.
- **Other fixatives**—merthiolate iodine-formalin (MIF) or sodium acetate acetic acid formalin (SAF) can also be used for preservation.

Examination of Blood Samples

Peripheral blood smear (PBS)

PBS can be used to demonstrate *Plasmodium* in red blood cells, the amastigote forms of hemoflagellates (e.g., *Leishmania donovani*) inside monocytes; and the microfilariae of *W. bancrofti* and *B. malayi*.

- The timing of blood collection is very important for the detection of parasites. In malaria, parasites are visualised during the febrile phase, whereas in filariasis, microfilariae can be seen in blood collected between 10 p.m. and 4 a.m. (nocturnal periodicity).
- Blood is also collected for the detection of antibodies in serum.

Blood smear preparation

- Smears can be prepared by the finger-prick method or from blood collected in EDTA tubes.
- Peripheral blood film examination is the most useful, easy, and economical method to diagnose blood parasites.
- Both thick and thin films can be made from blood.

Thick film

- This method is useful as a screening tool in **malaria**, to visualise **Leishman–Donovan (LD) bodies** in leishmaniasis, and to detect microfilariae in **filariasis**.
- A big drop of blood is placed on a grease-free slide. Using the edge of a second clean slide, the drop is spread into a circle with a diameter of about 1 cm (Fig. 7.9). The smear should be around 20–30 times thicker than a thin smear in an equal area.
- The blood is continuously mixed for 30 seconds to prevent fibrin deposition.
- It is then allowed to air dry before staining.
- Dehemoglobinisation of the thick film is carried out by placing the slide in distilled water in a glass cylinder or Koplin jar for 5–10 minutes.

Thin film

- It is used to detect *Plasmodium* spp. in suspected cases of malaria.
- The finger index or heel (for babies) is pricked, and the first drop of blood is wiped away. The next drop is touched with a clean, grease-free glass slide. With the help of another slide held at 45° to the first, the drop is spread out to form a tongue-shaped smear (Fig. 7.9).
- The smear should be one cell thick and occupy the central area, ending before the edge of the slide.
- The film is air-dried and fixed with methanol before staining.

DEC/Hetrazan provocation test

- Blood is examined after administering diethylcarbamazine (DEC) 2 mg/kg body weight to stimulate nocturnal microfilariae to appear in the peripheral blood in the daytime.
- The sample is collected 30–60 minutes after administering the drug.

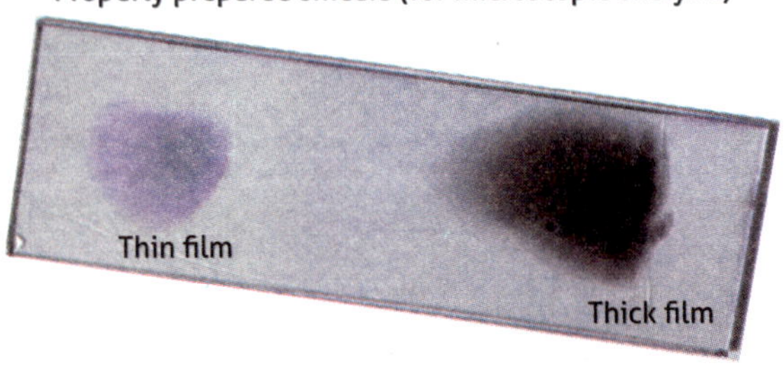

Fig. 7.9 Thick and thin films (*Source:* Centers for Disease Control and Prevention [CDC], Public Health Image Library [PHIL], Image ID: 5905)

Staining of blood smear

The following stains are used for blood smears:

- **Leishman stain:** The staining process takes about 10–15 minutes and provides good visualisation of blood parasites.
- **Giemsa stain:** The smears are fixed by dipping the slides in methanol for a minute, air drying them, and staining them with Giemsa stain for 10–30 minutes. The slides are washed and examined under oil immersion after drying.
- **Wright stain:** This method is used in hematology and is not considered optimal for blood parasites. It stains the smear within 10–12 minutes and is considered a good rapid alternative to Giemsa stain.

Microfilariae can be detected by examining thick films or by making a wet mount. For preparing a wet mount, a drop of blood is placed on a slide and covered with a coverslip. The larval form of the parasite can be seen wriggling in the blood.

Culture

The culturing of parasites requires special media. Some protozoa such as *Leishmania*, *Trypanosoma*, *Entamoeba histolytica*, *Trichomonas vaginalis*, free-living amoeba, and *Plasmodium* can be cultured in the laboratory.

Three types of culture systems can be used to grow parasites in the laboratory:

i. **Xenic cultivation:** The parasite is grown in the presence of an undefined flora.
ii. **Monoxenic cultivation:** The parasite is grown in the presence of a single species of associate flora.
iii. **Axenic cultivation:** The parasite is grown in the absence of any other metabolising cells.

Trichomonas vaginalis and *Giardia lamblia* can be grown in axenic cultures, whereas the culturing of *Entamoeba histolytica* is laborious, time-consuming, and complex.

- A non-nutrient medium with *E. coli* is used to culture free-living amoeba.
- The **diamond medium** is used for the cultivation of *Trichomonas vaginalis*.
- **NNN (Novey, Nicoll, and McNeal) medium** is used for the growth of *Leishmania* species. It contains one part of defibrinated rabbit blood and one part of salt agar and is dispensed as a slant in a tube or Bijou bottle. The amastigote forms transform into promastigotes within four weeks. The water of condensation is examined weekly by staining.
- **RPMI 1640** with human RBC in 7% carbon dioxide and 1–5% oxygen is used to cultivate *P. falciparum*. This is a system with a layer of RBCs, over which the RPMI medium flows gently in a continuous manner. The culture system is used to prepare antigens.

Antigen Detection

ELISA-based antigen detection kits are available to detect *Giardia lamblia* and *Cryptosporidium* species in stool samples and *T. vaginalis* in urogenital specimens. Rapid kits based on lateral flow assays are available to detect antigens of *Plasmodium* and filariasis in blood.

- **Serological diagnosis:** Antibodies and antigens can be detected by ELISA, immunochromatographic tests such as rapid diagnostic tests (RDTs; Fig. 7.10), indirect hemagglutination assays, and complement fixation tests.
- **Skin hypersensitivity tests:** The immediate hypersensitivity test is performed to diagnose hydatid disease (**Casoni test**), and the delayed reaction test is used to detect leishmaniasis and trypanosomiasis.
- **Molecular diagnostic method:** DNA probes, PCR, and real-time PCR tests are being increasingly used for the detection of parasites. These can be used to diagnose malaria, filariasis, and trypanosomiasis.

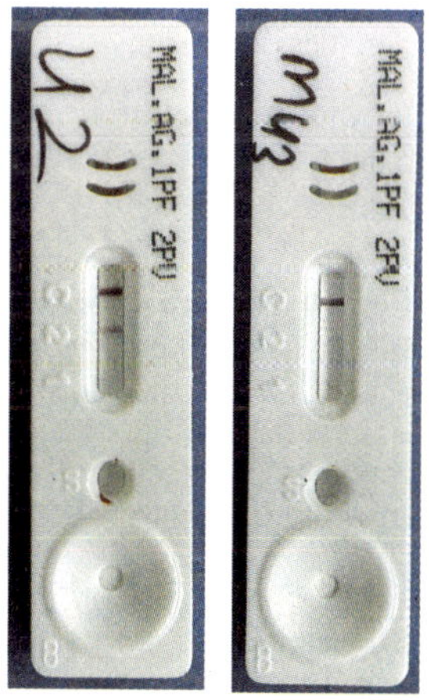

(a) (b)

Fig. 7.10 Rapid antigen detection test for malaria parasite: (a) Positive for *Plasmodium vivax* and (b) negative test (*Source*: Dept of Microbiology, Maulana Azad Medical College, New Delhi and Pondicherry Institute of Medical Sciences)

Key Points

- ✓ Parasites are eukaryotic organisms with a true nucleus. They are considered to be higher organisms than prokaryotic bacteria. Parasites may be unicellular, e.g., protozoa such as *Entamoeba histolytica*, or multicellular, e.g., helminths such as roundworms.
- ✓ Some parasites require two or more hosts to complete their life cycles.
- ✓ Parasites can enter the host and establish infection only at specific stages in their life cycle such as larvae, embryonated eggs or other stages. These stages are known as the infective forms.

- ✓ A host is an entity that provides nutrition and harbours the parasite.
- ✓ A definitive host is one in which the adult stage of the parasite lives (as is the case of helminths) or in which the parasite undergoes sexual reproduction (as in the case of protozoa).
- ✓ An intermediate host harbours the larval stage of the parasite.
- ✓ Parasites that live on the external body surfaces of the host are known as ectoparasites. Endoparasites live inside the host's body.
- ✓ Organisms that cannot exist without a host under any circumstance are called obligatory parasites. Those that can also exist in the free-living state are known as facultative parasites.
- ✓ There are different kinds of host–parasite relationships, which are as follows:
  - The host and parasite may be mutually dependent on each other so that both benefit from the association and neither is harmed (symbiotic).
  - The parasite may be fully dependent on the host and may cause harm to the host (parasitism).
  - In commensalism, the parasite depends on the host, but the latter is neither harmed nor benefited.
- ✓ Protozoa are unicellular, eukaryotic, simple, microscopic parasites that are classified as Amoebozoa (e.g., *Entamoeba*), Metamonada (e.g., *Giardia*), Euglenozoa (e.g., *Leishmania*), or Sporozoan (e.g., *Plasmodium*).
- ✓ *Plasmodium*, which causes malaria, and *Leishmania* species, which cause kala-azar are the commonest protozoal infections.
- ✓ Helminths are classified into nematodes (roundworms), cestodes (tapeworms), and trematodes (e.g., *Schistosoma*).
- ✓ Nematodes are roundworms which can infest the intestine or tissues. For their life cycles, nematodes require a definitive host or additionally, an intermediate host or the environment where certain phases of their sexual or asexual cycles take place respectively.
- ✓ Most patients infected with intestinal helminths remain asymptomatic. Some nematodes may cause anemia, malnutrition, or vague abdominal symptoms.
- ✓ Tissue nematodes cause lymphatic, subcutaneous, or serous cavity filariasis. Filariasis is characterised by the enlargement of the feet, hydrocele, or nodule formations. *D. medinensis* causes guinea worm disease, which is now eradicated.
- ✓ Cestodes are flat, tape-like hermaphrodite worms. Tapeworms cause taeniasis which is asymptomatic. *T. solium* causes cysticercosis while *Echinococcus* can cause hydatid disease.
- ✓ Trematodes are leaf-shaped worms that cause liver or urogenital disease. *Paragonimus* causes lung cysts.
  The laboratory diagnosis of parasitic diseases is based on the site infected by them. Direct microscopy is used to diagnose most parasitic infections. Newer diagnostic methods such as rapid antigen detection assays or molecular diagnostic tools may also be used.

Questions

Short answer questions

Write short notes on the following:

1. Diagnostic outline of parasitic infections
2. Common characteristics of parasites
3. Symbiosis
4. Vector-borne parasitic infections
5. Life cycle of trematodes

MCQs

1. **Which of the following defines a paratenic host?**
   a) Does not allow the parasite to develop further
   b) Harbours the adult stage of the parasite
   c) Harbours the larval stage of the parasite
   d) Acts as a reservoir of the infection also
2. **Which of the following statements is NOT true regarding protozoa?**
   a) They are unicellular, prokaryotic organisms.
   b) They include the causative agents of malaria.
   c) They obtain nutrition by ingesting nutrients present in the host.
   d) They are motile by flagella, pseudopodia, or cilia.
3. **Which of the following best describes nematodes?**
   a) Long, cylindrical, bilaterally symmetrical, and unsegmented worms
   b) Segmented, long, and dorsoventrally flat worms
   c) Leaf-shaped, flat, unsegmented worms
   d) Can cause hepatitis and hepatocellular carcinoma
4. **In which of the following is an NIH swab used to collect samples?**
   a) Roundworm infection b) Enterobius infection
   c) Hookworm infection d) None of these
5. **A thick film is used for detecting which of the following?**
   a) Malarial parasites b) LD bodies in leishmaniasis
   c) Microfilariae in filariasis d) All of these
6. **An organism that cannot exist without a host under any circumstance is known as:**
   a) Obligate parasite b) Facultative parasite
   c) Ectoparasite d) Endoparasite
7. **Which statement is correct regarding the phylum Amoebozoe?**
   a) The members use pseudopodia as organ of feeding.
   b) All members are parasites.
   c) They are flagellated organisms that lack mitochondria.
   d) They possess kinetoplasts.
8. **Which of the following is a nematode?**
   a) *Hymenolepis diminuta* b) *Loa loa*
   c) *Fasciola hepatica* d) *Paragonimus*
9. **Which of the following is characteristic of cestodes?**
   a) Adult worms are long, cylindrical, bilaterally symmetrical, and unsegmented, with both ends pointed.
   b) A pseudocele (body cavity) is present between the body wall and intestine.
   c) The sexes are separate, and females are generally longer than males.
   d) These worms lack an alimentary canal; they absorb nutrients through their outer coat.
10. **Trematodes are:**
    a) Leaf-like, flat, unsegmented worms
    b) Viviparous in nature
    c) Need only one host to complete their life cycle
    d) A body cavity is present between the body wall and intestine
11. **Pigs or dogs are the source of human infection by each of the following parasites except:**
    a) *Ascaris lumbricoides* b) *Taenia solium*
    c) *Echinococcus granulosus* d) *Trichinella spiralis*
12. **In which of the following worm the infection occurs by piercing of intact skin by filiform larval forms?**
    a) *Ancylostoma duodenale* b) *Ascaris lumbricoides*
    c) *Echinococcus granulosus* d) *Trichinella spiralis*

13. Vertical transmission is seen in:
- a) *Toxoplasma gondii*
- b) *Trichomonas vaginalis*
- c) *Trichinella spiralis*
- d) None of the above

14. Modified acid-fast staining technique is used to visualise the oocyst of:
- a) *Cryptosporidium*
- b) *Cystoisospora*
- c) *Cyclospora*
- d) All of the above

15. Which of the following statements is correct regarding hetrazan provocation test?
- a) Hetrazan is given in dosage of 0.1 mg/kg body weight.
- b) It is performed to stimulate nocturnal microfilariae to appear in peripheral blood.
- c) The blood sample is collected 10 minutes after administering the drug.
- d) The test can be done in suspected cases of malaria to stimulate *Plasmodium* species.

16. Diamond medium is used for the cultivation of:
- a) *Trichomonas vaginalis*
- b) *Leishmania donovani*
- c) *Trypanosoma*
- d) *Plasmodium* species

17. Which of the following is true of NNN (Novey, Nicoll, and McNeal) medium?
- a) Is used for the growth of *Plasmodium* species.
- b) It has one part of defibrinated rabbit blood and one-part salt agar.
- c) It is poured in petridishes.
- d) The promastigotes transform into amastigote forms within 4 weeks.

18. Peripheral blood smear cannot be stained with:
- a) Giemsa stain
- b) Wright stain
- c) Gram stain
- d) Leishman stain

Answers: 1:a, 2:c, 3:a, 4:b, 5:d, 6:a, 7:a, 8:b, 9:d, 10:a, 11:a, 12:a, 13:a, 14:d, 15:b, 16:a, 17:b, 18:c

CHAPTER 8

Part I: General Microbiology

Microbial Genetics

(MI1.1)

Chapter Navigation

- Principles of molecular biology
- Structure of DNA and RNA
- Mutations
- Clinical applications
- Gene transfer in bacteria
- Gene sequencing
- Applications of genetic engineering in microbiology

Learning objectives

- Describe various terms used in microbial genetics
- Discuss the methods of genetic transfer in bacteria
- Enumerate the clinical implication of microbial mutation in clinical practice
- Describe various molecular techniques applied to detect microbial genes in diagnostic laboratories

Microbes, like all other organisms, obey the laws of genetics. The application of the principles of genetics to the study of bacteria and viruses led to the birth of a new branch of science—**microbial genetics**—which has numerous applications ranging from understanding the genetic mechanisms of microbial virulence and survival to bacterial resistance to antibiotics and the detection of hitherto unknown microorganisms.

Applications of microbial genetics

- Development of molecular techniques for diagnosis
- Production of genetically engineered vaccines
- Detection of congenital anomalies, genetic disorders, metabolic disorders, and cancer markers
- New theories of disease mechanisms are being understood by studying the molecular makeup of microorganisms and host cells

BASIC PRINCIPLES OF MOLECULAR BIOLOGY

- The central dogma of molecular biology rests on the **transcription of deoxyribonucleic acid (DNA) into ribonucleic acid (RNA)**, which in turn, is **translated into a particular polypeptide**.
- This polypeptide goes on to form proteins and other structural and functional elements of the cell. As the nature and functions of a cell are determined by the specific polypeptides that constitute its proteins and enzymes, it follows that the **essential material of heredity is the DNA**.
- DNA is the storehouse of all information for protein synthesis except in some viruses in which the genetic material is RNA instead of DNA (Chapter 5).

Structure of DNA

The DNA molecule is composed of two chains of nucleotides wound around each other in the form of a '**double helix**'. Each chain has a 'backbone' of alternating units of deoxyribose and phosphate residues. Attached to each deoxyribose is one of four nitrogenous bases:

- **Purines**
  - Adenine (A)
  - Guanine (G)
- **Pyrimidines**
  - Thymine (T)
  - Cytosine (C)
- The double-stranded nature of the molecule is stabilised by hydrogen bonding between the bases on the opposite strands (Fig. 8.1) such that adenine is always linked to thymine (A–T) and guanine to cytosine (G–C; Fig. 8.2).
- The ratio of each pair of bases (A–T)/(G–C), though constant for each species, varies widely from one bacterial species to another. **Bacteria are grouped based on their G–C content.**
- The DNA molecule replicates by first unwinding at one end to form a fork, each strand of the fork acting as a template for the synthesis of a complementary strand with which the newly synthesised strand then forms a double helix.

Structure of RNA

- RNA is structurally similar to DNA except for two major differences—it contains the sugar **ribose instead of deoxyribose** and the base **uracil instead of thymine.**
- Three distinct types of RNA can be distinguished on the basis of structure and function:
  - Messenger RNA (mRNA)
  - Ribosomal RNA (rRNA)
  - Transfer RNA (tRNA)
- DNA acts as the template for the synthesis of mRNA and, therefore, the bases in the two are complementary

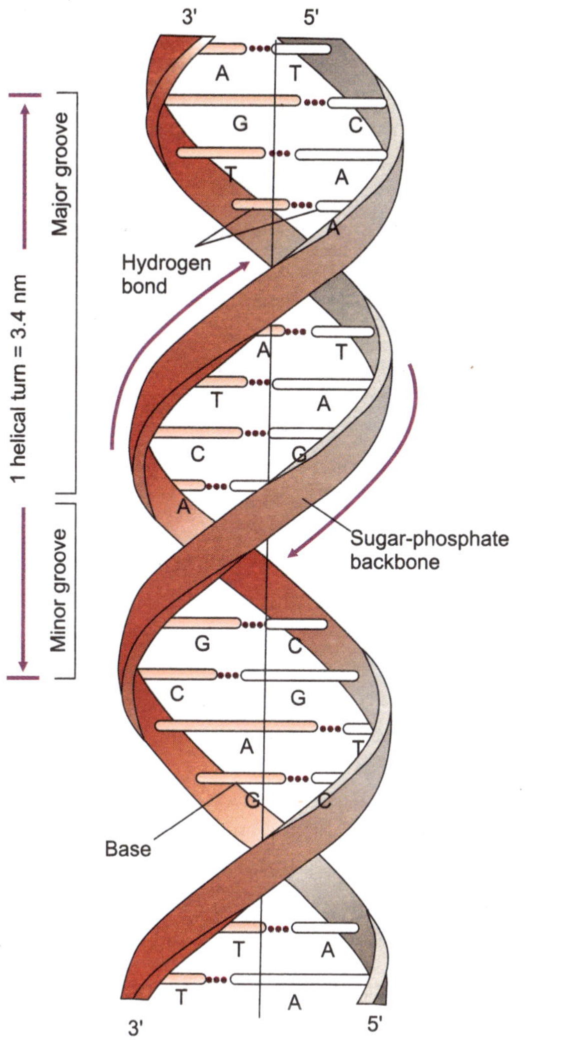

Fig. 8.1 A schematic drawing of the Watson–Crick structure of DNA, showing the helical sugar–phosphate backbones held together by hydrogen bonding between the bases

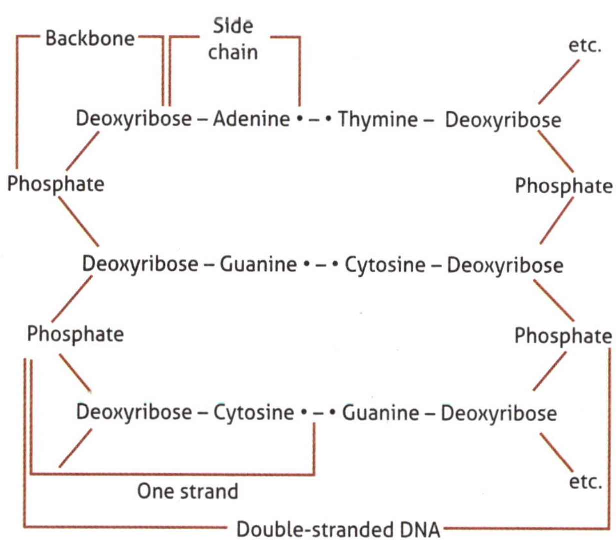

Fig. 8.2 The chemical nature of a segment of double-stranded DNA

to each other. The adenine, guanine, cytosine, and uracil units in mRNA pair with thymine, cytosine, guanine, and adenine, respectively, in DNA.

TERMS RELATED TO GENETICS

Gene

- A segment of DNA carrying **specific codons** for the synthesis of a particular polypeptide is called a **gene**.
- A DNA molecule comprises a large number of genes, each of which contains hundreds of thousands of nucleotides.
- The bacterial chromosome is a circular, double-stranded DNA.
- Bacterial DNA is about 4,000 kilobases (kb) long, whereas the human genome is about 3 million kb long.

Genome

This term refers to the sum of an organism's genetic material.

Phenotype

- The phenotype (*phaeno* meaning display) of a cell is the physical expression of its genotype in a given environment.
- A cell may exhibit different phenotypic characteristics in different situations.
- For example, *S.* Typhi is normally flagellated, but when it is grown in agar containing phenol, the flagella are not synthesised.

Genotype

- All the genes present in a cell (the **genome**) collectively constitute its **genotype**.
- The genotype is the hereditary constitution of a cell and is transmitted to its progeny.
- It includes the complete genetic potential of the cell, all of which may or may not be expressed in a given environmental situation.

Extrachromosomal Genetic Elements

In addition to chromosomal DNA, most bacteria possess extrachromosomal genetic elements. These are not essential for the normal life and functioning of the host bacterium but may confer properties such as drug resistance and toxigenicity on the bacterium, providing it with a survival advantage.

Plasmids

- These are extrachromosomal circular DNA molecules that are present in the cytoplasm of bacteria and yeasts.
- Plasmids are capable of autonomous replication (independent replicons and can be transferred from one cell to another).

- Plasmids have become important tools as vectors in genetic engineering.
- Plasmid DNA may sometimes be integrated with chromosomal DNA (episomes).
  There are two types of plasmids:
  - **Conjugative plasmids**, which are self-transmissible
  - **Non-conjugative plasmids**, which are non-transmissible

Transposable Genetic Elements: Transposons/ Jumping Genes

- Transposons were discovered by **Barbara McClintock** in plants. For this discovery, she was awarded the Nobel Prize for Medicine in 1983.
- These are structurally and genetically discrete segments of DNA that have the **ability to move around in a 'cut-and-paste' manner** between chromosomal and extrachromosomal DNA molecules within cells.
- A transposon is a segment of DNA with one or more genes in the centre and 'inverted repeat' sequences of nucleotides at the two ends. These are also known as **jumping genes**.
- Small transposons (1–2 kb) are known as '**insertion sequences**' or IS. They attach at certain regions of chromosomal, plasmid, or phage DNA.
- The insertion of a transposon leads to the acquisition of new characteristics by the recipient DNA molecule.
- Unlike plasmids, transposons are **not self-replicating** and depend on chromosomal or plasmid DNA for replication (Fig. 8.3).

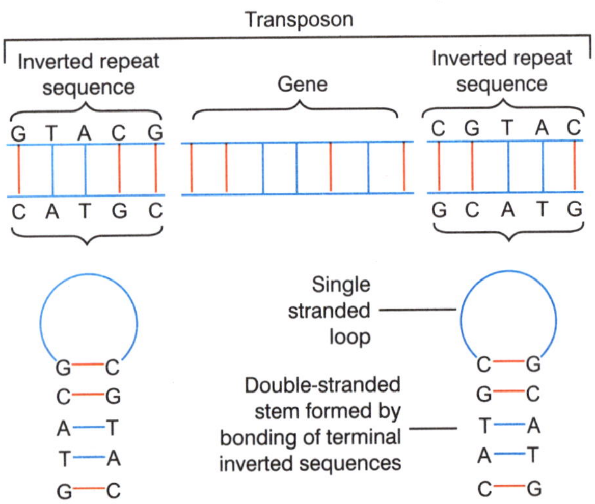

Fig. 8.3 Structure of a transposon

GENETIC VARIATIONS

These occur due to alterations in the genome and are stable and heritable. They may occur by:

- **Mutation**
- **Genetic transfer or exchange** (transformation, transduction, **lysogenic conversion**, or conjugation)

Mutation

- A mutation is a **heritable change in the DNA sequence** caused by an alteration in the nucleotide sequence.
- Mutations in genes are important in evolution.
- They can be beneficial or harmful to the cell.
- A mutation may spontaneously occur due to processes within the cell or may be induced by an extraneous substance such as a mutagen in the environment.
- Based on the type of change in the nucleotide sequences, mutations are categorised as follows (Fig. 8.4):
  - **Substitution of a nucleotide**
  - **Deletion of a nucleotide**
  - **Addition of a nucleotide**
  - **Inversion/duplication of a nucleotide**

  Some examples of mutations are listed in Table 8.1.
  Mutations can also be classified as follows:
- **Spontaneous mutations:** Each gene undergoes mutation with a fixed frequency. Mutation rates of individual genes in bacteria range from 10^4–10^{10} per bacterium per division.
- **Induced mutations**: Though mutations occur spontaneously, their frequency can be increased by several agents (mutagens) such as UV rays, alkylating agents, acridine dyes, 5-bromouracil, and 2-aminopurine.

Clinical applications of mutations

- A mutation can confer a **survival advantage**. For example, if a streptomycin-resistant mutant of the tubercle bacillus develops in a patient on treatment with the drug, it multiplies selectively and ultimately replaces the original drug-sensitive population of bacteria.
- It may modify any characteristic of the bacterium.
- Mutants may vary in properties such as nutritional requirements, biochemical reaction, antigenic structure, morphological features, colony form, drug susceptibility (conferring drug resistance in bacteria and virus), virulence, and host range.
- Some respiratory viruses like influenza and coronaviruses tend to mutate at a rapid rate, resulting

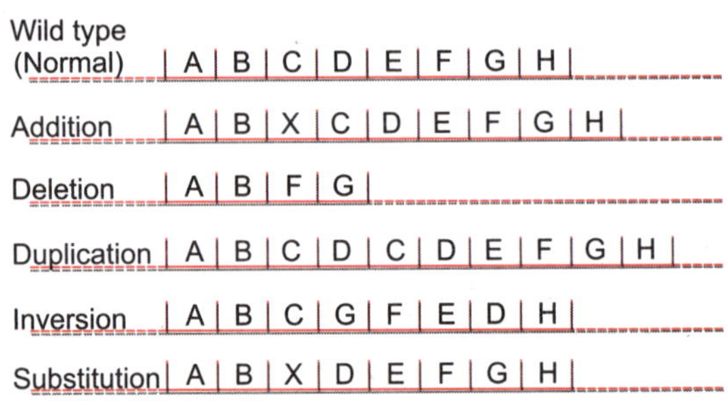

Fig. 8.4 Examples of the types of mutations

Table 8.1 Types of mutations in microorganisms

Type	Description
Missense mutation	A mutation where the triplet code is altered so as to code for an amino acid different from the one normally located at a particular position in the protein
Nonsense mutation	The deletion of a nucleotide within a gene may cause premature polypeptide chain termination by generating a nonsense codon
Transversion	The substitution of a purine for a pyrimidine and vice versa in base pairing
Suppressor mutation	The reversal of a mutant phenotype by another mutation at a position on the DNA distinct from that of the original mutation
Lethal mutation	Some mutations involve vital functions, rendering the mutant non-viable; a type of lethal mutation which is of great interest is 'conditional mutation'
Conditional lethal mutation	Confers the ability to be able to live only in certain conditions (permissive conditions)—the commonest type of a conditional mutant is the temperature-sensitive (ts) mutant, which can live at the permissive temperature (say, 35°C), but not at the restrictive temperature (say, 39°C)
Frameshift mutation	The deletion or insertion of multiple nucleotides
Multiple mutations	Extensive chromosomal re-arrangement
Point mutation	The deletion or insertion of one or more bases

in epidemics and resurgence of pandemics (e.g., the COVID-19 pandemic of 2020–21).

- The practical importance of microbial mutation lies in the fields of public health, drug resistance, the spread of disease (SARS-CoV-2 mutants), and in the development of vaccines.

Detection of mutations

Molecular techniques like gene probes and sequencing are used widely to detect changes in the bacteria due to mutation.

GENE TRANSFER IN BACTERIA

Genetic information is transferred from one bacterium to another by:

- **Horizontal transfer**, which occurs through transformation, transduction, and conjugation
- **Vertical transmission**, in which genetic information is passed down from a bacterium to its progeny via replication

Horizontal Transfer Mechanisms

Transformation

- This is the transfer of genetic information through free DNA.
- In 1928, Griffith demonstrated that the injection of a mixture of live, non-capsulated (type II) *S. pneumoniae* and heat-killed, capsulated (type I) *S. pneumoniae* (neither of which were individually fatal) could kill mice.
- The pneumococci that were isolated from the dead mice were found to be live and capsulated (type I), demonstrating that the gene (DNA) coding for the capsule had been transferred from the killed bacterium to the live, non-capsulated strain (Fig. 8.5).

Transduction (see Chapter 5)

The transfer of a portion of DNA from one bacterium to another by a **bacteriophage** is known as transduction. Transduction may be:

- **Generalised**, when it involves any segment of the donor DNA at random
- **Restricted**, when a specific bacteriophage transduces only a particular genetic trait

Conjugation

- It is a process whereby a **donor** ('male') bacterium conjugates or makes physical contact through a sex pilus with a **recipient** ('female') bacterium and transfers genetic elements to it (Fig. 8.6).
- Bacterial conjugation was first described by **Lederberg and Tatum** (1946) in a strain of *E. coli* called K12, in which conjugation has been most extensively studied. The steps involved in conjugation are as follows:
  - The donor status of a cell is determined by the presence of a plasmid that codes for a specialised fimbria (sex pilus) which projects from the surface of the cell.
  - The plasmid DNA replicates and a copy of it passes from the donor to the recipient cell through the sex pilus (which forms a conjugation tube).
  - The recipient attains donor status and can, in turn, conjugate with other recipient cells.
  - The ability to conjugate is a transmissible or 'infectious' characteristic located in a plasmid.

Along with plasmid DNA, portions of the host DNA may also be sometimes transferred to the recipient.

- The donor DNA then combines with the DNA of the recipient, effecting genetic recombination.
- The plasmid responsible for the transfer is called the **'fertility (F) factor'** and the bacteria possessing this factor are called **F⁺ bacteria**.

The F factor

- It is a transfer factor that contains the genetic information necessary for the synthesis of the sex pilus and self-transfer but is devoid of other identifiable genetic markers such as drug resistance.

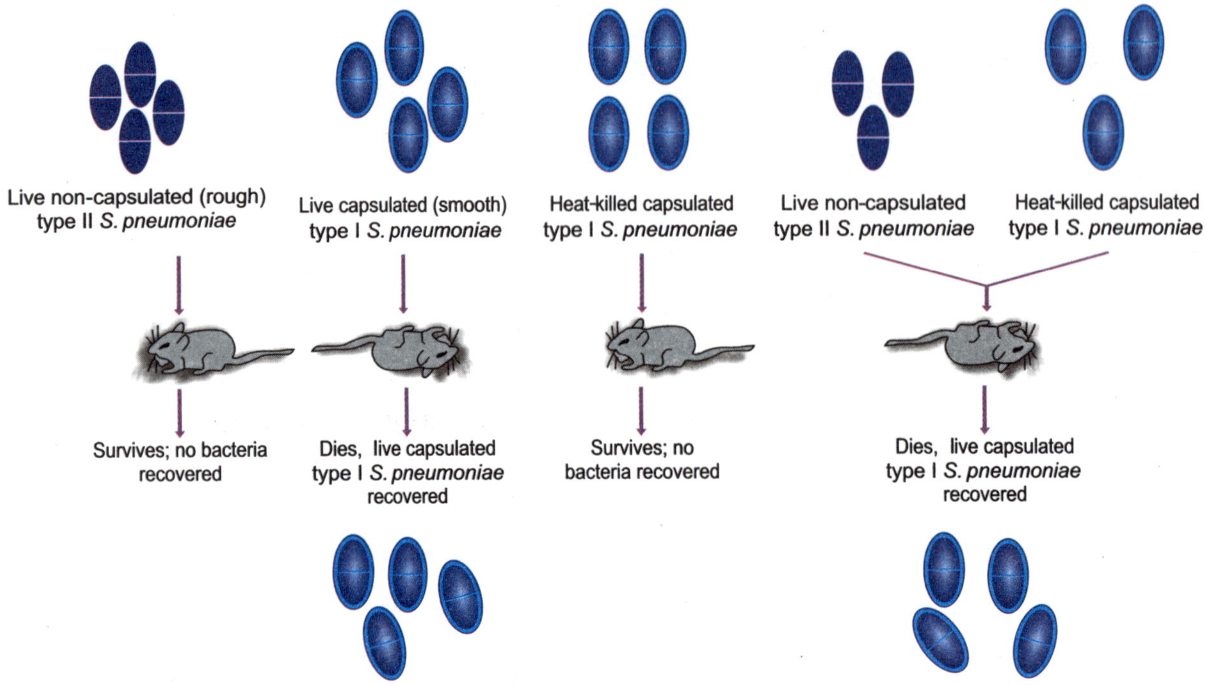

Fig. 8.5 Griffith's experiment to demonstrate genetic transfer by transformation—DNA is transferred from the dead *S. pneumoniae* to the live bacterium in the mouse, thus killing it

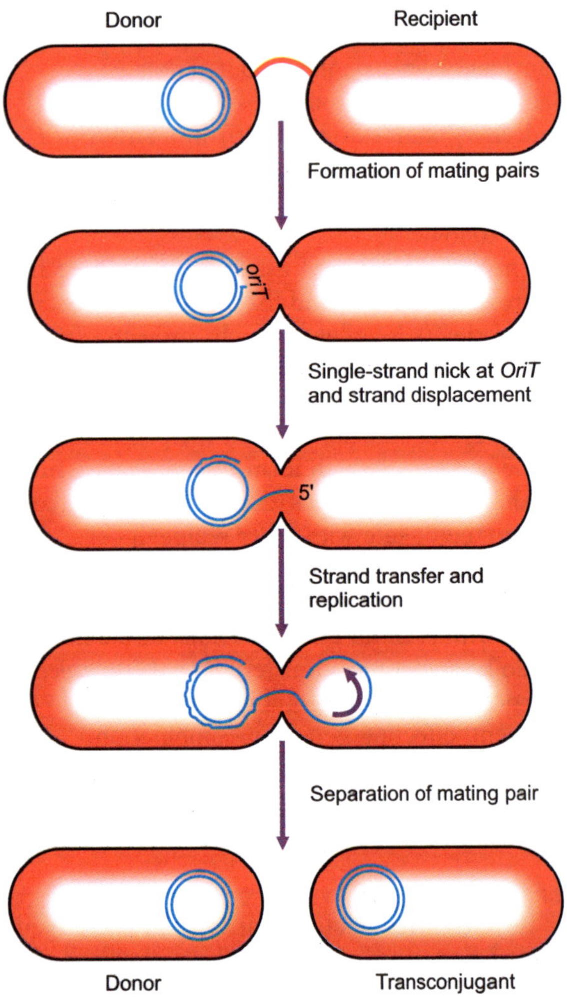

Fig. 8.6 Mechanism of DNA transfer through conjugation

- Cells carrying the F factor (F^+ cells) have no distinguishing features other than their ability to mate with F^- cells and render them F^+.
- Cells that can transfer chromosomal genes to recipient cells with high frequency are known as **Hfr cells**. Following conjugation with an Hfr cell, an F^- only rarely becomes F^+, even though it receives chromosomal genes from the donor.
- **F prime (F') factor:** The conversion of an F^+ cell into an Hfr cell is a reversible event. When the F factor reverts from the integrated state to the free state, it may sometimes carry with it some chromosomal genes from near its site of attachment. Such an F factor, which incorporates some chromosomal genes, is called an F prime (F') factor.
- **Sexduction:** When an F' cell mates with a recipient, it transfers, along with the F factor, the host genes incorporated with it. This process of the transfer of host genes through the F' factor resembles transduction and has, been named sexduction (Fig. 8.7).

Resistance transfer factor (RTF or R Factor)

- This plasmid contains genes for resistance to several antibiotics and permits the transfer of drug resistance between bacteria.
- It shares many genes with the F factor and is largely homologous to it.
- The transfer of multiple drug resistance has been demonstrated between *E. coli* and *Shigella* strains both in vitro and in vivo.

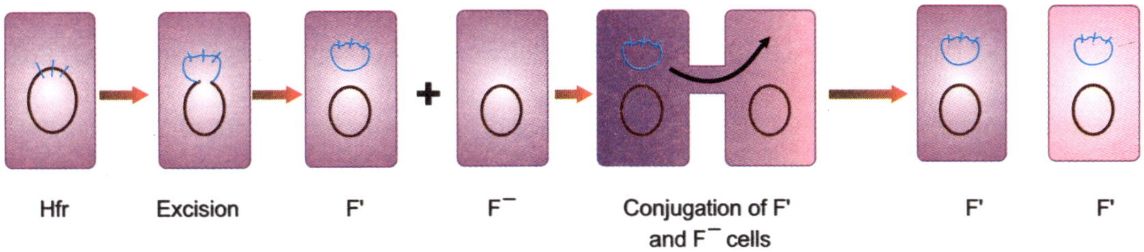

Fig. 8.7 Sexduction—the integrated F factor of an Hfr cell may revert to the cytoplasmic state; during excision, some host genes may be incorporated in the F′ factor (F)

- The resistance is plasmid-mediated and is transferred by conjugation. This mechanism of drug resistance is known as transferable, episomal, or infectious drug resistance.

The extrachromosomal mechanism of drug resistance was first reported by Japanese workers (1959) who were investigating a sudden increase in infections caused by *Shigella* strains that were resistant simultaneously to sulphonamides, streptomycin, chloramphenicol, and tetracycline. The workers observed that patients excreting such *Shigella* strains also shed *E. coli* strains resistant to the same drugs in their feces.

Components of RTF

The RTF has two components:

- The transfer factor called the **resistance transfer factor (RTF)**, which is responsible for conjugational transfer (Fig. 8.8)
- The **resistance determinant (r)** for each of the several drugs—the whole plasmid (RTF+*r* determinants) is known as the **R factor**; a plasmid may harbour up to 10 resistance determinants, which can be transferred simultaneously

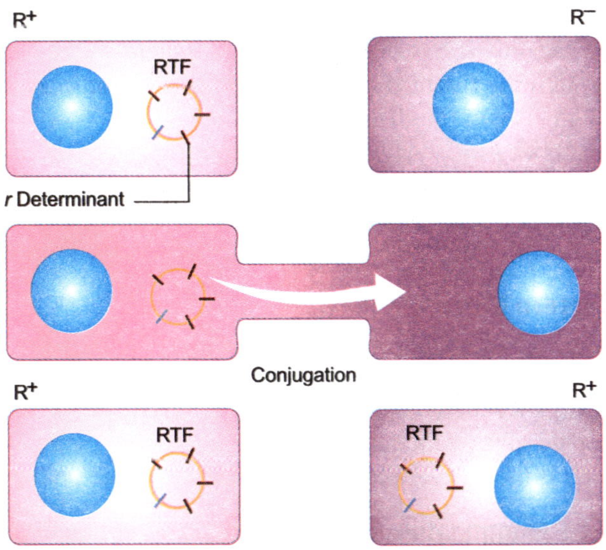

Fig. 8.8 Transferable drug resistance—the R⁺ cell carries the R factor, consisting of the RTF and *r* determinants; its transfer to a sensitive R⁻ bacterium converts the recipient into a resistant R⁺ cell

Sometimes, the RTF may dissociate from the r determinants, the two components existing as separate plasmids. In such cases, though the host cell remains drug-resistant, the resistance is not transferable.

Clinical implication of RTF

Transferable drug resistance is seen widely in various pathogenic and commensal bacteria in humans and animals.

- The transfer occurs easily in vitro but in the normal gut, it is inhibited by several factors such as anaerobic conditions, bile salts, alkaline pH, and the abundance of anaerobic gram-positive bacteria, which minimise the chance of contact between donor cells and suitable recipient cells.
- On the other hand, in the intestines of persons on oral antibiotic therapy, transfer occurs readily due to the destruction of the sensitive normal flora and the selection pressure produced by the drug.

Vertical Gene Transmission

- During the replication of bacteria, genetic information is passed on from the parent cell to its progeny.
- The parent DNA replicates into two double-stranded DNA, each containing a template strand from the parent and a newly synthesised strand.
- Small pieces of chromosomes replicate in this way and move to one end of the bacterial cell. These pieces are enclosed in a plasma membrane, around which a new cell wall develops, giving rise to daughter cells carrying copies of the parent genes.

Bacterial Recombination

- A piece of DNA (containing specific genes) that has been created artificially from two or more sources is incorporated into a single bacterial molecule. This results in the expression of that genetic information by the bacteria and its progeny.
- Recombination helps in generating large quantities of DNA carrying the specified genetic information.
- Bacterial recombination may be of the following types:
  - **Homologous recombination:** Occurs between DNA molecules that have very similar sequences

- Non-homologous recombination: Occurs between pieces of DNA among which there is no significant similarity in sequences
- **Site-specific recombination:** Occurs between short sequences present in dissimilar molecules
- **Replicative recombination:** Generates a new copy of a segment of the DNA

MOLECULAR TECHNIQUES APPLIED TO DISEASE DETECTION AND DIAGNOSIS IN CLINICAL MICROBIOLOGY AND EPIDEMIOLOGY

Techniques to extract nucleic acid molecules from cells and determine their structure and methods of replication have opened the door to the world of microbial genomics.

- DNA manipulation is based on the use of restriction endonucleases (restriction enzymes), which are microbial enzymes that cleave double-stranded DNA at specific oligonucleotide sequences.
- A number of such enzymes have been recognised, each of which acts at a different nucleotide sequence (for example, *Eco* RI, *Hind* III, and *Taq* I).
- Restriction enzymes split DNA strands into fragments of varying lengths, which can be separated by gel electrophoresis.

Nucleic Acid Probes

DNA probes

- DNA strands can be detected using specific DNA probes, which hybridise with the target DNA during DNA or RNA synthesis.
- The probes are radioactively labelled, biotinylated copies of cloned single-stranded DNA fragments, usually 20–25 nucleotides long and containing unique nucleotide sequences.
- DNA probes are used in the diagnosis of infectious diseases.
- Probes containing sequences unique to the microbe (strain, species, or group) to be detected can be added to microbial cultures or the sample suspected to contain the microbe or its DNA.
- If the sample or culture contains the suspected microbe, the DNA probe hybridises with the specific complementary sequences on the microbe's DNA.

Advantages

- High degree of specificity
- Ability to detect minute quantities of complementary DNA even in the presence of other microbes
- Capacity to recognise microbes that are either difficult or impossible to culture

Blotting Techniques

These techniques are used for the detection of DNA, RNA, and proteins of interest.

Southern blotting

This technique was named after **E. M. Southern**, who devised it. In this method, DNA–DNA hybridisation is used to **identify fragments of DNA**.

- DNA fragments are cut using restriction enzyme digestion.
- The fragments are separated and transferred on to the gel by blotting them onto nitrocellulose or nylon membranes that bind to the DNA.
- The DNA bound to the membrane is denatured (converted to the single-stranded form) and treated with radioactive single-stranded DNA probes.
- These probes hybridise with homologous DNA to form radioactive double-stranded segments, which can be detected on an X-ray film.
- Applications—to detect genetic disorders, cancers such as monoclonal leukemia, etc.

Northern blotting

- This is an analogous method for the **analysis of RNA** (named 'northern' in contrast to Southern blotting).
- The RNA mixture is separated by gel electrophoresis, blotted, and then identified using labelled DNA or RNA probes.
- Application—it is used for the detection of mRNA in a sample to study gene expression.

Western blotting

- A similar technique used for the identification of **proteins (antigens)** is called immunoblotting (or, in conformity with other blotting techniques, western blotting).
- The protein–antigen mixture is separated by SDS–**PAGE (sodium dodecyl sulphate–polyacrylamide gel electrophoresis)**.
- The mixture is blotted onto nitrocellulose strips and identified with the help of radiolabelled or enzyme-labelled antibodies used as probes.
- Applications—it is used for the diagnosis of HIV antibody in sera, in the TORCH panel (*Toxoplasma*, rubella, cytomegalovirus, and herpes antibodies), and for the diagnosis of several other infectious diseases.

Eastern blotting

- Eastern blot is a biochemical technique used to analyse protein **post-translational modifications (PTM)** including the addition of lipids, phosphates, and glycoconjugates.
- It is most often used to detect carbohydrate epitopes.

DNA Amplification Techniques

Polymerase chain reaction (PCR)

In 1983, **Kary B. Mullis** developed the PCR technique to amplify DNA; he was awarded the Nobel Prize in Chemistry in 1993 for his work.

Principle

- The amplification of DNA is a chain reaction in which a gene or a small fragment of the DNA of interest serves as the template for producing the primers that initiate the reaction.
- One DNA molecule is used to produce two copies, which then form the template for the next reaction in a thermocycler.
- Thus, it is possible to obtain several copies of the DNA of interest, and thereby, an abundant quantity of specific DNA from a minuscule source (Fig. 8.9).
- Post amplification, the number of copies of nucleic acid after n number of cycles = 2n.
- In the first step, the sample is heated to 94–95°C to **denature the DNA** to form a single strand.
- Then, a **primer** of 18–22 base pairs binds to the complementary regions of the single-stranded molecules at 50–56°C.
- In the third step, the **DNA fragments are extended** by the action of the **DNA polymerase** at 72°C.
- Thermostable DNA polymerase is used because it remains active at high temperatures.

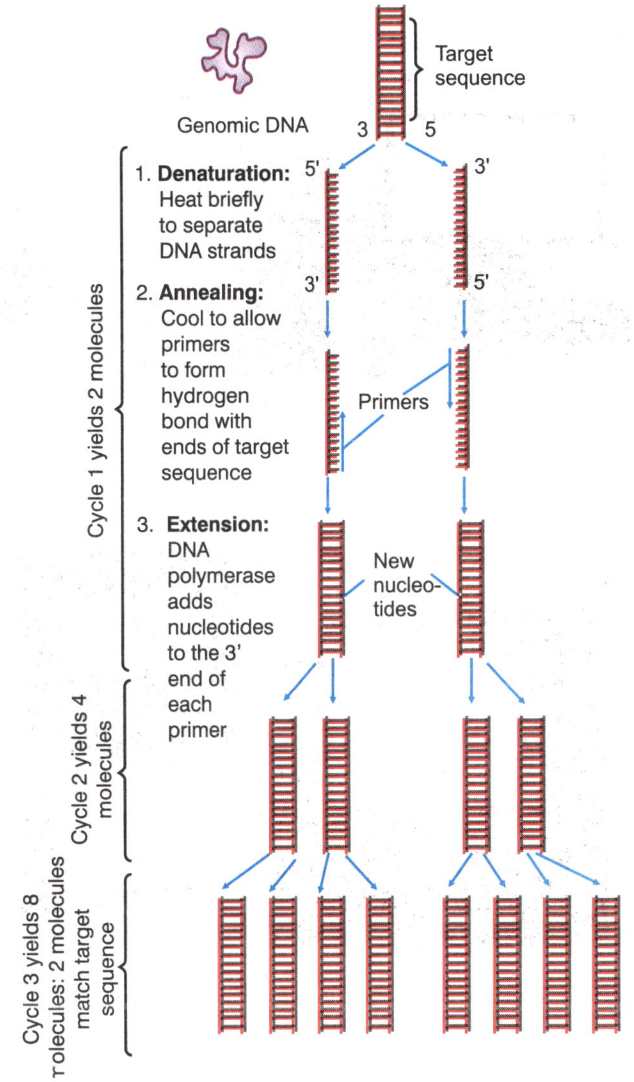

Fig. 8.9 Diagrammatic representation of the amplification of genomic DNA by PCR

> **Applications of PCR**
>
> With its enormous capacity to amplify DNA, PCR (Fig. 8.10a and b) is a versatile tool, useful in diverse areas such as:
>
> - Diagnosis of infectious diseases
> - Genetics of neoplastic diseases
> - Forensic investigations
> - Archeo-biological studies of ancient specimens
> - Examination of phylogenetic relationships in evolution

Modifications of PCR

Reverse transcriptase-PCR (RT-PCR)

- The conventional PCR technique can only amplify dsDNA (double-stranded DNA) sequences. Therefore, RNA viruses cannot be amplified by this method directly.
- **RT-PCR** is a modification of the PCR method in which RNA is used as a template to obtain a complementary DNA (cDNA) strand with the help of **reverse transcriptase (RT)** enzyme. This cDNA is then used as a template for PCR (Fig. 8.11).

Quantitative PCR (qPCR)

- This method was formerly called RT-PCR but renamed qPCR to avoid confusion with reverse transcriptase PCR.
- **Advantage**—the quantification of nucleic acid can be achieved by monitoring the amplification process while the PCR is ongoing, i.e., in real time (Fig. 8.12); hence the name.
- This method estimates the pathogenic DNA load in the test specimen. Samples can be processed for rapid detection of specific DNA load, e.g., hepatitis B DNA viral load.

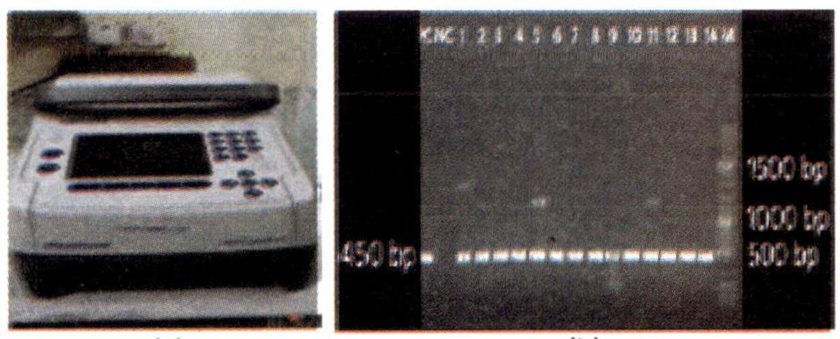

(a) (b)

Fig. 8.10 (a) Thermocycler (conventional PCR) to amplify the desired DNA and (b) gel documentation of amplified DNA bands of specific base pairs compared to a molecular weight ladder *(Source:* Department of Microbiology, Pondicherry Institute of Medical Sciences, Puducherry)

- The pathogenic DNA is detected using:
  - **Non-specific fluorescent** dyes like SYBR green, which intercalate with the dsDNA as it gets synthesised
  - **Specific DNA probes** labelled with fluorescent reporters that get detected only after the probe

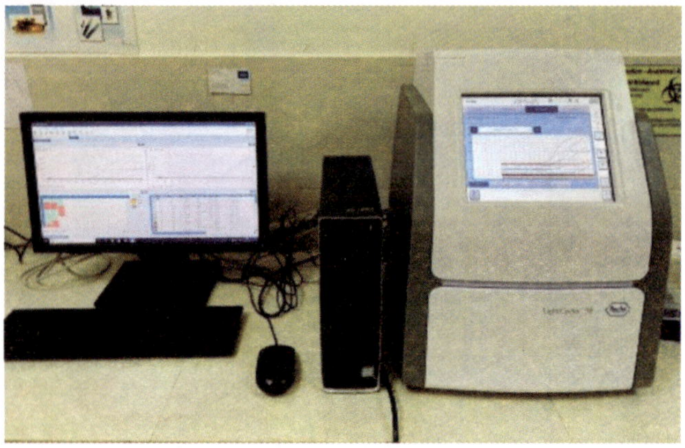

Fig. 8.11 Conventional RT-PCR (*Source*: Molecular Division, Dept of Microbiology, Pondicherry Institute of Medical Sciences, Puducherry)

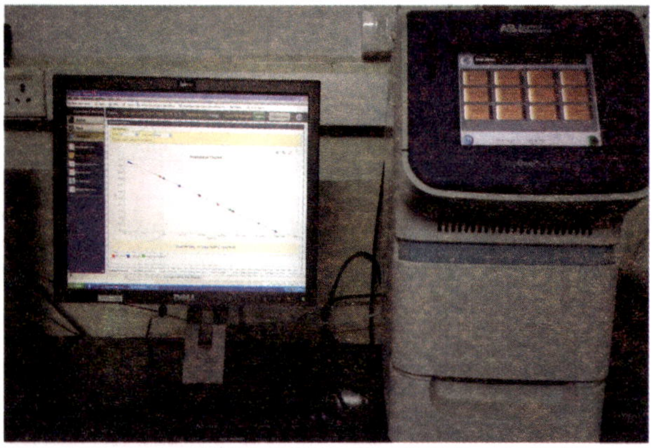

Fig. 8.12 Quantitative PCR (qPCR) (*Source*: Pushpagiri Institute of Medical Sciences and Research Centre, Thiruvalla, Kerala)

hybridises with the complementary DNA sequence; the fluorescent reporter signal strength is directly proportional to the number of amplified DNA molecules

Nested PCR

- In conventional PCR, sometimes, the primers bind to non-specific regions of DNA.
- To counter this non-specificity, a second set of specific primers is used in successive cycles to amplify a secondary target within the first target.

Multiplex PCR

Primers specific to more than one pathogen are used simultaneously in the same reaction to yield multiple PCR products.

Point-of-care PCR tests

PCR tests are intensive on equipment and require trained manpower. The following tests are used as rapid point-of-care tests:

- **Chip-based TrueNat**, which was recently used for detecting SARS-COV-2 during the COVID pandemic
- **Cartridge-based GeneExpert/CB NAAT**, which is used widely to detect *Mycobacterium*

Isothermal tests

Amplification requires the use of a thermocycler, temperature regulation, and several cyclical runs. To circumvent the need for these resources, non-amplification techniques have been developed.

Loop-mediated isothermal assay (LAMP)

- It is a simple technique, performed at a constant temperature. It does not require the reaction to be carried out at a series of different temperatures; therefore, it does not require a thermocycler.
- In the common format, four sets of primers are used to bind to six regions on the target DNA.
- Primers with a specific nature, called '**loop primers**', yield higher amounts of DNA.
- The polymerase enzyme used in this method is also special because it has a '**high strand displacement**' activity.
- This method has the potential for use as a screening test in the field or as a point-of-care test.

DNA microarray

- Several microscopic DNA 'spots' are coated on a solid surface, e.g., a silicone or glass chip.
- These spots contain specific DNA sequences that serve as probes.
- These probes hybridise with cDNA in the test sample. This method is useful in cases where the DNA in the sample does not have to be amplified.
- DNA microarray can simultaneously detect a large number of nucleic acid sequences in the sample.

As an important tool in a diagnostic laboratory, molecular diagnosis has become an integral part of the workflow for the following:

- Rapid detection of pathogens in clinical material (in viral meningitis, encephalitis)
- Detecting molecular markers (hepatitis B)
- Tracing the source of infection in an outbreak (gene sequencing to map phylogenetic relatedness)
- Identifying new microorganisms or mutants
- Detection of genes responsible for drug resistance

These methods have proven to be a major advantage over time-consuming, conventional culture and isolation methods and have greatly impacted patient management. Automated machines to perform these techniques have greatly improved disease detection.

MOLECULAR EPIDEMIOLOGY

Molecular techniques are used for the matching of microbial isolates for epidemiological purposes by:

- Plasmid profile analysis
- Genomic fingerprinting
- Typing of microorganisms

Sequence-based Assays

Analysing the nucleic acid sequence enables the detection of new pathogens and mutations in existing pathogens. These are also useful tools in molecular epidemiology.

Advantages and disadvantages of molecular methods in disease diagnosis

Advantages

- Highly sensitive and specific
- Can be applied directly to the specimen, even when only minute quantities are available
- Relatively rapid compared to culture
- Can detect the disease in the early stages
- Can detect several characteristics of the microorganism, e.g., antibiotic resistance and virulence
- Can be used for epidemiological purposes to trace the origin or source of infection

Disadvantages

- High cost (which has now come down due to wider application)
- Require specific technology and equipment

GENE SEQUENCING

Gene sequencing has made it possible to differentiate microorganisms at the genetic level. This has brought about a sea change in the following:

- Classification of microorganisms
- Tracing the lineage of microorganisms
- Establishing relatedness among microbes
- Discovery of several new microorganisms responsible for infections in humans and animals
- Detection of mutations within a cluster of microorganisms
- Study of factors leading to mutations

The cost of equipment and infrastructure like computer capacity and storage, the expertise required to analyse big data, etc., limit the wider use of gene sequencing methods at present.

APPLICATIONS OF GENETIC ENGINEERING IN MICROBIOLOGY

- The most important application of molecular genetics is genetic engineering or recombinant DNA (rDNA) technology.
- The DNA of a microorganism can be manipulated.
- Its most common application is cloning.

Method

- rDNA methods involve the isolation of the genes coding for desired proteins from microorganisms or cells of higher forms of life including human beings and their introduction into suitable microorganisms in which the genes would be functional, directing the production of the specific protein.
- Such cloning of genes in microorganisms enables the preparation of the desired protein in pure form, in large quantities, and at a reasonable cost.
- The target DNA segments are cleaved by specific enzymes called **restriction endonucleases**, and the fragments containing the desired genes are isolated.
- These fragments are incorporated into suitable vectors or carriers, such as plasmids or temperate bacteriophages, for insertion into microorganisms (most commonly, *E. coli* K12).
- These bacteria multiply quickly and produce a large amount of the desired gene product.

Commercial applications of genetic engineering

- Safer vaccines can be produced by cloning the protective antigens of pathogens, as has already been done in the case of foot-and-mouth disease, hepatitis B, and rabies viruses.
- Application of cloning for antigen preparation in kits has also paved the way for various sensitive and specific tests for the diagnosis of infectious diseases.
- Cloned human insulin, interferons, somatostatin, growth hormones, and many other biologicals have already been marketed.

Key Points

✓ Deoxyribonucleic acid (DNA) carries genetic information, which is transcribed into ribonucleic acid (RNA) and then translated into a particular polypeptide (DNA → RNA → polypeptide).

✓ Plasmids are circular DNA molecules present in the cytoplasm of bacteria, capable of autonomous replication (independent replicons).

✓ The total genetic constitution of an organism is called its genotype. The genotype of a cell is determined by the genetic information contained in its entire DNA content or genome.

✓ The characteristics displayed by an organism in any given environment comprise its phenotype.

✓ Mutation is a random, undirected, heritable variation caused by an alteration in the nucleotide sequence at some point in the DNA of the cell.

✓ Several agents (mutagens) such as UV rays, alkylating agents, acridine dyes, 5-bromouracil, and 2-aminopurine can induce mutations.

✓ Three important modes of genetic transfer are as follows:

- Transformation, in which there is direct uptake of donor DNA by the recipient cell, which may be natural or forced

- Transduction, in which donor DNA is carried in a bacteriophage and is transferred to the recipient by the mechanism used for phage infection
- Conjugation, in which the donor cell contributes energy and building blocks to the synthesis of a new DNA strand, which is physically transferred into the recipient cell, usually by a tube called a sex pilus

✓ Transferable drug resistance mediated by the R factor is the most significant way in which cells acquire drug resistance.

✓ Molecular diagnosis of infectious diseases and understanding the biology of the infectious agents have been made possible by studying the genetic elements of the microorganisms.

✓ PCR and its modifications are based on the amplification of the small amount of DNA sequences in a test sample to numbers that can be detected.

Questions

Short answer questions

1. Enumerate the mechanisms of transfer of genetic materials in bacteria. Explain the steps in any one of these mechanisms.
2. Describe the differences between the resistance acquired by mutation and transferable drug resistance.
3. Describe the mechanism of transformation.
4. Briefly discuss transduction.
5. What is conjugation? Describe its mechanism in bacteria.
6. Enumerate the uses of molecular diagnostic methods in a clinical laboratory.

MCQs

1. **The transfer of DNA from one bacterium to another by a bacteriophage is called:**
   a) Conjugation b) Transduction
   c) Transformation d) Transposition
2. **Which of the following is responsible for the transfer of drug resistance in bacteria?**
   a) Colicinogenic factor b) Hfr cells (high frequency)
   c) F factor d) None of these
3. **Which of the following refers to the transfer of genetic information through free DNA?**
   a) Transformation b) Transduction
   c) Conjugation d) Lysogenic conversion
4. **Which of the following does NOT APPLY to bacterial plasmids?**
   a) They are extrachromosomal.
   b) They are eliminated by radiation treatment.
   c) They cannot replicate independently.
   d) They are not essential for bacterial life.
5. **Genetic engineering has been used for infectious diseases in the following instances EXCEPT:**
   a) Cloning for vaccines
   b) Antigen preparation for diagnostic kits
   c) Using non-pathogenic organisms as vectors
   d) Inducing drug resistance in pathogenic bacteria
6. **Northern blotting is used for the analysis of:**
   a) DNA b) RNA
   c) Proteins d) All of these
7. **What is the function of a nonsense codon?**
   a) To act as punctuation marks terminating the message for polypeptide synthesis
   b) To bring the correct amino acid in line with a messenger RNA during protein synthesis
   c) It is the first codon of an mRNA from which the translation begins
   d) It helps in copying a particular segment of DNA into RNA
8. **Jumping genes are:**
   a) Episomes b) Plasmids
   c) Transposons d) Transfer factor
9. **The cell in which the F factor carries with it some chromosomal genes is called:**
   a) F-cell b) Hfr cell
   c) F'cell d) None of these
10. **All the following bacterial toxins are coded by lysogenic phages except:**
    a) Cholera toxin
    b) Diphtheria toxin
    c) *Streptococcus* pyrogenic exotoxin A
    d) *Streptococcus* pyrogenic exotoxin B
11. **Which of the following statement is true regarding mutation?**
    a) Missense mutation is the deletion of a nucleotide within a gene
    b) Point mutation is the substitution of a purine for a pyrimidine
    c) Spontaneous mutations occur on exposure to UV rays
    d) Frameshift mutation is the deletion or insertion of multiple nucleotides
12. **The technique for identifying DNA fragments by DNA-DNA hybridisation is:**
    a) Northern blotting b) Southern blotting
    c) Eastern blotting d) Western blotting
13. **The procedure of Southern blotting is:**
    a) Fragmentation → gel electrophoresis → blotting → hybridisation
    b) Gel electrophoresis → fragmentation → blotting → hybridisation
    c) Fragmentation → gel electrophoresis → hybridisation → blotting
    d) Fragmentation → blotting → gel electrophoresis → hybridisation
14. **The technique for the analysis of RNA is:**
    a) Northern blotting b) Southern blotting
    c) Eastern blotting d) Western blotting
15. **Immunoblotting is otherwise called:**
    a) Northern blotting b) Southern blotting
    c) Eastern blotting d) Western blotting
16. **Kary B. Mullis was awarded the Nobel Prize for:**
    a) PCR b) Northern blotting
    c) Southern blotting d) Western blotting
17. **Eastern blotting can be considered an extension of biochemical techniques of:**
    a) Northern blotting b) Southern blotting
    c) Western blotting d) None of these
18. **Reverse transcriptase is:**
    a) DNA-dependent RNA polymerase
    b) RNA-dependent DNA polymerase
    c) DNA-dependent DNA polymerase
    d) RNA-dependent RNA polymerase
19. **Which of the following is a phenotypic method for microbial characterisation?**
    a) Polymerase chain reaction
    b) Microbial array
    c) Ribotyping
    d) Serotyping
20. **The method of DNA amplification that is performed at a constant temperature is:**
    a) Nested PCR b) Hot-start PCR
    c) LAMP assay d) Microarray
21. **PCR in which a second set of specific primers is used is:**
    a) Nested PCR b) Hot-start PCR
    c) High-fidelity PCR d) Arbitrary primed PCR

22. Quantitative measurement of amplified product is carried out by:
a) Reverse transcriptase PCR (RT-PCR)
b) Real-time PCR
c) Nucleic acid sequence-based amplification (NASBA)
d) Ligase chain reaction (LCR)

23. Identify the correct sequence of steps involved in polymerase chain reaction (PCR):
a) Extraction, denaturation, annealing, extension
b) Denaturation, extraction, extension, annealing
c) Extension, annealing, denaturation, extraction
d) Annealing, extraction, denaturation, extension

24. All the following are true with regard to polymerase chain reaction except:
a) Conventional PCR can detect as few as 10 bacilli from a clinical sample
b) Nested PCR involves sequential use of two primer sets
c) RT-PCR can amplify dsDNA but not RNA
d) Multiplex PCR can amplify two or more target sequences simultaneously

25. All the following are true regarding molecular microbiological techniques except:
a) Rapid
b) High sensitivity
c) Low specificity
d) Lesser turn-around time

26. Which one of the following statements is false regarding molecular techniques?
a) Sensitive and specific as compared to the conventional techniques
b) Lesser turn-around time as compared to the conventional techniques
c) Require a more sophisticated laboratory than conventional techniques
d) Require less expertise

Answers: 1:b, 2:d, 3:a, 4:c, 5:d, 6:b, 7:a, 8:c, 9:c, 10:d, 11:d, 12:b, 13:a, 14:a, 15:d, 16:a, 17:c, 18:b, 19:d, 20:c, 21:a, 22:b, 23:a, 24:c, 25:c, 26:d

CHAPTER 9

Part I: General Microbiology

Epidemiology of Infectious Diseases

(MI1.3)

Chapter Navigation

Learning objectives

- Identify microbial factors causing infection
- Enumerate the types of infections
- Identify predisposing factors in the host leading to infection
- Enumerate environmental factors leading to infection
- Discuss the interaction between microbe, host, and environment that leads to infection

- The study of the epidemiology of an infection helps us to understand its occurrence, course, and transmission and to plan control measures.
- It defines the distribution and determinants of a disease in a population and the sociodemographic profile of those affected, the geographic patterns of its occurrence, and—in the case of epidemics and pandemics—the factors responsible for its occurrence and spread so that measures can be taken to control and prevent it.

TERMS RELATED TO INFECTIOUS DISEASES

- An **infection** occurs when an organism enters the body and establishes itself in the host, resulting in a disease process.
- **Pathogen:** An organism that causes a disease is called a pathogen and the process by which it produces the disease in the host is known as **pathogenesis**. A pathogen may be a bacterium, virus, parasite, fungi, or prion.
- **Infectious disease:** A disease resulting from an infection caused by the direct effect of a pathogen is known as an infectious disease.
- **Epidemiological triad:** The triad involved in the epidemiology of infections is as shown in Fig. 9.1.

Communicable Disease

An infection is termed communicable when the causative microbial agent has the ability to **spread from one host to another**.

- **Communicability:** This refers to the ability of the microbe to spread from one host to another. This property does not influence the production of a disease in an individual host but determines the survival and distribution of a microbe in a community. A correlation need not exist between virulence and communicability. The development of epidemic and pandemic diseases requires the pathogenic strain to possess a high degree of communicability, e.g., SARS-CoV-2, which causes COVID-19.
- **Infectivity:** This is the ability of an infectious agent to cause a new infection in a susceptible host.
- **Contagious disease:** Infections that are highly communicable, in which the pathogen is shed in secretions—such as respiratory and intestinal diseases (e.g., measles and cholera)—are known as contagious diseases.
- **Dead-end infections:** In some instances, such as rabies, human infection represents a dead-end, beyond which point, the infection does not spread from one host to another (though there may be rare

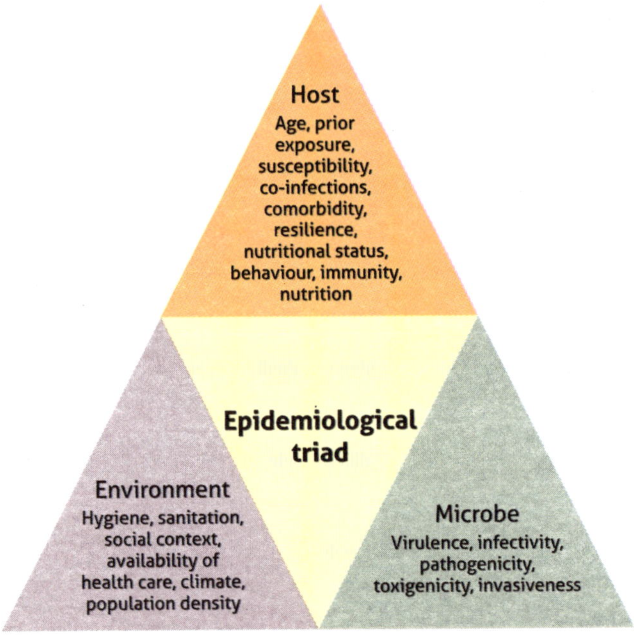

Fig. 9.1 Epidemiological triad of infection

cases of transmission through organ transplant or human bite).

- **Zoonotic diseases:** Certain infectious diseases can be transmitted from animals to humans. Such a disease is known as a zoonosis, e.g., rabies is transmitted by dogs.
- **Non-communicable infectious disease:** This is a disease that does not spread from one person to another, e.g., tetanus, which is caused by *Clostridium tetani*. Such a disease is only transmitted through contact with soil or objects containing spores of *C. tetani*, which enter the body through a skin wound.

Terms Related to the Epidemiology of Infectious Diseases

- **Primary infection:** This refers to the initial infection by a pathogen in a host.
- **Re-infection:** Subsequent infection by the same organism is called a re-infection, e.g., malaria can recur in an individual on repeated exposure.
- **Secondary infection:** When a new pathogen sets up an infection in a host whose resistance is lowered by a pre-existing infectious disease, it is called a secondary infection, e.g., fungal infection of the lung in a patient with pulmonary tuberculosis.
- **Focal infection:** More appropriately called focal sepsis, this indicates a condition caused by infection or sepsis at localised sites, e.g., appendicitis or tonsillitis.
- **Cross-infection:** When a patient who has an existing infection or disease acquires a new infection transmitted from another host or another external source, it is called a cross-infection.
- **Healthcare-associated infections/nosocomial infection:** Infections occurring in a hospital or healthcare setting after two days of admission are called healthcare-associated infections.
- **Iatrogenic infection:** These are infections resulting from investigative, therapeutic, or other procedures, which are physician-induced, e.g., abscesses resulting from injections. The use of unsterile syringes and surgical instruments or equipment or the transfusion of infected blood or organs could result in such an infection. Laboratory personnel handling infectious material are also at risk of acquiring infections if the biological material is not handled with care.
- **Endogenous or exogenous infection:** These terms refer to the source of infection—whether it originated within or is acquired from outside the host's own body, e.g., peritonitis following intestinal perforation (endogenous flora from the gut).

Based on the clinical effects they produce, infections may be described as follows:

- **Inapparent infection (subclinical infection):** This is an infection in which the clinical effects are not apparent.
- **Atypical infection:** This is an infection in which the typical or characteristic clinical manifestations of that particular infectious disease are not present (e.g., *Mycoplasma* causing atypical pneumonia).
- **Latent infection:** Some microorganisms, following infection, may remain in the tissues in a latent or hidden form, proliferating and producing clinical disease when the host's resistance is lowered (e.g., herpes zoster).
- **Colonisation:** Some organisms live on the host's body without causing infection but have the potential to become pathogenic if the host's immunity is compromised. For example, coagulase-negative staphylococci are colonisers of the skin but can cause bloodstream infection in a patient on chemotherapy who is on an intravenous catheter line.
- **Infestation:** This chiefly refers to ectoparasites like lice, fleas, and ticks, which live parasitically on the host, producing harmful effects. It also refers to intestinal parasitic infestations by helminths.

SOURCES OF INFECTION

Humans acquire infections from one of the following sources (Fig. 9.2):

1. Other humans
2. Animals
3. Insects
4. Inanimate objects
5. Environment

Humans as Sources of Infection

The commonest sources of infection in humans are humans themselves. The pathogen may originate from a patient or a carrier.

Carrier: A person who harbours the pathogenic microorganism without suffering any ill effect on account of it is known as a carrier. Carriers may be of the following types:

- **Healthy carrier**—one who harbours the pathogen but has not suffered from the disease caused by it.
- **Convalescent carrier**—one who has recovered from the disease and continues to harbour the pathogen in his body.
- **Temporary carriers**—convalescent carriers in whom the microorganisms persist for less than six months.
- **Chronic carrier**—a state that may last for several years and sometimes even for the rest of the person's life.
- **Contact carrier**—a person who acquires the pathogen from a patient.
- **Paradoxical carrier**—a carrier who acquires a pathogen from another carrier.

Animals as Sources of Infection

- A number of pathogens can infect both humans and animals.

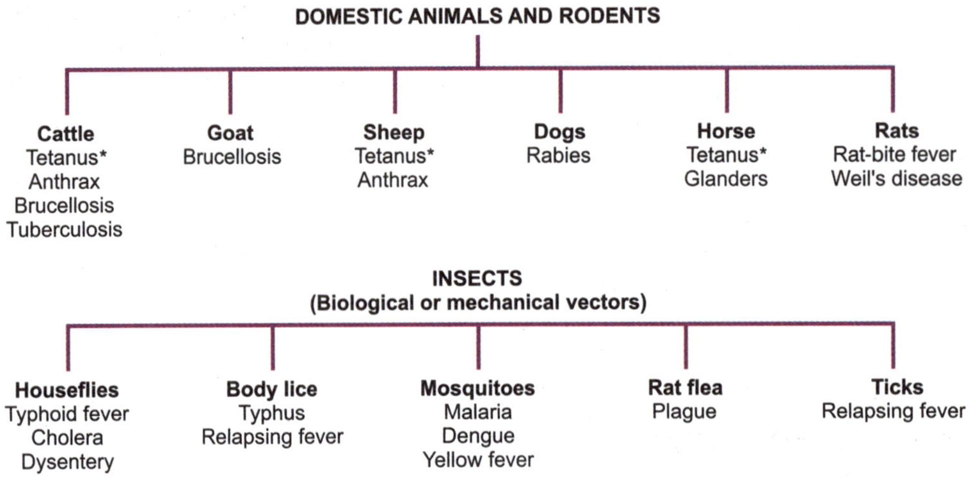

*Tetanus bacilli are shed in the feces of these animals

Fig. 9.2 Infections transmitted by animals and insects

- Animals may, therefore, act as sources of human infection.
- **Reservoir hosts**—in some instances, the infection in animals may be asymptomatic. Such animals serve to maintain the microbe in their system and act as reservoirs of human infections, e.g., brucellosis in cattle.

Zoonoses

Infectious diseases transmitted from animals to humans are called zoonoses. Zoonotic diseases may be of the following etiologies:
- Bacterial (e.g., plague from rats)
- Viral (e.g., rabies from dogs)
- Protozoal (e.g., toxoplasmosis from cats)
- Helminthic (e.g., hydatid disease from dogs)
- Fungal (e.g., zoophilic dermatophytes from cats and dogs)

Insects as Sources of Infection

- **Arthropod-borne diseases:** Bloodsucking insects may transmit pathogens to human beings, causing disease.
  Vectors are insects such as mosquitoes, ticks, mites, flies, fleas, and lice that transmit infections. Vectors may be of the following types:
  - **Mechanical vectors:** The pathogen is transmitted mechanically by the vector (e.g., transmission of dysentery or typhoid bacilli by the domestic fly).
  - **Biological vectors:** The pathogen multiplies in the body of the vector, often undergoing part of its developmental cycle in the vector's body (e.g., *Aedes aegypti* mosquito in dengue and female *Anopheles* mosquito in malaria).
- **Extrinsic incubation period:** The period between the entry of the pathogen into the biological vector and the vector becoming infective is known as the extrinsic incubation period. The infection is maintained in insects by the transovarial or transstadial passage (e.g., ticks in relapsing fever).

Transmission Through Inanimate Objects

The mechanical transfer of microbial agents can occur through fomites, medical equipment and instruments, and unhygienic, contaminated objects and surroundings.

Infections Acquired from the Environment

- **Soil:** Some pathogens can survive in the soil for very long periods. Spores of the tetanus bacilli remain viable in soil for several decades. Fungi (*Histoplasma capsulatum*, *Nocardia asteroides*) and parasites such as roundworm and hookworm survive in soil and cause human infection.
- **Water:** Water may act as a source of infection either due to contamination with pathogenic microorganisms (*Vibrio cholerae*, hepatitis A virus) or due to the presence of aquatic vectors (*Cyclops* in Guinea worm infection).
- **Food:** Food may be a source of infection if it contains pathogens. This may be due to external contamination (food poisoning by staphylococci) or pre-existing microorganism, e.g., infection in meat or other animal products (salmonellosis).

ROUTES OF TRANSMISSION OF INFECTION

Contact

- **Direct contact:** Infection may be acquired by direct contact, i.e., pathogens may be transmitted from person to person, e.g., sexually-transmitted diseases such as syphilis and gonorrhea.

- **Indirect contact:** Transmission of an infection may occur through **fomites**—inanimate objects such as clothing, bedding, towels, etc.
- **Inhalation:** Pathogens from the nose or throat of patients are shed during sneezing, speaking, or coughing. Large drops of secretions fall to the ground and dry there. Pathogens resistant to drying may remain viable in the dust and act as sources of infection. Small droplets, under 0.1 mm in diameter, evaporate immediately to become minute particles or **droplet nuclei** (usually 1–10 µm in diameter), which remain suspended in the air for long periods and act as sources of infection. Respiratory infections such as influenza and tuberculosis are transmitted by inhalation.
- **Ingestion:** Intestinal infections are generally acquired by the ingestion of food or drink contaminated by pathogens. Infections acquired by ingestion may be waterborne (e.g., cholera), foodborne (e.g., food poisoning), or may spread through unclean hands (e.g., dysentery).
- **Inoculation:** Pathogens may be introduced directly into the tissues through a breach in the skin and soft tissue. For example, tetanus spores are implanted in deep wounds following a penetrating injury contaminated by soil.
- **Congenital:** Some pathogens can cross the placental barrier and infect the fetus in utero. This is also known as **vertical transmission**, e.g., congenital syphilis and intrauterine infection with the rubella virus. Some of these infections disturb the development of the fetus and may result in birth defects; these are known as **teratogenic infections**.

HOST-PARASITE INTERACTION

When an individual is exposed to a pathogen, the outcome depends on the interaction between the host and the parasite/pathogen (Fig. 9.3).

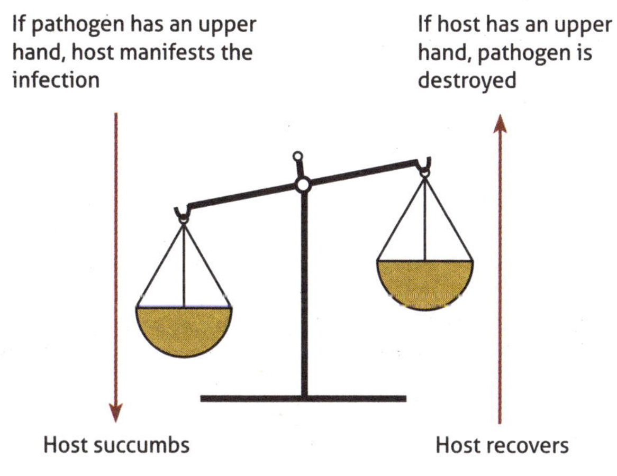

Fig. 9.3 Outcomes of host–pathogen interaction

MICROBIAL FACTORS PREDISPOSING TO INFECTION

Bacteria produce multiple virulence factors that permit them to cause disease. The genes encoding for these factors are often present on plasmids and may be transferred by bacteriophages. These include:

- **Pathogenicity:** This refers to the ability of a microorganism to produce disease.
- **Virulence:** This refers to the lethal property of the microorganism. Virulence is not constant and may undergo spontaneous or induced variation.
- **Exaltation:** The enhancement of virulence produced in vitro by serial passage in susceptible hosts is known as exaltation.
- **Attenuation:** The virulence of a microbe can be reduced by **passage or culturing**. Repeated cultures in artificial media, culturing in high temperature or in the presence of weak antiseptics, desiccation, or prolonged storage removes the virulence of a strain without affecting its infectivity. This process is used to prepare **live-attenuated vaccines**.

Factors Influencing Virulence

Virulence is determined by one or more of the following properties of microorganisms:

- **Adhesion:** The attachment of the bacteria/parasite to body surfaces is the first step in pathogenesis. This attachment is a specific reaction between surface receptors on host cells and adhesive structures (ligands) called **adhesins** on the surface of bacteria.
  - Adhesins may occur as organised structures such as fimbriae or pili or as colonisation factors. This specific adhesin may account for the tissue tropisms and host specificity exhibited by many pathogens.
  - Loss of adhesins renders the strain avirulent.
  - Adhesins are made up of protein and are antigenic in nature. Specific immunisation with adhesins has been attempted as a method of prophylaxis in some infections.
- **Invasiveness:** This is the ability of a pathogen to spread in the host tissues after establishing infection. Highly invasive pathogens spread from the site of infection to other parts of the body (e.g., streptococcal septicemia following localised wound infection). Less invasive pathogens cause localised lesions (e.g., abscesses due to staphylococci). Some pathogens, though capable of causing serious or even fatal diseases, lack invasiveness altogether (e.g., *C. tetani* remains confined to the site of entry and produces the disease by elaborating a potent toxin.
- **Toxigenicity:** Bacteria produce two types of toxins—exotoxins and endotoxins (Table 9.1).

Table 9.1 Differences between exotoxin and endotoxin

Property	Exotoxin	Endotoxin
Source	Largely, gram-positive bacteria, a few gram-negative bacteria (Shiga's strain of *Shigella dysenteriae, Vibrio cholerae,* and enterotoxigenic *E. coli*)	Gram-negative bacteria
Chemical nature	Protein, extracellular component of the cell Excreted by the cell	Lipopolysaccharide Integral part of the cell wall
Potency	High	Low
Action of heat	Labile at 60°C	Heat-stable even at 100°C
Diffusion into surrounding medium	Easily diffuse into the surrounding medium	Does not diffuse
Mechanism of action	Acts locally, away from the site of production by binding to specific receptors on host cell surface	Acts systemically to elicit an inflammatory response by induction of interleukin 1 (IL-1) production
Antigenicity	Highly antigenic	Poorly antigenic
Neutralisation	IgG neutralises toxins in tissues while IgA neutralises toxins at mucosal surfaces within the body	Not completely neutralised by the homologous antibodies
Specific receptors for binding to toxin	Binds to specific receptors	No specific receptors
Toxoid	Yes, by the addition of formaldehyde	Not possible
Effects produced	No fever or systemic effects	Can cause endotoxic shock marked by fever, leukopenia, thrombocytopenia, significant drop in blood pressure, circulatory collapse, and bloody diarrhea, leading to death
Genes responsible for encoding production	Genes located extra-chromosomally on plasmids and transferred by bacteriophages For example, in the diphtheria bacilli, the gene for toxin production is present in beta or other tox+ corynephages	Genes located on bacterial chromosomes

- **Microbial products:** Some bacterial products other than toxins (Table 9.2), though devoid of intrinsic toxicity, may contribute to virulence by inhibiting the mechanisms of host resistance.
- **Bacterial appendages:** Capsulated bacteria such as pneumococci, *K. pneumoniae*, and *H. influenzae* are not readily phagocytosed. Some bacterial surface antigens such as the Vi antigen of *S.* Typhi and K antigens of *E. coli* also help the bacteria to withstand phagocytosis and the lytic activity of complements.
- **Biofilms**
  - These are well-organised microcolonies of bacteria enclosed in self-produced extracellular polymer matrices known as **glycocalyx**.
  - Free-floating bacteria come in contact with medical devices and attach to them with pili.
  - They then aggregate, multiply, and secrete extracellular polymers and, in turn, get encased under the polymers. These polymers act as a barrier to the action of antibiotics.
  - The matrices are separated by channels that remove water and deliver nutrients.
  - They may be classified as **adherent, clinging, and submerged biofilms.**
  - Depending on the number of bacteria making them up, they classified as **monomicrobial biofilms and polymicrobial biofilms.**
  - Biofilm formation in medical devices is prevented by using sonication, antibiotics, catheter lock solutions, and catheter flushing.

Mechanism of Pathogenicity

- **Infecting dose:** The development of infections requires that an adequate number of bacteria enter the host.
- **Route of infection:** Some bacteria such as streptococci can initiate infection following entry by any mode. Others can survive and multiply only when introduced by the optimal routes. For example, *V. cholerae* is infective orally but is unable to cause infection when introduced subcutaneously. Bacteria also differ in their sites of selection in the host body after introduction into tissues (organ- or tissue-specific).

Table 9.2 Various enzymes produced by bacteria

Enzyme	Action	Bacteria producing it
Coagulase	Thrombin-like enzyme which prevents phagocytosis by forming a fibrin barrier around the bacteria and walling off the lesion	*Staphylococcus aureus*
Fibrinolysins	Breaks down the fibrin barrier in tissues and promotes the spread of infections	*Staphylococcus aureus*, Group A beta-hemolytic *Streptococcus* (*Streptococcus pyogenes*)
Hyaluronidases	Split hyaluronic acid, which is a component of intercellular connective tissue, and thereby facilitate the spread of infection along tissue spaces	*Streptococcus, Staphylococcus, Peptostreptococcus, Propionibacterium,* and *Clostridium*
Leucocidins	Damage polymorphonuclear leukocytes, dendritic cells, T lymphocytes, and natural killer cells	*Staphylococcus aureus*
Hemolysins	Capable of destroying erythrocytes	*Staphylococcus aureus, E. coli,* and many others

Note: Fungi and parasites also produce enzymes and toxins.

- **Bacteremia/viremia/parasitemia/fungemia:** These terms refer to the circulation of microbial agents in the blood. In bacterial infections, transient bacteremia is a frequent event even in healthy individuals and may occur during chewing, brushing of teeth, or straining at stools. Bacteria are immediately mopped up by phagocytic cells and are unable to initiate infection. However, bacteremia of greater severity and longer duration is seen during generalised infections such as typhoid fever.
- **Septicemia:** Bacteria circulate and multiply in blood, form toxic products, and cause fever.

The infecting dose is defined in terms of:

- **Minimum infecting dose (MID):** This is the minimum number of bacteria required to produce clinical evidence of infection (e.g., a minimum of 10 *Shigella* dysentery bacilli can cause an infection, whereas $>10^{5-6}$ non-typhoidal *Salmonella* are required to establish an infection).
- **Minimum lethal dose (MLD):** This is the minimum number of bacteria required to cause the death of a susceptible animal under standard conditions. As animals exhibit considerable individual variation in susceptibility, these doses are more correctly estimated in statistical expressions as **ID 50 (infectious dose 50) and LD 50 (lethal dose 50)**—the doses required to infect and kill 50 per cent of the animals tested under standard conditions.

EPIDEMIOLOGY OF INFECTIONS

- **Sporadic infection:** An infection that occurs in a single individual due to some predisposing factor in the individual is said to be sporadic, e.g., tetanus infection acquired due to exposure to *C. tetani* spores in the soil. These infections are managed with specific interventions.
- **Outbreak:** It starts with an **index case** (first person to be affected) and spreads to several persons who come into direct contact with the case. From them, it spreads to others by contact. A **cluster** of cases occur within the geographic area, e.g., the Nipah virus outbreak in Kerala, which was effectively controlled before it became an epidemic. Unless containment measures are taken, outbreaks may spread to become epidemics and pandemics.
- **Endemic infection:** An infection that is prevalent within a localised area due to the climatic condition or presence of suitable vectors in the area is said to be endemic to that area. For example, malaria is endemic in areas where *Anopheles* mosquitoes can breed and transmit the infection.
- **Prosodemic infection:** A slow-spreading infection transmitted from person to person, e.g., cerebrospinal fever, is said to be prosodemic.
- **Epidemic infection:** This is an infection that affects many persons simultaneously in a geographically defined area, e.g., dengue, Japanese encephalitis during the cooler seasons, or cholera outbreaks in areas of mass gatherings.
- **Pandemic infection:** A pandemic originates as a small outbreak of an infection, which soon spreads widely and affects populations in regions, countries, and continents across the globe. This occurs by human-to-human transmission and is accelerated by travel and migration. To label any outbreak as a pandemic, it must be recognised as such by an agency such as the World Health Organization based on set criteria with regards to number of active cases, countries affected, etc. For example, COVID-19 was declared a pandemic on 11 March 2020 by the WHO despite the fact that it started to spread in China in January 2020. Pandemic risk assessment evaluates the

potential risk of an infection to become a pandemic. It is focused on three aspects:

- The emergence of a strain
- Its ability to infect an individual
- Its public health impact

Factors essential to assess the risk of a pandemic are grouped into three categories:

i. **Properties of the pathogen (agent):** Genomic analysis, transmission in animal models, and antiviral treatment options
ii. **Attributes of the population (host):** Population immunity, disease severity and pathogenesis, and antigen relatedness
iii. **Epidemiology of the pathogen (external environment):** Global distribution and infection in animals and humans

Based on a scoring system, a pathogen is categorised as **low-, moderate-, or high-risk** with respect to its potential for human-to-human transmission and causing a pandemic.

MANAGEMENT OF PANDEMICS

- **Assessment of threat and preparedness:** An epidemiological survey to gather information about an outbreak or the occurrence of a reportable infectious disease follows a standardised process. This includes studying the population details (host factors) and environmental factors as mentioned below:
  - **Planning and coordination:** This is done to integrate pandemic preparedness into national emergency preparedness frameworks.
  - **Situation monitoring and assessment:** This is done in three phases:
    i. **Before the onset** of the pandemic, information about the risk of disease is disseminated.
    ii. **During the pandemic**, the pathogen is studied, the disease is monitored, and its spread is traced. Collecting data on genetic studies is also important at this stage.
    iii. The effect of **control measures** to stop the spread of infection is assessed.
- **Reducing the spread of disease:** This involves implementing measures to prevent the spread of the disease by influencing individual, household and societal behaviour to a specific pathogen (for example, in the case of SARS-CoV-2, wearing masks, sanitising, and maintaining physical distance). This also includes travel restrictions or other international travel measures to contain spread. Pharmaceutical interventions in the form of prophylactic therapy and vaccination programmes (targeted or mass vaccination) to increase herd immunity also form an important part of control.
- **Continuity of healthcare provision:** This is a crucial step in the management of an outbreak. It is imperative to ensure accessibility and availability of healthcare services with equitable distribution across the population. Further, non-pandemic healthcare systems should be functional with channels open to treat the non-pandemic illnesses too.
- **Communication:** This is necessary to disseminate relevant information early in the pandemic, to take appropriate preventive measures, provide health recommendations and interact with the public while maintaining transparency. Communication is vital to minimise economic and social disruption.

The steps involved in the management of a pandemic are as follows:

- The first case in a cluster confirmed to have the infection is known as the **index case**. Based on the clinical characteristics of signs and symptoms of the index case, a **working case definition** is made.
- These set criteria are used to **identify** or suspect individuals with similar findings.
- Such individuals are then **screened** for the disease by appropriate diagnostic tests to confirm the infection.
- Once confirmed, steps are taken to **isolate and treat** these patients.
- **Contact tracing** and identification are also important steps in pandemic management.
- **Awareness programmes** are required to effectively disseminate information to the community regarding preventive and control measures.
- **Genomic sequencing** plays an important role in the study of the spread and transmission and the detection of mutants. This is necessary to prepare policies for vaccine preparation.
- **Vaccines** are rolled out for either specific groups or as mass vaccination (as in the COVID-19 pandemic).

Ring vaccination: This is the process of vaccinating and monitoring a ring of people around each infected individual. This strategy controls an outbreak by forming a buffer zone of vaccinated individuals around an infected individual or cluster so that infection cannot cross this barrier. By doing so, the rest of the individuals in the area or community are protected. This was successfully done during the Ebola outbreak in 2014.

Key Points

✓ Triad of an infection includes the host → microbe → environment.
✓ Primary infection: This is the initial infection in a host by a parasite.
✓ Re-infection: Subsequent infections following primary infection are known as re-infections.

- ✓ Secondary infection: When a new parasite establishes an infection in a host whose resistance is lowered by a pre-existing infectious disease, it is known as a secondary infection.
- ✓ Focal infection: More appropriately called focal sepsis, this indicates a condition caused by infection or sepsis at localised sites, e.g., appendicitis or tonsillitis.
- ✓ Cross-infection: This is a new infection in a patient already suffering from one infection or disease, which is transmitted from another host or another external source. Most hospital-acquired infections are due to cross-infection. Cross-infections occurring in hospitals are called healthcare-associated infections (HAI)
- ✓ Infectious diseases, in relation to their epidemiological characteristics, are referred to as:
  - Endemic (localised to a particular area)
  - Epidemic (outbreak of a disease in a defined geographic area or community)
  - Pandemic (outbreak involving a global community across countries)
- ✓ Iatrogenic infection is one that results from investigative, therapeutic or other procedures which are physician-induced, e.g., abscesses resulting from injections.
- ✓ Endogenous or exogenous infection—categorisation depending on whether the source of infection is from within or outside the host's own body, e.g., peritonitis following intestinal perforation (endogenous flora from the gut).
- ✓ The commonest sources of infection in humans are humans themselves. The parasite may originate from a patient or a carrier.
- ✓ Factors affecting infectivity include virulence of the microorganism and invasiveness.
- ✓ Infections can occur sporadically, as outbreaks in clusters or as endemics, epidemics, or pandemics.
- ✓ Pandemic management includes epidemiological survey, contact tracing, enforcing preventive measures, dissemination of information and control through vaccination and through planning, situation monitoring, reducing spread of disease, and ensuring continuity of healthcare services.

Questions

Short answer questions

1. Describe the triage responsible for an infection and the factors influencing an outbreak.
2. Define the carrier state in humans. Enumerate the different types of carriers.
3. Define an infection. What are the different types of infections?
4. What are the microbial factors that influence an outbreak of infectious disease?
5. Define the different types of disease outbreaks.
6. What are the steps in the management of a pandemic?

MCQs

1. **Which of the following influence an infection?**
   a) Host factors
   b) Environmental factors
   c) Microbial factors
   d) All of these
2. **Secondary infection is when:**
   a) A new parasite sets up an infection in an individual already infected by another
   b) After recovery, the same microbe causes the infection a second time
   c) Infection resurges in the host due to non-response to a drug
   d) A carrier transmits the organism he is carrying to another individual
3. **An epidemic of an infection in a community is defined as:**
   a) Localised outbreak of a disease in a community
   b) An infection that occurs round the year due to presence of a vector in the area
   c) Large outbreak involving several countries
   d) None of these
4. **Which of the following is not a part of pandemic management?**
   a) Contact tracing
   b) Identifying clusters
   c) Maintaining confidentiality
   d) Implement isolation
5. **Attenuation of a microorganism means:**
   a) Enhancement of its virulence
   b) Reduction of its virulence
   c) Its ability to produce teratogenicity
   d) Ability of a microorganism to produce disease
6. **Which of the following statements is not true regarding commensals?**
   a) They live in complete harmony with the host.
   b) They can produce disease when host immunity is low.
   c) They are free-living microbes that sustain on dead or decaying organic matter.
   d) They behave as facultative pathogens.
7. **Cross-infection means:**
   a) Subsequent infection by the same parasite
   b) A new infection occurring in a patient already suffering from one infection
   c) Infection by a parasite in a host whose immunity is lowered by a pre-existing disease
   d) Initial infection in a host by a parasite
8. **Who is a paradoxical carrier?**
   a) A carrier who acquires the pathogen from a patient
   b) An individual who is a carrier for less than six months
   c) A carrier who acquires a pathogen from another carrier
   d) An individual who is a carrier for more than six months
9. **Which of the following is a viral zoonotic disease?**
   a) Rabies
   b) Plague
   c) Hydatid disease
   d) Toxoplasmosis
10. **Which of the following is not a property of exotoxins:**
    a) Heat-labile
    b) Protein
    c) Highly antigenic
    d) Lipopolysaccharide
11. **Which of the following statements is false regarding biofilms?**
    a) Well-organised microcolonies
    b) Mostly monomicrobial
    c) Contain glycocalyx
    d) Barrier to antibiotic action
12. **The term prosodemic disease refers to:**
    a) An outbreak of disease in bird population
    b) A disease constantly present at a high incidence
    c) A disease affecting large numbers of people in a wide geographic area
    d) A disease which spreads slowly
13. **Lysozyme is present in all the secretions except:**
    a) Tears
    b) Saliva
    c) Sweat
    d) Mucus
14. **Which of the following statement best describes the definition of iatrogenic infections?**
    a) These are infections that are physician-induced.
    b) These are infections in which the clinical effects are not apparent.

c) These are infections acquired by a patient after 48 hours of stay in the hospital.
d) These are infections in which the typical clinical manifestations are not present.

15. Animals that serve to maintain the microbe in their system are known as:
a) Biological vectors
b) Reservoir hosts
c) Mechanical vectors
d) Contagion

16. Which of the following is correct regarding exotoxins?
a) These are heat-stable lipopolysaccharides that form an integral part of the cell wall.
b) They are chiefly produced by gram-positive bacteria.
c) They cannot be toxoided.
d) They are poor antigens.

17. Endemic infection refers to:
a) A slow-spreading infection transmitted from person-to-person.
b) Infection that affects many persons simultaneously in a geographically defined area.
c) Infection that is prevalent within a localised area.
d) Infection that spreads worldwide in a short duration.

18. The minimum number of bacteria required to produce clinical evidence of infection is known as:
a) Minimum bactericidal dose
b) Minimum inhibition dose
c) Minimum lethal dose
d) Minimum infection dose

19. Bacteria such as *Klebsiella* are not easily phagocytosed. This is due to presence of:
a) Capsule
b) Fimbriae
c) Flagella
d) Thick cell wall

Answers: 1:d, 2:a, 3:a, 4:c, 5:b, 6:c, 7:b, 8:c, 9:a, 10:d, 11:b, 12:d, 13:d, 14:a, 15:b, 16:b, 17:a, 18:d, 19:a

CHAPTER 10

Part I: General Microbiology

Sterilisation and Disinfection

(MI1.4 and 1.5)

Chapter Navigation

Learning objectives

- Define sterilisation and disinfection and discuss their applications in healthcare settings
- Enumerate and discuss the principles and methods of sterilisation and disinfection in healthcare settings
- Discuss the impact of water, air, food, and milk on human health and the methods to contain related infections

Microorganisms are ubiquitous in nature. They live in abundance in the environment, on inanimate objects, and on the surfaces of the human body. Since they are responsible for infection, contamination, and decay, it is necessary to remove or destroy microorganisms in healthcare settings and other spaces from where they may be acquired by human beings.

DEFINITIONS

- **Sterilisation:** This term refers to any process by which an article, surface, or medium is freed of all microorganisms in the vegetative and spore states. It involves the elimination, inactivation, or killing of microorganisms. The method of sterilisation depends on:
  - The purpose for which it is carried out, e.g., for the sterilisation of medical instruments
  - Material to be sterilised, e.g., glass, media, medical equipment
  - Nature of the microorganisms to be removed or destroyed, e.g., bacteria, spores, viruses
- **Disinfection:** The process of destroying all pathogenic organisms is called disinfection. Some disinfectants may not be able to destroy spores.
- **Asepsis:** The complete absence of viable pathogenic microorganisms in any environment is referred to as aspepsis.
- **Antiseptics:** These are agents that can be safely applied on the skin or mucous membrane to prevent infection; they act by inhibiting the growth of bacteria.
- **Bactericidal agents (or germicides):** These are chemicals that can kill bacteria.
- **Bacteriostatic agents:** These are chemicals that prevent the multiplication of bacteria but do not necessarily kill them. With these agents, bacteria may remain viable for a certain amount of time. A chemical that is bacteriostatic at a particular concentration may become bactericidal at another concentration.
- **Decontamination:** The process of rendering an article, area, object, or body surface free of contaminants, including microbial, chemical, radioactive, and other hazardous materials is known as decontamination.
- **Bioburden:** The number of microorganisms that are present before disinfection or sterilisation.
- **Cleaning:** The process of reducing the bioburden by the physical removal of organic matter (washing and scrubbing).

Spaulding's Classification

- **Spaulding's classification** has been used to classify the devices used on the patient and determine the methods to be used for their disinfection or sterilisation.
- E. H. Spaulding classified the sterilisation and disinfection procedures for the intended use of the object as **critical, semi-critical, and non-critical** (Table 10.1).

The resistance of various microbes to disinfection and sterilisation and the required levels of sterilisation and disinfection are presented in Fig. 10.1.

Table 10.1 Spaulding's classification for sterilisation of items used in hospitals

Category	Definition	Items for sterilisation	Sterilising/Disinfecting agents
Critical items	Medical devices that enter sterile tissue or the vascular system and pose a high risk of infection if the article is contaminated	Surgical instruments, cardiac catheters, implants	Autoclaving, ethylene oxide or hydrogen peroxide gas, or plasma (if heat-sensitive)
Semi-critical items	Articles that come in contact with mucous membranes or non-intact skin	Endoscopes, bronchoscopes, thermometers, pressure cuffs for neonates	High-level disinfection with glutaraldehyde, alcohol
Non-critical items	Items that come in contact with intact skin but not mucous membranes	Bedpans, blood pressure cuffs, bed rails, bedside tables	Low-level disinfectants—water and detergent

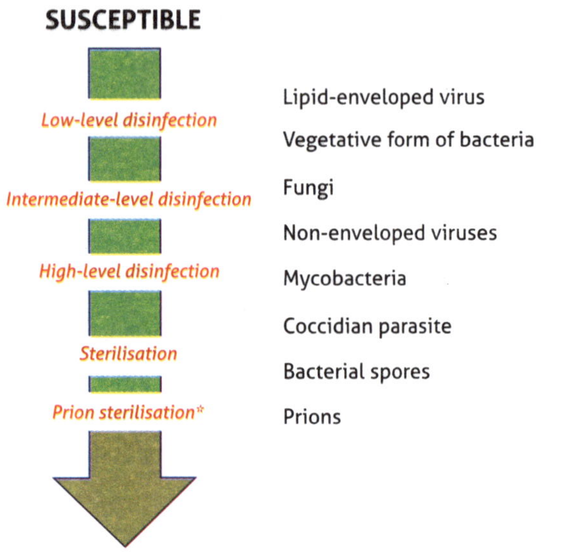

Fig. 10.1 Resistance of microorganisms to disinfection and sterilisation and the level of disinfection or sterilisation

*Combination of sodium hydroxide and autoclaving or autoclaving at 134°C × 18 minutes or 132°C × 1 hour

STERILISING AGENTS

Sterilisation can be achieved by either physical or chemical processes. Steam under pressure, dry heat, ethylene oxide (ETO) gas, hydrogen peroxide gas, plasma, vaporised hydrogen peroxide, and liquid chemicals are the principal sterilising agents that are used in healthcare facilities.

FACTORS THAT INFLUENCE STERILISATION

- **Bioburden and type of pathogen:** Spore-forming organisms are most resistant to sterilisation. The larger the number of microbes, the more is the time required to destroy all of them.
- **Cleaning:** Instruments, if not cleaned properly, can result in a higher bioburden, protein load, and salt concentration, thereby decreasing the efficacy of sterilisation.
- **Protein:** Residual protein decreases the efficacy of sterilisation. This can be reduced by cleaning.
- **Biofilm:** Biofilm formation reduces the efficacy of sterilisation as the exposure to the sterilant is impeded.
- **Contact period:** The sterilant must come into contact with microorganisms for a sufficient time to kill microbes. Device designs that prevent or inhibit this contact (e.g., sharp bends and blind lumens) can decrease the efficacy of sterilisation.
- **Product design:** The presence of a narrow lumen, as in an endoscope, impairs sterilant penetration. Forced flow through the lumen may be required to achieve sterilisation. Materials and design issues used in the construction of equipment can affect compatibility with different sterilisation processes and affect sterilisation efficacy.

I. PHYSICAL AGENTS

This includes the following:
- Sunlight
- Drying
- Heat
- Cold sterilisation
  - Filtration
  - Radiation

1. Sunlight

- It is a natural method of sterilisation.
- The **UV radiation** in sunlight is microbicidal.

2. Drying

- As 80% of a bacterial cell is constituted by water, the removal of water produces deleterious effects.
- However, **spores are not affected** by drying.

Sunlight and drying are not utilised in healthcare settings as the effectiveness of these methods is variable and cannot be measured.

3. Sterilisation With Heat

- This is the most reliable method of sterilisation.
- Materials that may be damaged at high temperatures can be sterilised by exposing them to medium heat for long periods or by repeated cycles.
- **Thermal death time** is the minimum time required to **kill a suspension of organisms at a predetermined temperature in a specified environment.** It does not include the time taken to reach the specified temperature. The time required for sterilisation is inversely proportional to the temperature applied for sterilisation.
- Factors influencing sterilisation by heat:
  - Nature of heat—dry or moist
  - Temperature and duration of exposure
  - Number of microorganisms present
  - Characteristics of the organisms such as species, strain, and the presence of spores
  - Type of material from which the organisms must be eradicated
  - The presence of organic substances, proteins, nucleic acids, starch, gelatin, sugar, fats, and oils, which increases the thermal death time
  - The presence of disinfectants and high acid or alkaline pH, which hasten bacterial killing
- The mechanism of action of dry and moist heat is presented in Table 10.2.

Sterilisation using dry heat

There are three methods in use in dry heat sterilisation—**flaming, hot air oven, and incineration.**

Table 10.2 Mechanism of action of dry and moist heat

Dry heat	Moist heat
• Protein denaturation • Damage to the cell due to the oxidation of molecules • Destruction of cell constituents • Elevation of electrolyte levels, which has a toxic effect	• Denaturation and coagulation of proteins • The latent heat liberated condenses on a cooler surface and raises the temperature of that surface, thereby denaturing its components • Steam condenses on spores, increasing their water content and resulting in the hydrolysis and breakdown of the bacterial protein
Examples: Flaming, hot air oven, incineration	Examples: Autoclave, steam steriliser, tyndallisation, boiling

Flaming and red heat

- Inoculating loops or wires, the tips of forceps, and searing spatulas can be sterilised by holding them over a Bunsen flame until they become red-hot.
- Inoculation loops carrying infective material may be dipped in a disinfectant before flaming to prevent spattering.

Incineration

- This is a method of terminal sterilisation used to destroy biomedical waste.
- Plastics such as PVC and polythene can be incinerated, whereas polystyrene materials cannot be incinerated as they emit clouds of dense toxic smoke which pollute the environment.
- The guidelines on 'Design and Construction of Biomedical Waste Incinerator' recommend various design features of an incinerator as well as air pollution control devices in a bid to regulate the quality of incinerators.

Hot air oven

- This is the most widely used method of sterilisation using dry heat.
- **Structure:** A hot air oven consists of a double-walled cabinet with insulation in-between the walls to keep in the heat and conserve energy (Fig. 10.2a and b). The oven has a heating element and is fitted with a thermostat to control the temperature and a fan that helps in the uniform distribution of heat. The cabinet is fitted with adjustable trays. There are indicators and controls for temperature and holding time.
- **Principle:** Sterilisation is achieved by conduction, convection, and radiation. The surface of the item to be sterilised absorbs the heat, which then radiates to the centre until the entire object reaches the desired temperature.
- **Working:** The oven is preheated to reach the desired temperature. Then, the items to be sterilised are

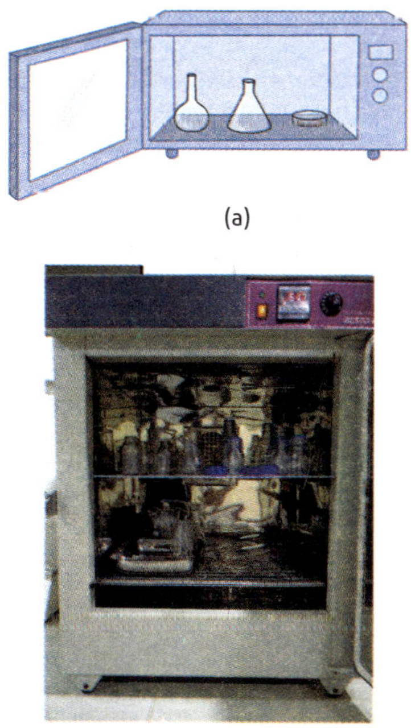

(a)

(b)

Fig. 10.2 Hot air oven: (a) diagrammatic representation and (b) a photograph (*Source*: Pushpagiri Institute of Medical Sciences and Research Centre, Thiruvalla, Kerala)

placed on the racks inside. The oven is switched off only after the holding time at the desired temperature is achieved. The holding times for the items commonly sterilised in hot air ovens are presented in Table 10.3.

- **Precautions:**
  - Glassware must be **completely dry** and wrapped in craft paper.
  - There should be **enough space between articles** to allow for efficient heat circulation.
  - The oven must be **allowed to cool slowly** for about two hours before opening the door to prevent glassware from cracking due to sudden or uneven cooling.
  - Rubber materials (except silicone rubber) cannot withstand the temperature inside a hot air oven, and hence, should not be kept in the hot air oven.
- **Sterilsation controls**
  - ***Biological***: Paper strips impregnated with 10^6 spores of *Bacillus atrophaeus* are placed in envelopes and inserted into proper packaging. After sterilisation, the strips are removed, inoculated in thioglycolate or cooked beef broth, and incubated at 37°C for five days for a sterility test under strict anaerobic conditions. When sterilisation has been effective, the spores will be eradicated, and there will be no growth.
  - ***Chemical***: The colour of **Browne's tube no. 3** changes from red to green after a holding time of 60 minutes at 160°C.
  - ***Physical***: Thermocouples and a temperature chart recorder can also be used as a sterilisation control.
- **Uses:** Hot air ovens are used for the sterilisation of glassware, petridishes, test tubes, flasks, oil, syringes, non-sharp surgical instruments, and pharmaceutical products such as liquid paraffin and dusting powder.

Table 10.3 Holding periods in a hot air oven for different items used in healthcare settings

Holding period	Items to be sterilised
Two hours at 160°C	Glassware, forceps, scissors, scalpels, and all glass syringes Swabs Pharmaceutical products such as liquid paraffin, dusting powder, and grease Materials made of silicone rubber
Two hours at 150°C	Sharp instruments such as those used in ophthalmic surgery
One hour at 150°C	Oils, glycerol, and dusting powder

Moist heat sterilisation

Moist heat sterilisation can be carried out by various methods (Fig. 10.3).

Sterilisation at temperatures below 100°C

- **Pasteurisation of milk:** Pasteurisation destroys all non-sporing pathogens such as mycobacteria, brucellae, and salmonellae.
  - ***Holder method:*** Milk is heated to 63°C and held at that temperature for 30 minutes. *Coxiella burnetii* is relatively heat-resistant and may survive the holder method.
  - ***Flash process***: Milk is heated to 72°C for 15–20 seconds and then rapidly cooled to 13°C or lower.
- **Vaccine bath:** Bacterial vaccines are heat-inactivated at 60°C for one hour in special vaccine baths. Serum or body fluids containing coagulable proteins can be sterilised by heating them for one hour at 56°C in a water bath on several successive days.
- **Inspissation:** Media are heated at 80–85°C for 30 minutes for three consecutive days. This is primarily a method of solidification rather than sterilisation—an inspissator (Fig. 10.4) is used for solidifying

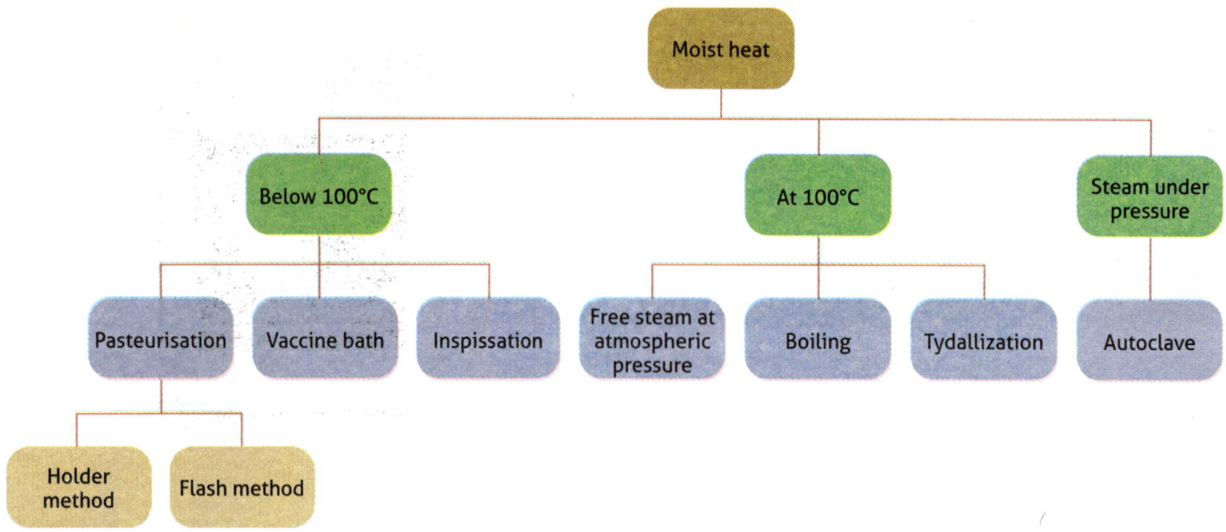

Fig. 10.3 Different methods of moist heat used for sterilisation

Fig. 10.4 Inspissator (*Source*: Intermediate Reference Laboratory for TB, SCB Medical College, Cuttack)

egg (in Lowenstein–Jensen medium) or serum-containing media.

Sterilisation at temperatures of 100°C

- **Boiling**
  - Vegetative bacteria are killed almost immediately at 90–100°C within 10–30 minutes.
  - Bacterial spores can withstand long periods of boiling.
  - Boiling is **NOT** recommended for the sterilisation of instruments used for surgical procedures as it kills only bacteria and not spores.
  - Hard water is not recommended for boiling—the addition of 2% sodium bicarbonate to water renders it soft and makes it suitable for sterilisation.
- **Free steam at atmospheric pressure (100°C)**
  - This method is useful for the sterilisation of **culture media** that may decompose if subjected to higher temperatures.
  - An autoclave with the valves left open can be used to generate free steam at 100°C.
- **Tyndallisation or intermittent sterilisation**
  Use: Media containing sugars or gelatin can be sterilised using this method.
  - **Method:** The media to be sterilised are heated at a temperature of 100°C for 20 minutes on three successive days. In between, the media is kept at 37°C to permit germination of spores.
  - **Principle:** The first exposure to heat kills all vegetative bacteria, while the spores, which germinate subsequently (during the resting period) because of the favourable medium they are in, are killed on subsequent exposure to 100°C for 20 minutes.
  - **Disadvantage:** Spores of certain anaerobes and thermophiles may not germinate and, as a result, may escape sterilisation.

Sterilisation using steam under pressure

Sterilisation by steam under pressure is carried out at temperatures between 108°C and 147°C.

Autoclave (steam steriliser)

- **Principle** (Fig. 10.5a and b)
  - When water boils, its vapour pressure equals that of the surrounding atmosphere. When the pressure inside a closed vessel increases, the temperature at which water boils also increases, generating steam at a higher temperature.
  - When this steam comes in contact with a cooler surface, it condenses and transmits its latent heat to that surface (1600 mL steam at 100°C and atmospheric pressure condenses into 1 mL of water at 100°C and releases 518 calories of heat).
  - The large reduction in volume results in the sucking in of more steam; this process continues until the temperature of that surface becomes equal to that of steam.
  - Condensed water ensures moist heat, which kills microbes by denaturing the macromolecules, particularly proteins.
- Two types of autoclaves are available:
  - **Gravity-displacement autoclave:** Steam enters the steriliser chamber and displaces the residual air through an open vent. This type of autoclave

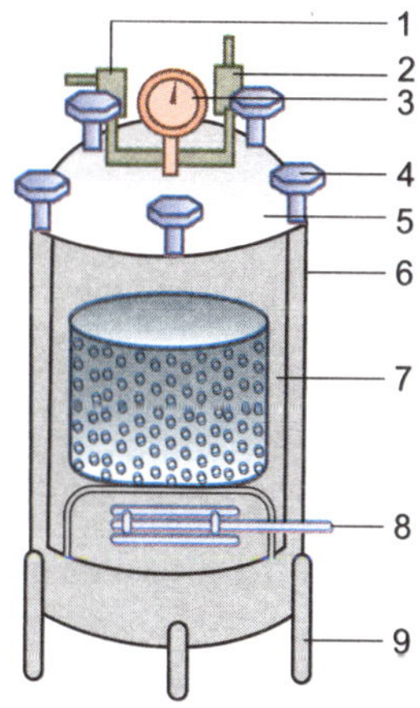

(a)

(b)

Fig. 10.5 (a) A line diagram of a simple vertical autoclave (1— release valve; 2—safety valve; 3—pressure gauge; 4—tightening screws; 5—cover; 6—body; 7—bin containing material; 8—heating element; 9—stand) and (b) a horizontal autoclave (*Source*: Department of Microbiology, Pondicherry Institute of Medical Sciences [PIMS], Puducherry)

can be vertical or horizontal. The drawback is that the method of air discharge is inefficient; materials remain moist after removal from the autoclave, a problem that is avoided with vacuum sterilisers.

- **High-vacuum autoclave:** The air is drawn out by vacuum, whereby steam penetrates rapidly. For this reason, the time required for sterilisation is less with this type of autoclave than it is with other methods. Once the sterilisation is complete, a post-cycle vacuum can be programmed for quick drying.

Table 10.4 lists the different types of autoclaves used in healthcare settings.

Table 10.4 Autoclaves used in healthcare settings

Type	Uses
Laboratory autoclave	For sterilisation of laboratory ware and media
Hospital dressing sterilisers	Best suited for parenteral solutions, injectables, and hospital dressings
Instrument sterilisers	Used to sterilise surgical and dental instruments
Rapid cooling sterilisers	May be used for hospital garments, parts of equipment, accessories, containers, culture media, laboratory waste, and bulk drugs

- **Structure**
  - The structure of an autoclave is similar to that of a pressure cooker. It is a cylinder made up of stainless steel and is fitted with a heating element, pressure gauze, and safety valve.
  - The function of the **safety valve** is to prevent accidents in case the pressure inside the chamber gets uncontrollable.
  - The lid is fitted with a rubber gasket and tightening screws.
  - The pressure **release valve** allows the release of pressure whenever required.
- **Working**
  - Sufficient water is added to the cylinder, and the material to be sterilised is placed on a tray within the bin inside (Fig. 10.5a).
  - The autoclave lid is screwed tight with the discharge tap open.
  - It is then heated by electricity. The steam–air mixture is allowed to escape freely till all the air inside has been displaced (tested by leading the escaping steam into a pail of water through rubber tubing). When all the air has been expelled, the discharge tap is closed.
  - When the steam pressure inside has risen to the desired set level, the safety valve opens, allowing the excess steam to escape. From this point, the **holding period** is calculated.
  - When the holding period is over, the heater is turned off and the autoclave is allowed to cool until atmospheric pressure is reached.
  - Finally, the discharge tap is opened slowly, allowing air into the autoclave.
- **Precautions**
  - The tap should be **opened only when the autoclave is precisely at atmospheric pressure**. If the tap is opened when the internal pressure is high, liquid media will boil violently, spill from the container, and even explode. If it is opened after the pressure inside has fallen below atmospheric pressure, an excessive amount of water will be lost from the media via evaporation.
  - Glassware should be wrapped in brown paper and tips covered with cotton before placing them in the autoclave.
- **Uses–autoclaves are used for the sterilisation of:**
  - Surgical instruments
  - Pharmaceutical products
  - Culture media
  - Hospital linen
  - Laboratory ware including glassware
  - Microbiological waste disposal
  - Heat-sensitive material
  - Aqueous solutions (between 108°C and 126°C)
- **Sterilisation control**
  - ***Biological*: *Geobacillus stearothermophilus*** is a thermophilic organism with an optimum growth temperature of 55–60°C; its spores require an exposure of 12 minutes at 121°C to be killed. This organism is used as the test organism for checking sterilisation by steam under pressure.
  - ***Physical***: Temperature and pressure logs and thermocouples are also used.
  - ***Chemical***: The **Bowie–Dick tape** is a heat-sensitive sheet placed in the middle of a linen load consisting of several layers of paper and foam. The sheet changes colour when exposed to a certain level of pressure from the saturated steam.

4. Cold Sterilisation

This term refers to methods in which sterilisation is performed at low temperatures by filtration or radiation.

Filtration

- Membranes are used to remove bacteria from **heat-labile liquids such as sera and solutions of sugars or antibiotics**.
- Since viruses pass through ordinary membrane filters (Fig. 10.6), the materials sterilised by this method are only free of bacteria.

Fig. 10.6 A membrane filter with filter pads (*Source*: Department of Microbiology, Pondicherry Institute of Medical Sciences, Puducherry)

- Generally, the pore size that prevents bacteria from passing through is estimated to be **0.2–0.22 µm**.
- Table 10.5 lists the different types of sterilisation by filtration.

Radiation

Two types of radiation are used for sterilisation—non-ionising radiation and ionising radiation.

Non-ionising radiation

- This type of radiation consists of **electromagnetic rays** with wavelengths longer than those of visible light. These rays produce low-energy radiation and are absorbed as heat.
- **Uses:**
  - ***Infrared radiation***: Rapid mass-sterilisation of pre-packed items such as disposable plastic syringes and catheters
  - ***Ultraviolet radiation***: Disinfecting enclosed areas such as biosafety cabinets in laboratories, entryways, and operation theatres

Ionising radiation

- Consists of high-energy radiations like **X-rays** and **gamma rays**, which are highly lethal to DNA and other vital constituents.
- These rays have a very high penetrative power.
- **Uses:** Gamma radiation is used by commercial plants for sterilising items like plastics, syringes, swabs, catheters, animal feeds, cardboard, oils, greases, fabric, and metal foils.

II. CHEMICAL STERILANTS

- Chemical sterilisation involves the use of chemical agents to achieve sterilisation.
- Chemical sterilisation is an important process for materials that cannot tolerate heat, high pressure, irradiation, or filtration.
- These agents are used for high level disinfection or sterilisation and have a wide range of antimicrobial activity.

Table 10.5 Types of filtration

Type	Pore size of filter	Properties	Uses
Membrane filtration	0.1–5 µm, 0.22 µm	Made up of cellulose esters or other polymers	Water purification and analysis, sterilisation and sterility testing, and for the preparation of solutions for parenteral use
Ultrafiltration (UF)	0.01–0.1 µm	Uses hydrostatic pressure to force liquids through a semipermeable membrane	Treating wastewater Concentrating proteins Chemical process separation Separating oil/water emulsions Removing pathogens from milk Clarifying fruit juices
Nanofiltration (NF)	0.001–0.01 µm	Removes monovalent ions such as chlorides or sodium salts, sugars, and dyes	Water treatment Pre-treatment for RO Pharmaceuticals Dairy
High-efficiency particulate air (HEPA) filtration	0.3 µm	Removes at least 99.97% of airborne particles measuring up to 0.3 µm in diameter Larger particles are filtered with high efficiency	In biosafety cabinets, operation theatres, and laboratories where there is a requirement for clean air
Ultra-low penetration air (ULPA) filtration	0.12 µm	Remove from the air at least 99.99% of dust, pollen, mould, bacteria, and any airborne particle measuring more than 120 nanometres	Clean rooms in precision industry, e.g., in semiconductor manufacturing units

- Chemical sterilants should be rapidly acting, compatible with materials, non-toxic, non-staining, and should have a good shelf life and be easy to use and cost-effective.
- There are two forms of chemical sterilisation methods: liquid sterilisation and gas sterilisation.
- Commonly used liquid sterilants are peracetic acid (PAA; 0.2%), glutaraldehyde (≥2.4%), ortho-phthalaldehyde (OPA; 0.55%), and hydrogen peroxide (7.5%) and a combination of peracetic acid (0.23%) and hydrogen peroxide (7.3%).
- Common gas sterilants include ethylene oxide (low temperature sterilisation technology), formaldehyde, nitrogen dioxide, and ozone. Details of these chemicals are covered in the section on disinfectants.

III. PLASMA STERILISATION

This is a relatively new method of sterilisation.

- Plasma is known as the **fourth state of matter and consists of ions, electrons, or neutral particles**.
- To generate plasma, radio frequency energy is applied to a gas, creating an electromagnetic field.
- Hydrogen peroxide vapours are introduced into this field, which generates plasma containing free radicals of hydrogen and oxygen. This state renders the articles in the environment of the plasma sterile by denaturing all microorganisms in it.

Applications

- **Heat-sensitive devices** such as arthroscopes and urethroscopes, and endoscopes (rigid and flexible) can be sterilised using plasma sterilisation.
- Plasma sterilisation is also used on electrocautery instruments, dopplers, laser probes, defibrilator paddles, thermometers, ophthalmic lenses, and harmonic cables.
- **Hollow loads** such as laryngoscopes and their blades, shaver handpieces, fiber optic light cables, and surgical power drills can also be sterlised by this method.

Table 10.6 presents the advantages and disadvantages of plasma sterilisation.

Table 10.6 Advantages and disadvantages of plasma sterilisation

Advantages	Disadvantages
• No chemical residues, so safe for environment • Easy and safe to handle • Aeration time is short	• Inability to sterilise liquids, powders, and strong absorbers • It requires specific synthetic packaging of the load • Sterilisation chamber is relatively small

IV. GASES

Ethylene Oxide (EtO)

- **Mode of action**—it acts by alkylating the amino, carboxyl, hydroxyl, and sulphydryl groups of protein molecules within microbes and spores. It also reacts with DNA and RNA (which makes it **virucidal**).
- **Advantage**—EtO is a colourless liquid with a sweet smell and a boiling point of 10.7°C. It has good penetrating ability at normal temperature and pressure (Fig. 10.7) and diffuses through several types of porous materials and readily penetrates some plastics.
- **Disadvantage**—it is highly inflammable, mutagenic, carcinogenic, and explosive at concentrations greater than 3%. The combination of ethylene oxide with 10% carbon dioxide or nitrogen makes it less explosive.
- **Uses**—EtO is used for the sterilisation of heat-sensitive equipment such as heart–lung machines, respirators, suture material, dental equipment, books, clothing, etc.

Beta-Propiolactone (BPL)

- Beta-propiolactone is an organic compound that reacts with water to produce **hydracrylic acid**.
- It is no longer used for fumigation as it is **carcinogenic**.

Hydrogen Peroxide

- It is bactericidal.
- **Fogging** with hydrogen peroxide has replaced fumigation.
- **Mode of action**—hydrogen peroxide produces destructive hydroxyl free radicals that can attack membrane lipids, DNA, and other essential cell

Fig. 10.7 A gas steriliser using ethylene oxide (*Source*: CSSD, Pondicherry Institute of Medical Sciences, Puducherry)

components and oxidises the cell wall of the organism.

- **Advantage**—it has **a short cycle time** and is non-toxic.

BIOLOGICAL INDICATORS

- Test systems that contain viable microorganisms with a defined resistance to a specific sterilisation process are used as controls.
- They are used to monitor whether the necessary conditions to kill a specified number of microorganisms for a given sterilisation process are met.
- Bacterial spores are some of the toughest microorganisms to kill. For this reason, **specific concentrations of spores are used to assess the efficacy of a sterilisation process.**

Table 10.7 lists some of the commonly used biological indicators and their uses.

Table 10.7 Biological indicators and their uses

Method	Spores of organisms used for monitoring (10^6 spores per indicator)
Dry heat	*Bacillus atrophaeus**
Moist heat (steam sterilisers)	*Geobacillus stearothermophilus* (for solids) *C. sporogenes, B. subtilis,* or *B. atrophaeus* (for liquids)
Ethylene oxide (EtO)	*Bacillus atrophaeus*
Ionising radiation	*Bacillus pumilus* *Bacillus cereus* *Lysinibacillus sphaericus*
Vapour-phase sterilisation	*G. stearothermophilus* *B. atrophaeus*
Filtration	*Serratia marcescens* *Brevundimonas diminuta*

*Previously known as *Bacillus subtilis* var niger

The commonly used methods of sterilisation are presented in Table 10.8.

Table 10.8 Commonly used methods of sterilisation in the hospital

Method	Types/Equipment	Agents for which it is used
Dry heat	Flaming	Inoculating loop/wire, tips of forceps, searing spatulas
	Hot air oven	Glassware, all glass syringes, sharp instruments, liquid paraffin, dusting powder, fats and grease

Table 10.8 Contd.

Method	Types/Equipment	Agents for which it is used
Moist heat	**Below 100°C**	
	Pasteurisation	Milk
	Inspissation	Lowenstein–Jensen medium, Loeffler's serum slope
	Vaccine baths	Vaccines and sera
	100°C	
	Boiling	Material used for domestic appliances, baby bottles, teats and caps
	Tyndallisation or intermittent sterilisation at 100°C for three consecutive days	Media containing sugars or gelatin
	Steam at atmospheric pressure	
	Steam sterilisers (Koch or Arnold steamer)	Culture media that are sensitive to high temperature
	Steam under pressure	
	Autoclave Gravity-displacement type High-vacuum type	Hospital linen, instruments, laboratory ware, media, and pharmaceutical products
Filtration	Sintered glass filters Asbestos filters Candle filters Membrane filters	Water purification and analysis, sterilisation, and sterility testing of liquids Solutions for parenteral use
Cold sterilisation	Ionising radiation X-rays Gamma rays	Plastics, syringes, swabs, catheters, oils, grease Industrial use—animal feeds, fabric and metal foils
	Non-ionising Infrared (not a method of sterilisation) Ultraviolet	Mass pre-packed materials—plastic syringes, catheters Laboratory cabinets, closed chambers
Liquid chemical sterilising agents	Glutaraldehyde Ortho-phthalaldehyde Peracetic acid Hydrogen peroxide Hypochlorous acid	Cystoscopes, endoscopes, tubings, and metal instruments
Plasma sterilisation	Free radicals of hydrogen and oxygen	Arthroscopes and urethroscopes

DISINFECTANTS

CHEMICAL AGENTS

Several chemical agents are used as **antiseptics** and **disinfectants**. These agents must possess certain properties or characteristics that make them suitable for use in healthcare settings. An ideal antiseptic or disinfectant should:

- Have a wide spectrum of activity and be effective against all microorganisms
- Be active in the presence of organic matter
- Be effective in acidic as well as alkaline media
- Have speedy action
- Have high penetrating power
- Be stable

Chemicals function on living tissues in the presence of organic matter, and hence, their activity can be affected by a number of variables. Factors that affect the potency of a disinfectant are as follows:

- Concentration of the substance
- Contact period
- pH of the medium
- Temperature
- Nature of the organism
- Presence of extraneous material

Chemicals act on bacteria, viruses, and spores by various mechanisms, which are presented in Fig. 10.8.

Alcohols

Ethyl alcohol (ethanol) and isopropyl alcohol

- This is the most frequently used sterilising agent.
- It is used widely as a **skin antiseptics** at concentrations of 60–90% in water.
- Mode of action—by **denaturing bacterial proteins**; however, they have **no effect on spores**.
- **Isopropyl alcohol** is preferred over ethyl alcohol as it is a better fat solvent and has potent bactericidal activity and is less volatile (as required for the disinfection of clinical thermometers).

Methyl alcohol

- It is effective **against fungal spores**.
- It is used to **clean cabinets and incubators** in the hospital.
- Methyl alcohol vapour is toxic and inflammable and has weak bactericidal action.

Aldehydes

Formaldehyde

- In aqueous solutions, it is markedly bactericidal, sporicidal, and virucidal.
- Mode of action—it acts **against the amino group in the protein molecule**, thereby destroying the protein.

Glutaraldehyde

- Effective against tubercle bacilli, fungi, and viruses.
- It has no deleterious effect on cement or the lenses of instruments.
- Used as a high-level disinfectant.

Ortho-phthalaldehyde

- It is a potent bactericidal agent.
- It is used to cleanse **endoscopes** between patients as it is quick, effective, and safe.
- It is used as a sterilant and a high-level disinfectant.

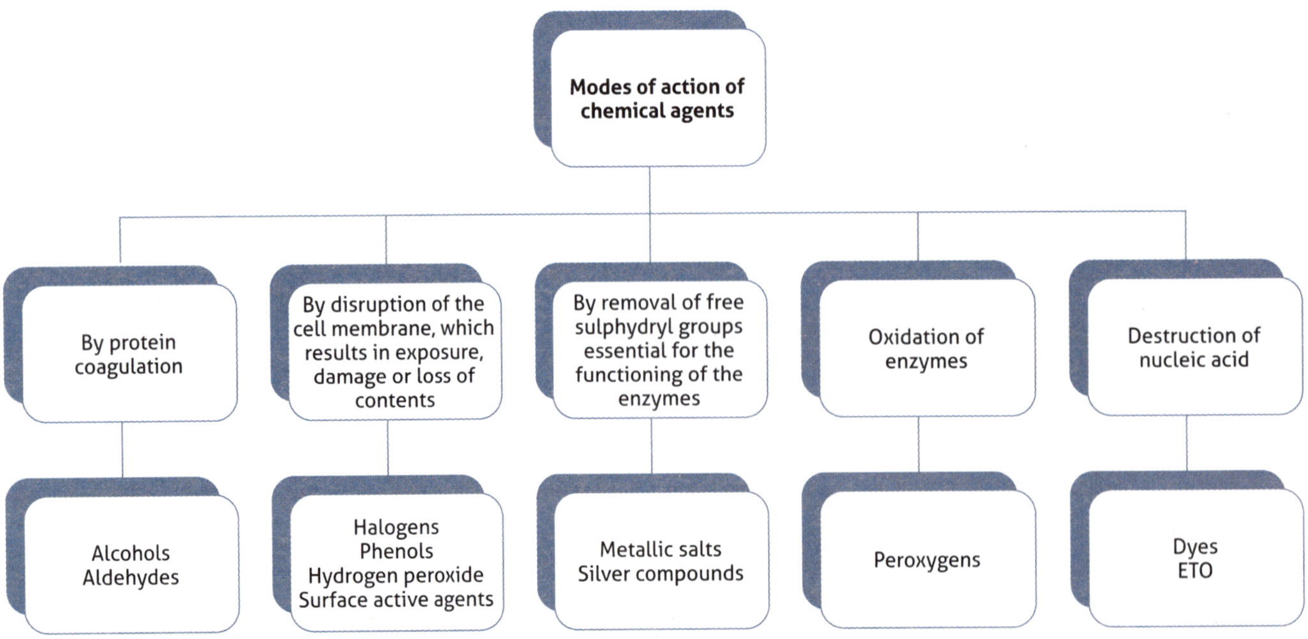

Fig. 10.8 Mechanism of action of chemical agents used for sterilisation and disinfection

Peracetic Acid

- This is a good bactericidal agent active against common antibiotic-resistant bacteria like methicillin-resistant *Staphylococcus aureus*, vancomycin-resistant *Enterococcus*, and *Clostridium difficile*.
- It is used as a sterilant and a high-level disinfectant. A combination of peracetic acid (0.23%) and hydrogen peroxide (7.5%) is used as a sterilant.

Hypochlorous Acid

- This acid is generated by the reverse reaction of sodium hypochlorite and hydrogen peroxide.
- It has bactericidal properties and is active against **biofilms**, killing microorganisms within them.

Dyes

- Two groups of dyes, **aniline** and **acridine**, are used extensively as **skin and wound antiseptics**.
- Both are bacteriostatic in high dilutions but have low bactericidal activity.
- They are more active against gram-positive than gram-negative bacteria.

Aniline dyes

- **Mode of action**—aniline dyes react with acid groups in bacterial cells.
- **Disadvantage**—they have no effect on tubercle bacilli and are inhibited by organic material.
- **Advantage**—they are non-irritant and non-toxic to tissues.
- **Examples**—brilliant green, malachite green, and crystal violet dyes.

Acridine dyes

- **Mode of action**—they **impair the DNA complexes** of the organisms and prevent replication.
- **Disadvantage**—they are not as selective as aniline dyes.
- **Advantage**—they are minimally affected by the presence of organic matter.
- **Examples**—proflavine, acriflavine, euflavine, and aminacrine.

Halogens

Iodine

- **Mode of action**—iodine disrupts oxidative phosphorylation in cells, causing the destruction of the cellular proteins, nucleic acids, and the cell walls or membranes of microorganisms.
- Compounds of iodine with non-ionic wetting or surface-active agents known as **iodophores** have better action than aqueous or alcoholic solutions of iodine.
- Idonine is used as a **skin disinfectant** (e.g., tincture iodine).
- **Advantage**—iodine is bactericidal, with moderate action against spores. It is active against mycobacteria and viruses.

Chlorine

- Chlorine and its compound **hypochlorite** have been used as disinfectants through the ages. These substances are markedly bactericidal and virucidal.
- It is used for the **disinfection of water** supply and swimming pools and in the food and dairy industries.
- **Organic chloramines are used as antiseptics** for dressing wounds.
- **Disadvantages**—some microorganisms have shown resistance to low doses of chlorine. Some parasites like *Giardia* may not get eliminated with chlorine. High doses of chlorine may cause diarrhea and other GI disturbances. Chlorine is also a skin irritant.

Phenols

- The bactericidal effect of phenols is due to their ability to damage the cell membrane and inactivate membrane-bound oxidases and dehydrogenases, leading to the lysis and death of the microorganism.
- Low concentrations of phenol precipitate proteins.

Phenol (carbolic acid)

- It is used as a common disinfectant in hospitals as **Lysol** and cresol.
- **Advantages**—phenols are not readily inactivated by the presence of organic matter; hence, they are good general disinfectants.

Chlorophenols and chloroxyphenols

- While they are not as toxic and irritant as other chemical agents, they are also less effective (inactive against *Pseudomonas*) and readily inactivated by organic matter.
- **Advantages**—they are active against most gram-positive organisms and fairly effective against gram-negative bacteria.
- **Examples**—chlorhexidine (hibitane) is a relatively non-toxic skin antiseptic; hexachlorophene is potentially toxic and should be used with care.

Surface-active Agents

- These alter the energy relationship at interfaces, producing a **reduction in surface tension** (Table 10.9) and are widely used as wetting agents, detergents, and emulsifiers.
- **Mode of action**—they act on the phosphate groups of the cell membrane and enter the cell. As a result, the membrane loses its semi-permeability, and the cell proteins are denatured. Surface-active agents act on bacteria but have no effect on spores, tubercle bacilli, and most viruses.

Table 10.9 Classification of surface-active agents

Category	Agents	Properties	Examples
Anionic	Sulphide, fluoride, bromide, iodide	Moderate activity Soaps prepared from saturated fatty acids (such as coconut oil) are more effective against gram-negative bacilli while those prepared from unsaturated fatty acids (oleic acid) have greater action against gram-positive bacteria and *Neisseria*	Alkyl sulphates, soaps
Cationic	Sodium, iron, lead	Most active at alkaline pH Organic matter reduces their action and anionic surface-active agents like ordinary soaps render them inactive	Quaternary ammonium compounds, acetyltrimethyl ammonium bromide (cetavlon or cetrimide), benzalkonium chloride
Non-ionic	Polyoxyethylene	Cause minimal skin and eye irritation	Tween, Triton™, Tergitol™
Amphoteric	Tego compounds	Used in personal care products	Betains, amphoacetates

Metallic Salts

- Though all salts have a certain amount of germicidal action, the salts of heavy metals are quite effective. Silver, copper, and mercury salts are used as disinfectants.
- **Mode of action**—salts are protein coagulants and have the capacity to combine with the free sulphydryl groups of cell enzymes.
- **Uses**—thiomersal, phenyl mercury nitrate, and mercurochrome are less toxic, and hence used as mild antiseptics. They have marked bacteriostatic, weak bactericidal, and limited fungicidal action. Copper salts are used as fungicides.

Based on their activity, disinfectants can be classified as follows:

- **High-level disinfectant:** It kills all microbial pathogens except spores—when they are present in large numbers. It may have some activity against a smaller number of spores if the contact time is increased. Examples include glutaraldehyde, peracetic acid (PAA), ortho-phthalaldehyde (OPA), and hydrogen peroxide.
- **Intermediate-level disinfectant:** It kills all microbial pathogens including mycobacteria and non-enveloped viruses but not spores. Examples include alcohol, phenolic compounds, and iodophores.
- **Low-level disinfectant:** It kills only vegetative bacteria, fungi, and lipid-enveloped viruses. This class inlcudes quaternary ammonium compounds.

The methods of antisepsis/disinfection are presented in Table 10.10.

Table 10.10 Methods of antisepsis/disinfection (by chemical agents)

Agents	Types	Uses
Alcohols	Ethyl alcohol (ethanol)	Skin antiseptics
	Isopropyl alcohol Methyl alcohol	Cleaning cabinets and incubators
Aldehydes	Glutaraldehyde	Sterilising instruments, heat-sensitive catheters, cystoscopes, bronchoscopes, rubber anesthetic tubes, plastic endotracheal tubes, and polythene tubing
	Formaldehyde (gas)	Fumigating OTs, wards, isolation rooms, and laboratories (not preferred in current practice) Under properly controlled conditions, clothing, bedding, etc., can also be sterilised
Dyes	**Aniline dyes** Brilliant green Malachite green Crystal violet	Used on skin and in some culture media (Lowenstein–Jensen medium)
	Acridine dyes Acriflavine Proflavine	Used as antiseptics on skin
Halogens	Iodine	Skin disinfectants
	Chlorine	Water supply, swimming pools
	Hypochlorites Chlorhexidine (Hibitane)	Disinfectants widely used in hospitals at different concentrations
	Hexachlorophene	Skin antiseptic, wound dressing
Quaternary ammonium compounds	Benzalkonium Chloride Cetrimide	Cystoscopes, cardiac catheters
Phenols	Lysol and cresol	Work benches, floors, and walls
Gases	Ethylene oxide	Heart–lung machines, respirators, suture materials, dental equipment, and clothing
	Hydrogen peroxide	Used for fogging

TESTING OF DISINFECTANTS

- There is no single reliable test available to determine the efficiency of a disinfectant, as several parameters influence its activity.
- Traditionally, phenol was taken as the standard, against which the action of the test agent was compared.

Kelsey-Sykes or In-use Test

Working solutions of the disinfectants used in hospitals are sampled at different times during their use.

Rideal-Walker Test

- Equal quantities of a microbe are challenged with varying concentrations of phenol or the disinfectant to be tested.
- The dilution of the disinfectant which does not show growth divided by the dilution of phenol that restricts growth gives the **phenol coefficient**.

Chick-Martin Test

This test is similar to the Rideal–Walker test except that it can be performed in the presence of organic matter.

Table 10.11 lists the concentrations of disinfectants routinely used in clinical practice.

Table 10.11 Routinely used concentrations of disinfectants

Name of the disinfectant	Concentration
Glutaraldehyde	1–2%, depending on the level of disinfection required
Phenol	5%
Sodium hypochlorite	0.5–5%*
Chlorhexidine	1–4%*
Povidone–iodine	10%
Alcohol	70–80%

*Household bleach

APPLICATIONS IN HEALTHCARE SETTINGS

DISINFECTION PROCESSES IN VARIOUS AREAS OF THE HOSPITAL

- **Operation theatres:** The processes employed for the disinfection of operation theatres are presented in Table 10.12.
- **Critical care and intensive care units:**
  - Standard precautions across the facility
  - Transmission-based precautions for droplets and airborne and contact precautions

Improved environment

- **Ventilation** with a focus on fresh air and natural ventilation or control systems with requisite air changes
- Cleaning, disinfection, and sanitation of the environment and frequently-touched surfaces with disinfectants like 1% sodium hypochlorite or 70% alcohol
- Safe water and food to prevent water- or foodborne diseases in hospital settings
- **Biomedical waste management** as per the central pollution control board (CPCB) guidelines

Table 10.12 Disinfection processes employed in operation theatres

Process	Agents used
Surface cleaning and disinfection	• Surfactant or detergent and water • Phenolic and iodophore compounds • Bacillocid (dihydroxy, 2–5 dioxahexane-glutaraldehyde benzalkonium chloride combination)
Air control by air handling units (AHU)	Positive pressure ventilation Air changes: 15–20 air changes per hour; reduced to 25% during non-operating hours Filtered air with HEPA filters Minimum positive pressure 2.5 Pascal (0.01 inches of water) (www.nabh.co)
Post-operative procedures for the disposal of soiled linen and contaminated instruments	Linen: Washed with detergents and autoclaved Instruments: Washed and autoclaved
Spills and discards management (blood/ body fluids)	Decontaminated with bleaching solution/chlorine solution (10% available chlorine)

STERILISATION AND DISINFECTION OF DENTAL INSTRUMENTS

- Since dental procedures include critical and non-critical interventions, surgical procedures and prosthetic implants, the infection prevention practices involved are as followed as in other areas of the hospital.
- Table 10.13 outlines some of the methods adopted for critical and non-critical areas as per the Indian Dental Association (https://www.ida.org.in/).

CENTRAL STERILE SUPPLY DEPARTMENT (CSSD)

CSSD plays an important role in the infection control process of all hospitals and healthcare settings. This facility is also called the **sterile processing department (SPD) or central supply department (CSD)**.

- Physical cleaning, washing, packing, and sterilisation of all hospital linen, medical devices, surgical instruments, and equipment are carried out in the CSSD.
- A unidirectional workflow is maintained (Fig. 10.9 and 10.10), with unsterile and soiled linen and instruments received at one end, followed by physical cleaning, washing, and drying. This is followed by packing and sterilisation either by autoclaves, ethylene oxide (EtO), or plasma sterilisers, depending on the articles to be sterilised.

Table 10.13 Methods of sterilising dental instruments

Level of contamination	Items	Method of sterilisation
Critical Items	Surgical instruments, periodontal knives, scaling instruments, forceps, burs, etc.	Steam sterilisation by autoclave
Non-critical items	Radiograph head/cone, blood pressure cuff, face bow	Cleaned and disinfected with alcohol-based disinfectants
Re-usable semi-critical items	Plastic impression trays, amalgam carriers, plastic instruments, etc.	Disinfected with 0.5% or 1% sodium hypochlorite
Contamination with blood and saliva	Surfaces and basins	Prompt removal and surface disinfection with 1% sodium hypochlorite or 5.25% household chlorine bleach to 1 gallon of water

- The packed materials are marked with a chemical indicator on the surface which indicates the status of sterilisation. Along with the date and time, the validity of the packed material is also indicated.

BACTERIOLOGY OF WATER, AIR, MILK, AND FOOD

BACTERIOLOGY OF WATER

- Drinking water is the most essential component of life. For this reason, it must be made acceptably safe, that is, free from physical, chemical, and microbiological contamination.
- Many major human diseases, for example, typhoid fever, cholera, and other diarrheal diseases, poliomyelitis, and viral hepatitis A and E are waterborne.
- These pathogens reach water sources through **fecal or sewage pollution**.
- Thus, it is essential to treat water suitably to remove and prevent such contamination.
- To ensure the safety of protected water supplies, **regular bacteriological surveillance** is recommended.
- Natural water sources, even when unpolluted, frequently contain bacteria such as species of *Pseudomonas*, *Serratia*, *Flavobacterium*, *Chromobacterium*, *Acinetobacter*, and *Alcaligenes*.
- Soil bacteria such as aerobic spore-forming bacilli, and those found on decaying vegetation such as the *Enterobacter* species may also be washed into natural waters during the rains; fortunately, these are harmless (Table 10.14).

Unsterile material → Physical cleaning & washing → Drying → Packing → Sterilisation autoclave/ETO/plasma sterilisation → Checking for sterility → Distribution to end users (OTs, wards, etc.)

Fig. 10.9 Workflow of CSSD

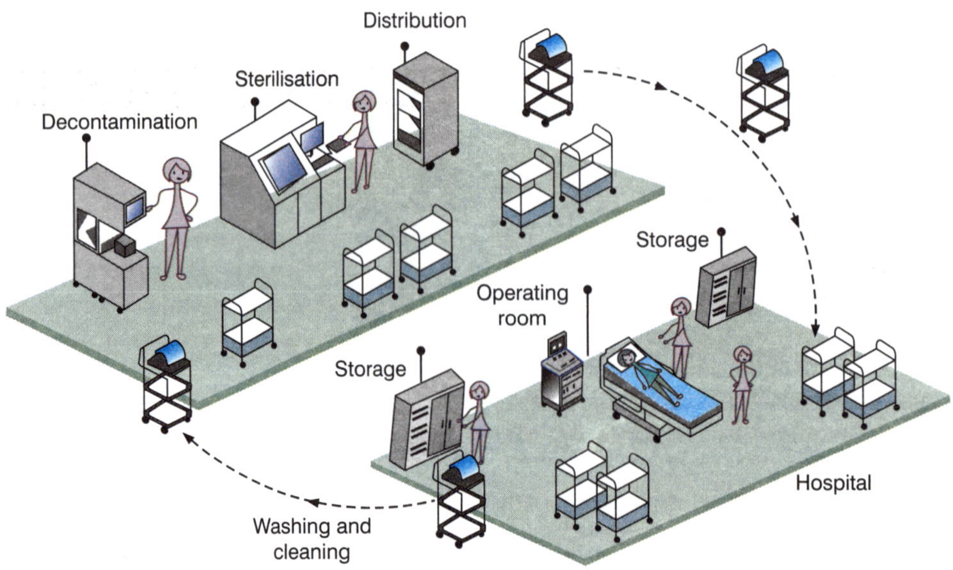

Fig. 10.10 Various steps in the CSSD

Table 10.14 Pathogens that spread through water

In the community		In the hospital		
Pathogens	**Disease caused**	**Pathogens**	**Found in**	**Disease caused**
Bacteria				
Shigella spp.	Bacillary dysentery	*Legionella pneumophila*	Hospital water, cooling towers	Pneumonia
Vibrio cholerae	Cholera			
Salmonella enterica serotypes Typhi and Paratyphi A, B, C	Enteric fever	*Serratia marcescens*	Water of humidifiers	Pneumonia
		Mycobacterium xenopi	Hot water taps	Pneumonia
Leptospira spp.	Leptospirosis	*Pseudomonas paucimobilis*	Water bottles for tracheal suction	Pneumonia
Parasites		*Mycobacterium chelonae*	Contaminated equipment and water tanks	Otitis and nasal septum cellulitis
Protozoa				
Giardia lamblia	Diarrhea, malabsorption	*Acinetobacter* spp.	Water bath used to thaw fresh plasma	Bacteremia
Entamoeba histolytica	Amoebic dysentery			
Helminths		*Pseudomonas aeruginosa*	Water bath used to thaw cryoprecipitate; tub water	Bacteremia; folliculitis and skin infections
Schistosoma haematobium	Hematuria			
Dracunculus medinensis	Skin lesions			
Viruses		*Clostridium difficile*	Bathwater	Diarrhea
Enteroviruses	Viral gastroenteritis			
Hepatitis A, E viruses	Hepatitis			
Polio viruses	Poliomyelitis			

Bacteriological Examination of Water

- The bacteriological analysis of water supplies should be performed at regular intervals and not as a random exercise.
- Drinking water should be free of any pathogenic microorganisms. Tests should be aimed at these pathogens, which are generally present in small numbers. Routine practice involves testing for fecal pollution; if fecal pollution is detected in a water sample, it is inferred that the water may be harbouring enteric pathogens.
- Bacteria that indicate fecal pollution of water:
  - Coliform bacteria (*Escherichia coli*)
  - *Enterococcus* species
  - *Clostridium perfringens*

The following sections elaborate on the tests that are generally performed for the routine bacteriological analysis of water.

Detection of coliform bacteria and *Escherichia coli* (Tables 10.15 and 10.16)

Detection of fecal streptococci

Fecal streptococci (such as the *Enterococcus* species) may be present in water samples but usually survive only for short durations. Hence, if fecal streptococci are detected in water after heating the sample up to 45°C (to detect enterococci, which survive this temperature), the water sample is said to have had recent fecal contamination.

Detection of Clostridium perfringens

C. perfringens may remain in water samples for a long time. Therefore, if *C. perfringens* alone is detected in a water sample (without *E. coli* or fecal streptococci), it suggests fecal contamination that is not recent.

Viruses in water

Enteroviruses and other cytopathogenic viruses can be isolated from water; however, such isolation is not routinely done. Generally, it is assumed that viruses in water are destroyed by chlorination. **The concentration of free residual chlorine is at least 0.5 mg per litre for a minimum contact period of 30 minutes at a pH below 8 and a turbidity of 1 nephelometric turbidity unit or less.**

Protozoa in water

Entamoeba histolytica, the *Giardia* species, and *Balantidium coli* may contaminate drinking water. Coliform counts do not reliably indicate protozoal contamination of chlorinated water as protozoa are more resistant to chlorine than coliforms.

BACTERIOLOGY OF AIR

In a healthcare setting, factors that may lead to healthcare-associated infections transmitted through air are as follows:

- Large number of patients housed in close quarters
- Patients with respiratory illnesses who produce aerosols
- Operative procedures performed on exposed body parts

Table 10.15 Tests to detect bacteria and *Escherichia coli* in water samples

Test	Principle	Method	Interpretation
Plate count	The greater the amount of organic matter present, the more likely the water is to be contaminated with parasitic and potentially pathogenic organisms Bacteria that grow at: • 37°C are most likely those associated with organic material of human or animal origin • 22°C are most likely to be saprobic—they normally inhabit water or are derived from soil and vegetables	For a given water sample, two pour plates are prepared on nutrient agar and incubated parallelly at 37°C for 1–2 days and at 22°C for 3 days, respectively The colonies on the two plates are then counted	A plate count at : • 22°C—provides an indication of the quantity of decomposing organic matter in the water • 37°C—provides an important index of dangerous pollution; an increase in the plate count should serve as an indication of some defect in the filter beds, requiring immediate correction
Presumptive coliform count—multiple tube technique	Deemed presumptive because the reaction observed may occasionally be due to the presence of organisms other than coliform bacteria	An estimate of the number of coliform organisms is usually made by adding varying quantities of water (0.1–50 mL) to bile salt–lactose–peptone water (with an indicator for acidity) and incubating at appropriate temperatures	The probable number of coliforms per 100 mL is read from the probability tables of McCrady This is known as the '**presumptive coliform count**' or the most probable number (MPN) of coliforms present in the water sample being tested as per Table 10.16
Differential coliform test (Eijkman test)	Performed to determine whether the coliform bacilli detected in the presumptive test are, in fact, *E. coli*	Following the presumptive test, subcultures are made from all the bottles showing acid and gas to fresh tubes of single-strength MacConkey medium These tubes are incubated at 44°C and examined after 24 hours	From the number of positive tubes obtained, the number of *E. coli* bacteria present in the water sample tested can be estimated by referring to McCrady's probability tables
Membrane filtration method	Bacteria, if present, are retained on the surface of the filter; these will grow upon subculturing on a suitable medium Specific pathogenic bacteria, such as *S.* Typhi or *Vibrio cholerae* may have to be looked for in this manner	A measured volume of water is filtered through a membrane filter with a pore size of 22 µm (bacteria-stopping filters) The membrane is then subcultured on a suitable medium	Colonies are counted and number extrapolated

Table 10.16 Classification of the quality of drinking water based on bacteriological tests

Classes	Presumptive coliform count per 100 mL	*E. coli* count per 100 mL
Class I—Excellent	0	0
Class II—Satisfactory	1–3	0
Class III—Suspicious	4–10	0
Class IV—Unsatisfactory	More than 10	0.1 or more

Infections transmitted through the air are classified into two broad groups, which are as follows:

- **Airborne infection:** Transmission of infections produced by respiratory droplets less than 5 µm in size
- **Droplet infection:** Transmission of infections produced by respiratory droplets larger than 5 µm in size; while coughing, sneezing, and talking, varying numbers of droplets are expelled from the body, ranging in size from less than 1 µm to 15 µm (Table 10.17).

Microbial Content of Indoor Air

Bacteria may be distributed through gross droplets and droplet nuclei from the nose and mouth of humans and through dust particles.

Dust

This consists of particles of varying sizes originating from animal, vegetable, or mineral sources. Heavy particles fall to the ground, while those 0.5–1 µm in diameter generally remain suspended in the air. The ultimate source of common pathogenic organisms is dust from human beings.

Table 10.17 Characteristics of droplet and airborne transmission and microorganisms involved in infections

Features	Droplet transmission	Airborne transmission
Size of droplet	>5 μm in size (respiratory droplet)	<5 μm in size (droplet nuclei)
Source of droplets	Coughing, sneezing, talking, invasive procedures (e.g., bronchoscopy)	Coughing, talking, sneezing, invasive procedures (bronchoscopy, suction aspiration)
Characteristics of droplets	Droplet transmission occurs due to evaporation Present in the air for a short time (≤1 minute) and travel only short distances Close contact is required for this mode of transmission	Remain suspended in air for long periods Travel several metres Susceptible individual may become infected even at some distance from infected persons
Microorganisms involved		
a) Bacteria	*Streptococcus pyogenes* *Neisseria meningitidis* *Corynebacterium diphtheriae* *Haemophilus influenzae* type B *Bordetella pertussis* *Yersinia pestis* (pneumonic plague) *Mycoplasma pneumoniae*	*Mycobacterium tuberculosis*
b) Viruses	Influenza viruses Rubella virus Mumps virus Adenovirus Parvovirus B19	Varicella-zoster virus Measles virus Influenza viruses SARS-CoV

Note:

- *S. aureus* and *S. pyogenes* are known to be shed and dispersed in the air in operating rooms and nurseries.
- Outbreaks of pneumonia due to *Legionella pneumophila* are associated with the presence of cooling towers close to the ventilation systems of hospitals.
- *Aspergillus* and other fungal spores are dispersed through the air during construction, renovation, and maintenance of buildings.
- Although carpets, linen, potted plants, and flowers are known to be reservoirs of opportunistic pathogens, epidemiological evidence linking these to nosocomial infections is lacking.

- Nasal secretions via the ala nasi and upper lip are carried by the hands to the skin, clothing, and bedding, from where they become detached as dust particles.
- Organisms may also get directly detached from the skin of different parts of the body including the perineum and septic wounds.
- Intestinal organisms, through dried particles of feces from napkins of infants, are also disseminated.
- The following organisms are found in the dust of wards where such patients are treated:
  - Hemolytic streptococci from patients or carriers
  - *M. tuberculosis*
  - *C. diphtheria*
  - Staphylococci

Under favourable conditions, these organisms may remain alive for many weeks. Bedclothes are an abundant source of bacteria-laden dust.

Air Quality of Operation Theatres

- Given the number of contaminants in the air, the air quality of operation theatres (OT) in hospitals has to be kept under control.
- The control measure employed depends on the type of OT—whether specialty OT, general OT, or minor (day-care) OT.
- Factors that are regulated in an OT through air-handling units (AHU) and high-efficiency particulate air (HEPA) filters are as follows:
  - Velocity
  - Air changes/hour
  - Positive pressure
  - Air filtration
  - Temperature and humidity

Measurement of Air Contamination

In critical care units or operation theatres, it is important to monitor the quality of air. This is done by the following methods:

- **Settle plate (sedimentation) method:** This is a means of estimating the number of bacteria present in the air by permitting bacteria to 'settle' on open petri dishes (containing culture media) over a fixed duration of time. Droplet nuclei require more time to settle than larger particles. This method provides an idea of the relative numbers and species of microorganisms present in air and is especially useful for testing the quality of air in surgical theatres and hospital wards.
- **Slit sampler:** This is a more elaborate method than the plate exposure and makes up for the many limitations of the latter. In this method, a known volume of air is directed onto a plate through a 0.25 mm wide slit, while the plate is being mechanically rotated so that the organisms are evenly distributed over it

BACTERIOLOGY OF MILK

Milk can be a potent vehicle for transmitting infections.

Types of Bacteria in Milk

- **Acid-forming bacteria:** The commonest bacteria in milk are lactic streptococci including *S. lactis* and *Enterococcus faecalis*. Lactobacilli are also found in

milk. These ferment the lactose in milk, producing acids—chiefly lactic acids—which leads to the formation of a smooth, gelatinous curd.
- **Alkali-forming bacteria:** These include the *Alcaligenes* spp., some aerobic spore bearers, and the *Achromobacter* species. These bacteria render the milk alkaline.
- **Gas-forming bacteria:** Coliform bacilli are the commonest in this group. Others are *C. perfringens* and *C. butyricum*. These organisms produce acid and gas, as a result of which, a smooth, gelatinous curd riddled with gas bubbles is formed. Coliform bacilli are responsible for the ropiness in milk.
- **Proteolytic bacteria:** Spore-bearing aerobes such as *Bacillus subtilis, Bacillus cereus, Proteus vulgaris*, staphylococci, and micrococci fall in this category.
- **Inert bacteria:** These are bacteria that produce no visible change in milk. These include some cocci of the udder, members of the *Achromobacter* group, and most of the pathogenic organisms in milk.
- **Human milk:** Breast milk contains small numbers of *S. epidermidis, S. mitis, Gaffkya tetragena,* and *S. aureus*. A few other species may also be found in some samples.

Milk-borne Diseases

- Occasionally, milk may be contaminated with *Streptobacillus moniliformis* from the nasal secretions of rats or *Campylobacter jejuni* from animal feces.
- *Yersinia enterocolitica* is not uncommon in milk and may give rise to gastroenteritis if present in large numbers (Table 10.18).

Table 10.18 Pathogenic microorganisms that spread through milk and cause human disease

Organism	Disease in humans
A) Bacteria	
Mycobacterium bovis	Tuberculosis
Bacillus anthracis	Anthrax
Brucella abortus	Undulant (remittent) fever
Streptococcus spp.*	Sore throat, scarlet fever
S. aureus	Diarrhea, vomiting
Salmonella spp.*	Enteric fever, food poisoning
*Coxiella burnetii***	Q fever
B) Viruses	
Cowpox virus	Lesions on the skin of the milker
Foot-and-mouth disease virus	Mild disease (fever, vesicles on skin and mucous membranes)

* May spread through milk in ice cream

** May not be the usual mode of spread of the infection

Sterilisation of Milk

Potential pathogens are eliminated from milk by the following methods:
- **Pasteurisation:** In this method, all vegetative pathogens are killed by heating milk to 63–66°C and maintaining it at this temperature for 30 minutes (the **holder method**). Alternatively, milk is heated to a temperature of 71°C and held at that temperature for at least 15 seconds before rapidly cooling it to 10°C or less. The **phosphatase test** is used to check the adequacy of pasteurisation.
- **Boiling:** The commonly-used household method of heating milk at or around boiling point (which destroys all but the most resistant spores) is adequate for short-term purposes. The efficacy of such treatment is tested by the **turbidity test**.
- **Ultraheat treatment:** In this method, milk is **heated to 132°C for one second** under specified conditions.

Bacteriological Examination of Milk

Routine bacteriological examination of milk is performed by the following methods:
- **Viable count method:** Raw milk always contains bacteria, varying in number from about 500 to several million per mL. The number of viable bacteria in a sample of milk is estimated by performing plate counts with serial dilutions of the milk sample.
  - **Significance:** The plate count gives a rough and direct assessment of the viable bacteria in a sample of milk.
- **Methylene blue reduction test:** This is a simple substitute for the viable count. It depends on the reduction of methylene blue by bacteria in milk when incubated at 37°C in complete darkness. The rate of reduction is related to the degree of bacterial contamination. Raw milk is considered satisfactory if it fails to reduce the dye in 30 minutes.
- **Resazurin test:** This is similar to the methylene blue reduction test, except that the dye resazurin, on reduction, passes through a series of colour changes—from blue to pink to colourless. The colour change is noted after incubating the milk for 10 minutes.
- **Phosphatase test:** This is an indicator of whether milk has been pasteurised. The enzyme phosphatase, which is normally present in milk, is inactivated if pasteurisation has been performed properly.

BACTERIOLOGY OF FOOD

Food-borne infections are a significant public health problem since they may be a major cause of morbidity, although, fortunately, an infrequent cause of mortality.
- Food is very easily contaminated and, in addition, is an excellent medium for the growth of various kinds of microorganisms.

- Etiological patterns of food-borne infections vary across the world. In the USA, *S. aureus* and *Salmonella* are responsible for more than half the food-borne outbreaks of disease, whereas in the United Kingdom, *Clostridium perfringens* is responsible for more than 90 per cent; in Japan, *Vibrio parahaemolyticus* is responsible for more than 50 per cent of such outbreaks. In India, several pathogenic Enterobacteriaceae cause food poisoning.

Definitions

- **Outbreak of food-borne disease:** When two or more persons experience a similar illness, usually gastrointestinal, after the ingestion of the same food and when epidemiologic analysis implicates the food as the source of the illness, it is referred to as an outbreak.
- **Food poisoning:** This refers to a group of diseases caused by the consumption of food contaminated with microbes or microbial toxins (usually caused by bacteria).

Sources of Food Contamination

Food should not normally contain microorganisms. If microbes are detected in food, particularly in cooked food, it indicates a breakdown of sanitary procedures and precautions. Food can be contaminated by:

- **Feces:** Food may get directly contaminated with feces through the hands of the infected person or indirectly through objects handled by an infected person.
- **Sewage:** Water used for cooking gets contaminated by sewage at some stage.
- **Flies:** These insects carry pathogenic organisms from feces or from waste in refuse dumps or dustbins to food; they also contaminate food with their excreta.
- **Hands of infected individuals:** Food may be contaminated by the hands of individuals who are either infected or carriers.
- **Air:** Uncovered food may be contaminated by organisms present in air and dust that settle on it.
- **Domestic animals and pets:** Food may inadvertently be contaminated by the excreta of domestic animals and pets, either directly or indirectly (through the hands of persons handling the animals).

Prevention of Food-borne Infections

Food may be contaminated prior to cooking. This contamination may be because the food has been taken from an infected animal or because the food may be contaminated at the time of the animal's slaughter in the abattoir or at the time of canning. To overcome this, care should be taken while cooking to ensure that heat penetrates all parts of the food so that all bacteria are killed. The following points should be borne in mind:

- **Refrigeration does not kill microorganisms**; it only serves to temporarily inhibit their growth.
- **Reheated foods** are often sources of food poisoning, particularly if they have been contaminated by enterotoxin-producing *S. aureus* since heat does not affect the heat-stable enterotoxin secreted by the bacterium.
- **Imperfectly or undercooked food**, for example, meat that has been broiled, is a potential source of danger.
- **Overall hygiene practices** must be maintained, and water used for cooking must come from protected or certified water supplies.

Key Points

- ✓ Sterilisation is the process by which an article, surface, or medium is freed of all living microorganisms, either in the vegetative or spore state.
- ✓ Disinfection is the destruction or removal of all pathogenic organisms or organisms capable of giving rise to infection.
- ✓ The factors that determine the type of sterilising or disinfecting process to be used include:
  - Purpose for which it is carried out
  - Material to be sterilised
  - Nature of the microorganisms to be removed or destroyed
- ✓ Physical methods of sterilisation are:
  - Heat (dry heat, moist heat)
  - Filtration
  - Radiation
- ✓ Chemical agents include alcohol, aldehyde, dyes, halogens, phenol, surface-active agents, and gases.
- ✓ Spaulding's classification categorises devices used in healthcare settings as follows—critical, semi-critical, and non-critical items.
- ✓ Bacteriology of water:
  - Drinking water must be visually acceptable, clear, and colourless, and without a disagreeable taste or odour.
  - Water is a potential source of microbial diseases and may contain potential pathogens (species of *Salmonella*, *Shigella*, and *Vibrio*, polioviruses, and hepatitis A virus), suggesting contamination with human excreta (less commonly, animal or bird droppings).
  - An increased presumptive coliform count suggests possible fecal contamination of a water supply.
  - The presence of fecal streptococci suggests recent contamination, whereas the presence of *C. perfringens* alone (without *E. coli* or fecal streptococci) suggests fecal contamination that is not recent.
  - Passing a known volume of the test water through a cellulose acetate membrane filter (22 μm pore size), which is then cultured, permits the simultaneous determination of total counts, counts of coliform bacilli, and counts of specific fecal *E. coli*.

✓ Bacteriology of air: The bacterial content of the air is important, particularly when pathogens are present. Heavy particles fall to the ground; particles <1 μm in size remain suspended in air.

✓ Bacteriology of milk:
- Human milk is seldom a vector of pathogens.
- Cow's milk may contain pathogens excreted in the milk or derived from the animal's udders (*Mycobacterium tuberculosis* and *Brucella abortus*) or from the animal's feces (salmonellae and campylobacters).
- To eliminate potential pathogens, milk is treated by:
  - Pasteurisation
  - Heating at or around boiling point
  - Ultraheat treatment, where milk is heated to 132°C for 1 second under specified conditions

✓ Bacteriology of food:
- Food-borne infections are serious causes of morbidity.
- Food may be contaminated by feces, sewage, flies, infected hands, air, and domestic animals and pets. Food poisoning may be caused by toxin-producing (*Clostridium botulinum*) and invasive (non-typhoid salmonellae) organisms.

Questions

Short answer questions

1. Mention the uses of an autoclave in a healthcare setting.
2. Enumerate the agents used for disinfection in hospitals.
3. Describe the process and use of pasteurisation.
4. List some antiseptic agents and their uses.
5. Mention the uses of gamma radiation for sterilisation.
6. What are the quality indicators for sterilisation?
7. Describe the principle of plasma sterilisation.
8. Describe the methods of cold sterilisation.
9. Describe milk-borne pathogens. How can they be eliminated?
10. What is the importance of air sampling in the OT?
11. Describe water analysis to check for waterborne infections in a hospital.

MCQs

1. Which of the following statements does NOT apply to ethyl alcohol?
a) It is used chiefly as a skin antiseptic.
b) It acts by denaturing bacterial proteins.
c) It is used in concentrations of 60–90% in water.
d) It is sporicidal.

2. The concentration of glutaraldehyde used in the disinfection of critical areas is:
a) 2% b) 0.5% c) 1.5% d) 1%

3. Which bacterial spores are used for the quality control of sterilisation in autoclaves?
a) *Clostridium tetani*
b) *Bacillus cereus*
c) *Geobacillus stearothermophilus*
d) *Bacillus subtilis*

4. Autoclaves are used for sterilising which one of the following?
a) Disposable gowns b) Liquid paraffin
c) Surgical instruments d) Central-line catheter

5. Which of the following are not sterilised in CSSD autoclaves?
a) Surgical instruments b) Dressing linen
c) Disposable syringes d) Rubber gloves

6. All the following statements are true about dry heat sterilisation except:
a) Mechanism of action is by protein denaturation
b) Incineration is a method of dry heat sterilisation
c) Efficacy is tested using spores of *Geobacillus stearothermophilus*
d) Glassware can be sterilised using dry heat sterilisation

7. Spores of *Bacillus atrophaeus* are used to test the efficacy of:
a) Moist heat sterilisation b) Ionising radiation
c) Dry heat sterilisation d) None of these

8. Endoscopes are sterilised by:
a) Hot air oven b) Autoclaving
c) Glutaraldehyde d) Lysol

9. All the following are chemical sterilants except:
a) Glutaraldehyde b) Soaps
c) Hydrogen peroxide d) Peracetic acid

10. The sterilisation conditions (temperature and holding time) required for autoclave are:
a) 121°C for 15 min. b) 120°C for 40 min.
c) 150°C for 1 hr d) 100°C for 2 hr

11. Steaming of a material at 100°C for 20 minutes for 3 consecutive days is called:
a) Inspissation b) Tyndallisation
c) Pasteurisation d) Incineration

12. Sterilisation of culture media containing serum is by:
a) Filtration b) Gamma radiation
c) Ethylene oxide d) Autoclave

13. Gamma rays are used for sterilising all the following except:
a) Plastic syringes b) Endoscopes
c) Infusion sets d) Metal instruments

14. Which of the following statements is true of ethylene oxide (EtO) sterilisation?
a) Heart-lung machines can be sterilised
b) Concentration of EtO required is 300 mg/L
c) Sterilisation is achieved in 8 hrs at 54°C
d) Mixing with 10% carbon dioxide makes it more explosive

15. Which of the following disinfectants act by disrupting the microbial membranes?
a) Halogens b) Cationic detergents
c) Aldehydes d) Heavy metals

16. Which of the following is used for the biological monitoring of ethylene oxide sterilisation?
a) *Geobacillus stearothermophilus*
b) *Serratia marcescens*
c) *Bacillus globigii*
d) *Bacillus pumilus*

17. Which is true of Spaulding's classification of medical devices?
a) Critical items are those that come in contact with mucus membranes
b) Semi-critical items require high-level disinfection
c) Cardiac catheters are semi-critical items
d) Non-critical items have to be sterilised with ethylene oxide

18. A postgraduate student working in a microbiology laboratory requires sterile serum for preparing media. What is the best method to obtain sterile serum?
a) Use aseptic precautions to collect blood
b) Autoclave
c) Candle filter
d) Membrane filter

19. A modern hospital is being constructed in a metro city. The architect has installed UV lamps in the corridors and entryways. What is the purpose of these installations?
a) These produce yellow light which looks beautiful.
b) They provide sterilisation of these areas.

c) They provide disinfection of these areas.
d) They are of no use.

20. Viruses in water can be effectively killed by chlorination (free chlorine 0.5 mg/L) for a period of:
a) 3 hrs, pH below 7, turbidity 10
b) 30 min, pH below 7, turbidity 10
c) 30 min, pH below 8, turbidity of 5 unit or less
d) 30 min, pH below 8, turbidity of 1 unit or less

21. Bacterial load in an operation theatre can be estimated by:
a) Settle plate
b) Slit sample
c) Open petriplate method
d) All of these

22. Milk is normally likely to contain any of the following organisms except:
a) *S. lactis*
b) *E. faecalis*
c) *S. moniliformis*
d) *S. epidermidis*

23. Effect of pasteurisation of milk is checked by:
a) Resazurin test
b) Phosphatase test
c) Methylene blue test
d) Kovac's test

24. Milk ring test is used to detect the presence of which of the following bacteria in milk?
a) *M. tuberculosis*
b) *M. bovis*
c) *Brucella* spp.
d) *E. faecalis*

25. Which of the following statement is true regarding differential coliform count?
a) It is done to detect the presence of coliforms in drinking water.
b) It detects whether the coliform bacilli detected in the presumptive test are *E. coli.*
c) It detects fecal streptococci in drinking water.
d) All of the above.

26. Which of the following is correct regarding infections that are spread by air?
a) Airborne infection refers to the transmission of infection produced by respiratory droplets less than 5 mm in size.
b) Droplet infection refers to the transmission of infection produced by respiratory droplets larger than 5 mm in size.
c) Droplet nuclei refers to varying numbers of droplets are expelled from the body that can range from 1–15 mm.
d) All are correct.

27. A farmer who is regularly involved in milking of cows developed small, ulcerating nodules on his hands. What is the likely diagnosis?
a) Brucellosis
b) Vaccinia
c) Cowpox
d) Yabapox

Answers: 1:d, 2:a, 3:c, 4:b, 5:d, 6:c, 7:c, 8:c, 9:b, 10:a, 11:b, 12:a, 13:b, 14:a, 15:b, 16:c, 17:b. 18:d, 19:c, 20:d, 21:d, 22:c, 23:b, 24:c, 25:b, 26:a, 27:c

CHAPTER 11

Part I: General Microbiology

Antimicrobial Agents

(MI1.6)

Chapter Navigation

Learning objectives

- Discuss the mechanisms of antimicrobial agents on bacteria, viruses, fungi, and parasites
- Describe the mechanisms of antibiotic resistance in bacteria and their transfer
- Enumerate strategies to contain antimicrobial resistance

The treatment of infectious diseases has progressed dramatically over the recent years with the discovery of new antibiotics and antiviral, antifungal, and antiparasitic agents. However, with increased use, resistance to these agents has developed exponentially. This has become a significant global problem, associated with increased morbidity and mortality and financial burden to the patient and hospital. Hence, it is important to understand the interaction of antimicrobial agents with microbes and hosts for effective and rational use.

- The first antibiotic was discovered accidentally by **Alexander Fleming** in 1920 when he found that staphylococci did not grow around colonies of fungi (*Penicillium*) that appeared as contaminants on a plate left at room temperature over a weekend.
- The active ingredient of the fungus, **penicillin**, was discovered by **Howard Walter Florey** and **Ernst Boris Chain** and went on to become one of the most widely used antibiotics.

ANTIBIOTICS

- '**Antimicrobial**' is a broad term that is used for all agents that act against different types of microorganisms, namely bacteria (antibacterial), viruses (antiviral), fungi (antifungal), and parasites (antiparasitic).
- **Antibiotics** refer to compounds that are produced by microorganisms and act against other microorganisms.

Antibiotics act by exerting selective toxicity, i.e., they selectively inhibit the growth of microorganisms without compromising the host cell. There are four major sites in the bacterial cell that antibiotics act on:

- Cell wall
- Cell membrane
- Nucleic acid
- Ribosome

Broad-spectrum antibiotics are active against different genera of bacteria (e.g., tetracyclines are active against gram-negative rickettsiae, mycoplasmas, and chlamydiae), while **narrow-spectrum antibiotics** only act against a single genus or limited genera (e.g., vancomycin is active against gram-positive bacteria, chiefly staphylococci). Based on their activity, antibiotics are classified as follows:

- **Bactericidal:** An antibiotic which, in optimum doses, kills the target bacteria
- **Bacteriostatic:** An antibiotic which, in appropriate doses, inhibits the growth of bacteria but does not kill them, i.e., the target bacteria can grow once the drug is withdrawn; the host's immune mechanism plays a role in eliminating the bacteria

Some examples of both types are presented in Table 11.1.

Table 11.1 Bactericidal and bacteriostatic antibiotics

Bactericidal agents	Bacteriostatic agents
Beta-lactam	Macrolide
Glycopeptide	Clindamycin
Cyclic lipopeptides	Tigecycline
Aminoglycoside	Tetracycline
Fluoroquinolone	Linezolid

MECHANISMS OF ACTION OF ANTIBIOTICS

Different classes of antibiotics have different modes of action. These mechanisms are summarised below and illustrated in Fig. 11.1 and Table 11.2.

1. Inhibition of Cell Wall Synthesis

i. Beta-lactam antibiotics

- These bind to **penicillin-binding proteins (PBP)**, which are **cell-surface receptors and inhibit transpeptidation**, thereby blocking the synthesis of the peptidoglycan layer of the cell wall.

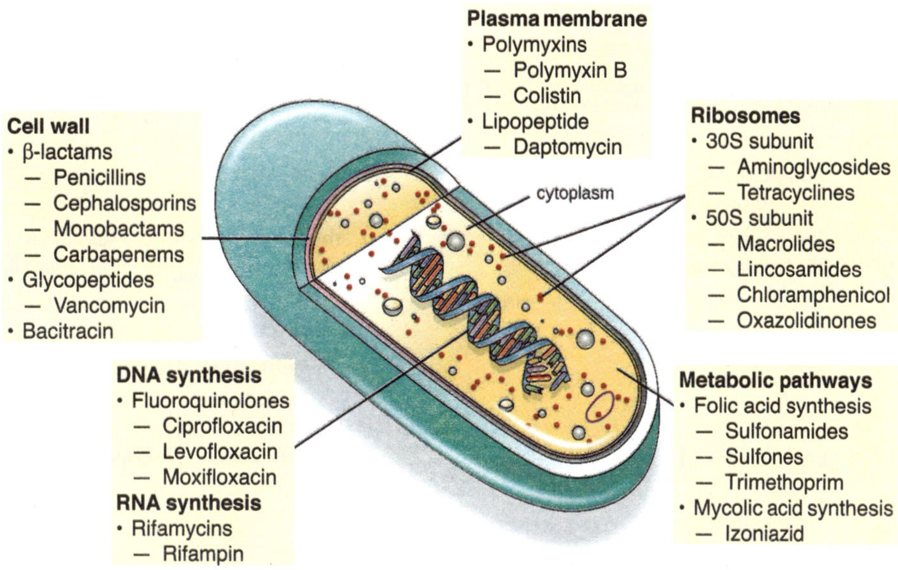

Fig. 11.1 Sites and targets of antimicrobial action on bacterial cells (*Source*: openstax.org/books/microbiology)

Table 11.2 Common antibiotics used in clinical practice

Class of antibiotics	Antibiotics	Spectrum of activity
1. Beta-lactams: Prevent cell wall synthesis		
Penicillins	Penicillin G, V (natural penicillins)	Gram-positive bacteria
	Cloxacillin, nafcillin	Gram-positive bacteria
	Ampicillin, amoxicillin	Gram-positive and gram-negative bacteria and spirochetes
	Carbenicillin, ticarcillin	Gram-positive bacteria
	Piperacillin	Gram-negative bacteria including *Pseudomonas* and *Klebsiella*
Cephalosporins	1st-generation: Cefazolin, cefadroxil, cephalexin, cephalothin, etc.	Gram-positive bacteria
	2nd-generation: Cefuroxime, cefaclor, cefamandole, etc.	Less gram-positive and more gram-negative bacteria
	3rd-generation: Cefotaxime, ceftriaxone, cefexime, cefoperazone, ceftazidime, etc.	Gram-negative and a few gram-positive bacteria Do not act against *S. aureus*
	4th-generation: Cefepime, cefpirome	Against MRSA
	5th-generation: Ceftaroline, ceftibiprole, cefedirocol	
Cephamycin	Cefotetan, cefoxitin	Gram-positive bacteria
Carbapenems	Imipenem, meropenem, ertapenem, doripenem	Gram-negative and gram-positive bacteria
Monobactam	Aztreonam	*Pseudomonas, Haemophilus,* and gram-negative bacteria
2. Prevention of protein synthesis		
i. Act on 30S subunit of the ribosome		
Aminoglycosides	Gentamycin, amikacin, tobramycin, netilmicin, streptomycin	Gram-negative bacteria; often used in combination with beta-lactams to broaden the spectrum of activity
Tetracycline	Tetracycline, doxycycline, minocycline	*Mycoplasma,* chlamydiae, and rickettsiae
ii. Act on 50S subunit of ribosome		
Chloramphenicol	Chloramphenicol	Gram-negative and gram-positive bacteria and anaerobes
Macrolides	Erythromycin Azithromycin	Gram-positive bacteria, *Campylobacter, Legionella,* rickettsiae, *Mycoplasma,* and chlamydiae (also gram-negative bacteria)
Lincosamide	Clindamycin	Anaerobes (both gram-positive and -negative)

(Table 11.2 Contd. . .)

Table 11.2 Contd.

Class of antibiotics	Antibiotics	Spectrum of activity
3. Inhibitors of nucleic acid synthesis		
Quinolones	1st-generation: Nalidixic acid	Gram-negative bacteria
	2nd-generation: Norfloxacin, ciprofloxacin, ofloxacin, lomefloxacin, levofloxacin	Gram-negative bacteria
	3rd-generation: Gatifloxacin, saparfloxacin, grepafloxacin	Gram-negative bacteria, *Pseudomonas* and *Staphylococcus aureus*
	4th-generation: Trovafloxacin, moxifloxacin	Gram-negative bacteria, *Pseudomonas, Staphylococcus,* and anaerobes
Rifamycin	Rifampicin	Antimycobacterial drug (indicator for resistance in TB)
		Prophylaxis for infections caused by *Haemophilus, Meningococcus, Staphylococcus, Pneumococcus* (not preferred as it is an anti-TB drug)
Sulphonamide	Sulphonamides, trimethoprim–sulphamethoxazole	Gram-negative bacteria that cause UTI
4. Inhibitors of function of cytoplasmic membrane		
Polymixin	Polymixin E, colistin	Gram-negative bacteria (except *Proteus*)
		Multidrug-resistant *E. coli, Klebsiella*
Anti-tuberculous drugs	Isoniazid, ethambutol, pyrazinamide	*M. tuberculosis* and atypical mycobacteria
Anti-leprosy drug	Dapsone	*M. leprae* (in combination with rifampicin due to resistance seen in monotherapy)

- As a result, cell wall synthesis is inhibited. Due to the absence of a rigid cell wall, the bacterial cell cannot withstand the osmotic pressure and hence, ruptures.
- Cefidirocol uses a Trojan horse mechanism to enter the periplasmic space using active iron transport by binding to iron and inhibiting cell wall synthesis.
- These antibiotics are bactericidal.

ii. Glycopeptides

- These substances **inhibit the polymerisation of peptidoglycans** in the bacterial cell wall.
- The drug binds to the acyl-D-ala-D-ala portion of the growing cell wall and prevents the cell wall from forming cross-linkages. Examples of drugs in this class are **vancomycin and teicoplanin.**

2. Inhibition of Protein Synthesis

- Bacteria have 30S and 50S ribosomes which differ from host cell ribosomes.
- Protein synthesis inhibitors take advantage of this difference.
- They act by **misreading messenger RNA and inhibiting the formation of the initiation complex.**

i. Drugs that act on 30S ribosomes

- **Aminoglycosides (gentamicin and amikacin): Promote mistranslation** by inducing codon misreading on the delivery of the aminoacyl transfer RNA.
- **Tetracycline:** Blocks the aminoacyl transfer RNA (tRNA).

ii. Drugs that act on 50S ribosomes

- **Chloramphenicol:** Blocks the **action of peptidyl transferase**, thereby preventing the formation of new peptides.
- **Macrolides (erythromycin, clarithromycin, and azithromycin):** Prevent translocation.
- Other antibiotics with similar action of inhibiting protein synthesis are clindamycin, streptogramin, and linezolid.

3. Inhibition of Nucleic Acid Synthesis

This may take place in three ways, which are as follows:

i. Inhibition of precursor synthesis

- **Sulphonamides:** These are structural analogues to **para-amino benzoic acid (PABA)**, which is required for folic acid synthesis.
- **Trimethoprim:** This drug **inhibits dihydrofolate reductase**, which is required for DNA synthesis.

Combining these two drugs (cotrimoxazole) provides the advantage of sequential blocking (Fig. 11.2).

ii. Inhibition of DNA synthesis

Fluoroquinolones: These inhibit the **DNA gyrase** (topoisomerase) enzyme.

iii. Inhibition of mRNA synthesis

Rifampicin: This drug blocks RNA polymerase and thereby, mRNA synthesis.

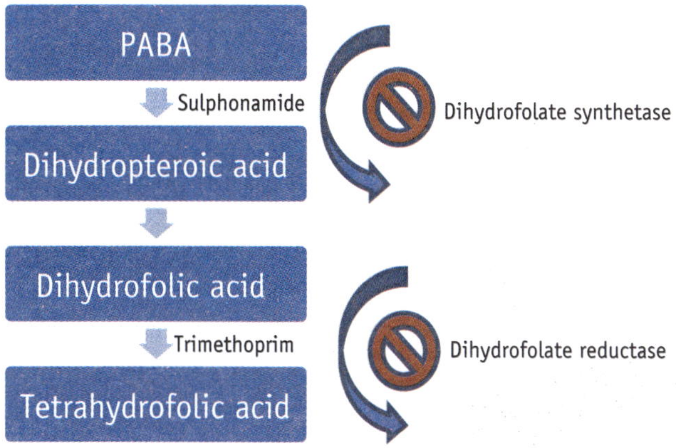

Fig. 11.2 Sequential blocking with the use of sulfonamides and trimethoprim

4. Action on Bacterial Cell Membrane

Polymixin: It destroys the **phospholipid component** of the cell membrane, e.g., daptomycin.

ANTIMICROBIAL RESISTANCE

- Bacteria may develop resistance to an antibiotic by several mechanisms and resist getting killed or inhibited by the antibiotic.
- A group of bacteria that are highly resistant to antibiotics on account of phenotypic variations are known as **persisters**.
- The phenotypic variation is observed when an **antibiotic is administered during the lag phase** of the bacterial growth cycle (when bacteria are not actively replicating). The antibiotics are ineffective at this time as they cannot act on dormant bacteria.
- Bacterial resistance follows one of two mechanisms—intrinsic and acquired.

Intrinsic Resistance

- Intrinsic resistance is a property shared universally within a bacterial species.
- It is independent of previous antibiotic exposure and is not related to horizontal gene transfer.
- It occurs due to **reduced permeability of the outer membrane and due to the natural activity of efflux pumps in bacteria**.

Bacteria inherently do not respond to certain antibiotics (Table 11.3).

Table 11.3 Intrinsic resistance in some commonly isolated bacteria

Bacteria	Drug
Gram-positive bacteria	Aztreonam
All gram-negative bacteria	Glycopeptides, lipopeptides
Klebsiella species	Ampicillin, amoxicillin-clavulanate, ticarcillin, cefazolin
Citrobacter species	Ampicillin, first- and second-generation cephalosporins, and cephamycins
Proteus species	Colistin, cefazolin, tetracycline, nitrofurantoin
Pseudomonas aeruginosa	Ampicillin, amoxicillin, amoxicillin-clavulanate, trimethoprim-suphamethoxazole, sulfonamides
	1st- and 2nd-generation cephalosporins: Cefotaxime, ceftriaxone, ertapenem, chloramphenicol, tetracycline
Acinetobacter species	Ampicillin, amoxicillin-clavulanate, aztreonam, chloramphenicol, fosfomycin, ertapenem
Enterococcus	Aminoglycosides, cephalosporins, clindamycin, many β-lactams, aztreonam, polymyxin B/colistin, and nalidixic acid

Acquired Resistance

Bacteria may stop responding to a drug to which they were originally sensitive by any of the following actions (Fig. 11.3):

- Production of **enzymes that destroy the antibacterial drug** (e.g., beta-lactamases in penicillin and cephalosporins)
- **Expression of efflux systems** that prevent the drug from reaching its intracellular target (e.g., efflux pump mechanism [fluoroquinolone resistance])
- **Reduction of permeability** of the drug through the mutation of porin proteins (as seen with aminoglycosides)
- **Modification of the drug's target site** (e.g., penicillin-binding protein for penicillins)
- Production of an **alternative metabolic pathway** that evades the action of the drug (e.g., folate metabolism)

Genetic Basis of Drug Resistance

Bacteria may acquire new genetic material that confers antibiotic resistance upon them by mutation or the acquisition of resistant genes.

Mutational resistance

i. **Stepwise mutation:** High levels of resistance are achieved only by a series of small step mutations, e.g., penicillin resistance.
ii. **'One-step' mutation:** These mutants differ widely in the degree of resistance, some exhibiting low resistance, while others may be highly resistant, and some even antibiotic-dependent. For example, if a patient is treated with streptomycin alone, bacilli initially die in large numbers; however, soon after, resistant mutants appear and multiply unchecked. If two or more anti-tuberculous drugs are combined, re-population by resistant mutants does not occur

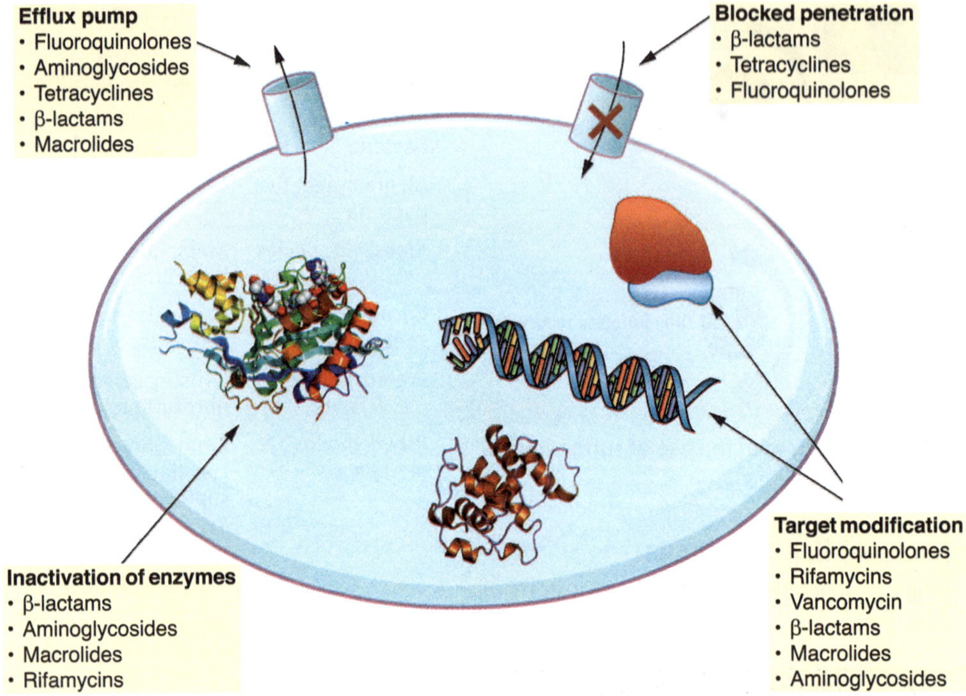

Fig. 11.3 Mechanisms of antibiotic resistance by a bacterium (*Source*: https://openstax.org/books/microbiology/pages/14-5-drug-resistance)

as a mutant resistant to one drug will be destroyed by the other drug. This is the rationale behind **combined treatment in tuberculosis**.

Transferable drug resistance

Transferable drug resistance is now universal and involves all antibiotics in common use.

1. **Vertical gene transfer:** Resistance genes are **transferred directly to all bacterial progeny** during DNA replication. A spontaneous mutation in the bacterial chromosome imparts resistance to a member of the bacterial population.
2. **Horizontal gene transfer:** Genetic material contained in small packets of DNA is transferred between individual bacteria of the same species or even between different species by one of the following mechanisms (see Chapter 8 for details):
   i. Conjugation
   ii. Transformation
   iii. Transduction
   iv. Transposons

Plasmid-mediated transferable drug resistance is mediated by the **R factor** and is the most important route of drug resistance. Some characteristics of plasmid-mediated antibiotic resistance are as follows:

- The acquisition of an R factor simultaneously confers resistance to several drugs.
- Resistance in such strains is due to the production of degrading enzymes, and the level of resistance is usually high.
- Resistance may be transferred between bacteria of different taxonomic groups.
- Resistant mutants usually have a lower growth rate and reduced virulence than the wild strains.
- In some cases, R factors may lead to enhanced virulence.

Bacteria carrying R factors can be transmitted from animals to humans. Hence, the indiscriminate use of antibiotics in veterinary practice or in animal feeds can also lead to an increase in multiple drug-resistant strains in the community. The features of mutational and transferable drug resistance are presented in Table 11.4.

Table 11.4 Comparison of mutational and transferable drug resistance

Mutational drug resistance	Transferable resistance
Bacteria are resistant to one drug at a time	Bacteria are resistant to multiple drugs
Low-degree of resistance is seen	High-degree of resistance is seen
It can be overcome with a high drug dose	High doses are ineffective
Drug combinations cannot prevent the development of resistance	Drug combinations can prevent the development of resistance
Resistance does not spread	Resistance spreads to the same or different species
Mutants may be defective	Mutants are not defective

Antibiotic Sensitivity Testing

Antibiotics must be judiciously used to treat and prevent infectious diseases. Treatment must be based on the susceptibility pattern of the bacterium (which is detected in vitro) and the pharmacokinetics and pharmacodynamics of the antibiotic and its safety for the patient. The results of AST are used to:

- Choose the most appropriate empirical antimicrobial agent
- Establish antimicrobial prescription policies at the institutional/state/national level
- Predict developing resistance
- Assess the efficacy of newly developed antimicrobial agents

MIC (minimum inhibitory concentration): This is the minimum concentration of an antibiotic that can inhibit visible growth of bacteria. To estimate this, the bacteria of interest are inoculated in a series of tubes with varying dilutions of the drug. The concentration of the drug in the tube that does not show any visible growth is said to be the MIC.

MBC (minimum bactericidal concentration): The lowest concentration of the drug that kills the bacterium is its MBC. It is estimated by subculturing the bacteria from liquid onto a solid medium.

Laboratory testing to detect susceptibility is carried out by **diffusion or dilution methods** (Fig. 11.4).

Disc diffusion method

Principle

The antibiotic is allowed to diffuse through a solid medium so that a gradient is established, the concentration being highest near the site of application of the drug and decreasing with distance (Fig. 11.5). The test bacterium is seeded on the medium, and the antibiotic discs are places on the culture. After overnight incubation, growth is inhibited around the discs containing antibiotics to which the bacterium is susceptible but not around those to which it is resistant.

The **diameter of the zone of inhibition** is influenced by a variety of factors such as the diffusibility of the drug, drug concentration, the nature, composition, and thickness of the medium, the presence of inhibitory or stimulatory substances, pH, and the time of incubation.

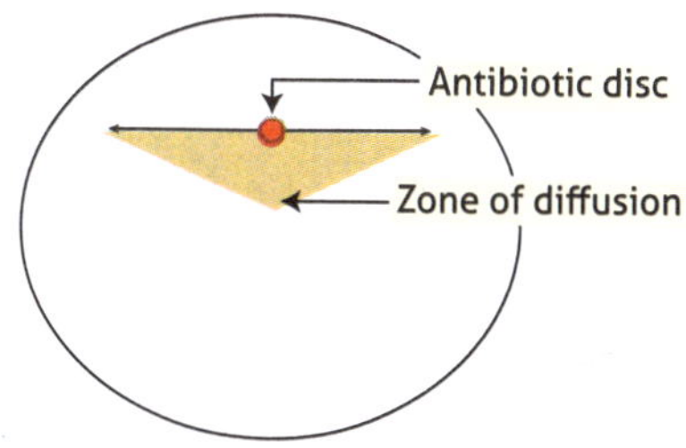

Fig. 11.5 Diffusion of an antibiotic disc in a medium

Requirements of disc diffusion

- Filter paper discs, 6.0 mm in diameter, impregnated with appropriate concentrations of the antibiotic, stored in cold and dry conditions
- Medium—**Muller–Hinton agar** (MHA) is cation–adjusted as calcium and magnesium can affect performance
- **Standard control strains**—ATCC (*S. aureus* ATCC 25923, *E. coli* ATCC 25922, or *P. aeruginosa* ATCC 27853) and McFarland standards for quality
- The choice of the antibiotic disc used is based on specific drug–bug combinations. e.g., chloramphenicol need not be tested against urinary pathogens as the drug is excreted in urine, usually in the inactive form

Methods of diffusion

Stokes method and Kirby–Bauer method are used in the routine testing of clinical strains, whereas the E-test is used to report MIC (Table 11.5).

- **Stokes method:** Test inhibition zones of the test strain are compared to the control strain on the same agar plate.

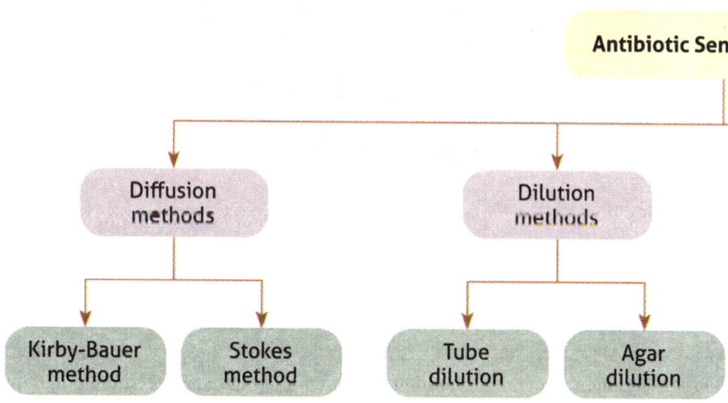

Fig. 11.4 Antibiotic sensitivity tests

Table 11.5 Salient differences between various methods of disc diffusion techniques

Characteristics	Stokes method Fig. 11.6	Kirby–Bauer method Fig. 11.7	E-test Fig. 11.8
Medium	MHA, NA*	MHA	MHA
Maximum number of discs/strips on one plate of 9 cm diameter	6 on the edge of test and control	6 in a circle	1
Interpretation	Test inhibition zones of the test strain are compared to the control strain	The test inhibition zone is measured and compared against a scale of standard inhibition zones	A gradient of concentrations, which cuts the zone of inhibition
Controls	On the same plate	Conducted separately daily or weekly	Conducted separately
Control stain source	NCTC	ATCC	ATCC
Measurement	Comparative zone size	Zone size	MIC
Utility	In resource-poor settings, for daily reporting of clinical isolates	Daily reporting of clinical isolates	Research Epidemiological studies Confirmation of borderline-resistant isolates MIC reporting

**MHA*—Muller-Hinton agar; *NA*—nutrient agar

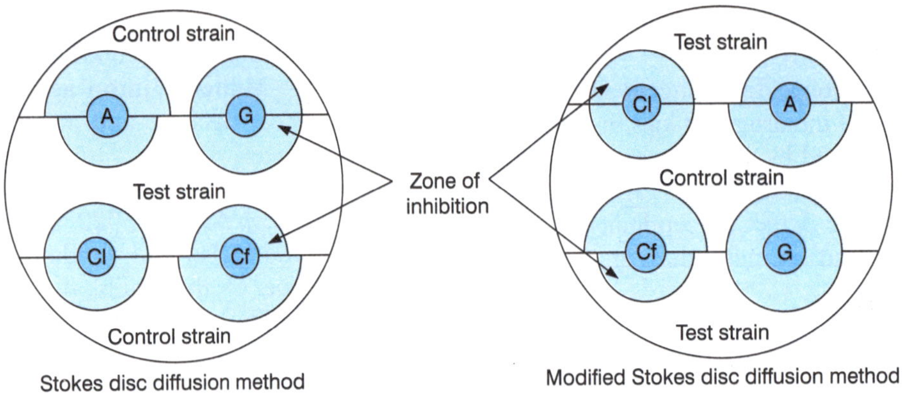

Fig. 11.6 Stokes method of testing for antibiotic susceptibility

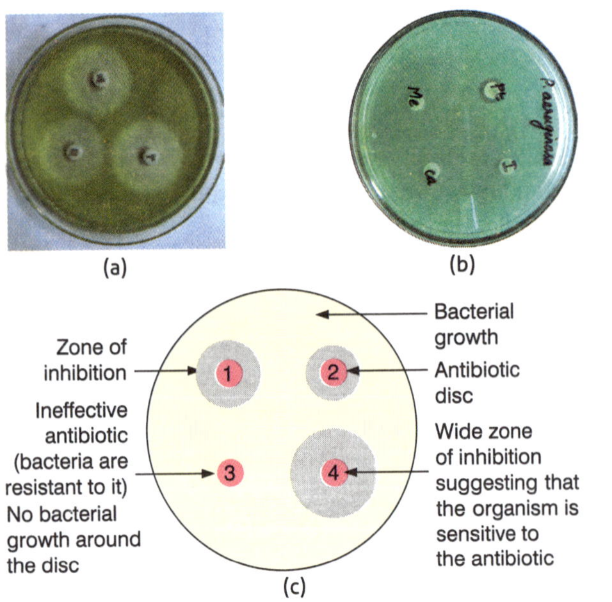

Fig. 11.7 Zones of inhibition around antibiotic discs on the lawn culture of the test bacterium: (a) Sensitive strain, (b) resistant strain (*Source*: [a and b] Department of Microbiology, Pondicherry Institute of Medical Sciences [PIMS], Puducherry) and (c) Diagrammatic representation of the same

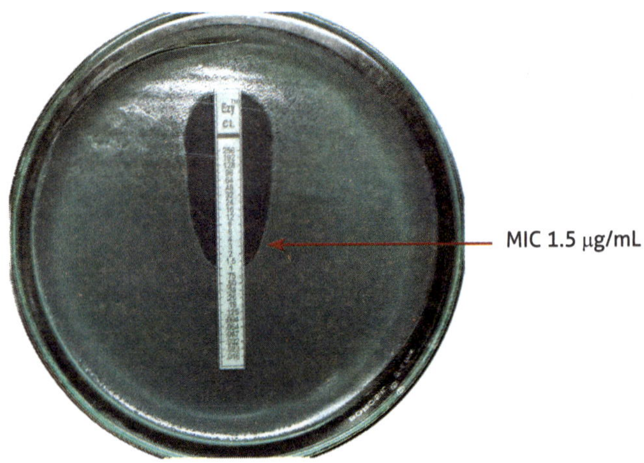

Fig. 11.8 MIC of colistin by E-test for *Pseudomonas aeruginosa*—the ellipsoidal zone of inhibition has a cut-off where the zone meets the strip incorporated with gradient concentrations of the antibiotic—MIC is 1.5 μg/mL (*Source*: Department of Microbiology, Pondicherry Institute of Medical Sciences, Puducherry)

- **Kirby–Bauer method:** The test inhibition zone is measured and compared against a scale of standard inhibition zone.
- **E-test:**
  - This is a modification of the agar diffusion test that employs a quantitative diffusion gradient known as the epsilometer (Fig. 11.8).
  - A non-porous plastic carrier strip impregnated with a predetermined gradient of antibiotic solution is applied to a plate of bacterial culture.
  - After overnight incubation, the concentration of the gradient at which the bacterial growth has been inhibited, marked by the tip of the ellipsoid, is taken as the MIC.

Dilution tests

- Serial dilutions of the antibiotic are prepared and inoculated along with the test bacterium.
- Dilution tests are generally employed when the therapeutic dose is to be regulated accurately as in the treatment of bacterial endocarditis, for slow-growing bacteria such as tubercle bacilli, and when small degrees of resistance are to be demonstrated.

Broth dilution

- In this method, serial dilutions of the antibiotic in a broth are tested against a standardised suspension of the test bacterium.
- After overnight incubation, the MIC is interpreted by noting the lowest concentration of the drug that inhibits growth.

Agar dilution

- This method is convenient when several strains are to be tested at the same time.
- Serial dilutions of the antibiotic are mixed in agar and poured into plates.
- Several strains can be inoculated on each of these plates to test for their susceptibility to the drug.

Automated methods to detect antibiotic susceptibility

Automated versions of sensitivity tests are available and are used in large laboratories for quick turnaround. Several automated systems have been developed to detect the MICs of antibiotics. Some of the commonly used ones are as follows:

- **Automated microtitre plates:** These use the fluorometric plate-reading system.
- **VITEK system:** This serves the dual purpose of identifying bacteria and determining the MIC levels of several antibiotics against the organism (Fig. 11.9).
- **Rapid methods:** These are used for the detection of inhibitory enzymes like the beta-lactamase produced by bacteria. Two types of rapid tests can be used:
  - **Colourimetric tests (lateral flow assays):** Rapid colour change if enzymes are present
  - **Molecular tests (chip- or cartridge-based):** Detect the gene responsible for antibiotic resistance

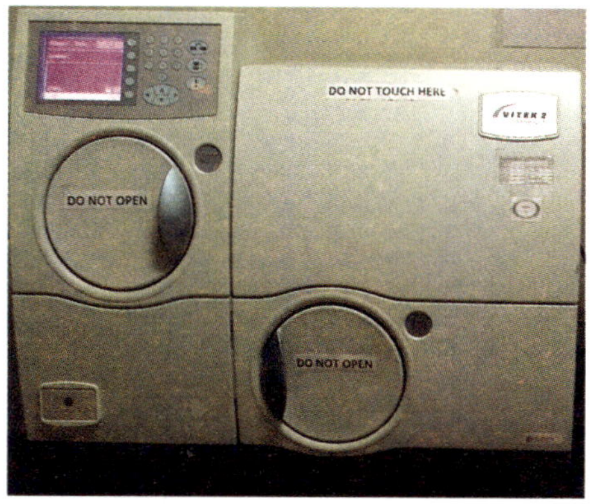

Fig. 11.9 VITEK system for the rapid identification of bacteria and determining the MIC levels of a set of antibiotics for an organism (*Source*: Centre for Virology and Molecular Biology, Pushpagiri Institute of Medical Sciences and Research Centre, Thiruvalla, Kerala)

Interpretation of antibiotic sensitivity reports

The interpretive criteria for drug susceptibility are based on standard dosing in a patient with normal renal function tests and no co-morbidities. The results of antimicrobial susceptibility testing (AST) are generally reported with the following interpretive categories.

- **Susceptible:** When the organism is inhibited by the concentration of the antibiotic usually achieved in the blood following administration at the recommended dosage, it is said to be susceptible.
- **Intermediate:** Such organisms are susceptible to the antibiotic at higher-than-normal dosages or when the drug has clinical efficacy at the body site where the drugs are physiologically concentrated, such as the urinary bladder.
- **Resistant:** These bacteria are not inhibited by the usually achievable concentrations of the agent achieved with normal dosage regimens and the clinical efficacy of the agent against the isolate may not be sufficient.
- **Susceptible-dose dependent (SDD):** The susceptibility of the bacteria depends on the dosage regimen that is used in the patient.
- **Non-susceptible (NS):** A susceptible breakpoint is designated because of the absence or rare occurrence of resistant strains. Also, this does not necessarily mean that the isolate has a resistance mechanism.

The interpretation of an antimicrobial sensitivity report is based on the recognition of the possible mechanism at the base of resistance, which can be extended to non-tested drugs or can lead to a change in the category obtained in vitro.

- **Indicator/surrogate drugs:** A member of the drug family to which the mechanism gives the most obvious resistance is chosen as the indicator. It is used to detect the presence of the mechanism that gives resistance not only to the indicator but also to related agents (Table 11.6).
- **Equivalent agents:** The results of one antimicrobial agent can be used to predict the results of another agent. For example, sensitivity to erythromycin can be used to predict sensitivity to clarithromycin or azithromycin.

The Global Threat of Antimicrobial Resistance

Antimicrobial resistance (AMR) develops when microorganisms such as bacteria, viruses, fungi, and parasites adapt and fail to respond to antimicrobials. The infections caused by resistant strains are harder to treat or control; the risk of the spread of infection to others is increased; and prolonged hospital stays—with added economic and social costs—cause an increase in complications and death. Factors contributing to the emergence of resistant microbial agents are as follows:

- Irrational and inappropriate use of antibiotics
- The use of the wrong antibiotics
- The administration of antibiotics at the wrong time
- The administration of antibiotics for the wrong infection
- The administration of antibiotics for the wrong pathogen

Antibiotic Policy

Given the growing global threat, guidelines are required for the rational use of antibiotics at the hospital level and at the national level. These guidelines are also a necessity to ensure appropriate antimicrobial treatment and to limit the inappropriate use of antibiotics in the treatment of infections. A national guideline (Treatment Guidelines for Antimicrobial Use for Common Syndromes, 2017) has been formulated by the Indian Council of Medical Research, Department of Health (Ministry of Health and Family Welfare, Government of India) based on national data.

- An antibiotic policy guides the treating physician on the **most appropriate antibiotic to be used empirically** (before the culture and sensitivity report) in a particular infection.
- It must also mention the need to switch to specific antibiotics following culture and sensitivity and the dose and duration of such treatment.

Table 11.6 Indicator/surrogate/equivalent antibiotics used in AST reports

Bacteria	Antibiotic tested	Inference (if indicator found resistant)
Staphylococcus spp.	Cefoxitin as a surrogate for predicting MRSA	A cefoxitin-resistant strain is inferred to be MRSA; such strains are resistant to all beta-lactam agents except ceftobiprole and ceftaroline
	Ciprofloxacin or ofloxacin	Acquisition of at least one target mutation that leads to resistance to all fluoroquinolones This may lead to the development of resistance during therapy with other quinolones
	Erythromycin	Can be used to predict the activity of azithromycin or clarithromycin
Enterococcus spp.	Ampicillin	Can be used to predict the activity of amoxicillin, amoxicillin-clavulanate, ampicillin-sulbactam, and piperacillin-tazobactam among non-β-lactamase producing enterococci
	Gentamicin (high level)	Determines loss of synergism of aminoglycosides with beta-lactam agents and glycopeptides irrespective of MIC value
Enterobacterales	Ciprofloxacin	Resistant to all fluoroquinolones
	Ceftriaxone or cefotaxime	Third-generation cephalosporin
Enterobacterales, *P. aeruginosa*, *Acinetobacter baumannii* complex	Colistin or polymyxin B	Any one can agent be used to predict the activity of another
Klebsiella/E. coli	Ceftazidime	Resistant to all cephalosporins except cephamycins (cefoxitin, cefotetan)
Gram-negative isolates from uncomplicated UTIs	Cefazolin	Predicts results for all oral cephalosporins such as cefaclor, cefdinir, cefpodoxime, cefprozil, cefuroxime, cephalexin, and loracarbef

- The antibiotic policy of a hospital should be **periodically reviewed and updated** to keep up with emerging resistance patterns in the hospital.
- It helps in the selection of an antibiotic and deciding on the following aspects of antibiotic therapy:
  - Dosing pattern
  - Route of administration
  - Duration of therapy
  - Emergence of resistance
  - Cost of therapy
- Such a policy should be based on the following:
  - **Data on the susceptibility patterns of common pathogens** causing infection in the hospital over a period of time are documented properly.
  - A **periodic antibiogram pattern** from the microbiology laboratory is reviewed by a team, and the policy is updated according to the current susceptibility pattern.
  - Information regarding the emergence of resistance to a particular antibiotic is also documented and disseminated.

Antibiotic Stewardship

Antibiotic stewardship is defined as the '**optimal selection, dosage, and duration of antimicrobial treatment** that results in the best clinical outcome for the treatment or prevention of infection, with minimal toxicity to the patient and minimal impact on subsequent resistance'. It refers to a set of coherent actions aimed at promoting responsible antimicrobial use. It includes the following components:

- Improving the diagnostic process (clinically, microbiologically, or in other ways)
- Prescribing appropriately according to treatment guidelines once there is a suspected or proven diagnosis
- Deploying strategies that cut across all levels (national, district, local) to promote responsible antimicrobial use

Hospitals draw up an **antimicrobial stewardship programme (AMSP)** for controlling antimicrobial resistance. The management must ensure the implementation of this programme by allotting money, personnel, and other facilities.

- The AMSP team must include **a physician** as the leader of the team to effectively monitor appropriate therapy, **a microbiologist**, **an infectious disease physician**, and **a pharmacist** to improve antibiotic dispensation and use and to monitor the quantity of antibiotics used per patient.
- Antimicrobial stewardship programmes have a direct responsibility to ensure prudent antibiotic prescribing. Antimicrobial stewardship programmes focus on the '**four Ds**' of prescribing antimicrobials (Fig. 11.10).

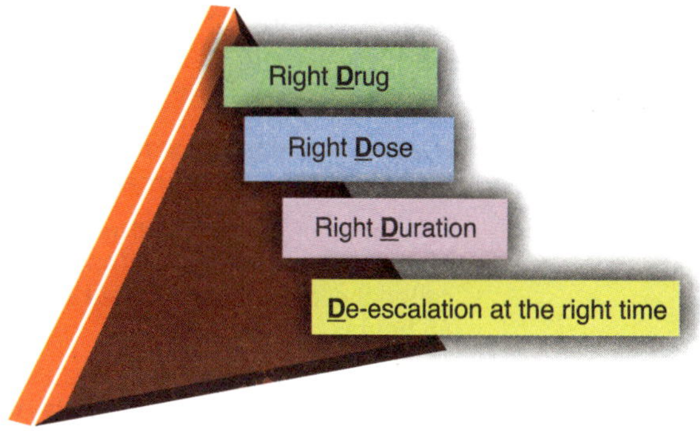

Fig. 11.10 The four Ds of prescribing antibiotics

- Interventions can be introduced in a phased manner or hospital-wide, depending upon the facilities available and the motivation. Broadly, the interventions that can be introduced are presented in Table 11.7.

Some of the core elements of the antibiotic stewardship programme are as follows:

- Limiting the adverse effects of antibiotics
- Reducing the cost to the patient or system

Strategies of antibiotic stewardship programmes

1. *Front-end or pre-prescription approach*

- This is a restrictive approach wherein certain antibiotics require prior authorisation or can be prescribed only by an authorised clinician.
- This list should be reviewed periodically, preferably every year, on the basis of antimicrobial usage data and rates of antimicrobial resistance.
- Prior authorisation is most commonly taken in hospitals and other inpatient settings. This intervention should be considered when antimicrobial stewardship teams and/or infectious diseases physicians have sufficient resources to dedicate to this activity, including experienced personnel and dedicated time.

2. *Back-end or post-prescription approach*

- Regular bedside reviews of patients should be conducted to analyse the prescriptions related to antimicrobial prescribing.
- After the review, feedback is provided to the prescriber advising changes (continue, adjust, modify, or discontinue), if required, on the optimal antimicrobial therapy.
- This approach has the advantage of advocating de-escalation of an antibiotic following empiric therapy.

Strategies for antimicrobial therapy

Empiric therapy

- It is initiated when the microorganism causing the infection is not known.

Table 11.7 Core and supplemental interventions in antimicrobial stewardship

Core interventions	
Prescription audit	Tracking—monitoring antibiotic prescription and resistance patterns
Antibiotic 'time-outs'	Evaluation of the need for treatment after a set period of initial treatment called the **'antibiotic time-out'**—after 48 hours
Antibiotic consumption analysis	Calculating the antibiotic utilisation/ consumption based on **patient days** or **defined daily dose (DDD)** in grams per adult patient as defined by WHO; this helps in estimating: • Antibiotic utilisation over time • Use of a particular class of drugs • Prescribing pattern
Formulary restriction	
Guidelines and clinical pathways	Antibiotic policy and rational empirical therapy guidelines
Supplemental interventions	
Dose optimisation and combination therapies	De-escalation can mean: • Changing a broad-spectrum antibiotic to one with narrower coverage • Changing from combination therapy to monotherapy • Stopping antibiotic therapy altogether when the drugs are no longer required
Streamlining or de-escalation of therapy	Targeted treatment using narrow-spectrum antibiotics (thereby preserving high-end, broad-spectrum antibiotics) based on susceptibility
Parenteral to oral conversion	
Laboratory surveillance and feedback	Reporting—regular dissemination of information on antibiotic use and resistance to doctors, nurses, and relevant staff by the laboratory
Information, education, and communication	Educating clinicians and patients regarding resistance and optimal prescription

- It is based on clinical signs and symptoms and knowledge of the most common infectious agent prevalent in the area and most likely to produce such an infection.
- The drug is administered only in serious/life-threatening infections. For example:
  - In a child with suspected pyogenic meningitis, a broad-spectrum antibiotic which covers the most common causes of pyogenic meningitis is started prior to laboratory confirmation.
  - In an area endemic for typhoid fever, a patient with signs and symptoms of this infection is given the specific antibiotic—ceftriaxone in this case—even before the blood culture report with the susceptibility pattern becomes available.

Definitive (targeted) therapy

Organism-specific therapy

This type of therapy is initiated following laboratory isolation and according to the antibiotic susceptibility pattern.

Prophylaxis

- The purpose of prophylaxis is to prevent an infection in a person who is expected to be exposed to certain known disease-causing microorganisms.
- **Surgical prophylaxis**, for example, is performing a surgery under antibiotic cover (through the duration) so that any microorganism gaining entry during the procedure will be eliminated.
- Another use of antimicrobial prophylaxis is for **persons travelling to an endemic area**, e.g., antimalarial drugs are prescribed for people travelling to malaria-endemic areas.

ANTIVIRAL AGENTS

Antiviral agents must be capable of selectively inhibiting viral functions without damaging the host. However, viruses being obligate intracellular parasites, are intimately associated with host cells. This presents the following challenges:

- Antiviral agents **cannot bypass cellular components**.
- **Nature of the replication cycle** of the virus—after the incubation period, the virus spreads to the target cell and starts producing features of the disease; viruses within the cells/organs are fairly inaccessible to drugs.
- **Latent stage**—the virus gets incorporated into the host cell and cannot be accessed by the drug.
- The **selective toxicity** of antiviral drugs usually depends on the greater susceptibility of viral enzymes to inhibitory actions than host cell enzymes.

Antivirals can help to reduce the morbidity and mortality and the economic loss resulting from viral infections. It is also vital for immunosuppressed patients, who are at a greater risk of infection. Some infections that are treated with antiviral agents are:

- Herpes
- HIV infections
- Influenza
- Hepatitis

TARGETS OF ANTIVIRAL ACTION

The target stages for antiviral therapy include the following (Fig. 11.11):

- Attachment of virus to host cells
- Uncoating of the viral genome
- Viral nucleic acid synthesis
- Synthesis of viral proteins
- Assembly of the virus
- Release of virus particles

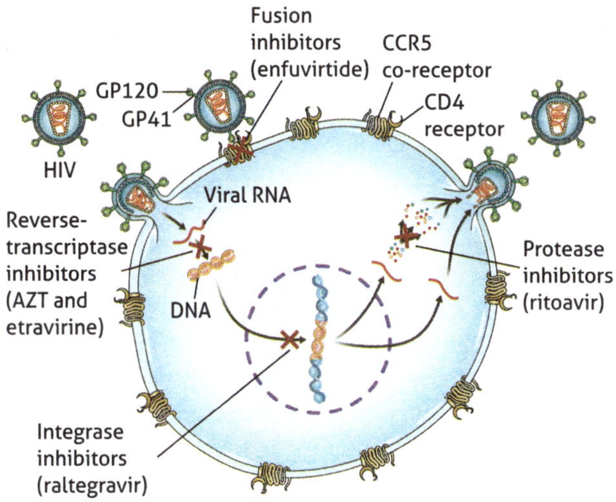

Fig. 11.11 Target sites of the action of antiviral drugs on the viral cell (*Source*: Openstax/books/microbiology)

Antiviral agents are currently available against herpes, influenza, HIV, and hepatitis.

MECHANISMS OF ACTION (TABLE 11.8)

Antiviral agents act by the following mechanisms:

- Inhibition of nucleic acid synthesis
- Viral nucleoside and nucleotide analogues inhibit nucleic acid synthesis
- Reverse transcriptase
- Inhibition of protein synthesis by protease inhibitors
- Prevention of the entry of the virus into the cell
- Inhibition via interferons, which prevent viral replication
- Other mechanisms

Often, the drug needs to be activated by enzymes in the cell before it can act as an inhibitor of viral replication. Most selective drugs are activated by a virus-encoded enzyme in the infected cell.

INTERFERONS (IFNs)

- These are proteins that are members of the cytokine family coded for by the host and which inhibit viral replication.
- Interferons are introduced very quickly (within hours) in response to viral infection or other inducers and are one of the body's first responders in the defense against viral infection.
- They play an important role in the host's innate antiviral immune response.

Table 11.8 Common antiviral agents and their sites of action

Target virus	Class of drug	Antiviral drug	Mechanism of action
Herpes virus	Nucleoside analogue	Acyclovir, ganciclovir, valacyclovir, penciclovir, cidofovir, foscarnet, vidarabine	Inhibit DNA polymerase, gets incorporated in viral DNA and stops elongation
Influenza virus	Receptor antagonist	Amantadine	Inhibits viral M2 protein and prevents the uncoating of the viral genome within the infected cell
	Neuraminidase inhibitor	Oseltamivir	Inhibits neuraminidase enzyme, thereby blocking the entry of viral nucleic acid into the host cell
Hepatitis B virus (HBV)	Nucleoside analogue	Adefovir, entecavir, tenofovir, telbivudine, lamivudine	Inhibit DNA polymerase
Hepatitis C virus (HCV)	Nucleotide analog inhibitor	Sofosbuvir, dasabuvir, semiprevir, ledipasvir	Inhibits RNA-dependent RNA polymerase
	Protease inhibitors	Telaprevir, boceprevir	Block the activity of the viral-encoded protease that is essential in the post-translational modification of the viral polypeptide
Antiretroviral drugs	Nucleoside reverse transcriptase inhibitors (NRTI)	Zidovudine, diadanosine, stavudine, lamivudine, abacavir, tenofavir	Inhibits reverse transcriptase and terminates proviral DNA
	Non-nucleoside reverse transcriptase inhibitors (NNRTI)	Nevaripine, efavirenz, delavirdine, ertavirine, rilpivirine	Non-competitive inhibitors of reverse transcriptase that do not require phosphorylation for activity
	Protease inhibitors	Saquinavir, ritonavir, indinavir, nelfinavir, lopinavir	Bind to protease enzyme and interfere with their cleaving function
	Integrase inhibitors	Raltegravir, dolutegravir (DTG)	Prevent the integration of proviral DNA into the host chromosome
	Viral entry blockers	Enfuvirtide	Binds to HIV-1 envelope transmembrane gp41 preventing fusion of viral-host cell
	CCR-5 receptor inhibitor	Maraviroc	CCR5 receptor is blocked so viral attachment and entry are prevented

- They modulate humoral and cellular immunity and have a broad range of cell growth regulatory activities. Their antiviral effects are as follows:
  - They provide a **non-specific defence** against viral infections.
  - They do not protect the virus-infected cell that produces them; nor are they antiviral agents independently. Rather, an IFN moves from the cell where it is produced to other cells, where it induces an antiviral agent.
  - They are almost always host species-specific but are not specific to a given virus.
  - They inhibit the replication of **a wide variety of DNA and RNA viruses.**
  - They are **extremely potent** and are active in very small amounts.

Virus mechanisms to counteract interferons

Viruses demonstrate different mechanisms that block the inhibitory activities of IFNs on virus replication in the following ways:

- They block the induction of the expression of IFN (herpes virus, hepatitis C virus, papillomavirus, and rotavirus).
- They block IFN-included signal transduction (adenoviruses, hepatitis B virus).
- They neutralise IFN-l by acting as soluble IFN receptors (myxoma virus).

(Details of the functions of interferons are discussed in Chapter 12.)

RESISTANCE TO ANTIVIRAL AGENTS

- Resistance to antiviral agents is an emerging problem.
- The seriousness of the problem is amplified by the limited spectrum of available antiviral drugs.
- Resistance usually develops due to mutations and can be prevented by the use of multidrug therapy.
- Resistance can be indirectly determined by estimating and monitoring the viral load in the patient's blood. If the viral load remains static or goes up despite treatment, it is inferred that the antiviral agent is ineffective against that virus, e.g., HIV patients who are on antiretroviral therapy.
- Mutations associated with drug resistance can be detected by sequencing.

ANTIFUNGAL AGENTS

- Antifungals target specific parts of the fungi viz. (i) the plasma membrane, (ii) fungal cell wall components, or its (iii) biosynthetic pathways (Fig. 11.12; Table 11.9). Antifungals can be polyenes, pyrimidines, azoles, grisans, and echinocandins.
- Some medications are used only topically on the skin or mucous membranes (pimaricin for eye or skin infections) and some only orally (griseofulvin), while others (amphotericin B, some azoles) can be administered by a variety of routes.

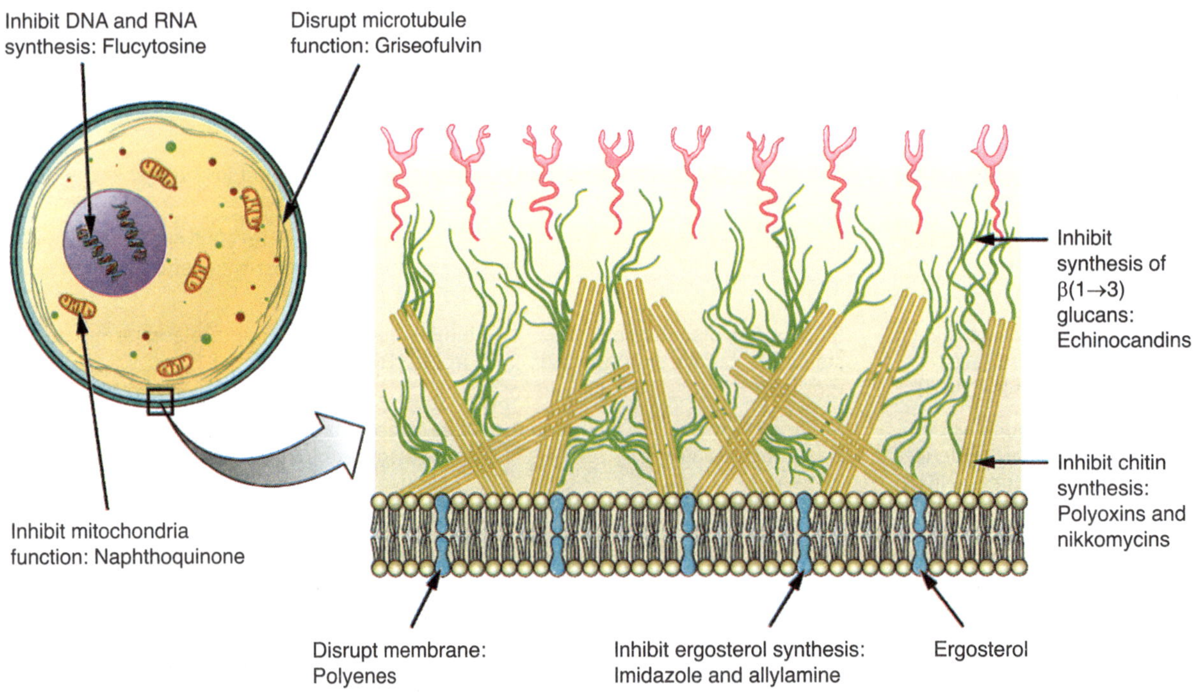

Fig. 11.12 Antifungal agents and their modes of action (*Source*: Openstax/books/microbiology)

Table 11.9 Common antifungal agents used to treat fungal infections

Class	Compounds	Mechanism	Uses
Imidazole and triazole (synthetic antifungal agents—azoles)	Itraconazole Voriconazole Ketoconazole	Inhibits ergosterol biosynthesis	Topical disease (ringworm) Systemic disease
Polyene	Amphotericin B Nystatin Pimaricin	Binds to sterols, causing perturbations in the cell membrane	Systemic disease Skin infection Eye infection
Pyrimidine	5-fluorocytosine (5FC)	Inhibits DNA and RNA synthesis	Used in association with amphotericin B
Echinocandins	Caspofungin Micafungin	Inhibits beta-glucan synthesis in the cell wall	Systemic diseases due to *Candida* and *Cryptococcus*
Grisans	Griseofulvin	Inhibits microtubule assembly	Dermatophytoses

RESISTANCE TO ANTIFUNGAL AGENTS

Resistance to antifungal agents has been on the rise in recent years, especially in the ***Candida* species**.

ANTIPARASITIC DRUGS

These are a group of drugs that are used for the treatment of infections caused by parasites. There are multiple classes of antiparasitic drugs, which can be broadly classified as **antiprotozoal and anthelminthic** (Table 11.10).

ANTIPROTOZOAL DRUGS

These drugs are used for the treatment of protozoal infections such as malaria, kala-azar, and toxoplasmosis.

1. Antiamoebic and Antigiardial Agents

These can be classified into luminal, systemic, and mixed amoebicides.

- Luminal amoebicides act on parasites in the lumen; iodoquinol, paromomycin, and diloxanide furoate.
- Systemic amoebicides (metronidazole and tinidazole) are used in the management of extra-intestinal amoebiasis.
  - **Metronidazole, ornidazole, and tinidazole:** These are 5-nitroimidazole derivatives that are effective against multiple protozoa such as *Entamoeba*, *Giardia*, *Trichomonas*, and anaerobic bacteria. They act by releasing toxic reactive intermediates which damage the parasitic DNA.
  - **Dehydroemetine:** It inhibits protein synthesis, but due to its high toxicity, it is used only in severe cases of intestinal and extra-intestinal amoebiasis.
  - **Diloxanide furoate:** It is postulated to interfere with protein synthesis. It is often used as a luminal amoebicide, especially for asymptomatic infection.
  - **Iodoquinol:** It is a halogenated hydroxyquinoline that acts by interfering with protozoal metabolism by increasing protein-bound serum iodine. It is a luminal amoebicide.

2. Antimalarial Agents

- **Chloroquine:** It acts by inhibiting nucleic acid synthesis and heme detoxification. It is the preferred agent used for the treatment of uncomplicated malaria caused by all species of *Plasmodium*. In recent years, chloroquine-resistant strains of *P. falciparum* have emerged. Derivatives of chloroquine include mefloquine and primaquine.
- **Quinine:** It is used for the treatment of complicated *P. falciparum* malaria. The mechanism of action of this drug is same as that of chloroquine.
- **Artemisinin derivatives:** These include artesunate, artemether, and dihydroartemisinin. After binding to heme, the drug decomposes endoperoxide bridges, releasing toxic free radicals. These drugs are used for treating chloroquine-resistant uncomplicated *P. falciparum* malaria.
- **Amodiaquine:** Its mechanism of action is similar to that of chloroquine. It is used for the treatment of chloroquine-resistant uncomplicated *P. falciparum* malaria.
- **Atovaquone-proguanil:** It is a synergistic quinone–folate antagonist combination. Atovaquone acts by interfering with the cytochrome electron transport system while proguanil inhibits dihydrofolate reductase. It is used to treat chloroquine-resistant uncomplicated *P. falciparum* malaria and mild babesiosis.
- **Pyrimethamine–sulfadoxine:** Pyrimethamine acts as a dihydrofolate inhibitor, while sulfadoxine is an inhibitor of dihydropteroate synthetase. The two compounds act synergistically to block sequential steps in the folate synthesis pathway. They are useful in the management of chloroquine-

resistant uncomplicated *P. falciparum* malaria. Pyrimethamine–sulfadoxine is also used as a first-line drug in the management of toxoplasmosis.

3. Antileishmanials and Anti-trypanocidal Agents

- **Sodium stibogluconate:** It is a pentavalent antimonial that inhibits the phosphorylation of ADP and guanosine diphosphate (GDP) to form ATP and GTP and thereby inhibits glycolysis and the citric acid cycle. This causes *Leishmania* to lose its viability. This drug also increases the production of cytokines and increases the concentrations of reactive oxygen radicals.
- **Meglumine antimoniate:** The mechanism of the action of this drug is similar to that of sodium stibogluconate but is more effective for intralesional applications and is also used as the first-line drug in the treatment of all types of leishmaniasis.
- **Benznidazole:** Chemically known as 2-nitroimidazole, benznidazole is a drug used to treat Chagas' disease, which is caused by *T. cruzi*.
- **Nifurtimox:** It is a 5-nitrofuran derivative that generates nitro anion radicals which are toxic to parasites. It is used as a second-line drug for treating Chagas' disease.
- **Pentamidine:** It is an aromatic diamidine that disrupts the membrane potential of the protozoan mitochondria, inhibits enzymes, and interacts with the parasitic DNA. It is used as the first-line drug for African trypanosomiasis or sleeping sickness. It can also be used to treat visceral leishmaniasis as an alternative to sodium stibogluconate.
- **Suramin:** It acts by inhibiting ATP production via glycolysis, which in turn, inhibits energy metabolism. It is the first choice for the treatment of the first stage of East African trypanosomiasis. It is also effective against onchocerciasis.
- **Eflornithine:** It acts by inhibiting the enzyme ornithine decarboxylase. It is used as the first-line drug for the second stage of West African trypanosomiasis.
- **Melarsoprol:** It is a trivalent organic arsenical compound that competitively inhibits the antioxidant enzyme trypanothione reductase. The drug can cross the blood–brain barrier and is ideal for the management of central nervous system symptoms in the second stage of *T. brucei* rhodesiense infections.

ANTHELMINTIC AGENTS (TABLE 11.10)

These drugs either kill the worm (**vermicide**) or expel it (**vermifuge**).

- **Albendazole and mebendazole:** These are benzimidazole derivatives that act by inhibiting microtubule synthesis, thereby blocking glucose uptake in the parasite, and depleting glycogen stores. These are the drugs of choice for the treatment of ascariasis, hookworm infestations, trichuriasis, trichinosis, cutaneous larva migrans, and pinworm infections. Albendazole is also the first choice for the management of hydatid disease and cysticercosis.
- **Diethylcarbamazine:** It acts as an opsonin, thereby sensitising the microfilariae to phagocytosis. It is the drug of choice for filariasis, loiasis, and tropical eosinophilia.
- **Pyrantel pamoate:** It is a neuromuscular blocking agent which acts on the nicotinic acetylcholine receptors of nematodes, causing their paralysis and leading to their expulsion.
- **Ivermectin:** It is a microfilaricidal agent that acts by interacting with glutamate-gated chloride channels on nematode motor neurons, causing hyperpolarisation and paralysis. It is the preferred agent for the treatment of onchocerciasis.

Table 11.10 Common parasitic infections and the preferred treatment

Disease	Drug of choice
Amoebiasis/ giardaisis/ trichomoniasis	Metronidazole, ornidazole, and tinidazole Paromomycin (giardiasis)
Uncomplicated malaria	Chloroquine
Complicated/ *P. falciparum* malaria	Artemisinin derivative/mefloquine/ pyrimethamine–sulfadoxine
Treatment of toxoplasmosis	Pyrimethamine–sulfadoxine
Toxoplasmosis prophylaxis	Cotrimoxazole Spiramycin (in pregnant women)
East African sleeping sickness	1st stage: Suramin 2nd stage: Melarsoprol
West African sleeping sickness	1st stage: Pentamidine 2nd stage: Eflornithine
Chagas' disease	Nifurtimox/Benznidazole
Leishmaniasis	Sodium stibogluconate/meglumine antimoniate/pentamidine/ amphotericin B Paromomycin (visceral)
Intestinal nematode infestations	1st line: Albendazole/mebendazole Alternate: Pyrantel pamoate Ivermectin (for strongyloidiasis)
Tissue nematode infestation (lymphatic filariasis)	1st line: Diethylcarbamazine Alternate: Albendazole/doxycycline Ivermectin (onchocerciasis)
Trematode infection	Praziquantel
Cestode infection	Praziquantel Albendazole (cysticercosis and hydatid disease)

- **Praziquantel:** It is a synthetic isoquinoline–pyrazine derivative that acts by disrupting ion transport in worms. It increases cell membrane permeability to calcium and binds to the actin and myosin light chain of helminths, altering their membrane fluidity and reducing the glutathione concentration in schistosomes. It is used as the drug of choice to manage infections caused by cestodes (tapeworms) such as *Taenia saginata*, *Diphyllobothrium latum*, and *Taenia solium*.

OTHER AGENTS WITH ANTIPARASITIC ACTIVITY

Amphotericin B, which is primarily an antifungal, has been used for the treatment of *Naegleria fowleri* infections and leishmaniasis. The use of this drug is limited due to its high toxicity. It acts by disrupting the cell membrane. Some antibacterial drugs are also used for the treatment of parasitic infections (Table 11.11).

Table 11.11 Antibacterial agents and their use in parasitic diseases

Drug	Disease
Paromomycin	Giardiasis, trichomoniasis, leishmaniasis
Clindamycin	Severe babesiosis
Doxycycline	Lymphatic filariasis
Furazolidone	Giardiasis in children
Cotrimoxazole	Prophylaxis for toxoplasmosis and opportunistic intestinal infections
Spiramycin	Toxoplasmosis in pregnancy

Key Points

✓ There are four major sites on the bacterial cell that antibiotics act on: (i) the cell wall, (ii) cell membrane, (iii) nucleic acid, and (iv) ribosome.

✓ Bacteria acquire resistance by mutation or acquisition of resistance genes from other organisms.

✓ Resistance mechanisms:
- Produce enzymes
- Express efflux systems
- Modify the drug's target
- Produce an alternative metabolic pathway

✓ The genetic transfer of resistance occurs by the following mechanisms:
- Conjugation
- Transformation
- Transduction
- Through plasmids

✓ Antiviral agents act by the following mechanisms:
- Inhibition of nucleic acid synthesis
- Inhibition of protein synthesis—protease inhibitors
- Preventing viral entry into the cell and interferon, which prevents viral replication

✓ Antifungals act on the following sites:
- Plasma membrane
- Fungal cell wall components
- Biosynthetic pathways

✓ Anti-protozoal drugs are of the following types:
- Antiamoebic and antigiardial agents
- Antimalarial agents
- Antileishmanials and trypanocidal agents

✓ Anthelmintics act by either of the following modes:
- Killing the worm (vermicide)
- Expelling the worm (vermifuge)

✓ The drugs used for helminthic infestation are chosen based on efficacy, safety profile, ease of administration, and cost.

✓ Antibiotic susceptibility testing by Kirby–Bauer disc diffusion and Stokes methods are key to rational antibiotic therapy and to know the status of susceptibility of bacteria.

✓ Guidelines and principles of appropriate antibiotics for different infections are documented in an antibiotic policy. It is specific to each hospital and healthcare setting.

✓ An antibiotic stewardship programme is an attempt to contain and rationalise antibiotic therapy in order to decrease and arrest antibiotic resistance in a hospital.

Questions

Short answer questions

1. Describe the mechanisms of action of antibiotics on bacteria.
2. Mention the importance of rational antibiotic use.
3. How can an antibiotic stewardship programme help to reduce antibiotic resistance?
4. What are the challenges of antiviral therapy?
5. What are the sites of action of antifungal agents?
6. List the antiprotozoal agents against specific protozoa.
7. Describe the important mode of action of anthelmintics.

MCQs

1. **All the following agents act on the cell wall except ________.**
   a) Penicillins b) Cephalosporins
   c) Fluoroquinolones d) Carbapenems
2. **Important targets for antiviral agents are the following except ________.**
   a) Attachment to host cells b) Peptidoglycan synthesis
   c) Uncoating of the genome d) Nucleic acid synthesis
3. **Antifungals target specific parts of the fungi except ________.**
   a) Plasma membrane
   b) Fungal cell wall components

c) Ergosterol synthesis
d) Lipopolysaccharide synthesis

4. **Anthelmintics act by the following methods except for ________.**
a) Killing the worm
b) Expelling the worm
c) Increasing cell membrane permeability
d) Selectively killing the ovum

5. **Which of the following statements is not true of antibiotic stewardship programmes?**
a) A physician should be the leader of the team to effectively monitor appropriate therapy.
b) A pharmacist should be on the team to improve antibiotic dispensation and monitor the quantity of antibiotics used per patient.
c) A defined daily dose is not an essential component of the stewardship programme.
d) Evaluating the need for treatment after a set period of initial treatment, called the 'antibiotic time-out', is a crucial part of stewardship programmes.

6. **Antimicrobial resistance has become a matter of grave concern globally due to the following reasons except for ________.**
a) Irrational and inappropriate use
b) Wrong antibiotics at the wrong time
c) Combination therapy
d) Antibiotics prescribed for the wrong infection

7. **An example for a narrow-spectrum antibiotic is:**
a) Vancomycin b) Tetracycline
c) Ampicillin d) Amikacin

8. **All the following are protein synthesis inhibitors except:**
a) Aminoglycosides b) Tetracycline
c) Macrolides d) Glycopeptides

9. **Which of the following drugs does not act on the 50S ribosome?**
a) Aminoglycosides b) Chloramphenicol
c) Macrolides d) Clindamycin

10. **Nucleic acid synthesis inhibitors include all the following except:**
a) Sulphonamides b) Fluoroquinolones
c) Rifampicin d) Monobactams

11. **Polymyxins and daptomycin are drugs that act on:**
a) Cell wall b) Cell membrane
c) Ribosome d) Nucleic acid

12. **Methods of acquired drug resistance in bacteria include all the following except:**
a) Production of enzymes
b) Expression of efflux pumps
c) Alternation of porin channels
d) Structural or functional characteristics

13. **Methods of horizontal gene transfer that help in the acquiring of drug resistance in bacteria include:**
a) Transduction b) Transformation
c) Conjugation d) All of these

14. **Antibiotic sensitivity testing methods include all the following except:**
a) Eijkman method b) Stokes' method
c) Epsilometer method d) Kirby–Bauer method

15. **Automated method of antibiotic sensitivity testing is:**
a) E-test b) VITEK™
c) Stokes' method d) BACTEC™

16. **The streaking method used for antibiotic sensitivity testing is:**
a) Streak method b) Stokes' method
c) Lawn culture d) Pour plate method

17. **The ideal culture media for performing disc diffusion method of antibiotic sensitivity testing (AST) is:**
a) Nutrient agar b) Blood agar
c) Nutrient broth d) Mueller-Hinton agar

18. **The method of antibiotic sensitivity testing where comparison of zone size between a test and control strain is done on the same plate is called:**
a) Kirby-Bauer method b) Stokes' method
c) E-test d) VITEK

19. **The most commonly used AST method is:**
a) Broth microdilution method
b) Agar dilution method
c) Kirby-Bauer method
d) E-test

20. **Standard ATCC strains of all the following organisms are used as quality control except:**
a) *E. coli*
b) *Klebsiella pneumonia*
c) *Pseudomonas aeruginosa*
d) *Staphylococcus aureus*

21. **When a small degree of resistance needs to be demonstrated for therapeutic purposes, which method of AST is better?**
a) Diffusion method b) Dilution method
c) E-test d) All of these

22. **The lowest concentration of a drug that inhibits growth in broth dilution method is called:**
a) Minimum bactericidal concentration
b) Minimum bacteriostatic concentration
c) Minimum inhibitory concentration
d) Trough concentration

23. **Agar dilution method of AST is better than broth dilution for all the following reasons except:**
a) It is easy to perform
b) It is easy to observe
c) Multiple strains can be tested simultaneously
d) There are no chances of contamination

24. **Guidelines on antibiotic usage in a hospital are known as:**
a) Antibiotic policy b) Antibiotic stewardship
c) Antibiotic audit d) Antibiogram

25. **The front-end or pre-prescription approach in antimicrobial stewardship is a:**
a) Non-restrictive approach wherein antibiotics are prescribed by all clinicians.
b) A strategy to ensure restricted prescription of reserved antibiotics.
c) Strategy based on reviewing prescriptions and advising clinicians to modify antibiotic based on the susceptible data.
d) Strategy to advocate de-escalation of an antibiotic following empiric therapy.

26. **Which of the following is an effective strategy for an antimicrobial stewardship programme?**
a) Back-end approach, in which the physician without authority cannot write an order for a restricted antibiotic
b) Front-end approach, in which prescription is audited after the antibiotic has been dispensed
c) Reporting on antibiotic use and resistance to doctors, nurses, and relevant staff
d) All of the above

27. **All of the following antivirals act by inhibition of nucleic acid synthesis except:**
a) Amantadine b) Acyclovir
c) Nevirapin d) Zidovudin

28. **Which one of following is a reverse transcriptase inhibitor?**
a) Nevirapin b) Acyclovir
c) Lamuvudin d) Indinavir

29. **The drug used in the treatment of influenza A is:**
a) Amantidine b) Zeduvidin
c) Zalcitabin d) Indinavir

30. **Which of the following is used in the treatment of hepatitis B infection?**
a) Zalcitabin b) Zeduvidin
c) Amantidin d) Adefovir

31. **Which one of the following is false with regard to interferons?**
   a) Protect the host against viral infections.
   b) They are some of the effective antiviral agents.
   c) They do not protect the cell which produced them.
   d) They are not specific for a given virus.
32. **All of the following diseases can be treated with antiviral drugs except:**
   a) HIV/AIDS b) Dengue
   c) Hepatitis C d) Hepatitis B
33. **What is the mechanism of action of amphotericin B on fungi?**
   a) Inhibits ergosterol biosynthesis
   b) Binds to sterols causing perturbations in the cell membrane
   c) Inhibits DNA and RNA synthesis
   d) Inhibits beta-glucan synthesis in cell wall
34. **Which of the following is correct regarding echinocandins?**
   a) Act by inhibition of microtubule assembly
   b) Drugs of choice for dermatophyte infections
   c) Useful in treatment of systemic infections caused by *Candida* and *Cryptococcus*
   d) Both b and c
35. **A patient is diagnosed as having a trematode infection. Lacking a more specific identification of the causative organism, which of the following drugs would most likely be effective?**
   a) Thiabendazole b) Praziquantel
   c) Diethylcarbamazine d) Tetracycline
36. **A 4-year-old girl is brought to the pediatric outpatient department with complaints of nighttime restlessness and itching around her anus. She is prescribed mebendazole. The mechanism of action of mebendazole is by:**
   a) Inhibition of microtubule synthesis
   b) Block glucose uptake in the parasite
   c) Depletion of glycogen stores in parasite
   d) All of the above
37. **Which of the following is correct regarding ivermectin?**
   a) It acts by causing hyperpolarisation and paralysis of the worm.
   b) It is the drug of choice for all nematodes.
   c) It can cause a disulfiram-like reaction with alcohol intake.
   d) All of the above.

Answer: 1:c, 2:b, 3:d, 4:d, 5:c, 6:c, 7:a, 8:d, 9:a, 10:d, 11:b, 12:d, 13:d, 14:a, 15:b, 16:c, 17:d, 18:b, 19:c, 20:b, 21:b, 22:c, 23:d, 24:a, 25:b, 26:d, 27:a, 28:a, 29:a, 30:d, 31:b, 32:b, 33:b, 34:c, 35:b, 36:d, 37:a

CHAPTER 12

Healthcare-associated Infections and Their Control

(MI8.5–8.7)

Chapter Navigation

- Types of healthcare-associated infections
- Standard precautions for hospital infection control
- Biomedical waste management
- Infection control practices in the hospital

Learning objectives

- Define healthcare-associated infections (HCAIs) and identify factors contributing to these infections
- Discuss measures to control HCAIs
- Explain standard precautions and describe steps of hand hygiene and parts of personal protective equipment
- Discuss the steps of biomedical waste management in a healthcare setting

INTRODUCTION

- The WHO defines a healthcare-associated infection (previously known as 'nosocomial' or 'hospital-acquired') as an infection occurring **in a patient in a hospital or other healthcare facility in whom the infection was not present or incubating at the time of admission.**
- Approximately 5–10 per cent of patients admitted to healthcare services in developed countries and more than 25 per cent of such patients in developing countries acquire **healthcare-associated infections (HCAI)**.
- Such infections increase morbidity, mortality, and the burden of cost to the patient and the hospital.
- It is believed that up to 20 per cent of healthcare-associated infections in developed countries and 40 per cent in developing countries can be prevented.
- Moreover, in developed countries, 5–10 per cent of such infections occur as outbreaks or in clusters; these figures are likely to be higher in developing countries. Therefore, concerted efforts are now being taken to implement infection control programmes in hospitals and health centres.
- The Centre for Disease Control's **National Health Safety Network (NHSN)** in USA tracks healthcare-associated infections. The definitions furnished by the NHSN are followed worldwide for the surveillance of HCAI.
- In India, the **Ministry of Health and Family Welfare** has laid down national guidelines for infection prevention and control.
- **CDC and AIIMS-ICMR** have also established a network for the surveillance of HCAI.
- Healthcare-associated infections are also defined as infections **occurring after 48 hours of admission/procedures in the hospital and up to 30 days after a procedure/discharge from the hospital**. The term also applies to the following:
  – An infection related to an operative procedure that occurs at or near the surgical incision within 30 days of the procedure or within 90 days in cases where prosthetic material is implanted at surgery (as defined by CDC)
  – Infections contracted in the hospital but not evident until after discharge

  HCAI does not apply to patients who are in the incubation period of an infection—such an infection is deemed to have been **acquired in the community**.
- Healthcare-associated infections are influenced by the interplay between the host, pathogen, and environmental factors. This interplay is depicted in Fig. 12.1.

Host (patient risk factors)

Microbes
Multidrug resistance
Ability to survive in adverse conditions
Resistance to disinfectants

Environment
Unclean, unsterile instruments
Healthcare personnel
Improper biomedical waste disposal

Fig. 12.1 Interplay of factors influencing HCAIs

TYPES OF HEALTHCARE-ASSOCIATED INFECTIONS

Common types of healthcare-associated infections, their risk factors, and organisms commonly associated with them are presented in Table 12.1.

Healthcare-associated Bacteremia (Bloodstream Infections [BSI])

While bacteremia may result from infections at any site, healthcare-associated bloodstream infections are commonly caused by infected intravenous cannulae (Fig. 12.2) such as the central venous catheter and peripheral lines. The longer the cannula is kept in situ, the greater is the risk of infection.

Table 12.1 Common types of healthcare-associated infections

Type of infection	Risk factors	Common organisms*	Method of prevention
Bloodstream infections: • Catheter-related bloodstream infection (CRBSI) • Central line-associated bloodstream infection (CLABSI)	• Central or peripheral intravenous catheters (femoral, subclavian, and jugular) • Immunocompromised patient • Total parenteral nutrition • End-stage renal disease	• ESKAPE pathogens: *E. coli, Staphylococcus* (MRSA), *Klebsiella, Acinetobacter, Pseudomonas* and *Enterobacter* • Other organisms: *Enterococcus, Candida*	• Proper preparation of the insertion site • Care to be exercised while using IV lines (bundle care)
Pneumonia (including ventilator-associated pneumonia)	• Patients on ventilator and those who have had a tracheostomy	• *Acinetobacter, Pseudomonas,* and *Klebsiella*	• Ventilator care • Respiratory physiotherapy
Urinary tract infection	• Catheter	• *E. coli, Proteus,* and *Enterococcus*	• Catheter care, appropriate types of catheters for long-term use
Surgical site infection	• Prolonged surgery • Surgical drains	• MRSA, coagulase-negative staphylococci, *E. coli*	• Surgical prophylaxis • Proper wound care and dressing
Post-antibiotic diarrhea	• Prolonged antibiotic therapy	• *Clostridioides difficile*	• Short duration of therapy (antibiotic monitoring) • Probiotics

*Colour-coded to depict the relative significance of causative organisms with respect to the organ system discussed in that chapter—**red denotes the most significant, orange indicates a slightly less etiological value and blue, lesser so.**

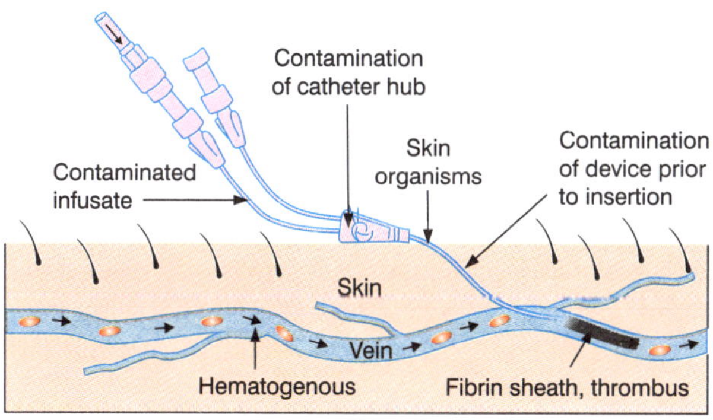

Fig. 12.2 Sources of catheter-associated bloodstream infections

The Centers for Disease Control and Prevention (CDC) define catheter-related bloodstream infection (CRBSI) and central line-associated bloodstream infection (CLABSI) as follows:

1. **Catheter-related bloodstream infection (CRBSI):** It is suspected if the following criteria are fulfilled:
   i. The same pathogen is isolated from a quantitative blood culture drawn through a central line and a second sample from a peripheral vein, such that the colony count is three times higher in the sample from the central line than that obtained from a peripheral vein **or**
   ii. A shorter time to positive culture (less than 2 hours) in the central line sample than the peripheral sample (**differential time to positivity [DTP]**)

   If, in a sample collected from the peripheral vein, an organism grows earlier than it does in a sample from the central line, or if the former has a higher colony count, the infection is not considered to be a CRBSI.
2. **Central line-associated bloodstream infection (CLABSI):** A central line or central venous catheter is a catheter inserted into a large vein in the neck, chest, or groin for parenteral management. This is different from intravenous catheters or IV lines, which are placed in a peripheral vein in the arm or hands for a short period.

   Bacteria can enter the bloodstream through the central line and cause infection.

   CLABSIs are defined based on the isolation of a pathogen from a blood culture (**a single isolation of an organism not commonly present on the skin or two or more blood cultures of organisms commonly present on the skin**) in a patient who had **a central line at the time of infection or within 48 hours prior to the development of the infection.** This is not related to any other source of infection the patient may have had when he was admitted to the hospital.
3. **Infection at the site of insertion of the catheter:** This is a third type of infection, which may develop due to improper catheter care. Such infections may be of two types:
   i. **Exit site infection:** Signs of inflammation are seen around the catheter exit site and are

confined to an area measuring <2 cm; a culture-positive exudate is present in such cases.

ii. **Tunnel infection:** This refers to an inflammation extending along the vein from the entry site for a length of more than 2 cm, which is associated with pain and tenderness along the subcutaneous track. The exudate from the exit site shows positive bacterial growth.

Healthcare-associated Pneumonia

This is a lower respiratory infection that presents clinically 48 hours or more after hospitalisation in a patient who was not incubating the infection at the time of hospital admission.

Ventilator-associated pneumonia (VAP)

Ventilator-associated pneumonia is defined as pneumonia that develops 48–72 hours after endotracheal intubation with the following criteria:

i. New or extensive infiltrates
ii. Fever and hyper or hypothermia
iii. Increased white blood cell count $<4,000/mm^3$ or $>11,000/mm^3$
iv. PO_2 or FiO_2 >240 or signs of acute respiratory distress syndrome (ARDS) and the detection of the causative agent from endotracheal or bronchoalveolar lavage (semiquantitative culture)

These findings are scored based on a **clinical pulmonary infection score (CPIS)**.

Catheter-associated Urinary Tract Infection (CAUTI)

- These are the commonest healthcare-associated infections (2 per cent of all admitted cases).
- CAUTI is defined as a UTI that occurs in cases where an indwelling urinary catheter was in place for >2 consecutive days in an inpatient location on the date of event or the day before, with the day of device placement being considered as day 1.

Healthcare-associated Wound Infections (Surgical Site Infections [SSI])

- Infections at the incision site or the subcutaneous tissue around the incision site or organ or space that occur within 30 days of surgery or within 90 days in cases where prosthetic material has been implanted in surgery are known as healthcare-associated wound infections or surgical site infections.
- The infections can range from superficial infections involving only the skin to infections of the deeper tissues under the skin, organs, or implanted material.
- Such an infection will have at least one of the following **features**:
  - Purulent drainage from the superficial incision
  - Organism(s) identified from an aseptically-obtained specimen from the site of infection
  - Localised pain or tenderness, localised swelling, erythema, or heat
- **Risk factors** for developing SSI—underlying disease, extremes of age, diabetes, contaminated wound, prolongation of surgery, poor nutritional status, tobacco use, type of surgery, inappropriate use of antibiotics.

Other Healthcare-associated Infections

Needlestick and sharps injuries—every percutaneous needlestick and sharps injury carries a risk of infection by blood-borne pathogens like hepatitis B, C, and D viruses (HBC, HCV, and HDV) and HIV as well as other infections. These infections can arise due to an injury (prick) with the following:

- Devices with hollow-bore needles, suture needles, and IV stylets
- Needle devices that are taken apart or manipulated (e.g., re-capping a needle) by the healthcare worker during or after drawing blood (as done in a phlebotomy)
- Syringes that retain an exposed needle after use
- Butterfly needles that are attached to tubing and cannot be disposed of in a sharps container

Healthcare-associated Infections Due to Hepatitis B and C Viruses (Transfusion-associated Infections)

- Patients receiving blood transfusions or undergoing renal dialysis are at risk of contracting hepatitis B and C.
- Hospital personnel working with patients in dialysis and cardiac surgery units and those who handle blood specimens are also at high risk of contracting these infections.
- Although screening for hepatitis B surface antigen (HbsAg) and anti-hepatitis C virus antibody has markedly reduced the risk of acquiring such infections, other predisposing factors still exist.
- Strict attention to aseptic technique greatly reduces this risk to hospital personnel.

Healthcare-associated Episodes of Acute Gastroenteritis

Outbreaks of *Salmonella* food poisoning, enterotoxic manifestations arising from staphylococcal contamination of cooked food, and outbreaks of diarrhea caused by *Escherichia coli*, and certain enteroviruses fall under this category of HCAIs.

Antibiotic-associated Diarrhea

- Diarrhea sometimes occurs as a result of the prolonged use of antibiotics like fluoroquinolones, cephalosporins, penicillins (ampicillin, amoxicillin), and clindamycin.

- In such cases, it occurs within a few days of starting the antibiotic due to the suppression of the normal flora of the gut. This leads to the overgrowth of *C. difficile*, which secretes exotoxins. These endotoxins act on the colonic epithelium and immune cells and induce events that result in fluid secretion, inflammation, and tissue damage, leading to diarrhea.

SOURCES AND RESERVOIRS OF HCAIs

Microbes that cause HCAIs can originate from endogenous or exogenous sources.

- **Endogenous sources:** This comprises infecting microorganisms that are not pathogenic and are often part of the body's normal flora under normal conditions. However, when there is an underlying (usually debilitating) disease or when invasive diagnostic and therapeutic procedures (including the use of immunosuppressives and antibiotics) have been performed, such a microbe spreads and localises at a site where it may produce an infection.
- **Exogenous sources:** Some microorganisms come from patients' environments—healthcare workers (HCW), visitors, patient care equipment, or medical devices. For any infection or colonisation to occur, a sequence of events occurs that transmits an infectious microorganism to a susceptible host. These interactions are more common in hospital environments and provide microorganisms with an opportunity to cause infection in susceptible hosts (Fig. 12.3).

Cross-infection

In this type of infection, the microorganism spreads by:

- **Direct contact** between an infected patient and another individual—by droplets of saliva or respiratory secretions or from the patient's hands (e.g., herpes)
- **Air**—respiratory pathogens through aerosols (e.g., TB, respiratory virus)
- **Hospital personnel** can directly transmit an infection to a patient through contaminated hands; alternatively, the organisms may be harboured in the respiratory tract, skin, or gut of the healthcare worker, from where it may get transmitted by direct contact or aerosols to patients (e.g., MRSA, bacterial or viral respiratory infections)

Infections from Environmental Sources

In these types of infections, the microorganisms concerned are derived from the following sources:

- Surfaces that are contaminated by the patient's secretions, excreta, blood or body fluids, animals, and insects (e.g., salmonellosis, hepatitis A infections)
- Inanimate objects contaminated by patients' secretions, the hands of healthy or unhealthy hospital personnel, visitors, food or infected water, animals, and insects

Contaminated needles

- The use of disposable syringes, needles, and other equipment has reduced the incidence of HAIs.
- Needlestick injury is a concern for healthcare personnel—a large proportion of healthcare workers experience percutaneous exposure to infectious diseases every year, a majority of which are due to needlestick injuries; the most common pathogens transmitted by needlestick injury are hepatitis B and C and HIV, in that order.

MICROBES CAUSING HCAIs

Almost any microbe can cause healthcare-associated infections. However, more than three-fourths of these infections are caused by the organisms listed in Table 12.2.

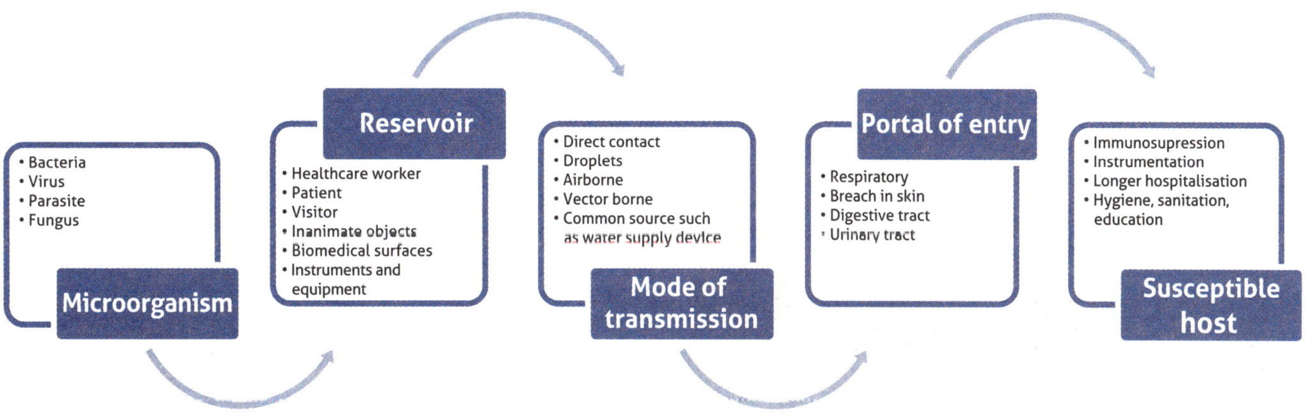

Fig. 12.3 HCAI spread in hospitals

Table 12.2 Common microorganisms that cause healthcare-associated infections

Bacteria	
• *Staphylococcus aureus*	• *Acinetobacter baumannii*
• *Enterococcus* species	• *Enterobacter* species
• *Escherichia coli*	• *Proteus* species
• Coagulase-negative staphylococci	• *Clostridium difficile*
• *Klebsiella* species	• *Bacteroides* species and other anaerobes
• *Pseudomonas aeruginosa*	• *Mycobacterium tuberculosis*
Viruses	
• Influenza • Noro virus	
Fungi	
• *Candida* species • Yeasts other than *Candida*	
Multidrug-resistant pathogens	
MRSA (methicillin-resistant *Staphylococcus aureus*) Vancomycin-resistant Carbapenem-resistant Extended-spectrum cephalosporin-resistant Multidrug resistant	*E. faecium* *Pseudomonas aeruginosa* *Klebsiella* species *Acinetobacter baumannii*
Novel organisms	
Burkholderia cepacia *Stenotrophomonas maltophila* *Candida auris* *Clostridium sordellii*	
Blood-borne infections	
Hepatitis B, C, and D and HIV	

INFECTION CONTROL

- Healthcare-associated infections can occur because of interactions that allow the infectious agent to enter a susceptible host. This chain of transmission favours healthcare workers who form a link between the hospital environment, host, and agent.
- To break this chain of transmission, certain interventions need to be instituted, which call for structural and cultural changes on the part of the health teams so as to demonstrate long-term effectiveness.
- Interventions for the prevention and control of HCAIs include standard specific precautions and proper biomedical waste disposal (Table 12.3).

Standard Precautions

Standard precautions are the minimum infection prevention practices that apply to all patient care, regardless of the suspected or confirmed infection status of the patient in any setting where healthcare is delivered (Fig. 12.4). It considers every person as potentially infectious and susceptible to infection.

Hand hygiene

- The hands of healthcare workers can have two types of microbial flora—resident and transitory ones. Therefore, healthcare workers can transmit microorganisms from one infected patient to another or through inanimate objects.
- Hand hygiene reduces the number of microorganisms present on the hands.

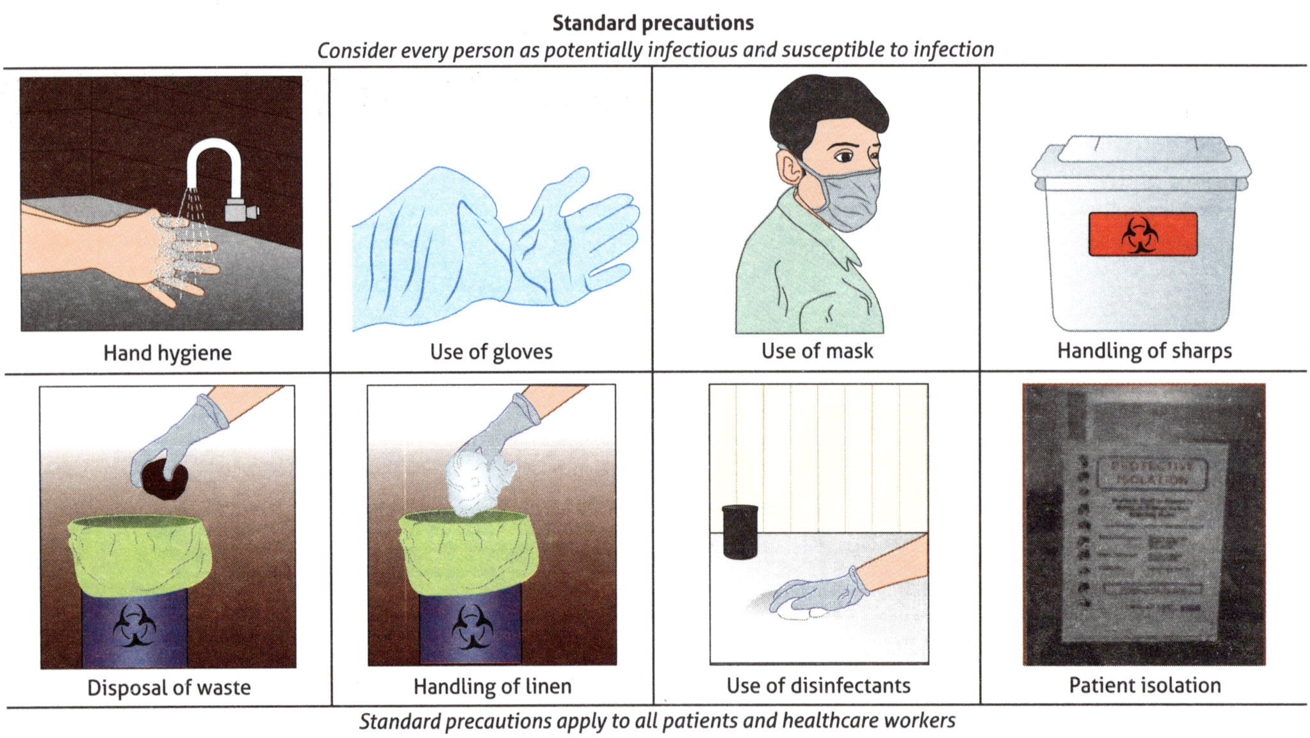

Fig. 12.4 Standard precautions (*Source*: Department of Clinical Microbiology, Christian Medical College and Hospital, Vellore)

Table 12.3 Standard and specific precautions for control of HCAI

Elements of standard precautions

- Hand hygiene
- Personal protection equipment (PPE)—gown, mask, face protection, gloves, goggles, etc.
- Safe injection practices
- Sharps management
- Proper disposal of biomedical waste
- Treatment of spills
- Patient care equipment/devices management
- Environmental control
- Respiratory hygiene/cough etiquette

Specific precautions

Source control	Patient to wear mask	Isolation of patient	Restriction of movement of patients	Use of PPE appropriately by healthcare workers	Disposable or dedicated patient equipment	Prioritise cleaning or disinfection of patient rooms
Contact precautions	No	No	Limit patient's movement outside his room Follow contact precautions if transfer is needed Covering colonised areas of the patient's body	Gloves and gown	Yes	Daily cleaning and disinfection No room and material should be allowed to be used by another patient prior to cleaning
Droplet precautions	Patient MUST wear a mask	In single room if possible Restrict movement of patients	Yes	Gloves, apron, and mask	Yes	To be done
Airborne precautions	Fit-tested, NIOSH-approved N95 or higher level respirator	In an airborne infection isolation room If not possible, then mask the patient and place him in a private room with the door closed	Yes	Full PPE with fit-tested, NIOSH-approved N95 or higher level respirator	Yes	To be done

- It is considered one of the most effective strategies for preventing cross-transmission of the microorganisms that cause HCAIs.
- Two methods can be used for hand hygiene: (1) washing hands with water and soap and (2) rubbing hands with an alcohol-based solution.
- While maintaining hand hygiene, attention must be paid to washing between the fingers, under the nails, and both sides of the hands. Drying is equally important as it increases compliance and removes moisture from the hands which is conducive to the growth of bacteria.
- The basic steps of handwashing take around 30–40 seconds in entirety (Fig. 12.5). Using a hand rub (Fig. 12.6) generally reduces the time to around 15–20 seconds.
- However, the process of a **surgical scrub** (prior to performing surgery) takes five minutes.

Five moments of hand hygiene are (Fig. 12.7):

- Before touching a patient, even if gloves are to be worn
- Before coming out of the patient care area after touching the patient or the patient's immediate environment
- After contact with blood, body fluids or excretions, or wound dressings
- Prior to performing any aseptic task (e.g., placing an intravenous line or preparing an injection)
- If the hands are likely to move from a contaminated body site to a clean body site during patient care and after the removal of gloves

Personal protective equipment (PPE)

- PPE refers to equipment intended to protect healthcare workers (HCWs) from exposure to or contact with infectious agents. Examples include gloves, gowns, face masks, respirators, goggles, and face shields.
- A full PPE is required while providing care to patients who have highly infectious diseases like COVID, Ebola, and Nipah virus infections, which require isolation and barrier nursing in containment areas of the hospital.

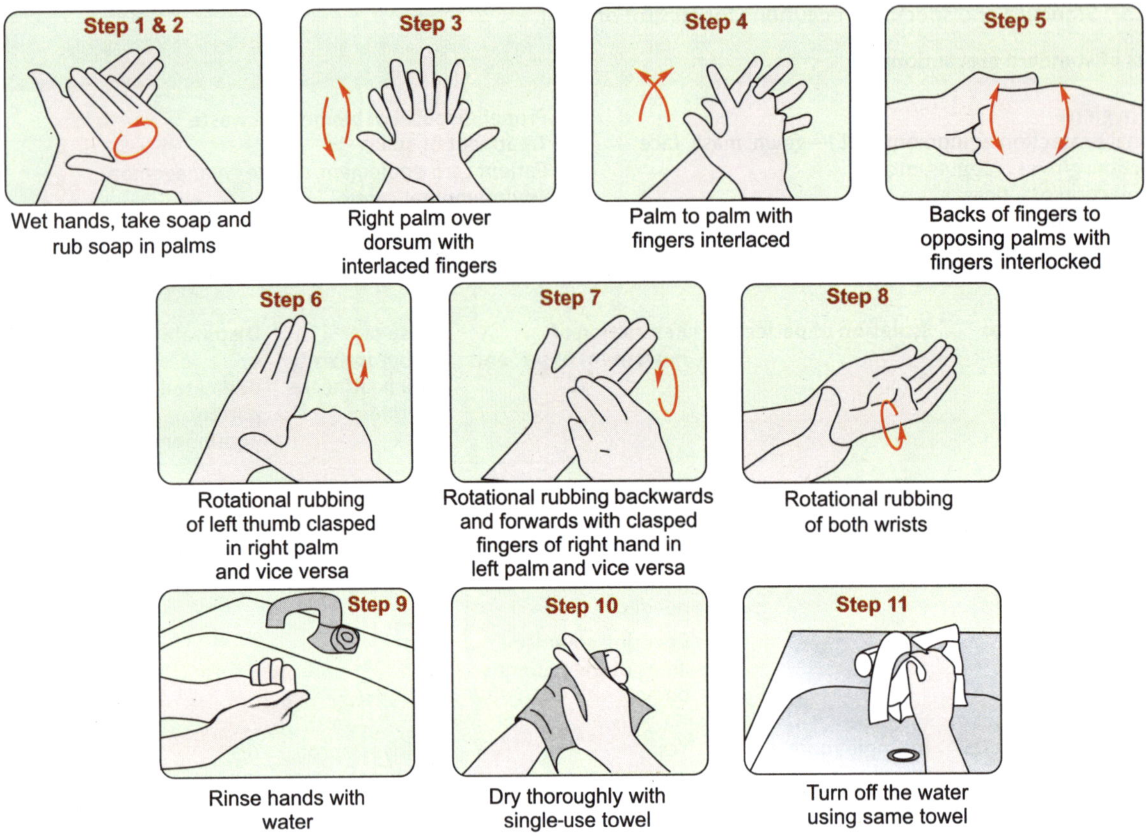

Fig. 12.5 Steps of hand washing followed by rinsing and drying

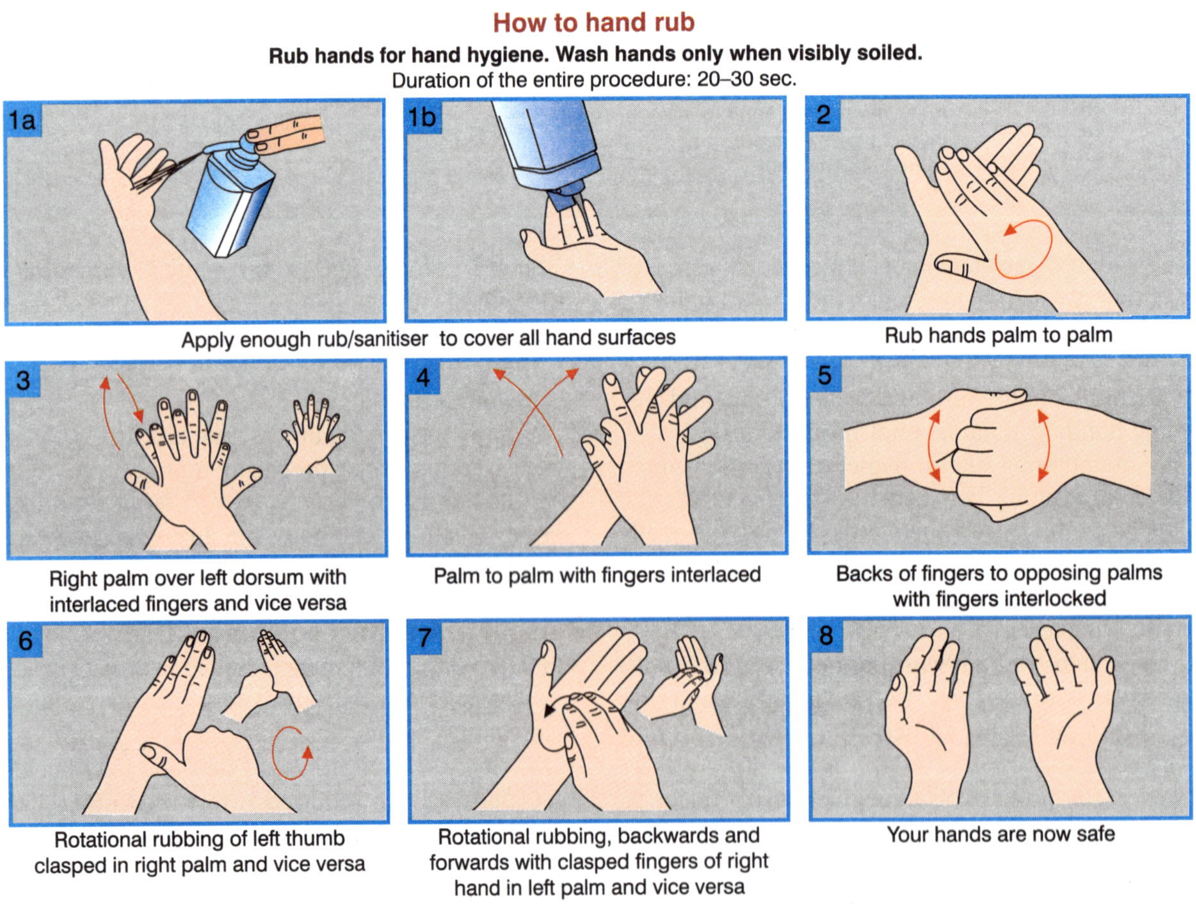

Fig. 12.6 Correct method of hand rubbing

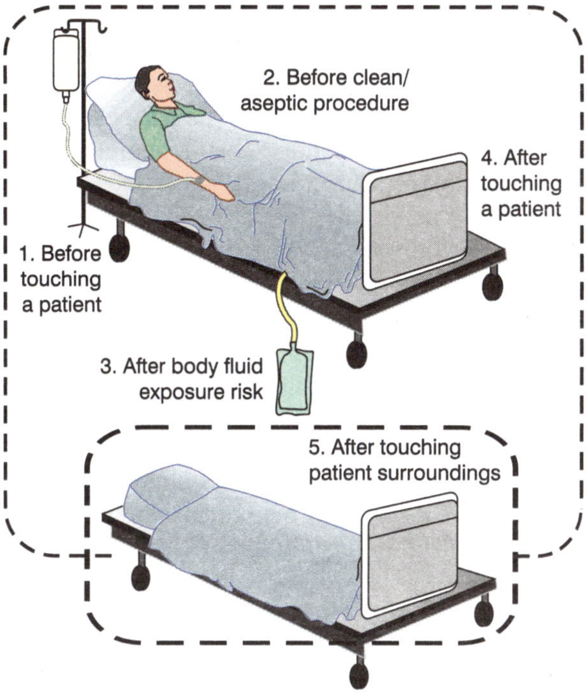

Fig. 12.7 Five instances where hand hygiene steps should be undertaken

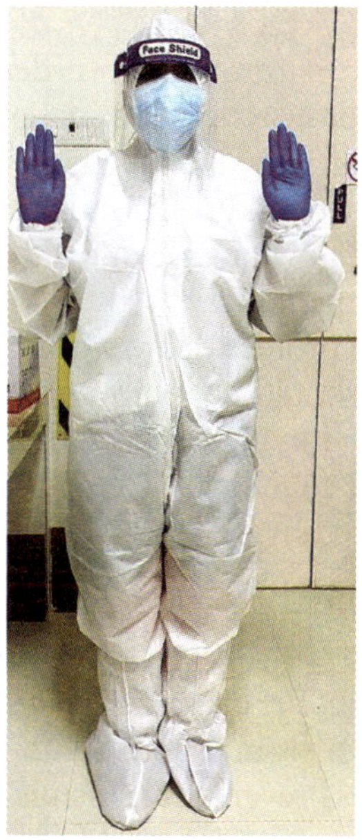

Fig. 12.8 Full PPE consisting of a gown, mask, face shield, gloves, and shoe-covers worn while caring for patients with highly infectious diseases like COVID, Ebola, etc. (*Source*: Dept. of Microbiology, Pondicherry Institute of Medical Sciences [PIMS], Puducherry)

- The same PPE should not be worn for the care of more than one patient.

Components of PPE (Fig. 12.8)

- **Gowns and aprons:** A gown is meant to cover the body from the neck to the knees and the arms down to the wrists with means to fasten and unfasten it, usually at the back. Gowns are meant to protect HCW's clothes/scrubs from getting contaminated while performing procedures that might create spatters of blood or body fluids. They may be disposable or non-disposable and are made of cloth, linen, or plastic. An apron is made of waterproof material and is worn over the gown as an additional barrier.
- **Safety eyewear such as glasses, wraparounds, and goggles:** The purpose of safety eyewear is to prevent aerosols, spatters, and droplets from coming in contact with the conjunctival mucous membrane.
- **Face protectors and face shields:** These cover the entire face so that the wearer does not require additional eye protection or a mask to guard against droplet-transmissible agents.
- **Masks:** Just like the conjunctiva, the mucous membranes of the nose and mouth are portals of entry for infectious agents. Hence, it is essential to protect them. The barriers can be in the form of masks or respirators. Masks are made up of impervious material and can be pleated or preformed. The efficiency of masks reduces with moisture and hence, they should be changed periodically.
- **Gloves:** These should be donned before touching blood, body fluids, secretions, excretions, contaminated items, mucous membranes, and non-intact skin. Gloves are intended to prevent contact of the hands with contaminated sources such as the skin of patients colonised or infected with multidrug-resistant microorganisms. They should be pulled up over the top of the cuff of the gown while using long-sleeved gowns. Gloves should be used only when absolutely indicated and they should always be changed between patients or if they develop breaks or tears. Hand hygiene should be performed immediately before and, even more importantly, after using gloves. Under no circumstance should glove use replace hand hygiene.
- **Boots, jumpsuits, overall, and hoods:** These do not provide any added protection but only prevent the soiling of clothes.

Donning of PPE

- Before donning PPE, the worker must tie up their hair and remove any jewelry or accessories on their body. The worker must ensure that the PPE is of the correct size before breaking open the seal.
- Hand hygiene should be performed before donning PPE.
- The PPE must be worn in the following order: gown → mask → eye protection → gloves.

Doffing of PPE

- PPE should be removed in an order that minimises the potential for cross-contamination.
- PPE should doffed in the following order:
  - Remove gloves and perform hand hygiene.
  - Remove the gown inside out without touching the exterior of the gown.
  - Remove the protective eye wear followed by mask. The mask should be removed by looping the ear loops or by untying cords. The front of the mask must not be touched.
  - Perform hand hygiene.

Sharps management

Injection safety (safe injection practices)

Every percutaneous needlestick and sharps injury carries a risk of infection from blood-borne pathogens. Safe injection practices prevent the transmission of infectious diseases in-between patients and between patients and healthcare workers (HCW) during the preparation and administration of parenteral medications. The following precautions are recommended:

- Aseptic techniques should be used while preparing and administering medications.
- Access diaphragms of medication vials should be cleaned with 70% alcohol before inserting a device into the vial.
- A single syringe should be used only for a single use in a single patient.
- A used syringe should not be used to draw medication from a vial or solution.
- Fluid infusion or administration sets (e.g., intravenous tubing) should not be used for more than one patient.
- Multi-dose vials should be dedicated to a single patient whenever possible.
- Used needles should never be re-capped.
- In the event of a needlestick injury/splash of blood or body fluid into the eye, the area should be washed with running tap water or an eye wash.
- Needles and sharps should be disposed of in puncture-proof containers and empty sharps containers before they are full.

Needlestick or sharp injury

In the event of a needlestick or sharp injury, the following protocol needs to be followed immediately to prevent the transmission of infection:

- The event should be immediately reported to the infection control nurse and medical officer. As a part of the injury reporting system, the source patient's status should be verified. (If the status is not known, and with the consent of the patient, the person may be tested for HIV, hepatitis B, and hepatitis C.)
- If the patient is known to be HIV-positive or if his status is unknown, post-exposure prophylaxis (PEP) should be initiated (according to NACO, CDC, and WHO guidelines; details in Chapter 22).
  - Hepatitis B—if the healthcare worker is vaccinated, no treatment is required, but if not vaccinated, HBIg and HB vaccine are initiated as described in Table 12.4.
  - Hepatitis C—no treatment is currently recommended.
- Needlestick injury is prevented by following safe practices—ensuring that re-capping is not done, disposing needles and sharps in puncture-proof containers, and emptying sharps containers before they are full.

Environmental cleaning

Cleaning refers to the removal of visible soil and organic contamination from a device or environmental surface using the physical action of scrubbing with a

Table 12.4 Risk of infection and management of a needlestick injury

Pathogen	Risk of infection from a known positive person	Steps in management	If source is NOT positive
HIV*	• 0.3%	• Wash the area immediately • Determine status of patient's infection (if not known) • Start post-exposure prophylaxis as per protocol • No post-exposure prophylaxis available for HCV	• Only local cleaning • No treatment
HBV**	• 0% if post-exposure prophylaxis (PEP) is immediately instituted • 6–30% without PEP		
HCV	• 1.8%		

*HIV protocol: The course is a four-week regimen of two drugs—zidovudine (ZDV) and lamivudine (3TC). If the patient tests negative, no PEP is required.

**HBV protocol: If the HCW has been vaccinated for HBV and the antibody titre is >10 mIU/mL, then the person is a responder, and hence, requires no treatment. If their titre is <10 mIU/mL or the HCW is not vaccinated, then hepatitis B immunoglobulin (HBIg) is administered immediately along with the full course of HB vaccine. If the source is unknown, HBIG and a full course of HB vaccine are administered.

surfactant or detergent and water or an energy-based process (e.g., ultrasonic cleaners) with appropriate chemical agents. This process removes large numbers of microorganisms from surfaces and must always be performed before disinfection.

General cleanliness

The hospital and its premises must be always maintained in a clean and hygienic state. This requires frequent washing with soap and water, mopping of floors (in a zig-zag manner), and good housekeeping practices. Biowaste management should be carried out to prevent harm from medical waste, minimise its volume, retrieve reusable materials, and ensure safe and economical disposal.

Medical equipment

- Medical equipment may be reusable or meant for single use.
- Reusable medical equipment (e.g., endoscopes) come with instructions for their cleaning and disinfection or sterilisation, as appropriate.
- Reusable medical equipment (e.g., blood glucose meters and other point-of-care devices, surgical instruments, endoscopes) is cleaned and reprocessed appropriately before being used on another patient.
- Single-use devices are labelled by the manufacturer for one-time use and come with reprocessing or disposal instructions.

Respiratory hygiene/cough etiquette

- Any individual with signs of illness including cough, congestion, rhinorrhea, or an increased production of respiratory secretions needs to be promptly identified to prevent the transmission of respiratory infections.
- They must cover their mouths/noses when coughing or sneezing, use and dispose of tissues properly, and perform hand hygiene following contact with respiratory secretions (Fig. 12.9).

Isolation

- All patients infected with tuberculosis, COVID-19, MRSA, and multi drug-resistant organisms need to be isolated and treated by barrier nursing.
- The nursing care is individualised so that the infection does not spread to other patients via the nurse.
- All personal protective equipment is dedicated to single-use.

Precautions in the Operating Theatre

Besides maintaining clean environments, equipment, and sterile instruments in an operation theatre, it is vital to control clean airflow.

- Conventional or plenum-type ventilation in operating theatres maintains approximately 20 air changes per hour; in laminar flow operating theatres, there may be as many as 300 air changes per hour.
- Airborne organisms typically exist in counts of between 150 and 300 CFU/m^3 (CFU = colony-forming units) in conventional operating theatres. With laminar flow ventilation, a count of 10 CFU/m^3 or less is achieved.
- Table 12.5 presents the precautions recommended for maintaining airflow in healthcare settings.

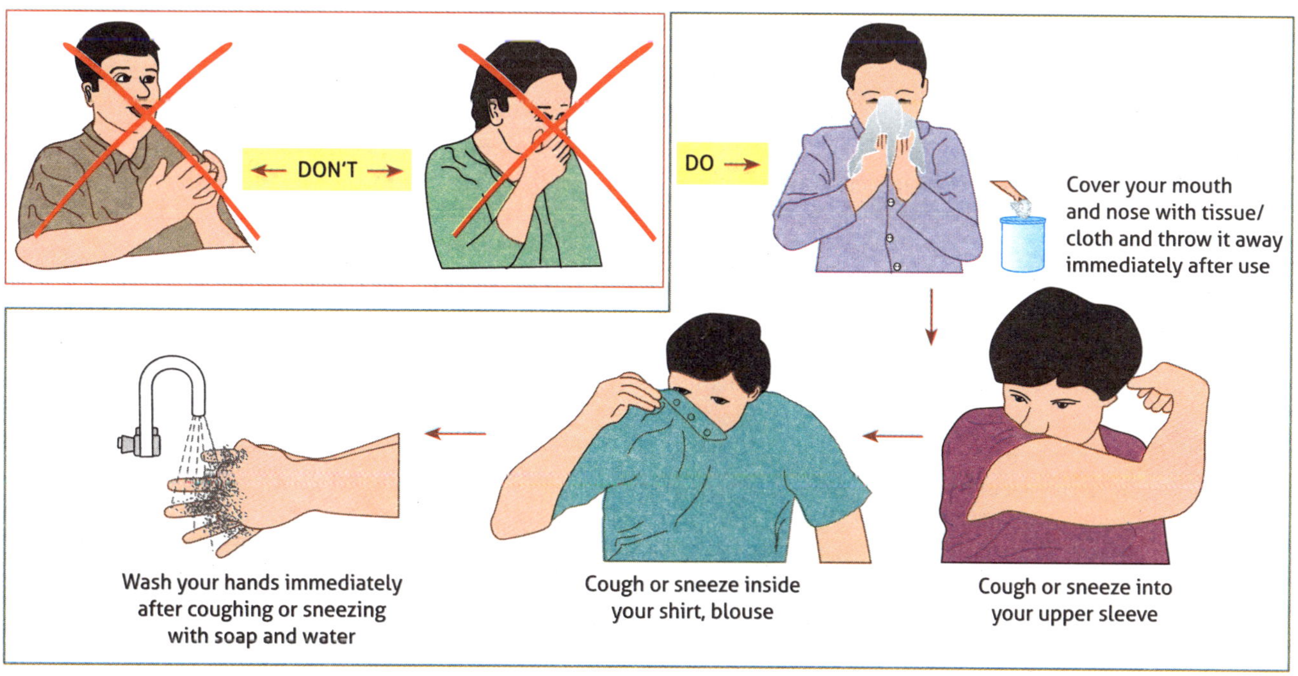

Fig. 12.9 Respiratory hygiene while coughing and sneezing

Table 12.5 Recommendations for maintaining airflow

Type of operation theatre	Acceptable bioload (CFU/m^3) in clean OT in the morning	Acceptable bioload (CFU/m^3) in OT during surgery
Conventional	<35	<180
OT with laminar flow	<10	<35
Ultraclean air OT	<1	<10 at the centre <20 at the periphery

CARE BUNDLE

This refers to a set of several individual procedures performed in a group (or bundle) in total and together. The individual components of a care bundle are an essential part of infection control practice. Insertion and maintenance bundles for devices such as CVP line, urinary catheter, and ventilators are presented in Table 12.6.

Table 12.6 Care bundles for prevention of HCAI

Care bundle	Insertion care	Maintenance/daily care
Central venous line care	Hand washing/hand hygiene Barrier precautions upon insertion (surgical mask, sterile gloves, cap, sterile gown, and large sterile drape) Skin antisepsis with chlorhexidine 2% w/v in 70% isopropyl alcohol Optimal selection of catheter site (avoid femoral site; prefer subclavian vein) Daily review of the necessity of central line and prompt removal when not needed Dedicated trained and competent staff for central line insertion Standardised insertion packs Availability of insertion guidelines and using checklists with trained observers Use of single lumen unless otherwise indicated Use of semipermeable dressing	Hand hygiene Chlorhexidine bath of ICU patients (>2 months of age) daily Monitor the catheter sites visually or by palpation through an intact dressing regularly Disinfecting the access port or hub immediately before each use or manipulation Cover catheter hubs and stopcocks with sterile protective cap at all times Change gauze dressings at least every 2 days and semipermeable dressings at least weekly Replace catheter site dressing if the dressing becomes damp, loosened, or visibly soiled or patient complains of pain at the site Change administration sets within 3 days
Urinary catheter care	Avoid unnecessary catheterisation Use sterile items/equipment Choosing a catheter of the appropriate size Identify the indication of insertion of the urinary catheter Hand hygiene Use an aseptic non-touch technique for insertion and proper maintenance after insertion Maintain unobstructed urine flow	Use antiseptic technique for daily catheter care Keep drainage bag above the floor but below the bladder level to prevent reflux or contamination Daily review of the need for catheter and prompt removal Maintain a closed drainage system Collect urine sample, if required, from the sampling port
Ventilation care	**Ventilator care:** Clean and disinfect after every patient Sterilisation or high-level disinfection of re-usable components **Suctioning care:** Use of single-use disposable gloves and hand wash before and after the procedure Use of sterile suction catheter and sterile fluid to flush catheter Change of suction tubing between patients Use of closed suction system if possible **Breathing circuits care:** Replacement of mechanical ventilator if soiled or malfunctioning Periodic drainage of breathing tube condensation traps Use of HME ventilator circuits if possible **Nebuliser care:** Filling nebuliser with sterile water only Change nebulisers between patients	Head end elevation at 30–45° Daily assessment of sedation and spontaneous breathing trials Gastric ulcer prophylaxis Deep vein thrombosis prophylaxis Oral hygiene Subglottic suctioning of respiratory secretions Daily assessment of readiness to extubate Oral care with chlorhexidine gluconate (0.12%–0.2%w/v) Control of endotracheal cuff pressure at 20–30 cm water Subglottic secretion drainage must be done every 2 hourly

SSI Core Prevention Bundles

- Soap or antiseptic agent bath a night before surgery
- Avoiding hair removal; only clipping of hair if indicated
- Selection of appropriate agent for surgical prophylaxis
- Prophylactic antibiotic to be given within 1 hour prior to incision (2 hours for vancomycin and fluoroquinolones)
- Use appropriate alcohol-containing antiseptic agent (unless contraindicated) and technique for skin preparation
- Maintenance of normothermia and normoglycemia in the patient throughout the surgery
- Administration of a higher fraction of inspired oxygen during surgery and after extubation in patients with normal pulmonary function

The implementation of care bundles has shown remarkable improvement in reducing critical care infections, especially in the intensive care units. Apart from the direct benefit to the patient, it leads to a reduction in the duration of hospital stay, reduced cost, and improved resource utilisation.

BIOMEDICAL WASTE MANAGEMENT

Hospitals generate large quantities of potentially hazardous waste daily. This includes infectious and non-infectious material. Proper disposal of hospital waste constitutes an important measure to prevent healthcare-associated infections. After all, the first rule in medicine as well as nursing is '*Primum non nocere*', which means 'First, do no harm'.

- **Biomedical or hospital waste** refers to any waste generated while providing healthcare, performing research, and undertaking investigations or related procedures on human beings or animals in hospitals, clinics, laboratories, or similar establishments (Management and Handling Rules, Government of India, 2016).
- Biomedical waste is a **potential hazard** as it is far more dangerous and offensive than domestic waste for the following reasons:
  - Infectious or other hazardous chemicals and radioactive materials may injure, infect, or otherwise harm patients, their visitors, hospital personnel, and the public at large.
  - 'Sharps' such as needles or broken glass can cause injury and infection.
  - If kept untreated, hospital waste may decompose and attract flies and other insects, birds and animals, rendering the site of generation and storage filthy and unhygienic.
  - It can attract rag pickers, who may repack disposables or drugs and sell them.
  - Since biomedical waste can be a serious pollutant of soil, water, and air unless carefully managed, governments across the world have been compelled to introduce legal restraints in this area. The Government of India had promulgated the **Medical Waste (Management and Handling) Rules**, 1998, amended in 2016, 2019, and 2020, under which, persons who are in charge of medical and other institutions where such wastes are generated (called '**occupiers**') are held legally responsible for maintaining the conditions prescribed in the rules. These rules came into effect on 1 January 2003.

Types of Biomedical Waste

The amount of waste generated under Indian conditions has been estimated as 1 to 2 kg per bed per day and is composed of different types of waste, of which, on average, about 85 per cent is harmless and only 15 per cent is hazardous, which includes infectious waste (Table 12.7).

Steps in Waste Management

1. **Segregation:** This refers to the separation of waste at the point of generation into various types with

Table 12.7 Categories of waste generated in healthcare settings

Category	Items included
Harmless waste	Paper, cardboard cartons, flowers, and ordinary office or kitchen waste (akin to domestic waste)
Infectious waste	• Human or animal tissues or organs removed during biopsy, surgery, or autopsy • Placenta and other products of conception • Any pathological fluid or discharges • Dressings, swabs, and other soiled items • Laboratory samples sent for microbiology, pathology, and biochemical tests
Non-infectious hazardous waste • Chemical • Radioactive • Pharmacological	Toxic, corrosive, inflammable, reactive and otherwise injurious waste, the handling and management of which are under the direction of the Bhabha Atomic Research Centre Surplus or time-expired drugs

respect to their category and mode of final disposal. Segregated waste is put into different coloured containers (Table 12.8) for appropriate treatment.
2. **Storage:** These are measures that are taken to ensure that biomedical waste is kept safely at the point of generation before being sent to the biomedical waste treatment facility.
3. **Transportation:** Biomedical waste needs to be moved safely from the point of generation to the treatment facility.
4. **Treatment:** These are procedures and processes intended to reduce the bulk of the waste and make it non-infectious and harmless.

Methods of Waste Management

The place of final disposal is always away from crowded areas. Some methods of waste treatment are listed below.

- **Chemical disinfection:** This is a very useful method for various items, particularly in small healthcare settings such as clinics. For some materials, it is also an important preliminary process before final treatment. For example, contaminated materials such as sputum or pus must be disinfected before being buried or autoclaved.
- **Deep burial:** Where large areas of uninhabited land are available, this is a convenient method of disposing of waste materials. After chemical disinfection, materials are placed in deep trenches, covered with lime, and buried under soil. This is also a safe method for the disposal of sharps.
- **Incineration:** This is a safe method of treating large, solid, and infectious wastes, particularly anatomical waste and amputated limbs, animal carcasses, and similar materials. An incinerator exposes the waste material to a very high temperature, converting it into ash and reducing the volume by a tenth of the original. However, this method is expensive and generally used only by very large establishments. The stack height of the incinerator should be more than 30 meters as per pollution control recommendations.

Table 12.8 Colour-coded bags for biomedical waste segregation (rules amended in 2019)

Colour of the bag	Type of waste	Waste treatment
Yellow 	a) Human anatomical waste	Incineration or plasma pyrolysis or deep burial
	b) Animal anatomical waste	
	c) Soiled waste	Incineration or plasma pyrolysis or deep burial
	d) Expired or discarded medicines	Returned to the manufacturer or supplier for incineration at >1,200°C
	e) Chemical waste	Incineration, plasma pyrolysis, deep burial, or encapsulation
	f) Chemical liquid waste	Pre-treatment and then disposal
	g) Discarded linen, mattresses, bedding contaminated with blood or body fluids	Non-chlorinated chemical disinfection followed by incineration or plasma pyrolysis
	h) Microbiology, biotechnology, and other clinical laboratory waste	Pre-treated to sterilise with non-chlorinated chemicals on-site as per NACO or WHO guidelines, thereafter incinerated
Red 	Contaminated waste (recyclable) like plastic bag, bottles, pipes, or containers	Autoclaving or microwaving/hydroclaving followed by shredding or mutilation Treated waste to be sent to registered or autoclaved recyclers or for energy recovery of plastics to diesel or fuel oil or for road-making
White, translucent 	Waste sharps including metals: Needles, syringes with fixed needles, needles from needle tip cutter or burner, scalpels, blades	Autoclaving or dry-heat sterilisation followed by shredding, mutilation, or encapsulation in metal containers or cement concrete Sent for final disposal to iron foundries (having consent to operate from the state pollution control committees) Alternatively, sent to a sanitary landfill or designated concrete pits for discarded sharps
Blue cardboard box with a blue label or blue leak- and puncture-proof container 	Glassware—broken or discarded and contaminated glass including medicine vials and ampoules except those contaminated with cytotoxic wastes; metallic body implants	Disinfection (by soaking the washed glass waste after cleaning with detergent and sodium hypochlorite treatment) or through autoclaving or microwaving or hydroclaving; then, it is sent for recycling

- **Autoclaving:** This method is widely used in laboratories and clinics for treating infectious waste before disposal.
- **Microwaving:** This is another useful method of sterilising small-volume waste at the point of generation. It cannot be used for animal or human body parts, metal items, or toxic or radioactive material.
- **Inertisation:** This process involves mixing waste with cement and other substances before disposal in order to minimise the risk of toxic substances contained in the wastes migrating into the surface water or groundwater.
- **Liquid waste:** Pathological, chemical, and toxic liquid waste should be appropriately treated with disinfectants or reagents and neutralised before flushing them into the sewer.

Spill Management

- Exposure to blood and other body fluids poses a risk of infection.
- The management of spillages is an essential part of the prevention and control of infections like HIV, hepatitis B, and hepatitis C which are transmitted through blood or other body fluids.
- All members of the hospital staff must wear PPE during this operation (gloves, mask, and gown). After completing the procedure, hand decontamination is carried out with a skin disinfectant or thorough hand washing.
- The procedure for the management of small spills is shown in Fig. 12.10.

Mandatory Practices for all Healthcare Settings

- Colour-coded, non-chlorinated bags for infectious, non-infectious, and general waste
- Puncture-proof containers for sharps and needles
- Pre-treatment autoclaving of infectious waste (blood and body products) prior to disposal
- Deep burial and incineration of appropriate hospital waste in a central facility

INFECTION CONTROL PRACTICES IN THE HOSPITAL

- Infection control programmes are mandated as a part of patient care and patient safety in all hospitals.
- Healthcare-associated infection control (HAIC) programmes are a mandatory requirement in all healthcare settings including hospitals that are accredited by the National Accreditation Board for Hospitals and Healthcare Providers (NABH).
- Every healthcare institution should have a properly constituted and functioning infection control team to monitor and prevent infections in the healthcare setting.
- Basic principles of personal and institutional hygiene should be strictly observed to ensure patient safety.

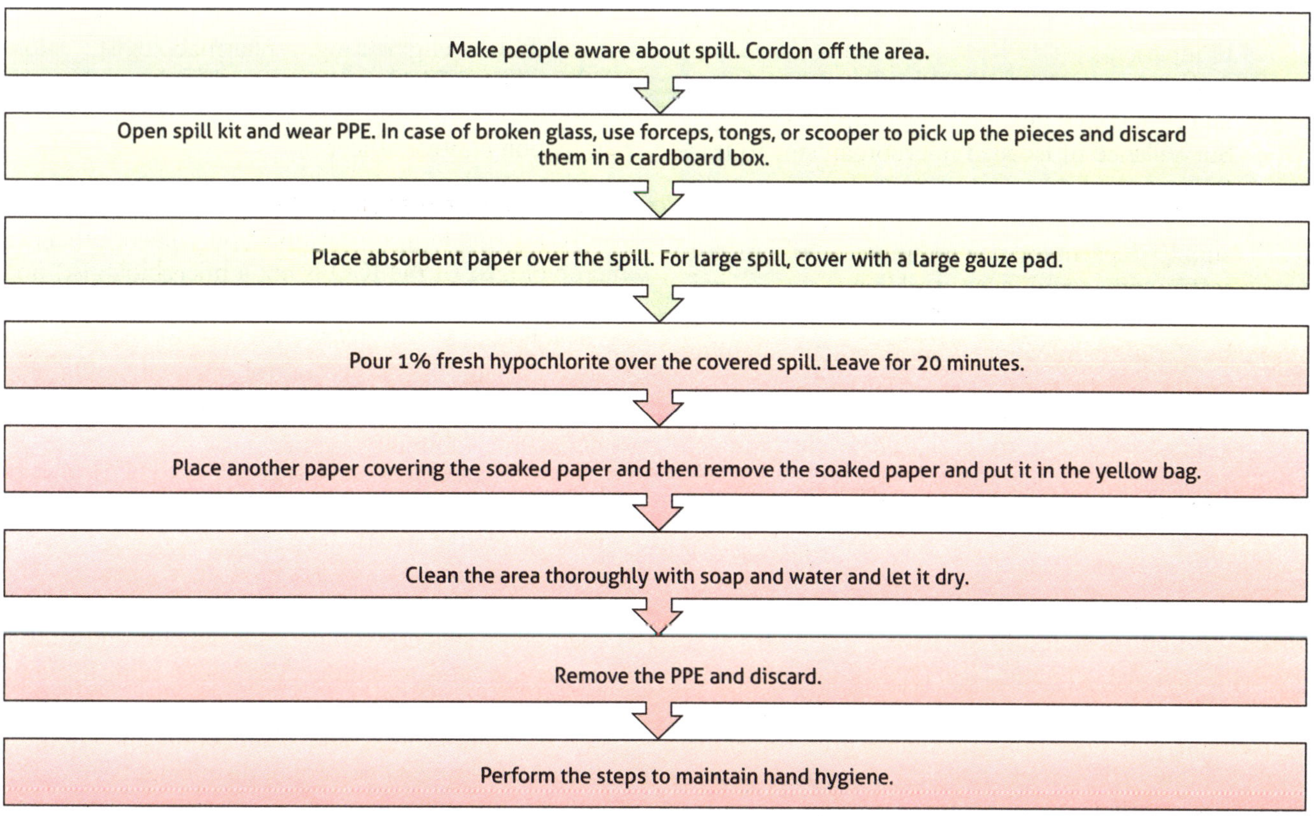

Fig. 12.10 Management of small spillages

- The activities for preventing HCAI may be described under the following broad groups:
  - Surveillance activities
  - Intervention
  - Preventive activities

Surveillance for HCAI

- It is important to establish a hospital surveillance programme in which infection patterns are monitored on a regular basis.
- An outbreak of HCAI can be detected and rapidly controlled through routine surveillance, review of the data, and prompt reporting so that control measures can be implemented.

i. **Passive surveillance:** The infection control team undertakes routine surveillance to monitor the rates of infection through daily review and tabulation of microbiology laboratory-generated reports. Healthcare workers and data from critical care units, general wards, and outpatient departments are also an important part of passive surveillance.

ii. **Active surveillance:** The surveillance team actively monitors for the following:
  - Presence of infections in inpatients—CAUTI, SSI, VAE (ventilator-associated event), and CRBSI
  - Environmental microbiological surveillance
  - Monitoring of water for microbial contamination (e.g., in dialysis units)
  - Surveillance of food handlers and food supplied to inpatients
  - Surveillance of air quality, e.g., in operation theatres and critical care units and other special areas
  - Surveillance of isolated microorganisms to trace the source of infection
  - Checking for quality control in the central sterile supply department for optimum sterilisation of instruments, equipment, and linen
  - Behavioural surveillance of healthcare personnel by auditing infection control practices among healthcare workers

iii. **Sentinel surveillance:** This comprises targeted surveillance for a specific disease (e.g., tuberculosis) which is of importance in terms of communicability, morbidity and mortality, or for characterising the agent.

Monitoring and Control of HCAIs: Hospital Infection Control Committee (HICC)

With the increasing incidence of HCAIs, hospitals must monitor and regulate the spread of such infections and implement and take preventive measures to control and contain HCAIs. A hospital infection control committee (HICC) is constituted to plan, monitor, evaluate, update, and educate healthcare professionals on standard infection control practices. The scope of the work of such a committee comprises the following:

- Meet periodically to evaluate the infection rates in the hospital
- Review audits of infection control practices
- Look at infrastructural corrections and personnel requirements to contain the spread of infection in the healthcare setting
- Review procedures to detect and investigate hospital outbreaks of infection
- Strategise the containment of outbreaks and implement preventive measures
- Review antibiotic resistance patterns in the hospital and update guidelines and antibiotic policy
- Review surveillance reports of critical areas in the hospital
- Advise the management on the improvement of structural and functional aspects to improve infection control measures
- Review and monitor the training of HCWs

Composition of HICC

- Chairperson: Head of the institute (preferably)
- Member secretary: Senior microbiologist
- Members: Representation from management/administration (dean/director of the hospital; nursing services; medical services; operations)
- Relevant medical faculties
- Support services: OT/CSSD, housekeeping/sanitation, engineering, pharmacologist, store officer/materials department
- Infection control nurse (s)
- Infection control officer

Infection control team (ICT)

The ICT is made up of an infection control officer, a microbiologist (if the ICO is not a microbiologist) and an infection control nurse. The functions of the ICT are as follows:

- Develop an infection control manual containing policies and procedures for aseptic isolation and antiseptic techniques
- Conduct targeted surveillance of HCAIs, data analysis for presentation in HICC meetings, and take corrective steps
- Train healthcare workers to ensure the implementation of infection control practices and maintain a safe environment for patients and staff
- Supervise and monitor cleanliness and hygienic practices
- Oversee sterilisation and disinfection and monitor the use and quality control of disinfectants
- Advise the management of at-risk patients and supervision of isolation procedures

- Investigate outbreaks of infection and take corrective measures for the control and prevention of outbreaks
- Biomedical waste management
- Audit infection control procedures and antimicrobial usage
- Monitors healthcare workers' safety programmes

Investigation and Follow-up of an Outbreak of HCAI

The occurrence of two or more similar infectious cases relating to a place and time is identified as a **cluster** or an outbreak and needs investigation to discover the route of transmission of infection and possible sources of infection in order to apply measures to prevent further spread.

Common source outbreak: If a large number of cases occur following a shared exposure, e.g., an operation, it is termed a common source outbreak, implying a common source of the infection.

In case of a suspected outbreak, the infection control team (consisting of the Infection Control Officer and ICN) acts after confirming that it is an outbreak, which is based on the following: (i) an increased number of cases over and above the usual number encountered in the hospital and (ii) a clustering of cases within close proximity/or common source.

Investigating an outbreak (Fig. 12.11)

Environmental surveillance and screening of healthcare personnel are carried out to detect the

Step 1: Recognition of the outbreak

- Develop a case definition, identify the site, pathogen and affected population
- Define the outbreak in time, person, and place

Step 2: Develop case defintion

- Is there an increase in the number of cases of a particular infection or a rise in prevalence of an organism?

Step 3: Risk assessment & confirmation

- Determine the magnitude of the problem
- Verification of the diagnosis
- Assess and institute emergency control measures

Step 4: Notify

- Determine the magnitude of the problem
- Verify the diagnosis
- Assess and institute emergency control measures

Step 4: Line listing

- Line listing for every case, patient details, place and time of occurrence and infection details
- Develop epidemic curve, analyse data

Step 5: Hypothesis

- Formulate hypothesis based on literature search and the features common to the cases

Step 6: Microbiological investigation

- Perform: (a) microbial culture of cases, carriers and environments (b) epidemiological typing of the isolates to identify clonal relatedness

Step 7: Test hypothesis

- Test hypothesis by reviewing additional cases in a case control study, cohort study, and microbiological study

Step 8: Control measures

- Implement specific control measures as soon as the cause of outbreak is identified

Step 9: Assess control measures

- Monitor for further cases and assess effectiveness of control measures

Step 10: Report preparation

- Prepare and shrare report with HICC for further training and learning

Fig. 12.11 Steps involved in the investigation of an outbreak

source and necessary containment is done by treating (decolonising) the source individuals. If the outbreak is caused by a notifiable disease (e.g., cholera), it is reported to the authorities as per the **Integrated Disease Surveillance Programme** (IDSP).

LABORATORY-ACQUIRED INFECTIONS

Infections acquired by personnel working in or having access to the laboratory through their laboratory-related activities are known as laboratory-acquired infections.

Biological Safety Levels (BSL)

A series of specific protections must be instituted in biological labs. Biosafety levels are individual safeguards designed to protect lab workers (Fig. 12.12) and the surrounding environment and community. The various categories of organisms and biosafety level laboratories (BSL) are presented in Table 12.9.

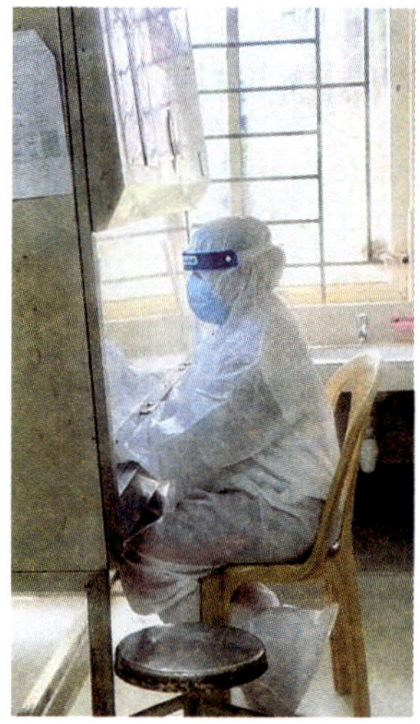

Fig. 12.12 Laboratory technologist in PPE working with hazardous material that has a risk of aerosol production (*Source*: Dept. of Microbiology, PIMS)

Table 12.9 Biosafety level laboratories

Biosafety level	BSL-1	BSL-2	BSL-3	BSL-4
Description	No containment Open lab bench used	Containment Biosafety cabinets class 1 or 2 used	High containment Biosafety cabinet class 2 used	Maximum containment in specialised laboratories
Pathogen characteristics	Agents that show minimum potential hazard to personnel and the environment and are unlikely to cause disease	Associated with human disease and pose moderate hazard to lab personnel and environment	Indigenous or exotic agents, agents that present a potential for aerosol transmission, and agents causing serious or potentially lethal disease	Dangerous and exotic agents that pose a high risk of aerosol-transmitted laboratory infections and life-threatening disease
Example of organisms	*E. coli*	Influenza, HIV, *Salmonella*, *Vibrio* and herpes	*Mycobacterium tuberculosis*, *Brucella*, *Bacillus anthracis*	Ebola, Marburg, Nipah virus

Key Points

- ✓ A healthcare-associated infection (hospital-acquired infection) is an infection that develops in a hospitalised patient, which was not present or was not in incubation at the time of admission. It usually becomes evident 48–72 hours after admission to the hospital but may sometimes manifest only after discharge.
- ✓ Hospital infections are exogenous, the source being any part of the hospital ecosystem (people, objects, food, water, air) or the patient himself (endogenous).
- ✓ Healthcare-associated infections manifest as wound infections, urinary tract infections, respiratory infections (nosocomial pneumonia), bacteremia, or septicemia.
- ✓ Healthcare-associated infections may occur sporadically or as outbreaks; possible sources of infection are investigated by routine bacteriological methods to identify the source.
- ✓ Standard precautions are the minimum infection prevention practices that apply to all patient care. They include hand hygiene, the use of personal protective equipment, safe injection practices, safe handling of potentially contaminated equipment or surfaces in the patient environment and respiratory hygiene/cough etiquette.
- ✓ For conventional operating theatres, the bioload (bacteria-carrying particles per cubic metre) should not exceed 35 in an empty theatre or 180 during an operation; for ultraclean air-operating theatres, the bioload should be less than 1.0 in the centre of an empty theatre and less than 10 during an operation and should not exceed 20 at the periphery.
- ✓ Biomedical or hospital waste refers to any waste generated during healthcare, research, testing or related procedures on human beings or animals, conducted in hospitals, clinics, laboratories, or similar establishments. Biomedical waste can act as a serious pollutant of soil, water, and air.
- ✓ To regularise the optimal management and disposal of biomedical waste, the Government of India issued legal restraints in this area by enacting Medical Waste Management and Handling Rules, 2016.
- ✓ In Indian conditions, about 1–2 kg of waste per bed per day is generated, 15 per cent of which is hazardous. The different steps in waste management are reduction, reuse, segregation, storage, transportation, and treatment. The methods of waste treatment, depending on the item of waste and the facilities available, include chemical disinfection, deep burial, incineration, autoclaving, and microwaving.

- ✓ The most essential part of hospital waste management is the segregation of biomedical waste, which should be performed within the premises of the institution where the waste is generated.
- ✓ BMW 2018 rules have changed the colour codes and categories of waste.
- ✓ Spill management requires cordoning off the area and placing absorbent paper over the spill. Fresh 1% hypochlorite should be poured over the covered spill and left for 20 minutes. Later, the soaked paper should be put in a yellow bag, and the area cleaned thoroughly with soap and water.
- ✓ Care bundles are a set of several individual procedures performed in a group (or bundle) in total and together as an essential part of infection control practice.
- ✓ Insertion and maintenance bundles are used for devices such as CVP line, urinary catheters, and ventilators.
- ✓ Insertion care bundle for the urinary catheter includes hand hygiene, choosing a catheter of the appropriate size, using of sterile items, avoiding unnecessary catheterisation, identifying the indication of insertion of the urinary catheter, using an aseptic non-touch technique for insertion, and proper maintenance after insertion and to maintain unobstructed urine flow.
- ✓ Ventilation care bundle includes ventilator care, suctioning care, breathing circuits care and nebuliser care.
- ✓ A hospital infection control committee (HICC) is constituted to plan, monitor, evaluate, update, and educate healthcare professionals on standard infection control practices.
- ✓ HICC is chaired by head of the institute. Its members include a microbiologist, representation from the management/administration (dean/director of the hospital; nursing services; medical services; operations), relevant medical faculties, support services, an infection control nurse, and an infection control officer.
- ✓ The occurrence of two or more similar infectious cases relating to a place and time is identified as a cluster or an outbreak. After recognising an outbreak, a case definition must be made, and risk assessment and confirmation should be performed before notifying it. A line list should be made, and the outbreak must be microbiologically investigated. The hypothesis must be tested by reviewing additional cases in a case control study, cohort study, and microbiological study, and specific control measures should be implemented.
- ✓ Infections acquired by personnel working in or having access to the laboratory through their laboratory-related activities are known as laboratory-acquired infections.
- ✓ Biosafety levels are individual safeguards designed to protect lab workers and the surrounding environment and community. BSL can range from 1 to 4; from no containment to maximum level of containment.

Questions

Short answer questions

1. Describe the types of healthcare-associated infections.
2. Detail the steps involved in hand hygiene.
3. Mention the need for and components of personal protective equipment (PPE).
4. Describe safe injection practices.
5. What are the roles and responsibilities of HICC in infection control in the hospital?
6. Discuss the types of biomedical waste generated in a hospital.

MCQs

1. All the following are true with regard to HCAI except ________.
   a) Patients do not present with symptoms at the time of admission.
   b) Patients are not in the incubation period of the disease at the time of hospital admission.
   c) Infection develops within 48–72 hours of hospital admission.
   d) Infection develops within 24–48 hours of hospital admission.

2. A 50-year-old man develops a surgical site infection on the eighth postoperative day. What pathogen is likely to have caused this infection?
   a) Staphylococci b) Clostridia
   c) Gram-negative bacilli d) Gram-negative cocci

3. The recommended number of air changes per hour in a conventional operation theatre is ________.
   a) 20 b) 40 c) 100 d) 300

4. Which of the following is the recommended time for the seven-step hand washing procedure?
   a) 5–10 seconds b) 15–20 seconds
   c) 25–30 seconds d) 35–40 seconds

5. Which of the following does biomedical segregation refer to?
   a) Separation of waste at the point of generation
   b) Categorising items at the site of disposal
   c) Collecting waste in a single, segregated bag to be kept separately in the hospital
   d) Weighing each segregated item separately prior to disposal

6. A man was hospitalised with second-degree burns. After five days of hospital stay, he developed an infection of the burn wounds. All the following bacteria are likely to be involved in this case except:
   a) *Pseudomonas* species b) *Bacillus* species
   c) *Proteus* species d) *Klebsiella* species

7. Which of the following is correct regarding the ESKAPE pathogens?
   a) It includes *E. coli*, *Staphylococcus* (MRSA), *Klebsiella*, *Acinetobacter*, *Pseudomonas*, and *Enterobacter*.
   b) It includes *Enterococcus*, *Staphylococcus* (MRSA), *Klebsiella*, *Acinetobacter*, *Pseudomonas*, and *Enterobacter*.
   c) It includes *E. coli*, *Streptococcus*, *Klebsiella*, *Acinetobacter*, *Pseudomonas*, and *Enterobacter*.
   d) It includes *E. coli*, *Staphylococcus* (MRSA), *Klebsiella*, *Acinetobacter*, *Pseudomonas*, and *Enterococcus*.

8. A 65-year-old ICU patient is suspected to have pneumonia 48–72 hours after endotracheal intubation. Which of the following criteria is not to be fulfilled before diagnosing ventilator-associated pneumonia (VAP)?
   a) Presence of hyper or hypothermia
   b) PO2 or FiO_2 >240 or signs of acute respiratory distress syndrome (ARDS)
   c) Detection of causative agent from the endotracheal or bronchoalveolar lavage
   d) Concurrent blood culture with growth of the same causative agent

9. A 22-year-old adult male is admitted for the past four days with fever and hypotension. A central line was inserted upon admission. Coagulase-negative *Staphylococcus* is isolated from two blood cultures taken on successive days. The patient should be diagnosed with:
   a) No healthcare-associated infection
   b) Community-acquired sepsis
   c) Central line-associated bloodstream infection
   d) Catheter-associated UTI with sepsis

10. Which of the following antibiotic is not associated with antibiotic-associated diarrhea?
   a) Penicillin b) Clindamycin
   c) Vancomycin d) Fluoroquinolones

11. Which of the following is a moment of hand hygiene?
a) Before touching a patient, even if gloves are to be worn
b) After contact with blood, body fluids, excretions, or wound dressings
c) Prior to performing any aseptic task on the patient
d) All of the above

12. Which of the following is not a part of care bundle required before putting in a central line?
a) Draping of area with sterile linen (barrier precaution)
b) Chlorhexidine skin preparation is not required if the site is clean
c) Optimal or best catheter site for insertion
d) Daily review of insertion site and need for a line

13. The recommended number of air changes in a laminar flow operation theatre is:
a) 100 air changes per hour
b) 200 air changes per hour
c) 300 air changes per hour
d) 400 air changes per hour

14. Common causative agents of catheter-associated UTI include all of the following except:
a) *Klebsiella pneumoniae*
b) *Escherichia coli*
c) *Pseudomonas aeruginosa*
d) *Streptococcus pyogenes*

15. Biosafety level 1 type of laboratories will handle the following type of organisms:
a) Coagulase-negative *Staphylococcus*
b) Hepatitis
c) Ebola
d) Mycobacteria

16. A laboratory has received a specimen from an animal which died of anthrax. What is the recommended level of biosafety (BSL) cabinet for handling this specimen?
a) BSL-1 b) BSL-2
c) BSL-3 d) BSL-4

17. District health authorities are planning to start investigating an outbreak of Marburg virus disease. What is the recommended level of biosafety (BSL) cabinet to be procured for handling this specimen?
a) BSL-1 b) BSL-2
c) BSL-3 d) BSL-4

18. In an ultraclean air operating theatre, the permitted bio-load during operation should be:
a) Less than one b) Less than ten
c) Between 10-20 d) More than 20

19. Yellow bags should be used as per BMW-2016 rules to dispose all of the following except:
a) Clinical laboratory waste
b) Expired medicines
c) Contaminated plastic ware
d) Contaminated body fluids

20. As per BMW-2016 rules, white, translucent containers should be used for the disposal of:
a) Sharp waste
b) Contaminated glassware
c) Contaminated plastic ware
d) Liquid chemical waste

21. Which of the following requires chemical treatment before disposal?
a) Contaminated recyclable plastic ware
b) Clinical laboratory waste
c) Human anatomical waste
d) Soiled dressings

22. Which is the best method for the disposal of human anatomical waste?
a) Chemical disinfection b) Autoclaving
c) Incineration d) Incarceration

23. Which is the recommended waste treatment for contaminated recyclable plasticware?
a) Incineration
b) Plasma pyrolysis
c) Chemical treatment and deep burial
d) Autoclaving and energy recovery

24. A student working in a clinical laboratory needs to discard some slides after examining wet mounts. How should he discard these slides in accordance with Bio-Medical Waste (Management and Handling) Rules, 2016?
a) Clear plastic/white translucent boxes
b) Cardboard boxes with blue marking
c) Red boxes with blue marking
d) Yellow boxes with blue marking

25. An orthopedic surgeon has removed an implant from the leg of a 70-year-old woman. The hospital does not have access to a common biomedical waste treatment facility. What is the correct procedure to discard this implant?
a) Autoclaving followed by washing and sent for recycling
b) Dipping in any disinfectant and sending for recycling
c) Simple washing and sending for recycling
d) Sending for recycling in a red, chlorine-free plastic bag

Answers: 1:d, 2:b, 3:a, 4:d, 5:a, 6:b, 7:a, 8:d, 9:c, 10:c, 11:d, 12:b, 13:c, 14:d, 15:a, 16:c, 17:d, 18:b, 19:c, 20:a, 21:b, 22:c, 23:d, 24:b, 25:a

Part II: Immunology

CHAPTER 13

Immunological Mechanisms in Health

(MI1.7)

Chapter Navigation

Learning objective

- Explain the immune mechanisms of the human body

INTRODUCTION AND GENERAL CONCEPTS

- The invasion and multiplication of microorganisms in a host constitute an **infection**. However, an infection does not necessarily result in disease. In fact, disease is a rare consequence of an infection in cases where the host's immunity is intact.
- The immune system in living organisms is responsible for protecting them from invading foreign substances including pathogens.
- The human immune system has evolved into a specialised system which facilitates protection by several mechanisms. It is highly adaptable, involving several different cells and molecules that help to specifically recognise foreign objects. This system has two important functions—first, to recognise a substance as foreign (to the host) and second, to respond to it.

IMMUNITY

Immunity is the host's response to injuries caused by microorganisms or any foreign substance. Protection against infectious diseases is only one of the many consequences of an immune response. The characteristic features of immunity are as follows:

- To **distinguish non-self** (that which does not belong to the body) **from self** (that which belongs to the body)
- Seek and destroy the **body's own cells that have gone rogue** (e.g., cancer cells)
- Recognise various types of pathogens and eliminate them by mounting a specific and selective response

Immunity depends on several factors including age, nutritional status, genetic makeup, underlying metabolic disorders, etc.

IMMUNE SYSTEM

- Human beings are protected from pathogens and harmful foreign substances by a three-tiered immune response.
- The lines of defence in different types of immunity are presented in Fig. 13.1 and 13.2.
- To cause an infection, microbes first have to cross **physical barriers** such as the skin, mucous membranes, and secretions of the digestive, respiratory, and reproductive tracts.

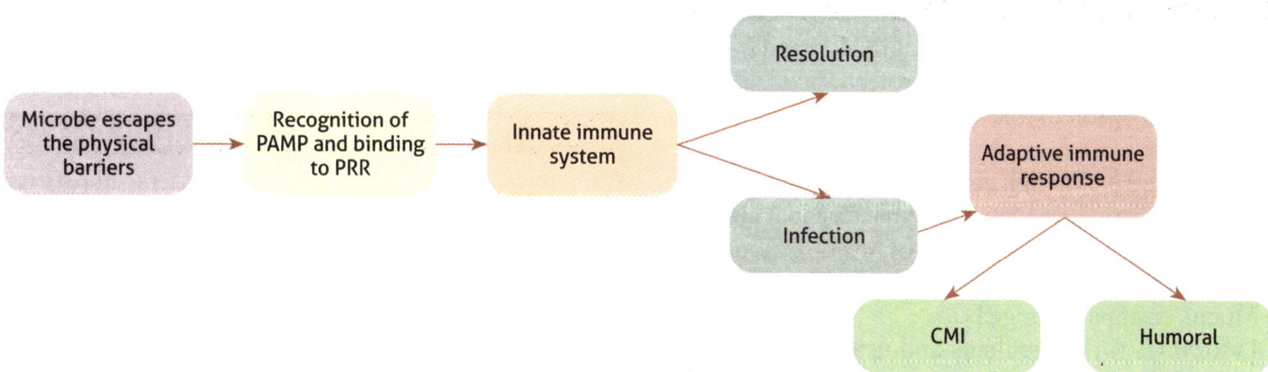

Fig. 13.1 Lines of defence in a host
PAMP, pathogen-associated molecular patterns; *PRR*, pathogen recognition receptors

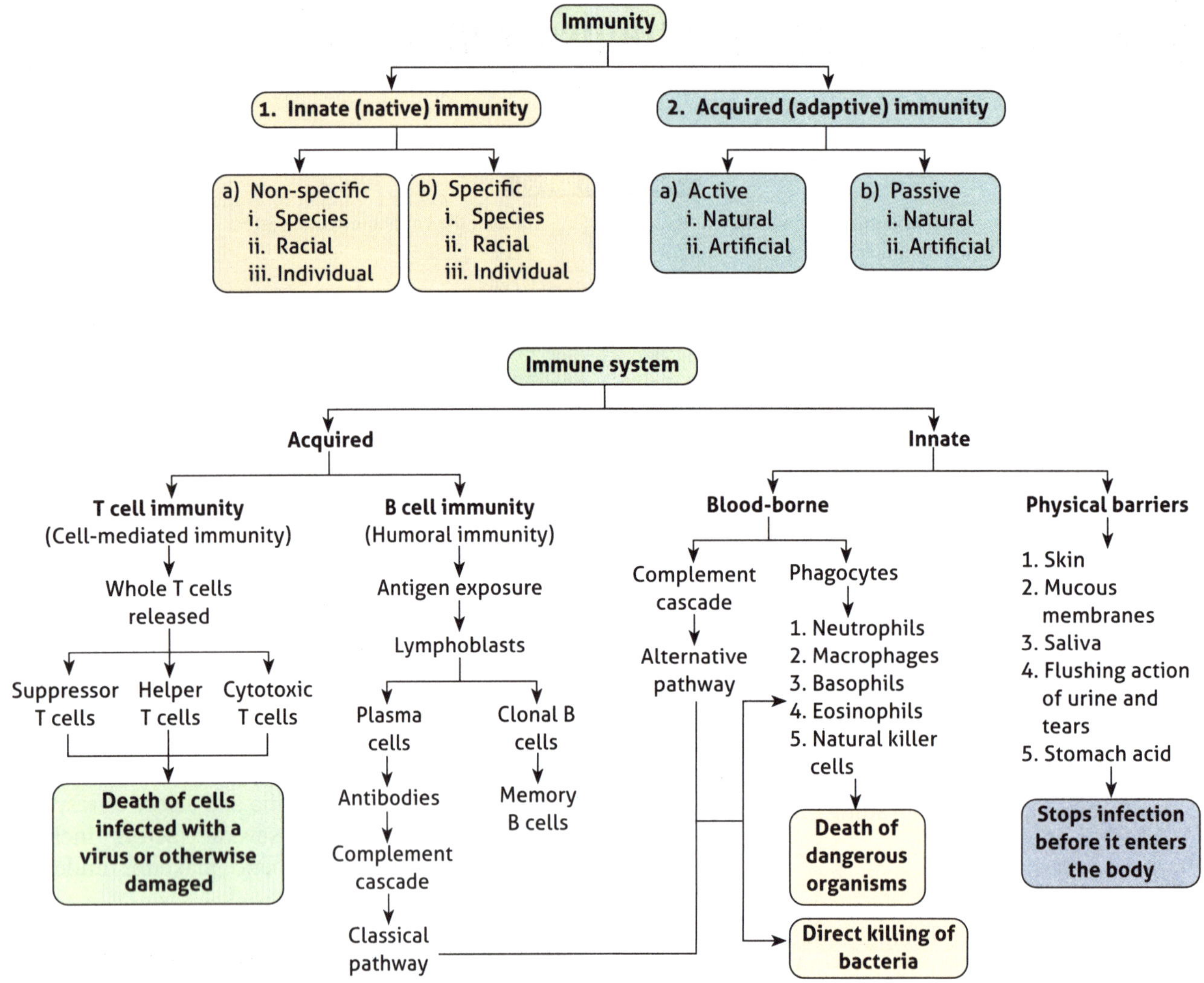

Fig. 13.2 Types of immunity in a human host

- The second line of defence is provided by the **innate immune system**. The third level of defence is **adaptive immunity**, which can be cellular or humoral.
- The differences between the innate and adaptive immune systems are presented in Table 13.1.

PHYSICAL BARRIERS: THE FIRST LINE OF DEFENCE (TABLE 13.2)

Epithelial surfaces including the intact skin and mucous membranes covering the body protect it considerably against the invasion of microorganisms by the following mechanisms:

- The presence of a high concentration of salt in sweat and sebaceous secretions has a bactericidal effect.
- Resident microbes on the skin prevent colonisation by exogenous pathogenic microorganisms.
- Mucus entraps pathogens.
- Lysozymes (in tissue fluids and nearly all secretions except cerebrospinal fluid, sweat, and urine) act by splitting certain polysaccharide components of the cell walls of susceptible bacteria.
- Phagocytic cells in mucous secretions trap particles that manage to reach the pulmonary alveoli and ingest them.
- Nasal and respiratory secretions contain mucopolysaccharides capable of combining with influenza and certain other viruses.
- Cilia on the respiratory epithelial cells propel particles and expel them through the nose and mouth.

INNATE IMMUNITY: THE SECOND LINE OF DEFENCE

After microbes cross the physical barriers, the second line of defence, innate immunity takes over. As the components of innate immunity are always in an active state, they respond immediately. Innate immunity may be considered at the level of the species, race, or individual.

- **Species immunity:** This refers to the total or relative refractoriness of a host to a pathogen, as exhibited

Table 13.1 Differences between innate and adaptive immune systems

Immune system	Innate	Adaptive
Response time	Immediate—within minutes to hours	Late—could take days to set in
Specificity	Limited	Highly specific
Memory	None	Yes
Diversity of response	Not seen	Diverse
Response to repeated infection	No enhancement	Rapid and more effective
Cellular mediators	Macrophages, phagocytes, NK cells, eosinophils, neutrophils	T and B lymphocytes
Soluble molecules	Interferons, complement proteins, LPS-binding protein, C-reactive protein and other acute-phase reactants, antimicrobial peptides and mannose-binding lectin	Immunoglobulins (secreted by B lymphocytes)
Types of activation—receptors	TLR, NLR (NOD-like receptors), and complement activation	T cell and B cell receptors
Basis	Resistance that an individual possesses by virtue of his genetic and constitutional make-up	Dependent on conditioning by the innate system and mediated via T and B lymphocytes
Prior contact with pathogen	Not required	Required

Table 13.2 Non-specific barriers that act as the first line of defence

Body systems and factors involved	Active component	Effector mechanism
Skin	Squamous cells, salt of sweat glands, sebaceous glands (long-chain fatty acids and lactic acid)	Desquamation, flushing, organic acids, and bactericidal pH (3–5) retard bacterial growth
Mouth	Saliva	Bathing, flushing, lysozyme activity
GI tract	Highly acidic gastric secretions, mucus	Peristalsis, entrapment of pathogens, low pH, flushing
Respiratory tract	Tracheal cilia, cough reflex, mucus	Mucociliary elevator, surfactants, forcible removal
Nasopharynx	Nasal hair, mucus	Mechanical barrier, entrapment, and swallowing
Eye	Tears	Flushing action of lachrymal secretions, lysozyme activity
Adult vagina	Acidic pH	Retards bacterial growth
Semen	Spermine and zinc	Antibacterial activity
Urinary system	Acidic pH and flow	Retards bacterial growth, flushing

by all members of the host species. For example, humans are not susceptible to plant pathogens. The mechanisms of species-specific immunity may be attributed to physiological and biochemical differences between the tissues of the different host species.

- **Racial immunity:** Within a species, races may show differences in susceptibility to infection. Such racial differences are known to be genetic in origin and are shaped by selection and inbreeding. For example, resistance to *Plasmodium falciparum* malaria is seen in some parts of Africa and the Mediterranean coast.
- **Individual immunity:** This is the difference in innate immunity exhibited by different individuals belonging to a race. It is documented that homozygous twins exhibit similar degrees of resistance or susceptibility to lepromatous leprosy and tuberculosis. Such correlation is not seen in heterozygous twins.

Factors Affecting Innate Immunity

1. **Age:** The very young and the very old are more susceptible to infectious diseases than the rest.
2. **Hormonal influences:** Endocrine disorders such as diabetes mellitus, hypothyroidism, and adrenal dysfunction are associated with increased susceptibility to infection. Elevated hormones during pregnancy may be related to the heightened susceptibility of pregnant women to many infections.
3. **Nutrition:** In general, both humoral and cell-mediated immune processes are reduced in malnourished individuals. For example, the Mantoux test is negative for cell-mediated immune response in severe protein deficiency, as in kwashiorkor.
4. **Drugs:** Immunosuppressants, anti-cancer drugs, antimetabolites, and corticosteroids significantly influence the host's response to infection. Corticosteroids depress the host's resistance due to their anti-inflammatory and anti-phagocytic effects and suppress antibody formation and hypersensitivity.

5. **Chronic stress:** Long-term psychological stress can result in elevated levels of cortisol and corticosteroids, which impair immunity. This can result in chronic infections, chronic inflammatory autoimmune diseases, cancers, and other physiological disorders.

Components of the Innate Immune System

The following cells constitute innate immunity (Fig. 13.3):

- **Phagocytic cells**, which were discovered by Metchnikoff in 1883, mediate natural defence against the invasion of blood and tissues. A major function of the reticuloendothelial system is the removal of foreign particles that enter the body. Phagocytic cells reach the sites of inflammation in large numbers, attracted by chemotactic substances, and ingest particulate material.
- **Macrophages**, which include monocytes, histiocytes, and fixed reticuloendothelial cells, detect, engulf, and destroy pathogens and apoptotic cells. Macrophages are produced by the differentiation of monocytes.
- **Mast cells** release granules and powerful chemical mediators such as histamine, cytokines, granulocyte macrophage colony-stimulating factor (GM-CSF), leukotrienes, heparin, and many proteases into the environment.
- **Dendritic cells** are antigen-presenting cells which are found in tissues and skin.
- **Granular leukocytes** (microphages)—neutrophils (polymorphonuclear leukocytes) get rid of possibly harmful microorganisms and other unwanted debris in the cells. An increase in the number of peripheral blood leukocytes is an indication of infection.
- **Mucosa-associated lymphoid cells** are abundant in the gut and the oral and rectal mucosa. They have specialised pathways for antigen uptake.
- **Natural killer cells (NK cells)** selectively kill virus-infected cells and tumour cells. NK cells are activated by interferons.

Mechanism of Innate Immunity

The innate immune system is able to identify pathogens based on **pathogen-associated molecular patterns (PAMPs)** found on microbial agents (or any foreign substance). PAMPs can be of the following types:

1. Nucleic acids (DNA, dsRNA, ssRNA)
2. Surface glycoproteins
3. Lipoproteins
4. Peptidoglycans
5. Lipoteichoic acid
6. Flagellin

- The immune system uses cell surface receptors and soluble receptors to sense threats known as **pathogen recognition receptors (PRR)** to bind to and mount an immune response.
- As the host does not produce PAMPs, particles bearing these patterns are immediately recognised as foreign and immobilised by soluble molecules of PRR.
- They are then ingested and phagocytosed or broken down by **complement-mediated lysis** (interaction of innate with adaptive immunity).

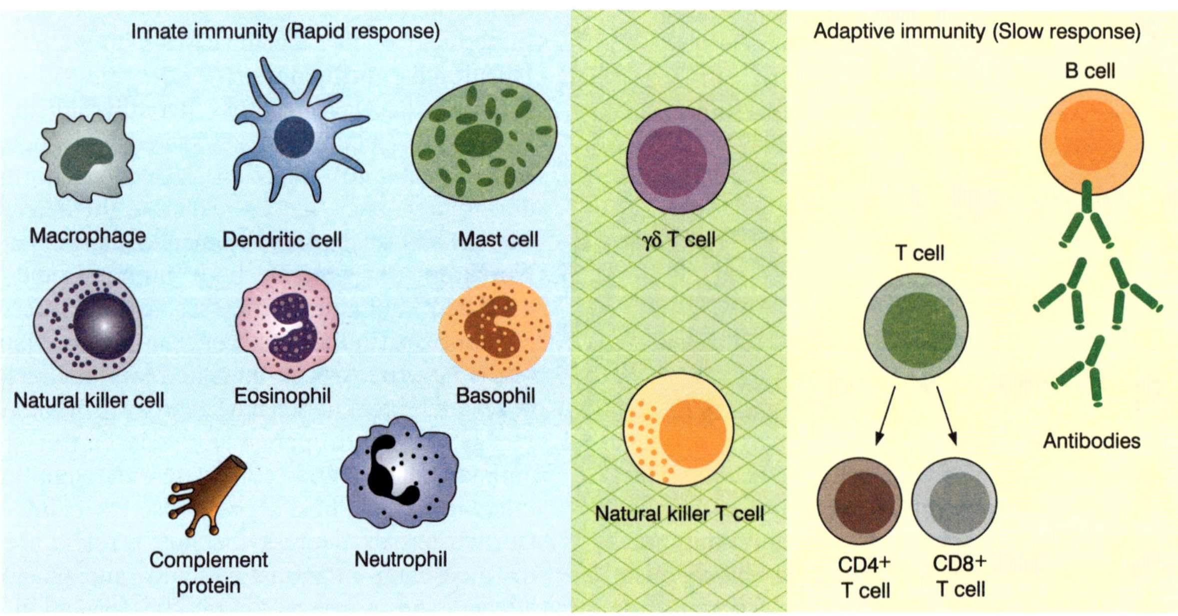

Fig. 13.3 Some of the cells involved in innate and adaptive immunity

- A PRR can recognise a PAMP that may be present on a number of different pathogens, thereby enabling the receptor to recognise a variety of different pathogens.
- PRRs are of the following types:
  - **Toll-like receptors (TLRs):** These are transmembrane receptors present on macrophages and dendritic cells. There are 13 different TLRs that recognise pathogens, enhance phagocytosis, and further lead to inflammation at the site.
  - **Scavenger receptors:** These include CD36, CD68, and SRB1 receptors. These can bind to bacterial cell wall components such as LPS, peptidoglycans, and infected, injured, or apoptotic cells and modified lipoproteins.
  - **Mannose receptors:** These bind to mannose-rich glycans and are present on the surface of phagocytes. They are short carbohydrate chains with mannose, N-acetyl glucosamine, or fucose as the terminal sugar. Mannose receptors help to internalise the bacterial antigen into the phagocyte.
  - **NOD-like receptors (NLRs):** These are nucleotide-binding oligomerisation domain (NOD)-like receptors. They recognise intracellular bacterial cell products.
  - **Opsonins:** These are molecules that bind to the surface of microbes. They are phagocytes that have receptors for opsonins, due to which, they are attracted to the opsonised pathogen and can engulf them. This is known as **opsonisation**. Capsulated bacteria such as *S. pneumoniae* are not readily phagocytosed except in the presence of opsonins. These organisms are readily phagocytosed when trapped by the opsonin against a firm surface such as the alveolar wall than when they are free in tissue fluids.

Phagocytosis

- Bacteria are phagocytosed into a vacuole (**phagosome**), which fuses with the lysosomes found in the cell to form the **phagolysosome** (Fig. 13.4a and b). The trapped bacteria are subjected to the action of the lytic enzymes in the phagolysosome and are destroyed.
- Some bacteria, such as *Brucella* and lepra bacilli resist intracellular digestion and may actively multiply inside the phagocytic cells. Phagocytosis in such instances may actually help to disseminate infection to different parts of the body.

Soluble molecules involved in innate immunity

- **Type I interferons:** IFN-α and IFN-β are rapidly produced in response to the detection of PAMP on viral cells.
- **Defensins:** These are cysteine-rich peptides secreted by neutrophils, epithelial cells, and macrophages on the skin and mucous membrane. They produce channels in the bacterial cell surface, causing an influx of chemicals and cell death.

Other Non-specific Mechanisms of the Innate Immune System

Inflammation

This is a non-specific defence mechanism that occurs as follows:

- The arterioles at the site of injury constrict initially and then dilate, leading to increased blood flow.
- This is followed by the slowing of blood flow and margination of leukocytes.
- Leukocytes escape into the tissues by diapedesis and accumulate in large numbers, attracted by the chemotactic substances released at the site of injury.

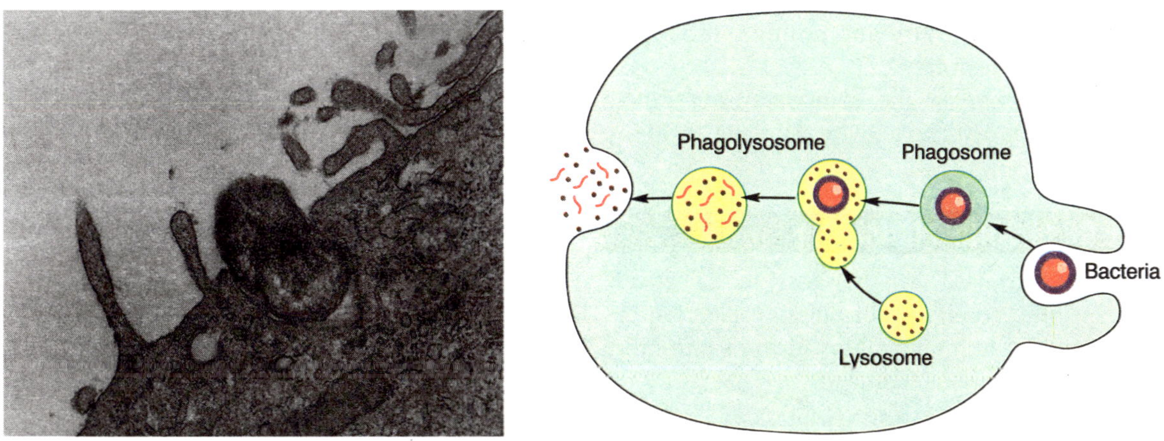

(a) (b)

Fig. 13.4 (a) Transmission electron microscopic (TEM) picture of *Rickettsia tsutsugamushi* getting phagocytosed by a mouse mesothelial cell and (b) diagrammatic representation of phagocytosis (*Source*: [a] Centers for Disease Control and Prevention [CDC], Public Health Image Library [PHIL], Image ID 923)

- As a result, microorganisms present at the site are phagocytosed and destroyed.
- There is an outpouring of plasma, which helps to dilute the toxic products present.
- A fibrin barrier is laid, serving to wall off the site of infection.

Fever

Fever acts as a natural defence mechanism. It not only helps to accelerate the physiological processes but may also, in some cases, destroy the infecting pathogens. Fever stimulates the production of interferons and aids in the recovery from viral infections.

Acute-phase reactant proteins

Infection or injury leads to a sudden increase in the plasma concentration of certain proteins, collectively called acute-phase proteins.

- **C-reactive protein (CRP):** An acute-phase reactant that accompanies inflammation and tissue injury, it is commonly used in clinical practice as an adjunct diagnostic and monitoring test in bloodstream infections, especially in the pediatric population.
- **Procalcitonin (PCT):** A 14.5 kDa peptide—its secretion is increased by IL-6, IL-1, and TNF-α; not activated by IFN-γ (which is chiefly produced in response to viral infections), making it a sensitive marker of bacterial infections.

Erythrocyte sedimentation rate (ESR)

ESR acts as an indirect acute phase reactant. It reflects an increase in plasma viscosity due to the presence of acute phase proteins.

Mechanisms of action of acute-phase reactants

- CRP and some acute-phase proteins **activate the alternative pathway** of complements.
- They **enhance host resistance**.
- They **prevent tissue injury** and **promote the repair of inflammatory lesions**.

Table 13.3 summarises the components of innate immunity, the sites of their activity, and the resulting outcomes.

Table 13.3 Factors contributing to innate immunity and their outcomes

Site	Components	Outcome
Circulation and lymphoid organs	Phagocytic cells, granulocytes, NK cells	Phagocytosis, intracellular killing, direct and antibody-dependent cytolysis
Serum	Lactoferrin and transferrin, interferons, TNFα/β, lysozyme, fibronectin, complement proteins	Iron binding, antiviral proteins, phagocyte activation, peptidoglycan hydrolysis, opsonisation and phagocytosis, enhanced intracellular killing
Cell-associated receptors	Toll-like receptors Scavenger receptors Opsonins	Pattern recognition of pathogen and elimination
Acute-phase proteins	CRP, mannose-binding protein, alpha-1-acid glycoprotein	Inflammation Fever
Breast milk	Antibodies, fibronectin, interferon, lactoferrin, lysozyme, mucin, oligosaccharides	Protection of newborn

ADAPTIVE OR ACQUIRED IMMUNITY

The immunity acquired by the host to selectively eliminate specific foreign molecules is known as acquired (adaptive) immunity. This form of immunity displays four characteristic features:

1. **Antigenic specificity:** The components of the acquired immune system—viz a viz antibodies—can distinguish antigens as separate entities from the host. They can detect the differences between two proteins (of an antigen) up to a single amino acid.
2. **Diversity:** It is capable of generating enormous antibody diversity in its recognition molecules, allowing it to reorganise billions of unique structures/patterns on foreign antigens. Genes form the basis of such diversity.
3. **Immunologic memory:** This type of immunity exhibits memory on the second encounter with the same antigen and generates a secondary response, which is more specific, heightened, and quick.
4. **Self/non-self-recognition:** Self-tolerance is one of the unique characteristics of the immune system, which prevents it from acting against the body's own molecules while still effectively eliminating foreign antigens. Multiple mechanisms ensure self-tolerance. Failure of these mechanisms may lead to autoimmunity.

Acquired immunity may be broadly categorised into two types—active immunity and passive immunity.

Active Immunity

This refers to the immune response mounted by a host in response to an antigenic stimulus. This type of immunity involves the active functioning of the host's immune apparatus, leading to:

- The synthesis of antibodies (**humoral immunity**)
- Production of immunologically active cells (**cellular immunity**)

Types of active immunity

i. **Natural active immunity:** Occurs in response to naturally-occurring infection.
ii. **Artificial active immunity:** Occurs after vaccination.

Features of active immunity

- **A latent period:** Required for the immunological machinery to be set in motion.
- **A negative phase:** The active immunity develops, and the level of measurable immunity may be lower than it was before the antigenic stimulus. This is because the antigen combines with any pre-existing antibody and lowers its level in circulation.
- **Secondary response:** Once developed, active immunity is **long-lasting**. If an individual who has been actively immunised against an antigen is exposed to the same antigen subsequently, the immune response is faster and more abundant than it was during the first encounter (Fig. 13.5).
- Besides the development of humoral and cellular immunity, active immunity is associated with **immunological memory**.

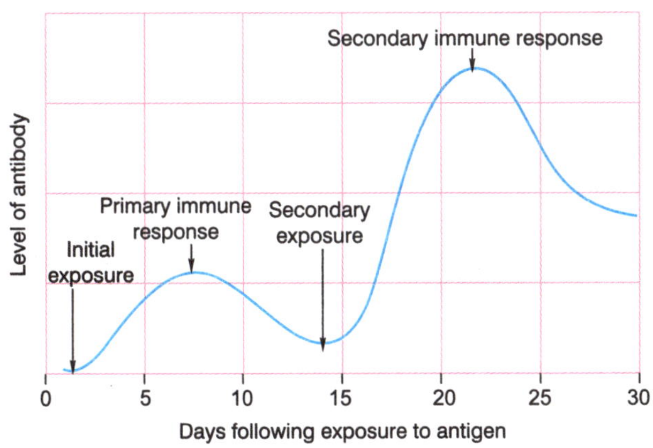

Fig. 13.5 Primary and secondary immune responses in active immunity

Herd immunity (also known as population immunity)

- This is a form of adaptive immunity seen in a population where the **majority of individuals develop immunity either due to an infection or immunisation** (universal or mass vaccination).
- Consequently, the transmission of the infection is minimised, and any individual within that community who has not been vaccinated has a lower chance of acquiring the infection.
- This type of immunity is seen during epidemics and pandemics and is one of the positive outcomes of the strategy for mass vaccination as was targeted during the COVID-19 pandemic.

Passive Immunity

- This type of immunity develops when a **preformed antibody is administered to an individual** to provide immediate protection against an infection.
- Since these antibodies (natural, in the form of convalescent sera) are preformed, their half-life is short. Thus, passive immunity is **short-lived**.

Types of passive immunity

There are two types of passive immunity—natural passive immunity and artificial passive immunity.

i. Natural passive immunity

This type of immunity is passively transferred from a **mother to her baby.**

- **Human colostrum**, which is rich in IgA antibodies and resistant to intestinal digestion, provides protection to the neonate.
- **Maternal antibodies** provide passive protection against infectious diseases to the infant. It is for this reason that most pediatric infections are more common after the age of three months than in younger infants.
- **Active immunisation of mothers** during pregnancy helps to improve the quality of passive immunity in infants. Immunisation of pregnant women with **tetanus toxoid** is recommended for this purpose as a part of routine antenatal medical care to prevent both maternal and neonatal tetanus.
- Though **fetuses synthesise antibodies (IgM) from about the twentieth week of life**, their immunological capacity is inadequate at birth. However, if there is an infection during intrauterine life, specific IgMs can be detected in the neonate's serum soon after birth.
- It is only by the age of about three months that the infant acquires some measure of immunological independence.

ii. Artificial passive immunity

This refers to the resistance passively provided to a recipient by the administration of preformed antibodies (discussed in detail in Chapter 19).

The differences between active and passive immunity are presented in Tables 13.4 and 13.5.

INTERACTION BETWEEN INNATE AND ADAPTIVE IMMUNITY

- The innate and adaptive immune responses often interact with each other and function simultaneously with overlapping activity of some cells like the natural killer cells.
- In summary, immunity is a multipronged protective mechanism in the human body that acts through innate and adaptive systems.

Immune mechanisms are not always beneficial. They may undergo disruption, which may result in hypersensitivity, autoimmunity, and immunodeficiency. These will be discussed in the subsequent chapters.

Table 13.4 Differences between active and passive immunity

Characteristic	Active immunity	Passive immunity
Mechanism	Immunocompetent cells are stimulated	Immunocompetent cells do not have any role
Trigger for induction	• Occurs naturally after an active infection • Artificially induced by vaccination	Maternal antibodies transferred from mother to fetus Preformed immunoglobulins (hyperimmune sera, natural or monoclonal) are injected into the host
Time to activation	Requires time to develop	Confers immediate protection (preformed antibodies do not have a lag time)
Duration of immunity	Long-lasting	Short-lived
Boosting immunity	Second/booster doses increase response in a short time and increase the level of immunity	Booster dose is not required (may result in a hypersensitivity reaction)
Memory cells	Memory cells are formed	Memory cells are not formed
Therapeutic role	Long-term protection by vaccination	Immediate protection by preformed antibodies

Table 13.5 Mechanisms of acquisition of immunity (*Source*: CNX OpenStax, CC BY 4.0, via Wikimedia Commons)

Type of immunity	Natural acquired	Artificial acquired
Passive	Immunity acquired from antibodies passed in breast milk or through the placenta 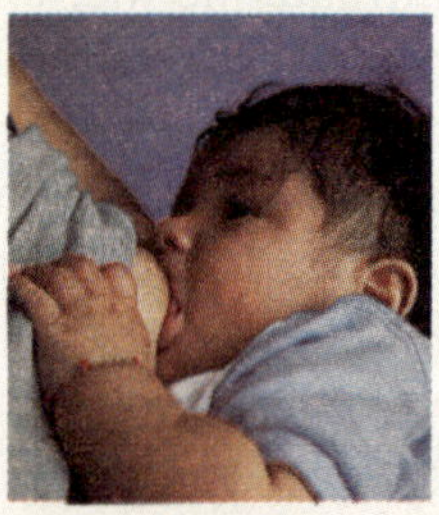	Immunity gained through antibodies harvested from another person or an animal 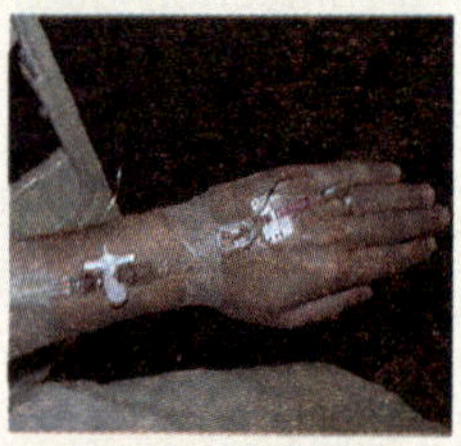
Active	Immunity gained through illness and recovery 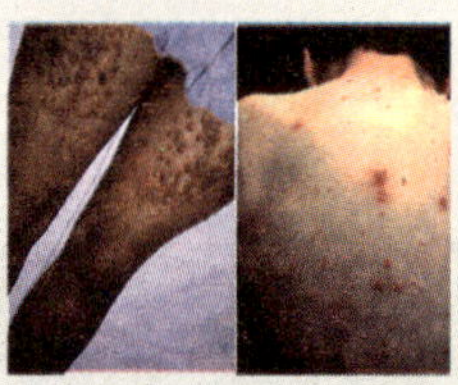	Immunity acquired through a vaccine 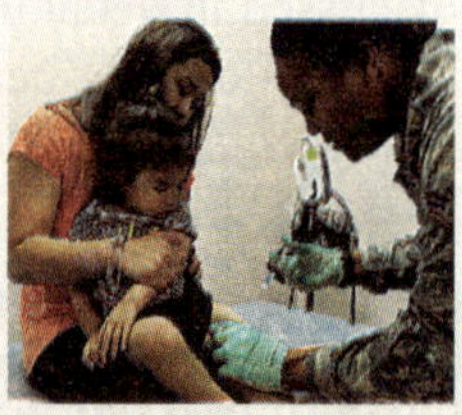

TYPES OF IMMUNE RESPONSES

There are two types of immune response to an antigen, which are as follows:

- **Humoral immunity (antibody-mediated):** It is mediated by antibodies produced by **plasma cells** present in blood and other body fluids ('humoral' is derived from the word 'humors', which means 'body fluids').
- **Cellular immunity:** This is mediated by **sensitised (activated) lymphocytes**. Cells for each of these components (T and B cells) develop through separate channels and remain independent, though they also interact in some instances.

CELLS INVOLVED IN THE IMMUNE RESPONSE (IMMUNOCOMPETENT CELLS)

ANATOMY OF THE IMMUNE SYSTEM

There are two major types of organs involved in the development and maturation of the cells of the immune system:

1. **Primary lymphoid organs,** in which lymphoreticular cells develop and mature—the bone marrow and the thymus
2. **Secondary lymphoid organs**, in which lymphoid cells and antigens react—the lymph nodes, the spleen, the tonsils, and certain tissues in various mucous membrane layers in the body

Lymphoreticular System (Mononuclear Phagocyte System)

- The lymphoreticular system is a complex organisation of cells of diverse morphology distributed widely in different organs and tissues of the body namely, blood, connective tissue, spleen, liver, lungs, bone marrow, and lymph nodes.
- These organs are made up of lymphoreticular cells which consist of lymphoid and reticular components.
- The mononuclear phagocyte system consists of cells with clearly demarcated functions:
  - **Lymphoid cells—lymphocytes and plasma cells**—are primarily concerned with specific immune response.
  - **Phagocytic cells—monocytes and macrophages** (otherwise known as **scavenger cells**)—are primarily concerned with ingesting and digesting exogenous antigens and foreign particles. They

contribute to non-specific immunity and remove microorganisms from blood and tissues. They also play a role in antigen processing and specific immunity in the afferent and efferent limbs of the immune response.

Hematopoiesis

All cells of immunological importance originate from a **hematopoietic stem cell** (HSC; Fig. 13.6).

Lymphoid System

The lymphoid system consists of lymphoid organs and cells (Fig. 13.7).

I. Central (primary) lymphoid organs

1. Thymus

- The primary function of the thymus is the production of **thymic or T lymphocytes**.
- It is the main site for lymphocyte proliferation in the body.
- Lymphocytes conditioned in the thymus are called **thymus-dependent lymphocytes or T cells**.
- T cells that cannot recognise the antigen–major histocompatibility complex (MHC) complexes and those that react with the self-antigen–MHC complex (and thereby cause autoimmune disease) are destroyed.
- More than 95 per cent of all thymocytes die by **apoptosis**.

Clinical significance of thymus dysfunction

- **DiGeorge syndrome:** Deficient cell-mediated immunity (CMI) is seen in congenital aplasia of the thymus in human beings and mice ('nude mice').
- **Runt disease:** The deficiency of CMI is evident from lymphopenia, deficient graft rejection, and the so-called 'runt disease' seen in neonatally thymectomised mice.
- **Post-thymectomy:** Thymectomy primarily affects the CMI. It also diminishes antibody response to many types of antigens (thymus-dependent antigens) such as sheep erythrocytes and bovine serum albumin. Humoral response to other antigens is unaffected.

2. Bone marrow

In humans and other mammals, the bone marrow is the **site of B cell origin and development**. All lymphocytes originate in the bone marrow. Immature B cells originating from lymphoid progenitors proliferate and differentiate within the bone marrow with the help of cytokines.

- In the human fetus, **Peyer's patches**, isolated lymphoid follicles (**gut-associated lymphoid tissue [GALT]**), develop and appear along with cells in

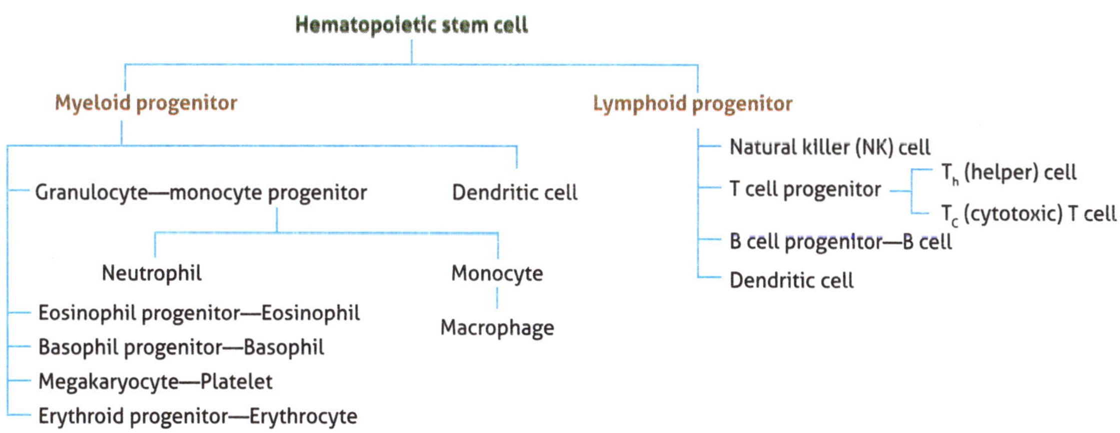

Fig. 13.6 Hematopoiesis

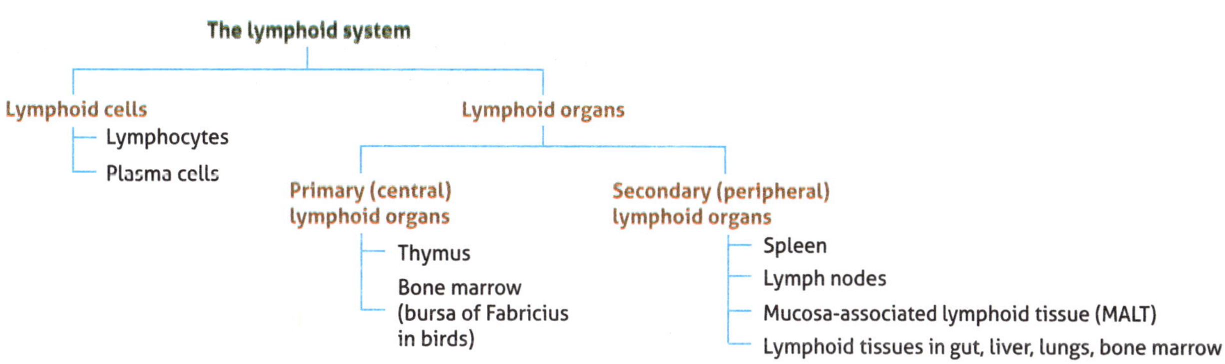

Fig. 13.7 Lymphoid system

the spleen and lymph nodes by the **20th week of gestation**.

- From then on, the fetus is able to produce IgM and IgD.
- IgA and IgE are not present at this stage.
- At birth, IgM production is enhanced, but the maternal IgG level falls steadily to reach minimum levels by the third month after birth.
- Production of IgG by the neonate then picks up and becomes adequate by 2–3 years. **Full immunocompetence is attained only after the first decade of life.**

Clinical significance of bone marrow dysfunction

- Depletion or suppression of the bone marrow leads to humoral immunodeficiency (hypo- or agammaglobulinemia); this leads to increased bacterial infection
- **X-linked (Bruton type) agammaglobulinemia**—children suffer from repeated pyogenic infections.
- **Common variable immunodeficiency** (IgA deficiency along with IgG_2) results from bone marrow dysfunction.
- **Transient hypogammaglobulinemia** occurs when maternal antibody wanes; it becomes severe in premature babies.

II. Peripheral (secondary) lymphoid organs

1. Lymph nodes

- Lymph nodes are distributed along the lymphatic vessels.
- They contain lymphocytes (B and T cells) in the cortex, dendritic macrophages, plasma cells, and interdigitating dendritic cells in the medulla.
- They enlarge following local antigenic stimulation (as seen in most infections, the enlargement of draining lymph nodes forms a diagnostic criterion).
- They act as a filter for lymph, each group of nodes draining a specific part of the body. Their functions are:
  - Phagocytosis
  - Proliferation and circulation of T and B cells

2. Spleen

It is the largest secondary lymphoid organ and has the following functions:

- **Clearance of effete or damaged cells** from the bloodstream
- Host resistance to bacterial infections, primarily directed against blood-borne antigens (pneumococcal and meningococcal infections)
- The **white pulp (WP)** of the spleen consists of the periarteriolar lymphoid sheath (PALS), which contains T cells and antigen-presenting cells (APCs)

Clinical significance of splenectomy

- The effects of a splenectomy on the immune response depends on the individual's age.
- In children, a splenectomy often leads to increased incidence of bacterial sepsis caused primarily by *Streptococcus pneumoniae*, *Neisseria meningitidis*, and *Haemophilus influenzae*.
- In adults, the effects are less adverse and may lead to bacterial infections (bacteremia).

3. Mucosa-associated lymphoid tissue (MALT)

The mucosa lining the alimentary, respiratory, and genitourinary tracts and other lumina and surfaces are constantly exposed to numerous antigens. They have a rich collection of specialised lymphoid cells, which are:

- **M cells (microfold cells)**, which transport the antigen from the lumen of the gut to the immune cells in specialised aggregates
- **Peyer's patches**
- **Gut-associated lymphoid tissue (GALT)**, which are scattered, isolated lymphoid follicles
- **Bronchus-associated lymphoid tissue (BALT)**, which is a mucosa-associated lymphoid tissue in the bronchus, where the M cells take up antigens (namely, mycobacterial antigen)

The MALT contains lymphoid as well as phagocytic cells. While the predominant immunoglobulin produced in the mucosa is secretory IgA, the other immunoglobulin classes—IgG, IgM, and IgE—are also formed locally.

Clinical significance of GALT

The existence of a common mucosal or secretory immune system may explain the superiority of oral or nasal immunisation over the parenteral route for many enteric and respiratory infections.

4. Bone marrow

- Besides being an important primary lymphoid organ, the bone marrow also has a significant role as a secondary lymphoid organ.
- It is capable of antigen presentation in an immunological reaction similar to the lymph nodes.

CELLS OF THE LYMPHORETICULAR SYSTEM

Lymphocytes

These cells constitute 20–40% of the body's white blood cells and 99% of the cells in the lymph. Based on their functions and cell-membrane components, lymphocytes can be broadly subdivided into three types:

I. T cells
II. B cells
III. Null cells

Table 13.6 presents the difference between T cells, B cells, and macrophages.

Table 13.6 Some distinguishing characteristics of T cells, B cells, and macrophages

Property	T cell	B cell	Macrophage
CD3 receptor	+	–	–
Surface immunoglobulins	–	+	–
Receptor for F_c piece of IgG	–	+	–
Erythrocyte antibody complement (EAC) rosette formation (C3 receptor; CR2; EBV receptor)	–	+	–
SRBC rosette (CD2; measles receptor)	+	–	–
Thymus-specific antigens	+	–	–
Blast transformation with			
Anti-Ig	–	+	–
Endotoxins	–	+	–
Phagocytic action	–	–	+

Lymphocytic circulation

- Constant traffic of lymphocytes through the blood, lymph, lymphatic organs, and tissues ensures their contact with antigens in any part of the body.
- Specific lymphocytes reach the site of introduction of the antigen during their ceaseless wandering and mount an immune response.
- A lymphocyte completes one cycle of recirculation in about one or two days.
- Recirculating lymphocytes can be recruited by the lymphoid tissues whenever necessary.

Functions

- Recognition of antigens
- Storage of immunological memory
- Immune response to specific antigens

Immune response to antigens: Lymphocytes have specific antigen-recognition mechanisms on their surface. Each clone of a lymphocyte is able to recognise only one antigen. The outcome of the antigen–cell interaction is either an immune response or immune tolerance. Stimulated T cells produce certain activation products (lymphokines) and induce CMI, while stimulated B cells divide and transform into plasma cells which synthesise immunoglobulins.

Surface markers

- A number of surface antigens or markers have been identified on lymphocytes and other leukocytes with the help of monoclonal antibodies.
- These markers, called **cluster of differentiation (CD) markers**, reflect the stage of differentiation and functional properties of the cells.
- Over 150 CD markers have been identified thus far.

Table 13.7 lists a few CD markers along with their cell association for CD4 (helper/inducer) and CD8 (suppressor/cytotoxic) cells.

Table 13.7 Some CD markers and their cell type association

CD number	Cell type association
CD1	Thymocytes, Langerhans cells
CD2	T cell sheep red blood cell (SRBC) receptor
CD3	T cell antigen–receptor complex
CD4	Helper T cell (receptor for HIV)
CD8	Suppressor/Cytotoxic T cells
CD19	B cells

I. T cells

- T cells are an important component of white blood cells.
- T cells are differentiated from other lymphocytes by the presence of a receptor on their cell surface, the **T cell receptor (TCR).**
- T cell precursors from the yolk sac, fetal liver, and bone marrow migrate to the thymus during the embryonic and postnatal stages.
- The earliest identifiable cells of the T lineage are the $CD7^+$ pro-T cells, which acquire CD2 on entering the thymus.
- They synthesise CD3 in the cytoplasm and become pre-T cells. T cell receptor (TCR) synthesis also takes place during this time (Fig. 13.8).

T cell receptors (TCR)

Under the influence of thymic hormones, progenitor T cells within the thymus acquire receptors called T cell receptors (TCR). Pre-T cells differentiate into two lineages, expressing either of the following:

1. αβ TCR chains
2. γδ TCR chains

- The majority of T cells carry αβ TCR. This TCR DNA region undergoes rearrangement to form functional alpha (α) or beta (β) TCR complexes, which then become positive for the expression of CD4, CD8, and other co-receptors, thereby becoming functional TCR.
- These TCR, in association with CD3, act as the antigen recognition unit, analogous to Ig on the surface of B cells.
- They are vital for the recognition of antigens by T cells.

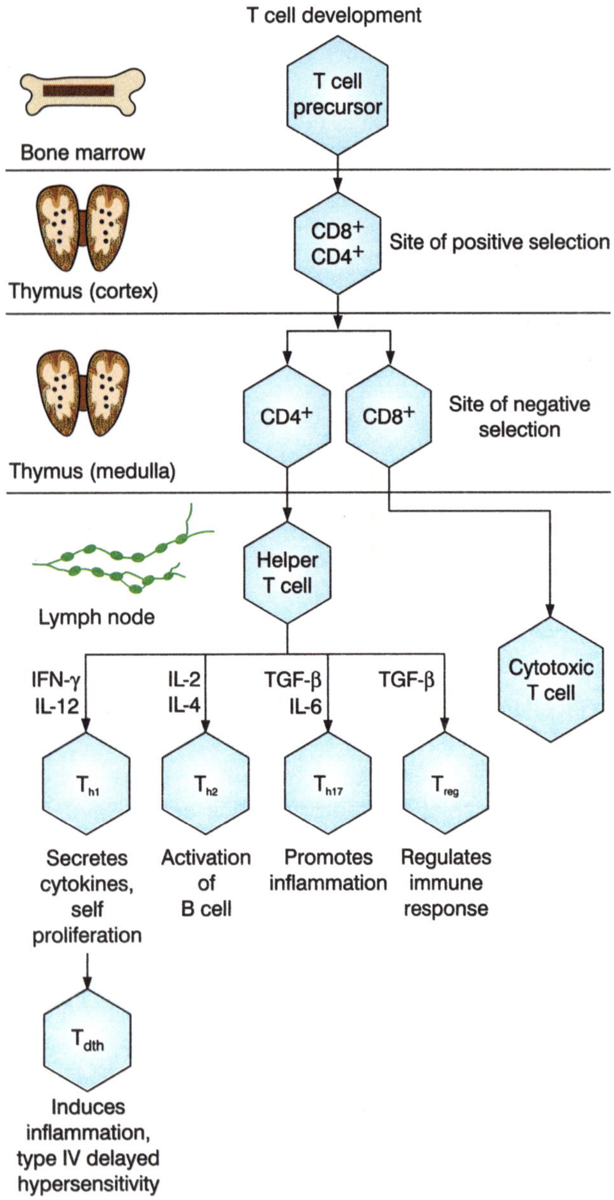

Fig. 13.8 T cell maturation

- As with immunoglobulins, TCR chains contain four separately encoded regions:
  - V or variable region
  - D or diversity region
  - J or joining region
  - C or constant region
- By reassortment of these regions (VDJC), a very wide repertoire of antigen specificities can be formed on the T cell surface.
- TCRs are capable of recognising host-specific antigen–major histocompatibility complexes (MHC).

Types of T cells

Various T cell categories have been identified based on their surface markers, target cells, and functions.

Effector T cells

- They interact with a target cell exhibiting a specific antigen and release proteins that recognise the antigens on the target cell.
- Different types of effector T cells are specialised to deal with different types of pathogens (antigens), which produce distinct and appropriate effects on the target cell (Fig. 13.9).

1. **Cytotoxic (suppressor) cells (Fig. 13.10)**
   - CD8 are suppressor/T cytotoxic cells (T_c).
   - They inhibit B cell antibody synthesis.
   - T_C cells are activated when they interact with an antigen–class I MHC complex on the surface of an altered self-cell (e.g., a virus-infected cell or a tumour cell) in the presence of appropriate cytokines. This activation results in proliferation, which causes the T_c cell to differentiate into an effector cell called **cytotoxic T lymphocyte (CTL)**, and thereby acquire the ability to recognise and eliminate altered self-cells.
   - They can kill and lyse target cells carrying new or foreign antigens, including tumours, allografts, and virus-infected cells.
2. **Helper/inducer (T_h) cells**
   - These cells have CD4 surface markers and MHC class II restriction.
   - They generally stimulate and promote the growth of T cells and macrophages.
   - They induce B cell differentiation.

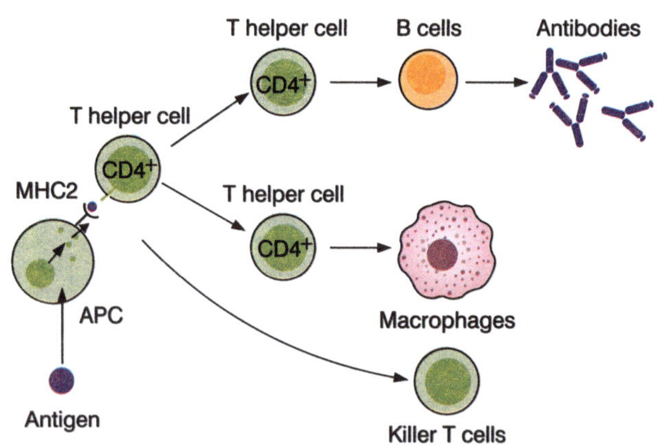

Fig. 13.9 Effector T cells

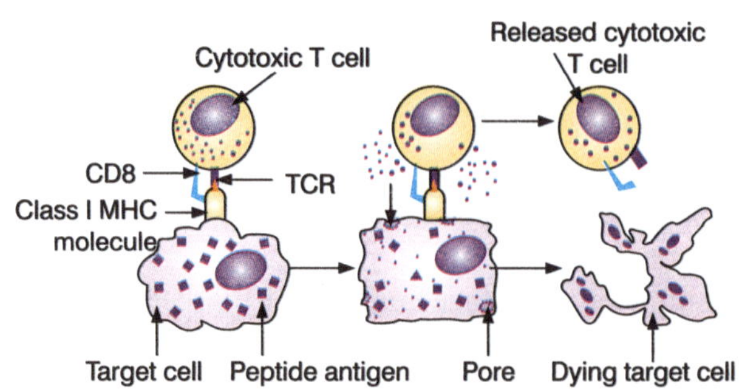

Fig. 13.10 Cytotoxic T cells

– They stimulate the proliferation of $CD8^+$ cytotoxic cells.
– They regulate certain stages of erythropoiesis.

$CD4^+$ cells can differentiate into T_h cells or T regulatory cells. T_h cells are further differentiated into the following types:

i. **T_{h1} cells:** These produce cytokines, interferon-gamma, and interleukin-2 (IL-2), which activate macrophage and T cells, promoting cell-mediated immunity, destruction of target cells and killing of intracellular pathogens (e.g., tubercle and lepra bacilli).
ii. **T_{h2} cells:** They produce cytokines IL-(interleukin) 4, 5, and 6, which in turn, stimulate B cells to form antibodies.
iii. **T_{h17} cells:** They produce cytokine IL-17 and promote inflammation, as seen in autoimmune diseases, systemic lupus erythematosus (SLE), rheumatic arthritis, and cancer.

3. **Memory cells (T_m)**
   – Both CD4 and CD8 provide memory and anamnestic immune responses.
   – These are antigen-specific T cells that persist for a long time after an infection has been eliminated.
   – As they are pre-stimulated to recognise the specific antigen, they quickly respond to a second stimulation by the same antigen, at which time, they are converted to effector T cells.
   – Subtypes of memory T cells are central memory T cells and effector memory T cells.

Regulatory T cells

- T_{reg} cells are $CD4^+$ cells that produce the cytokine TGF-β.
- These cells regulate the immune response and tolerance to self-reacting cells.

II. B cells

B cells are the major players in humoral immunity, responsible for producing immunoglobulins. B lymphocyte precursors, pro-B cells, develop in the fetal liver during embryonic life and in the bone marrow thereafter and continuously throughout life.

B cell maturation and synthesis of immunoglobulins (Fig. 13.11)

- Prior to the synthesis of immunoglobulins, the **rearrangement of immunoglobulin (Ig) genes** takes place in pre-B cells, which synthesise cytoplasmic IgM.
- In the next stage—**the immature B cell stage**—IgM is expressed on the cell surface. These cells undergo immunoglobulin isotype switching so that instead of IgM alone, the cell expresses on its surface IgD as well as one of the other Ig classes—IgM, IgG, IgA, or IgE.
- By re-assortment of Ig genes, B cells develop into mature B cells and then into plasma cells and memory cells with the capacity to produce Ig molecules that can react with all possible epitopes.
- By a process of **allelic exclusion**, each B cell is programmed to form only one class of Ig, with either kappa or lambda light chain, possessing specificity to a single epitope alone, and expressing it on its cell surface.
- Through contact with self-antigens during development, self-tolerance is developed by clonal deletion, leading to anergy or absence of response to antigens.

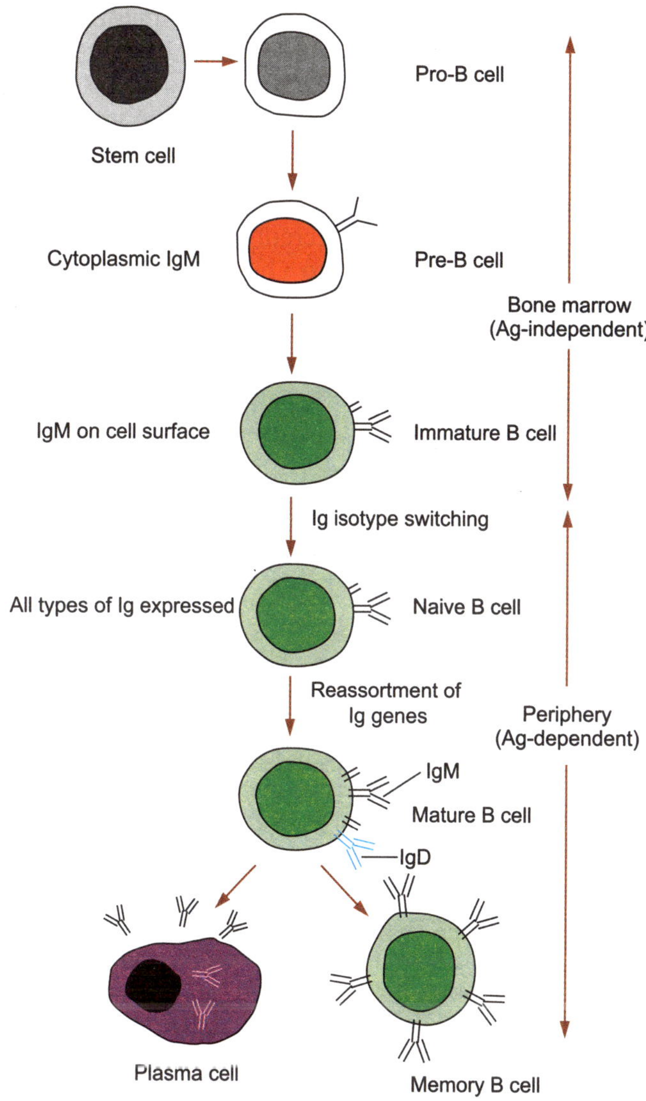

Fig. 13.11 B cell maturation

Types of B cells

- On coming into contact with its appropriate antigen, the mature B cell undergoes clonal proliferation.

- Some activated B cells become long-living **memory cells** responsible for the recall phenomenon seen on subsequent contact with the same antigen (Ag).
- Others are transformed into **plasma cells** (Fig. 13.12).

Plasma cells

- These are antibody-secreting cells that are structurally designed to be immunoglobulin-producing factories.
- They have a short lifespan of two or three days.
- One plasma cell makes an antibody of a single specificity, of a single immunoglobulin class and allotype, and a single light chain type only.
- An exception is seen in the primary antibody response, wherein a plasma cell which was producing IgM may switch to IgG production.

Memory B cells

- These express all isotypes—IgM, IgD, IgG, IgA, and IgE.
- They mediate a secondary immune response to subsequent encounters with the same antigen.
- **B1 cells** are a separate lineage of B cells, predominant in fetal and early neonatal life, express the T cell marker CD5 on their surface and have been named B1 cells. These are responsible for T-independent 'natural' IgM antibacterial antibodies, which appear in neonates, seemingly without antigenic stimulus.
- $CD5^+$ B cells may be relevant in the causation of autoimmune conditions.

III. Null cells

- Around 5–10 per cent of circulating lymphocytes lack features of T and B cells. These are called null cells.

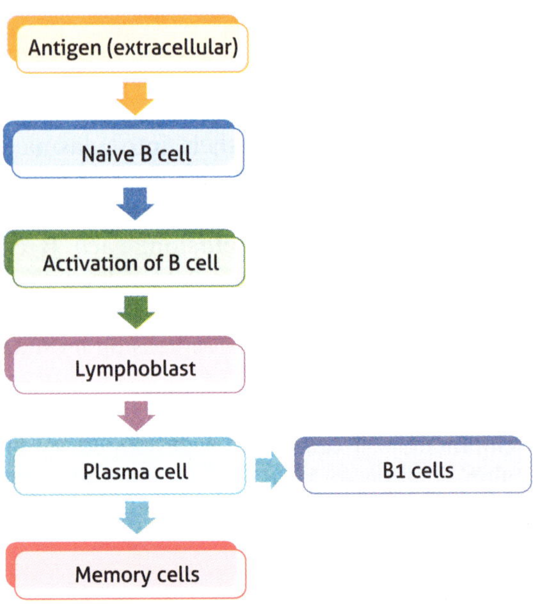

Fig. 13.12 Summary of B cell activation

- Because of their morphology, they are also known as **large granular lymphocytes (LGL).** LGL are a heterogeneous group of cells with different surface markers and functional characteristics.
- They are of three types:
  - **NK:** Natural killer cells
  - **ADCC:** Antibody-dependent cytotoxic cells
  - **LAK:** Lymphokine-activated killer cells

Natural killer cells (NK cells)

- They possess spontaneous cytotoxicity towards various target cells, chiefly malignant and virus-infected cells.
- They are considered to be important in immune surveillance and thought to be the natural defence against **virus-infected and malignant mutant cells.**
- Their cytotoxicity is not antibody-dependent or MHC-restricted.
- NK activity is 'natural' or 'non-immune' as it does not require sensitisation by prior antigenic contact; hence, they are considered a part of innate immunity.
- Their lineage is different from that of T and B cells.
- They are normally active in **severe combined immunodeficiency diseases (SCID),** in which mature T and B cells are absent.
- The cytolytic factor **perforin** of NK cells produces transmembrane pores through which cytotoxic factors such as the tumour necrosis factor-beta (TNF-β) enter the cell and destroy it by **apoptosis (programmed cell death)**.
- NK cell activity is augmented by interferons.

Antibody-dependent cytotoxic cells (ADCC)

- These are a sub-population of large granular lymphocytes (LGL), a group of NK cells possessing surface receptors for the F_c part of Ig.
- They are capable of killing target cells sensitised by IgG antibodies. This antibody-dependent cellular cytotoxicity is distinct from the action of cytotoxic T cells, which are independent of the antibody.

Lymphokine-activated killer (LAK) cells

- These are NK lymphocytes which, in conjunction with interleukin-2, are cytotoxic to a wide range of **tumour cells.**
- LAK cells have shown promise in the treatment of some tumours such as renal cell carcinoma.

Phagocytic Cells

- These cells are specialised in the removal of foreign and native particles.
- Phagocytic cells comprise mononuclear macrophages (of blood and tissues) and polymorphonuclear microphages.

1. Macrophages

They may be circulating or fixed cells.

- **Circulating monocytes (Fig. 13.13):** The blood macrophages (monocytes) are the largest of the lymphoid cells found in peripheral blood. Monocytes (10–15 μm) in circulation have an approximate half-life of three days. They leave the circulation and reach various tissues to be transformed into fixed macrophages.
- **Fixed tissue macrophages:** The tissue macrophages (histiocytes) are larger (15–20 μm) than circulating monocytes. They have morphological and functional features characteristic of the tissues in which they are localised and survive for months. These macrophages have different functions and are named according to their tissue location:
  - Alveolar macrophages
  - Histiocytes
  - Kupffer cells
  - Osteoclasts
  - Mesangial cells

Macrophages express several surface receptors including the following:

i. Receptor for the F_c part of IgG
ii. Receptor for activated complement components
iii. Receptor for various lymphokines
iv. M1 marker—a protein on human macrophages, which appears closely related to CR3, a cell receptor for C3 components

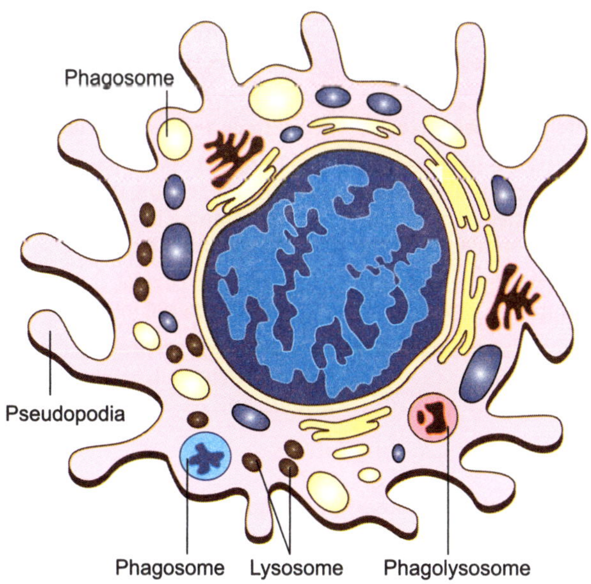

Fig. 13.13 Macrophage with phagosomes, pseudopodia, and lysosome granules

Mechanism of action (Fig. 13.14)

- Macrophages participate in several ways in the induction and execution of a specific immune response.

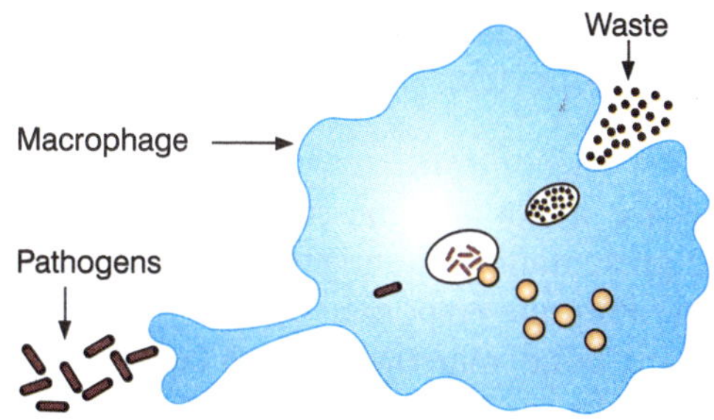

Fig. 13.14 Mechanism of action of macrophages

- As antigen-processing cells, they trap antigens, process them, and present them in optimal concentration to the lymphocytes for the induction of antibodies—too high a concentration of antigens may be tolerogenic and too low may be insufficient to be immunogenic.

2. Microphages

These are the polymorphonuclear leukocytes of the blood—neutrophils, eosinophils, and basophils.

Neutrophils

- These are non-specific active phagocytes that form the predominant cell type in **acute inflammation.**
- They are augmented by opsonins and do not have any role in specific immune processes.

Eosinophilic leukocytes

- These are found in large numbers in allergic inflammation, parasitic infection, and around antigen–antibody complexes.
- They primarily inhabit tissues rather than the bloodstream.
- They have granules containing a variety of hydrolytic enzymes that lead to the extracellular killing of large parasites.
- Eosinophils possess phagocytic activity but only to a limited degree.

Basophils

- These are found in the blood and **tissues (mast cells).**
- Their cytoplasm contains large numbers of prominent basophilic granules containing heparin, histamine, serotonin, and other hydrolytic enzymes.
- Degranulation of mast cells with the release of pharmacologically active agents constitutes the effector mechanism in **anaphylactic and atopic allergy**.

3. Antigen-presenting cells (APC) (dendritic cells)

- Activation of both the humoral and cell-mediated branches of the immune system requires cytokines produced by T_h cells.
- T helper cells can recognise only antigens that are displayed together with class MHC II molecules on the surface of antigen-presenting cells (APCs),

which are specialised dendritic cells, macrophages, and B lymphocytes.

Characteristics of dendritic cells

- These are major antigen-presenting cells involved in the presentation of antigens to T cells during the primary immune response.
- They are bone marrow-derived cells of a lineage different from the macrophages and T or B lymphocytes.
- They possess MHC class II expression.
- Co-stimulatory signals like B7 and CD28, which are necessary for T_h activation, are found on these cells.
- They are called **professional antigen-presenting cells**.
- Dendritic cells are **highly pleomorphic**, with a small central body and many long, needle-like processes (Fig. 13.15). They are present in peripheral blood and the peripheral lymphoid organs, particularly in the germinal areas of the spleen, lymph nodes, and skin.
- **Follicular dendritic cells (FDCs)** are present in the follicles of any secondary lymphoid organ and belong to the stromal cells of these organs. FDCs have important functions in the selection of memory B lymphocytes during **germinal centre reactions (GCR)**. FDCs present native antigens to potential memory cells, of which only B cells with a high affinity for B cell receptors (BCR) can bind.
- Ectopic FDC-formation is a feature of a number of autoimmune diseases and/or chronic inflammatory situations.
- The B cell is another antigen-presenting cell, which is particularly active during the secondary immune response.
- **Langerhans cells** in the skin possess features of macrophages and dendritic cells. They process and present antigens that reach the dermis.

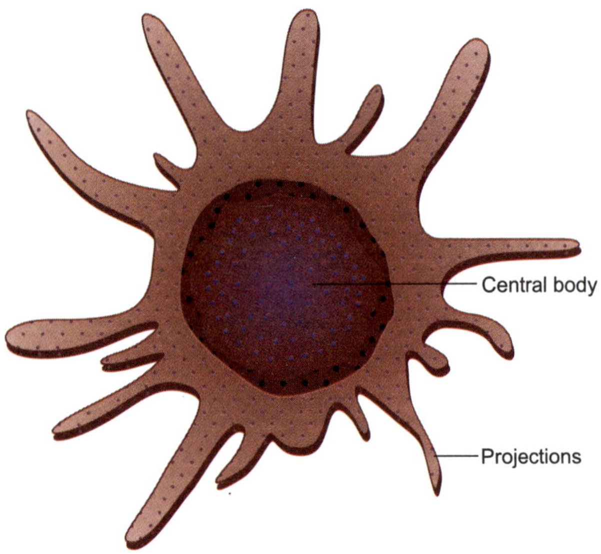

Fig. 13.15 Dendritic cell with needle-like projections

ABNORMALITIES OF IMMUNE CELLS

- Abnormalities in immune cells are generally inherited defects found in either specific or non-specific immune mechanisms.
- Individuals suffering from such abnormalities are susceptible to a variety of infections, depending on the nature of the cells involved.
- These abnormalities manifest as specific diseases as presented in Table 13.8.

Table 13.8 Abnormalities of the immune cells and the resulting diseases

Abnormality/Defective mechanism	Disease
Stem cell differentiation	Neutropenia–very few neutrophils
Lack of expression of CD18 molecule	LAD (leukocyte adhesion deficiency)
Defective phagocytosis	Chediak–Higashi syndrome
Defective encoding for NADPH oxidase results in defective intracellular oxygen-mediated killing	Chronic granulomatous disease leading to life-threatening bacterial and fungal infections
Defective IFN-γ and IL-12 receptor	Failure to control mycobacterial infections
Congenital aplasia of the thymus	DiGeorge syndrome, deficient CMI

Key Points

- ✓ The term immunity refers to the resistance exhibited by the host towards an injury caused by foreign antigens/microorganisms and their products.
- ✓ The two fundamental types of immunity are: innate and acquired. Each type is, in turn, composed of different subtypes.
- ✓ Innate (native) immunity is the resistance to infection that an individual possesses due to his genetic and constitutional make-up.
- ✓ Important non-specific antimicrobial factors found in blood, secretions, and tissues include lysozymes, beta-lysin, peroxidase enzymes, interferons, and lactoferrin.
- ✓ The resistance acquired by an individual during his life is known as acquired immunity. It displays four characteristic features—antigenic specificity, diversity, immunologic memory, and self/non-self recognition.
- ✓ Acquired immunity is of two types—active immunity, where the host produces its own response to foreign antigens to confer protection, and passive immunity, where preformed antibodies are introduced parenterally (not by the oral route but by the intramuscular or intravenous route) into the host to confer protection.
- ✓ Herd immunity is the immunity developed in a large proportion (80 per cent) of individuals in a population. This reduces the likelihood of epidemics arising in that community by the pathogen.

- ✓ Passive immunity is a type of immunity that develops when a preformed antibody is administered to an individual to provide immediate protection against infection. It also includes maternal antibodies in the neonate.
- ✓ The lymphoid system consists of the lymphoid cells and the lymphoid organs:
  - **Central (primary) lymphoid organs** include the thymus and the bone marrow in humans. All lymphocytes originate in the bone marrow, with T lymphocytes undergoing further differentiation in the thymus and B lymphocytes in the bone marrow.
  - **Peripheral (secondary) lymphoid organs** include the spleen, lymph nodes, and mucosa-associated lymphoid tissue (MALT).
- ✓ A lymphocyte that has been primed by the central lymphoid organs becomes an immunocompetent cell.
- ✓ Before they encounter antigens, mature T and B cells are called 'naïve cells', which subserve functions such as recognition of antigens and storage of immunological memory.
- ✓ Lymphocytes have antigen recognition mechanisms on their surfaces. When an immunocompetent cell reacts with its specific antigen, there may be induction of 'tolerance' or an immune response.
- ✓ Stimulated T cells produce lymphokines and affect cell-mediated immunity.
- ✓ A number of surface antigens or markers on lymphocytes reflect the stage of differentiation and functional properties of the cells, e.g., cluster of differentiation (CD) marker (CD4 in helper/inducer T cells, CD8 in suppressor/cytotoxic T cells). All T cells possess CD3 on their surface while B cells have Ig on their surface.
- ✓ T cells are classified as regulatory or effector cells. They have $CD4^+$ or $CD8^+$ markers on their surface.
- ✓ Helper/Inducer (T_h) cells, with a CD4 surface marker, stimulate the growth of T cells and macrophages in conjunction with major histocompatibility complex (MHC) class II to differentiate further.
- ✓ Cytotoxic T cells (T_c), with CD8 surface markers and MHC class I restriction, can kill and lyse target cells including tumour, allograft, and virus-infected cells.
- ✓ Memory T cells (T_m), both CD4 and CD8 cells, provide memory and anamnestic immune response.
- ✓ B cells undergo maturation in different stages as pro-, pre-, immature and mature B cells.
- ✓ On contact with its specific antigen, the mature B cell undergoes clonal proliferation to form long-living memory cells or plasma which function as antibody-secreting cells.
- ✓ Null cells are circulating lymphocytes which possess neither T cell nor B cell markers. They are of three types:
  - NK: Natural killer cells
  - ADCC: Antibody-dependent cytotoxic cells
  - LAK: Lymphokine-activated killer cells
- ✓ Phagocytic cells consist of mononuclear macrophages (of blood and tissues) and polymorphonuclear microphages and function as an effective defence against most microorganisms.
- ✓ Macrophages, dendritic cells, and B cells are the major antigen-presenting cells.

Questions

Short answer questions

1. Give an account of the types of innate immunity and the factors that affect them.
2. Describe the components of adaptive immunity.
3. List the differences between active and passive immunity.
4. List the differences between B and T cells and their process of maturation.
5. Enumerate the different types of T cells and list their functions.

MCQs

1. **The resistance to infections that an individual possesses by virtue of his genetic and constitutional make-up is referred to as ________.**
   a) Innate immunity b) Acquired immunity
   c) Artificial immunity d) Passive immunity
2. **Lysozyme is not found in which one of the following body fluids?**
   a) Tears b) Saliva c) CSF d) Sweat
3. **The immunity that an individual acquires during life by recognising and selectively eliminating specific foreign molecules is known as ________.**
   a) Innate immunity b) Non-specific immunity
   c) Adaptive immunity d) Immune tolerance
4. **Which of the following cells are not associated with the immune mechanism?**
   a) Basophils b) Dendritic cells
   c) Mast cells d) Epithelial cells
5. **Which of the following does not play a role in herd immunity leading to a decrease in the transmission of an infection?**
   a) Infection affecting a large population
   b) Universal immunisation
   c) Mass immunisation
   d) Antenatal immunisation of all pregnant women
6. **Which of the following is not a feature of acquired immunity?**
   a) Antigenic diversity
   b) No immunologic memory
   c) Non-recognition of self-antigen
   d) Non-self recognition
7. **Natural passive immunity in the newborn is due to maternal ________.**
   a) IgG b) IgM
   c) IgE d) IgD
8. **Active immunity is not acquired by:**
   a) Vaccination
   b) Immunoglobulin transfer
   c) Infection
   d) Sub-clinical infection
9. **Which of the following statements is true regarding passive immunity?**
   a) Confers immediate immunity
   b) Host immune system takes part
   c) Negative phase occurs
   d) Induced by infection
10. **All the following statements are true regarding the functions of the thymus except:**
    a) The thymus produces thymic lymphocytes.
    b) The thymus reaches its maximal relative size just before birth.
    c) The thymus functions best after puberty
    d) Thymus is the main site of lymphocyte proliferation.
11. **Toll-like receptors (TLRs) are:**
    a) Transmembrane receptors present on macrophages and dendritic cells
    b) Present on the surface of phagocytes to bind to mannose-rich glycans
    c) Nucleotide- binding oligomerisation domain like receptors to recognise intracellular bacterial cell products
    d) Molecules that bind to the surface of microbes

12. **Which of the following non-specific factor does not affect innate immunity?**
    a) Age of host
    b) Immune status of host
    c) Prior infection with same pathogen
    d) All of the above
13. **Differentiation between T cells and B cells is by all the following except:**
    a) T cells form sheep RBC rosette
    b) B cells form EAC rosette
    c) B cells have CD3 receptors
    d) B cells have surface immunoglobulins
14. **The earliest identifiable T cell linage is:**
    a) CD7 cells b) CD2 cells
    c) CD3 cells d) CD4 cells
15. **TCRs have the following features except:**
    a) TCR is a heterodimer of glycoprotein chains
    b) Majority of T cells carry $\gamma\delta$ TCR
    c) TCR chains contain four separately encoded regions
    d) TCR are not capable of recognising host-specific antigen–major histocompatibility complexes
16. **Which surface marker helps in the identification of chronic T cell malignancies?**
    a) CD4 b) CD7
    c) CD2 d) CD19
17. **Which of the following is true about memory B cells?**
    a) Not easily stimulated
    b) They are short-lived
    c) Express only IgM and IgD
    d) Express all isotypes of Ig
18. **B cells which may be associated with the causation of autoimmune diseases are:**
    a) Plasma cells b) B1 (CD5+) cells
    c) Memory cells d) Naive B cells
19. **Defective phagocytosis results in the following condition:**
    a) Chronic granulomatous disease
    b) DiGeorge syndrome
    c) Chediak–Higashi syndrome
    d) Common variable immunodeficiency
20. **Which is not a macrophage?**
    a) Monocyte b) Histiocyte
    c) Kupffer cells d) Lymphocytes
21. **MHC restriction is related to:**
    a) Antigen presentation by APCs to T cells
    b) Distribution of MHC in the population
    c) HLA typing done prior to transplantation
    d) Genetic predisposition to disease
22. **Which of the following is correct regarding lysozyme?**
    a) It is present in cerebrospinal fluid, sweat, and urine.
    b) It acts by splitting cell walls of susceptible bacteria.
    c) It is active against both gram-positive and gram-negative bacteria.
    d) It has no protective value.
23. **All are pattern recognition receptors except:**
    a) Toll-like receptors
    b) Mannose receptors
    c) Scavenger receptors
    d) B cell receptors

Answers: 1:a, 2:c, 3:b, 4:d, 5:d, 6:d, 7:a, 8:b, 9:a, 10:c, 11:a, 12:a, 13:c, 14:a, 15:b, 16:a, 17:d, 18:b, 19:c, 20:d, 21:a, 22:b, 23:d

CHAPTER 14

Part II: Immunology

Antigens, Antibodies, and the Complement System

(MI1.7 and 1.8)

Chapter Navigation

Learning objective

- Describe the structure and functions of antigens, antibodies, and complement systems

ANTIGENS

- Any **substance that induces an immune response in a host** is called an antigen.
- It stimulates the production of an antibody in a specific and observable manner.
- Antigens are substances that can be recognised by B cell receptors (Ig) and T cell receptors (along with the **major histocompatibility complex [MHC]**) and can evoke a specific immune response.
- Both the nature/molecular configurations and properties of an antigen and the host's immune system play an important role in eliciting an immune response.
- Both T and B lymphocytes recognise each antigen differently owing to the presence of different antigen recognition sites on each type of lymphocyte.
- Antigens can be of the following types:
  - **Complete antigens:** They are able to evoke an immune response independently.
  - **Partial antigens or haptens:** They require an additional carrier protein to be able to stimulate an antibody response (Fig. 14.1).

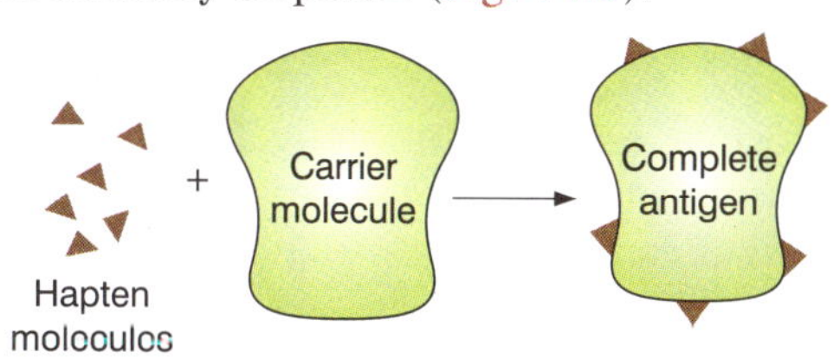

Fig. 14.1 Haptens when combined with a carrier particle form a complete antigen

TERMS RELATED TO ANTIGENS

- **Immunogen:** It induces a detectable immune response, resulting in either the production of antibodies or the activation of T cells, ultimately leading to either an immune response or immune tolerance.
- **Antigenicity:** It is the property that allows a substance to combine specifically with antibodies or T cell receptors (TCR), whether or not they are immunogenic. An immunogen can trigger an immune response and act as an antigen specific to that immune response. **All immunogens are antigens but not all antigens are immunogens.**
- **Immunogenicity:** The ability of a substance to induce humoral and/or cell-mediated immune response is called immunogenicity.
- **Epitope:** It is the smallest unit of an antigen responsible for antigenicity. It is also known as the **antigenic determinant**. The epitope consists of four or five amino acids or monosaccharide residues possessing a specific chemical structure, electrical charge, and steric (spatial) configuration. An epitope is capable of sensitising an immunocyte and reacting with its complementary site on the specific antibody or TCR. Epitopes may be present as sequential or conformational epitopes.
  - **Sequential or linear epitope:** A single, linear segment of the primary sequence. T cells recognise sequential epitopes.
  - **Conformational epitopes:** These are formed when surface residues from different sites of the peptide chain come together while folding into the tertiary structure. B cells identify the tertiary configuration of the conformational epitopes.
- **Paratope:** This is an area on the antibody molecule that corresponds to the epitope. Epitopes and paratopes determine the specificity of immunological reactions. Antigens such as bacteria or viruses carry many different types of epitopes,

presenting an antigenic mosaic. The presence of the same or similar epitopes on different antigens accounts for antigenic cross-reactions between unrelated microorganisms.

- **Adjuvants:** They enhance the immunogenicity of an antigen (when combined with it; details in Chapter 15).

Haptens

These are substances that are **incapable of inducing antibody formation by themselves but can react specifically with antibodies** (in Greek, *haptein* means 'to fasten').

- Haptens become immunogenic (capable of inducing antibodies) on **combining with a protein carrier**.
- Many drugs that cause allergic reactions act as haptens. For example, when **penicillin** is injected into a host, it combines with proteins in the body to form a neoantigen (newly-formed antigen) by forming a hapten–carrier complex. This may lead to a heightened immune response (anaphylaxis).
- Natural carriers include serum globulin, albumin, polysaccharides, and liposomes.
- Haptens can be categorised as follows:
  - **Complex haptens:** Polyvalent; can precipitate with specific antibodies
  - **Simple haptens:** Univalent and non-precipitating (it is assumed that precipitation requires the antigen to have two or more antibody-combining sites); they can inhibit the precipitation of specific antibodies by the corresponding antigen or complex hapten

DETERMINANTS OF ANTIGENICITY

1. **Foreignness:** Only antigens that are 'foreign' to the individual (non-self) induce an immune response. Individuals do not normally mount an immune response against their own cells (cell and tissue components). Tolerance to self-antigens is conditioned by contact with them during the development of the immune apparatus. The breakdown of this homeostatic mechanism results in the development of **autoimmune disease**. The antigenicity of a substance is related to the degree of its foreignness.

   Antigens from individuals of the same species are less antigenic than those from other species.
2. **Size:** Large molecules are highly antigenic, whereas particles with low molecular weight (less than 10,000 Da) are non-antigenic or feebly so. Low-molecular-weight substances may be rendered antigenic when they combine with tissue proteins.
3. **Chemical nature:** Most naturally-occurring antigens are proteins and polysaccharides. Lipids and nucleic acids, which are less antigenic, can become more antigenic by combining with proteins. However, not all proteins are antigenic.
4. **Susceptibility to tissue enzymes:** Antigens introduced into the body are degraded by the host into fragments of appropriate size containing the antigenic determinants. Phagocytosis and intracellular enzymes appear to play an essential role in breaking down antigens into immunogenic fragments. Substances that are not susceptible to tissue enzymes are not antigenic.
5. **Other factors:** These include the (i) dosage, (ii) route, and (iii) timing of the antigen administration. An antigen evokes the optimum response when administered in the right quantity. An excess dose produces immunological tolerance, while an inadequate dose fails to evoke an adequate response. Antigens administered intravenously evoke a rapid response while those administered by the subcutaneous or intradermal route elicit only a slow release.

During his lifetime, an individual encounters several antigens to which he may develop immunity, allergy, or hypersensitivity. The result may be innocuous (mild allergy), life-threatening (anaphylaxis), or a chronic, debilitating autoimmune disorder (Table 14.1).

Table 14.1 Examples of common antigens

Foreign antigens	Examples
Microbial agents	O antigen (cell wall), F antigen (pili) polysaccharides, H antigen (flagella), K antigen (capsule), nucleic acid, cell membrane, viral proteins, nucleic acids, fungal spores
Food	Egg protein, milk, fish, and crustaceans
Plants	Pollen
Animals	Fur, hair, snake venom
House dust	Mites
Heterologous transplants and infusions	Blood, blood products, tissues, and organs
Self-antigens	Cellular proteins, peptides, enzyme complexes, ribonucleoprotein complexes, and DNA

The positive aspect of an antigen is that it can be modified to form a vaccine to confer immunity against some communicable diseases.

TYPES OF SPECIFICITY OF ANTIGENS

- Antigenic specificity is based on stereochemistry and spatial configuration. Antigenic specificity is **not absolute**.
- Cross-reactions can occur between antigens that bear stereochemical similarities. In some instances, apparent cross-reactions may be due to the **sharing**

of **identical antigenic determinants** by different antigens.

- The specificity of natural tissue antigens is differentiated as follows:
  - **Species specificity:** The tissues of all individuals in a species contain species-specific antigens. There exists some degree of cross-reaction between antigens from related species. Species specificity has been used in: a) **tracing evolutionary relationships** between species and b) **forensic applications** for the identification of the species from blood and seminal stains.
  - **Isospecificity:** Isoantigens are antigens found in some but not all members of a species. A species may be grouped depending on the presence of different isoantigens in its members. The best examples of isoantigens are human erythrocyte antigens, based on which individuals can be classified according to their **blood groups**.
  - **Autospecificity:** Autologous or self-antigens are ordinarily non-antigenic, but there are exceptions. **Sequestered antigens**, which are not normally found free in circulation or tissue fluids (such as the eye lens protein, which is normally confined within its capsule) are not recognised as self-antigens. Similarly, antigens that are absent during embryonic life and develop later (such as the sperm) are also not recognised as self-antigens.
  - **Organ specificity:** Some organs such as the brain, kidneys, and lens protein of different species share the same antigen. Such antigens, which are found in different species, are called organ-specific antigens.
  - **Heterogenetic specificity (heterophile antigens):** Sometimes, the same antigens or closely related antigens may exist in different biological species, classes, and kingdoms. These are known as heterogenetic or heterophile antigens.
    - **Forssman antigen:** It is a heterophile antigen having a lipid–carbohydrate complex which is widely distributed in many animals, birds, plants, and bacteria. It is, however, absent in rabbits. For this reason, anti-Forssman antibodies can be prepared in these animals.
- **Heterophile reaction:** Some heterophile antigens are used for serological diagnostic tests. Antigens employed in these tests are unrelated to these etiological agents:
  - **Weil–Felix** reaction in typhus fever (antigens from *Proteus* species are used to detect antibodies against typhus fevers)
  - **Paul–Bunnell test** for infectious mononucleosis (human O group RBCs are used)
  - **Cold agglutinin test** for primary atypical pneumonia (human RBCs are used)

BIOLOGICAL CLASSES OF ANTIGENS

Depending on their ability to induce antibody formation, antigens are classified as:

1. T cell-dependent (TD) antigens
2. T cell-independent (TI) antigens
3. Superantigens

1. T Cell-Dependent Antigens (TD)

- Most natural proteins are T-dependent antigens.
- B cells cannot respond to these antigens without a co-stimulatory signal from T_h cells.
- Structurally, these antigens are characterised by a few copies of many different antigenic determinants.
- These antigens bind to the surface Ig on B cells and are internalised and processed to form smaller peptides, which are then expressed on the surface of B cells complexed with MHC-II and presented to T cells.
- Once these complexes are recognised by T_h cells, the latter secrete cytokines and start expressing the CD40 ligand, which interacts with CD40 on the B cells. The T–B interaction and cytokines provide the stimulus for B cell activation.

2. T Cell-Independent Antigens (TI)

- Some antigens can directly stimulate antibody production by B cells without the apparent participation of T cells. Such antigens are called TI antigens.
- These antigens react with B cell antigen receptors (BCRs) of innate immunity. Most microbial carbohydrates, lipids, and certain nucleic acids are T cell-independent antigens. They are of two types:
  - **Type 1 antigens** (endotoxin, lipopolysaccharide [LPS]) are directly mitogenic for B cells and cause polyclonal B cell activation.
  - **Type 2 antigens** are polymeric compounds like polysaccharides (bacterial cell wall lipopolysaccharide or pneumococcal capsular polysaccharides) or proteins (flagellar proteins). They activate B cells to generate specific antibodies with the help of cytokines, complement, and other cells like macrophages, dendritic cells, mast cells, and NK cells.

Some of the differences between T cell-dependent and T cell-independent antigens are listed in Table 14.2.

3. Superantigens

- These are antigens that evoke an **excessive immune response**.
- They are protein molecules (e.g., staphylococcal enterotoxins) that activate very large numbers

Table 14.2 Comparison between T cell-dependent and T cell-independent antigens

Property	T cell-dependent antigen	T cell-independent antigen
T cell involvement	Yes	No
Antigen interaction	Involves tertiary complex of T cell receptor, antigen, and MHC molecule	Involves binary complex of membrane Ig and antigen
Antigen processing by macrophages	Yes	No
Chemical nature of antigen	Mostly soluble protein	Type 1: LPS Type 2: Polymeric proteins and polysaccharide
Degradability	Easily degradable	Type 2: Poorly degradable
Complement activation	No	Type 2: Ag causes complement activation
Isotype switching	Yes	Type 1: No Type 2: Limited
Immunological memory	Yes	No
Polyclonal activation	No	Type 1: Yes, antigen in high doses
Effect on antibody production	Full range, IgM, IgG, IgA, and IgE	Limited to IgM and IgG_3

of T lymphocytes irrespective of their antigenic specificity.

- They do not get processed by an antigen-presenting cell (APC) but attach directly to the MHC-II α1 domain and β variable chain of TCR.
- There are two types of super-antigens—(i) **endogenous**, which are intracellular (e.g., virus) and (ii) **exogenous**, which extracellular (e.g., bacteria).
- Microbial superantigens are medium-sized proteins (MW 22–29 kDa) characterised by a high resistance to proteases.
- These are potent $CD4^+$ T cell activators.
- They initiate the release of cytokines (IL-2), which results in massive proliferation of T lymphocytes. This ultimately leads to the release of a variety of cytokines which have profound effects on the immune system.

Action of superantigens

- Superantigens (LPS, toxic shock syndrome toxin, and other bacterial cell wall factors) bind to the class II MHC antigens expressed on antigen-presenting cells (APCs) outside the peptide-binding groove and then sequentially bind to the T cell receptor (TCR) via the variable region of the TCR β-chain.
- High levels of IL-2 are released, which circulate in the blood, causing symptoms of fever, nausea, vomiting, diarrhea, and malaise.
- Excessive IL-2 leads to endothelial damage due to interleukin-1 (IL-1) and inflammatory chemokines such as IL-8 and platelet-activating factor (PAF), causing intravascular coagulation, shock, and multiple organ/system failure.
- Some human diseases associated with superantigens are presented in Table 14.3.

Table 14.3 Human diseases associated with microbial superantigens

Organism	Superantigen	Disease
Staphylococcus aureus	Enterotoxins A–E, G–J, K–R, U and V Toxic shock syndrome toxin (TSST-1, same as enterotoxin F)	Food poisoning Toxic shock syndrome Multiple sclerosis
Group A streptococci	Pyrogenic exotoxins (associated with streptococcal toxic shock syndrome and scarlet fever)	Shock Psoriasis Rheumatic heart disease
Mycobacterium tuberculosis	Not identified	Tuberculosis
HIV	Nef (negative regulatory factor)	AIDS
Rabies virus	Nucleocapsid protein	Rabies
Epstein–Barr virus	Not identified	B cell lymphoma Transactivates another endogenous virus HERV-K18
Cytomegalovirus (CMV)	Not identified	Causes enhanced HIV replication in Vb12 T cells, which are reservoirs for HIV

ANTIBODIES

- Antibodies are a group of proteins, collectively called **immunoglobulins** (glycoprotein molecules), that react specifically with the epitope on an antigen.
- Immunoglobulins are present on the B cell membrane and are secreted by plasma cells. Antibodies constitute 20% of plasma proteins.

- They circulate in blood and eliminate or neutralise the antigen by phagocytosis, antibody-dependent cell-cytotoxicity (ADCC), or opsonisation.
- An 'antibody' is a biological and functional model. **All antibodies are immunoglobulins, but all immunoglobulins may not be antibodies.**
- Immunoglobulins constitute 20–25 per cent of the total serum proteins.
- Based on physicochemical and antigenic differences, five classes of immunoglobulins have been recognised—IgG, IgA, IgM, IgD, and IgE.

ANTIBODY STRUCTURE

- Antibodies contain two identical light (L) chains weighing 22,000 Da each and two similar heavy (H) chains weighing 55,000 Da each (Fig. 14.2).
- **Papain digestion**—an IgG antibody, when digested by papain in the presence of cysteine, splits into two types of fractions:
  - **F_c (fragment crystallisable)**—an insoluble fraction that crystallises in the cold
  - **F_{ab} (fragment antigen binding)**—a soluble fragment which is unable to precipitate with egg albumin but can still bind with it
- Each molecule of immunoglobulin can be split by papain into one F_c and two F_{ab} pieces having a sedimentation coefficient of 3.5S.
- **Pepsin digestion**—when treated with pepsin, a 5S fragment is obtained, which is composed essentially of two F_{ab} fragments held together in position. This is the active part that precipitates with the antigen. This fragment is called $F_{(ab)2}$. The F_c portion can be digested by pepsin into smaller fragments.

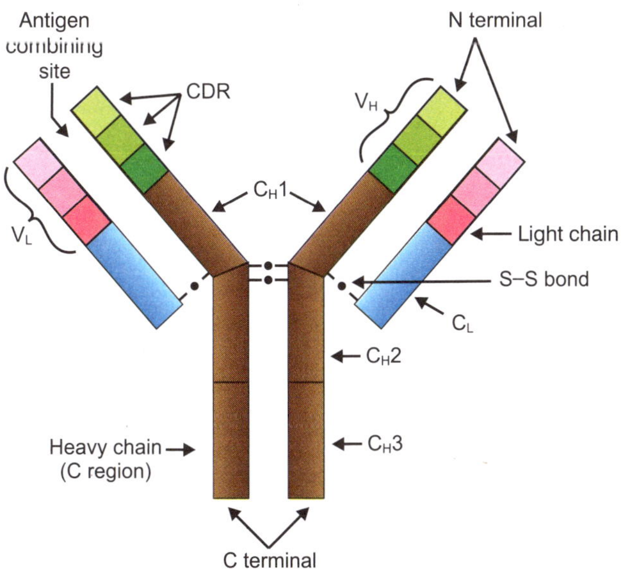

Fig. 14.2 Structure of an immunoglobulin molecule with light and heavy chains

Immunoglobulin Chains

- Antibodies or immunoglobulins are glycoprotein molecules consisting of **four polypeptide chains—two heavy and two light**.
- Each light chain is bound to a heavy chain by interchain disulphide bonds.

Heavy chains (H)

- Immunoglobulins are made up of two identical chains with a molecular weight of >55,000 Da each.
- The heavy chains are of five types—**alpha (α), gamma (γ), delta (δ), epsilon (ε), and mu (μ)** (Table 14.4).
- These are structurally and antigenically distinct for each class and are designated by the Greek letter corresponding to the immunoglobulin class.

Table 14.4 Immunoglobulin classes and their corresponding H chains

Immunoglobulin class	H chain
IgG	γ (gamma)
IgA	α (alpha)
IgM	μ (mu)
IgD	δ (delta)
IgE	ε (epsilon)

Light chains (L)

- Immunoglobulins possess two identical light (small) chains with a molecular weight of >22,000 Da each.
- There are two types of light chains, **kappa** (κ) and **lambda** (λ).
- A molecule of immunoglobulin may have either kappa or lambda chains but never both together.
- In humans, 60 per cent of L chains are kappa and 40 per cent are lambda.
- The L chains are similar in all classes of immunoglobulins.

Variable and Constant Regions

- The antigen-combining site of the immunoglobulin molecule is at its amino terminus. It is composed of both L and H chains (Fig. 14.2).
- The first 110 amino acids of the N terminal are variable in sequence. Hence, this region is called the **variable region**. It is made up of:
  - **VL—variable light chain**
  - **VH—variable heavy chain**
- The sequence beyond the variable region and spanning the rest of the molecule is relatively constant in all antibodies and is called the **constant region (C)**. It is made up of:
  - **CL—constant light chain**
  - **CH—constant heavy chain**

- The carbohydrate moieties of the variable region are linked to the constant region of the H chains.
- The carboxyl terminal constant region of the heavy chains mediates the effector functions.
- The C region of the light chains does not attach to the cell membrane and does not participate in its effector functions.

PROTECTIVE FUNCTION OF ANTIBODIES

- Antibodies are responsible for **humoral immunity**.
- The close structural complementary resemblance to the antigen, which provokes their synthesis, allows antibodies to bind with antigens in a non-covalent manner involving electrostatic Van der Waals and other weak forces.
- Antibodies protect against infection by the following mechanisms:
  - **Phagocytosis** by opsonisation
  - **Virus neutralisation** by preventing the virus from binding to the receptors on the cells, preventing their entry into the susceptible cell
  - **Complement-mediated lysis of cells**
  - **Antibody-mediated cell cytotoxicity** by attaching to the viral proteins on the cell surfaces of virus-infected cells, which leads to interaction with the killer cells, resulting in lysis
  - **Neutralisation** of toxins
  - Participate in **active immunisation** (vaccines induce their production) and **passive immunisation** (preformed antibodies injected into the host protect against toxins)

IMMUNOGLOBULIN CLASSES

Human sera contain the immunoglobulins IgG, IgA, IgM, IgD, and IgE in descending order of concentration. Table 14.5 lists the characteristics of these distinct classes.

Immunoglobulin G (IgG)

- IgGs are grouped into four subclasses—IgG_1, IgG_2, IgG_3, and IgG_4, which have γ_1, γ_2, γ_3, and γ_4 H chains respectively.
- Subtle differences are seen in the sequence of the constant regions of these subclasses.
- All four IgG subclasses are distributed in human serum in the approximate proportions of 65 per cent, 23 per cent, 8 per cent, and 4 per cent respectively.

Action of IgG

- IgG binds to microorganisms and enhances their phagocytosis (by a mechanism known as **opsonisation**).
- Extracellular killing of target cells coated with IgG antibody is mediated through the recognition of the surface F_c fragment by NK and K cells bearing the appropriate receptors.
- Interaction of IgG complexes with platelet F_c receptors probably leads to aggregation and vasoactive amine release.
- IgG participates in most immunological reactions such as complement fixation, precipitation, and neutralisation of toxins and viruses. It may be

Table 14.5 Characteristics of the five immunoglobulin classes

Characteristic	IgG	IgA*	IgM	IgD	IgE
Sedimentation coefficient (S)	7	7	19	7	8
Molecular weight (Da)	15,000	16,000	90,000	18,000	19,000
Serum concentration (mg/mL)	12	2	1.2	0.03	0.00004
Half-life (days)	23	6	5	2–8	1–5
Daily production (mg/kg)	34	24	3.3	0.4	0.0023
Intravascular distribution (per cent)	45	42	80	75	50
Carbohydrate (per cent)	3	8	12	13	12
Complement fixation—classical pathway	++	–	+++	–	–
Complement fixation—alternative pathway	–	+	–	–	–
Placental transport	+	–	–	–	–
Presence in milk	+	+	–	–	–
Selective secretion by seromucous glands	–	+	–	–	–
Heat stability (56°C)	+	+	+	+	–

* IgA may occur in 7S, 9S, and 11S forms

considered a general-purpose antibody, protective against infectious agents active in blood and tissues.

Clinical significance

- When its level is raised, as in chronic malaria, kala-azar, or myeloma, the IgG synthesised against that particular antigen is catabolised rapidly and may result in the deficiency of the particular antibody. Conversely, in hypogammaglobulinemia, the IgG administered for treatment is catabolised slowly.
- Passively administered IgG suppresses homologous antibody synthesis by a feedback process. This property is utilised for RhD isoimmunisation of women, which involves the administration of anti-Rh(D) IgG during delivery.
- With most antigens, IgG is a late antibody and makes its appearance after the initial immune response, which is mediated by IgM.

Immunoglobulin A (IgA)

IgA is the major immunoglobulin in colostrum, saliva, and tears. It occurs in two forms:

- **Serum IgA:** Monomeric 7S molecule (MW about 160,000 Da)
- **Secretory IgA** (dimeric SIgA):
  - Found on mucosal surfaces and secretions.
  - It is a dimer formed by two monomer units joined together at their carboxy terminals by a glycopeptide termed the **J chain** (J for joining).
  - The J chain is synthesised by plasma cells situated near the mucosal or glandular epithelium; the J chains are also present in polymeric IgM immunoglobulins.
  - SIgA binds to a receptor on the surface of the epithelial cells and is endocytosed and transported across the cells to the luminal surface. During this process, a part of the receptor remains attached to the IgA dimer. This part, which is a glycine-rich polypeptide, is known as the **secretory component or secretory piece** (Fig. 14.3). It is produced by mucosal or glandular epithelial cells and not by lymphoid cells.
  - The secretory piece is believed to **protect IgA from denaturation** by bacterial proteases in sites such as the intestinal mucosa, which have rich and varied bacterial flora.
  - SIgA is a much larger molecule than serum IgA (11S; MW about 40,000 Da).
  - SIgA is selectively concentrated in secretions and on mucous surfaces, forming an 'antibody paste', and is believed to play an important role in **local immunity against respiratory and intestinal pathogens.**

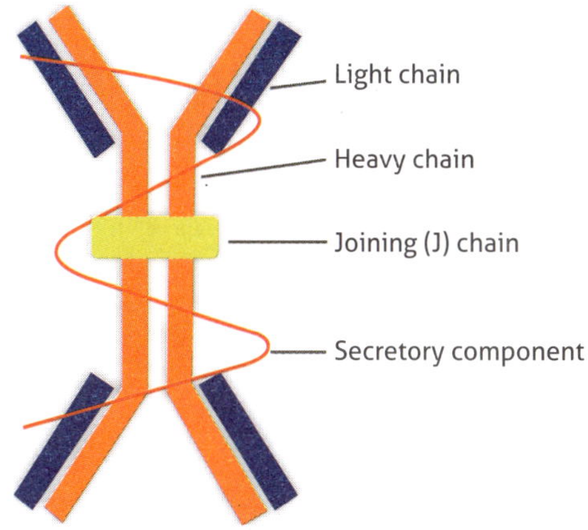

Fig. 14.3 Secretory IgA molecule

  - It is **relatively resistant to digestive enzymes and reducing agents**.
- IgA antibodies function by inhibiting the adherence of microorganisms to the surface of mucosal cells by engulfing organisms and thereby preventing their entry into body tissues. Cross-linkages form between the multivalent antigen and the polymeric IgA, and the pathogen is eliminated.
- **Breast milk**, which is rich in IgA, helps protect the newborn against infection during the first month of life. Colostrum contains large amounts of SIgA, which helps protect the mucous membranes in the throat, lungs, and intestines of the infant.
- IgA does not fix complement but can activate the alternative complement pathway.
- It promotes phagocytosis and the intracellular killing of microorganisms.
- Two IgA subclasses have been described: IgA1 and IgA2. IgA2 lacks interchain disulphide bonds between the heavy and light chains. Though IgA2 is a minor component of serum IgA, it is the dominant form in secretions.

Immunoglobulin M (IgM)

- It is a heavy molecule (19S; MW 90,000–100,000 Da; polymers of five four-peptide subunits on account of which, IgM is called '**the millionaire molecule**'; Fig. 14.4).
- Polymerisation of the subunits depends on the presence of the J chain.
- Though the theoretical valency of an IgM molecule is 10, this is observed only with small haptens. With larger antigens, the effective valency falls to five, probably due to steric hindrance.
- Monomeric IgM is the major antibody receptor on the surface of B lymphocytes for antigen recognition.
- Phylogenetically, IgM is the **oldest immunoglobulin class**. It is also the earliest immunoglobulin to be synthesised by the fetus, beginning at about **20 weeks of age**.

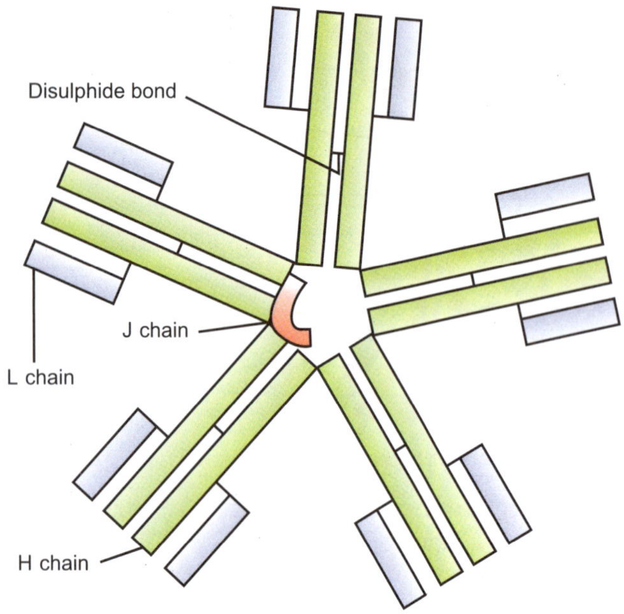

Fig. 14.4 IgM molecule

- IgM is not transported across the placenta; **the presence of IgM in the fetus or newborn indicates intrauterine infection** and its detection is useful in the diagnosis of congenital infections such as syphilis, rubella, HIV infections, and toxoplasmosis.

Role of IgM in infections

- A single molecule of IgM can bring about immune hemolysis, whereas 1,000 IgG molecules are required for the same effect.
- IgM is also 500–1,000 times more effective than IgG in opsonisation, 100 times more effective in bactericidal action, and about 20 times more effective in bacterial agglutination.
- In the neutralisation of toxins and viruses, however, it is less active than IgG.
- Being largely confined to the intravascular space, IgM is believed to be responsible for **protection against blood invasion** by microorganisms. IgM deficiency is often associated with septicemias.
- IgM is the **first antibody produced** as a primary response to an infection. It is seen in the early phase of the infection and appears to be involved in microbial infections, as opposed to IgG.
- IgM detection in the serum denotes an early infection and often, may be diagnostic in certain infectious diseases, importantly, viral infections for which specific detection methods are not available.
- Because of its pentameric structure, it is able to bind to many antigen particles and red blood cells. For this reason, smaller amounts of IgM are required to neutralise, opsonise, and activate complement than IgG.

Immunoglobulin D (IgD)

- IgD resembles IgG structurally.
- IgD and IgM are present on the surface of unstimulated B lymphocytes and serve as **recognition receptors for antigens**.
- The combination of cell membrane-bound IgD or IgM with the corresponding antigen leads to specific stimulation of the B cell—either activation and cloning to produce antibody or suppression.

Immunoglobulin E (IgE)

- It resembles IgG structurally.
- IgE is chiefly produced in the linings of the respiratory and intestinal tracts.
- IgE exhibits unique properties such as heat lability (inactivated at 56°C in one hour).
- It has an affinity for the surface of tissue cells (particularly mast cells) of the same species **(homocytotropism).**
- It does not cross the placental barrier or fix complement.
- Greatly elevated levels are seen in atopic (type 1 allergic) conditions such as asthma, hay fever, and eczema and in children living in unhygienic conditions with a high load of intestinal parasites.
- IgE deficiency has been associated with individuals with impaired immunity who present with undue susceptibility to infection.
- IgE mediates the **anaphylactic type of hypersensitivity**.
- It protects against pathogens by mast cell **degranulation and release of inflammatory mediators**.
- It is also believed to have a special role in defence against helminthic infections.

FACTORS INFLUENCING ANTIBODY PRODUCTION

- **Genetic factors:** The terms 'responder' and 'non-responder' are used to describe the individual's capacity to respond to a particular antigen. The *Ir* (immune response) genes control this property.
- **Age:** The embryo is immunologically immature. The capacity to produce antibodies develops only with the development and differentiation of lymphoid organs.
- **Immunocompetence:** This is not complete at birth but continues to develop as the infant grows. The infant becomes self-reliant for antibody production from the age of 3–6 months, by which time the maternal antibodies disappear. However, full competence is acquired only by 5–7 years for IgG and 10–15 years for IgA.
- **Nutritional status:**
  - Malnutrition affects the immune response adversely, though serum components necessary

for immunity are conserved selectively until the nutritional deficiency becomes marked.

– Protein calorie malnutrition suppresses both humoral and cellular immune response, the latter more severely.
– Deficiencies of amino acids (tryptophan, phenylalanine, methionine, glycine, and isoleucine) and vitamins (vitamins A and B, riboflavin, pyridoxine, pantothenic acid, and folic acid) have been shown to cause a decrease in antibody synthesis.

ABNORMAL IMMUNOGLOBULINS

Apart from antibodies, other structurally similar immunoglobulins are detected in the serum in many pathological processes, which are described in this section.

1. Multiple Myeloma

- Myeloma is a plasma cell dyscrasia in which there is unchecked proliferation of one clone of plasma cells, resulting in the excessive production of the particular immunoglobulin synthesised by the clone.
- Multiple myeloma may affect plasma cells synthesising IgG, IgA, IgD, or IgE.
- **Involvement of IgM-producing cells is known as Waldenstrom's macroglobulinemia**. In this condition, there is excessive production of the respective myeloma proteins (M proteins) and their light chains (Bence–Jones proteins).
- It is a low-grade B cell lymphoproliferative neoplasm characterised by small lymphocytes and IgM monoclonal gammopathy.

2. Heavy Chain Disease

- This is a form of paraproteinemia that causes lymphoid neoplasia and is characterised by the overproduction of the F_c parts of the immunoglobulin heavy chains.
- Three types of heavy chain disease (HCD) are recognised based on the class of immunoglobulin heavy chain produced (alpha, gamma, or mu) by the malignant cell.

3. Cryoglobulinemia

- A gel or precipitate is formed upon cooling the serum, which redissolves on warming.
- It may not always be associated with disease but is often found in myelomas, macroglobulinemias, and autoimmune conditions such as systemic lupus erythematosus.
- Most cryoglobulins consist of IgG, IgM, or their mixed precipitates.
- It is the presence of abnormal protein in the serum that tends to thicken as gel in serum in cold temperature.
- Cryoglobulins are antibodies that can form immune complexes that block blood vessels, causing vasculitis.

IMMUNOGLOBULIN SPECIFICITIES

The antigenic determinants or paratopes on immunoglobulin molecules fall into three main categories—isotype, paratype, and idiotype. These determinants are located in characteristic portions of the molecule.

Isotype

- Isotypic determinants refer to the genetic variations or **differences in the constant region of the heavy chain** of the Ig classes and subclasses within a species.
- Every healthy individual expresses all isotypes in the serum within the same species.
- Different species inherit different constant-region genes and, therefore, express different isotypes.
- When an antibody from one species is injected into another, the isotypic determinants are recognised as foreign, inducing an antibody response to the isotypic determinants.

Allotype

- Allotypic determinants refer to multiple alleles that exist for some genes and lead to subtle amino acid differences in some, but not all members of a species.
- In humans, allotypes have been characterised for all four IgG subclasses and one IgA subclass (IgA_2).
- Antibodies to allotypic determinants are sometimes produced by a mother during pregnancy in response to paternal allotypic determinants on the fetal immunoglobulins.
- Antibodies to allotypic determinants can also arise from a blood transfusion.

Idiotype

- Idiotypic determinants arise from the sequence of the heavy and light chain variable regions.
- Each **individual antigenic determinant of the variable region (paratope) is referred to as an idiotope**. The sum total of idiotopes on an Ig molecule constitutes its idiotype.
- By immunisation with F_{ab} fragments, anti-idiotypic antibodies can be produced. These resemble the epitopes of the original antigen. The anti-idiotypes provide protection against the original antigen (pathogen or tumour) in experimental animals.

ANTIBODY DIVERSITY

- An individual is capable of producing an enormous number of structurally distinct antibodies, each with a distinct specificity.
- The presence of a large number of antibodies that bind to different antigens/pathogens is referred to as antibody diversity, and the total collection of antibodies with different specificities is called the **antibody repertoire**.
- The genetic mechanisms that can generate such a large antibody repertoire occur exclusively in lymphocytes.

Clinical Significance of Antibody Diversity

The knowledge of immunoglobulin genes has made it possible to develop engineered antibodies for **therapeutic purposes to treat various lymphomas and autoimmune diseases**.

Genetic Basis of Antibody Diversity

- It is estimated that an individual has the capacity to produce 10^8 different antibody molecules. To have each such antibody molecule coded for by a separate gene would require millions of genes to be set apart for antibody production alone.
- The discovery of **split genes** for immunoglobulins demolished the longstanding dogma of 'one–gene–one–protein' with important implications in biology. For this discovery, **Susumu Tonegawa** was awarded the Nobel Prize for Medicine in 1987.

Class Switching

- In response to antigenic stimulation, plasma cells produce a single class of antibody (e.g., IgM).
- Later, gene arrangement allows for the production of the same antigen specificity with a different immunoglobulin class (e.g., IgG or IgA). This process is called class switching.

Clinical significance of class switching

- **Immunoglobulin class switching recombination deficiencies (Ig CSR deficiencies) or hyper IgM syndromes (HIGM)** are a group of primary immunodeficiency diseases characterised by defective CD40 signalling of B cells.
- Affected patients are characterised by low serum levels of IgG and IgA and normal or elevated levels of IgM, which lead to increased susceptibility to infections such as bacterial infections of the skin and respiratory tract, mucosal ulcers, and diarrhea.

GENETIC REGULATION OF IMMUNOGLOBULINS

B cells and their lineage express the immunoglobulin genes. Two major classes of regulatory sequences in the immunoglobulin DNA regulate the transcription of genes. They are as follows:

1. **Promoters**, which are a group of short nucleotide sequences that initiate transcription
2. **Enhancers**, which activate transcription from the promoter sequence

Genetic Characteristics of Immunoglobulins

- Immunoglobulins are unique in their property to respond to an array of foreign antigens.
- Variations in heavy and light chain proteins depend on the immunoglobulin genes.
- Multiple genes encode for a single immunoglobulin heavy or light chain.
- During the maturation of a B cell, the genes are reorganised in an ordered sequence (of Ig gene) to contribute to the diversity of the final product.
- The knowledge of antibody gene sequences and the structure of antibodies and the advent of molecular technology have made it possible to design and construct genes that code for the formation of immunoglobulins—artificial antibodies—using **hybridoma technology**.

MONOCLONAL ANTIBODIES

The discovery of **hybridoma technology** for the production of unlimited quantities of identical monoclonal antibodies of the same Ig class, possessing uniform specificity, affinity, and other properties, has revolutionised clinical immunology.

Principle

- A single antibody-forming cell or clone produces antibodies directed specifically against a single antigen or antigenic determinant only.
- When a clone of lymphocytes or plasma cells undergoes selective proliferation, as in multiple myeloma, antibodies with single antigenic specificity accumulate. Such **antibodies produced by a single clone and directed against a single antigenic determinant are called monoclonal antibodies**.
- Hybridomas are somatic cell hybrids produced by **fusing antibody-forming spleen cells with myeloma cells**. The resultant hybrid retains the antibody-producing capacity of the spleen cell and the ability of the myeloma cells to multiply indefinitely (Fig. 14.5).

Technique of Monoclonal Antibody Production by Hybridoma

- **Fusion:** Lymphocytes from mice spleens immunised with the desired antigen are fused with mouse myeloma cells that are deficient in the enzyme **hypoxanthine phosphoribosyl transferase (HPRT)**.

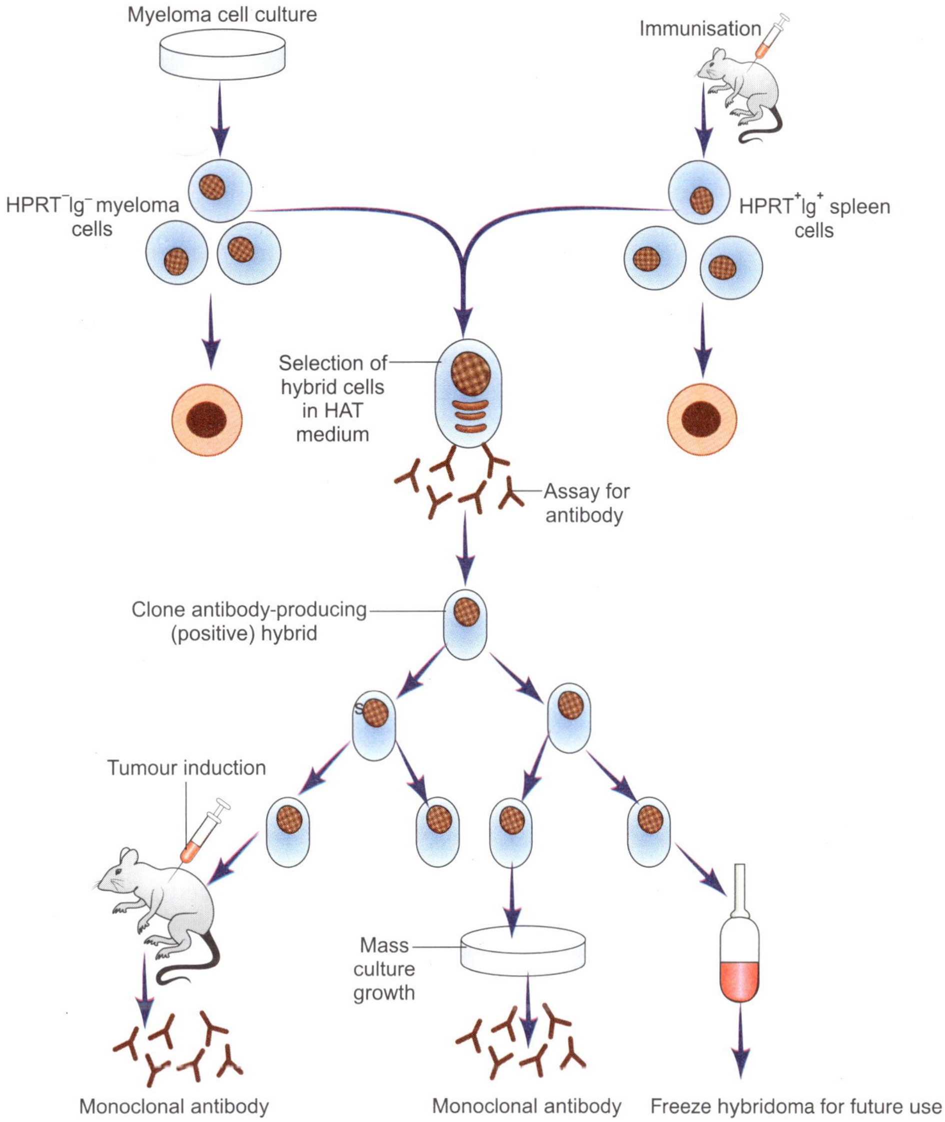

Fig. 14.5 Monoclonal antibody production by hybridoma technique
**HPRT*–hypoxanthine phosphoribosyl transferase, ***HAT*–hypoxanthine aminopterine thymidine

These hybrid cells are called hybridomas and are capable of replicating indefinitely.

- **Growth:** The fused cells are placed in a medium containing hypoxanthine, aminopterin, and thymidine (**HAT**). This medium does not permit the growth of enzyme-deficient myeloma cells. Only hybrid cells possessing properties of both the splenic lymphocytes and myeloma cells grow in the culture.
- **Cloning:** Hybridomas are cloned and examined for the production of antibodies.
- **Continuous cultivation:** Clones producing antibodies against the desired antigen are selected for continuous cultivation. These hybridomas can be maintained indefinitely in culture and will continue to form monoclonal antibodies.
- **Mice ascetic fluid:** They can also be injected intraperitoneally in mice; monoclonal antibodies may then be obtained by harvesting the ascitic fluid of mice and subsequently be frozen for prolonged storage.

Applications of Monoclonal Antibodies

Monoclonal antibodies have important clinical uses:

- **Therapeutic**—for the treatment of lymphomas, rheumatoid arthritis, and some serious infections. Monoclonal antibodies have been used as therapeutic agents in the treatment of COVID-19.
- **Diagnostic imaging**—radiolabelled monoclonal antibodies are used to detect tumour antigens.
- **Diagnostic reagents**—monoclonal antibodies are used for diagnosing infections (immunoassays) and measuring the blood levels of various drugs.
- **Research**—they are used for the in vitro production of immunoglobulins.

Humanised (Chimeric) Antibodies

- Mouse monoclonals have been humanised by genetic manipulation to make chimeric antibodies consisting of **murine variable regions and human constant regions**.
- Grafting of murine monoclonal complementarity-determining region (which is the antigen-binding surface of an antibody) on a human Ig framework provides a virtually human molecule.
- Humanised antibodies have been used therapeutically to treat patients with **refractory multiple myeloma**. Their use in other hematologic malignancies is under trial.

Antibody Engineering

- Entire libraries of antibodies have been built by fusing genes coding for antibodies with bacteriophage genomes.
- Large quantities of the desired human antibody can be prepared by infecting bacteria with the appropriate bacteriophage. Such antibody engineering holds great promise for immunotherapy.

COMPLEMENT

- The term complement (C) refers to a system of soluble and cell-bound proteins that occur in normal serum and are activated by antigen–antibody interaction. They subsequently mediate a number of biologically significant consequences.
- The complement system belongs to a group of biological effector mechanisms (called **triggered enzyme cascades**), which also include coagulation and the fibrinolytic and kinin systems.
- Each enzyme in the cascade can activate many molecules of the succeeding component, resulting in the amplification of the response at each step. Every step has its own control mechanisms so that the cascade can be regulated with precision.

PROPERTIES OF COMPLEMENT

- Complements are present in the sera of all mammals birds, amphibians, and fish.
- A complement from one species can react with antibodies from another species.
- It constitutes about 5 per cent of normal serum protein, and its quantity does not increase as a result of immunisation.
- Complement, as a whole, is heat-labile, though some of its components are heat-stable.
- It undergoes spontaneous denaturation slowly at room temperature and is destroyed in 30 minutes at 56°C. Serum deprived of its complement activity by heating at 56°C for 30 minutes is said to be 'inactivated'.
- It ordinarily **does not bind to free antigens or antibodies but only to antigen–antibody complexes**.
- The binding site (F_c) is expressed only when Ig is combined with its antigen.
- This site of C binding is located on the F_c piece of the Ig molecule (CH_2 domain on IgG and CH_4 on IgM).

COMPONENTS OF COMPLEMENT

- The complement system consists of at least 30 chemically and immunologically distinct serum proteins which make up the complement components.
- These include the **properdin system** and **control proteins**.
- The biological activities of this system affect both innate and acquired immunity. Interactions of cellular receptors with C proteins control B cell activation and play a crucial role in the highly developed acquired immune system.

Fractions

- Complement is a complex of nine different fractions, C1 to C9.
- The fraction C1 occurs in serum as a calcium ion-dependent complex, which on chelation with EDTA, yields three protein subunits called C1q, r, and s.
- Thus, C is made up of a total of 11 different proteins.
- C fractions are named C1 to C9 in the sequence of the cascading reaction, except that C4 comes after C1 and before C2.

> The C activity in an immune reaction is **cytolysis,** which is the lysis of erythrocytes sensitised by anti-erythrocyte antibody. The **erythrocyte–antibody complex is called EA**. When C components are attached to EA, the product is called **EAC**, followed by the components that have reacted with it (for example, EAC 14235 or EAC 1–5). When a C component acquires enzymatic or other demonstrable biological activity, it is indicated by a bar over the component number, for example, enzymatically activated C1 is shown as $\overline{C1}$. Fragments cleaved from C components during the cascade are indicated by small letters (e.g., C3a, C3b). Inactivated forms of C components are indicated by the prefix 'i' (e.g., iC3b).

FUNCTIONS OF THE COMPLEMENT

- **Cytolysis**—lysis of cells, bacteria, and viruses by the membrane attack complex
- **Chemotaxis**—the movement of phagocytic cells towards bacterial cells with the help of C5a (potent chemotactic agent)

- **Opsonisation**—C3b receptors on phagocytes render the process of phagocytosis more efficient
- **Binding of complement components** to cells trigger:
  - Inflammatory reactions
  - Specific cell functions
  - Secretion of immunoregulatory molecules
- **Immune clearance**—removes immune complexes and deposits them in the liver and spleen

COMPLEMENT ACTIVATION

- Complements are normally present in the body in an inactive form. When their activity is induced by antigen–antibody combination or other stimuli, C components react in a specific sequence as a cascade. This phenomenon is called **complement activation**.
- Various terms such as fixation, binding, and consumption are used to refer to the combination of C with bound immunoglobulin.
- The binding of the complement to the Ag–Ab complex leads to the activation of the classical C pathway.
- The fixation of C is not influenced by the nature of antigens but only by the class of immunoglobulins.
- Immunologically, complement is responsible for **cytolytic activity**.
- All classes of Ig do not fix C. Only IgM, IgG_3, IgG_1 and IgG_2 (in that order) fix C; IgG_4, IgA, IgD or IgE do not.
- Essentially, the **C cascade** is a series of reactions in which the preceding components act as enzymes on the succeeding components, cleaving them into dissimilar fragments.
- The larger fragments usually join the cascade. The smaller fragments often have biological effects, which contribute to defence mechanisms by various basic effector mechanism functions mentioned above.
- The resultant effects are the following (Fig. 14.6):
  - Amplification of the inflammatory process
  - Increase in vascular dilation and permeability (by anaphylatoxin, C3a and C5a)
  - Induction of smooth muscle contraction
  - Release of histamine from mast cells

COMPLEMENT PATHWAYS

The C cascade can be triggered by three parallel but independent mechanisms or pathways which differ only in the initial steps. Once C3 activation occurs, the subsequent steps are common to all pathways. The three pathways are as follows:

1. Classical C pathway
2. Alternative or properdin pathway
3. Lectin pathway

1. Classical Complement Pathway

- The chain of events in which complement components react in a specific sequence following

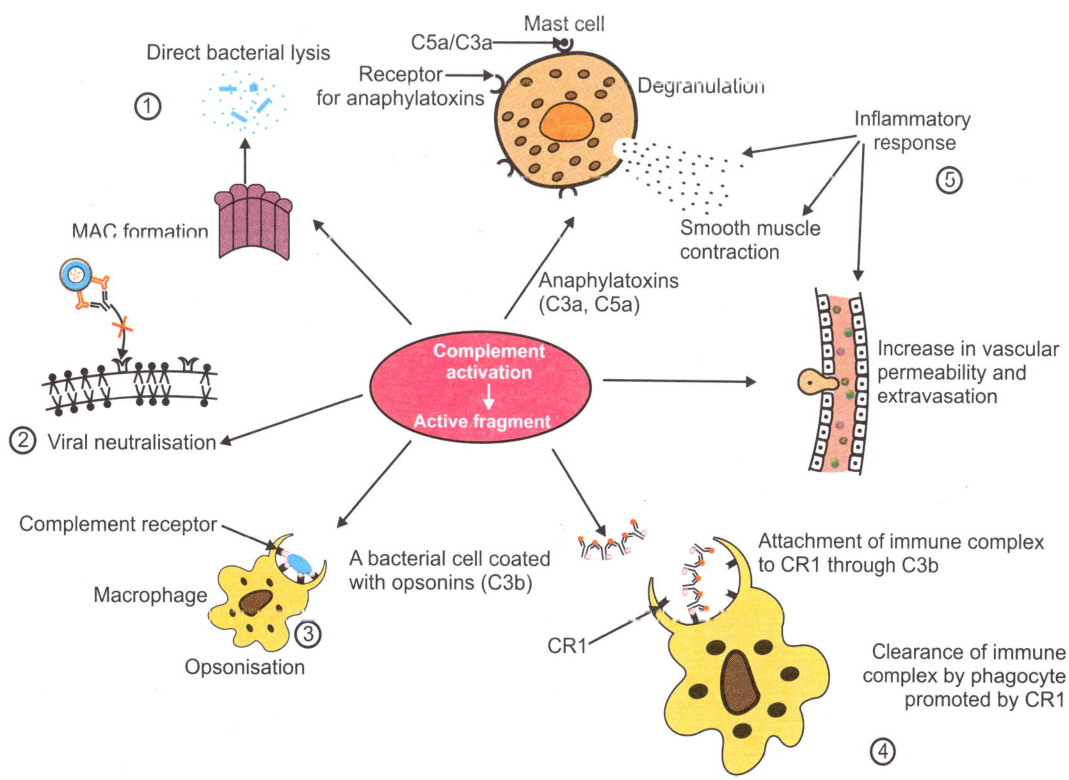

Fig. 14.6 Biological effects of complement activation

the activation of C1qrs, as mentioned below, and typically **culminate in immune cytolysis**, is known as the classical pathway.

- This pathway is generally activated by: (a) the antigen–antibody complex, (b) aggregated immunoglobulins, (c) DNA, (d) C-reactive protein, (e) trypsin-like enzymes, and (f) some retroviruses.
- The steps of the classical pathway are as follows:
  - **Step 1:** C1 is composed of C1q, C1r, and C1s. Of these, C1q reacts with the F_c piece of bound IgM or IgG of the antigen–antibody complex (traditionally represented as EA). C1q binding in the presence of calcium ions leads to the sequential activation of C1r and C1s, leading to the formation of the C1qrs complex.
  - **Step 2:** $C\overline{1}\overline{s}$ is an esterase ($C\overline{1}\overline{s}$ esterase), that cleaves C4 into C4a, (anaphylatoxin) and C4b, which binds to target cell membranes along with C1. C4b can cleave multiple molecules of C4.
  - **Step 3:** C4b, in the presence of magnesium ions, cleaves C2 into C2a, which remains linked to cell-bound $C4\overline{b}$, and C2b, which is released into the fluid phase. The resultant $C\overline{4}\overline{b}\overline{2}a$ has enzymatic activity and is referred to as the classical pathway C3 convertase.
  - **Step 4:** C3 convertase splits C3 into two fragments—C3a, (anaphylatoxin) and C3b, which remains cell-bound along with $C\overline{4}\overline{b}\overline{2}a$ to form a trimolecular complex—$C\overline{4}\overline{b}\overline{2}\overline{a}\overline{3}\overline{b}$. This complex has enzymatic activity and is called C5 convertase.
  - **Step 5:** C5 convertase cleaves C5 into C5a (an anaphylatoxin), which is released into the medium, and C5b which continues with the cascade. C6 and C7 then join together, forming a heat-stable trimolecular complex $C\overline{567}$, part of which binds to the cell membrane and prepares it for lysis by C8 and C9, which join the reaction subsequently.
    - Most of $C\overline{567}$ complexes escape and serve to amplify the reaction by adsorbing onto unsensitised 'bystander cells' and rendering them susceptible to lysis by C8 and C9. Unbound $C\overline{567}$ has some chemotactic activity.
    - **Complement-mediated cytolysis**—it occurs due to the production of 'holes/pores', approximately 100 Å in diameter, on the cell membrane. This disrupts the osmotic integrity of the membrane, leading to the release of the cell contents (Fig. 14.7).

2. Alternative Complement Pathway (Fig. 14.8)

- This pathway is **antibody-independent**.

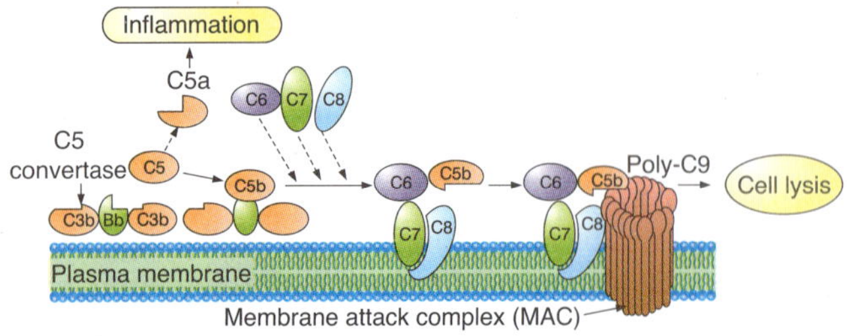

Fig. 14.7 Initiation and outcome of the classical complement pathway

- The activation of C3 without the prior participation of $C\overline{4}\overline{b}\overline{2}a$ is known as the alternative pathway.
- The activators of this pathway include the following:
  - i. Bacterial endotoxins and teichoic acid of gram-positive bacteria
  - ii. IgA and D
  - iii. Cobra venom factor and the nephritic factor (a protein present in the serum of glomerulonephritis patients)
  - iv. Zymosan (polysaccharide from the yeast cell wall)
  - v. Some viruses and virus-infected cells
  - vi. Parasites, e.g., trypanosomes

The steps of the alternative pathway are as follows:

- **Step 1:** C3b binds to an activator.
- **Step 2:** Bound C3b interacts with a serum protein called factor B (also known as 'C3 proactivator') to form a **magnesium-dependent complex 'C3bB'**.
- **Step 3:** The complex C3bB is cleaved by another serum protein, **factor D (also called 'C3 proactivator convertase')**, into two fragments, Ba and Bb. Fragment Ba is released into the medium. Fragment Bb remains bound to C3b, forming the esterase–C3b–Bb complex, which is the alternative pathway C3 convertase. The function of **properdin (also called Factor P)** is to stabilise the C3 convertase, which hydrolyses C3, leading to further steps in the cascade, as in the classical pathway.

3. Lectin Complement Pathway (Mannose-Binding Lectin or MB Lectin [MBL] Pathway)

- This involves the **host proteins binding to the microbial surface**.
- It represents an innate defence mechanism comparable to the alternate pathway while utilising classical pathway components except the C1 proteins.
- Pattern recognition proteins (PRPs), which are mannose-binding lectins, recognise and bind to specific carbohydrate targets.
- **Mannose-binding lectin (MBL)**, an acute-phase protein produced in inflammatory responses, binds to mannose residues (e.g., those present on the surfaces of microorganisms like *Salmonella*,

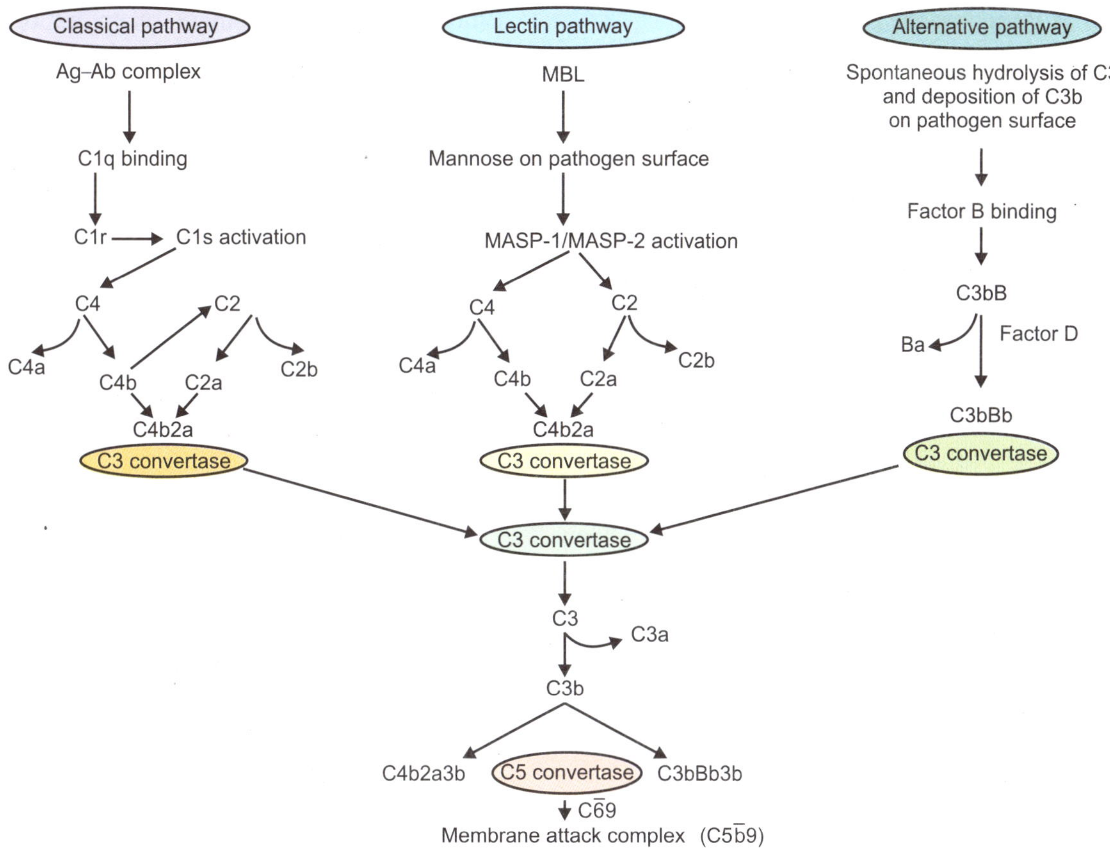

Fig. 14.8 Convergence of the three complement pathways

Neisseria, Listeria, Cryptococcus, and *Candida*) and activates the complement pathway.

- Hence, the lectin complement pathway is also called the MBL pathway. Its function is similar to that of C1q except that it does not depend on antibody for its activation.

The steps in the lectin complement pathway (Fig. 14.8) are as follows:

- **Step 1:** MBL binds to the foreign surface of the pathogen.
- **Step 2:** Proteases (MBL-associated serine protease-1[MASP-1] and MBL-associated serine protease-2 [MASP-2]) bind to it and form the C1-like active complex.
- **Step 3:** Cleavage and activation of C4 and C2 occur to produce C5 convertase without the need for specific antibody binding.

Convergence of the Three Pathways

- The three complement pathways converge at the **membrane attack complex (MAC)**, which contains C5b, C6, C7, C8, and C9 proteins, which form this macromolecular structure after sequential interaction.
- MAC forms large channels or pores through the membrane of a pathogen or target cell, enabling ions and small molecules to diffuse freely across the membrane. As a result, the cell cannot maintain its osmotic stability and is killed by an influx of water and loss of electrolytes.

An overview of the three pathways is presented in Fig. 14.8. The differences between the three pathways are listed in Table 14.6.

Table 14.6 Differences between the three complement pathways

Feature	Classical pathway	Alternate pathway	Lectin pathway
Activation	Antigen–antibody complex, aggregated immunoglobulin, DNA, C-reactive protein, trypsin-like enzymes, and some retroviruses	Endotoxins and teichoic acid, IgA and D, cobra venom factor, zymosan, some viruses and virus-infected cells, parasites (e.g., trypanosomes)	MBL-associated serine proteases (MASPs) binding to the mannose on the bacterial cell
Binding	C1q	Factor B and D	MASP1, MASP2
C3 convertase	C4b2a	C3bBb	C4b2a

Regulation of Complement Activation

Unchecked complement activity can cause not only exhaustion of the complement system but also serious damage to tissues. Several in-built control mechanisms regulate the complement cascade at different steps. There are chiefly two kids of control mechanisms, which are as follows:

1. **Inhibitors**, which bind to complement components and halt their further function; these include:
   - C1q inhibitors, which bind to and inactivate the serine protease activity of C1r and C1s.
   - Factor 1, which cleaves C3b and C4b, reducing the amount of C5 convertase.
   - Factor P (properdin), which stabilises C3 convertase and protects C3b.
   - S protein, which binds to $C\overline{67}$ and modulates the cytolytic action of the membrane attack complex.
2. **Inactivators**, which are enzymes that destroy complement proteins; these include:
   - Factor H (beta-globulin) which acts along with factor I, modulating C3 activation.
   - Anaphylatoxin inactivator (alpha-globulin), which enzymatically degrades C3a, C4a, and C5a, which are anaphylatoxins released during the C cascade.
   - TC4-binding protein, which controls the activity of cell-bound C4b.

BIOLOGICAL EFFECTS OF COMPLEMENT ACTION

The outcomes of complement-mediated immunological reactions are as follows (Fig. 14.9):

- Membrane damage (cytolysis, bacteriolysis)
- Amplification of inflammatory response
- Participation in the pathogenesis of certain hypersensitivity reactions (type III)
- Antiviral activity
- Promotion of phagocytosis and immune adherence

Listed below are some of the biological effects of complement activation on the host.

1. **Phagocytosis:** Complement receptors (CR) 1, 3, and 4 act as opsonins and stimulate phagocytosis. C3b induces phagocytosis of opsonised targets.
2. **Inflammatory response:** C fragments released during the cascade reaction help in amplifying the inflammatory response. For example:
   - C2 kinins are vasoactive amines and increase capillary permeability.
   - C3a and C5a are anaphylatoxic (histamine-releasing) and chemotactic.
   - $C\overline{567}$ is chemotactic and also brings about reactive lysis.
3. **Hypersensitivity reactions:** C participates in cytotoxic (type II) and immune complex (type III) hypersensitivity reactions.
4. **Autoimmune diseases:** The levels of serum C components are lowered in many autoimmune diseases such as systemic lupus erythematosus and rheumatoid arthritis.
5. **Endotoxic shock:** Endotoxin causes C3 fixation (via alternate pathway) and platelet adherence resulting in disseminated intravascular coagulation (DIC) and thrombocytopenia. Initial exposure to endotoxin or Ag–Ab complex produces fibrin thrombi, the clearance of which causes endothelial blockage. Second exposure to the endotoxin leads to tissue necrosis and hemorrhage.
6. **Immune adherence:** C bound to antigen–antibody complexes adheres to erythrocytes or non-primate platelets. This reaction, called immune adherence, contributes to defence against pathogenic microorganisms since such adherent particles are rapidly phagocytosed. C3 and C4 are necessary for immune adherence.

Conglutination

- Bovine serum contains an unusual protein called conglutinin (K) which causes the clumping of particles or cells coated with C, a process known as conglutination.

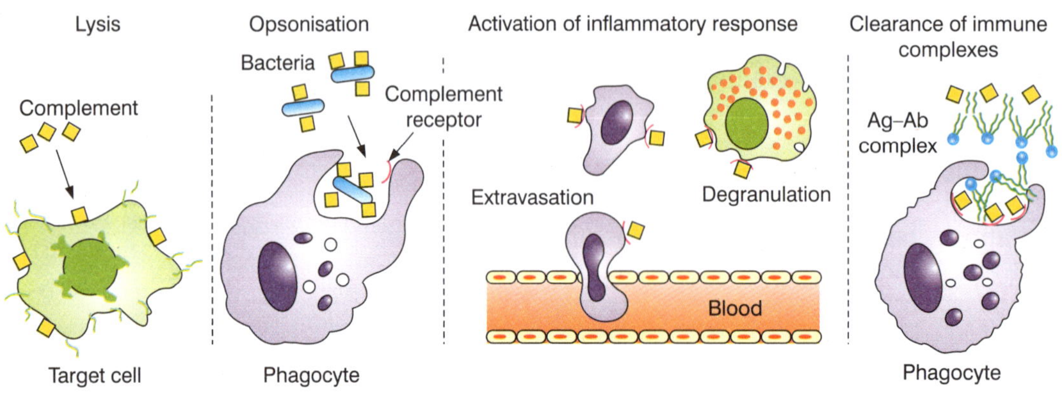

Fig. 14.9 Outcomes of complement-mediated immunological reactions

- Conglutinin reacts exclusively with bound C3.
- Though conglutinin behaves as an antibody to C, it is not an immunoglobulin and requires Ca^{++} for its activity.
- Antibodies with conglutinin-like activity (immunoconglutinin [IK]) can be produced by immunisation with complement-coated materials.
- They may also occur frequently in human beings and other mammals as autoantibodies to fixed C.

The complement system is generally effective in lysing gram-negative bacteria; however, some gram-negative bacteria and most gram-positive bacteria have mechanisms to evade complement-mediated damage, as shown in Table 14.7.

BIOSYNTHESIS OF COMPLEMENT

Complement components are synthesised at various sites in the body (Table 14.8).

DISEASES ASSOCIATED WITH ABNORMAL/DEFICIENT COMPLEMENT ACTIVITY

- Recurrent bacterial infections
- Autoimmune diseases (lupus and rheumatoid arthritis)
- Hereditary and acquired angioedema
- Glomerulonephritis, lupus nephritis, membranous nephritis, and IgA nephropathy, and transplant rejection of kidney
- Cirrhosis
- Hepatitis
- Malnutrition
- Septicemia, shock
- Serum sickness (immune complex disease)
- In acute or chronic inflammation, complement levels are increased along with other acute phase reactants like CRP

Deficiencies of the Complement System

- Complete or partial deficiencies of all components and complement inhibitors result in the host being unable to efficiently eliminate microbial antigens or circulating immune complexes.
- Some deficiencies are associated with severe diseases, while in others, clinical manifestations are mild.
- Complement deficiency results in recurrent bacterial and fungal infections and collagen diseases (Table 14.9).
- Other potential associations between complement deficiency and diseases include the following:
  - Rejection of transplanted organs
  - Implicated in diseases of the central nervous system such as Alzheimer's disease and other neurodegenerative conditions such as spinal cord injuries
  - Deficiencies in complement proteins produced in the liver can lead to a form of primary (congenital) immunodeficiency, in which the body is more susceptible to disease, particularly autoimmune diseases and severe bacterial infections

Complement-associated Diseases

The complement system plays a role in asthma, glomerulonephritis, various forms of arthritis,

Table 14.7 Microbial evasion of complement-mediated damage

Microbial components	Mechanism of evasion	Examples
	Gram-negative bacteria	
Cell wall lipopolysaccharides (LPS)	Side chains prevent insertion of MAC into bacterial membrane	Resistant strains of *E. coli* and *Salmonella*
Other membrane proteins	MAC interacts with membrane proteins and fails to enter the bacterial membrane	Resistant strains of *Neisseria gonorrhoeae*
Elastase	Anaphylatoxins C3a and C5a are inactivated by microbial elastase	*Pseudomonas aeruginosa*
	Gram-positive bacteria	
Peptidoglycan layer of cell wall	Thick layer prevents MAC from permeating the bacterial membrane	*Streptococcus*
Bacterial capsule	Provides physical barrier between C3b deposited on bacterial membrane and CR1 on phagocytic cells	*Streptococcus pneumoniae*
	Other microbes	
Proteins that mimic complement regulatory proteins	Proteins present in various bacteria, viruses, fungi, and protozoans inhibit the complement cascade	Vaccinia virus, herpes simplex, Epstein–Barr virus, *Toxoplasma*, *T. cruzi*, and *C. albicans*

Table 14.8 Sites of synthesis of C components

Complement component	Site of synthesis
C1	Intestinal epithelium
C2, C4	Macrophages
C5, C8	Spleen
C3, C6, C7, C9	Liver

Table 14.9 Clinical syndromes associated with genetic disorders resulting in complement deficiencies

Group	Deficiency	Syndrome
I	C1 inhibitor: Autocatalytic activation of C1 and the unrestrained breakdown of C4 and C2	Hereditary angioneurotic edema (episodic angioedema of subcutaneous tissues or of the mucosa of the respiratory or alimentary tracts)
II	Early components of classical pathway C1, C2, C4	SLE and other collagen vascular diseases
III	C3 and its regulatory protein C3b inactivator	Severe recurrent pyogenic infections
IV	C5 to C8	Bacteremia, chiefly with gram-negative diplococci, toxoplasmosis
V	C9	No particular disease

autoimmune heart disease, lupus erythematosus, multiple sclerosis, and inflammatory bowel disease. C is an acute-phase reactant; a rise in C levels (particularly C4, C3, C5, and C6) is observed during the acute phase of inflammation.

QUANTITATION OF COMPLEMENT AND ITS COMPONENTS

- Complement components need to be estimated in the following scenarios:
  - i. Acquired and inherited complement deficiencies
  - ii. Acquired and inherited C1 inhibitor deficiencies
  - iii. Disorders of complement activation
- C3 and C4 levels are the most frequently estimated, but the estimation of other components such as C1 inhibitors is also required when other deficiencies are suspected.

Indications for Complement Estimation in Specific Disease Conditions

- Unexplained inflammation or edema
- Autoimmune disorder (e.g., lupus)
- Immune complex-related diseases—here, the individual complement components need to be estimated when the total complement activity (CH50, which is sometimes called CH100) is abnormal; this is done to determine which components are deficient or abnormal
- To estimate the severity of the condition, when severity is linked to the decrease in complement levels
- To monitor the current activity of complement-related disease

Methods of Detection

Complement estimation can be based on any one of the following:

1. **Estimation of individual complement components:** In immune complex-mediated diseases, the individual complement components need to be estimated when the total complement activity (CH50, which is sometimes called CH100) is abnormal; this is done to determine which components are deficient or abnormal.
2. **Detection of activation products:** This is based on the identification of C3a or C3d.
3. **Estimation of complement function:** Hemolytic assays utilising sheep erythrocytes coated with rabbit antibodies (IgM with or without IgG) are used to estimate complement function.

The following methods are employed to detect complements:

- **Enzyme-linked immunosorbent assays**—performed to detect all three functional activities of complement
- **Hemolytic activity (HA) assays**—the complement activity of serum is measured by estimating the highest dilution of serum required to lyse sheep erythrocytes sensitised by anti-erythrocytic antibodies; the estimation of individual complement components is also performed based on hemolytic activity in systems containing an excess of all complement components except the one to be measured
- **Electroimmunoassay** (immunoelectrophoresis)
- **Radial immunodiffusion**
- **Quantitative estimations** can be made by turbidimetric and nephelometric methods

Commonly estimated complement levels

- **Total complement estimated by HA:** 41 to 90 hemolytic units or 30–75 units/mL
- **C3 and C4 levels in healthy individuals—C3:** 0.9–1.8 g/L and C4: 0.1–0.4 g/L

Key Points

- ✓ A substance that induces an immune response is called an antigen. It stimulates the production of an antibody, with which it reacts specifically, generally in an observable manner. Antigenicity refers to the ability of an antigen to induce an immune response.
- ✓ An immunogen is a substance that can induce an immune response, but which does not necessarily bind to its specific antibody.
- ✓ A hapten is a small molecule which, by itself, is not immunogenic, but which can form a complex with a large molecule to induce a specific immune response.
- ✓ Most antigens are foreign to the host. They are large molecules, such as proteins and polysaccharides. Small chemical groups on the antigen molecule, called epitopes, constitute the areas that are recognised by antibodies.
- ✓ Molecular properties like size, chemical complexity, and foreignness ultimately contribute to the activation of the immune system by the antigens.
- ✓ T cell-independent antigens can directly stimulate antibody production by B cells, without the apparent participation of T cells, whereas T cell-dependent antigens require a co-stimulatory signal and cytokines from T_h cells.
- ✓ Superantigens are protein molecules that bind outside the antibody-binding groove directly to the lateral aspect of the TCR β chain and activate very large numbers of T cells irrespective of their antigenic specificity.
- ✓ Toll-like receptors, scavenger receptors, and mannose receptors are components of the innate immune system and recognise unique molecular patterns which are shared by many related pathogens but not with their host; hence they recognise a variety of different pathogens/antigens.
- ✓ An antibody (immunoglobulin) is a glycoprotein molecule formed by the immune system in response to an antigenic stimulus; it is found in blood, bodily secretions, or on mucous surfaces, and binds to the specific antigen responsible for its production, thereby inactivating it.
- ✓ There are five classes of immunoglobulins: IgG, IgM, IgA, IgD, and IgE; all five classes can combine specifically with antigens.
- ✓ IgG constitutes 80–85 per cent of the total amount of antibodies; it can cross the placenta, fix complement, and specifically attach to phagocytes. It provides protection within the body, causes the lysis or removal of foreign antigens, and neutralises viruses and toxins.
- ✓ IgM constitutes 9–10 per cent of the total amount of antibodies. It fixes complement but is unable to cross the placenta or attach to phagocytes. It provides protection within the body and is especially effective in agglutinating antigens and in inactivating complement.
- ✓ IgA constitutes about 10 per cent of the total antibodies. It is secreted in external secretions (saliva, milk, and mucus) and guards the mucosal surfaces.
- ✓ IgD constitutes 1–3 per cent of the total amount of antibodies. It serves as an antigen receptor on B lymphocytes and promotes the development and maturation of the antibody response.
- ✓ IgE constitutes just 0.05 per cent of total antibodies. It specifically attaches to mast cells and basophils and is involved in type I hypersensitivity (allergic and atopic) reactions.
- ✓ Abnormal immunoglobulins include Bence–Jones proteins in multiple myeloma, overproduction of the F_c parts of the immunoglobulin heavy chains in heavy chain disease, and cryoglobulins consisting of IgG, IgM, or their mixed precipitates in cryoglobulinemia.
- ✓ Isotypic (genetic variations or differences in the constant region of the heavy chain), allotypic (multiple alleles for some genes), and idiotypic (sequences of the heavy and light chain variable regions) antigenic determinants on immunoglobulin molecules play an important role in antibody specificity.
- ✓ The presence of a large number of antibodies that bind different antigens/pathogens is called antibody diversity.
- ✓ The complement system consists of at least 30 distinct proteins. The major components are C1 through C9, which are numbered in the order of their discovery, not in the order in which they react.
- ✓ Some of the proteins of the complement system are inactive enzymes which, once activated, act in sequence, one after another in a cascade reaction. Other components perform specific biologic functions.
- ✓ The components released after activation often possess various biological effector functions which contribute to defence mechanisms such as lysis of cells and bacteria, promoting virus neutralisation, opsonisation (which promotes phagocytosis of particulate antigens), and immune clearance (which removes immune complexes from the circulation).
- ✓ Complement can be activated in three ways—by the classical pathway, which requires a specific immune reaction for activation, and by the alternative pathway and lectin pathway, which are antibody-independent. These pathways are activated by the reaction of complement proteins with surface molecules of microorganisms.
- ✓ The complement destroys invading bacteria and foreign cells by disrupting their cytoplasmic membranes and is also involved in inflammatory response since it contributes to vascular permeability, stimulates chemotaxis, and enhances phagocytosis.
- ✓ Complement proteins and protein fragments with receptors on the cells of the immune system control both innate and acquired immune response.
- ✓ The direct and indirect complement fixation tests are examples of in vitro diagnostic antigen–antibody reactions.
- ✓ Complement deficiencies result in the host being unable to efficiently eliminate microbial antigens or circulating immune complexes, causing recurrent bacterial and fungal infections.

Questions

Short answer questions

1. Define antigens. Describe the various determinants of antigenicity.
2. What are the differences between T cell-dependent and T cell-independent antigens?
3. Enumerate the factors responsible for the antigenicity of a molecule.
4. What are haptens?
5. What are the characteristics of superantigens? Give examples.
6. Define epitopes.
7. What are toll-like receptors?

MCQs

1. **Which one of the following is most immunogenic?**
   a) Proteins b) Lipids
   c) Polysaccharides d) Monosaccharides
2. **The main aim of an adjuvant is to increase ______________.**
   a) Distribution b) Absorption
   c) Antigenicity d) Metabolism
3. **Which of the following does isospecificity depend on?**
   a) Antigens present in some members of the same species
   b) Antigens present in all members of the same species
   c) Closely related antigens in biologically different species
   d) Self-antigens that are not ordinarily antigenic
4. **Weil–Felix reaction is based on ______________.**
   a) Autospecificity b) Isospecificity
   c) Species specificity d) Heterogenetic specificity
5. **Which of the following statements is not true regarding T cell-independent antigens?**
   a) They mostly include soluble proteins
   b) They are poorly degradable
   c) Antigen processing by macrophages is not required
   d) There is no immunological memory
6. **Which of the following is caused by superantigens?**
   a) Enhancement of phagocytosis
   b) Polyclonal activation of B cells
   c) Antigen presentation by macrophages
   d) Activation of complements
7. **An individual starts to produce IgG in adequate levels at ______________:**
   a) Birth b) 6 months
   c) 1 year d) 2 years
8. **The earliest Ig class to appear on the surface of B cells is ______________:**
   a) IgA b) IgM
   c) IgG d) IgE
9. **The alternate pathway is activated by all except:**
   a) Endotoxin b) IGA
   c) Teichoic acid d) C reactive protein
10. **All the following statements are true regarding hapten except:**
    a) They require additional carrier proteins
    b) Simple haptens have only one antibody binding site
    c) Complex haptens are non-precipitating
    d) Simple haptens can inhibit precipitation
11. **The conformational epitope is:**
    a) A product of a linear peptide sequence
    b) Produced by different sites of the peptide chain
    c) Recognised by T cells
    d) Also called a sequential epitope
12. **CD40 ligand is expressed by:**
    a) B cells b) T cells
    c) Macrophages d) Dendritic cells
13. **T cell-independent antigens:**
    a) Require antigen processing by macrophages
    b) All types are easily degradable
    c) Have no immunological memory
    d) Bind to the surface immunoglobulin on B cells and are internalised
14. **A single Ig molecule contains:**
    a) 1 light chain, 1 heavy chain
    b) 2 light chains, 1 heavy chain
    c) 2 light chains, 2 heavy chains
    d) 1 light chain, 2 heavy chains
15. **The earliest immunoglobulin synthesised by the fetus is:**
    a) IgG b) IgA
    c) IgE d) IgM
16. **The antigen combining site present on the antibody molecule is called:**
    a) Epitope b) Paratope
    c) Hapten d) Carrier protein
17. **Which immunoglobulin is normally transported across the placenta?**
    a) IgA b) IgM
    c) IgG d) IgE
18. **All the following statements are true regarding IgA except:**
    a) Has a monomeric structure
    b) Molecular weight is 400,000
    c) Glycopeptide bond
    d) Protects against polio virus
19. **Which of the following subclasses has the minimum serum concentration?**
    a) IgA b) IgG
    c) IgM d) IgE
20. **Which Ig class is involved in type I hypersensitivity?**
    a) IgA b) IgE
    c) IgM d) IgG
21. **Which immunoglobulin class has the maximum intravascular distribution?**
    a) IgA b) IgG
    c) IgE d) IgM
22. **Which of the following immunoglobulins is heat labile?**
    a) IgA b) IgE
    c) IgD d) IgM
23. **B cell receptors are constituted by:**
    a) IgG and IgM b) IgE and IgM
    c) IgD and IgG d) IgM and IgD
24. **Bence Jones proteins are best described as:**
    a) γ chain b) μ chain
    c) Fibrin split products d) Kappa and lambda chains
25. **Isotypic variation in Ig refers to:**
    a) Subtle changes in the amino acid sequence due to allelic difference
    b) Sequence variation of the heavy- and light chain-variable regions
    c) Sequence variation in the heavy chain of the constant region
    d) Sequence variation in the light chain of the constant region
26. **Defects in the process of immunoglobulin class switching may lead to increased serum level of:**
    a) IgG b) IgA
    c) IgM d) IgD
27. **What percentage of normal serum protein does complement constitute?**
    a) 5% b) 10%
    c) 15% d) 20%
28. **All the following immunoglobulins fix complement except:**
    a) IgM b) IgG3
    c) IgE d) IgA
29. **Serum can be 'inactivated' by exposing it to:**
    a) 45°C for 20 min b) 46°C for 30 min
    c) 50°C for 20 min d) 56°C for 30 min
30. **All the statements are true regarding classical pathway except:**
    a) It is activated by the antigen-antibody complex.
    b) Calcium ions are required for activation.
    c) C3 convertase has C4b2b.
    d) C5 convertase has C4b2a3b.
31. **Individuals are more prone to gonococcal infection if there is a deficiency of:**
    a) C1 inhibitor b) C1 and C2
    c) C5-C8 d) C3b inactivator
32. **In the alternate pathway, C3 pro-activator is:**
    a) Factor D b) Factor 1
    c) Factor H d) Factor B

33. Bovine serum contains an unusual protein which causes clumping of particles or cells by the process of:
a) Agglutination b) Conglutination
c) Precipitation d) Neutralisation

34. Biosynthesis of complements takes place at the following sites in the body except:
a) Macrophages b) Spleen
c) Liver d) Bone marrow

35. Which among the following is not a function of complement?
a) Phagocytosis b) Inflammatory response
c) Antigen presentation d) Immune clearance

36. Which of the following acts as an inhibitor of complement?
a) Factor P b) Factor H
c) C4 binding protein d) Anaphylatoxin inactivator

37. Collagen vascular diseases are associated with the deficiency of:
a) C1,C2,C3 b) C1, C2,C4
c) C1,C5,C8 d) C2,C3,C4

38. Which of the following statement is correct regrading IgD?
a) It resembles IgA structurally.
b) It has a half-life of about ten days.
c) It serves as recognition receptors for antigens on B cells.
d) Its detection in the serum denotes an early infection.

39. Which statement is correct regarding the participation of complement in hypersensitive reaction?
a) It is involved in cytotoxic (type II) and immune complex (type III) hypersensitivity reactions.
b) It has no role in any type of hypersensitivity reaction.
c) It is involved in type 1 hypersensitivity reaction.
d) It is involved in all types of hypersensitivity reactions.

40. A 12-year-old son of a farmer has malnutrition. He has a history of eating well and has frequently passed worms in stools. Investigation reveals eosinophilia and eggs of *Ascaris* in his stool. Which of the following immunoglobulins is likely to be elevated in such a case?
a) IgG b) IgE
c) IgD d) IgA

41. A 7-year-old boy presents with disseminated *Neisseria* infection. His other parameters are normal, but the pediatrician suspects complement deficiency. Such an infection is commonly associated with deficiency of:
a) Properdin b) Factor D
c) C1 inhibitory deficiency d) Membrane attack complex

42. The membrane attack complex consists of:
a) Colicins b) C3b3b,Bb
c) C5b,6,7,8,9 d) Properdin

43. Which of the following is not a C3 convertase?
a) C3bBb b) MBL/MASPC4b2a
c) C14b2a d) C14a2a

44. Which of the following acts as a anaphylatoxin?
a) C3a b) C3b
c) C4b d) C2a

45. A 27-year-old woman has had two miscarriages in past two years due to Rh incompatibility. This occurs due to complement-mediated lysis of fetal red blood cells. Which of the following is not true regarding complement system?
a) The classical pathway is antibody dependent.
b) Endotoxin is a potent stimulator of the alternate pathway.
c) Lectin pathway works independent of antibody.
d) The membrane attack complex is formed by C3 convertase.

Answers: 1:a, 2:c, 3:a, 4:d, 5:a, 6:b, 7:d, 8:c, 9:a, 10:c, 11:b, 12:b, 13:c, 14:c, 15:d, 16:b, 17:c, 18:a, 19:d, 20:b, 21:d, 22:b, 23:d, 24:d, 25:c, 26:c, 27:a, 28:d, 29:d, 30:c, 31:c, 32:d, 33:b, 34:d, 35:c, 36:a, 37:c, 38:c, 39:a, 40:b, 41:d, 42:c, 43:d, 44:a, 45:d

CHAPTER 15

Part II: Immunology

Mechanisms of Immune Response

(MI1.8)

Chapter Navigation

- Humoral immune response (antibody-mediated)
  - Primary and secondary responses
  - Antigen processing
  - Elimination of antigens
  - Factors influencing antigen–antibody reactions
  - Outcomes of antigen–antibody reactions
  - Adjuvants
  - Immunosuppressive agents
- Cellular immune response/cell-mediated immunity
  - Delayed-type hypersensitivity
  - Cytotoxic T lymphocyte-mediated response
  - Induction of CMI
- Paths of humoral and cellular immune responses
- Immunological tolerance
- Theories of immune response

Learning objective

- Describe the host immune responses in microbial infections

- The specific reactivity induced in a host by an antigenic stimulus is known as the immune response.
- In an infectious disease, it is generally equated with protection against invading microorganisms. However, immune response has a much wider scope and includes **reactions against any antigen, living or non-living**.
- It may lead to consequences that are beneficial, ineffective, or injurious to the host. It also includes the **state of specific non-reactivity (tolerance)** induced by certain types of antigenic stimuli.
- Immune response is classified into two types—**antibody-mediated immunity (AMI) and cell-mediated immunity**—depending on whether it is mediated by antibodies or immunocompetent cells (Table 15.1).

Table 15.1 Functions of AMI and CMI

Antibody-mediated immunity (AMI)	Cell-mediated immunity (CMI)
• Primary defence against most extracellular bacterial pathogens • Immunity against viruses that infect through the respiratory or intestinal tracts • Prevention of recurrence of virus infections • Involved in immediate (types 1, 2, and 3) hypersensitivity and certain autoimmune diseases	• Immunity against fungi, viruses, and facultative intracellular bacterial pathogens • Rejection of homografts and graft-versus-host reaction • Immunological surveillance and immunity against cancer • Involved in delayed (type 4) hypersensitivity and certain autoimmune diseases

ANTIBODY-MEDIATED (HUMORAL) IMMUNE RESPONSE

The mechanism of humoral or antibody-mediated immunity, which acts in response to an antigenic stimulus, is presented in Fig. 15.1.

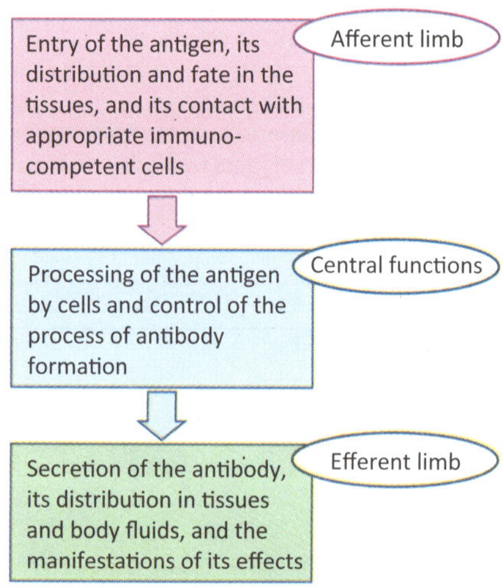

Fig. 15.1 Mechanism of AMI response

PHASES OF ANTIBODY PRODUCTION

Antibody production follows characteristic phases (Fig. 15.2), which are as follows:

i. **Lag phase**, the stage immediately following the antigenic stimulus, during which no antibody is detectable in circulation
ii. **Log phase**, in which the titre of antibodies rises steadily

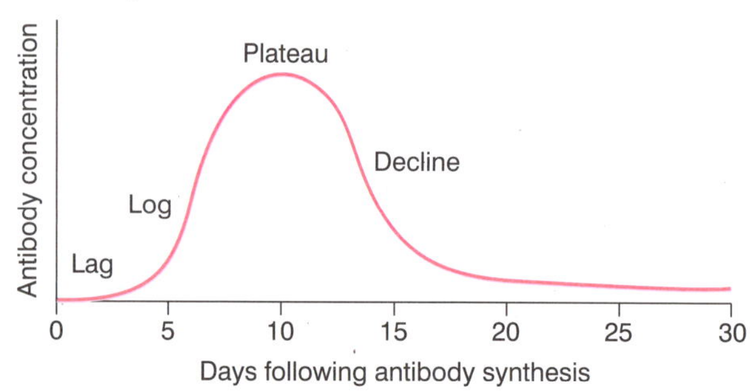

Fig. 15.2 Antibody production follows characteristic phases

iii. **Plateau or steady state**, wherein there is an equilibrium between antibody synthesis and catabolism
iv. **Phase of decline,** during which catabolism exceeds production and the titre falls

PRIMARY AND SECONDARY RESPONSES (Fig. 15.3)

- **Primary response:** This refers to the antibody response to an initial antigenic stimulus.
- **Secondary response:** This is a host's response to subsequent stimuli with the same antigen.

The primary response differs qualitatively and quantitatively from the secondary response. The differences between primary and secondary responses are presented in Table 15.2.

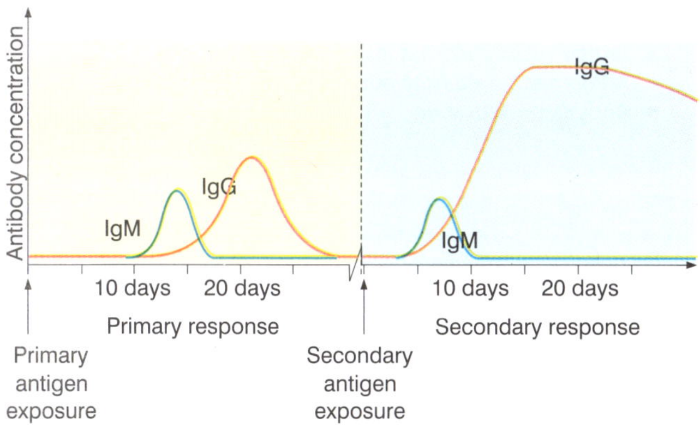

Fig. 15.3 Effect of repeated antigenic stimulus

Table 15.2 Differences between primary and secondary immune responses

Features	Primary response	Secondary response
Lag period after immunisation	Usually 4–10 days	Usually 1–3 days
Involving B cells	Naive B cell	Memory B cell
Peak response time	7–10 days	3–5 days
Peak antibody response	Depends on antigen	Usually 100–1,000 times more than the primary response
Isotype generated	In early response, IgM predominates	IgG predominates
Antigens	Both thymus-dependent and thymus-independent	Only thymus-dependent
Antibody affinity	Lower	Higher
Immunisation required	Comparatively high dose of antigens, preferably with adjuvants (protein antigens)	Adjuvants not required

The duration of the lag phase and the persistence of the antibodies vary with the nature of the antigen. With some antigens such as diphtheria toxoid, the lag phase in the primary response may be as long as 2–3 weeks, while with pneumococcal polysaccharide, antibodies can be detected as early as within a few hours.

Negative phase: When an antigen is injected into an animal already carrying the specific antibody in circulation, a temporary drop in the level of the circulating antibody occurs due to the combination of the antigens with the circulating antibodies. This is followed by an increase in the titre of the antibodies exceeding the initial level.

The appearance of different classes of antibodies from conception to adulthood varies with antigenic stimulation. The general trend of antibody rise and fall is presented in Fig. 15.4.

ANTIGEN PROCESSING AND FATE OF ANTIGENS FOLLOWING AN IMMUNE RESPONSE

Immune response to an antigen is brought about by three types of cells:

- Antigen-processing cells (APC—principally macrophages and dendritic cells)
- T cells
- B cells

There are several structural molecules that are critical components for antigen processing and presentation to the T or B cells. One such component is the **major histocompatibility complex (MHC)**, also known as **human leukocyte antigen (HLA)** (details in Chapter 18). Two of the complexes—MHC class I and MHC class II—help in presenting antigens to B and T cells respectively.

1. Antigen Processing and Presentation

Antigens need to be presented in combination with either MHC class II or class I complexes in order to activate T cells for further immunological action (Fig. 15.5).

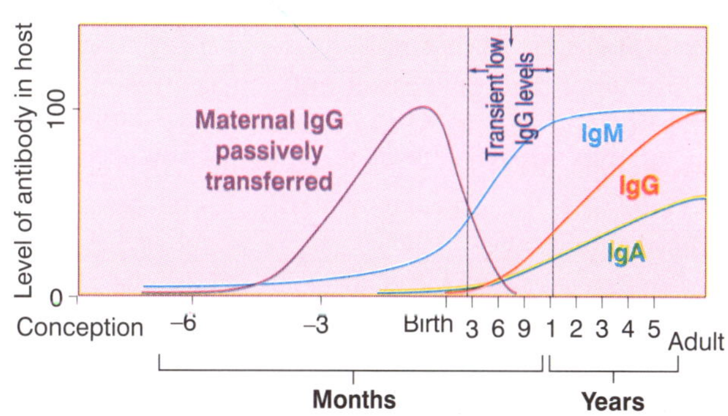

Fig. 15.4 Appearance of different classes of antibodies from conception to adulthood

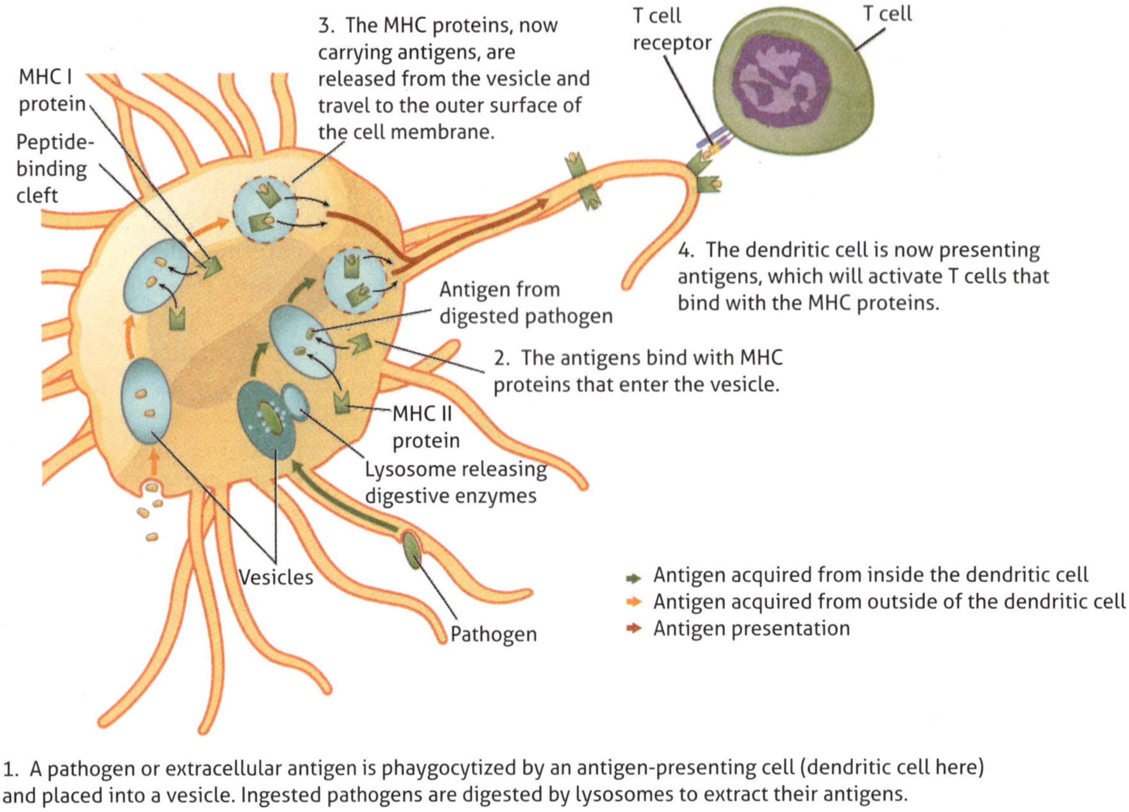

Fig. 15.5 Antigen processing and presentation by MHC class II and class I (*Source*: OpenStax College, https://commons.wikimedia.org/wiki/File:2216)

Presentation by MHC class II

Antigens introduced intravenously into the body rapidly get localised in the spleen, liver, bone marrow, kidneys, and lungs. Fig. 15.6 illustrates the mechanisms of antigen presentation by MHC class II complex and T helper cell (CD4) activation.

Presentation by MHC class I

Some antigens (viruses and intracellular bacteria) do not get phagocytised by macrophages. Instead, they bind to the cell membrane and directly enter the cell cytoplasm (Fig. 15.7).

2. T and B Cell Activation

- Only when the processed antigen is presented on the surface of the APC in association with MHC molecules to the T cell carrying the receptor (TCR) for the epitope, will the T cell be able to recognise it.
- In the case of CD4 (helper T/T_h) cells, the antigen has to be presented complexed with MHC class II, and for CD8 (cytotoxic T/T_c) cells, it has to be complexed with MHC class I molecules. B cells, which possess surface Ig and MHC class II molecules, can also present antigens to T cells, particularly during the secondary response.

The antigen is engulfed by macrophages and dendritic cells (professional APCs) present on the skin and mucous membrane

↓

Antigen enters the phagosome (endocytic vesicles) within the phagocyte

↓

Within the phagosome, the antigen is degraded to smaller peptides by the lysozyme

↓

Small antigenic peptides bind to the grooves of the MHC II within the endoplasmic reticulum of the APC

↓

The complex is transported to the cell surface to be exposed to $CD4^+$ T cells for recognition and activation

Fig. 15.6 Mechanism of presentation of antigens by MHC class II complex

3. Production of Antibodies Following an Antigenic Stimulation

- The T_h cell requires two signals for activation.
- The first signal is a combination of the T cell receptor (TCR) with the MHC class II-complexed antigen.
- The second signal is interleukin-1 (IL-1), which is produced by the APC.

Antigen (virus/bacteria)
↓
Binds to cell membrane
↓
Enters cytoplasm
↓
Binds to ubiquitin (76 amino acid protein in the cells)
↓
Antigen is broken down to peptides in the proteasomes
↓
Peptides bind to transporters associated with antigen processing (TAP)
↓
Enter endoplasmic reticulum and are loaded onto MHC class I molecule
↓
This complex is transported to the cell surface to be exposed to $CD8^+$ cells

Fig. 15.7 Mechanism of presentation of antigens by MHC class I molecules

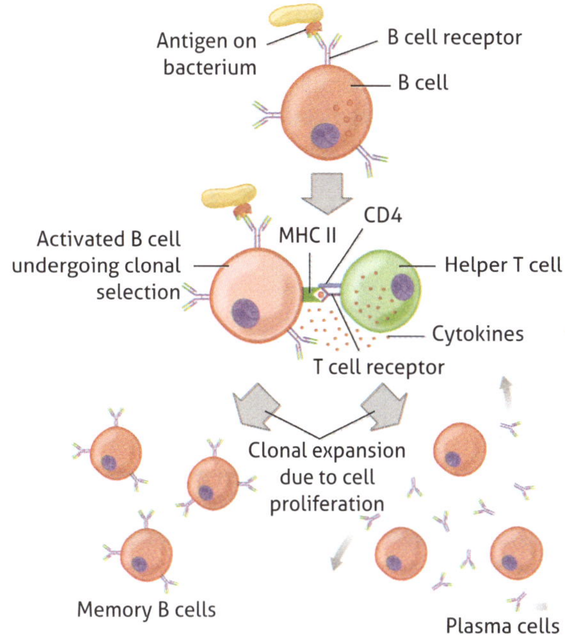

Fig. 15.8 T helper cells and cytokines stimulate B cells to form antibodies and interact with antigens (*Source*: https://openstax.org/books/microbiology/pages/1-introduction)

- The activated T_h cell forms interleukin-2 (IL-2) and other cytokines (IL-4, IL-5, and IL-6) required for B cell stimulation.
- These interleukins act as **B cell-activating factors (BCAF)** and **B cell-differentiation factors (BCDF)** to activate B cells that have combined with their respective antigens.
- Activated B cells clonally proliferate and differentiate into antibody-secreting plasma cells (Fig. 15.8).
- A small proportion of activated B cells, instead of being transformed into plasma cells, become **long-living memory cells**, capable of producing a secondary type of response to subsequent contact with the antigen.

4. Cytotoxic T Cell (CTL) or CD8/TC Activity

- Antigens that are generated within a cell (e.g., viral proteins in any infected cell) are degraded into fragments (e.g., peptides) by cytosolic **proteasomes**.
- These fragments are displayed on the surface of the cell within the grooves of the MHC I molecule, which in turn, are recognised by CTL ($CD8^+$ T cell).
- **Cytotoxic T cells (CTL)** are activated when they come into contact with antigens presented along with MHC class I molecules (target cells).
- IL-2 formed by activated T_h and the activated CTL release cytotoxins, which destroy the target cell (virus-infected or tumour cells).
- Some CTL cells also become memory cells.

ELIMINATION OF ANTIGENS

Removal of Particulate Antigens

Particulate antigens are removed from circulation in two phases—the non-immune phase and the immune phase.

Non-immune phase

- After the small, particulate antigen enters the body, it is engulfed by the phagocytic cells and subsequently broken down by lysozyme and eliminated.
- Degraded antigens are further broken down by the reticuloendothelial cells and excreted in urine; about 70–80 per cent of antigens are thus eliminated within one or two days.
- In contrast, antigens introduced subcutaneously are mainly localised in the draining lymph nodes. Only small amounts may be found in the spleen and are eliminated very slowly.

Immune elimination

- A specific antibody binds to the antigen to form an antigen–antibody complex.
- These complexes are rapidly phagocytised, resulting in the accelerated disappearance of the antigen from circulation.

Removal of Soluble Antigens

Soluble antigens are removed in three phases:

i. The **phase of equilibration** consists of the diffusion of the antigen to the extravascular spaces.
ii. The **metabolic phase** is the phase where the level of the antigen drops due to catabolic decay.

iii. **Immune elimination** consists of the rapid elimination of the antigen with the formation of antigen–antibody complexes. Such complexes may cause tissue damage and may be responsible for '**immune complex diseases**' such as serum sickness.

The rate of elimination of an antigen is related to the rate at which it is metabolised. Protein antigens are generally eliminated within days or weeks, whereas polysaccharides, which are metabolised slowly, persist for months or years. Pneumococcal polysaccharides, for instance, may persist for up to 20 years in humans following a single injection (pneumococcal polysaccharide vaccine).

FACTORS INFLUENCING ANTIGEN-ANTIBODY RESPONSES

Route of Administration

- Humoral immune response is better when the antigens are administered parenterally than through the oral or nasal routes.
- Large, particulate antigens such as bacteria or erythrocytes are more immunogenic when injected into tissues.
- The route of administration may also influence the type of antibody produced.
- The oral or nasal route is better suited for the production of IgA antibodies.
- The inhalation of pollen antigens induces IgE synthesis, whereas the same antigens introduced parenterally lead to the production of IgG antibodies.
- With some antigens, the route of administration determines whether tolerance or antibody response results. For example, injection of protein antigens either into the mesenteric vein or intrathymically usually induces tolerance.

Site of Administration

- The hepatitis B vaccine is less immunogenic following gluteal injection than following injection into the deltoid.
- This may be due to the paucity of antigen-presenting cells in gluteal fat, which results in the delaying of the presentation of the antigen to T and B cells.

Size and Number of Doses

- Antibody response is dependent on the dose of the antigen as a minimum critical dose is required to elicit an immune response.
- A single injection of an antigen does not elicit the generation of a high number of antibodies; rather, it helps in sensitising or priming the immunocompetent cells producing the corresponding antibody. Subsequent doses (optimal) enhance the intensity of antibody response.
- Beyond a certain level, an increase in the dose of antigen does not improve the antibody response; instead, such an increase may even inhibit the antibody response and induce tolerance. This phenomenon is called **immunological paralysis**.
- It is for this reason that killed vaccines are administered in multiple doses for active immunisation. The first injection is known as the **priming dose** and subsequent injections as **booster doses**.
- With live vaccines, a single dose is sufficient, as the multiplication of the organism in the body provides a continuing antigenic stimulus that acts as both the priming and the booster dose.

Multiple Antigens

- When two or more antigens are administered simultaneously, their effects may vary.
- Antibodies may be produced against the different antigens as though they had been given separately, or antibody response to one of the antigens may be enhanced and response to the other may be diminished (**antigenic competition**).

OUTCOMES OF ANTIGEN-ANTIBODY REACTIONS IN HUMORAL IMMUNITY (Fig. 15.9)

After forming antigen–antibody complexes, antibodies bring about a visible reaction like precipitation or agglutination, which culminates in the destruction or neutralisation of the antigen by:

- Augmenting the role of complement through the classical pathway
- Preventing the entry of the pathogen into cells by neutralising them (as in the case of viruses)
- Attaching themselves to the antigen and enhancing phagocytosis by opsonisation (Fig. 15.10)
- Stimulating natural killer cells to eliminate the antigen by antibody-dependent cell-mediated cytotoxicity
- IgE triggers inflammatory mediators from the mast cells

ADJUVANTS

- **Any substance that enhances the immunogenicity of an antigen** is known as an adjuvant.
- These substances may confer immunogenicity on non-antigenic substances.
- They are often used to boost the immune response when an antigen has low immunogenicity.
- They stimulate the antigen by helping it gain entry into the antigen-presenting cell.
- An antigen directly injected into a host undergoes rapid removal from the body, but an adjuvant helps to enhance the immune response against the antigen

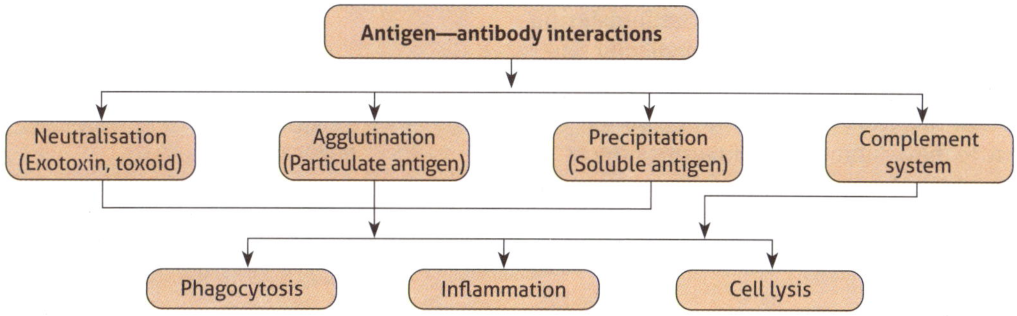

Fig. 15.9 Outcomes of antigen–antibody reactions in humoral immunity

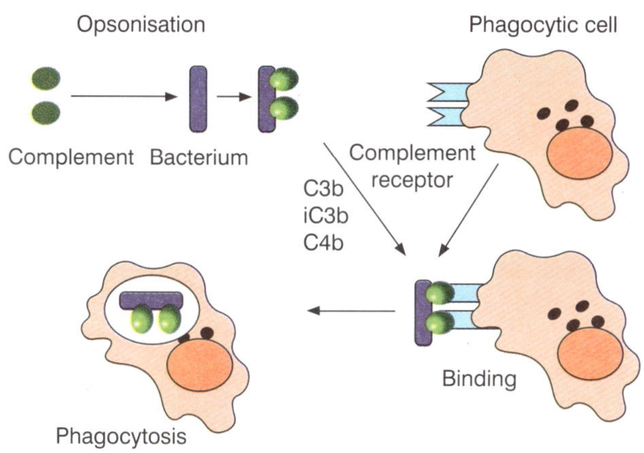

Fig. 15.10 Mechanism of opsonisation and phagcytosis

by maintaining a sustained release of the antigen and by producing a **depot (where the antigen is stored) effect**. This helps in maximising the immune response before the body clears the antigen.

- Killed vaccines and other non-living antigens do not induce CMI unless administered with the **Freund type of adjuvants.**

A few examples of adjuvants are as follows:

- **Freund's incomplete adjuvant:** An adjuvant incorporated in the water phase of a water-in-oil emulsion delays the release of the antigen from the site of injection and prolongs the antigenic stimulus, e.g., aluminium hydroxide or phosphate and protein antigens.
- **Freund's complete adjuvant:** A suspension of killed tubercle bacilli, when added to the incomplete adjuvant, increases both humoral and cell-mediated immune response. The adjuvant effect of the tubercle bacilli is due to the presence of a water-soluble peptide MDP (muramyl dipeptide). However, it is not fit for human use as it produces granuloma.
- **Other adjuvants:** Aluminum hydroxide and phosphate mineral salts are also used as adjuvants, e.g., aluminum hydroxide ('alum'), aluminum phosphate, and calcium phosphate. Aluminium salts or alum are used as adjuvants in diphtheria, tetanus, hepatitis, and HiB vaccines.

IMMUNOSUPPRESSIVE AGENTS

These are **substances that inhibit immune responses**. They are useful in certain situations like transplantation, in which it is necessary to prevent graft rejection, which is an immunological phenomenon. Some examples of immunosuppressive agents are as follows:

- **Corticosteroids** cause the depletion of lymphocytes from blood and lymphoid organs. They also stabilise the membranes of cells and lysosomes, inhibiting histamine release and the inflammatory response. Corticosteroids inhibit the induction and manifestations of delayed hypersensitivity in humans.
- **Antimetabolites** interfere with the synthesis of DNA, RNA, or both and thus inhibit the cell division and differentiation necessary for humoral and cellular immune responses. Examples include folic acid antagonists (methotrexate) and alkylating agents (cyclophosphamide).
- **Anti-lymphocyte serum (ALS)** is a heterologous antiserum against lymphocytes. The antibody prepared against thymus cells is called **anti-thymocyte serum (ATS).** The corresponding globulin preparations are called ALG and ATG. They are used to prevent graft rejection. While all other immunosuppressive agents have undesirable side effects, ALS is devoid of any action other than that on lymphocytes. ALS acts primarily against T lymphocytes in circulation and not cells in the lymphoid organs, and therefore, acts specifically on cell-mediated immunity.

Immunosuppression by Homologous Antibody

- The humoral immune response to an antigen can be suppressed specifically by passively administering homologous antibodies. The suppressive action here appears to be by a **feedback mechanism**.
- For example, the inhibitory effect of a passively administered antibody on the humoral immune response has been applied in the prevention of **Rh sensitisation in Rh-negative women** carrying an

Rh-positive fetus. This is achieved by administering anti-Rh globulin to the mother immediately after delivery (within 72 hours).
- Intravenous administration of immunoglobulin has been shown to have immunomodulatory effects. It has been used in the treatment of many diseases of presumed immunopathologic origin such as thrombocytopenias and autoimmune hemolytic anemias.

CELLULAR IMMUNE RESPONSE/ CELL-MEDIATED IMMUNITY (CMI)

- CMI is a specific immune response mediated by effector T cells generated in response to the antigen, primarily the intracellular pathogens.
- It is evoked by obligate and facultative intracellular microbial agents. These include bacterial (e.g., tuberculosis, leprosy, listeriosis, and brucellosis), fungal (e.g., histoplasmosis, coccidioidomycosis, and blastomycosis), protozoan (e.g., leishmaniasis and trypanosomiasis), and viral (e.g., measles and mumps) infections.

FUNCTIONS OF CELL-MEDIATED IMMUNITY: TRANSPLANTATION IMMUNITY AND GRAFT-VERSUS-HOST REACTION

- This type of immunity is involved in immunological surveillance and immunity against cancer.
- It is also involved in the pathogenesis of certain autoimmune diseases (e.g., thyroiditis and encephalomyelitis; discussed in detail in Chapter 17).
- Both activated T_h (T helper) cells and cytotoxic T lymphocytes (CTLs) serve as effector cells in CMI.
- Antibodies do not play any role in CMI.
- The two major pathways of CMI are:
  1. Delayed-type hypersensitivity (DTH) with the involvement of $CD4^+$ T cells
  2. Cytotoxic T lymphocyte-mediated response by $CD8^+$ T cells

1. Delayed-Type Hypersensitivity (DTH) with the Involvement of $CD4^+$ T Cells

The following events occur once macrophages get activated by $CD4^+$ T_{h1} (Fig. 15.11):
- There is an increase in phagocytic activity brought about by the release of enzymes (lysozyme) and superoxides (O_3).
- They not only kill the microorganism but also the cells harbouring them (e.g., macrophages harbouring *Mycobacterium tuberculosis*).
- Cytokines are released, which attract neutrophils to destroy the infected cells.

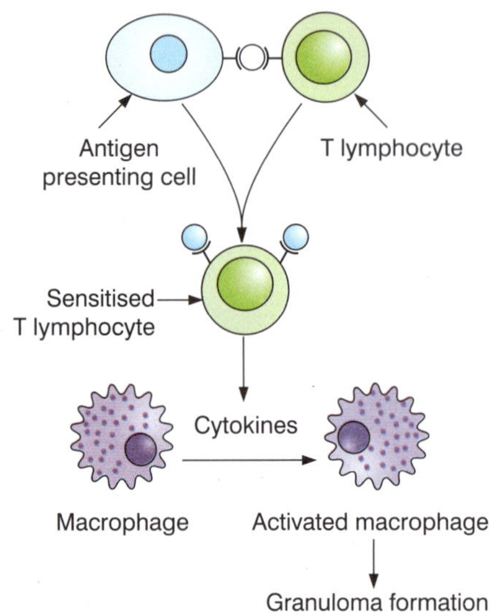

Fig. 15.11 Delayed-type hypersensitivity pathway

- The application of certain chemicals on the skin also induces delayed hypersensitivity. In such cases, a skin lesion appears 48–72 hours after the administration of the agent (antigen). The lesion appears as an indurated nodule infiltrated by mononuclear cells.

DTH is found to be immunologically specific but does not have any relation to antibodies and cannot be transferred passively by serum.

2. Cytotoxic T Lymphocyte (CTL)–$CD8^+$-mediated Response

This pathway consists of the following stages:
1. **Target cell recognition**—CTLs scan the body for cells with MHC class I target cells (in this case, cells infected with intracellular organisms) and bind to them.
2. **Target cell destruction**
   - CTL cells destroy the target cell by releasing perforins (which puncture the cells).
   - Alternatively, they release granzymes which induce the infected cell to destroy its own DNA by apoptosis (programmed cell death).

INDUCTION OF CELL-MEDIATED IMMUNITY (CMI)

- The nature of the antigenic stimulus is important in the induction of CMI. It is best developed following the infection of intracellular microbial agents. Only **T cell–dependent antigens elicit CMI**.
- Each T cell bears on its surface a specific receptor (T cell receptor) for a single antigenic epitope and combines only with antigens carrying that epitope. The chain of events that occur when T cells come in contact with the appropriate antigens is illustrated in Fig. 15.12.

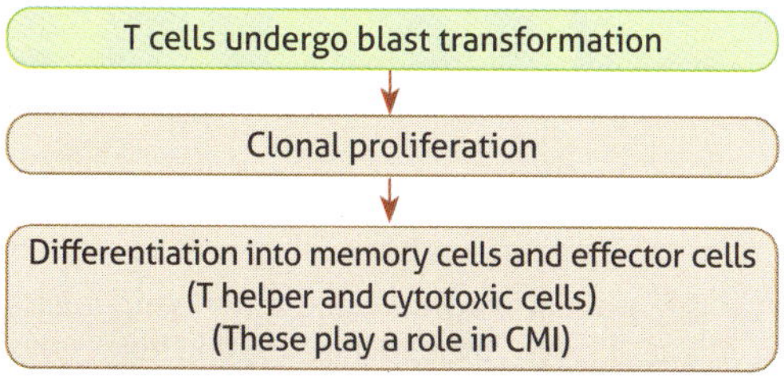

Fig. 15.12 Path of T cell proliferation on coming into contact with an antigen

- The outcomes of activated T cells are as follows:

a. T cells recognise antigens only when presented with MHC (major histocompatibility complex) molecules. Helper T cells react with antigens presented on the surface of macrophages or other cells, complexed with MHC class II molecules. They then release biological mediators (cytokines), which activate macrophages, enabling them to kill intracellular parasites.

b. Cytotoxic T cells (CTL) recognise antigens on the surface of cells (such as virus-infected tumour or allograft cells), and in association with MHC class I molecules, secrete cytokines and destroy the target cells.

Cytokines

Cytokines are biologically active substances released by activated T lymphocytes. They include interleukin products of leukocytes, which exert a regulatory influence on other cells, interferons, growth factors, and others with similar effects.

Functions of cytokines

I. **On macrophages**
   a. Migration inhibitory factor (MIF)
   b. Macrophage activation/aggregation factor (MAF)
   c. Macrophage chemotactic factor (MCF)

II. **On lymphocytes**
   a. Blastogenic/Mitogenic factor (BF/MF)
   b. T cell growth factor (TGF)
   c. B cell growth factor (BGF)

III. **On granulocytes**
   a. Chemotactic factor (CF)
   b. Colony-stimulating factor (CSF)

IV. **On cultured cells**
   a. Lymphotoxin (LT)
   b. Interferon (IFN)
   c. Tumour necrosis factor (TNF)

V. **Others**
   a. Skin-reactive factor (SRF)
   b. Transfer factor (TF)

Characteristics of cytokines

Cytokines are peptide mediators or intercellular messengers that regulate immunological, inflammatory, and reparative host responses.

- They are hormone-like substances, active even at femtomolar (10–15 M) concentrations.
- They are not produced by glands but by cells (such as lymphocytes, macrophages, platelets, and fibroblasts) and hence differ from endocrine hormones.
- They act on the cells producing them or on nearby cells.
- They are generally **pleiotropic**, i.e., having multiple effects on the growth and differentiation of various cell types.
- There is considerable overlap in the effects of different cytokines.
- The cloning of cytokines and the availability of monoclonal antibodies against them have helped to characterise them better (Table 15.3).

Regulation of cytokine production

- Cytokine production is regulated by exogenous stimuli such as antigens and mitogens as well as endogenous factors such as neuroendocrine hormonal peptides (corticosteroids and endorphins) and products of lipoxygenase and cyclooxygenase pathways. They also regulate each other by positive and negative feedback.
- With a better understanding of their properties, it is possible that many cytokines, their agonists, and antagonists could eventually be used in the management of inflammatory, infectious, autoimmune, and neoplastic conditions.
- A number of cytokines (e.g., IL-1, IL-2, IL-3, colony-stimulating factors, and interferons) have already found therapeutic application.

Transfer Factor (TF)

- CMI can be passively transferred by injecting viable leukocytes or their extract from a sensitised donor to a recipient. Such a leukocyte extract is known as the '**transfer factor**' (TF).
- Humoral immunity is not transferred by transfer factors.
- The transferred immunity is specific, as CMI can be transferred only to those antigens to which the donor is sensitive.
- TF is not antigenic.
- Chemically, TF appears to be a **polypeptide–polynucleotide complex**.
- TF is highly potent—an extract from 0.1 mL of packed leukocytes is sufficient for the transfer of specific immunity.
- Following TF injection, DTH and various in vitro correlates of CMI can be demonstrated in the recipient.
- TF transfers CMI to all the antigens to which the donor is sensitive as a whole. It is thus possible to transfer CMI from one recipient to another in a serial fashion.

Table 15.3 Sources and functions of cytokines

Cytokine	Chief sources and features	Major functions
A. Interleukins (IL)		
IL-1 (a and b)	Macrophages and other cell types Stable polypeptide that retains its activity at temperatures as high as 56°C and within a pH range of 3–11 Production stimulated by antigens, toxins, injury, and inflammatory processes Inhibited by cyclosporin A, corticosteroids, and prostaglandins	The proliferation and differentiation of T, B, and other cells Pyrogenic; induce acute-phase proteins; bone marrow cell proliferation Stimulates T cells for the production of IL-2 and other lymphokines, B cell proliferation and antibody synthesis, neutrophil chemotaxis, and phagocytosis Along with the tumour necrosis factor (TNF), causes hematological changes in septic shock and enhances initial meningeal inflammation in bacterial meningitis
IL-2	T cells Also called T cell growth factor (TCGF) produced by activated T cells.	Promote growth and differentiation of T and B cells, cytotoxicity of T and NK cells, and the secretion of other lymphokines It converts some null cells (large granular lymphocytes) into lymphokine-activated killer (LAK) cells which can destroy tumour cells; this property has been used in the treatment of certain types of cancer
IL-3	T cells	Growth factor for bone marrow stem cells Multi CSF—it induces growth of all types of hematopoietic cells (multi-colony-stimulating factor)
IL-4	T_h cells	Proliferation of B and cytotoxic T cells Increase IgG and IgE production Enhance MHC class II and IgE receptors Activates resting B cells and acts as a B cell differentiating factor Growth factor for T cells and mast cells
IL-5	T_h cells	Proliferation of eosinophils, stimulate IgA and IgM production
IL-6–IL-17	T cells, including T_h cells B cells, macrophages, fibroblasts spleen, bone marrow, and stromal cells	Promote B cell differentiation IgG production, acute-phase proteins B and T cell growth and proliferation Neutrophil chemotactic factor Inhibit IFN production and mononuclear cell functions Activate NK cells
B. Colony-stimulating factors (CSF)		
GM-CSF	T cells, macrophages, fibroblasts	Stimulate the growth and differentiation of pluripotent stem cells in the bone marrow T cell and macrophage growth stimulation
G-CSF	Fibroblasts, endothelium	Granulocyte growth stimulation
M-CSF	Fibroblasts, endothelium	Macrophage growth stimulation
C. Tumour necrosis factors (TNF)		
TNF-α (Cachectin)	Macrophages, monocytes, T helper cells	Tumour cytotoxicity Lipolysis Wasting Acute-phase proteins Phagocytic cell activation Antiviral and antiparasitic effects Endotoxic shock Immunomodulatory influence on other cytokines
TNF-β	T cells	Induce other cytokines
D. Interferons (IFN)		
IFN-α	Leukocytes	Macrophage activation Augmentation of monocyte and neutrophil function Anti-tumour activity
IFN-β	Fibroblasts	Antiviral activity
IFN-γ	T cells	Antiviral Macrophage activation MHC class I and II expression on cells

- The mechanism of action of TF is not known.
- TF activity was, until recently, demonstrable only in humans but it has now been reported in monkeys, guinea pigs, and mice.

Applications of transfer factor in clinical practice

- Used to restore immune capacity in patients with T cell deficiency (e.g., in Wiskott–Aldrich syndrome).
- They are useful in the treatment of disseminated infections associated with deficient CMI (e.g., lepromatous leprosy, tuberculosis, and mucocutaneous candidiasis).
- They are useful in the treatment of malignant melanoma, autoimmune diseases (like systemic lupus erythematosus and rheumatoid arthritis), and diseases of unknown origin (like sarcoidosis and multiple sclerosis).

PATHWAYS OF HUMORAL AND CELLULAR IMMUNE RESPONSES TO AN ANTIGEN

- Humoral immunity and cell-mediated immunity are the two arms of the immune system. Antigen elimination is the final outcome of both systems of immunity.
- While humoral immunity targets extracellular antigens, cell-mediated immunity targets intracellular pathogens.
- T helper cells ($CD4^{+}T_{h1}$) have a role to play in both arms of the immune system (Fig. 15.13).
- CMI also has the following additional functions:
  - Immune surveillance
  - Tumour immunity
  - The activation, pathway, and destruction of the antigen in these pathways are independent of each other.
  - Some of the salient differences between humoral and cell-mediated immunity are presented in Table 15.4.

IMMUNOLOGICAL TOLERANCE

- Immunological tolerance is defined as a condition in which contact with an antigen specifically abolishes the capacity to mount an immune response against that particular antigen when it is administered subsequently.
- While the immune system is non-reactive to that particular antigen, its reactivity to other antigens remains unaffected.

Mechanism of Immunological Tolerance

- The mechanism of tolerance is not clear. The most likely mechanism may be the elimination of T_h cells, which prevent B cell activation and humoral immunity.

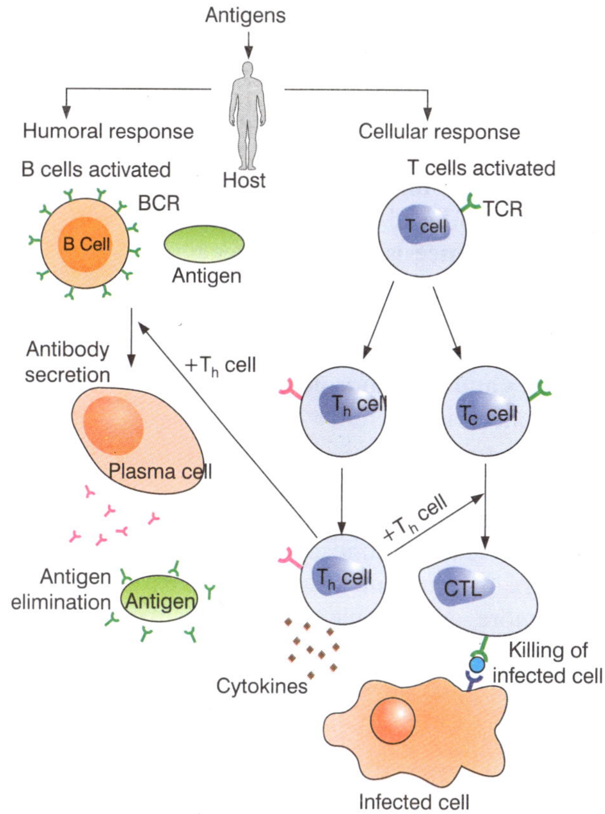

Fig. 15.13 Pathways of humoral and cellular immune responses to an antigen

Table 15.4 Differences between humoral and cell-mediated immunity

Feature	Humoral immunity	Cell-mediated immunity
Mediation	Mediated by antibodies	Mediated by cells
Cells involved	B cells (along with T helper cells) directly bind to soluble antigen	T cells ($CD4^{+}$ T_h, $CD8^{+}$ T_c cells)—T cell-mediated antigens can activate CMI
Mediators	Immunoglobulins IgG, IgM, IgE, IgA, and IgD	Cytokines (lymphokines and monokines) Interleukins and interferons
Role in complement pathway	IgG plays a role in the classical pathway of complement activation	Has no role
Hypersensitivity	Types I, II, and III	Type IV
Immunological surveillance	Limited role in surveillance and tumour immunity	Major player in immunological surveillance and tumour immunity
Transplant rejection	Minimal role in early rejection in pre-exposed person	Major role in homograft rejection and graft-versus-host reaction

- T and B lymphocytes appear to possess differing sensitivity to tolerance induction.
- High doses of antigens induce B cell tolerance.
- Repeated minute doses of antigens induce T cell tolerance.
- Tolerance to humoral and cellular types of immunity is usually induced simultaneously.
- **Split tolerance** is unresponsiveness by either B or T cell response.

Factors Affecting Immunological Tolerance

- **Species and immunocompetence of the host:** Strain differences in tolerance induction are seen within species (e.g., rabbits and mice can be rendered tolerant more rapidly than guinea pigs and chickens).
- **Age**: Embryos and newborns are particularly susceptible to immunosuppressive agents.
- **Nature of antigen:** Soluble antigens and haptens are more tolerogenic than particulate antigens. Solutions of serum proteins centrifuged at a high speed separate into tolerogenic supernatant and immunogenic sediment fractions.
- **Dose of the antigen:** There is a threshold dose, below which tolerance is not induced. Further increase in dosage increases the duration of tolerance.
- **Route of administration:** The best route to induce tolerance is the intradermal route, whereby the antigen equilibrates throughout the extra- and intravascular compartments. Certain haptens that are immunogenic in guinea pigs by the intradermal route are tolerogenic orally or intravenously.

Naturally-occurring tolerance is noted in certain viral infections such as congenital rubella and cytomegalovirus infections in which there is persistent viremia with decreased ability to produce neutralising antibodies (persistent tolerant infection).

When tolerance is interrupted by the induction of an antibody or an injection of sensitised lymphocytes, tolerance reverts, and disease occurs.

THEORIES OF IMMUNE RESPONSE

Several theories have been postulated on immune response and its specificity. Of these, the clonal selection theory stands out the most.

Clonal Selection Theory

This theory, which examined immunological specificity at the cellular level, was proposed by **Frank Macfarlane Burnet** (1957). The clonal selection hypothesis proposes the following:

- During immunological development, cells capable of reacting with different antigens are formed by a process of somatic mutation.
- Clones of cells that have immunological reactivity with self antigens are eliminated during embryonic life. Such clones are called **forbidden clones**.

Key Points

✓ T lymphocytes are responsible for cell-mediated immunity.

✓ When a T cell encounters an antigen recognised by the T cell receptors, a number of substances, collectively called lymphokines, are released; these activate macrophages and increase their ability to destroy the pathogens.

✓ Important cytokines are as follows:
- Interleukin
- Interferons
- Colony stimulating factors
- Tumour necrosis factor
- Transfer factors

✓ Immunological tolerance or unresponsiveness is a condition in which contact with an antigen specifically abolishes the capacity to mount an immune response against that particular antigen when it is administered subsequently.

Questions

Short answer questions

1. Briefly describe the types of immune responses and their functions.
2. List the factors that influence antibody production.
3. Discuss primary and secondary responses in antibody-mediated immunity.
4. Enumerate the functions of cytokines.
5. What are the clinical applications of transfer factors?
6. Describe immunological tolerance.
7. What are the differences between primary and secondary responses?

MCQs

1. **Which of the following is a function of antibody-mediated immunity?**
   a) Immunological surveillance
   b) Protection against fungi
   c) Acting against extracellular bacterial pathogens
   d) Involvement in graft-versus-host reaction
2. **Which among the following is true of primary immune response?**
   a) It is carried out by memory B cells.
   b) The predominant antibody is IgG.
   c) A high dose of antigen is required.
   d) Peak response is in 3–5 days.
3. **Which of the following statements is true for optimal antibody production?**
   a) Large particulate antigens should be administered in tissues.
   b) IgA production is better when antigens are administered parentally.
   c) Antibody production increases with increase in dosage.
   d) Simultaneous administration of multiple antigens has no effect.
4. **Immunological paralysis is seen when ____________.**
   a) Antibodies are administered multiple times
   b) Antigens are administered multiple times

c) Massive doses of antibodies are administered
d) Massive doses of antigens are administered

5. **Which one of the following effects does IL-1 produce?**
a) T lymphocyte activation
b) Delayed wound healing
c) Increased pain perception
d) Decreased PMN release from the bone marrow

6. **Which of the following are the chief sources of IL-4?**
a) Macrophages b) B cells
c) T cells d) Fibroblasts

7. **Which of the following is a pro-inflammatory marker?**
a) IL-4 b) IL-9
c) IL-11 d) IL-17

8. **Which of the following is an important endogenous pyrogen?**
a) IL-1 b) IL-2
c) IL-3 d) IL-4

9. **Which of the following statements is true for an interferon?**
a) It has an antiviral effect.
b) It was known to cause cachexia.
c) It causes macrophage activation.
d) It has antitumour activity.

10. **Which of the following induces the maturation of eosinophils?**
a) IL-5 b) IL-4
c) IL-3 d) IL-2

11. **Which of the following is a produced by T_h cells?**
a) IL-1 b) IFN-β
c) TNF-α d) IL-17

12. **All the following are true regarding IFN-γ except:**
a) Primarily produced by fibroblast
b) Has antiviral action
c) Activates macrophages
d) Increases MHC class I and II expression

13. **Which of the following does not affect immunological tolerance?**
a) Age b) Species
c) Dose of antigen d) Sex

14. **Which of the following is correct regarding the production of antibodies following an antigenic stimulation?**
a) The first signal is a combination of the T cell receptor with the MHC class II-complexed antigen
b) The second signal is interleukin-1, which is produced by the antigen processing cell
c) The activated Th cell forms interleukin-2 and other cytokines (IL-4, IL-5 and IL-6) required for B cell stimulation
d) All of the above

15. **Aluminium hydroxide is a:**
a) Freund's incomplete adjuvant
b) Freund's complete adjuvant
c) Not an adjuvant
d) None of the above

16. **Cytokines are:**
a) Hormone-like substances that are active even at minute concentrations
b) They are produced by endocrine glands
c) Regulated by exogenous stimuli such as antigens only
d) They act on the cells away from those producing them

17. **Which of the following is true regarding cell mediated immunity?**
a) It is involved in type 1 hypersensitivity.
b) It has minimal role in early rejection in pre- exposed person.
c) It plays a role in the classical pathway of complement activation.
d) It plays a major role in homograft rejection and graft-versus–host reaction.

18. **Interferon gamma, a type of cytokine has following function:**
a) It activates B cells to produce IgG
b) It has antiviral activity
c) It is used in the treatment of leukemia
d) It is secreted by macrophages

Answers: 1:c, 2:c, 3:a, 4:d, 5:a, 6:c, 7:a, 8:a, 9:b, 10:a, 11:d, 12:a, 13:d, 14:d, 15:a, 16:a, 17:d, 18:a

CHAPTER 16

Part II: Immunology

Laboratory Methods Used in the Detection of Immunological Response (Antigen–Antibody Reactions)

(MI1.10)

Chapter Navigation

Learning objective

- Discuss the laboratory methods used in immunology

INTRODUCTION

- Immunological responses in health and disease are a key indicator of the health status of an individual.
- A number of tests are performed in laboratories to detect both the humoral and the cell-mediated immune status of an individual. These tests are used to diagnose and monitor a patient's response to a disease.
- Antigens and antibodies combine/react with each other specifically and in an observable manner, i.e., the antigenic epitope binds to the variable domain of the antibody. This exclusive reaction forms the basis of several immunological assays to detect infectious diseases, monitor humoral immunity, and detect molecules of medical interest.
- These tests rely on the ability of the antibody to agglutinate a particulate antigen, precipitate a soluble antigen, or form an antigen–antibody complex to fix complement.
- The other methods of detection employ immunofluorescence or **chemically-modified antibodies to quantitate antigens by immunosorbent assays.**

STAGES IN ANTIGEN-ANTIBODY REACTIONS

The reaction occurs in three stages.

1. **Primary stage:** This is the initial interaction between antigens and antibodies without any visible effects. This reaction has the following characteristics:
   - **It is rapid,** occurs even at low temperatures, and is reversible.
   - Antigens and antibodies bind by **non-covalent bonds** namely:
     - *Electrostatic forces*: Attraction between opposite charges
     - *Hydrogen bonds*: Formed by the sharing of hydrogen between two electronegative atoms
     - *Van der Waal's forces*: Interaction between the outer electron clouds of two or more electrons
     - *Hydrophobic forces*: Forcing together of hydrophobic groups by water
2. **Secondary stage:** In almost but not all instances, the primary stage is followed by the secondary stage, leading to firmer covalent bonding.
3. **Tertiary stage:** Some antigen–antibody reactions occurring in vivo initiate chain reactions that lead to the neutralisation or destruction of injurious antigens or tissue damage. These are tertiary reactions and may be brought about either by a humoral or a cell-mediated immune response in the host. Tertiary reactions manifest as clinical allergies or other immunological diseases.

Laboratory methods based on antigen–antibody assay:

- Precipitation
- Agglutination
- Lysis of cells
- Killing of live antigens
- Neutralisation of toxins and other biologically active antigens
- Fixation of complement
- Immobilisation of motile organisms
- Enhancement of phagocytosis

- An antibody that causes agglutination is called an **agglutinin** while one that causes precipitation is called **precipitin**—the corresponding antigens are called **agglutinogen** and **precipitinogen** respectively.
- A single antibody can cause precipitation, agglutination, and most other serological reactions. An antigen can stimulate the production of different classes of immunoglobulins which differ in their reaction capacities as well as in other properties (Table 16.1).

GENERAL FEATURES OF ANTIGEN-ANTIBODY REACTIONS

- The reaction is **specific**—an antigen combines only with its homologous antibody and vice versa.
- The specificity, is not absolute. **Cross-reactions** may occur due to antigenic similarity or relatedness.
- Entire molecules, and not fragments, are involved in these reactions. When an antigenic determinant present on a large molecule or a 'carrier' particle reacts with its antibody, **whole molecules** or particles are agglutinated.
- There is **no denaturation** of the antigen or the antibody during the reaction.
- The combination occurs at the surface. Therefore, it is the surface antigens that are immunologically relevant. Antibodies to the surface antigens of infectious agents are generally protective.
- The antigen–antibody combination is **firm but reversible**. The firmness of the union is influenced by the affinity and avidity of the reaction.
- **Affinity** is defined as the quantitative intensity or strength of the binding between the epitope of the antigen and antibody molecules. Low-affinity antibodies bind weakly with antigens and tend to dissociate readily, whereas high-affinity antibodies bind to antigens more tightly and remain bound longer.
- **Avidity** is defined as the strength of the bond of the antigen and multivalent antibody complexes. It is a better measure of binding capacity within biological systems (e.g., the reaction of an antibody with antigenic determinants on a virus or bacterial cell). Pentameric IgM often has a lower affinity than IgG but high avidity due to its higher valence, which enables it to bind to antigens effectively.
- Both antigens and antibodies participate in the formation of agglutinates or precipitates.
- Antigens and antibodies can combine in varying proportions.
- Antibodies are generally bivalent, though IgM molecules may have five or ten combining sites. Antigens may have hundreds of valencies.
- In the body, antigen–antibody reactions form the basis of antibody-mediated immunity in infectious diseases or tissue injury in some types of hypersensitivity and autoimmune diseases.

Table 16.1 Comparative efficiency of the immunoglobulin classes in different serological reactions

Reaction	IgG	IgM	IgA
Precipitation	Strong	Weak	Variable
Agglutination	Weak	Strong	Moderate
Complement fixation	Strong	Weak	Negative
Lysis	Weak	Strong	Negative

USES OF ANTIGEN-ANTIBODY REACTIONS IN THE LABORATORY

Antigen–antibody reactions have great use in the diagnosis and study of infections. The following are some areas where they are applied:

- **Diagnosis of infections** by detecting antigens or early antibodies
- Detecting **exposure to an infection** by detecting late antibodies
- Detection of the **immune status** of an individual by quantifying the antibodies
- Detection and **quantitation of antigens**
- **Seroepidemiological surveys** to screen a population for the presence of antibodies to a particular infection

Parameters for Serological Tests

Antigen–antibody reactions in vitro are known as serological reactions. The parameters to define serological tests are as follows:

- **Sensitivity:** This is the ability of a test to detect even minute quantities of an antigen or antibody. When a test is highly sensitive, false-negative results are minimal to none.
- **Specificity:** The ability of a test to only detect reactions between homologous antigens and antibodies and no other. In a highly specific test, false-positive reactions are absent or minimal, and only the true-positive reactions are detected.
- In general, **sensitivity and specificity are inversely proportional**.
- **Qualitative assay:** Such a test can only detect the presence of an antigen or antibody (e.g., the test is reactive or non-reactive).
- **Quantitative assay:** This type of test can estimate the quantity of an antigen or antibody present in a sample. The simplest way to do this is by serially diluting the sample and detecting the antigen or antibody present in the highest dilution. There are other sensitive methods that estimate the exact quantity in micrograms or nanograms.

REACTIONS ON WHICH ANTIGEN-ANTIBODY ASSAYS ARE BASED

Some methods used in the laboratory are based on the antigen–antibody reactions listed below.

Precipitation Reaction

1. **Precipitation:** When a soluble antigen combines with its antibody in the presence of electrolytes (NaCl) at a suitable temperature and pH, the antigen–antibody complex forms an insoluble precipitate.
2. **Flocculation:** When, instead of sedimenting, the precipitate remains suspended as floccules, the reaction is known as flocculation.

Mechanism of precipitation

- Precipitation can take place in **liquid media** or gels such as agar, agarose, or polyacrylamide.
- Precipitation results when a large lattice is formed with alternating antigen and antibody molecules. This is possible only in the **zone of equivalence** (when antigens and antibodies are in equilibrium). In zones of antigen or antibody excess, the lattice does not enlarge, as the valencies of the antibody and the antigen respectively are fully satisfied.

Lattice hypothesis

The amount of precipitate formed is influenced by the relative proportions of antigens and antibodies. If increasing quantities of antigens are added to the same amount of antiserum in different tubes, precipitation/agglutination occurs most abundantly in one of the middle tubes in which the antigen and antibody proportions are optimal or equivalent. In the preceding tubes, in which the antibody is in excess, and in the later tubes, in which the antigen is in excess, there is weak precipitation or even lack thereof. This hypothesis also holds good for agglutination reactions.

If the amounts of precipitate in the different tubes are plotted on a graph, the resulting curve will have the following three phases (Fig. 16.1):

i. **Prozone (phenomenon):** This occurs in the presence of excess antibodies in the test system. The absence of a visible reaction is due to the inhibition of lattice formation by the excess antibody.

ii. **Zone of equivalence:** Here, the antigens and antibodies are in optimum proportions. Lattice formation and visible reactions are enhanced.

iii. **Postzone (phenomenon):** This is caused by the presence of an excess of antigens in the test system. No visible reaction occurs.

- The prozone and postzone phenomena may be corrected by making serial dilutions of serum, thereby reducing the concentration of antigen or antibody in the test serum and optimising the concentrations of the antigen and antibody.
- The prozone is of importance in clinical serology, as sera rich in antibody may sometimes give a false-negative precipitation or agglutination result unless several dilutions are tested. A classic example is the *Brucella* agglutination test.

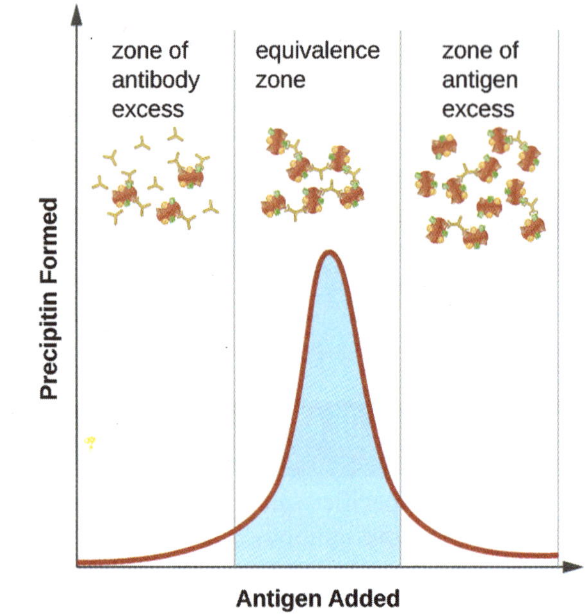

Fig. 16.1 Lattice hypothesis—as the antigen is added to a solution containing a fixed amount of antibody, the amount of precipitin increases as the antibody-to-antigen ratio approaches the equivalence zone and decreases once the proportion of antigen exceeds the optimal ratio (*Source*: https://openstax.org/books/microbiology/pages/20-2-detecting-antigen-antibody-complexes)

Applications of precipitation reactions

- The precipitation test may be carried out as a qualitative or quantitative test.
- It is sensitive in the detection of antigens—as little as 1 mg of protein can be detected.
- It is relatively less sensitive for the detection of antibodies. Precipitation tests have several applications:
  - Grouping of streptococci by Lancefield's technique
  - VDRL test for syphilis
  - Standardisation of toxins and toxoids
  - Testing of the toxigenicity of diphtheria bacilli
  - Forensic application in the identification of blood and seminal stains
  - Testing for food adulterants

Types of precipitation reactions

Immunodiffusion (precipitation in gel)

- The reaction is visible as a distinct band of precipitation, which is stable and can be stained for preservation, if necessary.

- As each antigen–antibody reaction gives rise to a line of precipitation, the number of different antigens in the reacting mixture can be readily observed.
- Immunodiffusion also indicates identity, cross-reaction, and non-identity between different antigens. Immunodiffusion is usually performed in a soft (1%) agar or agarose gel.

Applications

- Detection of antigens/antibodies
- Detection of the purity of an antigen
- Immunodiffusion techniques are employed in biochemistry laboratories to estimate globulins including immunoglobulins
- In diagnostic microbiology laboratories, immunodiffusion assays have been replaced by rapid methods like enzyme immunoassay methods

Modifications

- **Single diffusion in one dimension:** The antibody is incorporated into agar gel in a test tube and the antigen solution is poured over it. As the antigen moves towards antibody, a line of precipitation is formed in the gel.
- **Single diffusion in two dimensions (radial immunodiffusion):** The antiserum is incorporated into agar gel poured on a flat surface (slide or petri dish). When the antigen is added to wells cut into the surface of the gel, it diffuses radially from the well and forms ring-shaped bands of precipitation (halos) concentrically around the well (Fig. 16.2a and b). The diameter of the halo gives an estimate of the concentration of the antigen. This is used for the estimation of the immunoglobulin classes in sera.
- **Double diffusion in one dimension:** The antigen and antibody both move towards each other and form a line of precipitation at their meeting point in the agar block.
- **Double diffusion in two dimensions (Ouchterlony procedure):** Both the antigen and antibody diffuse independently through the agar gel in two dimensions, horizontally and vertically. An example is the Elek test for toxigenicity in diphtheria bacilli. When diphtheria bacilli are streaked at right angles to a filter paper strip carrying the antitoxin, and this paper is implanted on a plate of suitable medium, arrowhead-shaped lines of precipitation appear on incubation—these lines are indicative of a toxigenic bacillus (Fig. 16.2a and c). This test is used to compare different antigens and antisera directly. Cross-reaction or partial identity is indicated by **spur formation**.

Immunoelectrophoresis

- This involves the electrophoretic **separation of a composite antigen** (such as serum) **into its constituent proteins** and the subsequent immunodiffusion of these constituents against their antisera, resulting in separate precipitin lines, each indicating a reaction between an individual protein and its antibody.
- It is performed on agar or agarose gel on a slide. The resulting precipitin lines can be photographed

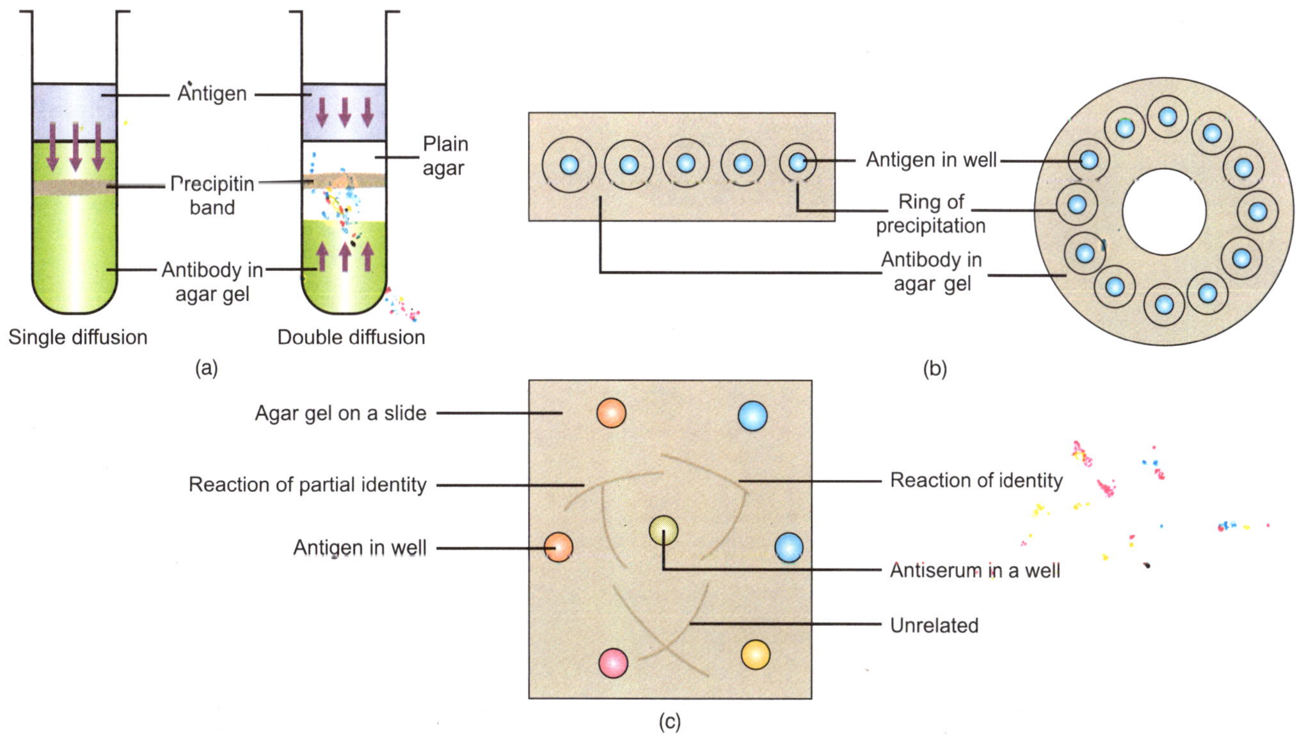

Fig. 16.2 Single diffusion in two dimensions and double diffusion in two dimensions (Ouchterlony technique)

and the slides dried, stained, and preserved for the record. Over **30 different proteins can be identified by this method in human serum**.

Applications

Immunoelectrophoresis is used for testing for normal and abnormal proteins in serum and urine (Fig. 16.3) and for detecting various antigens such as alpha-fetoprotein in serum and specific antigens of cryptococcus and meningococcus in cerebrospinal fluid (though advanced rapid immunological techniques have now replaced these methods).

Agglutination Reaction

- When a particulate antigen is mixed with its antibody in the presence of electrolytes at a suitable temperature and pH, the particles get clumped or agglutinated.
- Agglutination is more sensitive than precipitation for the detection of antibodies.
- Agglutination occurs optimally when antigens and antibodies react in equivalent proportions.
- The same principles govern agglutination and precipitation. The zone phenomenon may be seen when either an antibody or an antigen is in excess.

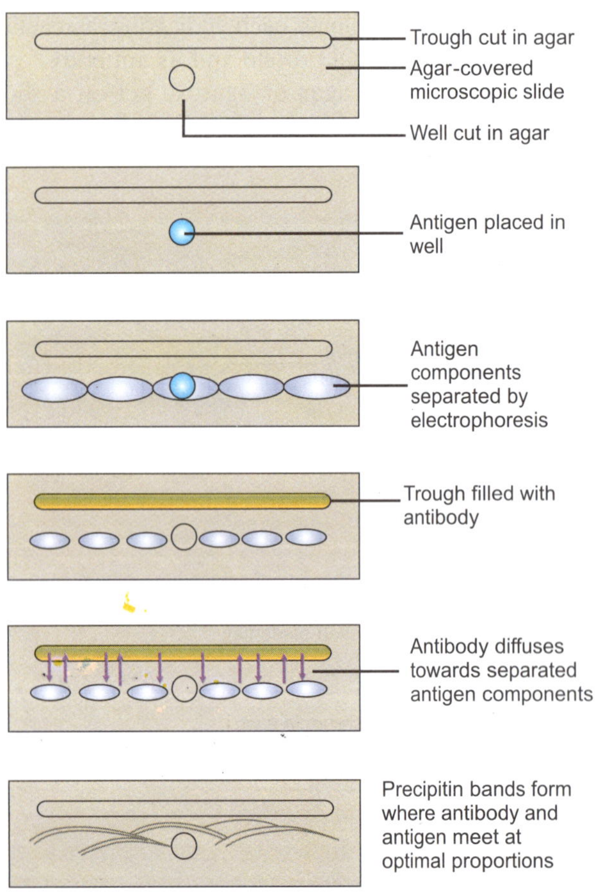

Fig. 16.3 Immunoelectrophoresis

- 'Incomplete' or 'monovalent' antibodies do not cause agglutination though they combine with the antigen. They may act as 'blocking' antibodies, inhibiting agglutination by the complete antibody added subsequently.

Agglutination tests may be performed on slides or in tubes. Table 16.2 presents the various applications of agglutination tests.

1. Agglutination on a slide

When a drop of the appropriate antiserum is added to a smooth, uniform suspension of a particulate antigen in a drop of saline on a slide or tile, agglutination takes place (Fig. 16.4).

Table 16.2 Clinical application of agglutination tests to detect antibodies in patient's serum

Diagnostic agglutination tests	Application
Identification of many bacterial isolates	*Salmonella, Shigella, Vibrio* typing
Blood grouping and cross-matching	Determining blood groups A/B AB/O
Febrile agglutination tests	Widal, *Brucella* agglutination test
Direct/Indirect Coombs test	Rh compatibility in newborn/Rh antibody in mother
Heterophile agglutination tests	
The *Streptococcus* MG agglutination test, cold agglutination test	Atypical pneumonia due to *Mycoplasma*
Weil–Felix (heterophile antibodies)	Rickettsial infections
Paul–Bunnell test	Infectious mononucleosis

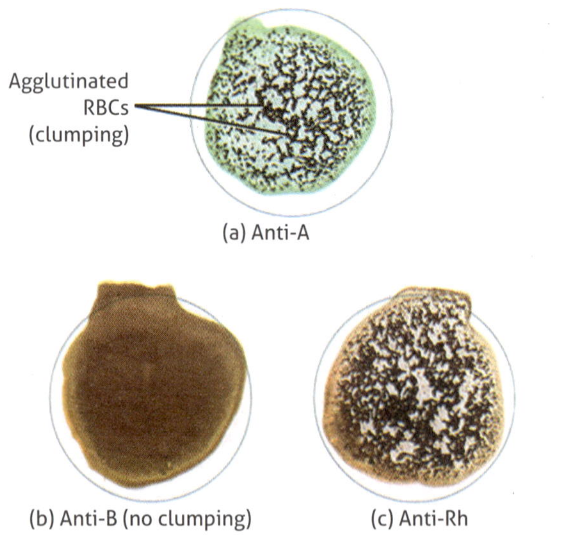

Fig. 16.4 Agglutination tests used for blood grouping (*Source*: Openstax.org/books/microbiology)

Uses of slide agglutination

- Identification of many bacterial isolates from clinical specimens (e.g., *Salmonella* sero-grouping or typing)
- Blood grouping and cross-matching (Fig. 16.4).

2. Agglutination in tubes

- An agglutination test is performed in tubes when serial dilutions of the serum need to be prepared to measure antibodies.
- A fixed volume of a particulate antigen suspension is added to an equal volume of serial dilutions of an antiserum in test tubes; the agglutination titre of the serum is then estimated.

Limitations of agglutination tests

- The agglutination test for brucellosis may be complicated by the prozone phenomenon.
- The presence of incomplete or 'blocking' antibodies (this may be detected by the antiglobulin **[Coombs] test**) may result in false-negative results.
- Cross-reacting or heterophile antigens, which share similar antigenic epitopes between unrelated microorganisms may result in false-positive results.

Antiglobulin (Coombs) test

This test detects anti-Rh antibodies that do not agglutinate with Rh-positive erythrocytes in saline.

Principle

- When sera containing incomplete anti-Rh antibodies are mixed with Rh-positive red cells, the incomplete immunoglobulin coats the surface of the erythrocytes but does not agglutinate.
- When such erythrocytes (anti-Rh antibody-coated) are washed free of all unattached protein and treated with a rabbit antiserum against human gammaglobulin (anti-red cell gamma globulin or Coombs serum), the cells agglutinate.

  Two types of the Coombs test are available:

  i. **Direct Coombs test:** The sensitisation of erythrocytes with incomplete antibodies takes place in vivo, as seen in hemolytic disease of the newborn due to Rh incompatibility. When the red cells of erythroblastotic infants are washed free of unattached protein and then mixed with a drop of Coombs serum, agglutination results. The direct Coombs test is often negative in hemolytic disease due to ABO incompatibility.

  ii. **Indirect Coombs test:** This is done to detect the presence of Rh factor antibodies in the mother's blood sample as part of the prenatal check. This is done to identify antigens that could cause problems in the newborn or the mother (in case transfusion is necessary). Sensitisation of red cells with the gammaglobulin is performed in vitro.

Uses of Coombs test

Originally employed for the detection of anti-Rh antibodies, the Coombs test is useful for demonstrating any type of incomplete or non-agglutinating antibody, as, for example, in brucellosis.

Passive agglutination test

- By attaching soluble antigens to the surface of carrier particles, it is possible to convert precipitation tests into agglutination tests. These are convenient and more sensitive for the detection of antibodies.
- The commonly used carrier particles are: (i) red cells, (ii) latex particles, and (iii) bentonite. Human or sheep erythrocytes adsorb onto a variety of antigens. Polysaccharide antigens may be adsorbed by simply mixing them with the cells. For adsorption of protein antigens, tanned red cells are used. Some types of passive agglutination tests are as follows:

  i. **Hemagglutination:** A special type of passive hemagglutination test is the **Rose–Waaler test.** In rheumatoid arthritis, an autoantibody (RA factor) appears in the serum, which acts as an antibody to gammaglobulin. The RA factor agglutinates sheep red cells coated with antibodies to sheep erythrocytes derived from rabbits.

  ii. **Latex agglutination test:** Polystyrene latex, which can be manufactured as uniform spherical particles measuring 0.8–1.0 µm in diameter, can adsorb several types of antigens on their surface. Latex agglutination tests (**latex fixation tests**) are widely employed in the clinical laboratory for the detection of anti-streptolysin O (ASO), C-reactive protein (CRP), RA factor, human chorionic gonadotropin (hCG), and many other antigens.

  iii. **Reversed passive agglutination:** Instead of the antigen, the antibody is adsorbed onto carrier particles to detect antigens. Passive agglutination tests are very sensitive and yield high titres, but may give false-positive results. This test is used to detect bacterial antigens in CSF, e.g., meningococcal, pneumococcal, and *Haemophilus* antigens in the CSF of patients with acute pyogenic meningitis.

  iv. **Co-agglutination test:** It is based on the agglutination of *Staphylococcus aureus* bearing a specific antibody-sensitised protein A with the soluble bacterial (e.g., *Legionella*) antigen in the clinical specimen.

Complement Fixation Test (CFT)

- This test can detect as little as 0.04 mg of antibody and 0.1 mg of antigen.
- In the presence of the appropriate antibodies, the complement lyses erythrocytes, kills and, in some

cases, lyses bacteria, immobilises motile organisms, promotes phagocytosis and immune adherence, and contributes to tissue damage in certain types of hypersensitivity.

Principle

- Complement plays a part in many immunological reactions and is absorbed during the specific reaction of antigens with their antibodies.
- The ability of antigen–antibody complexes to 'fix' complement is made use of in CFT (Fig. 16.5).

Procedure

- CFT is a complex procedure consisting of two steps and several reagents—antigen, antibody, complement, sheep erythrocytes, and amboceptor (rabbit antibody to sheep red cells). The antigen may be soluble or particulate.
- The antiserum is heated at 56°C (inactivated) for half an hour before the test to destroy any natural complement activity the serum may have and to remove some non-specific inhibitors of complement present in some sera (anti-complementary activity).
- The source of the complement is guinea pig serum. It can be preserved either in the lyophilised or frozen state or with special preservatives.

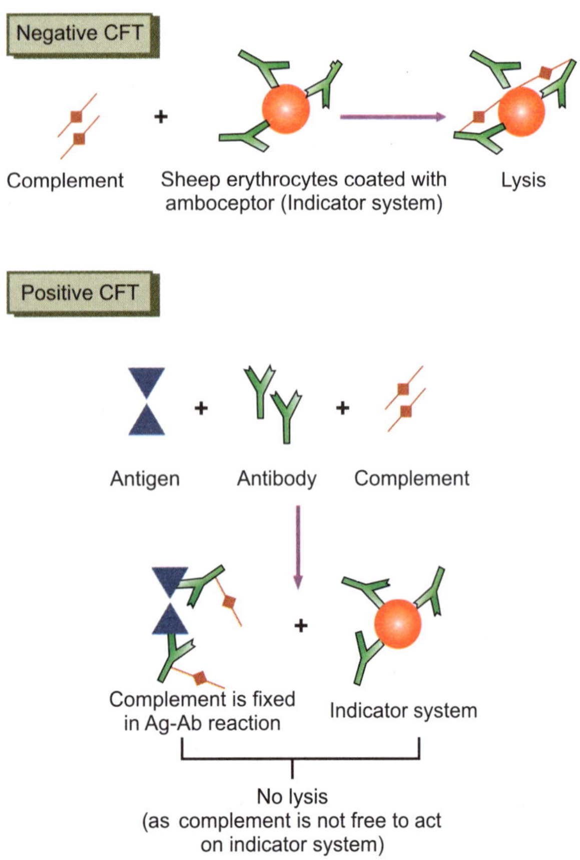

Fig. 16.5 Complement fixation test

Standardisation

- The guinea pig serum and the amboceptor should be titrated for complement activity.
- **Minimum hemolytic dose (MHD) or one unit of complement** is defined as the highest dilution of guinea pig serum that lyses one unit volume of washed sheep erythrocytes in the presence of excess hemolysin (amboceptor) within a fixed time (usually 30 or 60 minutes) at a fixed temperature (37°C).
- Controls included in the test procedure are:
  - Antigen and serum controls to ensure that they are not anti-complementary
  - Complement control to ensure that the desired amount of complement is added
  - Cell control to observe that sensitised erythrocytes do not undergo lysis in the absence of complement
- The **immobilisation test** is another complement-dependent reaction. In the *Treponema pallidum* immobilisation test, the test serum is mixed with a live, motile suspension of *T. pallidum* in the presence of the complement. On incubation, the specific antibody inhibits the motility of treponemes.

Applications of complement fixation test

- To detect and quantify an antibody that does not agglutinate or precipitate with its antigen
- To detect low levels of antibody (less than 1 µg per millilitre) during early infection
- Currently, CFT is used to screen viral or bacterial infections to detect IgG antibodies (for some respiratory viruses and *Coxiella burnetii* antibodies, and occasionally, antibodies from CSF)

Neutralisation Test

These are based on the action of antibodies to neutralise an end product or the outcome of the action of an antigen.

1. **Virus neutralisation test:** The neutralisation of viruses by their antibodies can be demonstrated in various systems such as the **plaque inhibition test**, in which specific antiviral serum inhibits plaque formation by the virus.
2. **Toxin neutralisation:** Bacterial exotoxins are good antigens and induce the formation of neutralising antibodies (antitoxins) which are clinically important in protection against diphtheria and tetanus. Toxin neutralisation can be tested in vivo.

Neutralisation tests in animals consist of injecting toxin–antitoxin mixtures and estimating the least amount of antitoxin that prevents death or disease in the animals.

The following are some examples of neutralisation tests:

- **Schick test:** This test is based on the ability of a circulating antitoxin to neutralise the diphtheria toxin given intradermally. It indicates either immunity or susceptibility to the disease.

- **Anti-streptolysin O (ASO) test:** This test demonstrates whether the antitoxin present in patient sera neutralises the hemolytic activity of the streptococcal O hemolysin (O is an immunogenic, oxygen-labile hemolytic toxin). In streptococcal infection, antibodies are produced against the various antigens of streptococci—ASO is one such antibody. Raised or rising levels indicate past or present infection.
- **Nagler's reaction:** This is an in vitro test for the identification of the alpha toxin of *Clostridium perfringens* in clinical specimens. The action of the toxin lecithinase is neutralised by the antitoxin.

Radioimmunoassay (RIA)

- This is a competitive binding assay in which fixed amounts of the antibody and radiolabelled antigen react in the presence of an unlabelled antigen.
- The labelled and unlabelled antigens compete for the limited binding sites on the antibody. This competition is determined by the level of the unlabelled (test) antigen present in the patient's serum samples.
- After the reaction, the antigen is separated into 'free' and 'bound' fractions and the radioactive counts of the two are measured. The concentration of the test antigen can be calculated from the ratio of the bound and total antigen labels using a standard dose–response curve.
- Hazards of radioisotope use have limited its applications.
- Applications—quantitation of hormones, drugs, tumour markers, IgE, and viral antigens.

Enzyme Immunoassay (EIA)

Antigens or antibodies labelled with enzymes were developed as serological reagents for the assay of corresponding antibodies and antigens.

- Their versatility, sensitivity, simplicity, economy, and absence of radiation hazards have made enzyme immunoassays (EIA) the most widely used procedure in clinical serology.
- The easy availability of a wide range of test kits and the scope for automation have added to the popularity of these assays.
- The term enzyme immunoassay includes all assays based on the measurement of enzyme-labelled antigen, hapten or antibody.
- Two types of EIAs are available:
  - **Homogeneous EIA** does not require the bound and free fractions to be separated; the test can thus be completed in one step, with all reagents added simultaneously. This type of EIA can be used only for the assay of haptens such as drugs and not for microbial antigens and antibodies. An example of homogeneous EIA is the **enzyme-multiplied immunoassay technique (EMIT)**, which is a simple assay for small-molecule drugs such as opiates, cocaine, barbiturates, or amphetamine in serum.
  - **Heterogeneous EIA** requires the separation of the free and bound fractions either by centrifugation or by adsorption onto solid surfaces and washing. It is, therefore, a multistep procedure, with reagents added sequentially. The main type of heterogeneous EIA is **enzyme-linked immunosorbent assay (ELISA)**. This is the most widely used technique in clinical laboratories.

Enzyme-linked immunosorbent assay (ELISA)

Principle

- The immunosorbent material, which may be particulate (for example, cellulose or agarose) or a solid phase (polystyrene, polyvinyl or polycarbonate tube, or microwell, membrane or disc of polyacrylamide, paper, or plastic), is loaded with a fixed antigen or antibody.
- The antigen or antibody in a patient's serum binds to the fixed component, forming an antigen–antibody complex.
- This test depends on the formation of enzyme–substrate conjugates. The enzyme converts a colourless substrate to a coloured product bound to the antibody, which is part of the antibody–antigen complex.
- ELISA is usually performed using 45/96-well microtitre plates suitable for automation or break-apart strips of eight for small sample loads.

Enzyme–substrate systems used in ELISA

- Horseradish peroxidase (HRP)-o-phenylenediamine dihydrochloride
- Urease–urea
- Beta-galactosidase–O-nitrophenyl-beta-D-galactopyranoside (ONPG)

Types of enzyme-linked immunosorbent assays

Several variations of the ELISA technique have been developed for simple diagnostic tests, including the card and dipstick methods, which are suitable for clinical laboratory and bedside applications (Table 16.3).

Chemiluminescence Immunoassay (CLIA)

- Chemiluminescence refers to a **chemical reaction that emits energy in the form of light**.
- Just as radioactive conjugates are employed in RIA, fluorescent conjugates in IFA, and enzymes in ELISA, chemiluminescent compounds (such as

Table 16.3 Principle, advantages, and disadvantages of various types of ELISA

Type of ELISA	Principle	Advantage	Disadvantage
Direct ELISA	The antigen is immobilised directly on the plate and a specific enzyme-conjugated detection antibody is added, which binds to the antigen The substrate is then added, producing a signal that is proportional to the amount of analyte in the sample	Fast and simple	Less specific Expensive because a specific antibody has to be enzyme-labelled No cross-reactivity
Indirect ELISA (Fig. 16.6)	The antigen is immobilised on a plate The antibody present in the patient's serum binds to the antigen A conjugated secondary antibody directed against the host species of the primary antibody is then added The substrate then produces a signal proportional to the amount of antigen bound in the well	Versatile Higher sensitivity Can be automated easily	Less expensive as non-specific goat anti-human immunoglobulin antibody is used Cross-reactivity caused by secondary antibody Time-consuming
Competitive ELISA (Fig. 16.7)	Similar to RIA Both the unknown antigen (sample) and the known antigen (standard) compete with each other for a fixed amount of antibody It yields an inverse curve, where higher values of the antigen in the samples/standards yield a lower amount of colour change	Ability to quantitate small molecules It is normally used for hapten detection	Less specific since only one antibody is used Requires a conjugated antigen
Sandwich ELISA (Fig. 16.7)	Direct sandwich—the antigen is sandwiched between two layers of antibodies (i.e., capture and detection antibodies) Indirect sandwich—the antibody is sandwiched between the antigen and a labelled antibody The microtitre plate is coated with goat anti-rotavirus antibody that binds to the antigen in the sample The antigen–antibody complex then binds to enzyme labelled antibody	Highest specificity and sensitivity Compatible with complex sample matrices	Long protocol Challenging to develop
Capture ELISA	Monoclonal IgM antibodies are coated onto wells The patient's serum (suspected to have IgM for a specific infectious agent) is added The coated anti-human IgM binds to the patient's IgM Then, the specific antigen (of the infectious agent) is added, which in turn, binds to the IgM of the patient's serum The antigen-specific mouse monoclonal antibody (Mab), which is conjugated with horseradish peroxidase (HRP), is then added The complex is detected based on the enzyme–substrate reaction	More specific	Expensive
Cylinder or cassette ELISA (Fig. 16.8)	The antigen is immobilised on a nitrocellulose membrane and the patient's serum is added and allowed to filter into the absorbent material placed below it in the cassette base The antibody, if present in the serum, will bind to the appropriate antigen A positive result is indicated by a coloured spot	Many samples can be tested Rapid (10–15 minutes) Does not require microplate washers or readers	–
Enzyme-linked immunospot assay (ELISpot)	It detects the secreted protein, e.g., cytokines from cytokine-secreting cells or Ig from immunoglobulin-secreting cells in a process similar to ELISA	Used for the detection of cytokine-secreting cells	Expensive
IgG avidity ELISA	A destabilising agent such as urea is used to dissociate the binding between the specific IgG and the antigen The result is expressed in terms of the avidity index: • Recent infection—weaker avidity as the antigen-antibody bond is weaker • Chronic infection—high avidity as there is no dissociation of antibody from the antibody	Discriminates acute from chronic infection	High cost
Enzyme-linked fluorescence assay (ELFA)	It utilises a substrate that yields a fluorescent rather than visible product on interaction with the immunoglobulin-linked enzyme	More sensitive	Background fluorescence can create detection problems

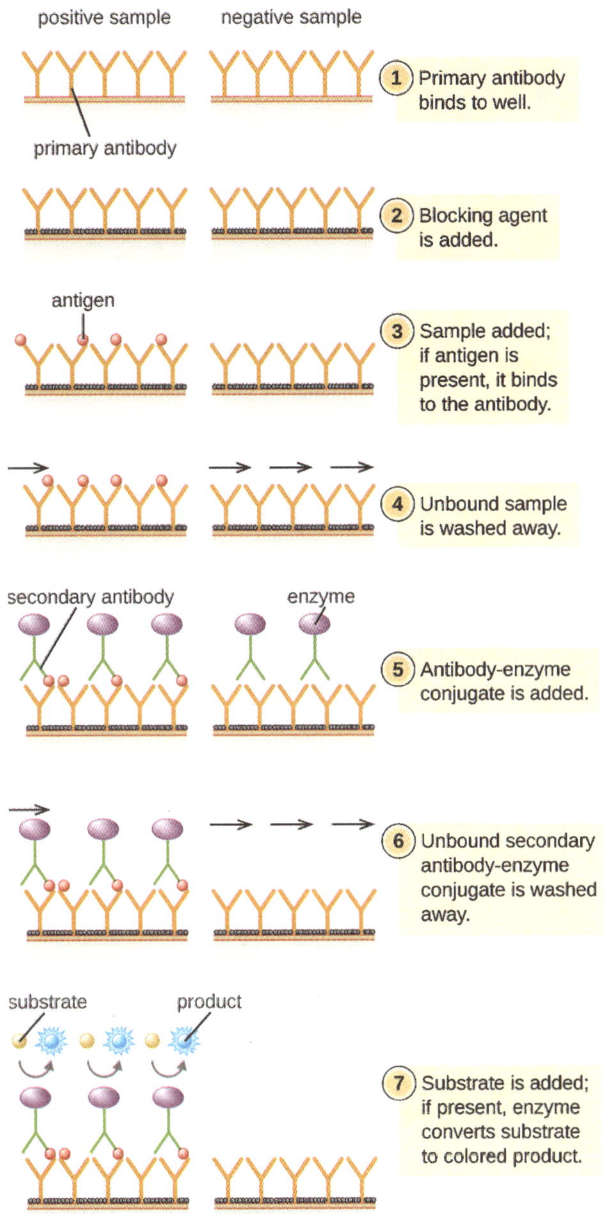

Fig. 16.6 The indirect ELISA is used to quantify antigen-specific antibodies in the patient's serum (*Source*: https://openstax.org/books/microbiology/pages/20-4-eias-and-elisas)

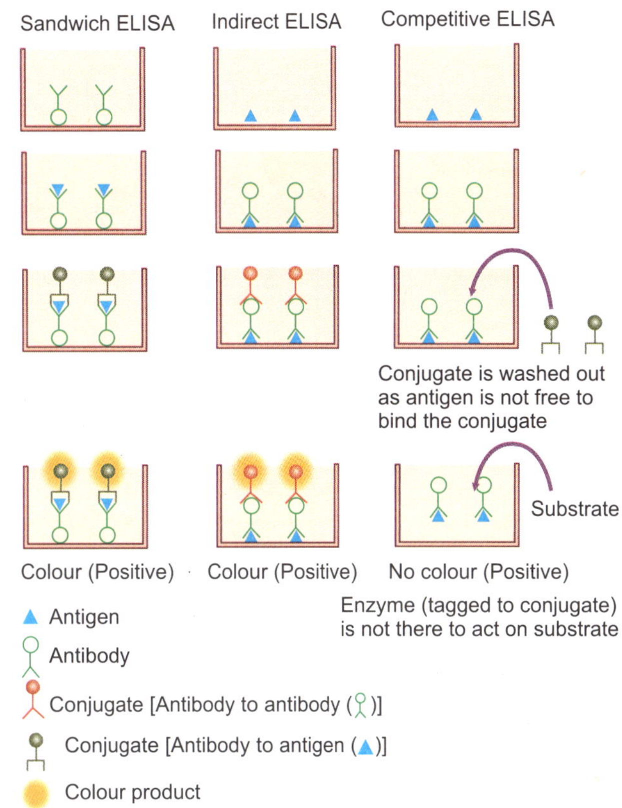

Fig. 16.7 Modifications of enzyme-linked immunosorbent assays (ELISA)

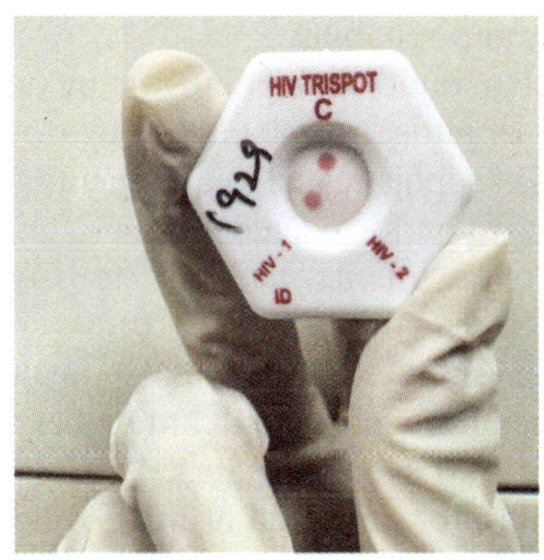

Fig. 16.8 HIV Trispot (*Source*: Maulana Azad Medical College, New Delhi)

luminol or acridinium esters) are used in CLIA as labels to signal the antigen–antibody reaction.

- The signal (light) can be amplified and measured and the concentration of the analyte calculated (Fig. 16.9).
- This method has been fully automated and is being increasingly used in laboratories where the volume of work is large. CLIA is used to detect HIV and hepatitis B/C.

Immunoelectroblot/Western Blot Techniques

Immunoelectroblot or western blot techniques combine the sensitivity of enzyme immunoassay with greater specificity. The technique is a combination of three separate procedures:

- Separation of ligand–antigen components by polyacrylamide gel electrophoresis
- Blotting of the electrophoresed ligand fraction onto nitrocellulose membrane strips
- Enzyme immunoassay (or radioimmunoassay) to:
  - Detect antibody in test sera against the various ligand fraction bands
  - Probe with known antisera against specific antigen bands

The **western blot test**, considered to be the definitive/confirmatory test for the serodiagnosis of

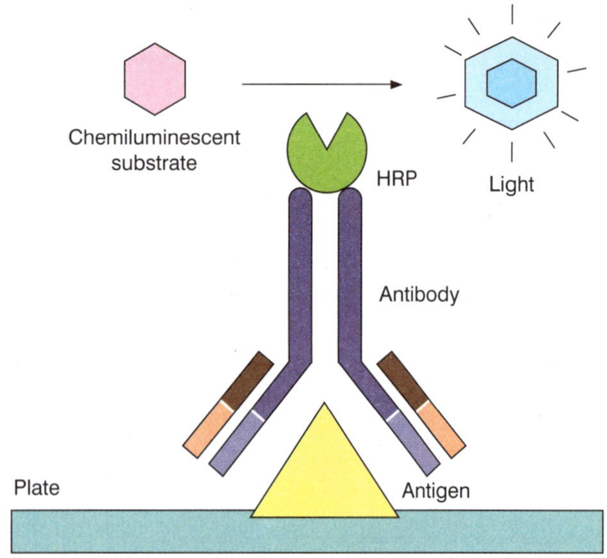

Fig. 16.9 Chemiluminescence

HIV infection, is an example of the immunoelectroblot technique.

Immunochromatographic (ICT) Assay/Lateral Flow Assay

- This is a one-step, qualitative immunochromatographic technique that has found wide application in serodiagnosis due to its rapidity, simplicity, economy, and reliability.
- These act a rapid and point-of-care tests for the detection of antigen/antibody. They do not require electricity, sophisticated equipment, or trained manpower to perform and read these tests.
- A description of its use for HBsAg detection illustrates the method (Fig. 16.10).
- The test system is a small cassette containing a membrane impregnated with antibodies tagged with a colloidal gold dye conjugate.
- The membrane is exposed at three windows on the cassette.
- The test serum is dropped into the first window. As the serum travels upstream by capillary action, a coloured band appears at the second window (test site) in a sample in which the serum antigen. This band appears due to the formation of the antigen–antibody conjugate complex and indicates a positive reaction.
- The absence of a coloured band at the test site indicates a negative reaction.
- A coloured band should appear in every case at the third window, which forms an in-built control. If the band does not appear in the third window, the test is considered invalid.
- This test is said to be nearly as sensitive and specific as EIA tests.

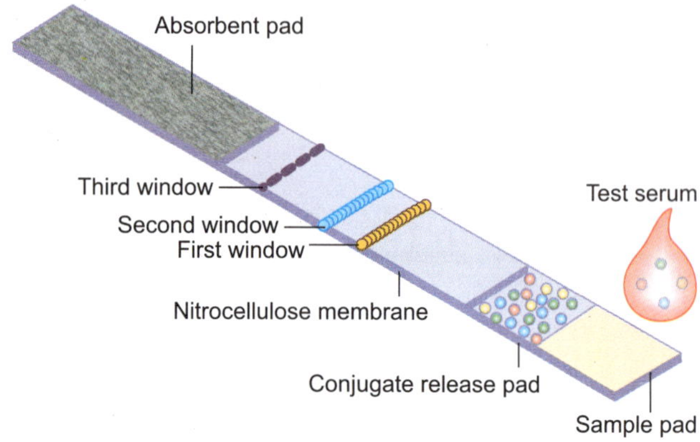

Fig. 16.10 Lateral flow immunoassay

Immunoelectron Microscopic Assays

1. **Immunoelectron microscopy:** When viral particles mixed with specific antisera are observed under the electron microscope, they appear clumped. This method finds application in the study of some viruses such as the hepatitis A virus and other viruses that cause diarrhea.
2. **Immunoferritin test:** Ferritin (an electron-dense substance from horse spleen) can be conjugated with an antibody. The reaction of such labelled antibodies with antigens can be viewed under the electron microscope.

Immunofluorescence Assay

Principle

- Fluorescence is the property of absorbing light rays of one wavelength and emitting rays of a different wavelength.
- Fluorescent dyes show up brightly under ultraviolet light because they convert ultraviolet light into visible light.
- Fluorescent dyes can be conjugated with antibodies and then used to locate and identify antigens in tissues. This 'fluorescent antibody' or immunofluorescence technique has several diagnostic and research applications (Fig. 16.11).

1. Direct immunofluorescence test

- The antigen-specific antibody labelled with a fluorescent dye is added to a slide containing the antigen (Fig. 16.12).
- A disadvantage of this method is that separate fluorescent conjugates have to be prepared against each antigen to be tested.
- Application—this method is used for the detection of rabies virus antigens in brain smears.

2. Indirect immunofluorescence test

- This test overcomes the difficulty mentioned above by using an **antiglobulin–fluorescent conjugate**.

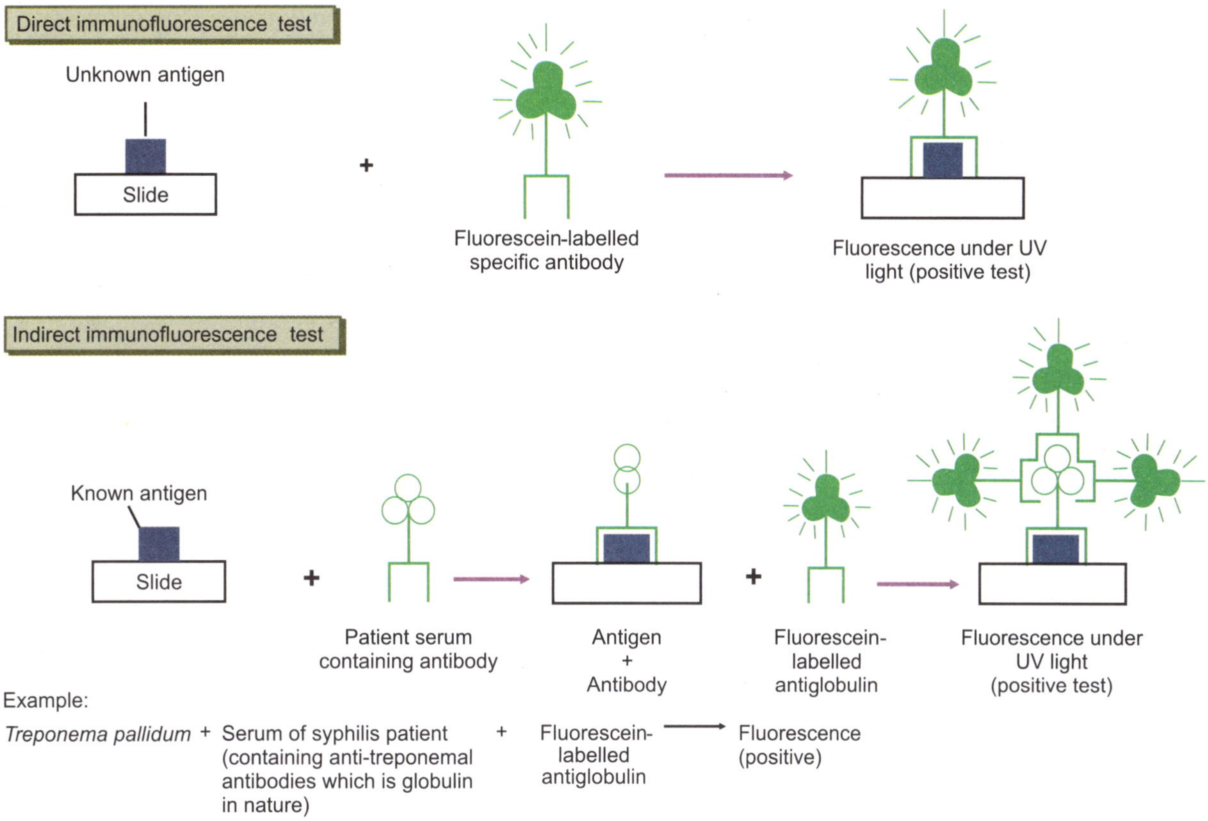

Fig. 16.11 Direct and indirect immunofluorescence tests

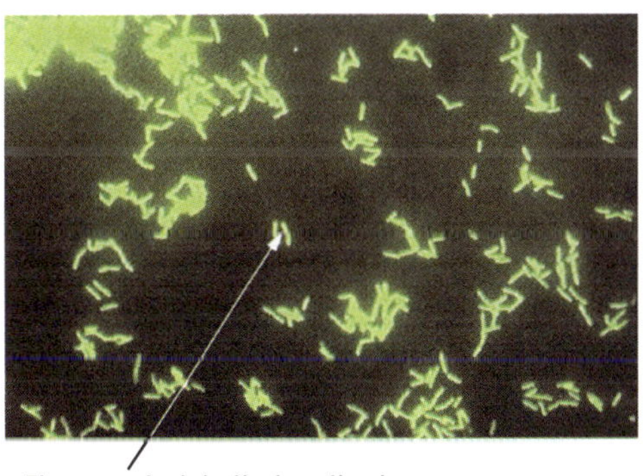

Fig. 16.12 A green fluorescent monoclonal Ab against *L. pneumophila* is used to visualise and identify bacteria from a smear of a sample from the respiratory tract of a pneumonia patient (*Source*: https://openstax.org/books/microbiology/pages/20-5-fluorescent-antibody-techniques)

- This antiglobulin conjugate binds to any antibody fixed to the antigen, thereby tagging it.
- Application—the fluorescent treponemal antibody (FTA) test is used for the diagnosis of syphilis.

Fluorescent dyes may also be conjugated with the complement. Labelled complement is a versatile tool and can be employed for the detection of an antigen or antibody. A sandwich technique of immunofluorescence is in use, in which the antibody is first allowed to react with an unlabelled unknown antigen. This complex is then treated with a fluorescent-labelled antibody. A sandwich is thus formed with the antigen in the middle and the labelled and unlabelled antibodies on its either side. This test is used to detect respiratory viruses from the throat and nasopharynx.

Immunohistochemical Techniques

- By combining the specificity of serology with the localising capacity of histology, immunofluorescence helps in the visualisation of antigen–antibody reactions in situ.
- The fluorescent dyes commonly used are fluorescein isothiocyanate and rhodamine, which exhibit blue-green and orange-red fluorescence respectively in tissue sections.

Flow Cytometry (Fig. 16.13)

- This is a fluorescence technique used to identify and enumerate cells bearing a particular antigen(s) or surface marker by suspending them in a stream of fluid and passing them through an electronic detection apparatus.
- It allows for the simultaneous, multiparametric analysis of the physical and/or chemical characteristics of up to thousands of particles per second.
- Different populations of molecules, cells, or particles can be differentiated based on size and shape using forward and right-angled light scatter.

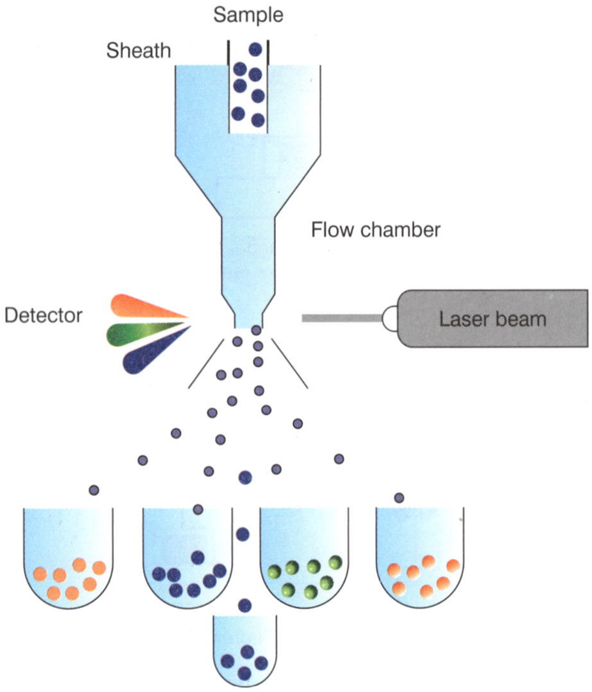

Fig. 16.13 Flow cytometry

Applications of flow cytometry

Multiple parameters can be gauged using flow cytometry, including the following:

- Size
- Granularity
- DNA or RNA content
- Cellular antigens
- Receptor levels of cells

This method is also widely used in research and diagnostics:

- To count blood cells including differential leukocyte count (DLC)
- To isolate T cell subsets (CD4 and CD8 counts in HIV patients)
- For the diagnosis, treatment, and prognosis of cancer (especially leukemias)
- To study the cell cycle and apoptosis

DETECTION OF CELL-MEDIATED IMMUNITY (CMI)

Some commonly used methods for detecting CMI are as follows:

- **T cell count** is an indirect method of detecting CMI. The absolute lymphocyte count is calculated by multiplying the total white cell count by the percentage of lymphocytes.
- T cell subsets are defined by a monoclonal antibody.
- The skin test is used for the detection of delayed hypersensitivity (e.g., the tuberculin test).
- The lymphocyte transformation test detects the transformation of cultured sensitised T lymphocytes on contact with the antigen to blast cells–blast cell transformation.
- Target cell destruction—involves the killing of cultured cells by T lymphocytes sensitised against them.
- The **migration inhibitory factor** test has been adapted for clinical use by incubating human peripheral leukocytes/macrophages in capillary tubes in culture chambers. When an antigen to which the individual has CMI is introduced into the culture medium, the leukocytes/macrophages are prevented from migrating. By comparing this with the control, it is possible to make a semi-quantitative assessment of the migration inhibition (Fig. 16.14).

Table 16.4 summarises the common immunological tests performed in the clinical laboratory.

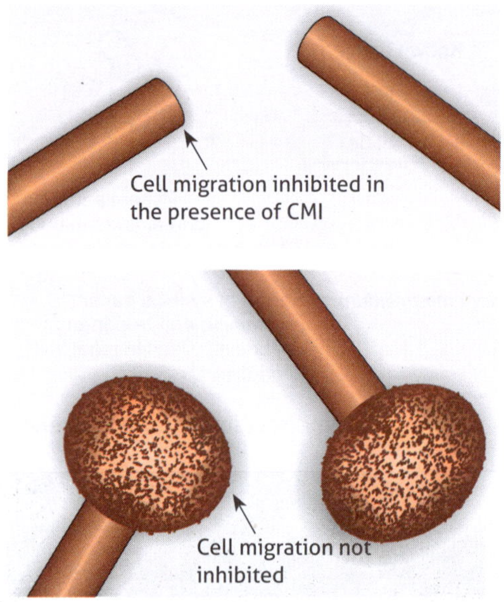

Fig. 16.14 Detection of cell-mediated immunity (CMI) by the leukocyte migration inhibition test

Table 16.4 Summary of the common immunological tests performed in a clinical laboratory

Test	Examples of applications
Precipitation	Typing and serogrouping of bacteria
Agglutination	Febrile agglutination tests Blood grouping
Immunoassays Immunochromatographic methods Lateral flow assays	Rapid antigen/antibody detection tests HIV, hepatitis B and C, malaria, dengue, SARS-CoV and several other infections
ELISAs direct/indirect/ sandwich/competitive (several other modifications)	Used to detect antigens and antibodies in most infectious diseases
Chemiluminescence immune assays (CLIA)	IgM/IgG antibodies in viral infections
Direct/Indirect immunofluorescence (FA/IFA)	Diphtheria/Respiratory virus
Western blot	HIV, TORCH panel (*Toxoplasma*, Rubella, CMV and Herpes)

Key Points

- ✓ Antigen–antibody reactions make it possible to detect, identify, and quantify (measure the concentration) antigens and antibodies. Since antibodies are present in serum, their study is known as serology.
- ✓ Antibodies may be demonstrated by various types of reactions:
  - **Precipitation reactions:** In these reactions, the antigen is in a soluble form and, on combination with the antibody, forms sediments or remains suspended in the form of floccules. Examples of diagnostic precipitation tests include the Kahn and VDRL tests.
  - **Agglutination reactions:** In these reactions, specific antibodies (agglutinins) formed in response to particulate antigens in host tissues combine with a homologous antiserum. Agglutination reactions are used for the diagnosis of infections due to salmonellae (Widal test) or brucellae (*Brucella* agglutination test).
  - **Complement fixation reaction:** In this type of reaction, an antigen combines with its (specific) antibody in the presence of complement, and the antigen–antibody complex adsorbs the complement. This reaction is not visible and requires the use of an indicator (sheep erythrocytes and the specific anti-erythrocyte antibody). Examples include the Wassermann test for syphilis and some tests for the detection of viruses.
  - **Neutralisation tests:** In these tests, the effect of the antigen, toxin, or virus is neutralised when it is mixed with its antibody. Examples include the Schick test for diphtheria toxin, anti-streptolysin O (ASO) test for streptococcal O infection, Nagler's reaction for the identification of alpha toxin of *Clostridium perfringens* and the *Treponema pallidum* immobilisation (TPI) test for *T. pallidum* in clinical specimens.
- ✓ In certain viral diseases, the antibody produced prevents the agglutination of certain red blood cells by the specific virus; this hemagglutination inhibition test can be used for the diagnosis of influenza.
- ✓ Tests based on the primary interaction between the antibody and antigen are very sensitive and specific. In radioimmunoassay, a radioisotope is used to indicate the presence of an antigen–antibody reaction.
- ✓ Enzyme immunoassay (EIA) is a detection system that involves the use of an enzyme-labelled antibody. ELISA is the most popular technique for the diagnosis of hepatitis, HIV, rotavirus, and many other viral and bacterial infections.
- ✓ Chemiluminescence refers to a chemical reaction emitting energy in the form of light.
- ✓ The immunofluorescence technique uses fluorochrome-labelled antibodies for the diagnosis of syphilis, treponemal infection, etc., in clinical samples and also for the visualisation of antigen–antibody reactions in situ by the immunohistochemical technique.
- ✓ Flow cytometry allows simultaneous multiparametric analysis of the physical and/or chemical characteristics of cells. Different populations of cells or particles can be differentiated by size and shape using forward and light scatter, where cells are made to flow in a single cell stream. It is widely used for CD4 and CD8 counts in HIV patients, the diagnosis and prognosis of cancer (especially leukemias), the study of cell cycle and apoptosis, etc.

Questions

Short answer questions

Describe the principles and applications of the following:

1. Precipitation reactions
2. Agglutination test
3. Complement fixation test
4. Neutralisation test
5. Immunofluorescence test
6. ELISA in clinical microbiology
7. Sandwich ELISA
8. Western blot technique

MCQs

1. **Which of the following antibodies have the highest avidity for binding to antigens?**
   a) IgA b) IgM
   c) IgG d) IgE
2. **Which of the following does not depict the general features of Ag–Ab reaction?**
   a) Reaction is specific
   b) Denaturation of the Ag or Ab during the reaction
   c) Entire molecule reacts
   d) Combination is firm but reversible
3. **Which of the following is an example of the immunoelectroblot technique?**
   a) Western blot b) Northern blot
   c) Southern blot d) All of these
4. **Chemiluminescence refers to a clinical reaction emitting energy in the form of ___________.**
   a) Electric current b) Chemical reaction
   c) Photons d) Ionising radiation
5. **Which of the following is not a component of ELISA?**
   a) Enzyme
   b) Substrate
   c) Immunoelectroblot
   d) Polystyrene platform
6. **All the following forces are involved in antigen–antibody reactions except:**
   a) Van der Waals forces b) Electrostatic bond
   c) Hydrogen bond d) Covalent bond
7. **Prozone phenomenon is due to:**
   a) Antigen excess
   b) Antibody excess
   c) Both antigen and antibody excess
   d) None of these
8. **All the following are true about precipitation reactions except:**
   a) Reaction of a soluble antigen with an antibody
   b) Very sensitive for the detection of antigens
   c) Elek test is an example
   d) Coombs test is an example
9. **Ouchterlony procedure is:**
   a) Single diffusion in one dimension
   b) Double diffusion in one dimension
   c) Single diffusion in two dimensions
   d) Double diffusion in two dimensions
10. **Coombs test is also known as:**
    a) Complement fixation test
    b) Agglutination test
    c) Neutralisation test
    d) Antiglobulin test
11. **Rose–Waaler test is an example of:**
    a) Latex agglutination test
    b) Reversed passive agglutination

c) Passive hemagglutination
d) Co-agglutination test

12. Which of the following parameters is not measured in flow cytometry?
a) Size b) Granularity
c) Cellular antigen d) Composition of cell wall

13. The Schick test is an example of:
a) Complement fixation test b) Co-agglutination test
c) Neutralisation test d) Passive agglutination test

14. Widal test is a type of:
a) Precipitation reaction b) Agglutination reaction
c) Neutralisation test d) Complement fixation test

15. The true statement regarding homogenous EIA is:
a) Needs separation of antigen and antibody
b) It is a multistep procedure
c) Can be used only for haptens
d) ELISA is an example

16. A 25-year-old man with history of multiple sexual partners is admitted with unexplained fever, loss of weight, and persistent diarrhea for past 6 months. His blood is drawn after counseling to detect antibody against HIV. The laboratory will test using ELISA. Which of the following statements is true regarding ELISA?
a) It uses horseradish peroxidase as a substrate.
b) It is performed in a microtiter plate having 46 wells.
c) Direct ELISA is used for detecting antibodies in test serum.
d) Indirect ELISA can be used to detect both antigen and antibody in serum.

17. Which of the following is the correct flow diagram to depict the principle of ELISA?
a) Ag–Ab complex-enzyme + substrate-----activates chromogen-----colour change----detection by spectrophotometry
b) Ag–Ab + substrate----activates enzyme-chromogen-----colour change----detection by spectrophotometry
c) Ag–Ab complex-substrate-----activates enzyme----add chromogen-----colour change----detection by spectrophotometry
d) Ag–Ab complex-enzyme-----activates chromogen-----colour change----detection by spectrophotometry

18. Another test used to confirm the findings of ELISA results in such a patient is the Western blot. Which of the following statements best describes the utility of Western blot?
a) It detects DNA fragments in patient sample.
b) It detects RNA in sample.
c) It detects specific proteins in sample.
d) None of the above.

19. Which statement is not correct regarding fluorescence?
a) It refers to the property of absorbing light rays of one particular wavelength and emitting rays of a different wavelength.
b) Fluorescent dyes convert ultraviolet light into visible light.
c) Fluorescent dyes can be conjugated to antibodies and then used to locate and identify antigens in tissues.
d) None of the above.

20. Lymphocyte transformation test is used to:
a) Measure antibodies in serum
b) Measure total B cell count
c) Detection of cell-mediated immunity
d) Identify viral infected cells.

21. Lateral flow assays are used to detect:
a) Antigen using fluorescent-labelled antibody
b) Antibody using fluorescent-labelled antigen
c) Antigen or antibody using nitrocellulose membrane
d) Antibody or antigen using chemiluminescence

Answers: 1:b, 2:b, 3:a, 4:b, 5:c, 6:d, 7:b, 8:d, 9:d, 10:d, 11:c, 12:d, 13:c, 14:b, 15:c, 16:d, 17:a, 18:c, 19:d, 20:c, 21:c

CHAPTER 17

Hypersensitivity

(MI1.10)

Chapter Navigation

- Type I (Immediate anaphylactic hypersensitivity)
- Type II (Cytolytic/cytotoxic hypersensitivity)
- Type III (Immune complex-mediated hypersensitivity)
- Type IV (Delayed-type T cell-mediated hypersensitivity)

Learning objective

- Describe immunological mechanisms of hypersensitivity and methods of laboratory diagnosis

INTRODUCTION

- Immune response is generally aimed at removing antigens using various effector molecules and without damaging the host tissue. However, in some instances, the immune system can have harmful effects like an **exaggerated inflammatory response or tissue injury, sometimes even leading to death**. This heightened response is known as hypersensitivity. Such a reaction generally occurs during the second contact with the specific antigen (**allergen**).
- Hypersensitivity is defined as a **set of undesirable injurious consequences in the sensitised host following contact with specific antigens**. The antigens are often innocuous substances such as serum proteins or pollen.
- Hypersensitivity is concerned with what happens to the host as a result of an inappropriate immune reaction.
- The host should have had **prior contact** with the antigen (allergen) for a hypersensitivity reaction to occur. The initial contact sensitises the immune system, leading to the priming of the appropriate B or T lymphocytes. This is known as the **sensitising or priming dose**. Subsequent contact (known as the **shocking dose**) with the allergen results in manifestations of hypersensitivity.

Table 17.1 Distinguishing features of immediate and delayed types of hypersensitivity

Features	Immediate hypersensitivity	Delayed hypersensitivity
Appearance and duration of reaction	Appears and recedes rapidly	Appears slowly, lasts longer
Mode of introduction	Induced by antigens or haptens by any route	Induced by an antigen or hapten introduced intradermally or with Freund's adjuvant or by skin contact
Mechanism	This is an antibody-mediated reaction involving circulating antibodies	The reaction is cell-mediated; circulating antibodies may be absent
Transference	Passive transfer is possible with the transfer of serum to a recipient	Transfer is possible with the transfer of T cells or transfer factor and cannot be transferred with serum
Desensitisation	Easy but short-lived	Difficult but long-lasting

CLASSIFICATION OF HYPERSENSITIVITY

Based on the Time Required for Manifestation

Based on how quickly a sensitised host develops clinical reactions on re-exposure to the antigen, hypersensitivity is classified as:

1. **Immediate hypersensitivity**
2. **Delayed hypersensitivity**

The chief differences between these types of hypersensitivity reactions are presented in Table 17.1.

Based on Mechanism

- Hypersensitivity reactions are classified into four types (Table 17.2) based on the different mechanisms of pathogenesis (originally described by **Philip Coombs and Robert Gell** in 1963).
- The four types of immunopathogenic mechanisms described here are not mutually exclusive. Any given hypersensitivity reaction may consist of components of more than one or all of these mechanisms.
- The pathology and clinical features of immunological diseases of hypersensitivity are also influenced by the humoral amplification system, which collectively contribute to many of the non-immune body mechanisms such as inflammation, complement involvement, coagulation, fibrinolytic, and kininogenic systems.

Table 17.2 Types of hypersensitivity reactions and their features

Type of reaction	Clinical syndrome	Time required for manifestation	Mediators
Type I: Immediate anaphylactic hypersensitivity (IgE-mediated)	1. Anaphylaxis 2. Atopy • Respiratory: Asthma, eczema, hay fever • GI: Allergy to shellfish (Familial or genetic predisposition)	Minutes Minutes	IgE, histamine, and other pharmacological agents
Type II: Cytolytic and cytotoxic	Antibody-mediated damage—thrombocytopenia—agranulocytosis, hemolytic anemia, etc.	Variable—hours to days	IgG, IgM, complement
Type III: Immune complex-mediated hypersensitivity	1. Arthus reaction 2. Serum sickness	Variable—hours to days Days	IgG, IgM, complement, leukocytes
Type IV: Delayed hypersensitivity (T cell-mediated)	1. Tuberculin 2. Contact dermatitis	Hours to days Hours to days	T cells, lymphokines, macrophages

TYPE I REACTIONS: IMMEDIATE ANAPHYLACTIC HYPERSENSITIVITY

Case 1. A 45-year-old farmer was brought to the emergency department following a bee sting with shortness of breath, a pounding sensation in his head, and tightness in his chest. He had a history of being stung by bees on two earlier occasions. The clinical diagnosis is of ***cutaneous anaphylaxis****. Intramuscular adrenaline and intravenous antihistamines were administered. The patient recovered completely by the following day.*

i. What is the mechanism of anaphylaxis?
ii. What are the mediators of anaphylaxis?
iii. How do antihistamines help in treatment?

The above case is classic example of type I hypersensitivity, which can occur in two forms:

- **Anaphylaxis:** Acute and potentially fatal with systemic manifestation
- **Atopy:** Chronic or recurrent, non-fatal, and typically localised

Principle of Type I Hypersensitivity

- These immunological reactions are rapid, occurring within minutes after exposure to the antigen.
- Anaphylactic hypersensitivity is **mediated by IgE** and involves the **degranulation of mast cells or basophils**.
- Antibodies ('cytotropic' IgE antibodies) are fixed on the surface of tissue cells (mast cells and basophils) in sensitised individuals. The antigen combines with the cell-fixed antibody, leading to the release of pharmacologically active substances (**vasoactive amines**), which produce the clinical reaction.

Anaphylaxis

- This is the typical immediate hypersensitivity reaction, which occurs within seconds of the antigen combining with IgE bound to mast cells, basophils, and possibly, eosinophils. The sequence of events involved in an anaphylactic reaction are illustrated in Fig. 17.1.
- There should have been prior exposure to the antigen with an interval of **at least 2–3 weeks between the sensitising dose and the shocking dose**. Once sensitised, the individual remains so for long periods.
- The shocking dose is most effective when injected intravenously, less effective intraperitoneally or subcutaneously, and least effective intradermally.

Signs and symptoms

- Tissues or organs predominantly involved in the anaphylactic reaction are known as '**target tissues**' or '**shock organs**'.
- The manifestations of anaphylactic shock begin with itching of the scalp and tongue, flushing of skin all over the body, and difficulty in breathing due to bronchial spasm. Nausea, vomiting, abdominal pain, and diarrhea, sometimes with blood in the stool, may be present.
- Other changes seen in anaphylaxis are edema, decreased coagulability of blood, drop in blood pressure and temperature, leukopenia, and thrombocytopenia.
- Acute hypotension, loss of consciousness, and death may follow.
- Human anaphylaxis, once commonly associated with heterologous serum therapy, is now chiefly

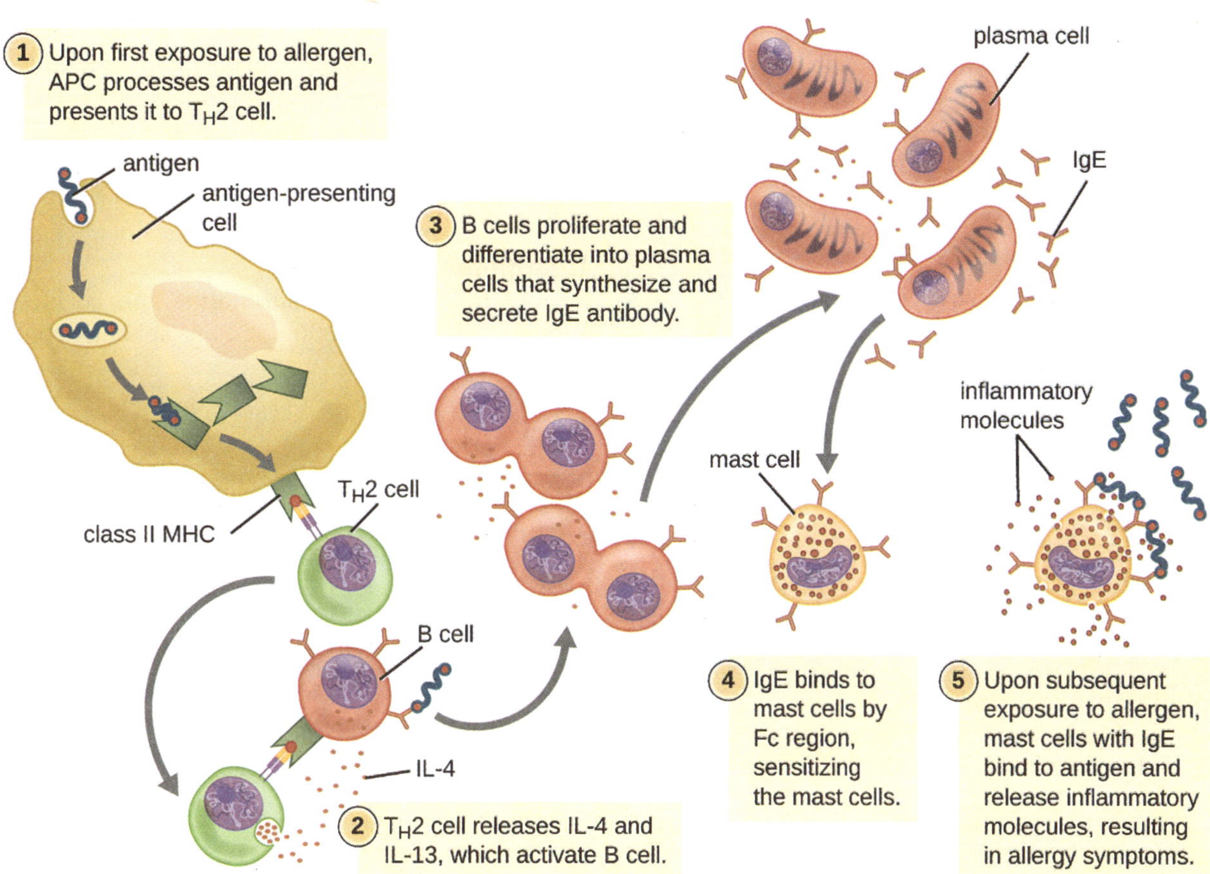

Fig. 17.1 Type I reaction—immediate anaphylactic hypersensitivity reaction mediated by IgE (*Source*: https://openstax.org/books/microbiology/pages/19-1-hypersensitivities-19-3)

seen following the **injection of antibiotics or other drugs**. **Insect stings** can also cause anaphylaxis in human beings.

Types of anaphylaxis

Cutaneous anaphylaxis

- A **wheal-and-flare response** (local anaphylaxis) occurs when an antigen is administered intradermally to a sensitised host.
- A wheal is a pale, central area of puffiness caused by edema, which is surrounded by a flare caused by hyperemia and subsequent erythema.
- This reaction is used to test for type I hypersensitivity and in identifying the allergen responsible for atopic diseases. In highly sensitised individuals, even the skin test may lead to serious and even fatal reactions.

Passive cutaneous anaphylaxis (PCA)

- This is an vivo method for the detection of antibodies that can passively lead to cutaneous anaphylaxis.
- A small volume of the antibody is injected intradermally into a healthy animal. If the antigen, along with a dye (such as Evans blue), is injected intravenously 4–24 hours later, there is immediate blue colouration at the site of intradermal injection of the antibody due to vasodilatation and increased capillary permeability (wheal-and-flare reaction).

Anaphylaxis in vitro

- Isolated tissues such as intestinal or uterine muscle strips from sensitised guinea pigs held in a bath of Ringer's solution will contract vigorously on the addition of the specific antigen to the bath. This is known as the **Schultz–Dale phenomenon**.
- Tissues from healthy animals can be passively sensitised by treatment with serum from sensitised animals.

Mechanism of anaphylaxis

- Anaphylaxis is an IgE-mediated hypersensitivity reaction.
- Cell-fixed antibody IgE is important since free IgE has no role.

Anaphylaxis occurs in the following manner (Fig. 17.1):

- **Step 1: Sensitisation phase**
  - During the first exposure to an allergen (antigen), plasma cells secrete IgE in response to activation by the allergen-specific T_{h2} cells.
  - These IgE molecules bind to the surface F_c epsilon receptors ($F_{cε} \rightarrow F_{cε}RI$) of tissue mast cells and blood basophils.

- $F_{c\varepsilon}$RI is a tetrameric membrane protein complex with high-affinity immunoglobulin receptors on its surface, which is present on the surface of mast cells and basophils. These receptors are analogous to the TCR receptors on the T cell surface.
- IgE molecules attach to these receptors by their F_c end.

- **Step 2: Effector phase**
  - Exposure to the same allergen causes crosslinking of the cell-bound IgE, which leads to the bridging of the gap between adjacent antibody molecules.
  - This crosslinking increases the permeability of the cells to calcium ions and leads to degranulation, releasing biologically-active substances contained in the granules.

Pharmacological mediators

The primary mediators of type I hypersensitivity are the preformed contents of mast cells and basophils, which are listed in Table 17.3.

Anaphylactoid reaction

The intravenous injection of peptone, trypsin, and certain other substances provokes a clinical reaction resembling anaphylactic shock. This is termed an anaphylactoid reaction. The only difference between this and anaphylaxis is that anaphylactoid shock has no immunological basis and is a non-specific mechanism involving the activation of complement and the release of anaphylatoxins.

Treatment

Prompt treatment with **adrenaline** can be lifesaving (0.5 mL of a 1-in-1,000 solution, subcutaneously or intramuscularly).

Atopy

Case 2. *A young girl was brought by her mother to the OPD with wheezing and itching of the nose and eyes. Her mother said that the girl had brought in a kitten from the streets four weeks ago. The symptoms appeared two weeks ago and kept worsening. The girl's brother too exhibited similar symptoms. The girl was prescribed antihistaminics and advised to limit physical contact with the kitten.*

i. What is the provisional diagnosis?
ii. How is the condition diagnosed and treated?

Table 17.3 Primary mediators of type I hypersensitivity

Mediator	Features	Function
Histamine	Most important vasoactive amine in human anaphylaxis Formed by the decarboxylation of histidine found in the granules of mast cells, basophils, and platelets	Stimulation of sensory nerves when released into the skin, producing burning and itching sensations Causes vasodilatation and hyperemia by an axon reflex (flare effect) and edema by increasing capillary permeability (wheal effect) Histamine-inducing smooth muscle contraction in diverse tissues and organs
Serotonin (5-hydroxytryptamine)	Base derived by the decarboxylation of tryptophan Found in the intestinal mucosa, brain tissue, and platelets	Smooth muscle contraction, increased capillary permeability, vasoconstriction
Chemotactic factors: Eosinophil chemotactic factors of anaphylaxis (ECF-A) Neutrophilic chemotactic factor (NCF)	Acidic tetrapeptides released from mast cell granules	Chemotactic for eosinophils contributing to the eosinophilia Attracts neutrophils
Leukotrienes (LTB4, C4, D4 and E4) Prostaglandin F2α and thromboxane A2	Formed from disrupted cell membranes of mast cells and other leukocytes	LTB4, C4, D4, and E4: Slow, sustained contraction of smooth muscles Powerful but transient bronchoconstrictors PG—affects secretion by mucous glands, platelet adhesion, permeability and dilatation of capillaries, and the pain threshold
Enzymatic mediators (proteases and hydrolases)	Released from mast cell granules	Catabolism of proteins and large molecules
Platelet-activating factor (PAF)	PAF is a low-molecular-weight lipid released from basophils	Aggregation of platelets and the release of their vasoactive amines

Note: Several biologically active substances such as the anaphylatoxins released by complement (C3a, C5a) activation, bradykinin and other kinins formed from plasma kininogens also mediate anaphylaxis.

- Case 2 is an illustration of atopy. **Atopy refers to naturally occurring familial hypersensitivities of human beings, typified by hay fever and asthma.**
- The antigens commonly involved in atopy—called **atopens**—are characteristically the following:
  - **Environmental inhalants** (e.g., pollen and house dust, animal furs)
  - **Ingestants** (e.g., eggs and milk)
  - **Contact allergens** to which the skin and conjunctiva may be exposed

Predisposition to atopy is genetically determined and probably linked to MHC genotypes. Atopy, therefore, runs in families. The tendency to produce IgE antibodies in unusually large quantities is inherited.

The atopic antibody **does not pass through the placenta**.

Mechanism

- Atopens induce IgE antibodies, formerly termed **'reagin' antibodies**.
- Atopic sensitisation develops spontaneously following natural contact with atopens. It is difficult to induce atopy artificially.
- All individuals are capable of forming IgE antibodies in small amounts. In atopics, however, IgE response is preponderant. Around 10 per cent of people have the tendency to overproduce IgE.
- Symptoms of atopy are produced by the release of pharmacologically active substances following the combination of the antigen and the cell-fixed IgE.

Atopy and IgA deficency

- Atopic sensitivity is also associated with the deficiency of IgA—this association has led to the suggestion that IgA deficiency may predispose a person to atopy.
- In healthy individuals, the inhalant and ingestant antigens are dealt with by the IgA that line the respiratory and intestinal mucosa and therefore, they do not come into contact with potential IgE-producing cells.
- When IgA is deficient, the antigens cause massive stimulation of IgE-forming cells, leading to the overproduction of IgE.

Clinical expression of atopic reaction

- This is usually determined by the portal of entry of the antigen.
- Manifestations include asthma, hay fever, conjunctivitis, rhinitis (these are precipitated by pollens and are usually seasonal), dust, fur from pet animals, gastrointestinal symptoms, and dermatitis after contact with the eyes, respiratory tract, intestine, or skin.

Detection of Type I Hypersensitivity

i. **Skin test:** Small amounts of the suspected allergen are introduced by intradermal injection or superficial scratching. A wheal-and-flare within 30 minutes confirms allergy (Fig. 17.2a and b). The local mast cells release histamine and other mediators.

ii. **Radioallergosorbent test (RAST):** The total levels of serum IgE can be determined by this test. IgE, the main antibody responsible for allergic reactions, can be estimated in serum by using the fluorescent-labelled anti-IgE by the ELISA method.

iii. **Radioimmunosorbent test (RIST):** This method is used to estimate the total circulating IgE antibodies.

Treatment of Type I Hypersensitivity

- **Acute anaphylaxis or allergy:** Epinephrine (adrenalin), antihistamines (block H1 and H2 receptors), and corticosteroids (reduce histamine levels) are effective in the treatment of anaphylaxis. The doses are adjusted according to body weight.

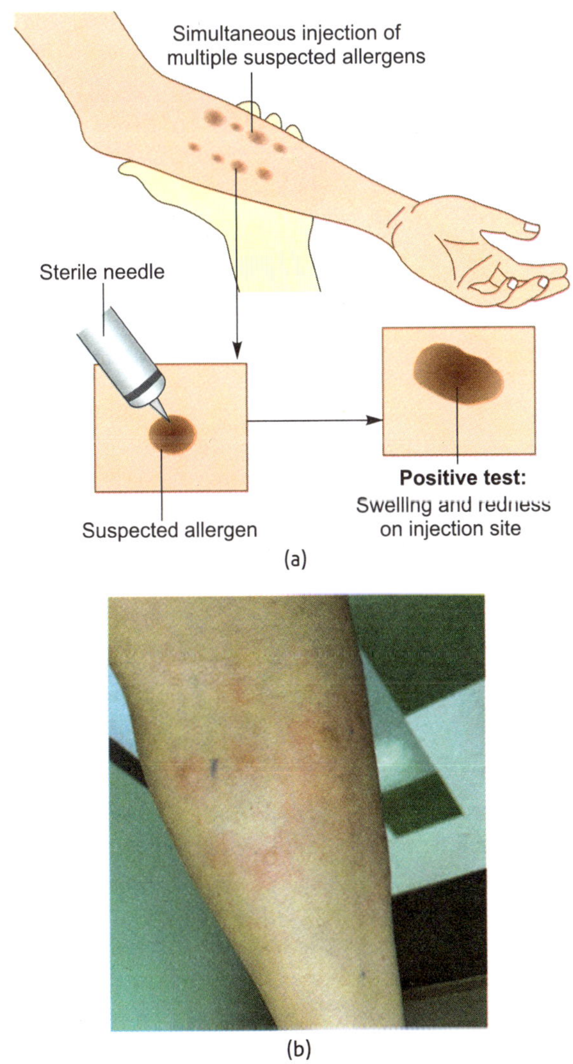

Fig. 17.2 (a) Skin test to demonstrate immediate hypersensitivity and (b) wheal-and-flare reaction on the forearm seen in an individual allergic to pollen and cat fur seen within 30 minutes of injecting minute doses of the allergens (*Source*: [b] Dr Reba Kanungo, personal collection)

- **Atopy:** Atopic persons can be treated by desensitisation, wherein small doses of the allergen are injected by the intradermal route over a period of time. The person develops anergy, and over time, tolerance.

TYPE II REACTIONS

> Case 3. *A 36-year-old man injured in a road traffic accident was brought to the casualty department. The patient was hemodynamically unstable and was transfused with one unit of blood. After a few minutes of starting blood transfusion, he became nauseated, experienced shortness of breath, sweating, and dizziness. On examination, he is febrile, his SpO_2 is 94% on room air, and he was hypotensive. The direct Coomb's test was positive.*
> i. Why did the patient deteriorate on being given blood transfusion?
> ii. What is the mechanism behind this condition?

- In the above case, the patient had a hemolytic transfusion reaction, which is a type II hypersensitivity reaction.
- It occurs within minutes to hours of exposure to a foreign red cell antigen.
- The mechanism of this reaction is antibody-mediated cell damage.
- This type of reaction is initiated by IgG (or rarely, IgM) antibodies that react either with cell-surface or tissue antigens. Cell or tissue damage occurs in the presence of complement or mononuclear cells.
- Type II reactions are intermediate, falling between hypersensitivity and autoimmunity.
- Combination with the antibody may, in some instances, cause stimulation instead of damage. In stimulatory hypersensitivity (referred to as type V), antibodies recognise and bind to the cell-surface receptors. This is a modified form of type II reactions.
- Type II reactions are **cytolytic and cytotoxic reactions** (antibody-mediated cytolytic and cytotoxic hypersensitivity).

Mechanism

- A combination of IgG and IgM antibodies, which are directed towards antigenic determinants on cell membranes (e.g., blood group antigens) or the extracellular matrix, leads to complement activation and consequently, cytotoxic or cytolytic effects.
- Examples: Lysis of red cells caused by anti-erythrocyte antibodies (e.g., in mismatched blood transfusion), autoimmune anemias, and hemolytic disease of the newborn (erythroblastosis fetalis).
- Alternatively, a free antigen or hapten may be absorbed onto cell surfaces. The subsequent reaction of the combined antigen or hapten with its corresponding antibody leads to cell damage. Many drugs may behave in this manner, leading to the complement-mediated lysis of red cells, leukocytes, and platelets, causing hemolytic anemias, agranulocytosis, and thrombocytopenic purpura.
- **Antibody-dependent cell cytotoxicity (ADCC)** occurs when cytotoxic cells are stimulated by the antigen–antibody complex. This may occur with ot without the help of complement (Fig. 17.3 and 17.4).

Some diseases involving type II reactions are as follows:

- **Transfusion reactions** are the commonest example of type II reaction-related complications. Antibodies to A, B, and O groups are called isohemagglutinins (IgM class). An A group individual possesses antibodies to antigenic epitopes on the B group. When the A group individual is transfused with B group blood, the anti-B antibodies cause a type II reaction. Symptoms of mismatched blood transfusion appear immediately. There is massive intravascular hemolysis, whereby hemoglobin is

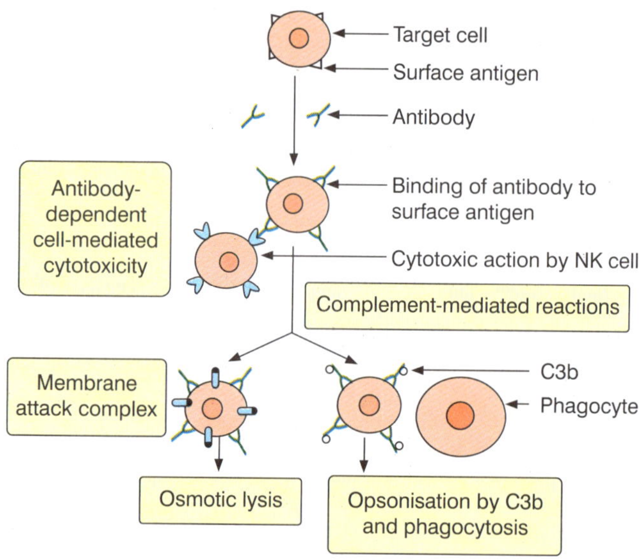

Fig. 17.3 A schematic diagram showing type II hypersensitivity

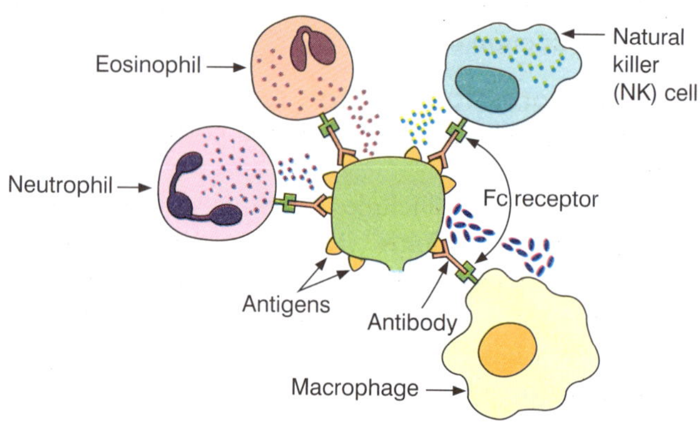

Fig. 17.4 Antibody-dependent cell-mediated cytotoxicity (ADCC)

released into plasma. Hemoglobulinuria also occurs in such a reaction.

- **Autoimmune hemolytic anemia and pernicious anemia** are caused by type II reactions.
- **Erythroblastosis fetalis**—in the hemolytic disease of the newborn, maternal IgG for the fetal blood group crosses the placenta to react with the fetal RBC when the latter has a different (Rh) antigenic epitope.
  - **Autoimmune thrombocytopenic purpura** involves type II reactions.
  - **Drug-induced hemolytic anemia** results from some drugs (such as penicillin, cephalosporins, and streptomycin), which act as haptens. They bind to the tissue or circulating proteins. By combining with carrier proteins, these haptens become complete antigens to react with antibodies, leading to hypersensitivity.
  - **Myasthenia gravis**—antibodies bind to the acetylcholine receptors on the motor end plates of muscles, causing the complement-mediated lysis of these cells and resulting in the weakening of skeletal muscles.

Type V Hypersensitivity

- An additional type of hypersensitivity, considered a subtype of type II, is a stimulatory type, sometimes known as type V hypersensitivity.
- In this type of reaction, IgG binds to the cell surface antigen receptors and has a stimulatory effect. This mechanism is seen in **Graves' disease**, in which antibodies are produced against receptors for the thyroid stimulating hormone.

TYPE III REACTIONS: IMMUNE COMPLEX-MEDIATED HYPERSENSITIVITY

Case 4. *A 19-month-old girl presented with marked leg swelling, pain, and inability to walk within six hours of receiving a booster of DPT. There was no significant history. She had received all her vaccinations on time and had no developmental delays. On examination, she was febrile with local tenderness, but no abscess was seen at the site of injection. The infant was given an antipyretic. Her condition improved in two days.*

i. What is the diagnosis in this case?
ii. What is the pathophysiological basis of this reaction?

In Case 4, damage is caused by antigen–antibody complexes, resulting in complement activation. This type of reaction is known as type III hypersensitivity reaction.

- The antibody in these cases is a high-affinity IgM or IgG that binds to the antigen.
- After combination, the antigen–antibody units form complexes with the complement.
- These complexes may precipitate in and around small blood vessels, causing damage to cells secondarily or on membranes, interfering with their function. They may relate to tissue deposition (**Arthus phenomenon**), e.g., vasculitis (skin), serum sickness (systemic), nephritis (kidneys), and extrinsic allergic alveolitis (lungs).

Mechanism

- In immune reactions, the antigen and antibody combine and take up the complement, thereby forming an immune complex.
- These complexes get deposited in the tissues and stimulate an inflammatory reaction.
- There are chiefly two types of systemic immune complex-mediated hypersensitivity—**serum sickness and local type II reactions**.

i. Serum sickness

- It is a systemic immune complex disease.
- It appears 7–12 days after a single injection of a high concentration of foreign serum such as the anti-diphtheritic serum in an acute case of diphtheria.

Pathogenesis

- Serum sickness occurs due to the formation of immune complexes (consisting of the foreign serum [antigen], the host antibody, and complement components [C5a, C4a, C3a]). This complex gets deposited on the endothelial lining of blood vessels in various parts of the body, causing inflammatory infiltration (Fig. 17.5).

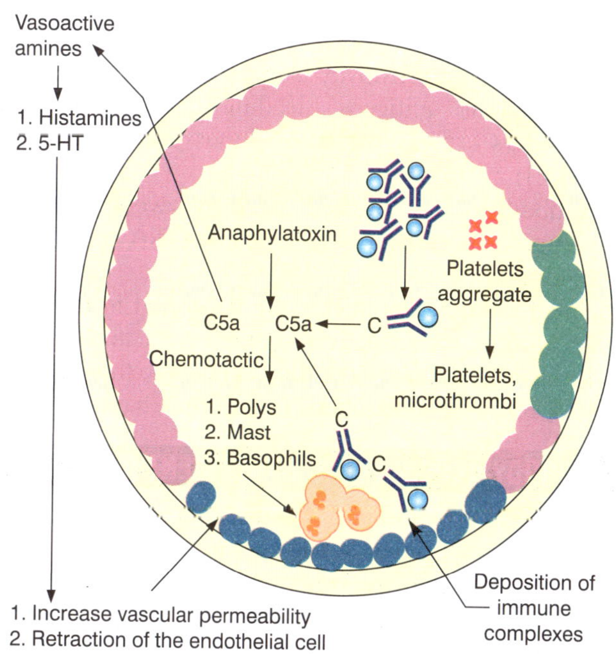

Fig. 17.5 Type III reaction (mechanism of immune complex-mediated hypersensitivity)

- It differs from other types of hypersensitivity reactions in that a single injection can serve both as the sensitising dose and the shocking dose. This syndrome is seen following the injection of penicillin or other antibiotics.

Clinical syndrome

- The typical features of serum sickness include fever, vasculitis, lymphadenopathy, splenomegaly, urticarial rashes, nausea, and vomiting.
- In addition, there may be arthritis, glomerulonephritis, and endocarditis.

ii. Local type III reaction

Arthus reaction

- Local manifestation of generalised type III hypersensitivity.
- Tissue damage is caused by the formation of antigen–antibody complexes, which cause complement activation and the release of inflammatory molecules. This leads to increased vascular permeability and infiltration of neutrophils.
- Leukocyte–platelet thrombi are subsequently formed, which occlude blood vessels, reducing blood supply and leading to tissue necrosis. Arthus reaction can be passively transferred with sera containing precipitating antibodies (IgG or IgM) in high titres.
- This reaction forms a pathogenic component of many clinical syndromes. For example, in an **intrapulmonary Arthus-like reaction** to inhaled antigens, thermophilic actinomycetes from mouldy hay or grain cause **farmer's lung** and other types of hypersensitive pneumonitis.

Autoimmune disorders

- Type III hypersensitivity also develops due to some endogenous antigens, leading to autoimmune disorders (e.g., systemic lupus erythematosus [SLE]).
- Antibodies are directed against several cellular components of host tissue such as DNA and DNA-binding proteins.
- Small immune complexes are deposited in the skin and joints, which activate complement and initiate local tissue damage including damage to the vascular bed, resulting in vasculitis (details in Chapter 18).

ANTIBODY-DEPENDENT ENHANCEMENT (ADE)

- In certain conditions, antibodies to a number of viral pathogens, formed during the process of infection, cause enhanced virus entry and replication in several of the cell types.
- Antibody-dependent enhancement occurs when virus–antibody immune complexes interact with cells bearing complement or F_c receptors.
- The internalisation of the virus is enhanced, leading to the progression of the infection and causes the exacerbation of the viral disease.
- ADE of infection may present obstacles to the prevention of viral diseases by vaccination, the primary aim of which is to induce a protective antibody response in individuals. It is postulated that vaccine-induced hypersensitivity may be a result of one such mechanism. However, a lot remains to be known about pathways and epitopes responsible for mediating ADE.

TYPE IV REACTIONS: T CELL-MEDIATED HYPERSENSITIVITY (DELAYED-TYPE)

> **Case 5.** *A 19-year-old college student developed a severe pruritic rash on the dorsal sides of both hands a week after applying* mehendi *(henna) for a wedding. The rash was erythematous and papulovesicular with ulceration and scabbing along the pattern of the* mehendi. *A diagnosis of* ***contact dermatitis*** *was made.*
>
> i. What is the pathogenesis of contact dermatitis?
>
> ii. What type of hypersensitivity is this?

- Type IV hypersensitivity reaction (Fig. 17.6) or delayed hypersensitivity is a cell-mediated response seen in tuberculin test and contact dermatitis.
- The antigen activates specifically sensitised CD4 and CD8 T cells, leading to the secretion of lymphokines and phagocyte accumulation, which in turn, manifest as delayed-type of hypersensitivity (DTH).
- It is a cell-mediated immune response.
- Type IV hypersensitivity is provoked by intracellular microbial infections or hapten-like chemicals applied on the skin.
- The reaction evolves slowly and consists of a mixed cellular reaction involving lymphocytes and macrophages. It is induced by sensitised T cells (T_{dth}, T_{h1}, T_{h2}, T_c) which, on contact with the specific antigen, release cytokines. These, in turn, produce biological effects on leukocytes, macrophages, and tissue cells.
- Delayed hypersensitivity cannot be passively transferred by serum (as it is not antibody-mediated) but can be transferred by sensitised lymphocytes or the transfer factor (TF).

Types of T Cell-Mediated Delayed Hypersensitivity

Two types of delayed hypersensitivity are recognised: **(i) the tuberculin (infection) type and the (ii) contact dermatitis type**.

i. Tuberculin (infection) type of delayed hypersensitivity

- When a small dose of tuberculin is injected intradermally in an individual sensitised to

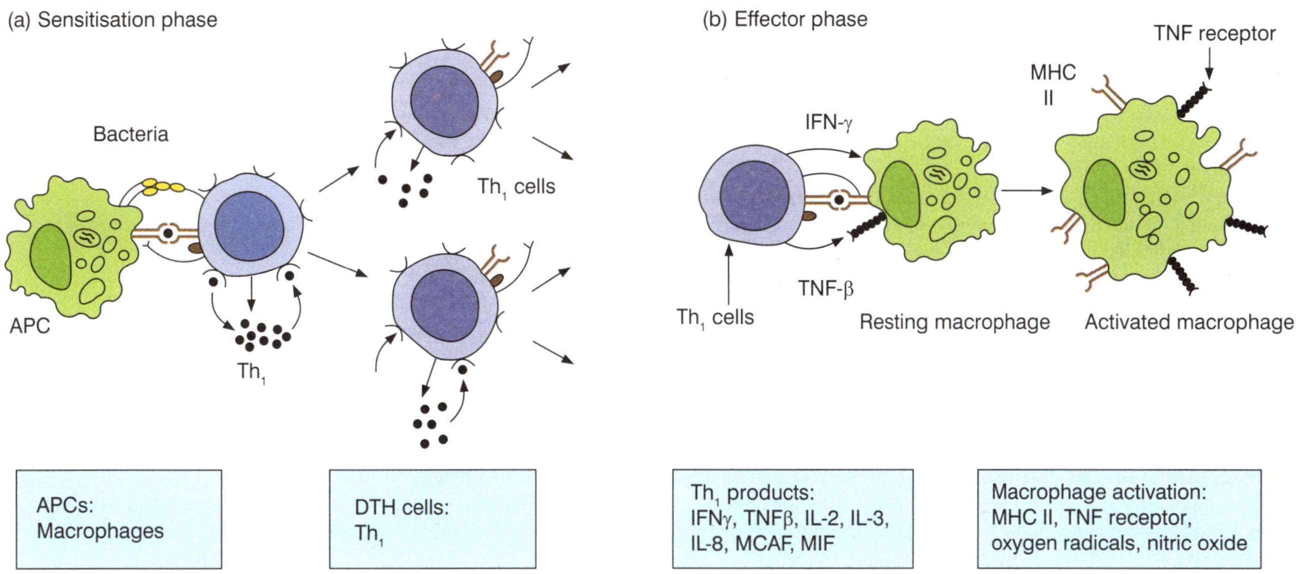

Fig. 17.6 Stages of type IV DTH

tuberculoprotein either by prior infection or immunisation, an indurated inflammatory reaction develops at the site within 48–72 hours.

- In unsensitised individuals, the tuberculin injection provokes no response.
- The tuberculin test, therefore, provides a useful indication of the state of delayed hypersensitivity (cell-mediated immunity) to the bacilli.
- Similar hypersensitivity is seen in several subacute or chronic bacterial, fungal, viral, and parasitic infections (intracellular pathogens), allograft reactions, and several autoimmune diseases.

Cutaneous basophil hypersensitivity

- A local reaction resembling the tuberculin response may be produced by the intradermal injection of some protein antigens.
- This is not a delayed hypersensitivity reaction as it can be passively transferred by serum.
- Its histology is different from that of the tuberculin response and is characterised by prominent **basophil infiltration**.
- This was formerly known as the Jones–Mote reaction but is now termed cutaneous basophil hypersensitivity.
- Its significance is not known.

ii. Contact dermatitis

- Delayed hypersensitivity sometimes results from skin contact with a variety of chemicals—metals such as nickel and chromium, dyes, dinitrochlorobenzene, drugs such as penicillin, plant allergens (poison ivy or parthenin from parthenium), and toiletries to which the individual has been sensitised earlier.
- Antibiotic ointments applied on patches of dermatitis frequently provoke sensitisation.
- The substances involved are in themselves not antigenic but may acquire antigenicity on combining with skin proteins.
- Sensitisation requires percutaneous absorption. As most of the substances involved are fat-soluble, passage along the sebaceous glands may be the method of entry of the allergens.

Mechanism of contact dermatitis

- Langerhans' cells of the skin capture the locally applied hapten along with the modified tissue proteins and migrate to the draining lymph nodes, where they present the processed antigen along with MHC molecules to T cells.
- The sensitised T cells travel to the skin site, where on contact with the antigen, they release various lymphokines. T_{h1} cells secrete IFN-γ and IL-2, which activate macrophages and other lymphocytes.
- Activated T_c cells mediate the killing of target cells.

Symptoms

Contact dermatitis lesions range from macules and papules to vesicles that break down, leaving behind raw, weeping areas, typical of acute eczematous dermatitis.

Detection using 'patch test'

- The allergen is applied to the skin under an adherent dressing.
- Sensitivity is indicated by the appearance of itchiness in 4–5 hours and local reaction, which may vary from erythema to vesicle or blister formation after 24–28 hours.

Key Points

✓ Hypersensitivity reactions are pathologic processes that result from exaggerated (very vigorous) specific interactions between antigens (exogenous or endogenous) and either humoral antibodies or sensitised lymphocytes, resulting in tissue injury.

✓ These reactions occur only if the host has had a previous instance of contact with the antigen (allergen). The initial or sensitising dose sensitises the immune system; subsequent contact (shocking dose) with the same antigen (allergen) causes manifestations of hypersensitivity.

✓ Hypersensitivity reactions can be classified as follows:
- Based on the time between exposure to antigen and the reaction, hypersensitivity is classified into immediate-type hypersensitivity and delayed-type hypersensitivity.
- Based on the component involved in the reaction, it is classified into antibody-mediated hypersensitivity and cell-mediated hypersensitivity
- Hypersensitivity is classified into type I, type II, type III, and type IV (based on Gell and Coombs classification).

✓ Type I (immediate, anaphylactic), IgE-mediated hypersensitivity reactions occur within minutes of exposure to the antigen. The crosslinking of mast cell-bound IgE with the allergen causes the release of vasoactive substances (histamine, leukotrienes, prostaglandins, etc.), which produce inflammation. Diagnostic tests include total IgE levels, direct skin test, radioallergosorbent test, leukocyte histamine release test, provocative challenge, and provocative food testing. Type I reactions include systemic anaphylaxis, allergic extrinsic asthma, seasonal allergic rhinitis, and reactions to insect stings.

✓ Type II (cytotoxic) hypersensitivity reactions, where antibody binding to cell surface antigens is followed by antibody-dependent cell-mediated cytotoxicity by K cells or complement-mediated lysis include transfusion reactions, hemolytic disease of the newborn, and certain drug allergies caused by cytotoxic T lymphocytes.

✓ Type III (immune complex-mediated) hypersensitivity reactions occur due to the activation of complement by immune complexes (through the classical pathway). These reactions are seen with malarial parasites and filarial worms; other examples include glomerulonephritis, alveolitis, and certain autoimmune diseases.

✓ Type IV (cell-mediated or delayed-type) hypersensitivity reactions manifest with symptoms at least 24–48 hours after exposure to the antigen. Allergic contact dermatitis (due to chemicals and some metals) and the skin test for exposure to tubercle bacilli (tuberculin test) are some examples.

Questions

Short answer questions

1. Classify hypersensitivity reactions with examples.
2. Define anaphylaxis and mention the mediators of anaphylaxis.
3. Describe atopy.
4. Describe Arthus reaction.
5. Define serum sickness.
6. What is tuberculin-type delayed hypersensitivity?
7. Describe contact dermatitis.

MCQs

1. Which of the following is not a step in the mechanism of anaphylaxis?
a) Activation by the allergen-specific T_{h2} cells
b) Secretion of IgE by plasma cells
c) Crosslinking of the cell-bound IgE
d) Activation of complements

2. Which of the following is RAST used to detect?
a) Type I hypersensitivity b) Type II hypersensitivity
c) Type III hypersensitivity d) Type IV hypersensitivity

3. Which of the following is not a primary mediator of anaphylaxis?
a) 5-HT b) Leukotrienes
c) Platelet activating factor d) Heparin

4. Which of the following statements is not true in relation to atopy?
a) It is genetically determined.
b) IgG is preponderant.
c) It is often associated with IgA deficiency.
d) Environmental inhalants are common.

5. Delayed hypersensitivity can be transferred by ___________.
a) Lymphocytes b) Neutrophils
c) Macrophages d) Eosinophils

6. Which type of enhanced immunological reaction mediates hypersensitive pneumonitis?
a) Type I hypersensitivity
b) Type II hypersensitivity
c) Type III hypersensitivity
d) Type IV hypersensitivity

7. Which antibody is predominantly involved in type II hypersensitivity reaction?
a) IgG b) IgM
c) IgE d) IgA

8. Delayed hypersensitivity can be transferred by:
a) Lymphocytes b) Neutrophils
c) Macrophages d) Eosinophils

9. Which type of hypersensitivity reaction occurs in the tuberculin test?
a) IgE type b) Cytolytic and cytotoxic type
c) Immune complex d) Delayed type

10. Patch test is done for which type of reaction?
a) Contact dermatitis b) Arthus reaction
c) Schwartzman reaction d) Atopy

11. One type of immune complex mediated hypersensitivity is:
a) Arthus reaction
b) Schwartzman reaction
c) Schultz–Dale phenomenon
d) Passive cutaneous anaphylaxis

12. Which among the following is reagin-dependent hypersensitivity?
a) Type I b) Type II
c) Type III d) Type IV

13. A 5-year-old girl presents with complaints of recurrent history of breathlessness and wheezing. Detailed history reveals similar episodes in the past few years usually around February and March. A clinical diagnosis of bronchial asthma is made. Which type of hypersensitivity reaction is this?
a) Type I b) Type II
c) Type III d) Type IV

14. Which of the following is not a primary mediator in case of bronchial asthma?
a) Histamine b) Prostaglandin
c) Heparin d) Serotonin

Answers: 1:d, 2:a, 3:c, 4:b, 5:a, 6:c, 7:a, 8:a, 9:d, 10:a, 11:a, 12:a, 13:a, 14:c

CHAPTER 18

Part II: Immunology

Autoimmune Disorders and Immunodeficiency States

(MI1.10)

Chapter Navigation

- Autoimmune disorders
  - Immunological tolerance
  - Autoimmunity
- Immunodeficiency
  - Primary immunodeficiency
  - Disorders of specific immunity
  - Secondary immunodeficiencies
  - Laboratory methods to detect immunodeficiency disorders

Learning objective

- Describe the mechanisms of autoimmune and immunodeficiency states

> **Case 1.** *A 36-year-old woman presented with lethargy, weight gain, cold intolerance, irregular menstrual periods, and dry/thinning hair for the past few months. On examination, her thyroid was mildly enlarged. An ultrasound scan of her neck revealed a left-sided thyroid nodule without any other significant finding. There was no cervical lymphadenopathy. The biochemical investigations revealed high thyroid-stimulating hormone (TSH), low free T4, and the presence of anti-thyroid antibodies.*
> i. What is the likely diagnosis?
> ii. What is the basis of such a condition?

- The above is a case of autoimmune or **Hashimoto's thyroiditis**.
- The immune system produces antibodies (known as **autoantibodies**) to endogenous (self) antigens, resulting in tissue injury and consequently, disease. In the past, the condition was known as **horror autotoxicus** (named by Paul Erlich, 1901).
- Autoimmune disorders occur due to the failure of the tolerance of the immune system to self-reactive lymphocytes.

IMMUNOLOGICAL TOLERANCE

Several innate mechanisms protect the body's cells from destruction by its own immune cells. This is referred to as **tolerance.** There are two types of immunological tolerance:

- **Central tolerance:** T and B cell clones that have any self-antigen-recognising receptors are deleted in the organ of maturation for the respective lymphocyte (thymus for T cells and bone marrow for B cells).
- **Peripheral tolerance:** If the central tolerance fails to delete autoreactive T and B cells, as a backup, the secondary lymphoid organs are equipped to render these cells inactive. This is known as peripheral tolerance. This can occur as follows:
  i) As anergy (state of non-responsiveness)
  ii) By the deletion of the clone of cells by apoptosis
  iii) By suppression by regulatory cells

  The failure of these mechanisms leads to autoimmunity.

AUTOIMMUNITY

There are several mechanisms of the immune system that lead to autoimmunity.

1. **Alteration of host tissue:** Sometimes, external factors alter host tissue and result in an inappropriate immune response to self-antigens. This can occur by the following mechanisms:
   - **Drug-induced reactions resulting in anemia, leukopenia, and thrombocytopenia** have an autoimmune basis.
   - **Infectious microorganisms**, particularly viruses and other intracellular pathogens, may induce alteration of cell antigens. Viral infections such as infectious mononucleosis are often known to precede autoimmune disease. Neuraminidases formed by myxoviruses and several bacteria act on erythrocytes, thereby releasing the T antigen.
   - **Neoantigens may also arise by mutation**. Such mutant cells may be immunogenic.
2. **Sequestered antigens:** Certain self-antigens are present in closed systems that are not accessible to the immune apparatus. These are known as sequestered antigens. Some examples of sequestered antigens are as follows:
   - The **lens antigen of the eye** is enclosed in its capsule. Hence, immunological tolerance against this antigen is not established during fetal life.

When the antigen leaks out of the capsule following a penetrating injury, it may induce an immune response, causing damage to the lenses of the eyes.
- Since **spermatozoa** develop only with puberty, tolerance towards the sperm antigen is not induced during fetal life. For this reason, this antigen is not recognised as 'self' and is immunogenic when it enters circulation. This is believed to be the reason for the pathogenesis of **orchitis** following mumps. The virus damages the basement membrane of seminiferous tubules, leading to the leakage of sperms and the initiation of an immune response, resulting in orchitis.

3. **Cross-reacting foreign antigens:** The similarity between some foreign and self-antigens is the basis of the 'cross-reacting antigen' theory of autoimmunity. Organ-specific antigens are present in several species. The injection of heterologous organ-specific antigens may induce an immune response, damaging the particular organ or tissue in the host.
   - The **antirabies neural vaccine** prepared from infected sheep brain tissue, which was used for immunisation in the past, resulted in neurological injury in humans. Its injection elicits an immune response against sheep brain antigens which may damage the individual's nerve tissue due to the cross-reaction between human and sheep brain antigens.
   - **Streptococcal M proteins** and the **heart muscle** share antigenic characteristics. The immune response induced by repeated streptococcal infection can therefore damage the heart.
   - **Nephritogenic strains of streptococci** possess antigens found in the renal glomeruli. Infection with such strains may lead to glomerulonephritis due to antigenic sharing.
4. **Molecular mimicry:** A related type of autoimmunisation is 'molecular mimicry', which occurs due to the presence of epitopes in some infecting microorganisms **with peptide sequences that are identical to those present on certain self-antigens** (in which aspect it differs from 'cross-reactions'). **Epitope spreading** is a phenomenon that occurs when the cryptic antigenic epitope of a self-antigen is exposed following tissue damage due to local inflammatory or cytokine action. Examples of such homologous sequences are seen in the following microorganisms and self-antigens:
   - Arthritogenic *Shigella flexneri* and HLA-B27
   - *Mycobacterium tuberculosis* and joint membranes
   - Coxsackie B virus and myocardium
5. **Polyclonal B cell activation:** It has been hypothesised that while an antigen generally activates only its corresponding B cell, certain stimuli non-specifically turn on multiple B cell clones. Such stimuli include the following:
   - Chemicals (e.g., 2-mercaptoethanol) and bacterial products (e.g., PPD and lipopolysaccharide)
   - Enzymes (trypsin) and antibiotics (nystatin)
   - Infections with some bacteria (*Mycoplasma*), viruses (EB virus) and parasites (*Plasmodium* in malaria)

   As a result, multiple non-specific antibodies are formed, such as antihuman erythrocyte cold antibodies in **atypical pneumonia** due to *Mycoplasma* and anti-sheep erythrocyte antibodies in **infectious mononucleosis**. These polyclonal antibodies are IgM in nature, similar to the 'natural antibodies' produced by $CD5^+$ B cells.
6. **Forbidden clones:** The breakdown of immunological homeostasis may lead to the cessation of tolerance and the emergence of forbidden clones of immunocompetent cells, which are capable of mounting an immune response against self-antigens. Autoimmunisation may result when tolerance to a self-antigen is eliminated. For example, the injection of the self-antigen with Freund's adjuvant leads to the stimulation of clones of antibodies to self-antigens.
7. **Altered T or B cell function:** Enhanced helper T cell and decreased suppressor T cell functions have been suggested as causes of autoimmunity. Defects in the thymus, in stem cell development, and macrophage function have also been postulated as causes.

Defects in the idiotype–anti-idiotype network have also been said to lead to autoimmunity. Genetic factors such as defective *Ir* or immunoglobulin genes have also been postulated. Despite so many different possible mechanisms proposed, their actual role in autoimmunity, if any, has not been established.

Features of autoimmune disorder
- Elevated levels of immunoglobulins
- Demonstrable autoantibodies
- Deposition of immunoglobulins or their derivatives at lesion sites (e.g., in renal glomeruli)
- Accumulation of lymphocytes and plasma cells at lesion sites
- The condition is diminished with corticosteroid or other immunosuppressive therapy
- The occurrence of more than one type of autoimmune lesion in an individual
- A genetic predisposition towards autoimmunity
- Higher incidence among females
- Chronicity, usually non-reversible

PATHOGENESIS OF AUTOIMMUNE DISEASES

- Many diseases are considered to be of autoimmune origin based on their association with the cellular or humoral immune response against self-antigens.
- The mere presence of autoantibodies during the course of a disease does not prove their causative role.
- Autoantibody formation may be a result of tissue injury, and the antibody may help in promoting the immune elimination of the damaged cell or tissue elements.
- Antibodies may cause damage by:
  - Cytolytic or cytotoxic (type II) reactions
  - Toxic complex (type III) reactions

It is likely that humoral and cellular immune responses may act synergistically in the production of some autoimmune diseases. For example, experimental orchitis can be induced only when both types of immune responses are operative.

CLASSIFICATION OF AUTOIMMUNE DISEASES

Based on the site of involvement and nature of the lesions, autoimmune diseases may be classified as follows (Fig. 18.1):

1. **Organ-specific**—hemocytolytic, localised
2. **Systemic** (non-organ-specific)

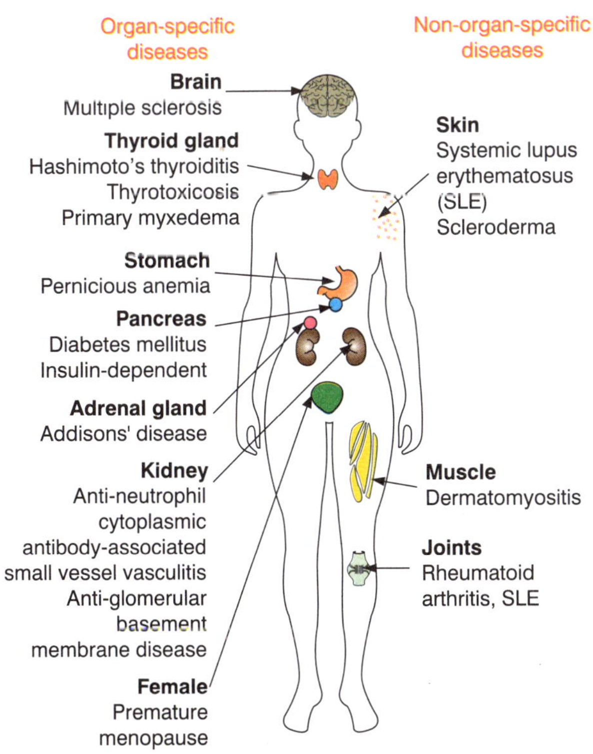

Fig. 18.1 Distribution of organ-specific and non-organ-specific autoimmune disorders

1. Organ-specific Autoimmune Diseases (Table 18.1)

Organ-specific autoimmune diseases are those in which a particular organ or tissue is preferentially targeted by the patient's immune system. For example, the **thyroid gland in Hashimoto thyroiditis or the beta cells of the endocrine pancreas in patients with type 1 diabetes**.

2. Systemic (Non-organ-specific) Autoimmune Diseases

> **Case 2.** *A 27-year-old woman was referred to a rheumatology clinic with a history of pain and swelling in the joints of her hand and a rash on her cheeks. An immunofluorescent test for antinuclear antibodies (ANA) showed a homogeneous pattern—the test is positive for anti-ds DNA. LE cells were demonstrated on the bone marrow biopsy. A diagnosis of **systemic lupus erythematosus** (SLE) was made. The patient was treated with low-dose corticosteroids and showed improvement.*
> i. What type of disease is SLE?
> ii. What is an LE cell?

This group includes conditions characterised by immune responses against a variety of self-antigens and damage to several organs and tissue systems (Table 18.2).

LABORATORY TESTS TO DETECT AUTOIMMUNE DISORDERS

The diagnosis of autoimmune diseases is based on symptoms and laboratory tests to capture autoantibodies from serum using immobilised autoantigens derived from cells. Some of the tests used to diagnose autoimmune disorders are discussed below.

- **Immunofluorescence tests:** Indirect immunofluorescence is commonly used to detect antinuclear antibodies (ANA) against antinuclear antigens, including cytoplasmic components.
- **Enzyme immunoassays (EIA, ELISA) or RIA:** These tests detect the presence or concentration of individual autoantibodies in biological fluids. They are also used to detect double-stranded DNA and other autoantigens including anti-DNA antibodies. Three major types of these antibodies are seen—(i) those reacting with single-stranded (ss) DNA, (ii) those reacting with double-stranded (ds) (DNA), and (iii) those reacting with both ss and ds DNA. Of these, high-titre anti-ds DNA antibody is relatively specific for SLE. Another SLE-specific antibody is the **anti-Smith (Sm) antibody**.
- **Multiplex immunoassay:** This enables the simultaneous detection of different autoantibodies. In this method, a large number of antigens are

Table 18.1 Organ-specific or localised autoimmune diseases

Disease	Features
Hemocytolytic autoimmune diseases	
Autoimmune hemolytic anemia	Autoantibodies against erythrocytes are demonstrable. • **Cold autoantibodies**: These are complete agglutinating antibodies of the IgM class. They agglutinate erythrocytes at 4°C but not at 37°C. Cold agglutinins are also seen in primary atypical pneumonia, trypanosomiasis, and blackwater fever (caused by *P. falciparum* infection). • **Warm autoantibodies**: These are generally incomplete, non-agglutinating antibodies, usually belonging to the IgG class and frequently seen in patients taking certain drugs such as sulphonamides, antibiotics, and alpha-methyldopa. RBCs coated with antibodies are prematurely destroyed in the spleen and liver. Complement-dependent intravascular hemolysis is rarely seen.
Autoimmune thrombocytopenia	Autoantibodies directed against platelets are seen in idiopathic thrombocytopenic purpura. Sedormid purpura is an example of immune response against drug-induced neo-antigens on platelets.
Autoimmune leukopenia	Non-agglutinating anti-leukocyte antibodies are seen in patients with systemic lupus erythematosus and rheumatoid arthritis.
Other organ-specific autoimmune diseases	
Hashimoto's disease (lymphadenoid goitre)	More common in females. It is characterised by the enlargement of the thyroid gland and symptoms of hypothyroidism or frank myxoedema. Histologically, the glandular structure is replaced by lymphoid tissue consisting of lymphocytes, histiocytes, and plasma cells. Autoantibodies that react with thyroglobulin, microsomal antigen, and a thyroid cell-surface component are seen.
Thyrotoxicosis (Graves' disease)	Most patients who have thyrotoxicosis possess antibodies to thyroglobulin. Long-acting thyroid stimulator (LATS), which is an IgG antibody to the thyroid membrane antigen, combines with the surface membrane of thyroid cells and seems to stimulate excessive hormone secretion.
Myasthenia gravis	There is an abnormal fatigability of muscles due to the malfunction of the myoneural junction. An antibody against the acetylcholine receptor on myoneural junctions of striated muscles is present in these patients. Infants born to affected mothers show symptoms of the disease but recover spontaneously by the age of two months, at which time maternal antibodies disappear. This suggests that neonatal myasthenia may be due to the autoantibody passively acquired from the mother.
Addison's disease	Antibodies are directed against the cells of the zona glomerulosa of the adrenal glands.
Autoimmune orchitis	Lymphocytic infiltration of the testes and circulating antibodies to the sperms and germinal cells can be demonstrated in this condition, which may follow mumps orchitis.
Eye	Cataract surgery sometimes leads to intraocular inflammation caused by an autoimmune response to the lens protein. This is known as **phacoanaphylaxis**. Perforating injuries of the eye, particularly those involving the iris or ciliary bodies, are often followed by sympathetic ophthalmia in the opposite eye.
Pernicious anemia	Two types of autoantibodies are present in this condition: • Antibodies against the parietal cells of the gastric mucosa, which probably cause achlorhydria and atrophic gastritis • Antibodies against the intrinsic factor prevent the absorption of vitamin B12 either by blocking its attachment to the gastric intrinsic factor or by binding to the B12 intrinsic factor complex and interfering with its uptake by the intestinal mucosa
Autoimmune diseases of the nervous system	'Neuroparalytic accidents' following rabies vaccination represent injury to the nervous system by the immune response against the sheep nerve tissue in the vaccine, which cross-reacts with human nerve tissue. Idiopathic polyneuritis (Guillain–Barré syndrome) is considered an autoimmune response against the peripheral nervous tissue.
Autoimmune diseases of the skin	Pemphigus vulgaris may be caused by the action of an antibody to the intercellular adhesion protein desmoglein (calcium-binding transmembrane glycoprotein). Bullous pemphigoid involves antibodies directed against the dermal epithelial junction.

Table 18.2 Non-organ specific autoimmune diseases and their features

Disease	Features
Systemic lupus erythematosus (SLE)	This is a chronic, multisystem disease with remissions and exacerbations, which terminate fatally. Patients have a variety of autoantibodies directed against cell nuclei, intracytoplasmic cell constituents, immunoglobulins, thyroid, and other organ-specific antigens. The lupus erythematosus (LE) cell is a neutrophil containing a large, pale, homogeneous body (LE body), which almost fills up the cytoplasm. The LE body is the immunologically damaged nucleus of a leukocyte. Sometimes, the LE body can be seen free, surrounded by a rosette of neutrophils.
Scleroderma	Cells are triggered by the immune system to produce excess collagen, which gets deposited in the skin and different organs, causing hardening and thickening.
Dermatomyositis	This is a chronic autoimmune disease affecting the skeletal muscles and the skin. It is characterised by muscle weakness and skin rash.
Rheumatoid arthritis	This is a symmetric polyarthritis with muscle wasting and subcutaneous nodules, commonly associated with serositis, myocarditis, vasculitis, and other disseminated lesions. It is more common in women than it is in men. The synovial membranes of the affected joints are swollen and edematous, with dense infiltration of lymphocytes and plasma cells. The rheumatoid factor (RF) is a circulating autoantibody that is usually a 19S IgM. RF acts as an antibody against the F_c fragment of immunoglobulins.
Polyarteritis nodosa	It is a necrotising angiitis involving medium-sized arteries, which ends fatally due to coronary thrombosis, cerebral hemorrhage, or gastrointestinal bleeding. Polyarteritis is seen as a component of serum sickness and other toxic complex diseases. Though it has been suggested that polyarteritis nodosa may be an autoimmune disease, the autoantibody responsible has not been identified.
Sjogren's syndrome (sicca)	It is characterised by a triad of conjunctivitis, dryness of the mouth with or without salivary gland enlargement, and rheumatoid arthritis. The syndrome may occur in association with other collagen diseases. Antinuclear antibodies and rheumatoid factor commonly occur in sera.

immobilised on solid carrier bead-based arrays. This is a relatively new technique.

- **Anti-CCP (cyclic citrullinated peptide) ELISA:** This test detects human autoantibodies against cyclic citrullinated peptide, which is a circular peptide containing the amino acid citrulline. IgG antibodies against this peptide are raised in rheumatoid arthritis (RA). A person positive for rheumatoid factor (RF) and raised anti-CCP has a higher likelihood of progressing to erosive disease. This test also helps to differentiate RA from other types of arthritis. If titres are raised in a healthy individual, the test predicts the development of RA in future.
- **Rheumatoid factor (RF) assay:** The RF factor can be detected by latex agglutination (LA) and ELISA.

MANAGEMENT OF AUTOIMMUNE DISORDERS

Once initiated, most autoimmune responses tend to be self-perpetuating. Their progress can be arrested by the following measures:

- **Immunosuppressive therapy** is used to treat autoimmune disorders, though the degree of response to such therapy varies in different diseases.
- **Vaccination with T cells** specific for a given autoantigen can be administered. These vaccines are synthetic blocking peptides that compete with the autoantigen for binding with the MHC molecules.
- Monoclonal antibodies to the self-antigen responsible for the autoimmune reactions can also arrest the progression of autoimmunity.

IMMUNODEFICIENCY

Case 3. *A nine-month-old male infant was brought to the hospital with symptoms of fever and difficulty in breathing. His mother gave a history of two similar episodes in the past two months for which he was managed symptomatically. At the age of 12 months, the child was again brought in for an episode of measles, from which he recovered after treatment. At 18 months of age, it was observed that the boy's height and* *weight were not appropriate for his age. The boy was the fourth child born to unrelated parents. The boy's three sisters were said to enjoy good health and did not have repeated infections.*

Tests on the boy showed his serum immunoglobulin IgG to be less than a tenth and IgA and IgM to be less than a hundredth of the normal level. A lymph node biopsy revealed a depletion of cells of the bursa-dependent areas. The child was diagnosed with ***Bruton's disease (X-linked agammaglobulinemia)*** *and was treated with an initial administration of 300 mg of gammaglobulin/kg body weight in three doses followed by monthly injections of 100 mg/kg. He responded well to treatment.*

i. What do you know about the etiology of this condition?
ii. Assuming there are no secondary causes, what would be the most likely cause of primary immunodeficiency in this infant?
iii. How should this infant be managed for the likely underlying condition? What other complications need to be monitored for?

- The immune system is a complex, multicomponent system. When one or more components of this system fail to function, it may lead to immunodeficiency. This results in repeated microbial infections of varying severity in the host, and sometimes enhanced susceptibility to malignancies.
- Deficiencies of defence mechanisms may involve specific immune functions—humoral immunity, cell-mediated immunity, or both—or non-specific mechanisms such as phagocytosis and complement activity.

Immunodeficiency is chiefly of two types:

i. **Primary immunodeficiency**, which results from abnormalities in the development of the immune system; the established types of primary immunodeficiency syndromes are listed in Table 18.3.

Table 18.3 Types of primary immunodeficiency syndromes

Primary immunodeficiency syndromes		Specific defect
I. Disorders of specific immunity	1. Humoral immunodeficiencies (B cell defects)	i. X-linked agammaglobulinemia (Bruton's disease) ii. Transient hypogammaglobulinemia of infancy iii. Common variable immunodeficiency (late-onset hypogammaglobulinemia) iv. Selective immunoglobulin deficiencies (IgA, IgM, or IgG subclasses) v. Transcobalamin II deficiency
	2. Cellular immunodeficiencies (T cell defects)	i. Thymic hypoplasia (DiGeorge syndrome) ii. Chronic mucocutaneous candidosis iii. Immunodeficiencies with hyper-IgM
	3. Combined immunodeficiency (B and T cell defects)	i. Cellular immunodeficiency with abnormal immunoglobulin synthesis (Nezelof syndrome) ii. Ataxia telangiectasia iii. Wiskott–Aldrich syndrome iv. Immunodeficiency with thymoma v. Immunodeficiency with short-limbed dwarfism vi. Episodic lymphopenia with lymphocytotoxin vii. Purine nucleoside phosphorylase (PNP) deficiency viii. Severe combined immunodeficiencies (SCID) a) 'Swiss-type' agammaglobulinemia b) Reticular dysgenesis of de Vaal c) Adenosine deaminase (ADA) deficiency
II. Disorders of complement	1. Complement component deficiencies 2. Complement inhibitor deficiencies	
III. Disorders of phagocytosis	1. Chronic granulomatous disease 2. Myeloperoxidase deficiency 3. Chediak–Higashi syndrome 4. Leukocyte G6PD deficiency 5. Job syndrome 6. Tuftsin deficiency 7. Lazy leukocyte syndrome 8. Hyper-IgE syndrome 9. Actin-binding protein deficiency 10. Shwachman disease	

ii. **Secondary immunodeficiency**, which is a consequence of disease, drugs, nutritional inadequacies, or other processes that interfere with the proper functioning of the mature immune system

PRIMARY IMMUNODEFICIENCIES

- Though primary deficiencies of specific immunity are classified as those affecting B or T cell responses or both, there is considerable overlap due to the intimate interaction between the B cell and T cell systems.
- For instance, T cell deficiencies involving helper or suppressor T cells will have a profound effect on antibody response.

I. Disorders of Specific Immunity

1. Humoral immunodeficiencies

i. X-linked agammaglobulinemia

This syndrome, which was first described by **Ogden Bruton** (1952), is the first immunodeficiency disease to have been recognised. It is seen only in male infants.

Manifestations

- It presents as **recurrent serious infections** with pyogenic bacteria, particularly with pneumococci, streptococci, meningococci, *Pseudomonas,* and *H. influenzae*.
- Patients respond normally to viral infections such as measles and chickenpox, though there have been reports of paralytic poliomyelitis and progressive encephalitis following immunisation with live virus vaccines or exposure to a wild virus.
- The disease is not apparent until about six months of age due to the passive protection provided by maternal antibodies.

As a rule, live microbial vaccines should not be given to children with any type of primary immunodeficiency.

Features

- All classes of **immunoglobulins are grossly depleted;** the IgG levels are less than a tenth, and IgA and IgM less than a hundredth of the normal level.
- Tonsils and adenoids are atrophic. Lymph node biopsy reveals depletion of cells of the bursa-dependent areas. Plasma cells and germinal centres are absent even after antigenic stimulation.
- There is a marked decrease in the proportion of B cells in circulation. Antibody formation does not occur even after injection of antigens.
- Cell-mediated immunity (CMI) is not affected. Delayed hypersensitivity of tuberculin and contact dermatitis types can be demonstrated. Allograft rejection is normal. Arthritis, hemolytic anemia, and atopic manifestations are frequently observed. However, the wheal-and-flare response of atopic hypersensitivity cannot be demonstrated.

Management

- This consists of the **maintenance of an adequate level of immunoglobulins.**
- The administration of 300 mg of gammaglobulin per kg of body weight in three doses is followed by monthly injections of 100 mg per kg are required.
- Commercial preparations of gammaglobulin contain only traces of IgA and IgM; therefore, whole plasma infusions are used.

ii. Transient hypogammaglobulinemia of infancy

- It occurs due to an **abnormal delay in the initiation of IgG synthesis** in some infants. Maternal IgG is slowly catabolised in the newborn and reaches a level of 200 mg per 100 mL by the second month of life. Ordinarily, an infant begins synthesising its own IgG by this age.
- A delay in the synthesis of IgG leads to immunodeficiency. **Recurrent otitis media and respiratory infections** are commonly seen in such cases. Spontaneous recovery occurs between 18 and 30 months of age. This disease may occur in infants of both sexes.

Management

Treatment with **gammaglobulin** may be required in some cases but is not recommended prophylactically as it may contribute to the prolongation of immunodeficiency by negative feedback inhibition of IgG synthesis.

iii. Common variable immunodeficiency

This common form of immunodeficiency is also known as **late-onset hypogammaglobulinemia** because it usually manifests only by 15–35 years of age. It is characterised by the following:

- Recurrent pyogenic infection
- Increased incidence of autoimmune disease
- Malabsorption
- Giardiasis
- Low level of total immunoglobulin
- Defective B cells in circulation—they fail to differentiate into plasma cells and secrete immunoglobulins.

Increased suppressor T cell and diminished helper T cell activity have been proposed as the causes of this disorder.

Management

Gammaglobulin preparations are administered intramuscularly or intravenously.

iv. Selective immunoglobulin deficiencies

- A selective deficiency of one or more immunoglobulin classes is seen, while the levels of the other immunoglobulins remain normal or elevated. These '**dysgammaglobulinemias**' are common and have been reported in about one per cent of all patients with recurrent infection.

- **Isolated IgA deficiency** is the most common condition in this group, with a reported incidence of about 0.2 per cent in normal populations. Patients exhibit increased susceptibility to respiratory infection and steatorrhea. IgA deficiency is often accompanied by atopic disorders. Anti-IgA antibodies are present in many of these patients.
- Selective IgM deficiency has been found to be associated with septicemia.
- Deficiencies of IgG subclasses have been observed in relation with chronic progressive bronchiectasis.

Management

- **Preventive antibiotics** may be used to help diminish the frequency of recurrent infections.
- Individuals with IgA deficiency often require a longer course of antibiotics for infections to clear.

v. Immunodeficiencies with hyper-IgM

- In this group of immunodeficiencies, some of which are X-linked and some inherited as autosomal recessive, low IgA and IgG levels are seen with elevated IgM.
- Patients show increased susceptibility to infections and autoimmune processes such as thrombocytopenia, neutropenia, hemolytic anemia, and renal lesions.
- Some patients develop malignant infiltration with IgM-producing cells. An elevated IgM level with immunodeficiency is sometimes seen in congenital rubella.

Management

Intravenous immunoglobulin (IVIg) therapy has reportedly decreased the frequency of severe and lower respiratory tract infections significantly.

2. Cellular immunodeficiencies

i. Thymic hypoplasia (DiGeorge syndrome)

This is a developmental defect involving the endodermal derivatives of the third and fourth pharyngeal pouches. The condition has the following characteristics:

- There is **aplasia or hypoplasia of the thymus** and parathyroid glands.
- It does not appear to be hereditary and does not show a familial incidence.
- It is probably caused by **intrauterine infection** or other complications.
- It is usually associated with **Fallot's tetralogy** and other anomalies of the heart and the great vessels, and a characteristic facial appearance.
- **Neonatal tetany** is present. Patients who survive the neonatal period show enhanced susceptibility to viral, fungal, and bacterial infections, which ultimately prove fatal.
- The immunodeficiency primarily involves cell-mediated immunity. The thymus-dependent areas of the lymph nodes and spleen are depleted of lymphocytes. Circulating T cells are reduced in number. Delayed hypersensitivity and graft rejection are depressed. The humoral immune mechanism is largely unaffected.
- Transplantation of fetal thymus tissue has been reported to restore the immunological function.

ii. Chronic mucocutaneous candidosis

- This constitutes an abnormal immunological response to *Candida albicans*.
- Patients develop severe chronic candidosis of the mucosa, skin, and nails. There is a selective defect in the functioning of T cells and B cells.
- They do not show increased susceptibility to other infections but often have endocrinopathies.
- CMI to *Candida* is deficient, and intracellular killing of *Candida* is defective in this condition.
- Delayed hypersensitivity to *Candida* antigens is absent, but circulating antibodies to them are found in high titres.
- Transfer factor therapy, along with amphotericin B, has been reported to be effective.

iii. Hyper-IgM globulinemia

- Similar to X-linked disease, it is caused by defective T cell CD40 ligand, which cannot signal B cells for class switching of IgM to IgE and IgA. As a result, there is an excess of IgM.
- **Clinical presentation** includes recurrent infections, diarrhea, and lymphoid hyperplasia.
- **Diagnosis**—serum IgM levels are high, whereas other immunoglobulins are absent. Flow cytometry to demonstrate the expression of the CD40 ligand on the T cell surface is diagnostic.
- **Management**—hyper-IgM globulinemia is managed by replacing IgG to prevent recurrent infections. Hematopoietic stem cell therapy may also be done.

3. Combined immunodeficiencies

i. Cellular immunodeficiency with abnormal immunoglobulin synthesis (Nezelof syndrome)

The term Nezelof syndrome has been rather loosely applied to a group of disorders, probably of varied origin, where depressed CMI is associated with selectively elevated, decreased, or normal levels of immunoglobulin. The consistent features of these disorders are as follows:

- A marked deficiency of T cell immunity
- Varying degrees of deficiency of B cell immunity
- Susceptibility to recurrent fungal, bacterial, viral, and protozoal diseases

- Thymic dysplasia with lymphoid depletion
- Autoimmune processes such as hemolytic anemia are common
- Despite normal levels of immunoglobulins, antigenic stimuli do not induce antibody formation
- Histocompatible bone marrow transplantation, transfer factor, and thymus transplantation have been used for treatment, with success in some cases; adequate antimicrobial therapy is essential.

ii. Ataxia telangiectasia

- This is a hereditary condition transmitted in the autosomal recessive mode, where combined immunodeficiency is associated with cerebellar ataxia, telangiectasia, ovarian dysgenesis, and chromosomal abnormalities.
- The majority of patients lack serum and secretory IgA (some have antibodies to IgA). IgE deficiency is also frequent. CMI is also defective, resulting in the impairment of delayed hypersensitivity and graft rejection.
- Transfer factor therapy and fetal thymus transplants have been attempted with some amount of success.
- There is a high incidence of malignancies, especially leukemias.

iii. Wiskott–Aldrich syndrome

- This is an X-linked disease characterised by eczema, thrombocytopenic purpura, and recurrent infections. Affected boys rarely survive the first decade of life; death is usually due to infection, hemorrhage, or lymphoreticular malignancy.
- CMI undergoes progressive deterioration associated with cellular depletion of the thymus and the paracortical areas of the lymph nodes.
- The humoral defect appears to be a specific inability to respond to polysaccharide antigens.
- Bone marrow transplantation and transfer factor therapy have been found to be beneficial.
- IgM concentrations are reduced, but IgG levels are normal. Both IgA and IgE levels are elevated.

iv. Severe combined immunodeficiencies (SCID)

- These include many syndromes with severe deficiency of both humoral and CMI response.
- They are inherited in the autosomal recessive mode.
- The primary defects are at the level of the early precursors of immunocompetent cells in the fetal liver and bone marrow.
- **Swiss-type agammaglobulinemia:** The basic defect is presumed to be at the level of the lymphoid stem cell.
- **Reticular dysgenesis of de Vaal:** Here, the defect is at the level of the multipotent hemopoietic stem cell.

II. Disorders of Complement

1. Complement component deficiencies

Genetic deficiencies have been detected for almost all the complement components in human beings. These defects are transmitted as autosomal recessive traits.

2. Complement inhibitor deficiencies

Hereditary angioneurotic edema

- It is caused by a genetic **deficiency of the C1 inhibitor** which is transmitted as an autosomal dominant trait. Asymmetric and mildly painful swellings on the face, lips, and/or tongue are seen. Swellings may also develop on the back of the hands or feet or on the genitals.
- **Management**—for acute attacks, C1 inhibitor, ecallantide, and tranexamic acid are administered. Attenuated androgens are used for prophylaxis.

III. Disorders of Phagocytosis

Phagocytosis may be impaired by the following:

- **Intrinsic disorders**, which may occur due to defects within the phagocytic cell, such as enzyme deficiencies
- **Extrinsic disorders**, which may occur due to a deficiency of an opsonic antibody, complement, or other factors promoting phagocytosis; they could also be caused by the effects of drugs or anti-neutrophil autoantibodies

Phagocytic dysfunction leads to increased susceptibility to infection, ranging from mild and recurrent skin infections to overwhelming systemic infection.

1. Chronic granulomatous disease

This familial disease manifests as recurrent infections with low-grade pathogens starting early in life. The characteristic features of this disease are as follows:

- Chronic suppurative granulomatous lesions develop in the skin and lymph nodes along with hepatosplenomegaly, progressive infiltration of the lungs and granulomatous septic osteomyelitis.
- Humoral and cellular immune responses are normal.
- The progress is chronic, and the outcome fatal.
- The bacteria involved are staphylococci and coliforms. The host's leukocytes are unable to kill catalase-positive bacteria following phagocytosis, as a result of which, the bacteria multiply intracellularly.
- Diminished H_2O_2 production leads to a defect in bactericidal activity.
- Delayed granule rupture and defective release of myeloperoxidase also contribute to inefficient bactericidal activity.
- Catalase-negative pathogens such as streptococci and pneumococci are handled normally.

- The disease shows two types of inheritance:
  - The more common **X-linked type**, which is seen in boys
  - The rare **autosomal recessive type**, which is seen in girls

2. Myeloperoxidase deficiency

- In this rare disease, leukocytes have reduced myeloperoxidase.
- Patients are particularly prone to *Candida albicans* infection.

3. Chediak–Higashi syndrome

- This is a genetic disorder characterised by decreased pigmentation of the skin, eyes, and hair and photophobia, nystagmus, and giant peroxidase-positive inclusions in the cytoplasm of leukocytes.
- The inclusions may be the result of autophagocytic activity. The leukocytes possess diminished phagocytic activity.
- Patients suffer from frequent and severe pyogenic infections.

4. Leukocyte G6PD deficiency

- Leukocytes are deficient in glucose-6-phosphate dehydrogenase and show diminished bactericidal activity after phagocytosis.
- This disease resembles chronic granulomatous disease in reduced myeloperoxidase activity and susceptibility to microbial agents.

5. Job syndrome

- It is characterised by multiple, large, **'cold' staphylococcal abscesses** containing large quantities of pus that develop repeatedly on the skin and in various organs, with little inflammatory response.
- Atopic eczema, chronic nasal discharge, and otitis media are common features.
- Serum immunoglobulins are normal, except for elevated IgE.
- The pathogenesis is probably a primary defect in phagocytic function.

6. Tuftsin deficiency

- 'Tuftsin' (discovered at Tufts University, Boston) is a leukokinin capable of stimulating phagocytosis.
- Individuals who have a deficiency of this enzyme are prone to local and systemic bacterial infections.

7. Lazy leukocyte syndrome

- The basic defect here is in chemotaxis and neutrophil mobility.
- Patients show poor leukocyte response to chemical and inflammatory stimulation and are susceptible to bacterial infection with recurrent stomatitis, gingivitis, and otitis.

8. Hyper-IgE syndrome

In this disease, which is seen in both sexes, patients have early-onset eczema and recurrent bacterial infections such as abscess, pneumonia, and secondary infection of eczema.

SECONDARY IMMUNODEFICIENCIES

- Secondary immunodeficiencies are much **more common** than primary deficiencies.
- A variety of **factors** such as malnutrition, malignancy, infection, metabolic disorders, and cytotoxic drugs may lead to deficits in specific and non-specific immunity.

Humoral Immune Deficiency

- B cells are depleted, as in lymphoid malignancy, particularly in chronic lymphatic leukemia.
- Immunoglobulin catabolism is increased, as seen in nephrotic syndrome.
- Excessive loss of serum protein occurs, as seen in exfoliative skin disease and protein-losing enteropathies.
- Excessive production of abnormal immunoglobulins occurs, as in multiple myeloma.

Cell-Mediated Immune Deficiency

- CMI is depressed in lymphoreticular malignancies, as in Hodgkin's disease.
- Lymph circulation is obstructed or lymphorrhea occurs.
- The thymus-dependent areas of the lymph nodes are infiltrated with non-lymphoid cells, as in lepromatous leprosy.
- CMI is suppressed transiently, following certain viral infections such as measles.
- Nutritional deprivation affects both types of immune response adversely. Ageing also causes a reduction in the efficiency of acquired immunity. Immunodeficiency follows the intentional or unintentional administration of immunosuppressive agents.

Acquired Immunodeficiency Syndrome

- A severe form of immunodeficiency occurs due to infection by the human immunodeficiency virus (HIV-1 and -2), which is transmitted through sexual contact, from mother to child, or through the transfusion of infected blood.
- It begins with cell-mediated immune deficiency and is followed by both humoral and CMI deficiency, leading to death (discussed in detail in Chapter 22).

Drug-induced Immunodeficiency

- Associated with the administration of steroids.

- T cells are more affected than B cells.
- Cytokine synthesis is inhibited.

Nutrient Deficiencies

- The immune system is impaired by protein and micronutrient deficiency.
- Malnutrition affects cell-mediated immunity, antibody production, phagocyte function, complement system, and cytokine synthesis.
- Many enzymes require zinc, iron, and other micronutrients.

Obesity

- It may cause impaired immune responses due to altered NK function.
- Cytotoxicity and phagocytosis are also compromised.

LABORATORY METHODS TO DETECT IMMUNODEFICIENCY DISORDERS (TABLE 18.4)

Table 18.4 Laboratory methods to detect immunodeficiency disorders

Test	Diagnostic utility
Complete blood count (CBC)	• Peripheral blood smear—granulocytes may have morphologic abnormalities. • Neutropenia (absolute neutrophil count <1,200 cells/µL) may be congenital or cyclic or may occur in aplastic anemia • Lymphopenia (lymphocytes <2,000/µL at birth, <4,500/µL at the age of 9 months, or <1,000/µL in older children or adults) suggests a T cell disorder because 70% of circulating lymphocytes are T cells • Leukocytosis that persists between infections may occur in leukocyte adhesion deficiency • Thrombocytopenia in male infants suggests Wiskott–Aldrich syndrome
B cell count	Humoral immunodeficiency is detected by estimating the B cells that may be absent in certain immune disorders associated with antibodies; estimation of CD19 and CD20
T cell count	TLC, CD3, CD4, and CD8 counts, CD4/CD8 ratio
Specific markers	Severe combined immune deficiency (SCID)—mutation analysis to test for certain typical mutations (e.g., in the IL-2 receptor gamma [IL-2RG or IL-2Rg] gene)
Complement assay	The standard screening test for deficiencies in the complement system is the total hemolytic complement assay or CH50 This is estimated in terms of the serum dilution required to lyse 50% of antibody-coated RBCs This test (called CH50) detects complement component deficiencies in the classical complement pathway but does not indicate which component is abnormal A similar test—AH50—can be performed to detect complement deficiencies in the alternative pathway
Quantitative serum immunoglobulins	Quantitative serum Ig levels—low serum levels of IgG, IgM, or IgA suggest antibody deficiency An IgG level <200 mg/dL usually indicates significant antibody deficiency and is also seen in nephrotic syndrome or protein-losing enteropathies IgM antibody—this is assessed by measuring isohemagglutinin titres (anti-A and anti-B) People with blood type AB have natural antibodies at a titre of ≥1:8 (anti-A) or ≥1:4 (anti-B) (except infants <6 months) IgG antibody—this can be assessed in immunised patients by measuring antibody titres before and after the administration of vaccine antigens, e.g., *Haemophilus influenzae* type B, tetanus, diphtheria, and conjugated or non-conjugated pneumococcal and meningococcal antigens Less than a two-fold increase in titre at 2 to 3 weeks suggests antibody deficiency
Flow cytometry for CD cell counts	Phagocytic cell deficiency—CD15 and CD18 are measured by flow cytometry, and neutrophil chemotaxis is tested A flow cytometric oxidative (respiratory) burst assay measured by nitroblue tetrazolium (NBT) can detect whether oxygen radicals are produced during phagocytosis; it is absent in chronic granulomatous disease

Key Points

✓ Autoimmune disorders are those in which the immune system produces auto/self-antibodies to endogenous (self) antigens, resulting in tissue injury.

✓ The body usually displays immunological tolerance, wherein the individual's immune system 'tolerates' its own ('self') antigens and is able to differentiate between 'self' and 'non-self' antigens. Autoimmunity occurs when this tolerance is lost, and the immune system attacks its own tissues. Autoimmune reactions may be mounted by antibodies and T cell responses.

- ✓ Self-antigens may be intracellular components, receptors, cell membrane components, extracellular components, plasma proteins, or hormones. Depending on the location of the self-antigen, the disease may be localised to a single organ or may affect multiple sites.
- ✓ Autoimmunity may involve antigenic alterations, sequestration, cross-reacting foreign antigens, molecular mimicry, or forbidden clones of self-antigens.
- ✓ Examples of autoimmune diseases are renal injury (nephritis) associated with systemic lupus erythematosus, orchitis following mumps virus infection, pernicious anemia, thyrotoxicosis (Graves' disease), and Hashimoto's thyroiditis.
- ✓ Treatment of autoimmune disorders is with immunosuppressants. Recently, other approaches like monoclonal antibodies and vaccination with specific T cells against self-antigens are also being studied.
- ✓ Immunodeficiency diseases are conditions in which the body's defence mechanisms are impaired, resulting in repeated microbial infections of variable severity and, sometimes, enhanced susceptibility to malignancies.
- ✓ Deficiencies of defence mechanisms may involve specific immune functions, namely humoral and/or cell-mediated immunity, or non-specific mechanisms (phagocytosis, complement), which augment specific immune processes. Immunodeficiencies may be primary or secondary.
- ✓ Primary immunodeficiencies result from abnormalities in the development of immune mechanisms and comprise the following:
  - Disorders of specific immunity, such as B cell defects or humoral immunodeficiencies (X-linked agammaglobulinemia, transient hypogammaglobulinemia of infancy, common variable immunodeficiency, and selective immunoglobulin deficiencies)
  - T cell defects or cellular immunodeficiencies (thymic hypoplasia and chronic mucocutaneous candidosis)
  - Combined B and T cell defects or combined immunodeficiencies (cellular immunodeficiency with abnormal immunoglobulin synthesis, ataxia telangiectasia, Wiskott–Aldrich syndrome, and severe combined immunodeficiencies)
  - Disorders of complement (deficiencies of complement components and deficiencies of complement inhibitor)
  - Disorders of phagocytosis (chronic granulomatous disease, myeloperoxidase deficiency, Chediak–Higashi syndrome, leukocyte glucose-6-phosphate deficiency, Job syndrome, and Tuftsin deficiency)
- ✓ Secondary immunodeficiencies occur due to certain infections (acquired immunodeficiency syndrome), nutritional deprivation (kwashiorkor), use of certain drugs (immunosuppressive agents, corticosteroids), metabolic disorders (protein-losing enteropathies), and malignancies (chronic lymphocytic leukemia, multiple myeloma and Hodgkin's disease). Secondary immunodeficiencies are more common than primary immunodeficiencies.

Questions

Short answer questions

1. What is molecular mimicry?
2. Describe the role of sequestered antigens in autoimmunity.
3. Enumerate localised (organ-specific) autoimmune diseases.
4. What is the immunological mechanism of Hashimoto's disease?
5. Describe the development of systemic lupus erythematosus (SLE).
6. Describe the laboratory tests to diagnose autoimmune disorders.
7. Describe primary immunodeficiency disorders.
8. Enumerate secondary immunodeficiencies.
9. List clinical conditions linked with humoral immunodeficiency disorders.
10. Describe cellular immunodeficiencies with examples.
11. List combined B and T cell defects with examples.
12. Describe chronic granulomatous disease.

MCQs

1. Which mechanism of immune response against self-antigens is responsible for arthritis seen in *Shigella flexneri* infection?
a) Sequestered antigens
b) Molecular mimicry
c) Cross-reacting foreign antigens
d) Antigenic alterations

2. Which of the following statements is not true regarding autoimmune disorders?
a) Has genetic predisposition
b) There is increased level of immunoglobulins
c) Seen in higher frequency in males
d) It is usually non-reversible

3. Which of the following autoimmune antibodies are relatively specific for SLE?
a) Anti-ds DNA antibody b) Anti-Sm antibody
c) Both (a) and (b) d) None of these

4. LE cell is a ____________.
a) Macrophage b) Dendritic cell
c) Neutrophil d) Lymphocyte

5. Anti-CCP is raised in which of the following conditions?
a) Polyarteritis nodosa b) Sjogren's syndrome
c) Rheumatoid arthritis d) SLE

6. All the following are features of Bruton's disease except ____________.
a) The disease is evident after six months of life.
b) Tonsils and adenoids are atrophic.
c) Live vaccines should not be given.
d) The patient presents with recurrent viral infections.

7. Which of the following statements is false about late-onset hypogammaglobulinemia?
a) Disease manifests after 15 years of age
b) Low level of total immunoglobulin
c) Recurrent bacterial infections
d) Also known as selective immunoglobulin immunodeficiency

8. Which selective immunoglobulin deficiency is associated with steatorrhea?
a) IgA deficiency b) IgG deficiency
c) IgM deficiency d) IgD deficiency

9. Which of the following diseases is not a severe combined immunodeficiency?
a) 'Swiss-type' agammaglobulinemia
b) Job syndrome
c) Reticular dysgenesis of de Vaal
d) Adenosine deaminase (ADA) deficiency

10. All the following are features of Wiskott–Aldrich syndrome except ____________:
a) Raised serum IgM
b) Absence of isohemagglutinins
c) Serum IgG levels are normal
d) Not genetically related

11. In myeloperoxidase deficiency:
a) There is an increased incidence of infections with pneumococci.
b) Patients are particularly prone to *Candida albicans* infection.

c) There is complement deficiency.
d) All of the above.

12. Which of the following is seen is DiGeorge syndrome?
a) Tetany
b) Eczema
c) Absent B and T cells
d) Low levels of immunoglobulins.

13. Hereditary angioneurotic edema is due to the deficiency of:
a) C1 protein
b) C8 protein
c) C1 inhibitor
d) C3 convertase

14. Which of the following is an example of organ-specific autoimmune disorder?
a) Hashimoto thyroiditis
b) Scleroderma
c) Pernicious anemia
d) All of the above

15. Which of the following is not true regarding cold autoantibodies?
a) These are complete agglutinating antibodies.
b) These belong to the IgM class.
c) These agglutinate erythrocytes at 4°C but not at 37°C.
d) These are seen in patients taking drugs as sulphonamides and antibiotics.

16. A child presents with frequent sinopulmonary infection and telangiectasia of the face. The deficiency of which of the following immunoglobulin classes should be investigated?
a) IgM
b) IgG1
c) IgG3
d) IgA

17. In lazy leukocyte syndrome:
a) There is a defect in neutrophil mobility
b) There is peripheral leukopenia
c) Both (a) and (b) are true
d) Both (a) and (b) are false

18. Chronic granulomatous disease is a:
a) B cell defect
b) T cell defect
c) Complement defect
d) Disorder of phagocytosis

19. Giant peroxidase-positive inclusions are a feature of:
a) Myeloperoxidase deficiency
b) Chronic granulomatous disease
c) Leucocyte G6PD deficiency
d) Chediak-Higashi syndrome

20. Mutational analysis is used to establish the diagnosis of:
a) B cell deficiency
b) T cell deficiency
c) Severe combined immune deficiency
d) Complement deficiency

21. An example of sequestered antigens responsible for an autoimmune condition is:
a) Drug-induced anemia
b) Mumps
c) Post-streptococcal glomerulonephritis
d) Neurological injury after neural vaccine administration

22. Cold autoantibodies are seen in all the following conditions except:
a) Drug-induced autoimmune anemia
b) Primary atypical pneumonia
c) Black water fever
d) Trypanosomiasis

23. Which among the following best characterises Bombay blood group?
a) Antigens A and B present, H absent on RBCs
b) Antigens A and B absent, H present on RBCs
c) Antigen A, B, and H present on RBCs
d) Antigen A, B, and H absent on RBCs

24. Antibodies present in a person with O blood group are:
a) Anti-A antibody only
b) Anti–B antibody only
c) Both anti–A and anti–B antibody
d) No antibody

25. The best method to detect Rh antibodies in serum is:
a) Direct Coombs test
b) Indirect Coombs test
c) Major cross-match
d) Minor cross-match

Answers: 1:b, 2:c, 3:c, 4:c, 5:c, 6:d, 7:d, 8:a, 9:b, 10:a, 11:b, 12:a, 13:c, 14:b, 15:d, 16:d, 17:c, 18:d, 19:d, 20:c, 21:b, 22:a, 23:d, 24:c, 25:b

CHAPTER 19

Part II: Immunology

Immunological Mechanisms of Transplantation, Immunohematology, and Tumour Immunity

(MI1.11)

Chapter Navigation

- Immunology of transplantation
  - Classification of transplants
  - Major histocompatibility complex
  - Histocompatibility antigens
  - Transplantation reaction
  - Graft-versus-host reaction
  - Laboratory tests required prior to transplantation
  - Immunosuppression
  - Privileged sites
  - Other sources of transplants
  - Post-transplant infections
- Immunohematology
  - ABO system
  - Rh blood group system
  - Medical implications of blood groups
- Immunology of malignancy
  - Tumour antigens
  - Immunological surveillance
  - Immunotherapy of cancer

Learning objective

- Describe the immunological mechanisms of transplantation and tumour immunity

IMMUNOLOGY OF TRANSPLANTATION

- **Transplantation** refers to the transfer of cells, tissues, or organs from one individual to another or from one site to another site in the same individual.
- **Graft** refers to the tissue or organ transplanted.
- **Donor** refers to the individual from whom the transplant is obtained and **recipient** refers to the individual to whom it is applied.

CLASSIFICATION OF TRANSPLANTS

Transplants may be classified in several ways based on several parameters:

1. Based on the **organ or tissue transplanted**, they are classified as kidney, heart, or skin transplants, and so on.
2. Based on the **anatomical site of origin of the transplant and the site of its placement**, grafts are classified as follows:
   i. **Orthotopic**—these grafts are applied in anatomically 'normal' sites, e.g., skin grafts.
   ii. **Heterotopic**—these grafts are placed in anatomically 'abnormal' sites, e.g., thyroid tissue transplanted into a subcutaneous pocket.
3. Based on whether transplants are from **fresh tissues and organs** or from **stored** ones, or whether they are from living or **dead materials (cadaveric transplants)**, they may be classified as follows:
   i. **Vital grafts**—live grafts such as the kidney or heart, which are expected to survive and function physiologically in the recipient
   ii. **Structural (static) grafts**—non-living transplants like bone or artery, which merely provide a scaffolding on which new tissue is produced by the recipient.
4. Based on the **genetic (and antigenic) relationship between the donor and the recipient**, they are classified as follows:
   - **Autograft**—an organ or tissue taken from an individual and grafted on the same person
   - **Isograft**—a graft taken from an individual and introduced into another individual of the same genetic constitution, e.g., grafts made between identical twins or between syngeneic members of highly inbred strains of animals
   - **Allograft (formerly called homograft)**—grafts between two genetically non-identical members of the same species
   - **Xenograft (formerly called heterograft)**—grafts between members of different species.

Organ/tissue transplantation is performed following organ failures due to various illnesses. Some of the common organ transplants are as follows:

- Kidney, liver, and cornea (e.g., kidney failure due to diabetes, liver damage due to poisoning, trauma, massive interstitial damage to the lungs due to fulminant infection, as was seen in the recent COVID-19 pandemic, irreversible pancreatitis, cornea, and heart transplants)
- Autologous skin transplant in burns and accidents

Although the frequency of cornea, kidney, skin, heart, lung, liver, and bone marrow transplants is increasing, the availability of organ donors is a challenge that might be solved in the future by innovations in transplantation and stem cell technology.

A major cause for concern is the availability of organs and tissues. This has been slightly alleviated since the advent of cadaver organ transplants.

MAJOR HISTOCOMPATIBILITY COMPLEX (MHC)

- The primary function of the immune system is to recognise and eliminate foreign cells and antigens that enter the body.
- Tissues and organs grafted from one individual to another member of the same species (**allografts**) are recognised as foreign and rejected.
- **Histocompatibility antigens** are cell surface antigens—also known as **alloantigens**—that induce an immune response leading to the rejection of allografts.
- The term '**major histocompatibility system**' refers to a system of cell antigens that exert a decisive influence on the fate of allografts.
- **Alloantigens** are characteristically found on the surface of leukocytes.
- Human MHC antigens are therefore synonymous with human leukocyte antigens (HLA), and the MHC complex of genes with the genes of the HLA complex.
- The genetic control of the immune response is carried out by specific genes called the immune response (*Ir*) genes.

Functions of Major Histocompatibility Complex (MHC)

- They are **cell surface markers** that help infected cells to signal cytotoxic and helper T cells.
- The polymorphism of MHC complexes helps **maximise protection** against microbial infections.
- By increasing the specificity of self-antigens, the MHC prevents microbes with related antigenic make-up from sneaking past host immune defences by molecular mimicry.
- The primary aim of the MHC may be defence against microbes and not against grafts. It plays an important role in **transplantation immunity**.
- The MHCs are regulated by genes.

MHC restriction

- Processing and presentation of an antigen by the macrophage to 'T' cells requires that both the cells possess **surface determinants coded for by the same MHC genes**. The T cell can accept the processed antigen only if it is presented by a macrophage carrying on its surface the self-MHC antigens. When the macrophage bears a different MHC antigen, it cannot cooperate with the T cells. This is known as MHC restriction.
- Cytotoxic T cells can recognise class I MHC antigens on target cells.
- Helper T cells can accept antigens presented by macrophages/dendritic cells only when they bear the same class II MHC molecules on their surface.
- In view of the importance of MHC restriction in immunological control, the Nobel Prize for Medicine was awarded to **Peter Doherty** and **Rolf Zinkernagel** in 1996 for their seminal work on how virus-infected cells are recognised by the immune system and the specificity of the cell-mediated immune defence.

Histocompatibility Antigens

- Antigens that participate in graft rejection are called transplantation or histocompatibility antigens.
- The major histocompatibility system in human beings is the **human leukocyte antigen (HLA) system**.
- The immune response against transplants depends on the presence of antigens in the grafted tissue which are absent in the recipient and hence recognised as foreign.
- If the recipient possesses all the antigens present in the graft, there will be no immune response, and consequently no graft rejection, even when the donor and recipient are not syngeneic.
- Some factors to note are as follows:
  - The first generation (F1) hybrids between two inbred strains possess antigens representative of both the parent strains and will, therefore, accept grafts from either of the parental strains. For example, if the two parental strains have genotypes AA and BB respectively, the F1 hybrid will be of genotype AB. It can, therefore, accept tissues with genotype AA as well as BB, as it possesses both alleles.
  - Transplantation in the reverse direction (from F1 to parent) will not succeed as strain AA will react against antigen B and strain BB against antigen A.

Eichwald–Silmser effect

- While transplants between members of a highly inbred strain of animals are successful, an exception is seen when the **donor is a male and the recipient is a female**.
- Such grafts are rejected as the grafted male tissue (XY) will have antigens determined by the Y chromosome which will be absent in the female (XX) recipient.
- On the other hand, grafts from the female to the male will succeed.
- This **unilateral sex-linked histoincompatibility** is known as the Eichwald–Silmser effect.

TRANSPLANTATION REACTION

- This is predominantly **cell-mediated**. It occurs as two sets of reactions—the first followed by the second set of reactions.

- The **first-set response** is brought about almost exclusively by T lymphocytes. Humoral antibodies are also produced during allograft rejection. This is a sensitisation phase, in which antigen-reactive lymphocytes of the recipient proliferate in response to the alloantigen on the graft.
- This is followed by an **effector stage (second-set response)** in which immune destruction of the graft takes place (Fig. 19.1).

Immunological Enhancement

- **Humoral antibodies may sometimes oppose cell-mediated immunity by inhibiting graft rejection.** This phenomenon is called immunological enhancement.
- If the recipient is pre-treated with one or more injections of killed donor tissue before the transplant is applied, the graft survives much longer than it does in control animals.
- The enhancing effect can be passively transferred to normal animals by an injection of serum from immunised animals, showing that the effect is due to humoral antibodies.
- The antibodies may bring about the enhancing effect in various ways:
  - They may combine with the antigens released from the graft so that they (antigens) are unable to initiate an immune response (**afferent inhibition**).
  - The antibodies may combine with the lymphoid cells of appropriate specificity and, by a negative feedback influence, render them incapable of responding to the antigens of the graft (**central inhibition**).
  - They may also coat the surface of cells in the graft so that sensitised lymphocytes are kept out of contact with them (**efferent inhibition**).

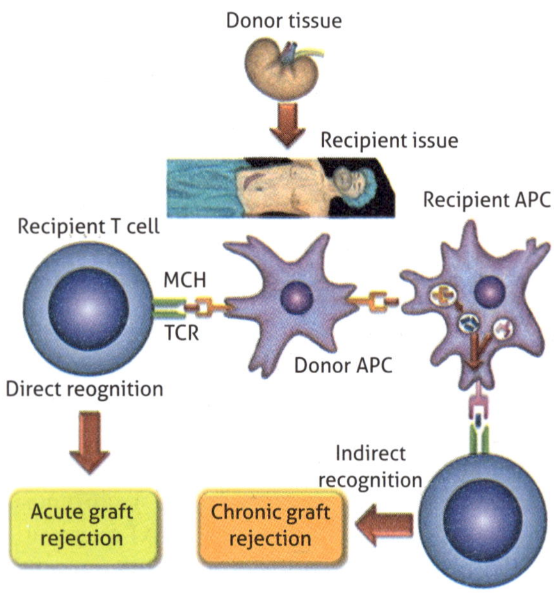

Fig. 19.1 Mechanism of graft rejection initiated by allogeneic immune response—sensitisation of the immune cells followed by effector mechanism leading to acute or chronic rejection (*Source*: Marco Antonio Ayala García et al. The Major Histocompatibility Complex in Transplantation. *Journal of Transplantation.* https://doi.org/10.1155/2012/842141. Creative Commons Attribution 3.0 Unported license)

Allograft immunity is a generalised response directed against all the antigens of the donor. A recipient sensitised by a skin graft will reject, by the second-set response, not only another skin graft but also any other organ or tissue graft from the same donor.

Immunological Basis of Allograft Rejection

> **Case 1.** *A 38-year-old man with end-stage renal failure due to chronic glomerulonephritis was given a cadaveric kidney transplant. His major blood group was A. The donor kidney was also of blood group A and was matched for one DR and four of the six ABC antigens. The patient was put on immunosuppressive therapy.*
>
> *On the third postoperative day, his urea and creatinine levels decreased appreciably. On the seventh postoperative day, however, his graft became tender and his creatinine level increased. A clinical diagnosis of **acute rejection** was confirmed on finding lymphocytic infiltration in the renal cortex biopsy. The patient was started on intravenous corticosteroids and improved 24 hours later. Oral steroids were continued.*
>
> i. What is the immunological basis of acute rejection?

The immunological basis of graft rejection is evident from the specificity of the response. Accelerated rejection is seen only if the second graft is from the same donor as the first.

1. First-set response

When a skin graft is applied on a genetically unrelated person, the graft appears to be accepted initially. It becomes vascularised and seems morphologically and functionally healthy during the first two or three days. Subsequently, the following reactions take place:

- By about the fourth day, inflammation becomes evident, and the graft is invaded by lymphocytes and macrophages.
- The blood vessels within the graft are occluded by thrombi.
- The vascularity of the graft diminishes.
- The graft undergoes ischemic necrosis.
- With extending necrosis, the graft assumes a scab-like appearance and sloughs off by the tenth day.

This sequence of events resulting in the rejection of the allograft is known as the 'first-set rejection or reaction' (Fig. 19.2).

2. Second-set response

If, in an individual who has rejected a graft by the first-set response, another graft from the same donor is applied, it will be rejected in an accelerated fashion. This is known as the second-set response. In this reaction, the following events take place:

- Antibodies are formed more rapidly and abundantly than during primary rejection.
- Antibodies are believed to participate in the second-set response along with cell-mediated immunity (Fig. 19.3).

3. Hyperacute rejection (white graft response)

When a graft is applied to an individual possessing the specific donor antibodies in high titres, a hyperacute rejection occurs, comprising of the following reactions taking place within a short span:

- Preformed anti-donor antibodies rapidly bind to antigens on donor tissue.
- This results in the activation of complement.
- Alternatively, they directly recruit host macrophages and neutrophils via their F_c region.
- This leads to microthrombi formation.
- Lastly, there is an occlusion of the blood vessels of the graft (Fig. 19.4a and b).

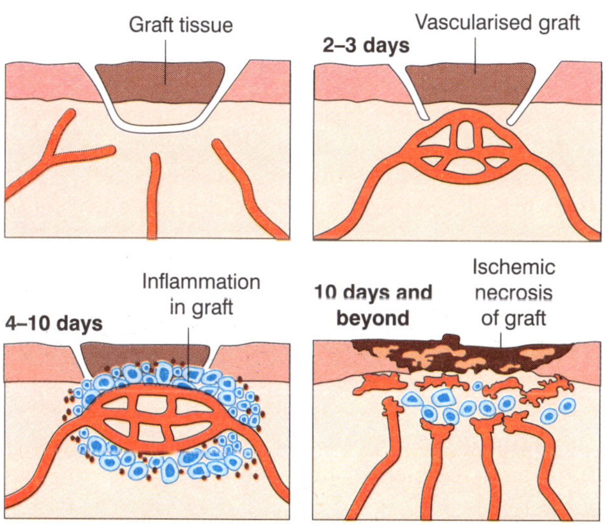

Fig. 19.2 First-set rejection—blood vessels within the graft are occluded by thrombi, the vascularity diminishes, and the graft undergoes ischemic necrosis

- The graft remains pale and is rejected within minutes to hours without even an attempt at vascularisation. This is known as the **white graft response**.
- It is sometimes seen in recipients of kidney transplants who may possess pre-existing antibodies as a result of prior transplantation, transfusion, or pregnancy.

4. Acute rejection

- It occurs within one week to ten days after a transplant has been grafted.
- It occurs due to the incompatibility of the allograft.
- The immune response against the graft increases in intensity, with the accumulation of lymphocytes, plasma cells, macrophages, and neutrophils, leading to endothelial damage. There is necrosis and edema, and finally, rejection of the graft.

5. Chronic rejection

- This is a late-onset rejection occurring after 60 days or more, often superimposed by acute changes.
- In kidney transplantation, there is mesangial cell proliferative glomerulonephritis and interstitial fibrosis.
- Such patients respond poorly to corticosteroids.

GRAFT-VERSUS-HOST REACTION

Case 2. *A 52-year-old man was diagnosed and acute myeloid leukemia and received an allogeneic bone marrow transplant 30 days ago. He also complains of abdominal pain, vomiting, and diarrhea.*

i. Based on the clinical findings, suggest a differential diagnosis.
ii. Describe the pathophysiology of the likely diagnosis.

- When a graft mounts an immune response against the antigens of the host (Fig. 19.5), the reaction is known as the **graft-versus-host (GVH) reaction**. This type of reaction is common in **stem cell therapy** in bone marrow transplants.
- Graft-versus-host disease is of two types—(i) **acute** and (ii) **chronic**. These are scored or graded on the basis of the tissue affected and the severity

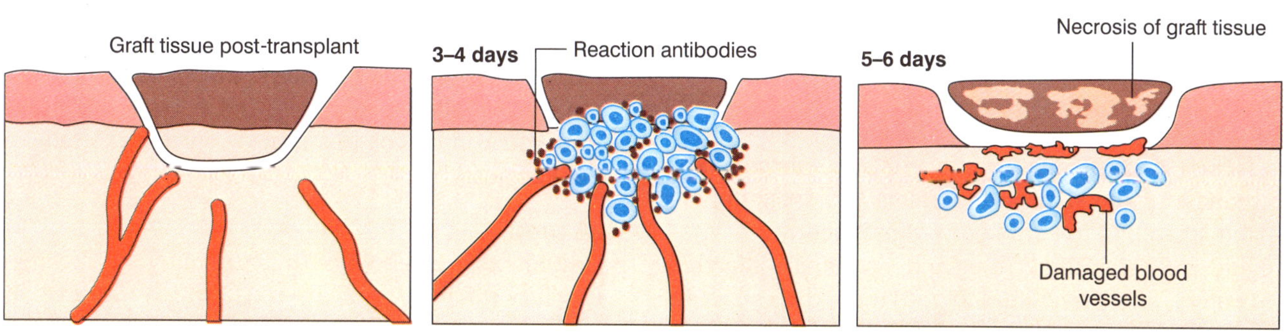

Fig. 19.3 Second-set rejection—the reactions are rapid with the involvement of antibodies and cell-mediated immunity, resulting in the rejection of the graft

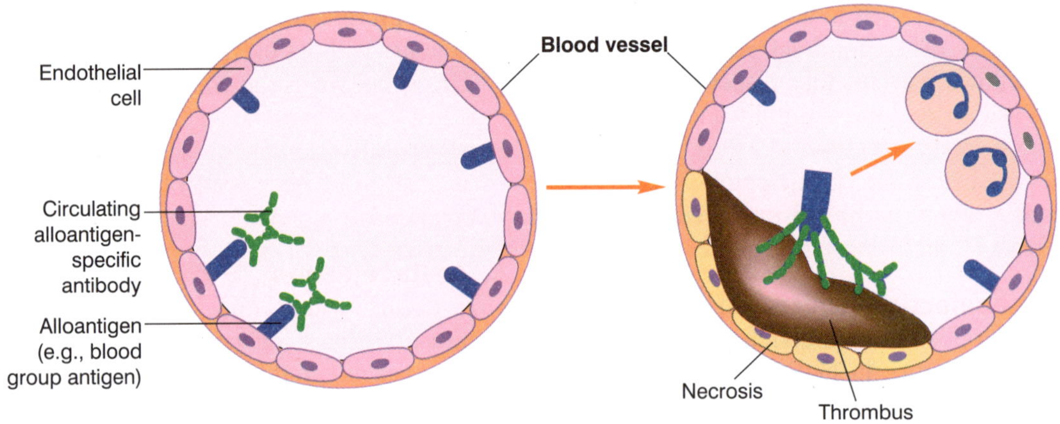

Fig. 19.4 (a) and (b) Vascular occlusion and thrombi formation in hyperacute rejection

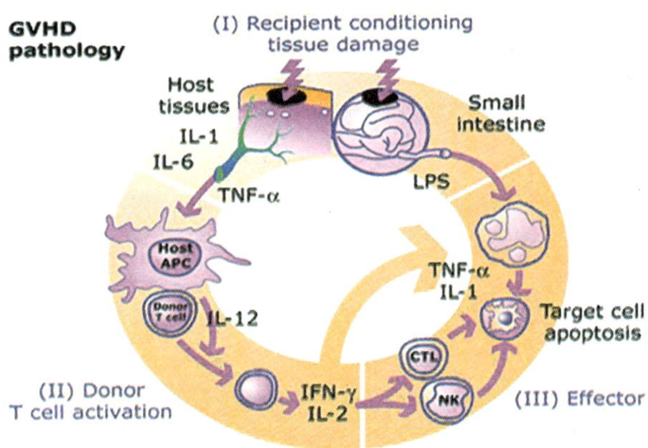

Fig. 19.5 Mechanism of graft-versus-host reaction in three phases of (i) recipients' conditioning, (ii) donor's T cell activation, and (iii) the effect (*Source*: http://www.stembook.org/node/548. 28 February, 2009. Creative Commons Attribution 3.0 Unported license)

of the reaction. GVH occurs under the following conditions:

i. The recipient possesses transplantation antigens that are absent in the graft (recipient's conditioning for tissue damage)
ii. The graft contains immunocompetent T cells (donor's T cell activation)
iii. The recipient is incapable of mounting an effective response to eliminate the graft immunocompetent cells (effect)

- The **acute or fulminant form** of the disease (GVH) is normally observed within the first 10 to 100 days post-transplant.
- This is accompanied by a rash and burning, and redness of the skin on the palms and soles initially and spreading to entire body subsequently.
- Nausea, vomiting, stomach cramps, diarrhea (watery and sometimes bloody), loss of appetite, jaundice, abdominal pain, and weight loss are other symptoms, which follow rapidly.
- Prognosis was poor in the past; however, in recent times, patients have been found to improve with intense immunosuppression with steroids.

Chronic GVH occurs over three months to two years after the transplant. The symptoms are prolonged with the GI, hepatobiliary, and respiratory systems getting affected over a period, affecting long-term survival.

Examples of situations leading to the GVH reaction

- Allograft in a recipient in whom specific immunological tolerance has been induced
- Adult lymphocytes injected into an immunologically-deficient recipient in whom the deficiency is due to immaturity (newborn) or immunosuppression
- F1 hybrid receiving a transplant from any one parental strain

LABORATORY TESTS REQUIRED BEFORE TRANSPLANTATION

In order to achieve a successful transplant, several tests and procedures must be performed to match the donor and recipient's immunological and antigenic parameters.

1. Histocompatibility Testing

- **Blood grouping:** ABO blood group antigen compatibility is important in transplantation.
- **HLA compatibility:** Second to ABO blood group compatibility, the most important factor in allograft survival is HLA compatibility. This is tested by HLA typing and tissue matching. HLA typing identifies the HLA antigens expressed on the surface of leukocytes.

HLA typing

- Monoclonal antibodies to HLA antigens are used in the **microcytotoxicity test** which detects the complement-mediated lysis of peripheral blood lymphocytes with a standard set of typing sera.

- However, serological typing is not possible for HLA-DR antigens, which are detected by the **mixed leukocyte reaction (MLR) and primed lymphocyte typing (PLT)**.

Microcytotoxicity test

- Lymphocyte suspensions are added to microwells of tissue typing trays predispensed with a panel of HLA typing sera, each containing alloantibodies to a specific HLA antigen.
- The plates are then incubated with complement. Cells carrying antigens corresponding to the HLA antiserum are killed by complement-mediated membrane damage.
- These cells can be detected by the addition of eosin or trypan blue, which stains only dead cells (antisera have now been replaced by monoclonal antibodies).

Molecular methods

- Genetic methods are being used increasingly for HLA typing in advanced centres. These employ restriction fragment length polymorphism (RFLP) and gene sequence-specific oligonucleotide probe typing.
- The HLA antigens coded for by the combination of alleles at each locus on one strand of a chromosome pair represent the haplotype. The complete HLA type of an individual comprises the antigens represented on both strands of the diploid chromosome and so will consist of two haplotypes (e.g., HLA-A1, -A2; -B7, -B12; -Cw3, -Dw8; Dw4; -Dw7; -DR1; -DR7; DQw1; -Qw3; -DPw4; -DPw6).

Applications of HLA typing

Due to the extreme pleomorphism of the HLA system, delineation of the HLA type provides a method of typing individuals that is far more discriminating than blood grouping.

- **Transplantation:** HLA typing is used primarily for testing compatibility between recipients and potential donors before tissue transplantation.
- **Paternity:** It also has applications in cases of disputed paternity.
- **Anthropological studies:** As the prevalence of HLA types varies widely between different human races and ethnic groups, HLA typing is used in anthropological studies. Population studies of HLA polymorphism suggest the origin of the human species in Africa and emigration as different subtypes to other continents.
- **Genetic predisposition to disease:** An association has been observed between HLA types and certain diseases. A strong association has been found between ankylosing spondylitis and HLA-B27, rheumatoid arthritis and HLA-DR4, and various autoimmune conditions and HLA-DR3.

2. Tissue Matching

Once a set of HLA-compatible donors is available (usually the patient's siblings), the best among them can be chosen by tissue matching. This is done by mixed lymphocyte reaction or culture.

Mixed lymphocyte reaction or culture (MLR and MLC)

- A culture of T lymphocytes, when exposed to HLA-incompatible antigens, will undergo blast transformation, the intensity of the reaction being a measure of the antigenic disparity between the donor and recipient lymphocytes.
- This is a one-way test in which donor lymphocytes are killed and only recipient lymphocytes are permitted to be transformed in response to the incompatible antigens on the donor cells.

IMMUNOSUPPRESSION FOR SUCCESSFUL TRANSPLANTATION

As allograft rejection is an immunological process, immunosuppression of the recipient is required to inhibit rejection. Clinical transplantation employs a combination of immunosuppressive drugs, including the following:

- Corticosteroids (prednisolone)
- Azathioprine
- Cyclosporine (fungal metabolite)—currently the most effective agent
- Cyclophosphamide
- Tacrolimus
- Antibodies against T lymphocytes, CD4 and CD8 cells, and anti-MHC-I/II antibodies

Immunosuppression in experimental animals is achieved by neonatal thymectomy, chronic lymphatic drainage, or the administration of ALS (anti-lymphocyte serum) procedures that inhibit cell-mediated immunity.

Tissue engineering, though still in the experimental stages, has a promising role in transplantation. The aim is to create artificial tissues with an objective of reducing immunogenicity, controling rejection, and increasing tolerance. However, to be fully effective, the application of tissue engineering tools to prepare synthetic cells and tissues requires a better understanding of the genes affecting acceptance and rejection.

PRIVILEGED SITES

- Any site that is **impenetrable to immunocompetent cells** (e.g., cartilage) is an immunologically privileged site. There appear to be certain privileged sites where allografts are permitted to survive, safe from immunological attack.

- Areas in which a lymphatic drainage system is absent (e.g., brain) or ineffective (e.g., testes) can accept allografts without rejection.
- Lack of vascularity at the site also prevents graft rejection. This is the reason for the success of corneal transplants.

- The **fetus can be considered an intrauterine allograft** as it contains antigens foreign to the mother. Why the fetus is exempt from rejection is not clear, though many explanations have been offered, which are as follows:
  - The placenta acts as an immunological barrier by generating a locally immunosuppressive hormone.
  - Major histocompatibility complex (MHC) antigens are present in low density on trophoblastic cells; the cell membranes are relatively resistant to attack by T or NK cells.
  - Antigen shedding by the fetus blocks the aggressive T cells or antibodies by exerting an enhancement effect.
  - An incomplete mucopolysaccharide barrier rich in sialic acid surrounds the trophoblastic cells, protecting them from cytotoxic lymphocytes.
  - The high concentration of alpha-fetoprotein in fetal blood may also be a factor, as it has immunosuppressive properties which may protect the fetus against immunological damage from any maternal leukocytes entering fetal circulation.

OTHER SOURCES OF TRANSPLANTS

- **Fetal stem cells** offer the benefit of generating new cells or tissues (in burns or spinal cord injury).
- **Adult stem cells** can also be used, e.g., hematopoietic stem cells in bone marrow transplants in hematological malignancies.
- **Non-human (xeno) transplants** are still in the experimental stages and pose several drawbacks (including social and ethical issues).

POST-TRANSPLANT INFECTIONS

The use of immunosuppresive agents to ensure the successful uptake of transplanted tissues or organs has a serious drawback—it makes the recipient prone to opportunistic (secondary) microbial infections.

The risk of infection in a transplanted patient depends on the following:

- Immunosuppression (determines susceptibility)
- Exposure to microbial agents

Types of Post-transplant Infections

Depending on the status of the recipient, there could be the following types of infections:

i. **Early infections:** These may be related to:
   - Healthcare-associated infection
   - Donor-associated infection
   - Recipient colonisers

   These infections are generally caused by bacteria such as staphylococci and yeasts such as members of the *Candida* species.

ii. **Second phase:** During the next 3–6 months, infections may be caused by the following:
   - **Viruses:** CMV, EBV, herpes simplex virus (HSV), human herpes virus 6, HBV, and HCV
   - **Fungi:** *Aspergillus*, *Pneumocystis jirovecii*
   - **Parasites:** *Toxoplasma* and *Strongyloides*

iii. **Late-onset (>6 months):** These infections can be grouped as follows:
   - **Community-acquired infections** occur in those with good allograft function—respiratory viruses, pneumococcal pneumonia, nocardiosis, or aspergillosis (occupational)
   - **Chronic or progressive viral infections**—HBV, HCV, CMV, EBV, and papillomavirus infections, leading to malignancies such as hepatocellular carcinoma after HBV or HCV and lymphoma due to EBV

Secondary Effects of Viral Infection

Infections may result in graft rejection and a susceptibility to opportunistic infections.

Prevention of Infections in a Transplant Patient

This is achieved by prolonged or lifelong antimicrobial prophylaxis (e.g., trimethoprim/sulfamethoxazole or antifungal prophylaxis) and careful attention to environmental exposures and strict infection control practices in day-to-day life.

IMMUNOHEMATOLOGY

ABO SYSTEM

- The ABO system consists of four blood groups and is determined by the presence or absence of two distinct antigens, A and B, on the surface of erythrocytes.
- The **four groups** are also distinguished by the presence or absence of two distinct isoantibodies in the serum. Serum contains the isoantibodies specific to the antigen that is absent in the red cell.
- The serum of a group A individual has anti-B antibodies, a group B individual has anti-A antibodies, and a group O individual has both anti-A and anti-B antibodies. In a person of group AB, both anti-A and anti-B antibodies are absent (Table 19.1).

- In the Indian population, the distribution of the blood groups is approximately 40% for O group, 22% for A group, 33% for B group, and 5% for AB group.
- Anti-A and anti-B isoantibodies appear in the serum of infants by the age of six months and persist thereafter. These are called **'natural' antibodies** because they seem to arise from genetic control without any apparent antigenic stimulation.
- Natural anti-A and anti-B antibodies are IgM (saline-agglutinating) antibodies that react optimally between 4°C and 18°C but are less active at 37°C.
- Immune isoantibodies may develop following an ABO incompatible pregnancy or transfusion.

H Antigen

- The red cells of all ABO groups possess a common antigen, the H antigen or H substance, which is a precursor to the formation of A and B antigens.
- Due to its universal distribution, the H antigen is not ordinarily important in grouping or blood transfusion.

RH BLOOD GROUP SYSTEM

- The antibody described by **Philip Levine and Rufus Stetson** and identified as the anti-Rh factor antibody is found in the red cells of the majority of persons tested.
- Rh factor is an **antigen that reacts with rabbit antiserum to Rhesus monkey erythrocytes**; hence it is called the 'Rhesus' or Rh factor.
- In 1941, Levine and his colleagues proved that Rh sensitisation was the cause of hemolytic disease of the newborn.

Rh Typing

- For routine purposes, the typing of persons as Rh-positive or -negative depends on the **presence or absence of antigen D (Rho) on red cells.**
- This can be accomplished by testing with anti-D (anti-Rh) serum.
- The distribution of Rh-positive individuals differs in different races. Among Indians, approximately 93 per cent are Rh-positive and 7 per cent are Rh-negative.
- A variant of D is known as **Du.** Red cells of the Du subtype react with some but not all anti-D sera. Though Du cells may not be agglutinated by anti-D sera, they absorb the antibody on their surface. For the purpose of blood donation, Du cells are considered Rh-positive. However, when a Du individual requires transfusion, it is advisable to use Rh-negative blood because such a person can be immunised by standard Rh-positive blood.
- There are no natural anti-Rh antibodies in the serum. They arise only as a result of Rh-incompatible pregnancy or transfusion.

MEDICAL IMPLICATIONS OF BLOOD GROUP ANTIGENS

Blood Transfusion

- The existence of several different blood group antigens makes it almost impossible to obtain perfectly matched blood for transfusion.
- In routine transfusion practice, only the ABO and Rh antigens are relevant. The other antigens are too weak to be of importance.

Choice of donor

Safety in blood transfusion requires that the following conditions be satisfied in choosing a donor:

- The recipient's plasma should not contain any antibody that will damage the donor's erythrocytes.
  - Due to the absence of isoantibodies in plasma, AB group persons are known as '**universal recipients**'.
  - Rh compatibility is important only when the recipient is Rh-negative. An Rh-positive person may safely receive either Rh-positive or Rh-negative blood, but an Rh-negative individual receiving Rh-positive blood may result in the formation of antibodies against the Rh antigen.
- The donor plasma should not have any antibody that will damage the recipient's red cells.

Table 19.1 Distribution of ABO antigens and antibodies in red cells and serum

	Red cells		Serum	
Blood group	**Antigen present**	**Agglutinated by serum of group**	**Antibody present**	**Agglutinates cells of group**
A	A	B, O	Anti-B	B, AB
B	B	A, O	Anti-A	A, AB
AB	A and B	A, B, O	None	None
O	None	None	Anti-A and anti-B	A, B, AB

- Due to the absence of antigens, the O group is known as the '**universal donor**'. The anti-A and anti-B antibodies in the transfused O blood group do not ordinarily cause any damage to the red cells of A or B group recipients. They are rendered ineffective by the dilution of the recipient's plasma. However, some O-group plasma may contain iso-antibodies in high titres (1:200 or above), which results in damage to the recipient's cells. This is known as the **dangerous O group**.

Cross-matching

Besides ABO grouping and Rh typing of the donor and recipient, it is invariably necessary to perform cross-matching before transfusion to ensure that the donor's blood is compatible with the recipient's blood.

Coombs cross-match

- The most discriminating method of cross-matching is the Coombs cross-match, wherein washed donor cells and recipient serum are incubated in a water bath at 37°C for two hours and then subjected to a direct Coombs test.
- This test detects all incompatibilities including incomplete antibodies.

Complications of incompatible blood transfusion

- Red cells may undergo clumping and intravascular hemolysis or they may be coated with antibodies, engulfed by phagocytes, removed from circulation, and subjected to extravascular lysis.
- This may be accompanied by symptoms such as shivering, tingling sensation, excruciating headache, constricting precordial discomfort, and severe lumbar pain. Hypotension, cold and clammy skin, cyanosis, feeble pulse, and other signs of collapse may be seen. Jaundice, hematuria, oliguria, and anuria may follow.
- Some transfusion reactions may be due to immunological processes other than blood group incompatibility. Rigor, urticaria, and other manifestations often occur when the recipient is hypersensitive to some allergen present in the donor's blood. Serious reactions follow when hemolysed or contaminated blood is transfused.

Hemolytic disease of the newborn

- When an Rh-negative woman carries an Rh-positive fetus, she may be sensitised against the Rh antigen by the passage of fetal red cells into maternal circulation. Minor transplacental leaks may occur at any time during pregnancy, but it is during delivery that fetal cells enter maternal circulation in large numbers.
- The mother is usually **sensitised only at the first delivery** and, consequently, the first child escapes harm. During a subsequent pregnancy, Rh antibodies of the IgG class pass from the mother to the fetus and damage its erythrocytes.
- The clinical features may vary from a mere accentuation of **physiological jaundice in the newborn to erythroblastosis fetalis or intrauterine death due to hydrops fetalis**.
- Hemolytic disease does not affect all the offspring of Rh-incompatible unions.

Detection of Rh antibodies

Most Rh antibodies are of the IgG class, and being '**incomplete antibodies**', they do not agglutinate Rh-positive cells in saline. A minority are complete (saline-agglutinating) antibodies of the IgM class. These are not relevant in the pathogenesis of hemolytic disease as they do not traverse the placenta.

IgG anti-D antibodies may be detected by the following techniques:

i. Using a colloid medium such as 20 per cent bovine serum albumin
ii. Using red cells treated with enzymes such as trypsin, pepsin, ficin, or bromelain
iii. By the indirect Coombs test

The last is the most sensitive method.

Identification of Rh-incompatibility

- Rh typing is performed as part of routine antenatal examination.
- When a woman is Rh-negative and her partner is Rh-positive, fetal complications should be expected.
- Women with Rh-incompatible pregnancies should be screened for Rh antibodies by the **indirect Coombs test at 32–34 weeks of pregnancy** and at monthly intervals thereafter.
- When hemolytic disease is diagnosed antepartum, intrauterine transfusion with Rh-negative blood may be indicated.
- Red cells introduced into the fetal peritoneal cavity will find their way into circulation and will survive normally.
- Premature delivery followed by transfusion may be necessary in some cases.
- When a baby is born with hemolytic disease, exchange transfusion with Rh-negative, ABO-compatible blood is the treatment of choice.

IMMUNOLOGY OF MALIGNANCY (TUMOUR IMMUNITY)

- In normal circumstances, the rates of cell growth and cell death are in equilibrium.
- However, sometimes, cells bypass checks in the rate of division and death to turn **cancerous or malignant.**

- When a cell undergoes malignant transformation, it acquires new surface antigens. It may also lose some normal antigens. This makes a tumour antigenically different from the normal tissues of the host.
- A tumour can, therefore, be considered an allograft and be expected to induce an immune response.
- **CMI is believed to be the mechanism of host defence against malignancy.** The humoral response may not be relevant, or may even be detrimental, as it may facilitate tumour growth by the process of enhancement.
- Several clinical observations indicate the presence of an immune response that prevents, arrests, and occasionally, cures malignancies:
  i. **Spontaneous regression:** Instances of spontaneous regression of established tumours have been reported, especially with neuroblastoma and malignant melanoma. Based on the role played by the immune response in recovery from infections, it is believed that recovery from malignancy may also represent an immune process.
  ii. **Chemotherapy with cytotoxic drugs:** Dramatic cure sometimes follows chemotherapy of choriocarcinoma and Burkitt's lymphoma. Even a single dose of a cytotoxic drug may, on occasion, result in a complete cure. Again, in some types of tumours, such as hypernephroma with pulmonary metastases, the removal of the primary tumour often leads to the regression of the metastases. These observations suggest that once a large mass of tumour has been removed, mopping up operations may be affected by the immune process. The immune response appears to be effective only when the tumour is below a 'critical mass'.
  iii. **Overcoming of defence mechanisms:** There is a higher prevalence of certain types of cancers observed unexpectedly at autopsy than their clinical incidence would suggest. This indicates that the immune system is able to deal with malignant cells as they arise and that only some of them overcome the defence mechanisms and develop into clinical cancer.
  iv. **Cellular response:** Histological evidence of immune response against malignancy is provided by the infiltration of tumours by lymphocytes, plasma cells, and macrophages infiltrating tumours. The cellular response resembles that seen in the allograft reaction. Tumours showing such cellular infiltration have a better prognosis than those that do not.
  v. **Immunodeficiency states:** If the immune system plays a natural role in preventing tumour development, a high incidence of malignancy should be expected in immune deficiency states. This is indeed so. Increased incidence of cancer, particularly lymphoreticular malignancies, is found in congenital immunodeficiency states, in AIDS patients, and in patients undergoing chronic immunosuppressive therapy.

TUMOUR ANTIGENS

1. Tumour-specific Antigens

- These antigens are present in malignant cells but absent in the corresponding normal cells of the host.
- They induce an immune response when the tumour is transplanted in syngeneic animals.
- Such antigens, which induce rejection of tumour transplants in immunised hosts, are termed **tumour-specific transplantation antigens (TSTA)** or **tumour-associated transplantation antigens (TATA).**
- In chemically-induced tumours, the TSTA are tumour-specific.
- Different tumours possess different TSTA, even though they are induced by the same carcinogen.
- In contrast, the TSTA of virus-induced tumours are virus-specific in that all tumours produced by one virus will possess the same antigens, even if the tumours occur in different animal strains or species.

2. Tumour-associated Antigens

This type of antigen is found in some tumours and may also be present in a few normal cells.

- **Oncofetal antigens**
  - Fetal antigens that are found in embryonic and malignant cells but not in normal adult cells are tumour-associated antigens, e.g., alpha-fetoprotein in hepatomas.
  - Their synthesis represents a de-differentiation of malignant cells into more primitive forms.
- **Carcinoembryonic antigen**
  - This is a glycoprotein that can be detected in the serum of many patients with carcinoma of the colon, particularly in the presence of metastases.
  - However, it also appears in some other conditions such as alcoholic cirrhosis, so its diagnostic value is limited.
- **Alpha-fetoprotein**
  - This is an alpha-globulin secreted by normal embryonic hepatocytes.
  - Its serum level drops sharply after birth and is barely detectable in adults.
- **Prostate-specific antigen (PSA)**
  - It is a differentiation antigen; its levels are higher in cases of prostate cancer, for which PSA serves as a diagnostic indicator.
- **CA125 (cancer/carbohydrate antigen 125)**
  - It is widely used as a diagnostic and prognostic marker for ovarian cancer.

DEMONSTRATION OF IMMUNE RESPONSE IN MALIGNANCY

- Both humoral and cellular responses can be demonstrated in malignancy (Fig. 19.6).
- Anti-TSTA antibodies can be demonstrated by indirect membrane immunofluorescence.
- Delayed hypersensitivity to tumour antigens can be detected by skin testing with tumour cell extracts.
- Cell-mediated immunity (CMI) can be demonstrated by the stimulation of DNA synthesis and lymphokine production by the patient's leukocytes on exposure to the tumour antigens.
- The lymphocytes from the patient are cytotoxic to the cultured tumour cells.

IMMUNOLOGICAL SURVEILLANCE

- This theory was developed by **Lewis Thomas** in the 1950s.
- The primary function of cell-mediated immunity is to 'seek and destroy' malignant cells that arise by somatic mutation. Such malignant mutations are believed to occur frequently and would develop into tumours but for the constant vigilance of the immune system.
- Inefficiency of the surveillance mechanism, either as a result of ageing or in congenital or acquired immunodeficiencies, leads to an increased incidence of cancer.

Mechanisms

If immunological surveillance is effective, cancer should not occur. The development of tumours represents a lapse in surveillance. Though the mechanisms of such lapses are not clear, several possibilities have been suggested, which are as follows:

- Due to the rapid rate of proliferation of malignant cells, they may be able to 'sneak through' before the development of an effective immune response. Once the tumour reaches a certain mass, it may be beyond the scope of immunological control.
- Circulating tumour antigens may act as a 'smokescreen', coating the lymphoid cells and preventing them from acting on tumour cells. The tumour antigens on malignant cells may be inaccessible to sensitised cells, being covered by some antigenically neutral substance.
- Humoral antibodies may cause immunological enhancement. 'Blocking' activity has been demonstrated in humoral factors. This may be due to the circulating antigen, antibody, or antigen–antibody complexes.
- Some tumours may be of low immunogenicity or may form cytokines, e.g., the **transforming growth factor (TGF)**, which suppresses CMI.
- Tumour cells frequently express low levels of class I MHC molecules and hence may not be recognised by $CD8^+$ CTLs for destruction. Further, poor immunogenicity and co-stimulatory signals may lead to clonal anergy.

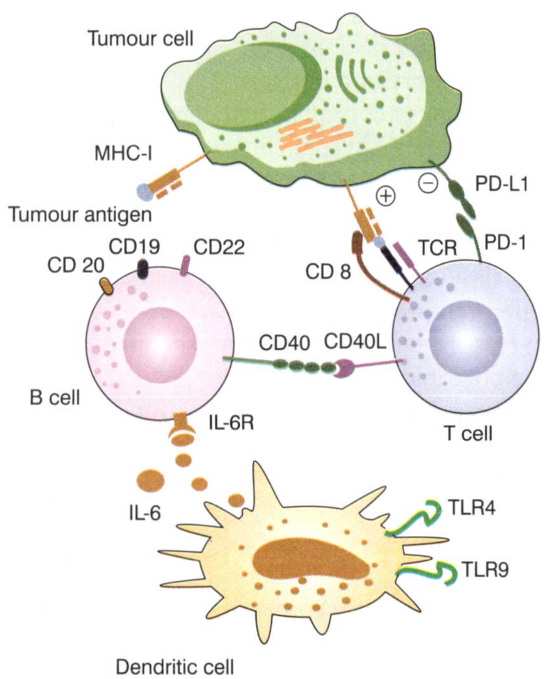

Fig. 19.6 Interaction of T and B cells with tumour cells

IMMUNOTHERAPY OF CANCER

Different approaches have been used for the immunotherapy of cancer—passive, active, and adoptive immunotherapy and specific and non-specific immunotherapy.

Passive Immunotherapy

- A special type of serotherapy has been found to be beneficial in experimental tumours.
- Appropriate antisera that possess 'de-blocking' activity in vitro have been found to cause regression of tumours, apparently by neutralising the circulating tumour antigens and permitting the sensitised lymphocytes to act on tumour cells.
- Monoclonal antibodies to tumour antigens may play a role as carriers in transporting cytotoxic or radioactive drugs specifically to the tumour cells.

Specific Active Immunotherapy

This method has been modified recently by using purified tumour cell membrane antigens and tumour cells treated with neuraminidase to increase their immunogenic potential.

Non-specific Active Immunotherapy

- This method uses BCG and non-living *Corynebacterium parvum*.

- **Georges Mathe**, the leading proponent of cancer immunotherapy, reported very good results in acute leukemia following combined treatment with BCG and allogeneic or autochthonous leukemia blast cells.
- **Intralesional BCG** in malignant melanoma has been reported to induce complete remission in a high percentage of patients. It has also been used against the intradermal recurrence of breast cancer following mastectomy.
- **Dinitrochlorobenzene** has been used in the treatment of squamous and basal cell carcinoma of the skin.
- **Glucan**, a pyran copolymer derived from microorganisms, and **levamisole**, originally introduced as an anthelmintic, have been used to stimulate CMI and macrophage function.
- **Interferons** have been employed in the treatment of leukemias.

Specific Adoptive Immunotherapy

- This has been attempted with lymphocytes, transfer factor, and 'immune RNA'. In these attempts, the donors were persons cured of their neoplasms or specifically immunised against the patient's tumour.
- Lymphokine-activated killer (LAK) cells obtained by the treatment of the natural killer cells with interleukin-2 have been found useful in the treatment of certain malignancies such as renal carcinomas. Immunotherapy is ineffective in the presence of a large mass of tumour cells.
- The role of this modality appears to be important in getting rid of the residual malignant cells after the gross tumour has been removed. The best results in the treatment of cancer apparently follow an integrated approach to therapy, combining surgery, radiotherapy, chemotherapy, and immunotherapy.

Immunomodulation

- Immunomodulators have the ability to stimulate natural and acquired defence mechanisms such as cytokines, which enable the immune system to help itself.
- Natural immunomodulators strengthen weak immune systems and moderate immune systems that are overactive.
- Plant sterols and sterolins are natural immunomodulators found in some raw fruits and vegetables. Immunomodulators may be either immunosuppressants or immunostimulants.
- Immunosuppressive agents are used to inhibit the immune response in transplant rejection and autoimmunity. Immunostimulants are used in immunodeficiency conditions, cancers, and in some infections (Table 19.2).

Table 19.2 Immunosuppressive and immunostimulating agents and their therapeutic uses

Agents		Therapeutic uses
Immunosuppressive agents		
Cytotoxic agents	Azathioprine, methotrexate, cyclophosphamide	
Antimetabolites, alkylating agents		
Cytokine inhibitors	Anticytokine-antibodies, TNFα inhibitors, IL-1 and IL-2 inhibitors	Organ transplantation Autoimmune disorders
Antibodies against specific immune cell molecules (e.g., antithymocyte globulin [ATG])	Polyclonal antibodies, monoclonal antibodies	
Immunostimulants		
Recombinant cytokines	Interferons (α, β, γ)	Malignant melanoma, follicular lymphoma, AIDS-related Kaposi's sarcoma
	Interleukins	Metastatic renal cell carcinoma and melanoma
	Colony-stimulating factors (CSF)	Leukopenia, ganciclovir-induced neutropenia
Bacillus–Calmette–Guerin (BCG)	Produces granulation tissue at the site of injection	Used for prophylaxis of primary and recurrent-stage papillary tumours after transurethral resection
Levamisole	Used as an adjuvant therapy with 5-fluorouracil	After surgical resection in patients with Duke's stage C colon cancer and agranulocytosis
Cancer vaccines	Several immunomodulants Still under research or in phases of clinical trials	Human papilloma virus vaccine to prevent cervical carcinoma

Key Points

- ✓ Transplants may be classified based on the organs or tissue transplanted, the anatomical site of origin of the transplant and its placement.
- ✓ Based on genetic relationships, transplants can be classified as follows:
  - Autograft (taken from an individual and grafted on the same person)
  - Isograft (between individuals of the same genetic constitution, like identical twins)
  - Allograft (between two genetically non-identical members of the same species)
  - Xenograft (between members of different species)
- ✓ Histocompatibility antigens are cell surface antigens that induce an immune response leading to the rejection of allografts coding for three different classes of proteins—I, II, and III.
- ✓ Human major histocompatibility complex (MHC) antigens are synonymous with human leukocyte antigens (HLA) and the MHC complex of genes with the HLA complex. The HLA complex of genes consists of three separate clusters of genes:
  - HLA class I, comprising A, B, and C loci, which are found on the surface of virtually all nucleated cells and are involved in graft rejection and cell-mediated cytolysis
  - HLA class II or the D region, consisting of DR, DQ, and DP loci are found only on cells of the immune system (macrophages, dendritic cells, activated T cells, and particularly on B cells)
  - HLA class III or the complement region, containing genes for complement components C2 and C4; they are heterogenous
- ✓ First-set response refers to the sequence of events resulting in the rejection of an allograft.
- ✓ In second-set response, there is accelerated rejection of another graft from the same donor.
- ✓ An allograft is accepted if the host is rendered immunologically tolerant.
- ✓ Transplantation immunity is predominantly cell-mediated. The first-set response is brought about almost exclusively by T lymphocytes, while in the second-set response, antibodies participate along with cell-mediated immunity.
- ✓ Histocompatibility (transplantation) antigens participate in graft rejection.
- ✓ The human leukocyte antigen (HLA) system is the major histocompatibility system in humans.
- ✓ The three classes of HLA antigens are I, II, and III. HLA class I antigens (A, B, C) are the principal antigens involved in graft rejection and cell-mediated cytolysis.
- ✓ The most important factors favouring allograft survival are the ABO blood group antigens and the HLA antigens.
- ✓ HLA compatibility is tested by HLA typing and tissue matching.
- ✓ In the graft-versus-host (GVH) reaction, the graft mounts an immune response against the antigens of the host.
- ✓ When a cell undergoes malignant transformation, it acquires new surface antigens and may also lose some antigens, making it antigenically different from the normal tissues of the host.
- ✓ Tumour-specific antigens, such as tumour-specific transplantation antigens (TSTA), are present in malignant cells but absent in the corresponding normal cells of the host.
- ✓ Tumour-associated antigens are found in some tumours and may also be present in a few normal cells. For example, fetal antigens are found in embryonic and malignant cells but not in normal adult cells. Some examples are alpha-fetoprotein in hepatomas, carcinoembryonic antigen in colonic cancers, and CA125 in ovarian cancers.
- ✓ Both humoral and cellular responses can be demonstrated in malignancy; cell-mediated immunity is the probable mechanism of host defence against malignancy.
- ✓ Lapses in immune surveillance allow tumours to emerge and may result in the proliferation of malignant cells, inaccessibility of tumour antigens on malignant cells to immunocompetent cells, and lowered cell-mediated and humoral immune responses.
- ✓ Different approaches have been attempted in the immunotherapy of cancer:
  - Passive immunotherapy, using antisera against TSTA, was found to be of no use.
  - Monoclonal antibodies to tumour antigens have been used as carriers to transport cytotoxic or radioactive drugs specifically to tumour cells.
  - Non-specific active immunotherapy uses BCG and non-living *Corynebacterium parvum*.
  - Specific adoptive immunotherapy has been attempted with lymphocytes, transfer factor, and 'immune RNA'. Lymphokine-activated killer (LAK) cells have been found useful in the treatment of certain malignancies. Immunotherapy is ineffective in the presence of a large mass of tumour cells.

Questions

Short answer questions

1. Describe the features of an allograft reaction.
2. Describe graft-versus-host reaction.
3. What are histocompatibility antigens?
4. What are the methods of HLA typing?
5. Describe the classes of the major histocompatibility complex (MHC) and their role in immunity.
6. What is HLA? Enumerate its classes and describe their functions.
7. What are the complications of mismatched blood transfusion?
8. Describe the method of Rh antibody detection in Rh compatibility testing.
9. Describe tumour antigens.
10. Discuss the role of immune surveillance in the prevention of malignancy.
11. Enumerate agents used in immunotherapy for cancer.

MCQs

1. **What is a graft transplant between two genetically non-identical members of the same species called?**
   a) Autograft b) Isograft
   c) Allograft d) Xenograft
2. **Which of the following are responsible for hyperacute rejection?**
   a) Preformed antibodies
   b) Lymphokines
   c) In situ antibody formation
   d) All of these

3. **Which one of the following is applicable to transplant reaction?**
   a) White graft response is due to delayed rejection
   b) Chronic graft rejection responds well to corticosteroids
   c) Acute rejection occurs within 10 days of transplantation
   d) All of these
4. **Which of the following defines Runt disease?**
   a) Graft rejection
   b) Graft-versus-host reaction
   c) Deficient T cell function
   d) Complement deficiency
5. **Which of the following is not used as an immunosuppressive agent?**
   a) Azathioprine b) Methotrexate
   c) Levamisole d) Cyclophosphamide
6. **Which is the best method to detect Rh antibodies in serum?**
   a) Direct Coombs test
   b) Indirect Coombs test
   c) Major cross-match
   d) Minor cross-match
7. **By what age do anti-A and anti-B isoantibodies appear in the serum of infants?**
   a) 4 months b) 6 months
   c) 8 months d) 10 months
8. **Which of the following antibodies does a universal donor have?**
   a) Anti-A b) Anti-B
   c) Anti-A and anti-B d) None
9. **Which of the following cells is not evidence of immune response playing a role in malignancies?**
   a) Lymphocytes b) Plasma cells
   c) Macrophages d) Eosinophils
10. **Which of the following statements is not true for tumour antigens?**
    a) Tumour-specific transplantation antigens induce an immune response when the tumour is transplanted in syngeneic animals.
    b) Tumour-associated antigens may also be present in a few normal cells.
    c) Oncofetal antigens are found in normal adult cells as well.
    d) Carcinoembryonic antigens are found in alcoholic cirrhosis.
11. **Which of the following factors does not lead to malignancy when there is a lapse of effective immunological surveillance mechanism ________?**
    a) Rapid proliferation of malignant cells allows them to 'sneak through'
    b) Circulating tumour antigens may act as a 'smokescreen'
    c) Some tumours may form cytokines, which suppresses CMI
    d) Tumour cells frequently express low levels of class I MHC molecules and hence they are recognised by $CD8^+$ and CTLs
12. **A common immunomodulant used in metastatic renal cell carcinoma therapy is ________.**
    a) Interleukin-2 (IL-2)
    b) Levamisole
    c) Colony-stimulating factor (CSF)
    d) 5-Fluorouracil (5-FU)
13. **Heterotopic graft refers to:**
    a) Graft placed in an anatomically normal site
    b) Graft placed in an anatomically abnormal site
    c) Graft that acts as scaffolding
    d) Graft from a cadaver
14. **Immunological enhancement results in:**
    a) Resistance to infections
    b) Hyperacute graft rejection
    c) Slowing down of graft rejection
    d) Enhanced functioning of thymus
15. **Eichwald-Silmer effect is:**
    a) Improved graft acceptance in genetically identical animals
    b) Unilateral sex-linked histoincompatibility
    c) Unilateral sex-linked histocompatibility
    d) Enhanced rejection in inbred animals
16. **Which of the following statement is not true regarding MHC complex?**
    a) The primary aim of the MHC is to give defence against grafts.
    b) The MHCs are regulated by genes.
    c) They are cell surface markers that help infected cells to signal cytotoxic and helper T cells.
    d) The T cell can accept the processed antigen only if it is presented by a macrophage carrying on its surface the self-MHC antigens.
17. **Antigens that participate in graft rejection are called:**
    a) Histocompatibility antigens
    b) Alloantigens
    c) Heterologous antigens
    d) Haptens
18. **Acute graft rejection is characterised by:**
    a) Rejection of graft within hours of implantation
    b) Appearance of 'white graft response'
    c) Damage that occurs due to preformed anti-donor antibodies
    d) Accumulation of lymphocytes, plasma cells, macrophages, and neutrophils, leading to endothelial damage

Answers: 1:c, 2:a, 3:c, 4:b, 5:c, 6:b, 7:b, 8:d, 9:d, 10:c, 11:d, 12:a, 13:b, 14:c, 15:b, 16:a, 17:a, 18:d

CHAPTER 20

Part II: Immunology

Immunoprophylaxis

(MI1.9)

Chapter Navigation

- Active immunisation
- Passive immunisation
- Combined active and passive immunisation
- Individual immunisation
- Vaccination during outbreaks and pandemics
- Newer vaccines

Learning objective

- Discuss the immunological basis of vaccines and describe the Universal Immunisation Schedule

INTRODUCTION

- The earliest evidence of the practice of vaccination against infectious diseases has been traced back to ancient civilisations in Greece, China, and India.
- Edward Jenner formally demonstrated the method in 1796 when he protected a young boy from smallpox by injecting him with material from cowpox blisters on the hands of a milkmaid. On subsequent injection of material from smallpox, the boy did not develop any symptoms of the disease, proving that he had been protected by the injected material. The disease, which killed millions of individuals in the pre-vaccine era, was finally eradicated with the help of the smallpox vaccine (a live-attenuated vaccine). This vaccine is no longer used except for laboratory infections.
- Robert Koch and Louis Pasteur expanded on this to develop vaccines for several diseases of animals and humans.
- Subsequently, some of the most dreaded and fearful infections were contained by immunising infants and young children.

Immunoprophylaxis

It involves the **administration of all or part of a microorganism** (live or inactivated) or a **product of that microorganism** (such as a toxoid) to the host to **stimulate an immunologic response** aimed at protecting the host against the infection. The process by which this protection is achieved is called **immunisation**.

Vaccination

It is the process of **inducing active immunity artificially**. It is aimed at inducing long-acting protective immunity in the form of antibody-producing plasma cells, memory B cells, and memory T cells.

The World Health Organization, Center for Disease Control and Prevention, and several countries including India developed universal immunisation programmes for children and other vulnerable populations to ensure protection from many fatal infectious diseases.

TYPES OF IMMUNISATION

Active Immunisation

- This refers to the induction of **artificially acquired immunity following the administration of a vaccine**, which may be a live-attenuated or killed microorganism or its components.
- The types of vaccines used to prevent infections (prophylaxis) are listed in Table 20.1.

Live-attenuated vaccines

- Organisms that have been made to lose their ability to cause disease (pathogenicity) while retaining their ability to induce an immune response (immunogenicity) and generate memory cells are said to be attenuated.
- This is achieved by growing organisms in altered culture conditions for prolonged periods.

Killed or inactivated vaccines

- Killed vaccines are relatively less efficacious as the pathogens are inactivated by heat or chemicals and hence do not cause disease.

Table 20.1 Types of vaccines

Vaccine type	Examples
Live-attenuated	Measles, mumps, rubella, varicella-zoster, BCG, rotavirus, oral polio (Sabin)
Killed vaccine	Rabies, pertussis, cholera, hepatitis A, killed polio (Salk)
Recombinant subunit	Hepatitis B, acellular pertussis
Toxoid	Tetanus, diphtheria
Protein conjugated	*H. influenzae* type-b (Hib), meningococcal, pneumococcal (13PCV)
Polysaccharide	Meningococcal and pneumococcal vaccines
DNA vaccines	Hepatitis B and C, HPV, SARS-CoV2 (malaria and HIV are under trial)
mRNA vaccines	Influenza, Ebola, Zika virus, rabies, and SARS-CoV-2
Edible vaccines	Cholera, measles, and hepatitis C and E (under trial)

- Killed vaccines require primary and booster doses.
- The differences between live and killed vaccines are presented in Table 20.2.

Toxoids

- These are purified **exotoxins** produced by bacteria (such as the organisms that cause tetanus and diphtheria) that are inactivated by **treatment with formaldehyde**.
- Antibodies generated against toxoids neutralise the toxin moiety produced during infection rather than acting upon the pathogen.
- Some toxoids are mixed with other vaccines to enhance their antigenicity. For example, in the DPT (diphtheria, pertussis, and tetanus) vaccine, diphtheria and tetanus toxoids are combined with the pertussis vaccine (which acts as an adjuvant).

Cellular fractions (bacterial polysaccharides)

- Certain vaccines are prepared from extracted cellular fractions of microorganisms.
- Polysaccharide vaccines induce poor-quality antibody responses, especially in children, as the carbohydrate antigens in general are not highly immunogenic. Immunogenicity can be increased by conjugating with a protein.

Conjugate vaccines

- Vaccines containing carbohydrate polysaccharide antigens require a protein conjugate to increase their immunogenicity, e.g., **heptavalent pneumococcal conjugate vaccine (7PCV)**.
- **Antigen valency** has been shown to influence the efficacy of a vaccine. High-valency antigens have more antibody-binding sites than low-valency antigens. High-valency antigens can stimulate the immune system to increase the number of antibody-producing B cells. However, the optimum valency for different vaccines may vary. Research has shown that a valency of 60 may be enough to boost B cell numbers in some while a valency of four might be enough in many.

Heterogenous vaccines

An allied pathogen is administered to protect against another, e.g., the administration of BCG vaccine made from *Mycobacterium bovis* protects against human tuberculosis caused by *Mycobacterium tuberculosis*.

Homologous vaccines

- They contain the same strain as the one causing the infection.
- For example, for the influenza epidemic caused by AH3N2, the vaccine strain used was also AH3N2.

Heterologous vaccine

- Due to antigenic drift (change in the antigenicity of the virus), the vaccine strain may differ from the strain causing the infection.
- In order to increase the cross-protection and breadth of response, a heterologous vaccine with an adjuvant is used.
- This is evidenced in the seasonal influenza epidemics (e.g., A/Panama/2007/99 [H3N2] with an adjuvant

Table 20.2 Differences between live-attenuated and killed vaccines

Live-attenuated vaccines	Killed vaccines
Produced by attenuating the live organism as a whole	Produced by inactivating pathogens by heat or chemicals
Primary course provides relatively long-lasting protection	Short-lasting immunity as they do not follow a natural infective course in the body
Generate memory cells	Do not generate memory cells
Highly effective in generating immune response	Less effective in generating immune response
They provide both humoral (antibody-mediated) and cell-mediated immunity	They do not result in infection and, therefore, the immune response produced is not as complete as that which is stimulated by a live vaccine
They exhibit wide antigenic diversity	Narrow antigenic diversity
Demonstrate good cross-reactivity against heterologous strains	Minimal cross-reactivity with heterologous strains
Generally, a single dose or two doses	Repeat doses are required as immunity is short-lasting (primary and booster dose)
Have the risk of reverting to the virulent strain	They are safer than live vaccines
Require cold chain facility to transport	Do not require cold chain facility to transport
Cannot be given to immunocompromised individuals	Can be given to immunocompromised individuals

was found to provide protection against A/Wyoming/3/2003 [H3N2], which is a drift variant).

Subunit vaccines

- These contain purified parts of the pathogen that are antigenic or are necessary to elicit a protective immune response.
- They are produced from purified macromolecules derived from immunogenic components of pathogenic microorganisms by recombinant DNA technology.

Recombinant vaccines

- Genes encoding antigens are cloned in bacteria, yeast, and some mammals using recombinant technology. The expressed antigenic proteins are purified and used as vaccines.
- Examples are the hepatitis B and pertussis vaccines.
- The surface antigen of the hepatitis virus was the first recombinant vaccine to be cloned in yeast.
- Since purified proteins are used, adverse reactions are minimal.

DNA vaccines

- These are made of small, circular pieces of the DNA (plasmid) of the microorganism that have been genetically engineered to produce one or more specific proteins of the pathogen against which immunity is desired.
- The gene is incorporated into a bacterial plasmid by recombinant technology. These bacteria replicate to produce large quantities of the plasmid, which are then purified.
- This forms the source of the required DNA which is delivered by intramuscular injection, dermal patches, or **electroporation (EP)**—the application of short electrical pulses to the delivery tissue—for DNA uptake by the dendritic cells (Fig. 20.1).

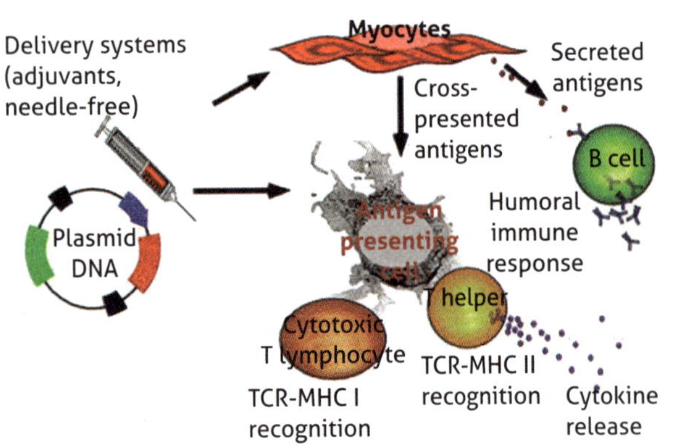

Fig. 20.1 DNA vaccine—mechanism of inducing immunity (*Source*: Moss R. B. 2009. Proposed mechanism of DNA vaccines. Prospects for control of emerging infectious diseases with plasmid DNA vaccines. *J Immune Based Ther Vaccines*, 7, 3. Creative Commons Attribution 2.0 Generic license)

- DNA vaccines can activate $CD8^+$ cytotoxic T cells (CTL) to evoke an immune response.

mRNA vaccines

- The mRNA of the infecting virus, which encodes for specific antigens like surface proteins or receptor-binding domains, is introduced into somatic cells to synthesise viral antigens (Fig. 20.2).
- These synthetic viral antigens can then induce the immune response.
- Being unstable, **mRNAs need carriers**—the commonly used ones are lipid-based, polymer-based, and peptide-based delivery systems.
- Compared to traditional vaccines, mRNA vaccine design needs virus gene sequences instead of virus strains and does not need cell culture.
- mRNA vaccines have been developed against some respiratory viruses like influenza, Ebola, Zika, and rabies.
- The development of mRNA vaccines takes far less time than other conventional vaccines, an attribute that expedited the development of SARS-CoV-2 mRNA vaccines in response to the coronavirus pandemic of 2020–21.

Edible vaccines

- Genes that code for the antigen of interest are introduced into plants by a process of **transformation**.
- These plants produce the encoded proteins (which are antigenic proteins and act like the subunit vaccines), which can stimulate an immune response in the human body.
- The altered plants are known as **transgenic plants**.

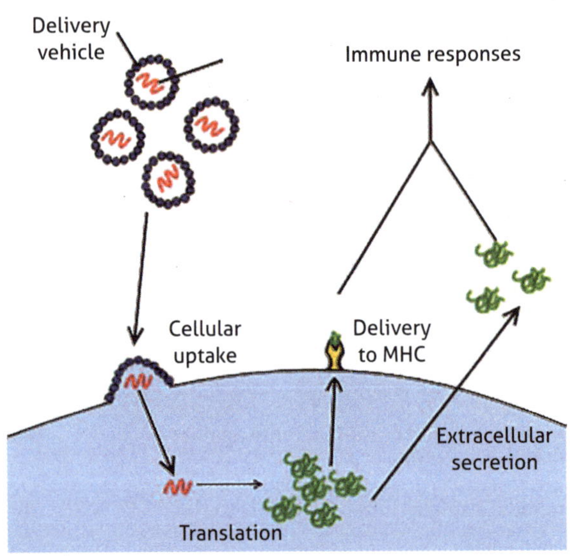

Fig. 20.2 mRNA vaccine—mRNA in a lipid nanoparticle (vehicle) which helps it to enter the cells (*Source*: Kuon Haku. 2020. Creative Commons Attribution-ShareAlike 4.0 International)

- Edible vaccines do not bear the pathogenic genes.
- The vaccine is given orally and acts through the intestinal mucosa.
- Edible vaccines have been tried for cholera, measles, and hepatitis C and E. Banana, tomato, potato, lettuce, and rice act as transgenic plants.

Newer vaccines in the schedule like HiB, rotavirus vaccine (Rotarix), pneumococcal vaccine, and 7PCV or 13PCV are available for the immunisation of children.

Passive Immunisation

- Passive immunisation refers to the use of **human or animal sera containing preformed antibodies**. This type of immunisation is administered when the need for immune intervention is immediate and protection cannot await the development of active immunity or when the recipient is not capable of mounting his own immune response.
- **Transplacental transfer** of immunity from the mother to the fetus is another example of passive immunity, an important source in infancy.

Passive immunisation is deployed using specific human immunoglobulins (Ig) for protection against different diseases, such as:

- Hepatitis B (HBIg)
- Tetanus (HTIg)
- Rabies (HRIg)
- Vaccinia (HVIg)

Human sera

Two forms of normal immunoglobulins can be injected into patients—pooled and specific.

i. **Pooled immunoglobulins** are used for short-term prophylaxis, e.g., in cases of exposure to hepatitis A or measles. Pooled sera are obtained from the plasma of a pool of healthy donors and contain a mixture of immunoglobulins, mainly IgG, with some amounts of IgM and IgA.
ii. **Hyperimmune immunoglobulins** are prepared from the plasma of patients in the convalescent phase of the infection against which immunity is sought.

Administration of human sera

- Hyperimmune sera are injected intramuscularly.
- In the case of rabies, half the dose is given around the bite wound and the other half, intramuscularly.

Plasma therapy

- Convalescent plasma from people who've recovered from an illness is used to passively help others recover.
- Plasma with high anti-COVID-19 antibody titres was used as therapy in the recent pandemic for patients who were in the early stages of the infection or those with low immunity.

Monoclonal antibodies

- These are derived from a single type or clone of antibody-producing cells (B cells) by hybridoma technology.
- These are specific to a single antigen or closely related antigens, e.g., monoclonal antibodies to the respiratory syncytial virus (RSV).

Animal sera

Equine sera were used in the past but have largely been replaced by human hyperimmune sera, which do not carry the risk of hypersensitivity.

Combined Active and Passive Immunisation

- A combination of active and passive immunisation is preferred when immediate passive immunity must be provided until active immunisation generates specific immune responses to the particular disease/infection.
- In this approach, a combination of vaccines (diphtheria/tetanus and rabies) is given simultaneously but injected at two different sites.

Individual Immunisation

The vaccines offered under national programmes are limited by economic considerations. For this reason, certain important vaccines may be omitted because of their high cost. These may be supplemented by individual initiative, whenever possible.

Varicella vaccine

- It is used for the prevention of **varicella and herpes zoster**.
- A live-attenuated **Oka strain of VZV** is used in the vaccine.
- Two doses are administered—the first at 12–15 months and the second at 4–6 years of age.
- This vaccine is **contraindicated in pregnancy**.

Typhoid vaccine

- This vaccine is recommended for the immunisation of children aged five years and above and may be employed at school entry.
- There are two types of typhoid vaccine—the **live oral Gal-E mutant vaccine** and the **injectable purified Vi polysaccharide vaccine.**

The immunoprophylaxis for individual diseases has been described in the respective chapters.

VACCINATION DURING OUTBREAKS AND PANDEMICS

- Population protection during outbreaks or epidemics of infections is an important measure of infection containment.

- It is aimed at producing **herd immunity.** Herd immunity refers to the indirect protection from an infectious disease that develops when a population is immune either through vaccination or immunity developed through previous infection.
- Some of the vaccines used for this purpose are flu vaccines (seasonal flu, H1N1 vaccines), cholera vaccines, Japanese encephalitis vaccine, etc.
- For example, a mass vaccination campaign was launched to control the 2020–22 COVID-19 pandemic; several mRNA vaccines, killed whole virus vaccines, and genetically engineered viral vector vaccines were deployed to bring down the caseload and protect the vulnerable population and healthcare workers from the highly transmissible virus. Some of the COVID-19 vaccines available for use are listed in Table 20.3.

Table 20.3 Vaccines used for adult immunisation against COVID-19 during the 2020–2022 pandemic

Vaccine platform	Name of the vaccine
Viral vector (non-replicating adenovirus)	COVISHIELD (made in the Serum Institute of India)
Protein subunit	Novovax
mRNA-based vaccine	Moderna
DNA-based vaccine	ZyCoV-D
Inactivated virus (whole virion, inactivated SARS-CoV-2 vaccine)	Covaxin (made in India)

ADULT VACCINATION

Adults need vaccinations to boost the efficacy of childhood vaccines, the effects of which may wane with time or get suppressed due to comorbidities and acquired illnesses. The different categories of adult vaccines are presented in Table 20.4.

Table 20.4 Categories of adult vaccination

Category of adult vaccination	Description
Booster	Boost protection acquired by immunisation earlier in life in the absence of 'natural' boosting from exposure to the infectious agent.
Accelerate control or elimination efforts	Disease control initiatives frequently aim at increasing herd immunity, interrupting transmission, and covering the non-immune cohorts.
Counter specific risks	These could be travel, high-risk behaviour, and immunocompromised states.

Though no national guidelines are available, the following vaccinations are recommended for adults.

- Tetanus, diphtheria, pertussis
- Human papillomavirus (details in chapter 34)
- Varicella
- Zoster (>60 years)
- Measles, mumps, rubella
- Influenza (>50 years)
- Pneumococcal vaccine (>65 years)
- Hepatitis A
- Hepatitis B
- Meningococcal vaccine

The vaccines recommended for pregnant women are listed in Table 20.5.

Table 20.5 Vaccines for pregnant women (*Source*: https://main.mohfw.gov.in)

Vaccine	When to give	Dose	Route	Site
TT-1	Early in pregnancy	0.5 mL	Intramuscular	Upper arm
TT-2	4 weeks after TT-1	0.5 mL	Intramuscular	Upper arm
TT-Booster	If the woman received 2 TT doses in a pregnancy within the last 3 years	0.5 mL	Intramuscular	Upper arm

NEWER VACCINES IN THE EXPERIMENTAL OR CLINICAL TRIAL STAGES

Several vaccines are in the developmental stages or under clinical trial. Some of the notable diseases for which vaccines are being developed are as follows:

- Dengue
- Nipah
- Chikungunya
- Influenza
- Zika
- Malaria (this has recently been released by the WHO for use in Africa for immunisation of the population)
- Influenza (universal influenza vaccine)
- Ebola
- Group B streptococci
- COVID-19

Key Points

✓ Immunoprophylaxis provides specific protection against infectious diseases by stimulating the immune system.

✓ Active immunisation involves both humoral and cell-mediated immunity and is associated with specificity and immunological memory.

- ✓ Vaccines used for immunisation may be:
  - Live-attenuated
  - Killed
  - Recombinant subunit vaccines
  - Toxoid
  - DNA vaccines
  - mRNA vaccines
  - Edible vaccines
- ✓ The transfer of preformed antibodies, human sera, or animal sera is used for passive immunisation.
- ✓ Immunoprophylaxis may be in the form of routine immunisation, which forms part of basic healthcare or immunisation of individuals exposed to the risk of specific infections.
- ✓ In the national immunisation schedule in India, BCG, oral polio, DPT, and measles vaccines are given to infants and children at different intervals according to the schedule.
- ✓ Newer vaccines like HiB, rotavirus vaccine, and pneumococcal conjugate vaccines are now included in the immunisation programme.

Questions

Short answer questions

1. Describe the mechanism of the following types of vaccines with examples:
   - Live-attenuated vaccines
   - Killed vaccines
   - Toxoids
   - Subunit vaccines
   - Four types of SARS-CoV-2 vaccines used during the pandemic
2. Define and describe the process of passive immunisation.
3. Describe the national immunisation schedule and its timeline.

MCQs

1. Which of the following is a live-attenuated vaccine?
a) BCG b) DPT
c) Hib d) Pertussis

2. Which of the following statements is not true regarding live vaccines?
a) The primary course provides relatively long-lasting protection
b) Provides both humoral and cell-mediated immunity
c) Indicated in an immunocompromised host
d) Good cross-reactivity against heterologous strains

3. Which of the following vaccines is administered at birth?
a) BCG b) Rotavirus vaccine
c) DPT d) Measles

4. Besides other mandated vaccines, which of the following vaccines should a child receive at the end of nine months?
a) BCG b) Measles
c) DPT d) HBV

5. Which of the following cells do DNA vaccines activate to evoke an immune response?
a) CD4 T cells b) CD8
c) Macrophages d) B cells

6. For which of the following viruses has an mRNA vaccine been attempted?
a) Influenza virus b) Measles virus
c) Rubella virus d) Rhinovirus

7. Which one of the following vaccines against SARS-CoV-2 is a live-attenuated vaccine?
a) COVISHIELD b) Covax
c) COVAXIN d) Sputnik

8. The following organisms have been tried in edible vaccine except:
a) HCV b) Cholera
c) HEV d) HPV

9. The following transgenic plants have been tried for the production of edible vaccines:
a) Banana b) Rice
c) Lettuce d) Beetroot

10. Hyperimmune immunoglobulins are used for passive immunisation in all the diseases EXCEPT:
a) Hepatitis B (HBIG) b) Vaccinia (HVIG)
c) Rabies (HRIG) d) Papilloma (HPIG)

11. HBV vaccine is an example of which type of vaccine?
a) Live b) Killed
c) Live attenuated d) Cloned subunit

12. The type of vaccine used against Japanese encephalitis is:
a) Killed b) Live attenuated
c) Live d) Subunit

13. Which one of the following is false with regard to attenuated live viral vaccines?
a) They act like the natural infection
b) Do not provide long-lasting immunity
c) There is a risk of the virus becoming more virulent
d) Limited shelf life

14. All of the following are true regarding killed vaccines except:
a) Provide brief immunity
b) No risk of reverting to virulent strain
c) Confer local immunity
d) Poor cell-mediated immune response

15. DNA vaccines are composed of:
a) Genes encoding antigens that have been cloned in bacteria or yeast.
b) Small, circular pieces of the DNA that has been genetically engineered to produce one or more specific proteins of the pathogen.
c) Purified macromolecules derived from immunogenic components of pathogenic microorganisms by recombinant DNA technology.
d) Messenger RNA in lipid nanoparticle.

16. Which of the following vaccine is not included in universal immunization program?
a) Japanese encephalitis vaccine
b) *Haemophilus Influenzae* vaccine
c) BCG
d) SARS-CoV2 vaccine

17. Which of the following statement is correct regarding killed vaccines?
a) Primary course provides relatively long-lasting protection.
b) They generate memory cells
c) They exhibit narrow antigenic diversity
d) They are highly effective in generating immune response

18. Which statement is correct regarding mRNA vaccines?
a) mRNA vaccine needs virus gene sequences and not virus in cell culture.
b) It takes far lesser time than other conventional vaccines.
c) It needs a carrier molecule as it is unstable.
d) All of the above

19. Which of the following is an example of polysaccharide vaccine?
a) *Haemophilus influenzae* type-b (Hib) vaccine
b) Hepatitis B vaccine
c) Meningococcal vaccine
d) Tetanus vaccine

Answers: 1:a, 2:c, 3:a, 4:b, 5:b, 6:a, 7:c, 8:c, 9:d, 10:d, 11:d, 12:a, 13:b, 14:c, 15:b, 16:d, 17:c, 18:d, 19:c

Part III: Microbiology as Applied to Infectious Diseases

The chapters in this section deal with the application of basic and clinical microbiology to infectious diseases. These chapters describe infectious diseases that commonly affect various systems of the human body and the microorganisms causing these syndromes. The microorganisms primarily involved in the disease spectrum are discussed in detail with respect to their virulence mechanisms in relation to the disease, epidemiology, specific laboratory diagnosis, treatment, and prevention.

Note: Some tables in this section have been colour-coded to depict the relative significance of causative organisms with respect to the organ system discussed in that chapter—**red denotes the most significant, orange indicates a slightly less etiological value and blue, lesser so**.

Part III: Microbiology as Applied to Infectious Diseases

CHAPTER 21

Cardiovascular and Bloodstream Infections

(MI2.1–2.6)

Chapter Navigation

Learning objectives

- Describe the etiologic agents, pathogenesis, and diagnosis of rheumatic fever and infective endocarditis
- Discuss the various causes for bloodstream infections including kala-azar, malaria, and filariasis

Infections affecting the cardiovascular system chiefly manifest in two forms:

1. Infections localised to the heart
2. Disseminated infections of the bloodstream

INFECTIONS LOCALISED TO THE HEART

- Infections affecting various layers of the heart may be classified as:
  1. **Endocarditis**
  2. **Myocarditis**
  3. **Pericarditis**
- Endocarditis, myocarditis, and pericarditis as well as vegetations on defective valves of the heart can lead to bacteremia and fungemia.
- Infections of the cardiovascular system can be life-threatening unless diagnosed and treated promptly.
- These infections occur when bacteria, fungi, or viruses from infectious foci (often the oral cavity) spread through the bloodstream and get attached to damaged areas in the heart.
- The infections involving the heart also include rheumatic fever.

Acute Rheumatic Fever (ARF)

Case 1. *A nine-year-old girl presented with a fever and sore throat. Group A streptococci were cultured from her throat. Three weeks later, she developed pain and tender swelling around both knees and ankles; a palpable effusion of the right knee was drained, which yielded straw-coloured fluid. Culture of the aspirate was negative. Subsequently, the girl began experiencing shortness of breath. Physical examination and chest X-ray suggested mild congestive heart failure. The girl was diagnosed with* ***acute rheumatic fever*** *with reactive post-streptococcal arthritis. On follow-up after one year, a murmur was audible at the cardiac apex, suggestive of* **rheumatic heart disease (RHD)**.

i. What are the criteria for diagnosing acute rheumatic fever?
ii. Describe the pathogenesis of ARF and RHD?

- Acute rheumatic fever (ARF) and rheumatic heart disease (RHD) are autoimmune manifestations of **group A streptococcal (GAS) infections.**
- These occur in **children between 5 and 14 years of age**.
- ARF occurs as an **autoimmune phenomenon**, which develops one to five weeks after a child has had pharyngitis due to GAS (which may sometimes be subclinical).
- The child develops features of rheumatic fever with polyarthritis and associated symptoms.

Agent—Group A streptococci (GAS)

- Group A streptococci are gram-positive cocci arranged in chains that produce beta-hemolysis on blood agar (Chapter 4).
- They can cause a diverse array of diseases, ranging from self-limited conditions (pharyngitis, impetigo, and erysipelas) to severe, invasive infections (bacteremia, necrotising fasciitis, myositis, and streptococcal toxic shock syndrome).
- Streptococcal pharyngitis or skin infections may have lasting sequelae such as post-streptococcal rheumatic disease or glomerulonephritis respectively.
- These bacteria produce a variety of toxins and enzymes (Chapter 4).

Pathogenesis of ARF

Acute rheumatic fever is a post-streptococcal sequel as depicted in Fig. 21.1.

Clinical features of ARF

- Fever
- Migrating polyarthritis (inflammation of multiple joints)
- Skin rashes (erythema marginatum)
- Multiple small, subcutaneous, and painless nodules may be seen on the bony prominences of the hands, legs, and back after 2–3 weeks
- **Chorea** manifests as a movement disorder that causes sudden, unintended, and uncontrollable jerky movements of the arms, legs, and facial muscles
- Symptoms and signs of carditis depend on the areas of the heart involved, such as the pericardium, myocardium, or heart valves

Laboratory diagnosis of ARF

The clinical diagnosis of ARF is based on a set of features known as the **Jones criteria**, which are further classified into major and minor criteria (Table 21.1).

Table 21.1 Jones criteria

Major criteria	Carditis Polyarthritis Chorea Erythema marginatum Subcutaneous nodules
Minor criteria	Fever (>38.5°C) Polyarthralgia History of rheumatic fever or rheumatic heart disease Prolonged PR interval Elevated acute phase reactant (ESR >60 mm and hsCRP >3 mg/dL)
Evidence of streptococcal infection	Group A streptococci grown on throat culture Anti-streptolysin-O (ASO) Anti-deoxyribonuclease B
Diagnostic criteria	Two major and one minor criteria **OR** Two major criteria with evidence of streptococcal infection

Treatment of ARF

The treatment of ARF is as presented in Table 21.2.

Rheumatic Heart Disease (RHD)

- Rheumatic fever, if left untreated, invariably damages the heart and leads to rheumatic heart disease in almost 60% of cases. Alternatively, RHD may even occur after a single episode of ARF.

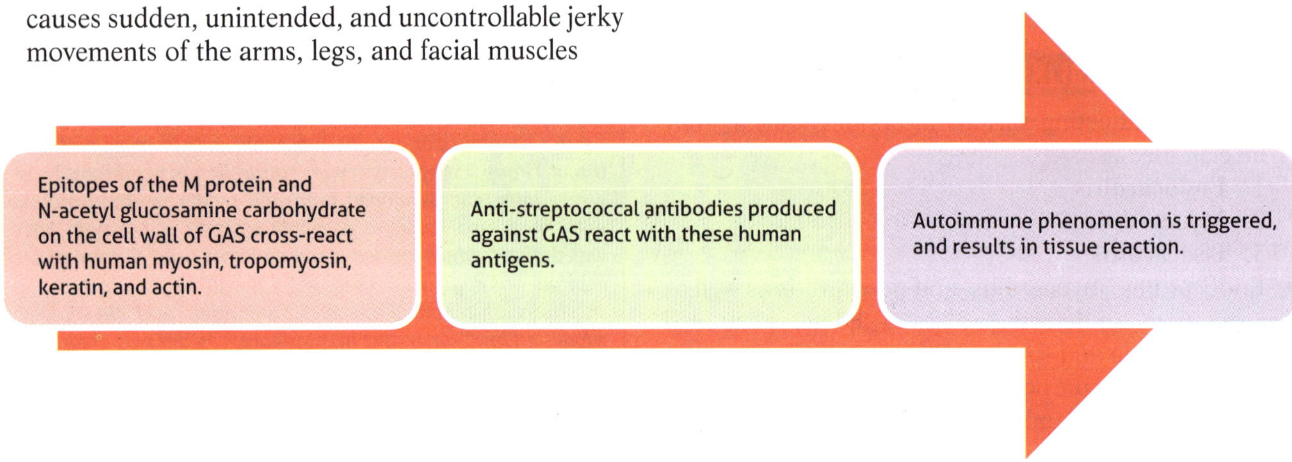

Fig. 21.1 Pathogenesis of ARF

Table 21.2 Treatment of ARF

Clinical entity	Drug of choice	Alternate treatment
Acute rheumatic fever	Penicillin oral 500 mg BDS for 10 days (250 mg for children) OR Intramuscularly (benzathine penicillin G) in a single dose 1.2 million units	Amoxicillin 50 mg/kg for 10 days
Arthritis	Salicylates and NSAIDs	-
Secondary prevention of ARF to prevent RHD	Long-term penicillin prophylaxis with benzathine penicillin 1.2 million units once in four weeks until adulthood	For individuals allergic to penicillin, erythromycin 250 mg is given twice daily

- RHD is characterised by a streptococcal antigenic cross-reaction with the **sarcolemma of the heart muscles.**
- Cross-reacting antibodies attach to the endothelium of cardiac valves, leading to **CD4$^+$ T cell-mediated inflammation.**
- Valvular damage commonly involves the **mitral valve**. Sometimes, the aortic valves are also affected, leading to regurgitation.
- Over the years, following repeated episodes, **valvular thickening and fibrosis** set in, leading to mitral and aortic valve stenosis.
- The pericardium, myocardium, and endocardium are also affected.

Infective Endocarditis (IE)

> Case 2. *A 29-year-old man was admitted to the hospital with cellulitis of the left arm following fever for two weeks. On examination, he was emaciated and had pallor. Physical examination revealed multiple needle prick marks on his upper and lower limbs. He had splinter hemorrhages and his spleen was marginally palpable. Auscultation revealed a systolic murmur on the left sternal border. Typical ECHO findings were also demonstrated. Two sets of blood cultures were performed, and he was started on empiric ceftriaxone. The blood culture revealed* ***methicillin-resistant Staphylococcus aureus (MRSA)****. He was diagnosed with infective endocarditis following IV drug abuse. The man was put on vancomycin and upon recovery, he underwent de-addiction and counselling.*
> i. What is infective endocarditis (IE)?
> ii. Enumerate the organisms causing IE.

Endocarditis is an endogenous infection involving the **endothelial lining of the heart and the valves** (Fig. 21.2a and b) that is caused by bacteria (Table 21.3) from various sites entering the bloodstream.

- The most common source of infection is the **oral cavity** (after dental extraction) or **IV drug abuse**, following foci of infection in the skin or elsewhere.
- It may present as an acute-onset infection or in a subacute form.

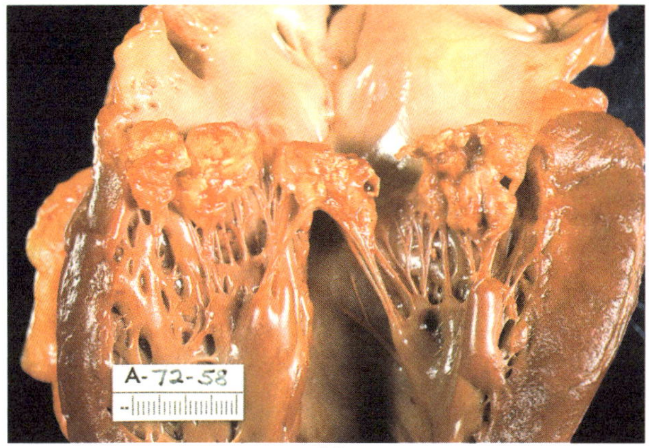

(a)

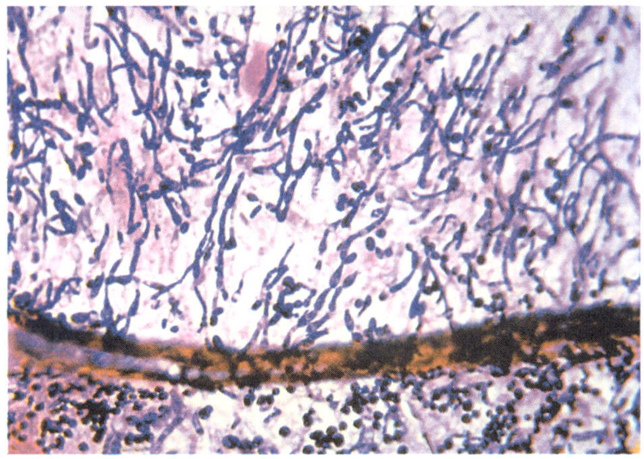

(b)

Fig. 21.2 (a) Sectioned human heart showing gross pathologic changes with large vegetations in a case of subacute bacterial endocarditis due to *Haemophilus parainfluenzae* and (b) histopathologic changes indicative of endocarditis caused by *Candida albicans*—bacterial colonisation of the mitral valve leaflets (*Source*: Centers for Disease Control and Prevention [CDC], Public Health Image Library [PHIL], Image IDs 851 and 2926)

Table 21.3 Microorganisms associated with endocarditis

Clinical entity	Causative organisms
Native valve endocarditis	*Staphylococcus aureus*, viridans streptococci, *Streptococcus bovis*, HACEK group (*Haemophilus*, *Aggregatibacter*, *Cardiobacterium*, *Eikenella*, and *Kingella*) and enterococci Rare pathogens—*Bartonella* spp., *Legionella* spp., *Tropheryma whipplei*, *Mycobacterium* spp., and fungi (*Candida* in prosthetic valve vegetations)
Prosthetic valve endocarditis (occurs days after valve surgery)	*Staphylococcus aureus*, coagulase-negative staphylococci, and *Candida* species
Endocarditis in IV drug abusers	*Staphylococcus aureus*, enterococci, and group B streptococci

Note: Some tables in this section have been colour-coded to depict the relative significance of causative organisms with respect to the organ system discussed in that chapter—**red denotes the most significant, orange indicates a slightly less etiological value and blue, lesser so.**

Risk factors for IE

- Underlying cardiac defect (congenital or a sequel of rheumatic fever)
- Heart valve disorders
- Prosthetic valve replacement

Pathogenesis of IE

Fig. 21.3 depicts the pathogenesis of IE.

Clinical features of IE

The clinical features of acute and subacute and endocarditis overlap and include the following:

- Fever, chills, and sweats
- Anorexia
- Myalgia and arthralgia
- Heart murmur

The **cutaneous lesions** of subacute bacterial endocarditis are a result of immune complex deposition. These may be of the following types:

- Petechiae
- **Splinter hemorrhages in the nail bed** (Fig. 21.4a)
- **Osler's nodes** (Fig. 21.4b)—small, tender, red–purple nodules on the pulp of the distal phalanges of the fingers and toes, the soles, and the thenar and hypothenar eminences of the palms
- **Janeway lesions**—small, painless hemorrhages with a macular or slightly nodular character on the thenar and hypothenar eminences of the palms and sole

Laboratory diagnosis of IE

Clinical diagnosis

This is based on cardinal findings known as the **modified Duke criteria** (Table 21.4).

- In cases of pericarditis, pericardial fluid can be cultured.
- Fastidious and non-cultivable microorganisms are identified by either antigen detection or molecular techniques like multiplex PCR.

Treatment of IE

- Infection is eradicated by eliminating microorganisms from the vegetations. Hence, treatment involves the use of bactericidal antibiotics for a prolonged period (Table 21.5).
- Specific antibiotics are prescribed based on the organisms isolated by culture and the sensitivity of the organism.

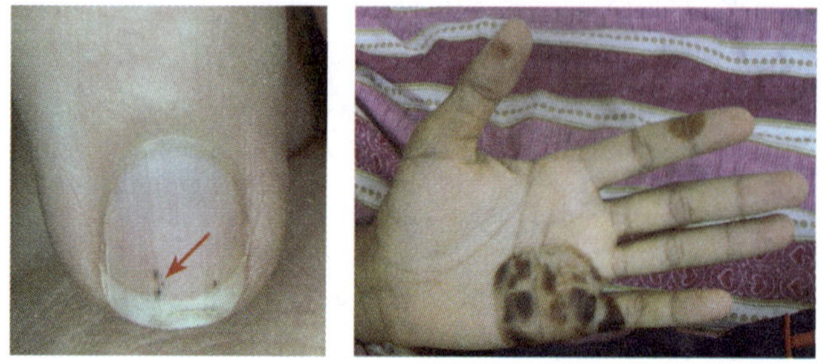

(a) (b)

Fig. 21.4 (a) Splinter hemorrhage and (b) Osler's nodes due to immune complex deposition (*Source*: [a] Splarka, commons.wikimedia.org and [b] Roberto J. Galindo, CC BY-SA 4., creativecommons.org)

Table 21.4 Modified Duke criteria

Major criteria	Minor criteria
i. **Isolation of microorganism consistent with IE** from at least two separate blood cultures drawn >12 h apart **OR** All three/majority of ≥four separate cultures of blood (with first and last sample drawn at least 1 h apart) **OR** Single blood culture or serology (phase I IgG titre >1:800) positive for *Coxiella burnetii* ii. **ECHO** showing evidence of endocardial involvement—intracardiac mass on valve, abscess or new partial dehiscence iii. New **valvular regurgitation**	i. Fever (>100°F) ii. Predisposing heart condition or IV drug abuse iii. Vascular phenomenon, e.g., arterial emboli, Janeway lesions iv. Immunologic phenomenon, e.g., Osler's nodes, Roth's spot, rheumatoid factor v. Microbiological—positive blood culture but does not satisfy major criteria (excludes single positive cultures for coagulase negative staphylococci and organisms that do not cause endocarditis) OR serological evidence of active infection with organism consistent with IE

Diagnosis of infective endocarditis:

i. Two major criteria **OR**
ii. One major and three minor criteria **OR**
iii. Five minor criteria

- Antibiotics must be given **parenterally** so that they reach the vegetations in adequate concentrations.
- If the vegetations are caused by ***Candida* species** (*albicans*), amphotericin B, echinocandins, or fluconazole may be given.

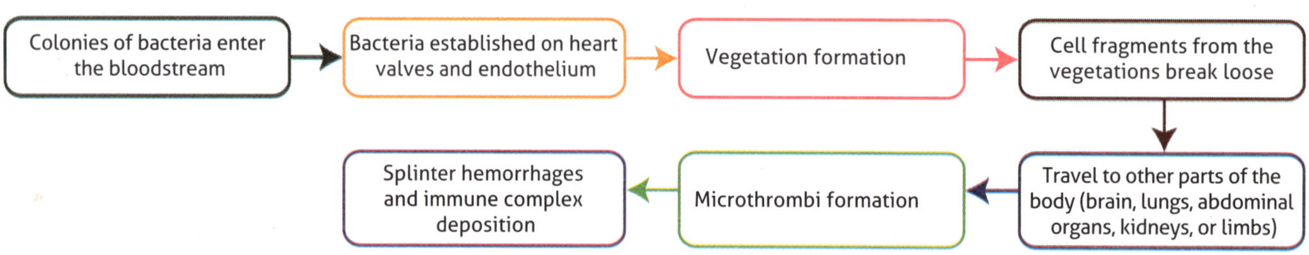

Fig. 21.3 Pathogenesis of infective endocarditis and associated lesions

Table 21.5 Drugs used to treat IE

Treatment	First choice	Alternate
Empirical treatment while awaiting cultures	Ampicillin–sulbactam plus gentamicin	Ampicillin plus ceftriaxone gentamicin
Native valve endocarditis (with strong suspicion of *S. aureus*)	Vancomycin	Daptomycin
Prosthetic valve endocarditis pending blood cultures or with negative blood cultures	Ceftriaxone, vancomycin, gentamicin, rifampicin	Use lower dose of rifampicin when there is severe renal impairment

Myocarditis and Pericarditis

- **Myocarditis** is an infection of the myocardium and is usually caused by a viral infection and can lead to heart failure, abnormal heartbeat, and sudden death.
- **Pericarditis** is the swelling and irritation of the thin, sac-like tissue surrounding the heart (pericardium) and often presents as chest pain.

The causative agents of myocarditis and pericarditis are listed in Table 21.6.

Clinical features

- Common features of infections resulting in pericarditis and myocarditis are **breathlessness, fever, and chest pain.**

Table 21.6 Causative agents of myocarditis and pericarditis

Causative agents	Myocarditis	Pericarditis
Viruses (most common)	Adenovirus, Coxsackievirus, Epstein–Barr virus, Cytomegalovirus, Hepatitis C virus, HIV, Parvovirus B 19, Influenza A virus	Echovirus, Coxsackievirus, Adenovirus, Influenza virus
Bacteria	*Borrelia, Streptococcus, Staphylococcus, Chlamydia, Mycoplasma, Treponema*	*Staphylococcus aureus, Haemophilus influenzae, Streptococcus agalactiae, Mycobacterium tuberculosis* *Mycobacterium avium intercellulare* *Legionella*
Parasites	*Trypanosoma cruzi* (Chagas' disease) Larva migrans Schistosoma	*Entamoeba histolytica* *Trypanosoma cruzi* (Chagas' disease/African trypanosomiasis)
Fungi	*Aspergillus* *Candida* *Cryptococcus* *Histoplasma*	*Candida* *Blastomyces* *Histoplasma*

- Pericarditis may be acute (<6 weeks), resembling a myocardial infarction, subacute (6 weeks to 6 months), or chronic and constrictive (>6 months).
- There may be **pericardial friction rub** or effusion.

Laboratory diagnosis

- The mainstay of diagnosis is imaging by **electrocardiography.**
- Microbiological tests include **blood culture** for bacterial etiology and **serology** to detect IgM antibodies or a paired sample to show a rising IgG titre in a case of viral etiology.

BLOODSTREAM INFECTIONS

- **Vasculitis:** Some bacteria, viruses, and fungi directly affect the endothelium of the blood vessels. Such infections involving the blood vessels result in vasculitis.
- **Septicemia:** This term refers to the presence of microbial antigens and endo- or exotoxins in the bloodstream.
- **Sepsis:** It is the response the host mounts against bacterial products. It occurs when there is active bacterial multiplication and a release of their products (toxins) in the bloodstream, causing harm to the host. The individual may develop **septic shock**.
- **Bacteremia:** This refers to the presence of a focus of disease or may represent the transient presence of bacteria in the bloodstream without multiplication. Bacteria may be recovered from 20–40% of patients by blood culture. Bacteremia may be transient, continuous, or intermittent.
  - **Transient bacteremia:** It occurs spontaneously following minor events such as brushing of teeth or chewing of food or because of procedures like manipulation of infected tissues, instrumentation of contaminated mucosal surfaces, and surgery involving non-sterile sites.
  - **Continuous bacteremia:** It is associated with bacterial endocarditis, endovascular infections, and septic shock, wherein organisms are released into the bloodstream at a fairly constant rate.
  - **Intermittent bacteremia:** This type is seen in meningitis, pneumonia, pyogenic arthritis, and osteomyelitis. It is associated with bacteria being recovered from blood early in the course of the disease. In the case of a sequestered focus of infection, such as an abscess, bacteria are released into the blood approximately 45 minutes before a febrile episode.
- **Catheter-related bloodstream infections**—central- and peripheral line-associated BSI (discussed in Chapter 11).
- **Bloodstream infections in an immunocompromised individual** (Table 21.7).

Table 21.7 Bloodstream infections in immunocompromised patients

Bacterial causes	Fungal causes
Pseudomonas aeruginosa *Serratia marcescens* *Acinetobacter baumannii* Coagulase-negative staphylococci (CoNS) Enterobacteriaceae (*E. coli, Klebsiella*, and *Enterobacter* species) *Staphylococcus aureus* *Enterococcus faecalis*	*Candida* species *Histoplasma* *Cryptococcus*

Etiology

The etiological agents of bloodstream infections are listed in Table 21.8.

Pathogenesis (Fig. 21.5)

- **Gram-negative septicemia** is a result of the endotoxin released by the organisms, which is a lipopolysaccharide (LPS). The core of the LPS, lipid A, mediates a systemic response which includes fever and the activation of complements and certain clotting factors.
- **Gram-positive bacteria** produce exotoxins which generate a similar response.

Laboratory Diagnosis

Fig. 21.6 outlines the investigations that are carried out to arrive at a diagnosis of bloodstream infection.

Blood culture

- **Two samples taken preferably from two different sites at two different times in a 24-hour period** must be collected for best results.
- 1–5 mL of blood is drawn for infants and small children, whereas for adults, 7–10 mL of blood per culture is recommended.
- Organisms like the HACEK bacteria, certain fungi, and *Brucella* are slow-growing and require either automated systems for culture or need to be incubated for up to two weeks by the conventional method.
- Fungal infections can be identified by inoculation on Sabouraud dextrose broth or automated blood culture bottles. These also support the growth of yeasts and yeast-like organisms.
- If malaria or leishmaniasis (presenting with other related symptoms) is suspected, the peripheral blood smear or bone marrow aspirate smear is prepared using Leishman or Giemsa stain.

Immunodiagnostic methods

- **Antigen detection:** This is a convenient and commonly used method for the identification of organisms that are difficult to cultivate, e.g., *Coxiella burnetii* and *Chlamydia*. **Immunofluorescent antibody (IFA)** and **enzyme immunoassay (EIA)** are the most routinely used tests.
- **Antibody detection:** Commonly used antibody detection tests are the **febrile agglutination tests** (e.g., Widal test), ***Brucella* agglutination test**, and IgM detection by **ELISA** for rickettsiae and chlamydiae.
- **Molecular detection:** As these infections are life-threatening, early diagnosis is of paramount importance for appropriate management. **Multiplex PCR** may be used, especially if viral or chlamydial etiology or non-cultivable organisms are suspected.

Table 21.8 Common etiological agents of bloodstream infections

Bacteria	Viruses	Fungi	Parasites
Staphylococcus aureus Beta-hemolytic streptococci *Enterococcus* species *Salmonella* spp. (enteric fever) Coagulase-negative staphylococci (device-associated) *S. pneumoniae* (associated with pneumonia or meningitis) *Viridans* streptococci (infective endocarditis) HACEK group of organisms (infective endocarditis) *Escherichia coli* *Pseudomonas* species (mainly HAI) *Brucella* species (brucellosis) *Klebsiella pneumoniae* *Enterobacter* species *Proteus* species Other organisms isolated from blood—*Leptospira*, *Burkholderia* (melioidosis), *Yersinia* (plague)	HIV Agents causing viral hemorrhagic fevers Dengue Kyasanur forest disease (KFD) virus Epstein–Barr virus Cytomegalovirus Crimean-Congo hemorrhagic fever (CCHF) virus	*Candida* *Cryptococcus* *Histoplasma capsulatum* *Coccidioides immitis* *Mucor* species *Blastomyces dermatitidis* *Aspergillus* species	*Plasmodium* *Trypanosoma* *Wuchereria* *Loa loa* *Babesia*

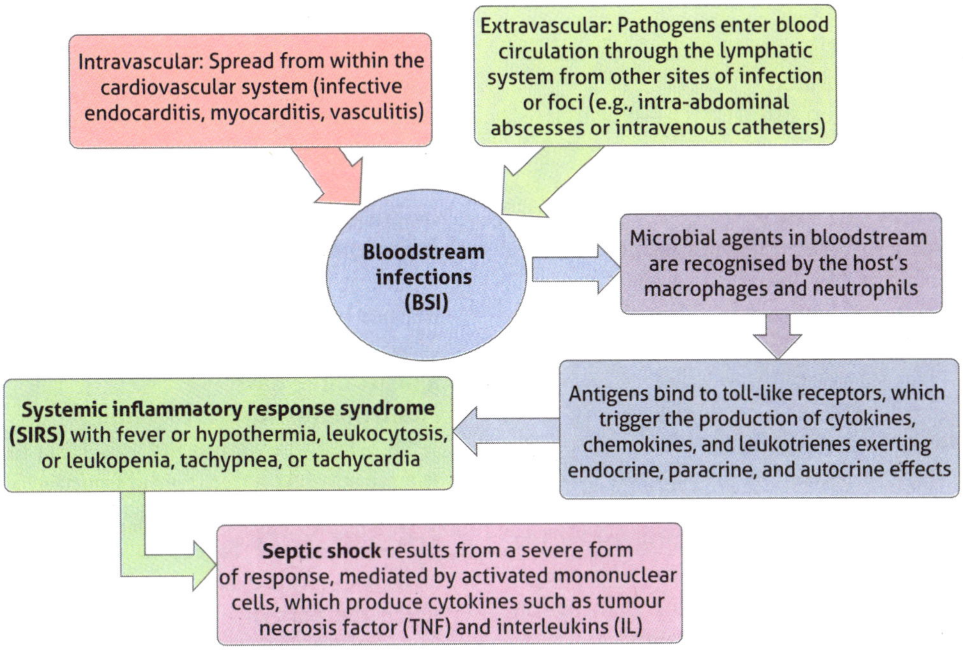

Fig. 21.5 Pathogenesis of bloodstream infections

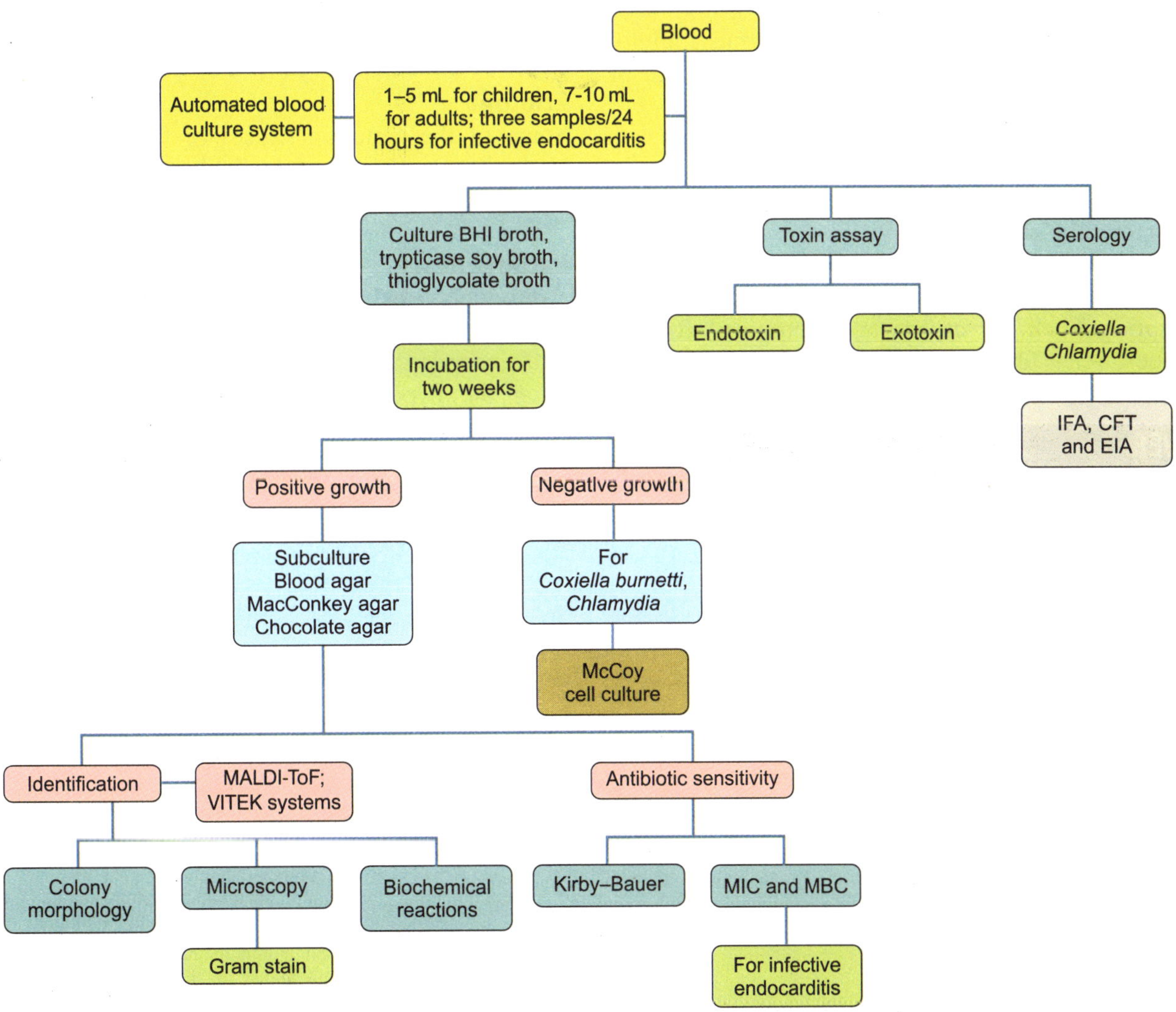

Fig. 21.6 Approach to the diagnosis of bloodstream infections

Other tests

- Total count—leukocytosis
- Troponin—levels are elevated
- Erythrocyte sedimentation rate (ESR)—elevated
- Procalcitonin—raised
- C-reactive protein—detectable
- Urine—culture may be positive
- Chest X-ray—lung infiltrates in pneumonia
- Echocardiography for the detection of vegetation
- Detection of bacterial endotoxin—**limulus lysate test**—is a sensitive test for the detection of bacterial endotoxin

SIRS and Sepsis

Systemic inflammatory response syndrome (SIRS) is an exaggerated defense response of the body to a noxious stressor (infection, trauma, surgery, acute inflammation, ischemia or reperfusion, or malignancy), while sepsis refers to a serious complication of infection that can lead to multiorgan failure and death.

Clinical features

- Fever
- Tachycardia (Heart rate more than 90 beats/minute)
- Respiratory rate more than 20 breaths/minute or partial pressure of CO_2 less than 32 mmHg
- Total leukocyte count greater than 12,000 or less than 4,000/microliter or over 10% immature forms or bands
- **Bacteremia/septicemia**—fever or hypothermia, chills, hyperventilation, and subsequent respiratory alkalosis
- **Septic shock**—fever, acute respiratory distress, shock, renal failure, and intravascular coagulation; can be initiated by exotoxins or endotoxins

FEVER

There are various types of fever, which are as follows:

- **Fever of unknown origin (FUO):** Fever in a patient who exhibits a temperature >38.3°C (101°F) on several occasions, has had the illness for more than three weeks, and for whom a diagnosis has not been established despite one week of in-patient investigations (blood/urine and sputum culture for common bacteria, serology, and complete blood count [CBC]) is said to be fever of unknown origin.
- **Nosocomial fever:** This refers to fever (>38.3°C) in a patient who has been admitted for ≥48 hours, who neither had fever not was incubating fever at the time of admission and who has undergone evaluation for at least three days.
- **Neutropenic fever:** In addition to a temperature >38.3°C and evaluation for at least three days, when the patient has a neutrophil count ≤500 per mm^3, he is said to have neutropenic fever.
- **HIV-associated fever:** In addition to temperature >38.3°C and evaluation for at least three days, when the patient has had fever for >4 weeks as an out-patient and >3 days as an in-patient and HIV infection has been confirmed, the fever is said to be HIV-associated.

Fever of Unknown Origin (FUO)

Etiology

Some of the common causes of fever of infectious etiologies are presented in Table 21.9. Some of the non-infectious causes are listed in Table 21.10.

Pathogenesis

- Products of bacterial breakdown including proteins (exotoxins) and lipopolysaccharide toxins released from bacterial cell membranes act on the hypothalamus either directly or through interleukin-1, causing fever.
- Fever due to non-infectious disorders such as myocardial infarction, pulmonary emboli, and neoplasms are caused by the injured or abnormal cells that incite the production of pyrogens.

Table 21.9 Infectious causes of FUO

Bacterial	Viral	Fungal	Parasitic
M. tuberculosis	Cytomegalovirus	*Candida albicans*	*Leishmania* spp.
Coxiella burnetii	Epstein–Barr virus	*Cryptococcus neoformans*	*Trypanosoma* spp.
Mycoplasma spp.	Arboviruses	*Histoplasma capsulatum*	*Toxoplasma* gondii
Atypical mycobacteria	Enteroviruses	*Coccidioides immitis*	*Plasmodium* spp.
Brucella spp.	HIV	*Aspergillus* spp.	*Wuchereria bancrofti*
Salmonella spp.			*Brugia malayi*
Chlamydia psittaci			
Leptospira spp.			
Rickettsia spp.			
Burkholderia pseudomallei			

Table 21.10 Non-infectious causes of FUO

Neoplasms	Connective tissue disorder	Granulomatous diseases
Lymphoma Leukemia Myeloma Renal cancer Colon cancer Liver cancer Metabolic disorder Gout Porphyria	Systemic lupus erythematosus (SLE) Polyarteritis nodosa	Crohn's disease Sarcoidosis Rheumatoid arthritis Polymyositis

Clinical features

FUO may present in varied forms (Table 21.11).

Laboratory diagnosis (Fig. 21.7)

A complete clinical history (including details of travel, immunisation, and exposure to diseases) should be obtained and followed by a detailed physical examination. Information from the patient's history and examination guide the choice of specimen collected for processing and the type of test prescribed.

Table 21.11 Various types of fever and causes

Type of fever	Fluctuation in body temperature	Causes
Intermittent fever	Returns to normal at least once every 24 hours	Gram-negative or gram-positive sepsis Abscesses Infective endocarditis
Remittent fever	Does not return to normal, although it varies a few degrees in either direction	Viral upper respiratory tract infection *Legionella* *Mycoplasma*
Sustained or continuous fever	Remains above normal with minimal variations (usually less than 1.5°F and 1°C)	Drugs Tuberculosis
Recurrent or relapsing fever	One or more episodes of fever, each as long as several days, with one or more days of normal temperature between episodes	*Borrelia* Chronic fungal infection
Undulant fever	Cyclical fluctuation like a wave	Brucellosis

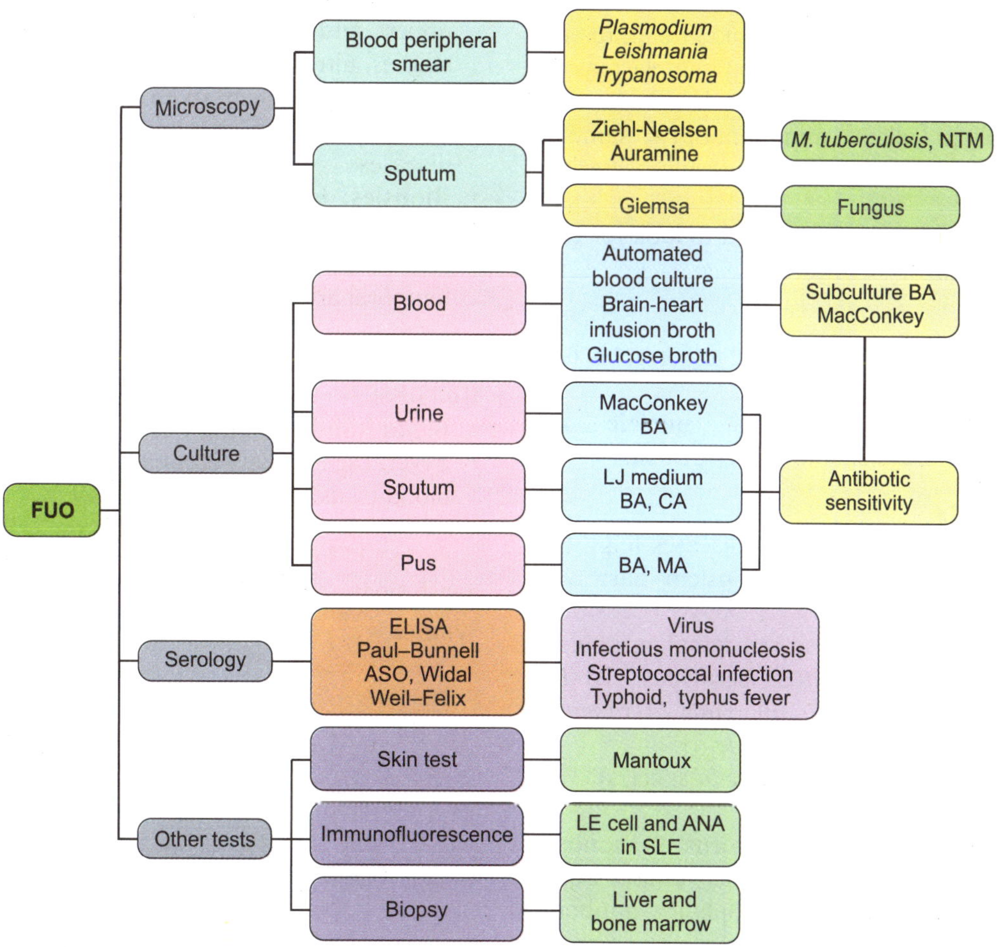

Fig. 21.7 Algorithm for the diagnosis of FUO (*BA*—blood agar; *MA*—MacConkey agar; *CA*—chocolate agar; *ASO*—antistreptolysin O; *NTM*—non-tuberculous mycobacteria; *SLE*—systemic lupus erythematosus)

- **Blood:** For culture and serology, complete blood counts (CBCs), erythrocyte sedimentation rate (ESR), microscopy (wet films and Leishman-stained smears), culture, and serological tests
- **Urine:** For microscopic urine analysis, antigen detection, and culture
- **Bone marrow:** For biopsy and culture
- **Others:** Material from abnormal tissues, lung, and abdominal viscera may be used for biopsy, if required

Diagnostic procedures for infective FUO

- **CBCs and ESR**—neutrophilic leukocytosis in CBC may suggest a pyogenic infection; markedly elevated ESR may suggest a tuberculous etiology
- **Microscopy of blood**—for example, a wet film may demonstrate microfilariae or trypanosomes; a stained smear may demonstrate malarial parasites, trypanosomes, and microfilariae
- **Wet mount** for *Entamoeba* (liver abscess)
- **Gram stain** for gram-negative bacilli, *Brucella* species
- **Ziehl–Neelsen stain** for *Mycobacterium* species
- **Giemsa stain** to detect fungi

Culture

- **Blood and aspirates**—multiple blood cultures may be required. Since a single blood sample may not yield a positive result due to transient bacteremia, it is advisable to collect a minimum of two samples.
- **Urine** is cultured to look for pyogenic bacteria or tubercle bacilli.
- Bone marrow culture is performed to look for fungi and *Leishmania*.

Molecular tests

- **Multiplex PCR** may be performed to detect bacterial or viral nucleic acids.
- If tuberculosis is suspected, **nucleic acid amplification assays** should be performed.

Serological tests

- ELISA for the diagnosis of viral diseases (e.g., HIV, cytomegalovirus [CMV], or Epstein–Barr virus [EBV] infections)
- CFT for certain bacterial diseases (lymphogranuloma venereum [*C. trachomatis* serotypes L1-L3], psittacosis [*C. psittaci*], tularemia [*Francisella tularensis*], Q fever [*Coxiella burnetii*])
- ELISA/SAT for brucellosis (*Brucella* species)
- IgM ELISA for *Rickettsia*
- Specific IgM antibodies to EB virus or monospot test (latex agglutination test, which utilises equine erythrocytes to detect heterophilic antibodies produced by the human immune system in response to EBV infection)
- ASO titre for rheumatic fever

Skin tests

Delayed hypersensitivity reactions may suggest that the FUO is due to tuberculosis (Mantoux skin test) or a fungal infection (e.g., the histoplasmin skin test for histoplasmosis).

Inflammatory markers

- CRP
- Procalcitonin
- IF/RA and other autoimmune markers for rheumatoid arthritis

Imaging techniques

- Chest X-ray and other imaging techniques may reveal lesions suggestive of a particular infection, e.g., lung infiltrates in pulmonary tuberculosis.
- Echocardiography may demonstrate findings indicative of rheumatic fever or infective endocarditis.

Diagnostic procedures for non-infective FUO

- **Serology:** Antinuclear antibody or anti-DNA antibody (suggestive of SLE), anti-thyroglobulin antibody (suggestive of subacute thyroiditis) or tests to detect other collagen vascular diseases
- **Imaging studies:** Chest X-rays, echocardiograms, upper and lower gastrointestinal tract barium studies, ultrasound, CT and MRI scans, intravenous urograms, and scans of lung, bone, and other tissues to detect neoplastic processes or collagen vascular diseases
- **Biopsies:** Examination of bone marrow (to detect leukemia), lymphoid tissue (Hodgkin's disease), gastrointestinal tract structures (by endoscopy), temporal artery, lung, and other tissues

Fever Due to Infections Transmitted by Blood Transfusion

- Blood transfusion is a possible source of infection.
- Several agents can potentially be transmitted through blood transfusions (Table 21.12).

Table 21.12 Infectious agents transmitted through blood transfusions

Viruses	Hepatitis B, C, and D virus HIV Cytomegalovirus Human T-cell lymphotropic virus
Bacteria	*Treponema* *Pseudomonas* *Staphylococcus* *Yersinia enterocolitica* *Serratia* *Flavobacterium*
Parasites	*Plasmodium*

INFECTIVE AGENTS ASSOCIATED WITH CARDIOVASCULAR AND BLOODSTREAM INFECTIONS

Microbial agents are not confined to one system or organ. Rather, they produce a syndrome which may involve one or more systems or organs in the body. The following sections discuss individual or groups of microorganisms that are primarily associated with bloodstream and cardiovascular infections. For clarity, organisms will be discussed as bacterial, viral, parasitic, and fungal agents involved in infections.

BACTERIAL BLOODSTREAM AND CARDIOVASCULAR INFECTIONS

- *Staphylococcus aureus*
- Coagulase negative staphylococci
- Viridans group of streptococci
- HACEK group

Staphylococcus aureus

Staphylococci are gram-positive cocci that appear in clusters. They produce two types of diseases:

- **Localised or disseminated infections:** These are characterised by localised pyogenic lesions, in contrast to the spreading lesions seen in streptococcal infections.
- **Toxin-mediated infections:** These are caused by enterotoxin-forming staphylococci. The toxin also exhibits pyrogenic, mitogenic, hypotensive, thrombocytopenic, and cytotoxic effects (details discussed subsequently). Examples include food poisoning and toxic shock syndrome.

Virulence factors

The four major virulence factors of staphylococci **include cell-associated and cell-surface proteins, extracellular enzymes and toxins (Chapter 4 for details).**

Regulation of virulence factors

- A number of staphylococcal factors may influence virulence, both cell-associated and extracellular.
- The production of virulence factors is regulated by complex regulatory mechanisms through a set of genes which are expressed by the binding of ligands to host cell receptors.

Pathogenesis of staphylococcal infections

Staphylococcal infections are among the most common bacterial infections and range from the mild to the fatal (see Chapter 4).

Localised infections

S. aureus gains access to damaged skin, mucosal, or tissue sites, colonises them by adhering to cells or extracellular matrices, evades host defence mechanisms, multiplies, and causes tissue damage.

Disseminated infections

A staphylococcal infection might disseminate from a focus and spread via the lymphatics and bloodstream to distant organs, causing osteomyelitis, empyema, pneumonia, meningitis, endocarditis, or sepsis.

Toxin-mediated infections

- Organisms may or may not be isolated from the sites of toxin action.
- The toxin also exhibits pyrogenic, mitogenic, hypotensive, thrombocytopenic, and cytotoxic effects.
- Some cases of post-antibiotic diarrhea are caused by enterotoxin-forming staphylococci.
- Toxin-mediated *Staphylococcus* infections are as follows:
- **Toxic shock syndrome (TSS)**
  - This potentially fatal multisystem disease presents with fever, hypotension, myalgia, vomiting, diarrhea, mucosal hyperemia, and an erythematous rash which desquamates subsequently.
  - It is caused by TSST-producing *S. aureus* strains, usually those belonging to bacteriophage group I.
  - Though tampon use was initially implicated in TSS, it is now rare among tampon users.
  - The syndrome occurs in other infections of the skin, mucosa, and also in some surgical wounds.
  - Staphylococcal enterotoxins and TSST-1 are superantigens which are potent activators of T lymphocytes. Being Vb-restricted T cell mitogens, they stimulate large numbers of T cells. This leads to an excessive and dysregulated immune response with the release of interleukins-1 and -2, tumour necrosis factor, and interferon gamma.
- **Kawasaki disease** clinically resembles TSS and is caused by the superantigen of *S. aureus*
- **Staphylococcal scalded skin syndrome** is discussed in Chapter 29.
- **Enterotoxin-mediated food poisoning** is covered in Chapter 25.

Epidemiology

Staphylococci are primary pathogens of human beings and animals, colonising the skin, apocrine glands, and mucous membranes.

Sources of infection

Human patients and carriers

- Patients with superficial infections and respiratory infections disseminate large numbers of staphylococci into the environment.

- It is estimated that 10–30% of healthy persons carry staphylococci in their nose and 10% carry them in their perineum and hair.
- Vaginal carriage is estimated to be 5–10%, which rises greatly during the menses, a factor relevant in the pathogenesis of TSS related to menstruation.
- Staphylococcal carriage starts early in life, colonisation of the umbilical stump being very common in babies born in hospitals.
- Some carriers, called '**shedders**', disseminate very large numbers of cocci for prolonged periods.

Inanimate objects

The cocci shed by patients and carriers contaminate fomites such as handkerchiefs, bed linen, and blankets and may persist on these objects for days or weeks.

Animals

Staphylococci may also be transmitted to humans from infected domestic animals such as cows.

Laboratory diagnosis of staphylococcal infections (Fig. 21.8)

Specimen

- The specimen to be collected depends on the type of lesion.
- For the detection of carriers, nasal swabs, axillary, and perineal swabs are collected.

Culture

- Staphylococci (Fig. 21.9a) grow readily on ordinary media within a temperature range of 10–42°C—the optimum being 37°C—and a pH of 7.4–7.6.
- Most *S. aureus* strains are hemolytic, especially when incubated under 20–25% carbon dioxide.
- **On nutrient agar**, golden-yellow pigment production occurs optimally at 22°C and only in aerobic cultures.
- **Selective media** may be needed for isolating *S. aureus* from specimens such as feces containing other bacteria such as media containing 8–10% NaCl (salt-milk agar, salt broth), lithium chloride, tellurite (Ludlam's medium), and polymyxin.
- **On blood agar**, dome-shaped colonies are formed with a golden-yellow pigment and a clear zone of hemolysis (Fig. 21.9b).

Coagulase test

- The identification of *S. aureus* is based on the coagulase test, wherein the enzyme coagulase can be detected from clinical isolates using rabbit plasma (Fig. 21.10a and b).

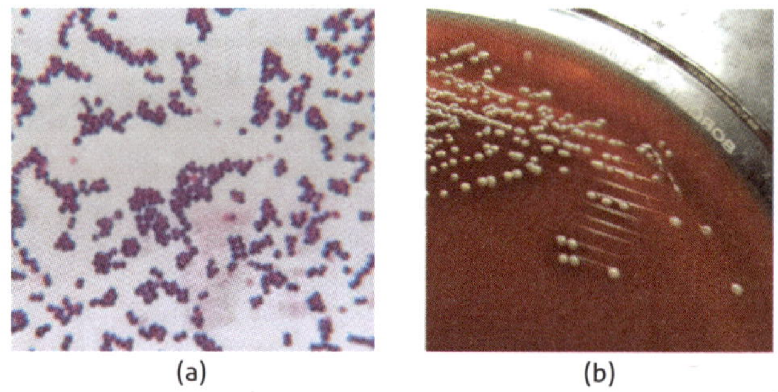

Fig. 21.9 (a) Gram-positive cocci (*Staphylococcus aureus*) in clusters and (b) colonies of *S. aureus* on blood agar showing golden-yellow hemolytic colonies (*Source*: Department of Microbiology, Pondicherry Institute of Medical Sciences, Puducherry)

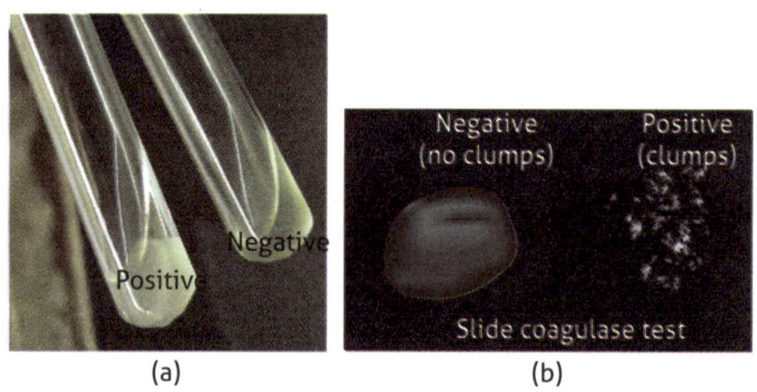

Fig. 21.10 (a) Free coagulase (tube coagulase) test—positive and negative and (b) clumping factor (slide coagulase) test—negative and positive

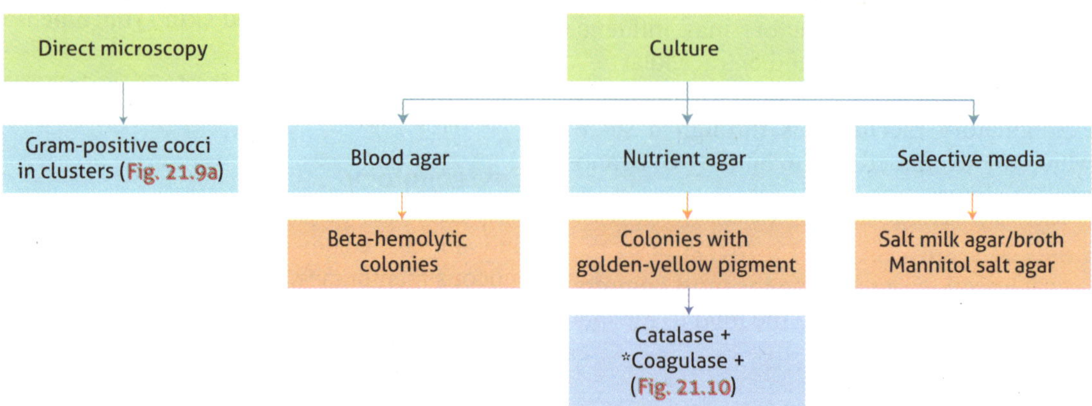

Fig. 21.8 Laboratory diagnosis of staphylococcal infections
* Differentiate *S. aureus* from *S. epidermidis*

- In the clinical laboratory, it is often necessary to differentiate *S. aureus* from *S. epidermidis* as the former is pathogenic, whereas the latter is a commensal which may contaminate the specimen.

Antimicrobial susceptibility testing

- Susceptibility testing is recommended for the following antibiotics:
  - Erythromycin
  - Cefoxitin
  - Gentamicin
  - Doxycycline
  - Clindamycin
  - Vancomycin
  - Linezolid
  - Ciprofloxacin
- **Any isolate resistant to cefoxitin is reported as an MRSA** and should be presumed resistant to all beta-lactam antibiotics.
- The current recommendation is to test using only the MIC for determining the susceptibility of *S. aureus* to vancomycin.

Molecular diagnosis

The *mecA* gene can be detected in MRSA infections.

Typing of *Staphylococcus aureus*

Typing is done when information is desired for epidemiological purposes. The two major techniques used for typing are as follows:

- **Phenotypic methods**—these are based on phenotypic characteristics and employ methods such as antibiograms, phage types, biotypes, etc.
  - *Bacteriophage typing*: Staphylococci may be typed based on their susceptibility to various bacteriophages.
- **Genotypic methods**—these are based on genetic composition and employ methods such as like **rRNA or pulse-field gel electrophoresis (PFGE) or random amplification of polymorphic DNA (RAPD).**

Treatment

- **Penicillin** is the drug of choice for community-acquired methicillin-sensitive disseminated staphylococcal infection.
- Mild infections caused by MSSA can be treated with first- or second-generation cephalosporins (cefazolin, cephalexin, or cefuroxime). However, most staphylococcal isolates from clinical material have acquired resistance to penicillin.

Antibiotic resistance

Beta-lactamase-mediated resistance

- This resistance to benzyl-penicillin is due to the production of beta-lactamases.
- Beta-lactamase or penicillinase inactivates penicillin by splitting its beta-lactam ring.
- These are inducible enzymes that are plasmid-mediated.
- They can be transmitted by transduction or conjugation.
- Penicillinase-producing strains remain sensitive to penicillinase-resistant penicillins such as methicillin and cloxacillin while the beta-lactamase-producing strains are sensitive to beta-lactamase inhibitor combinations like amoxicillin–clavulanic acid.
- Cloxacillins are used against penicillinase-producing strains to treat bone and skin infections.

Methicillin resistance

- **Methicillin-resistant *Staphylococcus aureus* (MRSA)** possess the *mecA* gene which encodes the 76 KDa **penicillin-binding protein, PBP2a** (also referred to as PBP2).
- Alteration at the target penicillin-binding site of PBP2a causes changes in bacterial surface receptors, thereby reducing the binding affinity of methicillin to cells.
- These sites are regulated by **staphylococcal cassette chromosomal mec genes (*SCC*$_{mec}$)**, especially the *mecA* gene.
- Based on the type of *SCC*$_{mec}$ gene, MRSA strains are classified as follows:
  - **Hospital-acquired MRSA (HA-MRSA):** The *SCC*$_{mec}$ is predominantly of the types I, II, and III. These strains are multidrug-resistant.
  - **Community-acquired MRSA (CA-MRSA):** These strains are less resistant, more likely to produce PVL toxin, and more transmissible.
- A majority of these strains also show resistance to other antibiotics like erythromycins, tetracyclines, and aminoglycosides and cause outbreaks of hospital infection.
- Vancomycin is the drug of choice to treat MRSA.
- Outbreaks of MRSA are seen in hospitals. Such events need to be identified early to undertake interventions for early control of spread (see Chapter 11).

Epidemic methicillin-resistant *S. aureus* (EMRSA)

These strains get easily disseminated in a hospital environment and are responsible for postoperative wound infections and other cross-infections in the hospital.

Detection of MRSA

- **Cefoxitin disc diffusion testing:** Cefoxitin resistance correlates with the presence of the *mecA* gene and is a surrogate marker to detect resistance to methicillin.
- **Oxacillin screen agar:** It is used to detect strains that are resistant to penicillins such as methicillin, oxacillin, and nafcillin. Some strains called borderline-resistant *S. aureus* (BORSA) have been found, which are missed by the cefoxitin susceptibility test and may need to be tested under

special incubation conditions using oxacillin screen agar.
- **PCR:** It can be performed to look for the *mecA* gene.
- Rapid assays are performed to detect MRSA.

Treatment of MRSA carriers
- For nasal carriers, treatment with local application of mupirocin is required.
- Chlorhexidine may be needed in some carriers or colonisers as a small number of strains have shown resistance to mupirocin.
- In resistant cases posing major problems, rifampicin along with another oral antibiotic may be effective in the long-term suppression or elimination of the carrier state.

Vancomycin resistance (VRSA)
- Strains showing resistance to vancomycin have started to emerge in some parts of the world.
- These strains may possess both *vanA* and *mecA* genes.
- Vancomycin resistance is a cause of major therapeutic challenges worldwide as these organisms remain susceptible only to linezolid as of now.

Linezolid
- This drug is the treatment of choice for patients infected with VRSA.
- It has good tissue penetration and hence can be used in cases of skin and soft tissue infections caused by MRSA.

Vancomycin-intermediate resistance (VISA)
- VISA strains have decreased susceptibility to vancomycin but do not carry any resistance genes. Instead, they have a thicker cell wall than usual.
- These organisms have been isolated from patients who are on prolonged vancomycin treatment but show clinical failure to treatment with vancomycin.

Coagulase-Negative Staphylococci (CoNS)
- Coagulase-negative staphylococci constitute a major component of the normal flora of the human body.
- *S. epidermidis*, *S. haemolyticus*, and *S. lugdunensis* are the important members of this group.
- Some species can produce human infections, especially in device-associated and healthcare-associated infections.

Staphylococcus epidermidis
- It is the **most common cause of CoNS infections** in humans.
- It is invariably present on healthy human skin and can cause disease when the host defences are breached.
- It is a common cause of the following disease conditions:
  - **Stitch abscess**
  - Colonisation of **implanted foreign bodies** such as artificial heart valves, shunts, intravascular catheters, and prosthetic appliances, leading to bacteremia
  - **Cystitis and central line-associated bloodstream infection** (CLABSI)
  - **Endocarditis**, particularly in drug addicts
- Hospital strains of *S. epidermidis* are usually multidrug-resistant.

Pathogenesis
- **Biofilm formation** is an important factor in the pathogenesis of infections caused by *S. epidermidis*.
- A biofilm is an extracellular polysaccharide matrix which protects bacteria from antibacterial agents and helps in colonisation and resistance of infections.

Staphylococcus saprophyticus
- These organisms may be present on normal human skin and the periurethral area and can cause **urinary tract infection**, particularly in sexually active young women.
- The infecting strains are usually sensitive to most common antibiotics, except nalidixic acid.
- *S. saprophyticus* is identified by its **novobiocin resistance**.

Viridans Group of Streptococci
- Formerly collectively called *Streptococcus viridans*, this group includes streptococci that normally reside in the mouth and upper respiratory tract and produce **greening (alpha-hemolysis) on blood agar**.
- Members include *S. mitis*, *S. mutans*, *S. salivarius*, and *S. sanguinis*.
- These bacteria are chiefly associated with **subacute bacterial endocarditis (SBE)**.
- *S. sanguinis* is most common cause of SBE.

Streptococcus mutans
- *S. mutans* is a gram-positive coccus.
- This alpha-hemolytic streptococcus is a part of the normal flora of the human oral cavity.
- From the mouth, it can spread to cause dental caries or endocarditis in individuals with risk factors (e.g., following dental extraction in individuals who have damaged heart valves).
- The bacterium has a **polysaccharide coat (glycocalyx),** which allows it to stick to teeth and damaged heart valves; it can invade the bloodstream.
- **Nutritionally-variant streptococci (NVS)**, also called nutritionally-deficient streptococci, need pyridoxal or cysteine for their growth on blood agar. They are part of the normal flora but may be responsible for infective endocarditis, especially culture-negative infective endocarditis or brain abscess.

Pathogenesis

- Following tooth extraction or other dental procedures, viridans streptococci cause transient bacteremia and get implanted on damaged or prosthetic valves or in a congenitally-diseased heart and grow to form vegetation.
- Prophylactic antibiotic cover is advisable before tooth extraction or similar procedures in such persons.

HACEK Group of Organisms

- This is a small group of heterogeneous bacteria which are fastidious, gram-negative organisms, located in the upper respiratory tract (mouth, throat, and lungs), gastrointestinal tract, and urinary tract.
- Although HACEK bacteria are usually not associated with problems, they have been known to cause infection in some cases, especially **prosthetic heart valve-associated infective endocarditis**.
- The HACEK group of bacteria includes:
  - *Haemophilus* species
  - *Aggregatibacter* species
  - *Cardiobacterium hominis*
  - *Eikenella corrodens*
  - *Kingella* species
- These organisms cause endocarditis, lung, and respiratory infections, gum infections, bone infections, urinary tract infections, and rarely, abscesses.
- These are fastidious, slow-growing organisms that need a carbon dioxide-enriched environment.
- Blood cultures take about 7–30 days to get a positive yield. However, with automated systems, VITEK and MALDI-ToF, their isolation and identification respectively have become much easier.

Haemophilus

- These are small, gram-negative, oxidase-positive coccobacillary species.
- They are primarily respiratory and CNS pathogens.
- Medically-important members include the following:
  - *H. aphrophilus*
  - *H. paraphrophilus*
  - *H. parainfluenzae*
  - Aggregatibacter

Aggregatibacter

- These are small, fastidious, gram-negative coccobacilli, which cause **infective endocarditis and periodontal infection**.
- They are part of the normal flora of the mouth, especially the gingival and supragingival crevices.
- The important members of this group are:
  - *Aggregatibacter actinomycetemcomitans* (formerly known as *Actinobacillus actinomycetemcomitans*)
  - Aggregatibacter *aphrophilus*

Cardiobacterium

- ***Cardiobacterium hominis*** is the chief species in this group.
- It is a gram-negative, pleomorphic bacillus, which is a common commensal in the human nose and throat and may cause endocarditis, particularly in those with pre-existing cardiovascular disease.
- It is sensitive to many antibiotics, **penicillin and streptomycin** being the recommended drugs.

Kingella

- It is a gram-negative bacillus that was formerly grouped under the genus *Moraxella*.
- It is an oxidase-positive, non-motile, facultative anaerobe with a tendency to occur as coccobacillary and diplococcal forms.
- Organisms belonging to this genus are part of the normal oral flora.
- They can cause endocarditis, especially in the **pediatric age group**, and infections of the bones, joints, and tendons.
- The drugs of choice are **ceftriaxone** or **ampicillin–sulbactam**.

Eikenella

- *Eikenella corrodens* is a gram-negative, oxidase-positive, capnophilic, facultatively anaerobic bacillus.
- The name '*corrodens*' refers to the characteristic **pitting or corroding of blood agar by colonies of this bacterium**.
- These organisms are present in the mouth, upper respiratory tract, and gastrointestinal tract of human beings.
- Infection follows **salivary or fecal contamination** and usually involves the skin and subcutaneous tissues. In rare cases, osteomyelitis, pneumonia, endocarditis, and meningitis may occur.
- They are sensitive to **penicillin and tetracycline**.

Moraxella (branhamella) catarrhalis

- These are gram-negative cocci, usually arranged in pairs, that grow readily on nutrient agar at 18–42°C producing non-pigmented colonies and not fermenting sugars.
- They were formerly known as *Neisseria catarrhalis*. They were reclassified as *Branhamella*, and again as *Moraxella catarrhalis*.
- They form part of the normal pharyngeal flora but can cause respiratory infections including otitis media, sinusitis, tracheobronchitis, and pneumonia.
- Since many strains produce beta-lactamases, penicillins are not usetul in treatment unless given in combination with clavulanate or sulbactam.

Moraxella lacunata

- These are short, plump, gram-negative bacilli usually arranged in pairs.
- They are non-flagellated but have been reported to be sluggishly motile and strictly aerobic. They grow on ordinary media.
- They are oxidase- and catalase-positive, indole- and H_2S-negative and non-fermentative.
- *M. lacunata* was first reported as the cause of **angular conjunctivitis** by Morax and Axenfeld. Hence, it is also known as the **Morax-Axenfeld bacillus**.

BACTERIAL INFECTIONS OF THE BLOODSTREAM WITH FEVER AS THE PRIMARY CLINICAL PRESENTATION

Enteric Fever

Case 3. *A 10-year-old boy was admitted to the pediatric ward with a history of remittent fever which increased gradually over 10 days. He was admitted with complaints of loss of appetite, pain in the abdomen, and general weakness. On examination, he was found to have a temperature of 102°F with pallor and coated tongue, mild splenomegaly, and abdominal tenderness. His blood culture was positive for* ***Salmonella Typhi*** *with a negative Widal test. The boy was diagnosed with enteric fever. The isolate was resistant to ciprofloxacin and so, he was treated with ceftriaxone. He responded after ten days of therapy.*

i. What is enteric fever?
ii. Detail the pathogenesis of enteric fever.
iii. How is the disease diagnosed?

The term enteric fever includes typhoid fever caused by *S.* Typhi and paratyphoid fever caused by *S.* Paratyphi A, B, and C.

Epidemiology

- Typhoid fever has been virtually eliminated in the developed countries over the past several decades, chiefly as a result of improvements in water supply and sanitation. However, it continues to be endemic in the developing world.
- Enteric fever is endemic in all parts of India.
- According to the WHO (2019), an estimated 9 million people develop typhoid and 1,10,000 people die from it every year.

Agent—*Salmonella*

- Members of the genus *Salmonella* cause enteric fever, gastroenteritis, and septicemia with or without focal suppuration.
- *Salmonella* are broadly implicated in two types of clinical entities:
  - **Typhoidal**—the **enteric fever group**, consisting of the typhoid and paratyphoid bacilli that are exclusively or primarily human pathogens
  - **Non-typhoidal**—these cause **food poisoning**; these are essentially animal parasites which can also infect human beings, producing gastroenteritis, septicemia, or localised infections
- DNA hybridisation studies have revealed two distinct species in the genus *Salmonella*—*Salmonella enterica* and *Salmonella bongori*.

Salmonella enterica

- *S. enterica* is further divided into six subspecies:
  - I–*S. enterica* subsp. enterica
  - II–*S. enterica* subsp. salamae
  - IIIa–*S. enterica* subsp. arizonae
  - IIIb–*S. enterica* subsp. diarizonae
  - IV–*S. enterica* subsp. houtenae
  - VI–*S. entericas* subsp. indica

  Most human infections are caused by subspecies enterica and rarely, by arizonae.
- All these species are further divided into more than 2,500 serovars or serotypes.
- *Salmonella enterica* subspecies enterica serovar Typhi is reffered to as *Salmonella* Typhi or *S.* Typhi (the serovar name is not expressed in italics and starts with a capital letter).

Antigenic structure

Salmonellae possess the following antigens, based on which they are classified and identified:

- **Flagellar antigen H:** It is present on the flagella and is **strongly immunogenic**. It induces antibody formation rapidly and in high titres following infection or immunisation.
- **Somatic antigen O:** This forms an integral part of the cell wall and is identical to the endotoxin. The antibody to the O antigen is cross-reactive with other enterobacteriaceae, while that to the H antigen is more reliable for the identification of *Salmonella*.
- **Surface antigen Vi:** Found in some species, the surface antigen Vi envelops the O antigen. This antigen is so named because of the enhanced virulence of strains that produce this antigen, which is associated with a **bacterial capsule**. Originally observed in *S.* Typhi, a Vi antigen with similar antigenic specificity is present in *S.* Paratyphi C and *S.* Dublin, as well as in certain strains of *Citrobacter*. The protective efficacy of the Vi antigen is demonstrated by the success of the purified Vi vaccine for typhoid, which is now in routine use.

Antigenic variations

The antigens of salmonellae undergo phenotypic and genotypic variations.

H–O variation

- This variation is associated with the **loss of flagella**. Rarely, salmonellae may lose their flagella by mutation.

- A stable, non-motile mutant of *S.* Typhi is the 901-O strain, which is widely employed for the preparation of O-agglutinable bacterial suspensions.

Phase variation

The flagellar antigens of most salmonellae occur in one of two phases:

- Phase 1 antigens, which are specific for a species and designated a, b, c, d, etc., and after z, as z1, z2, etc.
- Phase 2 antigens are widely shared and hence, called the non-specific or group phase, designated the numbers 1, 2, 3, etc.

Kauffman and White classification scheme (Table 21.13)

- This scheme permits serological varieties of the genus *Salmonella* to be differentiated from each other on the basis of the surface antigens that are produced by the bacterium.
- First, the somatic or 'O' antigen type is determined, followed by the flagellar or 'H' antigen.
- Different 'H' antigens are produced depending on the phase in which the *Salmonella* exists.

Table 21.13 Kauffman and White classification scheme of *Salmonella*

Serogroup/ Serotype name	O antigen	H antigen (phase1; phase2)
2/S Paratyphi A	1,2,12	a; 1–5
4/S Paratyphi B	1,2[5],12	b; 1,2
7/S Paratyphi C	6,7	c; 1,5
9/S Typhi	9,12 (Vi)	d

Host range

- Salmonellae are strict parasites of animals or humans.
- Infections caused by salmonellae range from asymptomatic conditions to fatal diseases, and sometimes epizootic disease.
- *S.* Typhimurium has a wide host range, affecting animals, birds, and humans.

Susceptibility

Salmonella survives in polluted water and soil for weeks and in ice for months. It is killed by the following:

- Heating at 55°C in one hour or at 60°C in 15 minutes
- Boiling or chlorination of water
- Pasteurisation of milk
- Mercuric chloride (1:500) or 5% phenol for five minutes

Carriers

- These bacilli **persist in the gallbladder or kidneys** and are eliminated in feces (fecal carrier) or urine (urinary carrier).
- **Food handlers or cooks** who become carriers are particularly of concern.
- Carriers exist for paratyphoid bacilli also.
- Depending on the amount of time for which they shed bacilli in their feces, carriers can be categorised as follows:
  - **Convalescent carriers:** Patients who continue to shed typhoid bacilli in feces for three weeks to three months after clinical cure
  - **Temporary carriers:** Those who shed the bacilli for more than three months but less than a year
  - **Chronic carriers:** Those who shed the bacilli for over a year

Typhoid Mary

The best-known typhoid carrier is Mary Mallon, a New York cook who, over a period of 15 years (in the 1900s), caused at least seven outbreaks of typhoid, affecting over 200 persons.

Pathogenesis (Fig. 21.11)

- *S.* Paratyphi A and B cause paratyphoid fever which resembles typhoid fever but is generally mild.
- *S.* Paratyphi C may also cause paratyphoid fever, but more often, it leads to frank septicemia with suppurative complications.
- Broadly, salmonellae cause the following clinical syndromes in humans:
  - Enteric fever
  - Septicemia with or without local suppurative lesions
  - Gastroenteritis or food poisoning

Clinical presentation of enteric fever

Typhoid fever occurs in two epidemiological types:

- **Endemic or residual typhoid**
- **Epidemic typhoid**, which may occur in endemic or non-endemic areas; typhoid epidemics are water-, milk- or food-borne

The disease is most common in the **age group of 5–20 years.**

Gastrointestinal lesions

- Salmonellae multiply abundantly in the gallbladder and are discharged continuously into the intestine.
- Intestinal lesions involve the **Peyer's patches and the lymphoid follicles of the ileum.**
- These become inflamed, undergo necrosis, and slough off, leaving behind the characteristic typhoid ulcers.
- Ulceration of the bowel leads to:
  - Intestinal perforation
  - Hemorrhage
- Intestinal lesions heal in 3–4 weeks.

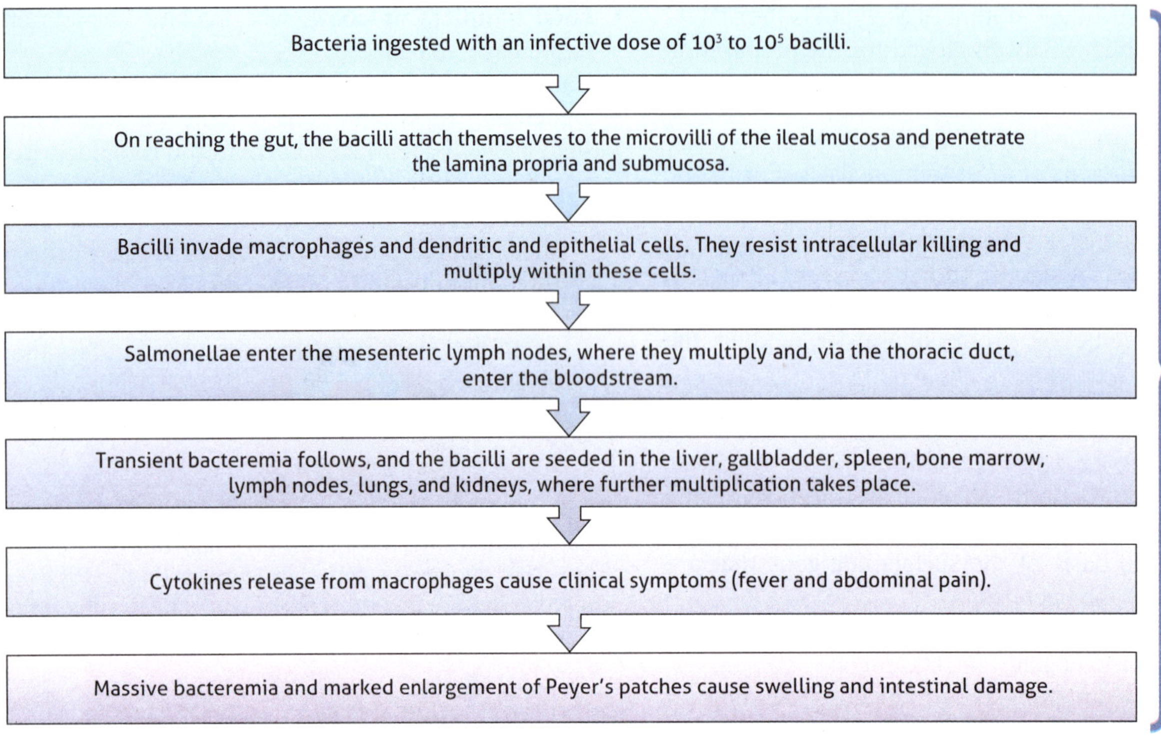

Fig. 21.11 The pathogenesis of *Salmonella* infection

- **Clinical features**
  - *Incubation period*: 10–14 days (may extend to 3–21 days); may vary from mild, undifferentiated pyrexia (ambulant typhoid) to a rapidly fatal disease (**Case 3**)
  - *Symptoms*: Stepladder pyrexia, myalgia, headache, anorexia, and diarrhea
  - *On examination*: Relative bradycardia and rose spots may be present
  - *Per abdomen examination*: Soft, palpable spleen, mild hepatomegaly
  - '*Rose spots*' (Fig. 21.12): These are skin spots that fade on pressure and reappear on the skin during the second or third week. However, these may not be apparent in the Indian population due to the darker hue of the skin.

Complications

- Intestinal perforation
- Hemorrhage
- Circulatory collapse
- Some degree of bronchitis or bronchopneumonia
- Some patients develop transient psychoses (coma vigil), deafness, or meningitis
- **Rare complications** include cholecystitis, arthritis, abscesses, periostitis, nephritis, hemolytic anemia, venous thrombosis, peripheral neuritis, and osteomyelitis
- Relapse occurs in 5–10 per cent of cases

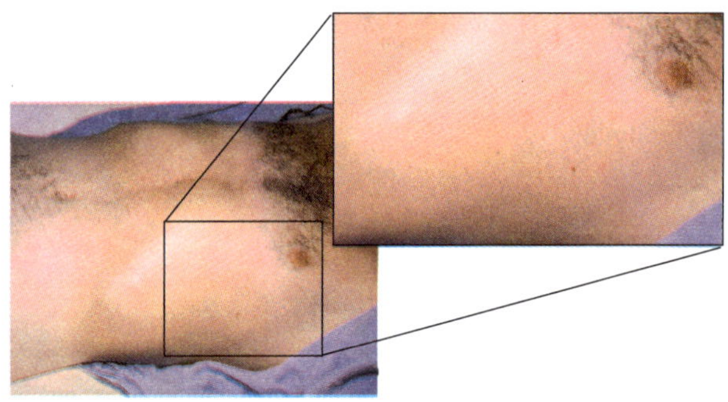

Fig. 21.12 Rose-coloured spots on the abdomen in a case of typhoid (*Source*: Centres for Disease Control and Prevention [CDC], Public Health Image Library [PHIL], Image ID 2214)

Laboratory diagnosis

Diagnosis depends upon the time of presentation of the individual. Bacteriological diagnosis of enteric fever consists of:

- The isolation of the bacilli from the patient
- Demonstration of antibodies in his serum after the first week of infection or depending on the time of presentation

The following laboratory tests are employed in the diagnosis of *Salmonella* Typhi infections:

Routine hemogram

- **CBC:** Low TLC, leukopenia with relative lymphocytosis
- **ESR:** Usually decreased

Culture

- **Blood culture**
  - A positive blood culture is diagnostic.
  - Bacteremia occurs early in the disease, and blood cultures are positive in approximately 90 per cent of cases in the first week, 75% in the second week, and 25% in the fourth week of fever.
- About 8–10 mL of blood is collected and inoculated into an automated blood culture bottle or into brain–heart infusion broth (manual) and incubated at 37°C. Subculture is done on MacConkey's agar and blood agar after 48 hours in manual mode until the indicator beeps in automated system.
- **Clot culture:** The blood is allowed to clot, and then lysed and cultured in broth. Clot cultures yield a higher rate of isolation than blood cultures.
- **Bone marrow culture:** It is valuable as it is positive in most cases, even when blood cultures are negative.
- **Feces culture:** Salmonellae are shed in feces throughout the course of the disease and even in convalescence with varying frequency.
- **Enrichment media:** Selenite F and tetrathionate broth are used for the culturing of *Salmonella* from feces.
- **Plating media:**
  - On XLD agar, *Salmonella* Typhi from feces grows as pink colonies that have a black center (Fig. 21.13).
  - On MacConkey's agar and deoxycholate agar, non-lactose fermenting, pale colonies are seen.
- **Urine culture:** Salmonellae are shed in urine irregularly and infrequently. Hence urine culture is less useful than the culture of blood or feces.

Identification and confirmation of colonies

- *S.* Typhi are **gram-negative, motile bacteria** that are **catalase-positive** and **oxidase-negative**.
- *S.* Typhi produce H_2S while *S.* Paratyphi do not. Both are negative for indole, urease, and citrate.

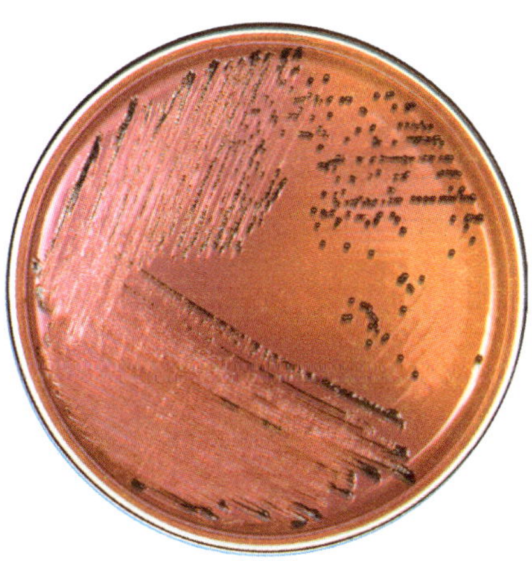

Fig. 21.13 A typical culture of *Salmonella* on xylose lysine deoxycholate (XLD) media (*Source*: CDC, PHIL, Image ID 6619)

- **Serotyping by agglutination**—the agglutination of a culture suspension with specific antisera indicates that the isolate belongs to that particular serotype, e.g., the identity of S. Typhi is established by agglutination with the flagellar antiserum (anti-D serum).
- The **identification of unusual serotypes** is done at the **National Salmonella Reference Centre at the Central Research Institute, Kasauli**. The reference centre for salmonellae of animal origin is the **Indian Veterinary Research Institute, Izatnagar**.

Serology

- Demonstration of antibodies is useful in 2nd or 3rd week of illness.
- It acts as as indirect evidence of infection.
- IgG or IgM antibodies can be detected by Widal or rapid tests.

Widal test: This is an agglutination test that detects antibodies against H and O antigens for *Salmonella* Typhi and *S.* Paratyphi. Since the somatic antigen is common to the two, only the O antigen of *S.* Typhi is used (Fig. 21.14). Antigens used in Widal test are:

- TO: Typhi O antigen (somatic)
- TH: Typhi H antigen (flagellar antigen)
- AH: Paratyphi A flagellar antigen
- BH: Paratyphi B flagellar antigen

Serial dilutions of the patient's serum are added to antigenic suspensions. Doubling dilutions are done to avoid prozone phenomeneon. The last dilution that produces agglutination is taken as the antibody titer.

The **interpretation—**the distribution of antibody levels in the 'normal population' in different areas should be taken into account to determine the cut-off titre.

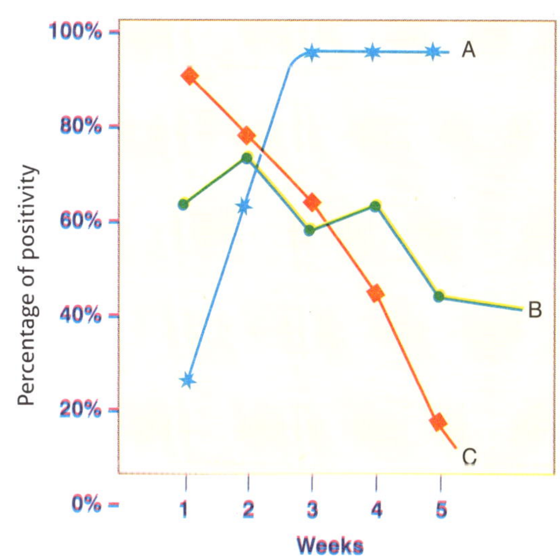

Fig. 21.14 Percent positivity of various laboratory diagnostic tests for typhoid fever during different stages of the disease (weeks 1–5); A. Widal test, B. feces culture, and C. blood culture

- In endemic areas like India, titres of **TO >100 and TH >200 are taken as indicative of early infection.** In non-endemic areas, lower titres may indicate infection.
- Alternatively, a four-fold rise in titre over one week is taken as positive.
- **Anamnestic response**—this is a false-positive result seen in non-typhoidal fever conditions such as malaria and scrub typhus and also in some immunological disorders, particularly active chronic hepatitis, rheumatoid arthritis, and ulcerative colitis.
- **False-negative results are also seen when the sample is collected early in disease** (Fig. 21.14).
- Vaccinated individuals may show positive response to all TH, AH, and BH antigens.
- Agglutinins usually appear by the end of the first week, so blood taken earlier may give a negative result. The titre increases steadily until the third or fourth week, after which it declines gradually.

Rapid tests

- **IgM detection kits:** These are useful in the diagnosis of infection in the initial weeks of a primary infection when a single serum sample is available. These kits detect IgM antibodies to lipopolysaccharides or outer membrane protein antigens.
- IgM O9 antibodies can be detected in patients within a few minutes by a tube test, e.g., IDL TUBEX.
- **Immunochromatographic tests:** These are used to detect specific IgM.
- **Widal slide test:** This test uses a coloured antigen that agglutinates with its specific antibody in the patient's serum.

Diagnosis of carriers

The detection of carriers is important for epidemiological and public health purposes and for the testing of food handlers and cooks. This is done by the following methods:

- Culture of stool, bile, and urine
- Detection of antibodies to Vi antigens
- Detection of the antigen in urine or serum by ELISA or PCR

Laboratory diagnosis of enteric fever (Table 21.14)

- **Specimen:** Blood for culture, bone marrow, feces, urine for culture, serum for serology
- **Routine hemogram:**
  - *CBC*: Low TLC, leukopenia with relative lymphocytosis
  - *ESR*: Usually decreased
- **Culture**
  - Enrichment media for feces: Selenite F and tetrathionate broth
  - MacConkey's agar and DCA: Non-lactose fermenting, pale colonies
  - XLD agar: Pink colonies that have a black center
- **Identification and confirmation of colonies:**
  - **Gram-negative, motile bacteria; catalase-positive, oxidase-negative**, negative for indole, urease, and citrate
  - *S.* Typhi produce H_2S while *S.* Paratyphi do not
  - **Confirmation by serotyping with slide agglutination** using specific antisera
- **Serology:**
  - Demonstration of antibodies is useful in 2nd or 3rd week of illness
  - IgG or IgM antibodies can be detected by Widal or rapid tests
- **Widal tube test:**
  - Titres of **TO >100 and TH >200 are taken as indicative of early infection**
  - Alternatively, a four-fold rise over one week is taken as positive
- **Rapid tests:**
  - *IgM detection kits*: Detection of IgM antibodies to the lipopolysaccharide or outer membrane protein antigens and IgM O9 antibodies
- **Diagnosis of carriers:**
  - Culture of stool, bile, and urine
  - Detection of antibodies to Vi antigens
  - Detection of the antigen in urine or serum by ELISA or PCR

Table 21.14 Percentages of positivity for tests conducted for enteric fever

Positivity %	1st week	2nd week	3rd week
Blood culture	90%	75%	25%
Widal	25%	60%	90%

Typing of *S.* Typhi

Bacteriophage typing

Phage typing helps in tracing the source of epidemics and provides information on the trends and patterns in the epidemiology of typhoid at the local, national, and international levels.

Among *S.* Typhi phage types, A and E1 are most common in India. Among the *S.* Paratyphi A phage, types 1 and 2 are the most common.

Molecular methods

Currently, more discriminating genotyping methods are being employed for epidemiological typing, which include the following:

- Pulse-field gel electrophoresis (PFGE)
- Multilocus enzyme electrophoresis
- IS-200 profiling
- Random amplified polymorphic DNA (RAPD) analysis

Salmonellae can be isolated from sewage by filtering the sample through millipore membranes and culturing the membranes on highly selective media such as the Wilson and Blair medium.

Prevention

- Typhoid bacilli are primarily intracellular parasites, hence cell-mediated immunity (CMI) develops during the course of infection.
- Typhoid fever can be effectively controlled by **improvements in sanitation and provision of safe water supply**. Many developed countries have been able to eliminate the risk of typhoid by implementing these measures; however, occasional outbreaks do appear due to unforeseen lapses.
- At present, vaccines provide short-term protection during outbreaks and are used as prophylaxis (Table 21.15).

Treatment (Table 21.16)

- For carriers, **a combination of antibacterial therapy and vaccination** has been attempted for the eradication of the carrier state. This combination has also been used to prevent relapses.
- Elimination of the persistent carrier state may, on rare occasions, require cholecystectomy, pyelolithotomy, or nephrectomy.

Antimicrobial resistance in *S.* Typhi

There are two categories of drug resistance in S. Typhi:

1. **Resistance to antibiotics such as chloramphenicol, ampicillin, and trimethoprim-sulfamethoxazole (MDR strains):** Quinolones such as ciprofloxacin are the drugs of choice to treat such infections.
2. **Resistance to the fluoroquinolone drugs, which may be total or partial; the so-called nalidixic-acid-resistant S. Typhi (NARST):** This strain was a marker of reduced susceptibility to fluoroquinolones compared to nalidixic acid-sensitive strains. Ceftriaxone is the best alternate in such infections.

Salmonella septicemia

- Certain salmonellae, *S. choleraesuis* in particular, may cause septicemic disease with focal suppurative lesions such as osteomyelitis, deep abscesses, endocarditis, pneumonia, and meningitis.
- The case fatality may be as high as 25 per cent.
- Salmonellae may be isolated from blood or pus from the suppurative lesion.
- Treatment is the same as that for enteric fever.

Non-typhoid salmonellosis

- This is a frequent infectious complication in **systemic lupus erythematosus (SLE).** The extraintestinal manifestations and sepsis are more frequent in SLE than in simple gastroenteritis.
- Vaccines do not protect against *Salmonella* gastroenteritis or *Salmonella* septicemia.

Table 21.16 Treatment regimen for typhoid fever

Condition	Optimal therapy	Alternative therapy
Empirical treatment	Ceftriaxone for 10–14 days Azithromycin for 5 days	–
Fully susceptible isolate	Ciprofloxacin for 5–7 days Azithromycin* for 5 days	Amoxicillin for 14 days Chloramphenicol for 14–21 days Trimethoprim sulphamethoxazole for 7–14 days
Multidrug-resistant (MDR) or quinolone resistant	Ceftriaxone for 10–14 days Azithromycin for 5 days	Ciprofloxacin for 5–14 days Ceftriaxone for 10–14 days Azithromycin for 5 days
As outpatient therapy in children	Cefixime for 7–14 days	–

* Azithromycin given in a dose of 1 g/day

Table 21.15 Typhoid vaccines

Vaccine	Features	Dosage	Recommendation
TAB vaccine	Contains 1,000 million *S.* Typhi and 750 million each of *S.* Paratyphi A and B per mL, which have been killed by heating at 50–60°C and preserved in 0.5% phenol Does not provide cell-mediated immunity	Two doses of 0.5 mL each, subcutaneously at an interval of 4–6 weeks	No longer used
Live oral vaccine (typhoral)	Stable mutant of *S.* Typhi strain Ty2 1a On ingestion, it initiates infection but 'self-destructs' after four or five cell divisions; therefore, cannot induce any illness Enteric-coated capsule containing 109 viable lyophilised mutant bacilli	One capsule is taken orally, an hour before food, with a glass of water or milk, on days 1, 3, and 5 No antibiotic should be taken during this period Booster after 3 years	Only for those over five years of age Protection starts 2–3 weeks after administration and lasts for at least three years
Vi vaccine (Typhim-Vi and VirEPA)	Contains purified Vi polysaccharide antigen from *S.* Typhi strain Ty2 Vi-rEPA is a newer conjugate vaccine in which Vi is bound to a non-toxic recombinant P. aeruginosa exotoxin A; it has high efficacy	Single subcutaneous or intramuscular injection Booster after 3 years	VirEPA is protective in children under 5 years

Plague

> **Case 4.** *A 45-year-old farmer from a village was brought with septic shock and features of pneumonia to the emergency department of a nearby referral hospital. His relatives informed the attending doctor that the farmer had a sudden onset of fever with difficulty in breathing two days ago and hemoptysis subsequently. X-ray on admission showed extensive bronchopneumonia. Immediate supportive therapy was started with antibiotics gentamicin and doxycycline. However, the patient succumbed to the infection.*
>
> ***Plague** was suspected, and cultures of the patient's sputum and blood were performed in a containment (biosafety level III) referral laboratory. With Wayson stain, his sputum showed bipolar staining bacilli, confirmed as* Yersinia pestis *on culture. While this patient was being treated, a close household contact also developed similar symptoms. He was immediately started on doxycycline with supportive therapy and recovered. The cases were notified to the public health authorities as part of the immediate control programme.*
>
> i. What is plague?
> ii. What are the various clinical forms of plague?
> iii. How is this condition diagnosed?

Plague is a zoonotic disease caused by *Yersinia pestis*.

Epidemiology

- Plague was known as **black death** in the Middle Ages.
- The plague bacillus is **parasitic in rodents**.
- The infection is transmitted among rodents by **rat fleas**. In the flea, the bacilli multiply in the stomach to such an extent that they block the proventriculus.
- The infection has an **extrinsic incubation period**. The interval between the ingestion of infected blood and the blockage of the proventriculus is usually about two weeks in *Xenopsylla cheopis* (Fig. 21.15a and b).
- When such a 'blocked' flea bites another rodent, blood mixed with the bacteria is regurgitated into the bite, transmitting the infection. Infection may also be transferred by the contamination of the bite wound with the feces of infected fleas.
- When a diseased rat dies (**rat fall**), the fleas leave the carcass and, in the absence of another rat, may bite human beings, causing **bubonic plague**.

Historical perspective in India

- India was one of the countries worst hit by the plague pandemic.
- Plague arrived in Bombay in 1896 and thereafter, spread throughout the country, causing more than 10 million deaths by 1918.
- It gradually receded afterwards, though occasional cases continued to be reported in endemic foci until 1967.
- Subsequently, no further plague cases were reported in India until 1994.

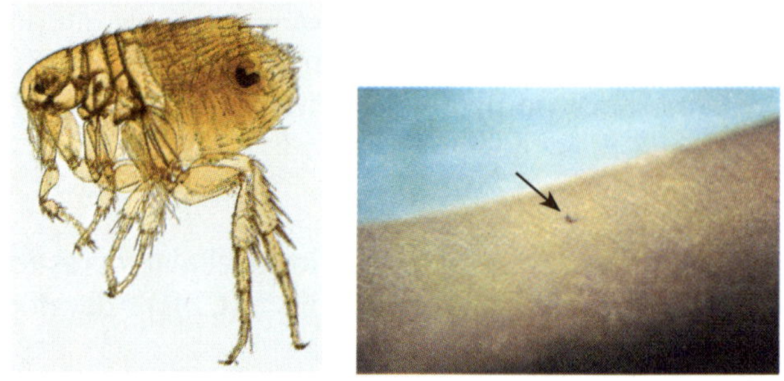

(a) (b)

Fig. 21.15 (a) Oriental rat flea, *Xenopsylla cheopis*, a well-known bubonic plague vector and (b) the flea biting its human host (*Source*: CDC, PHIL, Image IDs 22250 and 14286)

- Four foci of plague are known in India:
  1. The region near Kolar at the trijunction of Tamil Nadu, Andhra Pradesh, and Karnataka
  2. The Beed–Latur belt in Maharashtra, from where the Surat epidemic emanated
  3. Rhoru in Himachal Pradesh, where the 2002 outbreak took place
  4. A small pocket in Uttaranchal
- In the 1990s, there was a re-emergence of plague in India and China in Asia; Malawi and Zimbabwe in Africa; in the erstwhile USSR; in Europe; and in the USA.

Natural cycles of plague

- **Urban or domestic plague:** This refers to plague that is associated with human beings and the rodents living around them.
- **Wild or sylvatic plague:** This type occurs in nature and in wild rodents independent of human beings. In enzootic foci, plague may persist for long periods. Infected fleas may survive for over a year. The bacilli can remain alive and even multiply in the soil of abandoned rodent burrows. There, they infect new rodents that may reoccupy such burrows and humans who come in contact.

Agent—*Yersinia pestis*

- *Yersinia pestis* is a short, plump, ovoid, gram-negative bacillus about 1.5 × 0.7 μm in size, arranged singly, in short chains, or in small groups.
- It was discovered independently and simultaneously by **Alexandre Yersin and Kitasato Shibasaburo** (1894) in Hong Kong at the beginning of the last pandemic of the disease.
- It is easily destroyed by heat, desiccation, and chemical disinfectants, a temperature above 55°C, or with 0.5% phenol in 15 minutes.
- It remains viable for long periods in cold and moist environments.
- It can survive for several months and even multiply in the soil of rodent burrows.

Antigens, toxins, and other virulence factors

- **Protein envelope antigen (Fraction I or F-I):** This is a heat-labile antigen formed in cultures incubated at 37°C. It inhibits phagocytosis and is generally present only in virulent strains. This plasmid-encoded antigen has been considered a virulence determinant.
- **V and W antigens:** These inhibit phagocytosis and the intracellular killing of the bacillus. The production of these antigens is plasmid-mediated, and they are always produced together.
- **Pesticin I (bacteriocin), coagulase, and fibrinolysin:** These are produced by the virulent strains.
- **Endotoxin:** It is a lipopolysaccharide similar to the endotoxins of enteric bacilli.

Pathogenesis of plague

In human beings, plague occurs in three main forms: (i) bubonic, (ii) pneumonic, and (iii) septicemic.

Bubonic plague

- **Incubation period:** 2–5 days after the bite of an infected flea.
- The **draining lymph nodes of the site of entry of the bacillus become infected** (Fig. 21.16a).
- In some cases, only minor constitutional symptoms (**pestis minor**) occur.
- As flea bites are common on the legs, the inguinal nodes are most commonly involved, giving the disease its name, 'bubonic' (***bubon* meaning groin)** plague.
- Patients present with swollen, tender lymph glands (called buboes) with fever, headache, chills, and weakness.
- No interpersonal transmission is seen.

Pneumonic plague (Case 4)

- This type is seen during epidemics of bubonic plague. Rarely, primary pneumonic plague may occur in an epidemic form.
- It spreads by droplet infection. The bloody, mucoid sputum that is coughed out contains bacilli in enormous numbers. After inhalation, the bacilli spread through the lungs, producing **hemorrhagic pneumonia.**
- Patients present with fever, headache, weakness, and rapidly developing pneumonia with shortness of breath, chest pain, cough, and sometimes, bloody or watery sputum.
- The pneumonia progresses for 2–4 days and may cause **respiratory failure and shock**.
- **Cyanosis** is very prominent.
- Pneumonic plague is highly infectious.

Septicemic plague

- It is usually the terminal event in the bubonic or pneumonic variety but may sometimes occur primarily.
- Features of sepsis and rarely, signs of meningeal irritation may be seen.
- Clinically, fever, chills, prostration, abdominal pain, shock, and bleeding into the skin and other organs is seen.
- Septicemic plague does not spread from person to person.
- **Disseminated intravascular coagulation** is common and may lead to gangrene of the skin, fingers, and penis (Fig. 21.16b).
- Case fatality in untreated cases is 30–90 per cent.

Laboratory diagnosis

- Containment laboratories with biosafety levels III or IV are equipped to handle and process specimens from infected patients and rodents.
- In human bubonic plague, the bacilli may be readily demonstrated in buboes by microscopy, culture, or animal inoculation.
- In pneumonic plague, the bacilli can be demonstrated in sputum by microscopy, culture, or animal inoculation.

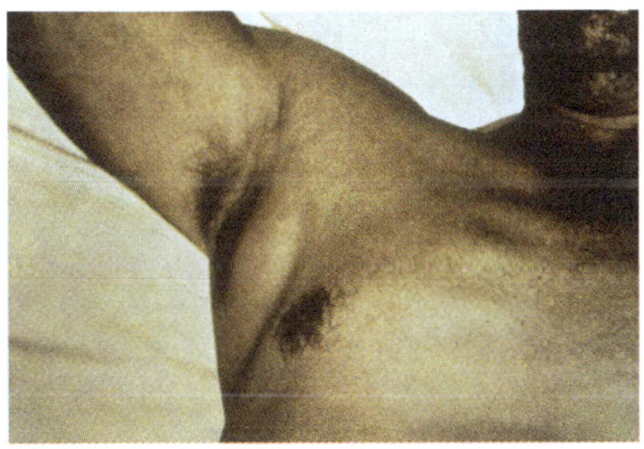

(a)

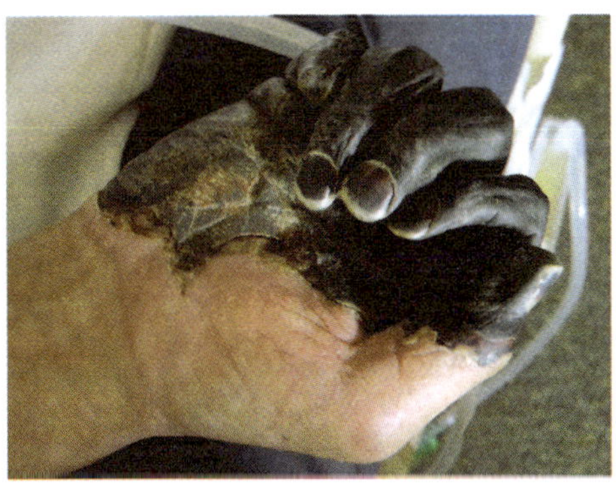

(b)

Fig. 21.16 Characteristic features of plague: (a) swollen axillary lymph node (bubo) and (b) gangrenous condition of the fingers and palm due to *Yersinia pestis* infection (*Source*: CDC, PHIL, Image IDs 2045 and 16551)

Specimen

- Buboes, sputum, and aspirates from the spleen, and blood are collected under strict sterile conditions.
- When an animal is suspected to have died from *Y. pestis* infection, the carcass is kept for investigations.

Microscopy

- Smears from the buboes, sputum, or blood stained with methylene blue (Wright or Wayson stain) show bipolar (**safety pin appearance**) stained bacilli (Fig. 21.17).
- Bacilli in buboes show considerable pleomorphism.

Culture

- **Primary isolation media:** These are blood agar, chocolate agar, and MacConkey's agar. The colonies are small, translucent, and **grey to dark brown** due to the **absorption of the hemin pigment. Stalactite growth** hanging from surface is seen in *ghee* broth.
- **Broth:** A flocculant growth occurs at the bottom and sides of tube.
- **Selective medium:** On **cefsulodin, irgasan, novobiocin (CIN) agar**, colonies have a deep-red centre surrounded by a transparent border, giving the appearance of a 'bull's eye'.
- **Incubation conditions:** *Y. pestis* are best incubated in ambient air at 28–30°C for 24–48 hours. The optimum temperature for the growth of *Y. pestis* is 27°C but the envelope develops best at 37°C.

Identification and confirmation

- *Y. pestis* are gram-negative, non-motile, catalase-positive, oxidase-negative, facultatively anaerobic straight rods or coccobacilli.
- They are negative for lactose fermentation, citrate, VP, and urease tests.

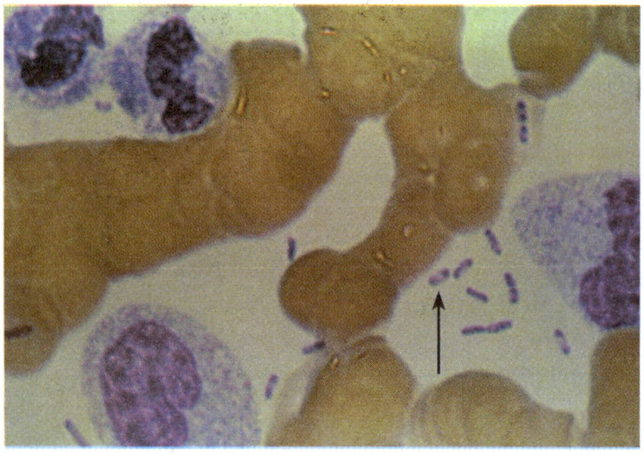

Fig. 21.17 Bubo smears stained with Wright stain showing bipolar staining (safety pin appearance) of *Y. pestis* (*Source*: CDC, PHIL, Image ID 2050)

Serological tests

- Antibodies to the F-I antigen may be detected by passive hemagglutination, IgG, and IgM ELISA or by a rapid dipstick assay.
- A rise in the titre of antibodies in paired sera or a titre of 128 or above in a single serum sample can be considered positive.
- The fluorescent antibody technique is performed on impression films of the affected tissues to demonstrate the F1 antigen.

Molecular method

PCR is a rapid and sensitive method for the presumptive diagnosis of plague from clinical material.

Isolation from rats

During an outbreak, the organism can be isolated from dead rats in the laboratory. The carcass is dipped in 3% Lysol to destroy the fleas before handling them for further procedures.

Laboratory diagnosis of plague

Essential—containment laboratories with biosafety levels III or IV.

- **Specimen:**
  - Buboes, sputum, aspirates from the spleen, and blood
  - Animal carcass
- **Microscopy:** Wright or Wayson (methylene blue) stained smears from the buboes, sputum, or blood; bipolar (**safety pin appearance**)
- **Culture:**
  - Optimum temperature: 27–37°C
  - *Blood agar, chocolate agar, and MacConkey's agar*: Small, translucent, and **grey to dark brown** colonies due to the **absorption of the hemin pigment**
  - *Ghee broth*: **Stalactite growth** hanging from surface
  - *Nutrient broth*: Flocculant growth occurs at the bottom and sides of tube
  - *Selective medium-CIN (cefsulodin, irgasan, novobiocin) agar*: Colonies with deep-red centre surrounded by a transparent border ('bull's eye')
- **Identification and confirmation:** Gram-negative, non-motile, catalase-positive, oxidase-negative, facultatively anaerobic straight rods or coccobacilli, negative for lactose fermentation, citrate, VP, and urease tests
- **Serological tests:**
  - Antibodies to the F-I antigen by passive hemagglutination, IgG, and IgM ELISA or by a rapid dipstick assay
  - F1 antigen detection: Impression films of the affected tissues
- **Molecular method-PCR**

Prophylaxis

General measures

- Control of fleas and rodents, sanitation, cleanliness, proper disinfection, and hand washing
- Isolation of pneumonic plague patients

Vaccines

- **Whole-culture antigen**
  - This is a killed vaccine (prepared at the Haffkine Institute, Mumbai) given subcutaneously in three doses at 1, 3, 6 months.
  - Vaccination provides some protection against bubonic plague but not against pneumonic plague.
  - Protection does not last for more than six months.
- **Recombinant rF1V plague vaccine**
  - It is used for pre-exposure prophylaxis in plague.
  - It was developed by scientists at the **US Army Medical Research Institute of Infectious Diseases (USAMRIID).**
  - The WHO recommends vaccination only for high-risk groups (those exposed occupationally or otherwise to infection, such as individuals who work at a plague laboratory, hospital personnel, and troops deployed in known plague-endemic areas).
  - The vaccine has no value in outbreaks or mass vaccination.

An episode of plague confers lifelong immunity.

Treatment

- Quinolones such as ciprofloxacin, levofloxacin, moxifloxacin, gentamicin, or streptomycin can be used.
- Doxycycline is useful in treating bubonic plague.

Leptospirosis

> **Case 5.** *A 45-year-old woman who worked in a rice farm (where she would stand in water during the planting season) developed fever off and on and did not seek medical advice. After two weeks, she presented with suffused conjunctiva, jaundice, and continuous fever. Her sera tested positive by ELISA for IgM antibodies to* ***Leptospira*** *antigens. A diagnosis of* ***leptospirosis*** *was made.*
>
> *For confirmation, the woman's serum sample was sent to the reference laboratory for microscopic agglutination test for* Leptospira*-specific antibodies. The test was positive at a titre of 1:10,000. The woman was treated with* ***doxycycline****.*
>
> i. What is leptospirosis?
> ii. What are the complications of leptospirosis?
> iii. How is this condition diagnosed?

Leptospirosis is considered the **most widespread of zoonoses** on account of its presence in all continents except Antarctica. It is caused by *Leptospira*.

Epidemiology

- *Leptospira* are transmitted by **rats and other rodents**, which are important reservoirs for the bacteria.
- Pathogenic leptospires survive for long periods in the **convoluted tubules of the kidneys** in natural hosts, where they multiply and are shed in urine, whereby they **contaminate water.**
- When people come into contact with contaminated water, these leptospires enter the body through **abraded skin or mucosa** and initiate infection.
- Occupational groups at risk include **agricultural workers** who work in rice or cane fields, **miners,** and **sewer cleaners.**
- Incidence rises following floods and waterlogging in urban areas. Outbreaks are often seen in pockets following rains that result in flooding.
- In India, leptospirosis occurs following the monsoons and floods.
- The **Andaman hemorrhagic fever**, which is also caused by *Leptospira* (serovar Portblairi and serogroup Sehgali), is endemic in the Andamans.

Agent—*Leptospira*

- Leptospires belong to the family of **spirochetes.**
- They are delicate, flexible, and helical rods, about 6–20 µm long and 0.1 µm thick.
- They have numerous coils set so close together that they can be distinguished only under dark-ground illumination in the living state or by electron microscopy.
- Their **ends are hooked** and resemble umbrella handles (Fig. 21.18).
- They are actively motile and stain poorly with aniline dyes.

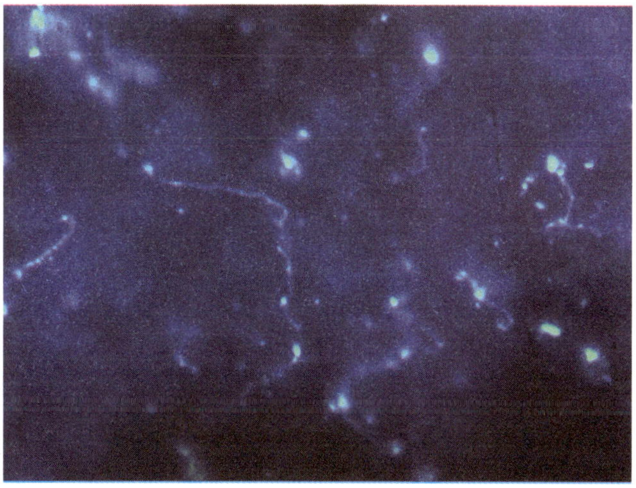

Fig. 21.18 Fluorescent antibody (FA) staining of a liver impression smear sample from a fatal case of leptospirosis shows corkscrew-shaped *Leptospira* (*Source*: CDC, PHIL, Image ID 1346)

Classification

The genus *Leptospira* is now classified into two species, and within each species are serogroups, which are further classified into serotypes (serovars).

1. *L. interrogans*: Pathogenic leptospires with about 22 serogroups having more than 200 serotypes (Table 21.17)
2. *L. biflexa:* Saprophytic leptospires found predominantly in surface waters; over 200 serovars have been identified and assembled into 22 serogroups

Antigenic properties

- Leptospires exhibit considerable antigenic cross-reaction.
- A genus-specific somatic antigen is present in all members of the genus.
- Classification into serogroups and serotypes is based on the surface antigens.
- Genetic methods such as restriction endonuclease analysis and DNA pairing are used for further classification into serotypes.

Resistance

- Leptospires are highly susceptible to heat (killed in 10 minutes at 50°C and in 10 seconds at 60°C), acids, gastric juices, and bile.
- They are readily destroyed by chlorine and most other antiseptics and disinfectants.
- They can survive for days in moist conditions at pH 6.8–8.

Pathogenicity (Fig. 21.19)

The **incubation period** of the infection is 2–26 days.

Clinical features

- Infection may be mild, with undifferentiated pyrexia, or it could manifest as **severe or fatal illness with hepatorenal damage (Weil's disease).**
- In severe cases, the onset is acute, with rigor, vomiting, headache, and conjunctival suffusion. The fever is irregular and usually subsides in about ten days.
- Jaundice occurs in about 10–20 per cent of cases by the second or third day.
- Purpuric hemorrhages sometimes occur on the skin and mucosa.
- Albuminuria is a constant feature (Case 5).
- There are two clinical types—**icteric and non-icteric.**
- It is a multisystem disorder.
- It may present as aseptic meningitis and, in some, abdominal symptoms predominate.

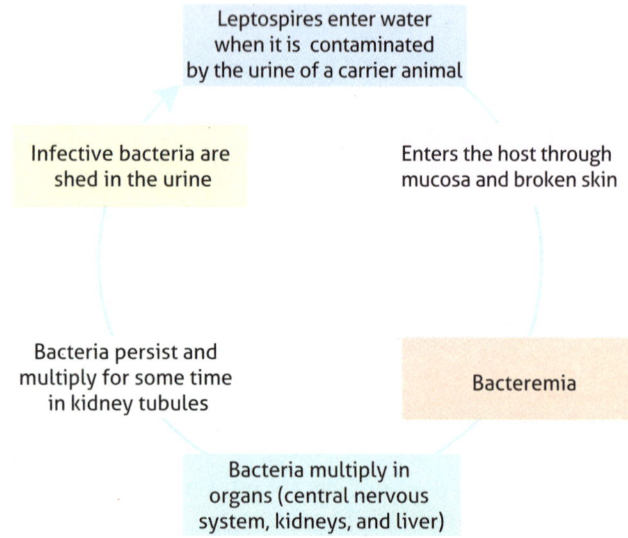

Fig. 21.19 The pathogenesis of *Leptospira*

Laboratory diagnosis

Faine's criteria to diagnose leptospirosis are as follows:

1. **Clinical features**—fever, headache, conjunctival suffusion, myalgia, dyspnea, meningism, and jaundice
2. **Epidemiological factors**—rainfall, contact with contaminated water or animals
3. **Laboratory findings**—isolation or PCR-positive for *Leptospira,* or positive for IgM by ELISA or MAT

Table 21.17 Important leptospiral infections (*L. interrogans*)

Serogroup	Disease	Clinical picture	Animal reservoir	Distribution
Icterohemorrhagiae	Weil's disease	Fever, jaundice, hemorrhages	Rat	Worldwide
Canicola	Canicola fever	Influenza-like aseptic meningitis	Dog	Worldwide
Grippotyphosa	Swamp or marsh fever	Fever, prostration, aseptic meningitis	Field mice	Europe, Africa, SE Asia, USA
Fort–Bragg	Pretibial fever Fort–Bragg fever	Fever, rash over tibia	Not known	Japan, SE Asia, USA
Pyrogenes	Febrile spirochetosis	Fever	Pig	SE Asia, Europe, USA
Bataviae	Indonesian Weil's disease	Fever	Rat	SE Asia, Africa, Europe

Microscopy

- **Blood:** *Leptospira* are seen during the acute phase of the disease (up to 8–10 days).
- **Urine:** *Leptospira* are seen in the later stages of the disease.
- Leptospires cannot be observed under the ordinary light microscope.
- They can be visualised by **dark-field microscopy** as thin, coiled, and rapidly moving microorganisms (Fig. 21.18). The sensitivity of dark-field microscopy is approximately 10^7 leptospires/L.
- Immunofluorescence, immunoperoxidase, silver staining, or Warthin–Starry staining may be done. However, these methods have a high risk of false-positive and false-negative results.
- Leptospires may be stained with Giemsa stain, but better results are obtained by the silver impregnation methods.

Culture

- *Leptospira* are aerobic and microaerophilic and grow optimally at **25–30°C** and at optimum **pH of 7.2–7.5.**
- On semi-solid media, growth occurs a few millimetres below the surface in the **form of a ring**.

Leptospires can be grown in media enriched with rabbit serum. Several liquid and semi-solid media are used:

- Korthof's, Stuart's, and Fletcher's media
- Semi-synthetic media such as EMJH (Ellinghausen, McCullough, Johnson, and Harris) are now commonly used for culturing leptospires
- Embryonated eggs
- Chorioallantoic membrane of chick embryos
- Demonstrated in blood drawn from the allantoic vessels after 4–5 days of infection

Identification

- The identification of isolates is a complicated procedure and is generally confirmed by either the WHO or FAO (global).
- The reference laboratory for leptospires in India is **ICMR's Regional Medical Research Center (RMRC) at Port Blair**.

Animal inoculation

On intraperitoneally introducing *Leptospira* in guinea pigs, the animals die within 8–12 days with jaundice and hemorrhage into the lungs and serous cavities.

Serological diagnosis

Antibodies appear in serum towards the end of the first week of the disease and increase until the fourth week, declining thereafter.

- **ELISA**—IgM and IgG antibodies are looked for to assess the stage of the infection
- **Rapid dipstick assay**—a *Leptospira*-specific IgM antibody test is performed.
- **Macroscopic agglutination test**—this is a slide agglutination test (SAT) that uses a suspension of a killed *Leptospira* serovar with serial dilutions of the test serum.
- **Cross-agglutinin absorption test (CAAT)**—the serum is absorbed to make it more specific.
- **Microscopic agglutination test (MAT)—**uses live cultures of different serotypes; agglutination is observed under a low-power, dark-field microscope.

Laboratory diagnosis of leptospirosis

- **Specimen:** Blood, urine
- **Microscopy:**
  - Seen during the acute phase in blood and in urine in the later stages
  - **Dark-field microscopy:** Thin, coiled, and rapidly moving bacteria
  - Immunofluorescence, immunoperoxidase, silver staining, or Warthin–Starry or Giemsa staining
- **Culture:**
  - *Aerobic and microaerophilic optimum temperature*: **25–30°C**, optimum **Ph 7.2–7.5**
  - *Semi-solid media*: Growth occurs a few millimetres below the surface in the **form of a ring**
  - *Selective media*: Korthof's, Stuart's, and Fletcher's media, EMJH (Ellinghausen, McCullough, Johnson, and Harris)
  - *Chorioallantoic membrane of chick embryos*: demonstrated in blood drawn from the allantoic vessels after 4–5 days of infection
- **Identification:** Done by agglutination with type-specific sera
- **Serological diagnosis:**
  - **ELISA**—IgM and IgG antibodies
  - **Rapid dipstick assay**—a *Leptospira*-specific IgM antibody test
  - **Macroscopic agglutination**
  - **Cross-agglutinin absorption test (CAAT)**
  - **Microscopic agglutination test (MAT)**
- **Molecular detection:** PCR to detect DNA in blood

Prophylaxis

- General measures for prevention such as rodent control, disinfection of water, and the use of protective clothing contribute to the prevention of leptospirosis.
- Vaccination has been attempted in persons at high risk such as agricultural workers.
- Doxycycline 200 mg can be orally given once a week for prophylaxis.

Treatment

- **Drug of choice:** Penicillin intravenous 1–2 million units six-hourly for seven days
- **Alternate drugs:** Ceftriaxone or cefotaxime

- A mild **Jarisch–Herxheimer reaction (JHR)** may occur in some patients: JHR is a transient clinical phenomenon that occurs in patients within 24 hours of antibiotic treatment. It is characterised by fever, chills, rigors, nausea and vomiting, headache, tachycardia, hypotension, hyperventilation, flushing, and myalgia

Brucellosis

> **Case 6.** *A 30-year-old man who worked at a cattle farm presents with complaints of intermittent fever, anorexia, profuse sweating, malaise, headache, muscle pains, and arthralgia for three days. He was seen by a physician, who prescribed antipyretics. The complaints, however, persisted on and off for the next one month and his condition worsened with significant weight loss.*
>
> *On examination, he had hepatomegaly and splenomegaly. The investigations revealed anemia, low neutrophil count, high lymphocytes, and a high erythrocyte sedimentation rate. His serum tested positive for antibodies to Brucella abortus. A diagnosis of **brucellosis** was made, and he was started on antibiotics.*
>
> i. What is brucellosis?
> ii. How is it diagnosed?

Brucellosis, also known as '**undulant fever**', '**Mediterranean fever**', or '**Malta fever**', is a zoonosis acquired by **direct or indirect contact with infected animals or their products**.

Epidemiology

- Human brucellosis is acquired from animals, either directly or indirectly. Goats, sheep, cattle, buffaloes, and swine are the common sources of infection.
- The most important vehicle of infection is **raw milk**.
- Milk products, meat from infected animals, and raw vegetables or water supplies contaminated by the feces or urine of infected animals may also be responsible.
- **Contact infection** is especially important as an occupational hazard in veterinarians, butchers, and animal handlers, and is particularly common during the calving season.
- In India, most human infections are caused by *B. melitensis* acquired from goats and sheep.

Agent—*Brucella*

Brucellae are gram-negative, coccobacilli or short rods that are non-motile, non-capsulated, and non-sporing (Fig. 21.20).

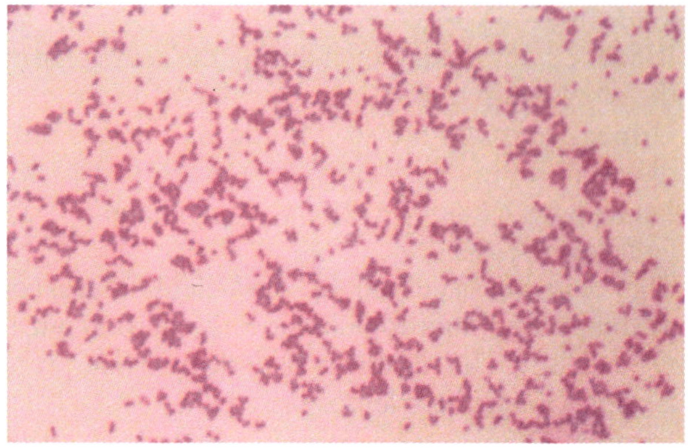

Fig. 21.20 *Brucella*—gram-negative coccobacilli (*Source*: Department of Microbiology, Pondicherry Institute of Medical Sciences, Puducherry)

Classification

Based on nucleic acid homology and gene sequencing, the genus *Brucella* is considered monospecific. The organisms that most commonly cause human infection are:

- *B. melitensis* (in goats and sheep)
- *B. suis* (in swine)
- *B. abortus* (in cattle)
- *B. canis* (in dogs)

Antigenic structure

- The somatic antigens of brucellae contain **two main antigenic determinants, A and M,** which are present in different amounts in the three major species.
- *B. abortus* contains predominantly A, while *B. melitensis* contains the M determinant. *B. suis* has an intermediate antigenic pattern.
- Antigenic cross-reactions occur between brucellae and *Vibrio cholerae*. Persons receiving the cholera vaccine may develop *Brucella* agglutinins lasting for about three years.
- Antigenic cross-reactions also exist with *Escherichia coli* O:116; O:157, *Salmonella* serotypes group N (O:30 antigen Kauffman and White), *Pseudomonas maltophilia*, *Yersinia enterocolitica*, and *Francisella tularensis*.
- A **superficial L antigen** resembling the *Salmonella* Vi antigen has been described.

Virulence factors

- Lipopolysaccharides on the cell walls of brucellae, especially the O-polysaccharides (O-PS) of the LPS—the binding and penetration of the organism into the host cells is regulated by the genes *BvrR*/*BvrS*
- Inhibition of fusion of phagosomes and lysosomes and replication in the phagosomes
- Intracellular bacterial intracellular growth in phagocytic and non-phagocytic cells

Pathogenesis (Fig. 21.21)

- *B. melitensis* is the most pathogenic while *B. abortus* and *B. suis* are intermediately pathogenic.
- *Brucella* are primarily **intracellular pathogens affecting the reticuloendothelial system** with a predilection for macrophages, dendritic cells (DCs), and trophoblasts.

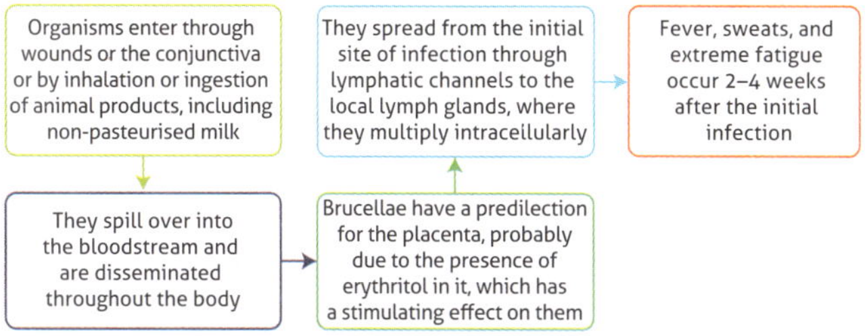

Fig. 21.21 Pathogenesis of brucellosis

- Their intracellular nature accounts for their **refractoriness to chemotherapy**.
- The ability of viable bacilli to co-exist with high levels of circulating antibodies also facilitates their pathogenesis.
- The incubation period is 10–30 days.

Clinical features

Human infection may be of three types:

i. Latent infection with only serological but no clinical evidence
ii. Acute or subacute brucellosis
iii. Chronic brucellosis

Acute brucellosis

- It is generally caused by *B. melitensis*.
- It is characterised by fever associated with muscular and articular pains, asthmatic attacks, nocturnal drenching sweats, exhaustion, anorexia, constipation, nervous irritability. and chills.
- **Undulant fever** (this term is misleading since only a few cases exhibit the undulant pattern) is associated with prolonged bacteremia and irregular fever.
- **Marked thrombocytopenia with low platelet counts**, which may result in bleeding or thrombocytopenic purpura, may be seen.
- Common complications are articular, osseous, visceral, or neurological.

Chronic brucellosis

- It may be non-bacteremic, with a low-grade infection and periodic exacerbations.
- Symptoms are generally related to a state of hypersensitivity in the patient.
- Sweating, lassitude, and joint pain with minimal or periodic pyrexia are seen.
- The illness lasts for years.
- Due to varied symptoms and prolonged fever, the infection is clinically considered as pyrexia of unknown origin (PUO).

Immunity

- Immunity is chiefly cell-mediated.
- Th_1 type of T helper cell response and cell-mediated immunity are required to eliminate brucellae through activated macrophages.
- Tissue reaction to *Brucella* consists of granuloma formation with epithelial cells, giant cells, lymphocytes, and plasma cells. Granulomas heal with fibrosis and sometimes become calcified.

Laboratory diagnosis

- A high index of suspicion is needed to presume brucellosis.
- Clinical diagnosis is almost impossible; laboratory confirmation is essential.
- Laboratory methods for diagnosis include culture, serology, and hypersensitivity tests.

Samples

- Blood culture is the most definitive method.
- Bone marrow cultures show higher yield and remain positive long after the blood culture has become negative.

Culture

- Brucellae are strict aerobes and do not grow anaerobically.
- *B. abortus* is capnophilic, requiring 5–10% CO_2 for growth.
- The optimum temperature is **37°C (range 20–40°C)** and pH is **6.6–7.4**.
- Solid media containing erythritol show small, moist, translucent, and glistening colonies (Fig. 21.22).
- Automated blood culture systems like **BacT/ALERT** and ID systems like **VITEK and MALDI-ToF** have greatly reduced the reporting time to 48–72 hours from the two to three weeks required for the conventional *Brucella* broth or Castaneda's method.
- **Castaneda's method of culture**: Blood is inoculated into a biphasic blood culture bottle containing both solid and liquid media (Fig. 21.23a and b).
  - Subcultures can be done simply by tilting the broth.

Fig. 21.22 *Brucella suis* colonies on chocolate agar (*Source*: CDC, PHIL, Image ID 20132)

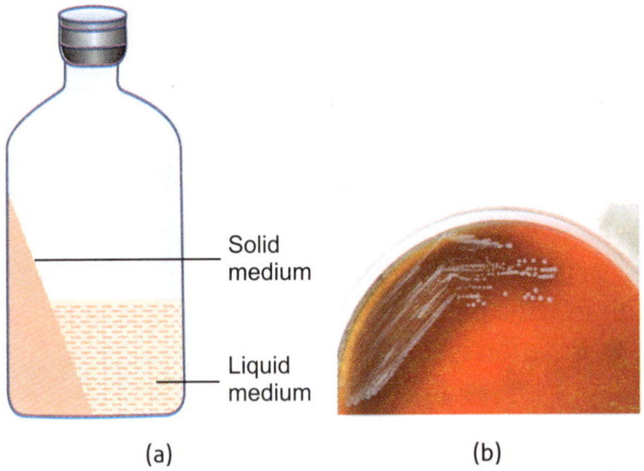

Fig. 21.23 (a) Castaneda's bottle for blood culture of *Brucella* and (b) subculture on sheep blood agar after 72 hours (*Source*: [b] CDC, PHIL, Image ID 12285)

- Cultures are positive only in about 30–50 per cent of cases, even when repeated samples are tested.
- *B. melitensis* and *B. suis* are isolated more readily than *B. abortus*.

Serological methods

ELISA

- It is sensitive and specific and can detect IgM and IgG antibodies separately.
- It is useful for differentiation between acute and chronic infection.

Standard agglutination test (SAT)

- Equal volumes of serial dilutions of the patient's serum and the standardised antigen (killed suspension of a standard strain of *B. abortus*) are mixed and incubated at 37°C for 24 hours or 50°C for 18 hours.
- A positive SAT is a titre of 160 or more.
- Most patients with acute brucellosis develop titres of 640 or more in 3–4 weeks of illness.
- Titres tend to decline after the acute phase of the illness.
- Both IgM and IgG antibodies appear in 7–10 days after the onset of clinical infection.
- As the disease progresses, IgM antibodies decline, while IgG antibodies persist or increase in titre.
- False-negative SAT may occur due to the presence of '**blocking' or 'non-agglutinating' antibodies** in sera. The blocking effect of these antibodies may sometimes be removed by prior heating of the serum at 55°C for 30 minutes or by using 4% saline as the diluent for the test.
- False-positive reaction may occur due to antigenic cross-reaction with gram-negative bacteria, including *V. cholerae* (in those receiving the cholera vaccine), *Escherichia coli* O:116; O:157, and *Salmonella* serotypes group N and Vi antigens.

Microagglutination test

The SAT is miniaturised in a microtitre plate form that uses small amounts of reagents and low serum volumes, permitting the simultaneous testing of multiple samples and producing results in much less time.

2ME test

- The IgG can be differentiated from IgM by treating the serum with 2-mercaptoethanol (2ME).
- IgM is destroyed by this agent.
- The titre following the treatment with 2ME reflects IgG.

Immunocapture agglutination test

Rapid test with strips of microtiter wells coated with anti-human immunoglobulins to detect IgM or IgG.

Dipstick/Lateral flow assay

It acts as a point-of-care test to detect IgM antibodies.

Diagnosis in animals

- Rapid methods such as rapid dipstick test and Rose Bengal card test have been employed for the detection of brucellosis in herds of cattle.
- The milk ring test is used to detect *Brucella* antibodies in milk.

Laboratory diagnosis of brucellosis

- **Samples:** Blood or bone marrow culture
- **Culture:**
  - Strict aerobes, capnophilic, optimum temperature: **37°C (20–40°C)**, pH: **6.6–7.4**
  - Automated blood culture systems like **BacT/ALERT** and ID systems like **VITEK and MALDI-ToF**
  - **Castaneda's method of culture:** Biphasic blood culture bottle
- **Serological methods**
  - *ELISA*
    - Sensitive and specific, can detect IgM and IgG antibodies separately
  - *Standard agglutination test (SAT)*
    - Positive: ≥160 or more
  - *Microagglutination test*
    - SAT is miniaturised in a microtitre plate
  - 2ME test
    - The IgG can be differentiated from IgM by treating the serum with 2-mercaptoethanol (2ME)
  - *Immunocapture agglutination test:* To detect IgM or IgG
  - Dipstick/Lateral flow assay: To detect IgM antibodies

Susceptibility

- Brucellae are destroyed by heat at 60°C in 10 minutes and by 1% phenol in 15 minutes.

- They are killed by pasteurisation and tend to die in buttermilk.
- They remain viable for 10 days in refrigerated milk, one month in ice cream, four months in butter and for varying periods in cheese depending on its pH. They may also survive for many weeks in meat.
- *B. melitensis* may survive for six days in urine, six weeks in dust, and ten weeks in water.

Prophylaxis

- As the majority of human infections are acquired by the consumption of contaminated milk, prevention consists of checking dairy animals for brucellosis.
- No suitable vaccine is available for humans.
- The *B. abortus* strain 19 vaccine is protective in cattle.

Treatment

- **Adults:** Doxycycline is prescribed for at least 45 days along with streptomycin IM daily for the first two weeks. A complete course is necessary to eradicate the infection.
- **Children:** Cotrimoxazole along with rifampicin or gentamicin.

Scrub Typhus

Case 7. *A 42-year-old woman presented with a history of fever, headache, and loss of appetite for past four days. She denied respiratory distress and other symptoms. She has a history of going hiking in the Uttarakhand mountains two weeks before presentation.*

On examination, purple-coloured, non-pruritic eruptions were observed on her face, extremities, and trunk. An eschar was observed in the left lower extremity (Fig. 21.24). Serology was positive for antibodies to Orientia tsutsugamushi. *She was diagnosed with* ***scrub typhus (tsutsugamushi disease)*** *and was started on 200 mg of minocycline. After two weeks of the treatment, her symptoms resolved, and the eruptions disappeared. She was discharged without complication.*

i. What is scrub typhus?
ii. How does it spread?
iii. How is it diagnosed?

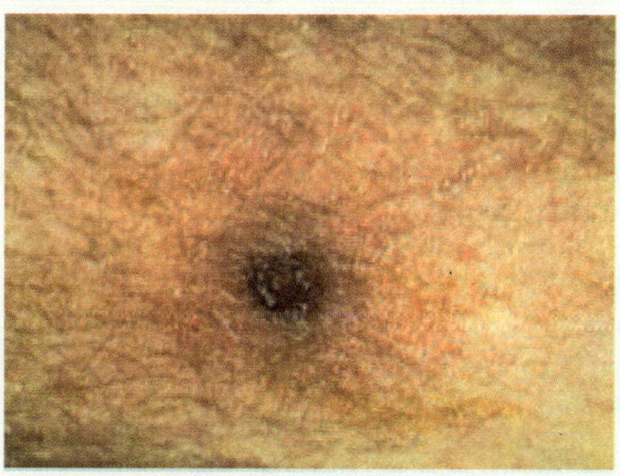

Fig. 21.24 Eschar of scrub typhus (*Source*: https://www.cdc.gov/typhus/scrub/index.html

Scrub typhus (also known as chigger-borne typhus) is an acute, febrile zoonotic illness, caused by bacteria belonging to family Rickettsiaceae *Orientia tsutsugamushi* (formerly known as *Rickettsia tsutsugamushi* and *R. orientalis*; details in Chapter 4).

Epidemiology

- In India, scrub typhus has been steadily increasing and has been reported from almost all states in the last few years.
- Epidemics have been reported from Pondicherry, Andhra Pradesh, and other southern states and Uttarakhand and other states in North India.
- Urbanisation and increased diversion of forest land for agricultural use and climate change have impacted the epidemiology, steering the disease transmission upward.
- **Zoonotic tetrad**—this refers to the co-existence and intimate relationship between *O. tsutsugamushi*, chiggers, rats, and vegetation (Fig. 21.25).

Vectors in India

- The disease is carried by trombiculid mites of the genus *Leptotrombidium deliensis*.
- These mites inhabit the soil and shrubs or bushes (low, bushy trees with woody branches close to the ground) where the microecosystem is favourable (mite islands).

Pathogenesis

- Humans are infected when they are bitten by the mite's larvae (**chiggers**). These larvae seek moist areas of the body like the groin and axilla.
- They bite and cause **eschars**, a pathognomonic finding in scrub typhus.
- Transovarian transmission occurs in mites.
- **Various rodents and birds act as reservoirs**.

Clinical features

- The **incubation period is 1–3 weeks**.
- Patients develop a characteristic eschar at the site of the chigger bite, with regional lymphadenopathy and maculopapular rash.
- The disease sets in with fever, headache, muscle pain, nausea, and vomiting.
- Encephalitis and pneumonia may be seen in severe cases.

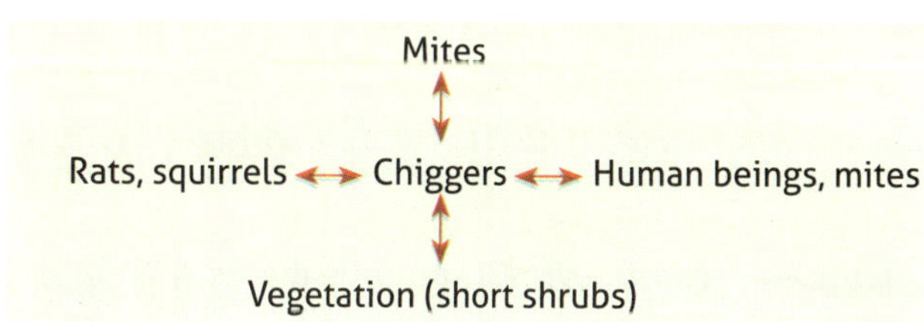

Fig. 21.25 Zoonotic tetrad of scrub typhus

- Complications include acute kidney injury, hepatitis, acute respiratory distress syndrome (ARDS), myocarditis, and septic shock.

Laboratory diagnosis

- Specific IgM antibodies are detected by ELISA.
- Weil–Felix—an agglutination test—uses *Proteus* strains as heterophile antigens.
- A rising titre is suggestive of scrub typhus.
- This is a non-specific test (see Table 21.18).

Treatment

- The treatment of choice is **doxycycline** (200 mg single dose) or tetracycline (100 mg four times a day for seven days).
- **Azithromycin** (500 mg as single dose) or levofloxacin can also be given in mild cases; however, it is not effective in severe cases with complications.
- No vaccines are available for the prevention of scrub typhus.

Ehrlichiosis

- Multiple disorders caused by *Ehrlichia*.
- *Ehrlichia* are small, gram-negative, obligate intracellular bacteria which have an affinity towards blood cells (details in Chapter 4).
- In the cytoplasm of infected phagocytic cells, they grow within phagosomes as mulberry-like clusters called **morula** (meaning mulberry).
- Unlike other rickettsiae, this bacterium affects the **granulocytes rather than mononuclear cells**.
- They are tick-borne.
- Doxycycline is recommended for the treatment of ehrlichiosis.

Pathogenicity

Three human infections caused by this group of organisms have been identified.

- **Glandular fever**—the patient shows serological response against the agent of canine ehrlichiosis. The causative agent has been named *Ehrlichia sennetsu*. It is endemic in Japan and parts of southeast Asia. It causes lymphoid hyperplasia and atypical lymphocytosis. No arthropod vector has been identified. Human infection is suspected to be caused by the ingestion of fish carrying infected flukes.
- **Human monocytic ehrlichiosis**—caused by *Ehrlichia chaffeensis*. It is transmitted by *Amblyomma* ticks. Deer and rodents are believed to be reservoir hosts. Human disease is associated with leukopenia, thrombocytopenia, and elevated liver enzymes. Multisystem involvement and fatality may occur.
- **Human granulocytosis anaplasmosis (HGA)**—caused by *Anaplasma phagocytophilum*. It is transmitted by *Ixodes* ticks. Dogs and rodents are important reservoirs. Leukopenia and thrombocytopenia are seen in patients. Giemsa-stained blood films may show the morula form of the ehrlichia.

Q Fever

Q fever is distributed worldwide as a zoonosis acquired from domestic livestock. It is transmitted among cattle, sheep, and poultry by *Ixodid* ticks. Transovarial transmission occurs in ticks. It is caused by rickettsia *Coxiella burnetii.*

Epidemiology

- *C. burnetii* is widely prevalent in birds and animals in India, as shown by serological surveys.
- Human disease, on the other hand, has been identified only rarely.

Agent—*Coxiella burnetti*

- *Coxiella burnetii,* the causative agent of **Q fever**, is a pleomorphic coccobacillary bacterium with a gram-negative cell wall and an ill-defined developmental cycle (Chapter 4).
- An obligate intracellular pathogen, it primarily infects monocyte–macrophage cells.
- It occurs as rods, 0.2–0.4 × 0.4–1.0 µm in size or as spheres, 0.3–0.4 µm in diameter.
- It is generally regarded as gram-negative, but is better stained with Gimenez and other rickettsial stains.
- *C. burnetii* shows phase variation. Fresh isolates are in phase I; they pass into phase II on repeated passage in yolk sacs but revert to phase I on passing through guinea pigs. Phase I is a more powerful immunogen than phase II and elicits a good antibody response to both I and II antigens.
- *Coxiella* differs from other rickettsiae in not needing a vector for transmission.
- The **Weil–Felix test cannot be used for the diagnosis of *Coxiella* infection.**
- Infection is endemic and largely subclinical in cattle, sheep, and goats, from where most human infections are acquired.

Pathogenesis

- Human infections occur occupationally through the handling of wool, hides, meat, or other animal products contaminated with the organism (not by the bite of ticks).
- The infection may also be transmitted by the consumption of infected milk.
- *Coxiella* may enter the body through abraded skin, mucosa, lungs, or the intestinal tract.
- Person-to-person transmission is rare.
- The human disease is an acute systemic infection characterised by **interstitial pneumonia**. The

clinical picture is quite variable, and asymptomatic infections very common.

- In **chronic Q fever**, the organism spreads to almost all organs and may cause hepatitis, meningoencephalitis, or endocarditis. Spontaneous recovery is the norm.
- *Coxiella* may remain latent in the tissues of patients for 2–3 years.

Epidemic Typhus

- This is a febrile disease caused by *Rickettsia prowazeki*, this infection is endemic in Eastern Europe, Africa, South America, and Asia. In India, the endemic spot is Kashmir.
- Humans are the only natural vertebrate hosts. Natural infection in flying squirrels has been reported from southeastern USA. They may act as reservoir hosts, infection being spread by the squirrel louse and flea.
- The human body louse *Pediculus humanus corporis* is the vector. The head louse may also transmit the infection but not the pubic louse.
- The rickettsia multiply in the gut of the lice and appear in their feces in 3–5 days. Lice remain infective till they die—in 2 weeks. Being sensitive to temperature changes in the host, they leave the febrile patient to parasitise other persons. Infection is transmitted when contaminated louse feces enters the body through the minute abrasions caused by scratching.
- The incubation period is 5–15 days.
- The disease starts with fever and chills followed by a characteristic rash on the fourth or fifth day on the trunk and limbs but not the face, palms, and soles. Towards the second week, the patient becomes stuporous and delirious.
- Case fatality may reach 40 per cent and increases with age.

Recrudescent Typhus (Brill-Zinsser Disease)

Epidemic typhus may remain latent and, at times, be reactivated, leading to recrudescent typhus. Such cases in louse-ridden communities may initiate epidemics.

Endemic Typhus (Flea-Borne Typhus)

- *Rickettsia typhi* (*R. mooseri*) causes endemic or murine or flea-borne typhus, which is worldwide in distribution.
- While rats are the reservoirs, the infection is transmitted by the flea *Xenopsylla cheopis*. The *Rickettsia* multiplies in the gut of the flea and is shed in its feces. Humans usually acquire the disease through:
  - The **bite of infected fleas**, when their saliva or feces is rubbed in
  - Through **aerosols of dried feces**
  - By ingesting **food recently contaminated with infected rat urine or flea feces**
- Human infection is a dead end.
- Person-to-person transmission does not occur.

In Kashmir and China, lice have been known to transmit endemic typhus to humans, producing smouldering outbreaks.

Indian Tick Typhus

- *R. conori* causes Indian tick typhus.
- The disease was first observed in India in the foothills of the Himalayas.
- The tick *Rhipicephalus sanguineus* is the most important vector. *Haemaphysalis leachi*, *Amblyomma*, and *Hyalomma* ticks can also transmit the infection.

Rocky Mountain Spotted Fever

- It is the most serious type of spotted fever and is the first to have been described.
- It is prevalent in many parts of North and South America and is transmitted by *Dermacentor andersoni* and related species of ticks.
- It is caused by *Rickettsia rickettsii.*

Trench Fever

- During the First World War, over a million cases of a disease known as trench fever or five-day fever occurred among soldiers fighting in the trenches in Europe (from quintana, meaning fifth, referring to 'five-day fever').
- The causative agent is *Bartonella quintana* (earlier known as *Rochalimaea quintana*).
- It is an exclusively human disease and has no known animal reservoir.
- It is transmitted by the body louse. The feces of lice become infectious 5–10 days after an infectious meal. The lice are unharmed and remain infective throughout their lives. Vertical transmission does not occur in lice.
- The disease frequently leads to chronic or latent infection. Recrudescence may occur as in Brill-Zinsser disease and relapses have been reported as long as 20 years after the primary disease. The chronic infection and late relapses help to maintain the *Bartonella* in the absence of animal reservoirs.

Oroya Fever (Carrion Disease)

- It is caused by *Bartonella bacilliformis*, which is found inside erythrocytes and in skin lesions.
- *B. bacilliformis* is a pleomorphic, gram-negative rod, which is motile on account of a tuft of polar flagella that can be cultivated in semi-solid agar with rabbit or human blood.
- The disease presents as **fever and progressive anemia** due to the bacterial invasion of erythrocytes.
- Mortality is high in untreated cases.

Verruga peruana

This is a late sequel to Oraya fever that presents with raised, reddish-purple skin nodules, usually on the limbs and face.

Laboratory diagnosis of rickettsial infections (trench fever and typhus)

Specimen

Blood and tissues samples are collected for culture, and serum for serological tests.

Direct microscopy

Aggregations of rickettsial particles (but not individual particles) are detected in cytoplasmic inclusion bodies under the light microscope by staining them with Giemsa (**purple-coloured basophilic inclusions**) or Macchiavello stain (**red-coloured inclusions**).

Culture

- **Isolation by culture** is performed in laboratories equipped with appropriate safety provisions (**biosafety class III cabinets**) as these organisms are highly infectious. For this reason, isolation is not used as a routine diagnostic technique.
- Cell and tissue culture are most widely used for isolation in containment laboratories.
- Rickettsiae grow well in 3–5 days on Vero cell or MRC-5 cell coverslip cultures, identified by immunofluorescence using group- and strain-specific antisera.
- The yolk sac of embryonated hens' eggs can also be used.

Serological tests

This is the mainstay in the diagnosis of rickettsial infections.

Detection of specific antibodies (generally IgM in acute infections) to rickettsial antigens by:

i. **Immunoassays**
   - Enzyme immunoassay (EIA/ELISA)
   - Immunofluorescence
   - Passive hemagglutination of red cells sensitised by ESS (**erythrocyte sensitising substance**)
   - Complement fixation test is performed with group-specific soluble antigen or the type-specific washed rickettsial antigen

ii. **Weil–Felix reaction (non-specific test):** This test detects antibodies against the O antigens of certain non-motile *Proteus* strains—OX 19, OX 2, and OX K.
   - *Principle*: An alkali-stable carbohydrate antigen of some rickettsiae is the same as that of *P. vulgaris* OX 19 and OX 2 and *P. mirabilis* OX K antigens.
   - *Interpretation*: The interpretation of the Weil–Felix test for rickettsial diseases is as presented in Table 21.18.

Table 21.18 Weil–Felix reaction in rickettsial diseases

Disease	Agglutination pattern with		
	OX 19	OX 2	OX K
Epidemic typhus	+++	+	–
Brill–Zinsser disease	Usually negative or weakly positive	–	–
Endemic typhus	+++	±	–
Tick-borne spotted fever	++	++	–
Scrub typhus	–	–	+++

Antibody response

- The antibody appears rapidly during the course of the disease, reaches peak titres of up to 1:1,000 or 1:5,000 by the second week, and declines rapidly during convalescence.
- A false-positive reaction may occur in some cases of urinary or other infections caused by *Proteus* and at times, in typhoid fever and liver diseases.
- Hence, it is desirable to demonstrate a rise in titre of antibodies for the diagnosis of rickettsial infection.
- After the introduction of specific ELISA tests to detect IgM and IgG, Weil–Felix is no longer preferred due to cross-reaction and the cumbersome process.

Molecular methods

Rickettsial DNA can be detected by polymerase chain reaction, which permits rapid, specific identification of the infecting agent in skin biopsy, necrotic tissue, and blood mononuclear cells.

Laboratory diagnosis of rickettsial infections

- **Specimen:** Blood and tissues samples and serum
- **Direct microscopy:** Giemsa (**purple-coloured basophilic inclusions**) or Macchiavello stain (**red-coloured inclusions**)
- **Culture: Not used for diagnosis**
- **Serological tests:**
  - Detection of specific antibodies (generally IgM in acute infections) to rickettsial antigens by:
    - Enzyme immunoassay (EIA/ELISA)
    - Immunofluorescence
    - Passive hemagglutination of red cells sensitised by ESS (**erythrocyte sensitising substance**)
    - Complement fixation test
  - **Weil–Felix reaction (non-specific test):** To detect antibodies against the O antigens of certain non-motile *Proteus* strains—OX 19, OX 2, and OX K.
  - In tick-borne spotted fever, both OX 19 and OX 2 are agglutinated. OX K agglutinins are found only in scrub typhus. The test is negative in rickettsial pox, trench fever, and Q fever.
- **Molecular methods:** Rickettsial DNA can be detected by polymerase chain reaction

Control

- The eradication of rickettsial diseases is virtually impossible because of the cycles maintained in rodents, wild animals, and vectors.
- Persons travelling to endemic areas should wear protective clothing to avoid infestation by vectors.

Treatment

- **Drugs of choice**—doxycycline, followed by tetracycline, chloramphenicol, and ciprofloxacin are preferred.
- **Penicillin and sulphonamides are ineffective.**
- In fact, **sulphonamides are contraindicated** as they worsen the condition by enhancing the growth of rickettsiae.

Besides the organisms and infections discussed heretofore, the following organisms can cause bloodstream infection and sepsis. However, they are primary pathogens of the gastrointestinal or urinary tract and hence, are dealt in the chapters dealing with those systems.

- *Escherichia coli*
- *Klebsiella pneumoniae*
- *Enterobacter* species
- *Proteus* species
- *Pseudomonas* species (mainly HAI)
- *Burkholderia* (melioidosis)

VIRAL BLOODSTREAM INFECTIONS

Some unrelated viruses cause infections localised to blood and reticuloendothelial cells. However, there is an initial viremic phase in most of the other systemic viral infections where the virus has a transient passage through blood before settling down in the target organs. At this time, they do not produce any pathological changes in the bloodstream and hence cannot be detected in blood or serum since this phase is transient and occurs during the incubation period.

Some of the common viral infections of blood and reticuloendothelial cells are:

- Arboviral hemorrhagic fever
  - Dengue
  - Chikungunya
  - KFD
  - West Nile fever
  - Crimean Congo hemorrhagic fever
  - Ebola
- Transfusion-transmitted bloodstream infections—hepatitis B, C, and D
- Human immunodeficiency disease

Viral Hemorrhagic Fever (VHF)

> **Case 8.** *A 40-year-old man presented to the medical OPD with high fever, severe headache, and pain around the eyes for the past three days. On examination, petechial lesions were noted on his forearm. His platelet count was 30,000/mm³. A provisional diagnosis of* ***viral hemorrhagic fever*** *was made. Later, dengue NS1 antigen by capture ELISA was positive. The patient recovered with supportive management.*
>
> i. What are the viral causes of hemorragic fever?
> ii. What is the pathogenesis of dengue hemorrhagic fever?

VHF is a severe febrile illness with an abnormal vascular regulation, vascular damage, and hemorrhagic manifestations. The viruses causing VHF share some similar characteristics, some of which are as follows:

- They are enveloped, single-stranded RNA viruses.
- They undergo cytoplasmic manifestations.
- They can cause gastrointestinal or neurologic manifestations.
- Many arthropods or rodents serve as effective reservoirs.
- They target monocytes, macrophages, or dendritic cells.
- They exhibit zoonotic cycles and are mostly geographically restricted.

Clinical features of VHF

- **Early symptoms:** Fever, muscle aches, fatigue, and dizziness
- **Late symptoms:** Bleeding gums, petechial hemorrhages, and bleeding from internal organs and orifices such as the mouth, eyes, and ears

Causative agents of VHF

Fig. 21.26 and Table 21.19 list the various viruses responsible for VHF.

Table 21.19 Reservoir hosts, geographic distribution, and vectors of some of the arboviruses and rodent-borne viruses associated with VHF in humans

Virus	Genus	Reservoir/hosts	Geographic distribution	Vector
Hemorrhagic fever				
Dengue types 1–4	Flavivirus	Not known	Widespread, especially in Asia and the Caribbean the tropics	Mosquito
Yellow fever	Flavivirus	Monkeys, humans	Africa, South America	Mosquito
Kyasanur Forest disease*	Flavivirus	Rodents, shrews, monkeys, and ticks	India	Tick

(Table 21.19 Contd. . .)

Table 21.19 Contd.

Virus	Genus	Reservoir/hosts	Geographic distribution	Vector
Crimean Congo hemorrhagic fever*	Nairovirus	Small mammals	The (former) USSR, Central Asia	Tick
Ebola virus	Filovirus	Chimpanzees, gorillas, fruit bats, monkeys, forest antelope, and porcupines	Mainly localised to African countries like Sudan, Zaire, Uganda, and Congo where it continues in several pockets Outbreaks have occurred in other Western countries—Italy, Spain, UK, and US	Not an arbovirus in the true sense as there are no arthropods incriminated in its transmission
Febrile illness with or without rash (non-hemorrhagic) and arthralgia				
Chikungunya	Alphavirus	Not known	Africa, Asia	Mosquito
O'nyong-nyong	Alphavirus	Not known	Africa	Mosquito
West Nile fever*	Flavivirus	Birds	Asia, Africa, USA	Mosquito
Sandfly fever	Bunyavirus	Not known	Mediterranean, Asia, tropical parts of America	Sandfly
Rift Valley fever	Bunyavirus	Sheep, cattle	Africa	Mosquito
Colorado tick fever	Orbivirus	Rodents	USA	Tick
Rodent-borne hemorrhagic fevers				
Hantavirus pulmonary syndrome	Hantavirus	Deer mouse	South America and the United States	Deer mice
Hemorrhagic fever with renal syndrome (HFRS)	Hantavirus	Rodents	China, Russia, and Korea	Rodents
Lassa fever	Lassavirus	Rodents	Western Africa	Rodents
South American Hemorrhagic fever	Junin and Machupo viruse	Rodents	South America	Rodents

*Chapter 24

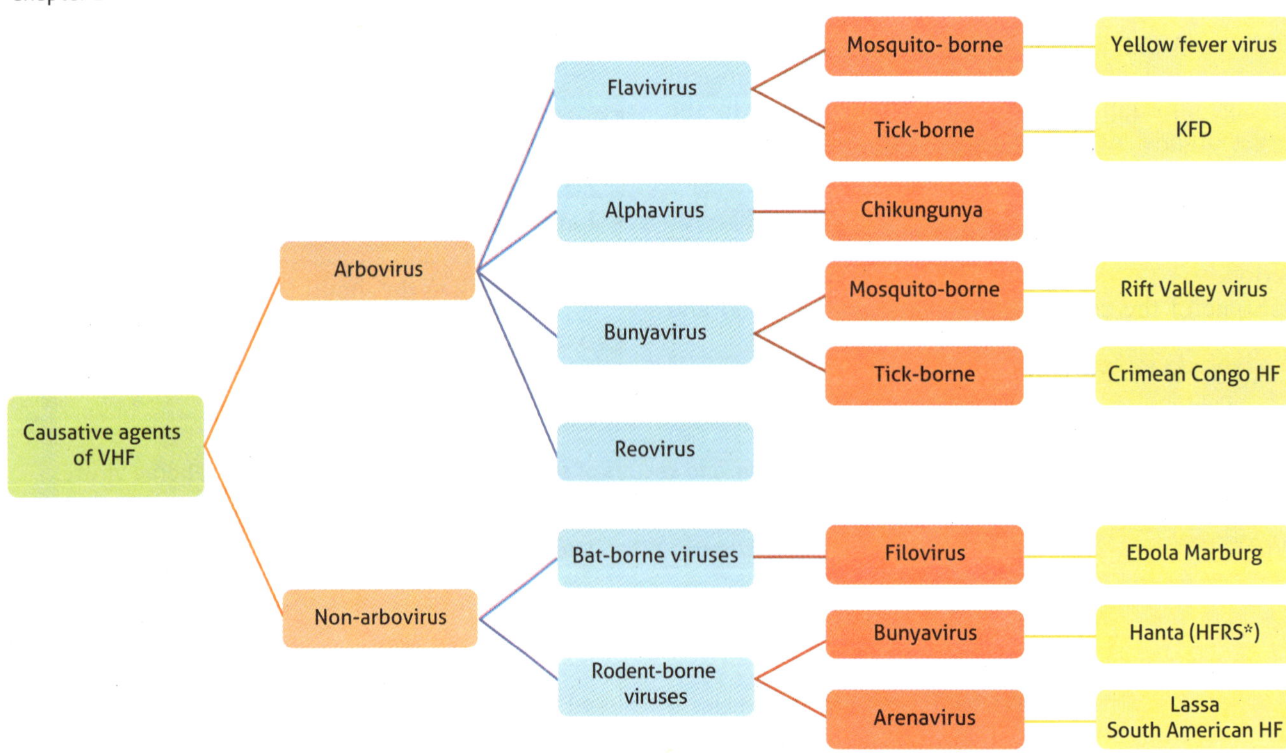

Fig. 21.26 Causative agents of VHF

**HFRS*—hemorrhagic fever with renal syndrome

Arthropod-Borne Hemorrhagic Fevers

Dengue

Epidemiology

- Dengue fever is an acute febrile condition caused by the dengue virus. It can result in life-threatening complications of **dengue hemorrhagic fever (DHF)** and **dengue shock syndrome (DSS).**
- Dengue came to be known as the **'break-bone fever'** during the Philadelphia epidemic in 1780.

Indian scenario

- Dengue virus was initially confined to the east coast of India and caused epidemics (sometimes along with the chikungunya virus).
- Subsequently, it spread westwards and, in the 1990s, Surat and Delhi saw major epidemics with several deaths due to dengue hemorrhagic fever (DHF) and dengue shock syndrome (DSS).

Global scenario

Dengue has been on the rise worldwide over the last few decades and today, ranks as the most important vector-borne disease, with over 2.5 billion people in 200 countries at risk.

Agent—Dengue virus

Dengue virus (DENV) is the cause of dengue fever. It is a mosquito-borne, single positive-stranded RNA virus of the family Flaviviridae; genus Flavivirus

Types

- Four types of the dengue virus exist—DEN 1–4.
- All four types of dengue virus are present in India.
- Occasionally, more than one type has been isolated from the same patient.

Distribution

The dengue virus is widely distributed throughout the tropics and subtropics (in Swahili, *Ki denga pepo* means a sudden seizure by a demon).

Mode of transmission

- **Vector:** *Aedes aegypti* mosquito
- **Extrinsic incubation period:** 8–10 days
- No vertebrate hosts other than humans have been identified
- A pregnant woman already infected with dengue can pass the virus to her fetus during pregnancy or around the time of birth. This **perinatal transmission** can cause harmful effects, low birth weight, premature birth or death of the fetus.

Pathogenicity

Fig. 21.27 illustrates the pathogenesis of dengue fever.

Clinical features

The incubation period of the infection is **3–14 days.**

Febrile phase

- Patients typically develop a **high-grade fever** (which may be **typically biphasic**) of sudden onset with headache, retrobulbar pain, and photophobia accompanied by facial flushing, skin erythema, and pain in the back and limbs (hence the name 'break-bone fever'), lymphadenopathy, and a maculopapular rash (Case 8).
- This acute febrile phase usually lasts 2–7 days and is often characterised by generalised body ache, myalgia, arthralgia, rubelliform exanthema, and headache.

Critical phase

- Patients become worse around the time of defervescence, when the temperature drops to 37.5–38°C or less and remains below this level—usually on days 3–8 of illness.
- Progressive leukopenia followed by a rapid decrease in platelet count usually precedes plasma leakage.
- An increasing hematocrit above the baseline may be one of the earliest additional signs.
- Complications with hemorrhagic manifestations (**dengue hemorrhagic fever**) or shock (**dengue shock syndrome**) can occur in persons who have non-neutralising heterologous antibodies from a previous infection or in infants with maternal antibodies.

> The clinical criteria for dengue fever, dengue hemorrhagic fever, and dengue shock syndrome are as follows (Fig. 21.28):
>
> - **Dengue fever (DF)** has three phases:
>   - *Febrile phase*: An acute febrile illness with two or more of the following manifestations—headache, retro-orbital pain, myalgia, arthralgia
>   - *Critical phase:* Rash and hemorrhagic manifestations
>   - *Recovery phase*
> - **Dengue hemorrhagic fever (DHF):** The clinical criteria of dengue fever plus hemorrhagic tendencies evidenced by one or more of the following:
>   - Positive tourniquet test
>   - Petechiae, ecchymoses, or purpura
>   - Bleeding from mucosa, gastrointestinal tract, injection sites, or other sites
>   - Thrombocytopenia—more than 20% drop in hematocrit following volume replacement treatment compared to baseline
> - In the severe form, there are signs of **plasma leakage (pleural effusion, ascites, hypoproteinemia)**
> - **Dengue shock syndrome (DSS):** This is evidenced by all the above criteria for DHF with evidence of **circulatory failure manifested by rapid and weak pulse and narrow pulse pressure** or hypotension for age, cold and clammy skin, and restlessness.

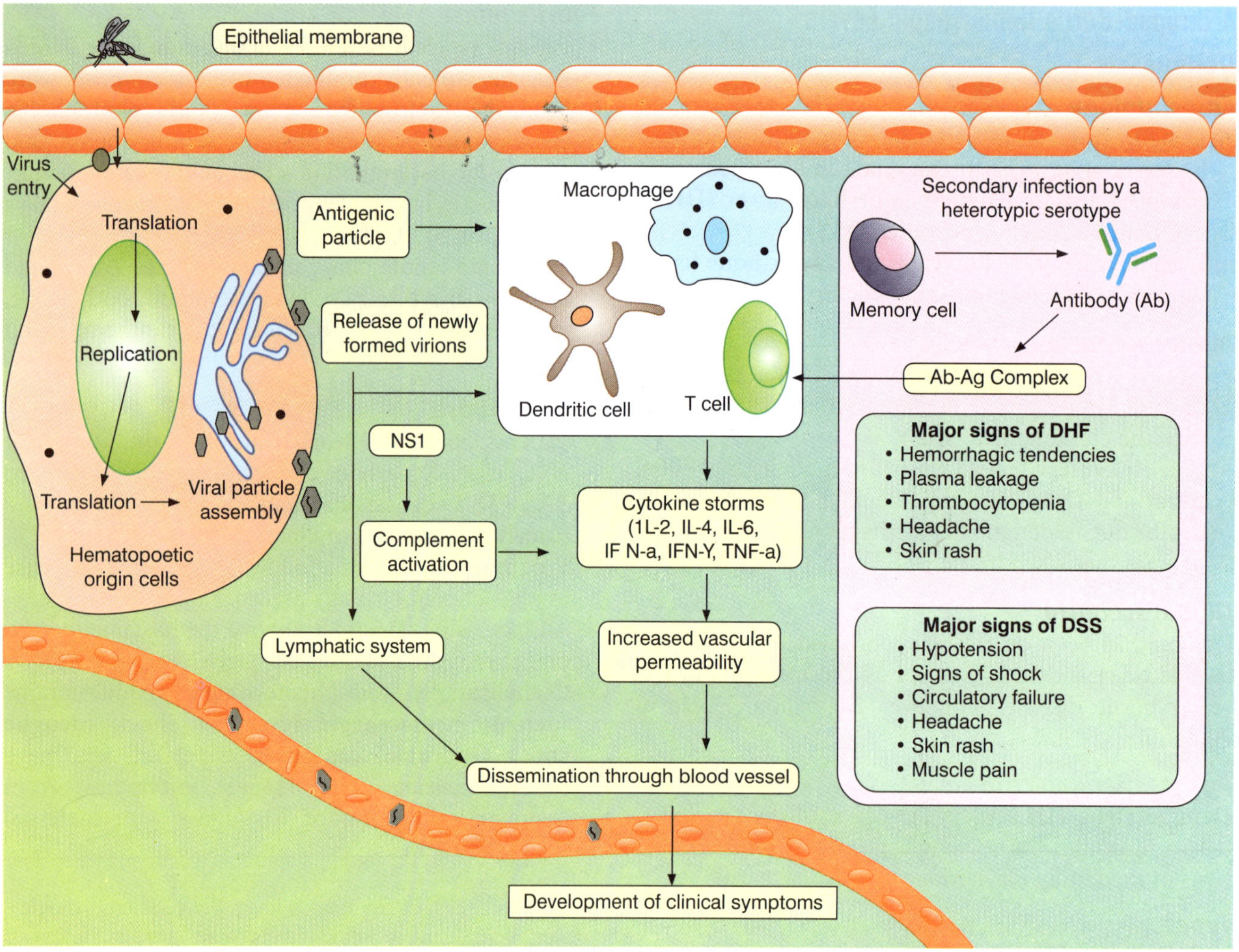

Fig. 21.27 Pathogenesis of dengue fever, DHF, and DSS (Ref: https://link.springer.com/article/10.1007/s12250-015-3624-1)

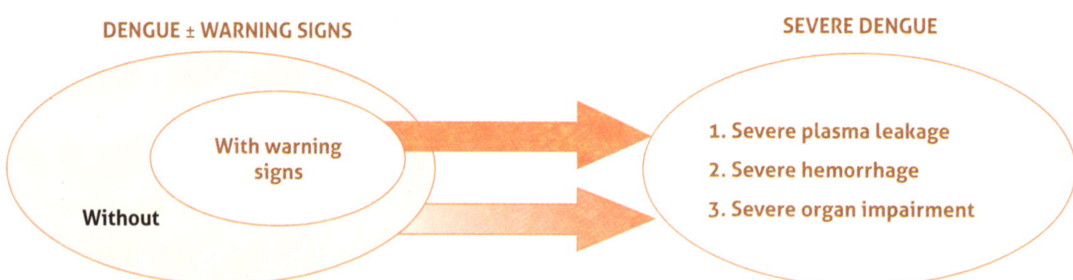

CRITERIA FOR DENGUE ± WARNING SIGNS

Probable dengue
Live in / travel to dengue-endemic area.
Fever and two of the following criteria:
- Nausea, vomiting
- Rash
- Aches and pains
- Tourniquet test positive
- Leucopenia
- Any warning sign

Laboratory-confirmed dengue
(Important when no sign of plasma leakage

Warning signs*
- Abdominal pain or tenderness
- Persistent vomiting
- Clinical fluid accumulation
- Mucosal bleed
- Lethargy, restlessness
- Liver enlargement >2 cm
- Laboratory: increase in HCT concurrent with rapid decrease in platelet count

*(Requiring strict observation and medical intervention)

CRITERIA FOR SEVERE DENGUE

Severe plasma Leakage
Leading to:
- Shock (DSS)
- Fluid accumulation with respiratory distress

Severe bleeding
as evaluated by clinician

Severe organ involvement
- Liver: AST or ALT ≥1000
- CNS: impaired consciousness
- Heart and other organs

Fig. 21.28 Clinical criteria for dengue fever, dengue hemorrhagic fever, and dengue shock syndrome

Laboratory diagnosis (Fig. 21.29)

Diagnosis in individuals presenting in the early phase of infection (<5 days) is based on the following.

Viral isolation

Though the virus can be isolated in the first week of illness, this is rarely done.

Serology

- **Diagnosis in the first week**
  - The **non-structural viral protein antigen (NS1)** can be detected in the blood up to 7 days after the onset of the disease.
- **Diagnosis after one week**
  - Demonstration of circulating IgM antibody helps in early diagnosis as it appears within 2–5 days of the onset of illness and persists for 1–3 months.
  - The **IgM ELISA test** offers reliable diagnosis.
  - A **strip immunochromatographic test** for IgM is available for rapid diagnosis. However, it has to be confirmed by ELISA.
  - The IgG antibodies are detected in a paired serum sample to show rising titres.

Molecular testing

- **Confirmatory diagnosis is by RT-PCR** to detect the genetic material of the dengue virus.
- Serotyping of the dengue virus is done for epidemiological purposes.

Control

- The control of dengue is limited to **vector control measures**.
- The WHO has approved a vaccine which has been licensed and launched in some South American countries. It is yet to be launched in India.
- Clinical trials of bivalent, trivalent, and quadrivalent dengue vaccines are being initiated to study the efficacy and safety of these vaccines before they can be put to use.

Treatment

No specific treatment is available. Current management is supportive and comprises the following measures:

- Cold/tepid sponging
- Paracetamol for fever (aspirin/NSAIDs like ibuprofen, etc., should be avoided since they may cause gastritis, vomiting, acidosis, platelet dysfunction, and severe bleeding), fluid and electrolyte replacement, and platelet infusion when counts are 10,000/mL and less
- Dengue shock is treated with whole blood transfusion and management of shock

Chikungunya

> **Case 9.** *A 35-year-old female presented with six days of fever, pruritic rash, and diffuse arthralgias. She complains of severe pain in the small joints of her hands, wrists, hips, knees, and ankles. There is no history of sore throat, bleeding, or bruising. Laboratory investigations showed a normal complete blood count with differential, metabolic panel, liver function tests, urinalysis, and negative NS1 antigen. The CHIKV IgM enzyme-linked immunosorbent assay (ELISA) test was positive. A diagnosis of* ***chikungunya virus infection*** *was made, and the patient was treated conservatively.*
>
> i. What are the clinical manifestations of CHIKV disease?
>
> ii. What are the criteria for diagnosing it?

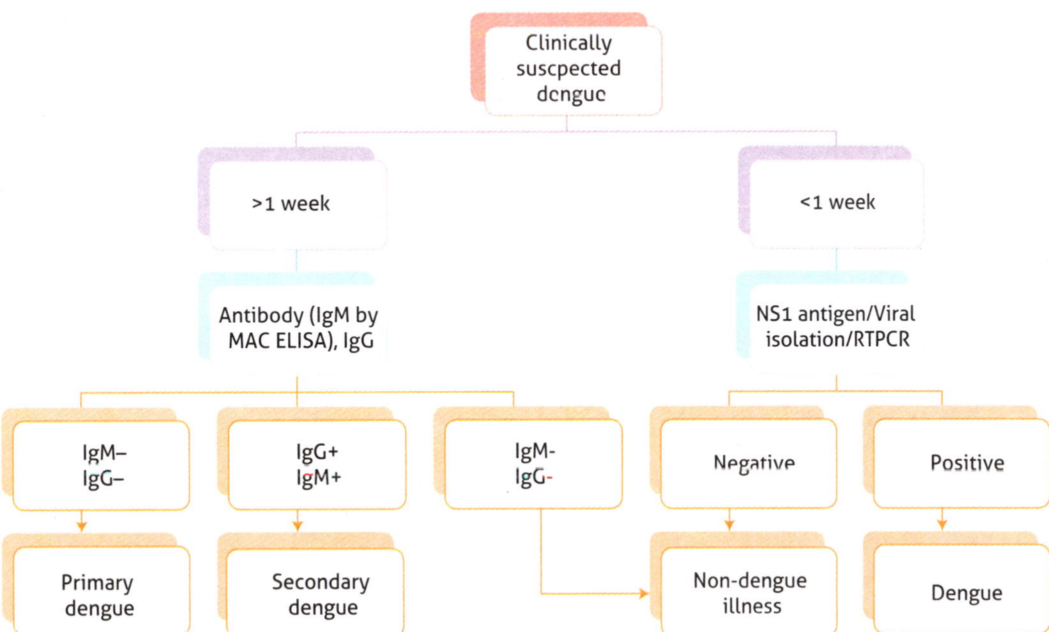

Fig. 21.29 Laboratory diagnosis of dengue

Chikungunya presents as acute febrile illness with joint pains and is caused by the **Chikungunya virus**.

Epidemiology

- The virus first appeared in India in 1963 when, along with dengue, it caused extensive epidemics in Kolkata, Chennai, and other areas.
- After that, chikungunya outbreaks occurred at irregular intervals along the east coast of India and in Maharashtra until 1973.
- Thereafter, the virus was quiescent for a few years, until it reappeared in 2004–2006, at which time, there was a large outbreak that began in Andhra Pradesh and Tamil Nadu and spread to Kerala, Karnataka, and Delhi, affecting nearly a million people.
- **Reasons for re-emergence**—the disease appears in epidemics after a gap of a decade or two; this is attributed to:
  - Mutation of the virus
  - Wider distribution of the vector *Aedes albopictus*
- During interepidemic periods, the virus is thought to be maintained in the sylvatic cycle in non-human primates.
- The chikungunya virus was first isolated from human patients in Tanzania.
- The name 'chikungunya' is derived from the native Makonde word for the disease in which the patient lies '**doubled up' due to severe joint pain**.
- The vectors *Aedes aegypti* and *A. albopictus* are found in endemic areas.
- *A. albopictus* has a wider geographical distribution than *A. aegypti*.

Agent—Chikungunya virus

Genotypes

There are three major chikungunya virus genotypes:

- West African
- East/Central/South African (ECSA)
- Asian

The genotypes are generally restricted to the geographic areas they are named after. However, a spread of ECSA to Asian regions has been reported.

Clinical features

- Chikungunya presents with a sudden onset of **fever and crippling joint pain**, especially in the ankles and wrists. Joint involvement is bilateral and symmetric.
- The fever is typically biphasic with a period of remission of 1–6 days.
- Associated symptoms include migratory polyarthritis, lymphadenopathy, and conjunctivitis.
- A **maculopapular rash** is common. The appearance of the rash often coincides with defervescence.
- Hemorrhagic manifestations are rare.
- A few patients have leukopenia with elevated aspartate aminotransferase (AST), and a positive C-reactive protein.
- Recovery takes weeks; some older patients may experience joint pain and stiffness for several years.
- Persistence may be seen in HLA B27 individuals.

Clinically, chikungunya cannot be differentiated from uncomplicated dengue.

Laboratory diagnosis

A patient is suspected to have CHIKV only if he meets the clinical criteria. A confirmed (definitive) case is defined as a patient who meets both the clinical and laboratory criteria.

Clinical criteria

- Acute onset of fever and severe arthralgia/arthritis with or without a skin rash
- A history of residence in an epidemic area or having been in and subsequently having left an epidemic area 15 days before the onset of symptoms

Laboratory criteria

- Virus isolation/the presence of viral RNA is detected by reverse transcriptase PCR
- The presence of virus-specific IgM antibodies in a single serum sample collected in the acute or convalescent stage or a four-fold increase in IgG values in samples collected at least three weeks apart by ELISA or other immune assays **confirms CHIKV**

Treatment

- No specific treatment is available.
- Supportive management includes antipyretics, analgesics like paracetamol and acetaminophen.
- Aspirin is avoided as it is difficult to differentiate dengue from chikungunya.
- Fluid and electrolyte replacement are also important supportive measures.

Prevention

- **Mosquito bites should be avoided, and mosquito breeding should be controlled diligently** (as is done for dengue).
- There is no vaccine yet for chikungunya.

Kyasanur Forest disease (KFD)

A hemorrhagic fever was reported from Karnataka (India) in 1957, when several dead monkeys were found in the Kyasanur Forest in the Shimoga district of Karnataka, and a severe prostrating illness was noticed in some villagers in the area.

Agent—KFD virus

- A **new arbovirus antigenically related to the RSSE complex** was isolated by investigators from the National Institute of Virology (then known as

the Virus Research Centre), Pune, from the patients and dead monkeys.
- It was named the Kyasanur Forest disease (KFD) virus after the place where it was first isolated.
- An outbreak of KFD was reported in 2015 from Wayanad and Malappuram districts of Kerala.
- **Vector:** Ticks of genus *Haemaphysalis spinigera*.
- **Reservoir host:** Forest birds and small mammals (Fig. 21.30); *Haemaphysalis* ticks may act as reservoirs to some extent as transovarial transmission of the virus has been demonstrated in them.
- **Amplifier hosts:** Monkeys are linked to the fatal disease. They are not the primary reservoirs but only amplifier hosts.

Clinical features

- KFD presents with abrupt onset of fever, headache, conjunctivitis, myalgia, and severe prostration.
- Some develop hemorrhages in the skin, mucosa, and viscera.
- The case fatality rate is about 5 per cent.

Laboratory diagnosis

General

Marked leukopenia, moderate thrombocytopenia, low erythrocyte sedimentation rate (ESR), albuminuria, and moderately elevated transaminases

Specific

- Real-time PCR to detect viral nucleic acid.
- Detection of IgM or rising IgG titres by ELISA.

Treatment and prevention

- Treatment includes **early hospitalisation and supportive therapy** such as maintenance of hydration and the usual precautions for patients with bleeding disorders.
- A killed KFD virus vaccine is available for use in endemic areas which was used in a small field trial and appeared to provide some degree of protection.

Crimean Congo hemorrhagic fever (CCHF)

- The disease is endemic in eastern Europe, Central Asia, and many parts of Africa. Cattle, sheep, goats, and other domesticated animals act as natural reservoirs.
- In India, the first case of CCHF was reported from Ahmedabad, Gujarat, in 1911 as a hospital outbreak. Subsequently, several cases have been reported from Gujarat, Rajasthan, and UP between 2012 and 2015.
- Antibodies to the CCHF viruses have been detected in human and animal sera from India.

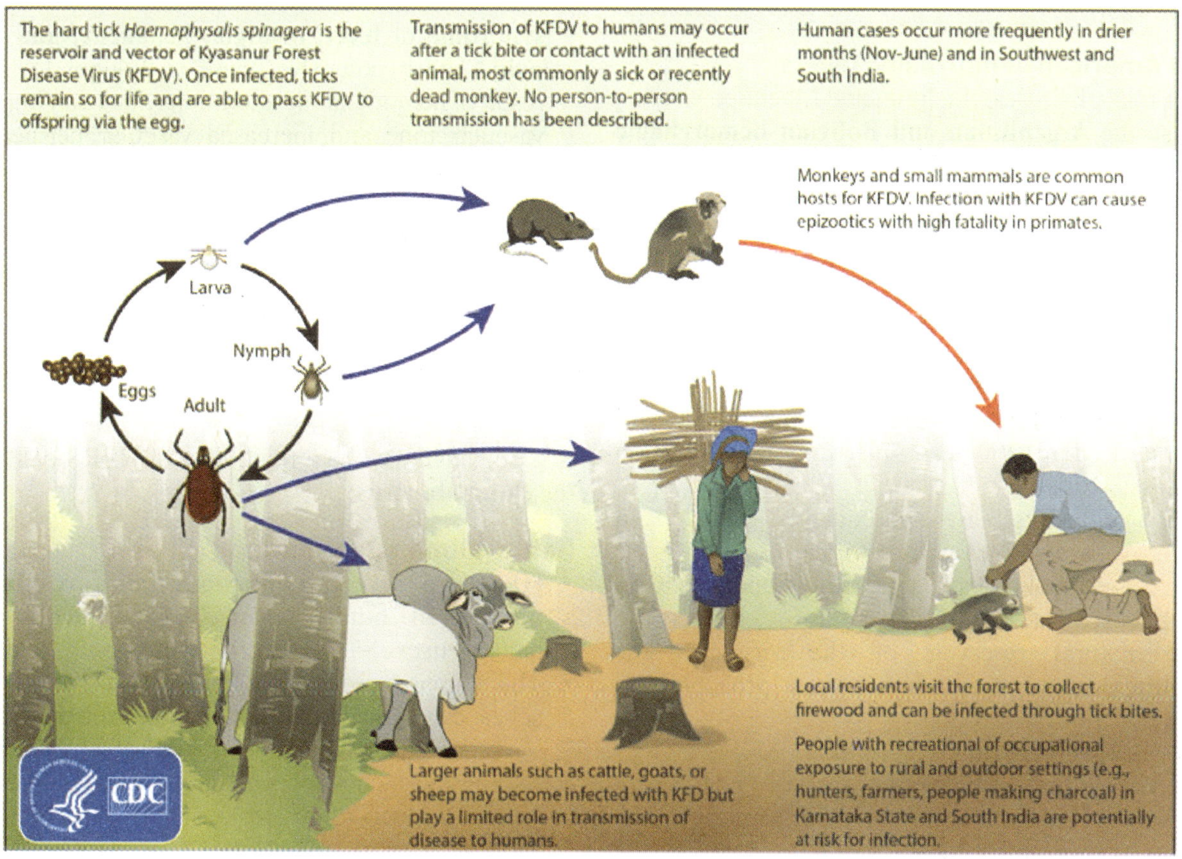

Fig. 21.30 Ecology of the Kyasanur Forest disease (*Source*: Https://www.insightsonindia.com/2019/03/25/in-news-kyasanur-forest-disease/)

Agent—CCHF virus

- It is an enveloped negative-sense tri-segmented RNA virus of the Bunyavirus family.
- It is transmitted by the bite or handling of *Hyalomma* ticks. It can also be transmitted via direct contact with blood or tissues of infected animals or humans.

Clinical features

- Clinical presentation is in four distinct stages of **incubation, pre-hemorrhagic, hemorrhagic, and convalescence**.
- During the acute phase of the disease, the blood of the patient is highly infectious, and direct transmission may occur through contact.
- High-grade fever, headache, body ache, nausea, vomiting, photophobia, petechia, ecchymosis, and visceral bleeding are the common features of CCHF.

Laboratory diagnosis

Diagnosis is based on the **detection of the viral genome by real time-reverse transcriptase PCR** or the **detection of rising antibody titres of IgM/IgG by ELISA**.

Rodent-borne Hemorrhagic Fevers

The zoonotic rodent borne hemorrhagic fevers include diseases caused by Arenaviruses (Lassa, Junin, and Machupo virus) and Bunyavirus (Haantan virus and Hanta virus).

South American hemorrhagic fevers

- Two related viruses, the Junin and Machupo viruses, cause the **Argentinian and Bolivian hemorrhagic fevers**, respectively. They belong to the Tacaribe group of arenaviruses.
- Rodents act as reservoirs for these viruses, and transmission is believed to occur through rodent excreta.

Lassa fever

- This is the most highly publicised of viral hemorrhagic fevers and is caused by another arenavirus. It was first noticed in 1969 in an American Mission station in Lassa, Nigeria.
- Many outbreaks have subsequently occurred in widely separated foci in West Africa.
- The case fatality rate was 35–70 per cent in hospitalised patients.
- The natural reservoir of the virus is the **multimammate rat**. **Rodent excreta** probably acts as the source of infection.
- The incubation period is 3–16 days.
- The virus is present in the throat, urine, and blood of patients. Hence person-to-person transmission may occur by droplet infection.
- **Nosocomial infection** has occurred frequently.
- **Ribavirin** has proved useful in treatment.

Agent—Arenaviruses

- This group includes the Junin virus, Machupo virus, and Lassa virus.
- These are enveloped viruses that have a negative-sense, single-stranded RNA genome and cause chronic, inapparent infection in rodents.
- Electron microscopy of thin sections shows characteristic **electron-dense granules resembling grains of sand within the virus** particles.
- The name of this group is derived from the Latin word *arena*, meaning sand.
- These sand-like particles are cellular ribosomes picked up by the virus presumably during maturation by budding from host cells.
- Arenaviruses are spherical or pleomorphic particles, ranging in size from **80 to 300 nm**.
- Arenaviruses have assumed considerable medical importance after the recognition that some members of the family cause hemorrhagic fevers in humans (Argentinian and Bolivian hemorrhagic fevers and Lassa fever).

Hemorrhagic fever with renal syndrome (HFRS)

- It is characterised by a **triad of fever, renal failure, and hemorrhage** caused by the Hanta virus. The disease can be divided into five phases: **febrile, hypotensive, oliguric, diuretic, and convalescent**.
- The patient initially presents with a **flu-like syndrome of fever, myalgia, headache, and flank pain**. The hypotensive phase is thought to be the result of hemodynamic alterations due to an impaired vascular tone and increased vascular permeability associated with pathological findings of pulmonary edema and retroperitoneal edema.
- Renal failure in HFRS often develops in the late acute stage.
- HFRS includes diseases such as Korean hemorrhagic fever, epidemic hemorrhagic fever, and nephropathia epidemica.

Nephropathia epidemica

It is considered to be a mild form of HFRS caused by the **Puumala virus**.

Hanta pulmonary syndrome (HPS)

- HPS is a **severe, sometimes fatal, respiratory disease** in humans caused by infection with hantaviruses.
- The patient presents with fatigue, fever, and muscle aches, especially in the large muscle groups—thighs, hips, back, and sometimes, shoulders. There may also be headaches, dizziness, chills, and abdominal problems, such as nausea, vomiting, diarrhea, and abdominal pain, followed by the acute onset of severe respiratory distress.
- Case fatality rate of up to 50% has been reported.

Agent—Hanta virus

- Hantaviruses belong to the family Bunyavirus.
- These viruses are enveloped, spherical, 90–100 nm in diameter with triple-segmented genome, negative-sense, single-stranded RNA.
- The members commonly associated with human disease are Hantaan virus, Seoul virus, Puumala virus, Prospect Hill virus, and Sin Nombre virus (Table 21.20).
- They replicate exclusively in the host cell cytoplasm.
- Entry into host cells occurs by attachment of virions to cellular receptors and subsequent endocytosis.

Transmission

- Rodents are natural reservoirs and shed the virus in their urine, droppings, and saliva. The virus is mainly transmitted via **respiratory route**.
- Infection can also rarely occur via **rat bite** or by eating food contaminated by urine, droppings, or saliva from an infected rodent.

Filoviral Hemorrhagic Fevers

- These are **long, thread-like viruses,** based on which they are so named (*filum* meaning thread).
- They range in size from 80 to 800–1,000 nm.
- Ebola viruses and Marburg viruses, which cause hemorrhagic fever, belong to the genus Filovirus.

Ebola virus disease

Ebola virus disease (EVD) is a rare and deadly disease in people and non-human primates. It is caused by the **Ebola virus** (Fig. 21.31).

Epidemiology

- The infection first occurred in a village near the Ebola River, after which the virus is named. The outbreak spread very rapidly within Africa, starting in Guinea and moving across land borders to Sierra Leone and Liberia.
- An outbreak that occurred between 2013 and 2016 in West Africa (Guinea, Liberia, and Sierra Leone) resulted in more than 11,000 deaths.
- The virus that caused this outbreak belonged to the Zaire Ebola virus species. It causes an acute illness which is serious and often fatal if untreated.

Table 21.20 Characteristics of some known hantaviruses

Hantaviruses	Reservoir	Pathology	Mortality
Hantaan	Field mouse	Renal, HFRS	5–15%
Seoul	Domestic rat	Renal, HFRS	1%
Puumala	Bank vole	Renal, nephropathia epidemica	1%
Prospect Hill	Meadow vole	No known human disease	N/A
Sin Nombre	Deer mouse	Pulmonary, HPS	50%

- The reservoir of the Ebola virus and its natural history still remain unclear, despite frequent outbreaks with high fatality.

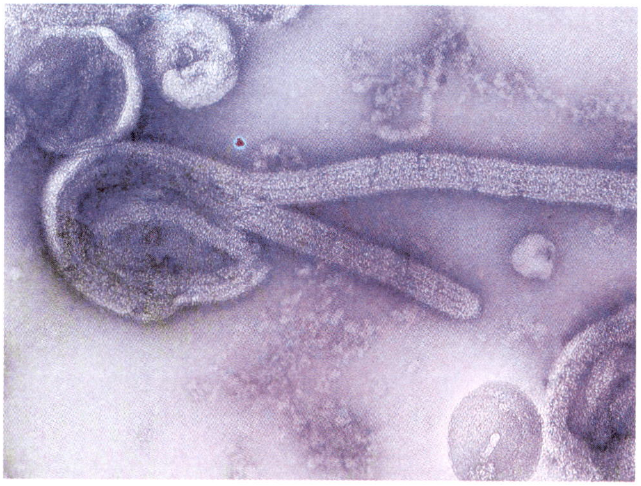

Fig. 21.31 Ebola virus—long, thin, and filamentous form (*Source*: CDC, PHIL, Image ID 23186)

Agent—Ebola virus

Ebola virus are filovirus and five distinct strains of Ebola virus have been recognised:

- Ebola-Zaire strain (EBO-Z), with a case fatality rate of up to 90 per cent
- Ebola-Sudan strain (EBO-S), with a case fatality rate of up to 50 per cent
- Ebola-Reston strain (EBO-R), isolated from quarantined monkeys imported from the Philippines and held at Reston, Virginia, and the US
- Ebola-Ivory Coast
- Ebola-Bundibugyo, which is found in Africa

Pathogenesis

- The high mortality of Ebola is attributed to virulence factors like the **glycoprotein** of the Ebola virus, which **kills endothelial cells and produces hemorrhage**.
- The virus also kills macrophages and lymphocytes.
- Viral proteins inhibit the action of interferons.

Clinical features

- The **incubation period is usually 5–7 days,** may go up to 21 days.
- The disease presents with fever, headache, sore throat, myalgia, arthralgia, vomiting, and diarrhea.
- Subsequently, **bleeding into the skin, marked thrombocytopenia, and lymphopenia** occur.
- **Multiorgan failure** leads to death.
- Mortality is 90%.
- When recovery occurs, the disease is followed by **post-Ebola syndrome (PES)** with blurring of vision, uveitis, arthralgia, and fatigue.

Laboratory diagnosis

- **RT-PCR** for detecting the viral RNA or ELISA to detect the viral antigen and rising antibody titres of IgG in serum have been employed to confirm Ebola.
- **Electron microscopy** may also be used to demonstrate long, filamentous viruses.
- The virus can also be **cultivated in monkey kidney cells** (Vero and MA104).
- This is done in biosafety level-4 containment laboratories as the patient's samples may contain viable virulent viruses.

Treatment

- No antiviral treatment is available currently.
- **Supportive therapy** is required with stringent infection-control practices to prevent secondary spread through body fluids with the use of personal protective equipment.
- A **monoclonal antibody (ZMapp)** was tried with little success.

Prevention

Two vaccines have been licensed:

- **rVSV-ZEBOV** was licensed in November 2019. It is based on the vesicular stomatitis virus which was genetically modified to express a surface glycoprotein of the Zaire Ebola virus.
- **Adenovirus 26 vectored glycoprotein/MVA-BN (Ad26.ZEBOV/MVA-BN)** was approved in 2020.

The current vaccination strategies for Ebola are as follows:

- **Pop-up vaccination:** To vaccinate all contacts in a protected vaccination site, i.e., at a healthcare facility, away from the residence of the contacts
- **Targeted geographic vaccination:** To vaccinate all contacts and their contacts in a particular geographic area; this is also done in a protected healthcare facility

A second revised vaccination strategy is to **vaccinate all contacts and their contacts**, including a third group of possible contacts at risk, to form a ring or barrier of immunised persons around the incident case.

Marburg disease

- Marburg disease is a hemorrhagic fever that occurred simultaneously in laboratory workers in Marburg, Frankfurt (Germany) and Belgrade (Yugoslavia) in 1967.
- The infection arose from the tissues of **African green monkeys** imported from Uganda, to which the laboratory workers had been exposed.
- The primary cases had a fatality rate of 30 per cent but the secondary cases were non-fatal.
- The **Marburg virus** was isolated in guinea pigs and in tissue culture from the blood and tissues of these patients.
- The virus appears to persist in the body and has been isolated after 80 days of the onset of the illness from semen and the anterior chamber of the eye. A case of sexual transmission has also been recorded.
- No further Marburg virus infections have been seen since, except for three cases identified in South Africa in 1975 and two in Kenya in 1980.

FUNGAL BLOODSTREAM INFECTIONS

Candidiasis

- **Candidiasis** refers to an infection of the skin, mucosa, and rarely, the internal organs caused by a yeast-like fungus, *Candida albicans.*
- It is an **opportunistic endogenous infection**, the commonest predisposing factor being diabetes.

Agent—*Candida*

- *Candida* species are normal inhabitants of the skin and mucosa.
- *Candida albicans* is an ovoid or spherical budding cell, which produces pseudomycelia both in tissues and in culture (Fig. 21.32a and b).

The *Candida* species of growing clinical importance are as follows:

- *C. glabrata*
- *C. tropicalis*
- *C. keyfr* (formerly, *C. pseudotropicalis*)

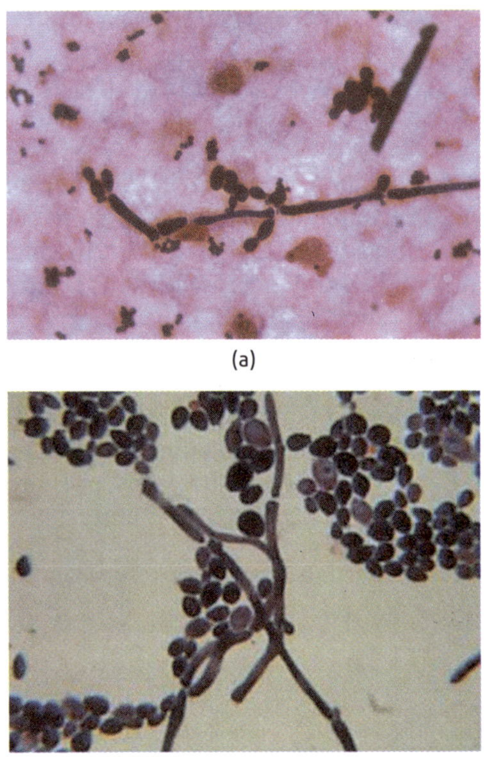

(a)

(b)

Fig. 21.32 Gram stain showing (a) gram-positive *Candida* yeast with pseudohyphae and pus cells in tissue and (b) *Candida* yeast with pseudohyphae in culture

- *C. krusei*
- *C. guilliermondii*
- *C. parapsilosis*
- *C. stellatoidea*

Candida auris is a recent finding globally that is resistant to many antifungal agents and can cause fatal infection.

Pathogenesis and clinical features

Cutaneous candidiasis

- May be **intertriginous or paronychial**.
- The **intertriginous type** is an erythematous, scaling, or moist lesion with sharply demarcated borders, where papular lesions are most prominent. The sites most often affected are those where the skin is macerated by perspiration—the groin, perineum, axillae, and inframammary folds.
- **Paronychia and onychomycosis** are seen in occupations that involve frequent immersion of the hands in water.

Mucosal lesions

- Common lesions are vaginitis, which is characterised by an acidic discharge and noted frequently in pregnancy, and oral thrush, which occurs commonly in bottle-fed infants and aged, immunocompromised, and debilitated persons (Fig. 21.33).

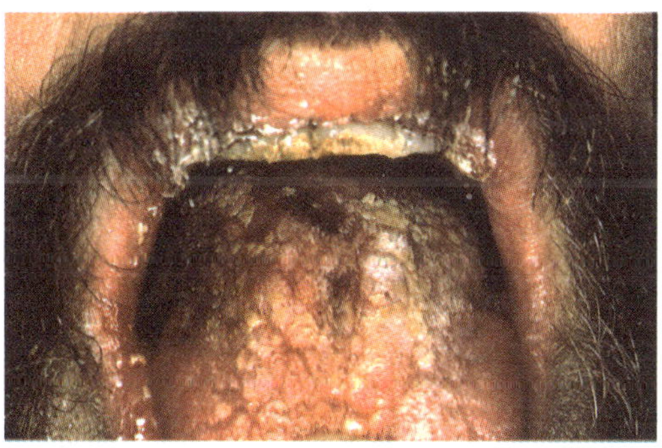

Fig. 21.33 Oral thrush due to *Candida albicans* in an immunocompromised patient (*Source*: CDC, PHIL, Image ID 6067)

- Creamy white patches appear on the tongue or buccal mucosa; following removal, these leave a red, oozing surface.
- These are also seen in individuals infected with HIV and in other immunocompromised individuals.
- **Esophageal candidiasis is now considered an AIDS-defining illness**.

Intestinal candidosis

This is a frequent sequel to excessive oral antibiotic therapy and may present as diarrhea which does not respond to antibacterial treatment.

Bronchopulmonary candidiasis

This is a rare complication of pre-existing pulmonary or systemic disease.

Nosocomial infections

- *Candida* has now emerged as a hospital-acquired pathogen in catheterised patients.
- Catheter-associated UTI or bloodstream infections occur due to catheter colonisation.

Systemic infections

- Septicemia, endocarditis, and meningitis may occur as terminal complications in severe generalised diseases such as leukemia and in persons on prolonged immunosuppression.
- Candida **granuloma and chronic mucocutaneous candidosis** are serious manifestations seen in immunodeficiencies.

Laboratory diagnosis

- **Microscopy:** Wet films or Gram-stained smears from lesions or exudates show budding, gram-positive cells (Fig. 21.34a). *Candida* can colonise normal skin or mucosa, but abundant presence is of significance. The **demonstration of mycelial forms** and pseudohyphae indicates colonisation and tissue invasion and is, therefore, of greater significance.

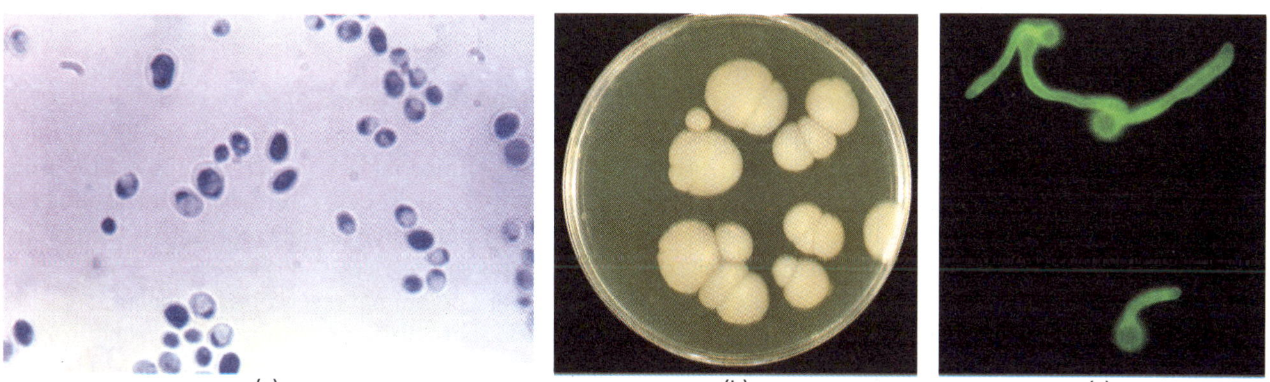

(a) (b) (c)

Fig. 21.34 *Candida albicans*—(a) oval budding yeast cells stained with Gram stain, (b) cream-coloured pasty colonies, and (c) germ tubes stained by calcofluor white (*Source*: CDC, PHIL, Image IDs 22204, 3102 and 205)

- **Cultures:** *Candida* grows readily on Sabouraud agar and on ordinary bacteriological culture media. Colonies are creamy white, and smooth, and have a yeasty odour (Fig. 21.34b). *Candida albicans* can be differentiated from other *Candida* species by growth characteristics, sugar assimilation and fermentation tests. *C. albicans* alone forms chlamydospores on cornmeal agar cultures at 20°C. A rapid method of identifying *C. albicans* is based on its ability to form germ tubes within two hours when incubated in human serum at 37°C (Reynold–Braude phenomenon) (Fig. 21.34c).
- **Serology:** There are no specific serological tests for *Candida*. However, the **serum (1–3)-β-D-glucan assay** may be done to detect invasive *Candida* infection. The test is used in conjunction with other methods for the diagnosis of invasive infection. Testing should be done prior to antifungal therapy.

Prevention

Candidosis is chiefly prevented by eliminating possible predisposing factors.

Treatment

- All *Candida* strains are susceptible to nystatin but since this drug is poorly absorbed from the gut, it is not useful in treating systemic diseases.
- **Amphotericin B, 5-fluorocytosine, imidazoles** (miconazole, ketoconazole), triazoles (itraconazole, fluconazole, voriconazole), and **echinocandins** (caspofungin, micafungin) may be used for disseminated candidosis.
- Some clinical isolates of *C. albicans* are **resistant to fluconazole** and *C. krusei* to **amphotericin**.

VECTOR-BORNE PARASITIC INFECTIONS OF THE BLOODSTREAM

- Parasites that infect cells of the circulatory system present clinically as a **febrile illness**. The patient may complain of fever, often with chills, malaise, rigors, sweating, myalgia, and headache.
- It is important to be aware of **local epidemiology and detailed history**, including that of recent travel. This should be accompanied by a complete physical examination to diagnose parasitic causes of fever. The most important parasites (hemoparasites) found in blood are:
- *Plasmodium*, which causes malaria
- *Leishmania*, which causes kala-azar
- *Trypanosoma*, which causes sleeping sickness
- Filarial worms, which cause filariasis
- *Babesia* and *Schistosoma*

Malaria

Case 10. *A 3-year-old girl presented to the emergency department with decreased oral intake, listlessness, and fever. She was febrile and had a depressed level of consciousness. Her peripheral blood smear showed multiple rings inside RBCs. A diagnosis of **malaria** was made, and she was started on antimalarial therapy.*

i. What causes malaria?
ii. How is the disease diagnosed?
iii. What are the complications of malaria?

Epidemiology

- Malaria has been known to exist since primordial times.
- The name was derived from the term '*mal*-air' or unhealthy air as it was once believed that the air during the rains became foul and caused the fever.
- Worldwide, more than 247 million cases of malaria were reported in 2021, with a whopping 6,19,000 deaths attributed to the disease. Africa accounts for more than 90% of malaria cases.
- The majority of malaria cases are caused by *Plasmodium falciparum* in Asia and Africa followed by *P. vivax*. In India, infections due to *P. malariae* and *P. ovale* are rarely seen.
- Recently, *Plasmodium knowlesi* has been reported to cause **zoonotic malaria in southeast Asia**. It primarily affects monkeys.
- The sporozoan *Plasmodium* was discovered by **Charles Lavern** in 1880 in the blood corpuscles of infected patients.
- In 1902, **Ronald Ross**, an army surgeon in India was awarded the Nobel Prize for his work on elucidating the life cycle of the malarial parasite.
- *P. falciparum* infections are associated with complications such as **cerebral malaria and blackwater fever**.
- The parasite has shown **resistance to chloroquine** in recent times.
- Despite several control programmes, malaria continues as an endemic infection in several parts of the world, including India.

Transmission to humans

- Malaria is naturally transmitted by the bite of female *Anopheles* mosquitoes, which require a blood meal to nourish their eggs. *Anopheles* mosquitoes are primarily night biters. Male mosquitoes feed on fruit juice and sap.
- There are no animal reservoirs for malaria.
- Malaria can also be acquired by transfusion of infected donor blood or by the sharing of needles amongst drug addicts.
- Rarely, congenital malaria has been reported in non-immune mothers.

Agent—*Plasmodium*

Malaria is caused by a protozoon, *Plasmodium*. Important members of this genus are:

- *Plasmodium vivax*
- *Plasmodium falciparum*
- *Plasmodium malariae*
- *Plasmodium ovale*

Life cycle (Fig. 21.35)

Intermediate host: Human being
Definitive host: *Anopheles* mosquito

Human phase

Two main stages of development are seen in humans—**the exoerythrocytic phase in the liver and erythrocytic schizogony** and **gametocytogenesis in red blood cells.**

i. *Exoerythrocytic or pre-erythrocytic cycle*
   - The human phase of the life cycle begins with the bite of a female *Anopheles* mosquito.
   - The **mature sporozoites** in the mosquito's saliva are injected into the bloodstream. Some of these sporozoites are phagocytosed and destroyed while others reach the liver and enter the hepatocytes.
   - The sporozoites become round (pre-erythrocytic schizont) and undergo division to develop into **merozoites.**
   - These are released into the bloodstream by the rupture of infected liver cells. This phase of growth is called the **exoerythrocytic or pre-erythrocytic stage** and can last from 8–25 days, depending on the *Plasmodium* species (Table 21.21).

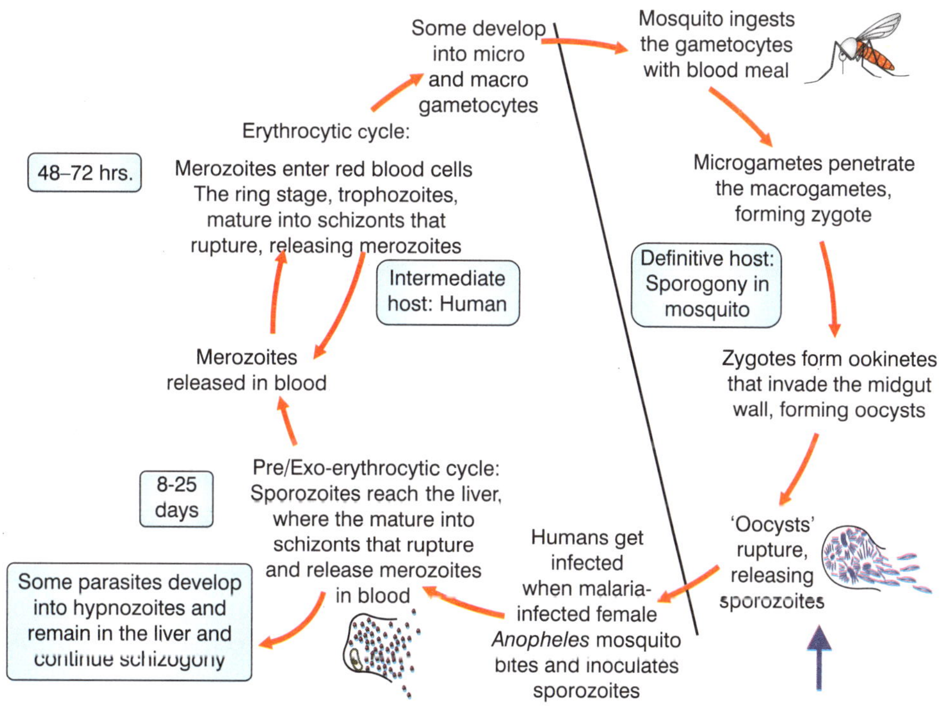

Fig. 21.35 Stages in the life cycle of *Plasmodium*

Table 21.21 Characteristics of *Plasmodium* species causing malaria

Features	*P. vivax*	*P. falciparum*	*P. malariae*	*P. ovale*
Incubation period	14 days	12 days	30 days	14 days
Relapse	Can occur up to five years after the primary infection	Not seen	Not seen	Rare
Level of parasitemia	Low	High	Very low	Very low
Duration of exoerythrocytic cycle (in days)	8	6	7–12	9
Number of merozoites in pre-erythrocytic schizont	12,000	40,000	2,000	15,000
Hypnozoites and relapses	Present	Absent	Absent	Present

(Table 21.21 Contd. . .)

Table 21.21 Contd.

Features	*P. vivax*	*P. falciparum*	*P. malariae*	*P. ovale*
Age of RBCs attacked	Young erythrocytes	All ages	Old erythrocytes	All ages
Peripheral blood smear (PBS) findings	All stages are seen Single ring seen in RBCs Low level of parasitemia (<2% RBC are parasitised)	Only ring forms and gametocytes are seen in PBS Multiple rings seen in RBC Heavy parasitemia is seen	Low parasitemia, predominantly ring forms are seen in PBS	Low parasitemia, all stages seen in PBS
Infected erythrocyte	Enlarged, pale Schuffner's dots present	Normal size Maurer's dots present	Normal size	Slightly enlarged Schuffner's dots present
Malaria pigment	Multiple, fine golden-brown particles	Few solid blocks of black pigment	Multiple, coarse, dark brown particles	Multiple black-brown particles
Trophozoite/Ring form seen in RBC	Single ring	Multiple delicate rings (2–6)	Same as *P. vivax*	Same as *P. vivax*
Schizont	10 μm, fills RBC completely	5 μm, fills 2/3rd of cell	6–7 μm	6–7 μm
Number of merozoites in a schizont	12–24 arranged in grape-like clusters	18–24 arranged in grape-like clusters	6–12 arranged in rosette forms	8–12 arranged irregularly
Gametocyte	Round	Crescentic and larger than red cell	Same as *P. vivax*	Same as *P. vivax*
Duration of sporogony	9–10 days	10–12 days	25–28 days	14–16 days

- In some species such as *P. vivax* and *P. ovale*, the sporozoites do not divide; instead, they remain dormant. These are known as **hypnozoites or sleeping forms**. The hypnozoites are responsible for causing relapses months to years after the initial disease by *P. vivax* and *P. ovale*.

ii. *Erythrocytic schizogony (Fig. 21.36)*

- This stage begins when the **merozoites released from the liver attach to specific receptors on the surface of erythrocytes and enter the cells,** thus initiating the erythrocytic cycle.
- The merozoite enters via endocytosis, and the membrane seals itself to form a vacuole.
- The, the merozoite develops into a **trophozoite (ring form), schizont, and gametocytes** within the red blood cell.

The various stages in the life cycle of *Plasmodium* are as follows:

Trophozoite

- After entry into an erythrocyte, the **merozoite becomes rounded with a vacuole in the center**, the cytoplasm pushed to the periphery and a nucleus at one pole.
- In a blood smear, these vacuoles appear as **thin rings of blue cytoplasm with a red nucleus and unstained central vacuole.**
- The parasite feeds on hemoglobin. Excess hematin combines with iron porphyrin to form the malaria pigment, which is seen in the late trophozoite and schizont stages. The timing of appearance differs according to the species (Table 21.21).
- Pigmented reticuloendothelial cells and pigment-laden macrophages are seen in histopathological sections as evidence of past infections. The ring form starts growing and becomes more irregular.
- In *P. falciparum*, a single red blood cell can be infected by multiple merozoites, giving the appearance of multiple rings in peripheral blood smear. **Maurer's dots** can be seen in *P. falciparum* as brick red pigments.
- In the late trophozoite stage of *P. vivax*, bits of the membrane from the parasite accumulate on the membrane forming '**Schuffner's dots**'.

Schizont

- The fully developed trophozoite becomes compact and loses its vacuole with the nucleus becoming large and getting pushed to the periphery. The nucleus starts dividing, followed by the cytoplasm; thus, **merozoites are formed**.
- A mature schizont contains **8–32 merozoites arranged in a rosette**.
- The erythrocyte ruptures, releasing merozoites, the malarial pigment, and toxins into the bloodstream. This gives rise to the **paroxysm of fever**. The released merozoites invade fresh erythrocytes and the cycle continues.

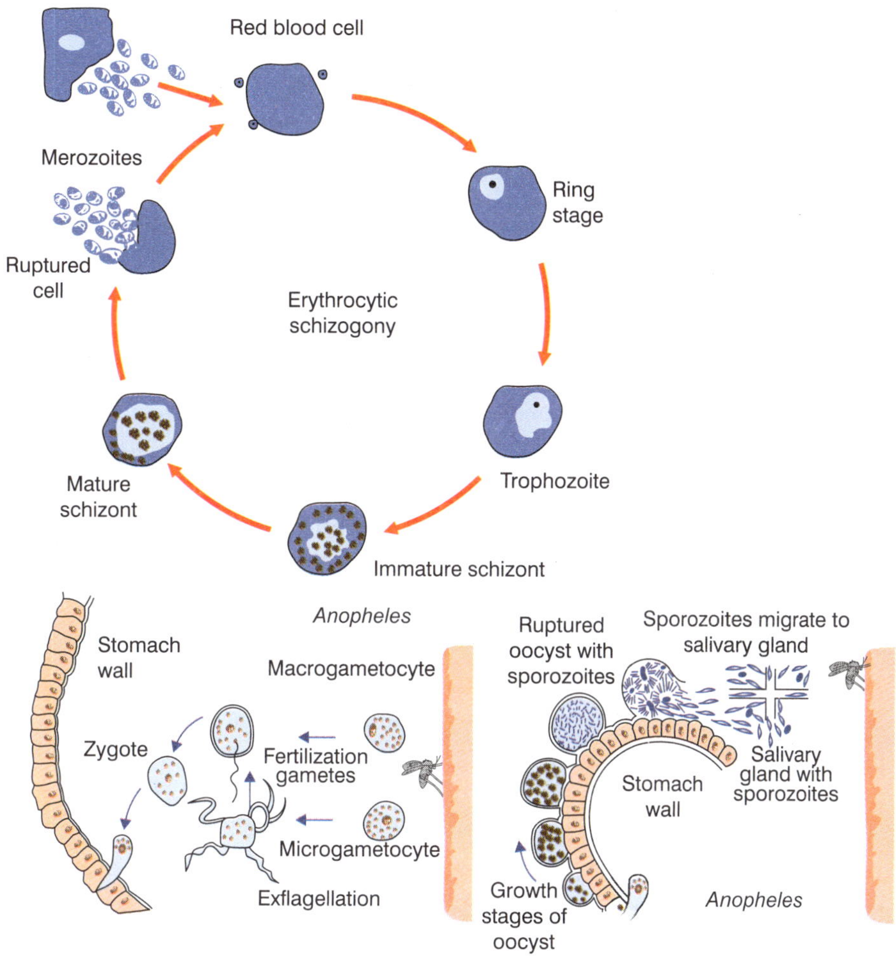

Fig. 21.36 Erythrocytic schizogony

- The erythrocytic stage takes 48 hours in *P. vivax*, *P. falciparum*, and *P. ovale* infections (called **tertian malaria**) and takes 72 hours in *P. malariae* infections (called **quartan malaria**).

Gametogony

- After several erythrocytic cycles, a few merozoites develop into gametocytes as the first step of sexual development. They grow in size, filling the entire red blood cell.
- Gametogenesis takes place in the capillaries of the spleen. Only mature forms are seen in peripheral blood.
- The female gametocyte, known as the **macrogametocyte**, has a dark blue cytoplasm with a deep red nucleus while the **microgametocyte** or male gametocyte has a pale blue cytoplasm with a large pale diffuse nucleus.
- Gametocytes are round in all species except in *P. falciparum*, where they are crescent- or sickle-shaped (or '*falx*'—hence, the name *falciparum*), stretching across the entire RBC.
- Both sexual and asexual forms circulate in blood.
- In *P. falciparum*, predominantly ring forms are seen in the peripheral blood.
- In *P. falciparum* infections, the parasite continues to grow in the erythrocyte, causing the cell membrane to become sticky and adhere to capillaries of internal organs. This results in **blockage and hypoxia**.
- Consequently, the blood vessels of internal capillaries rupture and produce complications as seen in cerebral malaria, algid malaria, or blackwater fever.

Mosquito phase

Sporogony

- The asexual forms ingested by the mosquito in its blood meal are digested while the sexual forms reach the stomach of the mosquito.
- In the gut, the microgametocyte divides into eight flagellated and highly motile bodies known as the **microgametocytes.**
- The macrogametocyte does not divide but develops into a **macrogamete**.
- After fertilisation with the microgamete, the macrogamete develops into a **zygote.**
- After 18–24 hours, the zygote elongates and develops into an **ookinete** which migrates to the stomach wall, secretes a thin wall, and grows into an **oocyst**.
- The oocyst continues to grow in size, and the nucleus divides. Each oocyst is filled with thousands

of sporozoites which rupture on the 10th day of ingestion.

- **Sporozoites** are released into the body cavity of mosquitoes, particularly into the salivary glands, from where they are released into the body of a human being by the bite of the mosquito.

Clinical features

- The **incubation period is 9–30 days**.
- **Benign malaria:**
  - The patient presents with **bouts of fever with chills and rigors followed by drenching sweats**.
  - The fever coincides with the release of merozoites from red blood cells.
  - Sweating is followed by an afebrile period of 48 hours in P. *vivax* and *P. ovale* and for 72 hours in infections with *P. malariae*.
- **Malignant falciparum malaria:**
  - The classic cycle of **cold–hot sweating** may not be seen in *P. falciparum* infections.
  - This is because several generations of parasites multiply at different intervals. Hence, the fever may be continuous or intermittent and periodic. In such cases, the patient may also complain of headache, nausea, vomiting, and epigastric pain.
  - Left untreated, multiple complications such as cerebral malaria, algid malaria, and septicemic malaria can develop.
- **Relapse:**
  - The hypnozoites get reactivated and initiate fresh erythrocytic schizogony.
  - This causes fresh episodes of malarial fever.
  - Such relapses can occur even up to five years after the initial infection with *P. vivax*.
- **Recrudescence:**
  - The primary attack subsides, making a person partially immune.
  - The malarial parasite persists in the capillaries of the internal organs at a subclinical level after an episode of malaria.
  - Erythrocytic schizogony probably occurs at a low level in such individuals.
  - After a period of clinical latency, features of clinical malaria are seen again.
  - This could be due to the waning immunity of the host or inadequate drug therapy.
  - This is called recrudescence and can be seen in all species, though most commonly in *P. falciparum* and *P. malariae* infections.

Complications of malaria

- **Cerebral malaria:** This is a fatal complication of *P. falciparum* malaria. The patient has a history of high fever with altered sensorium, confusion, and finally, coma. The pathogenesis is due to capillary plugging which occurs due to sticky red blood cells and malaria pigment causing anoxia, ischemia, and hemorrhage.
- **Algid malaria:** This syndrome is postulated to occur due to adrenal hemorrhage causing circulatory collapse. Clinically, it presents with low blood pressure, thread pulse, and cold, clammy skin. The patient may complain of abdominal pain, anuria, vomiting, and diarrhea.
- **Septicemic malaria:** It occurs due to multisystem involvement. The patient presents with complaints of high fever, pneumonia, and cardiac syncope.
- **Anemia:** More commonly seen in falciparum malaria, anemia develops due to the destruction of the infected red cells, dyserythropoiesis and destruction by the spleen, and erythrocytic sequestration.
- **Blackwater fever:** Patients present with bilious vomiting and passage of dark red urine—hence the name blackwater fever. This is seen in falciparum malaria, in patients with repeated infections, and those who have received inadequate quinine treatment. Intravascular destruction with the rapid destruction of red blood cells by anti-erythrocytic antibodies **causes hemoglobinuria.**
- **Quartan malaria nephropathy:** This syndrome involves immune complex injury to the kidney, resulting in nephropathy and is seen in infections with *P. malariae*.
- **Malaria in pregnant women:** This may lead to severe anemia, abortion in the early stage of pregnancy, or stillbirth. In endemic areas, the newborn may be protected by maternal antibodies, but as this protection wanes, they become susceptible to infection. In infants and children, malaria may lead to severe anemia and a fatal outcome. In African countries, the incidence of falciparum malaria has led to increased morbidity and mortality among children.
- **Malaria acquired by blood transfusion:** Malaria may be transmitted through blood transfusion. In such cases, the incubation period is reduced on account of the infusion of already parasitised blood. The incidence of transfusion-related malaria has, however, come down due to mandatory screening of blood prior to infusion.

Laboratory diagnosis (Fig. 21.37)

Demonstration of parasite

- **Peripheral blood smear (PBS)** examination is the most useful, easiest, and cheapest method to diagnose malaria. However, it requires a trained eye to identify **malarial parasites (mp)** in a PBS.
- A **thin film** is used to identify the species, whereas a **thick film** can be used as a screening tool to detect mp. The smeared slides are stained with Leishman or Giemsa or Wright stain.

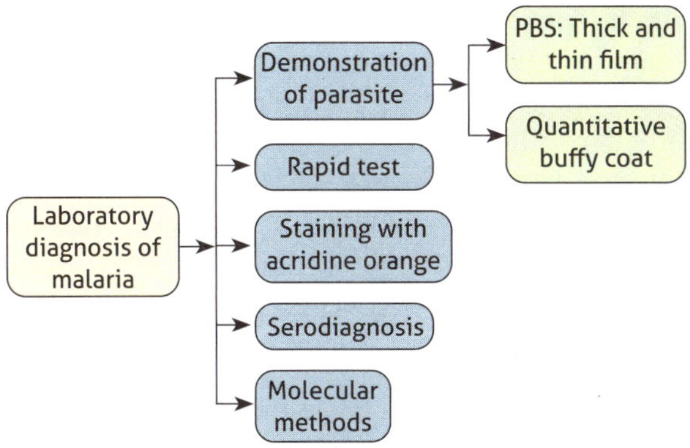

Fig. 21.37 Approach to the diagnosis of malaria

The quantification of parasites in a peripheral smear helps to monitor prognosis or gauge the severity of infection. 1% parasitemia may be roughly denoted by 1 parasite per 100 RBCs. However, there are other stringent calculations to estimate parasitemia.

Quantitative buffy coat (QBC)

- The blood sample is drawn in a small capillary tube which is pre-coated with acridine orange stain and potassium oxalate.
- This tube is centrifuged at 12,000 rpm for five minutes to separate the components of the blood and parasites by their density gradient.
- A plastic float in the tube increases sensitivity. The float settles on top of the centrifuged packed red cells which aids in the expansion of the centrifugally separated cell layers. This is surrounded by three discernible and now measurable layers of the buffy coat (Fig. 21.38).
- The parasite-laden RBC are less dense than normal RBCs and concentrate just below the leukocytes and above the erythrocyte column.

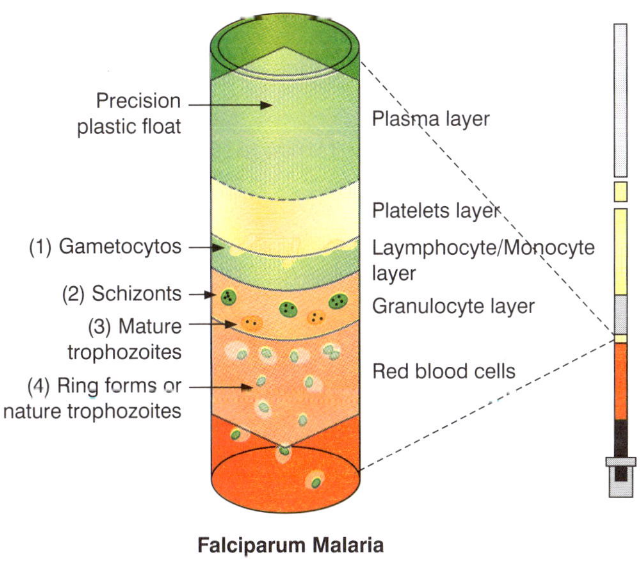

Fig. 21.38 Quantitative buffy coat

- The acridine orange binds to DNA and RNA. Malaria parasites bind to the acridine orange and fluoresces with green and red colors which, when excited to a frequency of 460 nm, allow the detection of *Plasmodium* morphology through fluorescent microscopy.
- This is detected as **bright specks of light under a fluorescent lens.**

Staining with acridine orange

- Blood smears can be stained with acridine orange and examined under a fluorescent microscope.
- **Nuclear DNA stains green** while the **cytoplasm is visible as a red-coloured structure**.

Rapid diagnostic test (RDT)

- This is a rapid, simple, sensitive, and specific test to detect the malarial antigen in lysed blood.
- **Monoclonal antibodies tagged with gold bind to the antigen and form a visible line.**
- Lateral flow assays are available commercially and are employed in emergency and field situations to screen fever cases.
- Different test formats like the dipstick, cassette, strip, well, card, and pad tests have been developed.

Plasmodium LDH

- Asexual and sexual forms of *Plasmodium* produce lactate dehydrogenase (pLDH). This binds to the antibodies in the test.
- The single band is genus-specific while the second band indicates *P. falciparum*. The test cannot distinguish between other species of *Plasmodium*.

Pan-specific Plasmodium aldolase test

- This test detects *Plasmodium* aldolase, an enzyme of the glycolytic pathway seen in the blood stages of all *Plasmodium* species.
- This test can detect all four species but cannot distinguish between species of *Plasmodium*.

P. falciparum-specific HRP-2 test

- This test detects histidine-rich protein-2 which is produced by trophozoites and young gametocytes of *P. falciparum*.
- It can remain positive for weeks but cannot detect infection due to any other species of *Plasmodium*.

Molecular methods

PCR kits are now available to detect *Plasmodium* DNA.

Serology

This has very limited value as it cannot distinguish between acute and chronic infections.

Treatment

- **Antimalarials** aim to destroy all asexual forms and gametocytes, eliminate sporozoites, and eradicate all exoerythrocytic forms to treat as well as prevent relapse and transmission.
- **Quinine** has been used since ancient times to treat malaria, but its use is now limited to the treatment of chloroquine-resistant malaria or complicated malaria (Table 21.22).

Prevention

Malaria vaccine

- Research on malaria vaccines is focused on *P. falciparum*. Multiple targets are being explored, and various clinical trials are ongoing.
- The WHO has recently recommended the **RTS, S/AS01 malaria vaccine** for children in sub-Saharan Africa and other falciparum-endemic areas. This vaccine is based on the *P. falciparum* circumsporozoite protein (CSP).
- Other vaccines in the pipeline:
  - **R21 vaccine:** The R21 vaccine uses a circumsporozoite protein (CSP) antigen used at a higher proportion than the RTS, S vaccine.
  - **Pf SPZ vaccine:** It is a candidate malaria vaccine developed using sporozoites attenuated by radiation.

Malaria control

- Malaria is a reportable disease. All smear-positive cases are reported to the local **IDSP programme** office or the **NVBDCP (National Vector-Borne Disease Control Programme)**.
- Mosquito control is one of the most important strategies for controlling malaria, filariasis, and other vector-borne infections.

Babesiosis

Babesia is a protozoan parasite that is transmitted by deer ticks (*Ixodes* species) and causes **babesiosis.** There are more than 100 species of *Babesia* that are known to cause infections in humans and animals. Of these, only *B. microti* and *B. divergens* cause human infections.

Agent—*Babesia*

Life cycle

Intermediate host: Human being
Definitive host: Hard tick

- In humans, the sporozoites enter red blood cells and transform into **trophozoites or ring forms**.
- Groups of **four rings are arranged back-to-back** in RBCs forming a cross (Fig. 21.39). This is known as the '**Maltese cross' appearance**.
- Some trophozoites develop into gametocytes, which are taken up by the tick during its blood meal.

Table 21.22 Treatment of malaria as per the National Vector-Borne Disease Control Programme

Infection	Specific antimalarial treatment
P. vivax	**Chloroquine:** 25 mg/kg body weight divided over three days **Primaquine:** 0.25 mg/kg body weight every day for 14 days (for prevention of relapse or prophylaxis for travellers to endemic areas)
Uncomplicated *P. falciparum*	**Artesunate:** 4 mg/kg body weight daily for three days plus sulphadoxine (25 mg/kg body weight)–pyrimethamine (1.25 mg/kg body weight) on the first day **Primaquine** 0.75 mg/kg body weight may be added for cases of mixed infections (*P. vivax* and *P. falciparum*)
Severe malaria (to be treated as a medical emergency)	**Artesunate:** 2.4 mg/kg body weight intravenous/intramuscular; then at 12 h, 24 h, and then once a day OR **Artemether:** 3.2 mg/kg body weight intramuscular as loading dose and then 1.6 mg/kg body weight per day OR **Arteether:** 150 mg intramuscular daily for three days in adults only OR **Quinine:** 20 mg/kg body weight either as intravenous infusion or divided intramuscularly followed by 10 mg/kg body weight eight-hourly
Short-term chemoprophylaxis (up to six weeks) for travellers to endemic areas	**Doxycycline***: 100 mg once daily for adults and 1.5 mg/kg once daily for children—to start two days before travel and be continued for four weeks after leaving the endemic area
Chemoprophylaxis for more than six weeks	**Mefloquine:** 250 mg weekly for adults Start two weeks before, at the time of, and for four weeks after exposure

*Not recommended for pregnant women and children under eight years of age

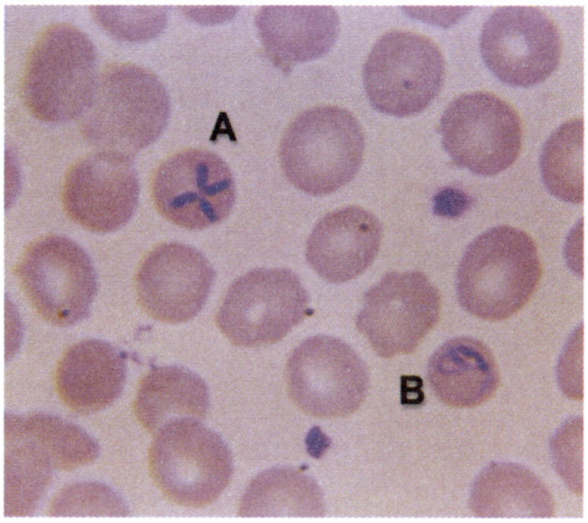

Fig. 21.39 Characteristic Maltese cross appearance of the trophozoites of *Babesia* in RBCs (*Source*: CDC, PHIL, Image ID 20166)

- The gametocytes in the tick's gut multiply sexually and develop into sporozoites which block the salivary glands.
- These are regurgitated with the blood meal.

Clinical features

- Babesiosis presents with fever, chills, headache, nausea, vomiting, and muscle aches (myalgia).
- Anemia is pronounced, and the patient may develop complications secondary to anemia.

Laboratory diagnosis

On a **peripheral blood smear**, the presence of the '**Maltese cross**' formed by groups of four rings arranged back-to-back in RBCs is diagnostic.

Treatment

Atovaquone plus azithromycin or **clindamycin plus quinine** are used to treat babesiosis.

Prevention

This involves the control of tick bites by applying anti-tick skin sprays, wearing protective clothing, and removing ticks from outdoor clothing.

Leishmaniasis and Trypanosomiasis

These are diseases caused by hemoflagellates.

Agent—Hemoflagellates

- These are flagellated protozoa found in peripheral blood.
- This group (Fig. 21.40) includes two important disease-causing pathogens in human beings—*Leishmania* and *Trypanosoma*.

Life cycle of hemoflagellates

The life cycle of hemoflagellates requires **two hosts—humans and insects**—and consists of various morphological stages.

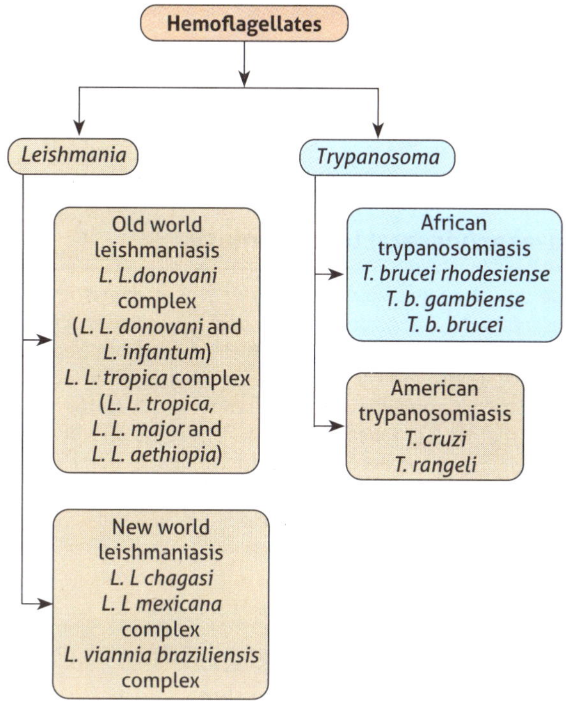

Fig. 21.40 Classification of hemoflagellates causing diseases in humans

Amastigote form

- Round/oval structure about 2–4 μm long and 1.5–3 μm wide with a delicate cell membrane and contains a nucleus, kinetoplast, axoneme, and vacuole.
- The **kinetoplast** comprises a parabasal body and blepharoplast, which are connected by delicate fibrils.
- The blepharoplast gives rise to a delicate filament called **axoneme**.
- A vacuole lies next to the axoneme.
- It is called the **Leishman–Donovan (LD) body** (named after the scientists who discovered it).
- On **Giemsa stain**, amastigotes appear with a pale blue cytoplasm, red nucleus, deep red parabasal body, and bright red kinetoplast.
- Amastogote forms are seen in the human body.

Promastigote form

- This is the lagellated, motile form, which is approximately 15–20 μm long and 1–2 μm wide.
- It is slender and spindle-shaped with flagella arising from the anterior end and extending without curving back.
- The promastigote form is seen in insect bodies and culture.

Epimastigote form

- This form of *Trypanosoma* is seen in insect vectors.
- The kinetoplast lies near the nucleus while the flagella emerge from the lateral side; a short, undulating membrane is seen.

Trypomastigote form

- It is the infective form of *Trypanosoma.*
- The kinetoplast is at the posterior end, and an elongated nucleus lies in the center with the flagella emerging at the posterior end and extending as an undulating membrane.

Kala-azar/visceral leishmaniasis

> **Case 11.** *A 45-year-old man working as a domestic helper presents with intermittent fever and a loss of weight and appetite over the last two months. He hails from Nepal and joined work around six months back. He has pallor and mild hepatomegaly and his spleen is palpable 10 cm below the costal margin. On investigation, he is anemic and his ESR is raised. Finding no other significant evidence, bone marrow aspiration is performed and is found positive for LD bodies. A diagnosis of* ***kala-azar*** *is made.*
> i. What is kala-azar?
> ii. How is it diagnosed?
> iii. Describe the pathogenicity of the disease?

Kala-azar is an anthropozoonosis caused by *Leishmania donovani* or *L. infantum* or *L. chagasi.* It is primarily a disease of the reticuloendothelial system.

Epidemiology

- The natural reservoirs include rodents, cats, and dogs.
- In endemic areas, the infection may be transmitted by a human–vector–human cycle.

Old world leishmaniasis (Table 21.23)

It can exist in three clinical forms—visceral, cutaneous, and mucocutaneous forms.

Visceral leishmaniasis

- It is the most serious form of leishmaniasis and is also known as **kala-azar (black fever) or Dum-Dum fever**.
- It can be fatal if not treated.
- Approximately 50,000–90,000 new cases are estimated to occur worldwide.
- In India, more than three-fourth of the total cases occur in Bihar, followed by West Bengal, Uttar Pradesh, Jharkhand, and the western coastal belt.

Mucocutaneous leishmaniasis

It occurs commonly in Ethiopia and South American countries (Brazil and Peru).

Cutaneous leishmaniasis

- This is more common globally, with more than 6,00,000 to 1 million new cases occurring annually.
- The majority of cases occur in the Mediterranean region, the Americas, Middle East, and Central Asia.

New world leishmaniasis

- It is a zoonotic disease confined to Central and South America.
- Visceral, cutaneous, and mucocutaneous leishmaniasis are caused by *Leishmania chagasi*, the *L. mexicana* complex, and *L. braziliensis* respectively.

Agent—Leishmania species

Different species of *Leishmania* causing disease are presented in Table 21.23.

Life cycle of *Leishmania*

Definitive host: Human being

Table 21.23 Features and characteristics of leishmaniasis

Agent	Clinical syndrome	Region affected	Vector	Reservoir
Old World leishmaniasis				
L. donovani (Asia and Africa)	Visceral leishmaniasis (kala-azar), post-kala-azar dermal leishmaniasis (PKDL)	South Asia, Sudan, Ethiopia, Kenya, Uganda, China, Middle East,	*Phlebotomus argentipes* (commonest)	Human beings
L. infantum (Mediterranean region)	Visceral leishmaniasis, PKDL (in immunosuppressed patients)	Mediterranean, Middle East, Central Asia, China	*P. perniciosus*	Wild animals (jackals, foxes)
L. tropica (Western India, North Africa and the Middle East)	Cutaneous leishmaniasis	China, India, Central Africa, Middle East, West Asia	*P. sergenti*	Human beings
L. major (India, China, Central and West Asia)	Cutaneous leishmaniasis	Western India, North Africa, Middle East	*P. papatasi*	Rodents
New World leishmaniasis				
L. chagasi (previously known as *L. infantum*)	Visceral and cutaneous leishmaniasis	Central and South America	*Lutzomyia* species	Foxes, dogs
L. (Viannia) *braziliensis*	Mainly cutaneous and mucocutaneous leishmaniasis (espundia)	South America	*Lutzomyia* species	Wild rodents

Intermediate host: Sandfly

- **Human phase:** The transmission of the infection from the sandfly to a human being is presented in Fig. 21.41.
- **Insect phase:** The sandfly (Fig. 21.42a) ingests both free as well as phagocytosed amastigotes (Fig. 21.42b) when it bites an infected person. Within 72 hours, the amastigote form develops into promastigote forms, which rapidly multiply and fill the gut of the insect. They migrate to the mouthparts of the sandfly in 4–15 days, causing a partial or complete blockage.

Pathogenicity and clinical features

Visceral leishmaniasis or kala-azar

- **Incubation period:** 2–6 months
- Irregular bouts of fever and loss of appetite and weight.
- Some patients also complain of pedal edema and ascites.
- The skin colour becomes dark due to **hyperpigmentation** (hence the name **kala-azar,** which **means 'black fever').**
- Examination reveals enlarged an spleen and liver, pancytopenia, hypogammaglobinemia, and anemia.

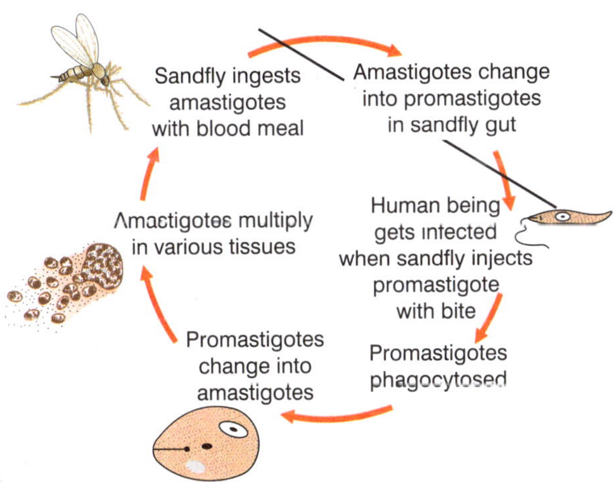

Fig. 21.41 Life cycle of *L. donovani*

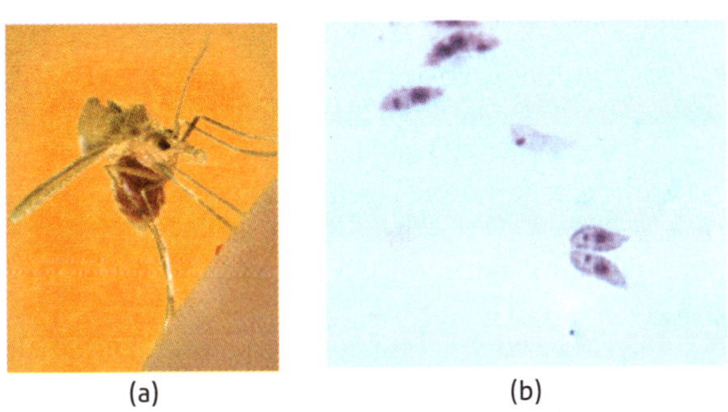

Fig. 21.42 (a) A sandfly taking its blood meal and (b) motile, flagellated promastigotes of *Leishmania donovani* (*Source*: CDC, PHIL, Image IDs 10274 and 544)

Post-kala-azar dermal leishmaniasis (PKDL)

- Seen in 5–10% of patients as a sequel to kala-azar, PKDL is common in East Africa and India.
- The patient presents with a **macular, papular, or nodular rash on the face, upper arms, and trunk.**
- It develops after six months to one or more years of recovery from kala-azar.
- Such people can also act as a potential source of *Leishmania* infection.

Cutaneous leishmaniasis

- **Incubation period:** 2–8 weeks.
- **Skin lesions** that occur mainly on the exposed parts of the body.
- The lesions begin as painless papules, become nodules, and finally ulcerate.
- They heal spontaneously, leaving lifelong scars and serious disfigurement.
- *L. tropica* lesions are singular and more swollen with a thick crust and less necrosis; these are also known as **Oriental sores, Delhi boils, Aleppo boils, and Baghdad buttons**.

Mucocutaneous leishmaniasis

- *L. mexicana* complex causes persistent ulcers on the pinna known as **chiclero ulcers**.
- The *L. braziliensis* complex causes **espundia**—the partial or total destruction of mucous membranes of the nose, mouth, and throat.
- It begins as nasal blockage, erythema, and mucopurulent discharge which later becomes ulcerative and leads to the erosion of the soft tissue and nasal cartilage.
- The destruction of the nasal septum can lead to nasal collapse.

Laboratory diagnosis

The approach to the diagnosis of leishmaniasis is detailed in Fig. 21.43.

The diagnosis of New World leishmaniasis is similar to that of Old World leishmaniasis. However, the amastigote form can be found only in the RE cells of the skin and mucous membrane.

Samples

- **Splenic aspirate**—for microscopy and culture, splenic aspirate is most sensitive; given the enlarged state of the liver, the procedure must be attempted with care to avoid hemorrhage
- Bone marrow aspirate
- Lymph node aspirate/liver aspirate/biopsy of the infected organ
- Peripheral blood—for routine investigations such as complete blood count, hemoglobin, etc.; PBS

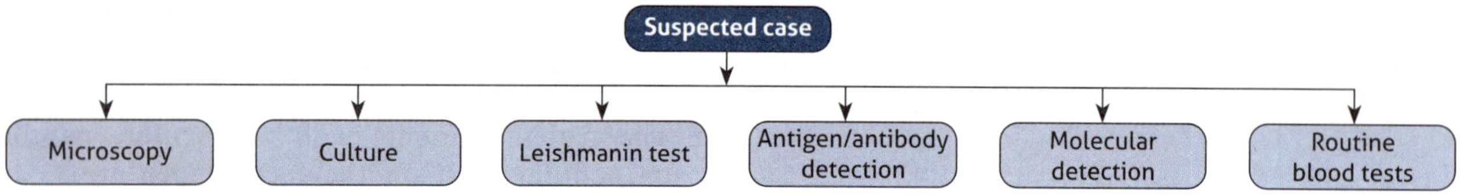

Fig. 21.43 Approach to the diagnosis of leishmaniasis

can be used for microscopy while serum is used for antibody detection

- Punch biopsy of the cutaneous lesion in cutaneous leishmaniasis

Microscopy

Smears made from bone marrow, spleen, aspirates, biopsy, and blood stained with Leishman, Giemsa, or Wright stains are examined to visualise the amastigote forms (also known as Leishman–Donovan or LD bodies) inside the macrophages.

Antigen detection

- A latex agglutination assay to **detect carbohydrate antigen** in urine is now available.
- It is a rapid method with good specificity but low sensitivity.

Culture

- Aspirates and buffy coats can be cultured in an NNN (described by Novy, MacNeal, and Nicolle) medium.
- Cultures are found positive in more than 75% of cases.

Non-specific tests

- Routine blood investigations reveal **pancytopenia, deranged liver function tests, and reversed albumin:globulin (A:G) ratio**.
- A:G ratio reversal is seen due to **hypergammaglobulinemia.**

Antibody detection

- **Immunochromatographic tests and direct agglutination** are more commonly used as these tests are rapid and easy to perform and do not require sophisticated equipment or trained manpower.
- Serological **tests fail to differentiate between recent and past infections**.

Immunochromatographic test

- The test uses an *L. chagasi* recombinant (rK39) antigen.
- A membrane is coated with antigen and protein A colloidal gold conjugate. When a drop of the patient's blood is placed in the well, the antibody (if present) binds to antigen–dye complex to produce a red line.
- It is a good screening tool, especially in field settings. However, this test **cannot detect PKDL**.

Direct agglutination test

- Microtitre plates are coated with *L. donovani* amastigote extracts. To these plates, serial dilutions of serum are added.
- Agglutination is seen as clumping/matting at the bottom of the tube; button formation is seen in negative cases.
- This test is useful in the **detection of PKDL**.

Leishmanin test

- This is a **delayed hypersensitivity skin test** to assess prior exposure to the *Leishmania* antigen.
- A suspension of *L. donovani* promastigotes is injected intradermally.
- An induration of more than 5 mm in 72 hours indicates a positive reaction.
- The test is positive only in cases of good CMI and is known to give false-positive results.
- The test is negative in cases of diffuse cutaneous leishmaniasis.

Molecular methods

Real-time PCR tests that **target mitochondrial DNA and 18sRNA** are available.

Treatment

The treatment regimen for leishmaniasis is presented in Table 21.24.

Prevention

- Vector control and blocking transmission by using nets and repellents are the only control measures available.
- Currently, no vaccines are available for the prevention of leishmaniasis. Both live attenuated and killed vaccines are at various stages of development.

HIV–leishmaniasis co-infection

- Visceral leishmaniasis is an important opportunistic infection associated with HIV infection.
- There exists a mutual reinforcement as the HIV infection predisposes to visceral leishmaniasis (VL) and the development of VL can cause the rapid progression of HIV to AIDS.
- Co-infected patients exhibit high parasitemia due to lower immunity. Hence, the diagnosis is made by direct examination or culture as serology is usually negative.
- These cases progress rapidly and may turn fatal in the absence of antiretroviral therapy.

Table 21.24 Treatment of leishmaniasis

	Kala-azar	PKDL	Cutaneous leishmaniasis	Mucocutaneous leishmaniasis
Drugs of choice	Liposomal amphotericin B intravenously as 10 mg/kg body weight Miltefosine 100 mg/day for 28 days—not recommended for pregnant and lactating women Pentavalent antimonials 20 mg/kg/day IM or IV for 20–28 days Miltefosine+paromomycin for 10 days	Miltefosine 150 mg/day × 12 weeks	Cryotherapy, thermotherapy with radiofrequency waves, and intralesional antimony leishmaniasis	***L. chagasi* complex** • Pentamidine • Imidazole • Quinolone • Amphotericin B ***L. mexicana* complex** • Miltefosine • Ketoconazole ***L. braziliensis* complex** • Pentavalent antimonials • Amphotericin B
Alternate therapy	Pentamidine 2–4 mg/kg/day, up to 15 doses IFNγ with pentavalent antimonials	Amphotericin 'B' deoxycholate infusion 1 mg/kg/day in 60–80 doses over 4 months Liposomal amphotericin B intravenously as 2.5 mg/kg/day for 20 days	Systemic therapy with testimonials and amphotericin B.	

Trypanosomiasis

Trypanosomes are limited to America and South Africa. Depending on the geographical location, the disease can present as the following entities:

- **African trypanosomiasis or sleeping sickness**
- **American trypanosomiasis or Chagas' disease**

African trypanosomiasis (sleeping sickness)

Trypanosomiasis is a chronic disease limited to the African continent.

Agent—*Trypanosoma* species

There are three species in the *Trypanosoma brucei* complex that cause disease.

- *T. brucei rhodesiense* causes East African sleeping sickness
- *T. brucei gambiense* causes West African sleeping sickness
- *T. brucei brucei* causes infection in cattle

Life cycle (Fig. 21.44)

Definitive host: Human being

Intermediate host: Tsetse fly (*Glossina* species)

- The infective **metacyclic trypomastigote forms** reside in the salivary glands of the tsetse fly and are inoculated in humans by the bite.
- After entering, the metacyclic trypomastigotes transform into long, slender forms, which divide by binary fission and change into short, stumpy forms.
- These invade the blood and reach various organs and are taken up by the tsetse fly during a blood meal to develop into metacyclic trypomastigotes once again.

Clinical features

- The disease begins as a self-limiting inflammatory lesion at the site of the bite, followed by an asymptomatic phase lasting for months.
- Subsequently, a systemic febrile illness is seen as **remittent irregular fever with night sweats** and lymphadenopathy. In the second stage, CNS invasion is seen.
- The patient develops **drowsiness in the daytime and insomnia and restlessness at night (hence the name sleeping sickness).**
- There is progressive meningoencephalitis leading to weakness, coma, and finally, death.

Laboratory diagnosis

The mainstay of diagnosis is the **detection of trypomastigote forms in the blood, CSF, or lymph node aspirates.**

Antigen detection in serum and CSF

- **Rapid tests and ELISA** to detect antibodies in serum and CSF; **antibody detection** is used widely in community screening of mass population and epidemiologic purposes.
- Newer molecular methods to detect parasitic DNA are also available.

Treatment

- **Pentamidine and suramine** are used for treatment.
- In the CNS stage, **eflorithine** may be prescribed.

American trypanosomiasis/Chagas' disease

Chagas' disease, also called American trypanosomiasis, is a potentially life-threatening illness caused by the protozoan parasite *Trypanosoma cruzi*.

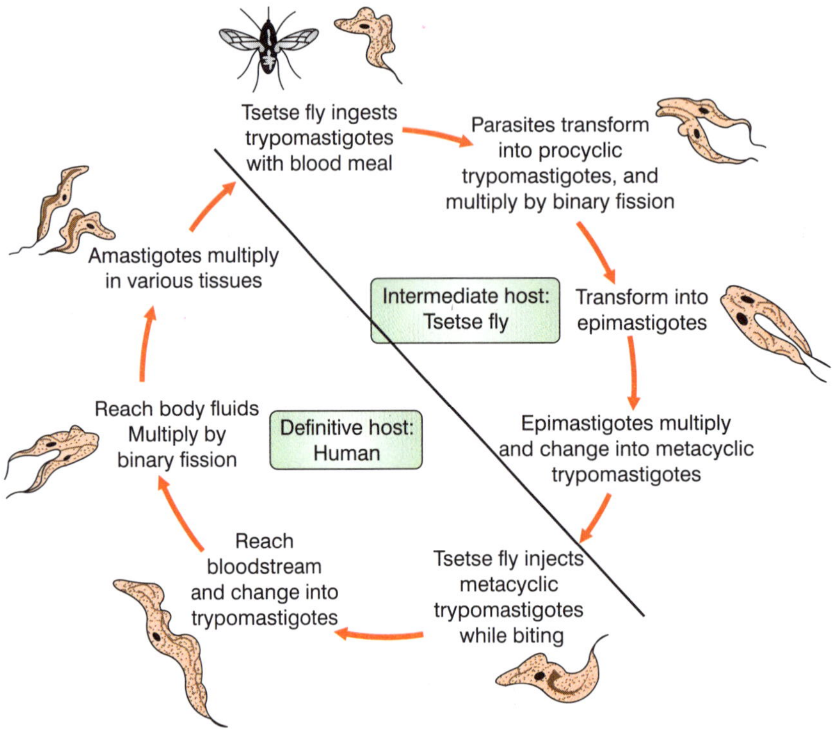

Fig. 21.44 Life cycle of the *Trypanosoma brucei* complex

Agent—*Trypanosoma* species

- *Trypanosoma cruzi* causes Chagas' disease or South American trypanosomiasis.
- *Trypanosoma rangeli* is non-pathogenic for humans.

Life cycle

Definitive host: Human being
Intermediate host: *Reduviid* bug (Fig. 21.45)

- The parasite exists in two forms—**amastigote and trypomastigote**.
- The bugs are nocturnal biters and take up blood meals containing trypomastigotes.
- These develop into epimastigotes, which are excreted in the bug's feces.
- From there, they **enter the human body through abrasions and bite wounds** on the skin during itching or rubbing.
- In humans, the metacyclic trypomastigotes enter tissues and transform into amastigote forms which transform into trypomastigotes in RE cells and back into amastigotes.
- The trypomastigotes are taken up by the insect vector in its blood meal.

Fig. 21.45 Triatomine bug, the vector of *Trypanosoma* cruzi, defecating on the wound after taking a blood meal (*Source*: CDC, DPDx)

Clinical features

Clinically, Chagas' disease develops in four stages.

1. **Early stage:** A subcutaneous nodule known as a **chagoma** is present at the site where the bug deposited its feces. The nodule is erythematous and painful and usually on the face. This resolves in 4–8 weeks. Unilateral swelling of the eyelid—the **Romana's sign**—may be present if the bug enters through the conjunctiva.
2. **Acute disease:** Non-specific symptoms like high fever, hepatosplenomegaly, and lymphadenopathy occur. Meningoencephalitis and acute myocarditis may also occur. The patient may succumb or recover within 4–8 weeks.
3. **Indeterminate stage:** The asymptomatic phase can last for years.
4. **Chronic stage:** The parasite invades the nerve plexuses in the gastrointestinal tract and cardiac muscles causing **megaesophagus and megacolon or features of cardiomyopathy**.

Laboratory diagnosis

- A **peripheral blood smear** is examined for amastigotes in the acute stage or in biopsy tissue of lymph node and subcutaneous nodule aspirate.

- The antibody can be detected in serum by **ELISA, ICT, or immunofluorescence assay.**
- **NNN medium** can be used for culture.

Treatment

- No effective treatment is known for Chagas' disease.
- Benznidazole has been used with some success.
- **Vector control** is more important than curing the disease.

Filariasis

Case 12. A 65-year-old man presents with a gradually increasing swelling of the left limb over many years, significant weakness, loss of appetite, and weight loss. On examination, he was found pale with left lower limb enlargement (Fig. 21.46). His inguinal lymph nodes were enlarged. A presumptive diagnosis of lymphatic filariasis was made.

i. What are the etiological agents of lymphatic filariasis?
ii. How is it diagnosed?

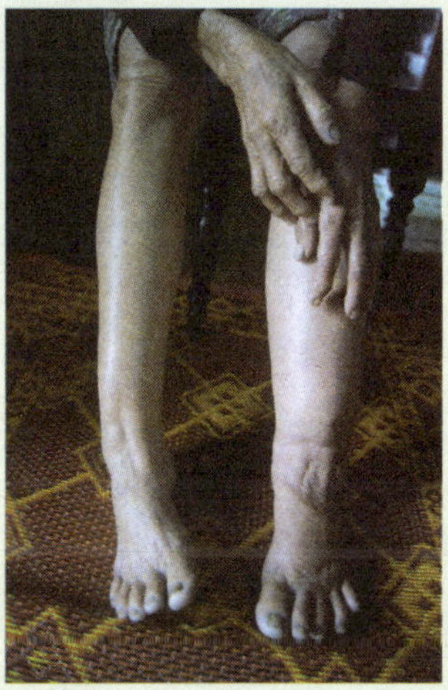

Fig. 21.46 Swelling in the limbs due to lymphatic filariasis (*Source*: CDC, PHIL, Image ID: 22458)

Filariasis is a chronic disease caused by a mosquito-borne parasitic worm, which results in the swelling of the extremities, hydroceles, and testicular masses.

Lymphatic filariasis

- It is caused by species of *Wuchereria bancrofti* (90% of the cases) and *Brugia malayi*.
- *Brugia timori* causes a disease localised to Timor Island and is uncommon in other parts of the world.
- It is reported from almost 50 countries in the tropical and subtropical belts.
- Lymphatic filariasis caused by *Wuchereria bancrofti* is confined to South-East Asia, India, China, the east coast of Africa, New Guinea, and some parts of South America.
- In India, more than 90% of cases are reported from nine states—Bihar, Uttar Pradesh, the coastal areas of Gujrat and Maharashtra, and coastal parts of Tamil Nadu, Andhra Pradesh, Orissa, and West Bengal.
- *B. malayi* is reported from South-East Asia, South India, and China.

Agent—Filarial nematodes

- **Filarial nematodes** cause blood and tissue infections.
- These are long, slender worms with tapering ends.
- The adult forms are found in tissues.
- Filarial worm infections can be divided into three groups based on the site of infections caused by them (Table 21.25).

Morphology

Adult (Fig. 21.47)

- Morphologically, adult worms of *Wuchereria* and *Brugia* are indistinguishable.
- The male and female adult worms are long, transparent, and filiform.
- The females are longer (8–10 cm) than the males (2.5 to 4 cm).
- Both ends taper with a slightly swollen head with two rows of 10 papillae each.
- Females are ovoviviparous and lay actively motile embryos.
- Male and females usually remain tightly coiled together in the lymph nodes.

Table 21.25 Clinical forms of filariasis and its causative agents

Infections	Causative agents
Lymphatic filariasis	*Wuchereria bancrofti*
	Brugia malayi
	Brugia timori
Subcutaneous filariasis	*Onchocerca volvulus*
	Loa loa
	Mansonella streptocerca
Serous cavity filariasis	*Mansonella ozzardi*
	M. perstans

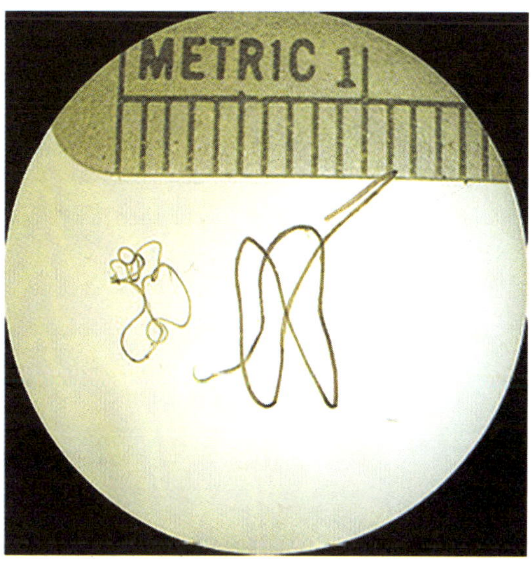

Fig. 21.47 Adults of *W. bancrofti* (*Source*: CDC, DPDx)

Larvae

- The microfilariae of various species are different.
- Their morphological features can be used to distinguish between species.
- They measure around 250 × 10 µm.
- The larvae retain a part of the chorionic envelope of the egg as a hyaline sheath and hence are called **sheathed microfilariae**. These sheathed larvae are seen in both *Wuchereria* and *Brugia*.
- The larvae have somatic cells known as nuclei which are distributed from the head to the tail end. In *Wuchereria bancrofti*, the tail tip is free of nuclei while in *Brugia malayi*, the tail has two nuclei—one at the extreme end and a second one just above the tip (Table 21.26).

Life cycle (Fig. 21.48)

Definitive host: Human being

Intermediate host: Mosquitoes (*Culex, Anopheles,* and *Aedes*)—the *Culex* mosquito is the vector in urban and semi-urban areas, whereas *Anopheles* plays this role in rural areas; *Aedes* is found mainly in the Pacific Islands

- Mosquitoes ingest microfilariae when they bite an infected host.
- Ex-sheathing of the larval stage occurs, and microfilariae undergo three moultings to mature into actively motile third stage larvae, which are the infecting forms (Fig. 21.49).

The characteristics of the worms that cause lymphatic filariasis are presented in Table 21.26.

Pathogenesis

- The worm causes lymphatic obstruction and inflammatory damage to the lymphatics.
- Antigen and toxic metabolites released by worms cause the infiltration of plasma cells, eosinophils, and macrophages, leading to valvular damage in the lymphatic vessels and resulting in lymphedema.
- After the death of worm, increased granulomatous reaction causes blockage and fibrosis of the lymph vessels leading to severe lymphatic obstruction.
- There is alteration of both CMI and humoral immunity.
- There is an increase in eosinophil count and IgE levels.

Clinical features

The incubation period is 8–16 months but can be longer. Clinically, the disease can present in the following forms.

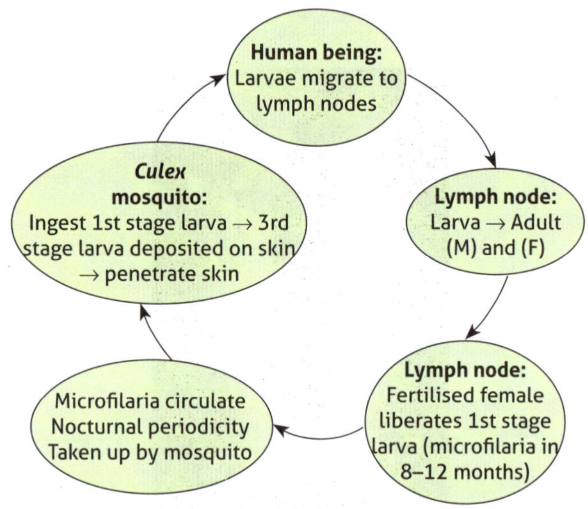

Fig. 21.48 Schematic diagram of the life cycle of *W. bancrofti*/*B. malayi*

Table 21.26 Characteristics of filarial worms causing lymphatic filariasis

Characteristics	*W. bancrofti*	*B. malayi*	*B. timori*
Vector	Worldwide—*Culex* Pacific islands—*Aedes*	*Mansonia, Anopheles*	*Anopheles*
Geographic distribution	South-East Asia, Africa, South America, Pacific Islands	South-East Asia	Indonesia
Microfilarial found in	Blood, may be seen in hydrocele fluid or urine	Blood	Blood
Microfilaria morphology	Sheathed	Sheathed	Sheathed
Somatic cell/nuclei	Coarse nuclei; well-separated	Large, coarse, overlapping nuclei	Large, coarse, overlapping nuclei
Tail end	No nuclei at tail tip	Two spaced nuclei at tail tip	Two spaced nuclei at tail tip
Diagram	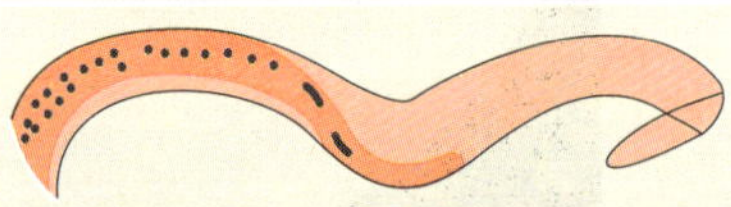 Tail end of microfilaria of *W. bancrofti*	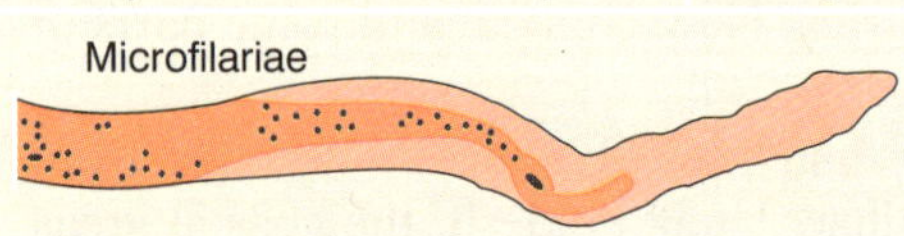 Tail end of microfilaria of *Brugia*	

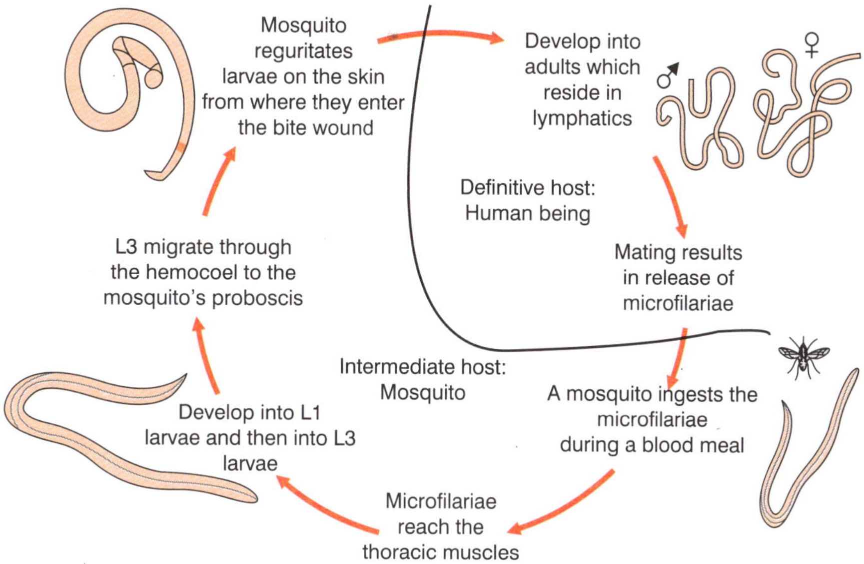

Fig. 21.49 The life cycle of filarial worms

- **Asymptomatic lymphatic filariasis:** Investigations may reveal microscopic hematuria, proteinuria, or microfilaria in peripheral blood.
- **Acute infection:** It is marked by high fever, lymphatic inflammation, and transient local edema. Upon examination, the regional lymph nodes are enlarged, and lymphatic channels appear thick and indurated. Scrotal tenderness and pain may be present.
- **Chronic infection:** Lymphatic damage over 10–15 years progressively leads to lymphangitis, lymphedema, and other complications. Severe lymphatic obstruction causes edema which progresses to fibrosis, chronic lymphedema, and elephantiasis. Initially, there is allergic lymphangitis with lymphadenitis, which progresses to dilatation of the lymphatic vessels, causing lymph retention. The patient may present with the following symptoms:
  - Lymph scrotum
  - Lymphuria or chyluria
  - **Hydrocele** is a common manifestation with the accumulation of a clear or straw-coloured fluid in the cavity of the tunica vaginalis of the testes
  - **Elephantiasis** occurs as a delayed sequel and commonly affects the lower limbs, genital organs, and breasts

Occult filariasis (tropical pulmonary eosinophilia)

- It is caused by a **hypersensitivity reaction to microfilarial antigens**.
- The patient presents with **dry cough, wheezing, fever, and loss of appetite**.
- Diffuse interstitial infiltrates are seen on chest X-ray with increased bronchoalveolar markings.
- Eosinophils counts can be raised up to 5,000 per mm^3 of blood, and IgE levels are elevated.

Laboratory diagnosis (Fig. 21.50)

- **Samples**: **Peripheral blood** is the best specimen, best collected at night or after administering a dose of diethylcarbamazine (DEC).
  - **DEC/hetrazan provocation test**—DEC is given in a dose of 2 mg/kg body weight to stimulate nocturnal microfilariae to appear in peripheral blood. A sample is collected after 30–60 minutes of giving the medicine.
  - **Urine or hydrocele fluid** is collected for microscopy or antigen detection.
- **Microscopy:**
  - Direct wet mount of **EDTA blood** shows **motile microfilariae**.
  - **Thick and thin films** are made and stained with Giemsa/Leishman or polychrome methylene blue. This method is inexpensive, easy, and feasible at individual and community lcvcls. It can also bc used to differentiate various species.

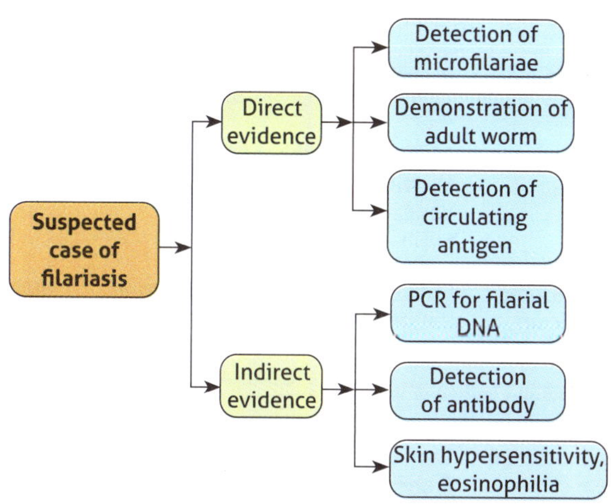

Fig. 21.50 Algorithm of the approach to the laboratory diagnosis of filariasis

- **QBC:** Acridine orange is used to stain the microfilariae after centrifuging blood in capillary tubes. It is then examined under a fluorescent microscope. Microfilariae are readily visible and seen close to the buffy coat layer.
- **Detection of circulating antigens:**
  - **ELISA** and **immunochromatographic (ICT) lateral flow assay** are available to detect the filarial antigen of *Wuchereria* (Og4C3 ELISA). These modalities are not yet available for *Brugia*.
  - Rapid point-of-care tests are more sensitive than microscopy and are used in filarial elimination programmes as a screening tool.
  - The antigen disappears after treatment, at which point, these tests are not effective.
  - Antigen detection can also be performed on urine samples.
- **Biopsy of lymph nodes:** The adult worm can be demonstrated in biopsy samples, but this modality is rarely used for diagnosis.
- **Molecular methods:** PCR for the detection of both types of filarial DNA are available.
- **Detection of antibody**
  - Antibody detection has little role in diagnosis in endemic areas.
  - It is used for **sero-epidemiological studies** or **for the detection of past infection** in non-endemic areas.
  - IgG antibody detection of bancroftian filariasis uses the **WbSXP-1 antigen**.
  - A *Brugia* rapid test, which uses recombinant Bm14 antigen to detect IgG antibodies, is also available.
- **Other non-specific diagnostic tests**
  - Skin hypersensitivity reaction
  - Blood tests—eosinophilia and elevated IgE are present; calcified worms may be seen on X-rays and moving larvae may be seen on ultrasound examination

Treatment

- The treatment of lymphatic filariasis does not depend on the type of worm causing the infection.
- **Diethylcarbamazine (DEC)** is the preferred drug and is given in doses of 6 mg/kg body weight for 12 days.
- **Albendazole** is given as 400 mg twice daily for 21 days.
- Both DEC and albendazole kill the adult as well as the larval form of the parasite.
- **Ivermectin (400 mg as a single dose)** has no effect on the adult worm but kills microfilariae. It is not used in India.

Prevention

- Under the National Vector-Borne Control Programme, the strategy employed is mass drug administration (MDA) along with vector control in endemic areas.
- DEC and albendazole tablets are administered annually.
- MDA repeated over 5–6 years eliminates microfilariae from blood so that mosquitoes cannot take them up in their blood meal.

Schistosomiasis

- Schistosomiasis is caused by **schistosomes**. These parasites are known as **blood flukes** as they reside in the venous plexus of the intestine and bladder of humans.
- Schistosomiasis is common in tropical and subtropical areas.
- The WHO estimates that more than 90% of cases of the disease are from Africa.
- It is the **second most common devastating parasitic disease after malaria**.
- Five species of schistosomes cause disease in humans (Table 21.27).

Table 21.27 Blood flukes pathogenic to humans

Species	Distribution	Habitat	Definitive host	Intermediate host	Eggs	Disease caused
Schistosoma mansoni	Africa, the Caribbean, Middle East, Brazil, Venezuela, and Suriname	Intestine	Human being, monkey, dog	Snail	Elliptical with sharp, lateral spine Eggshell is acid-fast	Intestinal and hepatosplenic disease
Schistosoma japonicum	China, Indonesia, Philippines	Intestine	Human being, dog, cat, rodents	Snail	Oval to round with rudimentary knob Eggshell is acid-fast	Intestinal, hepatosplenic disease and brain involvement
Schistosoma haematobium	Africa, the Middle East, Corsica, in France	Venous plexus of urinary bladder and ureter	Human being, monkey, chimpanzee	Snail	Elliptical with a sharp, terminal spine with larva Eggshell is not acid-fast	Urogenital disease (bilharziasis) Bladder carcinoma

Agent—*Schistosoma*

Morphology (Fig. 21.51a–d)

Schistosomes are similar in morphology to trematodes but show certain differences.

Adult

- The adults are cylindrical with a thick syncytial tegument.
- The suckers have delicate spines.
- The sexes are separate.
- Males measure 12–15 mm; female worms are slightly longer.
- Male and female worms coexist in the venous plexus.

Egg

- The eggs are non-operculated with a projection or spine.
- The size and position of the spine vary across species.

Larva

- Three larval forms are seen—**miracidium, sporocyst, and cercaria**.
- The infective form is the cercaria larva, which is oval and elongated with a bifurcated tail.
- The larva lives for about 24–72 hours.

Life cycle

The life cycle of schistosomes is detailed in Fig. 21.52.

Clinical features

Acute schistosomiasis

- **Cercarial dermatitis:** In acute conditions, the symptoms are related to the skin. The cercaria penetrates the skin, causing a maculopapular, itchy rash. This is seen in all *Schistosoma* infections and is known as the swimmer's itch.
- **Katayama syndrome:** This is seen in *S. mansoni* and *S. japonicum* infections 4–8 weeks after the infection. Presenting complaints include nocturnal fever, headache, cough, myalgia, and abdominal tenderness. Lymphadenopathy and hepatosplenomegaly may be present.

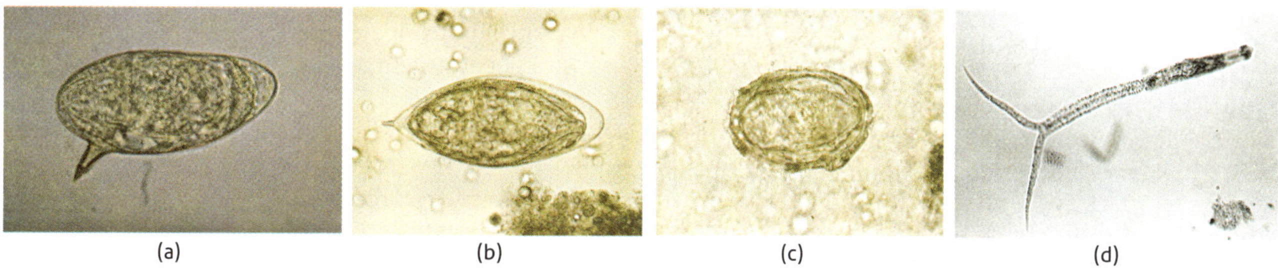

Fig. 21.51 *Schistosoma:* (a) egg of *S. mansoni*, (b) egg of *S. haematobium*, (c) egg of *S. japonicum*, and (d) cercaria larva (*Source*: CDC, PHIL, Image IDs 4841, 4842, 4843, and 8556)

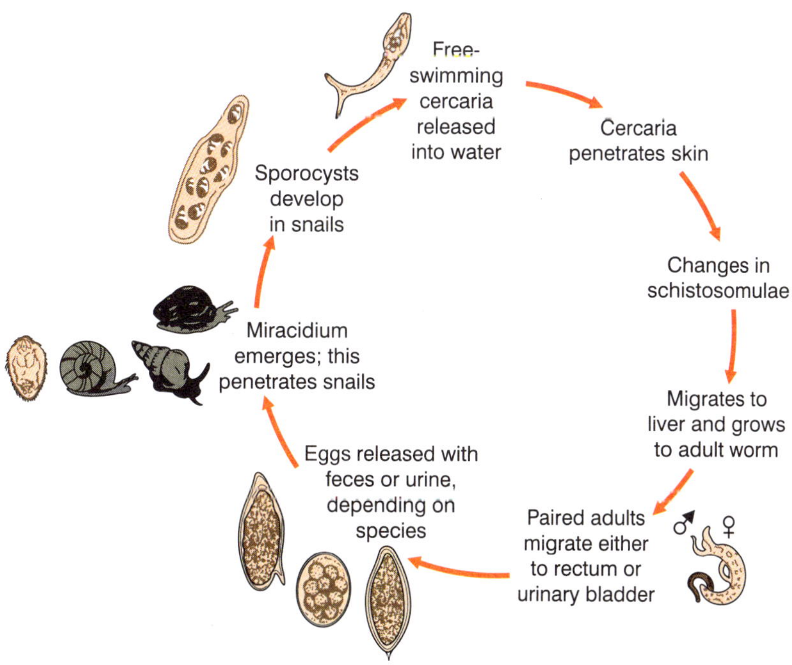

Fig. 21.52 Life cycle of schistosomes

Laboratory diagnosis

- Eosinophilia
- Chest X-ray shows diffuse pulmonary infiltrates
- Serum contains antibodies to *Schistosoma*
- Schistosome egg in feces

Treatment

- The drug of choice is praziquantel with or without steroids.
- **Artemisinin treatment** after exposure to water may decrease the risk of contracting the disease.

Chronic schistosomiasis

The features of the chronic stage vary with regard to the worm and depend on the site involved. The worms themselves do not cause much damage; the clinical features are chiefly due to tissue reaction:

- Reaction to mucosal injury
- Allergic reaction to eggs
- Granuloma due to eggs, which may be carried to sites like the spinal cord, brain, liver, lungs, and intestine, causing granuloma formation
- Complications such as nephrosclerosis, liver abscess, pulmonary hypertension, or encephalitis may occur

Hepatosplenic schistosomiasis

This condition is caused by chronic infection with *Schistosoma* parasites residing in the mesenteric plexus.

Laboratory diagnosis

- **Samples:** Urine, stool, biopsy specimen, and blood are collected.
- **Microscopy:** The eggs of *S. haematobium* can be detected in urine or bladder biopsy. The eggs of *Schistosoma* infecting the intestines can be detected in stool samples and intestinal biopsy. All eggs except those of *Schistosoma haematobium* are acid-fast.
- **Serology:**
  - *Antigen detection*: ELISA-based assays to detect circulating *Schistosoma* antigens (CCA and CAA) are available. Dipstick assays are also used to detect the antigen in urine.
  - *Antibody detection*: Antibodies to the *Schistosoma haematobium* **adult microsomal antigen (HAMA)** can be detected by ELISA and enzyme-linked immunetransfer blot assay. Elevated IgE and IgG may be present.

Treatment and control

- The drug of choice is **praziquantel**.
- Alternately, metrifonate can be used.
- Control measures include the correct disposal of human feces, eradication of intermediate hosts (snails) from water bodies, and proper treatment of infected persons.

ANEMIA DUE TO INFECTIOUS DISEASES

The World Health Organization defines anemia as a reduced number of red blood cells or lower than normal hemoglobin concentration in the cells, resulting in an inability carry an optimum concentration of oxygen to tissues. When this capacity reduces, symptoms such as fatigue, weakness, dizziness, and shortness of breath and other complications occur. The most common causes of anemia include the following:

- Nutritional deficiencies, particularly iron deficiency, though deficiencies in folate and vitamins B12 and A are also important causes
- Hemoglobinopathies
- Infectious diseases

Infections

The infectious causes of anemia are presented in Table 21.28.

Pathogenesis of Anemia in Infections

Anemia occurs due to various mechanisms:

- Some microorganisms affect the reticuloendothelial system and impair the release of iron by the endothelial system.
- Shortened red cell survival creates an increased demand for red cell production by the bone marrow, but the marrow cannot respond adequately because of impaired erythropoietin (EPO) production.
- The ability of the erythroid progenitors to respond to EPO is impaired.
- There is impaired mobilisation of reticuloendothelial system iron stores.
- Inflammatory mediators such as IL-1 and TNF decrease red cell survival.
- There is an inhibition of erythropoiesis by interferons.
- There is hemolysis with the destruction of RBCs (e.g., malaria).

Table 21.28 Microorganisms causing anemia in various infective syndromes

Group	Organisms
Bacteria	*Mycobacterium tuberculosis, Helicobacter pylori, Bartonella bacilliformis*
	Endotoxemia, exotoxins (hemolytic toxin of *C. perfringens*)
	Any bacteria causing severe sepsis may lead to anemia
Parasites	*Plasmodium* (*P. falciparum* causes blackwater fever), *Babesia, Leishmania*
	Hookworm
Virus	HIV (dysregulation and red cell destruction and ineffective production of RBC), Epstein–Barr virus, parvovirus B19

- Anemia may develop secondary to the inhibition of erythropoiesis (e.g., parvovirus).
- Anemia may develop due to iron deficiency (e.g., *H. pylori*).
- There is blood loss at intestinal attachment (e.g., hookworm).
- Chronic infection due to loss of appetite leading to malnutrition (e.g., tuberculosis).
- Immune-mediated lysis of RBCs occurs in some reactions (e.g., *Treponema pallidum*).
- Microangiopathic hemolysis/thrombotic thrombocytopenic purpura (e.g., HIV).
- Hemophagocytosis occurs in some infections (e.g., *Leishmania donovani*).
- Anemia may be a result of direct infection of hematopoietic progenitors (e.g., Epstein Barr Virus).

The pathogenesis and diagnosis of these infections are discussed in their specific sections.

Questions

Case review

1. A 10-year-old boy was admitted with complaints of loss of appetite, pain in the abdomen, and general weakness. On examination, he was found to have a temperature of 102°F with pallor and coated tongue, mild splenomegaly, and abdominal tenderness. His blood culture was positive for *Salmonella Typhi* with a negative Widal test.
   i. Give reasons for the negative Widal test.
   ii. How will you choose the appropriate antibiotic to treat this patient?
   iii. What are the complications of this infection?
2. A migrant worker from the Middle East came back to India on vacation. A few days later, he visited the outpatient department with low back, joint, and muscular pains. He also complained of on-and-off fever for the past two months, exhaustion and night sweats, which often left him drenched. He had a history of tending to goats and camels as part of his job. On examination, there was tenderness on his sacral prominence, but no other specific abnormalities could be detected.
   i. What are the important clues in this case that would help you narrow down your provisional diagnosis?
   ii. List the investigations you would order to arrive at a diagnosis?
   iii. Suggest the appropriate treatment and preventive measures you would prescribe to this patient.

Short answer questions

1. Discuss the association of Gr A streptococci with acute rheumatic fever.
2. Describe the etiopathogenesis of infective endocarditis.
3. Enumerate common viral hemorrhagic fevers prevalent in India and discuss the pathogenesis of DHF.
4. Draw up an algorithm of the microbiological investigations to diagnose FUO.
5. Describe the various laboratory methods to confirm malaria during the acute phase of the illness.

MCQs

1. **A nine-year-old girl presented with a fever and sore throat. Group A streptococci were cultured from her throat. Three weeks later, she developed pain and tender swelling in both her knees and ankles; a palpable effusion of the right knee was drained, which yielded straw- coloured fluid. Culture of the aspirate was negative. Subsequently, the girl began experiencing shortness of breath. Which of the following is a minor Jones diagnostic criteria used by the pediatrician to diagnose acute rheumatic fever?**
   a) Carditis b) Chorea
   c) Subcutaneous nodules d) Prolonged PR interval
2. **Which of the following is a non-suppurative complication of *Streptococcus pyogenes* infection?**
   a) Pharyngitis b) Impetigo
   c) Necrotising fasciitis d) Acute rheumatic fever
3. **A suspected case of endocarditis was admitted to the ICU. The consultant was asked to send the suitable sample for culture. What is the recommended protocol in this case?**
   a) Three blood samples from three different sites
   b) One blood sample during the peak of fever
   c) One blood and one urine sample
   d) Peripheral blood for smear, CRP, and ESR
4. **Classically, fever of unknown origin (FUO) is defined as fever in a patient who exhibits:**
   a) Temperature >101°F on several occasions for more than three weeks with no significant findings in one week of in-patient investigations (blood/urine and sputum culture for common bacteria, serology, and complete blood count).
   b) Temperature >101°F and evaluation for at least three days, at which time, the patient has a normal neutrophil count.
   c) Temperature >101°F on several occasions for less than three weeks with no significant findings on investigation (blood/urine and sputum culture for common bacteria, serology and complete blood count).
   d) Temperature <101°F on several occasions for more than three weeks with no significant findings in one week of in-patient investigations (blood/urine and sputum culture for common bacteria, serology and complete blood count).
5. **Septicemia refers to the presence of :**
   a) Bacteria in the blood
   b) Parasites in the blood
   c) Virus in blood
   d) Microbial antigens and toxins in the blood
6. **Which of the following statement is not correct regarding Duke's criteria?**
   a) It is used to diagnose rheumatic heart disease.
   b) It requires evidence of endocardial involvement by ECHO.
   c) It is used to diagnose infective endocarditis.
   d) It includes fever as a minor criterion.
7. **Penicillin resistance in *Staphylococcus aureus* due to penicillinase production is acquired by:**
   a) Transposition b) Mutation
   c) Transduction d) Transformation
8. **Methicillin resistance in staphylococci is due to:**
   a) Mutation
   b) Increase in cell wall thickness
   c) Alteration in penicillin-binding protein
   d) Tolerance to the drug
9. **D-test is used to detect which drug resistance in *Staphylococcus aureus*?**
   a) Vancomycin b) Clindamycin
   c) Ciprofloxacin d) Penicillin

10. **Which of the following is true of community-acquired MRSA?**
    a) Has SCC type ll
    b) Less transmissible
    c) Produces Panton-Valentine toxin
    d) Less virulent
11. **Which of the following is true of toxic shock syndrome toxin?**
    a) Produced by *Staphylococcus aureus* strains belonging to bacteriophage type II
    b) Is a super antigen
    c) Causes toxic shock syndrome only in females
    d) Anti-TSST antibodies are not protective
12. **Pitting type of growth is typical of:**
    a) *Kingella kingae* b) *Eikenella corrodens*
    c) *Moraxella catarrhalis* b) *Aggregatibacter* spp.
13. **HACEK group includes all the following except:**
    a) *Haemophilus* species b) *Aggregatibacter* species
    c) *Cardiobacterium* species d) *Elizabethkingia* species
14. **A typhoid carrier is best diagnosed by:**
    a) Detection of Vi antigen b) Blood culture
    c) Clot culture d) Stool culture
15. **A constant finding in enteric fever is:**
    a) Stepladder fever b) Rose spots
    c) Soft, palpable spleen d) Hepatomegaly
16. **Which of the following is associated with stepladder pyrexia?**
    a) Enteric fever b) Dengue fever
    c) Malaria d) Septecemia
17. **The culture medium used for isolating *Salmonella* from sewage is:**
    a) MacConkey agar b) Blood agar
    c) XLD d) Wilson and Blair medium
18. **The schedule for the live oral vaccine for typhoid, typhoral is:**
    a) Doses on days 1, 3, and 5
    b) Doses on days 1, 2, and 3
    c) Doses on days 1, 2, and 4
    d) Doses on days 1, 7, and 14
19. **All of the following are true of the Widal test except:**
    a) In endemic places, the baseline titre should be known to interpret the Widal test.
    b) In typhoid fever TH, AH, and BH titre will be elevated .
    c) Anamnestic response results in false-positive Widal test.
    d) Antibodies appear first following infection.
20. **In which organ does *Salmonella* Typhi exist in carriers?**
    a) Blood b) Kidney
    c) Gallbladder d) Spleen
21. **A 45-year-old presented with fever that he had for the past 12 days. He was suspected of having typhoid fever. Which of the following investigations is recommended in this case?**
    a) Blood culture b) Widal test
    c) Urine culture d) Stool culture
22. **Stalactite growth of *Yersinia pestis* will be seen in:**
    a) Nutrient broth b) Mueller-Hinton broth
    c) BHIB d) Ghee broth
23. **Which bacterium exhibits safety pin appearance on Gram stain?**
    a) *Klebsiella pneumoniae* b) *Proteus mirabilis*
    c) *Yersinia pestis* d) *Escherichia coli*
24. **The invariably fatal type of plague is:**
    a) Bubonic b) Pneumonic
    c) Septicemic d) All of these
25. **All the following are clinical types of plague except:**
    a) Bubonic b) Septicemic
    c) Meningitic d) Pneumonic
26. **Vectors of plague are:**
    a) Mosquito b) Flea
    c) Sandfly d) Housefly
27. ***Yersinia pestis* requires biosafety level:**
    a) I b) II
    c) III d) IV
28. **The optimum temperature for growth of plague bacilli is:**
    a) 35°C b) 37°C
    c) 27°C d) 45°C
29. **All the following are cultural characteristics of *Yersinia* except:**
    a) Pink colonies on MacConkey agar
    b) Dark brown colonies on blood agar
    c) Stalactite growth in ghee broth
    d) Dark brown colonies on chocolate agar
30. **Which of the following bacteria do not exhibit bipolar staining?**
    a) *Pasteurella* b) *Yersinia*
    c) *Francisella* d) *Legionella*
31. **Virulence factors of *Yersinia pestis* include all except:**
    a) Protein envelope antigen b) Hemolysin
    c) Pesticin I d) Endotoxin
32. **The causative agent of plague is:**
    a) *Yersinia pestis* b) *Yersinia pseudotuberculosis*
    c) *Pasteurella multocida* d) *Bordetella brochoseptica*
33. **Which of the following is true about killed vaccines for plague?**
    a) It is effective against pneumonic plague.
    b) It is effective against bubonic plague.
    c) The protection lasts lifelong.
    d) It is given intramuscularly
34. **Which of the following is used in the chemoprophylaxis for plague?**
    a) Cotrimoxazole b) Tetracycline
    c) Both (a) and (b) d) No antibiotic is effective
35. **A 45-year-old woman who worked in a rice farm where she would stand in water during planting season, developed fever off and on and did not seek medical advice. After two weeks, she presented with suffused conjunctiva, jaundice, and continuous fever. Her sera tested positive by ELISA for IgM antibodies to *Leptospira* antigens. Which of the following antibiotics is she likely to respond to?**
    a) Penicillin b) Vancomycin
    c) Gentamicin d) Ciprofloxacin
36. **Which of the following diseases is not related to *Leptospira*?**
    a) Swamp fever b) Five-day fever
    c) Pretibial fever d) Fort Bragg fever
37. **The false statement about leptospirosis is:**
    a) Incidence rises following floods and waterlogging in urban areas.
    b) *Leptospira* survive in convoluted tubules of the kidneys in natural hosts and is shed in urine.
    c) *Leptospira* enter the body by drinking contaminated water.
    d) Doxycycline 200 mg orally once a week is effective in prophylaxis
38. **Which of the following is included in Faine's criteria to diagnose leptospirosis?**
    a) Culture of *Leptospira* b) Dark-ground microscopy
    c) IgM ELISA d) PCR
39. **Which of the following is used for the culture of *Leptospira*?**
    a) Blood agar b) LJ medium
    c) Lysed horse blood agar d) Korthof's media
40. **Which of the following is a non-pathogenic *Leptospira*?**
    a) *L. icterohaemorrhagiae* b) *L. interrogans*
    c) *L. copenhageni* d) *L. biflexa*
41. **Brucellae are:**
    a) Motile, gram-negative, coccobacilli
    b) Non-motile, gram-negative coccobacilli
    c) Non-motile, gram-positive coccobacilli
    d) Motile, fermentative, gram-negative coccobacilli
42. **The term used to describe the cyclic rise and fall of body temperature associated with brucellosis is:**
    a) Relapsing fever b) Stepladder fever
    c) Undulant fever d) Shanghai fever

43. Modes of transmission of *Brucella* include all the following except:
a) Ingestion of contaminated raw vegetables
b) Inhalation of infected dust
c) Accidental inoculation
d) Person-to-person transmission

44. The factor that stimulates the growth and multiplication of *Brucella* is:
a) Erythritol b) Pyridoxal
c) Ferritin d) Interleukin-2

45. Brucellae have a special predilection for:
a) Brain b) Testes
c) Eye d) Placenta

46. 2-mercaptoethanol test is used for the detection of:
a) IgM b) IgG
c) Both IgM and IgG d) IgA

47. One rapid method used for diagnosing brucellosis in cattle is:
a) Milk ring test b) Whey agglutination test
c) Rose Bengal card test d) ELISA

48. Brucellosis can be treated with all the following drugs except:
a) Streptomycin b) Erythromycin
c) Doxycycline d) Rifampicin

49. A 56-year-old farmer from India presented with on and off fever and joint pains. He was closely associated with animals and gave a history of consuming raw milk. He subsequently developed an abscess in the right iliac fossa. Both his blood and pus culture grew gram-negative bacilli. What is the likely causative agent?
a) *Mycobacterium tuberculosis*
b) *Brucella melitensis*
c) *Mycobacterium haemophilum*
d) *Mycobacterium bovis*

50. The blood culture of a veterinary worker grew short gram-negative coccobacilli. He suffered from chronic fever, malaise, and joint pains. What is the likely etiological agent in this case?
a) *Mycobacterium tuberculosis*
b) *Brucella* species
c) *Listeria monocytogenes*
d) *Kingella* species

51. Eschar is a pathognomonic finding in:
a) Spotted fever b) Typhus fever
c) Q fever d) Scrub typhus

52. Q fever is not transmitted by:
a) Tick bite
b) Handling animal wool or hide
c) Drinking milk of infected animals
d) Aerosols

53. A patient with high fever shows a strongly positive agglutination with OXK antigen. The probable diagnosis is:
a) Endemic typhus b) Epidemic typhus
c) Spotted fever d) Scrub typhus

54. Oroya fever is caused by:
a) *Bartonella henselae* b) *Bartonella bacilliformis*
c) *Coxiella burnetti* d) *Bartonella quintana*

55. Verruga peruana is caused by:
a) *Bartonella henselae* b) *Bartonella quintana*
c) *Bartonella bacilliformis* d) *Coxiella burnetii*

56. Genus *Coxiella* differs from rickettsiae in which of the following features?
a) No vector for transmission
b) No obligate intracellular parasitism
c) Does not have a peptidoglycan cell wall
d) Does not have both RNA and DNA

57. A 36-year-old woman presented with fever, headache, and rashes since the last four days. Her platelet count was 25,000/mm^3 and the tourniquet test was positive. What is the disease she is likely to be suffering from?
a) Yellow fever b) Chikungunya
c) Dengue fever d) Kyasanur Forest disease

58. Crimean Congo virus belongs to the genus:
a) Flavivirus b) Alphavirus
c) Hantavirus d) Nairovirus

59. Break-bone fever is the common name for:
a) Chikungunya b) Dengue fever
c) Yellow fever d) Japanese encephalitis

60. *Haemophysalis spinigera* ticks can transmit:
a) Venezuelan equine encephalitis
b) Western equine encephalitis
c) Kyasanur Forest disease
d) Japanese encephalitis

61. Chikungunya virus belongs to the family:
a) Flaviviridae b) Reoviridae
c) Togaviridae d) Filoviridae

62. The vector that transmits Kyasanur Forest disease is:
a) Tick b) Mosquito
c) Sandfly d) Rodent

63. Ebola virus belongs to the family:
a) Flaviviridae b) Bunyaviridae
c) Reoviridae d) Filoviridae

64. The name 'Ebola' virus is associated with:
a) The name of the first person who succumbed to the disease
b) The scientist who first demonstrated the virus in the blood of patients
c) The river along the side of which the first few cases were reported
d) The city in South America where the initial case was reported

65. Which of the following viruses causes rodent-borne hemorrhagic fever?
a) Lassa fever virus b) Dengue virus
c) West Nile virus d) Western equine encephalitis

66. Reynold–Braude phenomenon is shown by:
a) *Candida albicans* b) *Cryptococcus neoformans*
c) *Candida auris* d) *Rhodotorula* species

67. Which test is used to confirm *Candida albicans* in culture?
a) Formation of chlamydospores on cornmeal agar at 37°C
b) Formation of chlamydospores on Sabouroud's dextrose agar at 37°C
c) Formation of chlamydospores on Sabouroud's dextrose agar at 20°C
d) Formation of chlamydospores on cornmeal agar at 20°C

68. A central reserve police officer posted in a forest in Jharkhand presented with fever, chills, and rigors. His blood picture showed neutropenia, and his peripheral smear was positive for multiple ring forms in the RBCs. Ophthalmologic examination revealed retinal whitening, and CT revealed a mass in the bilateral thalami. Which malarial species can give rise to such a picture?
a) *P. malariae* b) *P. falciparum*
c) *P. ovale* d) *P. vivax*

69. Malaria is naturally transmitted by the bite of which female mosquito?
a) *Culex* species b) *Anopheles* species
c) *Aedes* species d) Any of the above

70. Which species of *Plasmodium* causes zoonotic malaria?
a) *P. vivax* b) *P. falciparum*
c) *P. nowlesi* d) *P. ovale*

71. Schuffner's dots are seen in
a) *P. vivax* b) *P. falciparum*
c) *P. malariae* d) *P. knowlesii*

72. Maurer's dots are present in:
a) *P. vivax* b) *P. falciparum*
c) *P. malariae* d) *P. ovale*

73. The infective forms in malaria are:
a) Sporozoites b) Merozoites
c) Microgametocytes d) Macrogametocytes

74. The definitive host in malaria is:
a) Human being b) Mosquito
c) Sandfly d) Tick

75. A 40-year-old woman presented to the emergency room with a 3-day history of high-grade fevers, chills, abdominal pain, and generalised muscle aches. After a few hours, she became anuric and hypotensive and stopped responding to intravenous fluid resuscitation. What is the likely finding to be seen in peripheral thick film smear?
a) Ring forms of *Plasmodium falciparum*
b) Ring forms of *Plasmodium vivax*
c) Ring forms of *P. malariae*
d) No finding

76. The condition in previous question is known as:
a) Benign tertian malaria b) Malignant tertian malaria
c) Algid malaria d) Uncomplicated malaria

77. Which of the following statement is not true about relapse of malaria?
a) It is seen in *P. vivax* infection
b) It can occur even after 3 years of malaria
c) It is seen due to the persistence of parasites in the capillaries of internal organs
d) It occurs due to the reactivation of hypnozoites

78. *Plasmodium* LDH test can detect:
a) All *Plasmodium* species b) Only *P. falciparum*
c) Only *P. vivax* d) Both *P. vivax* and *P. falciparum*

79. As per the National Vector-Borne Disease Control Programme, the drug of choice to treat uncomplicated *P. falciparum* malaria is:
a) Chloroquine plus primaquine
b) Doxycycline plus sulphadoxine
c) Mefloquine
d) Artesunate plus sulphadoxine

80. Which of the following is true regarding old world leishmaniasis?
a) It is seen only in Central America.
b) The clinical presentation is limited to skin lesions only.
c) The vector is *Phlebotomus argentipes.*
d) All of the above

81. In India, the majority of kala-azar cases are reported from:
a) The southern part of the country
b) Jammu and Kashmir
c) Rajasthan and Maharashtra
d) Bihar and Jharkhand

82. The amastigote form develops into promastigote forms in insect body within:
a) One week b) 12 hours
c) 24 hours d) 48 hours

83. Amastigote forms are seen in:
a) The body of insects b) NNN medium
c) The human body d) Any of the above

84. Which of the following best describes the promastigote form?
a) It is spindle-shaped, flagellated and motile form.
b) It is commonly seen in the human body.
c) It is also known as the Leishman–Donovan (LD) body.
d) It has a delicate cell membrane with a nucleus, kinetoplast, axoneme, and vacuole.

85. Which of the following drug is not used to treat visceral leishmaniasis?
a) Liposomal amphotericin B b) Miltefosine
c) Ivermectin d) Pentavalent antimonials

86. *Trypanosoma cruzi* causes:
a) Chagas' disease or South American trypanosomiasis
b) East African sleeping sickness
c) West African sleeping sickness
d) No human disease

87. Chagoma is a pathognomonic finding seen in:
a) Visceral leishmaniasis
b) Post-kala-azar dermal leishmaniasis
c) South American trypanosomiasis
d) African trypanosomiasis

88. The drug of choice to treat trypanosomiasis is:
a) Liposomal amphotericin B
b) Mebendazole
c) Ivermectin
d) Pentamidine

89. Maltese cross appearance is seen in:
a) Babesiosis b) Malaria
c) Trypanosomiasis d) Leishmaniasis

90. Lymphatic filariasis is caused by:
a) *Onchocerca volvulus* b) *Loa loa*
c) *Brugia timori* d) *Mansonella streptocerca*

91. Which of the following is not a characteristic feature of occult filariasis?
a) It is a hypersensitivity reaction to microfilarial antigens.
b) The patient presents with dry cough, wheezing, fever, and loss of appetite.
c) Diffuse interstitial infiltrates are seen on chest X-ray with increased bronchoalveolar markings.
d) Eosinophil counts are normal.

92. Which of the following drug is capable of killing microfilariae but has no effect on the adult worm?
a) DEC b) Albendazole
c) Ivermectin d) None of the above

93. The egg of *Schistosoma mansoni* is:
a) Oval to round with a rudimentary knob
b) Elliptical with a sharp terminal spine with larva
c) Elliptical with a sharp lateral spine
d) Oval to round with a sharp terminal spine

94. Which statement is true regrading the larvae of *Schistosoma*:
a) Three larval forms are seen—miracidium, sporocyst, and cercaria.
b) The infective form is the cercaria larva, which is oval and elongated with a bifurcated tail.
c) The larva lives for about 24–72 hours.
d) All of the above

95. Warthin–Starry staining is used for demonstrating:
a) *Leptospira* b) *B. quintana*
c) *B. bacilliformis* d) *Ehrlichia sennetsu*

96. The *Trombiculid* mite is the vector for:
a) Endemic typhus b) Scrub typhus
c) Spotted fever d) Rickettsial pox

97. Which of the following bacteria cause glandular fever?
a) *Ehrlichia sennetsu* b) *Ehrlichia chaffeensis*
c) *Ehrlichia equi* d) *Anaplasma phagocytophilum*

98. Carrion disease is also known as:
a) Oraya fever b) Glandular fever
c) Q fever d) Trench fever

99. The selective medium used to isolate the causative agent of plague is:
a) Wright agar b) MacConkey agar
c) CIN agar d) CIS agar

100. The whole culture antigen vaccine of *Yersinia pestis* provides protection against:
a) All forms of plague
b) Bubonic plague but not against pneumonic plague
c) Pneumonic plague, but not bubonic plague
d) All zoonotic diseases except plague

Answers: 1:d, 2:d, 3:a, 4:a, 5:d, 6:a, 7:c, 8:c, 9:b, 10:c, 11:b. 12:b, 13:d, 14:d, 15:c, 16:a, 17:d, 18:a, 19:b, 20:c, 21:b. 22:d, 23:c, 24:b, 25:c, 26:b, 27:d, 28:c, 29:a, 30:d, 31:b, 32:a, 33:a, 34:c, 35:a, 36:b, 37:c, 38:c, 39:d, 40:d, 41:b, 42:c, 43:d, 44:a, 45:d, 46:b, 47:c, 48:b, 49:b, 50:b, 51:d, 52:a, 53:d, 54:b, 55:c, 56:a, 57:c, 58:d, 59:b, 60:c, 61:c, 62:a, 63:d, 64:c, 65:a, 66:a, 67:d, 68:b, 69:b, 70:c, 71:a, 72:b, 73:a, 74:b, 75:a, 76:c, 77:c, 78:a, 79:d, 80:c, 81:d, 82:d, 83:c, 84:a, 85:c, 86:a, 87:c, 88:d, 89:a, 90:c, 91:d, 92:c, 93:c, 94:d, 95:a, 96:b, 97:a, 98:a, 99:c, 100:b

Part III: Microbiology as Applied to Infectious Diseases

CHAPTER 22

Acquired Immunodeficiency Syndrome (AIDS)

(MI2.7, 8.15, and 8.16)

Chapter Navigation

Learning objectives

- Describe the structure and viral genes and antigens of the human immunodeficiency virus
- Discuss the pathogenesis of HIV infection
- Describe the clinical spectrum and the laboratory diagnosis of AIDS and outline the testing strategies
- Explain the epidemiology and preventive strategies employed for HIV infections

The emergence and pandemic spread of the acquired immunodeficiency syndrome (AIDS) has posed the greatest challenge to public health in modern times. A brief timeline of the history of AIDS is presented in Fig. 22.1.

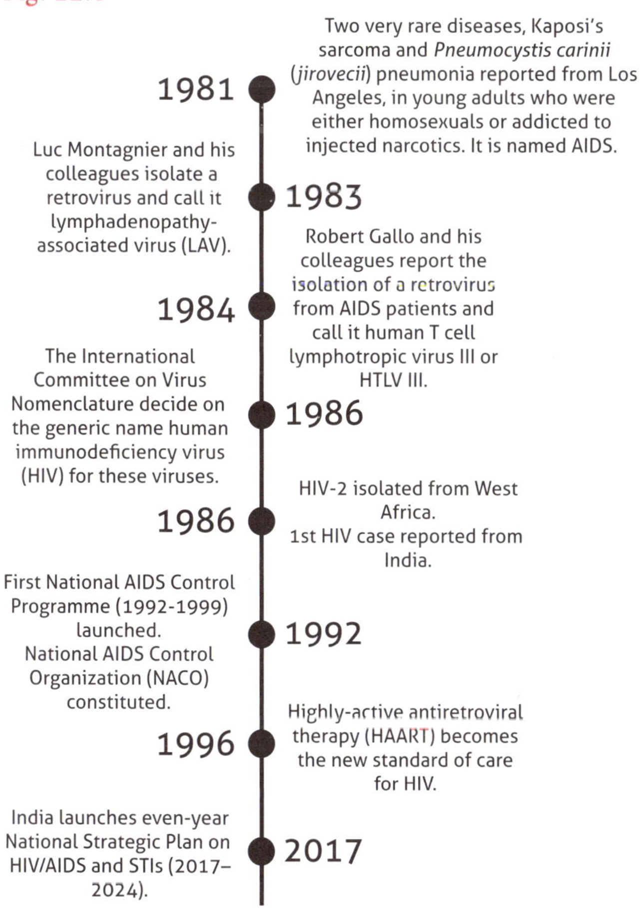

Fig. 22.1 An overview of the history of AIDS

Human immunodeficiency virus (HIV), the causative agent of AIDS, belongs to the lentivirus subgroup of the family Retroviridae, which comprises two genera:

- Genus **Lentivirus**: HIV-1 and -2
- Genus **Deltaretrovirus**: HTLV-1

Members of the lentivirus group that cause immunodeficiency

I. **In primates**
   1. Human immunodeficiency viruses (HIV) types 1 and 2
   2. Simian immunodeficiency viruses (SIV), which causes simian AIDS (SAIDS):
      a. Isolated from sooty mangabeys (SIV-SM) and rhesus macaque (SIV-MAC); closely related to HIV type 2
      b. Isolated from chimpanzee (cpz); closely related to HIV type 1

II. **In non-primates**
   1. Feline T lymphotropic virus (FTLV), which causes feline AIDS

EPIDEMIOLOGY

- The WHO reports a total of 38 million people currently living with HIV with a global prevalence rate of 0.7% amongst adults in 2021.
- The most affected region in the world is Sub-Saharan Africa.
- In India, there are more than 24 lakh people living with HIV, and the incidence rate is 0.05 in every 1,000 people.
- 1 December is celebrated as World AIDS Day across the world.

MORPHOLOGY OF HIV

HIV is a **spherical, enveloped RNA virus**, about 90–120 nm in size (Fig. 22.2). Its nucleocapsid has an outer **icosahedral shell** and an **inner conical core** enclosing the ribonucleoproteins.

Genome

- The genome of the HIV comprises two identical single-stranded, positive-sense RNA copies and the **reverse transcriptase enzyme** (a characteristic feature of retroviruses).
- When the virus infects a cell, the viral RNA is transcribed by the enzyme, first into single-stranded DNA, and then double-stranded DNA (**provirus**), which is integrated into the host cell chromosome.
- The provirus can remain latent for long periods, influencing host cell function.
- In response to viral promoters, the provirus initiates viral replication by directing the synthesis of viral RNA and other components.

Lipoprotein Envelope

- The lipoprotein envelope is acquired when the naked virus buds out through the host cell surface during viral replication.
- It consists of lipids derived from the host cell membrane and glycoproteins coded by the virus.
- The **glycoprotein spikes** constitute the chief surface component of the virus. These bind to the CD4 receptors (along with co-receptors CXCR4 and CCR5) on susceptible host cells.
- **Transmembrane pedicles** facilitate cell fusion.

VIRAL GENES AND ANTIGENS

Structural Gene

- The ***gag* gene** determines the core and shell of the virus, including the major antigens, p24, p18 (matrix or shell antigen), and p15 (core antigen). P24 is detected in serum in the early stages of HIV infection, before antibodies appear. The decline of free anti-p24 antibody and reappearance of p24 antigen in circulation indicates exacerbation of the illness.
- The ***pol* gene** codes for **polymerase reverse transcriptase** and other viral enzymes.
- The ***env* gene** determines the synthesis of the envelope glycoprotein gp160, which is cleaved into the two components—**gp120, which forms the surface spikes, and gp41**.
- Other non-structural and regulatory genes specific to the virus—the products of these genes, both structural and non-structural, act as antigens and play a role in antigen detection, which is useful for the diagnosis and prognosis of HIV infection.

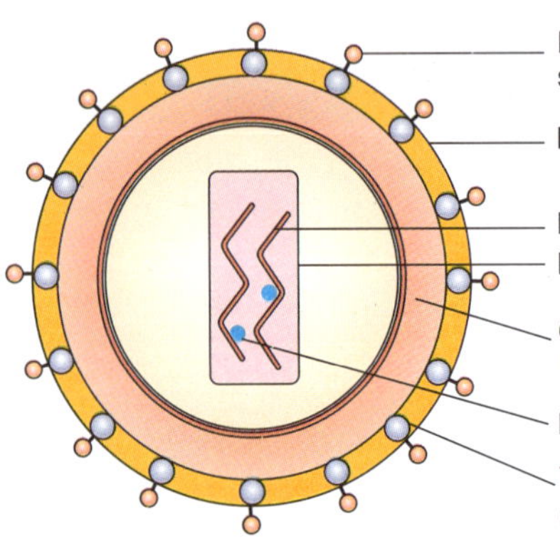

Fig. 22.2 Structure of HIV (diagrammatic representation)

Non-structural and Regulatory Genes

- ***tat*** (trans-activating gene) enhances the expression of all viral genes.
- ***nef*** (negative factor gene) downregulates viral replication.
- ***rev*** (regulator of virus gene) enhances the expression of structural proteins.
- ***vif*** (viral infectivity factor gene) influences the infectivity of viral particles.

Major Antigens of HIV

- Envelope antigens
  - **Spike antigen—gp120 (principal envelope antigen)**
  - **Transmembrane pedicle protein—gp41**
- Shell antigen
  - **Nucleocapsid protein—p21**
- Core antigens
  - **Principal core antigen—p24**
  - **Other core antigens—p15, p55**
- Polymerase antigens—**p31, p51, p66**

ANTIGENIC VARIATION AND DIVERSITY OF HIV

- HIV is a highly mutable virus.
- Antigenic variation exists in isolates of HIV from different places or persons and also between sequential isolates from the same person.
- This variability of HIV is due to the **error-prone nature of reverse transcription**, which may help the virus to evade the host's immune mechanisms.
- Antigenic differences between HIV strains **may be important in serodiagnosis.**

Subtypes: HIV-1 and HIV-2

Based on molecular and antigenic differences, two types of HIV have been recognised—HIV-1 and HIV-2.

- Infection by HIV-1 or -2 may not be identified unless the corresponding type is represented in the test antigen.

- Although HIV-1 and HIV-2 are related, there are important structural differences between them.
- Accurate diagnosis and differentiation of HIV-1 and HIV-2 is crucial for treatment, as HIV-2 is intrinsically resistant to **NNRTI (non-nucleoside reverse transcriptase inhibitors), the pillar of the national first-line ART regimen.**

HIV-1

- The original isolates of HIV and the related strains prevalent all over the world are HIV-1.
- HIV-1 strains have been classified into at least **ten subtypes** based on sequence analyses of their *gag* and *env* genes. These subtypes are designated **A to J.**
- All these subtypes constitute **group M (for 'major')** and cause most of the HIV-1 infections worldwide.
- Subtype A is the most prevalent worldwide, while B is the most common in the Americas and Europe.
- The common subtypes in **Africa** are **A, C, and D.**
- In **Asia,** the common subtypes are **E, C,** and **B.**
- Type E is presently considered a recombination of A and E and hence, is also called the AE type (also known as **CRFs or circulating recombinant forms**).
- The subtypes common in Asia and Africa (C and E) are more readily transmitted by heterosexual contact than the American strains (subtype B), which are preferentially spread through blood, i.e., through injection, and by homosexual contact.
- In India and China, subtype C is the most prevalent.

HIV-2

- This strain has only 40 per cent genetic similarity with HIV-1.
- It is more closely related to the simian immunodeficiency virus than to HIV-1
- HIV-2 is less virulent than HIV-1.
- It is largely **confined to West Africa** with a few reports from other areas, including western and southern India.

SUSCEPTIBILITY OF HIV

- HIV is **thermolabile,** being inactivated in 10 minutes at 60°C and in seconds at 100°C.
- At room temperature (20–25°C) and in dried blood, it may survive for up to seven days.
- On autopsy, HIV has been isolated from various tissues up to 16 days after death.
- It **withstands lyophilisation.** In lyophilised blood products, the virus can be inactivated by heating at 68°C for 72 hours, and in liquid plasma at 60°C for 10 hours.
- HIV is inactivated in 10 minutes by treatment with 50% ethanol, 35% isopropanol, 0.5% Lysol, 0.5% paraformaldehyde, 0.3% hydrogen peroxide, 1% nonidet p40, or 10% household bleach.
- It is also inactivated at the extremes of pH (1 and 13).
- Bleaching powder and household bleach are effective for surface decontamination.
- For the treatment of contaminated medical instruments, a 2% solution of glutaraldehyde is useful.

PATHOGENESIS

Infection is acquired when the virus enters the blood or tissues of a person and comes into contact with a suitable host cell, principally, the **CD4 lymphocyte** (Fig. 22.3).

Mode of Transmission

The various modes of transmission of HIV are presented in Table 22.1.

Table 22.1 Modes of transmission of HIV and risk of transmission

Route	Risk of transmission (%)	Percentage of total transmission
Sexual intercourse	0.1–10	>80 (mostly heterosexual)
Blood	>98	1–5
IV drug use	0.67	2–10
Needlestick injury*	0.3	<1
Mucous membrane, splash to eye, oro-nasal	0.09	-
Perinatal (without any intervention)	20–40	-

*Comparative risk after needlestick injury for HBV is 9–30% and for HCV is 1–1.8%. (*Source*: https://naco.gov.in/sites/default/files/National_Guidelines_for_HIV_Care_and_Treatment_2021.pdf)

HIV spreads by the following modes:

- **Sexual transmission:** Unprotected sexual contact with infected persons (heterosexual or homosexual) carries a 0.1–10% risk of transmission. Although the risk of transmission is low, sexual contact accounts for the maximum cases in the world.
- **Blood, blood products, and organ transplantation:** The incidence of transmission is very high but is now

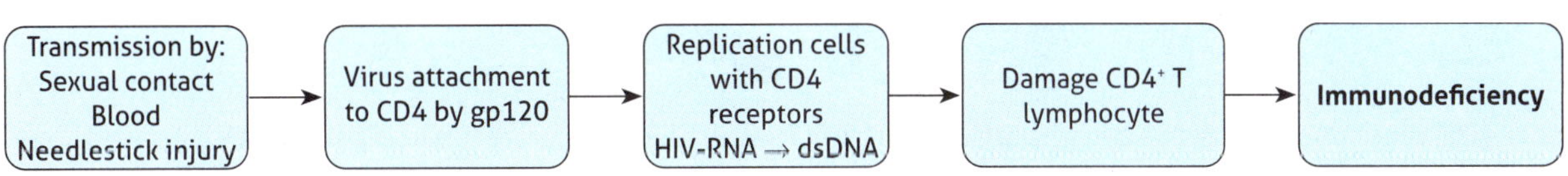

Fig. 22.3 Salient steps in the pathogenesis of HIV Infection

reduced due to mandatory screening of blood, blood products, and donor organs.

- **Contaminated needles:** This mode of transmission is seen among intravenous drug users who share syringes and needles. It also results from accidental needlestick injuries among healthcare workers. The use of disposable syringes, needles, and other equipment has reduced the incidence of such transmission.
- **Transmission from infected mother to child:** This can take place before, during, or after birth. HIV may be present in breast milk and may be transmitted through breastfeeding.

There is no evidence of HIV transmission by other means including casual contact, kissing, or through insect bites.

Virus Attachment

- Any cell bearing a CD4 antigen can act as a receptor for the virus; the following are suitable host cells for the HIV:
  - $CD4^+$ (helper/inducer)
  - T lymphocytes
  - Around 5–10 per cent of B lymphocytes
  - 10–20 per cent of monocytes and macrophages, including alveolar macrophages in the lungs and Langerhans cells in the dermis
  - Glial cells and microglia
  - Follicular dendritic cells from the tonsils
- The virus binds specifically to the CD4 receptor via the envelope glycoprotein gp120.
- Cell fusion is brought about by transmembrane gp41 (Fig. 22.4).

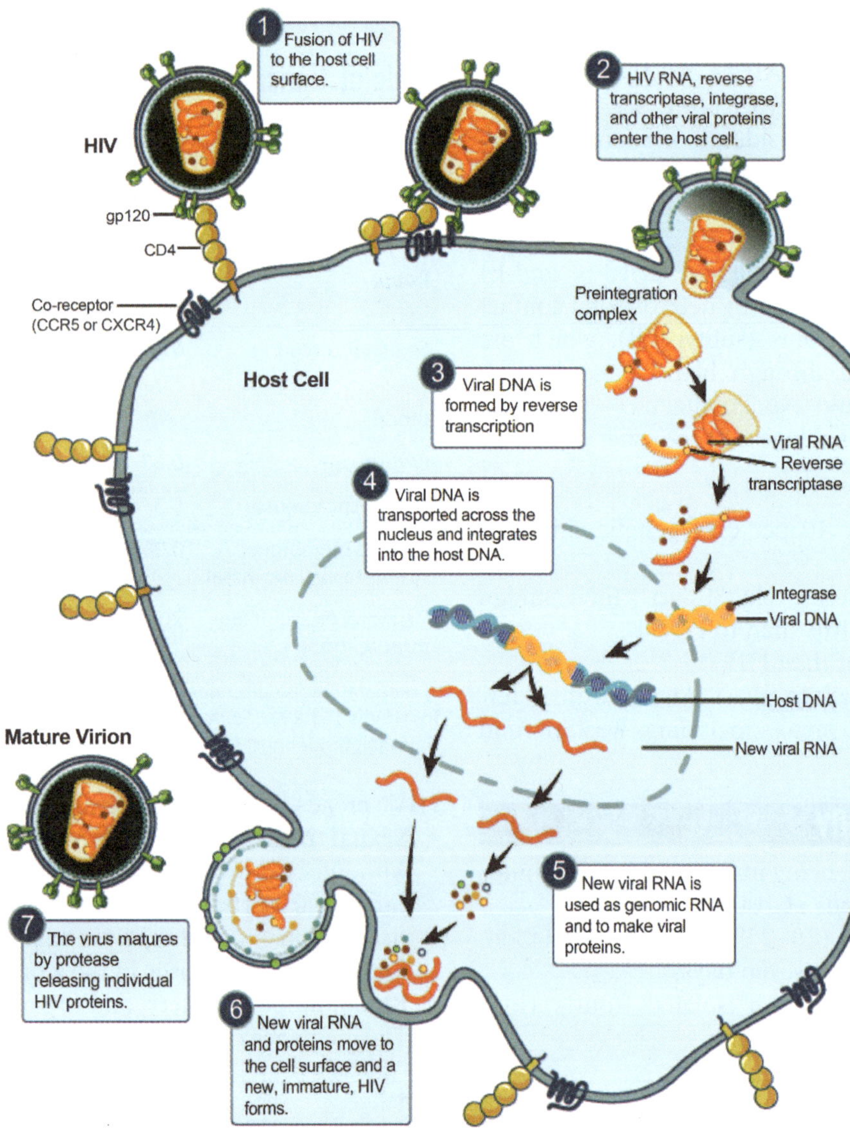

Fig. 22.4 The replication of HIV in a susceptible cell with CXCR4 and CCR5 receptors (*Source*: Centres for Disease Control and Prevention [CDC], Public Health Image Library [PHIL], Image ID 21162)

- Binding to the CD4 receptor requires the participation of a co-receptor molecule—CXCR4 for T cell-tropic HIV strains and CCR5 for macrophage-tropic strains.

Replication

- After the virus fuses with the host cell membrane, its genome is **uncoated and internalised** into the cell.
- Viral reverse transcriptase mediates the **transcription of HIV RNA into double-stranded DNA**, which is integrated into the genome of the infected cell through the action of the viral enzyme integrase. This results in a **latent infection**.
- The long and variable incubation period of HIV infection is because of this latency.
- The primary pathogenic mechanism in HIV infection is the **damage caused to the $CD4^+$ T lymphocyte.**
- The T4 cells decrease in number with the reversal of the T4:T8 (helper:suppressor) cell ratio.
- Infected T4 cells do not release normal amounts of interleukin-2, gamma interferon, and other lymphokines. This results in the suppression of cell-mediated immune response.
- The cellular immunity is chiefly damaged; humoral mechanisms are also affected.
- As a result, persons who develop AIDS are unable to mount a response to new antigens.
- There is polyclonal activation of B lymphocytes, leading to hypergammaglobulinemia of all classes of immunoglobulins, particularly IgG and IgA.
- In infants and children, IgM levels are also elevated.
- The hypergammaglobulinemia may be responsible for allergic reactions due to immune complexes (type III hypersensitivity).
- Monocyte–macrophage function is affected, apparently due to the lack of secretion of activating factors by the T4 lymphocytes.
- Chemotaxis, antigen presentation, and intracellular killing by monocytes/macrophages are diminished.
- The activity of NK cells and cytotoxic T lymphocytes is affected.
- The virus load in the plasma is of prognostic value.
- Viral killing of cells goes on throughout the illness.
- The steady state of the virus (**virus set point**) in a patient varies with individuals. A high set point correlates with rapid disease progression.
- The $CD4^+$ T cell count decreases steadily, from over 1,000 per microlitre to about 500 or less in the stage of acute infection. When the count falls to <200, clinical AIDS usually sets in.

Immunological abnormalities in HIV infection

- *Features that characterise AIDS:*
  - Lymphopenia
  - Selective T cell deficiency—reduction in the number of T4 (CD4) cells, inversion of T4:T8 ratio
  - Decreased delayed hypersensitivity on skin testing
  - Hypergammaglobulinemia—predominantly IgG and IgA; in children, IgM is also elevated
  - Polyclonal activation of B cells and increased spontaneous secretion of Ig
- *Other consistently observed features*:
  - Decreased in vitro lymphocyte proliferative response to mitogens and antigens
  - Decreased cytotoxic response by T cells and NK cells
  - Decreased antibody response to new antigens
  - Altered monocyte/macrophage function
  - Elevated levels of immune complexes in serum

CLINICAL FEATURES

There are three stages of HIV infection:

1. **Acute HIV infection:** This is the earliest stage of HIV infection, that develops within 2 to 4 weeks after infection with HIV. The patient may present with flu-like symptoms such as fever, headache, and rash. During this phase, HIV multiplies rapidly and spreads throughout the body and the person is highly infective. The CD4 counts are normal though the virus is steadily attacking and destroying CD4 cells.
2. **Chronic HIV infection:** This is also known as asymptomatic HIV infection or clinical latency. The HIV continues to multiply in the body but at very low levels. The patient may remain asymptomatic but can transmit HIV. This latency can be prolonged by ART.
3. **AIDS:** This is the final, most severe stage of HIV infection. The immune system is severely damaged. Dementia and other degenerative neurological lesions are due to the action of the virus on the central nervous system (CNS) cells.

- The Centers for Disease Control and Prevention, USA, had classified the clinical course of HIV infection under various groups ranging from acute infection (group I) to full-blown AIDS (group IV).
- Currently, however, it is the WHO classification that is followed (Table 22.2).

Opportunistic Infections

The clinical manifestations of HIV infections are not primarily due to viral cytopathology but are secondary to the failure of immune response. This renders the patient susceptible to opportunistic infections and malignancies (Table 22.3 and 22.4).

Table 22.2 WHO classification of the stages of HIV infection

Stages	$CD4^+$ cell count per microliter	Clinical features
Stage 1 (HIV infection)	At least 500	Asymptomatic Persistent generalised lymphadenopathy*
Stage 2 (HIV infection)	350–499	Moderate unexplained weight loss (<10% of presumed or measured body weight) Recurrent respiratory tract infections Herpes-zoster Angular cheilitis Recurrent oral ulceration Papular pruritic eruption Fungal nail infections Seborrheic dermatitis
Stage 3 (Advanced HIV disease [AHD])	200–349	Unexplained severe weight loss (>10% of presumed or measured body weight) Unexplained chronic diarrhea for longer than 1 month Unexplained persistent fever (intermittent or constant for longer than 1 month) Persistent oral candidiasis Oral hairy leukoplakia Pulmonary tuberculosis Severe bacterial infections (such as pneumonia, empyema, pyomyositis, bone or joint infection, meningitis, bacteremia) Acute necrotising ulcerative stomatitis, gingivitis, or periodontitis Unexplained anemia (<8 g/dL), neutropenia (<0.5×10^9/L) and/or chronic thrombocytopenia (<50×10^9/L)
Stage 4 (AIDS—full-blown AIDS which denotes the complete breakdown of the immune system)	<200 or the per cent of $CD4^+$ cells is less than 15% of all lymphocytes	HIV wasting syndrome Opportunistic infections *Pneumocystis jirovecii* pneumonia Recurrent severe bacterial pneumonia Chronic herpes-simplex infection Esophageal candidiasis Extrapulmonary tuberculosis Kaposi sarcoma Cytomegalovirus infection Central nervous system toxoplasmosis HIV encephalopathy Extrapulmonary cryptococcosis, including meningitis Disseminated non-tuberculous mycobacterial infection Progressive multifocal leukoencephalopathy Disseminated mycosis Lymphoma Invasive cervical carcinoma Atypical disseminated leishmaniasis

*The presence of enlarged lymph nodes in two or more non-contiguous extrainguinal sites that persist for at least three months in the absence of any current illness or medication that may cause lymphadenopathy

Long-term Survivors or Long-term Non-progressors

- Around 5–10 per cent of the infected appear to escape clinical AIDS for 15 years or more.
- The mechanisms for such prolonged survival are not clear, though many viral and host determinants may be responsible.

Pediatric AIDS

- About a third to half of the number of babies born to infected mothers are infected with HIV.
- Virus transmission to the fetus may occur as early as in the first trimester, but infection is more common perinatally.

Table 22.3 Major opportunistic infections and malignancies commonly associated with untreated AIDS patients

Etiology	Specific agents/infections
Bacterial	Mycobacterial infections—tuberculosis and non-tuberculous infections *Nocardia* and actinomycetes infections Salmonellosis *Campylobacter* infection Legionellosis
Viral	CMV infection Herpes simplex infection
Mycotic	*Pneumocystis jirovecii* infection Candidosis Cryptococcosis Histoplasmosis Aspergillosis
Parasitic	Cryptosporidiosis Isosporiasis Generalised strongyloidiasis Toxoplasmosis
Malignancies	Kaposi sarcoma Lymphomas—Hodgkin and non-Hodgkin types

Table 22.4 Effect of HIV/AIDS on various systems of body

System	Disease
General	Generalised wasting is an AIDS-defining feature with >10% involuntary weight loss, fever, chronic diarrhea, and fatigue **(wasting syndrome)**
Respiratory system	Dry cough, dyspnea and fever, tuberculosis, pneumonia caused by CMV or *Pneumocystis jirovecii*, *Cryptococcus*, or *Histoplasma*
Ophthalmic diseases	Cotton wool spots on the retina due to microvascular disease, CMV retinitis, herpes simplex, and varicella-zoster retinitis
Gastrointestinal system	Oral and esophageal thrush (candidiasis), herpetic stomatitis, gingivitis, hairy leukoplakia and Kaposi sarcoma, esophageal candidosis, infections caused by *Cystoisospora*, *Cryptosporidium*, atypical mycobacteria, salmonellae, CMV, or adenoviruses, disseminated strongyloidiasis
Central nervous system	Toxoplasmosis, CMV, herpes simplex, papovaviruses, mycobacteria, *Cryptococcus* and *Candida* infections Direct cytopathogenic damage in the central nervous system HIV can cross the blood–brain barrier and cause encephalopathy, leading to the loss of higher functions, progressing to dementia CNS lymphomas
Cutaneous diseases	Herpes lesions, candidosis, xeroderma, seborrheic dermatitis, prurigo, folliculitis, impetigo, and molluscum contagiosum
Malignancies	Kaposi sarcoma Lymphomas—Hodgkin and non-Hodgkin types

- A number of infected children fail to survive the first year of life.
- Children may also acquire the infection from blood transfusion or blood products.
- Children develop humoral immunodeficiency early, leading to recurrent bacterial infections, and as a result, they fail to thrive.
- Chronic diarrhea is more common among children than it is in adults.
- Lymphadenopathy is more pronounced than in adults.
- Tuberculosis and opportunistic bacterial infections are common manifestations in pediatric AIDS.
- Lymphocytic interstitial pneumonia is seen mostly in children.
- Kaposi sarcoma, toxoplasmosis, and cryptococcosis are less common in children than in adults.

Early infant diagnosis

- Early infant diagnosis was introduced by NACO to detect HIV infection in newborns.
- Infants born to HIV-positive mothers must be screened early after birth in order to start treatment or monitor for mother-to-child transmission.
- The strategy to diagnose infants/children below 18 months is by DNA polymerase chain reaction (DNA PCR), which detects HIV-1 proviral DNA integrated with the human genome.
- Whole blood or spots of the infant's blood dried on filter paper (DBS) are collected for this test. The blood is eluted out of the filter paper, and the sample is used to extract the DNA for PCR.
- Tests for antibodies in infants below six months of age yield false-positive results on account of the presence of maternal antibodies.

LABORATORY DIAGNOSIS OF HIV AND AIDS

Case 1. *A 35-year-old commercial sex worker presented to the medical OPD with history of weight loss, decreased appetite, and cough with expectoration for one month. Her total WBC count was low; sputum AFB was positive and graded 2+;* ***HIV serology was reactive*** *with the two-test (IIB) strategy of the National AIDS Control Organisation (NACO); and her CD4 count was 350 cells/mL. The patient was referred to an ART centre for antiretroviral and antitubercular chemotherapy. She was advised to return for a follow-up after two weeks and to use a condom during sexual contact.*

i. What are the various strategies recommended by NACO to diagnose HIV infection?
ii. Why is a follow-up test required in this case?

The indications for the laboratory confirmation of HIV are as follows:

- Diagnosis of clinically suspected cases
- Monitoring of treatment
- Screening of blood and organs
- Antenatal screening of mothers

- Early infant diagnosis
- Screening of high-risk groups

Policies established by the National AIDS Control Organisation (NACO)

- To establish a reliable set of standardised tests in India, the NACO is credited with implementing a strategy for testing different categories of individuals thought to have been exposed to HIV.
- To ensure quality and uniformity in reporting the incidence and prevalence of the disease, a set of guidelines has been provided for conducting serological tests and their interpretation. It has been made mandatory for all testing laboratories to follow these guidelines.
- Pre- and post-test counselling to educate the patient have been made mandatory.
- Further, it has been established that no test can be carried out without the prior consent of the patient.

Surveillance in India

Since its appearance in India in 1980, HIV has affected several millions of individuals, impacting the social, financial, and mental fabric of affected individuals. To contain the infection and treat and rehabilitate these individuals, a National AIDS Control Programme was initiated by the Government of India—spearheaded by the **National AIDS Control Organisation**, Ministry of Health. Details of the programme are discussed in Chapter 34.

Tests Used for Screening

- Screening tests are highly sensitive, have a broad spectrum of reactivity, are simple to perform and can be automated to handle large numbers of samples at a time.
- As they are not highly specific, they may give a few false-positive results.
- All sera testing positive on a screening test are to be confirmed before the sample is declared reactive.

1. Antigen detection

- The major core antigen, p24, is the earliest virus marker to appear in blood and hence, is tested for early diagnosis (Fig. 22.5).

p24 capture assay

- Antibody-bound p24 antigen may be demonstrable after dissociation.
- The p24 antigen capture assay (ELISA), which uses anti-p24 antibody as the solid phase, can be used.
- It is useful for diagnosis in persons recently exposed to infection, in whom the antibody test is negative in the first few weeks of infection and in the terminal phase of the illness.

Seroconversion

- This refers to the appearance of IgM antibody in the patient's serum following the initial period of p24 antigenemia and viremia.
- Later, free p24 antigen disappears from circulation and remains absent during the long asymptomatic phase, to reappear only when severe clinical disease sets in.

Uses of p24 antigen assay

- Diagnosis of HIV during window period
- Diagnosis of immune collapse stage
- Monitor progress of HIV infection
- Resolve unequivocal results of western blot
- Diagnosis of HIV encephalopathy

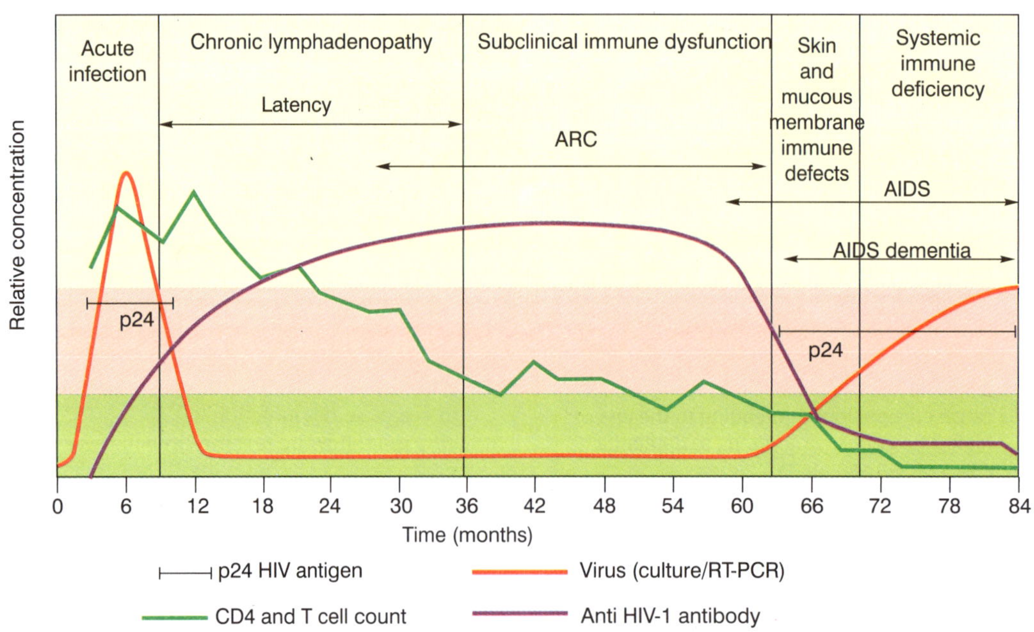

Fig. 22.5 Illustration of the usual course of immune response, viremia, and disease resulting from untreated HIV-1 infection

2. Antibody detection

- The demonstration of antibodies is the simplest and most widely employed technique for the diagnosis of HIV infection.
- **IgM antibodies appear in about 4–6 weeks, followed by IgG antibodies (Fig. 22.6).** Infection can be detected during the window period by p24 antigen assay.
- When immunodeficiency becomes severe following clinical AIDS, some components of the anti-HIV antibody (anti-p24) may disappear.
- Anti-HIV antibody tests are used for either screening or confirming the infection (Table 22.5).
- **Window period**—it may take 2–8 weeks to months for antibodies for appear in the patient's serum. The patient is highly infectious during this seronegative stage. Hence, antibody testing is not totally dependable for detecting infectious persons, e.g., from among blood donors.
- The window period is considerably shortened with the use of fourth-generation tests.

Table 22.5 Screening and confirmatory/supplemental tests used for antibody detection of HIV

Screening tests	Confirmatory or supplemental tests
ELISA: Third-generation ELISA detects only antibodies, whereas the fourth-generation test can detect both p24 antigens and antibodies **Rapid tests:** These yield results within 30 minutes and do not require specialised equipment **Cassette ELISA** **Immunochromatographic tests (ICT)** **Coated particle agglutination** **Immunoperoxidase or dipstick tests**	**Western blot test):** Detects antibodies to: • p24 (*gag* gene, core protein) • p31 (*pol* gene, reverse transcriptase) • gp41, gp120, or gp160 (*env* gene, surface antigens) A positive reaction with proteins representing the three genes is conclusive **Line immunoassays (LIAs):** Recombinant and synthetic peptide antigens on a plastic support strip like an immunoblot assay LIAs are second- or third-generation assays and have the potential to be used as supplemental tests Combination assays and differentiation between HIV-1 and HIV-2 can also be done in LIA

Methods to detect antibodies (Table 22.5)

NACO strategies of HIV testing comprise the detection of antibodies by ELISA, rapid or supplemental tests.

NACO strategies for HIV testing

- NACO has laid out guidelines and strategies for screening, testing, and monitoring HIV-infected/suspected individuals (*Source*: National Guidelines for HIV-Testing, July 2015; National AIDS Control Organisation, Ministry of Health and Family Welfare, Government of India).
- ELISA/rapid tests/supplemental tests (E/R/S) are used in strategies I, II, and III (Table 22.6).
- Supplemental tests are employed in cases of indeterminate/discordant result of E/R/S.

3. Detection of viral nucleic acids

This is the most sensitive and specific method for the detection of HIV. **Viral proviral** DNA or RNA can be detected by the following methods:

- **DNA PCR:** The amplified DNA is detected by probes that bring about nucleic acid hybridisation. The test can detect HIV proviral DNA at a frequency of **one copy per 10,000 cells**.
- **RT-PCR (reverse transcriptase PCR):** This test is used to detect disease progression and monitor response to therapy.

Uses of nucleic acid detection

- Assessment of viral load in HIV patients on antiretroviral therapy (ART)
- Detection during window period
- Differentiate between HIV-1 and HIV-2 infections
- Detection of specific genotype or subtype
- Detection of drug resistance genes in HIV

4. Virus isolation

- The virus is present in circulation in body fluids, within lymphocytes. Alternatively, it is cell-free, and its titres parallel those of p24.
- The virus is isolated only in containment laboratories from the peripheral lymphocytes of infected persons by co-cultivation.

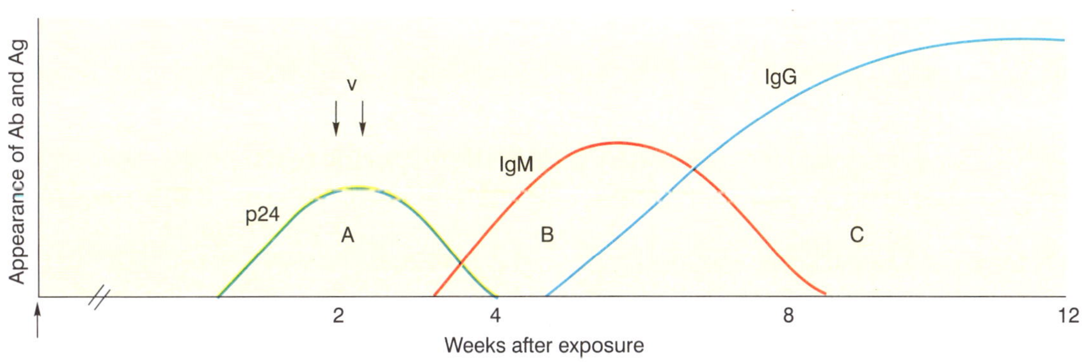

Fig. 22.6 The sequence of the appearance of p24 antigen and antibodies after a massive HIV infection

Table 22.6 NACO strategies for HIV testing

Strategy	Salient features	Use
I (Fig. 22.7)	The sample is subjected once to ELISA/Rapid (E/R) for HIV NACO recommends the use of E/R kits with a sensitivity of ≥99.5 per cent and specificity of ≥98 per cent and rapid kits with a sensitivity of ≥99.5 per cent and specificity of ≥98 per cent under strategy I If negative, the specimen is considered to be negative for HIV	To screen blood/blood products, organs, tissues, sperm, etc. A negative test means that the blood/organ is fit for donation
IIA (Fig. 22.8)	If a serum sample is positive in the first E/R test, it is subjected to a second E/R, which utilises a system different from the first one If the second E/R is positive, it is reported as positive If it is negative (the second E/R), the result is considered negative	Surveillance
IIB (Fig. 22.9)	Reported positive only if two E/R tests are positive and negative if two tests are negative If two tests out of three are positive, it is reported as indeterminate and repeated after 2–3 weeks (Case 1)	Diagnosis of symptomatic patients
III (Fig. 22.10)	It is similar to strategy II, with an additional third positive ELISA test being required for a sample to be reported HIV-reactive The first ELISA has the highest sensitivity and the second and third ELISAs have the highest specificity	Diagnosis of asymptomatic cases

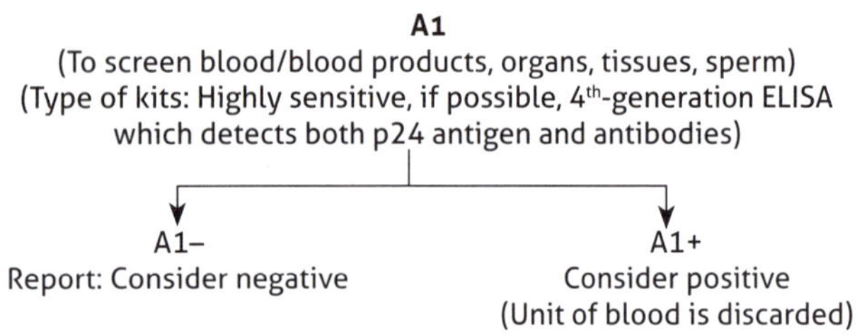

Fig. 22.7 Strategy I testing for HIV screening

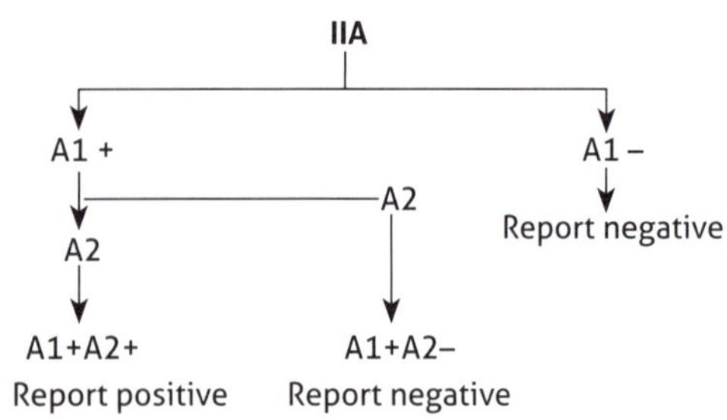

Fig. 22.8 Strategy IIA (for surveillance—anonymous and unlinked—two tests are required)

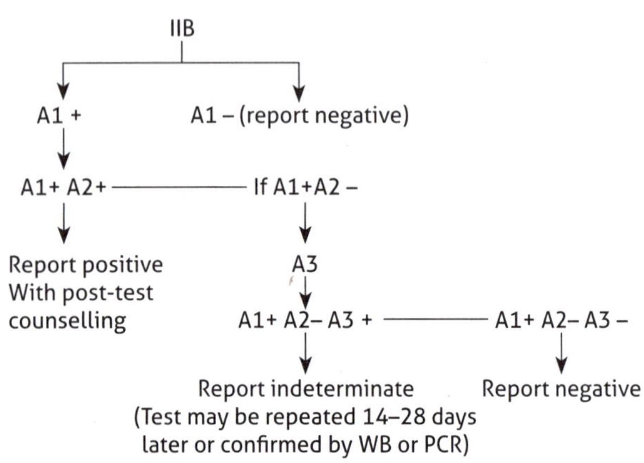

Fig. 22.9 Strategy IIB (for diagnosis of symptomatic or high-risk individuals)

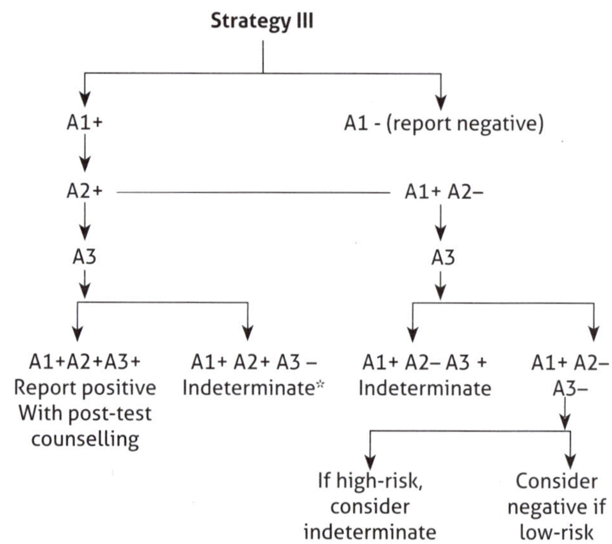

*All indeterminates must be counselled and advised for follow-up testing after a specified period.

Fig. 22.10 Strategy III—to detect HIV infection in asymptomatic individuals presenting to integrated counselling and testing centres (ICTC) or prevention of parent-to-child transmission (PTCT) centres

- This method is labour-intensive, expensive, and time-consuming.
- It is not used for diagnosis, but for research purposes only.

5. Non-specific tests

- Total leukocyte and lymphocyte count to demonstrate leukopenia (count usually below 2,000/mm^3)
- T cell subset assays—absolute CD4$^+$ T cell count is usually less than 200/mm^3; T4:T8 cell ratio is reversed
- Thrombocytopenia (low platelet count)
- Raised IgG and IgA levels
- Diminished CMI as indicated by skin tests
- Lymph node biopsy shows abnormalities

Significance of HIV screening
- Epidemiological surveillance using unlinked anonymous HIV testing
- Transfusion and transplant safety
- Diagnosis of HIV infection in symptomatic and asymptomatic individuals
- Prevention of parent-to-child transmission (including early infant diagnosis)
- For post-exposure prophylaxis (PEP)
- Research

The patient's lymphocytes are cultivated with uninfected healthy lymphocytes in the presence of interleukin-2.

Table 22.7 presents the serological markers that are detectable during various stages of HIV infection.

Laboratory Monitoring of HIV Infection

Some laboratory tests are important in monitoring the course of HIV infection.

- **CD4$^+$ T cell count:** A count below 500 indicates disease progression and the need for antiretroviral therapy. Counts below 200 denote risk of serious infection.
- **Viral load:** It is necessary to monitor HIV RNA during the course of treatment. This is usually done by RT-PCR and DNA assay.
- **Beta-2-microglobulin and neopterin:** These have a predictive value on the progression of HIV disease as they rise with advancing disease.

PROGNOSIS OF HIV INFECTION

In a person infected with HIV, the loss of detectable anti-p24 antibody indicates clinical deterioration. This is also associated with HIV antigenemia and increased virus titre in circulation.

PROPHYLAXIS

The prevention of AIDS depends on general measures such as health education, identification of sources, and decrease in high-risk behaviour.

Vaccines

Till date, no safe and effective vaccine against HIV infection has been released. Some problems with designing an HIV vaccine are as follows:

- High frequency of mutation of the virus
- Integration of the viral DNA into the host cell's genome, whereby the virus may remain in the latent form for years
- Lack of understanding of the mechanism of protective immunity
- Some infected persons are non-progressors and some remain uninfected despite being exposed several times

Several vaccines are under trial for the prevention of HIV infection. Some of these are:

- Engineered HIV envelope proteins delivered through messenger RNA
- Passive immunisation against HIV via broadly neutralising antibodies
- Immunogens that are designed to stimulate B cells of the immune system to generate **broadly neutralising antibodies (bnAbs)** against the virus
- An mRNA vaccine and a chimeric vaccine **HIVconsvX** that can target a wide range of HIV mutants are in phase-1 trials.

ANTIRETROVIRAL TREATMENT (ART)

The National AIDS Control Organisation (NACO) mandates that all persons diagnosed with HIV infection should be initiated on ART regardless of their CD4 count or WHO clinical stage or age group or population sub groups. ART should not be started in the presence of an active opportunistic infection.

The defined goals of ART are as follows:

- Rational sequencing of drugs in a manner that achieves clinical, virological, and immunological goals while maintaining future treatment options, limiting drug toxicity, and facilitating adherence
- Increased survival and improvement in the quality of life
- Greatest possible sustained reduction in viral load

Table 22.7 Serological markers detected during various phases of HIV infection

Stage of infection	Antigens/antibodies	Anti-HIV IgM	Anti-HIV IgG	Western blot pattern
Early	p24 antigen +	–	–	–
Acute (seroconversion)	p24 antigens +; gp120 + –	– +	– +	+
Partial illness	gp120, gp41, gp160 antibodies gp120	+ –	+	+
Carriers and asymptomatic individuals	P24 Ag–; gp120, gp41, gp160 antibodies +	–	+	+
PGL	P24 antibodies –	–	+	+
AIDS	P24, P17 and P55 antibodies decline	–	+	+

- Immune reconstitution (quantitative and qualitative)
- Reduction of HIV transmission by suppression of viral load

A number of effective drugs have become available in recent years (Table 22.8). The sites of action of major antiretroviral drugs are presented in Fig. 22.11.

Table 22.8 Antiretroviral drugs

Mechanism of action	Drugs
Entry inhibitors i. Fusion inhibitor ii. Co-receptor CR5 antagonist	 Enfuvirtide Maraviroc
Integrase inhibitors	Raltegravir Dolutegravir
Nucleoside reverse transcriptase inhibitors (NRTI)	Abacavir*, zidovudine* (azidothymidine [AZT]), didanosine, zalcitabine, lamivudine*
Non-nucleoside inhibitors (NNRTI)	Nevirapine*, etravirine, efavirenz*
Nucleotide reverse transcriptase inhibitors (NtRTI)	Tenofovir disoproxil fumarate (TDF)* Tenofovir alafenamide (TAF)
Protease inhibitors	Saquinavir, ritonavir*, indinavir, lopinavir*, atazanavir*, darunavir*

*ART drugs available in the NACO programme

Virological Monitoring of Patients on ART

- CD4 testing should be done every 6 months
- Plasma viral load estimation at 6 months and 12 months after ART initiation and then every 12 months

Management of AIDS

The approaches to the treatment of AIDS include the following measures:

- The treatment and prophylaxis of infections and tumours
- General management, which includes rehabilitation, and counselling both the patients and their family members
- Immunorestorative measures—protein-rich food and nutritional supplements
- Specific anti-HIV agents
- Prompt diagnosis, counselling, and appropriate treatment of opportunistic infections and tumours in the early stage of AIDS can be very useful and may help the patient to resume normal life in-between episodes of illness
- The general management of the patient requires the understanding and cooperation of the health staff in the hospital and of caregivers at home
- Fears about imaginary risks must be allayed and reassurance given that the patient can be kept at home or treated in the hospital without danger to contacts if proper precautions are taken

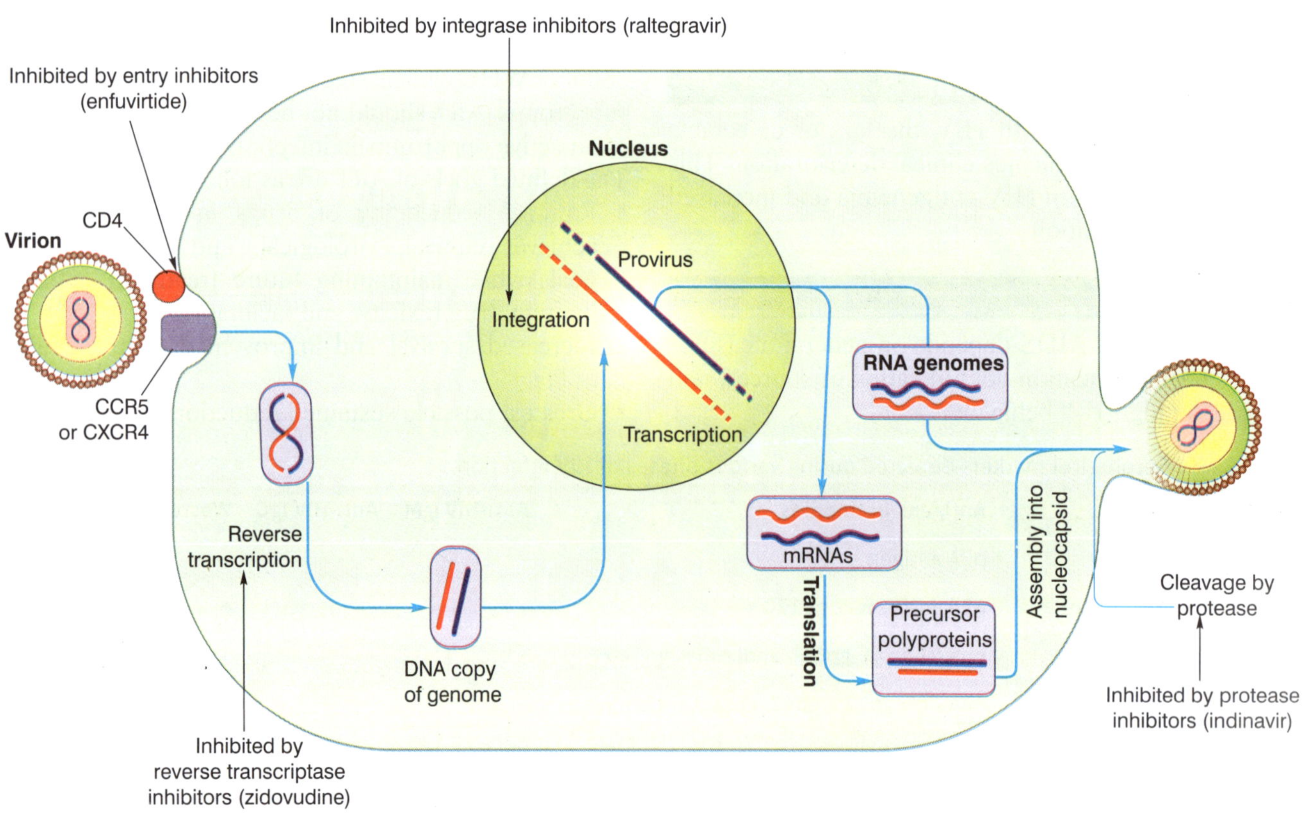

Fig. 22.11 Sites of action of antiretroviral drugs

Post-exposure Prophylaxis (PEP)

PEP is the comprehensive management given to minimise the risk of infection following potential exposure HIV. **'Exposure' (expressed as exposure code [EC])** to the risk of developing bloodborne infections is defined as (Table 22.9):

- Percutaneous injury (needlestick injury)
- Contact with mucous membrane of eye or mouth
- Contact with non-intact skin
- Contact with intact skin only

Although the risk of HIV transmission by these routes is less than 1%, PEP is recommended within 72 hours of exposure.

- Prophylaxis depends on the exposure category of the HCP (healthcare personnel) and the HIV status code of the patient (Fig. 22.12 and 22.13).

Table 22.9 Risk of exposure from different body fluids

Exposure to body fluids considered 'at risk'	Exposure to body fluids considered 'not at risk', unless these fluids contain visible blood
Blood Semen Vaginal secretions Cerebrospinal fluid Synovial, pleural, peritoneal, pericardial fluid Amniotic fluid Any body fluids contaminated with visible blood, serum	Tears Sweat Urine and feces Saliva Sputum Vomitus

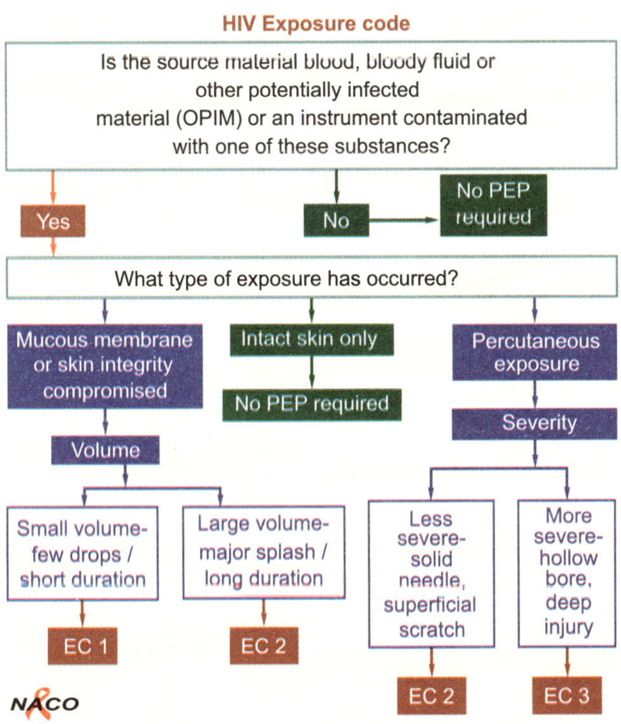

Fig. 22.12 Guidelines of the National AIDS Control Organisation (NACO)

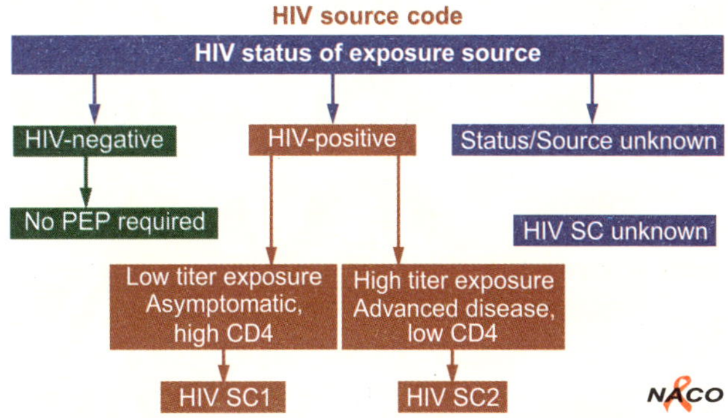

Effectiveness of PEP depends on...

- Efficacy of PEP is best if administered within two hours of exposure
- PEP needs to be given within 72 hours of exposure
- Do not delay PEP while waiting for result of HIV testing
- Informed consent must be obtained before testing a source as per National guidelines
- Baseline rapid HIV testing before PEP
- Negative result doesn't exclude HIV infection
- Positive HIV result helps in stopping the PEP

NACO

Determining PEP recommendation

EC	HIV SC	PEP recommendation
1	1	May not be warranted
1	2	Consider basic regimen (Two drugs)
2	1	Recommend basic regimen (Two drugs)
2	2	Recommend expanded regimen (Three drugs)
3	1 or 2	Recommend expanded regimen (Three drugs)
2/3	Unknown	Consider basic regimen if HIV prevalence is high in the given population

NACO

Fig. 22.13 Post-exposure prophylaxis (PEP) as recommended by NACO

- The first dose of PEP should ideally be administered within two hours (but preferably within the first 72 hours) of exposure.
- The risk is evaluated as soon as possible. If the risk is insignificant, PEP could be discontinued if already commenced.
- The duration of PEP is 28 days, regardless of the PEP regimen.

Drug regimens for PEP

Any of the following regimens can be administered as per the NACO guidelines:
Tenofovir + Lamivudine + Dolutegravir OR
Tenofovir +Lamivudine + Lopinavir/Ritonavir OR
Tenofovir +Lamivudine + Efavirenz

Questions

A 29-year-old HIV positive woman is admitted at 39 weeks gestation in active labour. While delivering her child an intern is injured by the needle. The needleprick was deep enough to draw out blood.

1. What is the category of exposure in this needlestick injury?
2. What is the recommended PEP in this case?

Short answer questions

1. Describe the antigens of HIV and their relevance.
2. Describe the steps in the pathogenesis of HIV.
3. Discuss the significance of viral markers in HIV infection.
4. Describe the applications of NACO strategies for HIV testing in India.

MCQs

1. Which of the following is a transmembrane pedicle protein of HIV?
a) gp120 b) gp41
c) p24 d) p18

2. Which of the following is a principle envelope antigen of HIV?
a) gp120 b) gp41
c) p24 d) p18

3. Which of the following is the major core antigen of HIV?
a) gp120 b) gp41
c) p18 d) p24

4. All the following are true regarding HIV-2 except:
a) First isolated from West Africa in 1986
b) More virulent than HIV-1
c) 40 per cent genetic similarity with HIV-1
d) Closely related to the simian immunodeficiency virus

5. Which of the HIV 1 subtypes is most prevalent in India?
a) HIV-1 subtypes A b) HIV-1 subtypes B
c) HIV-1 subtypes C d) HIV-1 subtypes D

6. All of the following immunological abnormalities are seen in HIV infection except:
a) Lymphopenia
b) Reversal of CD4:CD8 ratio
c) Decreased delayed hypersensitivity response
d) Decreased spontaneous secretion of immunoglobulins

7. All the following are specified infectious diseases listed in the CDC surveillance definition of AIDS, except:
a) *P. carinii* pneumonia b) Cryptosporidiosis
c) Coccidioidomycosis d) Cryptococcosis

8. As per the CDC case definition of HIV, what is the recommended $CD4^+$ T cell count?
a) $\leq$100 b) $\leq$200
c) $\leq$300 d) $\leq$400

9. A 42-year-old homosexual man presents with chronic diarrhea and fever. He had tested seropositive for HIV six months back. He was diagnosed as a case of gay bowel syndrome. All the following are likely causative agents of this condition except:
a) Rotavirus b) *Cryptosporidium*
c) *Isospora belli* d) *Giardia*

10. All of the following fungi are common causative agents of opportunistic infections in HIV/AIDS patients except:
a) *Aspergillus* species b) *Blastomyces dermatitidis*
c) *Candida* species d) *Histoplasma capsulatum*

11. What is the common manifestation of pediatric AIDS?
a) Cryptococcosis b) Interstitial pneumonia
c) Kaposi's sarcoma d) Toxoplasmosis

12. A paramedical worker gets a needle prick while handling a known HIV seropositive patient. His serum was sent for testing 15 days after exposure. What is the best testing method?
a) Western blot b) P24 capture assay
c) RT-PCR d) Viral culture by co-cultivation

13. What is the best laboratory method to test for HIV infection during the window period?
a) p24 antigen assay b) IgM capture ELISA
c) IgG ELISA d) Western blot test

14. The formation of a band against which one of the following HIV proteins is NOT considered significant in the Western blot test?
a) p18 b) p24
c) p31 d) gp41

15. Which one of the following is false regarding HIV testing in infants?
a) Suggested sample is a dried blood spot
b) DNA PCR test is preferred
c) Antibody detection in infants below six months is diagnostic
d) Early detection is possible by Western blot test

16. A 32-year-old man wants to donate blood at a local blood bank. Which strategy as per NACO should be implemented to test for HIV?
a) Strategy A1 b) Strategy IIA
c) Strategy IIB d) Strategy III

17. A 31-year-old man tested positive for HIV on routine screening. He conceals his contact history and has no HIV-associated symptoms. Which of the following is not correct about his phase of HIV infection?
a) Infective
b) CD4 below 200/μL
c) Actively replicating
d) Sequestrating in lymphoid tissue

18. Fatal fungal infection in AIDS patients is commonly caused by:
a) *Cryptococcus* b) *Histoplasma*
c) *Mucor* d) *Rhodotorula*

19. An HIV-infected person presented with fever and felt tired all the time. His radiologic findings were normal. CD4 counts were around 80 cells/mm³. His bone marrow smear showed granulomas and it grew budding yeast cells with hyphae and chlamydospores. Which of the following fungi is likely to be associated with this case?
a) *C. albicans* b) *H. capsulatum*
c) *C. immitis* d) *Mucor*

20. Acute HIV infection is characterised by:
a) Low CD4 counts
b) Fever, headache, and rash
c) Asymptomatic presentation
d) All of the above

21. An intern gets a needle prick injury while collecting a sample from an HIV-positive patient. The needle has barely penetrated skin, and the patient is asymptomatic with CD4 count >500/mm³. As per the NACO strategy, what regimen of ART is best recommended for this intern?
a) Single drug PEP b) Expanded three-drug regimen
c) Basic two-drug regimen d) PEP may not be warranted

Answers: 1:b, 2:a, 3:d, 4:b, 5:c, 6:d, 7:c, 8:b, 9:a, 10:b, 11:b, 12:b, 13:a, 14:a, 15:c, 16:a, 17:b, 18:a, 19:b, 20:b, 21:d

CHAPTER 23

Infections of the Respiratory System

(MI6.1–6.3)

Chapter Navigation

Learning objectives

- Discuss the pathogenesis of upper respiratory tract infections.
- List the agents responsible for URIs!
- Enumerate steps in the laboratory diagnosis of URIs
- Describe the pathogenesis of lower respiratory tract infections
- Describe the etiopathogenesis of LRIs
- Enumerate various diagnostic modalities to confirm LRIs

The respiratory tract, which comprises the upper and lower tracts, forms a single continuous entity from the nose to the alveoli. The respiratory tract is divided into upper and lower regions (Fig. 23.1).

1. **Upper respiratory tract:** This comprises the anatomically contiguous areas of the nasopharynx, oropharynx, middle ear, and paranasal sinuses and extends up to the pharynx and larynx.
2. **Lower respiratory tract:** It extends from the trachea and bronchi up to the alveoli.

Infections of the respiratory system are best discussed as upper respiratory tract infections and lower respiratory tract infections.

- However, the organisms that cause upper respiratory tract infection may descend to involve the lower tract.
- Commensal bacteria of the upper tract may also descend to cause lower tract infections, including those in the lungs. A common example of this is pneumococcal pneumonia following viral infection in infants and young children.

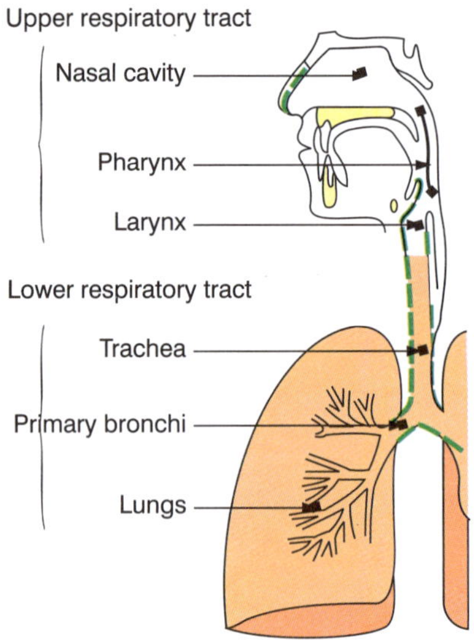

Fig. 23.1 Upper and lower respiratory tracts

PROTECTIVE MECHANISMS OF THE RESPIRATORY TRACT (Table 23.1)

Table 23.1 Protective mechanisms of the upper and lower respiratory tracts

Upper respiratory tract	Lower respiratory tract
• Ciliary movement of the epithelial cells • Hairs in the nostrils are a part of the innate immune system and prevent microbes from lodging in the tract • Natural secretions, cough reflex, and saliva also help in getting rid of the pathogens • The normal microbiota* of the upper respiratory tract competes with pathogens	• Local mucociliary mechanisms involve the integration of the ciliated epithelium, periciliary fluid, and mucus • Mucus acts as a physical and chemical barrier onto which organisms adhere • Cilia lining the respiratory tract propel the overlying mucus to the oropharynx where it is either swallowed or expectorated • Alveolar macrophages protect the alveoli

*Viridans group of streptococci, *S. pneumoniae*, coagulase-negative staphylococci, *Haemophilus influenzae*, *Moraxella*, commensal *Neisseria* species, diphtheroids, *Borrelia*, and anaerobes (*Fusobacterium* and *Bacteroides*) that colonise the mouth

PATHOGENS CAUSING RESPIRATORY TRACT INFECTIONS

Organisms infecting the tract are common, with some showing a special predilection for a particular site (e.g., rhinovirus to the nasopharynx). Based on the pathology of the infections caused by these organisms, they can be categorised as follows:

1. **Primary pathogens:** These organisms adhere to the normal mucosa, resist mucociliary interference and mucosal macrophages, damage mucosal and submucosal tissues, and establish an infection. For example, the influenza virus.
2. **Secondary pathogens:** These organisms have the ability to establish infection only on a damaged mucosal cell. For example, the influenza virus infects healthy respiratory epithelial cells and damages them, thereby predisposing them to secondary infection by pneumococci, which are otherwise commensals of the upper respiratory tract.
3. **Respiratory pathogens in the immunocompromised:** Opportunistic infections of the respiratory tract can be caused by commensal bacteria, e.g., oral thrush (candidiasis in an HIV-infected individual).

UPPER RESPIRATORY TRACT INFECTIONS (URIS)

- The upper respiratory tract (Fig. 23.2) is the most common site of infection by pathogenic microorganisms because it is in direct contact with the environment and is continuously exposed to microorganisms suspended in the air.
- URIs occur frequently and have a seasonal predilection.

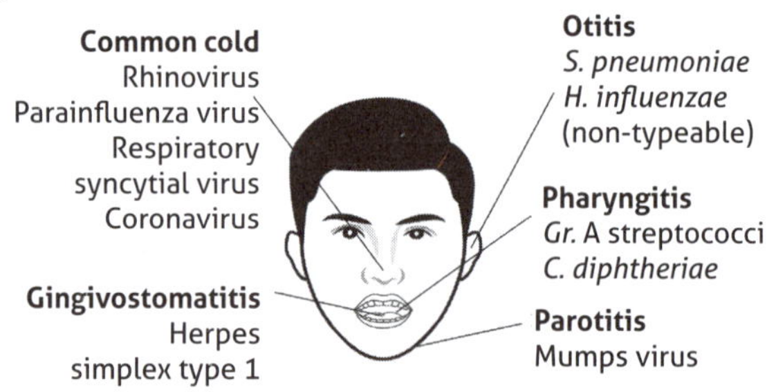

Fig. 23.2 Common microbial agents associated with URIs (*Source*: Anurupa Badrinath)

PATHOGENESIS OF URIs

Infections limited to the **nose, pharynx, sinuses, and larynx are caused by viruses** in 70% of cases.

- URIs are usually **self-limiting** illnesses. Generally, of viral origin, they may spread to the lower respiratory tract and predispose to bacterial complications (pneumococcal pneumonia).
- Common cold and pharyngitis of viral etiology can be transmitted easily, especially among school children on account of their close interactions.
- In children, the most common complication is **acute otitis media (AOM)** and in adults and the elderly, the most frequent complications are **paranasal sinusitis and pneumonia.**
- Microbial agents causing infections in different anatomical regions of the upper respiratory tract vary depending on the predilection of certain organisms to infect these sites.

- Their pathogenesis depends on the virulence of the organism and comorbidity of the individual, e.g., malignant external otitis, caused by *Pseudomonas aeruginosa*, can be fatal in diabetics.
- The agents associated with upper respiratory tract infections are listed in Table 23.2.

DIAGNOSIS OF URIs

- **Upper respiratory tract infections are chiefly viral and are diagnosed clinically.**
- Laboratory confirmation is required only for diphtheria, streptococcal pharyngitis, and tonsillitis. An algorithm for the diagnosis of URIs is presented in Fig. 23.3.

Collection and Transport of Specimens for Infections of Bacterial Etiology

- For infections of **bacterial etiology:**
  - **Two throat swabs** should be collected—one for direct microscopic examination and another for culture.
  - Sterile cotton or calcium alginate tipped swabs are used for collecting specimens from the affected area.
  - If **diphtheria** is suspected, a part of the pseudomembrane must be collected for processing.
- For infections of **viral etiology**:
  - Very rarely does a viral upper respiratory tract infection require laboratory confirmation, unless it is during an epidemic or pandemic.
  - Nasopharyngeal swabs, aspirates, or throat washings are collected in viral transport media.
  - Calcium alginate or Teflon-coated swabs are used for this purpose.
  - As these are aerosol-generating procedures, the clinical staff collecting the samples must take necessary precautions and follow infection control practices like wearing a mask, gloves, and visors before collecting respiratory samples.
  - The swab should be transported within one hour or the bacteria may dry up and impact their isolation.
- Serum is collected when streptococcal pharyngitis or infectious mononucleosis is suspected to look for antibodies.

Microscopy

- Gram staining is not very useful in the diagnosis of URIs, which are generally of viral etiology. Further, this method does not differentiate the pathogen from normal flora.

Utility of Gram stain in special situations

- **Diphtheria**: Abundant gram-positive rods are seen in cuneiform arrangement. The diagnosis is confirmed by Albert stain. Immunofluorescence may help in the rapid identification of diphtheria.
- **Vincent's angina:** Gram-negative spirochetes (*Borrelia vincentii*) and gram-negative fusiform bacilli (*Fusobacterium* spp.) are seen.

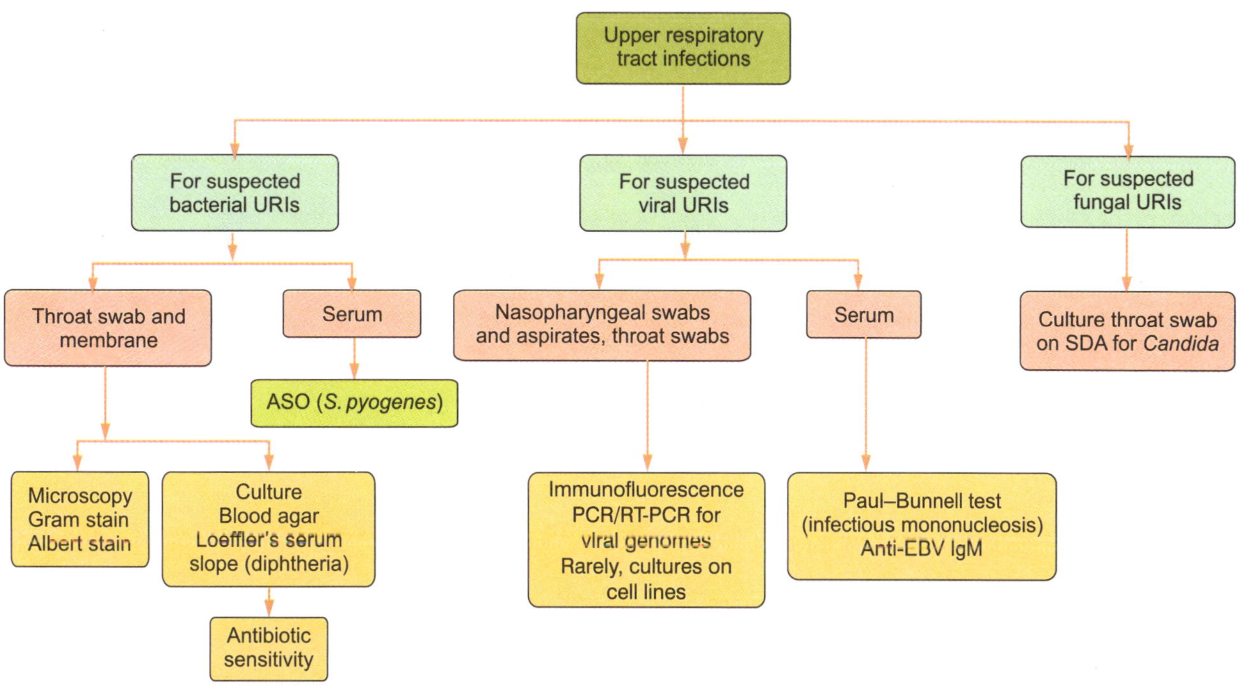

Fig. 23.3 Diagnostic algorithm for upper respiratory tract infections (*ASO*—antistreptolysin O; *EBV*—Epstein–Barr virus; *HSV*—herpes simplex virus; *SDA*—Sabouraud dextrose agar)

Table 23.2 Common etiological agents of acute upper respiratory tract infections

Clinical syndromes	Clinical features	Agents associated with upper respiratory tract infections (URIs)		
Site of infection		**Bacteria**	**Viruses**	**Fungi**
Infections of the oropharynx Pharyngitis Tonsillitis Common cold	**Pharyngitis (sore throat):** High-grade fever with cervical lymphadenopathy and the presence of exudates on the tonsillar area. **Common cold:** Running nose or nasal discharge, nasal obstruction and sneezing, malaise, hoarseness, mild burning, and lacrimation of the eyes. Subsequently, the nasal discharge becomes viscous and thick and subsides within a few days. **Peritonsillar abscess (quinsy):** Pus from a tonsillar abscess in-between the tonsils and posterior pharyngeal wall.	*Streptococcus pyogenes* *Corynebacterium diphtheriae* *Mycoplasma pneumoniae* *Staphylococcus aureus*	Adenovirus (3, 4, 7, 14, 21) Rhinovirus Parainfluenza virus (also causes croup) Respiratory syncytial virus Coronavirus Coxsackie virus (commonly A7) Influenza virus Rhinovirus and parainfluenza viruses Herpes simplex type I virus Epstein–Barr virus	*Candida albicans*
Infections of the sinus Rhinosinusitis	Often follows rhinitis and leads to rhinosinitis. Stuffiness, pain, and discomfort.	*Streptococcus pneumoniae* Non-typeable *H. influenzae* *Moraxella catarrhalis* *Staphylococcus aureus*	Rhinovirus Adenovirus Parainfluenza virus Influenza virus	*Rhizomucor* *Rhizopus* *Mucor* *Cunninghamella* *Absidia* *Aspergillus* *Fusarium*
Infections of the ear Acute suppurative otitis—externa and media Malignant otitis externa Mastoiditis (Details in Ch 33) Acute suppurative otitis media (ASOM)	Pain in ear, fever, ear discharge	*Pseudomonas aeruginosa* *Staphylococcus aureus* *S. pneumoniae* Non-typeable *H. influenzae* *Actinomycetes* *S. epidermidis*	Respiratory syncytial virus Rhinovirus Influenza virus	*Aspergillus* spp. *Fusarium* spp.
Infections of the salivary glands	Pain, swelling, fever	*Staphylococcus aureus*, anaerobic bacteria	Mumps virus	—
Infection of the epiglottis	Organism spreads from the nasopharynx to the epiglottis. Inflammation and edema of the epiglottis, leading to obstruction and respiratory distress. May be a life-threatening condition requiring intubation and antibiotic treatment.	*H. influenzae* type b	—	—
Croup	Inflammation of the larynx, trachea and large bronchi (laryngotracheobronchitis).	—	—	Parainfluenza

- ***Candida***: Budding yeast cells with pseudohyphae are seen in infections caused by members of the *Candida* species.

Culture

Culturing is recommended to confirm the diagnosis of diphtheria and streptococcal infections and for taking preventive measures. The media used for culture are:

- **Blood agar** (5% sheep BA); incubated under 5% CO_2 overnight
- **Loeffler's serum slope and potassium tellurite agar** (PTA)—for isolation of *C. diphtheriae*
- **Sabouraud dextrose agar (SDA)**—for the isolation of *Candida albicans*
- Identification and confirmation is done by characteristic colony morphology, Gram staining and biochemical reactions. Automated systems such as MALDI can be used instead of biochemical reactions.

TREATMENT OF URIs

- Specific antibiotics are prescribed for bacterial URI.
- For streptococcal sore throat, penicillin G, amoxicillin, and erythromycin are used (dose is dependent on age and body weight).
- If viral URI is suspected, antibiotics should not be given.

BACTERIAL CAUSES OF URIs

Streptococcal Throat Infection

> **Case 1.** *A seven-year-old girl presented with severe sore throat and fever up to 39°C for three days with difficulty in swallowing. On examination, she had bilateral tender submandibular lymphadenopathy, enlarged tonsils, and pharyngeal exudates. A culture of her throat swab was performed and, pending results, she was started on oral penicillin. After 24 hours, beta-hemolytic colonies grew on blood agar, which were gram-positive cocci in short chains and sensitive to penicillin. The diagnosis was confirmed as **streptococcal pharyngitis**.*
>
> i. What are the non-suppurative sequelae of such infections?
> ii. Describe the pathogenesis of the sequalae of streptococcal infections.

Epidemiology

- Streptococcal throat infections are caused by *Streptococcus pyogenes*.
- *S. pyogenes* colonises the human upper respiratory tract—throat and nasopharynx—of patients and carriers (up to 20%).
- Streptococcal infections of the respiratory tract are more frequent in **children between 5 and 8 years** of age than in children below the age of two years or adults.
- The **transmission** of infection is either by direct contact or through contaminated fingers, dust, or fomites.
- **Crowding** is an important factor in the transmission of infection. Outbreaks of infection may occur in **closed communities** such as **boarding schools or military camps**.

Agent—*Streptococcus pyogenes* or Group A *Streptococcus* (GAS)

- These are catalase-negative gram-positive cocci arranged in chains.
- They are human pathogens that produces a diverse array of diseases, ranging from **self-limited conditions** (pharyngitis, impetigo, and erysipelas) to **severe, invasive infections** (bacteremia, necrotising fasciitis, myositis, and streptococcal toxic shock syndrome).
- Streptococcal pharyngitis or skin infections may have lasting sequelae such as **post-streptococcal rheumatic disease** or **glomerulonephritis** respectively.

Virulence factors (*Table 23.3*)

Typing of GAS

S. pyogenes can be typed based on their virulence factors:

i. **M proteins (Griffith's M types)** act as virulence factors by inhibiting phagocytosis and are antigenic. About 80 M protein types have been recognised so far. Some are associated with the pathogenesis of rheumatic fever and others are associated with pyoderma.
ii. **T proteins** are present in many serotypes of *S. pyogenes*.
iii. **R protein** is a third antigen present in some types of *S. pyogenes*—types B, C, and G.
iv. **M-associated protein** is a non-type-specific protein.

Immunity

- The immune response elicited by *S. pyogenes* is **type-specific** and appears to be associated with the antibody to the M protein.
- **Re-infections** occur because of the multiplicity of serotypes.

Pathogenesis and clinical features

Streptococcal infections may be (i) localised, (ii) diffuse (invasive) or (iii) may manifest as sequelae to the original infection.

Localised infections

Clinical features of infections caused by GAS are as detailed below.

Streptococcal sore throat

- It is the most common streptococcal disease.

Table 23.3 Virulence factors of GAS

Virulence factor	Features
Cell wall-associated factors (Fig. 23.4)	
Capsule (hyaluronic acid)	Inhibits phagocytosis.
Carbohydrate antigen	This antigen shows cross-reactivity with some human tissues, leading to post-streptococcal sequelae. On the basis of the C carbohydrate antigen, *S. pyogenes* is classified under Lancefield group A.
Protein antigen	Several protein antigens have been identified in the outer part of the cell wall which help in typing the organisms for epidemiological purposes (see section on typing).
Pili (fimbria)	Attachment of streptococci to epithelial cells.
Toxins	
Hemolysins	i. **Streptolysin O**: Streptolysin is predominantly produced by group A, though groups C and G may also produce them. This is an antigenic and oxygen-labile toxin. **Anti-streptolysin O (ASO)** appears following streptococcal infection. The estimation of this antibody (ASO titre) is a standard serological procedure for the diagnosis of past infection with *S. pyogenes*. An ASO titre of more than 200 units is considered significant and suggests either recent or recurrent infection with streptococci. ii. **Streptolysin S:** This is an oxygen-stable hemolysin responsible for hemolysis around streptococcal colonies on blood agar plates. It is soluble (S) in serum and not antigenic. iii. **Streptococcal pyrogenic exotoxin (SPE):** Three types of SPEs—A, B, and C—have been identified. Like staphylococcal enterotoxins and the TSS toxin, SPEs are also superantigens. They are T cell mitogens that induce a massive release of inflammatory cytokines, causing tissue damage, fever, and shock.
Enzymes	
Streptokinase (fibrinolysin)	Promotes the lysis of human fibrin clots by activating a plasma precursor (**plasminogen**), thereby causing the spread of infection. Neutralising antibodies appear in convalescent sera. Anti-streptokinase antibodies provide retrospective evidence of streptococcal infection. Streptokinase is administered intravenously for the treatment of early myocardial infarction and other thromboembolic disorders.
Deoxyribonucleases (streptodornase, DNase)	Cause the depolymerisation of DNA and help to liquefy the thick pus in streptococcal exudate. Streptokinase and streptodornase are used therapeutically for liquefying the thick exudates as in empyema. Four antigenically distinct DNases-A, B, C, and D exist. Of these, type B is the most antigenic in human beings. Anti-DNase B antibody helps in the retrospective diagnosis of *S. pyogenes* skin infections, where ASO titres may be low.
Hyaluronidase	Breaks down the hyaluronic acid of the tissues, favouring the spread of infection along the intercellular spaces. Antibodies to this antigen appear in convalescent sera.
Other enzymes	Proteinase, phosphatase, esterase, amylase, N-acetyl glucosaminidase, and neuraminidase. **Serum opacity factor (SOF-a virulence determinant):** Its role in pathogenesis is unclear.

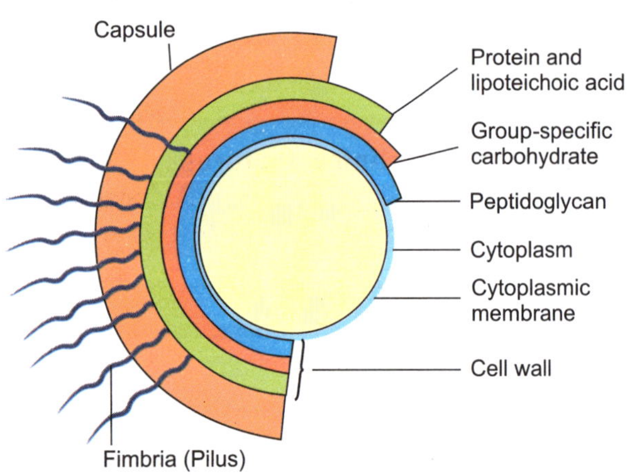

Fig. 23.4 Cell wall-associated antigens of *S. pyogenes*

- It may be localised as **tonsillitis**, which is more **common in older children** and adults than in younger children.
- It may involve the pharynx more diffusely (**pharyngitis**; as described in Case 1). This type of infection affects younger children, who commonly develop diffuse pharyngitis (Fig. 23.5a and b).
- Virulent group A streptococci adhere to the pharyngeal epithelium by means of the lipoteichoic acid covering the surface pili.
- From the throat, streptococci may spread to the surrounding tissues, leading to suppurative complications such as **otitis media, mastoiditis, quinsy, Ludwig's angina, and suppurative adenitis.**

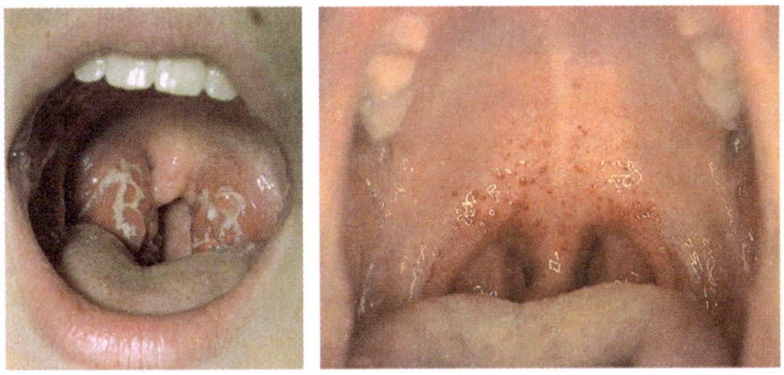

(a) (b)

Fig. 23.5 (a and b) Streptococcal pharyngitis with visible pus and exudate (*Source*: Wikimedia Commons)

Other serious group A streptococcal infections are as follows:

- Infections of the **skin and soft tissues** (discussed in Chapter 29)
- **Bloodstream infections** resulting in sepsis (discussed in Chapter 21)
- **Toxin-mediated infections**—toxic shock syndrome and scarlet fever
- **Scarlet fever—**it is caused by pyrogenic exotoxins A–C and is characterised by rashes on the trunk in combination with sore throat (Fig. 23.6); localisation is believed to be favoured by hypersensitivity due to prior contact

In the pre-antibiotic era, *S. pyogenes* was an important cause of death due to **puerperal sepsis**, with the infection usually being exogenous. In 1847, Ignaz P. Semmelweis demonstrated that hospital outbreaks of puerperal fever could be prevented by a simple measure of ensuring that those attending the labour wards wash their hands carefully—a finding that remains a landmark in clinical microbiology.

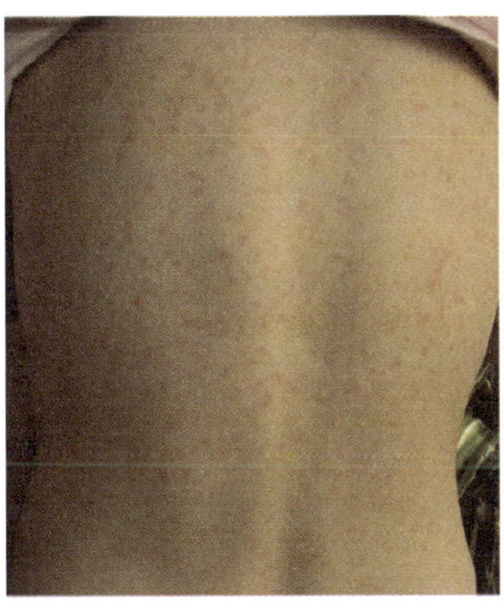

Fig. 23.6 Scarlet fever with rashes on the skin (*Source*: https://openstax.org/books/microbiology/pages/22-2-bacterial-infections-of-the-respiratory-tract)

Non-suppurative sequelae

S. pyogenes infections lead to important non-suppurative post-streptococcal sequelae, which occur one to four weeks after acute streptococcal infections.

- **Acute rheumatic fever (Chapter 21):** This is often preceded by streptococcal sore throat. This is an important sequel of hypersensitivity to cross-reacting antigens, antibodies to which can damage the myocardial tissue and heart valves.
- **Post-streptococcal glomerulonephritis:** This usually follows streptococcal skin infection with nephritogenic strains. The antibodies against streptococcal antigens cross-react with the glomerular basement membrane resulting in damage due to complement activation.

Pediatric autoimmune neuropsychiatric disorders associated with streptococcal (PANDAS) infections

In a subset of children with rapid onset of obsessive–compulsive disorder (OCD) or tic disorders, these symptoms are caused by Group A beta-hemolytic streptococcal (**GABHS**) infections. This has been attributed to autoimmune response to GABHS.

Laboratory diagnosis

In acute streptococcal infections, diagnosis is established by culture, whereas in the non-suppurative complications, diagnosis is mainly based on the demonstration of antibodies.

- **Specimen:**
  - **Swabs or aspirates** from the affected site are collected aseptically. If the sample cannot be transported immediately, **Pike's transport medium** is used.
  - **Serum** is collected for serology in rheumatic fever and glomerulonephritis.
- **Microscopy:** Gram-stained films of pus show gram-positive cocci in chains. However, smears are of no value in infections of the throat or genitalia, where streptococci may form part of the resident flora.
- **Culture**
  - The specimen is inoculated on 5% sheep blood agar and incubated at 37°C (range 22–42°C) anaerobically or under 5–10% CO_2 after 24 hours, show small (0.5–1.0 mm), circular, semi-transparent colonies with clear hemolysis around them (Fig. 23.7).
  - **Virulent strains,** on fresh isolation from lesions, produce '**matte**' (finely granular) colonies.
  - **Avirulent strains form 'glossy' colonies.**
- **Screen for group A streptococci carriage:** PNF medium **(polymyxin, nalidixic acid, and fusidic acid)**, which inhibits gram-negative bacteria and staphylococci, is used.

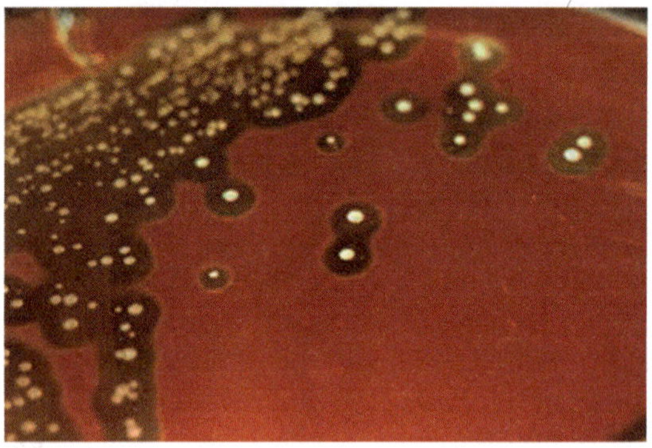

Fig. 23.7 Blood agar with streptococcal colonies showing beta-hemolysis (*Source*: Department of Microbiology, Pondicherry Institute of Medical Sciences, Puducherry)

- **Identification**
  - **Antigen detection:** Rapid diagnostic test kits using specific antisera are available commercially for the detection of streptococcal group A antigen from throat swabs. The tests can be completed in 1–4 hours and are nearly as specific as cultures, though less sensitive.
  - **Bacitracin sensitivity:** A wide zone of inhibition is seen around a 0.04-unit bacitracin with *S. pyogenes* but not with other streptococci.
  - **Typing**: Lancefield grouping uses group-specific antisera; streptococci are typed in the laboratory. **MTR protein-based typing** of *S. pyogenes* is required only for epidemiological purposes and may be done by precipitation or agglutination techniques using specific antisera.
  - **Antimicrobial susceptibility:** *Streptococcus* group A organisms are uniformly susceptible to penicillin. However, clinical failures are being reported following penicillin therapy. Hence, monitoring the minimum inhibitory concentration of penicillin is advisable in referral centres. Antimicrobial therapy is also required for cotrimoxazole and erythromycin, to which resistance is being reported
- **Serology**
  - In rheumatic fever and glomerulonephritis, a retrospective diagnosis of streptococcal infection may be established by demonstrating high levels of antibody to streptococcal toxins.
  - **Anti-streptolysin O (ASO) titration** is the standard test. ASO titres higher than 200 are indicative of prior streptococcal infection. The upper limit may vary with age, being higher in children than it is in adults. High levels are usually found in acute rheumatic fever; levels are often low in glomerulonephritis.
  - **Anti-deoxyribonuclease B (anti-DNase B)** is an antibody to DNase. Titres higher than 300 are considered significant.
  - **Anti-DNase B and antihyaluronidase tests** are very useful for the retrospective diagnosis of streptococcal pyoderma, for which ASO is of much less value.
- **Rapid streptozyme test:** This is a passive slide hemagglutination test using erythrocytes sensitised with a crude preparation of extracellular antigens of streptococci. It is a convenient, sensitive, and specific screening test. It is positive in nearly all types of streptococcal infections, whether of the throat or the skin.

Treatment

- All beta-hemolytic group A streptococci are sensitive to **penicillin G**, and most are sensitive to erythromycin.
- In patients allergic to penicillin, erythromycin or cephalexin may be used.
- Tetracyclines and sulphonamides are not recommended.
- Antimicrobial drugs have no effect on established glomerulonephritis and rheumatic fever.
- Bacitracin has been used for local application on skin lesions.

Laboratory diagnosis of streptococcal sore throat

Acute infection is diagnosed by culture while chronic infection (ARF/AGN) is diagnosed by serology.

- **Specimen:**
  - **Throat swabs or aspirates**
  - **Serum** in rheumatic fever and glomerulonephritis
- **Transport medium:** Pike's transport medium
- **Microscopy** (Gram-stain): Gram-positive cocci in chains, but difficult to differentiate from the resident flora
- **Culture:**
  - Aerobic conditions with 5–10% CO_2, optimum temperature is 37° C
  - **5% sheep blood agar:** Small (0.5–1.0 mm), circular, semi-transparent colonies with clear zone of hemolysis
  - **Screen for group A streptococci carriage**: PNF medium (polymyxin, nalidixic acid, and fusidic acid)
- **Identification:**
  - **Antigen detection:** Rapid diagnostic test kits to detect streptococcal group A antigen
  - **Bacitracin sensitivity**
  - **Typing** by precipitation or agglutination: **Lancefield grouping** using group specific antisera
  - **MTR protein-based** typing of *S. pyogenes*
- **Serology:**
  - **Anti-streptolysin O (ASO)** titres >200 in acute rheumatic fever; levels are often low in glomerulonephritis
  - **Anti-deoxyribonuclease B (anti-DNase B)** titres >300 are significant
  - **Rapid streptozyme test:** This passive slide hemagglutination test

Group C Streptococci

- Streptococci of this group are **predominantly animal pathogens**.
- Group C strains isolated from human sources usually belong to the *S. equisimilis* species.
- They can cause upper respiratory infections, endocarditis, osteomyelitis, brain abscess, pneumonia, and puerperal sepsis.
- Strains are often **tolerant to penicillin**. The addition of gentamicin is recommended in serious cases.
- *S. equisimilis* is the source of streptokinase used for thrombolytic therapy in patients.

Group F Streptococci

- These are also called the **minute streptococci**.
- They can cause suppurative lesions.
- *Streptococcus* MG (belonging to Group F) is isolated from cases of primary atypical pneumonia.

Group G Streptococci

- These are **commensals in the throats** of human beings.
- May occasionally cause **tonsillitis, endocarditis, and urinary infections** in human beings.

Group D Streptococci

- They reside in the **gastrointestinal tract**.
- These streptococci may cause urinary infection or endocarditis.
- Members of this group are generally susceptible to penicillins.
- They are associated with **bacteremia, infective endocarditis, colonic neoplasms**, and other gastrointestinal lesions.

Other Groups

- Groups H and K sometimes cause infective endocarditis.
- Group O is isolated chiefly from the healthy human throat and may cause acute tonsillitis and endocarditis.

Diphtheria

> **Case 2.** *A five-year-old child presented to the outpatient ward with a history of pain in the throat and difficulty in swallowing accompanied by low-grade fever for the past two days. He had enlarged cervical lymph nodes, and his tonsillar pillars were covered with a greyish-white, elevated membrane with a well-defined edge and surrounded by a zone of inflammation. His vaccination card revealed non-adherence to the immunisation schedule. Albert's stain of his throat swab smear showed rod-shaped bacteria, which were green in colour with bluish-black granules. A presumptive diagnosis of* ***diphtheria*** *was made, and the child was immediately started on penicillin—50 mg/kg (maximum 1.2 g) a day and passive immunisation with diphtheria antitoxin.* ***Prophylactic erythromycin*** *was prescribed to the child's siblings along with vaccination depending on the status.*
>
> i. What is the pathogenesis of diphtheria?
>
> ii. Describe the laboratory diagnosis of diphtheria.

Diphtheria is a vaccine-preventable but potentially lethal infection of the upper respiratory tract. The word is derived from the Greek word *'diphtheria'* which means hide or leather. The name draws on the presence of a characteristic pseudomembrane produced by the organism over the site of colonisation.

Epidemiology

Diphtheria has been virtually eradicated from most Western countries by effective prophylaxis and mass immunisation.

- In some developing countries where childhood immunisation programmes have not been implemented effectively, it continues to be a serious problem.
- In India, diphtheria falls under the ambit of the universal immunisation programme for infants. However, sporadic cases are still reported. This could be due to non-adherence to childhood immunisation schedules.
- Several adult diphtheria cases have also been reported in India, which may be attributed to **waning antibody levels or insufficient childhood immunisation.**
- **Asymptomatic carriers** outnumber active cases by a hundred-fold or more in endemic areas and are the most significant sources of infection.
- In temperate regions, the bacteria are carried predominantly in the nose and throat. Nasal carriers harbour the bacilli for longer periods than pharyngeal carriers.

Agent—*Corynebacterium diphtheriae*

The most important member of the genus *Corynebacterium* is *C. diphtheriae*, the causative agent of diphtheria.

- It is a **pleomorphic, gram-positive bacterium** measuring approximately 3–6 μm × 0.6–0.8 μm.
- These bacteria possess **polymetaphosphate granules or metachromatic granules** (details in Chapter 4).

Antigenic structure

- *C. diphtheriae* have been typed as **(i) gravis, (ii) intermedius, and (iii) mitis** based on their characteristics.
- The gravis and intermedius types are associated with high case fatality rates, while infections due to mitis are less lethal.

- The gravis and, to a lesser extent, the intermedius strains can spread more readily than mitis in populations that are naturally immune or artificially immunised.

Pathogenicity

The pathogenicity of diphtheria bacilli is mediated by their toxins (Fig. 23.8).

Diphtheria toxin

- Virulent strains of diphtheria bacilli produce a **powerful exotoxin** (Table 23.4; Fig. 23.9) known as the diphtheria toxin.
- Almost all strains of *C. diphtheria* gravis and intermedius (95–99 per cent) are toxigenic, while only about 80–85 per cent of mitis strains are toxigenic.
- The diphtheria toxin consists of two fragments, A and B (Table 23.4).

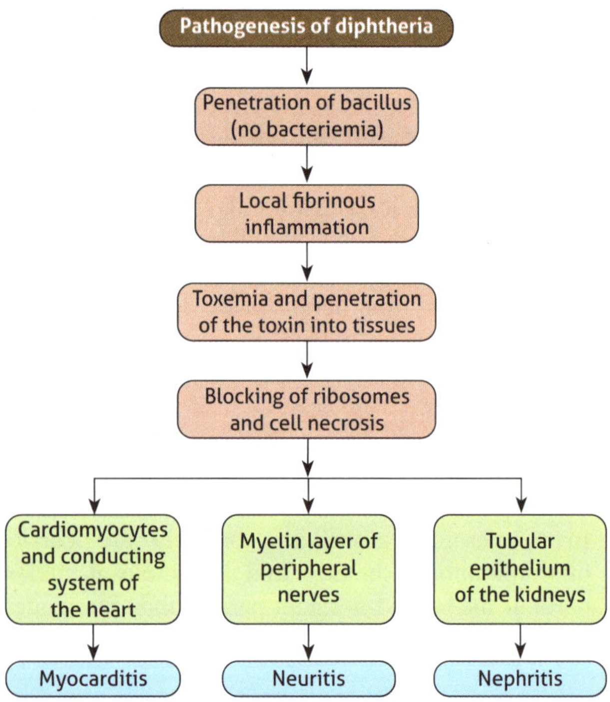

Fig. 23.8 Pathogenesis of diphtheria

Table 23.4 Characteristics of the diphtheria toxin

Lethal dose	0.1 µg/kg
Structure	Two subunits a) A—active domain, responsible for action b) B—binding domain, triggers entry into host cell
Host cell receptor	CD9 and HBEGF (heparin-binding epidermal growth factor)-like precursor
Mechanism	Entry is by attachment to receptor (endocytosis) Action—inhibits protein synthesis by inactivating elongation factor-2 (Fig. 23.9)

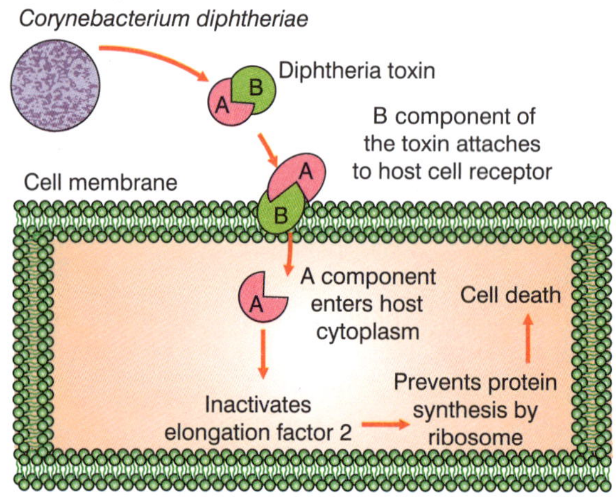

Fig. 23.9 Mechanism of the pathogenesis of diphtheria

- Toxin production is controlled by the tox **gene** which is present on bacteriophages (**tox+**) inside the diptheria bacilli.
- **Lysogenic or phage conversion** happens when non-toxigenic strains are rendered toxigenic by infecting them with beta or other phages. The toxigenicity remains only as long as the bacillus is lysogenic.

Toxoid

- This is a toxin that has lost its toxicity but not its antigenicity.
- It is capable of inducing antitoxin antibodies and hence is used as a vaccine candidate.
- The standard strain almost universally used for toxin production is the Park–Williams 8 strain.

Clinical features

- The incubation period is 3–4 days. In carriers, the incubation period may be very prolonged.
- The site of infection may be:
  - **Pharyngeal**—this is the most common type of infection and may vary from mild catarrhal inflammation to widespread involvement (Case 2/ Fig. 23.10)
  - Laryngeal
  - Otitic
  - Nasal
  - Genital—vulval, vaginal, or preputial
  - Conjunctival
  - Cutaneous

Diphtheria may be classified based on clinical severity into the following types:

i. **Malignant or hypertoxic**—It is charactersed by severe toxemia with marked adenitis (**bull neck**). Death is due to circulatory failure. There is a high incidence of paralytic sequelae in those who recover.

ii. **Septic**—This type leads to ulceration, cellulitis, and even gangrene around the **pseudomembrane.**

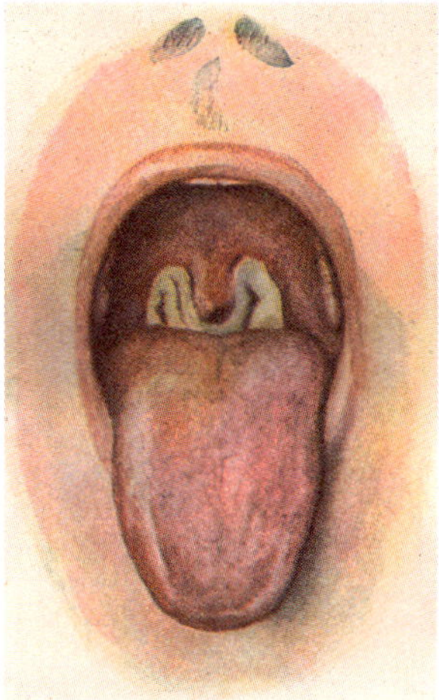

Fig. 23.10 Pharyngeal (faucial) diphtheria with a dirty white membrane bilaterally (*Source*: https://commons.wikimedia.org/wiki/File:Diphtheria.jpg)

iii. **Hemorrhagic**—It is characterised by bleeding from the edges of the pseudomembrane, epistaxis, conjunctival hemorrhage, purpura, and generalised bleeding tendency

Cutaneous diphtheria

- It is caused by non-toxigenic strains *of C. diphtheriae.*
- *Corynebacterium* infects the skin more often than the respiratory tract in the tropics.
- Toxigenic corynebacteria may persist in the skin for over three years.
- Cutaneous infections may stimulate natural immunity to diphtheria but can also lead to pharyngeal diphtheria in non-immune contacts.
- Cutaneous infections are usually secondary to pre-existing skin lesions.
- Sometimes, **diphtheritic whitlow or ulcer** may occur.

Common complications

- **Asphyxia** may occur due to mechanical obstruction of the respiratory passage by the pseudomembrane, for which an emergency tracheostomy may become necessary.
- **Myocarditis** usually occurs within two weeks. ECG changes are common, and there may be cardiac failure.
- **Acute circulatory failure** may be peripheral or cardiac.
- **Post-diphtheritic paralysis** may occur due to polyneuritis affecting the 9th cranial nerve and resulting in palatal palsy with regurgitation of fluids, leading to aspiration; this occurs typically in the third or fourth week of the disease.
- **Toxemia** occurs when the toxin is absorbed and produces toxic damage to the heart (myocarditis), kidney (tubular necrosis), liver, and adrenal glands.
- Local necrotic changes may lead to **fibrinous exudates**. These, together with the disintegrating epithelial cells, leukocytes, erythrocytes, and bacteria constitute the pseudomembrane, which is characteristic of diphtheritic infection.
- Non-toxigenic strains may cause infection even in immunised individuals, as immunity with the toxoid does not confer antibacterial immunity. Such an infection is mild, though pseudomembrane formation may sometimes occur, leading to a fatal outcome.

Typing

Typing methods are used to determine the transcontinental spread. The following methods are used for typing:

- **Ribotyping:** This is presently considered to be the most useful method to type the strains of *C. diphtheriae*.
- **Molecular methods**: **Pulsed-field gel electrophoresis (PFGE), random amplification of polymorphic DNA (RAPD), or amplified fragment length polymorphism (AFLP)** are the other molecular methods used for typing.
- **Bacteriophage typing:** Around 15 bacteriophage types have been described.

Laboratory diagnosis

Specific treatment should be instituted as soon as there is a suspicion of diphtheria without waiting for the results of laboratory tests.

Specimen

- Preferably, two dacron swabs are collected from the lesion under vision, using a tongue depressor.
- The area under the visible membrane should be sampled.

Microscopy

- **Neisser's or Albert's stain:** The granules made up of polymetaphosphates are often situated at the poles of the bacilli and are called **polar bodies.**
- **Gram stain:** The granules are more strongly gram-positive than the rest of the bacterial cell and appear slightly **swollen and club-shaped** (Fig. 23.11). They are often arranged in cuneiform fashion, which differentiates them from diphtheroids.
- **Loeffler's methylene blue:** The granules take up a bluish-purple colour and hence are called **metachromatic granules.**
- **Leishman stain:** This stain is used to rule out Vincent's spirochetes and fusiform bacilli responsible for Vincent's angina, which clinically resembles diphtheria.
- **Immunofluorescence:** Diphtheria bacilli may be identified in smears by the **direct immuno-**

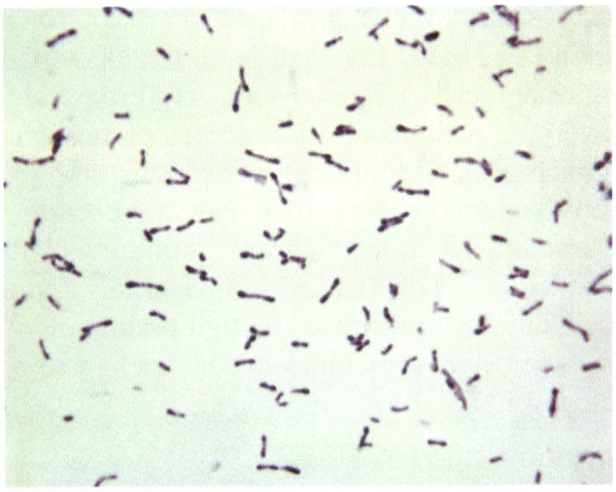

Fig. 23.11 Gram stain showing club-shaped bacilli (*Source*: Centers for Disease Control and Prevention [CDC], Public Health Image Library [PHIL], Image ID 12163)

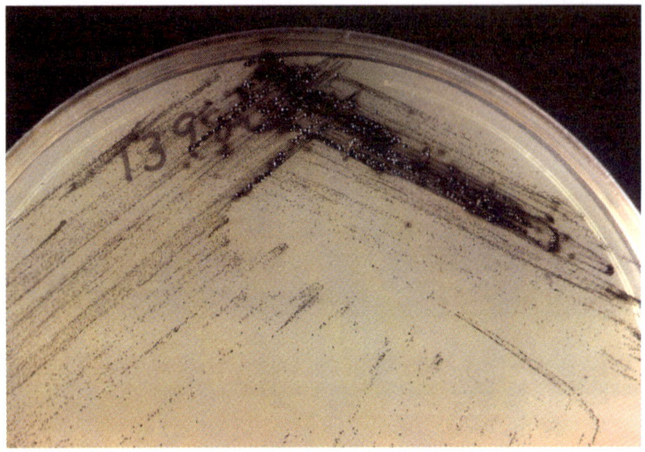

Fig. 23.12 Black colonies of corynebacteria on tellurite agar (*Source*: CDC, PHIL, Image ID 16561)

fluorescence test (DFA). This is most specific as the smear stained with specific antibody is tagged with fluorescent dye.

Isolation by culture

- **Enrichment with blood, serum, or egg** is necessary for good growth.
- The optimum temperature for growth is **37°C** (range 15–40°C) and the optimum **pH is 7.2**.
- *C. diphtheriae* are **aerobes and facultative anaerobes**. Specific media used for their growth are as follows:
  - **Loeffler's serum slope:** Growth is rapid on Loeffler's serum slope, but colonies are difficult to visualise. A smear is made after 6–8 hours of the water of condensation in the test tube and stained with Albert's stain to demonstrate granules.
  - **Tellurite agar or tellurite blood agar:** McLeod's and Hoyle's media yield grey or black colonies of diptheria bacilli (Fig. 23.12). Potassium tellurite in the medium is reduced to metallic tellurium, which produces the black colour of the colonies.

As mentioned before, *C. diphtheriae* is classified into the three biotypes—gravis, intermedius, and mitis—based on colony morphology on tellurite medium. While this differentiation may have no relevance in the clinical scenario, it may have epidemiological significance in an outbreak in a closed community.

Demonstration of toxicity (virulence testing)

Any isolate of *C. diphtheriae* should be tested for toxigenicity for the bacteriological diagnosis to be complete.

In vitro tests

- **Elek's gel precipitation test** (Fig. 23.13)
  - A filter paper disc impregnated with diphtheria antitoxin (10 IU/mL) is placed on a horse serum agar plate.

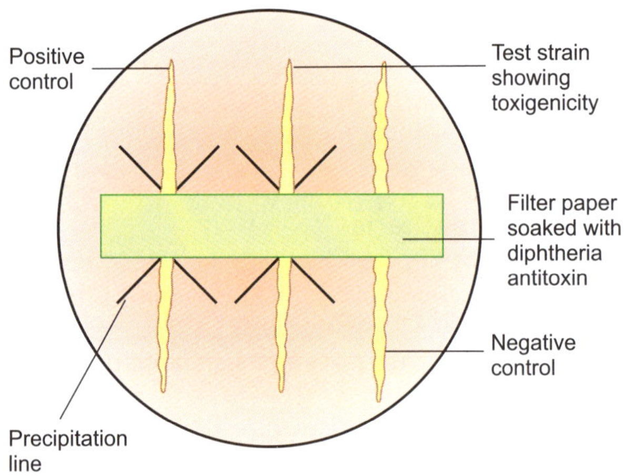

Fig. 23.13 Elek's gel precipitation test

  - At least 10 colonies are spot-inoculated 7–9 mm away from the disc and incubated for 48 hours.
  - The antitoxin diffuses out of the disc and forms precipitin bands between the disc and the bacterial growth at the point where it meets the diphtheria toxin (diffusing out of the organisms).
- **Tissue culture method:** In an agar overlay of cell culture monolayers, the toxin produced by the organism diffuses into the cells below and causes lysis of the cells.
- **PCR:** A polymerase chain reaction has been developed to detect the toxin gene in clinical isolates. This can also be used to detect the toxin from clinical material. This method is useful in patients already on antibiotics, in whom the organisms cannot be isolated.
- **ELISA:** The test may be performed to detect the toxin from the specimen using an antitoxin and enzyme–substrate system.
- **Immunochromatography:** This method can be employed to demonstrate the toxin within hours.

In vivo tests on animals

This method is not recommended anymore because they are prohibited by animal protection acts. They have been replaced by in vitro tests, which are more specific.

> **Laboratory diagnosis of diphtheria**
>
> - **Specimen:** Two dacron swabs collected from under the pseudomembrane
> - **Microscopy:**
>   - **Neisser's or Albert's stain:** Bluish-purple **metachromatic granules** on uniformly stained bacilli
>   - **Gram stain:** Dark purple stained granules in gram-positive bacilli arranged in cuneiform fashion (this differentiates them from diphtheroids, which are arranged in palisades)
>   - **Direct immunofluorescence assay:** The most specific test
> - **Culture:**
>   - Aerobe and facultative anaerobe, optimum temperature—37°C; optimum pH-7.2
>   - Enrichment required with blood, serum, or egg
>   - **Loeffler's serum slope:** Growth is rapid (within 6–8 hours); smear of the water of condensation is stained with Albert's stain to demonstrate granules
>   - **Tellurite agar or tellurite blood agar (McLeod's and Hoyle's media):** Grey or black colonies
>   *C. diphtheriae* are classified into the three biotypes—gravis, intermedius, and mitis—based on colony morphology on tellurite medium
> - **Identification:** Colonies identified by Gram stain and morphology
> - **Demonstration of toxicity (virulence testing)**
>   - ***In vitro tests***
>   - **Elek's gel precipitation test**
>   - **Tissue culture method**
>   - **PCR** for toxin gene
>   - **ELISA**
>   - **Immunochromatography:** Rapid method
>   - ***In vivo tests on animals:*** Not in use; replaced by in vitro tests

Prophylaxis

Immunisation is aimed at increasing the protective levels of antitoxins (antibodies) in the circulation.

Active immunisation

Immunisation against diphtheria is achieved using the toxoided form of the diphtheria toxin as a vaccine. Currently, two preparations are available for active immunisation:

1. **Formol toxoid (also known as fluid toxoid),** which is prepared by incubating the toxin with formalin
2. **Adsorbed toxoid** adsorbed onto insoluble aluminium—it is much more immunogenic than formol toxoid

Dosage: Diphtheria toxoid is usually given to children as a trivalent preparation which also contains tetanus toxoid and pertussis vaccine—the **DPT or triple vaccine**—as prescribed by the Universal Immunisation Programme (UIP). For young children, diphtheria toxoid is given in a dose of 10–25 Lf (limit of flocculation) units.

Treatment

The specific treatment of diphtheria consists of **antitoxin and antibiotic therapy**.

Antitoxin

- It should be administered as soon as diphtheria is suspected, as the fatality rate increases with any delay in initiating antitoxic treatment.
- Dosage—20,000–100,000 units are recommended for serious cases; half of this dose is given intravenously.
- Antitoxin treatment is generally not indicated for cutaneous diphtheria as the causative strains are usually non-toxigenic.

Antibiotic therapy

- *C. diphtheriae* is **sensitive to penicillin** and can be cleared from the throat within a few days of penicillin treatment.
- Diphtheria patients are given a course of penicillin as a supplement to and not instead of antitoxin therapy.
- In cases of penicillin allergy, erythromycin or azithromycin may be given.

Chemoprophylaxis

- Susceptible close contacts of a diphtheria patient are sometimes given **erythromycin or azithromycin** along with the booster dose of the vaccine.
- Erythromycin is more active than penicillin in the treatment of **carriers**.

Other Pathogenic Corynebacteria

Based on changes in growth upon adding lipid to the medium, the other corynebacteria can be divided into two groups—lipophilic and non-lipophilic corynebacteria.

Non-lipophilic corynebacteria (e.g., *C. pseudotuberculosis* and *C. ulcerans*)

- *Corynebacterium ulcerans* is closely related to *C. diphtheriae* and can cause **diphtheria-like lesions.**
- Human infections may be transmitted through **cow's milk.**
- *C. ulcerans* is sensitive to erythromycin. Diphtheria antitoxin is protective against infections caused by *C. ulcerans*.
- **Erythrasma**, a localised infection of the stratum corneum usually affecting the axilla and groin,

is caused by *C. minutissimum*, which produces superficial skin lesions. This is a lipophilic corynebacterium and can be grown readily in media containing 20% fetal calf serum.

Lipophilic corynebacteria (e.g., *C. jeikeium* and *C. urealyticum*)

- *Corynebacterium jeikeium* is one of the most important members of this group and can cause cutaneous and bloodstream infections with a high mortality rate.
- Such infections are often seen in immunocompromised hosts.
- This organism is usually **multidrug-resistant**, responding only to vancomycin.
- It is increasingly being associated with infective endocarditis.
- *Arcanobacterium haemolyticum* (formerly *Corynebacterium haemolyticum*) can cause **pharyngitis and skin ulcers**.

Diphtheroids

- Some corynebacteria occur as normal commensals in the throat, skin, conjunctiva, and other areas.
- These may sometimes be mistaken for diphtheria bacilli and are called diphtheroids.
- In general, diphtheroids stain more uniformly than *C. diphtheriae*, possess few or no metachromatic granules, and tend to be arranged in parallel rows (palisades) rather than in a cuneiform pattern.
- The common diphtheroids are ***C. pseudodiphtheriticum*** (*C. hofmannii*), which is found in the throat, and ***C. xerosis***, which is found in the conjunctiva.

Other Coryneform Bacteria

- Besides the genus *Corynebacterium*, a number of other genera of coryneform bacteria have been defined.
- Among them, the genus ***Propionibacterium*** is of medical interest as three species of this genus are constantly present on human skin—*P. acnes*, *P. granulosum*, and *P. avidum*. They are anaerobic and aerotolerant, growing well in media containing lipids. *P. acnes* is often isolated from lesions of acne, but its pathogenic role is uncertain.

Important *Corynebacterium* species and the diseases caused by them and are summarised in Table 23.5.

Table 23.5 Important diseases and species of *Corynebacterium*

Corynebacterium species	Disease produced
C. diphtheriae	Toxigenic strains—diphtheria Non-toxigenic strains—bacteremia, endocarditis
C. ulcerans	Toxigenic strains—diphtheria
C. pseudotuberculosis	Toxigenic strains—diphtheria
C. jeikeium	Infective endocarditis and bacteremia in immunocompromised hosts
C. urealyticum	Urinary tract infections
C. striatum	Healthcare-associated infections
Arcanobacterium haemolyticum	Pharyngitis, wound infections, and septicemia

Miscellaneous Bacterial URIs

Vincent's angina

- *Borrelia vincentii* is a motile spirochete, about 5–20 μm long and 0.2–0.6 μm wide, with 3–8 coils of variable sizes.
- It is easily stained with dilute carbol fuchsin and is gram-negative.
- It is a normal **mouth commensal** but may, under predisposing conditions such as malnutrition or viral infections, give rise to ulcerative gingivostomatitis (**Vincent's angina**; also known as trench mouth, due to the foul-smelling odour of the mouth; Fig. 23.14a) or laryngopharyngitis.
- In these cases, *B. vincentii* is always associated with **fusiform bacilli** (*Fusobacterium fusiforme*; Fig. 23.14b). This symbiotic infection is known as fusospirochetosis.
- **Diagnosis** may be made by demonstrating spirochetes and fusiform bacilli in stained smears

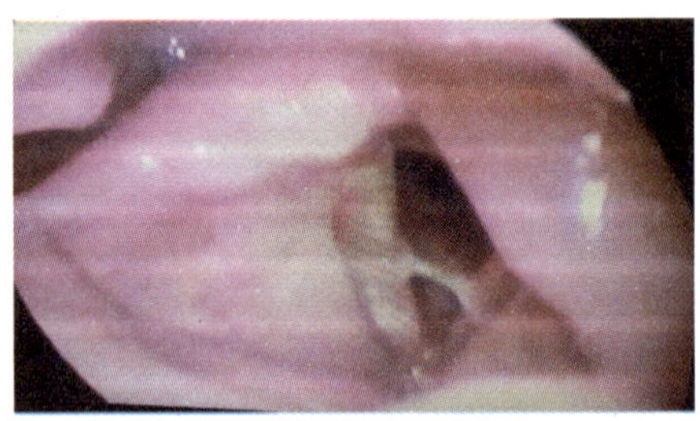

(a)

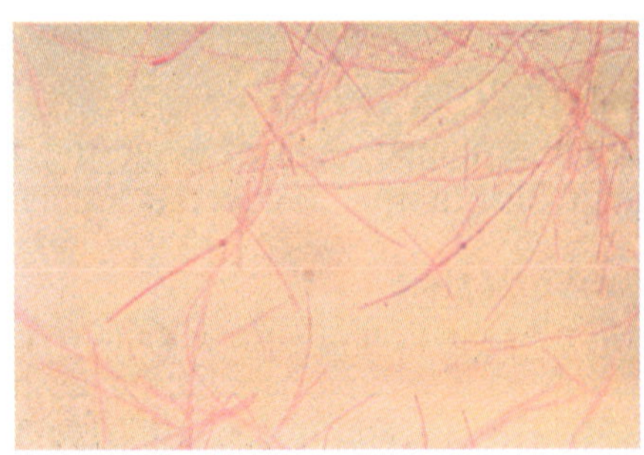

(b)

Fig. 23.14 (a) Oropharyngitis (Vincent's angina); patient recovered dramatically with penicillin and (b) fusiform bacilli from Vincent's angina (*Source*: [a] Department of ENT, Pondicherry Institute of Medical Sciences, Puducherry and [b] CDC, PHIL, Image ID 3044)

of exudates from the lesions. *B. vincentii* may be anaerobically cultivated with difficulty in enriched media.
- Penicillin and metronidazole are effective in **treatment**.

Moraxella catarrhalis

- *Moraxella catarrhalis* is an **exclusively human** opportunistic pathogen and is a common cause of **otitis media** in infants and children.
- It resembles commensal *Neisseria* species in culture. It is a gram-negative diplococcus and is catalase- and oxidase-positive.
- The tendency of the organism to colonise the upper respiratory tract is high in **infants and children** but decreases in adulthood.
- It can cause **sinusitis, laryngitis, bronchopneumonia, and meningitis**.
- In a significant proportion of **elderly patients** with chronic obstructive pulmonary disease and chronic bronchitis, it causes lower respiratory tract infections.
- These organisms are inert biochemically (asaccharolytic).
- *Moraxella* grow on ordinary media unlike the pathogenic *Neisseria*.

VIRAL URIs

- Almost 70% of upper respiratory tract infections are caused by viruses.
- They are responsible for endemic, epidemic, and pandemic outbreaks across the globe affecting all age groups.
- However, certain susceptible populations are at a greater risk—those at the extremes of age groups and immunocompromised individuals.

Common Cold

Epidemiology

- The common cold is a **mild infection of the upper respiratory tract** that is mostly viral in etiology.
- It is reported from across the globe.
- It is **transmitted by droplets**. Hand-to-hand contact followed by self-inoculation of conjunctival or nasal mucosa appears to be an important mode of transmission.
- Multiplicity of serotypes makes **vaccination impossible**.
- Common cold is a syndrome produced not only by rhinoviruses but also by a variety of other groups such as respiratory syncytial, corona, coxsackie, echo, adeno, influenza, and parainfluenza viruses.
- Most common viruses that can cause common cold are given below:
  - Rhinovirus
  - Respiratory syncytial virus
  - Influenza A viruses
  - Adenoviruses
  - Parainfluenza virus
  - Coronavirus
  - Metapneumovirus
  - Parainfluenza virus
  - Influenza virus
  - Adenovirus
  - Metapneumovirus
  - Bocavirus

Clinical features

Many cases may be asymptomatic or may with just some mild symptoms, which are as follows:

- Sneezing
- Stuffy or runny nose
- Sore throat
- Cough
- Headache
- Myalgia (body pain)

Respiratory viruses causing common cold may cause complications such as otitis media, sinusitis, bronchiolitis, pneumonia, and exacerbation of respiratory chronic diseases.

Pathogenesis of common cold (Fig. 23.15)

- Local inflammation and cytokines may be responsible for the symptoms of common cold.
- Interferon production occurs early, and specific antibodies appear in nasal secretions.

Adenoviral URIs

Agent—Adenoviruses

- Adenoviruses are **70–75 nm** in size, with a characteristic morphology.

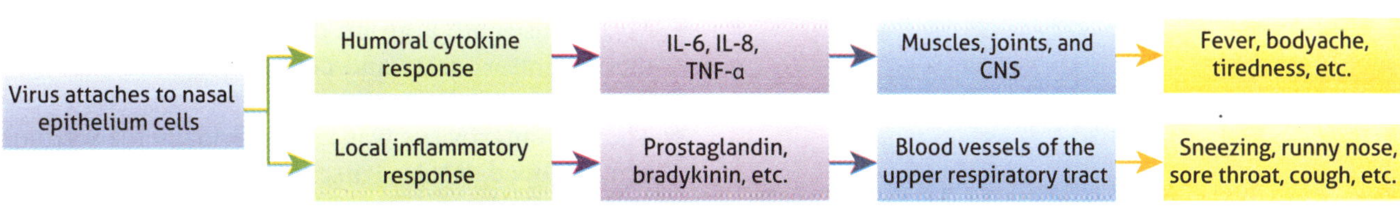

Fig. 23.15 Pathogenesis of common cold

- The capsid is composed of **252 capsomers arranged as an icosahedron with 20 triangular facets and 12 vertices**, giving the virion the appearance of a **space vehicle** (Fig. 23.16a and b).

Classification

- Over 50 serotypes of adenoviruses have been isolated from human sources and are divided into **groups A–F**.
- Human adenoviruses are classified into six groups (also called subgroups) based on properties such as hemagglutination, DNA fragment analysis, and oncogenic potential.
- Adenoviruses are relatively stable, remaining viable for about a week at 37°C. They are readily inactivated at 50°C. They resist ether and bile salts.

Pathogenicity

- Adenoviruses cause **infections of the respiratory tract, eye, bladder, and intestine**.
- More than one type of virus may produce the same clinical syndrome; on the other hand, a single type may cause clinically different diseases (Table 23.6).

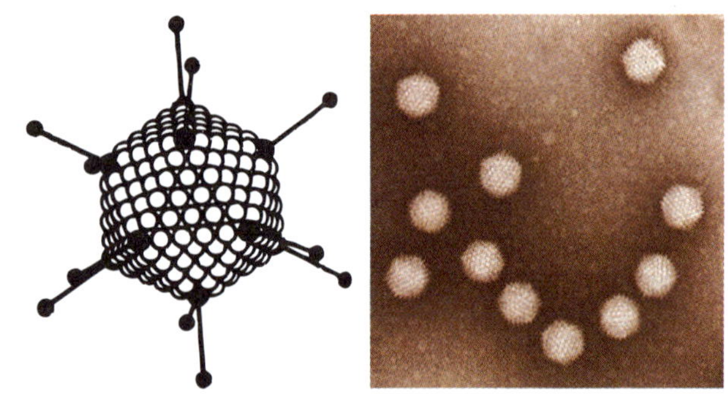

(a) (b)

Fig. 23.16 Adenovirus: (a) Diagrammatic representation of structure and morphology and (b) microscopic view

Table 23.6 Common syndromes associated with adenovirus infection

Syndrome	Principal serotypes
Respiratory disease in children	1, 2, 5, 6
Sore throat, febrile cold, and pneumonia	3, 4, 7, 14, 21
Acute respiratory disease (ARD) in military recruits	4, 7, 21
Follicular (swimming pool) conjunctivitis	3, 7
Epidemic keratoconjunctivitis (shipyard eye)	8, 19, 37
Diarrhea	40, 41
Pneumonia, hepatitis, colitis, etc., in transplant recipients	34, 35
Acute hemorrhagic cystitis in children and generalised exanthema	11, 21

Laboratory diagnosis

Routinely, the diagnosis of adenoviral URIs is not done in the laboratory as these are often self-limiting illnesses and diagnosis is time-consuming.

- **Tissue culture:** Human adenoviruses grow only in tissue cultures of human origin such as **human embryonic kidney, HeLa, or HEp-2**. Cytopathic changes may take several days to develop and consist of **cell rounding and aggregation** into grape-like clusters. Intranuclear inclusions may be seen in stained preparations.
- **Shell vial technique:** The rate of isolation of the virus from the throat, eye, urine, or feces can be increased using this technique. The processed specimen is inoculated onto shell vials with monolayers of cell lines and centrifuged for 15 minutes at 4,300 rpm (3,500 × g). This increases the adsorption of viral particles onto the cells, thus facilitating adherence and penetration into the cells. The time to positivity decreases. Cytopathic effects can be seen by direct fluorescent or immunofluorescent staining. **Typing** is done by neutralisation tests in tissue culture or by molecular techniques.
- **Serodiagnosis:** A rising antibody titre in paired sera is diagnostic. The examination of a single sample of serum is inconclusive as adenovirus antibodies are so common in the population.
- **Electron microscopy:** This modality is used for fecal virus detection.
- **Immunofluorescence:** It is used for viral antigen detection in nasopharyngeal and ocular infections.
- **PCR-based tests:** These are gaining popularity in the detection of viral DNA.

Adeno-associated viruses (AAV)

Electron microscopy of adenovirus preparations have revealed **small, icosahedral viral particles, 20–25 nm in diameter, which are unable to replicate independently and are called defective viruses** or viruses dependent on helper viruses. They can multiply only in cells simultaneously infected with adenoviruses and hence are called adeno-associated viruses (AAV) or adenosatellite viruses. They can be detected by electron microscopy and complement fixation or immunofluorescence with specific antisera. Types 1, 2 and 3 are of human origin and cause natural infection.

Myxoviral URIs

- Myxoviruses are a group of viruses associated with the respiratory tract and comprise **orthomyxovirus and paramyxoviruses.**
- The term myxovirus was used originally for a group of **enveloped RNA viruses characterised by their ability to adsorb onto mucoprotein receptors on erythrocytes**, causing **hemagglutination**—the name

refers to the **affinity of the viruses to mucins** (from *myxa*, meaning mucus).

Classification

- **Orthomyxoviridae**
- **Paramyxoviridae**

Table 23.7 lists the important differences between orthomyxoviruses and paramyxoviruses.

Influenza

> **Case 3.** *A 60-year-old woman with a history of chronic obstructive airway disease was admitted to the emergency ward with a cough and fever of 39°C. Her fever and cough showed rapid progression and she later developed dyspnea with respiratory distress. Her blood gas analysis was consistent with hypoxia and her chest X-ray showed bilateral ground-glass opacities with areas of consolidation. Her nasopharyngeal aspirate was subjected to PCR, which was positive for the* ***influenza virus H1N1****. She succumbed to the infection despite supportive care and oseltamivir antiviral therapy.*
> i. What is influenza?
> ii. Differentiate between antigenic drift and shift.

Influenza is an **acute infectious disease of the respiratory tract** which occurs in sporadic, epidemic, and pandemic forms. There are four types of influenza viruses—types A, B, C, and D.

Table 23.7 Distinguishing features of orthomyxoviruses and paramyxoviruses

Property	Orthomyxovirus	Paramyxovirus
Size of virion	80–120 nm	100–300 nm
Shape	Spherical; filaments in fresh isolates	Pleomorphic
Genome	Segmented; eight pieces of RNA	Single, linear molecule of RNA
Diameter of nucleocapsid	9 nm	18 nm
Site of synthesis of ribonucleoprotein	Nucleus	Cytoplasm
Genetic reassortment	Common	Absent
Dependent RNA synthesis	Required for multiplication	Not required
Effect of actinomycin D	Inhibits multiplication	Does not inhibit
Antigenic stability	Variable	Stable
Hemolysin	Absent	Present
Virus included	Influenza virus	Parainfluenza viruses, Newcastle disease virus, mumps virus, measles virus, and respiratory syncytial viruses

Epidemiology

- Influenza occurs sporadically as epidemics or in pandemic form. The **source** of infection is an infected individual.
- Influenza virus **type C** is endemic throughout the world and causes very mild or unapparent infections.
- **Type B strains** cause sporadic as well as epidemic influenza, while **type A strains** can cause pandemics as well.
- Epidemic influenza is significant in **temperate regions** during the winter months, causing considerable mortality in the aged and in those with cardiopulmonary diseases. In the **tropics**, it tends to occur frequently in the monsoon season.
- Pandemics make influenza an important and challenging disease. Under the auspices of the WHO, worldwide surveillance is maintained on influenza. Influenza pandemics have been recorded at irregular intervals from 1173. The **H1N1 virus**, which emerged in China and Russia in 1977, has spread across most parts of the world. The H3N2 virus currently causes human influenza.
- Non-human influenza viruses play an important role in the emergence of pandemic influenza.
- **Birds**, particularly aquatic, appear to be the primary reservoirs of influenza viruses; natural infection has been identified in several avian species.
- In birds, the virus usually produces an asymptomatic intestinal infection. The cloaca of healthy wild birds is the best source for the isolation of avian influenza viruses.
- Pandemics are caused by virus **strains that have undergone an antigenic shift** with marked variation involving many different polypeptides simultaneously.
- It is now held that pandemic strains originate from some animal or avian reservoir, either spreading to humans directly by host range mutation or as a result of recombination between human and non-human strains.
- Hybrids can be produced by growing human and non-human strains together in eggs. Recombinants can also be obtained from experimental animals exposed to mixed infection (e.g., the 1957 Asian virus and the 1968 Hong Kong virus).

Current status of influenza in India

- In India, epidemics occurred in 2017 (38,811 cases with 2,270 deaths) and 2019 (28,798 cases with 1,218 deaths).
- In 2023, 2,793 confirmed cases of H1N1 with 52 deaths were reported as of 31 July (National Centre for Disease Control [NCDC], Government of India).

Agent—Influenza virus

Classification

- The classification of influenza viruses into the three serotypes—A, B, and C—is based on the antigenic nature of the 'internal' or **ribonucleoprotein (RNP) and the matrix (M) protein antigens**.
- These antigens are not cross-reactive amongst the three types.

Influenza A virus

- Type A influenza virus comprises animal and human strains.
- The influenza virus type A virus is further divided into subtypes based on the **hemagglutinin (HA) and neuraminidase (NA)** glycoproteins on the surface of the virus.
  - 18 HA subtypes: H1–H18
  - 11 NA subtypes: N1–N11
- More than 130 influenza A subtype combinations have been identified in nature.
- **Human isolates belong to H1–H3, and H5 and N1, N2 subtypes.** Currently, human strains are subtype A(H1N1) and A(H3N2) influenza viruses.
- Only influenza type A viruses are known to have caused pandemics.

Influenza B virus

- These are not classified into subtypes but can be broken down into lineages.
- Currently circulating influenza type B viruses belong to either **B/Yamagata or B/Victoria lineage**.
- Type B consists of exclusively human viruses (natural infection has not been identified in animals or birds).

Influenza C virus

- It is detected less frequently and usually causes mild infections, and thus, does not present public health importance.
- Influenza C contains human and swine viruses.

Influenza D viruses

These primarily affect cattle and are not known to infect or cause illness in humans.

Nomenclature systems

- The standard nomenclature systems for influenza viruses include the following information—serotype, host of origin (except for human isolates), geographic origin, strain number and year of isolation, followed by HA and NA subtypes in parenthesis.
- For a human isolate, it is as follows—A/Hong Kong/01/68 (H3N2); for other host origin, it is expressed as follows—A/swine/Iowa/15/30 (H1N1).

Morphology

- The influenza virus is a **spherical, enveloped virus with a diameter of 80–120 nm**; some strains show pleomorphism.
- **Filamentous forms,** up to several micrometres in length and readily visible under the dark-ground microscope, are frequently seen in **freshly isolated strains**.
- The core consists of a ribonucleoprotein in helical symmetry.

Genome

- The genome comprises a **negative-sense single-stranded RNA** which is segmented into eight pieces (13.6 kb overall size).
- The virus contains a viral **RNA-dependent RNA polymerase** for the transcription of the viral RNA in infected host cells.

Envelope

The surface membrane contains two matrix proteins ('M proteins'):

- M1 provides structural integrity
- M2 is the ion channel between the internal milieu and the exterior

Projecting from the lipid envelope are two types of spikes—**hemagglutinin (HA) spikes and neuraminidase (NA)** spikes, which are less numerous (Fig. 23.17).

Antigenic properties

The internal antigens of the influenza virus can be classified into the following types:

- **RNP (ribonucleoprotein) antigen**
  - It is an internal antigen which is a ribonucleoprotein.
  - Based on this antigen, influenza viruses are classified into types A, B, and C.
  - Anti-RNP antibodies develop after infection but not following the administration of killed vaccines.

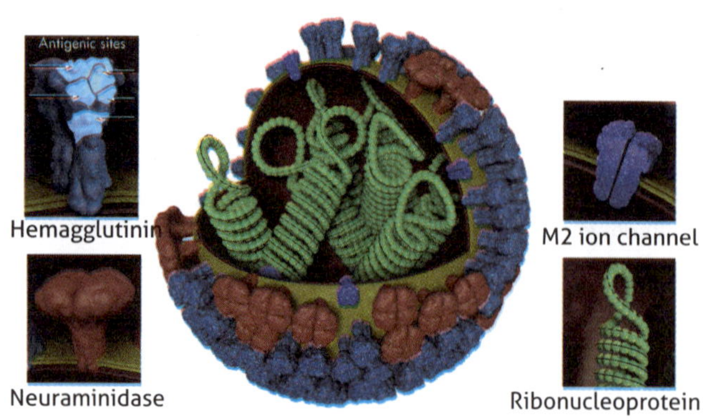

Fig. 23.17 Influenza virus showing surface protein hemagglutinin (HA), neuraminidase (NA) and the segments of ribonucleoprotein and M2 ion channels (*Source*: CDC, PHIL, Image ID 17346)

- **Hemagglutinin**
  - It is a glycoprotein composed of two polypeptides, HA1 and HA2.
  - It is responsible for hemagglutination and hemadsorption.
  - It enables the virus to adsorb onto mucoprotein receptors on red cells as well as respiratory epithelial cells.
  - Anti-hemagglutinin protective antibodies are produced following infection and immunisation.
  - Hemagglutination is an important characteristic of influenza viruses and is used to titrate the virus.
- **Neuraminidase**
  - It is a glycoprotein enzyme that destroys cell receptors by hydrolytic cleavage.
  - The anti-neuraminidase antibody is formed following infection and immunisation.
  - It is not as effective in protection as the anti-hemagglutinin antibody.
  - Of the 9 subtypes identified, only N1 and N2 have been recovered from humans.

Antigenic variation

- A unique feature of the influenza virus is its ability to undergo antigenic variation; this significantly impacts the epidemiology of the disease.
- Antigenic variability is **highest in influenza virus type A**; it is lesser in type B and has not been demonstrated in type C influenza viruses.
- The internal RNP antigen and M protein antigen are both stable, but these surface antigens undergo independent antigenic variations, either an **antigenic shift or drift** (Table 23.8).

Immunity

- An episode of influenza confers effective protection for one or two years. The apparent short duration of immunity is due to the antigenic variation that the virus exhibits frequently **(antigenic drift)**.
- It is the local concentration of anti-hemagglutinin and, to a smaller extent, of anti-neuraminidase antibodies (mainly IgA) in the respiratory tract that are more relevant in protection.
- Influenza virus infection induces cell-mediated immunity, but the role of CMI in protection has not been clarified.

> **'Original antigenic sin'**: When an individual experiences repeated infections with different antigenic variants of influenza virus type A, he responds by forming antibodies not only against each infecting strain but also against the strain that he first came into contact with. Hence, the dominant antibody response will be against the strain that caused the earliest infection. This is known as the doctrine of 'original antigenic sin'.

Table 23.8 Difference between antigenic drift and antigenic shift

Feature	Antigenic drift	Antigenic shift
Change in antigen structure	Gradual sequential change	Abrupt, drastic and discontinuous variation
Reason	Mutation and selection	Exchange of gene segments and re-assortment of segmented genome
Epidemiology	Periodic epidemics of influenza	Major epidemics or pandemics
New virus strain	Minor changes within subtype	Novel, unrelated antigenically to the predecessor strains
Occurs	Frequently	Rarely
Infectivity	New strain may infect the hosts of same species	New viral subtype may infect another host in a different species
Treatment	Easy	Difficult
Occurrence	In influenza type A, B, and C	In influenza type A only
Examples	A/Fujian (H3N2) replaced A/Panama (H3N2) in 2003–2004 Panama (H3N2) in 2003-2004	H3N2 replaced H2N2 in 1968

Susceptibility

- The influenza virus remains viable at 0–4°C for about a week.
- It can be preserved for years at –70°C or by freeze-drying.
- The virus survives slow drying and may remain viable on fomites for about two weeks.
- Ether, formaldehyde, phenol, salts of heavy metals, and many other chemical disinfectants destroy infectivity. Iodine is particularly effective.
- The hemagglutinating, enzymatic, and complement-fixing effects of the virus are more stable than its infectivity.

Pathogenesis

The disease is confined to the respiratory tract. Fig. 23.18 presents the sequence of events involved in an influenza virus infection.

Clinical features

- The incubation period is **1–3 days.**
- The disease varies in severity from **mild coryza** to **fulminating and rapidly fatal pneumonia**.
- Most infections are subclinical.
- It usually has an **abrupt onset** with fever, headache, and generalised myalgia.

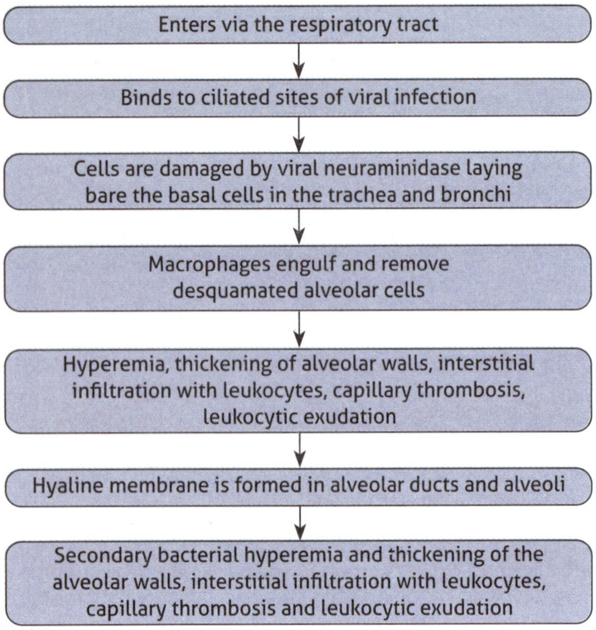

Fig. 23.18 Sequence of events involved in the pathogenesis of influenza

- Respiratory symptoms are prominent with **sore throat and cough**.
- Severe prostration is common.
- Abdominal pain and vomiting may occur, especially in type B infections in children, which may even present as an acute abdominal emergency (**gastric flu**).
- The uncomplicated disease resolves within about seven days.
- Very young and elderly individuals, pregnant women and immunocompromised individuals are at a greater risk of infection.

Complications

- **Pneumonia** may develop due to bacterial superinfection or, rarely, may be caused by the virus itself (Case 3).
- Cardiac complications such as **congestive failure or myocarditis** and neurological involvement such as encephalitis may occur rarely.
- **Reye syndrome (encephalopathy and fatty liver changes with raised transaminase)** may be a complication of influenza, with type B affecting young children.

Laboratory diagnosis

- **Specimen collection:** Necessary infection control measures are adopted for the collection of specimens in suspected cases of influenza. Throat garglings, nasal washes, or throat swabs (alginate swabs) are collected.
- **Transport:** Viral transport media or suitable buffered salt solution should be used. If the specimen is not processed immediately, it should be stored at 4°C, or if the delay is long, at –70°C.
- **Demonstration of the virus antigen:** A rapid diagnosis of influenza may be made by demonstrating the virus antigen on the surface of the nasopharyngeal cells by immunofluorescence.
- **Isolation of the virus**: Virus isolation is achieved readily during the first two or three days of the illness but less often in the later stages.
  - ***Primary monkey kidney cell cultures***—the virus grows well.
  - ***Continuous cell lines***—cytopathic effects are not prominent and virus growth is detected by hemadsorption or the demonstration of hemagglutinin in the culture fluid.
  - ***Amniotic cavity of chick embryos***—for viral isolation. After a few egg passages, the virus grows well in the allantoic cavity also, except for the type C.
  - ***Shell vial cultures***—the virus can be identified in 1–4 days on the basis of monoclonal antibodies.
- **Serology**: Several antibodies are produced against hemagglutinin, the matrix, neuraminidase, and nucleoprotein. Serodiagnosis is based on **hemagglutination inhibition (HAI) tests**, which offer a convenient and sensitive method for the detection and quantitation of the antibody for serological diagnosis of influenza. The highest dilution of serum that inhibits hemagglutination is its **HAI titre**. It is essential to examine paired sera in parallel to demonstrate a rise in the titre of antibodies.
  - ***Disadvantages***—as the anti-hemagglutinin antibodies are subtype-specific, it is necessary to use the strain currently causing infection as the antigen.
- **Hemagglutination:** This test provides a convenient method for the detection and titration of the influenza virus in egg and other culture fluids. The highest dilution of the virus suspension that produces agglutination of a fixed quantity of cells is known as its **hemagglutination (HA) titre**. Hemagglutination is used to titrate inactivated influenza viruses for the standardisation of killed influenza virus vaccines.
- **Hemadsorption:** Virus growth can be identified by hemadsorption (wherein RBCs are adsorbed onto the cell surfaces of the tissue culture) with human O group, fowl, or guinea pig erythrocytes.
- **Molecular methods:** Due to the cumbersome and time-consuming nature of tissue culture methods and lack of specificity with serological tests, laboratory confirmation now relies on molecular methods to detect the viral genomes. This also helps in monitoring mutations.

- **PCR-based diagnosis:**
  - Real-time reverse transcriptase PCR (RT-PCR) uses specific primers targeted at the viral RNA.
  - Multiplex PCR is used to detect subtypes of the virus as well as other respiratory viruses.
  - These techniques are sensitive, specific, and rapid (results are available within 3–4 hrs). The tests help in specific management and preventive strategies.

Laboratory diagnosis of influenza

- **Specimen collection:** Throat garglings, nasal washes, or throat swabs (alginate swabs)
- **Transport:** Viral transport media or suitable buffered salt solution
- **Storage:** In case of delay, store at 4°C, for 2 hours and at –70°C for more than 2 hours
- **Demonstration of the virus antigen using** immunofluorescence
- **Isolation of the virus:**
  - Primary monkey kidney cell culture
  - Continuous cell lines
  - Amniotic cavity of chick embryos
  - Shell vial cultures
- **Identification in cell lines done by hemadsorption and hemagglutination**
- **Serology**: Hemagglutination inhibition (HAI) test to detect antibody
- **Molecular methods:**
  - ***Real-time reverse transcriptase PCR (RT-PCR)*** to detect viral RNA
  - ***Multiplex PCR*** to detect subtypes of the virus as well as other respiratory viruses

Immunoprophylaxis

- The most significant difficulty in the immunoprophylaxis of influenza is the frequent change in the antigenic makeup of the virus.
- In colder countries, where it is necessary to protect the elderly and other high-risk individuals, the accepted practice is to immunise them with a vaccine containing the latest strains of type A and B viruses.
- Both live and killed vaccines are licensed for use (Table 23.9).

Indications for influenza vaccination

The World Health Organization recommends vaccination for the following groups:

- Those >65 years of age
- Children with asthma
- Immunocompromised persons
- Those with chronic obstructive pulmonary disease, bronchial asthma, heart disease, liver disease, kidney disease, blood disorders, diabetes, and cancer, as well as persons who might transmit to other high-risk groups (healthcare providers attending to persons with influenza)

National vaccine recommendation

- **Seasonal influenza** caused by HINI, H3N2, and influenza B affects the elderly (>65 years) and young children, who are at a higher risk. With this in mind, the Ministry of Health and Family Welfare, Directorate General of Health Services (National Centre for Disease Control), India, has issued 'Seasonal Influenza Guidelines' for vaccination with a quadrivalent influenza vaccine composed of A (H1N1 and H3N2) and B strains.
- **Chemoprophylaxis**—antiviral drugs **amantadine and rimantadine are** used. These drugs block the viral M2 protein, which functions as an ion channel. However, these drugs act only against the type A virus and not against the type B, which lacks the M2 components.

Table 23.9 Available vaccines for influenza

Type of vaccine	Feature	Adverse reactions	Approval
Egg-based inactivated influenza vaccine (egg-based IIV4s)	Made with influenza viruses that have been grown in eggs and contain very small residual amounts of egg proteins	Pain, redness, and swelling at the injection site Fever, malaise, myalgia, and other systemic symptoms Guillain–Barré syndrome (GBS)	Persons over six months of age Pregnant women Persons with chronic medical conditions High-risk groups with comorbidities
Cell culture-based inactivated influenza vaccine (ccIIV4)	Quadrivalent vaccine made with influenza viruses that have been grown in a cell culture		
Recombinant influenza vaccine (RIV4)	Vaccine is produced in an insect cell line using genetic sequences from cell-derived influenza viruses and is manufactured without the use of influenza viruses or eggs		
Live-attenuated (cold-adapted) influenza vaccine (LAIV)	Administered intranasally Made with a temperature-sensitive mutant (Ts) that can be readily isolated and is able to grow at the lower temperature of the nasopharyngeal mucosa (32–34°C) but not in the lungs at 37°C	Runny nose or nasal congestion	Persons aged 2–49 years who do not have underlying medical conditions

Treatment

- Four approved antiviral drugs are available:
  - Oseltamivir (Tamiflu)—designed to block viral neuraminidase when administered as a nasal spray
  - Zanamivir (Relenza)—designed to block viral neuraminidase
  - Peramivir (Rapivab)
  - Baloxavir (Xofluza)
- Amantadine and rimantadine reduce the average duration of the disease and cause symptomatic improvement, though virus shedding and antibody response are not affected. However, resistance to these drugs develops rapidly.
- Antivirals are prescribed within the first day or two of symptoms to reduce the severity of symptoms and possibly, the risk of complications.

Avian influenza

- A highly pathogenic avian influenza (bird flu) caused by the **H5N1 strain** was reported from some countries in 1914. This led to millions of birds being culled in several countries with great economic loss (Fig. 23.19).
- The virulence of H5N1 is greater than that of H1N1.
- Despite several poultry outbreaks, India has not reported a single human H5N1 case.
- These incidences indicate the value of influenza surveillance and the potential danger from avian strains.

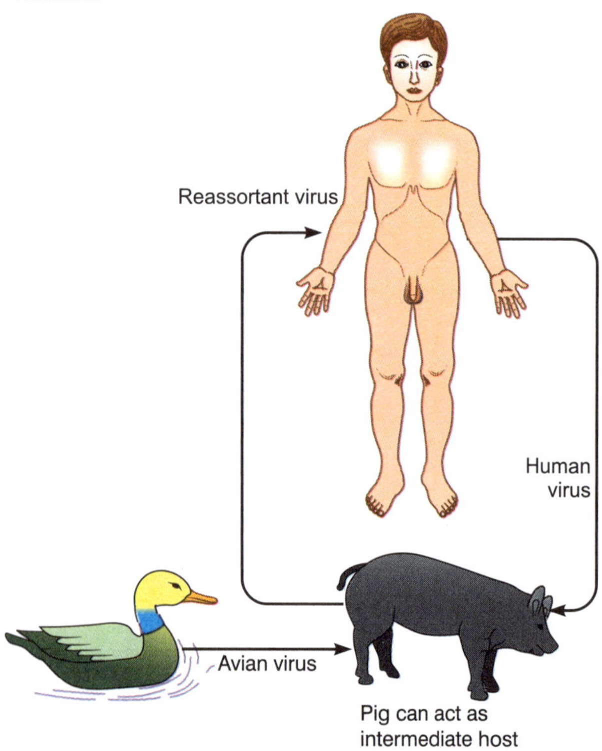

Fig. 23.19 Transmission cycle of avian influenza virus—human → pig → aquatic bird (waterfowl)

Transmission of novel strains

- Wild aquatic birds carry the full repertoire of the genes of all influenza strains, including old human pandemic strains.
- The viruses do not cause any disease in them or undergo any mutational changes.
- Birds shed the viruses abundantly in feces, which in turn, contaminate lakes and ponds.
- Recombination may take place in pigs, and such hybrid strains may cause human infection with a potential to reach pandemic proportions.
- The reappearance of old strains—like the H1N1 in 1977—is linked to avian reservoirs of strains.
- A new H7N9 influenza virus emerged in 2013, which caused infections in humans. In the past, this strain was known to infect only birds.

Swine influenza

- In March 2009, a new H1N1 virus was detected, which was a reassortant between previously circulating swine virus and a Eurasian swine virus.
- Human cases of influenza A (H1N1) have been reported globally since 2009 with several thousand deaths.
- This virus has now become the cause of seasonal influenza in various parts of the world, including India, and is expected to continue to do so for many years.
- Annually, several cases are admitted to the hospitals during the colder seasons.

Paramyxoviral Infections

- Paramyxoviruses resemble orthomyxoviruses in morphology but are larger and more pleomorphic. The members of this group and the infections caused by them are presented in Table 23.10.
- They are roughly spherical and range in size from 100–300 nm, sometimes with long filaments; some giant forms measure as much as 800 nm.
- The **helical nucleocapsid has a diameter of 18 nm** (except in the pneumovirus, in which it is 13 nm).
- Its genome comprises an unsegmented, linear, single-stranded RNA.

Classification

The family Paramyxoviridae is divided into two subfamilies:

i. Paramyxovirinae
ii. Pneumovirinae

All the genera in these two subfamilies are monotypic (one serotype) and antigenically stable.

Antigenic structure

- The nucleocapsid (Fig. 23.20) is surrounded by a lipid envelope that has the **matrix (M) protein** at its

Table 23.10 Important characteristics of Paramyxoviridae

Properties	Paramyxovirinae				Pneumovirinae	
	Respirovirus	**Rubulavirus**	**Morbillivirus**	**Henipavirus**	**Pneumovirus**	**Metapneumovirus**
	Parainfluenza 1–3	**Mumps Parainfluenza 2–4**	**Measles**	**Nipah Hendra**	**Resp. syncytial virus**	**Human metapneumovirus**
Diameter of nucleocapsid	18 nm				13 nm	
Fusion (F) protein	Fusion protein is present in all of them					
Hemagglutinin/ Hemadsorption	Possess HA and exhibit hemadsorption on specific RBCs			Do not have HA and do not exhibit hemadsorption		
Neuraminidase (NA)	Neuraminidase on the surface		NA is absent	Do not possess NA on the surface		
Intracellular inclusions in cytoplasm (C) / nucleus (N)	Intracytoplasmic inclusion bodies		Intranuclear and intracytoplasmic bodies	Intracytoplasmic inclusion bodies		Not known

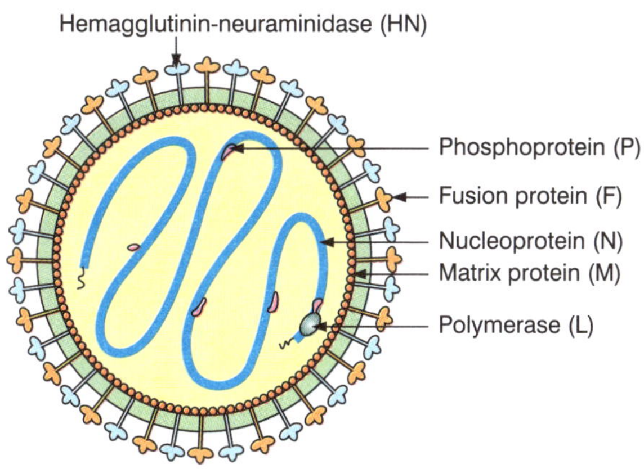

Fig. 23.20 Antigenic components of paramyxoviruses

base and **two types of transmembrane glycoprotein spikes on the surface.**

- The longer spike is **hemagglutinin (HA) and may also possess neuraminidase (NA) activity**; hence, it is known as the **H or HN protein.** It is responsible for the adsorption of the virus to the host cell surface.
- The second spike is the F (fusion) protein, responsible for the fusion of the viral envelope with the plasma membrane of the host cell, which is an essential early step of infection.
- It also brings about cell-to-cell fusion, forming large giant cells or syncytia, which are characteristic of paramyxovirus infections.
- The F protein also mediates the hemolytic activity of paramyxoviruses.

Parainfluenza (PI) Viral Infections

- Infection is confined to the respiratory tract, unlike mumps, which is a systemic disease, disseminating through blood and multiplying in various organs and tissues.
- The paramyxoviruses are responsible for about 10 per cent of respiratory infections in children needing hospitalisation.
- All human parainfluenza types are strongly correlated with specific clinical syndromes, ages, and times of the year.
- PI types 1 and 2 are the pathogens most commonly associated with **croup**, a serious clinical disease.
- PI type 3 is the pathogen most commonly associated with bronchitis, bronchiolitis, and pneumonia in infants and young children.
- Type 4 causes minor respiratory illnesses.
- In adults, parainfluenza viruses cause milder respiratory infections in which sore throat and hoarseness of the voice are common.
- Otitis media is a common complication in immunocompromised individuals.

Epidemiology

- Parainfluenza virus type 3 infection is often experienced in the first year of life with about 50 per cent of infants being seropositive by 12 months of age.
- Types 1 and 2 cause disease chiefly in preschool children.
- Type 3 infection is more endemic than types 1 and 2, which tend to occur as epidemics. Immunocompromised individuals have increased susceptibility to parainfluenza virus infections.
- First infections are more serious than re-infections, which are not infrequent.
- With the type 4 virus, even first infections are very mild. Infected children shed the virus in respiratory secretions for about a week.
- The infection spreads through air or contaminated fingers.
- Nosocomial spread is not uncommon.
- Currently, there is no vaccine against this group of viruses.

Laboratory diagnosis

- **Specimen:** Throat and nasal swabs are collected.

- **Virus isolation:** Primary monkey kidney cell cultures or continuous monkey kidney cell lines (LLC–MK2) with trypsin are used. Cytopathic changes are not readily apparent, except with the type 2 virus. Isolation may take ten days or more. Virus growth is detected by hemadsorption. Typing is by immunofluorescence, hemadsorption inhibition, or hemagglutination inhibition.
- **Serology:** Serological diagnosis is hampered by a wide range of antigenic cross-reactions. Paired sera can be tested by neutralisation, ELISA, HAI, or CF (complement fixation) to look for a rise in the titre of antibodies.
- **PCR:** Molecular diagnosis using reverse transcriptase PCR is gaining acceptance as a more specific and rapid technique.

Mumps

Case 4. *A 6-year-old unimmunised boy presented with acute, facial and neck swelling, fever, and malaise (Fig. 23.21). He had severe pain on the left side of his face and neck with a degree of trismus, and limitation of neck movement. He did not report any specific symptoms of otalgia or pharyngotonsillitis. Investigations were negative for EBV, neoplasms, and sepsis. The viral throat swab (tested for influenza, parainfluenza, respiratory syncytial virus, adenovirus, and rhinovirus) was negative, but* ***mumps PCR*** *on the parotid duct swab was positive. He was started on supportive management with intravenous fluids and analgesia. Four days later, the neck swelling had significantly reduced, trismus had resolved, and he was clinically much better.*

i. What are the characteristics of mumps?
ii. How can it be prevented?

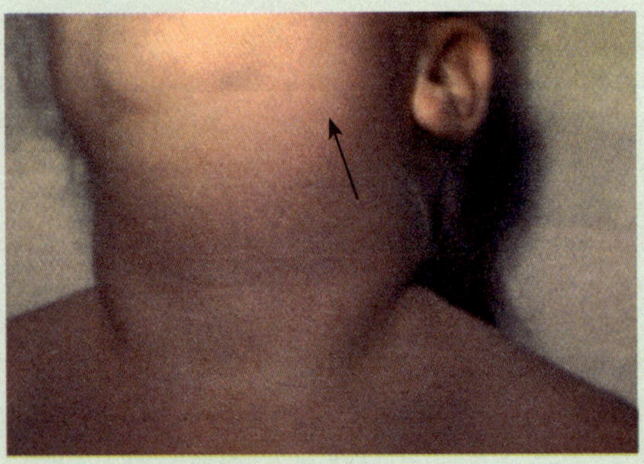

Fig. 23.21 Parotid gland swelling in a child with mumps (*Source*: CDC, PHIL, Image ID 3186)

Mumps or **acute parotitis** is caused by mumps virus. It is an infection commonly affecting children that is characterised by non-suppurative enlargement of the parotid glands.

Epidemiology

- **Distribution:** Mumps, which is endemic worldwide, has become less common in recent times due to immunisation.
- **Incidence:** It often occurs as epidemics in children between 5 and 15 years of age and in young people living in groups.
- **Hosts:** Humans are the only natural hosts. No human carriers or animal reservoirs exist.
- **Source of infection:** The infection spreads from a patient in the late incubation or early clinical stage of the illness.
- **Transmission** occurs by the following routes:
  - Direct contact—household spread is common
  - Airborne droplets
  - Fomites contaminated with saliva, and also possibly, urine
- The virus is detectable in saliva for about a week before and a week or two after the onset of parotitis.
- Peak infectivity is about a day or two before parotitis becomes evident and subsides rapidly thereafter. The virus is also shed in urine for up to two weeks after the clinical symptoms begin, though its role in the transmission of infection is not clear. One episode of mumps confers lasting immunity.

Agent—Mumps virus

- The mumps virus possesses HA, NA, and F proteins.
- It agglutinates the erythrocytes of fowls, guinea pigs, humans, and many other species. Hemagglutination is followed by hemolysis and elution at 37°C.
- As in influenza viruses, two complement-fixing antigens can be recognised—**the soluble (S) antigen and the viral (V) antigen**.

Immunity

- Infection leads to antibody response against both the internal (S) and surface (V) antigens.
- Antibodies to the S antigen appear early—within 3–7 days of the onset of symptoms—but disappear after about six months. The anti-hemagglutinin antibody correlates well with immunity to infection.
- Even subclinical infections lead to hemagglutinin antibody and resistance to infection.
- As antibodies are widespread in the population, passive immunity from the mother protects newborns. Mumps is, therefore, very rare before six months of age.

Susceptibility

- The mumps virus is labile, being rapidly inactivated at room temperature or by exposure to formaldehyde, ether, or ultraviolet light.
- It can be preserved at –70°C or by lyophilisation.

Clinical features

- The incubation period is **12–25 days**.
- The infection is acquired by **inhalation**.

- The virus replicates in the nasal and upper respiratory tracts and cervical lymph nodes and is disseminated through the bloodstream to various organs.
- The virus is shed from the saliva three days before to about nine days after salivary gland swelling.
- Almost one-third of infections are subclinical, including the majority in children below two years of age.
- **Parotid swelling** is usually the first sign of illness, though it may sometimes be preceded by prodromal malaise. It is unilateral to start with but may become bilateral (Fig. 23.21).
- The patient has **fever, local pain, and tenderness**, but the skin over the gland is not warm or erythematous.
- Parotitis usually resolves within a week. However, the involvement of the extraparotid sites may be more serious and may sometimes occur even in the absence of parotitis.

Complications

- **Epididymo-orchitis** is a complication seen in about a third of post-pubertal male patients.
- The testes become swollen and acutely painful, with accompanying fever and chills.
- **Orchitis is usually unilateral**; when it is bilateral and followed by testicular atrophy, sterility or low sperm counts may result.
- **Pancreatitis** may be a life-threatening complication of mumps.
- **Meningitis** is a symptom in only 10 per cent of cases even though the central nervous system is involved in about 60 per cent of cases, as indicated by pleocytosis in the CSF. The virus can be grown readily from the CSF in the early phase of meningitis.
- Mumps has been reported to cause about 10–15 per cent of cases of **aseptic meningitis**. It usually resolves without sequelae.
- **Meningoencephalitis** may also occur, but like meningitis, it usually resolves without sequelae; deafness may sometimes result. It may present 5–7 days after infection of the parotids or as parotitis presenting as atypical mumps. Other less common complications are arthritis, oophoritis, nephritis, thyroiditis, and myocarditis.

Laboratory diagnosis

A classic case of mumps does not need laboratory confirmation. However, it may be essential in atypical infection and where meningitis or other systemic involvement is the sole manifestation. Diagnosis may be established by **virus isolation and serological tests**.

- **Specimen:** The virus may be isolated from saliva (within 4–5 days), urine (up to two weeks), or CSF (8–9 days after the onset of the illness).
- **Virus isolation:** Cell cultures are better suited for isolation—**primary monkey kidney** being the preferred cell. Human amnion and HeLa cells also support the growth of the mumps virus. Its **cytopathic effect is slow and consists of syncytium formation** and the presence of acidophilic cytoplasmic inclusions. The virus can also be grown in **chick embryos**—in the amniotic cavity for primary isolation and the allantoic cavity after adaptation. It can be identified by hemagglutination inhibition using specific antisera. However, egg inoculation is less sensitive than cell cultures for isolation
- **Antigen detection:** Direct antigen detection by IFA is helpful in early diagnosis.
- **Serology:** Serological diagnosis depends on the demonstration of a rise in the titre of antibodies in paired serum samples. IgM-ELISA is useful in this respect because cross-reacting antibodies are IgG and do not interfere with IgM-ELISA.
- **PCR:** Molecular diagnosis using reverse transcriptase PCR is more rapid and sensitive than the other methods.

Prophylaxis

- An effective live virus vaccine is available against mumps.
- The **Jeryl–Lynn strain** of the mumps virus, which is grown in chick embryo fibroblast culture, is used as the vaccine.
- It is recommended for use only after one year of age as maternal antibodies may interfere with the multiplication of the vaccine virus if given earlier.
- Vaccination is contraindicated in cases of pregnancy, immunodeficiency, and hypersensitivity to neomycin or egg protein.
- The vaccine is given as a single subcutaneous injection, alone or in combination with the measles and rubella vaccines (MMR vaccine).
- Children should receive two doses of the MMR vaccine—the first dose at 12–15 months of age and the second dose between 4–6 years of age.
- The vaccine provides effective protection for at least ten years but may not prevent the disease if it is administered after exposure to the infection. However, there are no contraindications for its use in this situation.
- A **quadrivalent MMRV (measles-mumps-rubella-varicella) combination** vaccine has recently been launched. This vaccine is highly immunogenic, with seroconversion rates against all antigens.

Measles

Case 5. *A 5-year-old boy was brought to a primary health centre with complaints of fever, cough, running nose, and conjunctivitis. On examination of his throat, a white spot was noted on his buccal mucosa near the molars. He was* *given antipyretics and advised rest. He returned two days later with a maculopapular rash on his face, chest, and back. He tested positive for* ***measles virus*** ***IgM antibody by ELISA****. The patient improved with supportive care and aspirin.*

i. What are the complications of measles?
ii. How can measles be prevented?

Measles is a highly contagious airborne disease in children.

Epidemiology

In 2021, the WHO estimated 1,28,000 measles deaths globally, mostly among unvaccinated or under vaccinated children under the age of 5 years. After the immunisation coverage dipped in India, there was an upsurge in the cases in India.

- **Incidence:** Maximum incidence is seen in children between 1 and 5 years of age. It is uncommon in the first six months of life due to the presence of maternal antibodies.
- **Host:** Humans are the only natural hosts of the measles virus.
- **Route:** The virus enters the body through the respiratory tract and conjunctiva.
- **Spread:** It spreads by direct contact with respiratory secretions and aerosols created by coughing and sneezing.
- **Infectivity:** The infectivity of the measles virus is maximum at the prodrome (three days prior to the appearance symptoms) and diminishes rapidly with the onset of the rash.

Agent—Measles (Rubeola) virus

- The measles virus has the general morphology of paramyxoviruses. It is a roughly spherical but often pleomorphic particle, 120–250 nm in diameter.
- The **tightly coiled helical nucleocapsid is surrounded by the lipoprotein envelope**, which bears hemagglutinin (HA) spikes on its surface (Fig. 23.22).
- The envelope also bears the **F protein** which mediates cell fusion and hemolytic activities.
- It grows well on **human or monkey kidney and human amnion cultures** which are the preferred cells for primary isolation.
- Isolates can be adapted for growth on continuous cell lines (HeLa and Vero) and in the amniotic sac of hen's eggs.
- Cytopathic effects consist of **multinucleate syncytium formation** with numerous acidophilic nuclear and cytoplasmic inclusions.
- **Multinucleate giant cells (Warthin–Finkeldey cells)** are also found in the lymphoid tissues of patients.
- The virus is labile and **readily inactivated by heat, ultraviolet light, ether, and formaldehyde**.

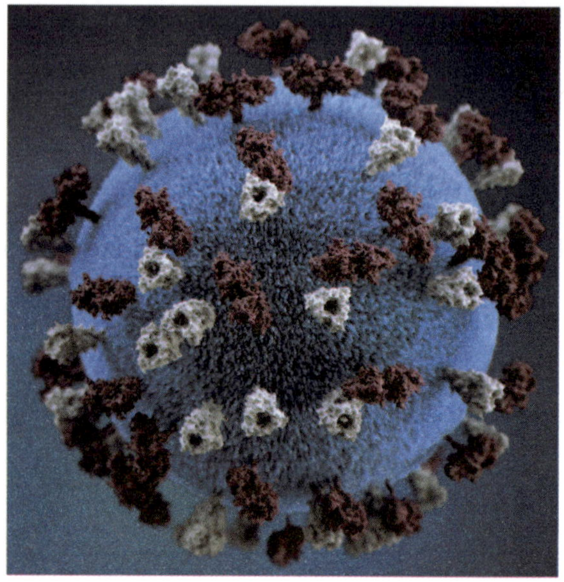

Fig. 23.22 Measles virus with H spikes and F protein (*Source*: CDC, PHIL, Image ID 21071)

- It can be **stabilised by molar $MgSO_4$ so that it resists heating at 50°C for one hour**.
- The measles virus is antigenically uniform. It shares antigens with the viruses of canine distemper and bovine rinderpest.
- A single episode confers **lifelong immunity**. In the immunodeficient, infection almost always results in clinical disease.

Pathogenesis

- The virus enters the body through the respiratory tract or the conjunctiva and multiplies locally and in the adjoining lymph nodes.
- It then enters the reticuloendothelial system through blood and after multiplying there, is transported by secondary viremia to the epithelial surfaces including the skin, mouth, respiratory tract, and conjunctiva.
- The pathognomonic **Koplik's spots** contain giant cells, cytoplasmic and intranuclear inclusions, and virion components, indicating local viral replication.
- Evidence of viral replication can also be seen in the vascular endothelial cells at the sites of the skin rash.
- The rash represents an immune reaction between T lymphocytes and cells in which viral replication is taking place.

Clinical features

The incubation period is 9–11 days.

Prodromal phase

- The patient presents with malaise, fever, conjunctival congestion, cough, and nasal discharge.
- After 3–4 days of prodromal illness, a **rash appears** (Case 5).

- A day or two before the rash begins, **Koplik's spots** develop on the buccal mucosa and occasionally, on the conjunctiva and intestinal mucosa.
- The prodromal illness subsides within a day or two of the appearance of the rash.

Eruptive phase

- The **red maculopapular rash** of measles typically appears on the forehead first and spreads (Fig. 23.23) downwards.
- It disappears in the same sequence 3–6 days later, leaving behind a brownish discolouration and finely granular desquamation.

Post-measles phase

- After the temperature subsides (3–4 days later) and the rash fades (in 5–6 days), the skin is discoloured (post-measles staining) and often peels off.
- It may continue to peel for weeks.

Complications

- Croup and bronchitis may develop due to the virus.
- Pneumonia or otitis media may occur due to secondary bacterial infection.
- Giant cell pneumonia may occur, particularly in children with immunodeficiencies or severe malnutrition.
- Meningoencephalitis is the most serious complication; many survivors have neurological sequelae.
- Subacute sclerosing panencephalitis (SSPE) is a rare late complication.
- Protracted diarrhea is often seen as a complication in children in the less developed nations; the virus may be recovered from the stools of patients with measles enteritis.

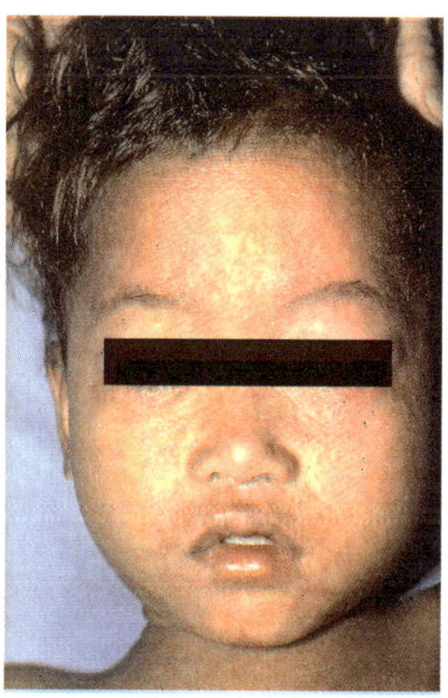

Fig. 23.23 Maculopapular rash on an infant's face; the hallmark of measles (*Source*: CDC, PHIL, Image ID 21074)

- Complications are more common and serious in the developing countries.
- There occurs a suppression of delayed hypersensitivity after measles infection, which may last for weeks or a few months. Mantoux and other allergic skin tests may be negative during this period.
- The underlying tuberculosis may become worse following an attack of measles. Recovery from measles may also be associated with an improvement of allergic eczema or asthma, Hodgkin's disease, or lipoid nephrosis.
- The virus may cross the placenta and infect the fetus in maternal measles but there is no evidence of teratogeny.
- Thrombocytopenia may develop, leading to purpura and bleeding from the mouth, intestines, and genitourinary tract.

Laboratory diagnosis

Diagnosis is self-evident and made clinically. In atypical cases and for differentiation from rubella, laboratory tests are useful.

- **Specimen:** Nasal secretions, throat, and conjunctival swabs, and blood can be used. CSF is collected in suspected cases of **subacute sclerosing panencephalitis (SSPE).**
- **Direct microscopy:** A simple diagnostic test, which can be performed even before the rash appears, is the demonstration of **multinucleated giant cells in Giemsa-stained smears of nasal secretions.**
- **Immunofluorescence assay:** The measles virus antigen can be detected in the cells of nasal secretions by immunofluorescence.
- **Virus isolation:** The virus can be isolated from the nose, throat, conjunctiva, and blood during the prodromal phase and up to about two days after the appearance of the rash. The virus may be obtained from urine for a few more days. Primary human or monkey kidney and amnion cells are most useful. Cytopathic changes may take up to a week to develop, but earlier diagnosis of viral growth is possible by immunofluorescence.
- **Serological diagnosis:** The demonstration of measles-specific IgM in a single specimen of serum drawn between one and two weeks after the onset of the rash is confirmatory. A four-fold rise in titre is looked for using paired sera collected during the acute phase and 10–21 days after.
  - Specific neutralisation, hemagglutination inhibition (HAI), and complement-fixing antibodies (in CFT) develop early. False-negatives may occur if the serum is collected earlier than one week before or later than two weeks after the onset of the rash.
  - The demonstration of high-titre measles antibody in the CSF is diagnostic of SSPE.

- **PCR:** Reverse transcriptase PCR for the *MeV* N gene RNA is a sensitive and specific method of diagnosis and may be positive even after 10–14 days of rash.

Prophylaxis

Passive protection

- Normal **human gammaglobulin or immunoglobulin (Ig) is given within six days of exposure.**
- This is useful in people who are at risk of severe illness and complications from measles such as infants younger than 12 months of age, pregnant women without evidence of measles immunity, and people with severely compromised immune systems.

Active immunisation

- The **Edmonston–Zagreb strain**, attenuated by passage in human diploid cells, is preferred because it is able to produce seroconversion even in infants who are 4–6 months old.
- The vaccine is given either by itself or in combination, like the MMR vaccine.
- In India, two doses of the vaccine are given, the first at 9 months and the second at around 16–24 months.
- **Contraindications** are immunodeficiency, untreated tuberculosis, and pregnancy.
- The original live vaccines used the **Edmonston strain** developed by multiple passages through human kidney and amnion and chick embryo cultures.
- The **Schwartz and Moraten** strains are safe but effective only in children older than 15 months.
- A live-attenuated vaccine has been developed which can be given as an intranasal aerosol to babies; this renders good protection, irrespective of the presence of maternal antibodies.

Measles eradication

- The WHO had aimed to eradicate measles by 2023 but this goal was impacted by COVID-19. India made substantial progress toward the goal of measles and rubella elimination.
- This will be achieved by providing high vaccination coverage with two doses of measles- and rubella-containing vaccines, by monitoring disease using effective surveillance, by being prepared for measles outbreak management, and by creating awareness among the public about the importance of vaccination.

Respiratory Syncytial Viral Infections

Agent—Respiratory syncytial virus (RSV)

- **Respiratory syncytial virus (RSV)** is so named because it causes cell fusion and the formation of multinucleated syncytia in cell cultures.
- RSV is pleomorphic and ranges in size from 150–300 nm.
- The viral envelope has two glycoproteins—the **G protein** by which the virus attaches to cell surfaces and the **fusion (F) protein** which brings about fusion between viral and host cell membranes.
- The F protein is also responsible for cell-to-cell fusion, which leads to the characteristic syncytial cytopathic changes in RSV infection.
- RSV does not possess hemagglutinin activity, nor does it possess neuraminidase or hemolytic properties, and thus differs from other paramyxoviruses. Another difference is that its nucleocapsid diameter (13 nm) is less than that of other paramyxoviruses (18 nm).
- It is propagated on heteroploid human cell cultures such as HeLa and HEp-2. It is highly labile and is inactivated rapidly at room temperature. It can be preserved by lyophilisation.
- It is antigenically stable, and only one antigenic type exists. However, studies using monoclonal antibodies have identified two subtypes, A and B.

Clinical features

- Most RSV infections are asymptomatic.
- The most common clinical presentation in RSV infection is an upper respiratory infection (common cold), but in young children, it can present as bronchiolitis, otitis media and can rarely progress to pneumonia, respiratory failure, apnea, and death.
- **Incubation period:** 4 to 6 days.
- Symptoms of RSV infection usually include:
  - Runny nose
  - Decrease in appetite
  - Coughing
  - Sneezing
  - Fever
  - Wheezing
- Most patients recover in 1–2 weeks, but those with immunodeficiency or cardiac defects may have a protracted illness and high death rates.

Epidemiology

- RSV is global in distribution.
- Outbreaks are common in children's wards, nurseries, and daycare centres.
- Infection is most common in children between six weeks and six months of age, with the peak at 2–3 months.
- Newborns are believed to be protected by high levels of maternal antibodies.

Immunity

- Secretory IgA is considered superior to circulating IgG in protection.
- Cell-mediated immunity appears more important than humoral antibodies in recovery from infection.

Laboratory diagnosis

It is rarely required due to the mild and self-limiting nature of the illness.

- **Specimen:** Nasopharyngeal swabs or nasal washings are collected for investigations. Unlike most viruses, RSV may be killed on freezing clinical samples.
- **Virus isolation:** Samples are inoculated onto cell cultures (HeLa or HEp-2) immediately after collection. In cultured cells, RSV causes characteristic giant cell and syncytial formation; cytopathic effects take about 10 days to appear. Earlier detection of viral growth in cells is possible by immunofluorescence tests.
- **Rapid diagnosis:** Immunofluorescence test on smears of nasopharyngeal swabs produces rapid results.
- **Serology:** Serological diagnosis is by the demonstration of rising antibody titres in paired serum samples by ELISA, CF, neutralisation, or immunofluorescence tests.
- **PCR:** Molecular diagnosis by reverse transcriptase PCR is sensitive and rapid.

Treatment

- Management is primarily by **supportive care.**
- Ribavirin administered as a continuous aerosol has been found beneficial in hospitalised patients, decreasing the duration of illness and of virus shedding.
- No effective vaccine is available yet.

Human metapneumovirus

- This respiratory pathogen was first described in 2001. The symptoms it produces resemble infection by RSV.
- Besides children, this virus can cause disease in adults and the elderly, manifesting as lymphomas or leukemias.

Rhinoviral Infections

- The **common cold** is probably the most common infectious disease caused by rhinoviruses in humans.
- The common cold virus was isolated by David Tyrrell and his colleagues (1960).
- Currently, more than 100 serotypes are known, classified through neutralisation tests.

Agent—Rhinoviruses

- Rhinoviruses resemble other picornaviruses in size and structure.
- They are inactivated below a pH of 6, inactivation being complete at a pH of 3.
- They are relatively stable at 20–37°C and may remain viable on fomites for days.
- Some serotypes survive for one hour at 50°C.

Pathogenesis

- The virus attaches to receptors on nasal ciliated epithelial cells and replicates within them, spreading to other cells. As a result, the cilia and cells are damaged, and the epithelium is subjected to secondary bacterial infection.
- Immunity to rhinovirus is type-specific.
- Rhinoviruses can cause serious complications like pneumonia and respiratory distress syndrome in the elderly with underlying co-morbid conditions.

Laboratory diagnosis

The virus can be isolated from nasal or throat swabs collected early in the infection and is achieved using the following measures:

- **Human cell cultures**, preferably MRC5 or W138 strains, are suitable for cell culture.
- Good oxygenation (achieved by **roller drum culture**), low pH (around 7), and low temperature (33°C) are required.
- Growth, as evidenced by cytopathic effect (CPE), may take two weeks to appear.
- Due to the large number of serotypes, serology is not feasible for diagnosis.

Newcastle Viral Disease

- The Newcastle disease virus **(NDV)** is a avian paramyxovirus type 1 and is a natural pathogen of poultry.
- Human infection with NDV is confined to **self-limited conjunctivitis in poultry workers** and others in contact with infected birds.
- In India, it is known as the **Ranikhet virus.**
- Control measures consist of vaccination and slaughter of infected birds.
- Other types of avian paramyxoviruses cause inapparent infection in many species of birds.

Rubella (German Measles)

Rubella or German measles is a **mild, exanthematous fever** characterised by a transient macular rash and lymphadenopathy. In itself, the disease is trivial but rubella in pregnant women may lead to congenital malformations in the baby.

Epidemiology

- Rubella is worldwide in distribution.
- Serological surveys in different countries have shown that **80–90 per cent of the population is immune to the virus by the age of 15 years.**
- The disease occurs **principally in children** but may affect all ages.
- Around 10–20 per cent of pregnant women are non-immune and, therefore, vulnerable. The infection is **transmitted to the fetus through the maternal bloodstream.**
- Infection is **acquired by inhalation.** The virus enters through the respiratory tract and disseminates to the skin to produce rashes.

Agent—Rubella virus

- **Morphology:** It is a pleomorphic, roughly spherical particle, 50–70 nm in diameter, with a single-stranded RNA genome and surrounded by an envelope carrying hemagglutinin peplomers.
- **Replication:** The viruses are released by budding from the host cell surface (Fig. 23.24).
- **Classification:** The rubella virus has been classified in the family Togaviridae as the only member of the genus Rubivirus.

Resistance

- The virus is inactivated by ether, chloroform, formaldehyde, beta-propiolactone, and deoxycholate.
- It is destroyed by heating at 56°C but survives for several years at –60°C.

Clinical features

- The **incubation period** is 2–3 weeks.
- A **generalised rash** develops on the face, spreading to the neck, trunk, and extremities, sparing the palms and soles. The rash is generally discrete and disappears by the third day.
- Viremia has been demonstrated as early as the seventh day before the rash and ceases shortly after the appearance of the rash.
- There is non-tender enlargement of the posterior cervical glands.
- Koplik spots, which are seen in measles, are absent.
- Some patients develop subclinical infection and are infectious to others.

Complications

- Arthralgia and arthritis may occur, more commonly in women and the elderly.

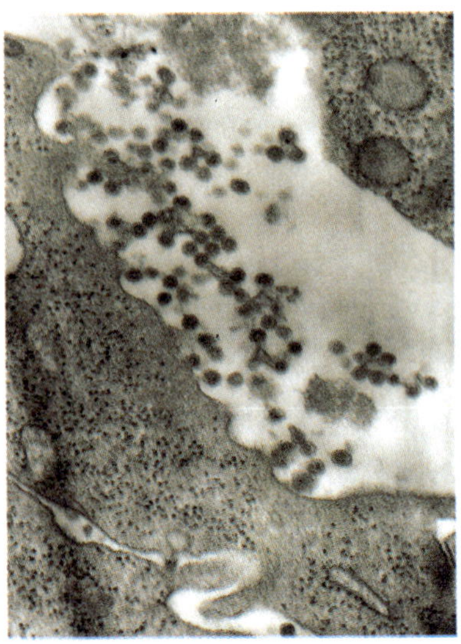

Fig. 23.24 Transmission electron microscopic (TEM) image shows rubella virions in the process of budding (arrow) from the host cell surface to be freed into the host's system (*Source*: CDC, PHIL, Image ID 10221)

- Based on the results of in vitro studies, rubella infection is presumed to cause chromosomal breakages and inhibition of mitoses in infected embryonic cells.

Congenital rubella

- Fetal damage caused by maternal rubella is related to the stage of pregnancy.
  - **Very early pregnancy:** Infection at this stage ends in an abortion.
  - **First trimester:** There is a 90 per cent risk of congenital malformations.
  - **Later in pregnancy:** The damage is subtle and manifests as communication defects or developmental retardation. These may not be apparent until the child grows older.
- The commonest malformations seen in congenital rubella syndrome are:
  - **Congenital heart disease**
  - **Suspicion of hearing impairment**
  - **Eye lesions**—cataract (Fig. 23.25), congenital glaucoma, and pigmentary retinopathy (salt and pepper)
  - **Blueberry muffin rash**—a distinctive rash in babies that presents as blue, purple, or dark spots on the face and body
  - **Expanded rubella syndrome**—hepatosplenomegaly, thrombocytopenic purpura, myocarditis, and bone lesions

Laboratory diagnosis

- Routine diagnosis of rubella is not required, but laboratory confirmation becomes important when rubella is suspected in pregnant women.
- Diagnosis can be established by serology or virus isolation.
- The virus can be recovered from the throat up to seven days before the rash appears.
- The virus can also be demonstrated in feces and urine.

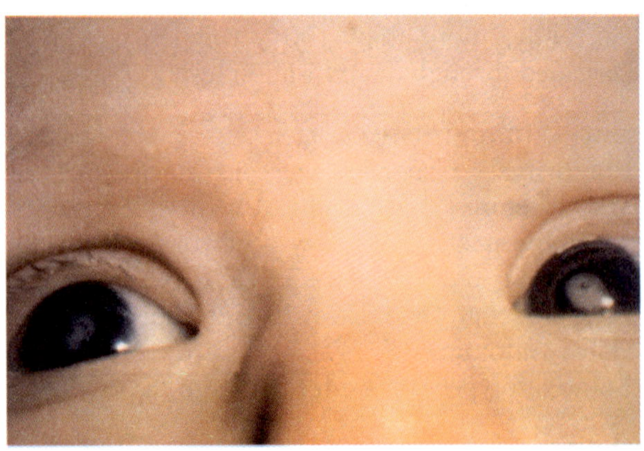

Fig. 23.25 Congenital rubella in a child—bilateral cataract (*Source*: CDC, PHIL, Image ID 4284)

Diagnosis in pregnant women (Fig. 23.26)

- **Serology:** Serological diagnosis by ELISA to detect IgM and IgG antibodies.
  - IgM antibodies alone, without IgG, indicate a current, acute infection.
  - IgG antibody alone without IgM indicates a past infection or vaccination and denotes immunity.
  - A paired serum sample is tested 10 days apart to demonstrate a significant rise or fall in the antibody titre.
  - A screening test for pregnant women using a **TORCH panel** (toxoplasmosis, rubella, cytomegalovirus, and herpes) is performed to rule out the possibility of congenital infections.
- **Cultivation:** The virus can be grown in many primary cell cultures and continuous cell lines. Virus growth can be identified by interference, using a challenge virus such as ECHO 11.

Diagnosis of congenital rubella (Fig. 23.27)

Serology: Since maternal IgM do not cross the placenta, IgM antibodies in the newborn indicate a fetal response to the infection, i.e., intrauterine infection.

- If a diagnosis is needed at a later time, paired serum from the mother and child can be used to demonstrate IgG antibodies.
- In a congenital infection, the IgG antibodies will be present in the newborn beyond six months. If they are passively transferred from the mother, levels will decline in the newborn while remaining the same in the mother.
- In congenital rubella, the virus may be isolated from the neonate from urine, throat swabs, leukocytes, bone marrow, or cerebrospinal fluid.

Prophylaxis

- Rubella infection confers **lasting immunity** as the virus has only one antigenic type. Re-infections have, however, been reported.
- A live-attenuated vaccine made from the RA 27/3 strain grown in human diploid cell culture is administered by subcutaneous injection, alone or in combination with measles and mumps components as the **MMR vaccine**.

Contraindication to the rubella vaccine

- **Immunocompromised individuals** should not be given the vaccine.
- **Pregnancy** is an absolute contraindication. Conception should be avoided for three months after vaccination (however, the vaccine virus is apparently not teratogenic; hence, the inadvertent administration of the vaccine to a pregnant woman may not lead to congenital defects in the baby).

Herpes Viral Infections of the Respiratory Tract

Herpesvirus includes a group of DNA viruses. Of the more than 100 known herpesviruses, 8 routinely infect only humans: herpes simplex virus types 1 and 2, varicella-zoster virus, cytomegalovirus, Epstein-Barr virus, human herpesvirus 6 (variants A and B), human herpesvirus 7, and Kaposi's sarcoma virus or human herpesvirus 8.

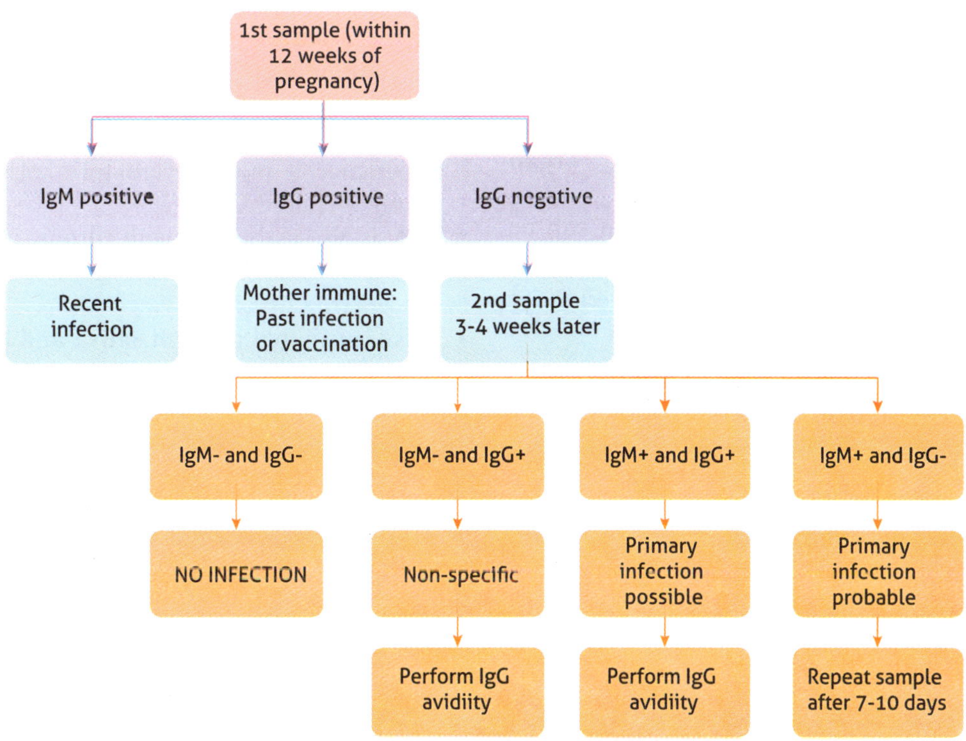

Fig. 23.26 Diagnosis in pregnant women

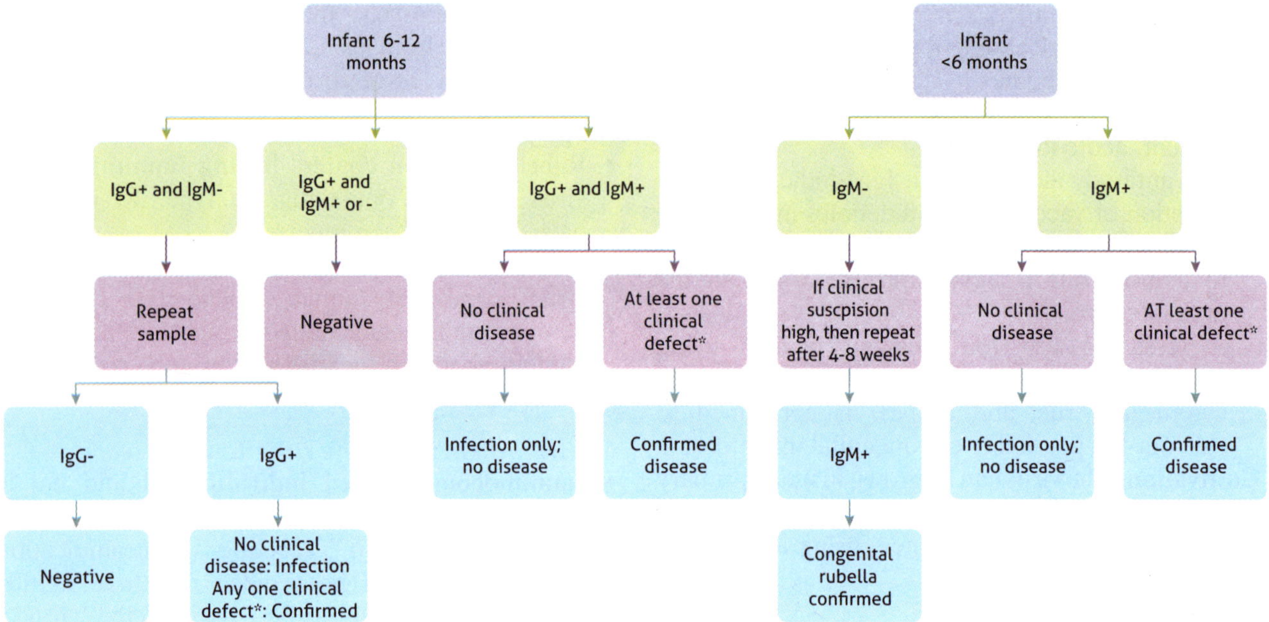

Fig. 23.27 Diagnosis of congenital rubella in the presence of clinical suspicion or in an infant born to a diagnosed mother

The family Herpesviridae consists of three subfamilies which are in Table 23.11.

Cytomegaloviral infections

- Infection caused by cytomegalovirus (CMV) is often asymptomatic. However, it may cause a florid infection leading to **pneumonia in organ transplant recipients and immunosuppressed individuals and congenital abnormality of the fetus in infected pregnant women**.
- CMV is a DNA virus belonging to the herpes group. It is a large virus which produces **multinucleated cells seen in salivary secretions**.
- Besides respiratory secretions, CMV can also be transmitted through **sexual contact, organ transplantation, and blood transfusion**.
- The diagnosis in suspected CMV infection can be made by the **detection of IgM in serum or viral DNA from bronchoscopic washings**. Biopsy of infected tissue may show multinucleate giant cells, which gives them an **'owl eye' appearance**.
- CMV can be treated with **foscarnet, ganciclovir, or cidofovir** (details in Chapter 30).

Epstein–Barr virus (EBV) disease

- Infectious mononucleosis or glandular fever is an acute infectious disease characterised by **fever, cervical lymphadenopathy and pharyngitis** accompanied by atypical large peripheral blood lymphocytes.
- It is caused by the Epstein–Barr virus (EBV).

Clinical features

- The patient presents with a sudden onset of sore throat, often with a swollen neck that reflects cervical lymph node enlargement. Another typical presentation is the slow development of malaise, myalgia, and fatigue.
- **Complications** include hepatitis, hepatosplenomegaly, cervical lymphadenopathy, and skin rashes.

Epidemiology

- The disease occurs worldwide with no seasonal predilection and is more **common in adolescents and young adults**.
- EB virus infection is associated with patients who are **immunocompromised, are on chemotherapy, or have had organ transplantation**.
- Healthy individuals, older children and young adults experience a more severe form of infection than younger children.
- EBV is also associated with chronic infections.

Mode of transmission

- It is transmitted through **saliva and oropharyngeal secretions** via close contact, on account of which, the infection it is often called the '**kissing disease**'. It can also be transmitted congenitally or via the sexual route (details in Chapter 32).
- EBV is a member of herpes group of viruses.

Pathogenesis

EB virus produces 'atypical lymphocytes' seen in the peripheral blood. The pathogenesis is shown in Fig. 23.28.

Laboratory diagnosis

EBV infection can be diagnosed by the serology.

Table 23.11 Classification and diseases caused by herpesviruses

Subfamily	Taxonomic name	Common name	Diseases caused
Alpha-herpesvirinae	HHV-1	Herpes simplex virus 1 (HSV-1)	Mucocutaneous infections such as gingivostomatitis, herpetic keratitis, herpes simplex labialis (cold sores) and dermal whitlows, neonatal infection and encephalitis
	HHV-2	Herpes simplex virus 2 (HSV-2)	Herpes genitalis, neonatal infection
	HHV-3	Varicella-zoster virus (VZV)	Chicken pox and shingles
Beta-herpesvirinae	HHV-5	Cytomegalovirus (CMV)	1. Congenital infections: Hepatosplenomegaly, retinitis, rash, and central nervous system involvement 2. Mononucleosis syndrome with fever, malaise, atypical lymphocytosis, and pharyngitis 3. Hepatitis, retinitis, esophagitis, colitis, pneumonitis, transverse myelitis and subacute encephalitis in immunocompromised hosts (transplant recipients and HIV-infected patients)
	HHV-6	HHV-6 variant A/B	Exanthem subitem (roseola or sixth disease) and with rejection of transplanted kidneys
	HHV-7	HHV-7	Exanthem subitem (less common than HHV-6)
Gamma-herpesvirinae	HHV-4	Epstein-Barr virus (EBV)	Infectious mononucleosis Hodgkins lymphoma Multiple sclerosis Gastric cancer Burkitt's lymphoma Nasopharyngeal cancers Inflammatory bowel disease
	HHV-8	Kaposi's sarcoma associated virus (KSHV)	Kaposi's sarcoma in AIDS patients, intra-abdominal solid tumors

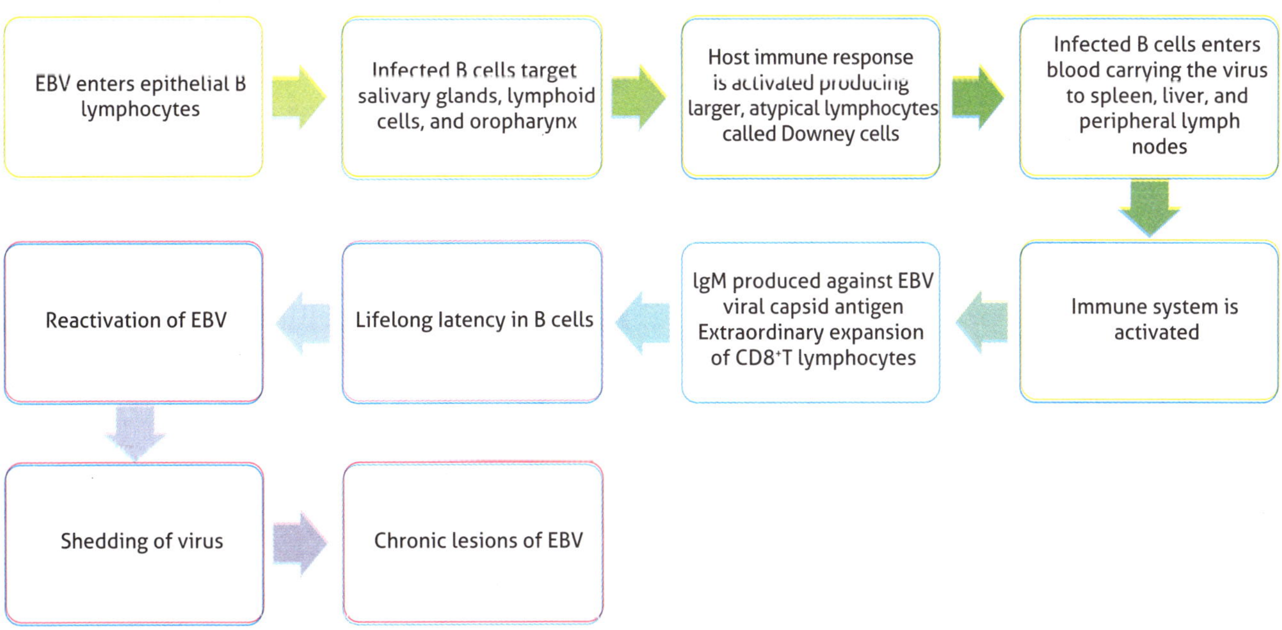

Fig. 23.28 Pathogenesis of EBV infection

- **Antibody to viral capsid antigen (VCA):**
  - Anti-VCA IgM appears early in EBV infection and usually disappears within four to six weeks.
  - Anti-VCA IgG appears in the acute phase of EBV infection, peaks at two to four weeks after onset, then persists for life.
- **Early antigen (EA):** Anti-EA IgG appears in the acute phase of illness and generally falls to undetectable levels after 3–6 months.
- **EBV nuclear antigen (EBNA):** Antibody to EBNA appears two to four months after the onset of symptoms and persists for the rest of a person's life. Detection is by the immunofluorescent test.
- **Monospot test:** The Monospot test detects heterophile antibodies. It is not recommended for diagnosis of EBV infection as it produces both false-positive and false-negative results. The antibodies detected by Monospot can be caused by conditions other than infectious mononucleosis.
- Antibodies to both VCA and EBNA suggests past infection.
- In the acute stage, anti-VCA IgM may be positive and atypical lymphocytes can be demonstrated in blood.

INFECTIONS IN THE ORAL CAVITY

Since the oral cavity is a part of the upper respiratory and gastrointestinal tract, a few oral cavity infections are mentioned here.

Oral Candidiasis

- *Candida* species (commonly, *C. albicans* and *C. tropicalis*) are normal commensals of the oral cavity.
- Prolonged antibiotic therapy or immunosuppressive diseases, e.g., HIV infection, may lead to oral candidiasis (**oral thrush**).
- It may also be seen in newborns or the elderly.
- In HIV patients, the infection may progress to cause esophagitis.
- The diagnosis of oral candidiasis is based on Gram staining, upon which budding yeast cells can be demonstrated.
- The treatment of oral candidiasis is with oral fluconazole, topical nystatin, or clotrimazole.

Periodontal Infection

- The space between the teeth and gingiva is a potential space for infection.
- Anaerobic bacteria like the *Actinomyces*, *Bacteroides*, and *Actinobacillus* interact with polymorphonuclear cells to produce inflammation and exudative lesions, leading to gingivitis.

Dental Caries

- Caries develop in the crevices of the teeth at low pH due to colonisation by the bacteria, which metabolise sucrose in the mouth to lactic acid and consequently, decalcify the tooth. The bacteria live in bacterial communities known as dental plaque which accumulates on the tooth surface.
- A common organism associated with dental carries is *Streptococcus mutans*. Other organisms such as *Streptococcus sobrinus*, and *Lactobacilli* are also involved.

LOWER RESPIRATORY TRACT INFECTIONS

- Of the infections affecting the lower respiratory tract, pneumonia is the most significant, being the most common cause of morbidity and mortality in children under the age of five years.
- Although rates have come down in the developing world due to effective vaccination and improved child health programmes, childhood pneumonia continues to be one of the leading causes of infant death in the developing world.
- Pneumonia also contributes to a large proportion of morbidity in the elderly.
- Infections of the lower respiratory tract can be clinically classified as follows:
  - Tracheobronchitis
  - Bronchitis (acute and chronic)
  - Bronchiectasis
  - Pneumonia
  - Lung abscess
  - Empyema
  - Epiglottitis
  - Laryngotracheobronchitis
  - Whooping cough

The common causes of lower respiratory tract infections are listed in Table 23.12.

Pathogens may gain entry to the lower respiratory tract through the following routes:

i. **Inhalation of aerosolised infectious particles**
ii. **Aspiration of colonisers** of the upper airway, which trickle down due to damaged mucociliary lining
iii. Through the **bloodstream**

TRACHEITIS AND LARYNGITIS

Because the larynx and trachea are non-expandable due to the cartilaginous rings, inflammation and edema give rise to hoarseness of voice, retrosternal pain, and in children, and inspiratory stridor, which might warrant hospitalisation.

Acute Bronchitis

- Viral infection leads to the destruction of the respiratory epithelial cells, predisposing to secondary bacterial infection.

Table 23.12 Common causes of lower respiratory tract infections

Clinical type	Bacteria	Viruses	Fungi	Parasites
Tracheitis Bronchitis	*Haemophilus influenzae* *Streptococcus pneumoniae* *Mycoplasma pneumoniae* Group B streptococci (in neonates and infants)	Parainfluenza virus, adeno, and sometimes, RSV Adenovirus Parainfluenza virus Influenza RSV	*Candida* *Aspergillus* spp. *Talaromyces* (*Penicillium marneffei*) *Penicillium*	
Bronchiolitis	*S. pneumoniae, H. influenzae, B. pertussis,* and *M. catarrhalis*	RSV Adenovirus Influenza		
Pneumonia	*Mycoplasma pneumoniae* *Bordetella pertussis* *Chlamydophila pneumoniae* *S. pneumoniae* *H. influenzae* *M. tuberculosis* *Coxiella burnetti* Atypical mycobacteria *Pseudomonas aeruginosa* *Klebsiella* species	Influenza A and B Coronaviruses (recent pandemic) Parainfluenza 1–4 RSV Adenoviruses CMV, VZV Human metapneumovirus BOCA virus	*Pneumocystis jirovecii* *Cryptococcus neoformans* *Blastomyces dermatitidis* *Coccidioides immitis* *Aspergillus* species	*Strongyloides stercoralis* (larvae in the immunocompromised) *Paragonimus westermani* *Ascaris* *lumbricoides* (larva) *Ancylostoma duodenale* (larva)

- *Mycoplasma* infection leads to the attachment of the bacteria to the bronchial mucosal epithelium, leading to necrosis and sloughing.
- Acute exacerbation of COPD is the commonest adult lower respiratory infection. It is generally caused by pneumococci or *Haemophilus*.

Chronic Bronchitis

- Along with infection, cigarette smoking and exposure to dust and fumes cause chronic bronchitis.
- This is aggravated by secondary infections by *S. pneumoniae* and non-capsulated *H. influenzae*.
- *Staphylococcus* and *Mycoplasma* are less commonly associated with exacerbation of chronic bronchitis.

BRONCHIOLITIS

- Commonly seen in children below two years of age, bronchiolitis is commonly caused by RSV.
- The narrow bronchioles get occluded by inflammation and edema due to the viral infection.
- Infants develop cyanosis associated with cough and rapid respiratory rate.

CROUP

- The infection of the upper airway causes tissues around the voice box (larynx) and the windpipe (trachea) to swell. Movement of air through the narrow passage produces a barking sound.
- It is commonly caused by parainfluenza virus, adenovirus, enterovirus, or RSV.

PNEUMONIA

- Pneumonia is an **infection of lung parenchyma, including the alveoli.**
- It affects all ages, but causes increased morbidity and mortality in infants, neonates, and the elderly.
- It is caused by a wide range of microorganisms including bacteria, viruses, fungi, and parasites, with indistinguishable clinical features.

Classification

Pneumonia can be classified in various ways (Table 23.13).

Clinical Features (Fig. 23.29)

- The patient generally presents with fever, weakness, chills, and sweating.
- There may be a cough, which may be productive or non-productive.
- Sputum may be mucoid, purulent, rusty or blood-tinged, and copious.
- There may be chest pain, which may be pleuritic with pain during inspiration.
- Severe cases may have dyspnea and shortness of breath.
- On **auscultation** of the chest, **abnormal breath sounds or crackling bronchial sounds and pleural friction rub** may be noted.
- Radiographic examination is an important adjunct to physical examination.

Table 23.13 Classification of pneumonia

Based on the site of acquisition pneumonia	Based on etiological agents	Based on the site of localisation
Community-acquired pneumonia (most viral agents, *S. pneumoniae, H. influenzae, K. pneumoniae, Mycoplasma, Chlamydophila*) Healthcare-associated pneumonia (*P. aeruginosa, A. baumanii, Klebsiella pneumoniae*) Pneumonia in the immunocompromised (*P. jirovecii*) Neonatal pneumonia (*S. pneumoniae, H. influenzae*)	Bacterial Viral Fungal	Lobar pneumonia: It involves a distinct lobe of the lung. The alveoli get filled with inflammatory infiltrates, which clots and renders the alveoli solid. Infection spreads to adjacent alveoli until the whole segment of the lung gets consolidated. Bronchopneumonia: It involves more regions of the lung with patchy consolidation. Interstitial pneumonia: The interstitium of the lung gets involved. It is often seen with viral etiology. Necrotising pneumonia (lung abscess): This is due to localised destruction of the lung parenchyma. It forms a cavity in the lung.

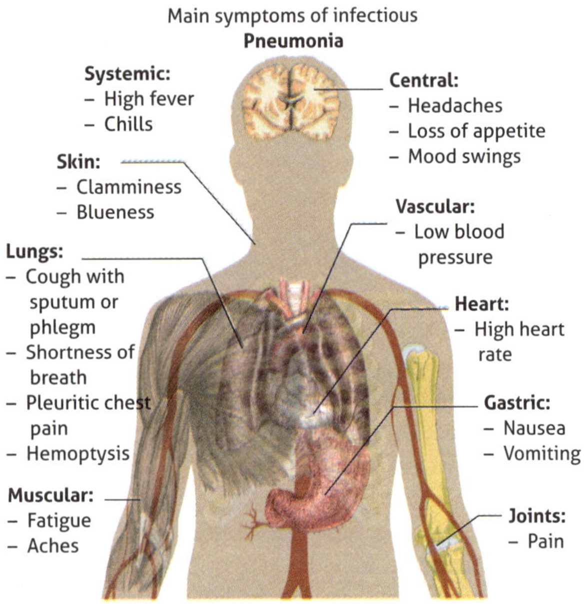

Fig. 23.29 Features of pneumonia caused by infectious organisms (*Source*: https://en.wikipedia.org/wiki/Pneumonia)

Community-acquired Pneumonia (CAP)

- Pathogens gain entry to the lower respiratory tract through the upper respiratory tract.
- Risk factors include the extremes of age (>70 and infants), tobacco smoking, alcoholism, and asthma. Other conditions may be concomitant HIV, chronic obstructive pulmonary disease, and seizure disorders (due to aspiration).
- CAP may be mild to severe, and sometimes fulminant.

Healthcare-associated Pneumonia

- Healthcare-associated pneumonia is defined as pneumonia that occurs **48 hours after admission to the hospital**.
- Generally, this kind of pneumonia is caused by **multidrug-resistant (MDR) organisms** and are associated with infections acquired within the hospital, e.g., *Pseudomonas*, *Acinetobacter*, and *Klebsiella*.
- Healthcare-associated pneumonia may be of two types:
  - Hospital-acquired pneumonia (HAP)
  - Ventilator-associated pneumonia (VAP)

Lung Abscess

- This condition may occur due to the seeding of organisms in the lung parenchyma, following aspiration or blood from a different site, e.g., endocarditis caused by *Staphylococcus aureus*.
- This condition is seen in unconscious patients who cannot cough out secretions.
- The organisms most commonly implicated are anaerobes, *Peptostreptococcus*, *Fusobacterium*, and *Prevotella*.
- **Clinical features** resemble pneumonia, with fever, cough and foul-smelling sputum.
- **Diagnosis** is based on X-ray showing lung infiltrates with cavities and on the air–fluid level.
- The condition is **treated** with clindamycin and an ampicillin–sulbactam combination.
- Surgical drainage is required if the patient does not respond to antibiotics.

Empyema

- This refers to a **collection of pus in the pleural space**.
- This is a sequel to pneumonia, with *S. pneumoniae* as the most common causative organism. Other causative organisms include *S. aureus*, enteric gram-negative bacteria. Anaerobes, *Fusobacterium* and *Prevotella* may also be involved.
- **Clinical features** include fever, chest pain, and cough.
- **Diagnosis** is by Gram staining and culture of the aspirated pus.
- **Treatment** is with antibiotics like clindamycin and third-generation cephalosporins.

Laboratory Diagnosis of Pneumonia

Fig. 23.30 is an algorithm of the approach to the diagnosis of pneumonia.

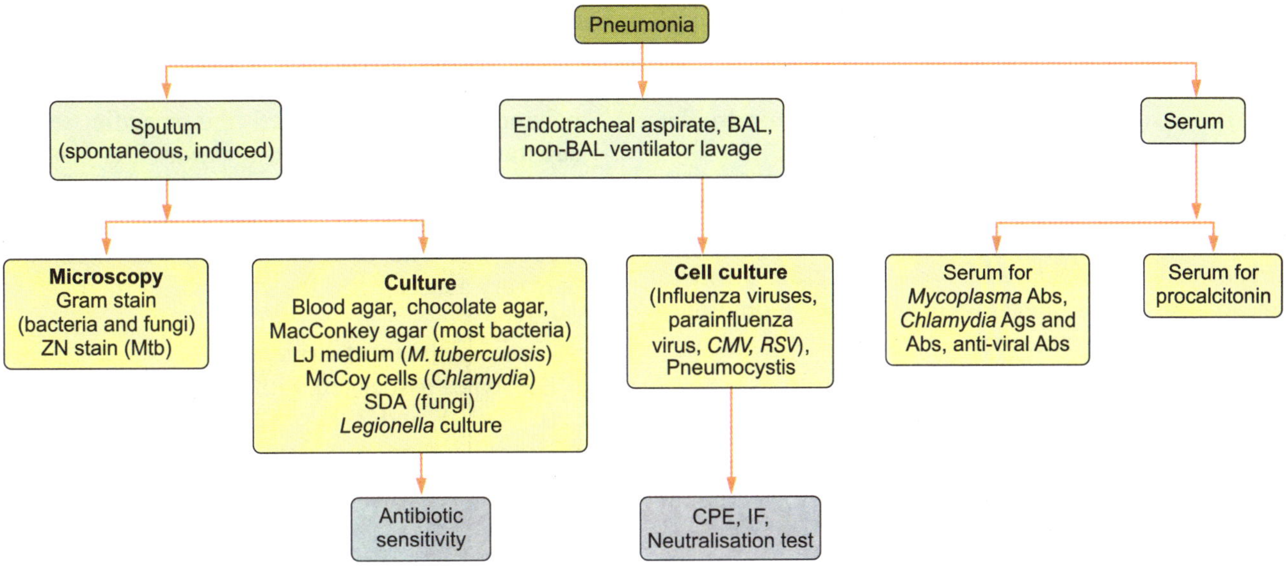

Fig. 23.30 Algorithm of the approach to the diagnosis of pneumonia (*BAL*—bronchoalveolar lavage; *CPE*—cytopathic effect; *IF*—immunofluorescence)

Radiological diagnosis

- It helps to differentiate bacterial, tubercular, fungal, viral, and atypical pneumonia..
- Chest X-ray may indicate patchy consolidation (Fig. 23.31), cavity, interstitial pneumonia (viral infections), and typical ground-glass appearance (*P. jirovecii* pneumonia).
- These findings are correlated with other laboratory and clinical parameters to confirm the etiology of pneumonia.

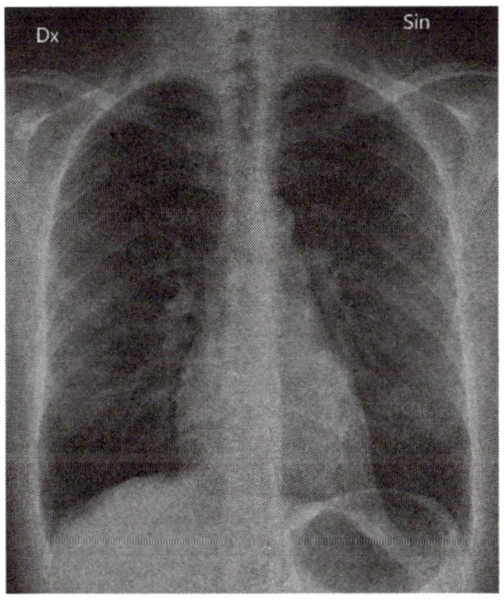

Fig. 23.31 Patchy consolidation located in the left lower lobe, characteristic of infection by *Mycoplasma pneumoniae* (*Source*: CDC, PHIL, Image ID 21493)

Bacterial Pneumonia

Pneumococcal pneumonia

> **Case 6.** *A 60-year-old man was admitted with a history of high-grade fever, chills, and rigors with mild chest pain and productive cough for the last two days. An X-ray chest showed consolidation in his right lower lobe (Fig. 23.32). The sputum smear observed after Gram-staining showed plenty of polymorphs with gram-positive, lanceolate diplococci. A sputum culture incubated for 24 hours produced greenish colonies on blood agar. The diagnosis was confirmed as **pneumococcal pneumonia**. Antibiotic susceptibility showed the organism to be sensitive to penicillin, to which the patient responded.*
>
> i. What are the infections caused by pneumococci?
>
> ii. Outline the diagnosis of pneumococcal pneumonia.

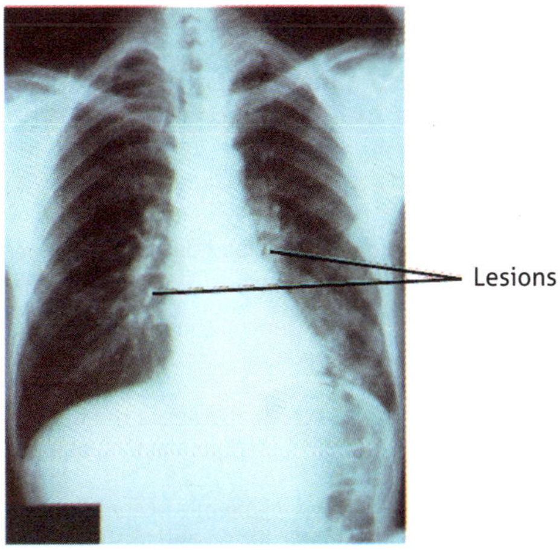

Fig. 23.32 Chest radiograph of a patient with pneumonia shows the consolidations (lesions) present as opaque patches (*Source*: https://openstax.org/books/microbiology/pages/22-2-bacterial-infections-of-the-respiratory-tract)

Epidemiology

- Community-acquired pneumonia is most commonly caused by *S. pneumoniae* worldwide. These are most commonly seen in winter and the early spring months.
- The source of human infection is the respiratory tract of carriers and, less often, of patients.
- *S. pneumoniae* is found in the throats of approximately half the population sampled at any time.

- They are the single most prevalent bacterial agents linked to pneumonia and otitis media in children.
- It is transmitted by contaminated droplets or droplet nuclei.
- Dissemination is facilitated by crowding.

Agent—*Streptococcus pneumoniae*

Streptococcus pneumoniae is a gram-positive, lanceolate, catalase-negative diplococcus (commonly referred to as pneumococci).

Morphology

- *S. pneumoniae* are typically small (1 μm), slightly elongated diplococci (Fig. 23.33) rounded at one end and pointed at the other, i.e., they are **flame-shaped or lanceolate**.
- The capsule may be demonstrated as a clear halo in India ink preparations or may be stained directly by special techniques (see Chapter 4).

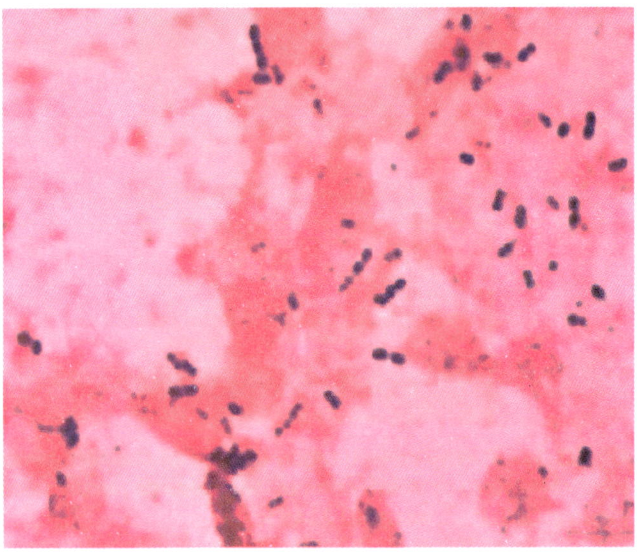

Fig. 23.33 Gram-stained smear of blood cultured in an automated system containing lanceolate gram-positive cocci (*Source*: Department of Microbiology, Pondicherry Institute of Medical Sciences, Puducherry)

Antigenic properties

1. **Capsule:** The most important antigen of *S. pneumoniae* is its type-specific capsular polysaccharide, also called the **specific soluble substance (SSS)**. *S. pneumoniae* are classified based on the antigenic nature of the capsular polysaccharide. 100 different serotypes are now recognised. The case fatality rates of pneumonia may vary according to the virulence of the infecting serotype.
   - Type 3 is the most virulent.
   - Serotypes associated with invasive infections are 6, 14, 19F, and 23F.
   - In adults, types 1–8 are responsible for about 75 per cent of the cases of pneumococcal pneumonia and bacteremia.
   - In children, types 6, 9V, 14, 19, and 23 are frequent causes of meningitis.
2. **Other antigens:** *S. pneumoniae* contains other antigens like the nucleoprotein deep inside the cell and a somatic 'C' carbohydrate antigen, both of which are species-specific.

Toxins and other virulence factors

- **Capsular polysaccharide** protects the cocci from phagocytosis because of its acidic and hydrophilic properties.
- Non-capsulated strains are avirulent. The antibody to the capsular polysaccharide affords protection against infection. **This is the basis of vaccination.**
- **Pneumolysin,** a membrane-damaging toxin produced by *S. pneumoniae*, may also contribute to its virulence.

Pathogenesis

The pathogenesis of pneumococcal pneumonia is described in Fig. 23.34.

Clinical features

Infections caused by pneumococci:

- Sinusitis
- Otitis media

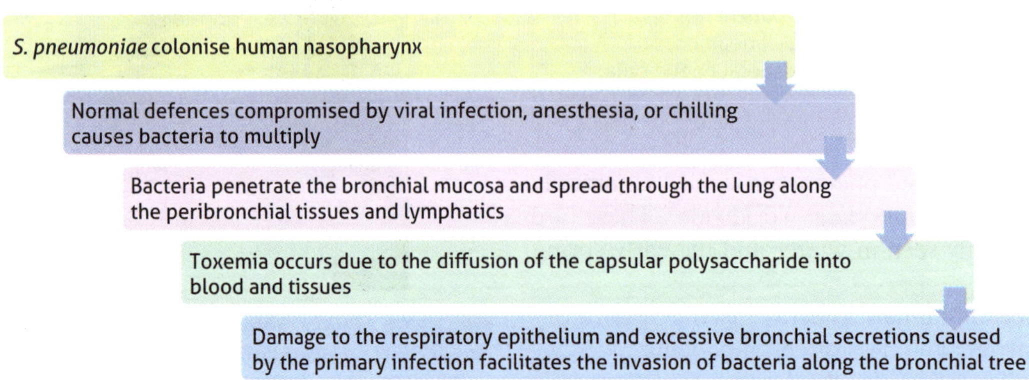

Fig. 23.34 Pathogenesis of pneumococcal pneumonia

- Mastoiditis
- Lobar pneumonia
- Bronchopneumonia
- Acute exacerbation of chronic bronchitis
- Joint infections
- Endocarditis
- Meningitis
- Bactremia
- Septicemia
- Abscesses in organs following septicemia
- Conjuctivitis

1. *Invasive infections that spread through the bloodstream*

- **Lung infections**—lobar pneumonia, bronchopneumonia, acute empyema, acute exacerbations of chronic bronchitis, and tracheobronchitis can occur.
- **Bacteremia** is common during the early stage of lobar pneumonia.
- **Bronchopneumonia** is almost always a secondary infection and may be caused by any serotype of *S. pneumoniae.* It is frequently a terminal event in aged and debilitated patients.
- **Meningitis** (discussed in Chapter 24) may occur.
- **Satellite infections**—pneumococcal bacteremia may also lead to distant infections in the heart, peritoneum, or joints.

2. *Suppurative lesions*

- Otitis media, sinusitis, pericarditis, conjunctivitis, suppurative arthritis, and peritonitis may occur.
- Ocular infections like keratitis and dacryocystitis may occur.

Laboratory diagnosis

- **Specimen:** Sputum, blood, and aspirated fluid are collected for direct demonstration, culture, and antigen detection.
- **Microscopy:** In the acute phase of lobar pneumonia, the rusty sputum contains *S. pneumoniae* in large numbers, with hardly any other kind of bacterium. They are demonstrated by Gram stain (Fig. 23.35a and b). In acute otitis media, *S. pneumoniae* may be demonstrated in the fluid aspirated from the middle ear. Gram-positive diplococci can be seen both inside polymorphs (which are often scanty) and extracellularly.
- **Culture:** In the acute stage of bacteremia, meningitis, pneumonia, and other localised infections, the organism may be isolated from blood (in automated blood culture systems or conventional brain–heart infusion broth), CSF, sputum, and body fluids respectively.
  - *S. pneumoniae* has complex growth requirements and grows only in enriched media such as sheep blood agar and chocolate agar with 5–10% CO_2.
  - Growth appears as 0.5–1 mm, dome-shaped, glistening colonies showing alpha-hemolysis, which subsequently become flat with raised edges, central umbonation and concentric rings (**Draughtsman** or **carom coin appearance**; Fig. 23.36).
  - The sensitivity of *S. pneumoniae* to **optochin** (ethyl hydrocuprein – 1/500,000), differentiates it from streptococci (Fig. 23.37).

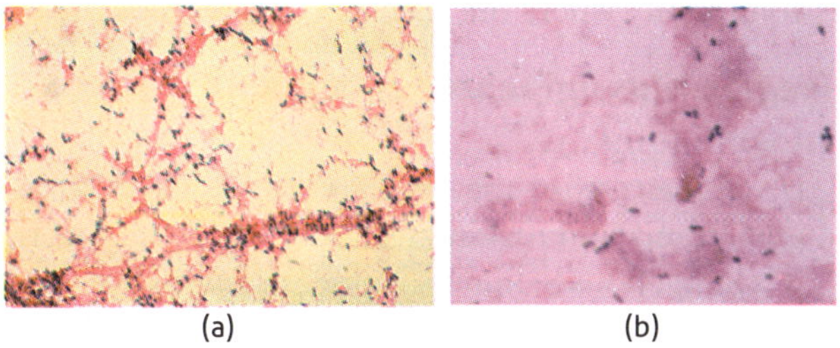

Fig. 23.35 Gram-positive diplococci with polymorphs: (a) in smear of sputum and (b) in CSF (*Source*: [a] CDC, PHIL, Image ID 2896 and [b] Department of Microbiology, Pondicherry Institute of Medical Sciences, Puducherry)

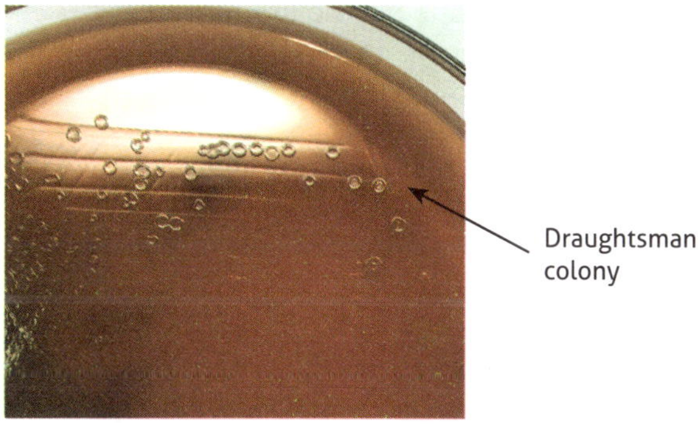

Fig. 23.36 Draughtsman or carrom coin appearance of pneumococcal colonies on sheep blood agar (*Source*: Department of Microbiology, Pondicherry Institute of Medical Sciences, Puducherry)

Fig. 23.37 *S. pneumoniae* colonies sensitive to optochin (*Source*: Department of Microbiology, Pondicherry Institute of Medical Sciences, Puducherry)

- **Antigen detection:** Rapid diagnosis of pneumococcal meningitis can be achieved by antigen detection in CSF with a latex agglutination test using polyvalent antisera.
- **Biomarkers:** CRP and procalcitonin are elevated in invasive pneumococcal disease. These levels are monitored to determine prognosis and response to treatment.
- **Molecular methods:** PCR-based methods have significant potential in cases where the patient has taken antibiotics and for genotyping isolates. The technique also has research potential for detecting virulence and resistance genes.

Serotyping of pneumococci

Currently, typing methods are based on (i) Quellung reaction, (ii) latex agglutination, and (iii) PCR-based methods.

Quellung reaction

- This method of typing based on capsule swelling.
- A suspension of *S. pneumoniae* is mixed on a slide with a drop of the type-specific antiserum and a loopful of methylene blue solution.
- In the presence of the homologous antiserum, the capsule becomes apparently swollen, sharply delineated and refractile (Fig. 23.38).

Latex agglutination

Typing of cocci is done using type-specific antiserum.

PCR-based typing

- Typing of pneumococci is done using real-time multiplex PCR, multilocus sequence typing (MLST), multilocus enzyme electrophoresis (MLEE), and DNA fingerprinting methods such as pulse-field gel electrophoresis (PFGE) and amplified fragment length polymorphism (AFLP).
- Typing by molecular methods is slowly replacing conventional methods.

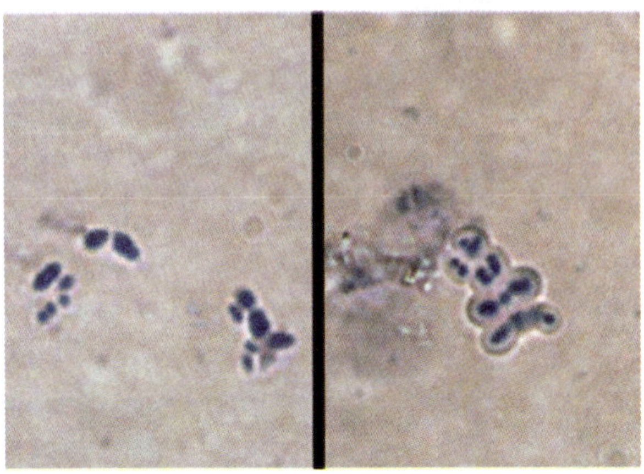

Fig. 23.38 Quellung reaction (a clear halo can be seen around pneumococci stained with methylene blue)

> **Laboratory diagnosis of pneumococcal infections**
>
> - **Specimen:** Sputum, blood and aspirated fluid, CSF in meningitis
> - **Antigen detection:** CSF with a latex agglutination test
> - **Biomarkers**: CRP and procalcitonin are elevated
> - **Microscopy**: Gram-positive diplococci both inside polymorphs and extracellularly
> - **Culture:**
>   - Aerobic, capnophilic, fastidious, optimum temperature: 37° C, pH: 7.8
>   - Blood agar and chocolate agar: Colonies are 0.5–1 mm, dome-shaped, glistening with alpha-hemolysis and later develop the Draughtsman or carom coin appearance
> - **Identification:**
>   - **Optochin** sensitive
>   - **Bile solubility test** is positive
> - **Serotyping of pneumococci**
>   - **Quellung reaction:** To demonstrate the capsule
>   - Latex agglutination using type-specific antiserum
>   - **PCR-based methods**

Treatment

- **Penicillin** is the drug of choice for penicillin-sensitive isolates and amoxycillin in mild cases.
- Penicillin resistance is noted in several strains, which are also resistant to other antibiotics like erythromycin and tetracycline. The development of resistance is due to alteration in the penicillin-binding protein (PBP2a and PBP2×).
- **Third-generation cephalosporins** are indicated in cases with penicillin-resistant isolates.
- Vancomycin is to be reserved for life-threatening illnesses with highly resistant strains.

Prophylaxis for pneumococcal diseases (pneumococcal vaccines)

- Immunity against pneumococcal infections is type-specific and associated with antibodies to the capsular polysaccharide.
- The existence of nearly 100 serotypes makes a complete polyvalent vaccine impracticable.
- Protection would depend on whether the serotypes included in the vaccine are also prevalent in the community where the vaccine is used (Table 23.14).

Haemophilus influenzae pneumonia

Epidemiology

- *H. influenzae* is the primary cause of childhood pneumonia and meningitis in the developing world
- It is indigenous to human beings, primarily colonising the upper respiratory tract.
- Carriage in the upper respiratory tract is common, particularly in young children, but such strains are usually non-capsulated and not responsible for acute invasive infection.

Table 23.14 Some currently available pneumococcal vaccines

Current pneumococcal vaccines	Serotypes
23 valent pneumococcal—polysaccharide (PPSV-23) vaccine Not recommended for children below 2 years	Capsular antigens of the 23 (1, 2, 3, 4, 5, 6B, 7F, 8, 9N, 9V, 10A, 11A, 12F, 14, 15B, 17F, 18C, 19F, 19A, 20, 22F, 23F, and 33F) most prevalent serotypes give 80–90 per cent protection; used for adults and children above 5 years of age
7-valent conjugate vaccine targeting types (PCV7) Conjugated to CRM197 protein of *C. diphtheriae*	4, 6B, 9V, 14, 18C, 19F, 23F Can be used in children from two months to two years
10-valent pneumococcal conjugate vaccine (PHiD-CV) conjugated to NTHi protein	1, 4, 5, 6B, 7F, 9V, 14, 18C, 19F, 23F
13-valent pneumococcal conjugate vaccine (PCV 13)	1, 3, 4, 5, 6A, 6B, 7F, 9V, 14, 18C, 19A, 19F, and 23F
15-valent PCV conjugated to CRM197	1, 3, 4, 5, 6A, 6B, 7F, 9V, 14, 18C, 19A, 19F, 22F, 23F, and 33F

- Immunity is **type-specific.**
- **Maternal antibodies are protective** and can be transmitted from mother to child.
- Non-immunised children between **six months and two years** of age become susceptible when maternal antibody wanes.
- With the increased use of the Hib vaccine, infections due to *H. influenzae* type B are decreasing.

Agent—*H. influenzae*

- The genus *Haemophilus* includes a number of species of organisms that are **small, pleomorphic, gram-negative, oxidase-positive, fastidious, non-motile, and non-sporing.**
- They require two accessory growth factors (**X or hemin and V or NAD**) that are present in blood.
- *H. influenzae* is the first free-living organism whose complete genome has been sequenced.

Morphology

- *H. influenzae* is a small (1.0 × 0.3 μm), gram-negative bacillus, exhibiting considerable pleomorphism.
- In sputum, it usually occurs as clusters of coccobacillary forms, while in CSF from meningitis cases, long, bacillary, and filamentous forms predominate.
- Cells from young cultures (18–24 hours) are usually coccobacillary, while older cultures are distinctly pleomorphic. Strains isolated from acute infections are often capsulated.

Antigenic properties and virulence factors

- The major antigenic determinant of capsulated strains is the **capsular polysaccharide.**
- *H. influenzae* strains have been classified by Pittman into six capsular types—A to F. Typing is done by the agglutination of type-specific antisera.
- *H. influenzae* type B (Hib) is the most common type associated with infection. Almost 95 per cent of *H. influenzae* isolates from acute invasive infections such as meningitis belong to type B.
- The **polyribosyl ribitol phosphate (PRP) antigen** of Hib capsule induces IgG, IgM, and IgA antibodies, which are bactericidal, opsonic, and protective. Hib PRP is used for immunisation.
- *H. influenzae* strains lacking a capsule cannot be typed and are called **non-typeable strains**. After Hib, the non-typeable strains are the most relevant in clinical infections.
- The **outer membrane protein antigens (OMP)** show considerable variation. OMP antigens of Hib have been classified into at least 13 subtypes.
- **Lipooligosaccharides (LOS)** are antigenically complex. OMP and LOS subtyping may be of epidemiological value.

Pathogenesis (Fig. 23.39)

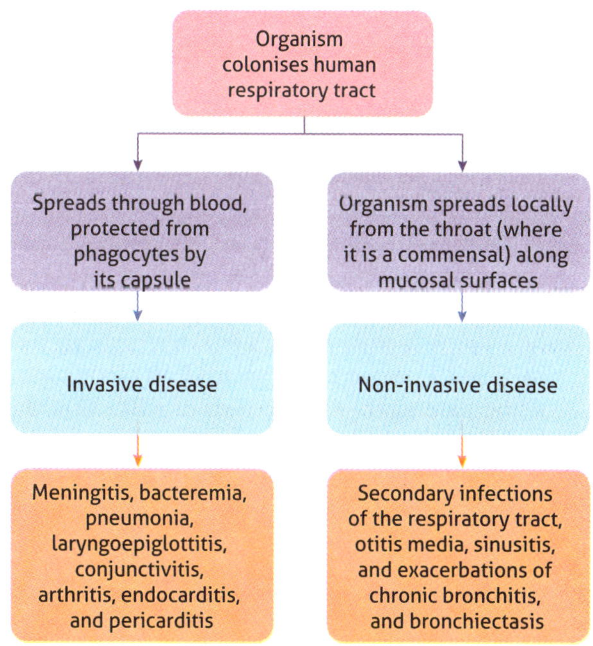

Fig. 23.39 Pathogenesis of invasive and non invasive disease caused by *H. influenzae*

Clinical features

- **Laryngoepiglottitis (croup)**
- **Bronchitis**
- **Meningitis**
- **Pneumonia**—occurs in infants and is accompanied by empyema and sometimes, meningitis as well.

In older children and adults, lobar pneumonia is seen. While these are primary infections caused by capsulated strains, bronchopneumonia may occur as a secondary infection with the non-capsulated strains.

- **Suppurative lesions**—arthritis, endocarditis, and pericarditis may result from hematogenous dissemination. Otitis media occurs by direct spread from the nasopharynx. Cellulitis, particularly in the buccal and periorbital areas, is seen in young children.

Laboratory diagnosis

- **Specimen:** CSF, blood, or sputum is collected depending on the site of infection. Since the bacillus is very sensitive to low temperatures, specimens should not be refrigerated before inoculation. Blood cultures are often positive in cases of laryngoepiglottitis and pneumonia. Sputum is not an ideal specimen to isolate the organism in cases of pneumonia.
- **Microscopy:** Direct sputum smear examination may not be very specific unless typical pleomorphic slender bacilli with abundant polymorphs are seen.
- **Direct antigen detection:** The capsular polysaccharide antigen may be present in CSF in meningitis and in urine in systemic infection. Latex particle agglutination for the detection of type B antigen is useful in rapid diagnosis.
- **Culture:** *Haemophilus* requires accessory growth factors to facilitate growth of the organism from clinical specimen.
  - **X factor**—hemin or other porphyrins required for the synthesis of cytochrome and other heme enzymes such as catalase and peroxidase involved in aerobic respiration
  - **V factor**—a co-enzyme, nicotinamide adenine dinucleotide (NAD,) or NAD phosphate (NADP);
  - **Blood agar and nutrient agar** with supplements of X and V factors (Fig. 23.40). Alternatively a disc of X and V factor can be placed on blood agar.
  - **Chocolate agar**—due to lysis of RBC, the X and V factor are released in the medium and extra supplementation may not be required.

(a) (b)

Fig. 23.40 *H. influenzae*: (a) on blood agar showing satellitism around *S. aureus* streaks and (b) around discs impregnated with XV factor (*Source*: Department of Microbiology, Pondicherry Institute of Medical Sciences, Puducherry)

  - **Filde's agar**—best for the primary isolation of *H. influenzae*.
  - Levinthal's medium is also used.
  - **Incubation:** Incubate in an environment containing 5–10% CO_2 and high humidity. After overnight incubation at 37°C, **small, opaque colonies of gram-negative, short, oxidase-positive cocci appear.**
- Typing may be done by slide agglutination using specific antisera.
- **Biochemical tests:** These are not routinely used for identification.

Satellitism

- A streak of *Staphylococcus aureus* is made on blood agar with culture of *Haemophilus* species.
- Growth of tiny colonies of *Haemophilus influenzae* are seen near the *Staphylococcus* streak.
- *Staphylococcus* lyses the red blood cells, releasing hemin (X factor) and NAD (V factor) for growth of *Haemophilus* spp. Therefore, *Haemophilus* spp. grow adjacent to the streak line where the nutrients are available.
- This demonstrates the dependence of *H. influenzae* on the V factor, which is available in high concentrations near staphylococcal growth.
- Another differentiating test is growth around discs incorporated with X, V, and XV factor.
- This helps in differentiating the species of *Haemophilus* (Fig. 23.40).

Laboratory diagnosis of *H. influenzae* infections

- **Specimen:** CSF, blood, sputum, or urine
- **Transport and storage of sample:** Transport immediately or store at room temperature or in incubator as the bacillus is very sensitive to low temperatures
- **Microscopy:** Direct sputum smear examination may not be very specific
- **Antigen detection**: Detection of capsular polysaccharide antigen in CSF and in urine
- **Culture**
  - Fastidious, capnophilic (5–10% CO_2) requires accessory growth factors (X factor or hemin and V factor or nicotinamide adenine dinucleotide [NAD] or NAD phosphate [NADP]
  - Blood agar with a disc of X and V factor, nutrient agar with supplements of X and V factors
  - Chocolate agar
  - Filde's agar—best for the primary isolation of *H. influenzae*
  - Levinthal's medium
- **Identification**
  - Typing may be done by slide agglutination using specific antisera
  - Catalase- and oxidase-positive
  - **Satellitism**

Treatment

- Cefotaxime or ceftriaxone are the drugs of choice for the treatment of *Haemophilus* meningitis.
- Ampicillin and cotrimoxazole were popular for respiratory infections, but as plasmid-borne resistance to these drugs is now common, amoxicillin–clavulanate or azithromycin is now considered more effective.

Prophylaxis

- **Active immunisation:** Hib PRP vaccine is indicated as serious infections are caused by type B strains.
- **Conjugate vaccines:** Purified PRP is immunogenic in older children and adults. However, as with other polysaccharide antigens, PRP is poorly immunogenic in children below the age of two years. Its immunogenicity has been improved by coupling it with protein carriers like diphtheria and tetanus toxoids or meningococcus outer membrane protein. Such conjugate Hib–PRP vaccines are available for use in young children and can be given along with DPT immunisation.

Other species of *Haemophilus*

Haemophilus aegyptius

- Recent DNA studies have shown that the bacillus is identical to non-capsulated *H. influenzae*. Therefore, it is now named *H. influenzae* biotype *aegyptius*.
- It is worldwide in distribution and causes a highly contagious form of conjunctivitis ('pink eye'). It is especially common in the tropics and subtropics and may occur in epidemic form.
- It responds to local sulphonamides or gentamicin.
- It has also been identified as the causative agent of **Brazilian purpuric fever (BPF)**, an endemic in South America in which conjunctivitis proceeds to fulminant septicemia in infants and children with high fatality.

Haemophilus parainfluenzae

- It is a commensal in the upper respiratory tract and has been reported to cause acute pharyngitis, subacute bacterial endocarditis, and urethritis.
- It requires only the V factor and not the X factor.

Whooping cough

Case 7. *A four-month-old child was brought from a peripheral centre in a remote rural area with complaints of severe cough ending with a typical whoop for the past 10–12 days. The boy had apnea during the bouts followed by vomiting but was said to be sleeping comfortably between the bouts. Blood counts were raised with lymphocytosis. There was no history of immunisation since birth. The laboratory reported growth of* ***B. pertussis*** *from the nasopharyngeal swab. A diagnosis of* ***whooping cough*** *was made. The child responded well to erythromycin and supportive care.*

i. How is whooping cough diagnosed?
ii. Describe the virulence factors of the *B. pertussis*.

Whooping cough (pertussis) is an infection of the respiratory system caused by *Bordetella pertussis*. It is characterised by severe coughing spells, which can sometimes end in a 'whooping' sound due to deep breath.

Epidemiology

- Whooping cough is predominantly a pediatric disease, the incidence and mortality being highest in the first year of life.
- **Maternal antibodies do not protect against the disease**. Immunisation should, therefore, be started early.
- The disease is **more common in females** than in males at all ages.
- It is worldwide in distribution.
- It occurs in an **epidemic form** periodically; the disease is never absent from any community.
- Whooping cough is one of the **most infectious** of bacterial diseases; non-immune contacts seldom escape the disease.
- With universal immunisation, childhood pertussis is on the decline. However, adolescent pertussis is on the rise due to waning of immunity at that age.
- Adults become prone to infection unless they receive booster immunisation.

Agent—*Bordetella pertussis*

- These are small, ovoid or rod-shaped, gram-negative organisms.
- The genus *Bordetella* comprises three human pathogens:
  - *B. pertussis*
  - *B. parapertussis*
  - *B. bronchiseptica*
- *B. pertussis* causes 95 per cent of whooping cough cases. Around 5 per cent are caused by *B. parapertussis* and infrequently, whooping cough may be caused by *B. bronchiseptica*.

Virulence factors

Agglutinogens

- Bordetellae carry surface agglutinogens associated with fimbriae.
- Strains that cause infection belong to **types 1, 2, and 3**; vaccine strains, therefore, contain these factors.
- Agglutinogens promote virulence by helping bacteria to **attach to respiratory epithelial cells**.
- They are useful in **serotyping strains and in epidemiological studies**.

Filamentous hemagglutinin (FHA)

- FHA **adheres to the cilia** of the respiratory epithelium and to erythrocytes.
- Antibodies to FHA are protective.
- FHA is used in **acellular pertussis vaccines** along with the pertussis toxin and pertactin.

FHA and PT hemagglutinins promote secondary infection by coating other bacteria such as *Haemophilus influenzae* and *S. pneumoniae* and assisting their binding to the respiratory epithelium. This phenomenon has been termed **piracy of adhesins.**

Pertactin

- This is an outer membrane protein (OMP) antigen present in all virulent strains of *B. pertussis*.
- Antibodies to pertactin can be seen in the blood of children after infection or immunisation.
- Pertactin is included in acellular pertussis vaccines.

Toxins

Pertussis toxin (PT)

- This is present only in *B. pertussis*.
- PT is expressed on the surface of the bacillus and secreted into the surrounding medium.
- The toxin is a diverse **lymphocytosis producing factor** (LPF) and causes profound lymphocytosis in pertussis patients.
- The **histamine-sensitising factor (HSF)**—responsible for heightened sensitivity to histamine— and **islet-activating protein (IAP)**, which induce excessive insulin secretion by the pancreatic islet cells, are seen only in experimental animals and not in patients.
- PT, with a molecular weight of 117,000 Da, is a **hexameric protein** composed of six subunits with an A–B structure (A being the enzymatically active moiety and B the binding component). It can be toxoided. PT toxoid is the major component of acellular pertussis vaccines.
- *Bordetella pertussis* strains carry a novel allele for the pertussis toxin promoter, which confers **increased pertussis toxin (Ptx) production**. Epidemiologic data suggest that these strains are more virulent in humans.

Adenylate cyclase (AC)

Known as the **AC toxin (ACT)**, it acts by catalysing the production of cAMP by various types of cells.

Heat-labile toxin (HLT)

- It is a **cytoplasmic protein** present in all bordetellae.
- It is dermonecrotic and lethal in mice.
- Its pathogenic role is not known.

Tracheal cytotoxin (TCT)

- It is a low-molecular-weight peptidoglycan that induces ciliary damage in hamster trachea.
- Its role in disease is not known.

Lipopolysaccharide (LPS)

- LPS or the heat-stable toxin is present in all bordetellae and exhibits features of gram-negative bacterial endotoxins.
- It is present in the whole-cell pertussis vaccine but is not considered to be a protective antigen.

Susceptibility

- *B. pertussis* is a delicate organism, killed readily by heat (55°C for 30 minutes), drying, and disinfectants.
- It survives at 0–4°C.
- Outside the body, *B. pertussis* survives for five days in dried droplets, three days on cloth, and a few hours on paper.

Pathogenesis

- *B. pertussis* is an obligate human parasite.
- Infection occurs through aerosols and close contact.
- In the initial stages, the bacilli are confined to the nasopharynx, trachea, and bronchi.
- Clumps of bacilli may be seen enmeshed in the cilia of the respiratory epithelium.
- As the disease progresses, inflammation extends into the lungs, producing diffuse bronchopneumonia with desquamation of the alveolar epithelium (Case 7).
- The infection is limited to the respiratory tract; the bacilli do not invade the bloodstream.

Clinical features

The incubation period of the disease is 1–2 weeks. The disease takes a protracted course, comprising three stages, each lasting approximately two weeks.

i. **Catarrhal stage:** This stage is insidious in onset, with low-grade fever, catarrhal symptoms, and a dry, irritating cough. Clinical diagnosis in this stage is difficult. This is also the stage of maximum infectivity. (If suspected, treatment at this stage can arrest the infection.)
ii. **Paroxysmal stage:** There is an increase in the intensity of the cough, which occurs in distinctive bouts. During the paroxysm, the patient experiences violent spasms of continuous coughing, followed by a long in-rush of air into the almost empty lungs, with a characteristic whoop (hence the name whooping cough; Fig. 23.41).
iii. **Convalescent stage:** The frequency and severity of cough gradually decrease. The disease usually lasts 6–8 weeks, though in some, it may be very protracted.

Complications

- **Pressure effects** (subconjunctival hemorrhage, subcutaneous emphysema)—these occur during the violent bouts of coughing.
- **Respiratory complications** (bronchopneumonia, lung collapse)—these are self-limited, with the atelectasis resolving spontaneously.
- **Neurological complications** (convulsions and coma)—these may result in permanent sequelae such as epilepsy, paralysis, retardation, blindness, or deafness.

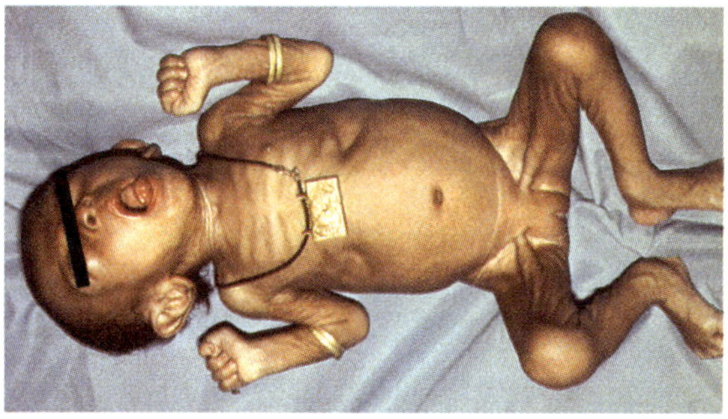

Fig. 23.41 Pertussis in a female infant who presented with paroxysms of cough—notice the indrawn chest muscles and protruding tongue (*Source*: CDC, PHIL, ID 6379)

Laboratory diagnosis

The bacilli are present in the upper respiratory tract most abundantly in the early stages of the disease. They may be demonstrated by microscopy, or more reliably, by culture.

- **Specimen:** Respiratory samples can be collected using a per-nasal swab, a post-nasal swab (dacron or calcium alginate swabs), or using the cough plate method.
  - *Per-nasal or nasopharyngeal swab:* A flexible nichrome wire is passed along the floor of the nasal cavity and material is collected from the pharyngeal wall. Nasopharyngeal aspirate collected through a soft catheter attached to a syringe is a better source. It can be used for PCR also.
  - *Post-nasal (per-oral) swab*: Secretions from the posterior pharyngeal wall are collected with a cotton swab on a bent wire passed through the mouth.
  - *Cough plate method:* A culture plate is held about 10–15 cm away from the patient's mouth during a bout of spontaneous or induced coughing so that droplets of respiratory exudates impinge directly on the medium.
- **Transport medium:** The swabs should be plated without delay or transported in a 0.25–0.5 mL Casamino acid solution, at pH 7.2, in modified Stuart's medium or Mischulow's charcoal agar.
- **Microscopy:** Microscopic diagnosis depends on the demonstration of the bacilli in respiratory secretions by Gram staining or the fluorescent antibody technique.
- **Culture:** Useful in the first 2 weeks of cough.
  - *B. pertussis* is an obligate aerobe. It grows best at 35–36°C. Complex media are necessary for primary isolation. The media commonly used are:
  - Bordet–Gengou glycerine-potato-blood agar
    - Regan–Lowe media with horse blood and charcoal and cephalosporins to inhibit the normal upper respiratory flora
    - Charcoal blood agar—charcoal or ion-exchange resins are incorporated in culture media to neutralise the inhibitory substances in the respiratory specimen
- **Incubation:**
  - Plates are incubated in high humidity at 35–36° C. Growth is slow.
  - After incubation for 48–72 hours, colonies are **small, dome-shaped, smooth, opaque, greyish-white, refractile and glistening, resembling 'bisected pearls' or 'mercury drops'.**
  - Confluent growth bears an '**aluminum paint**' appearance.
- **Identification:**
  - Bacterial growth can be confirmed by direct immunofluorescence using specific antisera or biochemical tests.
  - *B. pertussis* is biochemically inactive.
- **Polymerase chain reaction (PCR):** PCR-based tests are more sensitive and more commonly used in current practice since the culture yield is poor in pertussis. PCR has optimal sensitivity during the first 3 weeks of cough when bacterial DNA is still present in the nasopharynx.
- **Serology:** Serological diagnosis is not helpful and is not used routinely for diagnosis. Vaccination can be a cause of increase in antibodies. ELISA kits to detect IgG antibodies to pertussis toxin are available.
- **Other laboratory parameters:** Marked leukocytosis is seen with relative lymphocytosis (total leukocytic counts 20,000–30,000 per mm^3 with 60–80 per cent lymphocytes).

Laboratory diagnosis of whooping cough

- **Specimen:**
  - *Per-nasal or nasopharyngeal swab*
  - *Cough plate method*
- **Transport**: Transport without delay or in transport media.
  - Casamino acid solution
  - Modified Stuart's medium
  - Mischulow's charcoal agar
- **Microscopy**: Gram staining or the fluorescent antibody technique
- **Culture**:
  - Obligate aerobe, fastidious, requires complex media, slow growing, need high humidity, optimum temperature: 35–36°C
  - Bordet–Gengou glycerine-potato-blood agar
  - Regan–Lowe media (horse blood and charcoal and cephalosporins)
  - Charcoal blood agar (charcoal or ion-exchange resins)

- **Identification:**
  - Colonies—**small, dome-shaped, smooth, opaque, greyish-white, refractile and glistening, resembling 'bisected pearls' or 'mercury drops'.**
  - Confluent growth as an **'aluminium paint'** appearance.
  - Direct immunofluorescence using specific antisera
  - Catalase and oxidase positive, otherwise *B. pertussis* is biochemically inactive.
- **Polymerase chain reaction (PCR)**
- **Serology:** Not used routinely for diagnosis. ELISA kits to detect IgG antibodies to pertussis toxin
- **Other laboratory parameters:** Marked leukocytosis with relative lymphocytosis

Treatment

- Antimicrobial therapy is beneficial only if initiated within the first ten days of the disease.
- Erythromycin or clarithromycin and azithromycin are the drugs of choice.

Prophylaxis

- Non-immunised contacts should receive erythromycin prophylaxis for 10 days after contact with the patient has ceased.
- Pertussis vaccine is given as a part of the immunisation programme, in combination with tetanus and diphtheria as DT(w)P—diphtheria, tetanus, and whole cell pertussis.
- *B. pertussis* also acts as an adjuvant for the toxoids, producing better antibody response.
- Both whole-cell and acellular vaccines have a protection rate of about 90 per cent.
- With whole-cell vaccines, the protection declines to 50 per cent in about five years and is absent after 12 years. Even fully immunised subjects may develop pertussis, but the disease is mild.

Whole cell killed vaccine

- The smooth, phase I strain is used for vaccine production.
- It is a suspension of whole *B. pertussis* that has been inactivated, usually with formalin.
- Children under four years who are contacts of patients should receive a booster even if they have been previously immunised.

Complications

- Post-vaccinal encephalopathy, seizures, shock, or hyperpyrexia seen rarely (1 in 5–10 million vaccinees).
- **Hypotonic-hyporesponsive episode (HHE)** is seen as a reversible adverse event after 48 hours. There is decreased sensory awareness or loss of consciousness accompanied by pallor and muscle hypotonicity. Patients recover with supportive management.

Contraindications

- If severe complications such as encephalopathy, seizures, shock, or hyperpyrexia develop following pertussis vaccination, subsequent doses of the vaccine are contraindicated.
- Routine pertussis vaccination is not advisable after the age of seven years since adverse reactions are likely.

Acellular vaccine

- It contains the protective components of the pertussis bacillus (PT, FHA, agglutinogens 1, 2, 3, and pertactin).
- It has fewer side effects than the whole-cell vaccine.

Agent—*Bordetella parapertussis*

- This is an infrequent cause of whooping cough and causes mild disease.
- The pertussis vaccine does not protect against *B. parapertussis* infection.

Agent—*Bordetella bronchiseptica*

- This is motile by peritrichate flagella, but is antigenically related to *B. pertussis* and *Brucella abortus*.
- It has been found to cause a very **small proportion (0.1 per cent) of cases of whooping cough.**

Nosocomial Bacterial Pneumonia

Case 8. *An adult male was admitted to the ICU following a road traffic accident. He was put on a ventilator due to severe respiratory distress. On the third day, his condition worsened with symptoms of high fever, and features of pneumonia seen on chest X-ray. His blood culture yielded lactose-fermenting mucoid colonies, identified as **Klebsiella pneumoniae**. The strain was multidrug-resistant, but sensitive only to imipenem and piperacillin-tazobactam. He was treated with imipenem. However, he succumbed to the infection on the tenth day despite aggressive management.*

i. What are the probable causative agents in this situation?
ii. Describe laboratory diagnosis of such infections.

The common bacteria causing pneumonia in nosocomial are:

- *Pseudomonas aeruginosa*
- *Escherichia coli*
- *Klebsiella pneumoniae*
- *Enterobacter* spp.
- *Acinetobacter* spp.
- *Staphylococcus aureus*
- *Enterococcus*

Many organisms cause only healthcare-associated pneumonia while some can cause both healthcare-associated and community-acquired pneumonia.

Pneumonia caused by *Klebsiella pneumoniae*

Epidemiology

- Pneumonia due to *Klebsiella* have high case fatality (Case 8).

- It occurs in **middle-aged or elderly persons with risk factors** such as alcoholism, chronic bronchopulmonary disease, or diabetes mellitus. It can be community-acquired or hospital-acquired.
- The disease is characterised by **massive mucoid inflammatory exudates of lobar or lobular distribution**, involving one or more lobes of the lung.
- Necrosis and abscess formation are more frequent than in pneumococcal pneumonia.
- Serotypes 1, 2, and 3 are usually responsible for pneumonia.
- In recent years, a new hypervirulent, hypermucoviscous variant of *Klebsiella* has emerged. This is associated with community-acquired infections including liver abscess, pneumonia, meningitis, and endophthalmitis and the ability to metastatically spread.

Agent—*Klebsiella pneumoniae*

- Klebsiellae, which are member of Enterobacterales, are widely distributed in nature, occurring both as commensals in the intestines and as saprophytes in soil and water (Chapter 4).
- Three species are pathogenic to man—*K. pneumoniae*, *K. ozaenae*, and *K. rhinoscleromatis*. *K. oxytoca* is rarely isolated from clinical infections.
- The species *K. oxytoca* may rarely be isolated from clinical specimens.
- The common infections associated with *K. pneumoniae* are presented Table 23.15.

Table 23.15 Common infections caused by different species of *Klebsiella*

Species	Infection
K. pneumoniae	Pneumonia (community- or healthcare- or ventilator-associated), urinary tract infection, including catheter-associated UTI, septicemia- and several pyogenic infections
K. ozaenae	Ozena, a disease characterised by foul-smelling nasal discharge
K. rhinoscleromatis	Rhinoscleroma, a chronic granulomatous hypertrophy of the nose, prevalent in south-eastern Europe, India, and Central America

Morphology

- *Klebsiella* is a non-motile rod, which is capsulated, short, plump, and straight, measuring about 1–2 μm × 0.5–1.8 μm in size.
- The mucopolysaccharide capsule is often prominent and can be made out even in Gram-stained smears as **halos around the rods**.

Pathogenicity and virulence factors

The following factors are responsible for the pathogenesis of *Klebsiella* infections:

- **Capsule and slime layer:** Protect bacteria from phagocytosis by polymorphonuclear granulocytes, and also prevents killing of the bacteria by bactericidal serum factors.
- **Pili (Fimbriae):** These promote attachment to the host cell (adherence).
- **Lipopolysaccharide**
- **Siderophores:** These make iron available to bacteria.
- **Hypervirulent and hypermucoviscous (hvKp) strains of *Klebsiella*:** These are more virulent and possess virulence factors like siderophores and increased capsular production. These bacilli are carried in the gut and easily spread in the community and in hospitals, causing hospital-acquired infections. The K1 and K2 serotypes of hvKp are resistant to antibiotics.

Laboratory diagnosis

Diagnosis is made by culturing appropriate specimens and identifying the isolate by biochemical reactions.

- **Culture:** Klebsiellae grow well on ordinary media, forming large, dome-shaped, mucoid colonies with varying degrees of stickiness. On MacConkey agar, they grow as lactose-forming colonies. Fresh clinical isolates are often highly mucoid, almost flowing down the surface of the medium (Fig. 23.42).

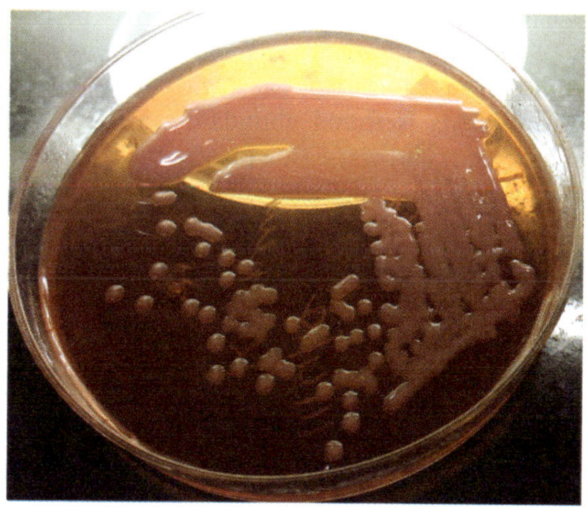

Fig. 23.42 Mucoid lactose-fermenting colonies of *Klebsiella pneumoniae* (*Source*: Department of Microbiology, Pondicherry Institute of Medical Sciences, Puducherry)

Treatment

Many *Klebsiella* strains carry plasmids that determine multiple drug resistance by producing carbapenamases and metallo-beta-lactamases.

Several strains have shown extensive drug resistance, including resistance to carbapenems like imipenem and meropenem. These are known as **XDR strains**. Colistin or polymixin (liposomal) must be used for these drugs as the last option.

Pseudomonas aeruginosa pneumonia

> **Case 9.** *A three-year-old boy presented with a history of repeated respiratory tract infections with productive cough. Further investigations showed that the sweat test was positive for high sodium chloride concentration, suggestive of cystic fibrosis. Culture of his sputum showed the growth of transparent, colourless colonies with a mucoid character. The organisms were oxidase-positive and motile. These were identified as* ***P. aeruginosa****. Based on the antibiotic susceptibility, the boy was started on a combination of piperacillin–tazobactam and amikacin along with supportive therapy of airway clearance and nutritional supplement.*
>
> i. What are the infections caused by *Pseudomonas* species?
>
> ii. Describe the virulence of *Pseudomonas*.

- Pseudomonads are hardy organisms; they are mostly saprophytic and frequently colonise hospital environments, equipment and moist areas.
- Some members of this group are pathogenic to plants, insects and reptiles.

Epidemiology

- Community-acquired infections by *P. aeruginosa* are **suppurative otitis and respiratory tract infections in cystic fibrosis patients**.
- Healthcare-associated infections by *P. aeruginosa* are observed in ICUs, especially with immunocompromised patients.
- Occasional opportunistic infections may be caused by a few other species such as *P. fluorescens*, *P. putida*, and *P. stutzeri*.
- *Pseudomonas* is found abundantly in the hospital environment and causes a wide variety of infections.
- It survives on equipment such as respirators and endoscopes, medicines such as lotions, ointments and eye drops, and even stocks of distilled water and articles such as bedpans.
- It is also present on the skin of the axilla and perineum in some individuals.
- The bacterium can establish its pathogenesis if it is introduced into devitalised tissue through disrupted skin or mucous membrane. The process is facilitated in neutropenia along with bacterial factors.
- The role of *P. aeruginosa* in hospital infections is due to its resistance to common antibiotics and disinfectants and its ability to establish itself widely in hospitals. It also has the ability to survive in a wide temperature range

Agent—Pseudomonas

Morphology

- *Pseudomonas* is a slender, gram-negative bacillus, 1.5–3 × 0.5 μm in size, which is actively motile by a polar flagellum.
- Mucoid strains, particularly isolates from cystic fibrosis patients, have an abundance of **extracellular polysaccharides** composed of alginate polymers.
- These form a loose capsule (**glycocalyx**), in which microcolonies of the bacillus are enmeshed and protected from host defences.

Virulence factors

- **Exotoxin A**—it has a mechanism of action similar to that of the diphtheria toxin. The toxin has two subunits—A (active) and B (binding)—and inhibits protein synthesis. Strong antibody response to exotoxin A is considered a favourable sign in severe infections with *P. aeruginosa*.
- **Other extracellular enzymes and toxins**—these include proteases, elastases, hemolysins, and enterotoxin.
- The **slime layer** and **alginate glycocalyx** and the **ability to form biofilm**s.
- **Pili** to enhance adhesion.
- **Pigment production** causes tissue injury.
- **Pyocyanin** is a bluish-green phenazine pigment that is soluble in water and chloroform (pyocyanin is produced only by *P. aeruginosa* (Fig. 23.43). This is enhanced by King A medium.
- **Fluorescein** is a greenish-yellow pigment soluble in water but not in chloroform. King B medium enhances fluorescein production. This pigment may be produced by many other species.
- **Pyorubin (red) and pyomelanin (brown)** are produced in various combinations by some species and are sometimes seen in clinical isolates.

It is not known whether the pigments have any role in pathogenesis. Some pigments, particularly pyocyanin, inhibit the growth of many other bacteria and may, therefore, contribute to *P. aeruginosa* emerging as the dominant bacterium in mixed infections.

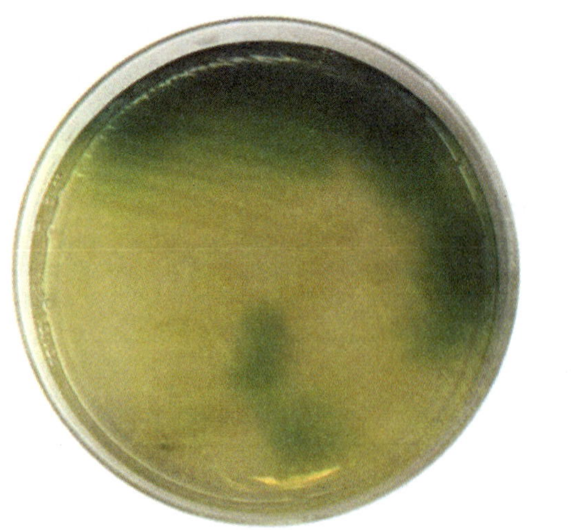

Fig. 23.43 *P. aeruginosa* on nutrient agar showing green pigment production (pyocyanin)

Multidrug resistance

- *Pseudomonas* are intrinsically resistant to many antibiotics such as ampicillin, amoxicillin-clavulanate, trimethoprim-suphamethoxazole, 1st- and 2nd-generation cephalosporins, cefotaxime, ceftriaxone, chloramphenicol, and tetracycline.
- Many hospital strains of *Pseudomonas* have developed acquired resistance to antimicrobials. This restricts the antibiotic treatment options as strains in the hospital are multidrug-resistant and even pan-drug-resistant.

Clinical features

Multiple types of localised and disseminated infections are associated with *Pseudomonas* (Table 23.16).

Laboratory diagnosis

As *P. aeruginosa* is a frequent contaminant in the hospital setting, the isolation of the bacillus from a specimen should not always be taken as proof of its etiological role. Repeated isolation and clinical correlation helps to confirm the diagnosis.

- **Culture:** *P. aeruginosa* grows well on ordinary media, producing **large, translucent, irregular colonies with a distinctive musty or earthy smell.** The colonies often produce a diffusible pigment (green colour). **Iridescent patches with a metallic sheen** may also be seen in cultures.
  - *Blood agar*: Many strains are hemolytic on blood agar.
  - *Selective media*: Identification of pigmented strains of *Pseudomonas* from clinical specimens is easy. It may be necessary to use selective media such as cetrimide agar for isolation from samples with mixed flora.

Table 23.16 Types of localised and disseminated infections associated with *Pseudomonas*

Superficial infections	Generalised/Organ-specific infections
Wounds and bed sores Secondary infection in burns patients Ecthyma gangrenosum Necrotising fasciitis and cellulitis Nail infections	Septicemia IV line-associated bloodstream infections Endocarditis Ventilator-associated pneumonia Chronic obstructive pulmonary lung disease (COPD) Catheter-associated urinary tract infections (CAUTI) Community-acquired enteric illness, leading to sepsis and necrotising enteritis (Shanghai fever) Iatrogenic meningitis (following lumbar puncture) Eye infections, ear infections (malignant otitis externa)

- **Typing:** For epidemiological purposes and to trace the source of infection, *Pseudomonas* isolates can be typed by several methods.
  - *Phage typing*: This method is dependent on specific bacteriophages which lyse only certain strains of *Pseudomonas*. This method was used to clone hospital isolates. Some laboratories still use it.
- *P. aeruginosa* produces a number of pigments but some strains may be non-pigmented (see above).

Treatment

While *Pseudomonas* are generally resistant to many antibiotics, they may be susceptible to colistin.

Acinetobacter pneumonia

Case 10. *A 30-year-old male was admitted to the emergency department following a road traffic accident with polytrauma and head injury and a Glasgow Coma Scale (GCS) of 5/15. He was sedated, intubated, mechanically ventilated and hemodynamically unstable. His baseline laboratory parameters were normal. On the fourth day of admission, he developed high-grade fever, increased tracheal secretions with leukocytosis, and a procalcitonin level of 4 ng/mL. In addition, his chest X-ray was suggestive of bronchopneumonia. He was diagnosed with* ***ventilator-associated pneumonia (VAP)****. A broncho-alveolar lavage (BAL) culture grew gram-negative coccobacilli which was oxidase-negative and did not ferment sugars; this was identified as* Acinetobacter baumannii. *It was resistant to all antibiotics except carbapenems and colistin.*

i. What are the infections caused by *Acinetobacter baumannii*?

Agent—*Acinetobacter baumannii*

Acinetobacter is a complex group of coccobacillary bacteria which are subdivided as follows:

- **Baumannii**, in which glucose is broken down oxidatively (causes clinically important infections)
- **Lwoffii**, which does not metabolise carbohydrates, and may be non-hemolytic
- **Haemolyticus** is the hemolytic strain

These bacteria are ubiquitous in nature and are found within the hospital environment, commonly associated with infections acquired in the hospital. *A. baumannii* is more often associated with infections, whereas *A. lwoffii* is found in the environment.

Laboratory diagnosis

Acinetobacter baumannii forms pinkish colonies (not clearly lactose-fermenting; often referred to as **query LF colonies**) on MacConkey medium.

Treatment

- These organisms have acquired resistance to cephalosporins, quinolones, aminoglycosides, and in some cases, carbapenems (Fig. 23.44).

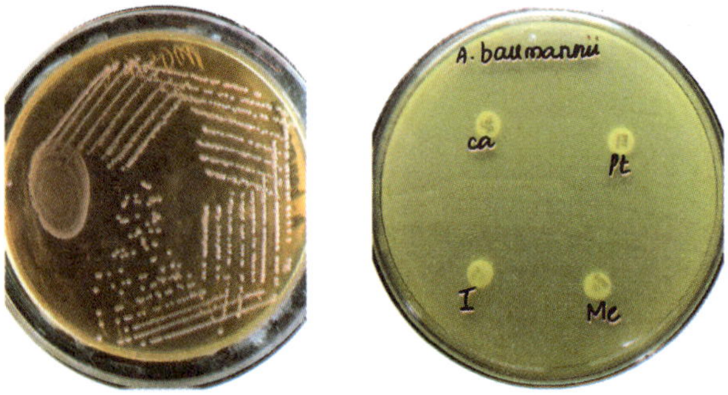

(a) (b)

Fig. 23.44 *Acinetobacter baumannii*: (a) light pink colonies due to partial breakdown of carbohydrate (lactose) on MacConkey agar plate and (b) multidrug-resistant *A. baumannii* on an antibiotic susceptibility plate (*Source*: Department of Microbiology, Pondicherry Institute of Medical Sciences, Puducherry)

- *Acinetobacter baumannii* has become problematic in intensive care units (ICUs) because it has acquired multiple resistance genes, resulting in treatment failures.

Atypical bacterial pneumonia

- It was noticed that some pneumonia patients had different characteristics than those with 'typical' pneumonias.
- These are caused by agents that are **not identified by routine staining and culture**.
- They presented with milder symptoms and the characteristic patch on chest X-ray was not seen.
- The causative agents of atypical pneumonia are specific respiratory pathogens such as *Mycoplasma pneumoniae*, *Chlamydia pneumoniae*, *Chlamydia psittaci* (psittacosis), *Francisella tularensis* (tularemia), *Coxiella burnetii* (Q fever), or *Legionella* species.
- The patient presents with atypical symptom such as mental confusion, headache, bodyaches, pleuritic pain, or abdominal pain. The pulmonary symptoms are often absent or mild.

Mycoplasmal pneumonia

- *Mycoplasma pneumoniae* infection occurs worldwide and is more prevalent in colder months.
- It chiefly affects children who are 5 to 9 years old.
- It is spread by close personal contact and has a long incubation period.
- *Mycoplasma* cause **primary atypical pneumonia (PAP).** The disease is typically tracheobronchitis.
- Acute pharyngitis is uncommon, and only a third of patients develop pneumonia.
- **Incubation period:** 1–3 weeks.
- The **onset** is gradual, with fever, malaise, headache, and sore throat. Paroxysmal cough may occur with blood-tinged sputum.
- The disease is characterised by minimum respiratory signs on physical examination but marked radiological evidence of consolidation, which is usually patchy, involving one of the lower lobes, starting at the hilum and fanning out to the periphery (Fig. 23.45).
- The disease is usually self-limiting, with recovery in 1–2 weeks.
- **Bullous myringitis and otitis** are common complications. Meningitis, encephalitis, and hemolytic anemia are other complications.
- The disease is found worldwide and in individuals of all ages.
- **Transmission** is by droplets of nasopharyngeal secretions.
- *Mycoplasma* may remain in the throat for two or more months after recovery from the disease.

Agent—Mycoplasma pneumoniae

Mycoplasma are cell wall-deficient bacteria that may be spherical or filamentous, which are associated with infections of the respiratory and urogenital tract (details in Chapter 4).

Antigenic properties

- Mycoplasmal surface antigens are mainly glycolipids and proteins.
- Serological tests such as complement fixation, agglutination, passive hemagglutination, ELISA, and immunofluorescence have been used to detect antibodies in sera:
  - Glycolipid antigens are identified by complement fixation.
  - Protein antigens are detected by ELISA.
- The **growth inhibition test** is based on the ability of antisera to specifically inhibit the growth of the homologous species on solid media.

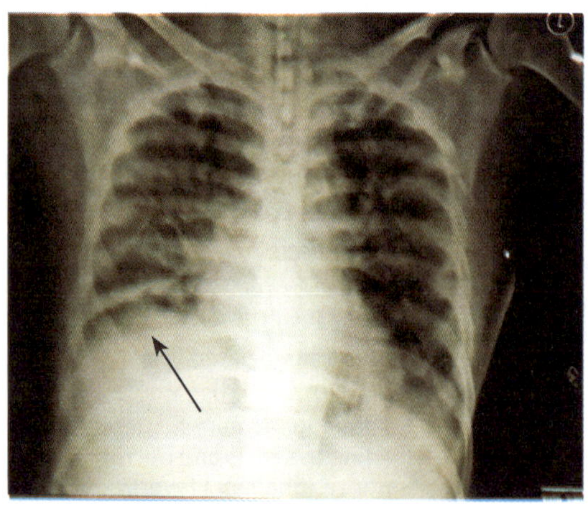

Fig. 23.45 Primary atypical pneumonia due to *M. pneumoniae* showing consolidation involving the right lower lobe (*Source*: Dr. Aneesh Basheer, Department of Medicine, Pondicherry Institute of Medical Sciences, Puducherry)

Pathogenicity

- Pathogenic *Mycoplasma* exhibit host specificity.
- They generally produce infections by adhering to the mucosa of the respiratory, gastrointestinal, and genitourinary tracts.
- *Mycoplasma* does not produce any toxin but probably cause disease by production of hydrogen peroxide, the end product of respiration in mycoplasma.
- *Mycoplasma* cause two types of disease in humans—**pneumonia and genital infections**.

Genital infections

These are caused by *Mycoplasma hominis*, which has a similar spectrum as *Ureaplasma urealyticum* (these are discussed in Chapter 27).

- Mycoplasmas tend to cause severe and prolonged infections in HIV-infected and other immunodeficient subjects.

Laboratory diagnosis

- **Specimen:**
  - For isolation, throat swabs or respiratory secretions are inoculated into viral or mycoplasma transport media to prevent drying and bacterial overgrowth.
  - *Blood sample for serology*—the patient's blood sample should not be refrigerated prior to separation of the serum as the agglutinins are readily absorbed by the homologous erythrocytes at low temperatures.
- **Culture:**
  - *Mycoplasma* may be cultivated in liquid or semi-solid media.
  - *Selective media*—PPLO (pleuropneumonia-like organism) medium and Hayflick medium. PPLO medium is agar or broth that contains beef heart infusion, peptone, yeast extract, salts, glucose or arginine, and horse or fetal calf serum.
  - The colonies may be embedded in the agar. They can be seen with a hand lens.
  - They have a '**fried egg appearance**' (Fig. 23.46) with the raised center and flat, spreading periphery. They are best studied after staining by the Dienes method and appear as pleomorphic forms under a light microscope.
  - In a liquid medium, no turbidity can be appreciated; pleomorphic forms are found in smears from the medium.
  - *Molecular methods of detection*—as isolation is difficult and delayed, PCR assays such as multiplex PCR are used to detect *Mycoplasma*; these methods are rapid and specific.
- **Serology:** Serological diagnosis may be made by specific and non-specific tests. A sustained high antibody titer may not be significant as the antibodies amy persist fro sometime after infection.

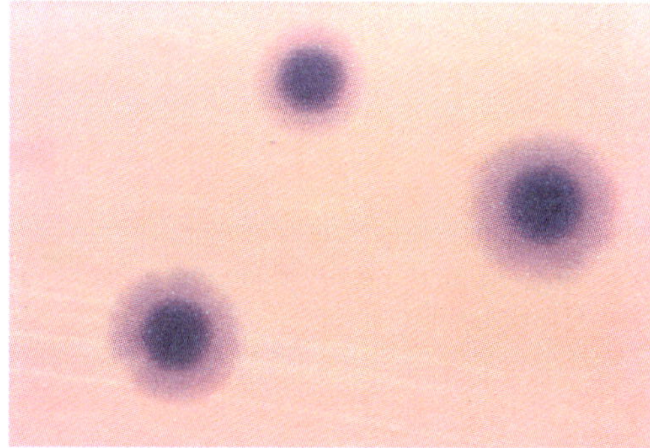

Fig. 23.46 'Fried egg appearance' of *Mycoplasma hominis* and T-strain *Mycoplasma* isolates, which have been grown on agar medium (*Source*: CDC, PHIL, Image ID 11024)

  - **Specific tests:** Tests using mycoplasmal antigens, which include immunofluorescence, hemagglutination inhibition and growth inhibition are the most sensitive tests.
  - **Non-specific tests:** The *Streptococcus* MG test is carried out by mixing serial dilutions of the patient's unheated serum and a heat-killed suspension of *Streptococcus* MG *intermedius* (a strain of *S. milleri* anginosus group) and observing agglutination after overnight incubation at 37°C. A titre of 1:20 or over is considered suggestive of mycoplasmal infection.
  - **Cold agglutination test:** This is based on the presence of macroglobulin antibodies in primary atypical pneumonia, which agglutinate human group O cells at a low temperature. Serial dilutions of the patient's serum are mixed with an equal volume of 0.2% washed human O group erythrocytes. Clumping is observed after incubation at 4°C overnight. The clumping dissociates at 37°C. A titre of 1:32 or above is suggestive. Demonstration of a rise in titre in paired serum samples is more reliable.
  - The **indirect Coombs test** may also be positive in some cases.

Laboratory diagnosis of mycoplasmal pneumonia

- **Specimen:**
  - For isolation, throat swabs or respiratory secretions
  - Blood sample is collected for serology
- **Transport medium:** *Mycoplasma* transport medium and viral transport medium are used
- **Culture**
  - Fastidious, very slow growing, microaerophilic, optimum temperature: 35–37°C, optimum pH: 7.8±0.2
  - Selective media—PPLO (pleuropneumonia-like organism) medium and Hayflick medium
  - Colonies may be embedded in the agar and seen with a hand lens; **'fried egg appearance'** with the raised center and flat, spreading periphery

- **Identification:**
  - Liquid medium: No turbidity; pleomorphic forms seen in smears
  - Solid medium: Fried egg colonies stained by the Dienes method
  - Pleomorphic forms by Gram stain
  - Staining directly on agar with homologous fluorescein-conjugated antibody
  - Specific antiserum to *M. pneumoniae*; inhibits their growth on agar
- **Molecular methods:** Multiplex respiratory panel PCR
- **Serology:**
  - *Specific tests*: Immunofluorescence, hemagglutination inhibition and growth inhibition using mycoplasmal antigens
  - *Non-specific tests*:
    1. Streptococcus MG test
    2. Cold agglutination test
- **Indirect Coombs test** may also be positive in some cases

Treatment

- As these are cell-deficient forms, beta-lactam antibiotics have no role in treatment.
- Severe infections can be treated with doxycycline or tetracycline.
- Erythromycin, clarithromycin, or azithromycin can also be used.
- As *Mycoplasma* infections usually resolve on their own, antibiotic treatment of mild symptoms is not always necessary.

Chlamydial pneumonia

- The pneumonia caused by *Chlamydia* is atypical with mild features.
- *Chlamydia* produce a number of diseases including respiratory, ocular, and sexually transmitted diseases in humans.
- Several species are implicated in human infections, as presented in Table 23.17 (for morphological details, refer to Chapter 4).

Table 23.17 Human diseases caused by chlamydiae

Species	Serotype*	Disease
C. pneumoniae	Only one serotype TWAR strain	Acute respiratory disease Outbreaks in young populations living in close quarters and in long-term care facilities
C. trachomatis A, B, Ba, C		Endemic blinding trachoma
C. trachomatis D, E, F, G, H, I, J, K		
C. trachomatis L1, L2, L3		
Inclusion conjunctivitis (neonatal and adult)		
Genital chlamydiasis		
Infant pneumonia		
Lymphogranuloma venereum		
C. psittaci	Complex serological groups	Pneumonia in humans (psittacosis)

* Predominant types associated with the disease

Agent—*Chlamydia*

Growth cycle

- Infection is initiated by the attachment of the elementary body to the surface of a susceptible epithelial cell, followed by its endocytosis (Fig. 23.47).

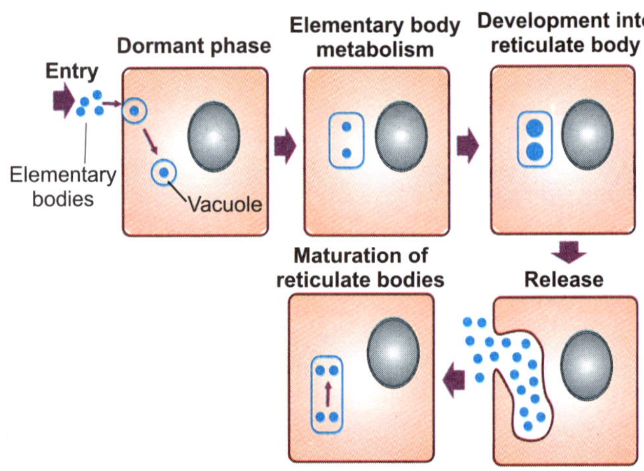

Fig. 23.47 Reproductive cycle of *Chlamydia*

- The mature inclusion body contains 100–500 elementary bodies, which are ultimately released from the host cell.
- During active intracellular growth, the *Chlamydia*-specific lipopolysaccharides accumulate on the host cell surface. This highly antigenic material induces inflammatory and immunological responses, which contribute to the pathogenesis of chlamydial diseases.

Antigenic properties

Chlamydiae possess three main kinds of antigens:

1. **Genus-specific antigen:** This is a heat-stable antigen, common to all chlamydiae. It is a lipopolysaccharide similar to LPS of enteric gram-negative bacilli and is present in all stages of the developmental cycle and can be identified by the complement fixation test.
2. **Species-specific protein antigen:** The second type is present on the envelope surface. These antigens are present in all strains of chlamydial species. They help in classifying chlamydiae into species—*C. trachomatis*, *C. psittaci*, *C. pneumoniae*, and *C. pecorum*.

3. **Serotype/var-specific antigen:** The third kind helps in intraspecies typing, as it is found only in some members of a species. These antigens are located on the **major outer membrane proteins (MOMP)** and can be demonstrated by microimmunofluorescence (micro-IF) and are designated serotype or serovar.

A number of serotypes of *Chlamydia* have been identified. Human diseases associated with *Chlamydia* are listed in Table 23.17.

Chlamydophila pecorum, a recently defined species, includes some *Chlamydia* strains that affect ruminants. *C. psittaci*, which causes mild pneumonia in humans, is transmitted by birds.

Suceptibility

- Chlamydiae are **heat-labile**, being inactivated within minutes at 56°C.
- They are susceptible to ethanol, ether and low concentrations of phenol and formalin.
- Infectivity is maintained for several days at 4°C.
- They can be preserved frozen at –70°C.

Laboratory diagnosis

Four approaches are available for the laboratory diagnosis of chlamydial infections:

i. Microscopic demonstration of inclusion or elementary bodies
ii. Isolation of *Chlamydia*
iii. Demonstration of chlamydial antigen
iv. Demonstration of antibodies or hypersensitivity

Clinical specimen

- The type of specimen depends on the clinical type of infection.
- Blood and sputum are collected for microscopy and culture, and serum is collected for serology in suspected psittacosis.

Microscopy

- Chlamydiae are gram-negative but are stained better by the Giemsa, Castaneda, Macchiavello, and Gimenez stains.
- Immunofluorescence (IF):
  - More sensitive and specific method.
  - IF can identify not only inclusions but also extracellular elementary bodies.
  - Also useful in the examination of cervical or urethral specimens, which may contain elementary bodies but few intact intracellular inclusions.

Tissue culture

- Cell culture is the preferred mode of isolation.
- Many cell lines are susceptible, but the McCoy and HeLa cells are the most commonly used.
- *C. trachomatis* strains vary in their infectivity to cell cultures. **Lymphogranuloma venereum (LGV)** strains grow well, while *Trachoma* inclusion conjunctivitis (TRIC) strains are less infective.
- Cell cultures used for isolation are pre-treated by irradiation or chemicals such as 5-iodo-2-deoxyuridine or cycloheximide to enhance chlamydial replication and facilitate the detection of inclusion bodies.
- Pre-treatment of cells with diethylaminoethyl (DEAE) dextran or centrifugation after inoculation promotes contact between chlamydial particles and the cell monolayer in shell vial cultures, thereby increasing the chances of isolation.

Embryonated chick egg inoculation

- This can be done by inoculating the specimen material into embryonated eggs.
- *Chlamydia* can grow in the yolk sac of 6–8-day-old chick embryos.
- However, this method is not widely applied now.

Antigen detection

- Chlamydial antigens are detected by immunofluorescence.
- ELISA is the method preferred for screening as it enables rapid testing for the LPS antigen in a large number of specimens.

Molecular methods

- DNA probes, nucleic acid amplification techniques (NAAT), and ligase chain reaction have greatly increased the sensitivity and specificity of antigen detection.
- These tests are slowly replacing enzyme immune assay and cultures (which need viable organisms in the specimen).

Antibody detection

- Diagnosis by the demonstration of antibody in serum may be done by the group specific. During chlamydial infection, specific antibodies are produced. First, the IgM antibodies (signifying primary infection) are formed, which persist for approximately two months, followed by IgG antibodies.
- CF test or type-specific micro-IF—**micro-IF can test IgG and IgM antibodies separately**. Titres of 1:8 or greater are usual in infected persons. The initial antibody response is IgM, which is replaced by IgG after about a month. Recurrent infection with the same serotype induces only IgG response. Four-fold rise in IgG titre or the presence of IgM antibody is suggestive of past or recent infections. High-titre antibodies are usually seen only in infant pneumonia, salpingitis and LGV.
- **Complement fixation test**—used mainly in invasive chlamydial infections such as psittacosis and LGV. A four-fold rise in titres is diagnostic. A single CF

antibody test is not diagnostic of psittacosis or LGV unless the titre is high (1:64 or greater).

Laboratory diagnosis of chlamydial infection

- **Specimen:**
  - Depends on the clinical type of infection
  - Blood and sputum for microscopy and culture, and serum is for serology
- **Microscopy:**
  - Giemsa, Castaneda, Macchiavello, and Gimenez stains
  - Immunofluorescence (IF) to visualise inclusions and extracellular elementary bodies
- **Tissue culture:**
  - McCoy and HeLa cell culture
  - Embryonated chick egg inoculation: *Chlamydia* can grow in the yolk sac of 6–8-day-old chick embryos
- **Antigen detection:**
  - Immunofluorescence
  - ELISA for the LPS antigen for rapid testing
- **Molecular methods:**
  - DNA probes, nucleic acid amplification techniques (NAAT), and ligase chain reaction
- **Antibody detection:**
  - CF test or type-specific micro-IF—micro-IF can test IgG and IgM antibodies separately; titres of >1:8 are usual in infected persons
  - Past infection: Four-fold rise in IgG titre
  - Recent infection: IgM antibody
  - Complement fixation test: A four-fold rise in titres; a single CF antibody test is not diagnostic of psittacosis

Treatment

Chlamydia are sensitive to tetracyclines like doxycycline and macrolides, erythromycin, and azithromycin.

Legionellosis

- Two distinct patterns have been identified, namely Legionnaire's disease and Pontiac fever—together known as **legionellosis**.
- Risk factors for pneumonia by *Legionella* are smoking, alcohol, advanced age, intercurrent illness, hospitalisation, and immunodeficiency.
- Men are more often affected than women.
- Its prevalence in the developing countries is not adequately known.

Agent—Legionella pneumophila

- The original isolate in this genus is designated *L. pneumophila* serogroup 1 (SG1), which accounts for nearly all severe infections.
- Other species that cause human infection less often are *L. micdadei*, *L. bozema*, *L. dumoffii*, and *L. gormanii.*

Morphology

- Legionellae are thin, non-capsulated bacilli, 2–5 × 0.1–0.3 μm in size.
- They are coccobacillary, assuming longer forms in culture.
- Most are motile with polar or subpolar flagella.
- They are gram-negative but stain poorly. They stain better by silver impregnation but are best visualised by direct fluorescent antibody (DFA) staining with monoclonal or polyclonal sera.

Habitat

- *Legionella* exists in environmental water as a symbiotic relationship with free-living amoebae.
- Also, *Legionella* is relatively resistant to the effects of chlorine and heat and can survive relatively inhospitable environments such as hot water tanks.

Pathogenesis (Fig. 23.48)

- Person-to-person transmission has never been demonstrated, and *Legionella* is not a member of the bacterial flora in humans.
- Following entry into the alveoli through aerosols, legionellae multiply inside the monocytes and macrophages.
- Dissemination occurs by endobronchial, hematogenous, lymphatic, or contiguous spread.
- **Immunity**—because of their intracellular location, humoral antibodies are ineffective. Cellular immunity is responsible for recovery.

Clinical features

- All age groups are susceptible, though a greater number of cases have been reported in the elderly.
- **Incubation period:** 2–10 days.
- **Legionnaire's disease:** Fever, non-productive cough, and dyspnea, rapidly progressing to pneumonia. Diarrhea and encephalopathy are common. Case

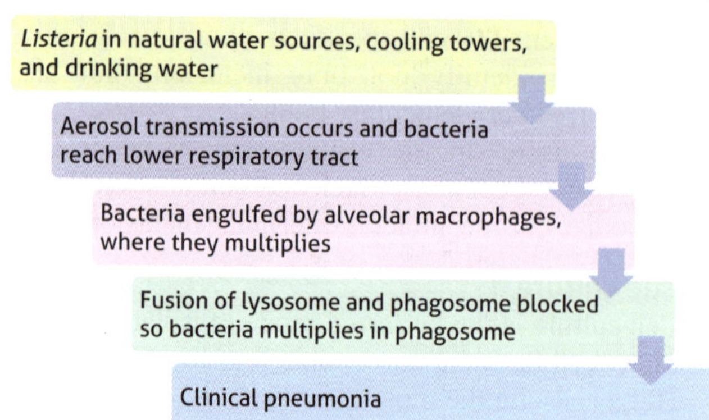

Fig. 23.48 Pathogenesis of legionellosis

fatality may be 15–20 per cent, the cause of death being progressive respiratory failure and shock.

- **Pontiac fever:** It is a milder, non-fatal, 'influenza-like' illness with fever, chills, myalgia, and headache. Outbreaks with high attack rates may occur.

Laboratory diagnosis

Culture characteristics

- Sputum, bronchial aspirates, and lung biopsies are examined by direct fluorescent antibody test and culture.
- They are fastidious organisms that grow slowly on complex media—buffered charcoal yeast extract (BCYE) agar—with L-cysteine and antibiotic supplements and 5% CO_2.
- *Legionella* antigens in urine can be detected by latex agglutination. Alternatively, ELISA antibodies are detected by ELISA or indirect immunofluorescent assay.

Treatment

- Newer macrolides, ciprofloxacin, and tetracyclines are effective.
- Rifampicin is used in severe cases.
- Beta-lactamase antibiotics and aminoglycosides are ineffective.

Viral Pneumonia

The etiology of viral pneumonia includes all the agents that are responsible for upper respiratory infections. Viruses that cause pneumonia are discusses below.

Agent—Coronavirus pneumonia

- The human coronavirus was first isolated in vitro from cases of common cold in organ cultures of human embryonic trachea by inoculating nasopharyngeal washings.
- Coronaviruses are common viruses that infect most people at some time in their lives.
- They appear to be the second most common cause of the common cold, particularly in the winter, next only to rhinoviruses.
- Human coronaviruses were thought to cause mild-to-moderate upper respiratory tract illnesses until the 2020 pandemic.
- In the recent past, they have been associated with severe lower respiratory infections like MERS-CoV, SARS and COVID, which threw up many challenges.

Morphology

- Coronaviruses are a group of spherical or pleomorphic, enveloped, single-stranded RNA viruses, carrying club-shaped peplomers on their surface.
- The name refers to the fringe of surface projections (**spike proteins**) surrounding the virus resembling the solar corona (Fig. 23.49 and 23.50).
- The spike proteins of the coronavirus help the virus to bind to the receptors on the target cell membrane.
- There are two groups of coronaviruses, which are:
  - **Acid-labile viruses** (associated with common cold-like illnesses)
  - **Acid-stable viruses** (associated with human and animal gastroenteritis)

Classification

Coronaviruses are divided into four main subgroups based on their genomic structure as follows:

- Alpha
- Beta
- Gamma
- Delta

  Of these, only alpha and beta infect mammals.

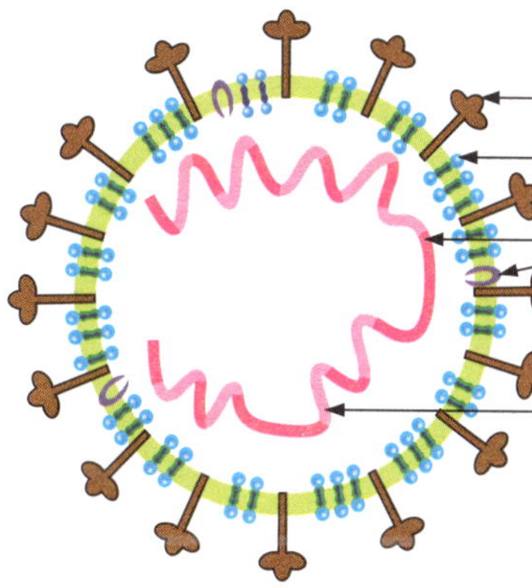

Fig. 23.49 Coronavirus with its structural proteins: S (spike protein), M (membrane protein), N (nucleocapsid), and E (envelope) protein and the genome (RNA)

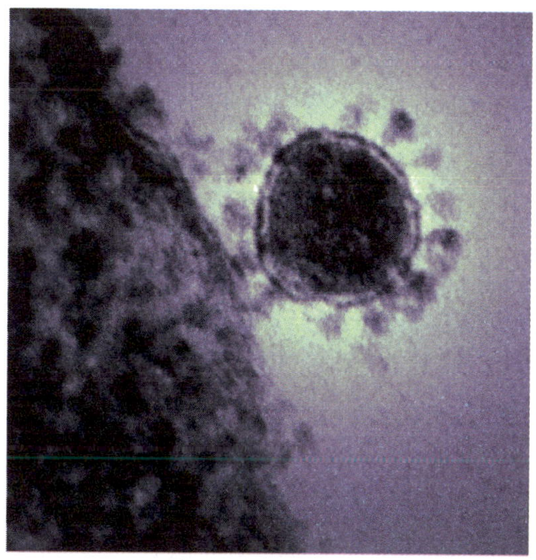

Fig. 23.50 Coronavirus—the spike protein of the virus binding to the cell membrane (*Source*: CDC, PHIL, Image ID 18114)

Coronavirus types infecting humans

- Human coronavirus 229E
- Human coronavirus OC43
- SARS-CoV (severe acute respiratory syndrome–coronavirus)
- Human coronavirus NL63 (HcoV-NL63, New Haven coronavirus)
- Human coronavirus HKU1
- Middle East respiratory syndrome coronavirus (MERS-CoV), previously known as novel coronavirus 2012 and HcoV-EMC (human coronavirus-Erasmus Medical Center)
- SARS-CoV-2—the coronavirus strain implicated in the 2019–21 pandemic was initially called the 'novel coronavirus' (NcoV-2); it was subsequently named COVID-19 (coronavirus disease-19) as it first appeared in December 2019 in the Wuhan District of China

Common human coronavirus—HcoV-OC43, HcoV-HKU1, HcoV-229E, and HcoV-NL63—cause common colds and self-limiting upper respiratory tract infections in immunocompetent individuals. However, in immunocompromised subjects and the elderly, lower respiratory tract infections can occur due to these viruses.

Coronaviruses causing severe acute respiratory syndrome (SARS)

The coronavirus responsible for SARS appears to be distinct from other coronaviruses, which have been classified into three types:

- **CoV-SARS**—mammalian viruses are types 1 and 2
- **Avian virus** is type 3
- The **present SARS-CoV** is coronavirus type 4

Agent—SARS-CoV

In November 2002, the Guangdong province in South China experienced an outbreak of an unusual respiratory infection with many deaths. The world outside knew about it only in February 2003, when a physician from Guangdong visited Hong Kong, fell ill and died there, after infecting 12 persons who had stayed in the same hotel. They, in turn, fell ill and upon returning to their respective countries, initiated outbreaks there. By July, when the pandemic was controlled, it had affected over 30 countries, with many thousands of cases and over 800 deaths. India escaped the SARS epidemic; however, a few suspected cases were detected and quarantined. In 2003, the WHO reported 6,521 cases of SARS with 461 deaths in 30 countries.

Evidence suggesting SARS-CoV as a new virus:

- There is no antibody against it in human and animal sera collected in the past.
- It may be a recombinant of some animal and human viruses.
- The virus has been isolated from Chinese wild civets and raccoon dogs but not from pigs, dogs, cattle, or poultry.

Spread

- SARS spreads by the inhalation of the virus present in droplets or aerosols of respiratory secretions of patients.
- Fecal aerosols also may be infectious.

Clinical features

- The incubation period is <10 days.
- The disease starts as a fever with cough or other respiratory symptoms.
- Diarrhea is sometimes seen.
- The chest radiograph shows pneumonic changes.
- Death is due to respiratory failure.

Laboratory diagnosis

- The virus was identified by **electron microscopy**.
- The virus can be grown in Vero cell culture, animal inoculation, cloning, sequencing, and histology.
- Real-time reverse transcription PCR (real-time RT-PCR) has been used for early diagnosis.
- A demonstration of rise in titre of antibodies by ELISA or indirect immunofluorescence test in paired serum samples is useful later in the infection.

Treatment

- No specific therapy or prophylaxis has been identified.
- The virus is highly mutable.
- Vaccine prophylaxis has not been possible due to the high rate of mutation.
- Control has been achieved by strict isolation and quarantine.

Agent—MERS-CoV

- The Middle East respiratory syndrome (MERS) is a viral respiratory disease caused by Middle East respiratory syndrome coronavirus (MERS-CoV).
- The first report of MERS-CoV was from a 60-year-old man from Bisha, the Kingdom of Saudi Arabia) in 2012, who died of rapidly progressive community-acquired pneumonia and acute renal failure. Samples analysed revealed a variant of corona virus (MERS-CoV).
- According to the WHO, 27 countries have reported 2,605 laboratory-confirmed cases since 2012, leading to 936 known deaths due to the infection and related complications at a case-fatality ratio (CFR) of 36%. The majority of these cases were reported from Saudi Arabia, with 2,196 cases and 855 related deaths (CFR: 39%).
- MERS-CoV is a zoonotic virus.

- MERS-CoV has been identified and linked to human infections in **dromedary camels** in several countries in the Middle East, Africa, and South Asia.
- Infection spreads through **respiratory secretions**.
- Another name for the virus is human coronavirus-emerging in the Middle East (HcoV-EMC).

Clinical features

- Persons infected with MERS-CoV develop severe respiratory illness.
- Symptoms include fever, cough, and shortness of breath.
- Severe cases may have acute renal failure.

Treatment

- Currently, there is no vaccine to prevent the infection.
- Strict isolation and barrier nursing are the only ways to prevent spread.

COVID-19

- In December 2019, the first few cases of pneumonia with severe acute respiratory syndrome were admitted to a hospital in Wuhan in the Hubei province of China.
- The early COVID-19 cases were traced to the Huanan seafood market with a possibility of human-to-human transmissions.
- The disease spread rapidly within Wuhan and the neighbouring districts with high infectivity and mortality.
- From China, the disease spread to other parts of the globe, including Europe, the USA, the Middle East, Africa, Australia and parts of Asia including India.
- On 11 March 2020, the World Health Organization (WHO) declared the novel coronavirus (COVID-19) outbreak a global pandemic.
- In India, the first case was confirmed in January 2020 by the National Institute of Virology, Pune, in a medical student.
- **As on 6 December 2023, there have been 772,138,818 confirmed cases of COVID-19, including 6,985,964 deaths, reported to the WHO (https://covid19.who.int/).**

Epidemiology

- The infectivity of the virus is very high although the case fatality is lower than that of MERS-CoV.
- **Human-to-human transmission** is the main mode of spread.
- When a virus circulates widely in a population, causing widespread infections, the chances of the virus mutating increases manifold as it replicates that much faster.
- This results in the continuous evolution of the virus with its **single nucleotide polymorphism (SNP) variants**, resulting in many lineages. In the COVID-19 pandemic, it is speculated that it kept the caseload high, besides other human factors.
- Mutants in the spike protein of SARS-CoV-2 across several countries (adapted from https://www.cdc.gov/coronavirus / 2019-ncov / variants / variant.html accessed on 20.07.2021) are as follows:
  - **B.1.1.7 (Alpha):** This variant was first detected in the United States in December 2020. It was initially detected in the United Kingdom.
  - Pathogenesis
  - **B.1.351 (Beta):** This variant was first detected in the United States at the end of January 2021. It was initially detected in South Africa in December 2020.
  - **P.1 (Gamma):** This variant was first detected in the United States in January 2021. P.1 was initially identified in travellers from Brazil, who were tested during routine screening at an airport in Japan in early January.
  - **B.1.617.2 (Delta):** This variant was first detected in the United States in March 2021. It was initially identified in India in December 2020. This virus was declared a cause for concern as it spread rapidly.
  - **B.1.1.529 (Omicron):** This was first detected in South Africa and was declared a virus of concern. This variant was found to have high transmissibility but produced less severe symptoms and fewer cases of hospitalisation.

Some of these mutated viruses were designated **VoC (variants of concern)** as they were transmitted at a faster rate and were able to infect a large number of people. The other concern was escape mutation, where the virus would escape the immunity conferred by vaccination. However, most vaccines appear to provide protection against the mutants. Post-vaccination infections were noted, but they were mild and often asymptomatic, not requiring hospitalisation.

Agent—SARS-CoV-2

The strain implicated in this pandemic is a new β-coronavirus belonging to the subgenus of coronaviridae.

- It is closely related to the bat SARS-CoV.
- Its reservoir host is thought to be the **pangolin**, an ant-eating mammal.
- This new coronavirus was found to share 79.5% of its genetic identity with the SARS-CoV, which caused the 2002–2003 epidemic.
- The virus responsible for the current COVID-19 is considered the third major coronavirus outbreak in the last 20 years (the other two being SARS and MERS-CoV).

Pathogenesis

The pathogenesis of SARS-CoV-2 is described in Fig. 23.51 and 23.52.

Clinical features

- The **incubation period** is 2–14 days.
- Individuals of all ages are at risk of contracting this infection and severe disease.
- However, **patients aged ≥60 years and patients with underlying comorbidities** like obesity, cardiovascular disease, chronic kidney disease, diabetes, chronic lung disease, smoking habits, and malignancies have an increased risk of developing severe COVID-19 infection.
- The percentage of COVID-19 patients requiring hospitalisation is higher among those with pre-existing medical conditions than those without such history.
- Symptoms include **fever** (>38°C) with **cough and dyspnea accompanied by chest discomfort**. Patients also experience fatigue and muscle pain.
- A small number of patients also develop headaches, hemoptysis, and diarrhea.
- Patients with mild-to-moderate symptoms recover spontaneously.
- SARS-related damages to alveolar cells can, in turn, trigger a series of systemic reactions.
- Respiratory distress appears seven days after the onset of symptoms and worsens within the next 2–3 days, requiring mechanical ventilation. Patients manifest signs and symptoms of systemic inflammatory response and develop acute respiratory distress syndrome which may end fatally.
- The timely detection of persons with decreased oxygen saturation and breathlessness and individuals with comorbidities helps in specific management and reduces mortality.

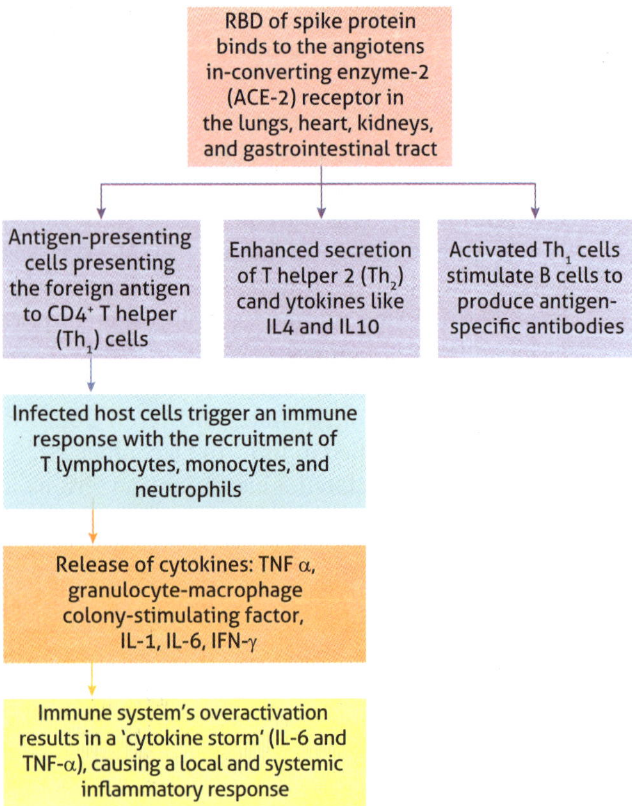

Fig. 23.51 Pathogenesis of SARS-CoV-2

Laboratory diagnosis

- **Imaging techniques:** Abnormalities in computed tomography (CT) images of the chest are seen in nearly 100% of the patients. **Bilateral ground glass-like appearance and consolidation** are also seen.

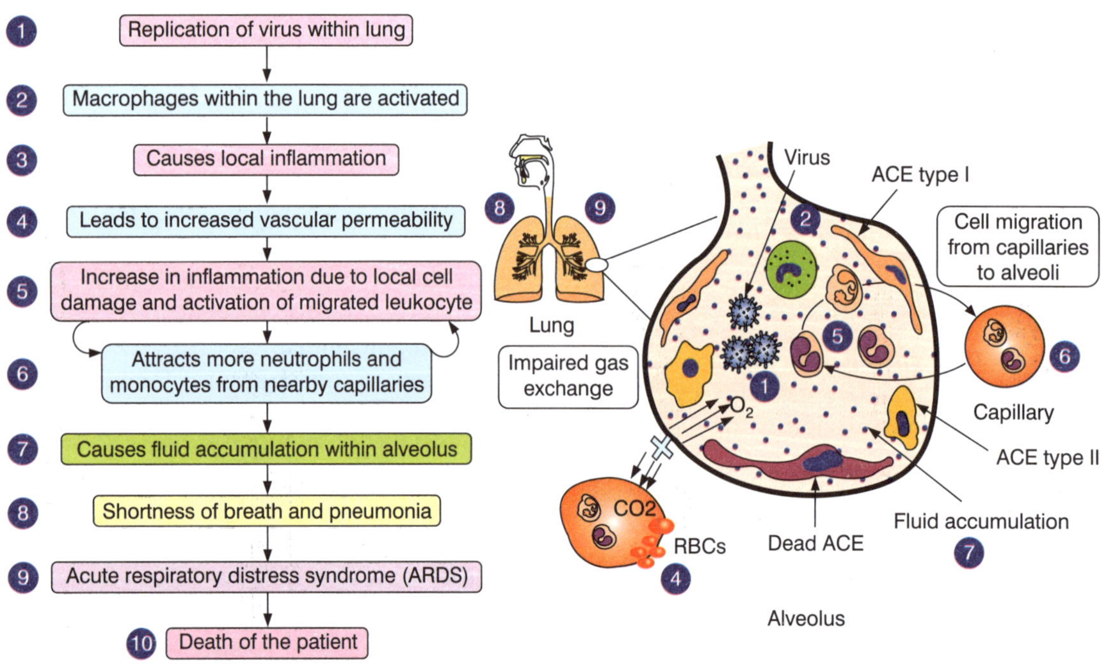

Fig. 23.52 Transmission and pathogenesis of SARS-CoV-2

Laboratory confirmation is achieved with the following methods:

- **Collection of specimens:** The recommended samples for the molecular detection of viral genomes are oropharyngeal or nasopharyngeal swabs (Fig. 23.53). Due to the highly transmissible nature of the virus, **strict procedures to prevent aerosols and personal protection in the form of PPE including facial visors, are necessary**. The samples are sealed in biohazard containers prior to transportation to the laboratory.
- **Molecular methods:** The diagnosis is confirmed by nucleic acid amplification techniques (NAAT). The viral genomes are tested by real-time RT-PCR or by reverse transcriptase quantitative PCR (RT-qPCR). The genes coding for the following proteins are detected:
  - Nucleoprotein (N)
  - Envelope protein I
  - Membrane protein (M)
  - Spike protein (S)
- Other confirmatory genes for detection are: (i) *ORF* (open reading frame 1a/b) and (ii) RNA-dependent RNA polymerase (RdRp).
- Rapid antigen detection: **Kits demonstrating SARS-CoV-2** nucleocapsid antigen have also been used to screen populations.
- **Serological tests:** IgG and IgM antibodies start to increase around 10 days after the onset of symptoms with seroconversion within the first three weeks. Antibodies against the nucleoprotein and the surface spike receptor-binding domain correlate with neutralising activity.
- **Other ancillary tests:** These include the estimation of D-dimer, CRP; and interleukin-6. All three are elevated in those infected and useful to monitor the patient.

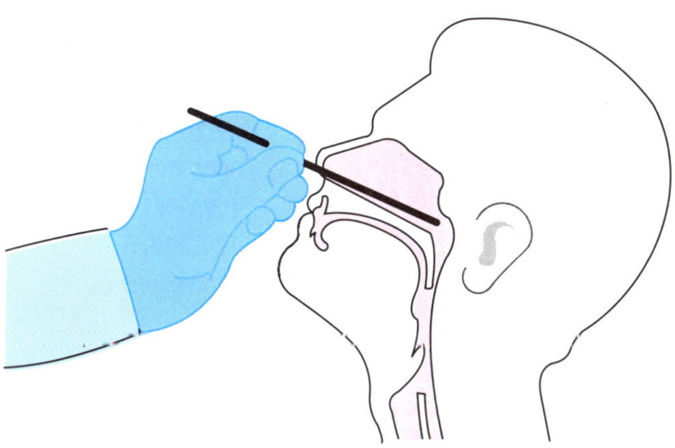

Fig. 23.53 Collection of nasopharyngeal swab for detection of SARS-CoV-2 viral genomes

Laboratory diagnosis of COVID-19

- **Specimens:** Oropharyngeal or nasopharyngeal swabs
- **Transport:** Swabs with viral transport medium
- **Molecular methods:** RT-PCR for genes coding for:
  - Nucleoprotein (N)
  - Envelope protein I
  - Membrane protein (M)
  - Spike protein (S)
- **Other confirmatory genes:** (i) ORF (open reading frame 1a/b) and (ii) RNA-dependent RNA polymerase (RdRp)
- **Rapid antigen detection:** Kits with SARS-CoV-2 nucleocapsid antigen
- **Serological tests:** IgG and IgM antibodies start to increase around 10 days after the onset of symptoms with seroconversion within the first three weeks
- **Other ancillary tests:**
  - D-dimer, CRP, and interleukin-6 are all elevated in infected patients
  - **CT scan:** Bilateral ground glass-like appearance and consolidation

Prophylaxis

The main strategy for prevention is **blocking transmission by wearing masks and practicing social distancing and personal hygiene measures** like handwashing and cough etiquette. The vaccine depends on the spike protein found on the surface of the virus, which is the most important component for invading human cells. The following types of vaccines are available:

- **Whole virus:** Inactivated or live-attenuated (e.g., COVAXIN)
- **Protein subunit:** Fragments of spike protein subunit to trigger an immune response; this requires an adjuvant (Novovax)
- **Nucleic acid:** Genetically engineered mRNA (Moderna & BNT162b2) or DNA (nCoV) to generate a protein that itself safely prompts an immune response
- **Viral vector:** A virus that cannot cause disease but serves as a platform (vector) to encode coronavirus spike proteins (S glycoprotein) to generate an immune response; this is done using recombinant techniques, e.g., recombinant adenovirus with a SARS-COV-2 gene which codes for the spike protein incorporated into its genome (COVISHIELD)

Prevention of transmission of infection

- The spread of SARS-CoV can be prevented by infection prevention measures.
- **Wearing masks, keeping a distance of at least six feet, and maintaining hand hygiene** are three key factors for preventing transmission.
- **Quarantining** of individuals testing positive for SARS-CoV-2 is another key factor.

- In COVID care hospitals, the following preventive measures are undertaken:
  - Designated sample collection centers with barriers to avoid close contact
  - Biosafety level II laboratories to perform molecular tests
  - Strict isolation policies, with dedicated COVID care wards and critical care units which are set up to prevent hospital outbreaks and transmission by healthcare workers (HCW)
  - Dedicated teams of doctors, nurses, and other ancillary staff who wear personal protective gear (PPE) while on duty in the wards or ICU
  - Biomedical waste handling and disposal are given special attention and carried out according to standard protocols to avoid spread of infection

Treatment

- **Remdesivir**, an adenosine analogue, has been used with some benefit.
- Treatment has also been attempted with **lopinavir and ritonavir**, two antiretroviral drugs.
- Maintaining oxygen saturation (by mechanical ventilation) is of utmost importance.
- Preventing lung damage (due to complications produced by heightened immunological response with steroids) is important.
- Sepsis should be prevented (with antibiotic prophylaxis).

Fungal Pneumonia

- Lung infections due to fungal agents occur **due to immunosuppression or underlying disease**.
- Several fungi are associated with lower respiratory tract infections.
- Two notable fungi associated with such infections are ***Aspergillus fumigatus*** **and** ***Pneumocystis jirovecii*** (Table 23.18).
- Other fungi like *Penicillium* spp., *Talaromyces marneffei*, and *Fusarium* may be associated with infections in immunocompromised individuals.

Aspergillosis

It is primarily a pulmonary disease caused by fungi *Aspergillus*.

Agent—*Aspergillus*

- *Aspergillus* species are the commonest mould seen on damp surfaces or almost any other organic matter, soil, and dust; their spores are ubiquitous.
- Of the 300 species of *Aspergillus*, ***A. fumigatus*** **is the chief pathogen**. It causes an invasive disease in immunocompromised human beings.
- Other species associated with infection are *A. niger*, *A. flavus*, and *A. nidulans*.

Pathogenesis and clinical features

Aspergillus species can cause a variety of clinical syndromes, which are:

1. Allergic bronchopulmonary aspergillosis

- Inhalation of spores by atopic persons with raised IgE levels can provoke a **hypersensitivity reaction**, which may be:

Table 23.18 Respiratory infections of fungal etiology

Disease	Fungus associated	Signs and symptoms	Diagnostic tests	Antifungal agents
Aspergillosis	*Aspergillus fumigatus*	Fever, chills cough, wheezing, bloodstained sputum, breathlessness Specific lesions: Aspergilloma (*Aspergillus*) Disseminated lesions elsewhere in the body Meningitis (*Histoplasma*) Facial and orbital lesions (Mucormycoses)	Chest radiograph Microscopy of sputum, aspirates—fungal hyphae, yeast cells Direct fluorescence for specific fungi Tissue biopsy Culture with appropriate media Antibody tests: EIA PCR for specific genes	Itraconazole Voriconazole
Pneumocystis pneumonia (PCP)	*Pneumocystis jirovecii*			Trimethoprim suphamethoxazole
Cryptococcosis	*Cryptococcus neoformans*			Amphotericin B Flucytosin
Histoplasmosis	*Histoplasma capsulatum*			Amphotericin B Ketoconazole Itraconazole
Mucormycosis	*Rhizopus arrhizus* *Mucor* spp.			Amphotericin
Coccidioidomycosis Paracoccidioidomycosis	*Coccidioides immitis* *Paracoccidioides brasiliensis*			Amphotericin B
Blastomycosis	*Blastomyces dermatitidis*			Amphotericin Ketoconazole

- Type I hypersensitivity (asthma), which occurs in atopic individuals following sensitisation to inhaled *Aspergillus* spores
- Type III hypersensitivity (extrinsic alveolitis)
- Combined type I and type III hypersensitivity reactions

- In bronchopulmonary aspergillosis, the fungus grows within the lumen of the bronchioles, which may be occluded by fungal plugs.
- The patient presents with asthmatic symptoms with eosinophilia and pulmonary infiltrates.

2. Aspergilloma

- A **fungal ball** grows within, and is usually restricted to, an existing lung cavity, for example, due to old tuberculosis or bronchiectasis.
- Fungal mycelia form a compact mass surrounded by dense fibrous wall.
- The patient may be asymptomatic or have moderate degree of cough and sputum production. Occasionally, hemoptysis may occur.
- Treatment is by the surgical removal of the fungal ball followed by antifungals.

3. Invasive aspergillosis

- The fungus first causes pneumonia and later disseminates to involve other organs such as the brain, kidneys, or heart.
- Patients are usually immunocompromised or debilitated.
- The disease has poor prognosis and may prove fatal.

4. Superficial infections

- *Aspergillus* species can cause infections of the external ear **(otitis externa)—otomycosis**; *niger* and *flavus* are the species involved.
- Secondary infections also affect the **eye (mycotic keratitis) and nasal sinusitis**.

Laboratory diagnosis

The demonstration of septate hyaline hyphae from tissues or specimen by direct microscopy is suggestive of *Aspergillus* infection but not diagnostic.

- **Specimen:**
  - Exudates, bronchial washings, and bronchoalveolar lavage
  - Tissue sections of biopsy or post-mortem material
- **Direct microscopy:**
  - Samples are first examined for the presence of fungal elements by wet mount preparation in 10% potassium hydroxide.
  - Tissue sections are stained by the **periodic acid Schiff** method. The hyphae are poorly stained by hematoxylin and eosin (H and E staining). Typical dichotomous branching at acute angles is characteristic of *Aspergillus* infection (Fig. 23.54).
- **Culture:** Colonies on **Sabouraud dextrose agar incubated at 25–37°C** grow after 48 hours, but longer incubation may be required before characteristic morphological features develop. After 3–4 days' incubation, the colonies have a velvety or powdery surface and are characteristically coloured: *A. fumigatus*—dark green, *A. niger*—black, and *A. flavus*—yellow-green (Fig. 23.55a, b, and c).
  - *Microscopic appearance of the colony*: A wet preparation stained with lactophenol cotton blue demonstrates **septate hyphae and conidiophores** (specialised aerial hyphae that bear spores or conidia). The conidiophores have swollen, rounded ends ('vesicles') with chains of conidia borne on elongated cells called **sterigmata**. The general morphology is characteristic of the genus. There are also inter-species differences that are useful in identification (Fig. 23.56a, b, and c).

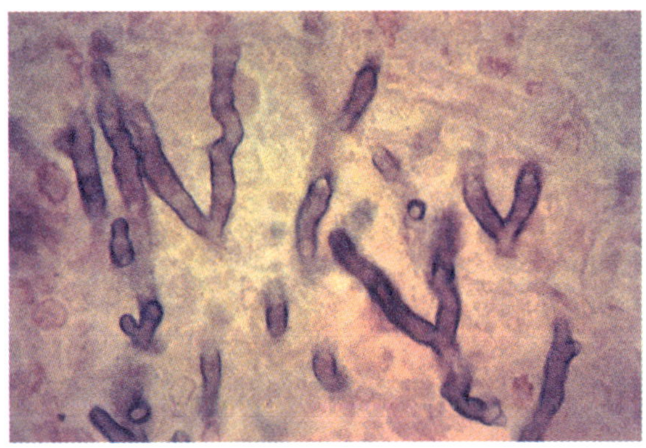

Fig. 23.54 H and E staining of a human lung tissue showing dichotomous, branching hyphae (*Source*: CDC, PHIL, Image ID 15630)

(a) (b) (c)

Fig. 23.55 *Aspergillus* colonies on Sabouraud agar: (a) *A. fumigatus*, (b) *A. niger*, and (c) *A. flavus* (*Source*: Department of Microbiology, Pondicherry Institute of Medical Sciences, Puducherry)

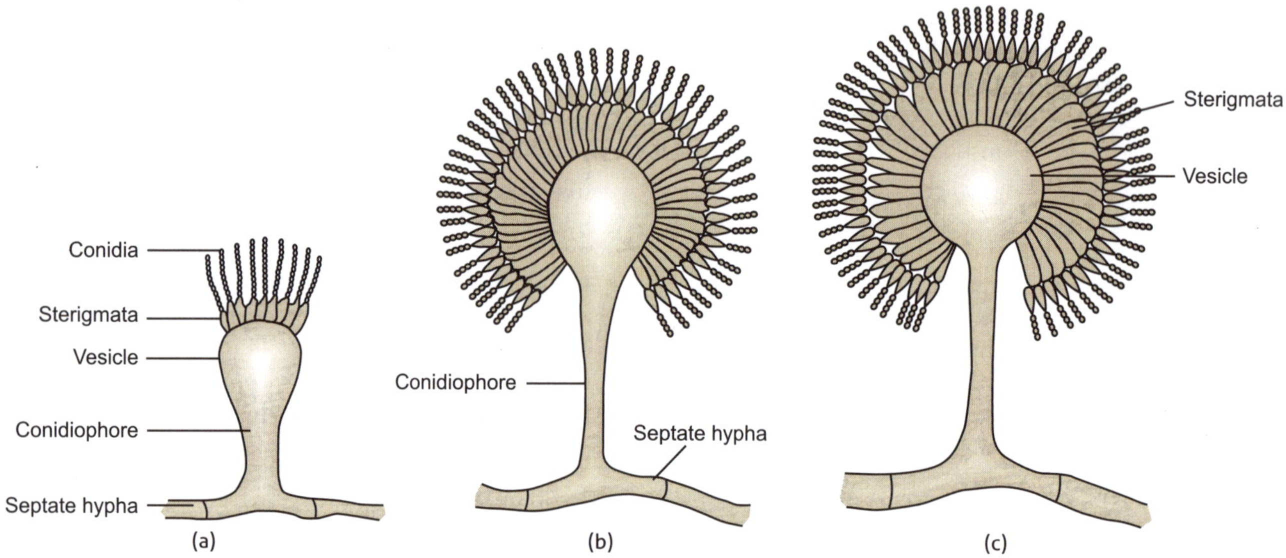

Fig. 23.56 *Aspergillus spp.: (a) A. fumigatus, (b) A. flavus, and (c) A. niger*

- **Serology:**
  - *Detection of the galactomannan antigen*: Galactomannan is a polysaccharide antigen found on the *Aspergillus* cell wall. It may be released into the blood and other body fluids even in the early stages of *Aspergillus* invasion, and the presence of this antigen can be sustained for 1–8 weeks. Its detection in serum or bronchoalveolar lavage fluid by ELISA acts as a marker for the diagnosis of invasive aspergillosis.
  - *Precipitating antibodies to Aspergillus antigen*: Can be demonstrated by countercurrent immunoelectrophoresis, immunodiffusion, and enzyme-linked immunosorbent assay (ELISA).

Antibodies are usually absent in the sera of healthy individuals—they can be detected in the majority (70 per cent) of patients with allergic aspergillosis and approximately the same proportion of those with pneumonia or invasive disease.

- **Polymerase chain reaction and nucleic acid sequence-based amplification:** These are advanced methods for establishing the diagnosis of aspergillosis tracheobronchitis.
- **Skin tests:** Dual immediate (after 15 minutes) and Arthus type (after 4–6 hours) skin test reactions which develop after the intradermal injection of *Aspergillus* spp. antigens are important criteria to establish a diagnosis of allergic bronchopulmonary aspergillosis.

Laboratory diagnosis of aspergillosis

- **Specimen:**
  - Exudates, bronchial washings, and bronchoalveolar lavage
  - Tissue sections of biopsy or post-mortem material
- **Direct microscopy:**
  - Wet mount preparation in 10% potassium hydroxide: Septate hyaline hyphae
  - Tissue sections stained by the periodic acid Schiff method: Hyphae with typical dichotomous branching at acute angles, poorly stained by hematoxylin and eosin (H and E staining)
- **Culture:**
  - Sabouraud dextrose agar incubated at 25 and 37°C:
  - Velvety or powdery surface and are characteristically coloured: *A. fumigatus*—dark green, *A. niger*—black, and *A. flavus*—yellow-green (Fig. 23.55a, b, and c)
- **Identification:**
  - Lactophenol cotton blue wet mount: Septate hyphae and conidiophores—specialised aerial hyphae that bear spores or conidia
- **Serology:**
  - **Detection of the galactomannan antigen**
  - **Precipitating antibodies to *Aspergillus* antigen:** By countercurrent immunoelectrophoresis, immunodiffusion, and enzyme-linked immunosorbent assay (ELISA)
  - Antibodies can be detected in the majority of patients with allergic aspergillosis pneumonia or invasive disease
  - **Polymerase chain reaction and nucleic acid sequence-based amplification**
  - **Skin tests:** Dual immediate (after 15 minutes) and Arthus type (after 4–6 hours) skin test reactions
  - Antigens are helpful in the diagnosis of allergic bronchopulmonary aspergillosis

Treatment

- Invasive aspergillosis is treated with **intravenous amphotericin B/liposomal amphotericin B**. However, the mortality is high.
- **Intravenous formulations of azoles** (voriconazole, itraconazole, and posaconazole) are currently used.

Talaromycosis (formerly, Penicilliosis)

- *Penicillium marneffei*, now known as *Talaromyces marneffei*, is endemic in South-East Asia, southern China, and North-Eastern India.
- It affects **immunocompromised patients**—those with HIV hematological malignancy and those undergoing immunosuppressive therapy.
- The natural reservoirs are **bamboo rats**.
- Infection is associated with high humidity and exposure to plants and farm animals.
- This fungus is unique among *Penicillium* in that it is a true pathogen.

Agent—*Talaromyces* (formerly, *Penicillium*)

- *Talaromyces* is associated with **respiratory infection in diabetics** and opportunistic mycoses.
- It is present in the environment and grows on various substrates such as bread, jam, fruit, and cheese.
- In the laboratory, they are common airborne contaminants of culture media.
- **Colonies** are blue–green in colour with a white border and a powdery surface (Fig. 23.57a).
- Microscopy demonstrates septate hyphae with branched conidiophores with two rows of sterigmata bearing chains of spores; **their appearance is like a brush or broom** (Fig. 23.57b).
- A **red diffusible pigment is seen on the reverse** of a *T. marneffei* colony.

Clinical features

- Symptoms range from mild-to-moderate infection with localised disease to disseminated infection with multiple organ involvement or severe disease including respiratory failure and shock.
- The fungi may invade the gastrointestinal tract, causing **oropharyngeal ulcerations and diarrhea**.
- Occasionally, the central nervous system is involved, leading to **meningoencephalitis**.

Laboratory diagnosis

- Microscopy of sputum smears or biopsies shows binary fission of yeasts on fungal staining.
- The yeast forms are small, oval, and 2–4 µm in diameter.
- The mycelial form produces a red, diffusible pigment (Fig. 23.57c) and morphologically resembles other members of the *Penicillium* species.

Treatment

The infection can be treated with **amphotericin B**, followed by oral itraconazole.

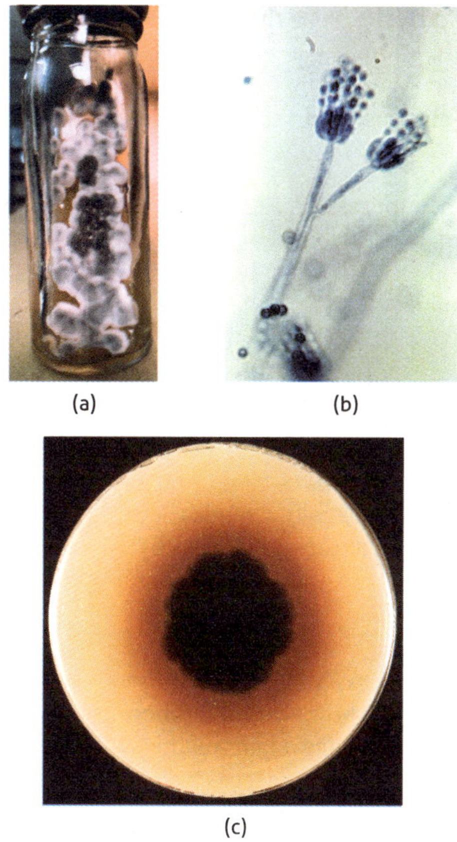

Fig. 23.57 (a) Blue–green colonies on SDA, (b) chains of conidia produced by phialides, and (c) red diffusible pigment on the reverse of a *P. marneffei* colony (*Source*: [b] Dept of Microbiology, PIMS; [a] and [c] CDC, PHIL, Image IDs 8398 and 11967)

Mucormycosis

Mucormycosis (previously known as **zygomycosis**) is an invasive disease caused by members of the order Mucorales that includes 55 genera and 261 species, of which 38 species are known to cause mucormycosis infection in humans.

Etiological agents of mucormycosis

- *Rhizopus*
- *Mucor*
- *Absidia*
- *Cunninghamella*
- *Saksenaea*
- *Rhizomucor*

These fungi are ubiquitous in soil and their spores are present in air and dust.

Pathogenesis and clinical features

- Infection is acquired by the **inhalation of spores**.
- Zygomycosis develops as a **systemic infection** following dissemination from a primary focus in the upper respiratory tract or nasal cavity, where the spores germinate.
- Subsequently, the mycelia invade the adjacent tissues.
- **Angio-invasiveness** or invasion of blood vessels by the fungus results in a profound loss of lung parenchyma.

- **Small arteries are thrombosed** with hemorrhage and infarction.
- Invasion may occur via the blood vessels into the **orbit, sinuses, and the brain** and may prove fatal.

Risk factors

- Immunosuppression
- Uncontrolled diabetes or diabetic ketoacidosis
- Neutropenia
- Malignancy

Clinical features

The clinical features of mucormycosis are presented in Table 23.19.

Laboratory diagnosis

Diagnosis is usually made during histological examination of autopsy material by demonstration of broad and bizarre, non-septate (or sparsely septate) hyphae in tissues.

- **Specimen:**
  - Exudate: A wet mount in 10% potassium hydroxide shows **broad, ribbon-like hyphae.**
  - Tissue: Hyphae stain readily with hematoxylin and eosin (unlike *Aspergillus*) **showing non-septate, broad hyphae, often invading the blood vessels** (Fig. 23.58a).
- **Culture**
  - Isolation is by culture on Sabouraud dextrose agar with an antibacterial agent but without cycloheximide, on which the fungi grow easily (however, growth may sometimes be difficult to achieve from necrotic material even when abundant hyphae are seen).
  - Culture characteristics of the main genera are similar.
  - After incubation for 3–4 days on Sabouraud agar at 30–37°C, the colonies are **grey-white with a thick, cottony, fluffy surface** (Fig. 23.58b). Under the microscope, non-septate, broad hyphae with aerial sporangiophores which end in a sporangium (a sac containing spores) are seen.
  - *Mucor* shows branched sporangiophores arising randomly along aerial mycelium. Rhizoids are absent. *Rhizopus* bears rhizoids and sporangiophores, which arise in groups directly above the rhizoids (Fig. 23.60a and b).

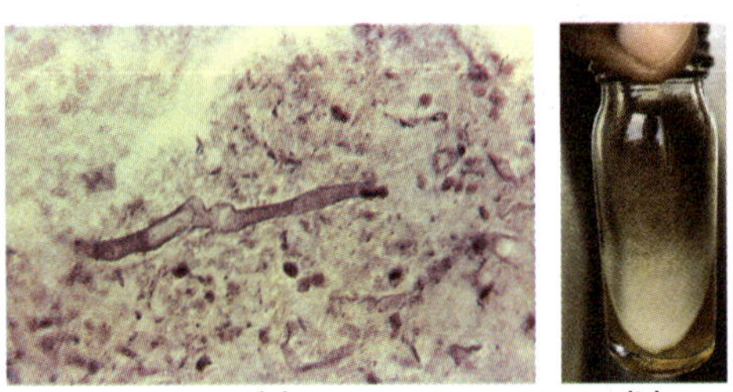

(a) (b)

Fig. 23.58 Mucormycosis: (a) Broad, ribbon-like hypha and (b) cottony appearance of colony (*Source*: [a] CDC, PHIL, Image ID 15440 and [b] Dept. of Microbiology, PMS)

Table 23.19 Clinical presentation of mucormycosis

Type	Features	Clinical features
Rhinocerebral mucormycosis (Fig. 23.59)	Spreads from the nasal mucosa to the paranasal sinuses, orbit, and cribriform plate and to the brain Destruction of the bones, invasion of blood vessels with infarction	Facial pain, nasal congestion, difficulty breathing, headache and anosmia, and brownish nasal discharge
Pulmonary mucormycosis	Invasion of tissues with profound destruction of lung parenchyma, infarction, and hemorrhage into lung tissues	Non-specific symptoms with chest pain, dyspnea, and hemoptysis
Gastrointestinal mucormycosis	Seen in extremely malnourished persons Infection acquired by inhalation of fungal spores or following surgery Terminal ileum, caecum, and colon develop ulcers and infarction	Non-specific with abdominal pain, diarrhea, hematemesis, and melena
Cutaneous mucormycosis	Seen in severe burn patients Infection can be primary or secondary to disseminated mucormycosis Focal necrosis of parenchyma and adipose tissue is seen in biopsy	Painful lesions that enlarge with nodular eccymotic areas of cutaneous infarction
Disseminated mucormycosis	Widely disseminated infection with involvement of skin, lungs, heart, and brain	Depend on the area affected

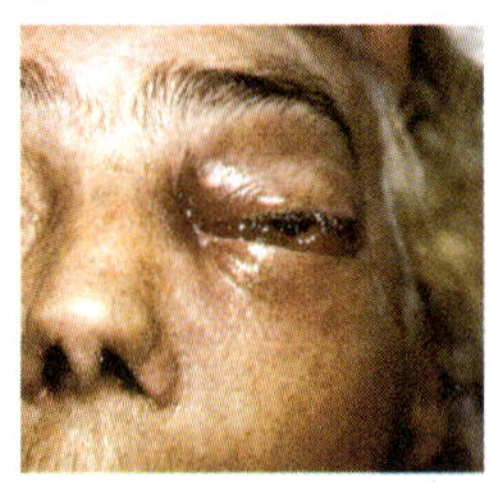

Fig. 23.59 Mucormycosis of the upper eyelid (*Source*: CDC, PHIL, Image ID 2831)

Note: These common environmental moulds are frequent contaminants of culture plates.

Treatment

- Intravenous amphotericin B combined, where appropriate, with surgical drainage is the recommended treatment.
- Good medical control of diabetes is required.

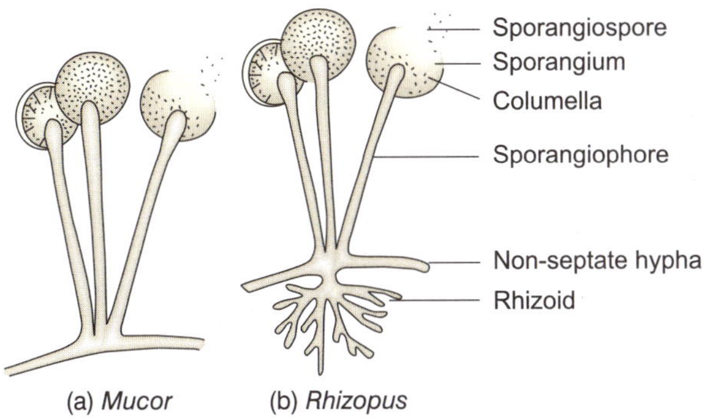

Fig. 23.60 (a) *Mucor* and (b) *Rhizopus*

Cryptococcosis (Torulosis)

Cryptococcosis (torulosis) is a subacute or chronic infection caused by the yeast *Cryptococcus neoformans* and, less frequently, other species.

Distribution

- Cryptococcosis occurs worldwide since the fungus is a **soil saprophyte** and is particularly abundant in the feces of pigeons and other birds.
- Since cryptococcosis was originally reported from Europe, it was formerly referred to as European blastomycosis.
- Cryptococcosis has been identified as a common opportunistic infection in AIDS patients.

Agent—*Cryptococcus*

Cryptococcus is a round or ovoid budding cell, 4–20 μm in diameter, with a prominent polysaccharide capsule. Species of importance include the following:

- *C. neoformans* (*C. neoformans* var. neoformans is most common)
- *C. neoformans* var. gatti
- *C. albidus*
- *C. laurentii*

Four serological types of the cryptococcal capsular polysaccharide—A, B, C, and D—have been identified.

Pathogenesis and clinical features

- Infection is usually acquired by **inhalation** but may sometimes be acquired through the skin or mucosa.
- Most infections are asymptomatic.
- Pulmonary cryptococcosis may lead to **mild pneumonitis**.
- Dissemination of infection leads to visceral, cutaneous, and meningeal disease.
- Visceral cryptococcosis may simulate tuberculosis and cancer clinically; bones and joints may be involved.
- Cutaneous cryptococcosis ranges from small ulcers to large granulomata.
- Cryptococcal meningitis and meningoencephalitis, the most serious type of cryptococcal infections, can mimic tuberculous or other chronic types of meningitis.
- Its onset is insidious and the course slow and progressive. It is often seen in AIDS.

Laboratory diagnosis

The diagnosis of cryptococcosis is established by direct microscopic examination and by culture.

- **Microscopy:** Microscopic examination of India ink-stained CSF and other material from lesions reveals **capsulated, budding yeast cells**; the capsules are prominent in the India ink preparations (Fig. 23.61a).
- **Culture:** The fungus grows readily on Sabouraud agar, forming **smooth, mucoid and cream-coloured colonies** (Fig. 23.61b). The ability to grow at 37°C and hydrolyse urea differentiates *C. neoformans* var. neoformans from non-pathogenic cryptococci.

The pathogenicity of a cryptococcal isolate can be demonstrated by intracerebral or intraperitoneal inoculation into mice, which develop a fatal infection.

- Capsulated budding yeast cells can be demonstrated in the brain of infected mice (this method is not routine practice).
- **Antigen detection tests:** Demonstration of the **capsular antigen by precipitation and latex agglutination can be of value in diagnosing cryptococcal meningitis**.

> **Laboratory diagnosis of cryptococcosis**
>
> - **Microscopy**
>   - India ink-staining of CSF and other material from lesions: Capsulated, budding yeast cells; with prominent capsules.
> - **Culture**
>   - Sabouraud dextrose agar: Smooth, mucoid and cream-coloured colonies.
> - **Identification**
>   - Hydrolyses of urea to differentiate *C. neoformans* var. neoformans from non-pathogenic cryptococci.
>   - **Antigen detection tests**: Demonstration of the capsular antigen by precipitation and latex agglutination in cryptococcal meningitis.

Treatment

Amphotericin B, 5-fluorocytosine, imidazoles (miconazole, ketoconazole), triazoles (itraconazole, fluconazole, voriconazole), and echinocandins (caspofungin, micafungin) may be used.

Pneumocystis pneumonia (PCP)

- *Pneumocystis carinii* pneumonia is now referred to as *Pneumocystis jirovecii* pneumonia.
- It is a fungal infection that most commonly affect the immunocompromised and, in some cases, can be severely life-threatening.

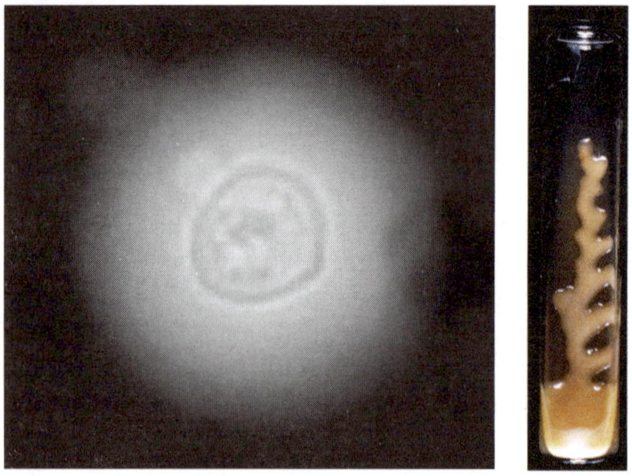

(a) (b)

Fig. 23.61 *Cryptococcus neoformans* var. neoformans: (a) India ink preparation of spinal fluid showing yeast cells surrounded by a large capsule and (b) mucoid colony (*Source*: [b] CDC, PHIL, Image ID 15677)

Agent—*Pneumocystis jirovecii*

- *Pneumocystis* was, until recently, thought to be a protozoan. However, nucleic acid sequencing has conclusively shown that the organism is a fungus related to *Ascomycetes*.
- *P. jirovecii* is found in humans.
- *Pneumocystis* pneumonia is one of the common opportunistic infections in AIDS.
- *P. jirovecii* has three morphological forms:
  1. **Trophozoite**—thin-walled, irregularly shaped, 1–5 µm in size
  2. **Pre-cyst**—an intermediate stage of the sexual phase, 5–8 µm in size
  3. **Cyst**—thick-walled and spherical, containing up to 8 intracystic bodies and up to 8 µm in size

Pathogenesis and clinical features

- Infection is transmitted by **respiratory droplets** and is asymptomatic in immunocompetent individuals.
- In **immunocompromised patients, life-threatening pneumonia** develops.
- X-ray shows **interstitial lung infiltrates** (Fig. 23.62a).

Laboratory diagnosis

- **Specimen:** Induced sputum, bronchoalveolar lavage, or lung biopsy are collected for investigations.
- **Microscopy:** Trophozoites can be demonstrated by Giemsa, toluidine blue, Grocott's (Gomori's) methenamine silver stain (GMS; Fig. 23.62b) and calcofluor white stains. The cyst wall stains black with methenamine silver stain.
- **Culture:** The organism cannot be cultured.
- **Serology:** Complement fixation test with titres of 1:4 or above indicates active disease. A latex agglutination test can also be used.

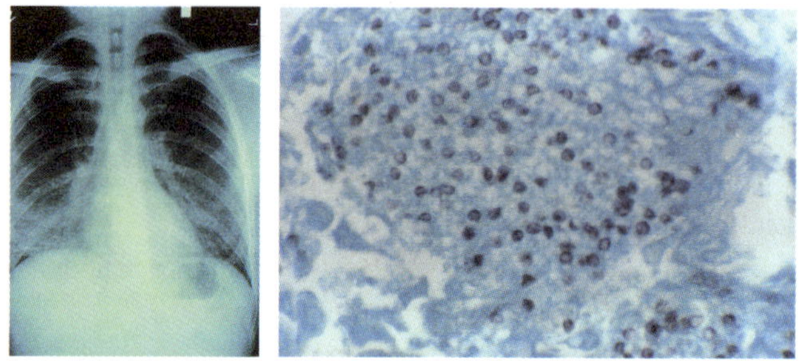

(a) (b)

Fig. 23.62 *Pneumocystis jirovecii* (a) X-ray shows interstitial lung infiltrates and (b) Grocott's methinamine silver stain—fungal elements in the alveolar spaces (*Source*: CDC, PHIL, Image IDs 14372 and 15734)

Treatment

Trimethoprim–sulphamethoxazole (TMP–SMZ) and pentamidine isothionate are the drugs of choice for the treatment of acute cases. TMP–SMZ is also used as prophylaxis in the management of HIV-infected individuals.

Histoplasmosis

- It is an infection of the reticuloendothelial system caused by the thermally dimorphic fungus *Histoplasma capsulatum*.
- The disease was originally described by Samuel T. Darling (1905) and came to be known as **Darling's disease.**

Distribution

- The disease occurs worldwide; however, it is most common in the USA where it is endemic in many central and eastern states.
- In endemic areas, the fungus is present in the soil and decaying trees and is particularly abundant in bird droppings.

Agent—*Histoplasma*

- *Histoplasma capsulatum* is differentiated into two varieties:
  - *Histoplasma capsulatum* var. capsulatum, which causes the classical, ubiquitous form of histoplasmosis
  - *Histoplasma capsulatum* var. duboisii, which causes 'African histoplasmosis'

Pathogenesis

- Infection is acquired by **inhalation**.
- It affects the reticuloendothelial system and manifests as lymphadenopathy, hepatosplenomegaly, fever, and anemia with a high rate of fatality.
- *H. capsulatum* produces an intracellular granulomatous lesion that is ulcerative.
- *Histoplasma* may also develop on the skin and mucous membranes (Fig. 23.63).

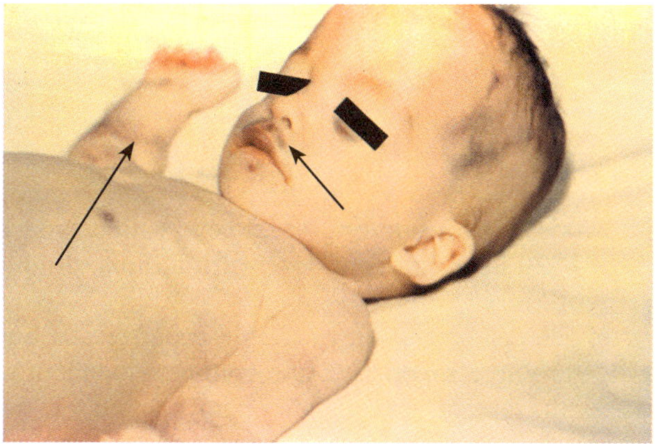

Fig. 23.63 Disseminated histoplasmosis in a child showing purpura and surface markings for enlarged liver and spleen (*Source*: https://phil.cdc.gov/Details.aspx?pid=20335)

Clinical features

- Most infections of classical histoplasmosis are asymptomatic and heal spontaneously, leaving behind an area of miliary calcification
- Some infected individuals develop a pulmonary disease that resembles tuberculosis.
- Disseminated histoplasmosis develops only in a minority of infected individuals.
- African histoplasmosis involves chiefly the skin, subcutaneous tissues and bones.
- The lungs are not commonly affected, and disseminated disease is infrequent.

Laboratory diagnosis

- **Microscopy:**
  - Giemsa- or Wright-stained smears of blood, bone marrow, scrapings from lesions or biopsies of lymph nodes show small intra- and extracellular yeast cells.
  - In tissues, the yeast phase occurs within phagocytic cells. The yeast appears as an oval, budding cell measuring 2–4 μm (Fig. 23.64a and b), while in African histoplasmosis, much larger yeast-like cells (7–15 μm) are seen.
- **Culture:**
  - On blood agar at 37°C, yeast cells seen.
  - On Sabouraud (Fig. 23.64c) agar at room temperature (25°C), **white, cottony, mycelial growth** is seen.
  - Microscopy of culture—large (8–20 μm) thick-walled, spherical spores with tubercles (finger-like projections) are seen. These '**tuberculate**' **spores** are of diagnostic relevance; the mycelial phases of both varieties are indistinguishable.
- **Serodiagnosis:**
  - Antibodies formed during the course of systemic mycoses increase in titre in progressive disease. Hence, serological tests such as **latex agglutination, complement fixation, and precipitation tests** are useful in the diagnosis of histoplasmosis.
  - The titre is considered positive at reciprocal dilutions greater than 1:8. A titre with dilutions greater than 1:32 suggests an active histoplasmosis infection.
  - Antigen detection—antigens can be detected in urine and serum and are useful in immunocompromised individuals in whom antibody formation is impaired.
- **Skin tests:**
  - **Delayed hypersensitivity** develops following infection and can be demonstrated by testing skin with 'histoplasmin', which is analogous to the tuberculin test for tuberculosis.

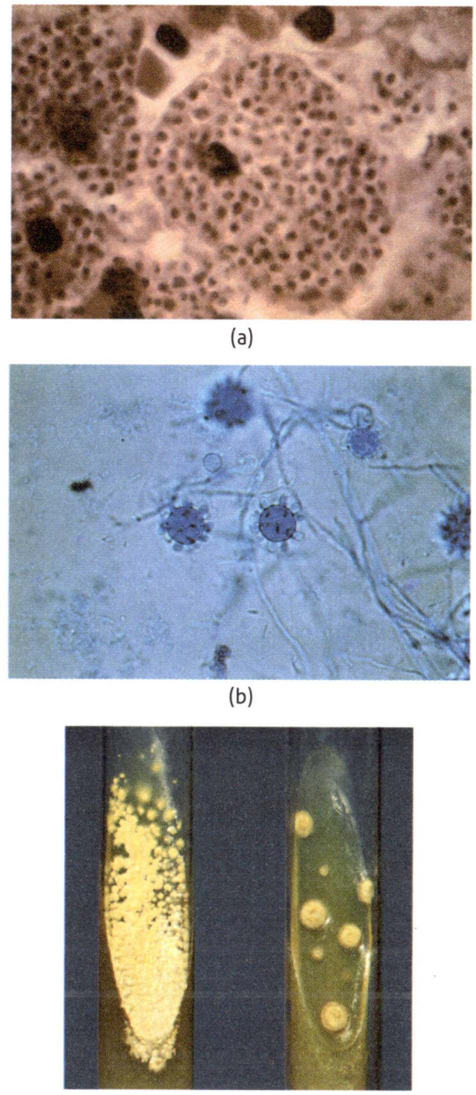

(a)

(b)

(c)

Fig. 23.64 *Histoplasma capsulatum*: (a) yeast form (intracellular), (b) mycelial form at 37°C, and (c) slant cultures of *H. capsulatum* colonies on Sabouraud agar (*Source*: CDC, PHIL, Image IDs 14422, 15364 and 4189)

– In histoplasmosis, skin tests are more specific than serological methods.

Blastomycosis

- This is a chronic infection characterised by the formation of **suppurative and granulomatous lesions with a marked predilection for the lungs and skin.**
- It can also occur in any other part of the body.
- Blastomycosis is caused by the thermally dimorphic fungus *Blastomyces dermatitidis* (its **teleomorph** [perfect state] is *Ajellomyces dermatitidis*).

Distribution

- The infection is largely confined to North America; hence it is known as **North American blastomycosis** (to differentiate it from paracoccidioidomycosis, which is also known as South American blastomycosis).
- In recent years, several cases have been reported from Africa and India.
- The fungus has been isolated in Delhi from the bronchial aspirates of a patient and from the lungs of insectivorous bats.

Agent—*Blastomyces dermatitidis*

Pathogenesis and clinical features

- Infection is acquired by the **inhalation of contaminated soil** which is the habitat of these fungi.
- Primary infection of the lungs may resemble tuberculosis or histoplasmosis.
- It may be asymptomatic or may lead to focal or diffuse consolidation, miliary lesions, or abscess formation.
- Disseminated disease has a high fatality rate and may develop through the bloodstream to form multiple abscesses in various parts of the body.
- Cutaneous disease usually affects the skin of the face or other exposed parts of the body.
- The initial lesion is a papule, around which secondary nodules develop and coalesce.

Laboratory diagnosis

- **Culture**
  – In tissues and in culture at 37°C, the fungus appears as a **budding yeast cell, which is large (7–20 μm) and spherical, with a thick, double-contoured wall.**
  – Each cell carries only a **single broad-based bud.** At room temperature, the culture is filamentous with septate hyphae and many round or oval conidia, and, in older cultures, chlamydospores are also seen (Fig. 23.65a and b). In tissues, single budding yeast cells with broad bases are seen (Fig. 23.65c).

Coccidioidomycosis

Coccidioidomycosis is a primary pulmonary infection that may be inapparent, benign, severe, or even fatal.

Distribution

The disease is endemic in the dry, arid regions of southwestern USA, where the fungus is present in the soil and in rodents.

Agent—*Coccidiodes immitis*

The thermally dimorphic fungus *Coccidioides immitis* (Fig. 23.66) causes coccidiodomycosis.

Pathogenesis and clinical features

- Infection is acquired by the **inhalation of dust containing arthrospores of the fungus.**
- In most cases, the respiratory infection is asymptomatic and leads to lifetime immunity.
- A self-limited influenza-like fever (known as valley fever or desert rheumatism) may occur in some individuals.
- Less than one per cent of infected persons develop the chronic, progressive, disseminated disease,

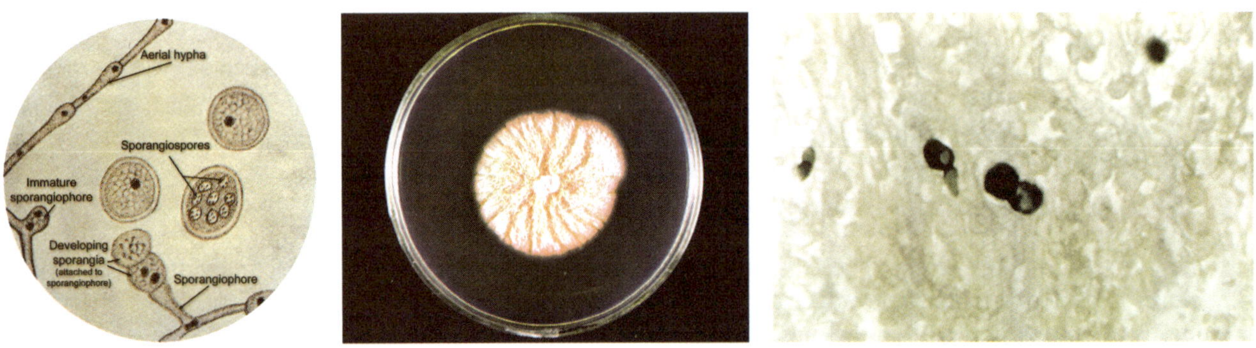

(a) (b) (c)

Fig. 23.65 (a) Mycelial and yeast forms, (b) culture of *Blastomyces dermatitidis*, and (c) histopathology of tissue section in blastomycosis by methenamine silver stain; shows broad-based, single budding cell (*Source*: CDC, PHIL, Image IDs 14355, 19914, 4226)

coccidioidal granuloma (Fig. 23.66a), which is highly fatal.

Laboratory diagnosis

- **Culture:**
  - The fungus is **thermally dimorphic**, occurring as a spherule in tissues and in culture at 37°C and as the mycelial form in soil and in culture at room temperature.
  - The spherule is 15–75 µm in diameter, with a thick, doubly refractile wall and filled with endospores (Fig. 23.66a).
  - The mycelial phase consists of hyphae, which fragment into **arthrospores** that are **highly infectious** (Fig. 23.66b and c).
  - Culture confirmation can be performed rapidly and with less risk to staff by DNA probe or exoantigen testing in a biological safety cabinet.

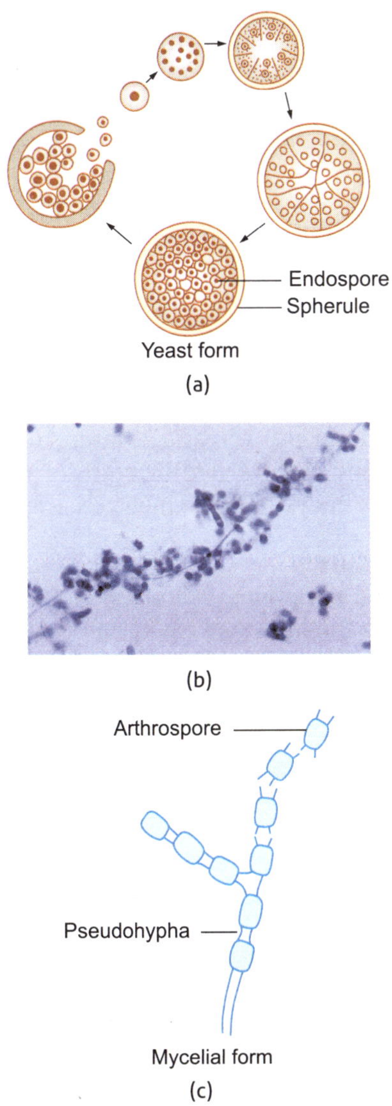

Fig. 23.66 *Coccidioides immitis*: (a) spherules, (b) arthroconidia, and (c) thick-walled arthroconidia (*Source*: [b] CDC, PHIL, Image ID 12201)

- **Skin test:** Delayed hypersensitivity can be demonstrated by a positive skin test with coccidioidin (analogous to the tuberculin test).

Treatment

- **Itraconazole, fluconazole**, and, more recently, **voriconazole**, are used for less severe infections.
- For paracoccidioidomycosis, **oral itraconazole** 50–100 mg/day is the drug of choice; ketoconazole and fluconazole are less effective.
- Fluconazole is useful for coccidioidal meningitis.
- Corrective surgery may be used for pulmonary and cutaneous lesions.

Paracoccidioidomycosis

This is a chronic granulomatous disease of the skin, mucosa, lymph nodes, and internal organs.

Distribution

As the disease is confined to South America, it is called South American blastomycosis. It occurs in rural areas.

Agent—*Paracoccidiodes brasiliensis*

Paracoccidioides brasiliensis is a thermally dimorphic fungus and one of the two species that cause paracoccidioidomycosis (the other is Paracoccidioides lutzii).

Pathogenesis and clinical features

- The **spores are inhaled** and result in a primary pulmonary infection that spreads by the hematogenous route to the mucosa of the nose, mouth, gastrointestinal tract, skin, and the lymphatics.
- This leads to **ulcerative granulomas of the buccal and nasal mucosa**, which are a prominent feature of this disease (Fig. 23.67a).

Laboratory diagnosis

- **Microscopy:** The yeast phase occurs in tissues and in cultures at 37°C as large, globose, or oval cells with multiple buds encircling the mother cell; this **mariner's wheel or pilot's wheel** appearance is characteristic (Fig. 23.67b).
- **Culture:** The mycelial phase is found in nature and can be isolated in culture after incubation at 25–30°C for two to three weeks (Fig. 23.67c).

PARASITIC LUNG INFECTIONS

Primary pathogenesis of the respiratory system caused by parasites is uncommon. However, several parasites can invade the lung, causing verious diseases including invasive amoebiasis, paragonimiasis, hydatid disease, ascariasis, hookworm infection, and filariasis as a part of their life cycle within the human host. Of these,

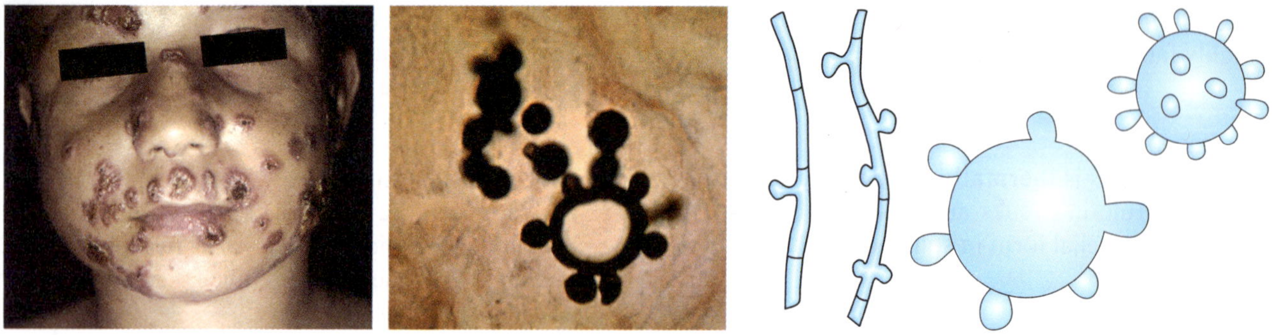

(a) (b) (c)

Fig. 23.67 (a) Lesions on the face due to paracoccidioidomycosis, (b) budding yeast form in methenamine silver stain of tissue, and (c) mycelial and yeast forms (*Source*: [a] and [b] CDC, PHIL, Image IDs 12156 and 527)

Paragonimus westermanii is the primary infection, and is dealt with in detail here.

- The three modes of infection by parasites are as follows:
  - **Primary infection** caused by *Paragonimus westermanii* (lung fluke)
  - **As part of the migratory path of the larval forms** or by producing ectopic lesions in the respiratory system—Loeffler's syndrome and hydatid cysts
  - **Contiguous spread**—amoebic lung abscess by spread from the liver abscess, through the diaphragm

Paragonimiasis

- Paragonimiasis caused by *Paragonimus westermanii* (also known as the **oriental lung fluke**).
- More than 40 species of *Paragonimus* are known, but only 10 species cause infection in humans.
- *Paragonimus westermanii* is found in Far East-Asian countries.
- In India, it has been reported from North-East states, particularly from Manipur.

Agent—*Paragonimus westermanii*

Morphology (Fig. 23.68)

- **Adult:**
  - The adult is oval, thick, fleshy, dark red, and a hermaphrodite like other trematodes.
  - It measures 16 × 8 mm in size and resides in the lung parenchyma.
  - It has an **oral sucker at the anterior end and a ventral sucker in the middle.**
- **Eggs:**
  - These are oval, operculated, brown and about 80–120 µm long; they are un-embryonated when laid.

Life cycle (Fig. 23.68)

- The **definitive hosts** of *Paragonimus* are humans, cats, and dogs.
- The **intermediate hosts** are snails and crabs or crayfish.
- **Mode of transmission**—by the consumption of undercooked, raw, pickled crab or crayfishes, which carry the metacercaria of the fluke.
- The larva hatches and pierces the peritoneal cavity to reach the diaphragm, and then the lung parenchyma and the bronchus, where the larvae mature into adults.
- These adults, in turn, release eggs, which leave the body in sputum. Swallowed sputum causes eggs to reach the intestine, in which form, they are released with stools.
- Eggs develop further in water and release miracidium after 2–3 weeks. This stage of the fluke enters the snail and multiples in it asexually to form sporocysts, redia, and finally, cercaria. Mature cercaria leave the snail and are ingested by crabs or crayfishes, in which they encyst to form metacercaria.

Clinical features

- **Abdominal symptoms:** Migration of the metacercaria larvae into the intestinal cavity can cause abdominal tenderness, nausea, and vomiting.
- **Pulmonary symptoms:** Cysts or eosinophilic granuloma may form surrounding the worm in the lungs. The cysts are 1 cm wide and enclose one or more flukes along with eggs, blood, and thick, purulent fluid. The cysts break into bronchioles.
  - The patient may present with a cough with rusty, blood-tinged sputum with a fishy odour.
  - Sometimes, frank hemoptysis may be present. In chronic conditions, bronchitis, pneumonia, or lung abscess may be present.
- **Extrapulmonary symptoms:** The migration of worms through the liver, spleen, and brain can result in satellite lesions in these areas.
  - Cysts in the brain can cause **cerebral paragonimiasis** characterised by features of space-occupying lesions.

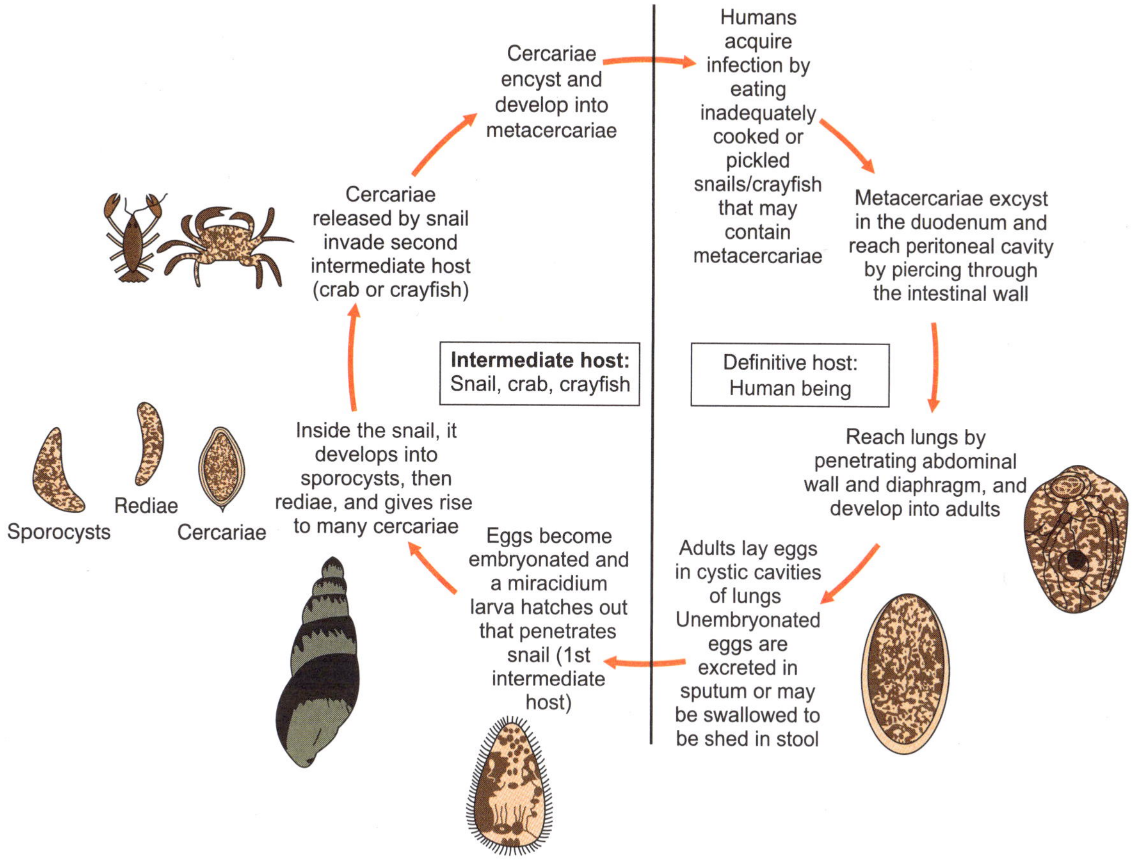

Fig. 23.68 Morphology and life cycle of *Paragonimus*

– **Cutaneous lesions** are seen as tender nodules. Features of **cutaneous larva migrans** may be present if the infesting worm is a non-human species of *Paragonimus*.

Laboratory diagnosis

- **Microscopy of sputum:** Multiple sputum samples may need to be examined to detect characteristic eggs.
- **Serological diagnosis:** This is mainly used for epidemiological purposes.
- **ELISA:** Antibodies can be detected by ELISA. Antigen detection can be done by dot ELISA and indicates active infection.

Treatment

Praziquantel, bithionol, and niclofolan can be used for treating such conditions.

Secondary Parasitic Infections of the Lung

Amoebic lung abscess

- This occurs as a complication of invasive amoebiasis caused by *Entamoeba histolytica*.
- Lung amoebic abscess occurs due to the penetration of the diaphragm to involve the right lung by a large liver abscess.

Schistosomiasis

- The lungs may be involved in schistosomiasis.
- The eggs or ectopic adult worms in the lungs can result in fibrosis, causing **bilharzial cor pulmonale**.

Echinococcosis

- *Echinococcus granulosus* cysts may, in rare cases, reach the lungs and develop in the lung parenchyma.
- Rupture of the pulmonary cysts can cause a severe **anaphylactic reaction and metastatic spread**, which can be fatal.
- The liver and other sites must be investigated for the presence of hydatid cysts.

Loeffler's syndrome

- Migration of intestinal nematodes like *Ascaris*, *Strongyloides*, or the human hookworm through the lungs can result in Loeffler's syndrome.
- Presenting complaints include **fever and cough with or without bloody sputum**.
- The chest X-ray in such cases shows **mottling of the lungs**; marked eosinophilia may be present.
- Larvae may be demonstrated in sputum.

Tropical pulmonary eosinophilia (occult filariasis)

- It is caused by *Wuchereria bancrofti* or *Brugia malayi*.
- Clinically, the patient complains of fever and cough.
- **Lung consolidation** is visible on X-ray; very marked **eosinophilia** is present.

- An **antibody test** may be positive for *W. bancrofti* and *B. malayi*.
- **Treatment** is diethylcarbamazine or ivermectin.
- The condition may also be seen in infection with *Dirofilaria immitis* (dog heartworm).
- The worms do not develop into adults producing microfilariae.
- A **chest X-ray may show a 'coin-lesion'** in the lung, which may be misdiagnosed as tuberculosis.
- Diagnosis is based on serology.
- The infection responds to antifilarial drugs.

Alveolar echinococcosis (AE)

- This is caused by *Echinococcus multilocularis* (see Chapter 25).
- It is characterised by parasitic tumours in the liver, which may spread to other organs including the lungs and brain.
- Liver failure can occur.
- The disease is fatal in 50–75% cases.

TUBERCULOSIS

Case 11. *A 50-year-old man presented with a history of low-grade fever with an evening rise in temperature and productive cough for the past two months. More recently, he had expectorated blood-tinged sputum. He also gave a history of loss of weight (around 10 kg over the past four months) and appetite. A chest X-ray revealed a nodular infiltrate in the apical area of the right upper lobe.*

i. What is the provisional diagnosis?
ii. How is such a patient investigated?
iii. What is the treatment strategy for such a patient?

Tuberculosis is a potentially fatal infection, caused predominantly by the *Mycobacterium tuberculosis* complex (MTC), which can affect any part of the body but most commonly involves the lungs.

Epidemiology

It is estimated that a third of the world's population (two billion) is infected with the tubercle bacilli.

- Every year, between eight and nine million new cases of tuberculosis appear, and three million people die from the disease. India is one of the worst affected countries, with half a million deaths attributed to tuberculosis every year.
- **TB–HIV coinfection** has caused further rise in morbidity and mortality.

Factors responsible for the spread of tuberculosis

- Poverty and overcrowding go hand-in-hand with tuberculosis.
- HIV infection reactivates latent tuberculosis and increases the severity of the disease; it also renders treatment ineffective. Tuberculosis may, in turn, hasten the development of HIV infection into the active disease.
- Multiple drug resistance among tubercle bacilli also contributes to the spread of the disease.

Agent–*Mycobacterium tuberculosis*

- These are aerobic bacilli, 0.2–0.6 µm wide and 1.0-10 µm long, that are generally non-motile. They stain poorly with Gram stain but resist decolorisation with acid after they are stained.
- *Mycobacterium tuberculosis* and *M. leprae* are pathogenic, but other mycobacteria are opportunistic pathogens.

Classification of *Mycobacterium*

The classification of mycobacteria is presented in Fig. 23.69.

Staining characteristics

- Mycobacteria are slender rods that **do not stain readily, but once stained, resist decolourisation** with dilute mineral acids due to the presence of mycolic acid in their cell walls.
- *Mycobacterium tuberculosis* are **acid-fast** (with 20% sulphuric acid) and **alcohol-fast** (3% hydrochloric acid and alcohol).
- Electron micrographs of thin sections show a thick cell wall composed of three layers enclosing a **trilaminar plasma membrane**. They transform into spheroplasts and L-forms when grown in the presence of lysozymes.
- Saprophytic mycobacteria stain uniformly without a **barred or beaded appearance** and are usually only acid-fast.

Antigenic properties

1. Cell wall antigens

i. **Lipids:** The mycobacterial cell wall is rich in long-chain fatty acids called mycolic acids. Mycolic acids, when complexed with peptidoglycan, are responsible for granuloma formation. The cord factor is also responsible for the virulence of bacteria.
ii. **Lipoarabinomannan:** It is a cell wall-associated glycolipid and a virulence factor that can inhibit macrophage activation, leading to the persistence of mycobacteria within mononuclear phagocytes.
iii. **Proteins:** These induce delayed-type hypersensitivity and elicit tuberculin reaction. There is also some antigenic relationship between lepra and tubercle bacilli.
iv. **Polysaccharides:** Group specificity is due to polysaccharides. Their role in pathogenesis is not clear but they can induce immediate-type hypersensitivity.

Antibodies against the polysaccharide, protein, and phosphatide antigens of tubercle bacilli have no protective or diagnostic relevance.

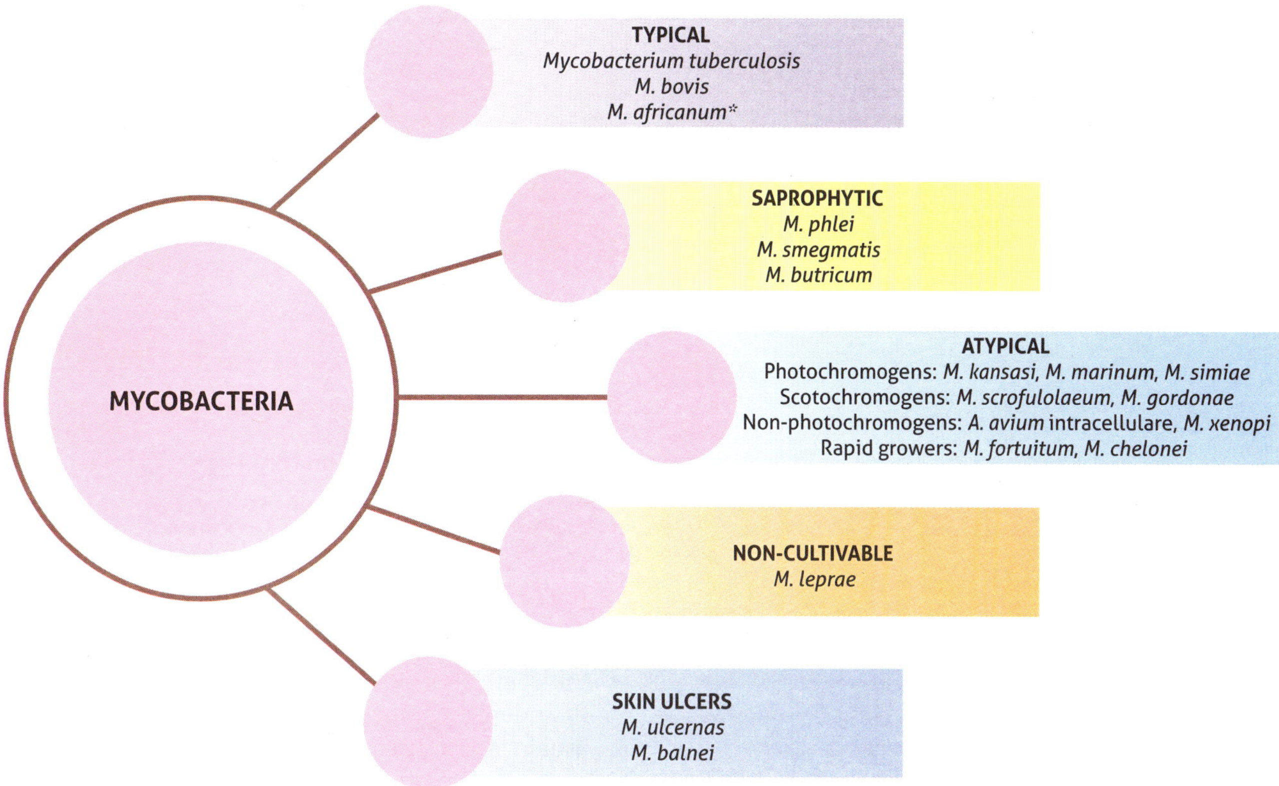

Fig. 23.69 Classification of mycobacteria
*Includes some non-human pathogens: *M. caprae* (cattle), *M. microti* (voles and other small mammals), *M. canetti* (similar to *M. africanum*), and *M. pinnipedii* (seals)

2. *Cytoplasmic antigens*

These are protein antigens that stimulate dendritic cells and increase the secretion of cytokines.

Host range

- *M. tuberculosis* causes natural infection in humans and other primates and animals that come into close contact with humans.
- Experimentally, guinea pigs and hamsters are highly susceptible to the infection (mice are only moderately susceptible).
- *M. bovis* produces tuberculosis in cattle, humans, other primates, carnivores including dogs and cats, badgers, swine, parrots, and some birds of prey.
- Experimentally, guinea pigs are highly susceptible to this pathogen while rats are moderately susceptible.
- Bacillus–Calmette–Guerin (BCG), the tuberculous vaccine, is an attenuated strain of *M. bovis*.

Susceptibility

- Mycobacteria are killed at 60°C in 15–20 minutes.
- Bacilli in sputum may be viable for 20–30 hours and in droplet nuclei and up to 8–10 days under suitable conditions.
- Cultures remain viable at room temperature for 6–8 months and may be stored for up to two years at –20°C.
- Mycobacteria can survive exposure to 5% phenol, 15% sulphuric acid, 3% nitric acid, 5% oxalic acid, and 4% sodium hydroxide.
- They are killed by **formaldehyde and glutaraldehyde**.
- They are destroyed by tincture of iodine in five minutes and by 80% ethanol in 2–10 minutes. Ethanol is a suitable disinfectant for skin, gloves, and clinical thermometers.

Pathogenesis

- Transmission follows the person-to-person route from an active case of tuberculosis by way of **aerosols produced during coughing** (Fig. 23.70).
- Aerosols produced by an infected person may contain as many as 3,000 infectious nuclei per cough.
- The disease spreads most often among members of the household or close and prolonged contacts of open cases (whose sputum may contain a minimum of 10,000 bacilli per mL).
- Infection also occurs infrequently by ingestion, through contaminated milk, and rarely, by inoculation.
- The inhaled bacilli reach the lungs and are phagocytosed by the alveolar macrophages.
- The bacilli escape being killed by the phagosomal vacuole and modulate macrophage apoptosis.

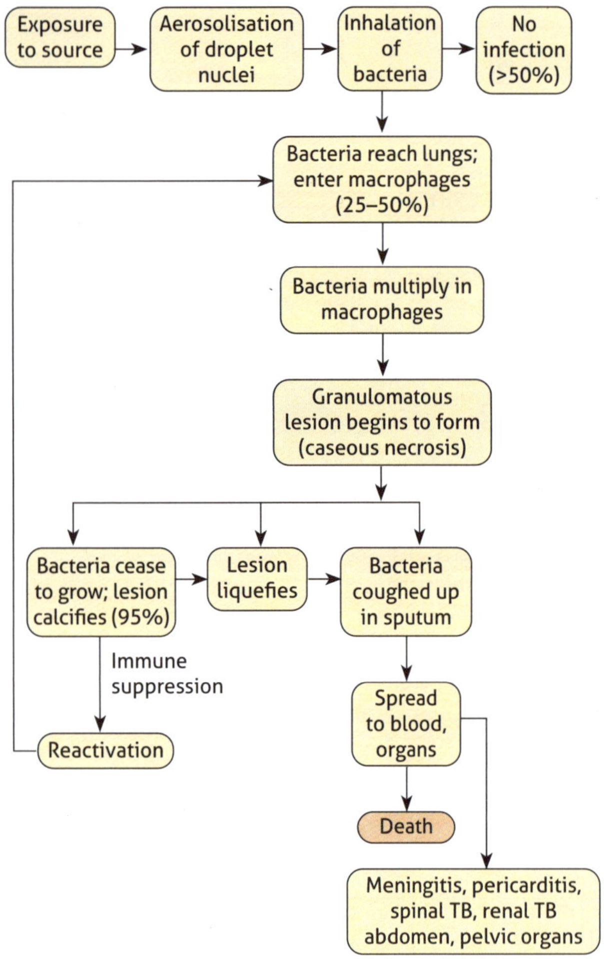

Fig. 23.70 Steps in the pathogenesis of tuberculosis

- Alternatively, they remain quiescent during the initial infection and subsequently transform into a highly replicative and invasive phenotype, migrate across the alveolar barrier and into the circulation, resulting in dissemination.
- Mycobacteria produce the following two proteins, which are thought to play a role in macrophage destruction:
  - **Tuberculosis necrotising toxin**, which cleaves the NAD within the macrophage, resulting in cell death
  - **Early secreted antigen-6 (ESA toxin-6),** which decreases immunity by decreasing production the of interferon

Two types of lesions are produced in tuberculosis:

i. A **granulomatous lesion** is produced with central giant cells (called the Langhans giant cells) containing the bacilli surrounded by epithelioid cells. A tubercle is produced, which consists of granuloma surrounded by fibrous tissue. This tubercle undergoes **caseation necrosis**. The granulomatous lesion is characteristic of tuberculosis (Fig. 23.71).

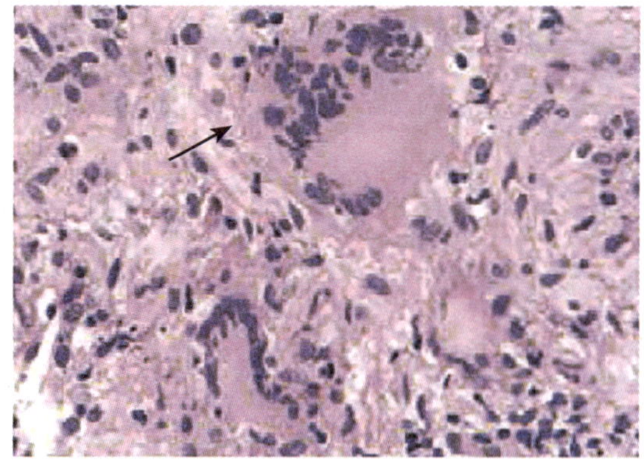

Fig. 23.71 Tubercle or granuloma with caseation

ii. The **exudative lesion** consists of the acute inflammatory response seen in the early stage of tuberculosis.

Depending on the time of infection and the type of response, tuberculosis may be classified as follows.

Primary tuberculosis (primary complex)

- Initial infection in a susceptible host by mycobacteria.
- In endemic countries like India, young children are usually more susceptible.
- Alveolar macrophages engulf the bacilli, which then multiply intracellularly.
- This gives rise to a subpleural focus of tuberculous pneumonia, commonly located in the lower lobe or the lower part of the upper lobe (**Ghon focus**).
- The hilar lymph nodes are involved.
- The Ghon's focus together with the enlarged hilar lymph node constitutes the **primary complex**. In most cases, the lesion heals spontaneously in 2–6 months, leaving behind a calcified nodule.
- In children with impaired immunity, the primary lesion may enlarge and cause miliary, meningeal, or other forms of disseminated tuberculosis.

Post-primary (secondary or adult) tuberculosis

- This involves the reactivation of a latent infection (post-primary progression, endogenous reactivation) or exogenous re-infection and differs from the primary type in many respects.
- It chiefly affects the upper lobes of the lungs, the lesions undergoing necrosis and tissue destruction, leading to **cavitation** (Fig. 23.72).
- The necrotic materials break out into the airways, leading to expectoration of bacteria-laden sputum, which is the main source of infection for contacts.
- In the immunodeficient, there is widespread dissemination of lesions to the lungs and other organs from a primary lesion elsewhere, via the bloodstream.
- The post-primary lesion heals with scarring (named **Simons' focus** after Georg Simon, a German pediatrician).

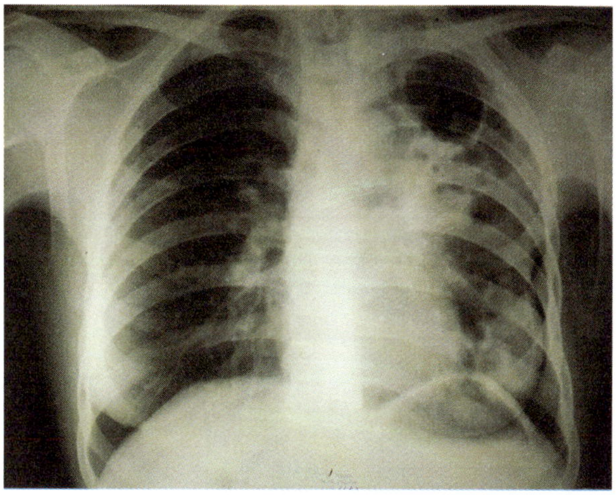

Fig. 23.72 Chest radiograph showing upper lobe cavity (*Source*: Giller Boris, CC0, via Wikimedia Commons)

Tuberculous lesions are primarily of two types:

- **Exudative type**—an acute inflammatory reaction with accumulation of edema fluid and polymorphonuclear leukocytes
- **Productive type**—predominantly cellular and associated with protective immunity

Extrapulmonary tuberculosis (EPTB)

This is generally associated with HIV, other immunocompromised conditions, and diabetes. The incidence of extrapulmonary tuberculosis has been increasing in the recent years. Almost all organs of the body may be affected, including the following:

- Lymph nodes: Lymphadenitis
- GI tract: Diarrhea/constipation
- Urogenital organs: Infertility
- CNS, skeletal system (generally the spine): Chronic pain and immobility
- Ocular system: Blurring of vision, photosensitivity

Symptoms and clinical features depend on the organ affected. As a result, the condition may not get detected easily in the absence of cough, hemoptysis, and abnormal chest X-ray. Appropriate investigations including **histopathology of the tissue involved**.

Immunology

- Cell-mediated immunity plays a major role in tuberculosis; humoral immunity, on the other hand, has little or no role in protection or pathogenesis.
- The key cell is the activated CD4$^+$ helper T cell, which induces delayed-type hypersensitivity, leading to tissue destruction and progressive disease.
- The role of delayed hypersensitivity is evidenced by using a purified protein derivative (PPD) of *M. tuberculosis* as an intradermal skin test. A positive test is evidence that the person is infected but does not necessarily indicate active disease.

Koch's phenomenon

Robert Koch demonstrated that when virulent tubercle bacilli are injected into a healthy guinea pig, it develops a nodule at the site of inoculation, which develops into an ulcer that persists, with caseation of draining lymph nodes. On the other hand, if the bacilli are injected into an already infected guinea pig (infected 4–6 weeks earlier), the lesion develops rapidly—within one or two days—into a shallow ulcer that heals completely without involving the draining lymph nodes.

Laboratory Diagnosis of Pulmonary Tuberculosis (Fig. 23.73)

Biosafety

- All laboratory procedures are mandated to be performed in class IIB biosafety cabinets (Fig. 23.74).
- The use of personal protective gear like masks and gloves is mandatory.
- In the disposal of laboratory waste, biomedical waste guidelines have to be adhered to.

Specimen

- **Sputum**
  - Bacillary shedding in the sputum is abundant in cases of caseation but relatively scanty in organised lesions that do not communicate with the airways.
  - Sputum is best **collected in the morning**, before any meal. Repeated testing of samples increases the chance of detection.
  - **Two sputum sample examinations** are currently recommended by the National TB Elimination Programme (NTEP):
    - Spot sample examination at the time of the first visit has been recommended. Spot sample refers to a sample collected during a patient's first visit to a healthcare setting.
    - The second sample is an early morning one collected before any meal (CDC and WHO, however, recommend three sputum smear examinations).
- **Other samples**
  - If sputum is not available, laryngeal aspirates or bronchial washings (with appropriate precautions) may be collected.
  - In small children who tend to swallow their sputum, gastric lavage can be examined.
- **Decontamination and concentration of specimens**
  - Specimens from non-sterile sites and sputum need prior treatment so that organisms other than mycobacteria are inhibited.
  - Liquefaction, decontamination, and concentration improve the yield. NaOH kills the contaminating bacteria.

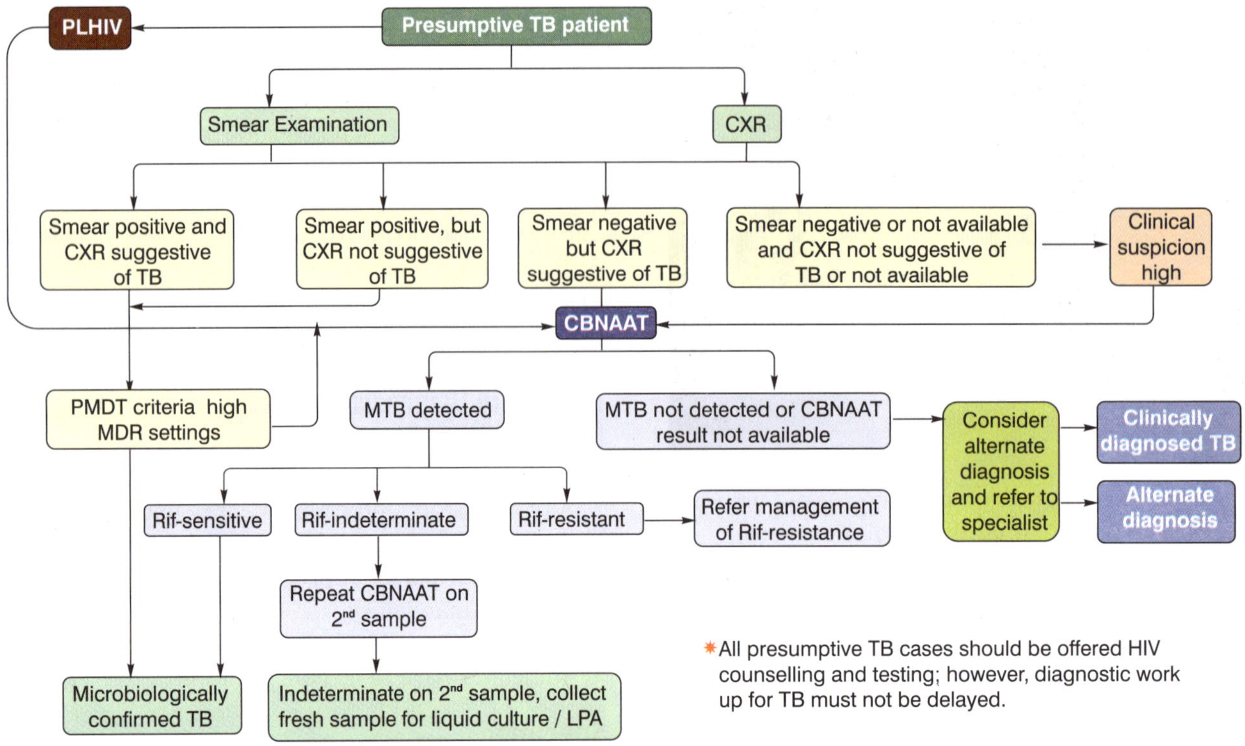

Fig. 23.73 Algorithm for the diagnosis of pulmonary tuberculosis (Updated in 2016 in accordance with the Revised National Tuberculosis Control Programme [now known as NTEP])

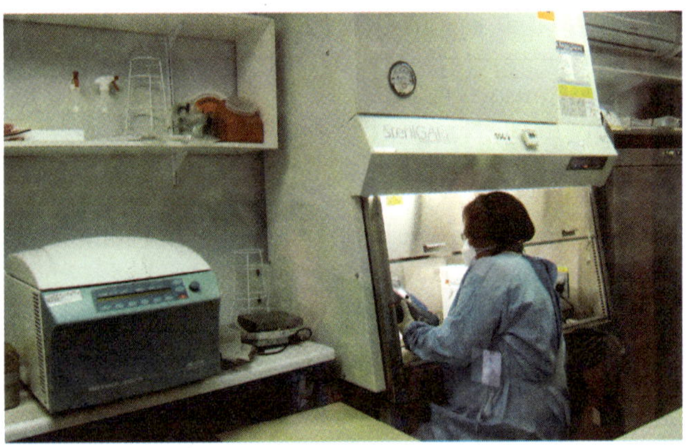

Fig. 23.74 Laboratory technician in the process of conducting tuberculosis sample evaluations in a biosafety cabinet, which prevents laboratory contamination from pathogenic samples (*Source*: CDC, PHIL, Image ID 19753)

Petroff's method

Fig. 23.75 describes the steps involved in Petroff's method.

NALC (N-acetylcysteine) combined with 2% NaOH

- This method is considered better than Petroff's.
- N-acetylcysteine is used for the liquefaction of sputum.
- The sample is then neutralised with a buffer and concentrated by centrifugation.
- This method is also compatible with culture in automated systems.

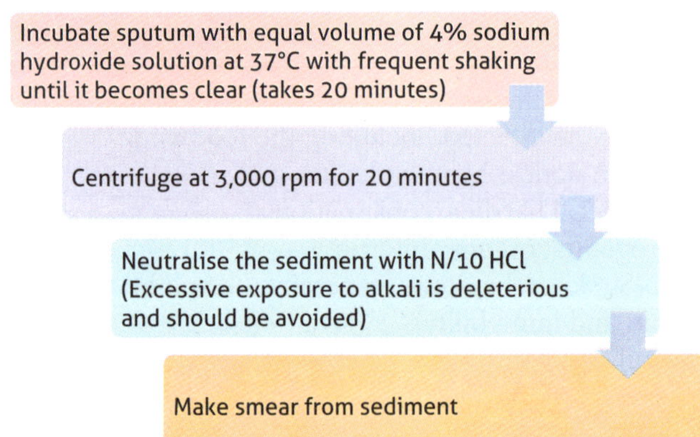

Fig. 23.75 Petroff's method of decontamination

Microscopy

- Sputum microscopy is the single most reliable method for the diagnosis and control of tuberculosis. Smears should be prepared from the **thick, purulent part of the sputum**.
- A minimum of 10,000 acid-fast bacilli should be present per mL of sputum for them to be readily demonstrable in direct smears. Routinely used staining methods are:
  - Ziehl–Neelsen technique (ZN)
  - Acid-fast staining
- Mycobacteria, when stained with strong carbol fuchsin by the Ziehl–Neelsen method, resist decolourisation by 20% sulphuric acid and hence, are called **acid-fast**.

- Acid-fastness has been ascribed to the presence of an **unsaponifiable lipid-rich (mycolic acid) material in the mycobacterial cell wall or a semipermeable membrane around the cell.**
- Staining may be **uniform or granular**.
- **Beaded or barred forms** are frequently seen in *M. tuberculosis*, whereas *M. bovis* stains more uniformly.
- Mycobacteria are seen as slender, pink bacilli in a smear (Fig. 23.76).
- At least 100 fields should be examined by Ziehl–Neelsen technique over 10 minutes before reporting the test as negative. The report is said to be positive if two or more typical bacilli are detected.
- Smears are graded based on the NTEP guidelines (Table 23.20).

Alcohol-fastness

In addition to being acid-fast, *M. tuberculosis* is also alcohol-fast (resists decolourisation with 3% hydrochloric acid in 95% alcohol), which differentiates it from saprophytic mycobacteria, which are only acid-fast.

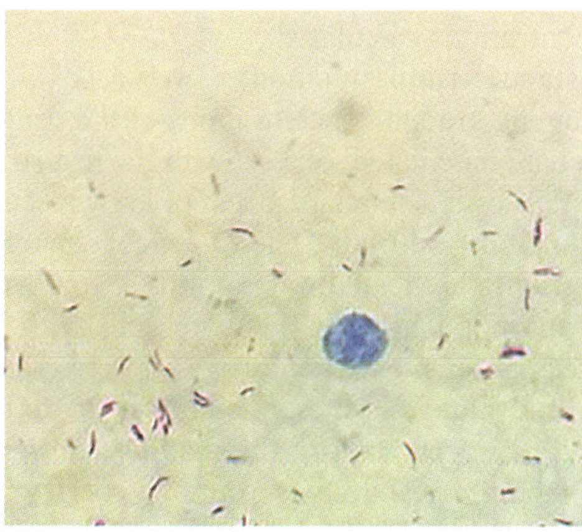

Fig. 23.76 *Mycobacterium tuberculosis*—thin, pink bacilli in sputum smear—acid-fast bacilli on ZN stain (*Source*: Department of Microbiology, Pondicherry Institute of Medical Sciences, Puducherry)

Table 23.20 ZN smear evaluation for sputum smear as per NTEP guidelines

Number of bacilli	NTEP	
	Result	Grading (Report)
>10/field	+	3+
1–10/field	+	2+
10–99/100 field	+	1+
1–9/100 field	+	Scanty*
No. of AFB in 100	–	Negative

*Record actual number of bacilli seen in 100 fields

Kinyoun's modification of acid-fast staining

Acid-fast bacilli are seen as bright pink rods while the background is blue. Since a high concentration of phenol and basic fuschin is used, heating is not required.

Auramine–rhodamine or fluorochrome staining

- This is the current recommendation for sputum smear examination using a fluorescence microscope.
- Bacilli appear as bright rods against a dark background (Fig. 23.77). Because of the contrast, the bacilli can be seen even under the dry objective, enabling large areas of the smear to be screened rapidly.
- When several smears are to be examined daily, it is convenient to use fluorescence microscopy.
- Smears are stained with auramine phenol or auramine rhodamine fluorochrome dyes and examined under ultraviolet illumination or LED (light-emitting diode).

Culture

- This is the gold standard for the diagnosis of tuberculosis, making it possible to detect as few as 10–100 bacilli per mL.
- *Mycobacterium tuberculosis* has a generation time of 14–15 hours, i.e., it grows slowly. **Colonies appear in 2–8 weeks**.
- The optimum temperature is 37°C; temperatures below 25°C or above 40°C do not favour growth. The optimum pH is 6.4–7.0.
- *M. tuberculosis* is an obligate aerobe.

Conventional culture methods

Solid media—Löwenstein–Jensen (LJ) medium

- It consists of coagulated hen's eggs, mineral salt solution, asparagine, and malachite green, the last acting as a selective agent that inhibits other bacteria (Fig. 23.78).

Fig. 23.77 Under a magnification of 450X, this photomicrograph of green, glowing *Mycobacterium tuberculosis*, stained using fluorochrome stain, and viewed through an oil-immersion condenser (*Source*: CDC, PHIL, Image ID 22206)

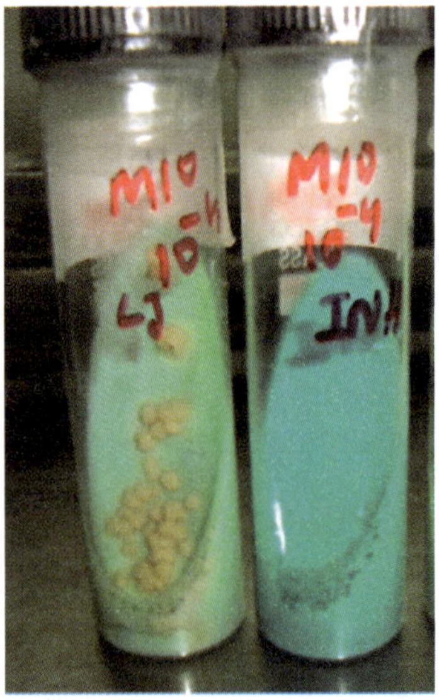

Fig. 23.78 *M. tuberculosis* on LJ medium (left); LJ medium without growth (right)

- The growth appears after **four weeks due to slow generation time**. The culture is incubated for up to eight weeks and examined weekly for growth.
- **Buff-coloured colonies**, which are rough in consistency and not easily emulsified, are subjected to acid-fast staining to identify acid-fast bacilli.

Liquid media commonly used are:

- Dubos' medium
- Middlebrook's medium
- Proskauer and Beck's medium
- Sula's and Sauton's medium

Virulent strains tend to form **long, serpentine cords** in liquid media.

Newer methods of cultivation

BacT/ALERT

This method uses colourimetry for growth detection, which is measured on the basis of the production of CO_2 as a result of bacterial metabolism during growth.

BACTEC–MGIT

- Automated mycobacteria growth indicator tube (MGIT; Fig. 23.79) is a rapid growth detection method, which uses 7H9 Middlebrook medium with fluorometric technology for the detection of O2 consumption.
- An added advantage is that the incorporation of **pyrazinamide (PZA)** in the medium detects the organism's resistance to this drug as well.

Drug sensitivity tests

As drug resistance is a significant problem in tuberculosis, it is desirable to test the sensitivity of isolates as an aid to treatment. Sensitivity tests for *M. tuberculosis* are carried out using the following methods:

Fig. 23.79 Bactec–MGIT for the culture of *M. tuberculosis* (*Source*: Intermediate Reference Laboratory for TB, SCB Medical College, Cuttack, Odisha)

Conventional methods

i. **Absolute concentration method:** A number of media containing serial concentrations of the drugs are inoculated, and the minimum inhibitory concentrations calculated.
ii. **Resistance ratio method:** Two sets of media containing graded concentrations of the drugs to be tested are inoculated, one set with the test strain and the other with a standard strain of known sensitivity.
iii. **Proportion method:** The average sensitivity of the strain is indicated, taking into account the fact that any population will contain cells with varying degrees of sensitivity to a drug.

The disadvantage of these methods is the time taken for growth to occur before the reading can be taken. Hence, laboratories do not routinely perform these tests.

Automated systems

Molecular methods (to detect MTB and drug resistance genes) including nucleic acid amplification methods by polymerase chain reaction (PCR) and ligase chain reaction (LCR) are replacing culture methods, especially for extrapulmonary tuberculosis (e.g., **cartridge-based nucleic acid amplification**—CB–NAAT) because they are rapid and can detect drug resistance genes simultaneously.

i. **Gene Xpert MTB/RIF assay** (Fig. 23.80) is a fully automated molecular diagnostic test for tuberculosis. This method is used more commonly now because it can simultaneously detect MTB and rifampicin resistance and because of the shorter turnaround time.
ii. **Truenat (and Truenat MTB plus for the detection of drug resistance)** is an indigenous molecular

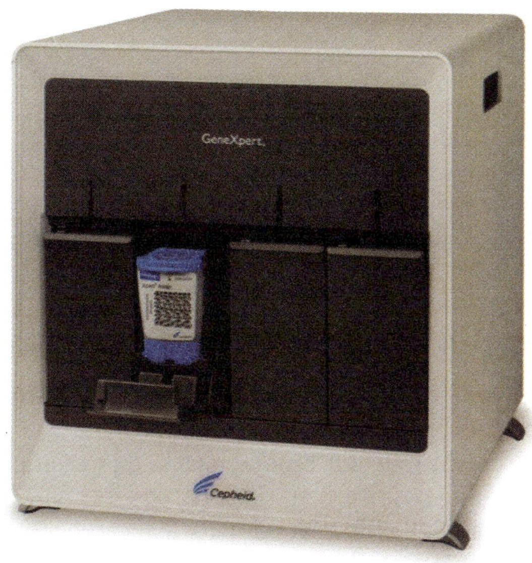

Fig. 23.80 Gene Xpert—rapid growth and resistance to rifampicin detected in a few days (*Source*: Wikimedia Commons)

detection method that was developed in India. It is similar in principle to CB–NAAT and has now been approved by the WHO for use as a point-of-care tool for the detection of tuberculosis.

iii. **Transcription-mediated amplification (TMA)**, which targets ribosomal RNA, has been introduced as an improvement of PCR-based DNA amplification.

iv. **Line probe assay (LPA)** is used both for the identification of MTB and detection of mutations associated with drug resistance genes.

Typing methods

- **Molecular typing:** This is based on DNA fingerprinting, which is a powerful epidemiological tool for differentiating between strains of tubercle bacilli.
- **IS6110 restriction fragment length polymorphism (RFLP) typing:** The IS6110 is a target sequence in several methods currently used for molecular typing of *M. tuberculosis*. The patterns are strain-specific and can be used for fingerprinting (IS6110-based PCR fingerprinting).

Immunodiagnosis

Serological tests are not useful in the diagnosis of tuberculosis, especially in endemic areas, and are not recommended.

Tuberculin test (also known as Mantoux test)

- Demonstration of hypersensitivity (delayed) to tuberculoprotein by tuberculin testing is a standard procedure to detect exposure to the bacilli.
- 0.1 mL of PPD containing 5 TU (tuberculosis unit) is injected intradermally.
- After **48–72 hours, an induration of 10 mm or more at its widest point is taken as positive.**
- Induration of 5 mm or less is considered negative and 6–9 mm equivocal. A PPD dose of 1 TU is used when extreme hypersensitivity is suspected.
- An increased dose of 10 or 100 TU is used when the 5 TU test is negative.

Interpretation

- **Positive:** A positive tuberculin test indicates **hypersensitivity to tuberculoprotein, denoting infection or prior immunisation with BCG**. The test becomes positive 4–6 weeks after infection or immunisation and wanes gradually and disappears after 4–5 years. In endemic areas, the allergy is maintained by repeated contact with the bacilli.
- **Negative:** Persons who have not had contact or been exposed to the tubercle bacilli test negative. **False-negative tests (anergy)** may be seen in miliary tuberculosis, convalescence from some viral infections like measles, lymphoreticular malignancy, sarcoidosis, severe malnutrition, immunosuppressive therapy or impaired cell-mediated immunity.

Gamma interferon assays

- Interferon-gamma release assays (IGRAs) are in vitro blood tests of cell-mediated immune response; they measure T cell release of interferon-gamma following stimulation by antigens unique to *Mycobacterium tuberculosis* and a few other mycobacteria.
- In most individuals infected with *M. tuberculosis*, mononuclear cells in the blood release interferon-gamma when stimulated with antigens derived from *M. tuberculosis*.
- This test is useful to diagnose latent tuberculosis.

Diagnosis of extrapulmonary tuberculosis

- The general procedure is as for pulmonary tuberculosis.
- The **specimen** depends on the site of infection; urine, CSF, joint fluid, biopsy material, blood, or any other body fluid may be collected as a sample. CSF from **tuberculous meningitis often develops a spiderweb clot on standing**, the examination of which may be more successful than that of the fluid.
- **Microscopy and culture** are used for the diagnosis of extrapulmonary tuberculosis, though it is difficult to obtain conclusive results. This is because the bacilli are present in fewer numbers in these lesions than in pulmonary disease.
- **Molecular techniques** can add to the diagnosis of extrapulmonary tuberculosis by conventional smear and culture but cannot replace the latter methods. The use of PCR and DNA probes may be more efficient in detecting the bacilli.

- Bone marrow and liver biopsy specimens from miliary tuberculosis and blood from those with HIV co-infection are useful for culture.
- Pus from tuberculous abscesses often yields positive results in smear and culture.
- Pleural effusion and other exudates may be collected with citrate to prevent coagulation. They may be directly cultured after centrifugation.
- Urinary excretion of bacilli in renal tuberculosis is intermittent. Hence, it is advisable to test at least 2–3 first, whole-voided morning samples for culture. Each sample is centrifuged at 3,000 rpm for 30 minutes and the sediment is used for culture after concentration.

Immunoprophylaxis

- The **BCG (Bacillus–Calmette–Guerin) vaccine**, administered by intradermal injection of the live-attenuated vaccine (developed by Albert Calmette and Camille Guerin in 1921). This vaccine is prepared from a strain of *M. bovis* which has been attenuated by 239 serial subcultures in a glycerine–bile–potato medium over a period of 13 years.
- The resulting immunity may last for 10–15 years and is similar to the immunity following natural infection.
- **In endemic countries such as India, the BCG vaccine is administered to neonates by intradermal injection on the deltoid, immediately after birth or as early as possible, before the age of 12 months.**

Advantages of BCG vaccine other than protection from tuberculosis

- BCG induces non-specific stimulation of the immune system, providing some protection against leprosy and leukemia.
- Multiple injections of BCG have been attempted as adjunctive therapy in some malignancies.
- Some workers have reported that BCG is superior to PPD for tuberculin testing.

Complications of BCG vaccine

- **Local:** Abscess, indolent ulcer, keloid, tuberculids, confluent lesions, lupoid lesions, and lupus vulgaris
- **Regional:** Enlargement and suppuration of draining lymph nodes
- **General:** Fever, mediastinal adenitis, erythema nodosum, tendency to keloid formation, and very rarely, non-fatal meningitis; very few cases of progressive tuberculosis have been reported and are believed to have developed in immunodeficient subjects
- **Efficacy:** BCG may not offer protection from the risk of tuberculosis infection but gives protection to infants and young children against the more serious types of the disease, such as meningitis and disseminated tuberculosis

Contraindication to BCG

- The vaccine is contraindicated for infants and children with active HIV disease (may be given with benefit to asymptomatic HIV-positive cases) and babies born to mothers with AFB-positive sputum.
- However, in the latter case (babies born to mothers with AFB-positive sputum), it may be given after a course of preventive chemotherapy.

Indications for chemoprophylaxis or preventive chemotherapy

- Those with latent tuberculosis (asymptomatic, tuberculin-positive)
- Those at high risk of developing active tuberculosis
- Uninfected persons who are at a high risk of infection through exposure
- Infants of mothers with active tuberculosis
- Children living in the same house as a person with active tuberculosis
- HIV-infected contacts of persons with active tuberculosis

Drug and dosage

Isoniazid 5 mg/kg daily for 6–12 months is recommended. Trials have shown that this reduces the risk of developing active disease by 90 per cent.

Anti-tuberculosis Therapy (ATT)

Bactericidal drugs

- **Rifampicin (R) and pyrazinamide (Z)** are called sterilising drugs because they effectively kill the bacilli in the lesions.
- **Isoniazid (H)** is effective only against replicating bacilli.
- **Streptomycin (S)** acts only against extracellular bacilli unable to sterilise the lesions on their own.

Bacteriostatic drugs

- These are combined with bactericidal drugs for effective clearance of bacteria.
- Ethambutol (E) is the bacteriostatic drug of choice.

Alternate methods of grouping ATT are presented in Table 23.21.

Combination therapy of ATT

- Combination therapy prevents drug resistance.
- A short-course regimen of 6–7 months is effective and convenient.
- A combination of four drugs (HRZE) given three times a week during the initial intensive phase of two months, followed by 4–5 months of the continuation

Table 23.21 Anti-TB agents

Group	Drugs
Group 1: First-line oral anti-TB agents	Isoniazid (H); rifampicin (R); ethambutol (E); pyrazinamide (Z)
Group 2: Injectable anti-TB agents	Streptomycin (S); kanamycin (Km); amikacin (Am); capreomycin (Cm); viomycin (Vm)
Group 3: Fluoroquinolones	Ciprofloxacin (Cfx); ofloxacin (Ofx); levofloxacin (Lvx); moxifloxacin (Mfx); gatifloxacin (Gfx)
Group 4: Oral second-line anti-TB agents	Ethionamide (Eto); prothionamide (Pto); cycloserine (Cs); terizadone (Trd); para-aminosalicylic acid (PAS)
Group 5: Agents with unclear efficacy (not recommended by the WHO for routine use in MDR-TB patients)	Clofazimine (Cfz); linezolid (Lzd); amoxicillin/clavulanate (Amx/Clv); thioacetazone (Thz); imipenem/cilastatin (Ipm/Cln); high-dose isoniazid (high-dose H); clarithromycin (Clr)

phase with only two drugs (HR), three times a week has been found to be effective.

- All forms of TB except TB meningitis and bone and joint TB (osteoarticular TB) are treated with rifampicin, isoniazid, pyrazinamide, and ethambutol in the intensive phase followed by a continuation phase of four months of rifampicin and isoniazid.
- In TB meningitis and bone and joint TB (osteoarticular TB), the continuation phase is for ten months with rifampicin and isoniazid.
- In cases of drug-resistant TB, the guidelines discussed in the section on NTEP are followed under intense monitoring and supervision.

Drug Resistance

- Drug-resistant tuberculosis has become a problem in high-TB-burden countries, including India.
- Factors that have led to the emergence of MDR-TB (multidrug-resistant tuberculosis) are:
  - Lapses in prescribing practices, drug delivery and patient compliance
  - High frequency of mutations with an approximate rate of 1 in 108 cell divisions

Classification of drug resistance

1. **Resistance in newly treated:** This occurs when the patient is infected with a strain of the tubercle bacilli that is already resistant.
2. **Resistance in previously treated (secondary, post-treatment):** This occurs when the infecting strain, which was initially sensitive, becomes resistant. This is usually a result of improper or inadequate treatment and is the more common type of resistance.

Multidrug-resistant tuberculosis (MDR-TB)

- This refers to resistance to rifampicin (R) and isoniazid (H), with or without resistance to one or more of the other drugs.
- This is because R and H form the sheet anchor of short-term chemotherapy and any strain resistant to both these drugs is unlikely to respond to treatment.
- MDR-TB is a global problem, menacing the developed and developing nations alike.
- It may be primary or acquired.
- MDR-TB is more dangerous for those with a concomitant HIV infection.
- When first-line drugs become ineffective, second-line drugs must be tried.
- Large numbers of old and new drugs are being used—quinolones, aminoglycosides, macrolides, para-aminosalicylic acid, thiacetazone, cycloserine, capreomycin, and others. However, these drugs are unsatisfactory, being much less effective, costlier, more toxic, and requiring prolonged treatment schedules.
- In view of the poorer outcomes with kanamycin and capreomycin, the WHO recommends avoiding these drugs.

Pre-extensively drug-resistant tuberculosis (MDR-TB)

Pre-XDR TB is now defined as multidrug-resistant/rifampicin-resistant (MDR/RR) TB with any fluoroquinolone resistance.

Extensively drug-resistant tuberculosis (XDR-TB)

XDR-TB is now defined as pre-XDR TB with additional resistance to bedaquiline and/or linezolid.

National TB Elimination Programme (NTEP)

- It was formerly known as the Revised National Tuberculosis Control Programme (RNTCP), and was initiated from 1 January 2020 by the Govt. of India.
- The revised programme encompasses strategies to promote quality of life and health by preventing, controlling, and finally, eliminating TB.
- In NTEP, the principle of TB treatment (except confirmed DR-TB) is to administer a daily **fixed dose combinaton (FDC)** of first-line ATT in appropriate weight bands, under direct observation.
- For patients being treated in the private sector, FDCs may be provided by the NTEP upon request.
- RNTCP was implemented in India in 1992 with the objectives of detecting new cases through **designated microscopy centres (DMC)** and providing standardised treatment and proper diagnostic facilities. This was based on the **Directly Observed Treatment, Short course (DOTS) strategy** of the WHO.

- The treatment is initiated following diagnosis made primarily by morning and spot sputum microscopy.
- DOTS was made available free of cost to patients at DMCs. Treatment is provided under direct observation by a DOTS provider at the DOTS centre near the patients' home.
- DOTS is the fastest expanding programme in the world.
- For the purpose of treatment, based on the sputum smear positivity and the seriousness of the disease, regimens are categorised as category I, II, and III of treatment.
- The algorithms for laboratory diagnosis and treatment strategies are standardised.
- DOTS relies on treatment with the first-line drugs rifampicin and INH (isoniazid).
- DOTS-Plus refers to DOTS programmes that add components for MDR-TB diagnosis and treatment using quality-assured culture and drug susceptibility testing (DST)

Standardised treatment regimen for drug-sensitive TB as per NTEP

Under the NTEP, the following changes have been made in the DOTS strategy:

i. Injectable drugs have been removed from Cat II
ii. Newer drugs bedaquiline and delamanid have been included in DOTS Plus regimen

   Treatment is given in two phases:

   1. **Intensive phase:** Therapy for 8 weeks (56 doses) of isoniazid (H), rifampicin (R), pyrazinamide (Z), and ethambutol (E) given under direct observation in daily dosages as per weight band categories.
   2. **Continuation phase:** It consists of 16 weeks (112 doses) of isoniazid, rifampicin, and ethambutol in daily dosages. Only pyrazinamide will be stopped in the continuation phase. It has to be extended to 24 weeks in certain forms of TB like CNS TB and skeletal TB. TB meningitis requires the addition of steroids. In disseminated TB or in cases of slow response, the treating physician may extend the regimen on a case-to-case basis.

Treatment of drug-resistant TB

- **The 6-month treatment regimen** is composed of bedaquiline, pretomanid, linezolid, and moxifloxacin (BPaLM).
- **The 9-month regimen** comprises bedaquiline (used for 6 months) in combination with levofloxacin/moxifloxacin, ethionamide, ethambutol, isoniazid (high dose), pyrazinamide, and clofazimine (for four months, with the possibility of extending to 6 months if the patient remains sputum smear positive at the end of four months), followed by treatment with levofloxacin/moxifloxacin, clofazimine, ethambutol, and pyrazinamide (for five months).

National Strategic Plan, India—2017–2025 (NTEP)

The Govt. of India plans to make India TB free by 2025. The strategy is as follows:

- Including private healthcare settings
- Tracing people with TB who have been missed by the care provided through DOTS
- Finding active cases among key populations
- Preventing the development of active TB in people with latent TB in high-risk groups like HIV

PREVENTION OF RESPIRATORY INFECTIONS

- Mortality and morbidity due to respiratory infections across the globe are high, especially in middle- and low-income countries.
- To decrease these rates, the WHO and member countries have instituted immunisation programmes.
- These programmes include the immunisation of vulnerable populations (infants, children, the elderly, and immunocompromised individuals) against common respiratory pathogens and seasonal viral respiratory infections as an important strategy to prevent lower respiratory infections (Table 23.22).
- Besides these, awareness programmes for the prevention of smoking and discouraging of cooking with firewood, which are risk factors to respiratory infections, have also been implemented by governments.

Table 23.22 Common vaccines to prevent respiratory infections

Pathogens	Vaccines
C. diphtheriae	DtaP, Tdap, DT, Td, DTP
M. tuberculosis	BCG
H. influenzae	HiB PRP vaccine
S. pneumoniae	Pneumococcal polysaccharide vaccines—PPSV-23, PCV-7, PCV-13
B. pertussis	DTaP, Tdap
Viral respiratory pathogens	
Influenza	Inactivated, FluMist
Measles	MMR
Rubella	MMR
SARS-CoV	mRNA whole cell vaccine, genetically engineered vector vaccines

Questions

Case reviews

1. A seven-year-old girl presented with severe sore throat and fever up to 39°C for three days, with difficulty in swallowing. On examination, she had bilateral tender submandibular lymphadenopathy, enlarged tonsils and pharyngeal exudates.
   i) List the differential diagnosis.
   ii) What investigations will you perform to confirm the etiology of this condition?
   iii) Discuss the indications for antibiotics in treating pharyngitis.
2. A 22-year-old male was admitted to the respiratory unit of a hospital with symptoms of cough with expectoration, thoracic pain, weight loss, and low-grade fever. On examination, he had laterocervical lymphadenopathy. There was a suspicion of tuberculosis.
   i) List the investigations you will carry out further to arrive at a conclusive diagnosis.
   ii) How will you categorise the patient based on the smear report?
   iii) What histopathological features will you find in the lymph node biopsy specimen?
   iv) What precautions will you advocate for the laboratory personnel processing the specimen?

Short answer questions

1. Discuss the reasons for the epidemic and pandemic potential of influenza virus.
2. Outline the algorithm for the laboratory confirmation of community-acquired pneumonia.
3. Describe the pathogenesis of SARS-CoV-2 infection.
4. What are the laboratory tests to confirm bronchopulmonary aspergillosis?
5. What are the current vaccines in use to prevent respiratory tract infections?
6. Describe the procedure of Zeihl–Neelson staining for a sputum smear.
7. Describe the pathogenesis of primary tuberculosis.
8. Classify the mycobacteria of clinical importance.
9. Draw a flowchart of the pathogenesis of pulmonary tuberculosis.
10. Discuss various methods to culture *M. tuberculosis.*

MCQs

1. **A 45-year-old known HIV seropositive patient was admitted in the ICU with pneumonia. His CD4 counts were 80 cells/mm^3. GMS stain smear from his broncho-alveolar lavage (BAL) fluid showed cysts measuring 4–6 microns. What is the likely organism associated with this disease?**
   a) *T. gondii* b) *P. jirovecii*
   c) *H. nana* d) *C. sinensis*
2. **A 55-year-old farmer, a known asthmatic, was admitted for evaluation of chronic productive cough. On admission, he had high grade fever and his chest X-ray showed lobular infiltrates. On Gram, his sputum showed numerous eosinophils and fungal filaments with septate hyphae at 45° branching. Name the organism associated with this infection.**
   a) *Aspergillus fumigatus* b) *Candida albicans*
   c) *Mucor* d) *Cryptococcus*
3. **An infant was admitted with cough and fever with moderate respiratory distress. Chest examination showed generalised wheeze, and the infant was diagnosed as a case of bronchiolitis. Bacterial culture from the respiratory fluid did not yield any growth. What is the likely causative agent in this condition?**
   a) Varicella zoster virus b) Respiratory syncytial virus
   c) Mumps virus d) *Candida albicans*
4. **A 35-year-old chronic smoker with chronic cough consulted a physician after he noticed red-coloured sputum that morning. His sputum was negative for acid-fast bacilli. However, sputum culture also yielded red-coloured colonies on nutrient agar. What is the likely organism associated?**
   a) *Serratia marcescens* b) *Pseudomonas. aeruginosa*
   c) *Klebsiella pneumoniae* d) *Streptococcus pneumoniae*
5. **A 52-year-old HIV seropositive patient was suspected of pulmonary tuberculosis and was admitted for evaluation. His sputum sample was submitted for investigation. Which of the following does not apply to *M. tuberculosis* culture?**
   a) Colonies on LJ medium are produced after four days
   b) Colonies are typically rough, tough, and buff on LJ medium
   c) BACTEC culture gives rapid results
   d) Concentration technique yields better results
6. **A 56-year-old chronic smoker presented with fever and chronic cough with rusty sputum from the past month. His chest X-ray showed dense infiltration in the lower lobe. His sputum growth on blood agar showed alpha-hemolytic colonies. What is the likely pathogen in this case?**
   a) *Klebsiella pneumoniae*
   b) *Staphylococcus aureus*
   c) *Streptococcus pneumoniae*
   d) *M. avium*-intracellulare complex
7. **A two-year-old girl presented with high fever, severe throat pain, and difficulty in swallowing since the past two days. Her tonsils were enlarged and covered with a white patch. Albert staining of a specimen from the patch showed bacteria that resemble *Corynebacterium diphtheriae* morphologically. What is the recommended treatment in this case?**
   a) Antibiotics and supportive therapy
   b) Diphtheria antitoxin
   c) DPT vaccine booster dose
   d) Antibiotics, diphtheria antitoxin, and DPT vaccine
8. **A 50-year-old diabetic man developed fever, cough, and respiratory distress, and flu-like syndrome. He was diagnosed as a case of pneumonia and admitted. On the third day, his condition deteriorated and he was put on ventilator. A suitable clinical sample was collected in a viral transport medium and screened for H1N1 influenza. Which of the following is true with regard to better case management?**
   a) Viral culture is recommended
   b) ELISA is a good test for screening
   c) Immunochromatography test yields rapid results
   d) Real-time PCR is the best diagnostic test
9. **A 33-year-old smoker was admitted with pneumonia. His sputum was subjected to serological test and was positive for *Mycoplasma pneumoniae*. What is the best method to confirm the diagnosis?**
   a) Bacterial culture
   b) Gram stain microscopy
   c) Complement fixation test with paired sera
   d) ELISA for cell wall antigens
10. **A 53-year-old woman presented with fever and chronic cough with hemoptysis for more than six months. She was treated for these symptoms by the local practitioner but it did not help. She gave a history of consuming improperly cooked crabs. Her sputum was negative for acid-fast bacilli but showed parasitic ova. What is the likely pathogen associated with this condition?**
    a) *Loa loa* b) *Wuchereria bancrofti*
    c) *Hymenolepis nana* d) *Paragonimus westermani*
11. **The streptococcal enzyme responsible for the serous character of exudates is:**
    a) Streptokinase b) Streptodornase
    c) Hyaluronidase d) Lipoproteinase
12. **Quellung reaction is:**
    a) Used for the presumptive identification of *Streptococcus pneumoniae*

b) Based on antigen–antibody reactions
c) Can be used for serotyping
d) All of the above

13. The transport medium used for throat swabs in cases of streptococcal pharyngitis is:
a) Pike's medium b) Stuart's medium
c) Buffered glycerol saline d) Cary-Blair medium

14. Lancefield grouping of streptococci is based on:
a) Protein antigen b) Carbohydrate antigen
c) Capsular antigen d) Teichoic acid antigen

15. *Streptococcus pyogenes* can be differentiated from other hemolytic streptococci by:
a) Optochin sensitivity b) Bacitracin sensitivity
c) Novobiocin sensitivity d) Erythromycin sensitivity

16. Individuals suffering from *Mycoplasma pneumoniae* infection have agglutinins to:
a) *Streptococcus mutans* b) *Streptococcus* MG
c) *Streptococcus equisimilis* d) *Streptococcus agalactiae*

17. A 12-year-old boy developed acute glomerulonephritis a week after he had pyoderma. The etiological agent can be identified by the serotyping of:
a) Capsule b) M proteins
c) Pili d) Carbohydrate

18. All the following are true about the M protein of *Streptococcus pyogenes* except:
a) It is responsible for virulence
b) It is heat and acid-stable
c) It is used for typing streptococci
d) M protein is not antigenic

19. Alpha-hemolysis on blood agar is produced by:
a) *Streptococcus pyogenes*
b) *Staphylococcus aureus*
c) *Streptococcus pneumoniae*
d) *Streptococcus agalactiae*

20. Scarlet fever is due to:
a) Production of endotoxin by *Streptococcus pyogenes*
b) Production of exotoxin A & C by *S. pyogenes*
c) Lipoteichoic acid of *S. pyogenes*
d) Capsule of *S. pyogenes*

21. The beaded appearance of *Corynebacterium diphtheriae* on staining indicates:
a) Live bacilli b) Dead bacilli
c) Storage of energy d) Presence of exotoxin

22. The toxigenicity of *Corynebacterium diphtheriae* is due to:
a) Beta phage b) Lambda phage
c) Gamma phage d) Delta phage

23. The receptor for diphtheria toxin is:
a) CD3 b) CD8
c) CD9 d) CD16

24. In its action, diphtheria toxin resembles the toxin of:
a) *Staphylococcus aureus* b) *Bacillus cereus*
c) *Clostridium perfringens* d) *Pseudomonas aeruginosa*

25. Diphtheria toxin attacks all the following tissues except:
a) Myocardium b) Skin
c) Adrenals d) Liver

26. The organ that is maximally damaged by the diphtheria toxin is:
a) Heart b) Kidney
c) Liver d) Brain

27. The best method for typing *Corynebacterium* is:
a) Phage typing b) Bacteriocin typing
c) Biotyping d) Ribotyping

28. Post-diphtheritic paralysis is seen usually at:
a) 2–3 weeks b) 3–4 weeks
c) 4–5 weeks d) 5–6 weeks

29. The stain not used for *Corynebacteria* is:
a) Albert b) Kinyoun
c) Ponder d) Neisser

30. Isolation of *Corynebacteria* requires enrichment with all the following except:
a) Egg b) Starch
c) Blood d) Serum

31. A lipophilic *Corynebacterium* species is:
a) *C. jeikeium* b) *C. ulcerans*
c) *C. pseudotuberculosis* d) *C. minutissimum*

32. A 7-year-old unvaccinated child presents with complaints of throat pain and fever for the past two days. On examination, the child has a dirty white membrane in his throat. A throat swab is collected and sent for processing. Which medium gives rapid growth for early identification?
a) Tellurite blood agar b) Loeffler's serum slope
c) Tinsdale medium d) Lowenstein–Jensen medium

33. *C. diphtheria* and *C. ulcerans* have similar colony morphology on tellurite blood agar. Which biochemical test differentiates the two?
a) Catalase test b) Oxidase test
c) Urease test d) Nitrate test

34. A multidrug-resistant *Corynebacterium* species was isolated from a case of prosthetic valve endocarditis. The species will be:
a) *C. ulcerans* b) *C. xerosis*
c) *C. parvum* d) *C. jeikeium*

35. The drug of choice for treating diphtheria carriers is:
a) Penicillin b) Ampicillin
c) Erythromycin d) Doxycycline

36. A *Corynebacterium* species that causes healthcare-associated infections is:
a) *C. striatum* b) *C. parvum*
c) *C. pseudotuberculosis* d) *C. pseudodiphtheriticum*

37. The sputum culture of an 80-year-old male patient with respiratory infection yielded a large number of gram-negative diplococci. His sputum also showed plenty of pus cells. The pathogen is likely to be:
a) *Neisseria meningitidis* b) *N. bovis*
c) *Moraxella catarrhalis* d) *Moraxella lactamica*

38. 'Hockey puck' sign is typically seen with:
a) *N. gonorrhoeae* b) *N. meningitidis*
c) *M. catarrhalis* d) *N. sicca*

39. Which non-sporing anaerobic gram-negative bacilli is responsible for Vincent's angina?
a) *Bifidobacterium* b) *Fusobacterium*
c) *Leptotrichia* d) *Prevotella*

40. Human adenoviruses are grown in:
a) HeLa cell lines
b) Monkey kidney cell lines
c) Rhesus cell lines
d) Baby hamster kidney cell lines

41. Which one of the following adenovirus serotypes is involved in the causation of tumours in humans?
a) 12 and 18 b) Serotypes 12,18, 31
c) All serotypes d) None

42. All of the following clinical features are characteristic of adenoviral infection except:
a) Diarrhea b) Hemorrhagic fever
c) Hemorrhagic cystitis d) Keratoconjunctivitis

43. Which one of the following is false regarding the shell vial technique?
a) Virus is grown on monolayer cell lines
b) Cannot be performed directly on clinical samples
c) Centrifugation decreases the time to culture positivity
d) Direct immunofluorescence method is used for reporting

44. Which one of the following is false regarding adeno-associated viruses?
a) They are defective viruses
b) They are oncogenic

c) They can grow only in cells already infected by adenoviruses
d) Belong to the Parvoviridae family

45. Which non-sporing anaerobic gram-negative bacilli is responsible for Vincent's angina?
a) *Bifidobacterium* b) *Fusobacterium*
c) *Leptotrichia* d) *Prevotella*

46. Genetic reassortment is a common natural event of:
a) Influenza A virus b) Influenza B virus
c) Influenza C virus d) Influenza D virus

47. The name Myxovirus refers to:
a) Ability of the virus to undergo point mutations
b) Affinity of the virus towards mucin
c) Ability of the virus to integrate
d) Mixed antigenic structure of the virus

48. The family Paramyxoviridae includes all of the following viruses except:
a) Mumps virus b) Rhinovirus
c) Newcastle disease virus d) Respiratory syncytial virus

49. Gradual sequential change in the viral antigenic structure at regular intervals is known as:
a) Antigenic shift b) Antigenic drift
c) Point mutation d) Antigenic recombination

50. The incubation period in human influenza is:
a) 1–3 days b) 8–10 days
c) 10–12 days d) 12–14 days

51. Which of the following viral infections is associated with Reye's syndrome?
a) Influenza type A b) Influenza type B
c) Influenza type C d) Adenovirus

52. Which of the following neuraminidase inhibitor drugs is effective against influenza?
a) Acyclovir b) Adefovir
c) Oseltamivir d) Gancyclovir

53. An eight-year-old school-going girl developed fever, malaise and non-suppurative right parotid gland swelling. During the next two days, the left parotid gland was enlarged, and the whole episode got resolved within a week. What is the likely causative agent?
a) Measles virus b) Mumps virus
c) Newcastle disease virus d) Respiratory syncytial virus

54. Which of the following occurs as a complication of mumps?
a) Giant cell pneumonia
b) Acute hepatitis
c) Epididymo-orchitis
d) Subacute sclerosing panencephalitis

55. Jeryl-Lynn strain is used for vaccination against:
a) Influenza b) Measles
c) Rubella d) Mumps

56. Which of the following was originally isolated from children with acute laryngotracheobronchitis?
a) Influenza virus
b) Parainfluenza virus
c) Respiratory syncytial virus
d) Human metapneumovirus

57. All are true regarding Koplik's spots except:
a) They appear one to two days after the rash appears
b) They are bluish-white ulcerations on the buccal mucosa
c) They indicate local viral replication
d) They contain cytoplasmic and intranuclear inclusion bodies

58. Presence of Warthin-Finkeldey cells in lymphoid tissue is characteristic of:
a) Measles b) Mumps
c) Rubella d) Newcastle disease

59. All of the following are complications of measles except:
a) Giant cell pneumonia
b) Meningoencephalitis
c) Epididymo-orchitis
d) Subacute sclerosing panencephalitis

60. All of the following are true regarding measles except:
a) Koplik's spots develop one-two days before the appearance of a rash
b) Maculopapular rash first appears on the forehead
c) Mantoux and other allergic skin tests may be positive
d) Can result in spontaneous abortion or premature delivery

61. Which strain of virus is preferred for active immunisation in measles?
a) Edmonston b) Edmonston-Zagreb
c) Jeryl-Lynn d) Schwartz Moratena

62. Which of the following statement is not correct regarding parainfluenza viruses?
a) The paramyxoviruses are responsible for about 10% of respiratory infections in children needing hospitalisation.
b) In adults, parainfluenza viruses cause milder respiratory infections in which sore throat and hoarseness of the voice are common.
c) Otitis media is a common complication in immunocompetent individuals.
d) Currently, there is no vaccine against this group of viruses.

63. Which of the following is a clinical feature of rubella?
a) Koplik spots are seen on buccal mucosa
b) Rash develops on the face, spreading to the neck, trunk, and extremities
c) Subacute sclerosing panencephalitis is a common complication
d) Croup, bronchitis is seen in adults

64. Owl's eye appearance is associated with:
a) HSV-2 b) Herpes zoster infection
c) CMV d) Tana pox

65. Shell vial culture is used for which one of the following organisms?
a) HIV b) Rabies virus
c) H1N1 d) CMV

66. Which of the antivirals can be used to treat CMV?
a) Foscarnet b) Ganciclovir
c) Cidofovir d) All of the above

67. A common organism associated with dental caries is:
a) *Streptococcus pyogenes* b) *Streptococcus pneumoniae*
c) *Streptococcus viridans* d) *Streptococcus mutans*

68. Lanceolate appearance is shown by:
a) *Streptococcus pyogenes* b) *Staphylococcus aureus*
c) *Enterococcus faecalis* d) *Streptococcus pneumoniae*

69. All of the following are virulence factors produced by *Streptococcus pneumoniae* except:
a) Pneumolysin b) Capsule
c) Autolysin d) Exfoliative toxin

70. The capsule of *Streptococcus pneumoniae* in CSF can be demonstrated by:
a) India ink
b) Latex agglutination test
c) Precipitation with antisera
d) All of these

71. Alpha-hemolysis on blood agar is produced by:
a) *Streptococcus pyogenes*
b) *Staphylococcus aureus*
c) *Streptococcus pneumoniae*
d) *Streptococcus agalactiae*

72. Draughtsman colony on blood agar is shown by:
a) *Streptococcus pyogenes*
b) *Streptococcus pneumoniae*
c) *Streptococcus viridans*
d) *Streptococcus mutans*

73. All the following are characteristics of *Streptococcus pneumoniae* except:
a) Optochin sensitivity b) Bile solubility
c) Inulin fermentation d) Oxidase positive

74. Penicillin resistance in pneumococci is due to:
a) Beta-lactamase production
b) Alteration in penicillin-binding proteins
c) Acquisition of *mecA* gene
d) All of these

75. Which of the following pneumococcal vaccines is not recommended for children under 2 years of age?
a) 7-valent conjugate vaccine
b) 23-valent polysaccharide vaccine
c) 13-valent conjugate vaccine
d) All of these

76. Which of the following is causative agent of pink eye?
a) *H. aegyptius* b) *H. ducreyi*
c) *H. influenzae* d) *H. parainfluenzae*

77. Brazilian purpuric fever is caused by:
a) *H. aegyptius* b) *H. aphrophilus*
c) *H. paraprophilus* d) *H. haemolyticus*

78. Pittman has classified *Haemophilus* into six types—a–f. This classification is based on:
a) Protein antigens b) Carbohydrate antigens
c) Capsular antigens d) Flagellar antigens

79. The culture medium that is not used for isolating *Haemophilus influenzae* is:
a) Chocolate agar
b) MacConkey's agar with a streak of *Staphylococcus aureus*
c) Levinthal's medium
d) Filde's agar

80. The species of *Haemophilus* that does not require both X and V factors for its growth is:
a) *H. influenzae* b) *H. aegypticus*
c) *H. ducreyi* d) *H. parainfluenzae*

81. *Haemophilus parainfluenzae* differs from *H. influenzae* in that it:
a) Requires only the V factor for growth
b) Does not require X or V factor
c) Requires only X factor and not V factor
d) Does not cause respiratory tract infections

82. The false statement about *H. influenzae* is:
a) Direct antigen detection is possible with CSF in meningitis
b) Cefotaxime is the drug of choice for *Haemophilus meningitis*
c) Characteristic iridescence can be demonstrated in Filde's agar
d) Capsule is formed of polyribosyl phosphate

83. The best specimen for the culture of *Bordetella* is:
a) Tracheal aspirate b) Sputum
c) Nasopharyngeal aspirate d) Endobronchial aspirate

84. The toxin responsible for the destruction of ciliated respiratory epithelial cells in whooping cough is:
a) Pertussis toxin
b) Tracheal cytotoxin
c) Filamentous hemagglutinin
d) Pertactin

85. The major component of acellular pertussis vaccine is:
a) Pertussis toxin
b) Tracheal cytotoxin
c) Filamentous hemagglutinin
d) Pertussis toxin promoter

86. The appearance of *Bordetella* in a smear resembles:
a) Palisading b) Cuneiform
c) Thumbprint d) Box car

87. The maximum infectivity in whooping cough is during:
a) Catarrhal stage b) Paroxysmal stage
c) Convalescent stage d) Acute stage

88. *B. pertussis* swabs may be transported in the following except:
a) Casamino acid solution
b) Stuart's medium
c) Mischulow's charcoal agar
d) Lacey's medium

89. All the following are true about *Bordetella pertussis* except:
a) Bipolar metachromatic granules can be demonstrated with toluidine-blue staining
b) Ability to grow on nutrient agar
c) The source of infection is always a patient
d) Confluent growth presents an aluminium paint appearance

90. The false statement about the pertussis vaccine is:
a) Adverse reactions are less with acellular vaccine
b) Hypotonic hyporesponsive episodes are a serious reaction to pertussis vaccination
c) Rough phase-1 strains are used for the vaccine
d) Encephalitis may occur as a complication

91. A four-month-old unimmunised child was brought with complaints of severe cough ending with a typical whoop for the past 10–12 days. The boy had apnea during the bouts, followed by vomiting, but was said to be sleeping comfortably between bouts. The culture of the nasopharyngeal swab shows typical colonies of *B. pertussis* on blood agar. The colonies are known as:
a) Bisected pearls colonies b) Draughtsman colonies
c) Pitting colonies d) Punched out colonies

92. Common cause of spurious haemoptysis is:
a) *E. coli* b) *Klebsiella pneumoniae*
c) *Serratia marcescens* d) *Enterobacter cloacae*

93. Which of the following is not a non-fermenter:
a) *Pseudomonas* b) *Acinetobacter* spp.
c) *Burkholderia* d) *Kingella*

94. Which of the following is true about non-fermenters?
a) Most non-fermenters utilise carbohydrates by the oxidative pathway
b) They are resistant to common antibiotics used in hospitals
c) *Acinetobacter* and *Pseudomonas* are both non-fermenters
d) All of these

95. Pseudomonads are motile by:
a) Polar flagella b) Lophotrichous flagella
c) Peritrichous flagella d) Amphitrichous flagella

96. Exotoxin A of *Pseudomonas aeruginosa* has a mechanism of action resembling:
a) Cholera toxin b) Diphtheria toxin
c) Staphylococcal exotoxin d) Listeriolysin

97. The pigment that is unique to *Pseudomonas aeruginosa* is:
a) Pyorubrin b) Pyoverdin
c) Pyocyanin d) Pyomelanin

98. Necrotising enteritis caused by *Pseudomonas* is called:
a) Shanghai fever b) Ecthyma gangrenosum
c) Erythema marginatum d) Malignant enteritis

99. The capsule of *Pseudomonas* is made up of:
a) Glycoprotein b) Hyaluronic acid
c) Glutamate d) Alginate

100. Tracheal aspirate of a patient with ventilator-associated pneumonia grew a non-fermenter in culture. The least likely organism will be:
a) *Pseudomonas aeruginosa*
b) *Acinetobacter baumannii*
c) *Burkholderia pseudomallei*
d) *Stenotrophomonas maltophilia*

101. All the following are virulence factors of *P. aeruginosa* except:
a) Exotoxin A b) Alginate
c) Pili d) Protein A

102. Most common agent causing respiratory infection in cystic fibrosis patients is:
a) *Pseudomonas aeruginosa*
b) *Burkholderia cepacia*
c) *Burkholderia mallei*
d) *Stenotrophomonas maltophilia*

103. The false statement about *Mycoplasma pneumoniae* is:
a) Sputum is not an ideal sample for culture
b) Treatment is with erythromycin

c) Diagnosed by serum cold antibody
d) Growth requires anaerobic environment

104. Urogenital infections may be caused by all except:
a) *Chlamydia. pecorum*
b) *Ureaplasma urealyticum*
c) *Mycoplasma hominis*
d) *Mycoplasma genitalium*

105. The cell wall of *Mycoplasma* is largely made up of:
a) Muramic acid b) Diaminopimelic acid
c) Sterols d) Lipoteichoic acid

106. Morphology of mycoplasma is:
a) Cocci c) Star shape
b) Rods d) All of these

107. Mycoplasmas are resistant to all the following except:
a) Tetracycline b) Penicillins
c) Cephalosporins d) Lysozymes

108. All the following are human pathogens except:
a) *Mycoplasma urealyticum*
b) *M. genitalium*
c) *M. hominis*
d) *M. pneumoniae*

109. All the following are requirements for mycoplasma culture medium except:
a) Serum b) Yeast extract
c) Penicillin d) Starch

110. Which of the following is not true of *Mycoplasma*?
a) They form poached-egg colonies
b) Colonies stain by Diene's method
c) Colonies can be seen by tetrazolium reduction test
d) No turbidity in liquid medium

111. The extracellular infective form of *Chlamydia* is:
a) Elementary body b) Inclusion body
c) Reticulate body d) Initial body

112. All the following conditions are associated with *Chlamydia trachomatis* except:
a) Trachoma b) Psittacosis
c) Infant pneumonia d) Inclusion conjunctivitis

113. All are methods used for staining *Chlamydia* except:
a) Castaneda b) Macchiavello
c) Gimenez d) Gomori

114. *Chlamydia* can be grown in the following sites except:
a) McCoy cells
b) HeLa cells
c) Chorioallantoic membrane
d) Yolk sac

115. TWAR agent is:
a) *C. trachomatis* b) *C. pneumoniae*
c) *C. psitacci* d) *C. pecorum*

116. Which of the following is true regrading Pontiac fever?
a) It is a milder, non-fatal, 'influenza-like' illness
b) It can cause outbreaks with high attack rates
c) It is caused by *Legionella pneumophila*
d) All of the above

117. The selective medium used for isolation of *Legionella pneumophila* is:
a) Pike's medium
b) Buffered charcoal yeast extract agar
c) Chocolate agar
d) Blood agar with colistin

118. Infection transmitted by aerosols from air conditioners, cooling towers, etc., are:
a) Legionnaires disease b) Listeriosis
c) Tularemia d) Donovaniosis

119. Coronaviruses are divided into four main subgroups based on their genomic structure. Of these, which are associated with human infections?
a) Alpha and beta b) Beta and gamma
c) Alpha and gamma d) Alpha and delta

120. Which animal is thought to be a reservoir of SARS-CoV2?
a) Bats b) Cats
c) Pangolin d) Man

121. The receptor-binding domain of the spike protein of SARS-CoV 2 binds to:
a) ACE2 receptors present in the lungs only
b) ACE1receptors present in the lungs only
c) ACE1 receptors present in the lungs and other organs such as the heart, kidneys, and gastrointestinal tract
d) ACE2 receptors present in the lungs and other organs such as the heart, kidneys, and gastrointestinal tract

122. Match the following COVID-19 vaccines with their principles correctly:

1. Covaxin	i. Genetically engineered mRNA
2. Novovax	ii. Recombinant adenovirus with a SARS-COV-2 spike protein gene
3. BNT162b2	iii. Fragments of spike protein subunit with adjuvant
4. Covishield	iv. Inactivated or live-attenuated

a) 1-iii, 2-ii, 3-i, 4-iv b) 1-iv, 2-iii, 3-i, 4-ii
c) 1-ii, 2-iii, 3-iv, 4-i d) 1-iv, 2-i, 3-ii, 4-iii

123. Which of the following clinical presentations is not seen in *Aspergillus* infection?
a) Otomycosis
b) Allergic bronchopulmonary aspergillosis
c) Zygomycosis
d) Mycotic keratitis

124. Detection of galactomannan in serum is diagnostic of:
a) Invasive aspergillosis b) Mucormycosis
c) *Candida* infection d) Cryptococcosis

125. A 56-year-old man with uncontrolled type 2 diabetes presented with complaints of chills, fever, cough, and sore throat. Her chest X-ray showed a left lower lobe consolidation and diagnosis of pneumonia was made. Culture of bronchoalveolar lavage in wet mount showed typical hyphae with dichotomous branching at acute angles. This is suggestive of:
a) Rhizopus b) Mucor
c) Aspergillus d) Cryptococcus

126. Zygomycosis is caused by:
a) *Aspergillus* and *Penicillium*
b) *Coccidiodis* and *Histoplasma*
c) *Mucor* and *Rhizopus*
d) *Candida* and *Cryptococcus*

127. High case fatality due to vascular invasion in an immunocompromised host is the characteristic feature of:
a) Candidiasis b) Mucormycosis
c) Blastomycosis d) Sporotrichosis

128. A 55-year-old uncontrolled diabetic man recovered after suffering from COVID-19 and was discharged from ICU. Three weeks later he developed fever, nausea, with tenderness and swelling of the eyelids. A diagnosis of rhino-orbito-cerebral mucormycosis was made. Wet preparation of the exudate is most likely to show:
a) Hyphae with dichotomous branching at acute angles
b) Broad ribbon-like aseptate hypha
c) Capsulated yeast cells
d) Non-capsulated yeast cells

129. Which of the following fungus is capsulated?
a) *Candida* b) *Rhodotorula*
c) *Pneumocystis* d) *Cryptococcus*

130. Which statement is not true regarding *Cryptococcus neoformans*?
a) Grows at 37°C
b) Grows on SDA

c) Possesses polysaccharide capsule
d) Negative urease test

131. Which of the following statement is true regarding morphology of *Pneumocystis jirovecii*?
a) Round or ovoid budding cell with a polysaccharide capsule
b) Round or ovoid budding cell, without capsule
c) Exists in three forms namely trophozoite, pre-cyst, and cyst
d) Both hyphal and yeast forms are seen

132. Which stain can be used to stain trophozoites of *Pneumocystis jirovecii*?
a) Lactophenol cotton blue b) Giemsa
c) India ink d) Gram

133. Which of the following is a dimorphic fungi?
a) Histoplasma capsulatum
b) Cryptococcus neoformans
c) Candida albicans
d) *Mucor* species

134. Dimorphic fungi are characterised by:
a) Hyphal forms at 37°C and yeast form at 25°C
b) Hyphal forms at 25°C and yeast form at 37°C
c) Both hyphal and yeast forms at 37°C
d) Both hyphal and yeast forms at 25°C

135. Tuberculate spores are characteristic of:
a) Cryptococcus neoformans
b) Candida albicans
c) *Mucor* species
d) Histoplasma capsulatum

136. Which fungus gives typical mariner's wheel or pilot's wheel appearance after the tissue is stained with methenamine silver stain?
a) *Histoplasma capsulatum*
b) *Blastomyces dermatitidis*
c) *Paracoccidioides brasiliensis*
d) *Coccidioides immitis*

137. All of the following fungi are dimorphic except:
a) *H. capsulatum* b) *B. dermatitidis*
c) *P. marneffei* d) *C. albicans*

138. South American blastomycosis is caused by:
a) *Histoplasma capsulatum*
b) *Blastomyces dermatitidis*
c) *Paracoccidioides brasiliensis*
d) *Coccidioides immitis*

139. A 50-year-old man living in Manipur regularly eats crab. He occasionally noticed small brownish specks in his sputum. Microscopic examination of these brown specs revealed brown coloured oval eggs with operculum. What is the most probable diagnosis?
a) Paragonimiasis b) Ascariasis
c) Echinococcosis d) Schistosomiasis

140. The definitive hosts in paragonimiasis is:
a) Snail b) Crab
c) Crayfish d) Man

141. The infective form of paragonimiasis in humans is:
a) Miracidium larva b) Cercaria larva
c) Metacercaria larva d) Sporocysts

142. Which statement best describes the parasite *P. westermani*?
a) The adult is oval, thick, fleshy and dark red
b) Sexes are separate
c) It can measure up to 1 meter in length
d) It has no suckers

143. Which of the following drugs is commonly used to treat paragonimiasis?
a) Praziquantel b) Albendazole
c) Metronidazole d) DEC

144. *Mycobacterium tuberculosis* complex includes all the following except:
a) *M. tuberculosis* b) *M. bovis*
c) *M. africanum* d) *M. intracellulare*

145. The virulence factor that helps in the persistence of mycobacteria in phagocytosis:
a) Lipid b) Lipoarabinomannan
c) Cord factor d) Cytoplasmic antigen

146. Decontamination of sputum prior to culture for mycobacteria is done using:
a) Sodium hydroxide b) Sodium chloride
c) Sodium citrate d) Sodium bicarbonate

147. The following methods are used for staining *M. tuberculosis* except:
a) Ziehl–Neelsen b) Kinyoun's
c) Seller's d) Auramine O

148. The lowest detection limit of culture is:
a) 10,000 b) 1,000-10,000
c) 100-1,000 d) 10-100 mycobacteria/mL

149. Identification of *M. tuberculosis* is confirmed by:
a) Growth on Lowenstein Jensen (LJ) medium
b) Growth on LJ with paranitrobenzoic acid
c) Growth in MGIT (mycobacteria growth indicator tube)
d) Growth on LJ followed by smear

150. Indications for the tuberculin test include the following except:
a) Diagnosis of active infection in all individuals
b) Diagnosis of latent TB infection
c) Studying the prevalence of infection
d) Studying the effect of vaccination

151. The true statement regarding interferon gamma release assay (IGRA) is:
a) It is the assay used for diagnosing active TB
b) It is an in vivo test
c) Antigens used are purified protein derivatives
d) In children, it is inferior to the tuberculin skin test

152. Which of the following is true about the BCG vaccine?
a) It is a live-attenuated vaccine derived from *M. tuberculosis.*
b) Protective efficacy of the vaccine is 80–90%.
c) The size of skin test reaction after vaccination predicts the degree of protection afforded.
d) There is decreased efficacy in populations with prior exposure to environmental mycobacteria.

153. The best sample for diagnosing urinary TB is:
a) Early morning mid-stream urine
b) Early morning whole voided sample
c) 24-hour urine
d) Early morning terminal urine sample

154. The molecular test that detects both rifampicin and INH resistance in *Mycobacterium tuberculosis* is:
a) MGIT b) BACTEC
c) CBNAAT d) Line probe assay

155. Among the anti-tuberculous drugs, the one which is bacteriostatic is:
a) Rifampicin b) Pyrazinamide
c) Ethambutol d) Streptomycin

156. The anti-TB drug given for chemoprophylaxis is:
a) Rifampicin b) Isoniazid
c) INH and rifampicin d) INH and ethambutol

157. MDR-TB is a *Mycobacterium tuberculosis* strain resistant to:
a) INH and rifampicin
b) INH and fluoroquinolone
c) INH, rifampicin and fluoroquinolone
d) INH, rifampicin, fluoroquinolone, and aminoglycoside.

158. A 65-year-old man in the US attended a camp in a forest area. Soon he developed cough, fever and fatigue. His blood sample grew fungus and the sputum showed plenty of spherules. Which of the following is likely to be associated in this case?
a) *Histoplasma capsulatum* b) *Candida albicans*
c) *Coccidioides immitis* d) *Cryptococcus neoformans*

159. The other name for Darling's disease is:
a) Histoplasmosis b) Sporotrichosis
c) Chromoblastomycosis d) Rhinosporidiosis

160. A 35-year-old woman developed pneumonia. She was seropositive for HIV, her chest X-ray showed interstitial lung infiltrates, and sputum examination revealed black cyst walls on methenamine silver stain. What is the most likely causative agent?
a) *P. jirovecii* b) *C. albicans*
c) *H. capsulatum* d) *B. dermatitidis*

161. A tourist from North America who is a known diabetic develops pneumonitis. What is the likely causative agent in this case?
a) *Candida albicans* b) *Blastomyces dermatitidis*
c) *Coccidioides immitis* d) *Paracoccidioides brasiliensis*

162. Which organism is associated with foul-smelling nasal discharge, atrophic changes in the nasal mucosa, and resorption of the nasal bone?
a) *Citrobacter freundii* b) *Pantoea agglomerans*
c) *Enterobacter cloacae* d) *Klebsiella ozaenae*

Answers: 1:b, 2:a, 3:b, 4:a, 5:a, 6:c, 7:d, 8:d, 9:c, 10:d, 11:b, 12:d, 13:a, 14:b, 15:b, 16:b, 17:b, 18:d, 19:c, 20:b, 21:c, 22:a, 23:c, 24:d, 25:d, 26:a, 27:d, 28:b, 29:b, 30:b, 31:a, 32:b, 33:c, 34:d, 35:c, 36:a, 37:c, 38:c, 39:b, 40:a, 41:d, 42:b, 43:b, 44:b, 45:b, 46:a, 47:b, 48:b, 49:a, 50:a, 51:b, 52:c, 53:b, 54:c, 55:d, 56:b, 57:a, 58:a, 59:c, 60:c, 61:b, 62:c, 63:b, 64:c, 65:d, 66:d, 67:d, 68:d, 69:d, 70:d, 71:a, 72:b, 73:d, 74:b, 75:c, 76:a, 77:a, 78:c, 79:c, 80:c, 81:a, 82:c, 83:c, 84:b, 85:a, 86:c, 87:a, 88:d, 89:b, 90:c, 91:a, 92:c, 93:d, 94:d, 95:a, 96:b, 97:c, 98:a, 99:d, 100:c, 101:d, 102:a, 103:d, 104:a, 105:c, 106:d, 107:a, 108:b, 109:d, 110:a, 111:a, 112:b, 113:d, 114:c, 115:b, 116:d, 117:b, 118:a, 119:a, 120:c, 121:d, 122:b, 123:c, 124:a, 125:c, 126:c, 127:b, 128:b, 129:d, 130:d, 131:c, 132:b, 133:a, 134:a, 135:d, 136:c, 137:d, 138:c, 139:a, 140:b, 141:c, 142:a, 143:a, 144:d, 145:b, 146:a, 147:c, 148:d, 149:b, 150:a, 151:d, 152:d, 153:b, 154:d, 155:c, 156:b, 157:a, 158:c, 159:a, 160:a, 161:b, 162:d

Central Nervous System Infections

(MI5.1–5.3)

Chapter Navigation

- Meningitis
- Encephalitis
- Brain abscess
- Acute bacterial (pyogenic) meningitis
  - Meningococcal meningitis—*Neisseria meningitidis*
  - Neonatal meningitis—*Streptococcus agalactiae* (Group B streptococci)
  - Listeriosis—*Listeria monocytogenes*
- Chronic bacterial meningitis
  - Tetanus—*Clostridium tetani*
  - Botulism—*Clostridium botulinum*
  - Aseptic viral meningitis
  - Enteroviral infections
    - Poliomyelitis—Polio virus
    - Coxsackievirus diseases—Coxsackie viruses A and B
    - Echoviral diseases—Echoviruses
  - Rabies—Rabies virus
- Slow viral diseases
  - Subacute sclerosing panencephalitis (SSPE)
  - Progressive multifocal leukoencephalopathy (PML)
  - Prion diseases
- Arboviral diseases
  - Encephalitis
    - Japanese encephalitis (JE) virus
    - Tick-borne encephalitis—Russian spring–summer encephalitis (RSSE) complex
    - West Nile virus
    - Zika virus
- Fever with arthritis—Chikungunya virus
- Parasitic CNS infections
  - Taeniasis—*Taenia solium*
  - Neurocysticercosis (NCC)
  - Primary amoebic meningoencephalitis (PAM)—*Naegleria fowleri*
  - Granulomatous amoebic encephalitis (GAE)—*Acanthamoeba* species
  - Toxoplasmosis—*Toxoplasma gondii*
  - Trypanosomiasis—*Trypanosoma*
- Fungal CNS infections
  - Candidiasis—*Candida*
  - Cryptococcal meningitis—*Cryptococcus*
  - Histoplasmosis—*Histoplasma*
  - Microsporidiosis—*Encephalitozoon*

Learning objective

- Describe the etiopathogenesis and clinical course and discuss the laboratory diagnosis of infections of the central nervous system, including meningitis and encephalitis

Central nervous system (CNS) infections are common medical emergencies. They may first present with non-specific prodromal symptoms such as fever, headache, and lassitude and progress to acute neurological manifestations like seizure and other focal features.

- These infections affect all age groups. However, they are of special concern in neonates, immunocompromised hosts, and the elderly, often requiring emergency intervention.
- All forms of CNS inflammation are serious and can lead to blindness, deafness, coma, and death. Prompt diagnosis and appropriate management are the key to reducing morbidity and mortality associated with CNS infections.

Table 24.1 presents a brief outline of infections of the central nervous system.

MENINGITIS

Case 1. *A five-year-old child was brought to the emergency department with a history of seizures following high-grade fever with 1–2 episodes of projectile vomiting. On examination, he had altered sensorium, and neck rigidity with a* ***positive Kernig's sign****. A lumbar puncture was performed after fundus examination for CSF cytology and biochemical examination. It showed increased polymorphs with low glucose and raised proteins. A Gram smear showed the presence of gram-positive diplococci.*

i. Enumerate the etiological agents of meningitis.
ii. Describe the laboratory diagnosis of meningitis.

Table 24.1 Outline of infections of CNS

Infections of the CNS nervous system	Clinical definition	Most common etiological agent
Meningitis	Inflammation of the meninges	Acute: a) Pyogenic—bacterial origin b) Aseptic—viral origin Chronic: Tubercular or fungal
Encephalitis	An inflammation of the brain tissue	Usually viral
Meningoencephalitis	Inflammation affecting both the meninges and the brain tissue	Free-living amoebae, bacteria, and viruses
Brain abscess and subdural empyema	Focal infections in the brain	Bacteria

- **Meningitis** is inflammation of the meninges and is a life-threatening condition.
- It is an inflammatory response elicited in the meninges by a microorganism (usually a bacterium or virus) that enters the subarachnoid space.
- It is the most frequently encountered CNS infection.
- Prior to the advent of childhood vaccination, bacterial meningitis was a common cause of morbidity and mortality in neonates and infants.

Predisposing conditions associated with meningitis:

i. Sinusitis (pneumococcal)
ii. CSF rhinorrhea
iii. Chronic alcoholism
iv. Diabetes
v. Hypogammaglobulinemia
vi. Splenectomy
vii. Trauma to the skull, including fracture of the base of the skull

- Meningitis may be classified as (i) acute or (ii) chronic depending on the speed of the initial presentation and subsequent progress.
- Most bacteria cause acute pyogenic meningitis, while *Mycobacterium* and *Cryptococcus* produce chronic meningitis.
- The etiological agents of meningitis can be bacterial (pyogenic or acute bacterial meningitis), fungi, or viruses (aseptic meningitis). These agents are listed in Table 24.2.

Pathogenesis (Fig. 24.1)

Bacterial meningitis is usually preceded by the following events:

- **Nasopharyngeal or middle ear colonisation**, leading to microbial invasion of the tissue and intravascular space.
- **Bacteremia,** following which, the infection is disseminated via the bloodstream.

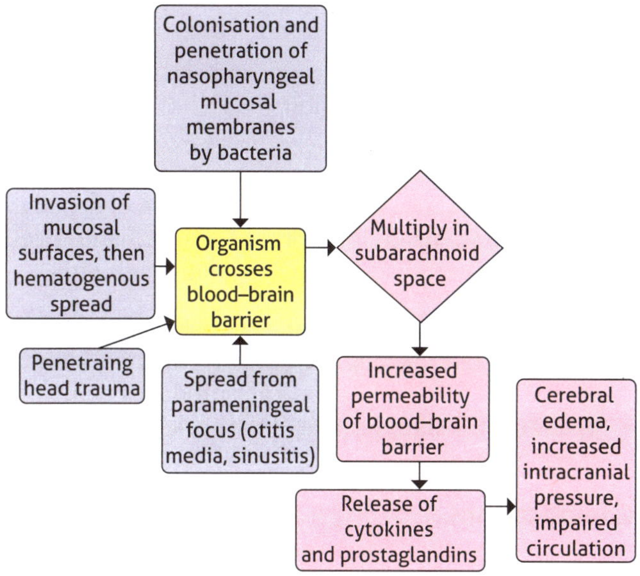

Fig. 24.1 Pathogenesis of meningitis

Table 24.2 Microorganisms that cause meningitis

Bacteria	Viruses (aseptic meningitis)	Fungi	Parasites
In neonates • *E. coli* • Group B streptococci (*Streptococcus agalactiae*) • *Haemophilus influenzae* • *Streptococcus pneumoniae* • *Staphylococcus aureus* • *Klebsiella* species • *Listeria monocytogenes* • *Mycobacterium tuberculosis*	• Enteroviruses (ECHO, coxsackie, polio) • Arboviruses (flavivirus, bunyavirus) • Paramyxoviruses (mumps, measles) • Herpes viruses (herpes simplex, varicella–zoster) • Adenoviruses Affects all ages	• *Cryptococcus neoformans* • *Histoplasma capsulatum* • *Coccidioides immitis* • *Candida albicans* • *Aspergillus* species	• *Naegleria* • *Acanthamoeba* • *Entamoeba histolytica* • *Toxoplasma gondii* Affects the immunonocompromised and swimmers exposed to contaminated water
In children (1 month onwards) • *Haemophilus influenzae* • *Neisseria meningitidis* • *Streptococcus pneumoniae* • *Listeria monocytogenes*	Arboviral infections may occur as epidemics and affect infants and young children		
In adults • *Neisseria meningitidis* • *Streptococcus pneumoniae*			
In the elderly • *Streptococcus pneumoniae* • *Staphylococcus aureus* • *Listeria monocytogenes* • Gram-negative enteric bacilli			

- On autopsy, the **brain parenchyma** shows features of dilated cerebral veins associated with the thick, **purulent exudate within the leptomeninges**.
- There may be cerebral infarcts, hemorrhages, abscesses, and inflammatory cell infiltrations (Fig. 24.2a and b).

Clinical Features

Symptoms

- Fever and severe headache
- Photophobia
- Altered sensorium, confusion, and sleepiness
- Neck rigidity and confusion
- Nausea and vomiting
- In neonates and infants—seizures or altered mental symptoms irritability, drowsiness, and lack of appetite; the **classical triad of fever, headache, and neck rigidity characterises acute pyogenic meningitis**

Signs of meningitis

- **Kernig's sign**: Flexing the patient's hip and knee at 90° and then extending the knee produces pain.
- **Brudzinski's sign**: Flexion of the neck causes flexion of the knees.

Complications

About 20% patients with bacterial meningitis may have long-lasting after-effects. The complications are as follows:

- **Hydrocephalus**
- **Arterial or venous infarcts** due to inflammation and thrombosis of arteries and veins in superficial and, sometimes, deep areas of the brain
- **Abducens palsy** due to compression or inflammation of the 6th cranial nerve
- **Deafness** due to inflammation of the 8th cranial nerve or structures in the middle ear

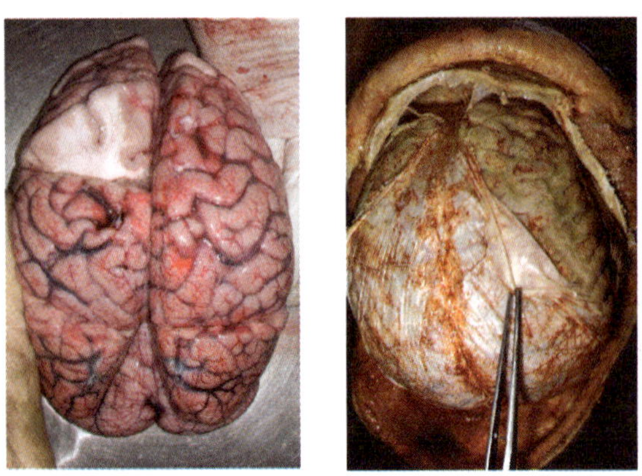

(a) (b)

Fig. 24.2 (a) A normal human brain removed during an autopsy and (b) the brain of a patient who died from bacterial meningitis—hemorrhage and pus under the dura mater can be seen (*Source*: https://openstax.org/books/microbiology/pages/26-2-bacterial-diseases-of-the-nervous-system)

- **Subdural empyema**
- **Increased intracranial pressure** due to cerebral edema
- **Brain abscess**
- **Brain herniation**
- **Systemic complications** such as septic shock, disseminated intravascular coagulation (DIC), or hyponatremia

Laboratory Diagnosis

An algorithm of the approach to the diagnosis of meningitis is presented in Fig. 24.3.

Samples

- Cerebrospinal fluid (CSF)
- Blood for culture and viral serology
- Urine (for antigen detection)

Collection and transport of specimens

- CSF is obtained by making a lumbar **puncture at the L3–L4 interspace**.
- **Lumbar puncture is NOT performed if there is a rise in intracranial pressure.**
- Under strict aseptic conditions, CSF is collected in three sterile containers, one each for (i) culture (ii) cell count, and (iii) biochemical examination.
- CSF samples collected for culture must not be refrigerated as the fastidious organisms are temperature-sensitive.
- For virus isolation, the sample must be transported on ice; for bacterial culture, it should be transported at room temperature.
- If there is a delay, the CSF sample should be kept in an incubator at 37°C and not in a refrigerator.

Routine (non-microbiological) tests

- The appearance of the CSF, total WBC and lymphocyte count, and protein and glucose concentrations are noted (Table 24.3).

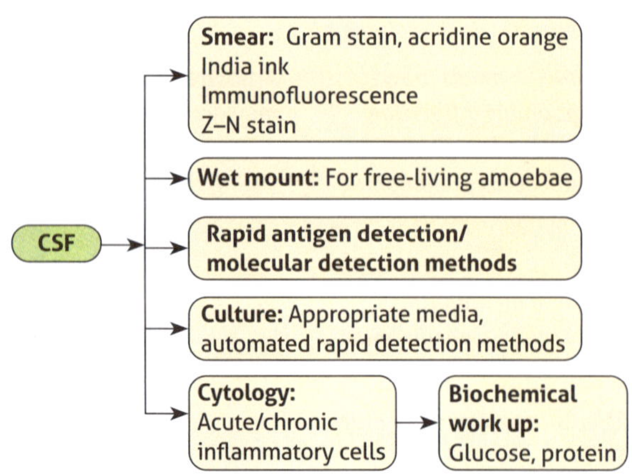

Fig. 24.3 Algorithm of the approach to the diagnosis of meningitis

Table 24.3 Interpretation of microscopic and biochemical assays of CSF

Condition	White cell count		Biochemistry	
	Neutrophils ($\times 10^6$/L)	Lymphocytes ($\times 10^6$/L)	Protein (g/L)	Glucose (CSF:blood ratio)
Normal	0	≤5	<0.4	≥0.6 (or ≥ 2.5 mmol/L) (>1 month of age)
Normal neonate	0*	<20	<1.0	≥0.6 (or ≥ 2.5 mmol/L)
Bacterial meningitis	100–10,000 (but may be normal)	Usually <100	>1.0 (but may be normal)	<0.4 (but may be normal)
Viral or aseptic meningitis	Usually <100 (but may be normal)	10–1,000 (but may be normal)	0.4–1	Usually normal
TB meningitis	Usually <100 (but may be normal)	50–1,000 (but may be normal)	1–5 (but may be normal)	<0.3

* In neonates, there may sometimes be up to 5% neutrophils in the CSF in the absence of meningitis

- In bacterial meningitis, CSF appears turbid with a large number of pus cells

Examination of CSF (Table 24.3)

The CSF is centrifuged, and the deposit is used for microscopy and culture while the supernatant can be used for antigen detection.

Microscopic examination

- A **heaped drop** (one over the other) of the supernatant is placed on a dry, grease-free slide without spreading.
- **Staining**
  - *Gram stain*: May show gram-positive and gram-negative bacteria and yeast cells. Meningococci are often seen extracellularly but a few are present intracellularly in the polymorphs.
  - *Acridine orange*: May show brightly fluorescing microorganisms.
  - *Ziehl–Neelsen stain*: Tubercle bacilli may be seen.
  - *India ink preparation*: Capsulated budding yeast cells of *Cryptococcus neoformans* (Fig. 24.4) or capsulated pneumococci may be seen.

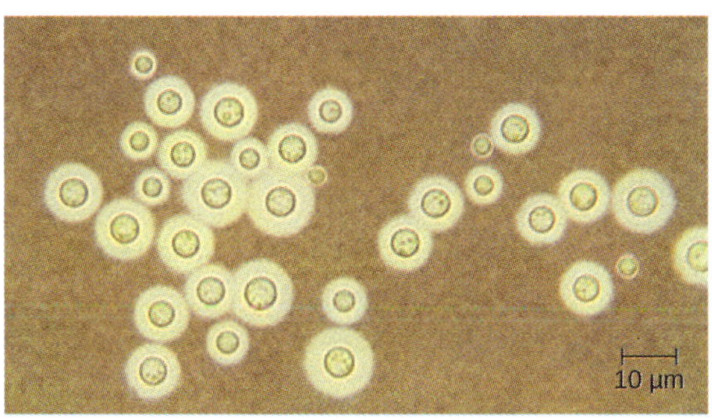

Fig. 24.4 India ink staining to demonstrate capsulated budding yeast cells of *Cryptococcus neoformans* in CSF (*Source*: https://openstax.org/books/microbiology/pages/26-4-fungal-and-parasitic-diseases-of-the-nervous-system)

- **Unstained wet mount**
  - Motile, free-living amoebae may be seen (Fig. 24.5).
  - Inflammatory cells and fungal elements are seen.
  - Bacteria in pairs/chains may be visualised.

Antigen detection

The supernatant may be used to demonstrate antigen by **latex agglutination** using meningococcal antisera.

- Commercially available kits containing antisera of *S. pneumoniae*, *H. influenzae* type B, and group B streptococci are available for rapid diagnosis.
- Antigen detection is particularly useful in partially treated patients in whom smear and culture tests may be negative.

Culture

Bacterial culture is done on the following media:

- **Chocolate and blood agar**
- **Enriched liquid medium**—BHI broth or an automated blood culture bottle (both aerobic and anaerobic)

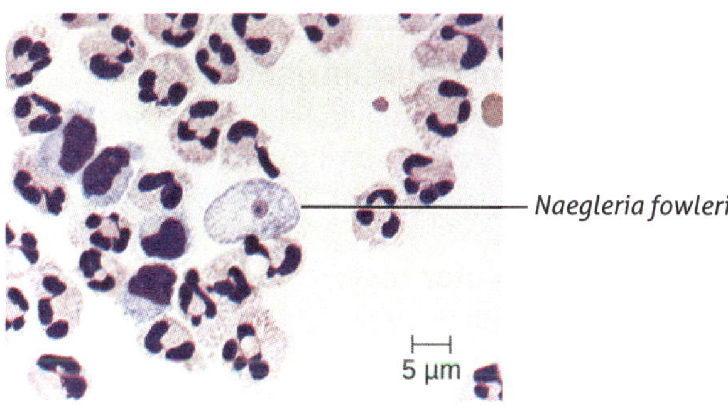

Fig. 24.5 *Naegleria fowleri* in human brain tissue collected from a patient suffering from primary amoebic meningoencephalitis (*Source*: https://openstax.org/books/microbiology/pages/26-4-fungal-and-parasitic-diseases-of-the-nervous-system)

- **Fungal culture is done on:** Sabouraud dextrose agar/broth or brain–heart infusion agar—two sets of media are inoculated and incubated at 25°C and 37°C respectively
- **Birdseed agar**—for the culture of *Cryptococcus* species

- A **non-nutrient agar medium** is used for the culture of free-living amoebae.
- The **automated *Mycobacterium* culture system** is used to cultivate *M. tuberculosis.*
- Viral culture is performed via **tissue culture, egg inoculation, or animal inoculation**.

However, these methods are **time-consuming and cumbersome**; instead, molecular methods like conventional and real-time PCR are performed.

Identification and confirmation

- This may be done by traditional biochemical tests or automated systems such as **VITEK or MALDI-TOF.**
- Antibiotic sensitivity testing is performed on confirmed isolates.

Serology

- **Antigens** can be detected using latex agglutination tests for *S. pneumoniae, H. influenzae,* and *N. meningitidis.*
- In neonates, antigens of group B streptococci (*S. agalactiae*) and *Listeria monocytogenes* are looked for.
- **Antibody detection**
  - Serological tests to detect antigens/antibodies are the mainstay of the diagnosis of viral meningitis, e.g., tests to detect IgM antibodies or glycoprotein antigens of different viruses.
  - **IgM ELISA** is the preferred test as it detects IgM, which is an acute phase antibody.
- **Detection of bacterial endotoxin**
  - The ***Limulus* lysate test** is extremely sensitive for the detection of bacterial endotoxin.
  - This test depends on the coagulation of horseshoe crab (*Limulus polyphemus*) amoebocytes (blood cells) with blood containing endotoxin.

Tests to detect microbial antigens in urine

Tests are available to detect *S. pneumoniae* (immunochromatography-based) and *Cryptococcus neoformans* (latex agglutination-based) antigens in urine.

PCR-based molecular tests

- Nucleic acids can be detected in CSF by PCR or real-time PCR.
- PCR-based tests can also be performed on blood samples if CSF-based tests do not yield results.
- Rapid molecular platforms such as cartridge-based RT-CPR can be used to detect *M. tuberculosis.*
- Syndromic panels (such as film array-based systems) or multiplex PCR can detect multiple pathogens.

Treatment

- **Empiric treatment and supportive management** are started as soon as a provisional diagnosis based on clinical presentation is made.
- A broad-spectrum antibiotic is started promptly if bacterial etiology is suspected (Table 24.4).
- Antibiotics must be given parenterally.
- Only antibiotics that can cross the blood–brain barrier should be used.
- For **viral meningitis and encephalitis**, supportive therapy and antivirals such as acyclovir IV are given.
- **Malaria (cerebral) and encephalitis** due to free-living amoebae are treated with chloroquine or natamycin and metronidazole respectively.
- **Fungal meningitis** due to *Cryptococcus* is treated with liposomal amphotericin B.

Table 24.4 Antibacterial therapy in case of meningitis

Age	First-line therapy	Alternate agents
Neonates	Meropenem and vancomycin	Cefotaxime and gentamicin
Children	Ceftriaxone and vancomycin	Cefotaxime and vancomycin
Adults	Ampicillin, ceftriaxone, and vancomycin	Meropenem and vancomycin

ENCEPHALITIS

The variable symptoms of encephalitis make it difficult to make a provisional diagnosis. The encephalitis may be acute or chronic encephalitis, subacute sclerosing panencephalitis (SSPE), or meningoencephalitis.

Acute Encephalitis

- This is an **acute inflammation of the brain tissue.**
- The etiology may be infection or immunological (Table 24.5).

Table 24.5 Some of the common microbial agents causing encephalitis

Type of encephalitis	Agents
Viral	Rabies virus, herpes virus, measles virus, Epstein–Barr virus Arbovirus group—JE, West Nile, western equine encephalitis, eastern equine encephalitis
Bacterial	*Bacillus anthracis*
Parasitic	*Acanthamoeba, Naegleria,* and *Balamuthia*

Subacute Sclerosing Panencephalitis (SSPE)

- This is a non-permissive infection by the **measles virus**. The disease may occur **after measles** in 1 in 100,000–500,000 cases.
- It is a **progressive, demyelinating disease**.
- The majority of patients are between the ages of 5 and 23 years of age with a history of contracting measles at an early age (around 2 years).
- **Symptoms:** Patients exhibit intellectual deterioration, generalised seizures, and ataxia and progressively become quadriplegic and spastic.

Chronic Encephalitis

- It is caused by **progressive multifocal leukoencephalopathy (PML).**
- It is a slowly progressing demyelinating disease of the brain affecting all areas except the spinal cord and optic nerves.
- It is caused by the **Jakob Creutzfeldt (JC) virus.**
- Associations have also been with AIDS and other immunosuppressive disorders.

Meningoencephalitis

- This is an **inflammation of the meninges and the brain tissue.**
- **Symptoms**: Patients present with fever, headache, photophobia, altered sensorium, behavioural changes, and lowered consciousness. Later, focal to generalised seizures may occur.

Laboratory Diagnosis

- **Cerebrospinal fluid examination**: Elevated protein and white blood cells are found on CSF examination. Viral agents may be detected by specific amplification methods by conventional PCR or RT-PCR.
- **Neuroimaging by MRI and CT**: Focal or diffused encephalitic lesions are seen.

Treatment

- There are limited treatment options available.
- Early laboratory confirmation helps in the treatment of herpes encephalitis by acyclovir (10 mg/kg IV every 8 hours by slow infusion in adult patients).
- Management is supportive with the maintenance of fluids and respiratory airway.

BRAIN ABSCESS

> **Case 2.** *A 65-year-old woman who had been a diabetic for several years presented to the neurosurgery OPD with symptoms of focal seizures. Computed tomography of her brain showed a 'space-occupying' lesion in the left cerebral hemisphere with 'ring-enhancement', which was consistent with a brain abscess. The pus was sent in a sealed gassed-out vial filled with carbon dioxide. Microscopic examination showed the presence of gram-positive cocci and a few fine gram-negative bacilli. Culture on anaerobic blood agar incubated in GasPak was positive for* ***peptostreptococci and Bacteroides fragilis****. Surgical drainage was done, and she was treated with cefotaxime and metronidazole, to which she responded.*
> i. What are the common causes of brain abscess?
> ii. Which organisms are responsible for brain abscess?

A **brain abscess** is a localised collection of pus in the brain parenchyma caused by a suppurative lesion. It may occur due to:

i. **Trauma** or any procedure within the cranium (commonly caused by MRSA, *Pseudomonas,* and other gram-negative bacteria)
ii. **Spread from a focus** in the paranasal sinus, otitis media, mastoiditis, or a dental infection (*Pseudomonas*, *Bacteroides*, *Haemophilus)*
iii. **Hematogenous spread** from any other focus in the body *(Staphylococcus aureus*, viridans group of streptococci)

The common bacterial causes of brain abscesses are the following:

- Anaerobes—*Bacteroides* and *Fusobacterium* species
- *Pseudomonas* species
- Enterobacteriaceae
- *Haemophilus* species
- *Staphylococcus* and viridans streptococci (may follow infective endocarditis)

Clinical Features

A brain abscess presents as a space-occupying lesion with a triad comprising the following:

i. **Focal neurological symptoms** such as paralysis, loss of muscle tone, and weakness or convulsions
ii. **Fever**
iii. **Headache,** which is a hemicranial, dull ache, which later becomes severe and is refractile to treatment

Other features depend on the location of the abscess.

Laboratory Diagnosis

- Neuroimaging techniques like **MRI and CT scans** are used for diagnosis.
- Specimens obtained by **CT-guided stereotactic needle aspiration of the purulent material** are stained with Gram stain and cultured (in aerobic and anaerobic media) to confirm the etiological agent.
- Blood culture may help in a small percentage of cases.
- **Lumbar puncture must not be performed** when there is a suspicion of brain abscess.

Management

- Broad-spectrum antibiotics are prescribed to cover both aerobic and anaerobic organisms.

- Neurosurgical drainage of the abscess and other supportive therapy constitutes the management of brain abscess.

Other localised infections of the brain include the following:

- **Neurocysticercosis** (caused by *Cysticercus cellulosae*)
- **Toxoplasmosis** (*Toxoplasma gondii*)
- **Rhinocerebral mucormycosis/aspergillosis** (*Mucor, Rhizopus, Aspergillus* species; dealt with in Chapter 23)
- Infections due to the microsporidium of *Encephalitozoon* (*E. cuniculi*)

MICROORGANISMS ASSOCIATED WITH CENTRAL NERVOUS SYSTEM INFECTIONS

ACUTE BACTERIAL (PYOGENIC) MENINGITIS

- Acute bacterial meningitis is a rapidly progressive bacterial infection of the meninges and subarachnoid space.
- Most bacteria cause acute pyogenic meningitis (Table 24.1).
- Some bacterium are described as per chapter numbers given below:
  - *Streptococcus pneumoniae* and *Haemophilus influenzae* (Chapter 23)
  - *Pseudomonas* and *Klebsiella* (Chapter 23)
  - Other Enterobacteriaceae (Chapter 25)

Meningococcal Meningitis

Case 3. *A seven-year-old boy presented to the emergency department with high-grade fever and a headache for two days and a single episode of convulsion and projectile vomiting the previous day. On examination, he was found to be disoriented, and his is neck rigid. The boy tested positive for the* ***Kernig's sign*** *and had a non-blanching petechial rash on his back.*

Following funduscopic examination, a lumbar puncture was performed along with a complete blood count and serum biochemistry. Cytology yielded the following results: polymorphs—1,000/mm3, protein—250 mg/dL, and glucose—20 mg/dL. His blood glucose levels were normal. On Gram stain, gram-negative cocci were seen in pairs, most of them intracellular with plenty of polymorphs. Latex agglutination test was positive for ***N. meningitidis*** *antigen, confirming the diagnosis of* ***meningococcal meningitis****. The CSF culture was sent to the laboratory for the isolation of the organism and antibiotic sensitivity testing. Empirically, ceftriaxone 100 mg/kg IV was started. Vaccination was advised for the patient's siblings along with ciprofloxacin chemoprophylaxis.*

i. What is the pathogenesis of meningococcal disease?
ii. How is it diagnosed?

Meningococcal disease can present as:

- Meningitis
- Meningococcal septicemia
- Nasopharyngeal carrier state (usually asymptomatic)

Epidemiology

- Meningitis is **common in children** between the ages of three months and five years and in the young adult population with a **male preponderance**.
- Epidemics usually occur in semi-closed communities living in crowded conditions.
- Meningococcal disease outbreaks have been documented periodically in India and have been found to affect adolescents and **young adults** in addition to the pediatric population.
- Transmission is essentially **airborne**—by droplets—or less often, by **fomites**.
- Natural infection by meningococci is **limited to human beings**.
- The human nasopharynx is the only reservoir of meningococci; there is no animal reservoir.
- Asymptomatic nasopharyngeal **carriers rarely contract the illness** but serve to infect their contacts.
- During inter-epidemic periods, the carrier rate is about 5–10 per cent.
- An increase in carrier rate heralds the onset of an epidemic. During epidemics, the carrier rates in closed communities may go up to 90 per cent.

Agent—*Neisseria meningitidis*

- *Neisseria meningitidis* (meningococcus) was first described and isolated in 1887 by **Anton Weichselbaum** from the spinal fluid of a patient.
- Meningococci have been associated in several outbreaks of meningitis and meningococcal disease in both young children and adults.

Morphology

- Meningococci are **gram-negative, oval or spherical cocci** which are typically **arranged in pairs with the adjacent sides flattened** (Fig. 24.6).
- In smears from lesions, the cocci are more regular and generally intracellular.

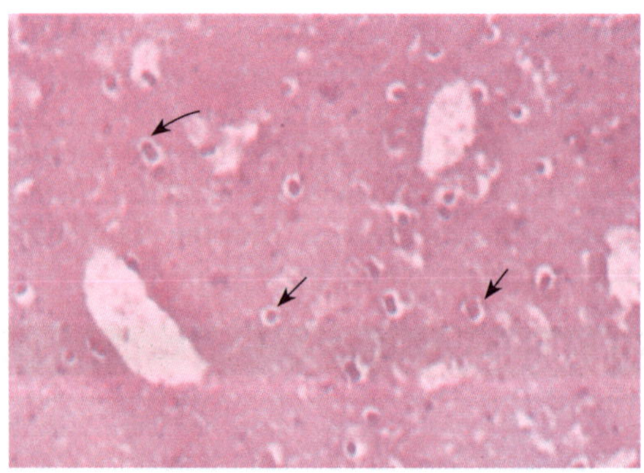

Fig. 24.6 *N. meningitidis* in cerebrospinal fluid—intracellular (within a polymorph), gram-negative diplococci with a large number of polymorphs (*Source*: Department of Microbiology, Pondicherry Institute of Medical Sciences, Puducherry)

Antigenic properties and classification

Serogroups

- Based on their **capsular polysaccharide antigens**, meningococci are classified into **13 serogroups**, of which groups A, B, C, X, Y, and W-135 are the most important.
- **Group A** is usually associated with epidemics and **group C** with localised outbreaks; **group B** causes both epidemics and outbreaks.
- **Groups 29-E, W-135, and Y** also frequently cause meningitis.
- Any serogroup may colonise the nasopharynx, but these six groups account for most cases of meningitis.

Serotypes

- Serogroups are further classified into serotypes and subtypes based on the **outer membrane proteins**.
- **Twenty** such serotypes have been identified.

Virulence factors

Meningococci have four virulence factors:

- **IgA protease**—which helps in binding to the mucous membrane of the upper respiratory tract by cleaving local IgA.
- **Factor H-binding protein**—which binds to factor H—a C3b inhibitor that impairs complement activation.
- **Endotoxin—**which causes fever and shock.
- **Capsular polysaccharide**—which resists phagocytosis.

Pathogenesis

- The endotoxin lipopolysaccharide released by autolysis causes injury to vascular endothelium.
- All major inflammatory cascade systems as well as cytokines and nitric oxide are triggered and upregulated. This results in a **systemic inflammatory response** syndrome, leading to multiorgan failure.
- **Vascular microthrombotic disease** occurs, causing adrenal hemorrhage and leading to profound shock and death.

Clinical spectrum of meningococcal infections

Meningitis

- Case fatality is variable but may be **as high as 80 per cent** in untreated cases.
- Survivors may experience sequelae such as blindness and deafness.
- Some cases develop **chronic or recurrent meningitis**.

Meningococcemia

- Presents as acute fever with chills, malaise, and prostration.
- A **petechial rash** occurs early in the disease. Meningococci may be isolated from the petechial lesions.
- Metastatic involvement of the joints, ears, eyes, lungs, and adrenals may occur.
- About 10 per cent develop pneumonia.

Fulminant meningococcemia

- This was formerly called **Waterhouse–Friderichsen syndrome**.
- It is a serious and usually fatal condition.
- It is characterised by high fever, shock, disseminated intravascular coagulation, and multisystem failure.

Complement deficiency (C5–C9) predisposes to meningococcal disease.

Laboratory diagnosis

Specimens

- Cerebrospinal fluid
- Blood
- Nasopharyngeal swabs (in carriers)
- Skin scrapings from petechial lesions in meningococcemia

Microscopy

- Meningococci are often seen extracellularly, but a few are present intracellularly in the polymorphs.
- Meningococci may sometimes be demonstrated in aspirates from petechial lesions by microscopy and culture.

Antigen detection

- Antigen detection is performed by latex agglutination using meningococcal antisera in CSF supernatant.
- Commercially available kits containing antisera of *S. pneumoniae*, *H. influenzae* type B, and group B streptococci are available for rapid diagnosis.
- Antigen detection is particularly useful in partially treated patients in whom smear and culture tests may be negative.

Culture

- In meningococcemia and early cases of meningitis, blood culture is often positive.
- Cultures should be incubated under 5–10% carbon dioxide. Meningococci do not grow in anerobic conditions.
- Meningococci have exacting growth requirements and do not grow on ordinary media. Growth occurs on media enriched with blood or serum (blood or chocolate agar).
- Selective media such as the modified Thayer–Martin medium (contains vancomycin, colistin, and nystatin) are used for skin or nasopharyngeal swabs in cases of meningococcemia.

Identification

- Meningococci must be differentiated from morphologically similar organisms such as *N. flavescens*, *N. flava*, and *Acinetobacter.*

- Rapid identification of suspected meningococcus isolates can be made by the VITEK systems, which also provide the MIC values in the antibiotic susceptibility panel.
- The isolated meningococcus is serogrouped, if required, by agglutination with the appropriate antisera.

Serology

Meningococcal antigens can be found in the blood in active disease.

Biochemical tests

- Conventional methods to identify meningococci are based on biochemical tests based on carbohydrate fermentations.
- Meningococci are catalase- and oxidase-positive; they can utilise glucose and maltose, but not sucrose and lactose.
- Identification can be confirmed by automated systems.

Molecular diagnosis

Group-specific diagnosis of infection can be made based on the detection of meningococcal DNA sequence in CSF or blood by PCR amplification.

> **Laboratory diagnosis of meningococcal infections**
>
> - **Specimens:**
>   - Cerebrospinal fluid
>   - Blood
>   - Nasopharyngeal swabs (in carriers)
>   - Skin scrapings from petechial lesions in meningococcemia
> - **Microscopy:**
>   - Gram stain: Gram positive diplococci often intracellularly in polymorphs
> - **Antigen detection:** Latex agglutination in CSF supernatant, urine, and blood
> - **Culture:**
>   - Aerobic, fastidious capnophilic, slow growing; optimum temperature: 37°C, grows on blood and chocolate agar
>   - Colonies: Greyish, nonhemolytic, round, convex, smooth, moist, and glistening with a clearly defined edge
> - **Identification:**
>   - Meningococci must be differentiated from morphologically similar organisms such as *N. flavescens*, *N. flava*, and *Acinetobacter*
>   - Catalase- and oxidase-positive; can utilise glucose and maltose, but not sucrose and lactose, can be confirmed by automated systems
>   - Serogrouping by agglutination with the appropriate antisera
> - **Serology:** Meningococcal antibody detection has no value
> - **Molecular diagnosis:** Detection of meningococcal DNA in CSF or blood by PCR

Treatment

- The drug of choice is **intravenous penicillin G.**
- Alternately, in persons who are hypersensitive to penicillin, ampicillin, ceftriaxone, or cefotaxime can be used.
- Ciprofloxacin or ceftriaxone may be used to decolonise a meningococcal carrier.

Prophylaxis

- **Ciprofloxacin** is recommended for chemoprophylaxis.
- As infection rates are very high in the household or among close contacts of meningococcal patients, they should be provided with chemoprophylaxis.
- Immunity following meningococcal infection is serogroup-specific, so vaccines are also serogroup-specific.

Vaccines

i. Polysaccharide vaccines

- Either bivalent (serogroups A and C), trivalent (A, C, and W), or tetravalent (A, C, Y, and W-135)
- Not effective before two years of age
- Protect for three years but do not induce herd immunity

ii. Conjugate vaccines

These are specific polysaccharides conjugated with a protein carrier. They can be of the following types:

- **Monovalent C**
- **Monovalent A**—Available in Africa and has potential to control outbreaks caused by serogroup A
- **Tetravalent** (serogroups A, C, Y, and W)
- **Outer membrane vesicle (OMV) vaccines—**Some countries like Cuba, New Zealand, and Norway recommend the OMV vaccines, which are parenteral licensed vaccines against meningococcal B disease

Neonatal Meningitis

Neonatal meningitis is a potentially devastating condition with dire long-term consequences. The immature immune system of neonates, especially preterm neonates, puts them at a high risk of contracting bacterial meningitis. Infection in the newborn is classified as follows:

i. **Early-onset type**—It occurs within a week of birth. This type, which is more common, presents as meningitis or septicemia and is often fatal. Infection is acquired from the maternal vagina during birth.

ii. **Late-onset type**—It develops between the second and twelfth weeks of life. Infection is often acquired from the environment and presents more often as septicemia and not meningitis.

Bacteria responsible for neonatal meningitis are presented in Table 24.6.

Table 24.6 Bacteria causing early- and late-onset meningitis in neonates

Early-onset meningitis	Late-onset meningitis
Streptococcus agalactiae (GBS)	*Escherichia coli* Coagulase-negative *Staphylococcus*
Escherichia coli	*Streptococcus agalactiae* (GBS) *Staphylococcus aureus*
Listeria monocytogenes	*Listeria monocytogenes*
Enterococcus species	*Streptococcus pneumoniae*
Gram-negative enteric bacteria	*Neisseria meningitidis* *Candida* species
Streptococcus pneumoniae	*Enterococcus* species Gram-negative enteric bacteria

Agent—*Streptococcus agalactiae* (Group B streptococci)

- It is a gram-positive bacteria that colonises the gastrointestinal and genitourinary tract. **Colonisation of the maternal genital tract by group B streptococci** during labour is a major risk factor for infection in the newborn.
- It is also an important pathogen of cattle, producing **bovine mastitis.**
- Nine capsular serotypes have been identified, antibodies to which confer type-specific protection.
- Human pathogenic group B strains possess a **polysaccharide capsule** which appears to confer virulence.

Clinical spectrum

Human infections include the following:

- **Neonatal infections** such as meningitis, arthritis, osteomyelitis, conjunctivitis, respiratory infections, peritonitis, omphalitis, and endocarditis
- **Infections in adults** such as sepsis, skin and soft tissue infections, respiratory tract infections, and urinary tract infection in women

Laboratory diagnosis

- On **blood agar**, tiny colonies with a wide zone of beta-hemolysis are seen.
- **Identification and confirmation**
  - *CAMP test (Christie, Atkins, and Munch–Peterson)*: An accentuated zone of hemolysis appears when *S. agalactiae* are inoculated perpendicular to a streak of *Staphylococcus aureus/Staphylococcus plazen* to produce an **arrowhead type of enhanced hemolysis** due to the beta-lysis of *Staphylococcus* on blood agar (Fig. 24.7).

Treatment

Empiric treatment is with a combination of ampicillin and penicillin for up to 21 days.

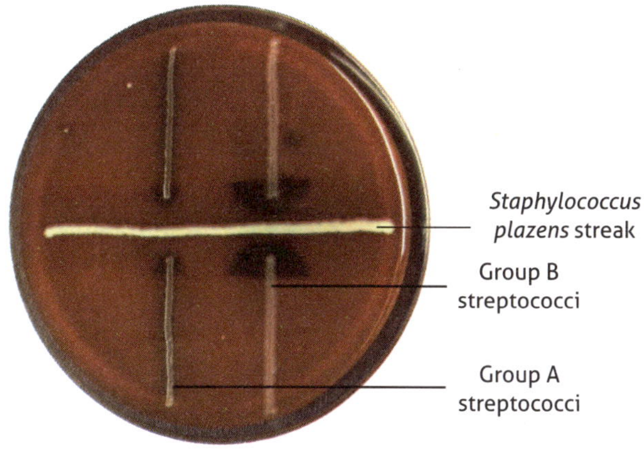

Fig. 24.7 CAMP test: Arrowhead shows hemolysis by *S. agalactiae*

Listeriosis

Epidemiology

- Listeriosis in humans occurs by (i) contact with infected animals, (ii) by inhalation of contaminated dust, or (iii) by the ingestion of contaminated milk or food.
- Outbreaks of food-borne listeriosis occur due to the ability of the organism to grow at refrigeration temperature, low pH, and high salt concentration (adopted for food preservation).
- Monocytosis is a predominant feature of human listeriosis.

Agent—*Listeria monocytogenes*

- The causative agent *L. monocytogenes* is widely distributed in nature. It has been isolated from mammals, birds, fish, ticks, and crustacea.
- It occurs as a saprophyte in soil, water, and sewage.
- It is a **short, non-sporing, gram-positive** bacillus.

Pathogenicity

- The bacteria are invasive and produce an important virulence factor, **listeriolysin O,** which allows the bacterium to escape from the cell membranes, and hence contributes to septicemia.
- These organisms can cross the blood–brain barrier to cause meningitis and encephalitis.
- Cell-mediated immunity involving macrophages activated by T lymphocytes confers protection.

Clinical features

Listeriosis in human beings may present as follows:

- **Meningitis or sepsis in neonates** occurs due to perinatal transmission, which can be of early or late onset; infants acquire infection in utero or by inoculation through an infected birth canal.
- **Meningoencephalitis in the immunocompromised and the elderly** is transmitted by the ingestion of contaminated food.

- Pregnant women who get infected may have an abortion or a stillbirth.
- Asymptomatic infection of the **female genital tract** may cause **infertility**.
- Listeriosis may also present as abscess, conjunctivitis, pharyngitis, urethritis, pneumonia, infectious mononucleosis-like syndrome, or endocarditis.

Laboratory diagnosis

- Appropriate specimens, depending on infection, may be examined for diagnosis.
- **Blood and CSF, cervical and vaginal secretions, lochia, meconium, or cord blood** may be cultured on blood agar and incubated in an environment containing 5% CO_2.
- It has a slow, **tumbling motility when grown at 25°C but is non-motile at 37°C.** This is because peritrichous flagella are produced optimally at 20–30°C but not at 37°C.
- It is **aerobic or microaerophilic** and grows in 5–10% CO_2 between 30°C and 37°C, but slow growth occurs even at 4°C.
- Colonies are **hemolytic** on blood agar.
- **Antibodies to listeriolysin O** can be detected.

Treatment

- **Ampicillin, cotrimoxazole, and gentamicin** are the first-line drugs.
- Cephalosporins are not recommended.
- Meningitis is treated with **ampicillin.**

Prevention

- Proper food handling and personal hygiene are important.
- Early diagnosis in pregnancy is important.

Chronic Bacterial Meningitis

Chronic meningitis is defined as **an inflammatory cerebrospinal fluid profile** that persists for **at least 1 month**. It presents clinically as headache, nausea, vomiting, cranial neuropathies, symptoms of elevated intracranial pressure, or focal neurologic deficits.

Etiology

1. **Infectious**

- Bacterial
  - *Mycobacterium tuberculosis* (Chapter 23)
  - *Treponema pallidum* tertiary stage as neurosyphilis (Chapter 27)
  - Lyme's disease (Chapter 29)
- Fungal (Chapter 23)
  - *Cryptococcus*
  - *Histoplasma*
  - *Coccidioides*
  - *Blastomyces*
- Viral
  - HIV (Chapter 22)
  - Varicella Zoster virus (Chapter 23)

2. **Autoimmune**
3. **Neoplastic**

Tetanus

Case 4. *A 50-year-old villager presented with slurring of speech and disorientation. On examination, he had board-like abdominal rigidity, muscle spasms, and risus sardonicus. These symptoms were easily induced by minor stimuli like noise or lights. On neurological examination, his bilateral plantar reflexes were deteriorating, but his jaw reflex was found to be increased. The man had sustained a non-healing and increasingly painful penetrating wound 7 days ago. The patient was admitted on strong suspicion of* ***tetanus****.*

i. Describe the pathogenesis of tetanus.
ii. How can it be prevented?

- Tetanus is a potentially fatal **nervous system disorder characterised by muscle spasms that are caused by the toxin-producing anaerobe *Clostridium tetani*.**
- The incidence of tetanus has reduced greatly due to immunisation and infection control practices.

Clinical features

- The **incubation period** is 6–12 days.
- It is characterised by tonic muscular spasms with **painful muscle rigidity**.
- Pain often begins in the jaw muscles (**trismus or 'lockjaw'**) and neck, shoulder, and abdominal muscles.
- As the disease progresses, muscle groups throughout the body are affected, and spontaneous generalised seizure-like spasms develop.
- **Autonomic dysfunction**, including hypertension and tachycardia alternating with bradycardia occurs.
- In the absence of ventilatory support, **death** is usually **due to respiratory failure**.
- Untreated tetanus has a very high mortality rate, especially in cases of neonatal tetanus.

Complications

- Airway obstruction
- Respiratory arrest
- Heart failure
- Pneumonia
- Damage to muscles
- Fractures
- Brain damage due to lack of oxygen during spasms

Maternal and neonatal tetanus (MNT)

- In rural India, tetanus used to be a common cause of death, particularly in the newborn. However, universal immunisation of infants and expectant mothers has reduced the incidence of maternal and neonatal tetanus to a large extent.

- Maternal and neonatal tetanus are the most common life-threatening consequences **of unclean deliveries and poor umbilical cord care practices.**
- Neonatal tetanus has been an indicator of inadequate access to immunisation and other maternal, newborn, and child health services.
- Neonatal tetanus develops by the eighth day of life.
- The child becomes **irritable, refuses feeds, and exhibits a facial grimace** with stiff muscles (Fig. 24.8).
- Mortality rates are extremely high, especially when appropriate medical care is not available. The disease has higher incidence in low-income countries. According to the WHO, as of March 2022, 47 countries including India had eliminated maternal and neonatal tetanus.

Agent—*Clostridium tetani*

- *C. tetani* is widely distributed in soil and in the intestines of humans and animals.
- It is ubiquitous and has been recovered from a wide variety of sources, including street and hospital dust, cotton wool, plaster of Paris, bandages, catgut, talc, wall plaster, and clothing.

Morphology

- It is a **gram-positive, slender bacillus**, generally measuring about 4–8 × 0.5 μm in size, though there may be considerable variation in length.
- The spores are spherical, terminal, and bulging, giving the bacillus a **characteristic 'drumstick' appearance** (Fig. 24.9).
- It is **non-capsulated and motile** due to the presence of peritrichate flagella.

Serotypes

Ten serological types have been recognised based on agglutination (types I to X).

Susceptibility

- **Spores** can survive in soil for years.
- Though most of them are killed by boiling for 10–23 minutes, some resist boiling for up to three hours.

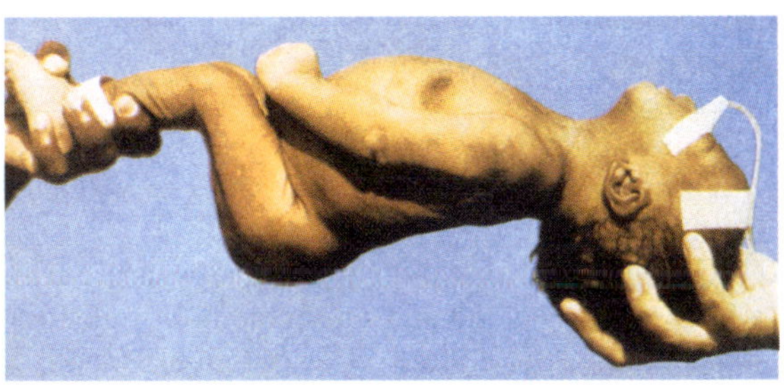

Fig. 24.8 Neonatal tetanus caused by *C. tetani* showing muscle stiffness (*Source*: Centers for Disease Control and Prevention [CDC], Public Health Image Library [PHIL], Image ID 6374)

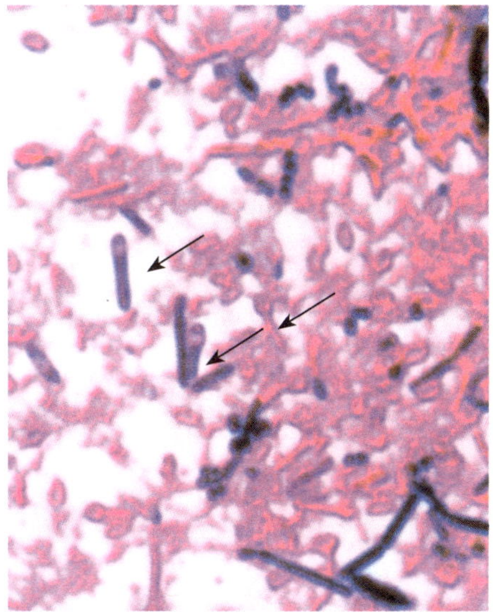

Fig. 24.9 *C. tetani*—drumstick appearance on Gram stain (*Source*: Department of Microbiology, Pondicherry Institute of Medical Sciences, Puducherry)

- **Autoclaving** at 121°C for 20 minutes is recommended to kill the spores.
- Disinfectants such as **iodine** (1% aqueous solution) and **hydrogen peroxide** (10 volumes) kill the spores within a few hours.
- They are not destroyed by 5% phenol or 0.1% mercuric chloride solution.

Pathogenesis

- **Route of entry:** Spores enter through the **skin or mucosa** through a breach of continuity following injury (puncture wounds), surgical operations (due to lapse in asepsis), unhygienic practices (septic abortion), and unsterile injections.
- **Transmission**—It is not transmitted from person to person.
- **Toxin**—*C. tetani* has little invasive power, but produces toxin locally.
  - *C. tetani* produces at least two distinct toxins, **tetanospasmin (neurotoxin) and tetanolysin (hemolysin)**.
  - Both are antigenically and pharmacologically distinct and their production is mutually independent.
  - A **third toxin**—a non-spasmogenic, peripherally active neurotoxin—has been identified. It is not known whether this plays any role in the pathogenesis of tetanus.
  - The toxin is absorbed by the motor nerve endings and transported to the central nervous system intra-axonally.
  - The toxin is specifically and avidly fixed by the gangliosides of the grey matter of the nervous tissue.

Tetanospasmin

- It is responsible for tetanus.
- It is **oxygen-stable but relatively heat-labile**, being inactivated at 65°C in five minutes.
- Its production is **plasmid-coded**.
- It is a simple protein composed of a single polypeptide that gets toxoided spontaneously or in the presence of low concentrations of formaldehyde.
- It is antigenic and is specifically neutralised by the antitoxin.
- It resembles strychnine in its effects.

Mechanism of action

- The toxin acts pre-synaptically (Fig. 24.10a and b).
- The abolition of spinal inhibition causes uncontrolled spread of impulses initiated anywhere in the central nervous system.
- This results in muscle rigidity and spasms due to the simultaneous contraction of agonists and antagonists in the absence of reciprocal inhibition.
- The **minimum lethal dose (MLD)** is 50–75 × 10–6 mg in mice.

Tetanolysin

- This is a heat-labile, oxygen-labile hemolysin, which is antigenically related to the oxygen-labile hemolysins of *C. perfringens*, *C. novyi*, and *S. pyogenes*.
- It is not relevant in the pathogenesis of tetanus.

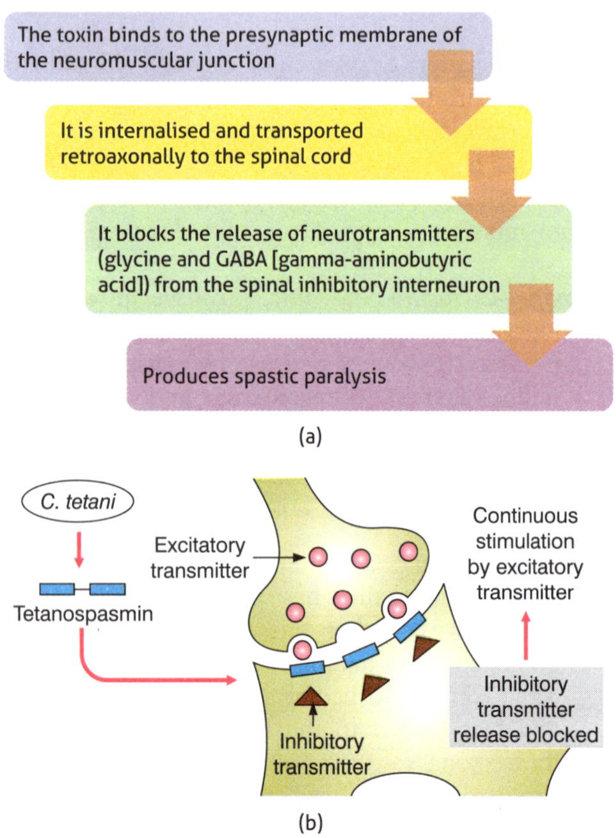

Fig. 24.10 (a and b) Mechanism of action of the tetanus toxin

Laboratory diagnosis

- The diagnosis of tetanus should always be made on clinical grounds and may be made by the demonstration of *C. tetani* by microscopy, culture, or toxigenicity tests. Laboratory tests are not usually helpful.
- **Specimen:** Necrotic tissue is collected from the site of injury.
- **Microscopy:**
  - It is unreliable as the demonstration of the typical 'drumstick' bacilli in wounds in itself is not diagnostic of tetanus. The bacilli may be present in some wounds without tetanus developing.
  - When present, it may be indistinguishable from non-pathogenic *C. tetanomorphum* and *C. sphenoides*.
- **Culture:**
  - *C. tetani* is an **obligatory anaerobe** that grows only in the absence of oxygen.
  - The optimum temperature for growth is 37°C; the optimum pH is 7.4.
  - It grows on ordinary media. Growth is improved by blood and serum but not by glucose.
  - *C. tetani* grows well in **Robertson's cooked meat broth** with **turbidity** and **some gas formation**. The meat is not digested, but turns black on prolonged incubation.
  - On blood agar, **beta-hemolysis** is produced due to the production of hemolysin (tetanolysin).

Toxigenicity testing

- **In vitro**: The inhibition of hemolysis on blood agar incorporated with tetanus antitoxin (1,500 units/mL) denotes neutralisation of hemolysin (tetanolysin) in contrast with hemolytic colonies on blood agar without the antitoxin.
- **In vivo:** Toxigenicity can also be tested in rats. Injection in rats causes rigidity that proceeds to the leg on the inoculated side, the opposite leg, trunk, and forelimbs, in that order. The animal dies within two days

Laboratory diagnosis of tetanus

Chiefly clinical

- **Specimen:** Necrotic tissue
- **Microscopy:**
  - Unreliable
  - Gram stain shows 'drumstick' bacilli in wounds
- **Culture:**
  - **Obligatory anaerobe**, non-fastidious; growth is improved by blood and serum but not by glucose
  - Optimum temperature 37°C; optimum pH: 7.4
  - *Robertson's cooked meat broth:* **Turbidity** and **some gas formation;** meat is not digested, but turns black on prolonged incubation
  - *Blood agar:* **Beta-hemolysis**

- *Toxigenicity testing:*
  - In vitro: Neutralisation of hemolysin (tetanolysin)
  - In vivo: In rats , it causes rigidity and death

Prophylaxis, treatment, and prevention

- The nature of prophylaxis depends largely on the type of the wound and the immune status of the patient.
- If the vaccination status is not known or if the patient was vaccinated >5 years ago, an intramuscular injection of tetanus toxoid is advised immediately after a penetrating injury or laceration.

Antibiotics

- **Metronidazole** 500 mg is given every six hours either intravenously or orally depending on the extent of infection.
- **Penicillin G** (100,000–200,000 IU/kg/day intravenously) is given in 2–4 divided doses.
- Tetracyclines, macrolides, clindamycin, cephalosporins, and chloramphenicol can also be used.
- Bacitracin or neomycin may be applied locally.

Surgical management

- Foreign bodies, necrotic tissue, and blood clots can be removed surgically in order to remove any anaerobic environment favourable for *C. tetani*.
- The extent of surgical treatment ranges from simple cleansing to radical excision, depending on the type of the wound.

Supportive management

- Spasms are controlled by benzodiazepine and autonomic dysfunctions by magnesium sulphate.
- Airway and respiratory control are mandatory.

Prevention by immunisation

Passive immunisation

This was originally achieved by the following methods:

1. **Anti-tetanus serum (ATS) extracted from hyperimmune horses**
   - The drawbacks of ATS are immune elimination and hypersensitivity.
   - Prior sensitisation also leads to type III hypersensitivity reactions, which may range from mild, local reactions to serum sickness.
   - Infrequently, fatal anaphylaxis (type I hypersensitivity) may also occur.
   - It is compulsory that a test for hypersensitivity be done prior to the administration of ATS.
2. **Tetanus immunoglobulin (TIG)**
   - This provides passive immunity without the risk of hypersensitivity.
   - This is effective in smaller doses (250 units) and has a longer half-life (3–5 weeks).
   - Passive immunisation is an emergency procedure to be used only once.
   - TIG 10,000 IU, suitably diluted, may be given by slow IV infusion, followed, if needed, by another 5,000 IU later.
   - Even though TIG may not neutralise the toxin already bound to the nervous tissue, it can inactivate the unbound toxin and any further toxin that may be produced.

Active immunisation

- Patients recovering from tetanus should receive a full course of active immunisation with tetanus toxoid, as an infection does not confer subsequent immunity.
- Second episodes of tetanus have been recorded.
- This is the most effective method of prophylaxis and is achieved by administering spaced injections of the following:
  - **Formol toxoid**, which is available as 'plain toxoid'
  - **Adsorbed toxoid**, which is adsorbed on aluminium hydroxide or phosphate
- The tetanus toxoid is administered alone or along with the diphtheria toxoid and the pertussis vaccine as a 'triple vaccine', in which the pertussis vaccine acts as an adjuvant as well.
- **Dosage**—three doses of tetanus toxoid are given intramuscularly, with an interval of 4–6 weeks between the first two injections and the third dose given six months later (or according to the recommendations of the National Immunisation Programme).
- A full course of immunisation confers immunity for a period of at least ten years. A **'booster dose'** of the toxoid is recommended after ten years. The booster dose is also given if a penetrating injury occurs three years or more after the full course of immunisation.
- ATS or TIG should not be given to an immunised individual.

Combined immunisation

This comprises administering to a non-immune person exposed to the risk of tetanus, a TIG injection at one site, along with the first dose of toxoid at the contralateral site, followed by a second and third dose of toxoid at monthly intervals. Ideally, combined immunisation should be used whenever passive immunisation is called for.

Table 24.7 presents the recommended integrated prophylaxis for tetanus following injury.

Botulism

Botulism is a rare but life-threatening condition caused by a toxin produced by *Clostridium*. Botulism can occur as the result of food or wound contamination.

Table 24.7 Tetanus prophylaxis in the wounded

Nature of wound	Immune status of the person		
	Immune	Partially immune	Non-immune
Clean (wound toilet performed within six hours)	Toxoid × 1*	Toxoid × 1	Toxoid × 3
Contaminated (soil or other foreign or necrotic material present)	Toxoid × 1*	Toxoid × 1 TIG + antibiotics	Toxoid × 3 TIG + antibiotics
Infected	Toxoid × 1* + antibiotics	Toxoid × 1 TIG + antibiotics	Toxoid × 3 TIG + antibiotics

Note:
- **Immune**: Patient has had a full course of three injections of toxoid
- **Partially immune:** Patient has had two injections of toxoid
- **Non-immune**: Patient has had one or no injection of toxoid, or immunisation status is not known
- **TIG:** Tetanus immunoglobulin

* The toxoid needs to be given only if three years or more have elapsed after active immunisation or the last booster injection.

Agent—*Clostridium botulinum*

Clostridium botulinum has similar staining characteristics, morphology, and growth requirements as *C. tetani*.

Classification

- **Eight types** of *C. botulinum* strains have been identified—A, B, C1, C2, D, E, F, and G—based on **immunological differences**.
- These are antigenically and serologically distinct but structurally similar.
- Human botulism is caused mainly by types A, B, E, and (rarely) F.
- They have similar pharmacological activity but are neutralised only by the homologous antiserum.
- C2 toxin shows enterotoxic activity, while the others are neurotoxins.

Pathogenicity

- The pathogenicity of *C. botulinum* is due to the action of its **toxin,** the manifestations of which are collectively called botulism.
- The potent toxin produced by *C. botulinum* differs from other exotoxins in that it is not released during the life of the organism.
- It is produced **intracellularly as a non-toxic protoxin or progenitor toxin**.
- Trypsin and other proteolytic enzymes activate the progenitor toxin to produce the active toxin and appears in the medium only on cell death and autolysis.
- Toxin production appears to be determined by the presence of bacteriophages, at least in types C and D.
- The lethal dose for human beings is probably 1–2 mg.
- It is a neurotoxin, which acts slowly, taking several hours to cause death—it is one of the most potent toxins known to humankind.
- The toxin is relatively stable and is inactivated only after 30–40 minutes at 80°C or after 10 minutes at 100°C.
- Food suspected to be contaminated with the botulinum toxin can be rendered completely safe by pressure cooking or boiling for 20 minutes.

Mechanism of action

The toxin produced by *C. botulinum* acts by blocking the production or release of acetylcholine at the synapses and neuromuscular junctions.

Clinical uses of toxins

- Intramuscular injection of the toxin was first used to treat **strabismus**.
- It is now recognised as a safe and effective **symptomatic therapy for many neuromuscular diseases** like blepharospasm, muscle stiffness/spasms, cervical dystonia, and torticollis.
- It is used **cosmetically** to temporarily remove wrinkles and frown lines.

Clinical features

The common symptoms of botulism include the following:

- Diplopia, dysphasia, and dysarthria (due to cranial nerve involvement)
- Descending symmetric flaccid paralysis of the voluntary muscles
- Loss of deep tendon reflexes
- Constipation
- There are no sensory or cognitive deficits
- Respiratory muscle paralysis may lead to respiratory failure and death

i. *Food-borne botulism*

- This is caused by the **ingestion of the preformed toxin**.
- Human disease is usually caused by types A, B, E, and very rarely, F.

- Types C and D are usually associated with outbreaks in cattle and wild fowl.
- Type G has been associated with sudden death in a few patients.
- Symptoms usually begin 12–36 hours after the ingestion of food.
- **Coma or delirium** may supervene. No vomiting or diarrhea is present.
- Death is due to respiratory failure and occurs 1–7 days after onset.
- The case fatality is 25–70 per cent.
- The source of botulism is usually preserved food—**meat and meat products** containing spores.
- **Cans of spoiled food are often inflated** and show bubbles on opening. Non-proteolytic varieties leave food unchanged.
- Type E is associated with fish and other seafood.

ii. Infant botulism

- Occurs in **infants below six months** who are bottle-fed.
- *C. botulinum* spores ingested in baby food get established in the gut and produce the toxin.
- The manifestations are constipation, poor feeding, lethargy, weakness, pooled oral secretions, weak or altered cry, floppiness, flaccid paralysis, ptosis, and loss of head control (Fig. 24.11).
- Some cases of sudden infant death syndrome have been attributed to infant botulism.
- Honey used to sweeten milk or formula food has been incriminated as a likely food item through which the bacillus enters the gut.

iii. Wound botulism

- This is a very rare condition resulting from wound infection with *C. botulinum*.
- Type A was found to be responsible for most of the cases studied.

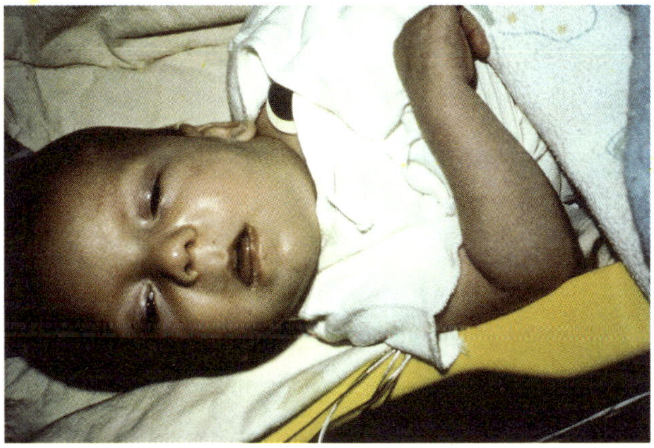

Fig. 24.11 Infant botulism—flaccid facial musculature and ptosis of the lids caused by *Clostridium botulinum* (*Source*: CDC, PHIL, Image ID 23681)

Laboratory diagnosis

Microscopy

- The **bacillus or its toxin** can be demonstrated in food or feces.
- **Gram-positive, spore-forming bacilli** may be demonstrated in smears made from the food.

Animal inoculation

- The food suspected to be contaminated is macerated in sterile saline and the filtrate inoculated into mice or guinea pigs intraperitoneally.
- Control animals protected by polyvalent antitoxin remain healthy.

Cultural characteristics

- *C. botulinum* is a **strict anaerobe**.
- The optimum temperature for its growth is 35°C, but some strains may grow even at 1–5°C.
- Colonies are large, irregular, and semi-transparent with a fimbriate border.

Prevention and treatment

- Since most cases of botulism follow the consumption of improperly canned or preserved food, control can be achieved by **proper canning and preservation**.
- When an outbreak occurs, a **prophylactic dose of antitoxin** should be given intramuscularly to all who consumed the food article.
- Active immunisation has been shown to be effective. Two doses of **aluminium sulphate adsorbed toxoid** are given ten weeks apart, followed by a booster a year later.
- Polyvalent antiserum to types A, B, and E may be administered as soon as a clinical diagnosis is made.
- **Supportive therapy with maintenance of respiration** is of great value.

VIRAL CAUSES OF CNS INFECTIONS

Aseptic Viral Meningitis

- This is meningitis that is accompanied by pleocytosis, normal or slightly elevated protein concentration, normal glucose, and normal or slightly elevated opening pressure with negative bacterial cultures of CSF.
- The most common etiology for aseptic meningitis is viral infection (Table 24.2), but it can also have non-infectious causes such as systemic diseases with meningeal involvement (sarcoidosis, Sjögren's syndrome, SLE), drugs (monoclonal antibodies, sulphonamides), or neoplasms.

Enteroviral Infections

- Most enteroviruses are host-specific, infecting only one or a few related species.

- There is no common group antigen for enteroviruses, though some of them show antigenic cross-reactions.
- They are found in the gastrointestinal or nasopharyngeal track, and hence are named Enterovirus and Rhinovirus respectively.

Classification

Enteroviruses of medical importance are as follows:

- Poliovirus types 1–3
- Coxsackievirus A types 1–24 (types not included: 23, 18, and 23)
- Coxsackievirus B types 1–6
- Echovirus types 1–34 (types not included: 8, 10, 22, 23, 28, and 34)
- Enterovirus types 68–123 (type not included: 72)

Retaining the enterovirus nomenclature, **coxsackie A** virus now falls under human enterovirus species A **(HEV-A) and HEV-C**, while **coxsackie B virus** falls under **HEV-B**, which also includes the **ECHO virus.** Newer enteroviruses discovered after 1969 have been assigned enterovirus (EV) type numbers, starting with 68. **Enterovirus 72**, which causes infectious hepatitis (hepatitis type A), has been reclassified as a separate genus—**Hepatovirus**.

Rhinoviruses were considered a separate genus in the past, but are now included in the genus Enterovirus. They have 123 antigenic types and are classified into three species of human rhinovirus (HRV)—A, B, and C.

Table 24.8 presents the clinical syndromes caused by enteroviruses.

Poliomyelitis

Case 5. *A 10-year-old boy from a slum presented to the emergency department with a history of mild fever, headache, and sore throat for the previous eight days. He also complained of vomiting for the previous two days. There was a brief asymptomatic period of two days. On the morning of presentation, he experienced pain in his right lower limbs, followed by weakness and inability to walk at the time of presentation.*

The boy had no history of vaccination in the first five years of his life and had missed the mass vaccination campaign. Throat and rectal swabs and cerebrospinal fluid (CSF) were sent for viral studies to a referral laboratory. The throat and rectal swabs showed cytopathic changes in primary monkey kidney cell line; the etiology was confirmed as ***poliovirus type 1*** *by the neutralisation test. Reverse transcriptase PCR in CSF was positive for poliovirus. The general condition of the child improved on supportive treatment.*

i. What is poliomyelitis?
ii. Describe the pathogenesis and control strategy of poliomyelitis.

Epidemiology

- Poliomyelitis is an exclusively human disease caused by poliovirus.
- The only sources of the virus are humans—the patient or, much more commonly, the symptomless carrier.
- Patients shed the virus in feces for varying periods—about 50 per cent for three weeks, and a small proportion for 3–4 months.
- There is no permanent carrier state. However, the virus may persist in the environment (sewage) for up to six months.
- Viruses shed in throat secretions during the early part of the disease may also be a source of infection.

Polio status—globally and in India

- The WHO has declared all the major continents free of polio except for a few African and Asian countries.
- India has also been declared polio-free as there have been no reports of wild poliovirus infection in the last five years.

Table 24.8 Clinical syndromes commonly caused by the serotypes of various enteroviruses

Clinical syndrome	Poliovirus	Coxsackie A	Coxsackie B	Echovirus	New enterovirus types
Paralysis	1, 2, 3	7	–	–	–
Aseptic meningitis	1, 2, 3	7, 9, 23	1–6	Several (2, 3, 4, 6, 7, 9, 11, 14, 23, 17, 18, 19, 25, 33)	71
Encephalitis	1, 2, 3	9	3–6	6, 9, 17, 19	71
Fever with rash	–	9, 23, 23	–	4, 6, 9, 23	–
Hand, foot, and mouth disease	–	5, 10, 23	–	–	71
Herpangina	–	1–6, 8, 10	–	–	–
Upper respiratory infection	–	21	–	11, 20	–
Pneumonitis, bronchiolitis	–	–	–	–	68
Bornholm disease	–	–	1.5	–	–
Myocarditis, pericarditis	–	–	1.5	–	–
Acute hemorrhagic conjunctivitis	–	24	–	–	70

- This has been possible due to active polio eradication programmes, sustained vaccination through pulse polio, and mop-up programmes.
- India has actively participated in the **Global Polio Eradication Initiative (GPEI)** to work towards a polio-free world.
- There are a few countries which still report cases, notably Pakistan and Afghanistan, with wild poliovirus infections being reported. The Democratic Republic of Congo, Nigeria, and Papua New Guinea reported cases of vaccine-derived poliovirus (cVDPV) infection in 2019 (http:/polioeradication.org/polio-today/polio-now/this-week/).

Agent—Poliovirus

- Poliovirus is a **non-enveloped, positive-sense, single-stranded RNA virus.**
- It is a member of the **Enterovirus** genus within the family Picornaviridae.

Morphology

- The virion is **a spherical particle, about 27 nm in diameter, composed of 60 subunits**, each consisting of four **viral proteins (VP1–VP4)** arranged in **icosahedral symmetry**.
- VP1, which faces outwards, carries the major antigenic site for combination with type-specific neutralising antibodies.
- The genome is a single strand of positive-sense RNA of 7.4 Kb.
- The positive-sense RNA genome is infectious.
- The virus can be crystallised, and arrays of virus crystals can be seen in the cytoplasm of infected cells (Fig. 24.12a and b).

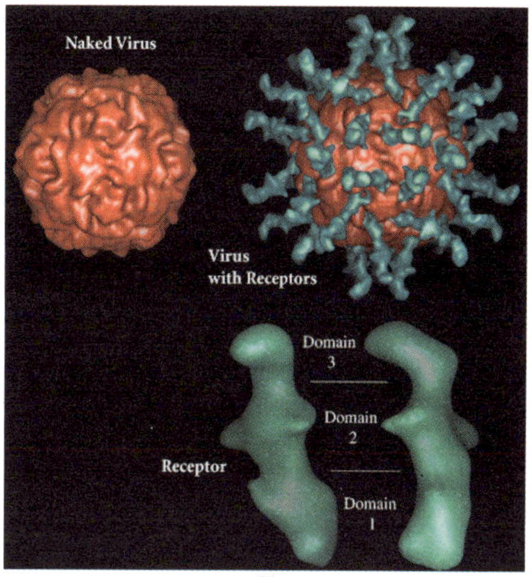

(a)

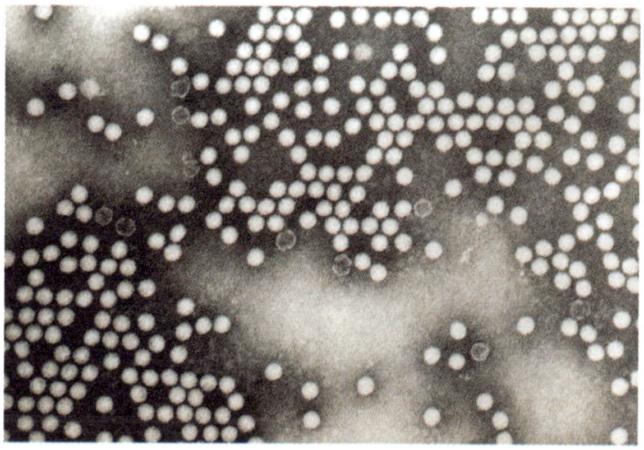

(b)

Fig. 24.12 (a) Structure of poliovirus and (b) TEM image depicting numerous, round, poliovirus type-1 virions, measuring 20–30 nm in diameter with icosahedral symmetry (*Source*: [b] CDC, PHIL, Image ID 235)

Antigenic types

There are three antigenic types, which are as follows:

i. **Type 1 (Brunhilde and Mahoney strain):** It is the most common and causes the most epidemics.
ii. **Type 2 (Lansing and MEFI strain):** It usually causes endemic infections and inapparent infections in western countries. In India, paralysis due to type 2 is common.
iii. **Type 3 (Saukett strain):** This type has also caused epidemics.

Two antigens, C and D, are recognised based on complement fixation, ELISA, or precipitation tests.

i. **The C antigen (coreless)**, also called the **heated or H antigen**, is associated with the 'empty' non-infectious virus. Anti-C antibody does not neutralise virus infectivity and is not protective.
ii. The **D antigen (dense)**, also called the **native or N antigen**, is associated with the whole virion and is type-specific. The D antigen is converted into the C antigen by heating the virus at 56°C. This antibody is protective. The potency of the injectable polio vaccine can be measured in terms of D antigen units.

Susceptibility

- Poliovirus is resistant to ether, chloroform, bile, proteolytic enzymes of the intestinal contents, and detergents.
- In feces, it can survive for days to weeks at room temperature, months at 4°C, and for years at –20°C.
- It is readily inactivated by heat (55°C for 30 minutes).
- Milk or ice cream protects the organism from getting killed by moderate heat.
- **Formaldehyde and oxidising disinfectants** destroy the virus.
- Chlorine at 0.1 ppm can destroy purified polio. However, higher concentrations are required to destroy the virus in sewage water, organic matter, and feces.

Pathogenesis (Fig. 24.13)

- Degenerative lesions generally occur in the anterior horns of the spinal cord, causing flaccid paralysis. However, the posterior horns and intermediate columns may also be involved to some extent.
- In some cases, encephalitis may occur, involving the brain stem and extending up to the motor and premotor areas of the cerebral cortex.

Immunity

- Immunity is type-specific but there is a significant amount of cross-protection between types 1 and 2, and types 2 and 3, and little or none between types 1 and 3.
- Humoral immunity provided by **circulating and secretory neutralising antibodies** is responsible for protection against poliomyelitis.
- The **IgM antibody** appears within a week of infection and lasts for about six months.
- The **IgG antibody** persists for life.
- Antibodies cannot protect once the infection spreads to the CNS. Hence, vaccination must be administered prior to the onset of CNS symptoms.
- **Secretory IgA** in the gastrointestinal tract provides mucosal immunity, preventing intestinal infection and virus shedding.
- **Breast milk containing IgA antibody** protects infants from infection.
- The virus also induces cell-mediated immunity (CMI) but its effect is unclear.

Clinical features

- Infection is **generally asymptomatic**. The ratio of subclinical to clinical infections has been stated to be 100 or 1,000 to 1.
- The outcome of infection is influenced by the virulence of the infecting strain, the infective dose, and the age of the individual (adults being more susceptible than children).
- Pregnancy carries an increased risk of paralysis, perhaps due to the associated hormonal changes.

Inapparent infection

- Following exposure to the poliovirus, 90–95 per cent of susceptible individuals only develop an inapparent infection, which causes seroconversion alone.
- It is only in 5–10 per cent that any sort of clinical illness results.
- The **incubation period is about 10 days** on an average, but may range from four days to four weeks.

Abortive poliomyelitis or minor illness

The earliest manifestations are associated with the phase of primary viremia consisting of **fever, headache, sore throat, and malaise** lasting 1–5 days.

Paralytic poliomyelitis or major illness

- If the infection progresses, the minor illness is followed 3–4 days later by the major illness.
- The fever returns (**biphasic fever**), along with headache, stiff neck, and other features of meningitis. This marks the stage of **viral invasion of the central nervous system** (Case 5).
- Paralytic polio is more common in older individuals, pregnant women, and persons who perform strenuous physical activity.
- Tonsillectomy predisposes a person to bulbar poliomyelitis.
- IM injection during the febrile polio phase increases the risk of paralysis.
- In those progressing to paralytic poliomyelitis, **flaccid paralysis** develops.
- Paralysis is focal in distribution initially, but spreads over the next 3–4 days.
- Depending on the distribution of paralysis, cases are classified as **spinal, bulbar, or bulbospinal**.

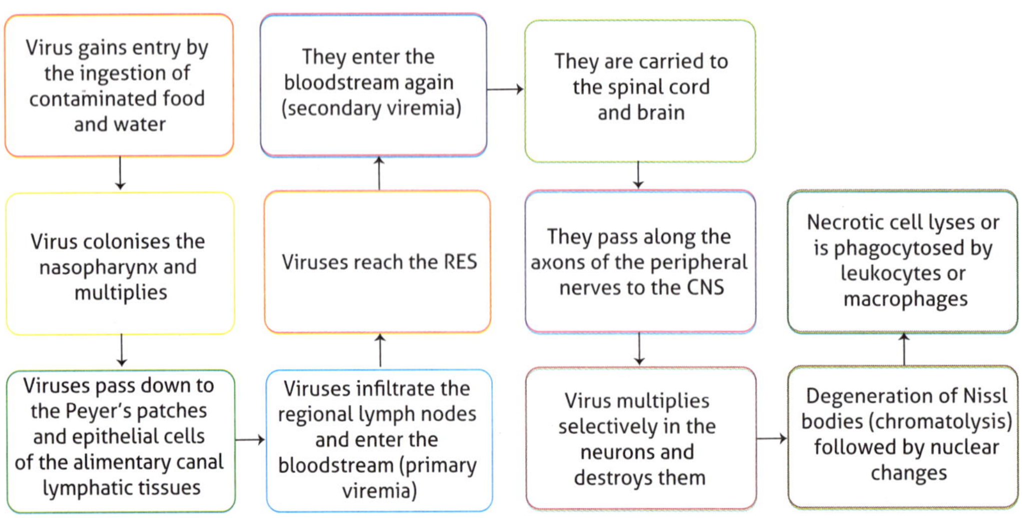

Fig. 24.13 Pathogenesis of poliomyelitis

- The mortality range is 5–10 per cent and is chiefly due to **respiratory failure**.
- Recovery of the paralysed muscles takes place in the next 4–8 weeks and is usually complete after six months, leaving behind varying degrees of residual paralysis (Fig. 24.14).
- Rarely, myocarditis or lymphatic hyperplasia may occur.

Non-paralytic poliomyelitis

Sometimes, the disease does not progress beyond the stage of aseptic meningitis.

Post-polio syndrome

- This occurs years (20–40 years) after the infection due to the loss of function of motor neurons which had compensated for the loss in the initial infection.
- This is seen more often in women and presents as pain, fasciculation, atrophy, fatigue, and weakness.

Laboratory diagnosis

Examination of CSF

Reveals lymphocytic pleocytosis, normal glucose and normal-to-slightly elevated protein.

Specimen

- Throat swabs, feces (rectal swabs), blood, and CSF are suitable samples.

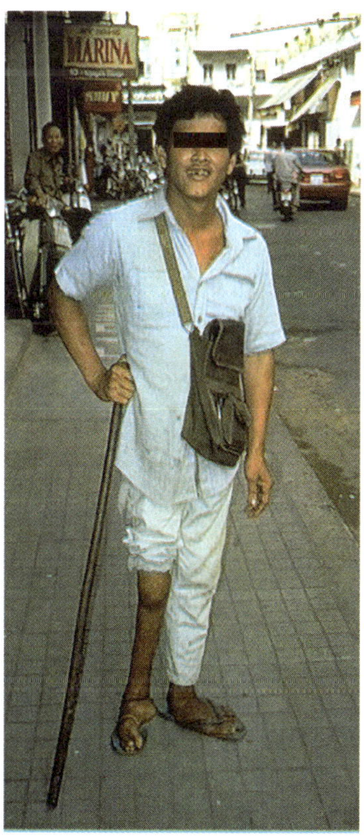

Fig. 24.14 Polio-afflicted individual with paralysis of the right lower limb (*Source*: CDC, PHIL, Image ID 134)

- As fecal excretion may be intermittent, the best results are obtained by testing fecal samples collected on two separate days, as early in the illness as possible.
- Unlike other enteroviruses, the poliovirus can seldom be isolated from cerebrospinal fluid (CSF) but can be obtained from the spinal cord and brain postmortem.

Isolation

- The virus can be isolated from blood during primary viremia, 3–5 days after infection, before neutralising antibodies appear. However, this is of little practical importance.
- The virus can be isolated from the throat within 14 days of the onset of the disease.
- Virus isolation from feces is usually possible in over 80 per cent of patients in the first week, 50 per cent till the third week, and 25 per cent till the sixth week.
- Prolonged fecal excretion may be seen in the immunodeficient. Carrier stages do not occur.

Viral isolation is the best method for specific diagnosis and typing of the strain, but virus isolation must be supported by clinical and serological evidence. The virus grows readily in tissue cultures of primate origin.

- Primary monkey kidney cultures or continuous cultures from human tissues are used.
- Detection of virus in cell lines
  - The infected cells round up and become refractile and pyknotic.
  - Eosinophilic intranuclear inclusion bodies may be demonstrated in stained preparations.
  - Well-formed plaques develop in infected monolayers.
  - Typical cytopathic effects seen in 2–3 days.
  - The viruses are identified by neutralisation tests with pooled and specific antisera.

Molecular diagnosis

- Viral RNA is demonstrated in CSF by reverse transcriptase PCR.
- Real-time reverse transcription PCR is used for intratypic differentiation using viruses isolated in culture as the starting material.

Sequencing

In a population, three types of strains may be in circulation—**the wild virus, the oral polio vaccine (OPV) virus strain, and the vaccine-derived polioviruses (VDPVs)**. These can be differentiated by sequencing.

- Partial genome sequencing is used to confirm the poliovirus genotype and determine its likely geographic origin.
- The use of monoclonal antibodies specific to vaccine strains, oligonucleotide fingerprinting, and nucleic acid sequencing help to differentiate the wild type from vaccine type strains.

Serodiagnosis

- Antibodies appear soon after the onset of paralysis.
- Neutralising antibodies appear early, and persist for life.
- In the CF test, antibodies to the C antigen appear first and disappear early, while anti-D antibodies take some weeks to appear after infection but last for about five years.
- Antibody rise can be demonstrated in paired sera by neutralisation or complement fixation tests.

> **Laboratory diagnosis of poliomyelitis**
>
> - **Examination of CSF:** Lymphocytic pleocytosis, normal glucose, and normal-to-slightly elevated protein
> - **Specimen:** Throat swabs, feces (rectal swabs), blood, and CSF
> - **Isolation:**
>   - From blood during primary viremia
>   - Virus isolation from feces possible in over 80 per cent of patients in the first week, 50 per cent till the third week, and 25 per cent till the sixth week
>   - Primary monkey kidney cultures or continuous cultures from human tissues are used to culture the virus
> - **Molecular diagnosis:** Viral RNA in CSF by reverse transcriptase PCR
> - **Sequencing:** Differentiation of **wild virus, OPV strain, and vaccine-derived polioviruses (VDPVs)**
> - **Serodiagnosis:** Antibody rise can be demonstrated in paired sera by neutralisation or complement fixation tests

Notification

- As polio is a notifiable disease, patient details and the details of household contacts must be obtained.
- In the case of a positive result, the local health officer and **Integrated Disease Surveillance Programme (IDSP) personnel** must be contacted for follow-up.

Immunoprophylaxis

- Passive immunisation by the administration of human gammaglobulin is of little value.
- Active immunisation forms the mainstay of prophylaxis for polio.

Active immunisation

Two types of vaccines are available. The salient differences between oral and injectable polio vaccines are presented in Table 24.9.

i. **Salk's killed vaccine (injectable polio vaccine [IPV])**

- This vaccine is prepared from formalin-inactivated types 1–3 poliovirus, grown in monkey kidney tissue culture.
- Inactivation with formalin (1:4,000) is done at 37°C for 12–23 days.

Vaccination schedule

- The Salk's vaccine is administered by injection and hence, it is also called the **injectable polio vaccine** (IPV).
- Three doses are given 4–6 weeks apart, followed by a booster six months later.

Table 24.9 Salient differences between the oral and injectable polio vaccines

Characteristics of polio vaccines	Oral polio vaccine	Injectable polio vaccine
Nature	Live-attenuated polioviruses grown in non-human cells—Sabin 1, 2, and 3 Bivalent OPV drops consist of attenuated strains of types 1 and 3	Inactivated (killed) poliovirus strains of all three poliovirus—types 1, 2, and 3 Combined with other routine vaccine antigens like diphtheria, tetanus, whole-cell or acellular pertussis, hepatitis B, or *Haemophilus influenzae* type B
Virulence and transmission	Have negligent neurovirulence and transmissibility	Non-transmissible and are avirulent as they are killed
Route of administration	Oral	IM or subcutaneous
Immunity	Mucosal immunity along with humoral immunity Produces herd immunity in the community through excretion by vaccinated children	Systemic humoral immunity Does not produce herd immunity
Advantages and drawbacks of oral and injectable polio vaccines		
Administration	Advantage: Easy to administer; additional doses of OPV on all pulse polio days for children up to 5 years of age	Drawback: Cannot be given as a mass schedule
Genetic change	Advantage: Cannot revert to virulent form	Drawback: Can genetically revert to iVDPV (immunodeficiency-related vaccine-derived poliovirus), aVDPV (ambiguous vaccine-derived poliovirus), or cVDPV (circulating vaccine-derived poliovirus)
Safety	Advantage: There is no danger of vaccine-associated paralytic polio (VAPP)	Drawback: Rarely, may give rise to VAPP

- The first dose should be given to **babies above six months of age** to ensure that maternal antibodies do not interfere with the vaccine.
- The second dose—an **enhanced potency IPV produced** in human diploid cells—induces better seroconversion following two subcutaneous **doses 4–8 weeks apart.**
- A third dose may be given **6–12 months later.**
- Booster doses help to sustain immunity; they can be administered every 3–5 years after the third dose.

ii. **Sabin's live-attenuated vaccine (oral polio vaccine [OPV])**

- The attenuated virus immunises by active multiplication.
- It stimulates systemic production of IgM, IgG, and secretory IgA in the intestine, producing local immunity.
- Very stringent precautions are taken to ensure freedom from extraneous agents like SV40 and the B virus in cell cultures used to prepare vaccines.
- **Monovalent or trivalent forms** are prepared in pleasantly-flavoured syrups. $MgCl_2$ or sucrose stabilises the vaccine against heat inactivation, particularly in tropical conditions.
- **Bivalent OPV** (bOPV containing types 1 and 3 Sabin viruses) has been licensed and used in some settings since December 2009. Trivalent OPV has been replaced with bivalent OPV from April 2016 as per the World Health Organization module. OPV containing serotype 2 monovalent is kept for emergency use.

Vaccination schedule

- The live polio vaccine is administered orally and is, therefore, known as the oral polio vaccine (OPV).
- Four doses are given at 4–8-week intervals.

Safety

- Both vaccines are safe.
- A few cases of vaccine-induced poliomyelitis have been reported, but the incidence of such cases is so low that the risk is negligible. However, it assumes importance where the disease has been eradicated by immunisation.
- OPV is not safe in immunodeficient or immunosuppressed subjects, in whom the killed vaccine is the safest alternative.

Vaccine-derived poliovirus (VDPV)

- On very rare occasions, OPV can lead to **vaccine-associated paralytic polio (VAPP).**
- VAPP may also be caused by a strain of poliovirus that has genetically changed in the intestine from the original attenuated vaccine strain contained in the OPV. This strain is known as **vaccine-derived poliovirus (VDPV).**
- In rare instances, the vaccine-derived virus can genetically change into a form that can paralyse the individual who gets infected—such a strain is known as a **circulating vaccine-derived poliovirus (cVDPV).**
- Two other VDPVs have been isolated:
  - **Immune-deficiency associated VDPV (iVDPV):** These is a VDPV isolated from persons with evidence of primary immunodeficiency.
  - **Ambiguous VDPV (aVDPV):** This is a VDPV isolated from individuals or from environmental samples without evidence of circulation in community or immunodeficiency (when isolated from individuals).

Efficacy

A full course of the killed vaccine induces a satisfactory immune response in a healthy child (Fig. 24.15).

Issues related to OPV failure in the tropics

- **Interference by other enteroviruses** is common in the tropics (experimentally, it has been shown that coxsackie B viruses may interfere with the poliovirus, while coxsackie A may be synergistic).
- **Frequent diarrheal diseases** prevent colonisation by the vaccine virus.
- **Breastfeeding** immediately before or after the vaccine may lead to the failure of the vaccine as the virus may be neutralised by antibodies in breast milk or other inhibitory agents in the intestinal secretions.
- Due to some drawbacks of OPV, as mentioned above, primary immunisation with killed vaccines (given along with the triple antigen) has been proposed. This may be followed by live oral vaccines for achieving intestinal immunity.

Key advantages of OPV

- **Ease of administration:** OPV is obviously preferable to killed vaccines given by injection. However, an advantage of the killed vaccine is that it can be administered along with the DPT vaccine as a quadruple vaccine.

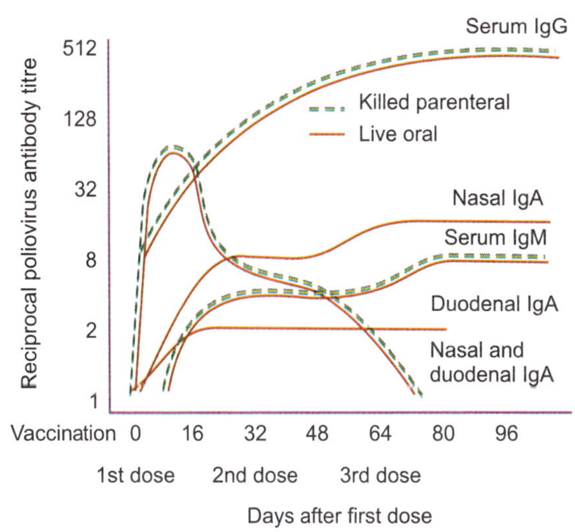

Fig. 24.15 Serum and secretory antibody response to three doses of OPV or three doses of IPV in a healthy child

- **Economy:** The live vaccine is much more economical. This is an important aspect of mass vaccination campaigns in the developing countries.
- **Nature of immunity:** This is perhaps the most important difference between the two types of vaccines. Live vaccines also induce local immunity in the gut so that wild viruses are unable to multiply in the intestines and subsequently be shed in feces. Hence, it protects the individual and the community, producing herd immunity.
- **Duration of immunity:** Immunity following a killed vaccine may need to be maintained with booster doses periodically, while immunity following a live vaccine resembles natural active immunity in lasting longer.
- **Use in epidemics:** Community-wide administration of OPV, ideally a monovalent vaccine of the same type as that causing the epidemic, early during an epidemic of paralytic poliomyelitis can stop the epidemic. This has been successfully practiced in different parts of the world.
- **Spread of vaccine virus in the community:** The tendency of the vaccine virus to spread naturally in the community, especially among children, is a disadvantage in developed countries.

Eradication of poliomyelitis

- The endgame strategic plan of the Global Polio Eradication Initiative (GPEI) was to deliver a sustained, polio-free world (implemented between 2013–2018).
- The GPEI has been updated to the **Polio Endgame Strategy 2019–2023** for the eradication, integration, certification, and containment of polio (http://polioeradication.org/wp-content/uploads/2019/06/english-polio-endgame-strategy.pdf).

Pulse polio immunisation programme (PPI)

- The strategy of administering the vaccine to all children in a region on the same day (**pulse immunisation**) has been found to be useful in developing countries.
- The Government of India has used PPI to achieve the goal of eradicating poliomyelitis in addition to the National Universal Immunisation Programme (UIP)—the PPI does not replace the UIP.
- Under this programme, the strategy was to administer **two rounds of vaccination at an interval of 4–6 weeks during the low polio transmission period in winter** (November to February).
- The World Health Organization declared India free of polio on 27 March 2014.

Coxsackieviral diseases

- Coxsackieviruses are typical enteroviruses. Two groups, **A and B,** are known.
- Based on neutralisation tests, group A viruses are classified into 24 types and group B into 6 types.
- The serotypes that comprise these types are presented in Table 24.8.
- Coxsackieviruses produce a variety of clinical syndromes in humans ranging from **trivial to fatal infections** (Table 24.8).
- Minor respiratory infections resembling the common cold may be caused by A10, A21, A24, and B3.

Agent—Coxsackie A virus

Aseptic meningitis may be caused by most group A and all group B viruses.

- A maculopapular rash may be present.
- The disease may sometimes occur as epidemics.

Herpangina (vesicular pharyngitis)

- It is a common clinical manifestation of Coxsackie group A infection in children.
- It is a severe febrile pharyngitis presenting with headache, vomiting, and pain in the abdomen.
- The characteristic lesions are small **vesicles on the fauces** and **posterior pharyngeal wall**, which break down to form ulcers.

Hand, foot, and mouth disease (HFMD)

- It is an exanthematous fever that chiefly affects young children.
- It is characterised by clusters of papulovesicular lesions on the skin and oral mucosa.
- It occurs as sporadic cases and as outbreaks.
- Coxsackie A23, A9, and B1–3 were the common causative agents initially. Subsequently, Enterovirus 71 caused extensive epidemics with serious complications like aseptic meningitis, encephalitis, flaccid paralysis, and pulmonary hemorrhage, with many fatalities, particularly in East Asia, from Taiwan to Singapore.
- HFMD is now an important emerging disease.

Agent—Coxsackie B disease virus

- **Epidemic myalgia or pleurodynia or Bornholm disease** (first described on the Danish island of Bornholm): It is a febrile disease with a sharp, piercing pain in the chest and abdomen, caused by group B viruses. The disease may occur sporadically or as epidemics.
- **Myocarditis and pericarditis in the newborn:** It is associated with high fatality. The disease may sometimes occur in older children and adults as well.
- **Juvenile diabetes:** It has been claimed to be associated with coxsackie B4 infection, but a causal role has not yet been established.
- **Orcihitis:** Coxsackie B has been associated with orchitis.

- **Congenital disease:** Transplacental and neonatal transmission have been demonstrated with coxsackie B viruses; this results in a serious, disseminated disease that may include hepatitis, meningoencephalitis, and adrenocortical involvement.
- **Post-viral fatigue syndrome:** Type B viruses have been associated with a condition called **post-viral fatigue syndrome** but neither the condition nor the association has been clearly defined.

Epidemiology

- Like other enteroviruses, the coxsackie virus primarily inhabits the alimentary canal and spreads by the fecal–oral route.
- Coxsackie B virus epidemics tend to occur every 2–5 years.
- Infants are most commonly affected.
- Vaccination is not practicable as there are several serotypes, and immunity is type-specific.
- Many enteroviruses bear similarities with coxsackieviruses.
- They may occur together in the environment, sewage, or in the gut of the human host.

Laboratory diagnosis

- **Virus isolation:** The virus may be isolated from lesions or feces by inoculation into suckling mice.
- **Identification:** The virus is identified based on histopathological studies on infected mice and by performing neutralisation tests.
- **Tissue culture:** Not all serotypes grow well in cell lines; hence, tissue culture is not useful. All coxsackie B viruses grow well in monkey kidney tissue cultures, while only types 7 and 9 of group A grow well in this type of culture. The A21 virus grows on HeLa cells.
- **Serodiagnosis:** Due to the existence of several antigenic types, serodiagnosis has not been of much use in diagnosis.
- **Animal inoculation:** Coxsackievirus can be isolated by inoculation in suckling mice. Inoculation is usually achieved by the intracerebral, subcutaneous, and intraperitoneal routes. Adult mice are not susceptible. Suckling hamsters can also be infected experimentally.

Echoviral diseases

- The introduction of tissue culture techniques in diagnostic virology led to the isolation of several cytopathogenic viruses from the feces of sick as well as healthy persons. They were called **orphan viruses** because they could not be associated with any particular clinical disease.
- They came to be known by the descriptive term enteric cytopathogenic human orphan (ECHO) viruses.
- Echoviruses resemble other enteroviruses in their properties. They have been classified into 34 serotypes based on neutralising antibodies.

Clinical spectrum

- Echoviruses are perhaps the most common cause of **aseptic meningitis.**
- **Fever with a rash and aseptic meningitis** occurs sometimes as epidemics caused by types 4, 6, 9, 23, 20, 28, and 30.
- They have also been implicated in **respiratory disease in children** (types 1, 11, 19, 20, and 22) and **gastroenteritis** (type 18), but their causative role has not been proved.
- Occasional cases of **paralysis and hepatic necrosis** have been reported.
- Most other infections by echovirus are asymptomatic or may resemble those produced by coxsackieviruses.

Epidemiology

- Like other enteroviruses, echoviruses primarily inhabit the alimentary tract and are spread by fecal transmission through contaminated drinking water.
- Epidemics may occur, especially in summer.
- Vaccination has not been attempted.

Laboratory diagnosis

- All echoviruses grow well in human and simian kidney cultures, producing **cytopathic effects**.
- Feces, throat swabs, or CSF may be inoculated into monkey kidney tissue cultures, and virus growth can be detected by cytopathic changes.
- Serological diagnosis is not practicable except in cases of epidemics where the causative serotype has been identified.

New enterovirus types

- New enteroviruses include types 68–71.
- **Type 68** was isolated from pharyngeal secretions of children with pneumonia and bronchitis.
- **Type 69** is not associated with any human disease.
- **Type 70** causes acute hemorrhagic conjunctivitis.
- **EV-71**, originally isolated from cases of meningitis and encephalitis, causes many other syndromes, including hand, foot, and mouth disease (HFMD). Enterovirus 71 has been associated with fatal meningitis, encephalitis, and paralysis.
- Acute hemorrhagic conjunctivitis caused by **EV-70** is discussed in Chapter 33. Radiculomyelopathy has been reported from India as a complication. Sometimes, it leads to paralysis resembling poliomyelitis.
- The most recently identified enteroviruses include EV-79–88, EV-97, and EV-100–123.

Rabies

Case 6. *A five-year-old boy who was bitten by a rabid dog and presented to the emergency department with multiple scratches on the upper arms and neck. His parents stated that the dog was a stray that lived in their locality (and hence could be observed). Wound toilet was performed, followed by the application of povidone-iodine. The wound was left open. The boy was given tetanus toxoid (as his booster was due), and a Vero cell culture vaccine was started. He completed the course (0, 3, 7, 14, and 28 days post-exposure). The dog died two days later. Its brain was sent to the veterinary lab and tested positive for* ***Negri bodies*** *on histopathology.*

i. What are Negri bodies?
ii. How can rabies be prevented in human beings?

- Rabies has been recognised since ancient times as a disease transmitted to humans and animals by the bite of 'mad dogs'.
- The name 'rabies' comes from the Latin word *rabidus,* meaning mad, which in turn is derived from the Sanskrit root *rabhas,* meaning frenzy.
- The disease in human beings is called **hydrophobia** because the patient exhibits a fear of water.
- It is caused by the rabies virus.

Louis Pasteur (in 1881) established that the rabies virus was present in the brain of infected animals. By serial intracerebral passage in rabbits, he obtained the fixed virus that could render dogs immune by a series of injections of fixed virus. This vaccine was prepared by drying pieces of spinal cord taken from rabbits infected with the fixed virus.

In July 1885, Joseph Meister, a nine-year-old boy, severely bitten by a rabid dog and in grave risk of developing rabies, was given a course of 13 inoculations of the infected cord vaccine by Pasteur. The boy survived. This dramatic event was a milestone in the development of medicine.

Epidemiology

- Rabies is endemic in India.
- It has been estimated that more than 30,000 people die of rabies in India every year and more than 700,000 receive the antirabies vaccine. In India, antirabies treatment is to be considered following the bite of any animal except rats.
- With the dog population in India estimated to be over 23 million, the problem is immense.
- The **National Rabies Control Programme (NRCP)** has both human and animal health components. The human health component is being implemented across the country.

Rabies in dogs (or other animals)

- About 60 per cent of rabid dogs shed the virus in their saliva.
- Rabid dogs usually die in 3–5 days.
- In dogs, rabies may be of two types:
  - **Furious rabies**: This is the more common type. The dog runs amok, biting without provocation and indiscriminately. The lower jaw droops and saliva drips uncontrollably from the mouth. Paralysis, convulsions, and death follow.
  - **Dumb rabies**: This is the paralytic form in which the animal lies huddled, unable to feed. The dog may not bite, but attempts to feed it are dangerous.

Types

Two epidemiological types of rabies exist:

i. **Urban,** transmitted by domestic animals like dogs and cats
ii. **Sylvatic,** involving animals in the wild, such as jackals, wolves, foxes, mongooses, skunks, and bats

Agent—Rabies virus

- The natural cycle of the rabies virus is seen in vampire bats. These bats may shed the virus as symptomless carriers over a period of several months. The virus also occurs in insectivorous and frugivorous bats.
- **Humans may be infected by aerosols** if they enter caves colonised by infected bats.
- Though the virus is present in the saliva of patients, direct person-to-person transmission of rabies has not been recorded. Therefore, there is no danger in examining or nursing hydrophobia patients, provided that suitable precautions are taken.

Morphology

- It is a bullet-shaped virus, 180 × 75 nm in size, with one end rounded or conical and the other planar or concave (Fig. 24.16).
- The lipoprotein envelope carries knob-like spikes, **composed of glycoprotein G**.
- Beneath the envelope is the membrane or **matrix (M) protein**.
- The membrane may project outwards from the planar end of some virions, forming a **bleb**.
- The core of the virion consists of a helically arranged ribonucleoprotein.

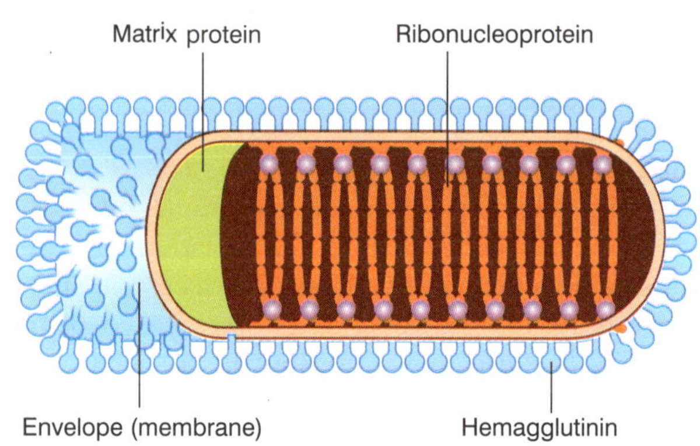

Fig. 24.16 Bullet-shaped rabies virus with its structural elements

- The genome is an unsegmented, linear, negative-sense RNA.
- RNA-dependent RNA transcriptase and some structural proteins are present in the nucleocapsid.

Antigenic properties

i. **Glycoprotein:** The surface spikes, which are composed of glycoprotein G, are important in pathogenesis, virulence, and immunity. The important properties of this glycoprotein are as follows:
- It mediates the **binding** of the virus to acetylcholine receptors in neural tissues.
- It induces **hemagglutination inhibiting (HI) antibodies**. As a result, the rabies virus possesses hemagglutinating activity, optimally seen with goose erythrocytes at 0–4°C and a pH of 6.2.
- It induces **neutralising antibodies (protective antibodies).**
- It stimulates cytotoxic T cell immunity.
- Purified glycoprotein may act as a safe and effective subunit vaccine.

ii. **Nucleoprotein:** This is a nucleocapsid protein, with the following properties:
- It induces **complement-fixing antibodies**, which are not protective.
- The antigen is **group-specific** and cross-reactions are seen with some rabies-related viruses.
- The antiserum prepared against the nucleocapsid antigen is used in **diagnostic immunofluorescence tests.**
- Other antigens identified include two membrane proteins—**glycolipid and RNA-dependent RNA polymerase.**

Susceptibility

- The virus is inactivated by ethanol, iodine preparations, quaternary ammonium compounds, soap, detergents, and lipid solvents such as ether, chloroform, and acetone.
- It is also inactivated by phenol, formalin, beta-propiolactone, and ultraviolet irradiation.

Host range and cultivation

- **All mammals** are susceptible to rabies infection, though differences in susceptibility exist between species.
- Cattle, cats, and foxes are highly susceptible, whereas skunks, opossums, and fowl are relatively resistant.
- Humans and dogs occupy an intermediate position.

Classification

The rabies virus has been classified into two types, **street and fixed virus**, based on the naturally infective strain and laboratory-maintained strain (Table 24.10).

Table 24.10 Differences between the types of rabies virus

Feature	Street virus	Fixed virus
Isolation	Natural infection in a human or animal	After several serial intracerebral passages in rabbits
Infectivity and fatal encephalitis	Inoculation via any route	More neurotropic, though it is much less infective by other routes
Incubation period	Long and variable incubation period of about 1–12 weeks	Short and fixed incubation period of 6–7 days
Negri body demonstration	Can be demonstrated in the brain of animals dying of street virus infection	Usually not demonstrable
Vaccine production	Not used	Used

Pathogenicity

Transmission

- Human infection is usually caused by the **bite of rabid dogs or other rabid animals**.
- The virus, present in the **saliva** of the animal, is **deposited in the wound**.
- Rarely, infection may occur following non-bite exposures, aerosols, transplantation of cornea, or other virus-infected tissues.
- Humans appear to possess a high degree of natural resistance to rabies.
- Rabies antibodies demonstrated in veterinarians without any history of antirabies vaccination suggests that inapparent infection may occur.
- The **incubation period in humans** is usually 1–3 months, though it may be as short as seven days or as long as three years.
- The incubation period is usually short in persons bitten on the face or head and long in those bitten on the legs. This may be related to the **distance the virus has to travel to reach the brain.**
- The incubation period is generally shorter in children than in adults (Case 6).

Sequence of events in rabies

- The virus appears to multiply in the **muscles, connective tissue, or nerves** at the site of deposition for 48–72 hours.
- It then spreads via the peripheral nerves to the **brainstem or spinal cord** at a rate of 3 mm/hour via **retrograde axonal transport** (Fig. 24.17a and b).
- It disseminates throughout the central nervous system along the neuroanatomical connections and selectively injures neuronal processes (axons and dendrites) without producing neuronal death.

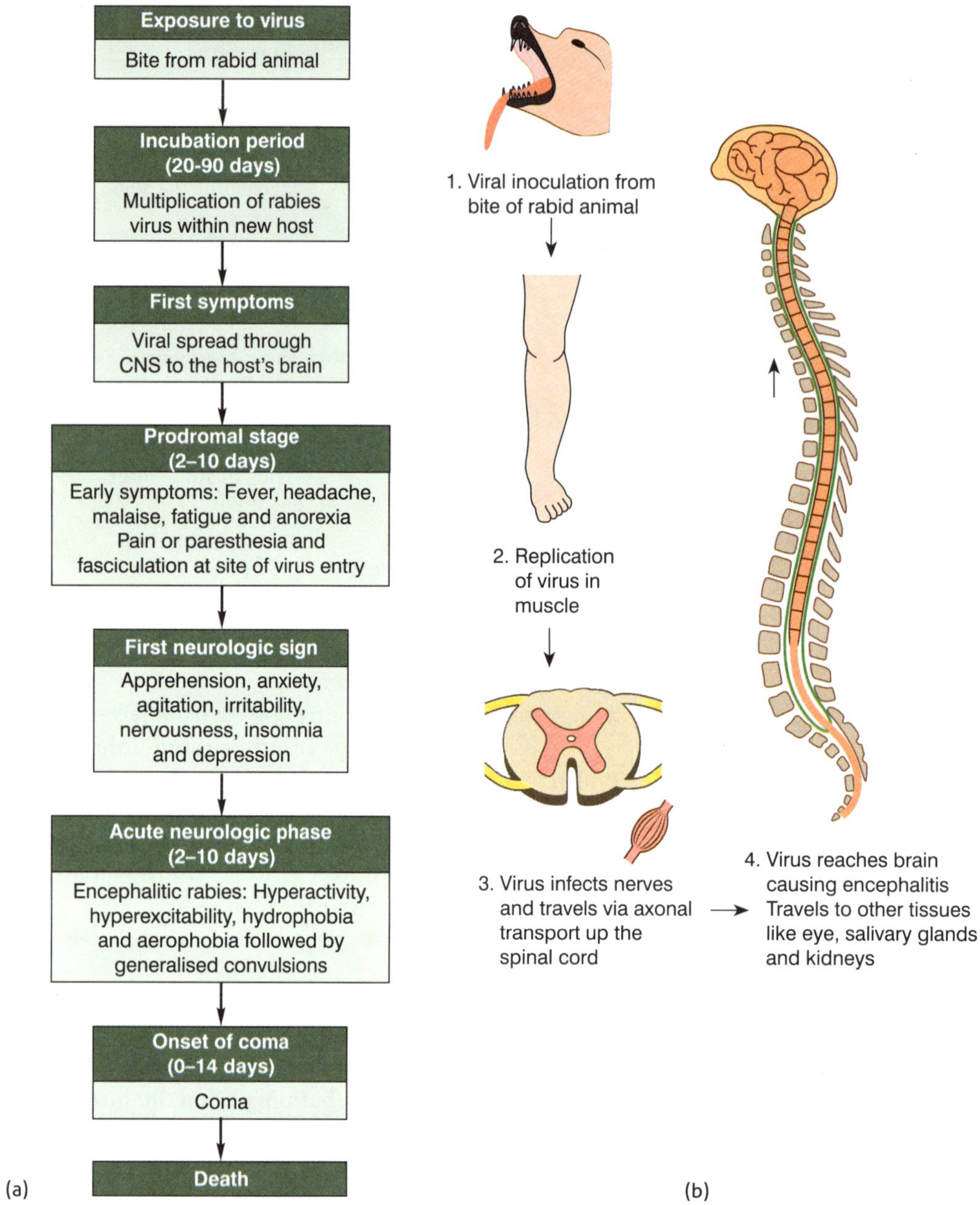

Fig. 24.17 (a) Flowchart depicting the pathogenesis and sequence of events in rabies and (b) diagrammatic representation of pathogenesis

- Histopathologic changes in encephalitis show **perivascular cuffing** due to the accumulation of inflammatory cell infiltrates, including lymphocytes and polymorphonuclear leukocytes (PMNs; Fig. 24.18).
- The virus also multiplies in the **salivary glands** and is shed in saliva.
- It ultimately reaches virtually every tissue in the body, though centrifugal dissemination may be interrupted at any stage by death.
- The virus is almost invariably present in the cornea and facial skin of patients, which provides a method for the antemortem diagnosis of human rabies.

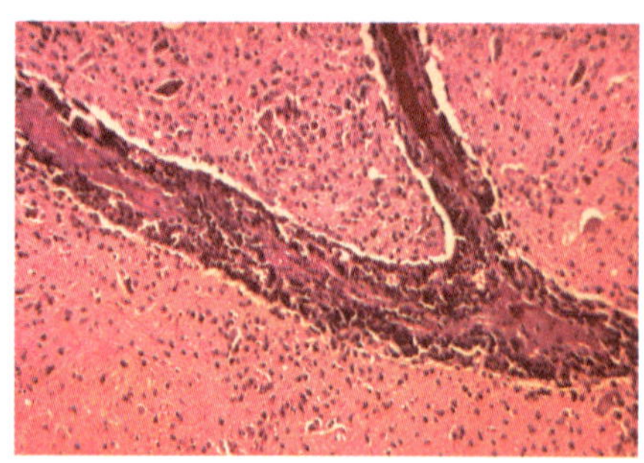

Fig. 24.18 Histopathologic changes associated with rabies encephalitis show perivascular cuffing by inflammatory cells like lymphocytes and polymorphonuclear leukocytes (*Source*: CDC, PHIL, Image ID 4069)

Clinical features

The course of the disease in humans can be classified into four stages, which are as follows:

Prodrome phase

This is characterised by:

- Fever, headache, malaise, fatigue, and anorexia
- Neuritic type of pain or paraesthesia and fasciculation at the site of virus entry
- Apprehension, anxiety, agitation, irritability, nervousness, insomnia, and depression
- Excessive libido, priapism, and spontaneous ejaculation may occur rarely
- Lasts 2–10 days

Encephalitic rabies

- The acute neurological phase usually begins with hyperactivity, which is characteristically intermittent with bouts of bizarre behaviour, agitation, or seizures appearing between apparently normal periods in 2–10 days.
- Such hyperactivity may be spontaneous or precipitated by external stimuli.
- The pathognomonic feature is **difficulty in drinking along with intense thirst**. Patients may be able to swallow dry solids but not liquids. Attempts to drink bring on such painful spasms of the pharynx and larynx with such violent choking or gagging that patients develop a dread of even the sight or sound of water (**hydrophobia**).
- Cardinal features are **hyperexcitability, hydrophobia, and aerophobia** followed by **generalised convulsions.**
- In about 20% of cases, encephalitic rabies does not develop.

Paralytic rabies

- **Muscle weakness** is predominant with flaccid weakness starting at the site of the bite, which spreads to produce **quadriparesis and facial weakness** in 2–10 days.
- These patients may survive a little longer than the encephalitic group, but ultimately, **multiorgan failure occurs, followed by death.**
- Paralytic rabies may be confused with Guillain–Barré syndrome.

Coma and death

- Some patients may enter into a comatose state.
- Death usually occurs within 1–14 days due to respiratory arrest during convulsions.

Laboratory diagnosis (Fig. 24.19)

- In humans, routine laboratory findings are often normal.
- CSF shows **mild mononuclear cell pleocytosis** with **slightly elevated protein**.
- CT and MRI are variable and non-specific.

Specific confirmatory diagnosis

Specimen

- Corneal impression smears are rarely used as their sensitivity is very low

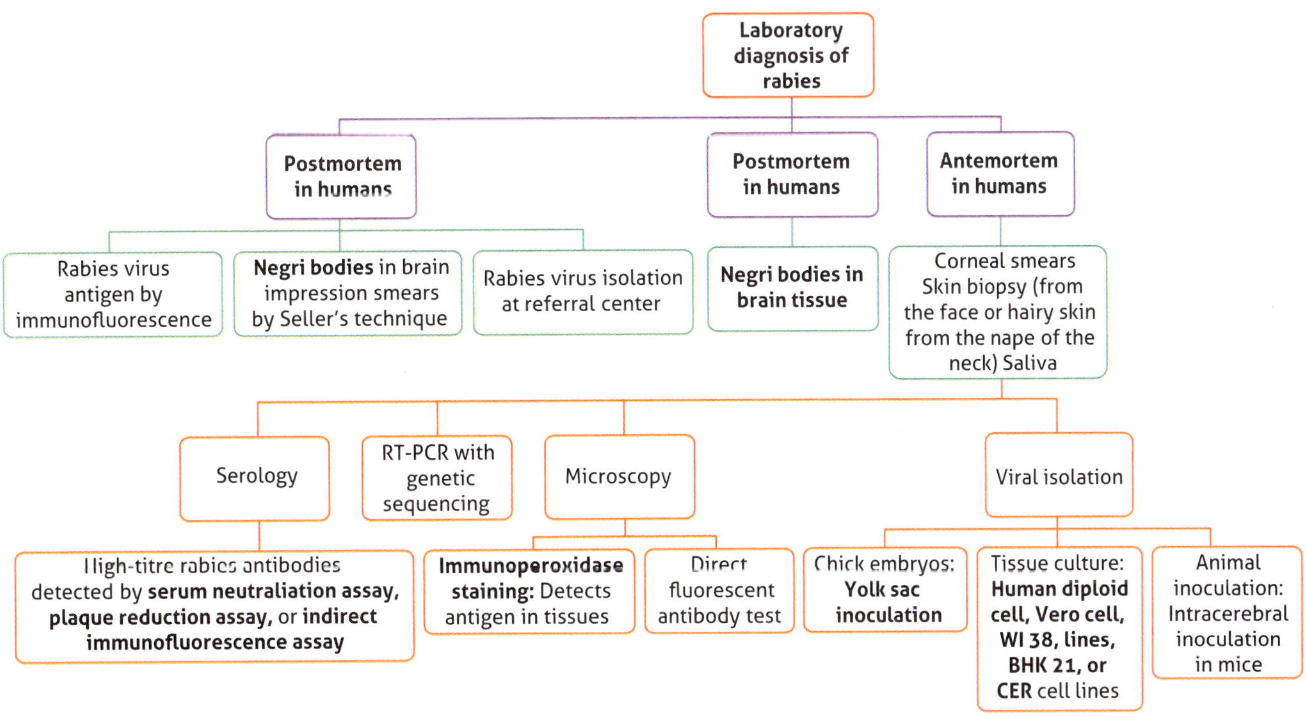

Fig. 24.19 Laboratory diagnosis of rabies

- Skin biopsy (from the face or hairy skin from the nape of the neck)
- Saliva
- Brain biopsy in postmortem cases

Direct fluorescent antibody testing (dFA)

- Skin biopsy, corneal smears, or fresh saliva are used to detect the rabies antigen.
- Impressions smears or frozen sections of the brain or nervous tissue are treated with antirabies serum or globulin conjugated with **fluorescein isothiocyanate** and examined under **ultraviolet light**.
- **Immunoperoxidase staining** can be used to detect antigen in tissues.

Antibody detection

- Antibodies can be detected in CSF in rabies encephalitis by **serum neutralisation assay**, **plaque reduction assay,** or **indirect immunofluorescence assay.**
- High-titre rabies antibodies are present in the CSF in rabies during infection but not after immunisation.

Molecular diagnosis

RT-PCR with genetic sequencing is highly sensitive and can detect the viral RNA from CSF, saliva, skin, and brain tissue.

Postmortem diagnosis

- **Negri bodies** can be demonstrated in the brain tissue from the **pyramidal cells of Ammon's horn** and the **Purkinje cells of the cerebellum**.
- They may be absent in about 20% of cases.

Virus isolation

- This is done only in reference laboratories with **biosafety level III facilities**.
- Viral isolation is required for **vaccine production** and **rarely for diagnosis**.

The following methods are used for the isolation of the rabies virus.

i. *Chick embryos*
   - **Yolk sac inoculation** is used.
   - Serial propagation in chick embryos has led to the development of **attenuated vaccine strains like Flury and Kelev.**

ii. *Tissue culture*
   - The fixed virus strains are adapted for growth in **human diploid cell** and **Vero cell cultures**.
   - Tissue culture cell lines such as **WI 38, BHK 21, or CER** are used.
   - Growth in cell lines can be detected by immunofluorescence from as early as 4–5 hours up to 5 days following inoculation.

iii. *Animal inoculation*
   - This is done by intracerebral inoculation in mice from the brain tissue, CSF, saliva, and urine.
   - Mice are examined for signs of illness, and their brains are examined at death or at 28 days post-inoculation for Negri bodies; immunofluorescence can also be used for identification.
   - Neutralising antibodies can be detected.

Diagnosis in animals

The diagnosis of rabies is of great importance in assessing the risk of infection and deciding post-exposure treatment.

- The portion of the animal's brain sent to the reference laboratory for confirmation should include the **hippocampus and cerebellum** since Negri bodies are most abundant in these areas.
- **Demonstration of inclusion bodies (Negri bodies)**—Impression smears of the brain are stained by **Seller's technique** (basic fuchsin and methylene blue in methanol). Negri bodies are seen as **intracytoplasmic, round or oval, purplish-pink structures with characteristic basophilic inner granules**. Negri bodies vary in size, ranging from 3–27 μm (Fig. 24.20). Failure to find Negri bodies does not rule out the diagnosis of rabies.
- Demonstration of rabies virus antigen by immunofluorescence—Examination of the **animal's salivary glands by immunofluorescence** is useful. It may indicate whether or not the animal was shedding the virus in saliva.
- Rabies virus isolation and other laboratory procedures are performed in reference laboratories equipped with biosafety cabinets and expertise to handle rabid tissues. The **National Centre for Disease Control (NCDC)** in Delhi is the designated reference laboratory in the country.

Prophylaxis for rabies

Active immunisation (antirabies vaccines)

- Licensed rabies vaccines for human use are based on inactivated purified rabies virus grown either in tissue culture or in embryonated duck or chicken eggs.

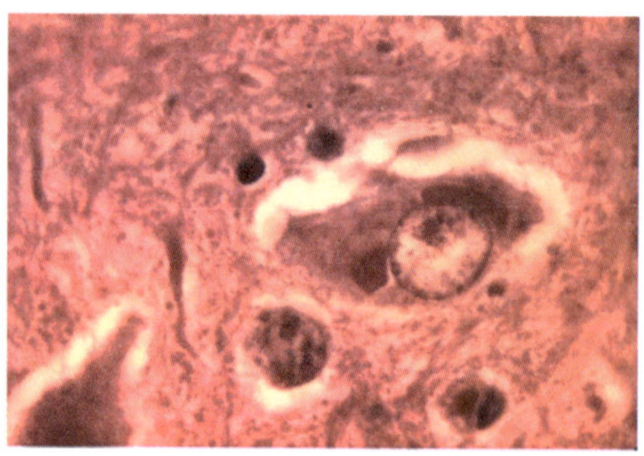

Fig. 24.20 Seller's stain showing intracytoplasmic inclusion body (**Negri body**), which is found most frequently in the pyramidal cells of Ammon's horn and the Purkinje cells of the cerebellum (*Source*: CDC, PHIL, Image ID 3372)

- Nerve-tissue grown vaccines, which are less costly but are also less immunogenic and can have serious side-effects, are no longer recommended by the World Health Organization and their use has largely been discontinued.

i. **Tissue culture vaccines**

- The **human diploid cell (HDC) vaccine** was the first cell culture vaccine. It was developed by Hilary Koprowski, Tadeusz Wiktor, and Stanley Plotkin.
- It is a purified and concentrated preparation of a **fixed rabies virus (Pitman–Moore strain)** grown on human diploid cells (WI38 or MRC5) and inactivated with beta-propiolactone or tri-n-butyl phosphate.
- **Advantage:** It is highly antigenic and free from serious side effects.
- **Disadvantage:** It is very expensive.
- **Dosage:** The HDC vaccine is given intramuscularly or at a 5–10-fold lower dose intradermally.

ii. **Chick-embryo cell vaccine and purified duck embryo vaccine**

- Purified duck embryo vaccine was the first vaccine developed for human use in 1957.
- It is made of dried, killed rabies viruses.
- Vaccination with PDEV involved a series of intradermal injections over several days.
- It is now replaced with cell culture vaccines.

iii. **Subunit vaccine**

- The **glycoprotein subunit on the virus surface**, which is the protective antigen, has been cloned and used to produce recombinant vaccines.
- This vaccine is still in the experimental stage.

iv. **Neural vaccines**

- These are **suspensions of nervous tissues of animals infected** with the fixed rabies virus.
- These are unsatisfactory as they are poor immunogens, largely containing the nucleocapsid antigen.
- They may contain infectious agents and hence, may be encephalitogenic.
- **Neural vaccines have now been abandoned in most parts of the world** since tissue culture vaccines are available at an affordable cost.

In India, the following rabies vaccines are available:

- **Purified Vero cell rabies vaccine (PVRV)**
- **Chromatographically purified rabies vaccine (CPRV)**
- **Human diploid cell vaccine (HDCV)**
- **Purified chick embryo cell vaccine (PCECV)**
- **Purified duck embryo vaccine (PDEV)**

Passive immunisation with antirabies serum

- Antirabies serum can be prepared by the hyperimmunisation of horses **(equine rabies immunoglobulin (ERIG) or humans (human rabies immunoglobulin (HRIG).**
- Immunisation achieved by administering human rabies immunoglobulin (HRIG) pooled from the sera of immunised human donors.
- HRIG is preferred over ERIG as it is free from the danger of sensitisation but should also be free from HIV and hepatitis viruses.
- Equine rabies immunoglobulin is no longer used due to adverse effects of type 3 hypersensitivity.

Prophylaxis

Pre-exposure prophylaxis

- Immunisation is recommended for persons at high risk such as **veterinarians and dog handlers**.
- **Cell culture vaccines** are used for pre-exposure vaccination.
- The vaccine is administered as one full dose of vaccine intramuscularly or 0.1 mL intradermally on days 0, 7, and either day 21 or 28.
- Neutralising antibody titres must be checked every six months during the initial two-year period after the primary vaccination. If it is less than 0.5 IU/mL, a booster dose of vaccine should be given **(National Guidelines on Rabies Prophylaxis)**.
- Rabies immunoprophylaxis is mandatory for all pet animals.

Post-exposure management

- Specific prophylaxis is generally used after exposure to infection and is therefore called antirabies treatment.
- This consists of local treatment, active immunisation with antirabies vaccines, and passive immunisation with antirabies serum as follows:
- **Local treatment**:
  - The wound should be scrubbed immediately with soap and water.
  - The wound is then treated with quaternary ammonium compounds (such as **cetavlon**), tincture or aqueous solution of **iodine**, or **alcohol** (40–70%).
  - In severe wounds, **antirabies serum may be applied topically** and infiltrated around the wound.
  - It is advisable to postpone suturing the wound.
  - Anti-tetanus measures and antibiotics to prevent sepsis may be used as necessary.
- **Post-exposure prophylaxis**
  - Antirabic vaccine should be administered when a person has been bitten, scratched, or licked by an animal that is rabid or cannot be apprehended.
  - When the animal can be observed, it should not be killed. Instead, it should be observed for 10 days.
  - This observation period is recommended because the virus may be present in saliva 3–4 days before

the onset of symptoms and the animal usually dies within 5–6 days of developing the disease.
- If the animal remains healthy after this period, there is no risk of rabies, and vaccination, if already started, may be discontinued.
- A rare possibility of the carrier state in dogs cannot be ruled out.
- In cases where the vaccine is started and the animal kept under observation, an alternative recommendation is to stop treatment after five days. The animal is observed for another five days, and the vaccine is started again if the animal becomes ill or dies during this period.
- The WHO guidelines on post-exposure prophylaxis are based on the risk category to which the patient belongs (Table 24.11).
- All three cell culture vaccines available in India **(HDC, PCECV, and PVRV)** have the same dosage schedule for adults as well as children.
- Exposure to rodents, rabbits, and hares seldom, if ever, requires specific antirabies post-exposure prophylaxis.
- If an apparently healthy dog or cat in/or from a low-risk area is placed under observation, the situation may warrant delaying initiation of treatment.
- Post-exposure prophylaxis should be considered when **contact between a human and a bat** has occurred, unless the exposed person can rule out a bite or scratch or exposure to a mucous membrane.
- This schedule is recommended by the Govt. of India in the National Guidelines on Rabies Prophylaxis on post-exposure regimen (updated Thai Red Cross Schedule [2-2-2-0-2]).
- **Dose**—0.1 mL of the reconstituted vaccine is administered per intradermal site and on two sites per visit (one on each deltoid area, an inch above the insertion of deltoid muscle) on days 0 (date of first dose and not animal bite), 3, 7, and 28.
- The duration of protection is five years, during which period, any further exposure may need only one or two booster doses (on days 0 and 3), depending on the degree of risk. After five years, it is advisable to give a full five-injection course on exposure to infection.
- The recommended route is IM or SC in the deltoid region, or in children, on the anterolateral aspect of the thigh. Gluteal injections are to be avoided as they have been found to be less immunogenic.
- **Vaccine for animals:** Antirabies immunisation in animals is to be performed as pre-exposure prophylaxis.

Treatment

- Rabies is considered to be an invariably fatal disease for which no serious treatment is available, apart from sedation.
- No specific antirabies agent is available.
- Very few cases have been reported to survive rabies in literature.

Slow Viral Diseases

- The term 'slow virus disease' is applied to a group of infections in animals and human beings characterised by a **very long incubation period and a slow but relentless course, terminating fatally just as is seen in rabies.**
- The recognition in recent years that some chronic degenerative neurological diseases of human

Table 24.11 Type of contact, exposure, and recommended post-exposure prophylaxis

Category	Type of contact with a suspected or confirmed rabid domestic or wild animal or animal unavailable for testing	Type of exposure	Recommended post-exposure prophylaxis
I	Touching or feeding of animals Licks on intact skin	None	None, if reliable case history is available
II	Nibbling of uncovered skin Minor scratches or abrasions without bleeding	Minor	Administer vaccine immediately Stop treatment if animal remains healthy throughout an observation period of 10 days or if animal is proven to be negative for rabies by a reliable laboratory, using appropriate diagnostic techniques
III	Single or multiple transdermal bites or scratches or licks on broken skin Contamination of mucous membrane with saliva (i.e., licks) Exposures to bats	Severe	Administer rabies immunoglobulin and vaccine immediately Stop treatment if animal remains healthy throughout an observation period of 10 days or if animal is found to be negative for rabies by a reliable laboratory using appropriate diagnostic techniques

beings may have a similar pathogenesis has led to considerable interest in this concept.

Characteristics of slow virus infections

- Incubation periods ranging from months to years
- Course of illness lasting for months or years, with remissions and exacerbations
- Predilection for the involvement of the central nervous system
- Absence of immune response or an immune response that does not arrest the disease but may actually contribute to pathogenesis
- Genetic predisposition
- Invariable fatal termination

Slow virus diseases may be caused by—(i) conventional viruses, (ii) slow viral diseases of animals, and (iii) prions.

i. Conventional viruses

a. Subacute sclerosing panencephalitis (SSPE)

- It is seen in **young adolescents**.
- It is a very rare, **delayed sequel to infection with the measles virus**.
- The disease sets in many years after the initial infection and is characterised by progressive deterioration of mental and motor functions.
- Death occurs 1–3 years after the onset of symptoms.
- Brain cells from patients show serological and electron microscopic evidence of measles virus infection. The virus cannot be isolated in routine cultures, but only by co-cultivation of infected brain cells with susceptible cells of non-neural origin.
- Measles virus strains isolated from SSPE are defective.
- Patients show very high levels of measles virus antibody in their serum.
- The **antibody is regularly found in CSF** and is pathognomonic.
- Cellular immune response to the measles virus is absent in SSPE.

b. Progressive multifocal leukoencephalopathy (PML)

- It is a rare, subacute demyelinating disease seen in **elderly persons** whose immune process is impaired as a result of malignancy or immunosuppression.
- There is **progressive deterioration of motor function, vision, and speech**.
- Death occurs in 3–4 months.
- **Human polyomavirus** has been demonstrated by electron microscopy and cultured from **brain biopsies** of patients.

ii. Slow viral infections of animals

These are slowly progressive infections of sheep and other ruminants, caused by serologically related, non-oncogenic retroviruses called **lentiviruses** (from the Latin *lentus*, meaning slow).

iii. Prion diseases

Prions are **protein-containing particles** that are infectious in nature. Some characteristics of prions are as follows:

- They are **devoid of DNA and RNA.**
- They are unusually resistant to physical and chemical agents such as heat, irradiation, and formalin.
- They can be transmitted to experimental animals by parenteral and oral challenge.

> In 1982, **Stanley B. Prusiner** identified certain proteinaceous infectious agents, which he named prions. He was awarded the Nobel Prize for Medicine in 1997 for his pioneering work in this area.

Pathogenic mechanism of prions

- The first step is the proliferation of an **abnormal prion protein (scrapie; PrPsc)**, which is derived from the normal cellular prion protein known as PrPc.
- This is followed by the accumulation of PrPsc in the central nervous system as diffuse deposits and in the form of plaques, which disrupts the architecture and function of the brain and causes cell death, and consequently, disease.

Human prion diseases

- Human prion-mediated diseases are called **transmissible spongiform encephalopathies (TSE)** or **subacute spongiform viral encephalopathies (SSVE).**
- They are **chronic, progressive, and degenerative diseases** of the CNS.
- The pathology of these diseases consists of progressive vacuolation in the dendritic and axonal processes of the neurons and extensive astroglial hypertrophy and proliferation, culminating in spongiform degeneration in the grey matter.
- There is no sign of any inflammatory or immune response.

a. Creutzfeldt–Jakob disease (CJD)

- CJD is characterised by subacute, presenile encephalopathy, with progressive incoordination and dementia, ending fatally in about a year.
- Both sporadic and inherited forms of the disease have been seen.
- Iatrogenic CJD has been known to occur after corneal transplantation and injection of the pituitary growth hormone, presumably from donors who had the infection.

b. Variant CJD

- The appearance of a new variant of CJD affecting younger persons (below 45 years) in Britain in 1996 raised fears of the infection getting transmitted through the consumption of beef from animals effected by **bovine spongiform encephalopathy (BSE)**.
- This aroused a panic reaction about the export of British beef, and many thousands of British cattle had to be slaughtered before the anxiety was allayed.

c. Gerstmann–Sträussler–Scheinker (GSS) syndrome and fatal familial insomnia

- These are two variants of CJD, the familial form of which is rare.
- These are caused by the inheritance of mutations of the *PrP* gene.

d. Kuru

- Identified in 1957, Kuru (meaning tremor) was a mysterious disease seen only in the Fore tribe inhabiting the eastern highlands of New Guinea.
- The disease had an incubation period of 5–10 years and led to progressive cerebellar ataxia and tremors, ending fatally in 3–6 months.
- The infection is believed to have been introduced through cannibalism and maintained by the tribal custom of eating the dead bodies of relatives after death as a part of the ritual.
- The disease disappeared following the abolition of cannibalism in New Guinea in 1958–59.
- **Carlton Gajdusek** was awarded the Nobel Prize for Medicine in 1976 for his important contributions to the study of Kuru.

Prion diseases of animals

- **Scrapie:** This is the prototype prion disease. It has been known as a natural disease of sheep for two centuries.
- **Mink encephalopathy:** This is a scrapie-like disease of mink. It is believed to have spread to mink through the ingestion of scrapie-infected sheep meat.
- **Bovine spongiform encephalopathy (BSE, 'mad cow disease'):** It has been enzootic in Britain since 1986. The infection is presumed to have spread to cattle through the ingestion of cattle feed contaminated with scrapie-infected meat and bone powder.

Laboratory diagnosis of prion diseases

- CSF is usually normal but may show protein elevation.
- The CT may be normal or may show **cortical atrophy.**
- **MRI** is more useful in differentiating CJD from other neurological diseases.
- The **measurement of PrPsc** is done by inducing proteolysis, which generates the PrP, which is then detected by immunoassay after denaturation.
- Sequencing of the *prnp* gene—this is done to detect the wild type gene sequence or mutations; it helps in diagnosing CJD or familial prion disease respectively.

Arboviral Diseases

- Arboviruses (arthropod-borne viruses) are a diverse group of RNA viruses that have common ecological and epidemiological attributes but diverse physical and chemical properties.
- They multiply in, and are transmitted by, hematophagous insect vectors to vertebrates including humans.
- Insect viruses and viruses of vertebrates that are sometimes mechanically transmitted by insects do not come under this category.
- Arboviruses are classified according to their physical and chemical features into taxonomical families named **Togaviridae, Flaviviridae, Bunyaviridae, Reoviridae, Arenaviridae, and Filoviridae** (Table 24.12).
- Within each family, viruses are classified into genera and antigenic groups based on serological relationships. Some viruses are not grouped.

Clinical spectrum of arboviral diseases

Arboviruses belong to a heterogenous group of viruses with a common mode of transmission—arthropods. However, they manifest as three distinct clinical entities as below:

i. **Encephalitis**
ii. **Fevers with or without rashes**
iii. **Hemorrhagic fevers**

- Primarily febrile illnesses and hemorrhagic episodes caused by arboviruses are considered infections of the cardiovascular system (Chapter 21).
- Arboviruses are worldwide in distribution but are far more numerous in the tropical zones than in the temperate zones.
- Over 500 viruses have been listed, the majority of which cause silent infections in rodents and other wild mammals; about 100 of them can infect humans.
- In India, over 40 arboviruses have been detected, of which more than 10 are known to produce human disease.
- The natural cycles of these viruses involve infected non-human vertebrate hosts including many species of animals and birds.

Arthropod vectors

The ability to multiply in arthropods is a special characteristic of arboviruses. The common arbovirus vectors are:

- Mosquitoes
- Ticks

Table 24.12 Taxonomy, characteristics, classification of some important arboviruses

Family	Characteristics	Genus	Important species	
Togaviridae	Spherical, enveloped viruses with a diameter of 50–70 nm with **single-stranded RNA genome that replicates** in the host cell cytoplasm and is released by budding through host cell membranes. All of them are **mosquito-borne (*Culex* and *Anopheles*).**	Alphavirus	**Fever with arthritis** Chikungunya,	**Encephalitis viruses group**: O'nyong-nyong, Mayaro, Semliki Forest, Sindbis, Ross River, Eastern, Western and Venezuelan equine encephalitis viruses
Flaviviridae	Smaller than alphaviruses, measuring about **40 nm in diameter**. The name Flavivirus refers to the **yellow fever virus** (*flavus* in Latin means yellow). They are worldwide in distribution. Classified into two groups (i) mosquito-borne and (ii) tick-borne viruses.	Flavivirus	**Tick-borne virus:** Russian spring-summer encephalitis complex, Kyasanur Forest disease, Omsk hemorrhagic fever viruses **Mosquito-borne virus**: Dengue types 1–4, Yellow fever, Zika virus	**Mosquito-borne encephalitis:** St. Louis encephalitis virus Ilheus virus West Nile virus Murray Valley encephalitis virus Japanese encephalitis (JE)
Bunyaviridae	Bunyaviruses cause fevers, sometimes with a rash, hemorrhagic hepatitis, encephalitis, or blindness. Bunyaviruses are spherical, enveloped particles 90 to 100 nm in diameter with three segments of single-stranded RNA combined with nucleoprotein. They replicate in the cytoplasm.	Bunyavirus	California encephalitis, Oropouche, and Turlock viruses	
		Phlebovirus	Sandfly fever and Rift Valley fever viruses	
		Nairovirus	Crimean Congo hemorrhagic fever, Nairobi sheep disease, and Ganjam viruses	
		Hantavirus	Hantan, Seoul, Puumala, Prospect Hill, and Sin Nombre viruses	
Reoviridae	Orbiviruses contain 10 segments of double-stranded RNA, exhibit icosahedral symmetry, and are spherical with a diameter of 60 to 80 nm. They replicate in the cytoplasm. They replicate in arthropods as well as in vertebrates and are transmitted by arthropods.	Orbivirus	Colorado tick fever, African horse sickness, and Bluetongue viruses	
Arenaviridae	Round, oval, or pleomorphic, enveloped viruses 110 to 130 nm in diameter. The genome consists of two distinct single-stranded viral RNA species, called L and S. These are generally spread by rodents, with each virus associated with one, or a few, closely related rodent species that may serve as the virus' natural reservoir. Arenaviruses are thought to multiply like typical antisense RNA viruses.	Arenavirus	Junin, Machupo, lymphocytic choriomeningitis, and Lassa fever viruses	
Filoviridae	Filoviruses can cause hemorrhagic fever in humans and primates. Filoviruses are filamentous enveloped particles with a negative-sense single-stranded RNA genome having a single glycoprotein spike on the surface and four structural proteins which include the virus-encoded polymerase.	Marburgvirus	African hemorrhagic fevers and Ebola viruses	

- *Phlebotomus*
- *Culicoides*
- *Cimicidae*

The rubella virus of the genus **Rubivirus** is also a togavirus but is not arthropod-borne and differs in its epidemiology and other features (discussed in Chapter 23). Rotavirus belongs to the family Reoviridae and causes diarrhea (Chapter 25).

Antigenic structure

- Three antigens are important in serological studies—hemagglutinins, complement-fixing, and neutralising antigens; these are all integral parts of the virus particle.

- Considerable antigenic cross-reactions occur among arboviruses.
- The **plaque reduction neutralisation test (PRNT)** shows the greatest specificity for the identification of these viruses.

Hemagglutination

- Most arboviruses agglutinate the red cells of day-old chicks or geese.
- Hemagglutination is influenced by pH and temperature, the optimal requirements varying with different viruses.
- Spontaneous elution does not occur.
- **Hemagglutination is inhibited specifically by antibodies;** this property can be used in diagnostic assays and non-specifically by lipoprotein inhibitors in the serum, brain, and other tissues.

Epidemiology

- Arbovirus infections are zoonoses (with a few possible exceptions such as dengue and O'nyong-nyong) maintained in nature by animals in a vector cycle, with human infection occurring when individuals come into contact with the vector either accidentally or on account of living in close proximity with the animal–vector cycle.
- In some areas, the arbovirus infection appears to be seasonal, and outbreaks are known to occur periodically.
- The reason for this could be an increase in the vector population or reintroduction of the virus to the area through migratory birds or other reservoirs.

Patterns

The epidemiology of arbovirus infections has two patterns:

i. It is linked with the ecology of the arthropod vectors and vertebrate hosts. Most arboviruses exist in nature in animal or avian species, in which infection is asymptomatic. The vector arthropod gets infected by biting a viremic vertebrate but becomes infective only after the incubation period (extrinsic incubation period). Human disease results when the virus is accidentally transmitted to humans, either directly by the vector or through the intermediary animal reservoirs.

ii. A second epidemiological pattern is seen in diseases like dengue where no non-human vertebrate host has been identified. Here, the virus is maintained in a cycle composed of humans and the domestic mosquito. Transovarial transmission in ticks maintains the virus in the vector.

Susceptibility

- In general, arboviruses are labile, being readily inactivated at room temperature and by bile salts, ether, and other lipid solvents.
- Infectivity may be retained at –70°C or by lyophilisation.

Pathogenesis

The life cycle of the arbovirus is based on its ability to replicate in the arthropod vector and the primate host (Fig. 24.21).

- The infected, viremic, bloodsucking vector is responsible for the transmission of the virus.
- Humans are generally dead-end hosts of arboviruses as the concentration of the virus in the blood is low, and the viremic period is short.
- Table 24.13 lists the reservoir hosts, geographic distribution, and vectors of some of the arboviruses and rodent-borne viruses associated with different clinical syndromes in humans.

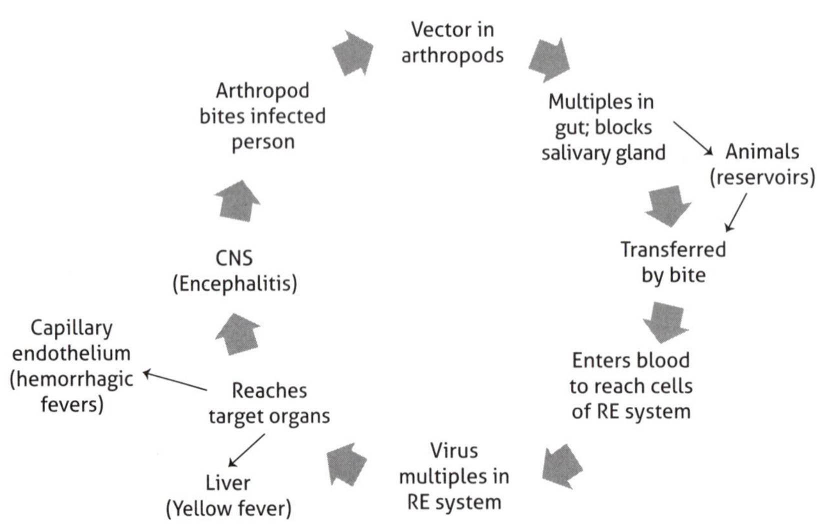

Fig. 24.21 Pathogenesis of arboviruses

Table 24.13 Reservoir hosts, geographic distribution, and vectors of some of the arboviruses and rodent-borne viruses associated with different clinical syndromes in humans

Virus	Genus	Reservoir/hosts	Geographic distribution	Vector
Encephalitis				
Eastern equine encephalitis	Alphavirus	Birds	Americas	Mosquito
Western equine encephalitis	Alphavirus	Reptiles	Americas	Mosquito
Venezuelan equine encephalitis	Alphavirus	Rodents	Americas	Mosquito
St. Louis encephalitis	Flavivirus	Birds	Americas	Mosquito
West Nile encephalitis	Flavivirus	Birds	Africa, Europe, USA, West Asia	Mosquito
Japanese encephalitis*	Flavivirus	Birds	East and Southeast Asia	Mosquito
Murray Valley encephalitis	Flavivirus	Birds	Australia	Mosquito
Hemorrhagic fever				
Chikungunya	Alphavirus	Not known	Africa, Asia	Mosquito
Dengue types 1–4	Flavivirus	Not known	Widespread, especially in Asia and the Caribbean the tropics	Mosquito
Yellow fever	Flavivirus	Monkeys, humans	Africa, South America	Mosquito
Kyasanur Forest disease*	Flavivirus	Rodents, shrews, monkeys and ticks	India	Tick
Crimean Congo hemorrhagic fever*	Nairovirus	Small mammals	The (former) USSR, Central Asia	Tick
Febrile illness with or without rash (non-hemorrhagic) and arthralgia				
Ebola	Filovirus	Chimpanzees, gorillas, fruit bats, monkeys, forest antelope and porcupines	Mainly localised to African countries like Sudan, Zaire, Uganda, and Congo, where it continues in several pockets Outbreaks have occurred in other Western countries: Italy, Spain, UK, and US	Not an arbovirus in the true sense as there are no arthropods incriminated in its transmission
O'nyong-nyong	Alphavirus	Not known	Africa	Mosquito
West Nile fever	Flavivirus	Birds	Asia, Africa, USA	Mosquito
Sandfly fever	Bunyavirus	Not known	The Mediterranean, Asia, tropical parts of America	Sandfly
Rift Valley fever	Bunyavirus	Sheep, cattle	Africa	Mosquito
Colorado tick fever	Orbivirus	Rodents	USA	Tick
Rodent-borne hemorrhagic fevers				
Hantavirus pulmonary syndrome	Hantavirus	Deer mouse	South America and the United States	Deer mice
Hemorrhagic fever with renal syndrome (HFRS)	Hantavirus	Rodents	China, Russia and Korea	Rodents
Lassa fever	Lassavirus	Rodents	Western Africa	Rodents

*Commonly reported from India

Veterinary diseases caused by arboviruses

- Eastern, Western, and Venezuelan equine encephalitis in horses in America
- Rift Valley fever in sheep and cattle in Africa
- Bluetongue in donkeys in India, Africa, and America
- Ganjam disease of sheep in India
- African horse sickness in horses and mules in Africa and Asia

Laboratory diagnosis

Diagnosis may be established by virus isolation or serology. Strict biosafety measures are required to carry out virus isolation.

- **Specimen:**
  - As all arbovirus infections are viremic, blood collected during the early phase of the disease may yield the virus.
  - Arboviruses can also be isolated from cerebrospinal fluid (CSF) in some encephalitic cases and tissues.
- **Virus isolation:**
  - Intracerebral inoculation in suckling mice is the most sensitive method for their isolation. The animals develop fatal encephalitis, though serial blind passages may be necessary in some cases.
  - The yolk sac or chorioallantoic membrane of a chick embryo can also be used.
  - The virus can be cultured on primary cells like chick embryo fibroblasts or continuous cell lines like Vero or HeLa and cultures of appropriate insect tissues (e.g., mosquito cell lines).
- **Identification of isolates:** Isolates can be identified by hemagglutination inhibition, complement fixation, immunofluorescence, immunochromatography, ELISA, or neutralisation with appropriate antisera. Isolation of the virus from insect vectors, birds or animals in reservoir areas indicates arbovirus activity in the area.
- **Serology:**
  - Demonstration of IgM antibodies in the acute phase by ELISA and immunochromatographic methods. A rise in IgG titres in paired serum samples (first sample taken soon after onset and a second sample 2–3 weeks later) to demonstrate a four-fold rise in titre is suggestive of ongoing or recent infection.
  - Serological diagnosis is often complicated due to the antigenic cross-reaction between related viruses (e.g., cross-reaction between Alphavirus and Flavivirus groups).
- **Molecular methods:** The detection of viral proteins, antigens or RNA by molecular methods (conventional or real-time PCR) are more specific and rapid. These methods have replaced conventional CFT, HAI, and neutralisation tests.

Control

- Control measures are indicated only in infections that lead to epidemics or epizootics.
- These consist essentially of **vector control**.

Vaccine

- A vaccine against yellow fever has been successful and is routinely used for travellers to Africa and other endemic areas.
- Vaccines against Japanese encephalitis (JE) have also been used during outbreaks or in endemic areas. However, they have not been used in the control of other arbovirus diseases.
- Vaccines have been used against Japanese encephalitis in Asia and the US. A **live-attenuated SA-14-14-2** vaccine against JE is used in some endemic areas in Asia including India.
- Due to large epidemics, the WHO has initiated vaccine trials against dengue. Trials with tetravalent dengue vaccine are ongoing in India.

Encephalitis

Table 24.14 lists the clinical spectrum of encephalitis caused by arboviruses. Japanese encephalitis (JE) is described in detail due to its importance in India.

Japanese encephalitis (JE)

- This virus is **widespread**—from East Asia to South Asia; from Korea and Japan in the north to India and Malaysia in the south. JE was first recognised in Japan in 1935 and was named **Japanese 'B' encephalitis** to distinguish it from 'encephalitis A' **(encephalitis lethargica or von Economo disease).**
- Several large-scale epidemics of JE have occurred since then.
- Epidemics show seasonal incidence (summer–autumn) in the temperate regions, though this is not evident in the tropical areas.

Clinical features

- The disease has an **abrupt onset** with fever, headache, and vomiting.
- After 1–6 days, signs of encephalitis set in with nuchal rigidity, convulsions, altered sensorium, and coma.
- The fever is high and continuous.
- Neutrophil leukocytosis is noted in peripheral blood, and pleocytosis with normal or raised sugar levels and slightly raised protein levels are noted in the CSF.
- The mortality rate in some epidemics has been as high as 50 per cent.

Table 24.14 Clinical spectrum of the arboviruses causing encephalitis

Virus	Features
Flavivirus	
St. Louis encephalitis virus	Prevalent in North and Central America and is the most important mosquito-borne disease in the USA Wild birds act as the reservoir and *Culex tarsalis* as the vector
Ilheus virus	South and Central America and maintained in forests by a cycle of mosquitoes, wild birds, and monkeys Human infection is largely subclinical Encephalitis is rare
West Nile virus	African countries, Israel, Cyprus, France, and India Dengue-like illness in humans and encephalitis in horses The virus is maintained in nature in wild birds In India, the virus has been isolated from *Culex* mosquitoes and from febrile as well as encephalitic patients from Rajasthan to Karnataka In 1999, West Nile fever appeared in New York, and since then, the virus has spread across much of the USA, becoming a major public health problem
Murray Valley encephalitis virus	Australia and New Guinea Occurs normally in an enzootic cycle involving wild birds and mosquitoes, and to break out only occasionally into epidemics *Culex annulirostris* is the vector
Japanese encephalitis (JE)	Widespread from East Asia to South Asia; from Korea and Japan in the north to India and Malaysia in the south
Alphavirus	
Eastern equine encephalitis (EEE)	Several species of *Culex* and *Anopheles* mosquitoes act as vectors for these viruses; wild birds act as the reservoirs Occurs along eastern Canada, USA and the Caribbean, causing sporadic cases and small epidemics
Western equine encephalitis (WEE)	More widely distributed in America and causes large epidemics Formalinised vaccines have been developed for EEE and WEE
Venezuelan equine encephalitis (VEE)	Prevalent in Central and South America A live-attenuated vaccine for VEE is available for horses but not humans

- Convalescence may take several weeks. Up to 50 per cent of survivors may have neurological or psychiatric sequelae. Infection in pregnant women during the first and second trimesters has led to fetal death.
- The large majority of infections are, however, asymptomatic, and it has been estimated that 500–1,000 inapparent infections occur for every case of clinical disease.

Clinical case definition: Acute onset of fever <7 days duration associated with changes in mental status (may include irritability, somnolence, or abnormal behaviour greater than that seen with usual febrile illness) and/or new onset of seizures (excluding simple febrile seizures).

Epidemiology

Reservoir hosts: Herons (Fig. 24.22)
Amplifier hosts: Pigs
Vertebrate hosts: Humans, cattle, and buffaloes (besides pigs)
Vector: *Culex tritaeniorhynchus*

- Human infection is a 'dead-end' event.
- Natural infection has been demonstrated in *Ardeid* birds (herons and egrets); bird-to-bird transmission also takes place through *Culex tritaeniorhynchus*.
- Other birds such as ducks, pigeons, and sparrows may also be involved.
- The mosquito *Culex tritaeniorhynchus* has a predilection for cattle over humans or pigs. However, cattle do not develop viremia and hence do not spread the virus.

JE in the Indian setting

In India, Japanese encephalitis was first recognised in 1955 in Pondicherry and Vellore and from mosquitoes of the *Culex vishnui* complex.

- Since 1976, the highest rates of human disease have been reported from the states of Andhra Pradesh, Assam, Bihar, Haryana, Goa, Karnataka, Kerala, Tamil Nadu, Pondicherry, Uttar Pradesh, and West Bengal.

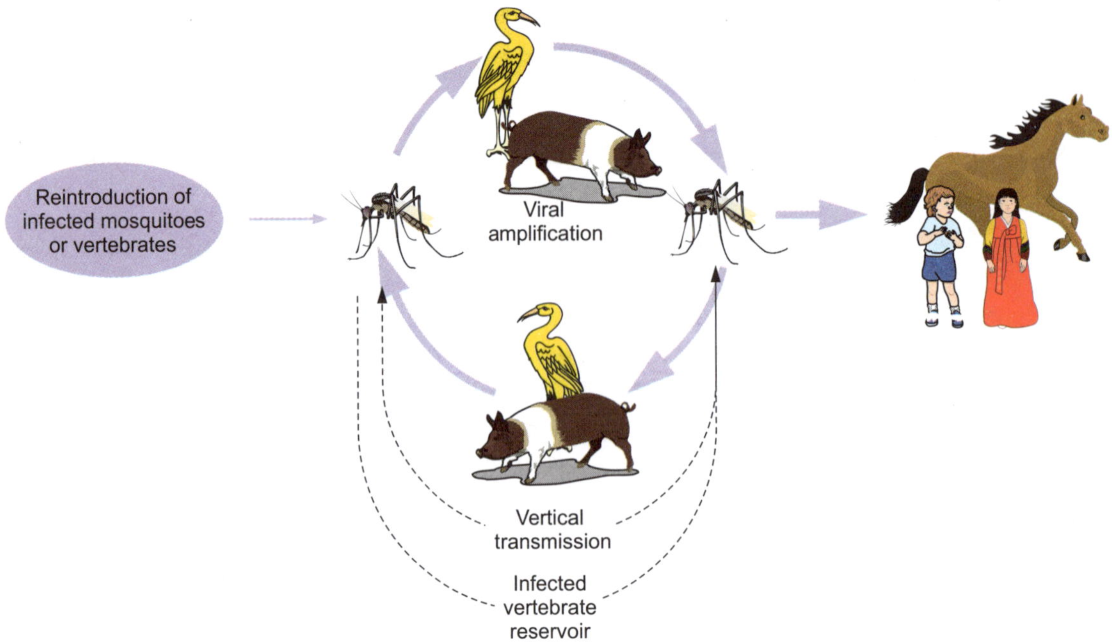

Fig. 24.22 Transmission cycle of Japanese encephalitis virus

- Sporadic cases have been reported from different parts of the country, excepting Dadra, Daman, Diu, Gujarat, Himachal Pradesh, Jammu and Kashmir, Lakshadweep, Meghalaya, Nagar Haveli, Punjab, Rajasthan, and Sikkim.
- Given its prevalence, Japanese encephalitis has become a major public health problem of national importance in India.
- The high cattle–pig ratio in India has been suggested to be a factor contributing to human infection.
- The virus continues to be active in Tamil Nadu, Andhra Pradesh, and Uttar Pradesh, causing illness chiefly in children, indicating the endemic nature of the virus. Cases are reported throughout the year in India, more so during the rainy season.

Laboratory diagnosis

- **Diagnosis is largely serological, however, isolation can be done from the CSF or brain tissue of animals or vectors for epidemiological purposes.**
- **Antibody detection:**
  - **Serological diagnosis** is often complicated due to the antigenic cross-reaction between related viruses.
  - It is based on the demonstration of IgM antibodies in the acute phase by ELISA and immunochromatographic methods.
  - IgM antibodies are usually detectable 3 to 8 days after the onset of illness and persist for 30 to 90 days, though longer persistence has been documented.
  - A rise in IgG titres in paired serum samples (first sample taken soon after onset and a second sample 2–3 weeks later) to demonstrate a four-fold rise in titre, which is suggestive of ongoing or recent infection.
  - Detection can be done by IM antibody capture ELISA, hemagglutination inhibition test, complement fixation test or plaque reduction neutralisation test.
- **Antigen detection in CSF**
  - Monoclonal antibody/immunogold/silver staining (M-IGSS)
  - Reverse passive hemagglutination
  - Immunofluorescence
- **Culture**
  - Intracerebral (I.C.) inoculation is carried out on 1–2 day-old mice pups, which are then observed for clinical features such as paralysis, followed by death. Virus isolation can be carried out from infected mice brains.
  - Susceptible mammalian cell lines such as Vero, BHK, and PS cells are infected and observed for cytopathic effects (CPE).
  - **Molecular techniques:** Detection of viral genomes by RT-PCR has largely replaced conventional cell culture methods can confirm diagnosis.

Laboratory diagnosis of Japanese encephalitis

Clinical case definition: Acute onset of fever <7 days duration associated with changes in mental status and/or new onset of seizures.

Confirmatory test:

- IgM antibody in serum and/or CSF
- Four-fold difference in IgG antibody titre in paired sera

- Virus isolation from brain tissue
- Antigen detection by immunofluorescence
- Nucleic acid detection by PCR

Control measures

Preventive measures include mosquito control and locating piggeries away from human dwellings.

Prophylaxis

- A formalin-inactivated mouse brain vaccine using the Nakayama strain has been used successfully for immunisation in Japan and, on a small scale, in India as well.
- Two doses are administered two weeks apart, followed by a booster 6–12 months later.
- The immunity produced by the vaccine is short-lived.
- Vaccines licensed in India for Japanese encephalitis:
  - Inactivated Vero cell culture-derived SA-14-14-2
  - Inactivated Vero cell culture-derived Kolar strain, 822364XY, JE vaccine

  These are recommended for children living in endemic areas.

West Nile fever

- The **West Nile virus (WNV)** belongs to the Flaviviridae family of viruses.
- ***Culex* mosquitoes** act as vectors for this virus.

Epidemiology

- WNV has been reported from various countries. In India, antibodies to the virus have been reported in human sera in various parts of the country including Tamil Nadu, Karnataka, Andhra Pradesh, Maharashtra, Gujrat, Madhya Pradesh, Odisha, and Rajasthan.
- Serologically confirmed cases have been reported from the Vellore district of Tamil Nadu.

Clinical features

- WNV usually causes a **mild, non-fatal, dengue-like illness** in humans.
- It can also produce a febrile illness in an epidemic form and clinically overt encephalitis.

Laboratory diagnosis

- The diagnosis of WNV infections is based on the detection of IgM antibodies in the acute phase of the illness or rising IgG titres.
- Real-time PCR may be used to confirm WNV infections.

Zika virus infection

- Zika virus has recently been recognised as a global threat.
- Outbreaks have occurred in French Polynesia and the South Pacific islands.
- Its vector is the *Aedes* mosquito.
- Zika is a sylvatic disease that is maintained in mosquitoes and non-human primates.
- The incubation period is 3–14 days.
- Initial symptoms are mild, such as rashes, fever, conjunctivitis, muscle and joint pain, malaise, and headache that usually last for 2–7 days.
- Complications include neurological complications such as **Guillain–Barré syndrome**. Zika virus infection during pregnancy is a cause of microcephaly and other congenital malformations in the infant, including limb contractures, high muscle tone, eye abnormalities, and hearing loss.
- Diagnosis is by detecting viral RNA by **RT-PCR** in serum and urine. Zika IgM antibodies can persist for years following infection, and so, detecting Zika IgM antibodies might not indicate a recent infection.

Tick-borne viral encephalitis

- A number of viruses belonging to the **Russian spring–summer encephalitis (RSSE) complex** cause encephalitis with high rates of fatality and permanent paralytic sequelae in some survivors.
- These infections are prevalent along a wide area of the northern landmass from Scotland to Siberia.
- Biphasic meningoencephalitis may be transmitted to human beings when they drink the milk of infected goats.
- Control of infection by the RSSE complex depends on avoiding tick bites. A formalin-inactivated RSSE vaccine has been found useful.

Fever with Arthritis

Chikungunya (CHIKV)

Epidemiology

- The name 'chikungunya' is derived from the native Makonde word for the disease in which the **patient lies 'doubled up' due to severe joint pain**. This virus was first isolated from human patients in Tanzania.
- **Vectors:** *Aedes aegypti* and *A. albopictus*.
- A mutation of CHIKV in the E1 protein enabled this virus to eliminate its cholesterol dependence. This resulted in better viral uptake, replication, and transmission by *A. albopictus*. This mutation is believed to have been instrumental in causing the outbreak of CHIKV infection in India in 2004–2006.
- *A. albopictus* has a wider geographical distribution than *A. aegypti*.
- No animal reservoir has been identified yet.
- The virus first appeared in India in 1963, when along with dengue, it caused extensive epidemics in Kolkata, Chennai, and other areas. Subsequently, chikungunya outbreaks occurred at regular intervals along the east coast of India and in Maharashtra until 1973.

- After 1973, the virus remained quiescent for several years, only to reappear in 2004–2006, at which time, there was a large outbreak that began in Andhra Pradesh and Tamil Nadu, which spread to Kerala, Karnataka, and Delhi, affecting nearly a million people.

Genotypes

- West African
- East/Central/South African (ECSA)
- Asian genotypes

These genotypes are generally restricted to the geographic areas after which they are named. However, the spread of ECSA to Asian regions has been reported.

Reasons for re-emergence

- The disease appears in epidemics after a gap of a decade or two. This is attributed to mutation of the virus and the wide distribution of the vector *Aedes albopictus*.
- During interepidemic periods, the virus is thought to be maintained in the sylvatic cycle in non-human primates.

Clinical features

- There is a sudden onset of fever, which is typically biphasic with periods of remission of 1–6 days.
- There is crippling joint pain, especially in the ankles and wrists. Joint involvement is bilateral and symmetric.
- Migratory polyarthritis is also seen in some individuals.
- Lymphadenopathy and conjunctivitis may occur.
- A few patients have leukopenia with elevated aspartate aminotransferase (AST); a positive C-reactive protein may be seen in some patients.
- A maculopapular rash may be seen. Appearance of the rash often coincides with defervescence.
- Hemorrhagic manifestations are rare.
- Recovery takes weeks; some older patients experience joint pain and stiffness for several years. Persistence may be seen in HLA B27 individuals.
- Clinically, chikungunya cannot be differentiated from uncomplicated dengue.

Laboratory diagnosis

A patient is suspected to have CHIKV only if she meets the clinical criteria. A confirmed (definitive) case is defined as a patient who meets both the **clinical and laboratory criteria**.

- **Clinical criteria:** The patient presents with an **acute onset of fever** and **severe arthralgia/arthritis** with or without a skin rash and a **history of residence in an epidemic area** or having **been in and subsequently having left an epidemic area within 15 days** before the onset of symptoms.
- **Laboratory criteria:**
  - The presence of **virus-specific IgM antibodies** in a single serum sample collected in the acute or convalescent stage or four-fold increase in IgG values in samples collected at least three weeks apart by ELISA or other immune assays confirms CHIKV.
  - Virus isolation/the presence of viral RNA is detected by **reverse transcriptase-PCR**.
  - Antigen detection kits based on ELISA, immunochromatographic tests, and IFA are also available.

Diagnosis of chikungunya

- **Clinical criteria:** Acute onset of fever >38.5°C **and** severe arthralgia/arthritis not explained by other medical conditions
- **Epidemiological criteria:** Residing in or having visited epidemic areas having reported transmission within 15 days prior to the onset of symptoms
- **Laboratory criteria:**
  - Virus isolation
  - Presence of viral RNA by RT-PCR
  - Presence of virus-specific IgM antibodies in a single serum sample collected in acute or convalescent stage
  - Four-fold increase in IgG values in samples collected at least three weeks apart

Treatment

- Supportive management and physiotherapy for relief of joint pains are the only options.
- No vaccine is yet available to prevent the infection.

O'nyong-nyong virus, Semliki Forest virus, and Sindbis virus disease

- These viruses also cause fever with arthritis.
- Antibodies to these viruses have been detected in some human sera in India.

PARASITIC CNS INFECTIONS

- Many parasitic infections can affect the human central nervous system (including the brain, the spinal cord, and the eyes).
- Some parasites infect the CNS as their primary infection site. For many others, the disease is a complication or an ectopic site of infection (Table 24.15).
- Many of these infections are prevalent worldwide, but most are restricted to or more prevalent in tropical areas and countries with poor hygiene and sanitation.
- Ectopic infections have been covered in Chapter 24. Their primary infections, however, are dealt in this chapter.

Table 24.15 Parasites infesting the CNS and their disease spectrum

Parasites infesting the CNS	Disease caused
Naegleria fowleri	Primary amoebic meningoencephalitis (PAM)
Acanthamoeba spp.	Granulomatous amoebic encephalitis (GAE), ulcerative keratitis, cutaneous infection, sinusitis
Balamuthia mandrillaris	GAE, cutaneous infection
Sappinia species, *Hartmannella*	Keratitis, meningitis
Entamoeba histolytica	Brain abscess
Trypanosoma gambiense *T. rhodesiense*	Sleeping sickness
T. cruzi	Chagas' disease
Plasmodium falciparum	Cerebral malaria
Toxoplasma gondii	Encephalitis, brain calcifications, blindness
Paragonimus spp. *Fasciola* spp. *Trichinella* spp. *Schistosoma* spp.	Brain cyst, neurological complications
Taenia solium (cysticercus cellulosae)	Neurocysticercosis
Echinococcus granulosus	Cerebral hydatid disease
Ascaris lumbricoides *Toxocara* spp.	Epilepsy
Strongyloides stercoralis	Meningitis, brain cysts
Loa loa *Onchocerca volvulus*	Worm in eye, blindness
Mansonella perstans	Coma

Taeniasis

- Taeniasis is caused by ***Taenia solium***, the pork tapeworm, which belongs to the cestode family **Taeniidae**.
- It is found throughout the world and is most common in countries where pork is eaten. In humans, the **larval form** of this tapeworm causes cysticercosis (see Chapter 25).
- These parasites show a predilection for the central nervous system (causing **neurocysticercosis** [NCC]), subcutaneous tissue, skeletal muscles, and eyes.
- **Human cysticercosis** is commonly seen in developing countries and is the cause of seizures in about one-third of cases. The endemic areas are Southeast Asia including India, Central and South America, and Africa.
- **Modes of infection:**
  - Consumption of green and leafy vegetables contaminated with the eggs of the tapeworm
  - Autoinfection, wherein the eggs are ingested either from the contaminated hands of the infected human or by reverse peristalsis (Fig. 24.23)

Agent—*Taenia solium*

Morphology

The adult worm is about **2–10 m long and 5 mm wide** and is segmented with each segment complete with a reproductive system.

Larvae

- **Cysticerci** are the larval forms of *T. solium* and are visible as milky-white bladder-like cysts in the muscles or viscera of humans.
- This larva has an **inner chamber** containing a small, **solid, eccentric scolex** (head) surrounded by an outer bladder wall.
- The **outer wall is filled with vesicular fluid.**
- The **bladder** may be round or lobulated. It may be large—up to 10–20 cm in size—and is often seen in the more spacious areas of the body or brain.

Eggs

- These are passed out from the gravid segment of adult worm.
- They are **oval and have the hexacanth embryo** embedded within the tissue.

Pathogenesis

- Eggs enter the intestine by reverse peristalsis, auto-infection, or by the consumption of contaminated water.
- These eggs hatch to release several larvae in the **duodenum and jejnum**.

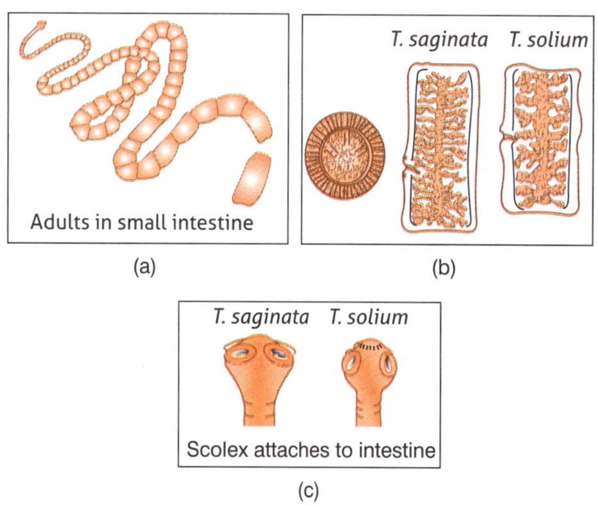

Fig. 24.23 Morphology of *Taenia*: (a) adult, (b) egg, and (c) scolex

- The larvae **(oncospheres) penetrate the intestinal wall** and travel via the mesenteric bloodstream to several parts of the body (chiefly, subcutaneous tissue, muscles, and CNS), where they lodge, **forming cysts.**
- This is an **end-stage disease**, in which the larvae cannot travel further or gain access to the exterior to continue their life cycle.
- In the tissue, they establish **space-occupying lesions**, with inflammatory cell infiltration and granuloma formation.
- As the lesions age, they get fibrosed and calcified.
- They may be discovered incidentally during radiological examination.

Clinical features

- The clinical presentation depends on the site of the cyst in the body.
- The most common sites of cyst formation are the brain, skeletal muscles, and eyes.

Neurocysticercosis (NCC)

- Almost 60–80% of cases of cysticercosis show neural involvement (Fig. 24.24). It is the most common parasitic infection of the central nervous system in humans.
- In some cases, the worm may remain in the brain without causing any symptoms.
- In others, seizures are the most common feature, followed by hydrocephalous, meningeal irritation, and dementia.
- Almost 50% of adult-onset epilepsy is due to NCC.

Ocular cysticercosis

- Ocular cysts are often found in the posterior chamber of the eye and then in the subretinal chamber, orbit, or anterior chamber.
- The symptoms depend on the site of cyst formation and may present as diplopia, proptosis, loss of vision, or slow-growing inflammatory nodule.
- The eye may be permanently damaged and may have to be surgically removed.

Subcutaneous cysticercosis

These infections are generally asymptomatic and may just present as palpable nodules.

Muscular cysticercosis

- These cases may be asymptomatic or may lead to myositis with accompanying fever and eosinophilia.
- Sometimes, muscular cysticercosis may lead to myositis with accompanying fever and eosinophilia.

Laboratory diagnosis

Careful history along with multispeciality radiological and histological investigations are required.

- **Radiology**—CT scan and MRI are performed to demonstrate cystic lesions, the scolex of the worm, and the surrounding inflammatory reaction.
- **Histopathology** of brain biopsies and surgically removed cysts from the muscles, subcutaneous tissues, and eyes may show the scolex.
- Stool microscopy has no diagnostic role as most patients do not harbor the adult worm at the time of larval infestation.
- **Fundoscopic examination** of the eye may show the larvae in the eye.
- **Serology**—the detection of specific antibodies in serum or CSF by immunodiagnostic assays such as ELISA is available. The antigen can be detected in serum or CSF by ELISA.

Treatment

- The treatment should be individualised based on the number and location of cysts and their viability.
- For parenchymal cysts, medical treatment is effective.
- For racemose cysts and those located in the intraventricular or subarachnoid space, medical treatment has no value.
- Dead or calcified cysts are not affected by medical treatment.
- The prescribed mode of treatment is with **albendazole for 10–14 days**.
- **Praziquantel** can be combined with albendazole if cysts are viable.
- Supportive medical therapy must be given to control seizures.
- High-dose glucocorticoids should be given before anti-parasitic drugs to reduce the acute inflammatory response which can occur following the degradation of viable cysts.
- Surgical removal of the cyst may be required in cases where there is extensive spread of infection.
- **Control measures**—Good food hygiene practices such as thoroughly washing fruits and vegetables,

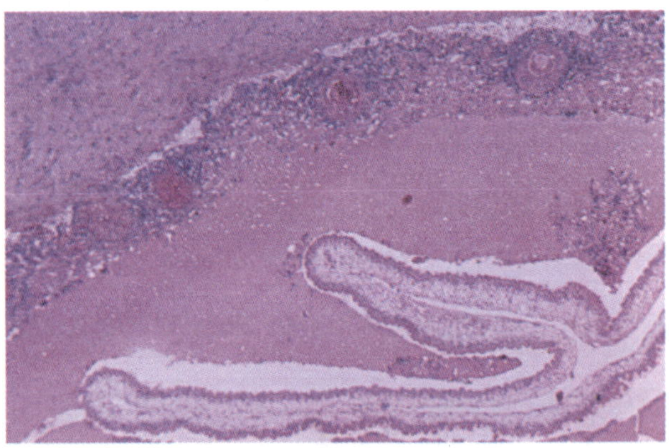

Fig. 24.24 Photomicrograph of a brain tissue specimen shows histopathologic features in a case of neurocysticercosis (*Source*: CDC, PHIL, Image ID 22813)

inspecting beef and pork for cysticerci, and destroying infected meat in slaughterhouses, avoiding raw or undercooked food, and improving sanitation and hygiene.

Primary Amoebic Meningoencephalitis (PAM) and Granulomatous Amoebic Encephalitis (GAE)

Case 7. *A 43-year-old swimmer visited the emergency department with a 2-day history of persistent occipital excruciating and intolerable headache, fever, myalgia, and fatigue. Physical examination revealed the presence of positive meningeal irritation sign and* ***left Babinski sign****. The brain CT scan demonstrated hydrocephalus, brain swelling, and signs that confirmed cerebral hernia and meningoencephalitis. Subsequently, external ventricular drainage was placed, and CSF samples were collected and examined. Trophozoites of N. fowleri were identified in the CSF via Wright-Giemsa staining and a diagnosis of PAM was made.*

i. What is primary amoebic meningoencephalitis?
ii. What is the pathogenesis of this condition?

Epidemiology

- Primary amoebic meningoencephalitis (PAM) and granulomatous amoebic encephalitis (GAE) have been reported worldwide.
- These infections are predominantly seen in **immunosuppressed patients and in divers, swimmers, and those who bathe in freshwater ponds.**
- The infection may also be acquired after **irrigation of nasal sinuses with contaminated** water using a neti pot.
- Both conditions are caused by free-living amoebae. Primary amoebic encephalitis (PAM) is caused by ***N. fowleri,*** while ***Acanthamoeba*** causes granulomatous amoebic encephalitis (GAE).
- Free-living amoebae are widely distributed in nature in soil and water. They are known to cause opportunistic infections that can be fatal in human beings (Table 24.1).

Primary amoebic meningoencephalitis (PAM)

Clinical features

- The incubation period is **1–14 days**.
- It is rapidly progressive and results in the **fatal destruction of brain tissue**.
- Initially, headache, fever, lethargy, nausea, vomiting, rhinitis, and disorientation are present.
- The disease is indistinguishable from acute bacterial meningitis at this stage.
- Later, the patient declines rapidly, becomes comatose, and dies within 7–10 days.
- It is important to look for a history of diving or bathing in freshwater sources in the last 10–14 days.

Agent—Naegleria fowleri

- It has a worldwide distribution and is commonly found in soil and freshwater lakes and rivers.
- Both trophozoite and cyst forms are seen in nature. Two forms of trophozoites—**amoeboid and flagellate**—are seen.

Morphology (Fig. 24.25)

- The amoeboid form can be seen in humans.
- It has lobular pseudopodia (**lobopodia**) and measures about **20 µm** in size.
- It has a **granular cytoplasm** and is the only replicating form.
- This form changes into the flagellated form in water at 25–27°C.
- The flagellated form is about 10–23 µm long and possesses two flagella at its broad, anterior end. These forms are **motile and exhibit a jerky or spinning motility**. The loss of flagella leads to reversion to amoeboid forms.
- The cysts are round, about 1–23 µm in diameter and are surrounded by a thick, smooth double wall. This form is not seen in humans.

Life cycle (Fig. 24.25)

- The infective form is the **amoeboid form**.
- It enters the body via the nose and reaches the brain tissue directly through the cribriform plate of the ethmoid bone, traveling along the olfactory nerve.
- The parasite causes extensive damage to olfactory bulbs and nasal mucosa.
- It multiplies there and rapidly forms nests of amoebae.
- It results in extensive hemorrhage and damage, leading to meningoencephalitis.
- Basilar portions of the cerebrum and the cerebellum are most commonly affected.
- Humans are the dead end in the life cycle of free-living amoebae.

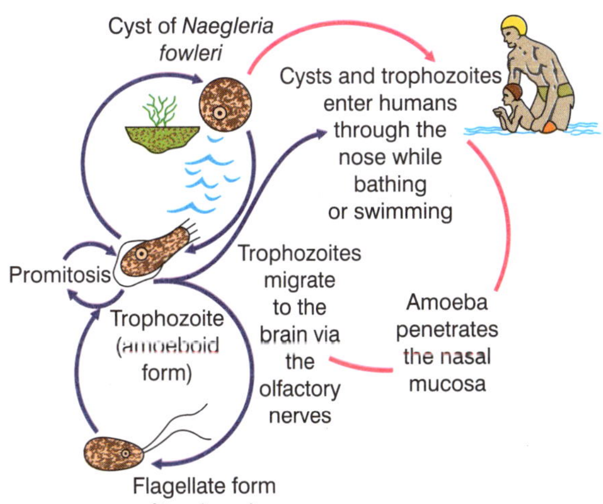

Fig. 24.25 Morphology and life cycle of *Naegleria fowleri*

Laboratory diagnosis

The diagnosis of *Naegleria* is dealt with under *Acanthamoeba*.

Treatment and control

- The disease shows rapid progression, and treatment is not effective.
- **Amphotericin B and miltefosine** have been used in some cases with success.
- Prevention is difficult because of the widespread distribution of the parasite.

Granulomatous amoebic encephalitis (GAE)

Clinical features

- Granulomatous amoebic encephalitis is caused by *Acanthamoeba* and *Balamuthia* in **immunocompromised persons.**
- It has an **insidious onset** with an **incubation period** that can be **as long as months**.
- The pathology is **focal granulomatous lesions** in the brain (Fig. 24.26a).
- The clinical presentation is **subacute to chronic**.
- Patients present with confusion, nausea, dizziness, and headache.
- Signs and symptoms suggestive of meningitis may be present.

Agent—*Acanthamoeba* species

- Multiple species of *Acanthamoeba* exist in nature.
- Human infections have been reported to be caused by *A. castellanii*, *A. astrocytes*, *A. polyphaga*, and *A. culbertsoni*.
- These organisms affect the CNS, sinuses, eyes, and skin.
- *Acanthamoeba* species are known as **Trojan horses** in the context of microbiology because they can harbour bacterial pathogens such as *Legionella* species, *Mycobacterium avium*, and *Listeria*.
- They can be present in **water cooling systems or in contact lens fluid.**
- Granulomatous amoebic encephalitis (GAE) caused by *Acanthamoeba* has been reported chiefly from the USA; very few cases have been reported from India.

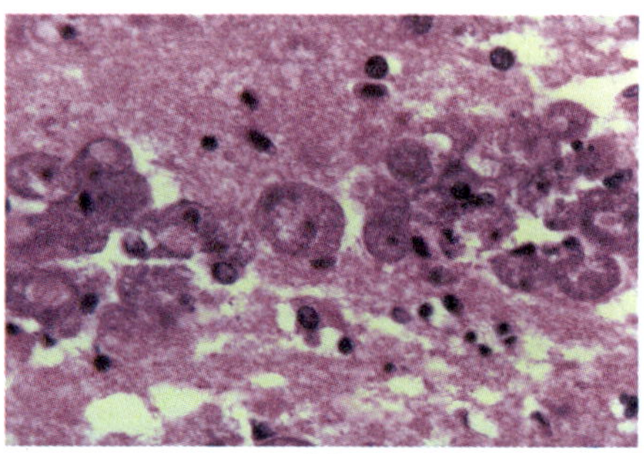

(a)

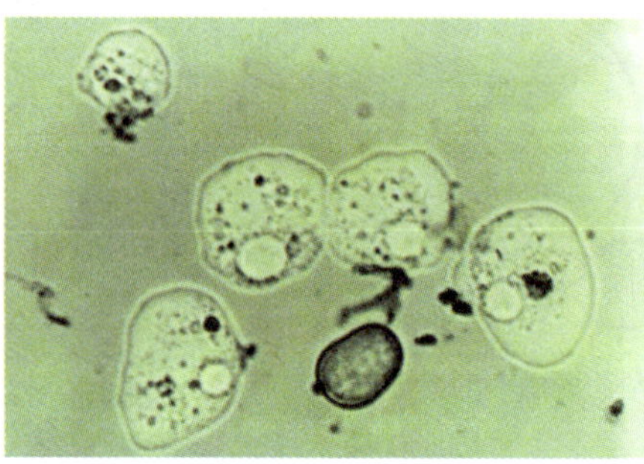

(b)

Fig. 24.26 (a) Histopathological changes associated with an infection found in a brain tissue specimen and (b) vegetative cells of *Acanthamoeba* (*Source*: CDC, PHIL, Image IDs 18941 and 18945)

Morphology (Fig. 24.26b and 24.27)

- The trophozoite is larger than that of *Naegleria* and has **spine-like pseudopodia (acanthopodia)**.
- The cyst form is 10–25 μm in diameter and has a double wall.
- The outer wall or ectocyst is wrinkled, while the inner wall is polyhedral.

Life cycle (Fig. 24.27)

- Humans acquire the infection by **inhaling the cyst or trophozoite forms with aerosols.**
- Trophozoites can reach the CNS from the lungs.
- *Acanthamoeba* can also enter the central nervous system via **skin ulcers or trauma wounds**.
- The parasite travels via the hematogenous route.
- The wounds caused by keratitis from the puncture of the corneal surface or ulceration from contaminated saline used with contact lenses can act as a conduit for the organism to enter the eyes.
- The parasite primarily lives in the lungs and sinuses.

Other infections caused by *Acanthamoeba*

Amoebic keratitis

- *Acanthamoeba* enters the cornea through minor abrasions present on the cornea; the resulting infection is known as amoebic keratitis.

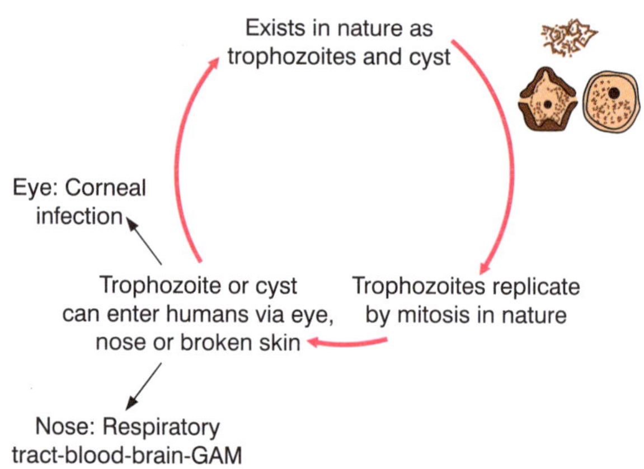

Fig. 24.27 Morphology and life cycle of *Acanthamoeba*

- The parasite is known to exist on **contact lenses** or in **contact lens solution** and hence, often affects contact lens users.
- Amoebic keratitis produces corneal ulceration, iritis, scleritis, and hypopyon, and can lead to the loss of vision.
- It is a painful condition.
- **Skin lesions**—Cutaneous ulcers, musculoskeletal abscesses, sinusitis, and nasal ulcers are seen in HIV patients.

Diagnosis of Naegleria fowleri and Acanthamoeba infections

- **Routine CSF examination:**
  - Lymphocytosis is seen in the CSF of a case of GAE, while neutrophilia, elevated proteins, and sugar are features of PAM.
  - CSF and corneal scrapings are examined for characteristic trophozoites and cysts of *Acanthamoeba*.
  - **Brain biopsy tissue and centrifuged deposit of CSF** can be stained with periodic-acid Schiff (PAS) and H and E stain.
  - **Calcofluor stain** is a better stain to visualise the cysts.
- **Culture:**
  - It is difficult to culture *Acanthamoeba* in the laboratory.
  - CSF specimens suspected of containing *Naegleria* can be inoculated on a non-nutrient medium such as Page saline with 1.5% agar.
  - Non-nutrient agar with bacterial supplements can be used.
  - Trophozoites (Fig. 24.23a) are seen in **furrows in the sloppy agar**.
  - *Naegleria* is cultured with *E. coli*, upon which it feeds; this is indicated by the **formation of tracks on the agar surface.**
- **Molecular methods:** Nucleic acid detection targeting 18S rRNA is available only in reference laboratories. Multiplex PCR is available for the simultaneous detection of *Acanthamoeba*, *Balamuthia*, and *N. fowleri* in CSF.
- **Radiology:** Imaging techniques such as MRI and CT scan in cases of PAM show non-specific signs like **diffuse enhancement around the cerebrum and obliteration of cisterns**. In GAE, space-occupying lesions may be seen in imaging of the brain.
- **Others:** Isoenzyme analysis is also available in specialised laboratories for the detection of *Naegleria fowleri*.

Diagnosis of PAM and GAE

- **Specimen:** CSF, corneal scrapings, brain biopsy tissue
- **Routine CSF examination:** Lymphocytosis in GAE while neutrophilia, elevated proteins, and sugar in PAM
- **Microscopy:**
  - Trophozoites and cysts of *Acanthamoeba* seen after periodic-acid Schiff (PAS) and H and E stain
  - **Calcofluor stain** to visualise the cysts
- **Culture:**
  - Difficult to culture *Acanthamoeba* in the laboratory
  - Non-nutrient medium such as Page saline with 1.5% agar, non-nutrient agar with bacterial supplements
  - Co-cultivation with *E. coli*: Na*egleria* indicated by the **formation of tracks on the agar surface**
- **Molecular methods:** 18S rRNA by multiplex PCR
- **Radiology** (MRI and CT scan):
  - PAM: non-specific signs like **diffuse enhancement around the cerebrum and obliteration of cisterns**
  - GAE: Space-occupying lesions
- **Others**: Isoenzyme analysis for *Naegleria fowleri*

Treatment

- The treatment of GAE is challenging, and fatality is high.
- Amoebic keratitis is treated with **topical application of chlorhexidine with or without pentamidine.**
- Surgical treatment may be required to restore vision.

Agent—Balamuthia mandrillaris

- It is a free-living amoeba that causes GAE.
- Its life cycle is similar to that of *Acanthamoeba*.
- The trophozoite form is **10 µm in size and irregular with extensive branching with a single nucleus** (Fig. 24.28).
- No flagellated form is seen.
- Its cyst measures 13–30 µm with a three-layered cell wall.
- It **enters through the respiratory tract or open wounds** and can cause GAE even in immunocompetent persons.
- The diagnosis of *Balamuthia* infections is based on the visualisation of trophozoites and cysts in

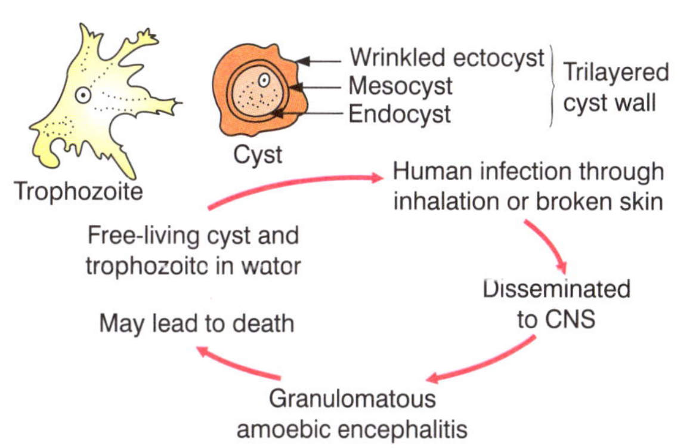

Fig. 24.28 Morphology and life cycle of *Balamuthia mandrillaris*

CSF and brain biopsy tissues. It cannot grow on laboratory media; cell lines are used for cultivation.

- An indirect fluorescent antibody test can also be used for diagnosis. A multiplex PCR assay is available for the detection of *Balamuthia* in clinical samples.
- Multidrug combinations of azithromycin, pentamidine, fluconazole and miltefosine have been used for treatment.

Other rare free-living amoebae

- *Sappinia* is a free-living amoeba found in the environment. Two species are known—*Sappinia diploidea* and *Sappinia pedata*.
- The latter causes amoebic encephalitis, which is similar to that caused by other free-living amoebae.
- *Hartmannella* is known to be a sporadic cause of keratitis in association with corneal trauma.

Toxoplasmosis

> **Case 8.** *A 32-year-old woman presented with a high-grade intermittent fever of 6 months' duration. She also complained of generalised arthralgia and significant weight loss. She had been extensively investigated and treated with multiple courses of broad-spectrum antibiotics, but fever continued. Physical examination revealed enlarged, non-tender cervical, and axillary lymph nodes. Her spleen was palpable 4 cm below the left costal margin. The bone marrow aspirate showed multiple oval cysts suggestive of T. gondii tachyzoites. Toxoplasma IgM was strongly positive, and Toxoplasma IgG was also positive. A diagnosis of probable* ***chronic disseminated toxoplasmosis*** *was made.*
>
> i. Describe the life cycle of *Toxoplasma*.
>
> ii. How is this infection diagnosed and treated?

Toxoplasmosis is caused by an obligate intracellular parasite, *Toxoplasma gondii*, which has a worldwide prevalence. A wide range of animals and birds are frequently infected. Cats are most commonly affected, with humans being the accidental hosts.

Clinical features

- Toxoplasmosis is **generally asymptomatic**.
- A **low-grade lymph node infection** resembling infectious mononucleosis may occur.
- The rupture of a tissue cyst releases numerous bradyzoites, causing a local hypersensitivity reaction and leading to inflammation, blockage of blood vessels and cell death near the damaged cyst.
- In **immunocompetent persons**, infection is usually asymptomatic or mild. The commonest presentation is **cervical lymphadenopathy**.
- In **immunocompromised patients**, the most common presentation of toxoplasmosis is encephalitis. It is a common opportunistic infection in HIV patients. In many such patients, active toxoplasmosis can occur from the reactivation of a latent infection when CD4 levels fall. Retinitis or chorioretinitis, encephalitis, or pneumonitis may be seen.

Toxoplasmosis in pregnant women

- When non-immune women become infected during pregnancy, the disease is severe. The classical triad of **chorioretinitis, hydrocephalous, and intracranial calcification** is pathognomonic of congenital toxoplasmosis.
- If the infection is acquired during the **first trimester**, it generally results in **stillbirth or major CNS anomalies.**
- **Second- and third-trimester infections** are more common but cause less severe neurologic damage (Table 24.16). In such cases, the clinical manifestations may be delayed until long after birth, even beyond childhood.
- **Neurologic problems or learning difficulties** may present late. Eye involvement can occur in the second or third decade due to the rupture of cysts in the eye.

Agent—*Toxoplasma gondii*

Morphology (Fig. 24.29)

T. gondii is an intracellular pathogen that exists in three forms—tachyzoites, bradyzoites, and oocysts. All three forms are infective.

- **Tachyzoites:**
  - These are rapidly multiplying crescent-shaped forms.
  - They are 4–8 μm long and 2 μm wide.
  - Tachyzoites can infect all nucleated cells. They divide asexually by internal budding.
  - In host cells, these are arranged as rosettes around the nucleus, surrounded by the vacuole.
- **Bradyzoites:**
  - These are slender, crescent-shaped forms present inside the tissue cysts.

Table 24.16 Relationship between time of acquisition of infection, disease transmission, and severity of congenital toxoplasmosis in pregnant women

Time of acquisition of infection by mother	Chances of transplacental transmission	Disease severity in neonate
1st trimester	23% (lowest)	High
2nd trimester	Medium	Less severe neurologic damage
3rd trimester	65% (maximum)	Asymptomatic infant at birth
Prenatal infection	Not infected	Not seen

- They multiply slowly and are present in chronic infection.
- They contain granules that are stained well with PAS stain.

- **Tissue cysts:**
  - Oval or round tissue cysts are present inside host cells, enclosing plenty of bradyzoites.
  - They range in size from 5–100 µm.
  - The cyst wall is eosinophilic and stains with PAS.
  - Tissue cysts are commonly found in the brain, muscles, and central nervous system.
- **Oocyst:** This is the sexual form found in cats and other feline hosts. It has double-layered colourless, refractile walls.

Life cycle (Fig. 24.29)

- The **definitive hosts are cats and other felines** while other mammals such as rodents, sheep, goat, pigs, and cattle are intermediate hosts.
- **Humans are accidental hosts**, and no further development takes place in them.

Mode of infection

- The infection is acquired by the **ingestion of sporulated oocysts with contaminated food and water.**
- It could be acquired by the ingestion of tissue cysts from the undercooked meat of pigs, goats, cattle, and other animals.
- The inoculation of tachyzoites via blood transfusion, needlestick injury, organ transplantation, placenta, and through the iatrogenic (laboratory-acquired) route can also result in infection.

Pathogenesis

- The oocyst releases sporozoites in the human being's or animal's duodenum.

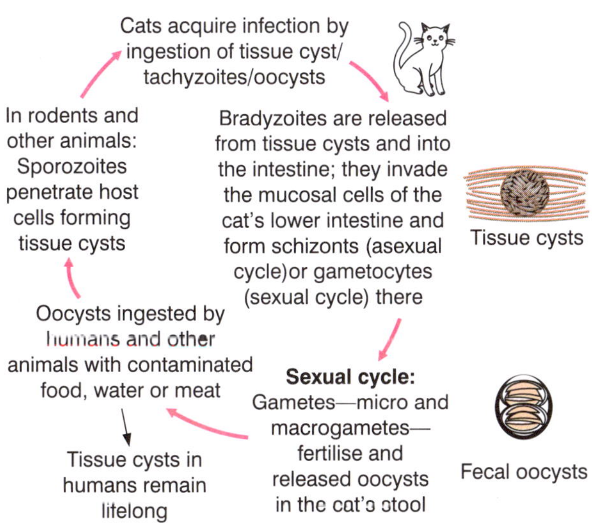

Fig. 24.29 Morphology and life cycle of *Toxoplasma*

- Tissue cysts (Fig. 24.30) release bradyzoites.
- These pass through the intestinal wall and reach the blood circulation.
- There, they transform into tachyzoites that invade various cells, especially macrophages, and reach lymph nodes and other organs. This marks the acute stage.
- Later, the tachyzoites penetrate nerve cells, especially those of the brain and eye, and transform into bradyzoites.
- There, the trophozoites multiply slowly, forming tissue cysts (chronic stage).
- The whole process takes about 8–10 days after infection.
- The tissue cysts in rodent flesh are eaten by cats, which get infected by oocysts or tachyzoites.
- Bradyzoites are released from tissue cysts in the intestine.
- They invade the mucosal cells of the cat's small intestine, where they form schizonts (asexual cycle) or gametocytes (sexual cycle).
- Male and female gametocytes mature to form gametes. Fertilisation occurs, and a zygote is formed, which changes into an oocyst.
- These oocysts are excreted in the cat's feces.
- The oocysts are ingested by humans who come in contact with the cat. The oocyst releases sporozoites, which mature into tachyzoites.
- These tachyzoites penetrate various cells in the body of the human host; this is the end-stage of the disease.
- The tachyzoite directly destroys cells, especially the reticuloendothelial and parenchymal cells.
- If the oocysts are ingested by a cat, the parasite continues its sexual and asexual cycles.
- If it is ingested by intermediate hosts such as birds, rodents, or mammals (including humans), the parasite, after establishing infection, can only reproduce asexually.

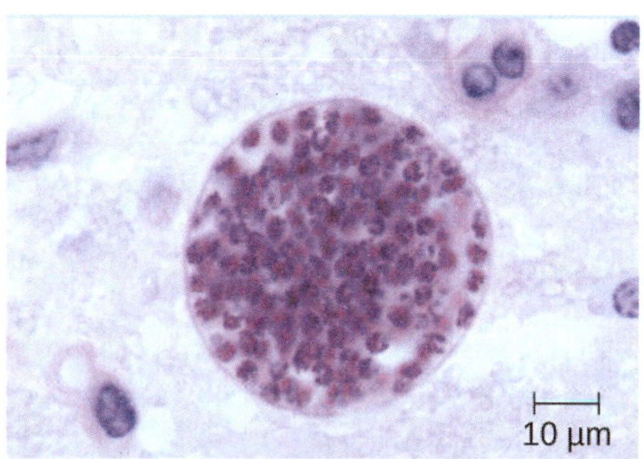

Fig. 24.30 *Toxoplasma gondii* cyst observed in mouse brain tissue, containing thousands of inactive parasites (*Source*: Modification of work by USDA- https://openstax.org/books/microbiology/pages/26-4-fungal-and-parasitic-diseases-of-the-nervous-system)

Laboratory diagnosis (Fig. 24.31 and 24.32)

- **Specimen:** Blood, body fluids, bone marrow, lymph node aspirate, bronchoalveolar lavage, CSF, aspirates, or biopsy tissues from the spleen, liver, and muscles tissue may be collected depending on the investigation.
- **Microscopy:**
  - Stains: Giemsa and periodic acid-Schiff's stain
  - H and E and PAS stain: For biopsy samples
  - **Crescent-shaped tachyzoites can be seen in blood smears, body fluids, and biopsy tissues**
  - Direct fluorescent antibody test: Fluorescein-conjugated antibody against the surface antigens of *Toxoplasma gondii* is added to the slides to detect tachyzoites
- **Histopathology:**
  - Tachyzoites in tissue sections or smears of body fluid (e.g., CSF or amniotic or BAL fluids) in acute infection by using the immunoperoxidase technique
  - Wright-Giemsa stain on centrifuged sediment of CSF or brain aspirate
  - In acute infection or reactivation of latent infection, the presence of multiple tissue cysts near an inflammatory necrotic lesion confirms the diagnosis
- **Culture:**
  - Cell lines and inoculation into mice are the two methods to propagate *T. gondii* from human tissue or body fluid.
  - Acute infection by *T. gondii* can be established by these two techniques.
  - Culture should only be attempted with all the precautions, as laboratory-acquired infections may occur.
- **Serology:**
  - Detection of specific antibodies to *T. gondii* is the mainstay of diagnosis.To establish diagnosis, a combination of serological tests is usually required.
  - Antibody detection assays such as ELISA, ELFA, and immunofluorescent technologies are available.

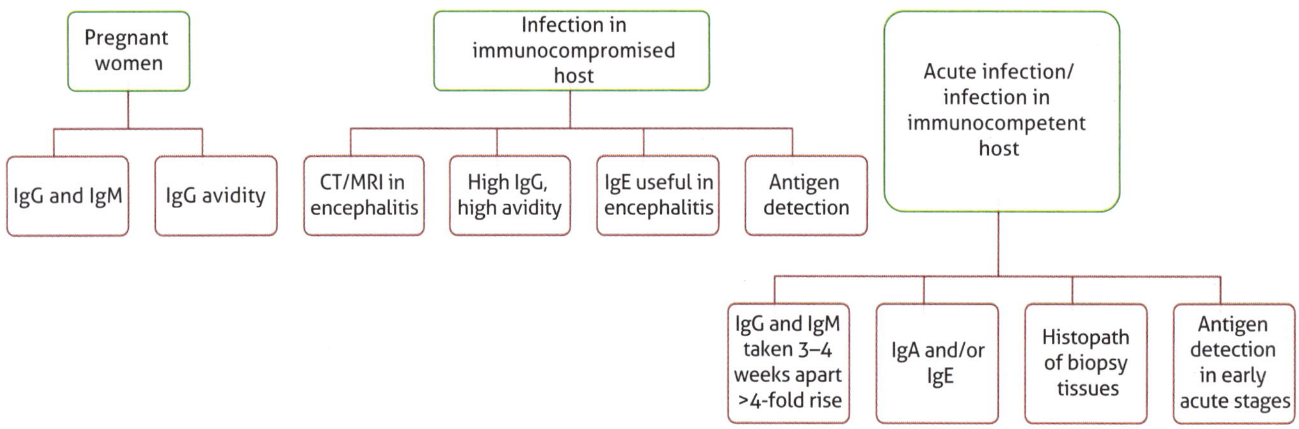

Fig. 24.31 Diagnosis of toxoplasmosis

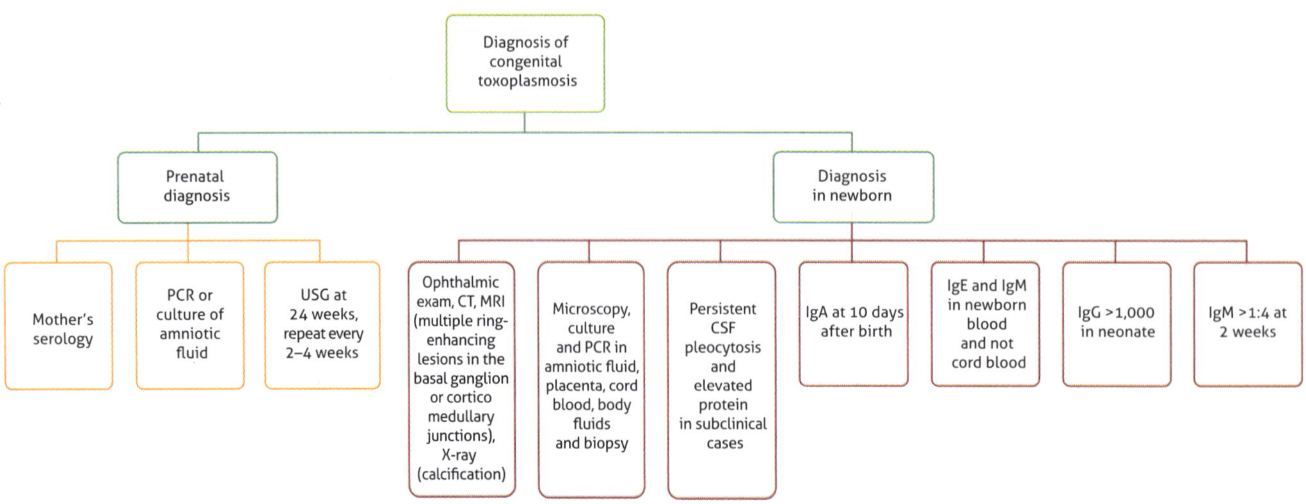

Fig. 24.32 Diagnosis of congenital toxoplasmosis

- **IgM antibody:** Develops within 1–2 weeks of infection, and can also persist variably until six months after the primary infection. In some patients, the persistence of IgM has been seen years after the primary infection.
- **IgG antibody:** It appears after four weeks. After eight weeks, the IgG levels reach baseline but persist lifelong. **Avidity** refers to the overall strength of binding between an antibody and an antigen. The avidity of IgG antibodies increases with time. In infections of less than 12 weeks duration, the avidity is low; and is high in past infections. Avidity testing of IgG is a good indicator to differentiate recent from past infections.
- IgG and IgM detection can be done for the initial evaluation of these patients. However, serial specimens after 3–4 weeks give the best diagnosis. **Negative results in both tests help to rule out toxoplasmosis**. These antibodies are detected in acute and congenital infections and may persist for one year but are not of much value in diagnosing adults.
- **IgA antibody:** It is more sensitive than IgM in detecting congenital infections. In the absence of IgM, diagnosis can be made by the detection of IgA and IgG antibodies.
- **IgE antibody:** It is detectable in the sera of acutely infected adults, congenitally-infected infants, and children with congenital toxoplasmosis. The seropositivity of IgE is shorter than that with IgM or IgA antibodies and hence is useful for detecting recently acquired infections.

- **Molecular assays:** PCR has been used for the detection of DNA of *T. gondii* in body fluids and tissues in congenital, cerebral, ocular, and disseminated toxoplasmosis, especially in immunosuppressed persons.

Sabin–Feldman dye test: This is a classical gold-standard test of historical importance. The patient's serum is first treated with tachyzoites and complement. After incubation, methylene blue is added. If antibodies to *Toxoplasma* are present in the patient's serum, these are activated by complement. As a result, the parasite membrane is lysed and hence, it does not take up the stain. On the other hand, tachyzoites with intact membranes are stained by methylene blue (negative result). It is a highly sensitive test but fails to differentiate between past and recent infections.

- **Imaging:** CT scan and MRI show multiple ring-enhancing lesions in the basal ganglion or corticomedullary junctions. X-rays may show calcifications.
- **CSF examination:** Lymphocytosis and a slight increase in protein and normal glucose levels are seen in suspected encephalitis cases.

The diagnostic strategies for various syndromes related to toxoplasmosis are presented in Table 24.17.

Treatment

Pyrimethamine, sulphadiazine, and clindamycin are the drugs used to treat toxoplasmosis.

Prophylaxis

Trimethoprim–sulfamethoxazole has prophylactic value.

Trypanosomiasis

Infection by trypanosomes is restricted to Africa. It primarily affects the CNS.

Agent—*Trypanosoma*

Two major species involved in the infection are as follows:

- ***Trypanosoma brucei* rhodesiense**—causes East African trypanosomiasis (EAT) and African sleeping sickness. It is transmitted to humans primarily by the bite of the tsetse fly (*Glossina* spp.).
- ***Trypanosoma cruzi* gambiense**—causes West African trypanosomiasis (WAT). The infection is transmitted by the *Reduviid* bug (infection usually occurs after bugs defecate on the bite site and the feces are rubbed into the wound upon scratching).

Pathogenesis

- The flagellated forms enter the host by the bite of a vector.
- These move into the circulatory system. Following a period of remission after initial fever, the trypanosomes move from the lymphatics into the CNS to manifest with neurological symptoms.
- *Trypanosoma* can change their surface proteins into over 100 antigenic types.
- Hence, protective immunological response is inadequate.
- However, there is some immune system response to the infection.
- As the number of parasites decreases, the clinical symptoms abate.
- Some evade the immune system.
- These antigenic variants rapidly proliferate and cause another bout of disease.
- If left untreated, the disease can be fatal.

Clinical features

- These include sleepiness during daytime leading to insomnia at night and mental deterioration.
- **East African type:** The disease lasts from weeks to months.

Table 24.17 Diagnosis and treatment strategies for various clinical syndromes of toxoplasmosis

Clinical syndrome	Diagnostic strategy	Treatment
Acute infection/ infection in immunocompetent host	Paired sera for IgG and IgM taken 3–4 weeks apart will show >4-fold rise IgA and/or IgE estimation Histopathological examination of biopsy tissues depending on lesion Antigen detection in early acute stages	Mild cases or asymptomatic ones do not require therapy Ocular toxoplasmosis—pyrimethamine and sulphadiazine/clindamycin for one month
Congenital infection Prenatal diagnosis	Mother's serology is suggestive PCR or culture of amniotic fluid Ultrasonography at 24 weeks, repeat every 2–4 weeks	Pyrimethamine 1 mg/kg once a day and sulphadiazine 100 mg/kg with folinic acid for 1 year
Congenital infection: Diagnosis in newborn	IgA at 10 days after birth; IgE and IgM detection in the newborn blood and not umbilical cord blood IgG >1,000 in neonate IgM >1:4 at 2 weeks Ophthalmological examination, CT, MRI, X-ray to detect characteristic lesions and calcifications Microscopy, culture and molecular diagnosis from amniotic fluid, placenta and cord blood or in body fluids and biopsy Persistent CSF pleocytosis and elevated protein content seen in subclinical cases	
Infection in immunocompromised host	CT/MRI in encephalitis High IgG, high avidity Brain biopsy if IgG negative or only single lesion is visible in MRI of suspected encephalitis IgE useful in encephalitis Antigen detection	HIV with CD4 <100/μL: Trimethoprim–sulfamethoxazole prophylaxis Can be stopped if ART compliant and CD4 levels >200/ μL HIV patients, after receiving treatment for toxoplasmosis, need long-term maintenance therapy Discontinue if asymptomatic and CD4 counts >200/ μL
Pregnant women	IgG and IgM IgG avidity	None

- **West African type:** It develops over a span of months to years.
- A pathognomonic sign is the **Winterbottom's sign,** which refers to the enlargement of lymph nodes on the back of the neck, indicating cerebral infections.

Laboratory diagnosis

- *Trypanosoma* can be directly observed in stained samples including blood, lymph node, and CSF.
- Microscopic detection of the parasite is sufficient to clinch the diagnosis (Fig. 24.33).
- Early diagnosis is important for treatment.
- Antibodies against the parasite are found in most patients with acute or chronic disease.

Treatment

- **Before the nervous system is involved:** Pentamidine (an inhibitor of nuclear metabolism) and suramin (mechanism unclear)
- **Once the sleeping sickness phase sets in:** Melarsoprol (an arsenic derivative) and eflornithine

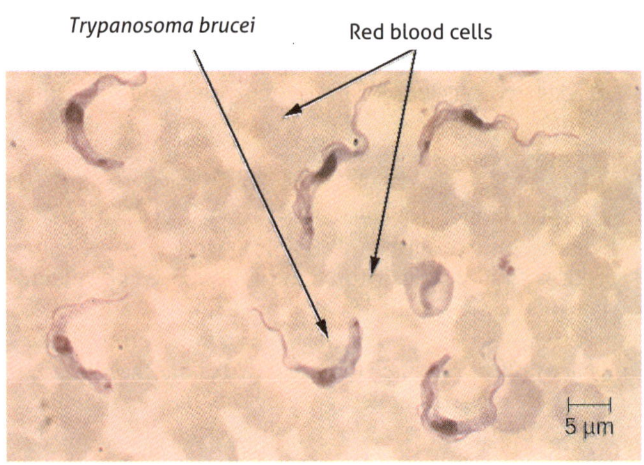

Fig. 24.33 *Trypanosoma brucei*, the causative agent of African sleeping sickness, in a human blood smear (*Source*: modification of work by the Centers for Disease Control and Prevention—https://openstax.org/books/microbiology/pages/26-4-fungal-and-parasitic-diseases-of-the-nervous-system)

Relapses of the disease may occur, and hence the patient's CSF should be examined periodically for up to two years. The most effective means of preventing these diseases is to control the insect vector populations.

FUNGAL CNS INFECTIONS

- Infections of the central nervous system caused by fungi are often seen in immunocompromised individuals.
- Fungi rarely enter the CNS of a healthy individual.
- The route of entry is through by the **hematogenous route from a primary site**—often the lungs or the sinus and mastoids or internal auditory meatus.
- Rarely, penetrating trauma and fracture of the skull may give rise to fungal infections of the CNS.

Candidiasis

- *Candida* infections occur invariably due to the hematogenous spread of the organism.
- Such infections are commonly caused by *Candida albicans*, which can produce acute or chronic meningitis or meningoencephalitis.
- More often, cerebral abscess, vasculitis leading to cerebral infarction and rarely, aneurysms are produced.
- *Candida* infections are generally seen in neonates, secondary to the implantation of ventriculoperitoneal shunts and other neurosurgical procedures, and in immunocompromised individuals.

Clinical features

Fever, meningismus, and localised neurological signs are present.

Laboratory diagnosis

- The microscopic examination of CSF shows **pleocytosis.**
- Gram staining may show **budding yeast cells**.
- Culture on blood agar or SDA and speciation confirms the diagnosis.

Treatment

Treatment is with **amphotericin B with flucytosine.**

Cryptococcal Meningitis

- *Cryptococcus* is the most common fungal agent causing meningoencephalitis, space-occupying lesions (cryptococcoma), and chronic meningitis in immunocompromised individuals.
- It is one of the most common opportunistic infections in HIV-infected patients.
- Two species of *Cryptococcus* involved in human infection are: (i) *Cryptococcus neoformans* var neoformans and (ii) *Cryptococcus neoformans* var gatti.

Clinical features

- The features of *Cryptococcus* infections are similar to those of tubercular meningitis.
- Fever, chronic headache, malaise, lethargy, nausea and vomiting, and altered mental status are the usual features.

Laboratory diagnosis

- **Microscopy:** Specimens are stained with India ink to visualise capsulated yeast cells.
- **Culture:** Specimens are inoculated on SDA for growth and speciation.
- **Serology:** The capsular polysaccharide antigen of *Cryptococcus* is detected by latex agglutination or enzyme immunoassays from CSF, serum, or urine.

Treatment

- **Mild to moderate infection**: Fluconazole
- **Severe infections:** Amphotericin B with flucytosine

Histoplasmosis

- *Histoplasma* causes histoplasmosis in immunocompromised patients.
- It can also cause acute or chronic meningitis.
- CNS infection by *Histoplasma* can manifest as a cerebrovascular accident due to infective emboli. Solitary lesions in the spinal cord can mimic a spinal cord mass or malignancy.
- Other lesions include military brain lesions and meningovascular lesions.

Clinical features

- These may be varied and include features such as fever, headache, confusion, and lethargy.
- Meningeal irritation and cranial nerve palsy may also be present.

Laboratory diagnosis

- *Histoplasma* infection is diagnosed based on CSF analysis for increased protein and culture, which often yields *Histoplasma capsulatum*.
- MRI and other imaging techniques also help.

Treatment

Amphotericin B, itraconazole, or fluconazole can be used.

Microsporiodiosis

Human microsporidiosis represents an important and rapidly emerging opportunistic disease, which occurs chiefly, but not exclusively, in severely immunocompromised patients.

Agent—Microsporidia

Microsporidia were initially considered obligate, intracellular, spore-forming protozoa that were parasitic in animals. Three distinct *Encephalitozoon* species have been detected in persons with human immunodeficiency virus (HIV) infection, in whom they cause causing cerebral, ocular and intestinal infections (Fig. 24.34):

- *E. cuniculi*
- *E. hellem*
- *E. intestinalis*

Clinical features

- In humans, *Encephalitozoon* species have been identified as agents causing neurologic disorders associated with fever, headache, vomiting, and loss of consciousness, sometimes associated with spastic convulsions.
- In immunocompromised patients such as HIV-infected individuals, associated renal failure, involvement of liver and pneumonitis have been documented.
- Infections with *E. hellem* include sinusitis and keratoconjunctivitis.
- Diarrhea is associated with *E. intestinalis.*

Laboratory diagnosis

- **Microsporidial spores** can be detected in CSF, stool, and urinary sediment. It may also be detected in sputum from cases of pneumonitis.
- The spores can be **stained by fluorochrome stains**. In tissue sections, hematoxylin and eosin stain or Gram stain shows intracellular spores by light microscopy.
- **Western blot** analysis of spore antigens and amplification of polymorphic DNA of the agent can be done.

Treatment

Albendazole is found to be effective against *Encephalitozoon*.

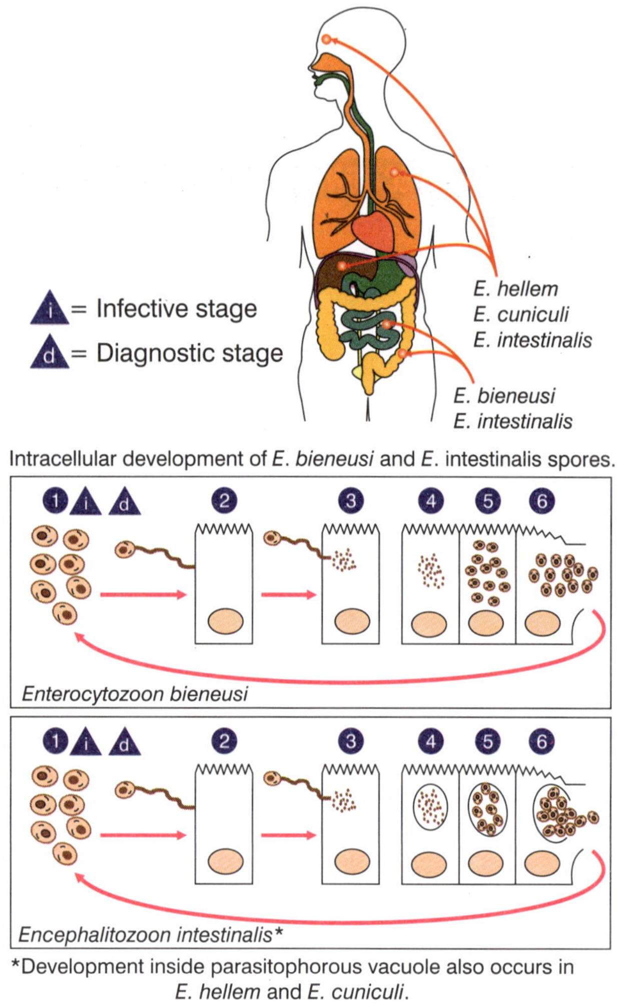

Fig. 24.34 Life cycle of the different species of microsporidia responsible for causing diarrheal illnesses and other infections, together known as microsporidiosis (*Source*: CDC, DPDx)

Questions

Case review

1. A 65-year-old woman who had been a diabetic for several years presented to neurosurgery OPD with symptoms of focal seizures. Computed tomography of her brain showed a 'space-occupying' lesion in the left cerebral hemisphere with 'ring-enhancement', which was consistent with brain abscess.
   i) Describe the pathogenesis of brain abscess.
   ii) List common bacteria causing brain abscess.
2. A 30-year-old woman was brought to a referral hospital from a rural area where she had been admitted with fever, anorexia, and vomiting for the past four days followed by delirium and hallucination and severe spasms on drinking fluids. Her relatives informed the doctor that she had been bitten by a stray dog about three months back. No medical help had been sought at the time, but wound care and local treatment had been provided. Following admission, she was started on IV fluids and other supportive care. The next day, she developed severe spasms in the pharyngeal muscles. This spasm increased in intensity and distribution, leading to generalised convulsions despite anticonvulsion therapy. She died during one of these attacks due to respiratory failure. On autopsy, her brain tissue was found positive for rabies virus antigen by direct fluorescent antibody test.
   i) Describe the pathogenesis of rabies.
   ii) What prophylaxis should have been provided three months back to the woman?

Short answer questions

1. Outline the algorithm for the laboratory confirmation of acute pyogenic meningitis.
2. Describe the epidemiology of Japanese encephalitis in India.
3. Outline the pathogenesis and laboratory diagnosis of primary amoebic meningoencephalitis.
4. Discuss prion disease.

MCQs

1. **A one-and-a-half-year-old non-vaccinated child presented with signs of meningitis; her CSF showed pleomorphic gram-negative bacilli. On blood agar, the organisms showed enhanced growth in the presence of X and V factors. What is the likely organism associated with this case?**
   a) *M. tuberculosis* b) *E. coli*
   c) *B. pertussis* d) *H. influenzae*
2. **A regular swimmer was admitted with diminished vision and neck stiffness. On examination, she was found to have keratitis and meningitis. She gave a history of using semisoft contact lenses. Her CSF culture on axenic media showed microbial growth. Which organism is commonly associated with such a condition?**
   a) *S. pneumoniae* b) *M. tuberculosis*
   c) *Acanthamoeba* spp. d) *T. gondii*
3. **A 23-year-old wrestler with a habit of consuming raw milk was admitted with signs of meningitis for the past three days. His CSF reveals gram-positive bacilli. Which organism is likely to be associated in this case?**
   a) *Mycobacterium tuberculosis*
   b) *Escherichia coli*
   c) *Listeria monocytogenes*
   d) *Bacillus cereus*
4. **A child with trigeminal neuralgia and vesicular painful ulcer around the mouth was admitted to the hospital. Her CSF sample inoculated on cell culture showed syncytia formation and ballooning. What is the likely organism associated with this condition?**
   a) Adenovirus b) Herpes virus
   c) Cytomegalovirus d) Varicella zoster virus
5. **A 20-year-old presented with projectile vomiting, headache, and fever. He is admitted for meningitis. The Gram staining of his CSF showed intracellular gram-negative diplococci. Which organism is responsible for this condition?**
   a) *Neisseria meningitidis* b) *Escherichia coli*
   c) *Streptococcus pneumoniae* d) *Klebsiella* spp.
6. **Meningococci are:**
   a) Non-capsulated c) Gram-positive diplococci
   b) Oxidase-negative d) Gram-negative diplococci
7. **Which serogroup of meningococci causes both epidemics and outbreaks?**
   a) A b) B
   c) W-135 d) Y
8. **Which of the following is not a virulence determinant in *Neisseria meningitidis*?**
   a) Capsule b) IgA protease
   c) Exotoxin d) Endotoxin
9. **One of the important risk factors in meningococcal disease is complement deficiency. Which complement deficiency favours the disease?**
   a) C1 deficiency b) C1 inhibitor deficiency
   c) C9 deficiency d) C5–C8 deficiency
10. **Primary agents of pyogenic meningitis include all the following except:**
    a) *Streptococcus pneumoniae* b) *Haemophilus influenzae*
    c) *Neisseria meningitidis* d) *Moraxella catarrhalis*
11. **The selective medium used for meningococci is:**
    a) Thayer–Martin medium b) Neomycin blood agar
    c) Mannitol salt agar d) MYPA agar
12. **The specimen that is not indicated for the diagnosis of severe meningococcemia is:**
    a) Blood b) Sputum
    c) Skin lesions d) CSF
13. **Treatment of meningococcal carriers is by:**
    a) Penicillin b) Erythromycin
    c) Ciprofloxacin d) Tetracycline
14. **A seven-year-old boy presented to the emergency department with high-grade fever and headache for two days and a single episode of convulsion and projectile vomiting the previous day. On examination, he was found to be disoriented, and his neck rigid; Kernig's sign was positive. On Gram stain, gram-negative cocci were seen in pairs, most of them intracellular with plenty of polymorphs. Which organism is suggested by this Gram stain?**
    a) *Neisseria meningitidis* b) *Streptococcus pneumoniae*
    c) *Haemophilus influenzae* d) *Streptococcus pyogenes*
15. **Which selective medium is used to isolate meningococci from carriers during epidemics?**
    a) Chocolate agar
    b) Cetrimide agar
    c) Modified Thayer–Martin medium
    d) Mueller Hinton agar
16. **Modified Thayer–Martin medium contains:**
    a) Vancomycin, colistin, and nystatin
    b) Vancomycin, imipenem, and nystatin
    c) Erythromycin, colistin, and nystatin
    d) Erythromycin, colistin, and naldixic acid
17. **Which of the following is used in the identification of *Streptococcus agalactiae*?**
    a) Bile solubility b) PYR test
    c) CAMP test d) Inulin fermentation
18. **The blood sample from a neonate suffering from fever grew beta-hemolytic colonies on blood agar. On Gram stain, gram-positive cocci in chains were seen under the microscope. Cervical swab from the mother also showed the same organism. Which is the likely causative agent?**
    a) *Streptococcus pyogenes* b) *Streptococcus pneumoniae*
    c) *Streptococcus agalactiae* d) *Streptococcus viridans*
19. **Which is true about *Listeria monocytogenes*?**
    a) Is non-motile at 25°C
    b) Major virulence factor is capsule
    c) Serovars associated with human infections are 1/2a, 1/2b, and 4b
    d) Infection is transmitted by respiratory droplets
20. **Tumbling motility is shown by:**
    a) *S. pneumoniae* b) *Legionella pneumophilla*
    c) *Listeria monocytogenes* d) *Neisseria meningitidis*
21. **Which organism shows drumstick appearance in Gram stain?**
    a) *Clostridium perfringens* b) *Clostridium botulinum*
    c) *Clostridium septicum* d) *Clostridium tetani*
22. **Which toxin (s) is responsible for pathogenicity of tetanus?**
    a) Tetanolysin
    b) Tetanospasmin
    c) Peripherally acting neurotoxin
    d) All of the above.
23. **Tetanospasmin acts by:**
    a) Inhibition of neurotransmitters such as GABA
    b) Blocking the production or release of acetylcholine
    c) Degeneration of Nissl bodies
    d) Combination of all of the above
24. **The method used for isolating *Clostridium tetani* from mixed cultures is:**
    a) Filde's technique c) Castenada's method
    b) Deep agar shake d) Dienes method
25. **A 28-year-old male was brought to the hospital in a state of altered mental status. He soon progressed to coma and respiratory paralysis. Family members informed that he had consumed preserved meat the previous day. Smear from the suspected meat showed gram-positive sporing bacilli. What is the likely agent associated?**
    a) Enterotoxigenic *E. coli* b) *Clostridium botulinum*
    c) *Campylobacter jejuni* d) *Yersinia enterocolitica*

26. **All the following bacteria produce neurotoxins except:**
   a) *Clostridium perfringens* b) *Clostridium botulinum*
   c) *Bacillus cereus* d) *Vibrio cholerae*
27. **The false statement about botulism is:**
   a) Wound botulism resembles food-borne botulism
   b) In infant botulism, spores are ingested and not the pre-formed toxin
   c) Type E strains are associated with fish and other sea food
   d) All the eight toxins produced by *Clostridium botulinum* are neurotoxins
28. ***Encephalitozoon* species have been identified as agents causing:**
   a) Renal failure and pneumonitis in HIV-infected persons
   b) Neurological disease only in immunocompetent persons.
   c) Haemorrhagic fever in HIV positive patients.
   d) No infections in humans.
29. **Which of the following statement correctly describes tetanospasmin?**
   a) It is heat-labile and oxygen-labile toxin.
   b) It is responsible for causing hemolysis on blood agar.
   c) It is plasmid-encoded toxin responsible for tetanus.
   d) It is similar to toxins produced by *Clostridium perfringens, Clostridium novyi*, and *Streptococcus pyogenes.*
30. **A 25-year-old man is brought to emergency after a road traffic accident in a state of altered sensorium. He has a contaminated lacerated wound on his legs. His vaccination status cannot be ascertained. What action should be taken immediately to provide him with tetanus prophylaxis?**
   a) Intramuscular injection of tetanus toxoid
   b) Intramuscular injection of tetanus toxoid with antibiotics
   c) Intramuscular injection of tetanus toxoid with anti-tetanus immunoglobulin and antibiotic
   d) Subcutaneous injection of tetanus toxoid with antibiotics
31. **Which of the following statements is not correct regarding the toxin produced by *Clostridium botulinum*?**
   a) The toxin is not released during the life of the organism.
   b) It acts by inhibiting neurotransmitters such as GABA.
   c) It is produced intracellularly as a non-toxic protoxin or progenitor toxin.
   d) Toxin production is regulated by the presence of bacteriophages.
32. **Aseptic meningitis is commonly caused by:**
   a) Enterovirus b) Adenovirus
   c) Poliovirus d) Echovirus
33. **All the following are members of Enterovirus except:**
   a) Echovirus b) Coxsackie virus
   c) Poliovirus d) Orf virus
34. **Which one of the following is false regarding the oral polio vaccine?**
   a) Induces only systemic antibody response
   b) Protects the individual and the community
   c) Induces long-lasting immunity
   d) Early administration can prevent an epidemic
35. **Degeneration of Nissl bodies is the characteristic feature of:**
   a) Poliovirus b) Rabies virus
   c) Adenovirus d) Vaccinia virus
36. **Pulse polio programme induces:**
   a) Active immunity b) Passive immunity
   c) Herd immunity d) Active and herd immunity
37. **Failure of the oral polio vaccine occurs because of all of the following except:**
   a) Interference with other enteroviruses
   b) Frequent diarrhea preventing the colonisation
   c) Intestinal enzymes
   d) Immediate breastfeeding after vaccination
38. **Sabin polio vaccine stimulates the production of the following immunoglobulins except:**
   a) IgM b) IgG
   c) IgA d) IgD
39. **Which of the following is true about the poliovirus?**
   a) Poliovirus type 3 is responsible for most epidemics of paralytic poliomyelitis.
   b) In India, poliovirus type 1 commonly causes paralysis.
   c) Immunity is not type-specific.
   d) The virus is destroyed by 0.1 ppm chlorine but higher concentrations are required to kill the virus in sewage water.
40. **Which statement is not true regarding the epidemiology of poliomyelitis?**
   a) Paralysis is seen in more than 90% of infections.
   b) Natural infection occurs only in humans as there are no animal reservoirs.
   c) The virus, after entry, colonises the nasopharynx and multiplies there.
   d) Primary viremia occurs when the virus enters the bloodstream from regional lymph nodes of the intestine.
41. **A 5-year-old boy presents with neck stiffness and fever for past two days. His CSF reveals lymphocytic pleocytosis, normal glucose, and slightly elevated protein. No organisms are seen on Gram stain. The most likely diagnosis in this patient is:**
   a) Pyogenic bacterial meningitis
   b) Viral meningitis
   c) Tubercular meningitis
   d) Can't be ascertained
42. **Coxsackie viruses have the characteristic ability to infect:**
   a) Suckling mice b) Adult mice
   c) Gnotobiotic mice d) Guinea pigs
43. **Coxsackie B viruses are associated with all of the following except:**
   a) Myocarditis b) Encephalitis
   c) Bornholm disease d) Hand, foot, and mouth disease
44. **Group A coxsackie virus is associated with all the following except:**
   a) Bornholm disease
   b) Hand, foot, and mouth disease
   c) Aseptic meningitis
   d) Herpangina
45. **Orchitis is a complication of which one of the following viral infections?**
   a) Varicella b) Herpes
   c) Coxsackie d) Polio
46. **Herpangina (vesicular pharyngitis) is caused by:**
   a) Coxsackie group A b) Coxsackie group B
   c) ECHO virus d) Poliovirus
47. **Which of the following viruses is associated with hand, foot, and mouth disease?**
   a) Enterovirus 71 b) Coxsackie virus type 5, 10
   c) Coxsackie type 23 d) All of the above
48. **A 25-year-old man reports to casualty with complaints of being bitten by a stray dog on the left leg. The dog cannot be traced. On examination, the leg shows lacerations with multiple bites. The nurse has washed the wound with soap and water. What is the immediate action required in casualty?**
   a) Take a swab from wound and send for culture.
   b) Administer rabies vaccine after proper wound toilet.
   c) Administer rabies immunoglobulin and vaccine immediately after proper wound toilet.
   d) Send the patient home.
49. **The staining technique used for the demonstration of Negri bodies is:**
   a) Giemsa stain b) Wright stain
   c) Field stain d) Seller stain
50. **Which part of the brain tissue will have Negri bodies in abundance?**
   a) Cerebrum b) Brainstem
   c) Hippocampus d) Corpus callosum

51. **Which vaccines are safe and effective for pre-exposure prophylaxis for rabies?**
   a) Semple vaccine
   b) Beta-propiolactone (BPL)
   c) Duck egg vaccine
   d) Human diploid cell vaccine (HDC)
52. **Post-exposure prophylaxis in rabies is recommended for all of the following except:**
   a) Minor abrasions without bleeding
   b) Licks on the intact skin
   c) Single, transdermal bite
   d) Exposure to bats
53. **Which one of the following is false regarding post-exposure prophylaxis for rabies?**
   a) Five-six doses of the vaccine at 3, 7, 14, 30 and 90 days after exposure
   b) The suspected animal needs observation for two days
   c) Stop treatment if the animal dies during the observation period
   d) Stop treatment if the animal is tested rabies-positive in a laboratory
54. **All the following statements are true regarding passive immunisation for rabies except:**
   a) It should be given after one to two days of the first dose of the vaccine
   b) Dose of human rabies immunoglobulin is 20 IU/kg body weight
   c) There is a risk of anaphylaxis with equine rabies immunoglobulin
   d) Both active and passive immunisation are preferred in high-risk cases
55. **Which one of the following is false regarding rabid dogs?**
   a) Bite without provocation and indiscriminately
   b) Usually die after 2 weeks
   c) Lie huddled, unable to feed
   d) 60% of viruses are shed virus in saliva
56. **Which of the following is false regarding Negri bodies?**
   a) Smears are fixed and stained by Seller's technique
   b) Have round to oval intranuclear bodies
   c) Possess basophilic inner granules
   d) Measure 3–27 microns
57. **All the following are true with regards to the treatment of an animal bite wound except:**
   a) Wash the wound thoroughly with soap and water
   b) Treat with Cetavlon/tincture iodine
   c) Infiltrate with anti-rabies serum
   d) Suture the wound immediately
58. **All of the following are examples of slow viral diseases except:**
   a) Eastern equine encephalitis
   b) Subacute sclerosing panencephalitis (SSPE)
   c) Progressive multifocal leucoencephalopathy (PML)
   d) Subacute spongiform viral encephalopathies
59. **Which of the following is not true regarding slow virus disease?**
   a) The course of illness lasts for months or years, with remissions and exacerbations.
   b) There is a predilection for the involvement of the central nervous system.
   c) The immune response is exaggerated.
   d) There is a genetic predisposition and usually, a fatal outcome.
60. **Prions are:**
   a) Protein-containing particles with either a DNA and RNA
   b) Viruses that are easily destroyed by physical and chemical agents such as heat, irradiation, and formalin
   c) Particles that can be transmitted to experimental animals by parenteral and oral challenge
   d) All of the above
61. **The only specific method of diagnosing human prion diseases is by:**
   a) Measuring an abnormal prion protein (PrPsc)
   b) CSF examination
   c) CT scan
   d) None of the above
62. **Progressive multifocal leukoencephalopathy (PML) is caused by:**
   a) Sindbis virus b) Lassa virus
   c) Polyomavirus d) West Nile virus
63. **Which one of the following is false with regard to prions?**
   a) They are proteinaceous infectious particles.
   b) They cause chronic progressive degenerative diseases of the CNS.
   c) They possess either DNA or RNA.
   d) Resistant to heat, irradiation, and formalin.
64. **The reservoir host for Japanese encephalitis is:**
   a) Monkey b) Bird
   c) Rodent d) Sheep
65. **The transmitting vector of Japanese encephalitis is:**
   a) Tick b) Mosquito
   c) Sandfly d) Rodent
66. **All of the following viruses belong to the genus Alphavirus except:**
   a) O'nyong-nyong b) Chikungunya
   c) Hantan d) Semliki
67. **Japanese encephalitis is transmitted by:**
   a) *Culex tarsalis* b) *Culex annulirostris*
   c) *Culex tritaeniorhynchus* d) *Aedes aegypti*
68. **Which of the following statements is not true regarding neurocysticercosis?**
   a) Stool microscopy has no diagnostic role.
   b) Medical treatment is effective only for parenchymal cysts.
   c) 60–80% of cases of cysticercosis show neural involvement.
   d) Meningitis is the most common presenting symptom.
69. **Primary amoebic encephalitis is caused by:**
   a) *Entamoeba histolytica* b) *Entamoeba coli*
   c) *Naegleria fowleri* d) *Acantheamoeba castelleni*
70. **Which of the following is an example of Trojan horses in microbiology?**
   a) *Acanthamoeba* species b) *Entamoeba histolytica*
   c) *Leishmania donovani* d) *Trypanosoma cruzi*
71. ***Acanthamoeba* species can cause:**
   a) Parasitic meningitis b) Primary amoebic encephalitis
   c) Amoebic keratitis d) All of the above
72. **Which of the following is true regarding the treatment of CNS infections caused by free-living amoeba?**
   a) It can be treated successfully with metronidazole
   b) The surgical treatment is often successful
   c) The condition is almost fatal
   d) All of the above
73. **Which of the following is a not a morphological form seen in toxoplasmosis?**
   a) Tachyzoites b) Bradyzoites
   c) Sporocyst d) Oocyst
74. **Which of the following statement is true regarding toxoplasmosis?**
   a) The definitive hosts are cats while the intermediate hosts are humans.
   b) The definitive hosts are rodents while cats are accidental hosts.
   c) The definitive hosts are cats and humans are accidental hosts.
   d) The intermediate hosts are cats while humans are the definitive hosts.
75. **Avidity testing of IgG against toxoplasmosis is a good indicator to:**
   a) Differentiate recent from past infections
   b) Diagnose past infections
   c) Diagnose recent infection
   d) Detect only congenital infections

76. Sabin Feldman test has been used to diagnose:
a) Toxoplasmosis b) Leishmaniasis
c) Trypanosomiasis d) Neurocysticercosis

77. HIV patients should be started on trimethoprim–sulphamethoxazole prophylaxis to prevent toxoplasmosis:
a) In all patients
b) In case of CD4 below 100/μL
c) In case of CD4 below 500/μL
d) In cases where IgG is high

78. Which of the following is a diagnostic criteria of toxoplasmosis in newborn?
a) Detection of IgA at 10 days after birth
b) Detection of IgE and IgM in newborn blood
c) Levels of IgG >1000 in neonates
d) All of the above

79. Winterbottom's sign is seen in:
a) Toxoplasmosis b) Sleeping sickness
c) Cryptococcosis d) Histoplasmosis

80. The vector responsible for the transmission of African sleeping sickness is:
a) Tse tse fly b) *Culex* mosquito
c) *Aedes* mosquito d) Rat flea

81. A 38-year-old smoker developed a severe pneumonia-like syndrome and was hospitalised. In next two days, he developed meningitis. He was seropositive for HIV and was on antiretroviral therapy. His CD4 counts were less than 50. His CSF sample was positive for capsular polysaccharide antigen by serology. What is the likely causative agent?
a) *Mycobacterium tuberculosis* b) *Cryptococcus* spp.
c) *Streptococcus pneumoniae* d) *Candida* spp.

82. Fatal fungal meningitis in patients with AIDS is commonly caused by:
a) Cryptococcosis b) Blastomycosis
c) Sporotrichosis d) Chromoblastomycosis

83. Which drug is given as prophylaxis against toxoplasmosis in HIV patients?
a) Trimethoprim–sulfamethoxazole
b) Clindamycin
c) Erythromycin
d) Pyrimethamine

Answers: 1:d, 2:c, 3:c, 4:b, 5:a, 6:d, 7:b, 8:c, 9:d, 10:d, 11:a, 12:b, 13:c, 14:a, 15:c, 16:a, 17:c, 18:c, 19:c, 20:c, 21:d, 22:b, 23:a, 24:a, 25:b, 26:d, 27:d, 28:a, 29:c, 30:a, 31:b, 32:a, 33:d, 34:a, 35:a, 36:d, 37:c, 38:d, 39:d, 40:a, 41:b, 42:a, 43:d, 44:a, 45:c, 46:a, 47:d, 48:c, 49:d, 50:c, 51:d, 52:b, 53:b, 54:a, 55:b, 56:b, 57:d, 58:a, 59:c, 60:c, 61:a, 62:c, 63:c, 64:b, 65:b, 66:c, 67:c, 68:d, 69:c, 70:a, 71:c, 72:c, 73:c, 74:c, 75:a, 76:a, 77:b, 78:d, 79:b, 80:a, 81:b, 82:a, 83:a

CHAPTER 25

Gastrointestinal Infections

(MI3.1–3.8)

Chapter Navigation

Learning objectives

- Describe the microbial agents that cause diarrhea and dysentery and the pathogenesis, clinical features, and diagnosis of these infections, including acid peptic disease
- Discuss the causative agents of food poisoning and the pathogenesis, clinical course, and laboratory diagnosis of the same

The gastrointestinal (GI) tract is protected by multiple defence mechanisms against invasion by pathogens:

- **Saliva in the mouth**: It is rich in lysozyme, which destroys bacteria by digesting their cell walls.
- **Colonisation by normal flora:** The intestinal commensal bacteria compete for space and energy resources with incoming pathogens. The balance between beneficial and harmful bacteria helps to maintain the mucosal integrity and modulates the immunological activity of the gut barrier.
- **Mucus layer:** The mucus produced by epithelial cells prevents direct contact between microorganisms and the cells beneath.
- **Gastric acidity**: Most bacteria are killed by the low pH of the stomach. Reduction of gastric acid secretion predisposes to infection.
- **Antibodies and epithelial and immune cells:** These are present in the gut barrier.
- **Gut-associated lymphoid tissue (GALT):** This comprises the tonsils, Peyer's patches, appendix, colonic and caecal patches, and a number of smaller, single follicular structures termed **isolated lymphoid follicles (ILF).**

INFECTIONS OF THE GI TRACT

- Infections are caused by the **ingestion of exogenous pathogens** in sufficient quantities for them to evade the immune mechanism of the host.

- Acute diarrheal diseases are the leading cause of **morbidity and mortality in children** in developing countries, next only to respiratory infections.
- Among the diseases that affect the gastrointestinal system, **parasitic infections** caused by protozoa and intestinal helminths are among the most prevalent, especially in the developing and lower middle-income countries.

Ingested pathogens produce disease by the following mechanisms:

- Production of toxins
- Invasion and multiplication

The **outcomes** of ingested pathogens may be as follows:

- The pathogens produce **primary pathology** and disease manifestation in the gut, as seen in infections caused by diarrheagenic *Escherichia coli*, *Vibrio cholerae*, and *Clostridium difficile*.
- A **systemic infection** is initiated in the gut, but pathology and disease are manifested elsewhere in the body, as seen in infections caused by *Salmonella* Typhi and Paratyphi A, Paratyphi B, hepatitis A and E and polio.

Pathogenesis of GI Infections

The sources of gastrointestinal infection are **contaminated food or potable** water, which in turn have been contaminated with human or animal waste/feces (or other environmental sources such as sewage).

Fig. 25.1 presents an algorithm of the pathogenesis of GI infections.

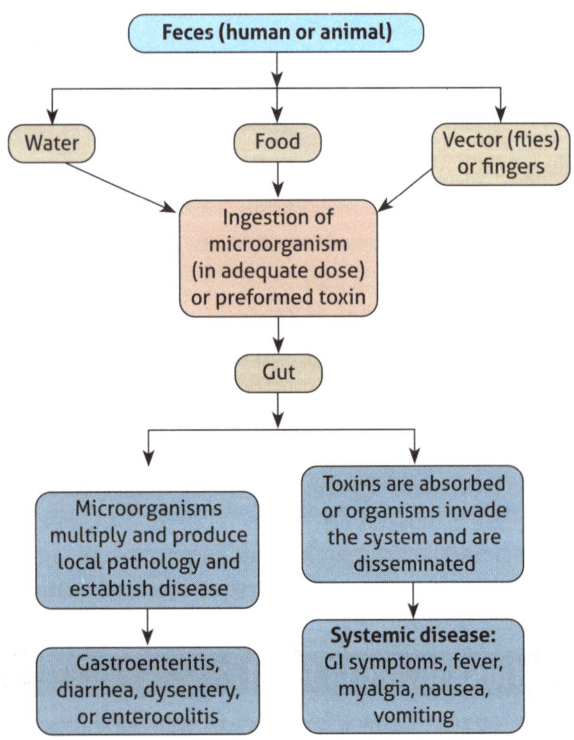

Fig. 25.1 Pathogenesis of GI infections

Clinical Presentations of Infections of the GIT

Diarrhea

- Diarrhea is defined as an increase in **fluid and electrolyte loss into the lumen,** causing **unformed liquid stool**, which is sometimes copious and watery.
- The frequency and volume of bowel movement are relative to the usual habits of the individual (generally, more than 3–5 episodes of liquid/loose stool per day).
- Diarrheal diseases are among the most common causes of morbidity and mortality in children in the developing nations. It is the second leading cause of death in children under five years of age.
- It can lead to **severe dehydration and death**. Long-term diarrhea leads to **malnutrition**.
- The most common etiological agents are **rotavirus** and ***E. coli***.
- Diarrhea may be of the following types:
  - **Acute diarrhea** with or without vomiting is the predominant symptom of infective gastroenteritis. Acute diarrhea may be watery or bloody (due to inflammation).
  - **Chronic diarrhea** is generally non-infectious (due to inflammatory bowel disease and pancreatic enzyme deficiency).
  - **Traveller's diarrhea** refers to the passing of watery stool three or more times in 24 hours and is often seen in travellers.

Dysentery

- It is a clinical condition of varying etiology, characterised by the frequent passage of **bloodstained, mucopurulent stools.**
- Clinical features include the **inflammation of the large intestine**, accompanied **by fever and abdominal cramps** associated with blood and mucus (pus) in stool.
- *Entamoeba histolytica* and *Shigella dysenteriae* are the common etiological agents.
- The two common types of dysentery are (i) **bacillary** and (ii) **amoebic**.

Enterocolitis

Enterocolitis is an inflammation of the digestive tract, including enteritis of the small intestine and colitis of the colon.

Gastroenteritis

- This refers to the inflammation of the stomach and intestines due to bacterial invasion or toxins or viral infection.
- It manifests with nausea, vomiting, abdominal discomfort, and loose motions.

The pathogenic mechanisms of various types of GIT infections are as follows (Table 25.1).

- **Small intestinal pathology**—manifests as large-volume, watery stools (enteritis), without pus or

Table 25.1 Pathogenic mechanism of gastroenteritis

Features	Toxins		
	Enterotoxins	Cytotoxins	Neurotoxins
Pathogenesis	Cause the outpouring of electrolytes and fluid into the lumen of the intestine, resulting in profuse, watery diarrhea	Disrupt the intestinal epithelium, leaving behind a raw, unprotected mucosa The resulting inflammatory response causes neutrophils and blood to be shed in stool	Result of ingesting toxins produced by microorganisms
Disease	Diarrhea	Dysentery	Food poisoning or intoxication
Organisms	*Vibrio cholerae* Non-cholera vibrios Enterotoxigenic *E. coli* *Salmonella* species, *Campylobacter jejuni, Clostridium perfringens, Clostridium difficile, Bacillus cereus, Giardia lamblia*	*Shigella dysenteriae* Enterohemorrhagic *E. coli* *Clostridium difficile* *Staphylococcus aureus* *Clostridium perfringens* *Entamoeba histolytica*	*Staphylococcus aureus* *Bacillus cereus* *Clostridium perfringens* *Clostridium botulinum*

blood (diagnosis is based on stool examination with the unaided eye and microscopy).

- **Large intestinal pathology**—usually manifests as frequent, small-volume stools with pus and/or blood (diagnosis is based on stool examination with the unaided eye and microscopy).

Etiology

The etiological agents responsible for various types of infections of the intestines are listed in Table 25.2.

Food poisoning

- Food poisoning is defined as **acute inflammatory diarrhea** that can be linked to the **consumption of contaminated food.**
- It may occur due to the ingestion of a **preformed toxin** (e.g., in clostridial infections) or due to a toxin produced as a **part of the pathogenesis** of the organism (e.g., staphylococcal food poisoning)
- A food poisoning outbreak is suspected when it follows the consumption of food from a common source, and the examination reveals non-inflammatory etiology of diarrhea.
- It may manifest as **diarrhea** (diarrheic type) or **vomiting** (emetic type).
- Bacterial toxin-mediated food poisoning has a short incubation period since the toxins in the food are preformed. The bacterial causes with symptoms and incubation period are listed in Table 25.3.
- Fungal toxins or mycotoxins, which are toxic metabolic products released by fungi, often contaminate food. The ingestion of such food results in disease. Examples of such diseases are as follows:
  - **Aflatoxicosis** results from the consumption of grains containing aflatoxins secreted by *Aspergillus flavus*, which contaminates groundnuts, corn and peas.

Table 25.2 Etiology of infective gastroenteritis

Diarrhea	Traveller's diarrhea	Dysentery	Food poisoning
Bacteria	**Bacteria**	**Bacteria**	**Bacteria**
Vibrio cholerae, Escherichia coli (EIEC, EPEC, EHEC), *Salmonella* species	*E. coli* (ETEC)	*Shigella* species	*Bacillus cereus*
Clostridium difficile, Yersinia enterocolitica	*Salmonella*	*Campylobacter* species	*V. parahaemolyticus*
Viruses	**Viruses**	*Escherichia coli* (EIEC, VTEC)	*Campylobacter jejuni*
Enterovirus, Rotavirus, Norwalk virus, Adenovirus	Norwalk virus	**Parasites**	*C. perfringens*
Calicivirus, Coronavirus, Astrovirus	Rotavirus	*Entamoeba histolytica*	*Staphylococcus aureus*
Fungus	**Parasites**	*Giardia lamblia*	*Clostridium botulinum*
Candida albicans (antibiotic-associated diarrhea)	*Giardia lamblia*	*Balantidium coli*	**Fungi**
Parasites		*Taenia saginata*	Mycotoxins
Cryptosporidium, Cystoisospora (previously known as *Isospora*)		*Taenia solium*	
Sarcocystis, Cyclospora		*Trichuris trichiura*	
		Fasciola hepatica	
		Hymenolepis nana	

Table 25.3 Bacterial food poisoning, incubation periods, symptoms, and sources

Bacteria causing food poisoning	Incubation period	Symptoms	Common food involved
Staphylococcus aureus	1–6 hours	Nausea, vomiting, diarrhea	Milk products (mayonnaise, cream, etc.), potato, ham
Bacillus cereus	1–6 hours	Emetic type—vomiting	Fried rice
Bacillus cereus	8–16 hours	Diarrheal type—cramps, diarrhea	Cereal, beans, meat
Clostridium perfringens	8–16 hours	Diarrhea, cramps, no vomiting	Beef, meat, canned food
Vibrio	>16 hours	Rice water stools	Water
Vibrio parahaemolyticus	>16 hours	Dysentery	Shellfish
Enterotoxigenic *E. coli*	>16 hours	Dysentery	Improperly cooked meat/beef or products, raw milk
Salmonella	>16 hours	Inflammatory diarrhea	Eggs, dairy products, beef
Campylobacter	>16 hours	Inflammatory diarrhea	Poultry, raw milk
Shigella	>16 hours	Dysentery	Water, unwashed vegetables

  - **Ergotism** results from the ingestion of rye contaminated with *Claviceps purpurea*.
- Non-infectious etiology of food poisoning may be attributed to the toxins of some fish and shellfish.

Laboratory Diagnosis of GI Infections (Fig. 25.2)

Specimen collection

- A clean, dry, screw-capped, wide-mouthed container is used to collect **feces** for culture.
- If the collection of feces is not possible, a **rectal swab** may be collected.
- For an outbreak of food poisoning, the **food implicated** and the **vomitus** of the patient are collected along with the feces sample.

Transport

- Though a freshly collected fecal sample is ideal for investigations, a transport medium such as **Cary–Blair or Venkatraman–Ramakrishnan medium** is used if a delay in transport to the laboratory is anticipated.
- For *Vibrio cholerae*, **alkaline peptone water**, which also serves as an enrichment medium, can be used for transport.

Microscopy

- **Wet preparation:** Microscopy of feces is performed to detect pus cells and RBCs and the ova, cysts, or segments of adult parasites.
- **Saline and iodine preparation:** This is done to detect helminth ova, protozoan cysts, and protozoan trophozoites.
- **Concentration techniques:** The **formol ether** or **salt floatation techniques** for ova and cysts may be performed to increase the positivity rate.
- **Gram-stained smear:** Given that the feces of healthy individuals contains several bacteria, Gram staining has limited use in the diagnosis of GI infections. However, it may be useful to show the following:
  - The presence of a **large number of yeast cells** in an immunocompromised host or infant, which is suggestive of ***Candida*** infection
  - The presence of **curved bacilli**, suggestive of ***Vibrio***
- **Modified acid-fast stain (Kinyon or ZN stain):** It can be used for the identification of coccidian parasites like ***Cryptosporidium***, ***Cystoisospora***, and ***Cyclospora***.
- **Electron microscopy:** It is used for the identification of **viruses** (only in referral laboratories).

Culture of feces

Direct culture

- **Non-selective media:** MacConkey agar
- **Selective media:** Xylose-lysine-deoxycholate (XLD) agar, deoxycholate citrate agar (DCA), or thiosulfate-citrate-bile salt-sucrose agar (TCBS)

Enrichment culture

- Selenite F broth, tetrathionate broth (incubated for 12–18 hours)
- Alkaline peptone water (6–8 hours)
- Following this, a subculture is made on the solid culture medium used for direct plating
- If the sample contains only a small number of pathogens, the enrichment medium allows selective growth of the pathogen, which then becomes easier to isolate

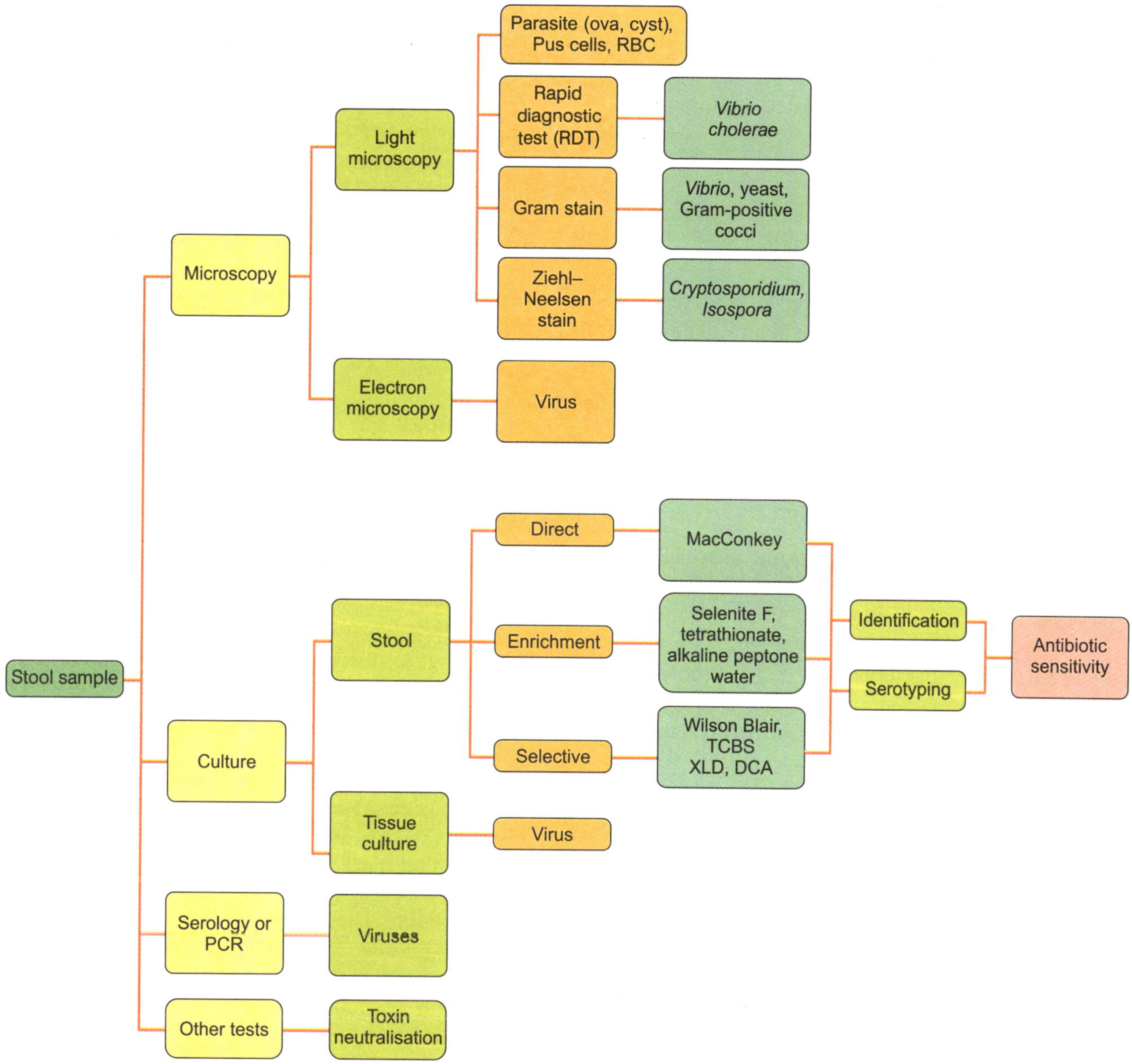

Fig. 25.2 Algorithm for the diagnosis of infective diarrhea (*TCBS*—thiosulfate-citrate-bile salts-sucrose agar; *XLD*—xylose lysine deoxycholate; and *DCA*—deoxycholate citrate agar)

Identification

Once isolated, the organism is identified by **biochemical tests** and **serotyping**.

Antibiotic sensitivity tests

- These are carried out only for selected bacterial pathogens—*Shigella*, *Vibrio*, and *Salmonella* isolates.
- *Shigella* and *Salmonella* may appear active in vitro to aminoglycosides, first- and second-generation cephalosporins and cephamycins but are not effective clinically.
- For stool isolates of *Salmonella* and *Shigella*, only ampicillin, a fluoroquinolone, and trimethoprim-sulfamethoxazole should be tested.

Tissue culture

- This method is used for the **culture of viruses** (it is not clinically important because viral diarrheas are self-limiting and regress spontaneously).
- Tissue culture may be done for epidemiological or research purposes, but not for diagnosis.
- ETEC penetrate HeLa and HEp-2 cells in culture. EHEC verotoxin (VT) can be detected by cytotoxic effect on Vero cells.

Serology

- ELISA is used for the detection of ***E. coli*** O157: H7 (EHEC), **Shiga toxin**, and ***C. difficile* toxins**.
- In addition, it can be used to detect rotavirus antigens.

PCR

- Specific primers can be used to detect the **Norwalk virus** and other viral agents.
- Multiplex PCR can be used to detect multiple targets in stool in viral diarrhea.

Detection of enterotoxin

Immunodiffusion, ELISA, neutralisation, and latex agglutination tests can be used for the detection of toxins.

Rapid diagnostic tests (RDT)

- *V. cholerae* O1 and O139 antigens are detected in stool specimens using monoclonal antibodies by the immunochromatographic method.
- A dipstick is dipped into a tube containing the stool sample and applied to the well of the immunochromatographic cassette. A coloured band denotes the presence of the antigen (O1 or O139).

Treatment of GI Infections

- The mainstay of the treatment of acute diarrhea is **fluid and electrolyte replacement** to maintain hydration.
- Viral diarrheas are self-limiting and may require only fluid and electrolyte correction.
- Some infective bacterial diarrheas (cholera) or dysentery (shigellosis) may require appropriate antibiotics like **ceftriaxone, ciprofloxacin, and tetracycline**.
- When parasites are the cause, **metronidazole and tinidazole** may be used to treat individuals.

Contraindication of antibiotics in GI infections

- A word of caution while treating diarrhea with antibiotics—they may increase the risk of **hemolytic uremic syndrome in children infected with *E. coli*** O157, H7 (Shiga toxin-producing *E. coli*).
- The prolonged use of antibiotics may also lead to the development of resistance among the normal gut flora and the pathogen.
- Antibiotics suppress the normal flora, allowing *C. difficile* or *Candida* to overgrow, invade, and cause necrotising enterocolitis.
- **Viral diarrheas** are NOT to be treated with antibiotics.

MICROORGANISMS CAUSING GASTROINTESTINAL INFECTIONS

BACTERIAL GASTROINTESTINAL INFECTIONS

Salmonellosis (Non-typhoidal Salmonellae)

- *Salmonella* gastroenteritis (more appropriately, enterocolitis) or food poisoning is generally a zoonotic disease, the source of infection being animal products.
- Non-typhoid salmonellosis is a frequent infectious complication in systemic lupus erythematosus (SLE).
- The extraintestinal manifestations and sepsis are more frequent in SLE than in simple gastroenteritis.
- There are several serovars of salmonellae associated with diarrhea such as:
  - *S. typhimurium*
  - *S. enteritidis*
  - *S. haldar*
  - *S. heidelberg*
  - *S. agona*
  - *S. virchow*
  - *S. seftenberg*
  - *S. indiana*
  - *S. newport*
  - *S. anatum*

Epidemiology

- The **mode of infection** is by the **ingestion of contaminated water and food**—chiefly poultry (including eggs and egg products), meat, milk, and milk products. Food contamination may also result from the droppings of rats, lizards, or other small animals.
- Outbreaks of gastroenteritis can occur without food poisoning, as in **cross-infection in hospitals**.
- The role of human carriers is minimal compared to the magnitude of infection acquired from animals.
- *Salmonella* food poisoning is a problem in **developed nations**. This is attributed to certain food habits (eating salads and undercooked meat and meat products), large-scale food production, packaging, storage, and marketing.

Pathogenesis

The pathogenesis of salmonellosis is depicted in Fig. 25.3.

Clinical features

- The incubation period of the infection is **less than 24 hours**.
- It manifests with frequent passage of **liquid or watery stool** with increased water and electrolyte loss with **abdominal pain** and may be associated with **vomiting and fever**.
- Though it is self-limiting in most cases, in **neonates and children**, fluid and electrolyte loss might lead to dehydration and electrolyte imbalance and may turn fatal unless treated immediately.
- The severity of the condition may vary from the passage of one or two loose stools to an **acute, cholera-like disease.**

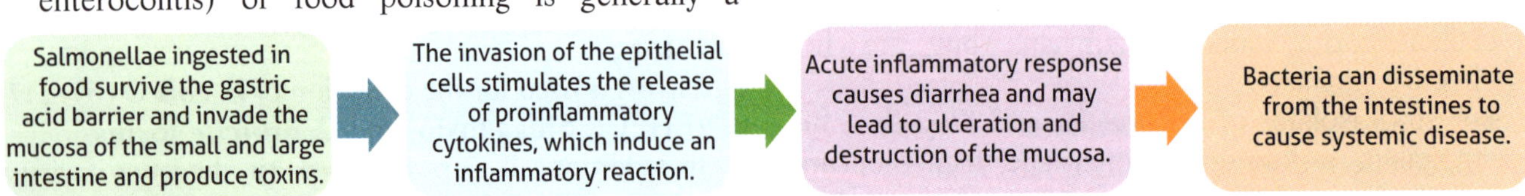

Fig. 25.3 Pathogenesis of salmonellosis

- *Salmonella* gastroenteritis usually subsides in 2–4 days, but in some cases, a more prolonged enteritis develops, with the passage of mucus and pus in feces, resembling dysentery. In a few, a typhoidal or septicemic type of fever may develop.

Laboratory diagnosis

- The etiological diagnosis is confirmed by the **culture** of stools.
- In outbreaks of food poisoning, the causative organism can be isolated from the article of food suspected to have been contaminated.

Treatment

- For uncomplicated, non-invasive *Salmonella* gastroenteritis, antibiotics should not be used as they may increase the period of fecal shedding of the bacilli.
- For serious, invasive cases, antibiotic treatment is needed.

Control

- Adherence to personal and food hygiene are the most effective control measures against *Salmonella* gastroenteritis.
- Vaccines do not protect against *Salmonella* gastroenteritis or *Salmonella* septicemia.

Antibiotic resistance in *Salmonella*

- **Multiresistant salmonellae** have now become important agents of hospital cross-infections.
- R factors conferring multiple drug resistance were first observed in *S.* Typhimurium phage type 29 in England in the 1960s. These resistant strains have become widely disseminated among salmonellae.

Reference centres for *Salmonella* in India

The **National *Salmonella* and *Escherichia* Centre (NSEC), Central Research Institute, Kasauli**, is the national reference centre for the serotyping of *Salmonella* and *Escherichia coli* for epidemiological studies. It was established in collaboration with Public Health Laboratories, Colindale, UK, and the WHO.

Diarrheal Diseases Caused by *Escherichia coli*

Multiple agents can cause diarrhea (Table 25.2). An important causative organism is *Escherichia coli* (EIEC, EPEC, EHEC) which can cause sporadic diarrhea in infants and children, traveller's diarrhea, dysentery, or bloody diarrhea. It can result in complications such as hemorrhagic colitis and hemorrhagic uremic syndrome. Uropathogenic strains of *E. coli* are found as causative agents of urinary tract infection.

Agent—*Escherichia coli*

- *E. coli* are **gram-negative, straight rods**, measuring 1–3 µm × 0.4–0.7 µm, which are arranged singly or in pairs (Fig. 25.4a and b).

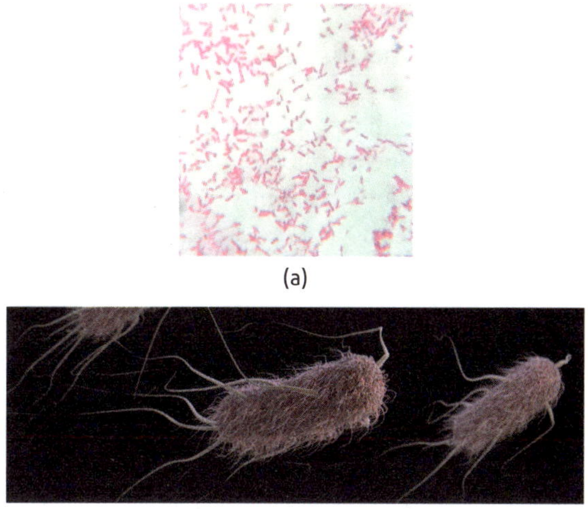

(a) (b)

Fig. 25.4 (a) Gram-stained smear of *E. coli:* gram-negative slender rods (*Source*: Dept of Microbiology, PIMS, Puducherry) and (b) a computer-generated 3D image of long, whip-like, peritrichous flagellae of *E. coli* (*Source*: Centres for Disease Control and Prevention [CDC], Public Health Image Library [PHIL], Image ID 21915)

- They are **motile** on account of the presence of peritrichous flagella.
- **Capsules and fimbriae** are found in some virulent strains.

Antigenic structure

- The serotyping or antigenic typing of *E. coli* is based on three antigens—**somatic antigen O, flagellar antigen H, and capsular antigen K**. In addition, *E. coli* also have **fimbrial or F antigens.**
- The antigenic pattern of a strain is based on the numerical type of the antigen it carries (e.g., O111: K58:H2).

Virulence factors

Two types of virulence factors have been recognised in *E. coli*—**surface antigens** and **toxins**.

1. Surface antigens

- **Somatic antigen O (somatic lipopolysaccharide):**
  - It exerts **endotoxic activity** and protects the organism from phagocytosis and the bactericidal effects of complements.
  - Around 170 types of heat-stable, lipopolysaccharide O antigens have been recognised.
  - O antigens are associated with the virulence of the organism.
  - The normal colon strains belong to the 'early' O groups (1, 2, 3, 4, etc.), while the enteropathogenic strains belong to the 'later' O groups (26, 55, 86, 111, etc.).
- **K envelope antigen:**
  - Strains causing **neonatal meningitis and septicemia** carry this antigen, which resembles the group B antigen of meningococci.

- The **envelope or K antigens** also offer protection against phagocytosis. Around 100 K antigens have been recognised to date.
- This is an acidic polysaccharide antigen located on the envelope or microcapsule.
- K antigens are currently classified into two groups, I and II.

• **H antigen (flagellar antigen):**
- H antigens are more specific since cross-reaction between other members is not common.
- 75 H antigens have been recognised so far.

• **F antigens:**
- Fimbriae are important virulence factors.
- They are **heat-labile** and get detached when the organism is heated to 100°C.
- Plasmid-coded fimbriae are found only in small numbers and mediate mannose-resistant hemagglutinins.
- **P fimbriae**, seen in uropathogenic strains, bind specifically to the P blood group substance on human erythrocytes and uroepithelial cells.
- Fimbriae are important in **initial attachment and colonisation**.

Certain **colonisation factor antigens (CFA)** present in enterotoxigenic *E. coli* cause human diarrhea.

2. Toxins

E. coli produce two kinds of exotoxins—**hemolysins** and **enterotoxins**.

- Hemolysins do not appear to be relevant in pathogenesis though they are produced more commonly by virulent strains than by avirulent strains.
- **CNF1 (cytotoxic necrotising factor-1)** and **siderophores** are virulence factors in uropathogenic *E. coli* and are important components of biofilm production and adhesion.
- Enterotoxins are important in the pathogenesis of diarrhea.
- Three distinct types of *E. coli* enterotoxins have been identified—**heat-labile toxin (LT), heat-stable toxin (ST), and verotoxin (VT)**, which is also known as **Shiga-like toxin (SLT).**

Heat-labile toxin (LT)

- *E. coli* LT resembles the cholera toxin in its structure, antigenic properties, and mode of action.
- The toxin has one A subunit (A for active) and five B subunits (B for binding).
- The toxin binds to the GM1 ganglioside receptor by means of subunit B.
- Subunit A is activated to yield two fragments—A1 and A2.
- The A1 fragment activates adenylyl cyclase in the enterocyte to form cAMP, which results in an increased outflow of water and electrolytes into the gut lumen, with consequent watery diarrhea.

Heat-stable toxin (ST)

It is a low-molecular-weight polypeptide, which is poorly antigenic. Two types of STs are known:

- **STA**, which acts by activating cGMP in the intestine
- **STB**, the mode of action of which is not entirely understood

Verocytotoxin (VT) (also known as Shiga-like toxin [SLT])

- It is so named because of its cytotoxic effect on Vero cells.
- It acts by inhibiting protein synthesis.

The methods used to detect these toxins are presented in Table 25.4.

Clinical features

Broadly, *E. coli* produce three types of clinical syndromes, which are as follows:

- Diarrhea
- Septicemia, neonatal sepsis, and neonatal meningitis (Chapters 21 and 24)
- Urinary tract infection (Chapter 28)

Diarrhea is caused by six types of *E. coli*, which are presented in Table 25.5.

Laboratory diagnosis

Since *E. coli* is part of the normal gut flora, mere isolation from stool samples does not confirm its role in the causation of disease. Specific tests are required to identify diarrheagenic *E. coli* (Table 25.5).

Table 25.4 Methods for detecting ETEC-labile (LT) and ETEC-stable (ST) enterotoxins

In vitro assays	LT	ST
1. Tissue culture tests		
Rounding of Y1 mouse adrenal cells	+	–
Elongation of Chinese hamster ovary (CHO) cells	+	–
2. Serological tests		
ELISA	+	ST ELISA with monoclonal antibody
Passive agglutination tests, passive immune hemolysis, precipitin (Eiken's) test	+	–
3. Genetic tests		
DNA probes	+	+

ETEC—enterotoxigenic E. coli

Table 25.5 Types of diarrheagenic *E. coli*, their clinical presentation, serotypes, pathogenesis, and laboratory diagnosis

Type of diarrheagenic *E. coli*	Clinical presentation	*E. coli* serotypes	Pathogenesis	Laboratory confirmation
Enteropathogenic *E. coli* (EPEC)	Sporadic diarrhea in infants and children Institutional outbreaks of diarrhea Disease seen less often in adults	*E. coli* 026:B6, 055:B5, 0111:B4	• Not fully understood • Non-invasive and do not produce enterotoxins • Infantile enteritis: Plasmid-encoded protein, EPEC adherence factor (EAF) involved in adherence to the enterocyte membrane	• Specific antisera for EEC types • Tissue culture adherence by IF • DNA-based detection of virulence gene
Enterotoxigenic *E. coli* (ETEC)	• Endemic diarrhea in developing tropical countries, in all age groups; with variable severity • Traveller's diarrhea in persons from non-endemic areas who visit endemic area	*E. coli* 06, 08, 015, 025, 027, and 0167	• Disease caused primarily by adherence to intestinal mucosa by fimbriae called colonisation factor antigens, of which several types have been identified (CFA I, II, III and IV) • ETEC produce enterotoxins which may be either LT or ST or both	• Gene probes for LT and ST from cultures
Enteroinvasive *E. coli* (EIEC)	Dysentery or bloody diarrhea	*E. coli* 028 ac, 0112 ac, 0124, 0136, 0143, 0114, 0152, and 0154	• Invasion of interstitial epithelial cells as seen in shigellosis	• Nucleic acid detection of invasive–associated genes or demonstration of invasiveness in tissue cultures
Enterohemorrhagic *E. coli* (EHEC)	Hemorrhagic colitis and hemorrhagic uremic syndrome (HUS), particularly in young children and the elderly that can be fatal	0157:H7 or rarely, 026:H1	• Two potent toxins, verocytotoxin (VT) and Shiga-like toxin (SLT) that damage the vascular endothelial cells • In HUS, a characteristic renal lesion of capillary microangiopathy is seen	• Shiga-like toxin detected from colonies by PCR based and immunological assays • Real-time PCR with specific DNA probes for the VT1 and VT2 genes • VT detected by its cytotoxic effects on Vero or HeLa cells • VTEC strains can be screened using sorbitol MacConkey screening medium for 0:157 VTEC
Enteroaggregative *E. coli* (EAEC)	Persistent diarrhea	Most are O-untypable; many are H-typable	• Bacteria aggregate in a 'stacked brick' formation on HEp-2 cells or glass	• Tissue culture assays for aggregate/adherence
Diffusely adherent *E. coli* (DAEC)	Diarrhea in children above 12 months of age		• Pattern of diffuse adherence is seen, in which the bacteria uniformly cover the entire cell surface • DEAC are less well-established as pathogens	• PCR or specific probe

Prevention and treatment

Personal hygiene and food safety are the two chief measures for preventing diarrhea in susceptible populations.

- Antibiotic treatment is not indicated for *E. coli* diarrhea.
- There are no vaccines available currently.

Cholera

Case 1. *A 12-year-old boy from an overcrowded colony presented to the pediatric emergency department during the monsoons with complaints of profuse and frequent (10–15 times), watery diarrhea since the previous day, accompanied by 2–3 episodes of vomiting. There was no history of fever,* *abdominal pain, or passage of blood in stools. The boy complained of decreased urine output, intense thirst, and leg cramps. On examination, he was found to be moderately dehydrated. He mentioned that some other residents of the slum were also suffering from a similar condition. The stool, on gross examination, had a characteristic **'rice water' appearance**. The stool culture was positive for **Vibrio cholerae**. Management included rehydration and a single dose of azithromycin 20 mg/kg of body weight.*

i. What is the provisional diagnosis?
ii. What is the pathogenesis of this infection?

Cholera is an acute diarrheal disease caused by *Vibrio. cholerae*. Natural infection with cholera occurs only in humans. *Vibrio cholerae* has multiple serogroups that are mentioned below.

Epidemiology

- Cholera can occur in many forms—sporadic, endemic, epidemic, or pandemic.
- It has been known to exist in the large deltaic area of the rivers Ganga and Brahmaputra in Bengal, India, since ancient times.
- The large-scale movement of persons, as occurs during fairs and festivals, has traditionally been associated with the spread of cholera.
- The persistence of the vibrio during the inter-epidemic periods may be attributed to the continuous transmission of subclinical or mild infection.
- The natural habitats of cholera vibrios are the **saline waters of coastal seas and brackish estuaries**, where they can persist for long periods, particularly in association with small crustaceans such as copepods, crabs, or plankton.

Historical perspective of cholera

Fig. 25.5 presents a timeline of the epidemiology of cholera.

El Tor vibrio

- The severity of illness is much less with the El Tor biotype than it is with classical cholera. A large proportion of the infections mild and asymptomatic.
- **Mortality is low and the carrier rate is high.**
- El Tor vibrios tend to remain endemic in many new geographic areas, causing periodic epidemics.
- In India, the classical vibrio has been replaced by El Tor; however, in Bangladesh, the classical vibrio has staged a comeback.

O-139 vibrio

- A new non-O-1 vibrio was isolated from a cholera outbreak in Chennai in 1992–93
- The new epidemic strain was designated serovar O-139 (or O-139 Bengal). Unlike the O-1 cholera vibrio, the O-139 vibrio is capsulated.
- As it possessed novel surface antigens, the O-1 strain vaccines could not protect against O-139 infection.

Agent—*Vibrio cholerae*

- Vibrios are **slender, gram-negative, curved bacilli** (Fig. 25.6a and b) of family Vibrionaceae.

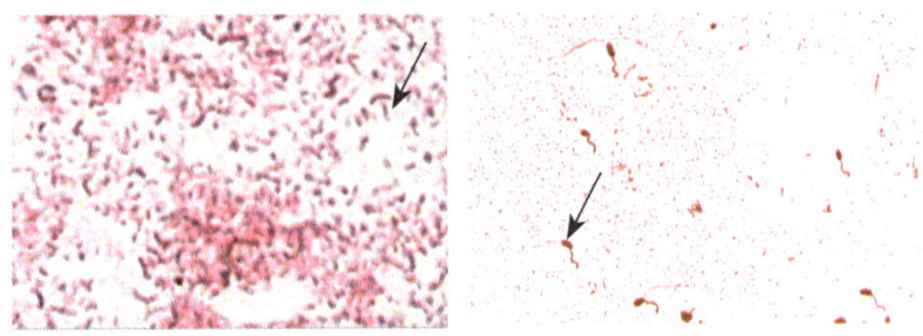

Fig. 25.6 *Vibrio cholerae:* (a) comma-shaped gram-negative rods and (b) digitally-enhanced photomicrograph of *Vibrio cholerae* stained by Leifson flagella stain visualised under 320 × magnification showing polar flagella (*Source*: [a] Department of Microbiology, Pondicherry Institute of Medical Sciences, Puducherry and [b] CDC, PHIL, Image ID 1034)

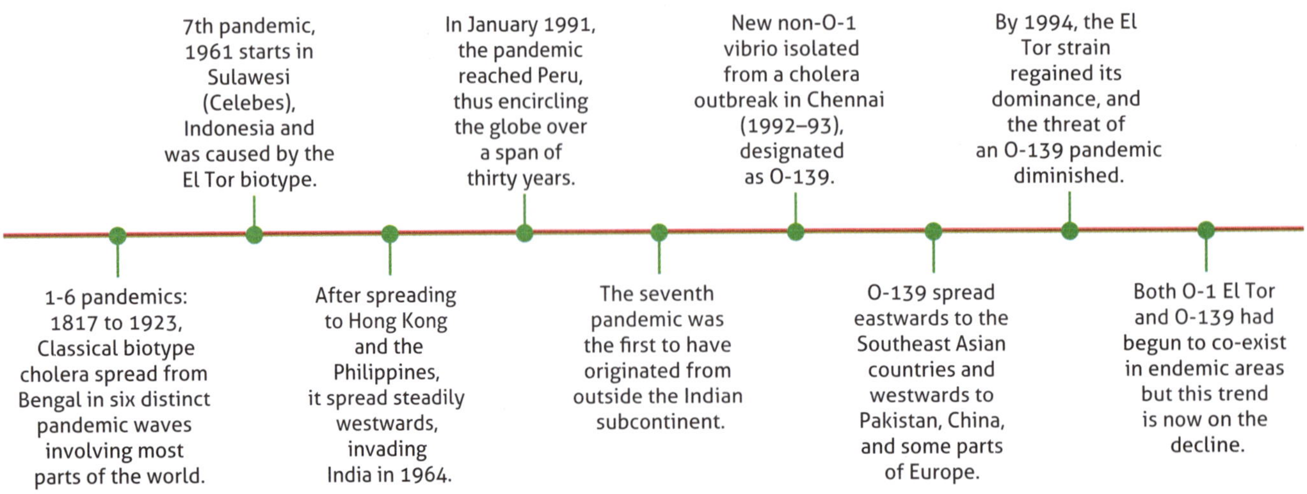

Fig. 25.5 Timeline of cholera

Classification

Based on their **requirement of sodium chloride**, vibrios are classified as halophilic and non-halophilic.

- **Halophilic:** *V. parahaemolyticus*, *V. alginolyticus*, *V. vulnificus*
- **Non-halophilic:** *Vibrio cholerae*

Another method of classifying vibrios is the **Gardner and Venkatraman's approach**, which is depicted in Fig. 25.7.

Biotypes of *Vibrio cholerae*

Based on a few biochemical properties, *Vibrio cholerae* has been classified into two biotypes—**classical and El Tor vibrios.**

- Vibrios isolated from Haj pilgrims at the Tor quarantine station on the Sinai Peninsula came to be called the El Tor vibrios.
- They were identical to cholera vibrios in all laboratory tests except that they lysed sheep erythrocytes and gave a positive Voges–Proskauer reaction.
- The incidence of mild and asymptomatic infections is more with El Tor vibrios than with the classical cholera vibrios.

Serotypes of *Vibrio cholerae*

- *V. cholerae* possess more than 200 O antigens. Based on these O antigens, it has been divided into different serogroups (Fig. 25.7). Non O1 vibrios are also known as NAG (non-agglutinable vibrios) as they are not agglutinated by O1.
- Of these, only the O1 and O139 serogroups have been found to cause epidemic cholera.
- Based on minor surface antigenic characteristics, classical and El Tor biotypes of cholera vibrios were classified as Ogawa and Inaba.

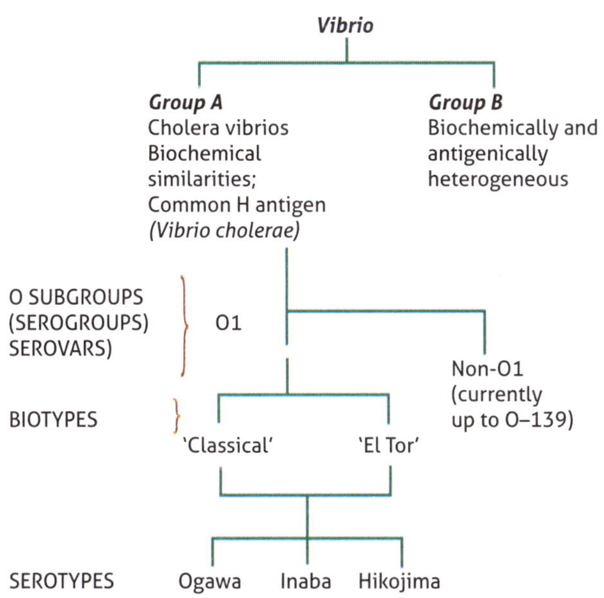

Fig. 25.7 Gardner and Venkatraman's classification of vibrios (updated)

- A third serotype, Hikojima, is agglutinated by both Ogawa and Inaba antisera (Table 25.6).
- The difference is only of epidemiological significance.

Typing

- **Phage typing:** Phage typing schemes have been standardised for classical and El Tor biotypes as well as for O-139 vibrios.
- **Molecular typing:** New molecular methods like ribotyping have added further refinements to strain typing.

Susceptibility

- *Vibrio cholerae* are killed easily by common disinfectants, drying, and pH <5.
- They can survive in tap water for up to one month and for a week at room temperature. They are killed by heating at 56°C for 30 minutes.
- Classical vibrios are more susceptible than El Tor.

Pathogenesis

- The action of *V. cholerae* on the intestinal epithelial cells is toxin (cholera toxin)-mediated.
- Cholera vibrios also possess the lipopolysaccharide O antigen (LPS and endotoxin) like gram-negative intestinal bacilli, but this antigen plays no role in the pathogenesis of cholera. It is, however, responsible for the immunity induced by killed vaccines.

Cholera toxin (CT)

- It is a protein that has a molecular weight of approximately 87 kDa and consists of six units—one A unit surrounded by five B subunits.
  - **B (binding) unit:** It attaches to the GM1 ganglioside receptors on the surface of jejunal epithelial cells.
  - **A (active) subunit:** It is transported into the enterocyte, whereupon it dissociates into two fragments: A1 and A2. The A2 fragment only links biologically active A1 to the B subunit.
  - CT is antigenic and induces the production of neutralising antitoxins and can be toxoided.
- The A1 fragment of the cholera toxin causes prolonged activation of cellular adenylate cyclase and the accumulation of cAMP.
- This results in the outpouring of intracellular electrolytes—Na, Cl, and K—and large quantities of water (Fig. 25.8) into the small intestinal lumen, leading to watery diarrhea.

Table 25.6 O serotypes of cholera vibrios

Serotype	O antigens
Ogawa	AB
Inaba	AC
Hikojima	ABC

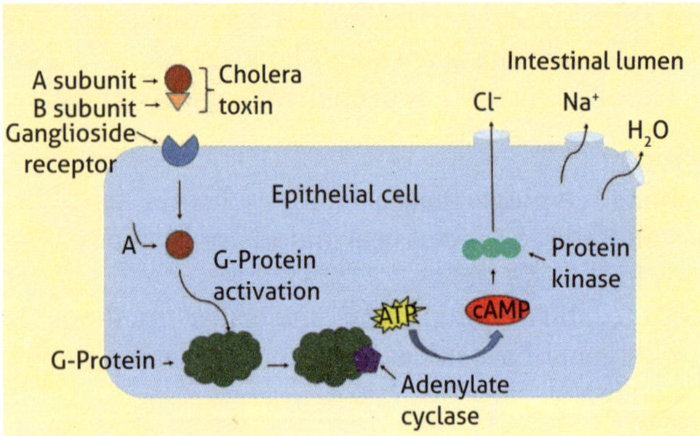

Fig. 25.8 Attachment of subunit B to GM_1 ganglioside on epithelial cells and entry of subunit A of the cholera toxin into the intestinal epithelial cells, leading to hypersecretion of sodium chloride and water (*Source*: Jessica Brochu, CC BY-SA 4.0, via Wikimedia Commons)

- Chloride (Cl) secretion occurs through **cystic fibrosis transmembrane conductance regulators** (CFTR) which are the transport regulators in the intestine.
- The fluid secreted is isotonic with plasma but contains much more of potassium and bicarbonate.
- The toxin also inhibits the intestinal absorption of sodium and chloride by the microvilli.

In-vitro **detection of activity of the cholera toxin:**

- It produces a characteristic **elongation of Chinese hamster ovary (CHO) cells** in culture and causes **histological changes in adrenal tumour (Y1) cell and Vero cell cultures**.
- It increases skin capillary permeability, also called the **'permeability factor'** (PF). This can be demonstrated by the **skin blueing test**—CT is injected intradermally into rabbits or guinea pigs; subsequently, pontamine sky blue is injected intravenously. The site of the injection of the toxin becomes blue.
- CT can also be estimated by **ELISA.**

Immunity

- Immunity may be localised to the intestine or systemic.
- In cholera, the vibrios remain confined to the intestine, where they multiply and elaborate the enterotoxin, which is responsible for the disease.
- Immunity, therefore, is directed against the bacterium or against the toxin—antibacterial or antitoxic.
- Natural infection confers some amount of immunity, but such, immunity does not last for more than 6–12 months.
- Re-infections after this period are known to occur.
- Local immunity is conferred by coproantibodies which consist of IgG, IgM, and IgA, which are local antibodies in the intestine.
- These antibodies appear in the feces. Their protective role is short-lived.

Clinical features

- The incubation period **less than 24 hours to five days.**
- The infection is acquired by the **consumption of contaminated water or food.**
- The clinical severity of cholera varies widely—from a **rapidly fatal disease to a transient, asymptomatic colonisation** of the intestine by the vibrios.
- **Rice water stool**—the cholera stool is typically a **colourless, watery fluid with flecks of mucus** (Case 1). It is said to resemble water in which rice has been washed (hence the name). It has a characteristic **inoffensive and sweetish odour**. In composition, it is a bicarbonate-rich isotonic electrolyte solution with little protein. Its outpouring leads to the diminution of extracellular fluid volume, hemoconcentration, hypokalemia, base-deficit acidosis, and shock.
- The disease begins slowly with mild diarrhea and vomiting in 1–3 days or abruptly with sudden, massive diarrhea.
- Severe illness manifests with profuse, painless, watery diarrhea and copious, effortless vomiting leading to hypovolemic shock and death in less than 24 hours.
- With fluid replacement and antibiotics, the disease may last 4–6 days (the patient may pass a total volume of liquid stool equal to twice his body weight).
- All the clinical features of severe cholera result from this massive loss of fluid and electrolytes.
- Complications include muscular cramps, renal failure, pulmonary edema, cardiac arrhythmias, and paralytic ileus.

Laboratory diagnosis

Diagnosis is based on the isolation and identification of *V. cholerae* from stool or rectal swabs.

Specimen

- **Stool is collected in the acute stage of the disease**, before the administration of antibiotics.
- **Rectal swabs** can also be used.
- Collecting samples from the pan is not recommended.
- Vomitus is not useful.

Transport or holding medium

- Since cholera vibrios die in a few hours in tropical temperatures, it is necessary to preserve the specimen at 4°C or in an appropriate holding medium.
- Stool samples may be preserved in VR fluid or Cary–Blair medium for long periods.
- **Venkatraman–Ramakrishnan (VR) medium** has a pH to 8.6–8.8.

- **Cary–Blair medium** has a pH 8.4 and is a buffered solution.
- **Autoclaved seawater** can also be used.

Microscopy

On direct microscopic examination of cholera stool, the **characteristic darting motility** of the vibrio is seen and its inhibition by specific antiserum can be demonstrated.

Culture

- The ideal temperature range is 16–40°C (**optimum temperature is 37°C**).
- The ideal pH is 6.4–9.6 (optimum being **8.2**).
- Growth is better in an **alkaline medium**.
- The enrichment medium (can also be used as transport media) used is **alkaline peptone water (APW)** with a pH of 8.6.
- On **blood agar**, colonies are initially surrounded by a **zone of greening, which later becomes clear due to hemodigestion.**

Selective media for culture and isolation

- **Thiosulphate–citrate–bile salt–sucrose (TCBS):** Cholera vibrios ferment sucrose, which is indicated by the appearance of **large, yellow, convex colonies** (Fig. 25.9).
- **Alkaline bile salt agar (BSA; pH 8.2):** Colonies are **translucent** like water or oil drops.
- **Monsur's gelatin taurocholate trypticase tellurite agar (GTTA):** Cholera vibrios produce **small, translucent colonies with a greyish-black centre and a turbid halo**. The colonies grow to a size of 3–4 mm in 48 hours.

Identification

String test (Fig. 25.10)**—**The mixing of colonies in 0.5% sodium deoxycholate saline on a slide creates a characteristic 'string' formed by DNA material that is released from the vibrios as a result of the lysis of the cells by sodium deoxycholate.

Fig. 25.9 Yellow colonies on TCBS (*Source*: Department of Microbiology, Pondicherry Institute of Medical Sciences, Puducherry)

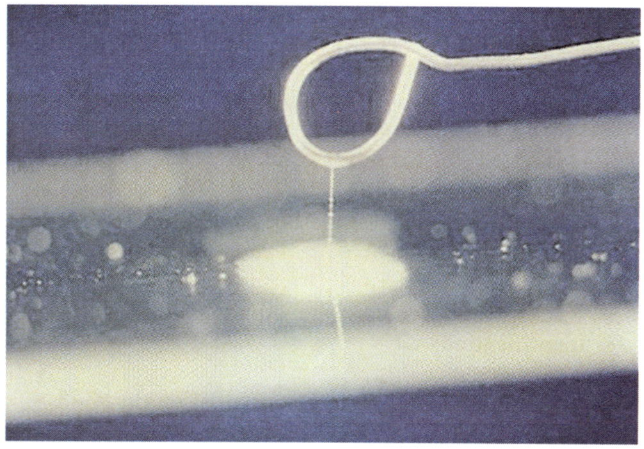

Fig. 25.10 String test used to differentiate *Vibrio* spp. from *Aeromonas* spp. and *Plesiomonas*

Biochemical reactions

- Cholera vibrios are **catalase- and oxidase-positive and produce indole.**
- They **reduce nitrates to nitrites** and produce **ornithine decarboxylase.**

Serogrouping

- Colonies suggestive of vibrios should be tested by agglutination with cholera O subgroup I serum (cholera non-differential serum). However, isolates of vibrios that are not agglutinated by the O subgroup I serum should not be ignored because non-O-1 vibrios are also known to produce a cholera-like disease.
- Specific antiserum against O-139 is now available.

The national reference laboratory for identification (including phage typing) and repository is the **International Reference Centre for Vibrio at the National Institute of Cholera and Enteric Disease (NICED)** in Kolkata.

Laboratory diagnosis of cholera

- **Specimen:** Stool or rectal swabs
- **Transport or holding medium:**
  - *Venkatraman–Ramakrishnan (VR) medium* (pH 8.6–8.8)
  - *Cary–Blair medium* (pH 8.4)
  - *Autoclaved seawater*
- **Microscopy: Characteristic 'darting' motility** that is inhibited by specific antiserum
- **Culture:**
  - Facultative anaerobe, **optimum temperature; 37°C (range:** 16–40°C), optimum pH: 8.2 (range 6.4–9.6)
  - Enrichment medium: **Alkaline peptone water** with a pH of 8

- **Blood agar:** Colonies are initially surrounded by a green zone that later clears due to hemodigestion
- **Selective media for stool culture:**
  - *Thiosulphate–citrate–bile salt–sucrose (TCBS):* **Large, yellow, convex colonies**
  - *Alkaline bile salt agar (BSA; pH 8.2):* Colonies are **translucent** like water or oil drops
  - *Monsur's gelatin taurocholate trypticase tellurite agar (GTTA)*: Small, translucent colonies with a greyish-black centre and a turbid halo
- **Identification:**
  - *String test*
  - Catalase- and oxidase-positive and produce indole, reduce nitrates to nitrites, and produce ornithine decarboxylase
  - **Serogrouping:** By agglutination with cholera O subgroup I serum (cholera non-differential serum) followed by Ogawa and Inaba

Testing water quality

Since water is the most significant vehicle for the spread of the infection, it is important to test water samples for *V. cholerae*. Two methods are employed for this purpose.

i. **Enrichment method:** In this method, 900 mL of the water to be tested is added to 100 mL ten-fold concentrated peptone water at pH 9.2. The mixture is incubated at 37°C for 6–8 hours and enriched a second time before it is plated on cholera-specific selective media to look for growth.
ii. **Filtration technique:** In this method, the water to be tested is filtered through a millipore membrane. The membrane is then placed directly on the surface of a selective medium and incubated. If vibrios are present, colonies appear after overnight incubation. Sewage should be diluted in saline, filtered through gauze, and treated as was described for water.

Prevention

Parenteral vaccines

- These are **suspensions containing 8,000 million killed *V. cholerae* per mL** with equal numbers of **Ogawa and Inaba serotypes.** These vaccines are given by **subcutaneous or intramuscular injection.**
- A vaccine for strain O-139 has also been prepared recently.
- Protection does not exceed 50–60 per cent and lasts only for 3–6 months.
- Also, injectable vaccines do not provide any local immunity in the intestinal mucosa. They are also unacceptably reactogenic.

Oral vaccines

There are three types of oral vaccines in use against cholera:

i. **Killed oral whole cell vaccines** with and without the inclusion of the B subunit of CT
ii. **Live oral vaccines with classical,** El Tor and O-139 strains, with their toxin genes deleted
iii. **Oral inactivated or non-live cholera vaccines—**prequalified by the World Health Organization (WHO) but not widely used

Dosage—Two doses of the vaccines are required for full protection (minimum 1 week and maximum 6 weeks apart). Three doses are required for children aged 2–5 years (minimum 1 week and maximum 6 weeks apart).

Treatment

- Prompt and adequate **replacement of fluids and electrolytes** using oral rehydration solution (ORS) is the first line of management. This is done either alone or along with supplementation of intravenous fluids.
- Cereal-based preparations are equally effective and are usually more acceptable.
- Antibacterial therapy is of secondary importance. **Doxycycline** is recommended as the first-line treatment for adults, while **azithromycin** is recommended as first-line treatment for children and pregnant women.

Other members of *Vibrio*

Vibrio mimicus

- *V. mimicus* infection is acquired by eating **seafood, especially oysters**.
- The disease is self-limiting.
- Clinical manifestations resemble those of infections caused by *V. parahaemolyticus.*
- *V. mimicus* has been responsible for many sporadic cases of diarrheal disease on the gulf coast of the USA.

Halophilic vibrios

- Vibrios that have a **high requirement of sodium chloride** are known as halophilic vibrios.
- Their natural habitats are seawater and marine life.
- Some halophilic vibrios have been known to cause human disease—*V. parahaemolyticus*, *V. alginolyticus*, and *V. vulnificus*.

Vibrio parahaemolyticus

- It is an **enteropathogenic, halophilic vibrio**, which was originally isolated in 1951 in Japan during an outbreak of food poisoning caused by sea fish.
- Infection is common in the coastal areas, where it is found in **fish and arthropods such as shrimps and crabs, and molluscs such as oysters.**
- It resembles the cholera vibrio except that it is **capsulated and shows bipolar staining**.
- Unlike other vibrios, it produces peritrichous flagella when grown on solid media and polar flagella in liquid cultures.
- It grows **only in media containing NaCl**.

- It tolerates salt concentrations up to 8% but not 10%. The optimum salt concentration for its growth is 2–4%.
- On TCBS agar, the colonies are green (as they do not ferment sucrose) with an opaque, raised centre and flat translucent periphery (note that *V. cholerae* colonies are yellow in colour). The string test is positive.
- It is differentiated from *V. cholerae* by various biochemical reactions.
- **Suceptibility**—*V. parahaemolyticus* is killed at 60°C in 15 minutes. It does not grow at 4°C but can survive refrigeration.
- Three antigenic components have been recognised—**somatic O**, **capsular K**, and **flagellar H antigens**. Serotyping is based on the O and K antigens; 12 O groups and 59 distinct K antigens have been recognised.
- Not all strains of *V. parahaemolyticus* cause disease in human beings.
- Strains isolated from environmental sources (such as water, fish, crabs, or oysters) are nearly always non-hemolytic when grown on a special high-salt blood agar (**Wagatsuma agar**).
- Strains from human patients are almost always hemolytic. This is called the **Kanagawa phenomenon** and is attributed to the presence of a heat-stable hemolysin.
- The significance of hemolysis is not known but is used as a laboratory test for pathogenicity; only the Kanagawa-positive strains are pathogenic. No enterotoxin has been identified.

Pathogenesis and clinical features

- *V. parahaemolyticus* causes **food poisoning associated with marine food**.
- It also causes **acute diarrhea** not associated with food poisoning.
- The vibrio is believed to cause enteritis by the invasion of the intestinal epithelium.
- Abdominal pain, diarrhea, vomiting, and fever are the usual signs.
- Feces contains cellular exudate and often also blood.
- Dehydration is of moderate degree, and recovery occurs in 1–3 days.
- Cases are more common in summer, and in adults more than in children.
- *V. parahaemolyticus* is found in sea fish from parts of India but human cases are much less frequent.
- Rarely, it may cause **wound infections**, and in more severe cases, sepsis.

Treatment

- The infection is usually self-limiting.
- The drug of choice is **doxycycline** in severe cases of gastroenteritis.
- For uncomplicated wound infections, minocycline or doxycycline can be used.

Vibrio alginolyticus

- This halophilic vibrio resembles *V. parahaemolyticus* in many respects and was formerly considered a biotype of the same.
- Its status as a human pathogen is uncertain.
- It has been associated with **infections of the eyes, ears, and wounds** in human beings exposed to **seawater**.

Vibrio vulnificus

- *V. vulnificus*, previously known as the L+ vibrio, is a marine vibrio of medical importance.
- It is VP-negative and ferments lactose but not sucrose. It has a salt tolerance of less than 8%.
- It causes two types of illnesses:
  - i. **Wound infection** may occur following contact of open wounds with seawater.
  - ii. In **immunocompromised hosts**, particularly those with liver disease, following the ingestion of the vibrio, usually in oysters, *V. vulnificus* penetrates the gut mucosa but does not cause gastrointestinal manifestations. Instead, it enters the bloodstream from the gut and rapidly leads to septicemia with high mortality.

Aeromonas and *Plesiomonas* Infections

- The family Vibrionaceae also contains the genera *Aeromonas* and *Plesiomonas*.
- *Aeromonas hydrophila*, originally isolated from frogs, in which it causes the **'red leg disease'**, has been reported in many cases of diarrhea and some pyogenic wound infections in humans.
- *Plesiomonas shigelloides* also has been reported from diarrheal disease.
- Both these are oxidase-positive, polar-flagellated, gram-negative rods and may be mistaken for vibrios.
- They may be differentiated from vibrios by biochemical tests such as utilisation of amino acids.

Bacillary Dysentery

Dysentery is **diarrhea that is associated with blood** (plus or minus mucus) and represents more invasive infection. It may be caused by bacteria or parasites. The bacterial causes may be:

- *Shigella* species
- *Campylobacter* species
- *Escherichia coli* (EIEC, VTEC)

Shigellosis

> **Case 2.** *A 16-year-old boy ate* pani-puri *made by a roadside vendor. Three days later, he developed abdominal cramps with mild fever followed by increased frequency of stools with blood and mucus.*
>
> *He visited the hospital, where his stool sample was sent for microscopic examination and culture. The sample showed plenty of pus cells, and the culture was positive for non-lactose-fermenting bacteria on MacConkey agar which agglutinated with antisera for* ***Shigella flexneri****. The boy was treated with ciprofloxacin 15 mg/kg of body weight, orally twice daily for 3 days and plenty of oral fluids as supportive therapy.*
>
> i. *What is dysentery?*
> ii. What are the various causative agents?
> iii. How is the condition diagnosed?

Shigellosis is an infection of the intestines caused by *Shigella*. The onset of the disease is marked by **diarrhea/dysentery, fever, and abdominal pain**. It may end with complications such as reactive arthritis, sepsis, seizures, and hemolytic uremic syndrome.

Epidemiology

- Human beings are the only natural hosts for shigellae.
- The infection is transmitted through contaminated fingers, food and water, flies, and fomites such as door handles, water taps, and lavatory seats.
- In young male homosexuals, transmission is a part of the gay bowel syndrome.

Clinical features

- The incubation period is 1–7 days, usually 48 hours.
- It is characterised by the frequent passage of loose, scanty feces containing blood and mucus, and abdominal cramps and tenesmus, fever, and vomiting (Case 2).

Complications

These are most often seen in **children and immunocompromised patients** like those affected by AIDS (mainly due to *S. dysenteriae* type 1).

- Some common complications are arthritis, toxic neuritis, conjunctivitis, parotitis and, in children, intussusception.
- Hemolytic uremic syndrome may occur as a complication in severe cases due to microangiopathic hemolytic anemia, thrombocytopenia, and renal failure.

Agent—*Shigella*

- These bacteria, which are frequently implicated in dysentery, were named after **Kiyoshi Shiga**, the scientist who isolated the first member of this genus from epidemic dysentery in Japan in 1896.
- These are **short, gram-negative bacilli, that are non-motile, non-sporing, and non-capsulated** and measure about 0.5 × 1–3 μm. Fimbriae may be present.
- They are classified into four species or subgroups based on a combination of biochemical and serological characteristics: *S. dysenteriae*, *S. boydii*, *S. flexneri*, and *S. sonnei*.
- *S. dysenteriae* type 1 forms a toxin (Shiga toxin), the earliest example of an exotoxin produced by a gram-negative bacillus (for classification, please refer to Chapter 4).

Antigenic structure

- Shigellae possess one or more 'major' antigens and a large number of 'minor' somatic O antigens.
- Some strains possess K antigens.
- The antigenic structure of *Shigella* is simple as compared to the complex structure of salmonellae. There is considerable antigenic sharing between some members of the genus and between shigellae and *E. coli*.
- Common fimbrial antigens may also occur, particularly in *S. flexneri*.
- It is, therefore, important that the identification of shigellae be made by a combination of antigenic and biochemical properties and not based on antigenicity alone.

Susceptibility

- Shigellae are killed at 56°C in one hour and by 1% phenol in 30 minutes.
- In ice, they last for 1–6 months.
- *S. sonnei* is, in general, more resistant than the other three species namely *S. dysenteriae*, *S. flexneri,* and *S. boydii*.

Classification

Shigellae are classified into four species or subgroups based on a combination of biochemical and serological characteristics. Serotypes are distinguished within the species. The four groups are given below, and the details are mentioned in Chapter 4.

- *S. dysenteriae* (subgroup A)
- *S. flexneri* (subgroup B)
- *S. boydii* (subgroup C)
- *S. sonnei* (subgroup D)

S. dysenteriae type 1 causes fulminant bacillary dysentery and hemolytic uremic syndrome (HUS).

Pathogenicity

Shigella is the most potent pathogen among the enteric bacteria.

- The infective dose is 10–100 bacilli.
- These organisms are capable of initiating the disease at such a low dose because they survive gastric acidity better than other pathogenic enteric bacteria.
- The pathogenic mechanisms of shigellae resemble those of enteroinvasive *E. coli* (Fig 25.9).

Invasion of cells by *Shigella*

Shigella causes cell injury and resulting complications by the direct invasion of colonic mucosa and production of enterotoxins.

- Invasiveness is attributed to the outer membrane protein, which is responsible for cell penetration. These proteins are called virulence marker antigens (VMA).
- The invasive property can be demonstrated by its ability to penetrate cultured HeLa or HEp-2 cells.
- The detection of VMA by ELISA serves as a virulence test for shigellae, similar to enteroinvasive *E. coli* toxins produced by *Shigella*.
- The endotoxin is responsible for diarrhea and ulcers in the intestine.
- The exotoxin is heat-labile protein that acts as an enterotoxin and neurotoxin.
  1. **Neurotoxicity:** The primary site of the action of the toxin appears to be not the nervous tissue but the blood vessels, mainly those of the central nervous system; neurological effects are secondary.
  2. **Enterotoxicity:** Two new *Shigella* enterotoxins have been identified, designated **ShET-1 and -2**, the former confined to *S. flexneri* 2a and the latter more widespread. *S. dysenteriae* type 1 forms an exotoxin, ShET-2, which has the ability to induce inflammation of the epithelial cells by decreasing the secretion of interleukin-8.
  3. **Cytotoxicity**: *Shigella* produces cytopathic changes in Vero cell lines. Shiga toxin appears to be the same as verotoxin (or the Shiga-like toxin of *E. coli*—VTEC). Two distinct verotoxins VT-1 and VT-2 are known.

Laboratory diagnosis

- **Specimen:** Stool or rectal swabs are collected.
- **Microscopy**: Plenty of pus cells with RBCs are seen along with a predominance of non-motile bacteria in wet mounts.
- **Enrichment media**: Selenite F broth or *Salmonella–Shigella* (SS) broth can be used.
- **Cultural characteristics:**
  - Shigellae are aerobes and facultative anaerobes.
  - Temperature range is 10–40°C (optimum of 37°C).
  - Optimum pH is 7.4.
  - MacConkey's agar: Pale, non-lactose fermenting colonies are seen.
  - Deoxycholate citrate agar (DCA) and xylose lysine deoxycholate (XLD): Red (alkaline) colonies (Fig. 25.11) develop.
- **Biochemical reactions:**
  - Colonies are catalase-positive and oxidase-negative and non-lactose fermenting.

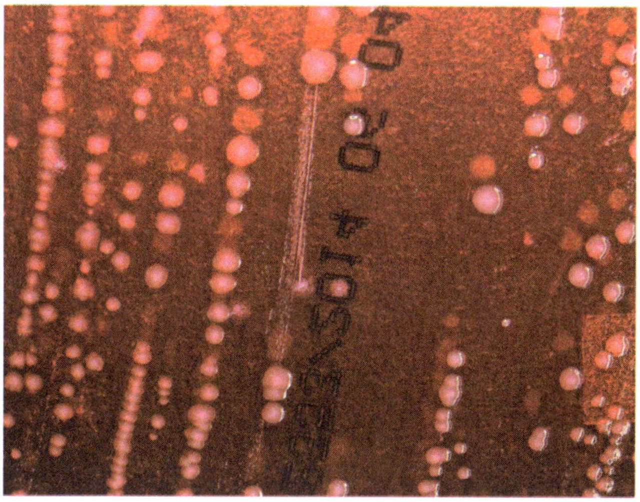

Fig. 25.11 Red colonies of *Shigella sonnei* on XLD agar (*Source*: CDC, PHIL, Image ID 17191)

  - Fermentation of mannitol—*S. dysenteriae* are non-fermenting while all the rest ferment mannitol.
- **Identification by specific antisera:** Identification is confirmed by slide agglutination with polyvalent and monovalent antisera.

Laboratory diagnosis of shigellosis

- **Specimen:** Stool or rectal swab
- **Microscopy**: Plenty of pus cells with RBCs are seen along with a predominance of non-motile bacteria in wet mount
- **Enrichment media**: Selenite F broth or *Salmonella–Shigella* (SS) broth
- **Culture:**
  - Aerobes and facultative anaerobes, optimum temperature 37°C (10–40°C), optimum pH 7.4
  - MacConkey's agar: Pale, non-lactose fermenting colonies
  - Deoxycholate citrate agar (DCA) and xylose lysine deoxycholate (XLD): Red (alkaline) colonies
- **Identification:**
  - Catalase-positive and oxidase-negative
  - Fermentation of mannitol (only *S. dysenteriae* are non-fermenting)
  - Agglutination by with polyvalent and monovalent antisera.

Treatment

- Uncomplicated shigellosis is a self-limiting condition and hence treatment with antibiotics is not recommended for adults.
- However, in acute cases, particularly in infants and young children, **dehydration should be corrected** promptly. Oral rehydration is adequate in most cases.

- Fluoroquinolones (ciprofloxacin) or trimethoprim-sulfamethoxazole may be given in severe or complicated cases.

Control

- Control consists essentially of improving personal and environmental sanitation.
- Antibiotics have no place in prophylaxis. No effective vaccine is available.

Campylobacteriosis

Campylobacteriosis can present as dysentery with abdominal cramps, fever, extraintestinal infection, or abscess. It is caused by *Campylobacter* species. *C. jejuni* is the most common species responsible for illness in humans.

Epidemiology

- The infection is zoonotic, the source being food of animal origin, especially raw milk, as the causative organisms are a part of the normal intestinal flora of domestic animals and birds and are shed in their feces.
- *Campylobacter* are important veterinary pathogens.
- *C. jejuni* is endemic in developing countries; asymptomatic infections are prevalent in humans as well as domestic animals and birds.
- The disease is generally confined to children, while adults are immune.

Agent—Campylobacter

- These are slender, spirally curved, typically comma-shaped, or spiral gram-negative rods, 0.2–0.5 μm thick and 0.5–5 μm long.
- These are motile with a single, unsheathed polar flagellum at one or both poles.
- Growth occurs under microaerophilic conditions, 5% oxygen concentration being optimal.
- Many pathogenic species are thermophilic, growing well at 42°C.
- Campylobacters do not ferment carbohydrates but are strongly oxidase-positive.
- Important species—*C. jejuni*, *C. coli*, and *C. lari*.

Medically, *Campylobacter jejuni* is the most important *Campylobacter* species as it causes diarrhea and food poisoning worldwide.

- *C. jejuni*, *C. coli*, and *C. lari* are thermophilic and do not grow at 25°C.
- Plates are incubated at 42°C in an atmosphere of 5% oxygen, 10% carbon dioxide, and 85% nitrogen.

Pathogenicity

- *C. jejuni* is the most common bacterial cause of diarrheal disease in many developed countries.
- The mode of infection is **ingestion.**
- The jejunum and ileum are the primary sites of colonisation, but the infection may spread to the colon and rectum.
- *C. jejuni* invades and damages the cells of the small intestine and consequently disrupts fluid absorption with ulceration of the mucosa.
- It is an invasive pathogen that causes the enlargement of the mesenteric lymph nodes.
- It also produces a heat-labile enterotoxin that causes a increase in intracellular levels of cAMP, resulting in watery diarrhea.

Clinical features (Table 25.7)

- The incubation period is 1–7 days.
- The illness starts with fever, abdominal pain, and watery diarrhea, often with blood.
- Stool contains leukocytes and blood.
- The disease is usually self-limiting, though *Campylobacter* shedding may continue for weeks after recovery.
- Sometimes, *Campylobacter* infections can cause complications such as irritable bowel syndrome, temporary paralysis, and arthritis. In immunosuppressed people, it can occasionally spread to the bloodstream causing a life-threatening infection.

Laboratory diagnosis

- **Microscopy:** Direct microscopic examination by **phase-contrast or dark-field microscopy** shows darting or tumbling motility of the spiral rods or small, curved rods in stained smears.
- **Culture:**
  - **Feces or rectal swabs** are plated on selective media.
  - **Transport media:** Campylobacters survive for 1–2 weeks at 4°C in **Cary–Blair transport medium.**
  - **Selective media**: Skirrow's, Butzler's, and Campy BAP are the most suitable selective media.
  - Colonies usually appear in 48 hours. They are non-hemolytic, grey or colourless, moist, and flat or convex.
- **Confirmation:** The organisms are identified on the basis of Gram staining, the appearance of darting motility on wet mount, and biochemical tests, the most common of which is **positive hippurate hydrolysis.**

Table 25.7 Clinical presentations of various *Campylobacter* infections

Clinical presentation	Species
Dysentery	*C. jejuni, C. coli, C. lari*
Extra-intestinal infection	*C. fetus*
Abscess	*C. sputorum, C. conciscus*

Treatment

- Fluid and electrolyte replacement is the mainstay of treatment.
- The drug of choice is erythromycin.

Other pathogenic species

Campylobacter fetus may infect humans, leading to bacteremia, sepsis, and meningitis in immunocompromised hosts.

Peptic Ulcer Disease

> **Case 3.** *A 35-year-old man presented to the hospital complaining of pain in the epigastrium along with nausea, flatulence, and bad breath for the previous four weeks. The pain typically increased in the middle of the night and decreased slightly after eating a meal. Antacids did not relieve the symptoms. On examination, there was epigastric tenderness. An endoscopic examination followed by gastric biopsy was done. The tissue bit was positive for rapid urease test. Histopathology of the tissue showed the presence of gram-negative, spiral bacilli in the submucosa. A diagnosis of* ***gastric ulcer with pylori gastritis*** *was made. The patient was relieved of his symptoms after treatment with a combination of amoxicillin, clarithromycin, and omeprazole along with other supportive care.*
>
> *i. What is the causative organism in this condition?*
>
> ii. Describe the pathogenesis of this disease.

- Peptic ulcer disease (PUD) is characterised by the **discontinuation in the inner lining of the gastrointestinal tract** because of gastric acid secretion or pepsin.
- It usually occurs in the stomach and proximal duodenum and may also involve the lower esophagus, distal duodenum, or jejunum.
- The only infectious cause of this disease is ***Helicobacter pylori.***
- *H. pylori* is a spiral, *Campylobacter*-like bacteria that was observed in close apposition to the gastric mucosa in several cases of gastritis and peptic ulcer by Robin Warren and Barry Marshall in Australia in 1983.

Epidemiology

- *H. pylori* is global, with a prevalence of 30–60 per cent, more so in developing countries than in the developed ones.
- The sole source of *H. pylori* is the human gastric mucus.
- The exact mechanism of transmission is not clear, but it is likely to be oral–oral or fecal–oral.
- Poverty, overcrowding, and poor hygiene favour transmission.
- With improvements in lifestyle, the prevalence of childhood infections has declined in developed countries.

Agent—*Helicobacter*

- *H. pylori* is a **gram-negative spiral rod,** motile by a **unipolar tuft of lophotrichous flagella**.
- It grows on **chocolate agar** or ***Campylobacter* media** under microaerophilic conditions.
- It produces abundant urease, which allows the organism to survive in gastric acidity; this property is used as a rapid test on gastric biopsy samples.
- It does not metabolise carbohydrates or reduce nitrate.
- *H. pylori* has adapted to the human gastric mucosa. The only other animals it infects are monkeys.
- *H. pylori* shows considerable genetic diversity, as is evident in molecular typing.

Pathogenicity (*Fig. 25.12*)

Pathogenicity is attributed to:

- **Urease production** allows these bacteria to tolerate the extremely low pH of 2.6 in the gastric mucosa.
- **Flagella** provide motility and allow movement toward the gastric epithelium.
- Interaction only with the overlying gastric mucosa (Fig. 25.13) and no invasion is seen; this protects the bacteria from the acidic pH of the stomach.

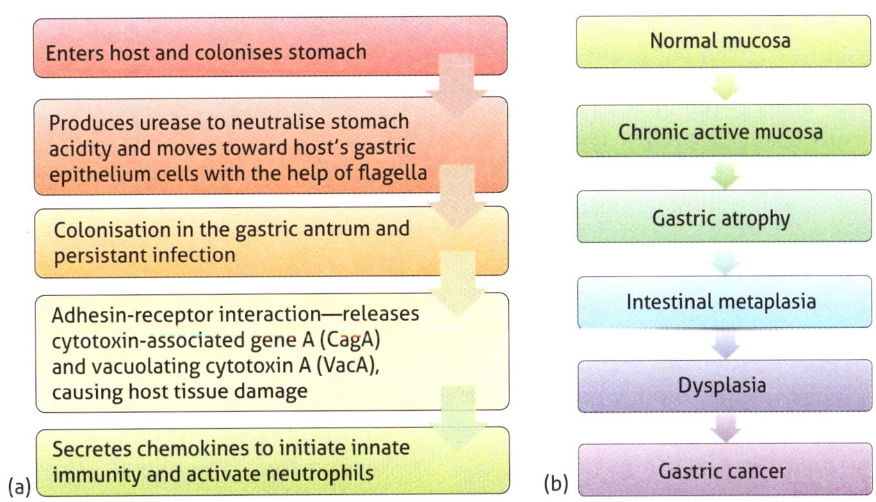

Fig. 25.12 (a) Pathogenesis of *H. pylori* infection and (b) sequence of events in the development of gastric ulcers

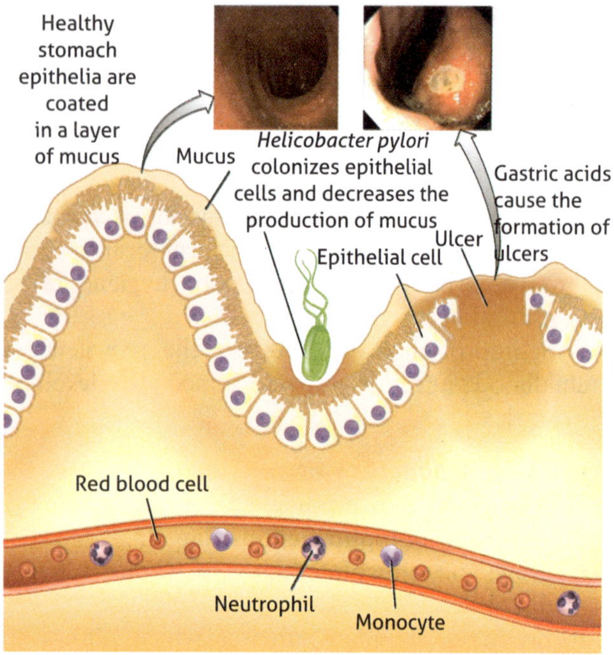

Fig. 25.13 *Helicobacter pylori* interact with gastric mucosa and colonise it (*Source*: https://openstax.org/books/microbiology)

- Toxins such as:
  - Vacuolating cytotoxin A
  - Cytotoxin-associated gene A (*cag*)

Clinical features

- The **incubation period** is 3–4 days.
- It begins with **mild acute gastritis** that may last for about two weeks.
- Once the symptoms subside, the organism persists as an asymptomatic coloniser, though chronic superficial gastritis may be demonstrable histologically.
- Peptic ulcer disease occurs in a proportion of those infected (Case 3).
- **Chronic atrophic gastritis** may be seen in the later stages.
- The infection is recognised as a risk factor for gastric malignancies such as adenocarcinoma and **'mucosa-associated lymphoid tissue' (MALT) lymphomas**. Such MALTomas appear to be antigen-driven and are found to regress after the elimination of *H. pylori* by treatment.
- Infection induces IgM, IgG, IgA, and cellular immune response, but these do not seem to be protective.

Laboratory diagnosis

1. **Invasive tests (endoscopic biopsy of gastric mucosa)**

- **Specimen:** Biopsy of gastric mucosa
- **Microscopy:** Of biopsy sections by silver staining or Gram staining shows spiral bacilli (Fig. 25.14).

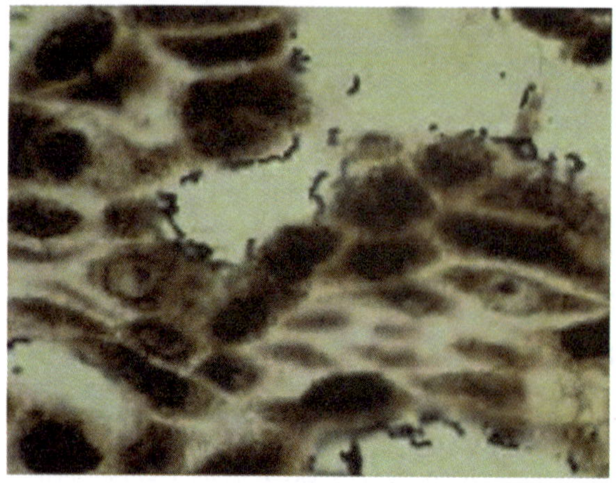

Fig. 25.14 *Helicobacter pylori* gastritis—silver stain

- **Culture:** This is more sensitive. It requires an enriched medium with horse blood (Skirrow *Campylobacter* medium or chocolate agar) in which the bacteria grow at 37°C in microaerophilic conditions within 3–7 days.
- **Identification:** Campylobacters are catalase- and oxidase-positive and produce abundant urease.
  - **Direct urease test:** The biopsy tissue in crushed in 0.5 mL urease solution with an indicator and incubated at 37°C. In a positive sample, the colour changes within a few minutes to 2 hours due to the production of ammonia.

2. **Non-invasive tests:**

- **Urease breath test**: The subject drinks a **carbon isotope-labelled urea solution**. This isotope can be detected in the breath. This method is sensitive and reliable but needs an isotope assay.
- **Serology:** IgM and IgG detection by ELISA is not considered useful for diagnosis. However, these methods may be used for seroepidemiological studies in the community.
- **Rapid tests:** Several rapid immunochromatographic cassette and dipstick methods have been developed for bedside testing for antigens or antibodies in stool, gastric aspirate, or blood.

Treatment

- Indications are an active peptic ulcer disease or past history of peptic ulcer without documented cure. Treatment is indicated only for *H. pylori*-related gastric or duodenal ulceration and not for asymptomatic colonisation. Recurrence is common after treatment.
- **First-line** treatment: Proton pump inhibitor with **clarithromycin (500mg), amoxicillin (1g), and metronidazole (500mg)** for 14 days,
- **Second-line** treatment: Proton pump inhibitor with **Bismuth salicylate, tetracycline, and metronidazole/nitroimidazole** for 14 days.

Pseudomembranous Colitis (PMC)

- Pseudomembranous colitis is severe **inflammation of the inner lining of the large intestine** characterised by elevated yellow-white plaques that coalesce to form pseudomembranes on the mucosa, which manifests clinically as an antibiotic-associated diarrhea.
- It most commonly results from a serious *Clostridium difficile* infection, usually in nosocomial settings.
- Infections usually **follow the use of broad-spectrum antibiotics** like clindamycin, ampicillin, or fluoroquinolones, to which the organism is resistant.

Agent—*Clostridioides difficile*

- *C. difficile* is a gram positive motile bacilli with oval, subterminal or terminal, non-bulging spores. It is an obligate anaerobe that grows easily on blood agar.
- It is the most common cause of healthcare-associated diarrhea in many developed countries.
- Antibiotic-associated diarrhea is seen in patients on prolonged antibiotic use and is mainly due to the alteration of the normal flora of the gut. Though 20–25% of cases are due to *C. difficile*, other anaerobic infections are also implicated in the causation.

Pathogenesis (Fig. 25.15 and 25.16)

- Two high-molecular-weight **exotoxins, A and B,** are involved in the pathogenesis of the condition.
- Toxin A is a potent **enterotoxin** that attaches to gut receptors; it may also be cytotoxic.

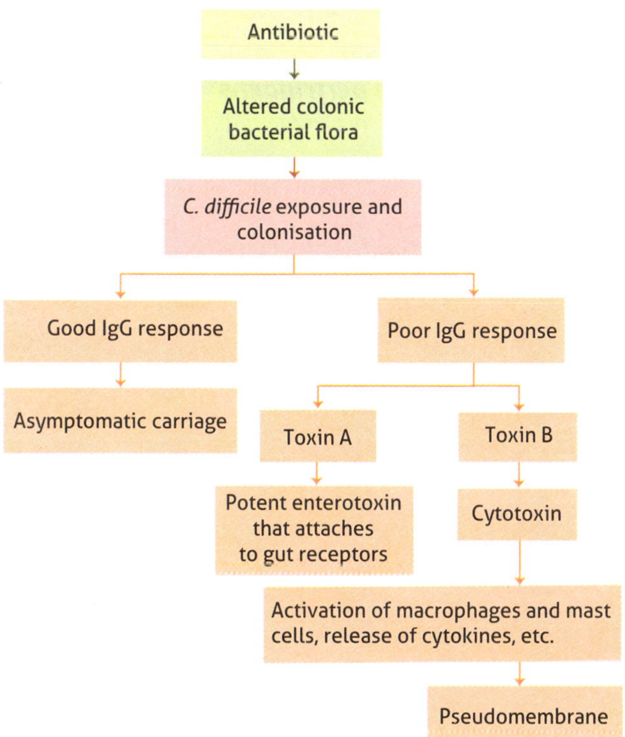

Fig. 25.15 Pathogenesis of PMC

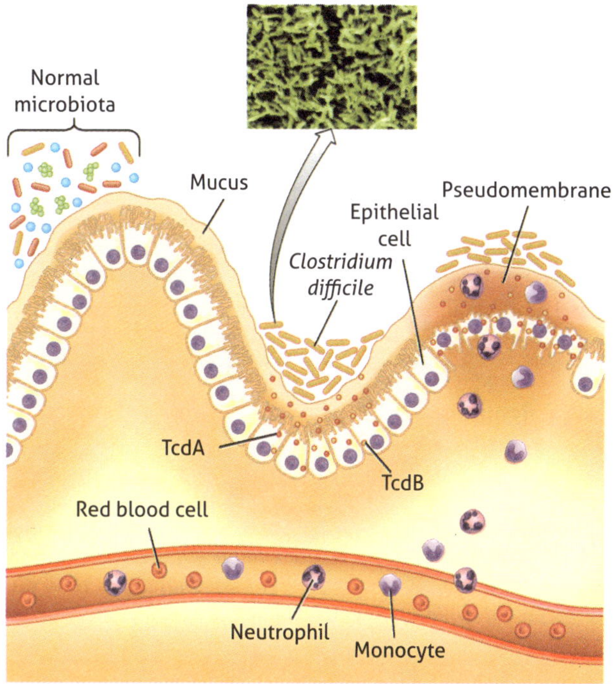

Fig. 25.16 Overgrowth of *C. difficile* and pseudomembrane formation (*Source*: https://openstax.org/books/microbiology); *TcdA*: Toxin A, *TcdB*: Toxin B

- Toxin B is a cytotoxin.
- *C. difficile* also bears the toxin genes on a chromosomal pathogenicity island.

Laboratory diagnosis

- Direct examination by **colonoscopy shows microabscesses.**
- **Specimen:** Stool samples are collected.
- **Culture:** Cycloserine, **cefoxitin fructose agar (CCFA)**, or cysteine cefoxitin yeast extract agar (CCYA) are used as selective media under anaerobic conditions.
- **Detection of toxins A and/or B in stool by:**
  - **ELISA** (most common method)
  - Cytopathic effect on **Hep-2 or/and human diploid cell cultures**7; also specifically **neutralised by the *C. sordellii* antitoxin**
  - **Rapid lateral flow assay to detect** *C. difficile* toxins and glutamate dehydrogenase in stool samples
- **Molecular methods: PCR** to detect genes for toxin A or B.

Prevention and treatment

- The disease is prevented by restricting the use of antibiotics associated with *C. difficile* outbreaks.
- The condition is treated by discontinuing the antibiotic causing the disease and instituting **vancomycin or metronidazole.**

Fecal transplant

Recently, interest has been generated in the treatment of ulcerative colitis with fecal transplants. Fecal transplantation (or bacteriotherapy) is the **transfer of stool from a healthy donor into the gastrointestinal tract of another for restoring their normal flora**. This is done to treat recurrent *C. difficile* colitis, which is associated with prolonged antibiotic therapy. Some success has been noted with this modality. Further studies are required to better understand the role of fecal transplants in restoring normal flora.

Food Poisoning

Case 4. *A 15-year-old boy began vomiting 6–7 hours after consuming a pineapple pastry at a birthday party. He passed loose stools but had no fever. Six other children who had attended the party reported a similar illness. On further questioning, it was found that the common dish that all the affected children had consumed was the pastry. The cream from the pastry tested positive for S. aureus. Further investigations found that a specimen obtained from the nose of the cook was positive for* ***S. aureus****. Pulse field gel electrophoresis ascertained both to be of the same type, which indicated a common source. The cook was treated for the carrier state with local antibiotic application. The children recovered within a day or two and did not require any antibiotics.*

i. What are the causative agents of food poisoning?
ii. What is the pathogenesis in the above-mentioned condition?

Food poisoning is an acute infection of the intestine that can be caused by the ingestion of preformed toxin or bacteria in food or water. The causative agents of food poisoning are mentioned of Table 25.2.

Agent—*Staphylococcus aureus*

- *S. aureus* produces a potent exotoxin (**enterotoxin**), which causes food poisoning resulting in nausea, vomiting, and diarrhea 2–6 hours after consuming food contaminated by the preformed toxin.
- **Meat, fish, and milk and milk products cooked and left at room temperature** after contamination with staphylococci are commonly responsible for the infection.
- The organisms in food form the toxin, which produces **inflammatory changes** throughout the gastrointestinal tract with severe lesions in the jejunum and ileum.
- The source of infection is usually a **food handler** who is a carrier.
- The illness is often self-limiting, with recovery in a day or two (Case 4).

Staphylococcal enterotoxin

- The enterotoxin is relatively **heat-stable**, resisting a temperature as high as 100°C for 10–40 minutes.
- Eight antigenic types of the staphylococcal enterotoxin are currently known—enterotoxins A, B, C1–3, D, E, and H.
- The toxin is believed to act **directly on the autonomic nervous system** to cause the illness, rather than on the gastrointestinal mucosa.
- The toxin is antigenic and is neutralised by the specific antitoxin.
- **Type A toxin** is responsible for most cases.
- Sensitive serological tests such as **latex agglutination and ELISA** are available for the detection of the toxin.

Botulism

Clostridium botulinum causes botulism, a paralytic disease that usually presents in the form of food poisoning.

Agent—C. botulinum

- The name botulism is derived from the Latin word *botulus*, meaning 'sausage'.
- *C. botulinum* is **motile** due to the presence of peritrichate flagella and produces **subterminal, oval, and bulging spores.**
- **Spores are heat- and radiation-resistant,** surviving several hours at 100°C and for up to 10 minutes at 120°C.
- Spores of the non-proteolytic types B, E, and F are much less resistant to heat.
- Although the infection occurs via the gastrointestinal tract or wounds (wound botulism), the disease manifests as a neurological condition (described in Chapter 24).

Agent—*Clostridium perfringens*

- Some strains of *C. perfringens* type A can produce food poisoning.
- They produce a **heat-labile enterotoxin** which, like the enterotoxins of *V. cholerae* and enterotoxigenic *E. coli*, leads to fluid accumulation in the rabbit ileal loop.

Pathogenesis

- Food poisoning is usually caused by a **cold or warmed-up meat dish.**
- When contaminated meat is cooked, the spores in the food may survive. During subsequent storage or warming, they germinate and multiply in the anaerobic environment of the cooked meat.
- Large numbers of clostridia may thus be consumed, which may pass into the intestines, unharmed by gastric acid due to the high protein content of the meat. In the intestines, their spores produce the enterotoxin.
- After an **incubation period of 8–24 hour**s, abdominal pain, diarrhea, and vomiting set in.

Laboratory diagnosis

- Isolating heat-resistant *C. perfringens* type A from feces is not of much value on account of its presence in the normal intestine.
- Isolation from food is attempted by direct plating on selective media, as the bacillus is generally present in food as vegetative cells.

Treatment

The illness is self-limiting, and recovery occurs in 24–48 hours.

Other infections caused by *C. perfringens*

- **Gangrenous appendicitis:** *C. perfringens* type A (and occasionally, type D) strains have been isolated from gangrenous appendicitis. It has been proposed that the toxemia and shock in some cases of intestinal obstruction and peritonitis may be due to the toxins of *C. perfringens.*
- **Necrotising enteritis:** This is a severe and often fatal enteritis. It is caused by *C. perfringens* type C strains with heat-resistant spores which germinate in the intestine, producing beta toxin and causing mucosal necrosis. Immunisation with the type C toxoid has been shown to protect against this condition.
- **Biliary tract infection:** This is characterised by acute emphysematous cholecystitis and post-cholecystectomy septicemia.
- **Endogenous gas gangrene of intra-abdominal origin:** This results when an endogenous organism derived from the gut contaminates the abdominal wall during surgery.
- **Brain abscess and meningitis:** These are also known to be caused by *C. perfringens*.
- **Panophthalmitis:** This may occur following penetrating eye injuries.
- **Thoracic infections:** They may develop following penetrating wounds of the thorax are often seen in battle casualties.
- **Urogenital infections:** These may develop following surgical procedures such as nephrectomy.
- **Septic abortion:** An infection of the uterus is a serious and not infrequent condition, commonly associated with septic abortion; septicemia is common in this condition.

Bacillus cereus food poisoning

- *B. cereus* is widely distributed in nature and may be readily isolated from soil, vegetables, and a wide variety of foods including **milk, cereals, spices, meat, and poultry.**
- *B. cereus* is generally motile, but non-motile strains may occur.
- It is not capsulated and not susceptible to gamma phages.
- Two types of food poisoning have been associated with *B. cereus*—the diarrheic and emetic types.

Diarrheal type

- It is associated with a wide range of foods including **cooked meat and vegetables**.
- It is characterised by diarrhea and abdominal pain, 8–16 hours after the ingestion of contaminated foods; vomiting is rare.
- *B. cereus* is **not found in large numbers in the fecal specimens** from these patients.
- The diarrheal disease is generally caused by serotypes 2, 6, 8, 9, 10, or 12.

Emetic type

- It is associated almost exclusively with the consumption of **cooked rice**, usually fried rice from Chinese restaurants.
- The illness is characterised by acute nausea and vomiting, 1–5 hours after the meal.
- Diarrhea is not common.
- *B. cereus* is present in large numbers in the cooked rice and fecal samples from these patients.
- This type is caused by serotypes 1, 3, or 5.
- The emetic toxin is produced only when *B. cereus* is grown in rice but not in other media.
- Two mechanisms of action have been described for the action of the enterotoxin of *B. cereus*, one involving the stimulation of the cyclic adenosine monophosphate (cAMP) system and the other independent of it.

Laboratory diagnosis

- Special selective media—**mannitol egg yolk phenol red polymyxin agar (MYPA)** and **polymyxin-pyruvate mannitol bromothymol blue agar (PEMBA)** are useful in isolating *B. cereus* from feces and food sources.
- The presence of 10^5 or more bacteria per gram of specimen is diagnostic.

Treatment

Both types of the illness are mild and self-limiting, requiring no specific treatment.

Table 25.8 summarises the bacterial infections of the GI tract.

FUNGAL AGENTS CAUSING FOOD POISONING

Many fungi form poisonous substances. Mycotic poisoning is of two types:

1. **Mycotoxicosis**, in which fungal toxins contaminate food, which on ingestion, cause infection
2. **Mycetism**, in which the ingestion of the fungus causes toxic effects

Table 25.8 Summary of bacterial infections of the GI tract

Disease	Pathogen	Signs and symptoms	Transmission	Diagnostic tests	Antimicrobial drugs
Bacillus cereus infection FP	*Bacillus cereus*	Nausea, pain, abdominal cramps, diarrhea, or vomiting	Ingestion of contaminated rice or meat, even after cooking	Testing stool sample, vomitus, or uneaten food for the presence of bacteria	None
Campylobacter jejuni gastroenteritis D/FP	*Campylobacter jejuni*	Fever, diarrhea, cramps, vomiting, and sometimes, dysentery; sometimes, more severe organ or autoimmune effects	Ingestion of unpasteurised milk, under-cooked chicken or contaminated water	Culture on selective medium with elevated temperature and low oxygen concentration	Generally, none; erythromycin or ciprofloxacin, if necessary
Cholera D	*Vibrio cholerae*	Severe diarrhea and fluid loss, potentially leading to shock, renal failure, and death	Ingestion of contaminated water or food	Culture on selective medium (TCBS agar); distinguished as oxidase-positive with fermentative metabolisms	Generally none; tetracyclines, azithromycin, or others, if necessary
Clostridium difficile infection D	*Clostridium difficile*	Pseudomembranous colitis, watery diarrhea, fever, abdominal pain, loss of appetite, dehydration; in severe cases, perforation of the colon, septicemia, shock, and death	Overgrowth of *C. difficile* in the normal microbiota due to antibiotic use; hospital-acquired infections in immunocompro-mised patients	Detection of toxin in stool, nucleic acid amplification tests (e.g., by PCR)	Discontinuation of previous antibiotic treatment; metronidazole or vancomycin
Clostridium perfringens gastroenteritis D	*Clostridium perfringens* (especially type A)	Mild cramps and diarrhea in most cases; in rare cases, hemorrhaging, vomiting, intestinal gangrene, and death	Ingestion of undercooked meats containing *C. perfringens* endospores	Detection of toxin or bacteria in stool or uneaten food	None
E. coli infection D	ETEC, EPEC, EIEC, EHEC	Watery diarrhea, dysentery, cramps, malaise, fever, chills, dehydration; in EHEC, possible severe complications such as hematolytic uremic syndrome	Ingestion of contaminated food or water	Tissue culture, immunochemical assays, PCR, gene probes	Not recommended for EIEC and EHEC; fluoroquinolones, doxycycline, rifaximin, and TMP/SMZ possible for ETEC and EPEC
Peptic ulcers	*Helicobacter pylori*	Nausea, bloating, burping, lack of appetite, weight loss, perforation of stomach, and blood in stools	Normal flora; can also be acquired via saliva; fecal–oral route via contaminated food and water	Breath test, detection of antibodies in blood, detection of bacteria in stool sample or stomach biopsy	Amoxicillin, clarithromycin metronidazole, tetracycline, lansoprazole; antacids may also be given in combination with antibiotics
Salmonellosis D	*Salmonella enterica* serotype Enteritides	Fever, nausea, vomiting, abdominal cramps, headache, and diarrhea; can be fatal in infants	Ingestion of contaminated food, handling of eggshells or contaminated animals	Culturing, serotyping, and DNA fingerprinting	Not generally recommended; fluoroquinolones, ampicillin, others for immunocompromised patients

(Table 25.8 Cont. . .)

Table 25.8 Contd...

Disease	Pathogen	Signs and symptoms	Transmission	Diagnostic tests	Antimicrobial drugs
Shigella dysentery DYS	*Shigella dysenteriae, S. flexneri, S. boydii,* and *S. sonnei*	Abdominal cramps, fever, diarrhea and dysentery; possible complications—reactive arthritis and hemolytic uremic syndrome	Fecal–oral route via contaminated food and water	Testing of stool samples for the presence of blood and leukocytes; culturing, PCR, immunoassay for *S. dysenteriae*	Ciprofloxacin, azithromycin
Staphylococcal food poisoning FP	*Staphylococcus aureus*	Rapid-onset nausea, diarrhea, vomiting lasting 24–48 hours; possible dehydration, and change in blood pressure and heart rate	Ingestion of raw or undercooked meat or dairy products contaminated with staphylococcal enterotoxins	ELISA to detect enterotoxins in uneaten food, stool, or vomitus	None
Typhoid fever D	*S. enterica,* subtypes Typhi or Paratyphi	Aches, headaches, nausea, lethargy, diarrhea or constipation; possible rash; lethal perforation of the intestine can occur	Fecal–oral route; may be spread by asymptomatic carriers	Culture of blood, stool, or bone marrow; serologic tests; PCR tests when available	Fluoroquinolones, ceftriaxone, azithromycin; preventive vaccine available
Yersinia infection D	*Yersinia enterocolitica, Y. pseudo-tuberculosis*	Generally mild diarrhea and abdominal cramps; in some cases, bacteremia can occur, leading to severe complications	Fecal–oral route, typically via contaminated food or water	Testing stool samples, tissues and body fluids	Generally, none; fluoroquinolones, aminoglycosides, others for systemic infections

(*FP*—food poisoning; *D*—diarrhea; *DYS*—dysentery)

Mycotoxicoses

- These are diseases caused by the toxic metabolic products released by fungi and result from the ingestion of food contaminated with mycotoxins (fungal toxins).
- Examples:
  - **Aflatoxicosis**—which results from the consumption of grains containing aflatoxins secreted by *Aspergillus flavus*, which contaminates groundnuts, corn, and peas.
  - **Ergotism**—which results form the ingestion of rye contaminated with *Claviceps purpurea*.
- The fungus does not necessarily have to be present in the tissues to exert its pathogenic effect as long as its toxic metabolites are present; moreover, there is **no invasion of tissues** by the fungus.
- Mycotoxins are natural products produced by fungi and found in some articles of food (Table 25.9).

1. **Aflatoxin:**

- *Aspergillus flavus* secretes aflatoxin B1 while *Aspergillus parasiticus* secretes aflatoxins B1, B2, G1, and G2, among others.
- Aflatoxins are frequently present in mouldy foods, particularly in groundnuts, corn, and peas.

Table 25.9 Important mycotoxins and their sources

Mycotoxin	Fungus	Contaminated food items
Aflatoxin	*Aspergillus* spp.	Groundnuts, corn, and peas
Fumonisin	*Fusarium*	Maize
Ochratoxin	*Aspergillus* and *Penicillium*	Cereals and bread
Ergotoxin	*Claviceps purpurea*	Rye
Zearalenone	*Fusarium graminearum*	Grains
Trichothecenes	Various species of *Trichoderma, Trichothecium* and *Myrothecium*	Livestock feeds or foodstuff

- These toxins are highly toxic to animals and birds as well as human beings.
- Aflatoxins are known to cause hepatomas in ducklings and rats, and their possible carcinogenic effect in human beings has caused great concern.
- There have been several reports of aflatoxicosis from India, involving human beings and animals.

2. **Ergot alkaloids:**

- **Ergotoxicosis (ergotism)** is caused by the toxic alkaloids produced by the fungus *Claviceps purpurea*, which grows on the fruiting heads of rye.
- **Trichothecenes** are toxins produced by certain species of *Fusarium.*
- **Zearalenone** is a toxin produced by *Fusarium graminearum*; animals that consume grains contaminated with this toxin may develop symptoms and signs mimicking an estrogenic disorder.

Mycetism

- Mycetism has been known to exist since ancient times, several varieties of poisonous mushrooms having been identified as inedible.
- Mycetism may cause gastrointestinal disease, dermatitis, or death.
- The **hallucinogenic agents** (d-lysergic acid, psilocybin) produced by the *Psilocybe* species and other fungi have attracted much attention in recent years.
- Some examples of fungi that produce this entity are as follows:
  - *Claviceps* species—ergot poisoning
  - *Coprine* species—coprine poisoning
  - *Inocybe* species—muscarine poisoning

VIRAL GASTROENTERITIS

Rotaviral Diarrhea

Epidemiology

- Rotavirus diarrhea is usually seen in children below the age of five years but is most frequent between 6 and 24 months of age.
- Rotaviruses are the commonest cause of diarrhea in **infants and children** the world over and account for about half the cases of children hospitalised for diarrhea.
- Infection is not infrequent in neonates, but they seldom develop diarrhea because of maternal passive immunity.
- By the age of five years, most children have had clinical or subclinical infection; hence, rotavirus diarrhea is very uncommon in older children and adults.
- Infections occur throughout the year but are predominant in the winter months when the virus may be detected in most patients. It sometimes produces **large epidemics of diarrhea in winter**.

Agent—Rotavirus

- Rotaviruses belong to the family Reoviridae.
- The family contains three genera:
  – Reovirus – Orbivirus – Rotavirus
- While the first two are not associated with human infections, rotavirus is a major viral agent responsible for diarrhea in children and adults. It accounts for several childhood deaths in developing countries.
- These are **double-shelled, non-enveloped, icosahedral viruses, 55–57 nm** in diameter.
- Their genome comprises a double-stranded RNA in 9–12 segments.
- The rotavirus was first identified by **Ruth Bishop** and colleagues (1973) in Melbourne.
- Outbreaks of rotavirus diarrhea in older children and adults have been reported from different parts of China. This rotavirus is called **adult diarrhea rotavirus (ADRV).**
- Rotaviruses resemble **cartwheels,** with short spokes radiating from a wide hub to a clearly defined outer rim. The name is derived from the Latin word *rota*, meaning 'wheel'.
- The incomplete or single-shelled virus is smaller, measuring about 60 nm, with a rough surface that has lost the outer shell (Fig. 25.17).
- The rotavirus genome comprises a double-stranded RNA in 11 segments.
- All rotaviruses share common antigens.
- Though the viruses are generally species-specific; interspecies infection can be induced experimentally.
- Human rotavirus infection has been transferred to piglets, calves, and monkeys.
- It is not known whether human infection can be caused by animal rotaviruses.

Classification

- The rotavirus is lassified into antigenic groups (A to G) based on the **antigenic epitopes of the internal structural protein (VP6).**

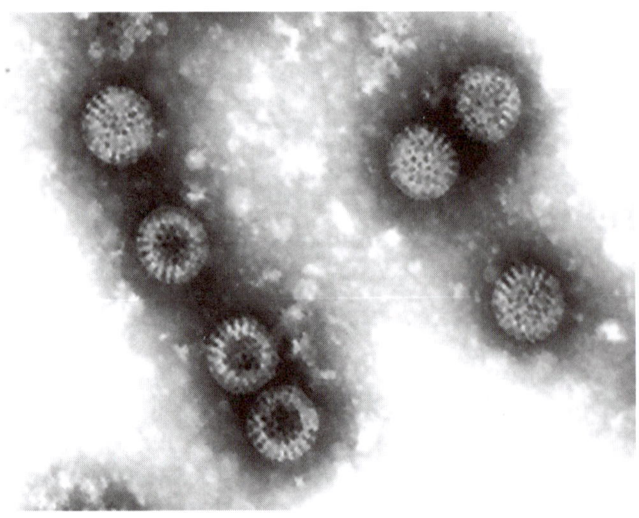

Fig. 25.17 Electron micrographic appearance of the rotavirus—double-shelled virus and single-shelled virus showing cartwheel appearance

- **Group A strains** cause the majority of human infections in children and have been classified into the following subgroups:
  - Subgroups I and II, which can be detected by ELISA and immune electron microscopy
  - Serotypes 1, 2, 3, etc., which can be detected by neutralisation tests

  Outer viral capsid proteins VP4 and VP7 carry epitopes important for neutralisation. The predominant antigen is the glycoprotein VP7.
- ADRV strains belong to **group B**.

Pathogenesis

- Infection is transmitted by the **fecal–oral route**.
- The virus infects the mature enterocytes in the upper and middle parts of the villi of the small intestine.
- Destruction of the villi leads to copious diarrhea.

Clinical features

- The **incubation period** is 2–3 days.
- **Vomiting and diarrhea** occur with little or no fever.
- Stools are usually greenish-yellow or pale, with no blood or mucus.
- The disease is self-limiting, and recovery occurs within 5–10 days.
- Mortality is low.

Laboratory diagnosis

- **Culture:** This is rarely used for diagnosis. Human rotavirus does not grow readily in cell cultures, but some strains have been adapted for serial growth in tissue cultures.
- **Serological techniques:** IgM and IgG antibodies can be demonstrated in the blood of infected children using ELISA.
- **Detection of virus:** The virus can be detected in stool filtrates by **enzyme immunoassays (EIA) and immune electron microscopy (IEM).**
- **Genotyping:** Polymerase chain reaction (PCR) is a sensitive detection method.

Treatment

Rehydration is all the treatment that is needed. There are no antiviral agents available currently.

Prophylaxis

Rotavirus vaccines are in use in many countries.

- **RotaTeq (RV5) vaccine** is commonly used. It is composed of five rotavirus strains, which are derived from a bovine rotavirus strain and contain a gene from rotaviruses of human origin. India has also recently introduced an indigenously developed rotavirus vaccine.
- **Rotarix (RV1):** Rotarix vaccine is an attenuated human rotavirus that contains representatives of the most frequently circulating rotaviruses.

Both vaccines are given orally.

Other Viral Causes of Diarrhea

Calicivirus

- Caliciviruses are a group of important viruses that cause gastroenteritis in humans.
- The most important virus in this group is the Norovirus whose prototype is the Norwalk virus.

Norwalk virus

- The Norwalk virus, which measures 27 nm, was shown to be responsible for an epidemic of gastroenteritis affecting school children and teachers in Norwalk, Ohio, in 1972.
- It has been included in the family **Caliciviridae, which consists of small, round RNA viruses, 22–30 nm in size,** many of which have been reported from diarrheal feces.
- These viruses bear **32 cup-shaped depressions** on their surface, based on which the family is named (derived from *calyx*, meaning cup).
- Serological surveys have shown that infection with the Norwalk virus is widespread in many countries.
- Epidemics of Norwalk virus diarrhea associated with the **consumption of raw oysters** have been reported from Australia and America.
- The virus can be demonstrated in feces by electron microscopy. Antibodies to the virus can be detected by immune electron microscopy and radioimmunoassay.

Adenovirus

- Several outbreaks of diarrhea in children have been associated with the presence of large numbers of adenoviruses in feces.
- These can only be grown with difficulty in tissue culture.
- They have been designated types 40 and 41.
- Adenovirus-associated diarrhea is seen more often in the summer months than in the winter.

Astrovirus

- These star-shaped, isometric particles measuring 28 nm have been associated with some epidemics of diarrhea in children.
- Similar viruses have also been identified in lamb and calf diarrhea.

Coronavirus

- These viruses are well-established causes of acute diarrhea in calves, piglets, and dogs and have been observed in human feces also.
- Though diarrhea is often observed in coronavirus respiratory infections, their relation to diarrhea is uncertain.

PARASITIC INFECTIONS OF THE GASTROINTESTINAL TRACT

- Parasitic infections of the intestine are responsible for significant morbidity and mortality in endemic countries.
- Several parasites can cause infections in humans. In immunosuppressed patients, these parasitic infections can be fatal.
- The majority of intestinal parasitic infections are asymptomatic.
- Clinically, these patients may present with diarrhea, dysentery, anemia, or malnutrition. Diarrhea may be persistent and may lead to malnutrition and death. In children, these infections may also cause failure to thrive. There may be complaints of eating mud or plaster (**pica**), poor appetite, or frank malnutrition.
- **Anemia** may be caused by heavy hookworm infections.
- Bleeding sometimes occurs from the esophageal varices.
- Intestinal helminths rarely cause death, but the burden of disease is related to the chronic and insidious effects on the health and nutritional status of the host.
- These infections impair the physical and mental growth of children.

Commensal Flagellates of the Intestine

- Most of these are found in the caecum as harmless commensals and are not known to cause any disease in humans.
- Examples include *Chilomastix mesnili*, *Retortamonas intestinalis*, and *Enteromonas hominis*.
- The cysts and trophozoites of these flagellates need to be differentiated from pathogenic causes of diarrhea.

Dientamoeba fragilis

- *D. fragilis* is a commensal that is known to cause intermittent **diarrhea, flatulence, and abdominal cramps** in some people, though the majority of cases are asymptomatic.
- It exists as **small (5–12 μm) trophozoites with 1–2 nuclei**; it **does not have a cystic stage**. The parasite's internal structure **resembles a flagellate, but it is motile with pseudopodia**.

Giardiasis

Case 5. *A 32-year-old male presented to the OPD with complaints of persistent diarrhea, abdominal cramps, bloating, and nausea with greasy and foul-smelling stools for the last two weeks. He complained of fatigue and weight loss. Physical examination revealed mild tenderness in the lower abdomen. His stool sample was collected and sent for microscopy and culture examination. The culture of the stool sample showed no growth after 24 hours of incubation. On microscopic examination of his stool, oval structures with four nuclei each and fibrils visible on both sides were seen.*

i. What is the provisional diagnosis?
ii. What are the probable etiological agents for the above conditions?
iii. Describe the life cycle of the parasite in this case.

Giardiasis is the most common parasitic disease in both developed as well as developing countries.

- The causative agent, *Giardia lamblia*, is a **flagellated intestinal protozoon** present worldwide.
- The disease is more commonly seen in children than in adults.
- It causes chronic diarrhea, often with malabsorption.
- In human hosts, *Giardia* resides in the duodenum and upper jejunum.

Agent—*Giardia duodenalis* (syn. *Giardia intestinalis* or *G. lamblia*)

Morphology

- **Cyst (Fig. 25.18):**
  - It is **oval** and measures 11–14 × 7–10 μm.
  - Immature cysts may have two nuclei, but a **fully mature cyst is quadrinucleate**—the four nuclei are seen at one end in pairs.
  - **Remnants of a flagella and a sucking disc** can also be seen inside.
- **Trophozoite** (Fig. 25.18):
  - It is a **pear-shaped, bilaterally symmetrical** structure that measures about 10 × 15 μm and exhibits a **'falling leaf' motility**.
  - It resembles a tennis racket and has a **sucking disc on its concave ventral surface**. This disc helps the parasite to attach to the intestinal mucosa.
  - It is highly motile with the help of **four pairs of flagella.**
  - The parasite has **four nuclei and two parabasal bodies** present on **axostyles.**

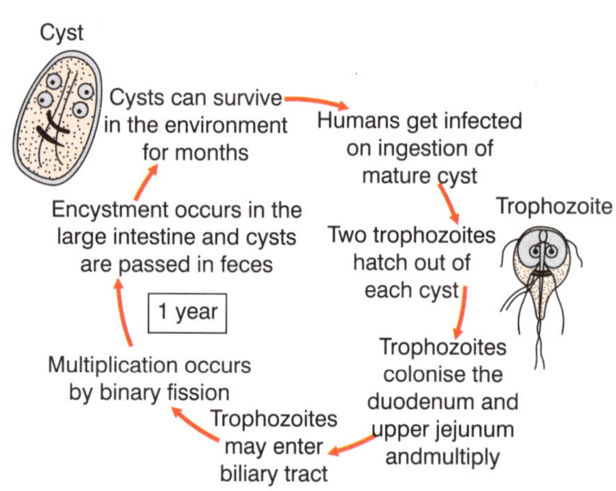

Fig. 25.18 Morphology and life cycle of *Giardia*

Life cycle (Fig. 25.18)

- The parasite lives its entire life in a **single host**.
- Humans acquire the infection by **ingesting mature cysts with food or water**.
- Excystation occurs in the duodenum, and two trophozoites are released from each cyst within 30 minutes of ingestion.
- The trophozoites multiply by binary fission and colonise the duodenum and upper parts of the jejunum.
- To avoid the acidity of the stomach, the parasite may localise in the biliary tract.
- The encystation occurs in transit in the colon. A large number of trophozoites may be excreted during the acute stage.
- Later, once stool loses moisture and becomes formed, cysts are excreted.

Clinical features

- The **incubation period** for giardiasis is 1–3 weeks.
- Infection is acquired by the **ingestion of water, food, or vegetables contaminated with the cysts.**
- The ingestion of as few as 10 cysts is sufficient to cause an infection in humans.
- Cysts are excreted by infected persons and are **resistant to chlorine disinfection**. They are infectious when excreted; hence, person-to-person transmission occurs in areas where fecal hygiene is poor.
- More than 50% of persons infected with *Giardia* are asymptomatic.

Acute presentation

- Clinically, the presentation of *Giardia* infection ranges from mild diarrhea to severe malabsorption syndrome.
- Patients develop diarrhea, abdominal pain, bloating, and vomiting.
- Loose motions are foul-smelling and contain mucus without blood.
- Most cases recover within 10–15 days; a few may develop a chronic condition.

Chronic presentation

- The presentation can be intermittent or recurring.
- Patients have foul-smelling diarrhea with weight loss due to malabsorption syndrome.
- Some patients present with extraintestinal manifestations such as urticaria, anterior uveitis, or arthritis.

Laboratory diagnosis

- **Specimens:**
  - Stool samples are usually collected for diagnosis.
  - **Entero string test** (Fig. 25.19)—secretions from the upper part of stomach are collected using an entero string test. A string with a weighted gelatin capsule is swallowed, and one end of the string is taped to the side of the patient's face. The capsule dissolves in the stomach, and the string, which is weighted at its distal end, passes into the duodenum by peristalsis. After about four hours, the string and any adsorbed gastrointestinal fluid is withdrawn through the mouth. This test is infrequently used now.
  - **Endoscopic biopsy** from the duodenum or upper jejunum may show trophozoites attached to microvilli.

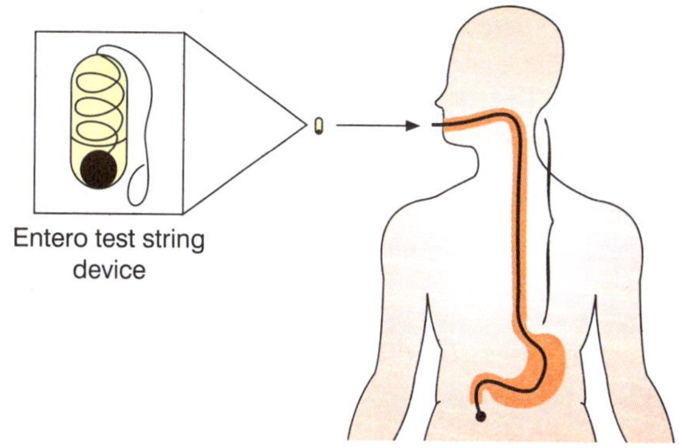

Fig. 25.19 Entero string test

- **Microscopy:**
  - **Stool microscopy by wet mount** shows the oval cysts of *Giardia* with four nuclei and fibrils. Trophozoites may be seen in acute infection.
  - The excretion of cysts is variable, and **repeated samples may have to be collected to make a diagnosis**.
  - Detection can be increased by stool concentration techniques.
  - Trophozoites are seen in fluid obtained by string test.
- **Antigen detection:** Giardial antigens can be detected in stool by ELISA or immunochromatographic test (ICT).
- **Serology:** Antibodies to *Giardia* can be detected by ELISA and immune fluorescence assay.
- **Molecular detection:** PCR can be used to diagnose giardiasis, but this method is expensive and hence rarely used.

Laboratory diagnosis of giardiasis

- **Specimens**
  - Stool samples
  - **Entero string test**
  - **Endoscopic biopsy** from the duodenum or upper jejunum
- **Microscopy:**
  - **Stool microscopy by wet mount** shows the oval cysts of *Giardia* and trophozoites in acute infection.
  - Trophozoites seen in fluid obtained by string test or biopsy.

- **Antigen detection**: In stool by ELISA or immunochromatographic test (ICT).
- **Serology**: ELISA and immune fluorescence assay to detect antibodies.
- **Molecular detection**: PCR is expensive and hence rarely used.

Treatment

- Metronidazole, tinidazole, or albendazole can be used.
- It is important to eradicate the carrier stage, particularly in food handlers.
- The cysts are resistant to chlorine and can be removed from drinking water by filtration.

Opportunistic GI Infections

- Parasites that otherwise cause self-limiting diseases in immunocompetent hosts can cause severe, life-threatening diseases in immunocompromised patients.
- Members of Sporozoa are known to cause intestinal, blood, and tissue infestations. The intestinal Sporozoa are known to cause chronic diarrhea in immunosuppressed conditions such as HIV infections.
- All members of the phylum exhibit a life cycle of **sexual (gametogony) and asexual (schizogony) reproduction**.
- Intestinal Sporozoa include the following organisms:
  - *Cryptosporidium*
  - *Cystoisospora* (previously known as *Isospora*)
  - *Sarcocystis*
  - *Cyclospora*

Cryptosporidiosis

Case 6. *A 31-year-old HIV-positive male presented with complaints of persistent diarrhea, abdominal cramps, and weight loss with severe dehydration and intermittent low-grade fevers for the past few weeks. His CD4 count was 150/μL. His stool sample was sent for microscopic examination and culture. Modified ZN stain of the sample revealed* ***multiple red spherules measuring approximately 4–6 μm.***

i. What is the provisional diagnosis?
ii. What are the probable etiological agents for the above condition?
iii. Describe its laboratory diagnosis.

Cryptosporidiosis generally develops in persons who are under two years of age, work with animals, or are immunosuppressed. It presents as diarrhea caused by *Cryptosporidium parvum*. Outbreaks of cryptosporidiosis associated with swimming pool or contaminated drinking water have been reported in many countries.

Agent—*Cryptosporidium parvum*

- *Cryptosporidium* is a zoonotic parasite occurring worldwide.
- It can cause severe, debilitating and chronic diarrhea in HIV-infected patients.
- Although there are more than 30 species of *Cryptosporidium*, only two species are known to cause infections in humans—*C. parvum* and *C. hominis*.
- *C. parvum* is zoonotic and is seen commonly in rural areas. The infection can spread from infected animals to humans.

Life cycle

- The parasite resides in the intracellular region, just **within the brush border cells** of the intestinal epithelium.
- Infection occurs by the ingestion of **sporulated thick-walled oocysts** (Fig. 25.20).
- These oocysts are immediately infectious when passed in feces and are known to be acquired through **contact with recreational water**.
- They are **resistant to chlorination**.
- The life cycle of *Cryptosporidium* is presented in Fig. 25.21.
- After ingestion, the oocyst excysts and releases four **sporozoites**, which develop into **trophozoites** or uninucleate **type 1 meronts**.
- Each type 1 meront develops into eight merozoites.
- Some merozoites re-enter the intestinal cells, as a result of which, the cycle continues.
- A few merozoites develop into type 2 meronts. From a meront, four merozoites are released. These enter the host cell and develop into sexual forms, namely **microgamonts and macrogamonts**.
- One microgamont releases 12–16 microgametes, whereas a single macrogamont develops into a single macrogamete.
- Fertilisation results in the formation of **unsporulated oocysts**.

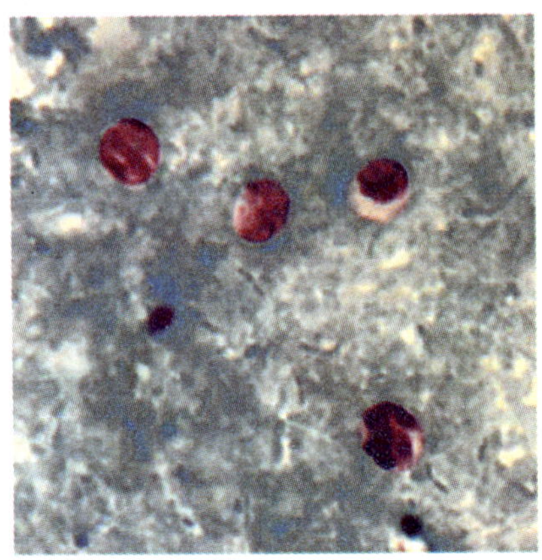

Fig. 25.20 Oocysts of *Cryptosporidium* (*Source*: CDC, DPDx)

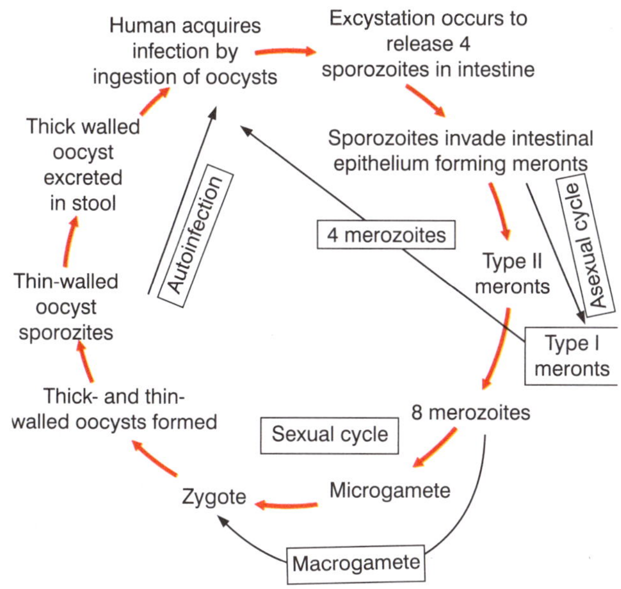

Fig. 25.21 Life cycle of *Cryptosporidium*

- Oocysts are colourless, spherical, or oval and measure about 2–5 µm in diameter.
- The thin-walled oocyst tends to re-infect the same host, while thick-walled oocysts pass in stools and infect new hosts.
- The oocyst does not stain with iodine and is acid-fast.

Clinical features

- The **incubation period** of the infection is 1–2 weeks.
- In an **immunocompetent host**, infection is often asymptomatic or may occur as **self-limiting, watery non-bloody, frothy diarrhea** a few times daily.
- In **immunosuppressed patients**, the disease presents as **chronic, persistent, and profuse diarrhea with abdominal pain.**
- This can lead to water and electrolyte loss, causing severe weight loss and wasting.
- The disease is more severe in patients with lower CD4 counts.
- Extraintestinal infections such as respiratory tract infections and pancreatitis develop in cases of severe immunosuppression.

Treatment

- **Nitazoxanide** 500 mg twice daily for three days is the drug of choice.
- Other drugs like azithromycin and clarithromycin have proven effective in some cases.

Cystoisosporaiasis

Agent—*Cystoisospora belli*

- *Cystoisospora belli* was formerly known as *Isospora belli*.
- It is a coccidian parasite that resides in the intestinal epithelium and causes tissue damage.
- The parasite occurs worldwide and is reported in stool specimens from healthy as well as immunosuppressed persons.

Life cycle

- *Cystoisospora* require a single host for their life cycle.
- Both sexual and asexual cycles take place in **intestinal epithelial cells**.
- Infection is acquired by the **ingestion of sporulated mature oocysts** (Fig. 25.22).
- These oocysts release sporozoites, which enter cells of the small intestine and multiply asexually.
- Immature oocysts bearing a single sporoblast are liberated. These sporulate outside the cell in 1–2 days to become mature, infective *C. belli*.
- These may enter intestinal cells to restart the cycle or develop into sexual forms (microgamete and macrogametes).
- Fertilisation results in the formation of oocysts.
- These oocysts are passed in the stool and become infectious after they sporulate. They are thin-walled and large, measuring 22×10 µm.
- Each sporulated mature oocyst contains two sporocysts, each of which contains four sporozoites.

Clinical features

- Patients develop acute-onset, watery diarrhea with fever and abdominal pain that can last for days or months.
- The infection is self-limiting in immunocompetent persons but could last for months in HIV-infected/ immunosuppressed persons.

Treatment

- **Trimethoprim–sulfamethoxazole** four times daily for 10 days is the first line of treatment.
- For HIV-infected patients, it is given three times daily for three weeks.

Sarcocystosis

Agents—*Sarcocystis suihominis* and *S. hominis*

- This parasite exists worldwide and is a rare cause of diarrhea in immunosuppressed persons.

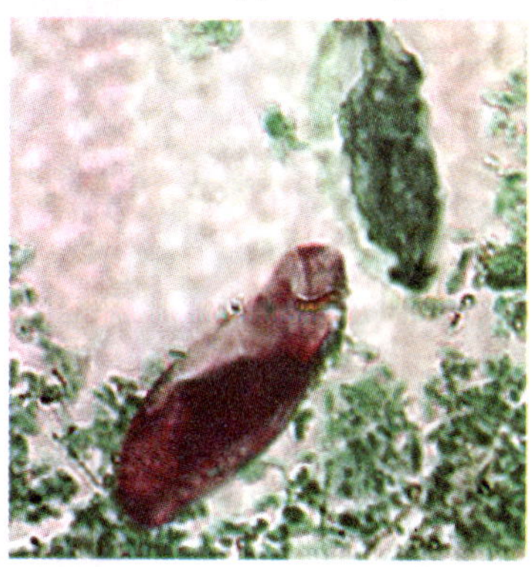

Fig. 25.22 Oocysts of *Cystoisospora* (*Source*: CDC, DPDx)

- Its **definitive host** is a **human being**.
- The **intermediate hosts** are pigs for *Sarcocystis suihominis* and cattle for *S. hominis*.
- Humans can be infected with non-human species of *Sarcocystis* after the accidental ingestion of oocysts and end up as **dead-end hosts.**

Morphology

Sarcocystis exists in three morphological forms:

- **Oocyst:** Contains two sporocysts, each of which contains four sporozoites
- **Sporocyst:** Small and oval, measuring about 10 μm in diameter, each containing four sporozoites; both oocysts and sporocysts can be passed in stool
- **Sarcocyst:** Found in the muscles of pigs and cattle; tiny compartments containing many bradyzoites are seen in the sarcocyst

Life cycle

- The intermediate host acquires the infection by ingesting oocysts with **contaminated food or water**.
- Sporocysts form schizonts in the endothelium and in the bloodstream; they subsequently release **merozoites**, which migrate to skeletal and cardiac muscles where they develop into sarcocysts.
- Humans (definitive hosts) get infected when they consume **improperly cooked meat or pork containing sarcocysts,** which release bradyzoites in the intestine.
- These bradyzoites develop into **microgametes and macrogametes**.
- A zygote is formed after fertilisation and matures into a sporocyst and sporulated oocyst.
- Both are released in the intestinal lumen and passed with stools.

Clinical features

- Disease in humans is usually self-limiting.
- The patient may present with fever, nausea, diarrhea, or pain abdomen.
- Muscle aches, myositis, and muscle weakness may be present in muscular sarcocystosis.

Cyclosporiasis

Agent—*Cyclospora cayetanensis*

- *Cyclospora cayetanensis* is globally prevalent.
- The parasite is transmitted by the ingestion of **oocysts with food and water**.
- It can cause diarrhea of varying severity in **travellers and immunosuppressed persons**.
- It can cause **inflammatory changes, villous atrophy, and crypt hyperplasia**.
- Most cases are asymptomatic.
- Some patients develop diarrhea, flu-like symptoms, flatulence, and belching.
- The illness may be self-limiting or it may wax and wane. Sometimes, *Cyclospora* may cause prolonged diarrhea, anorexia, and weight loss.
- The infective form is the oocyst, which is spherical and refractile and measures about 10 μm in diameter.
- The mature oocyst contains two sporocysts, each with two sporozoites.
- The life cycle of *Cyclospora* is very similar to that of *Cryptosporidium* species.

Laboratory diagnosis

- The patient must be investigated for HIV and CD4 counts, which are usually low in such patients. Compliance with antiretroviral therapy must be investigated.
- **Specimens**—Stool and biopsy samples of the small intestine are collected for diagnostic investigations.
- **Direct microscopy:** As oocysts are excreted in small numbers, stool concentration techniques may be performed on a stool sample to increase its yield (Table 25.10).
  - Wet mount shows the characteristic refractile cysts of the parasite.
  - Modified acid-fast staining is used to visualise oocysts.
  - Auramine, safranin, and lactophenol cotton blue stains can also be used to visualise cysts. Autofluorescence under UV light can be used for *Cyclospora* and *Cystoisospora* but is not useful for visualising *Cryptosporidium*.

Table 25.10 Microscopic characteristics of oocysts of intestinal sporozoa

Parasite oocyst	Size and shape	Ultrastructure on wet mount	Acid-fast staining
Cryptosporidium	4–5 μm, round	Refractile, double-walled, four sporozoites	1% sulphuric acid fast, Kinyoun stain shows red oocyst
Cystoisospora	20–30 μm, oval/ elliptical	Thin, twin-layered cyst wall contains two sporocysts, each with four sporozoites Both sporulated and non-sporulated cysts can be seen	Uniformly acid-fast
Cyclospora	8–10 μm, round	Two sporocysts, each containing two sporozoites	Variable acid-fastness

   - Direct fluorescent antibody staining can be used to detect *Cryptosporidium* oocysts and is more sensitive than acid-fast staining.
- **H & E stain of biopsy material:** *Cystoisospora* and *Cryptosporidium* can be detected in the small intestine.
- **Antigen detection in stool:** Both ELISA and immunochromatographic tests are available to detect *C. parvum*-specific coproantigen in stool.
- **Serology**: Antibody detection can be done by ELISA to detect IgG and IgM antibodies to *Cryptosporidium* in the patient's serum. The antibody can also be detected by an indirect fluorescent antibody test.
- **Molecular diagnosis:** PCR is available to detect *Cryptosporidium, Cyclospora,* and *Cystoisospora* in clinical as well as from environmental samples.

Laboratory diagnosis of opportunistic diarrhea

Always investigate patient for HIV and CD4 counts and check for compliance with antiretroviral therapy.

- **Specimens**—Stool and biopsy of the small intestine.
- **Direct microscopy**
   - Wet mount: Typical refractile cyst.
   - Modified acid-fast staining is used to visualise oocysts.
   - Auramine, safranin, and lactophenol cotton blue stains.
   - Direct fluorescent antibody staining: To detect *Cryptosporidium* oocysts and is more sensitive than acid-fast staining.
- **H & E stain of biopsy material**: *Cystoisospora* and *Cryptosporidium* can be detected.
- **Antigen detection in stool**: ELISA and immunochromatographic tests to detect *C. parvum*-specific coproantigen.
- **Serology**: Antibody detection by ELISA or fluorescent antibody test to detect IgG and IgM antibodies to *Cryptosporidium* in the patient's serum.
- **Molecular diagnosis**: PCR to detect *Cryptosporidium, Cyclospora,* and *Cystoisospora*.

Parasitic Dysentery

- *Entamoeba histolytica* and *Balantidium coli* are the most prominent causes of parasitic dysentery.
- Amoeba is a single-celled protozoa of phylum Amoebozoa.
- These organisms are motile with the help of pseudopodia.
- Amoeba can be classified based on their habitat.

Amoebic dysentery

Case 7. *A 38-year-old female presents with complaints of severe abdominal pain, bloody diarrhea, and low-grade fever. The patient's symptoms began three days ago with mild abdominal discomfort, which rapidly progressed to severe cramping accompanied by bloody diarrhea. She also experienced weakness and dehydration. The stool sample, on microscopic examination, showed spherical structures with four nuclei, and the chromatoid body was visible with presence of red blood cells.*

i. What is the provisional diagnosis?
ii. What are the probable etiological agents for the above conditions?
iii. What can be the complications of this infection?

This is a case of amoebic dysentery, an infection caused by *Entamoeba histolytica*.

Epidemiology

- Amoeba residing in the large intestine cause infections ranging from mild dysentery to life-threatening amoebic liver abscess. Infections in 90% of patients are asymptomatic.
- *E. histolytica* is distributed worldwide but is most common in tropical and sub-tropical areas.
- The infections are commonly seen in areas with poor sanitation and poor or contaminated water supply.

Agent—*Entamoeba histolytica*

Morphology

The parasite exists in three morphological forms (Fig. 25.23).

- **Trophozoite:**
   - This is the motile form of *E. histolytica*. It is 20–60 μm in diameter with vacuoles and a nucleus with a characteristic central nucleolus.
   - The outer layer of cytoplasm is called the **ectoplasm**.
   - The inner layer of cytoplasm or endoplasm may contain ingested erythrocytes, leukocytes, and debris.
   - The ectoplasm projects forward as pseudopodia, which act as organs of locomotion.

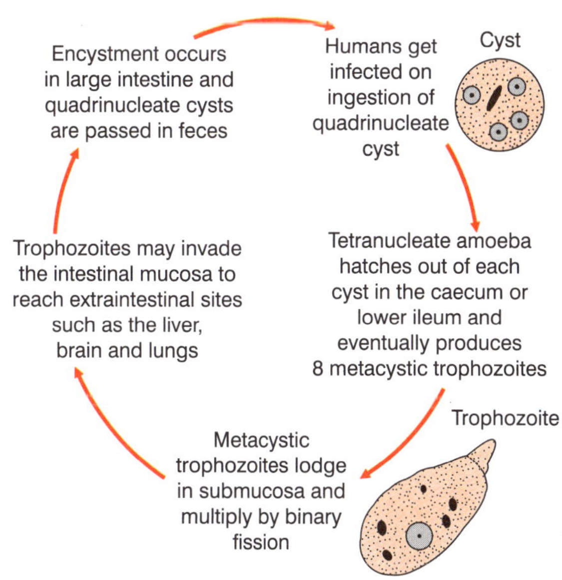

Fig. 25.23 Morphological forms and life cycle of amoeba

- **Precysts:** These are smaller than trophozoites and are uninucleate and oval with blunt pseudopodia.
- **Cysts:**
  - These are small, round structures measuring about 10–20 μm in diameter.
  - The cyst develops from uninucleate structures to those with 1–4 nuclei.
  - The cytoplasm contains round chromatid bars. The cyst wall is highly refractile.
  - The cysts can be seen in the fecal sample of an infected person, with either one, two, or four nuclei in various stages of development.

Life cycle (Fig. 25.23)

Entamoeba requires only one host for its life cycle.

- Cysts contaminate soil and water bodies and can be transmitted mechanically by flies and cockroaches or by the contamination of water sources by sewage water.
- The use of fertilisers containing human waste in agricultural areas can also contaminate fruits and vegetables.
- Upon excystation in the small intestine, a quadrinucleate trophozoite is released, known as the **metacyst.**
- The metacyst migrates to the large intestine and multiplies by binary fission to release eight metacystic **trophozoites.**
- The trophozoites attach to intestinal epithelial cells, producing **microulcerations in the large intestine**. This is characterised by acute amoebic dysentery. This condition must be differentiated from bacillary dysentery caused by bacteria such as *Shigella* (Table 25.11).

Pathogenesis

The mode of infection is by the ingestion of mature quadrinucleate cysts that are passed in the feces of infected persons including asymptomatic carriers.

- **Intestinal amoebiasis:**
  - Amoebae invade the colonic epithelial cells and lyse them.
  - The host's response to cell lysis causes the accumulation of erythrocytes, inflammatory cells, and epithelial cells.
  - The cellular destruction spreads horizontally in the submucosa, forming a classic '**flask-shaped' ulcer**.
  - The ulcer is found teeming with trophozoites at the margins and base.
  - Most ulcers heal without any scarring, but **occasionally, necrosis and perforation** can occur.
  - After some days, the intestinal lesions start healing, and the trophozoites transform into cysts which are released in stools upon maturation.

Table 25.11 Differentiation between bacillary and amoebic dysentery

Features	Bacillary	Amoebic
Clinical		
Onset	Acute	Gradual
Fever	Yes	No
Vomiting	Occasional	None
Gross stool findings		
Consistency	Watery, bloody, odourless	Blood and mucus, offensive odour
Appearance	Bright red blood	Dark red altered blood
Microscopy		
RBC	Discreet rouleaux	In clumps
Pus cells	Numerous	Scanty
Macrophage	Numerous	Nil/few
Ghost cells	Many	Nil
Trophozoite	Nil	Trophozoites
Charcot–Leyden crystals	Nil	Present

- **Amoeboma:**
  - Rarely, the infection results in the formation of a **tumour-like lesion** in the lumen of the large intestine.
  - The mucosa covering the lesion is very thin and ulcerated with thick, edematous walls and a large amount of granulation tissue.
- **Liver abscess:**
  - The trophozoites travel to the liver via the portal venous system.
  - The lysis of liver cells and thrombus formation leads to the formation of an abscess.
  - Inflammatory cells accumulate in the necrotic tissue, and the abscess is filled with necrotic material surrounded by a thin wall of congested liver tissue.
  - The pus is typically described as '**anchovy sauce pus'**. It contains necrotic cells and sterile granular debris and few trophozoites near the capsule of the abscess.
  - The abscess in one-third of cases can spread to involve the pleural and pulmonary areas.
  - Frank rupture into the pleural, pericardial, or peritoneal cavity can cause a medical emergency.
- **Satellite lesions**
  - Rarely, the infection can spread to the genitourinary system, skin, etc., to produce ulcers.
  - The ability of *Entamoeba* to cause satellite lesions in various organs can be attributed to the fact

that trophozoites are resistant to complement-mediated lysis.
- On the other hand, non-pathogenic *E. dispar* is lysed by complement, and thus remains restricted to the intestinal lumen.

Clinical features

- Most infected individuals remain asymptomatic. In such patients, the cyst may be seen in stool samples without any clinical manifestations.
- Initially, intestinal amoebiasis presents with symptoms ranging from **mild diarrhea to dysentery**. Later, systemic symptoms such as fever, rigors, and leukocytosis appear.
- Patients with liver abscess present with fever, right quadrant abdominal pain, hepatomegaly, and altered bowel movements. The presentation in such cases is slow and sub-chronic with mild symptoms like fever and weight loss lasting six months or more.
- Satellite symptoms develop due to effusions or spread from the liver. These can present with pleuritis or peritonitis.

Laboratory diagnosis

- **Samples:**
  - Stool collected for microscopy
  - Aspirate from liver abscesses
  - Biopsy specimen obtained after sigmoidoscopy
- **Microscopy for cyst or trophozoites:**
  - In stool, quadrinucleate cysts and trophozoites are seen.
  - The visualisation of trophozoites is pathognomonic of amoebic dysentery. However, these need to be differentiated from white blood cells.
  - It is important to differentiate cysts of *Entamoeba histolytica* from those of *E. hartmanii* and *E. coli*.
- **Culture:**
  - It is technically demanding and hence, performed only for research purposes.
  - *Entamoeba* species can be cultured in xenic and axenic media. Xenic culture is the growth of the parasite with unknown flora.
  - Examples of the xenic media used for the culture of *Entamoeba* spp. include the modified Boeck and Drbohlav medium and egg diphasic medium.
- **Isoenzyme analysis:**
  - Zymodemes are *Entamoeba* strains that have the same electrophoretic pattern for several enzymes.
  - Based on various enzymes, 24 different zymodemes are recognised. Zymodeme pattern analyses can differentiate *E. histolytica* from *E. dispar* (which is non-pathogenic).
  - In current practice, molecular techniques have superseded isoenzyme analysis.
- **Serology:**
  - The detection of antibodies by ELISA can help diagnose 90% of cases of colitis and amoebic liver abscesses.
  - Fecal antigen detection assays are also available.
- **Molecular diagnostics**: PCR or DNA probes can be used to detect infection. This helps to differentiate *E. histolytica* from *E. dispar*.

Treatment and control

- The drug of choice to treat amoebiasis is **metronidazole.**
- Other alternatives include iodoquinol and diloxanide furoate.
- Good hygiene, adequate sanitation measures, the availability of clean drinking water, and raising awareness about the importance of washing fruits and vegetables before consumption help in the control of infections.

Commensal, Non-pathogenic Amoebae

Some members of the genus *Amoeba* exist in the intestines of human beings without causing any disease (Table 25.12). Their cysts or trophozoites may be

Table 25.12 Amoebae causing disease in humans

Amoebae	Habitat	Species	Pathogenicity
Intestinal amoebae	Large intestine	*Entamoeba histolytica*	Amoebic dysentery, amoebiasis, amoebic liver abscess
		Entamoeba coli	Non-pathogenic
		E. dispar	
		E. hartmanii	
		Endolimax nana	
		Iodamoeba bütschlii	
	Oral cavity	*E. gingivalis*	Non-pathogenic
Free-living amoebae (Chapter 23)	Soil and water	*Acanthamoeba* species	Granulomatous amoebic meningoencephalitis Amoebic keratitis
		Naegleria fowleri	Primary amoebic meningoencephalitis
		Balamuthia mandrillaris	Granulomatous amoebic meningoencephalitis

passed in stool and need to be differentiated from those of *Entamoeba histolytica* as their life cycle and morphological forms are quite similar to that of *E. histolytica*.

- *Entamoeba dispar* is morphologically indistinguishable from *E. histolytica*.
- The cysts and trophozoites of *E. hartmanni* and *Endolimax nana* are slightly smaller than those of *E. histolytica*.
- The cyst of *Entamoeba coli* is double the size of *E. histolytica* cysts and has eight nuclei and filamented chromatin bars. The trophozoite of *Entamoeba coli* moves slowly, and the cytoplasm is not demarcated into ecto and endoplasm.

Balantidiasis

Agent—*Balantidium coli*

- *Balantidium coli* is a ciliated organism found worldwide, known to cause balantidiasis.
- Its **natural reservoirs are pigs**. Humans act as incidental hosts
- Most people infected with *Balantidium coli* remain asymptomatic. Others, especially the immunosuppressed or malnourished, present with persistent diarrhea, dysentery, abdominal pain, weight loss, nausea, and vomiting.
- Upon ingestion, the cyst excysts in the small intestine, releasing ciliated **trophozoites** (Fig. 25.24).
- These reach the large intestine and multiply in its **mucosal lining.**
- The trophozoite ranges from 50–200 µm in length and 40–70 µm in width. Its surface is covered with **cilia,** and the prominent internal structure is a **macronucleus and a smaller micronucleus**. The trophozoites are binucleate with a **cytosome (funnel-shaped mouth).** The small and large nuclei are involved in reproduction.
- Other organelles are the food vacuoles and two contractile vacuoles.
- The trophozoites encyst, and the **cysts are passed in feces**.

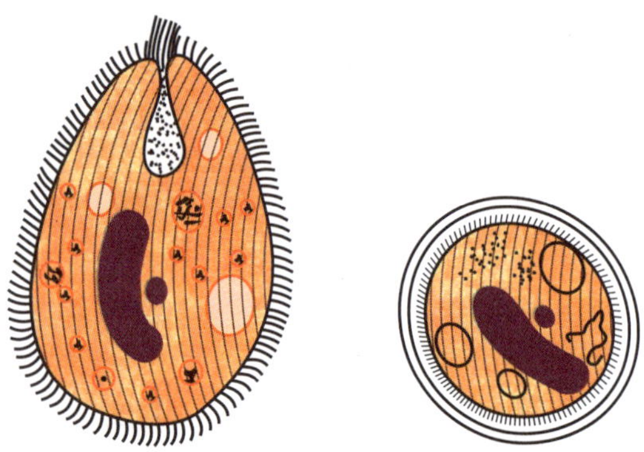

Fig. 25.24 Cyst and trophozoite of *Balantidium*

- These cysts are double-walled and round, measuring about 50 µm in diameter and containing a micro and a macronucleus.

Laboratory diagnosis

The microscopic examination of stool reveals characteristic trophozoites and cysts.

Treatment

The drug of choice is tetracycline; iodoquinol and metronidazole are the alternative antimicrobials. Prevention and control measures are similar to those recommended for amoebiasis.

Helminthiasis or Intestinal Nematode Infections

- **Helminthiasis** is an infestation with one or more intestinal parasitic roundworms or nematodes. They are transmitted by eggs present in human feces, which in turn, contaminate soil in areas where sanitation is poor.
- Intestinal nematode infections (Table 25.13) are among the most common infections worldwide, especially in low-income countries.

According to estimations made by the WHO, approximately 24% of the world's population is infected with roundworms.

- The freshly passed eggs of hookworms, roundworms, and whipworms need about **three weeks to mature in the soil** before they become infective. Hence, **no direct person-to-person transmission** of infection occurs from fresh feces.
- The eggs of these worms do not develop in the human host.
- Therefore, re-infection occurs only as a result of **contact with infective stages in the environment**.
- *Strongyloides stercoralis* can multiply in its host; in immunocompromised individuals, it can result in fatal hyperinfection.

Ascariasis

Ascaris lumbricoides, the causative agent of ascariasis, is the most common parasitic worm in humans.

Ascariasis can be asymptomatic, causing only malnutrition and growth retardation, or it may present with abdominal pain, nausea, vomiting, bloating, and diarrhea.

Table 25.13 Intestinal nematodes

Small intestine	Large intestine
Ascaris lumbricoides	*Trichuris trichuria*
Necator americanus	*Enterobius vermicularis*
Ancyclostoma duodenale	
Strogyloides spp.	

Agent—*Ascaris lumbricoides* (roundworm)

- It is the **largest nematode that resides in the small intestine**.
- Infestations are commonly seen in **tropical countries** like India. It is seen more commonly in **children**, especially those from the poor socio-economic backgrounds and those living in unhygienic conditions.

Morphology (Fig. 25.25)

The life cycle of *Ascaris* consists of the adult, larva (four stages), and egg stages.

- **Adult:**
  - It is round and around 20–40 cm long.
  - Female worms are nearly double the length of male worms.
  - The adult worm lives for 1–2 years.
  - The posterior end of the male worm is curved and pointed, with two spicules, whereas that of the female is straight.
  - The mouth opens anteriorly and has three-toothed lips, one dorsal and two ventral.
- **Egg:**
  - The fertilised egg is round or oval and around 90 × 45 μm in size.
  - It is surrounded by a thick, albuminous coat containing a large, unsegmented ovum of granular mass.
  - It floats in the saturated salt solution. The **fertilised egg** (Fig. 25.26a) **is the infectious form**. The unfertilised egg (Fig. 25.26b) is elongated and bears a thin and scanty coat which encloses a thin, unsegmented ovum. It does not float in the saturated salt solution. Both types of the egg are bile-stained and appear brown in saline mounts.

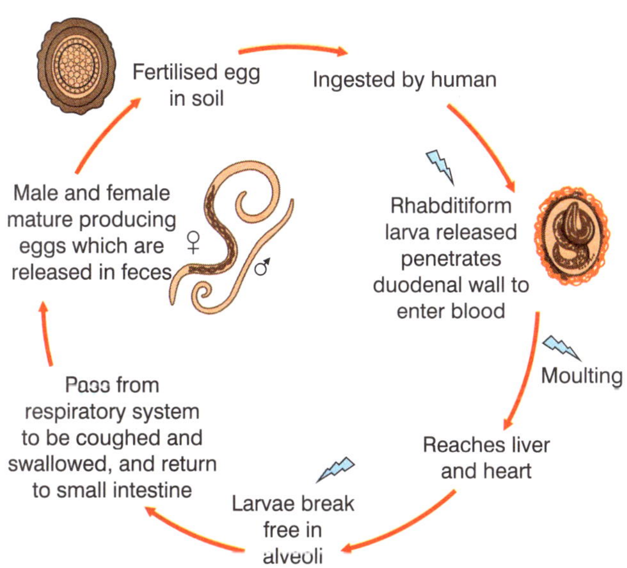

Fig. 25.25 Morphology and life cycle of *Ascaris*

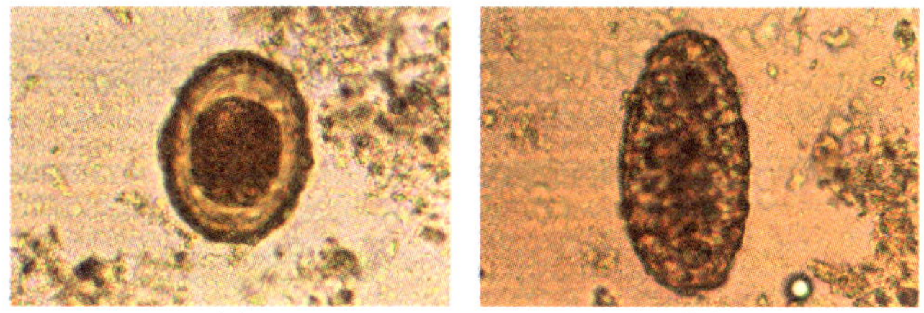

Fig. 25.26 A fertilised *Ascaris lumbricoides* egg and (b) an unfertilised egg (*Source*: CDC, PHIL, Image IDs 18741 and 18737)

Life cycle (Fig. 25.25)

- The mode of infection is by the ingestion of fertilised eggs containing stage two rhabditiform larvae.
- The eggs hatch to form larvae which moult to form stage 3 larvae and pierce the intestine to reach portal circulation and finally the lungs via pulmonary capillaries.
- In the lungs, they migrate into alveoli, then bronchi, trachea and finally, the pharynx.
- These larvae are swallowed by the host.
- The final moulting occurs in the small intestine.
- The worm then develops into a sexually mature adult.
- A mature female lays eggs that are immediately fertilised by the male worm.
- The life span of the adult female is around 1–2 years, during which time, it can lay up to 2 lakh eggs per day. Both fertilised and unfertilised eggs are released by the female and are passed in stools.
- The unfertilised eggs disintegrate, while the fertilised eggs develop further in soil.
- The rhabditiform larva is formed inside the egg, which moults to form the stage 2 larva.

Clinical features

- Most cases are asymptomatic.
- In children, malnutrition and growth retardation are seen as the worms feed on intestinal nutrition.
- The patient might have a history of passing the worm in stool or rarely, with coughing.

Complications

- These can occur either due to a high worm burden or due to the migration of larvae.
- Large entanglement by a bolus of adult worms can cause pain abdomen, intestinal obstruction, perforation, intussusception, or volvulus. The migration of adult worms through the biliary tract can cause biliary colic, cholecystitis, cholangitis, and pancreatitis.
- **Allergic manifestations** include fever, urticaria, angioneurotic edema, and conjunctivitis.
- **Löffler's syndrome** occurs due to the migration of larvae through lungs, usually 10–12 days after the

ingestion of eggs. It can result in an irritating, non-productive cough and burning substernal discomfort that is aggravated by coughing or deep inspiration. Mild fever with eosinophilia develops, which subsides slowly over weeks. Chest X-rays may reveal features of eosinophilic pneumonitis which may take several weeks to clear.

Treatment

- **Drug of choice:** Albendazole 400 mg once or mebendazole 100 g twice daily for three days or 500 mg once.
- **Alternatives:** Ivermectin (150–200 μg/kg), nitazoxanide, or pyrantel pamoate can also be used.
- Pyrantel pamoate can be given safely in pregnancy while other antibiotics are contraindicated.
- Control measures such as health education, improved sanitation, and not using human feces as a fertiliser are critical.

Hookworm infestation

Agents—*Necator americanus* and *Ancylostoma duodenale*

- Hookworm infection is one of the common parasitic infestations in developing countries. They are mainly causing iron deficiency anemia in exposed patients.
- Hookworm infections in humans are caused by two nematode species—*Necator americanus*, the **'New World' hookworm** and *Ancylostoma duodenale*, the **'Old World' hookworm.**
- Many hookworms are also known to infect animals and can cause cutaneous larva migrans (Chapter 29).
- Hookworm infections are present worldwide but are more common in **the tropics**.
- They are an important cause of **iron deficiency anemia**.
- In India, *N. americanus* is more common in southern India than in the north.

Morphology

- **Adult:**
  - The adult hookworm measures around 7–11 mm; females are slightly longer than males.
  - The **anterior end of the worm is bent like a hook** (hence the name 'hookworm').
  - The mouth parts of the worm are well developed with teeth in *Ancylostoma duodenale* and cutting plates in *Necator americanus* (Fig. 25.27).
  - Their large buccal cavity is lined with a cuticle and is meant for the blood-feeding parasitic life.
  - In males, the genital apparatus opens into the **cloaca**, while in females, the vulva is located at the latter one-third of the worm's body (posterior end).

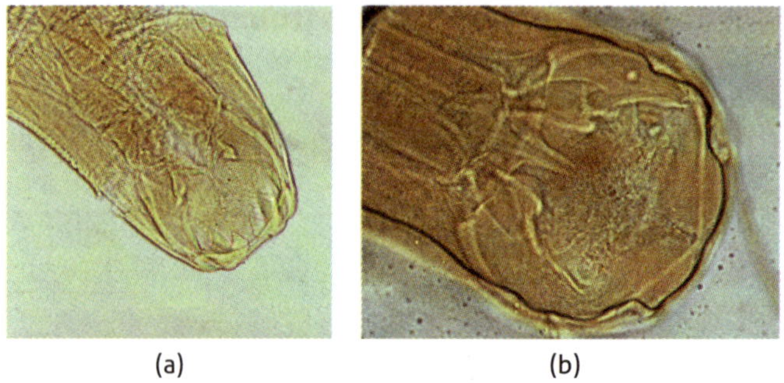

Fig. 25.27 Adult worm showing (a) cutting teeth for *Ancylostoma duodenale* and (b) cutting plates for *Necator* (*Source*: CDC, DPDx)

- **Eggs (Fig. 25.28):**
  - They are oval, about 60 × 40 μm in size and non-bile staining. They are visualised within a clear, hyaline shell containing a segmented ovum.
  - There is a clear space between the shell and the ovum.
  - Eggs of both species are the same, whereas the adult worms may show minor differences (Table 25.14).

Life cycle (Fig. 25.29)

- Hookworms require only one host for their life cycle.
- The immature hookworm eggs are excreted by infected persons in their stool. In soil that is moist, warm, and in a shaded area, the eggs release rhabditiform larvae (L1) in about 24 hours.
- The L1 larva moults twice to develop into a filariform larva (L3) within the next 5–10 days.
- This larva can remain viable in the soil for 3–4 weeks.
- It enters the human body by **penetrating skin**. Humans can get infected while **walking barefoot in fields, on grass and on damp soil. Autoinfection may also occur**.
- Upon entry, the larva travels **through the venous circulation** to reach the right side of the heart, and then the lungs through the alveoli.

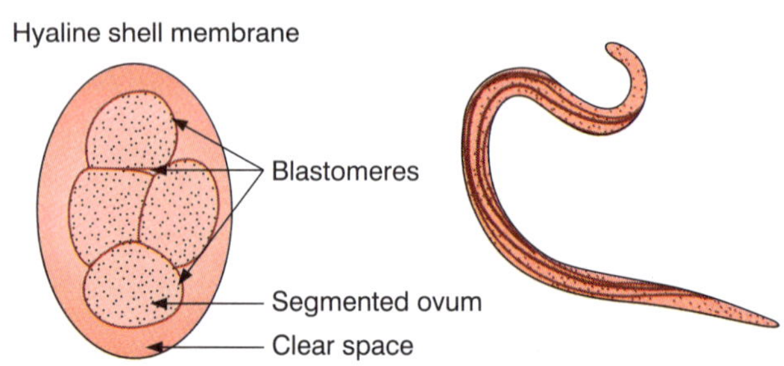

Fig. 25.28 Egg and larva of hookworm

Table 25.14 Differences between *Necator americanus* and *Ancylostoma duodenale*

Feature	*N. americanus* 'New World' hookworm	*A. duodenale* 'Old World' hookworm
Life span (years)	3–20	1–3
Size	Shorter	Longer
Mouth parts	Two cutting plates	Two pairs of teeth
Eggs released per day by single female	5,000–20,000	25,000–35,000
Blood loss/day/worm	0.04 mL	0.2 mL
Anemia	Less severe	More severe
Geographic distribution	South China, Southeast Asia, Americas, central and south Africa, southern India and Australia	North Africa, Mediterranean regions, North India, North China and Japan

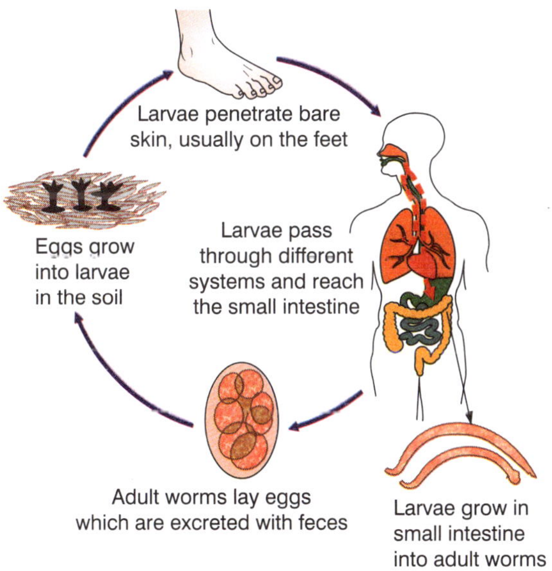

Fig. 25.29 Life cycle of hookworms

- From there, the larva reaches the trachea and pharynx to be swallowed.
- The **final moulting occurs in the migratory phase or in the intestine**.
- In the small intestine, the larva attaches to the mucosa using the teeth in its buccal cavity and develops into an adult worm.
- It feeds on capillary blood and secretes anticoagulants to keep feeding.
- Sexual maturity is attained in the next 5–6 months, following which, fertilisation occurs.
- The females lay eggs with segmented ovum, which are passed with feces.

Clinical features

- Patients are generally asymptomatic.
- **Acute infections:**
  - A maculopapular rash develops with itching at the site of entry of larva.
  - The patient may complain of **allergic symptoms** due to larval migration through the lungs.
  - Non-specific complaints include nausea, abdominal pain, and diarrhea.
  - Blood tests reveal eosinophilia and elevated IgE.
- **Chronic infections:** These are marked by signs and symptoms of iron deficiency anemia and malnutrition.
  - Children are more commonly affected by growth retardation and lethargy.
  - The child's parent may complain that the child has been eating mud, chalk, or wall plaster (a condition known as pica).
  - Complications of anemia such as edema and congestive heart failure may be reported.
  - Hematological investigations show features of iron deficiency anemia.
  - Stool may show the presence of occult blood.

Treatment and prevention

- Due to the ability of the hookworm to produce iron deficiency anemia and protein energy malnutrition, the infection must be treated promptly to prevent growth retardation in children.
- The drugs used to treat hookworms are albendazole 400 mg orally single dose, mebendazole 500 mg BD for three days or pyrantel pamoate 11 mg/kg for three days.
- Anemia needs to be corrected with iron and dietary supplements.
- As for most intestinal helminthic infections, prevention is by personal hygiene, the use of appropriate footwear and health education.
- Mass drug administration programmes also form a part of the preventive strategy.

Trichuriasis

- *Trichuris trichiura*, also known as the human whipworm, is a roundworm that causes trichuriasis in humans.
- Though *Trichuris trichiura* is found all over the world, its infections are more common in tropical areas with poor living conditions and low socioeconomic status.
- The infection is acquired during childhood and is usually associated with mixed infections with roundworm.

Agent—*Trichuris trichiura* (whipworm)

Morphology (Fig. 25.30)

- **Adult:**
  - The adult worm measures around 35–50 mm.
  - The posterior end is about 2/5th of the total length and is thick like the handle of a whip, while the anterior end is thin and slender (hence the name 'whipworm').
  - The anterior end remains burrowed in the mucosa of the large intestine (caecum and ascending colon) while the posterior end extends to the lumen of the gut.
  - The life span of the adult worm is 1–3 years.
- **Eggs:**
  - The eggs are bile-stained, brown, barrel-shaped and approximately 55 × 25 µm in size.
  - They have characteristic bipolar mucus plugs (Fig. 25.30).
  - The embryo is released after chitinases lyse the mucus plugs.
  - The eggs float in a saturated solution of common salt.

Life cycle (Fig. 25.30)

- Humans get infected after ingesting embryonated eggs present in contaminated food or water or via fomites or hands.
- The eggs release larvae in the small intestine which moult four times to develop into adult worms over the next three months.
- The adults mate in the large intestine and release a large number of eggs, which are excreted in stools.
- In soil, the eggs mature in 2–4 weeks and become infective.
- They can survive outside the human body for many months.

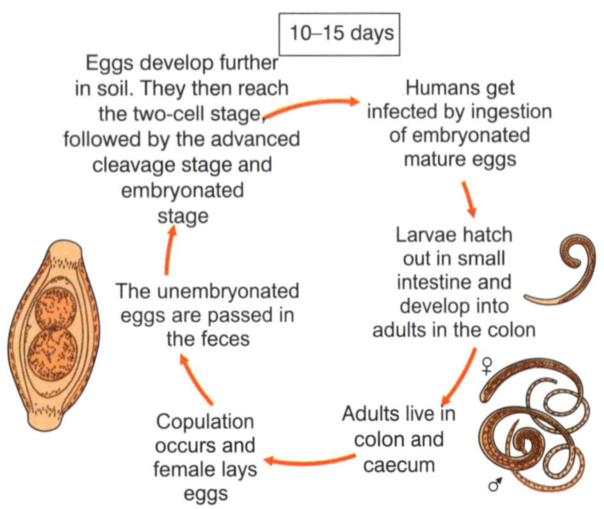

Fig. 25.30 Morphology and life cycle of whipworms

Clinical features

- Most infections are asymptomatic.
- In heavy infections, vague abdominal pain, anorexia, flatulence, fever, nausea, and vomiting may be reported.
- The worm burrows into the intestine; this can result in mucosal injury and bleeding.
- Such patients may present with dysentery syndrome characterised by abdominal cramps, tenesmus, and loose motions with large amounts of mucus and blood.
- Weight loss and pruritus are also seen.
- Complications—recurrent rectal prolapse, growth retardation, and cognitive impairment may occur in children with heavy infestation.

Laboratory diagnosis

Microscopic examination of stool shows **barrel-shaped ova** of *T. trichiura* which are bile-stained.

Treatment

The drug of choice is albendazole (400 mg single dose) or mebendazole 100 mg BD for three days.

Enterobiasis

Enterobiasis is a **common contagious parasitic infestation seen mainly in children.** It is caused by *Enterobius vermicularis*, which is one of the most common nematode infections in the world.

Humans are the only natural host for this infection.

Agent—*Enterobius vermicularis* (pinworm)

- *Enterobius vermicularis* is also known as **pinworm or threadworm.**
- Like other nematodes, the infection is common in tropical areas.

Morphology (Fig. 25.31)

- **Adult:**
  - The adult worms are tiny, thread-like, and whitish and reside in the large intestine.

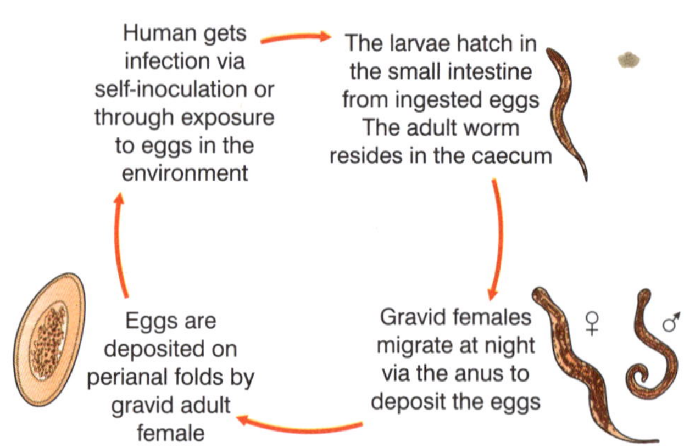

Fig. 25.31 Adult pinworm, egg, and life cycle of *Enterobius*

- Males are 2–5 mm in length, while the females are longer, about 8–10 mm long.
- **Eggs:**
  - These are typically planoconvex, colourless, and walled.
  - Each egg contains a tadpole-shaped larva inside and floats in the saturated salt solution.

Life cycle (Fig. 25.31)

- Humans are the only hosts; they acquire infection by ingesting the fully developed embryonated egg.
- Autoinfection is also common.
- The egg, upon reaching the caecum, releases the larva, which develops into an adult worm in the next 4–6 weeks.
- The male dies soon after fertilisation, while females can live for up to two months.
- The gravid female migrates to the anus to lay eggs on the perianal skin.
- The eggs are infective and can hatch into larvae, which can travel back into the alimentary canal.
- The eggs are usually laid at night and cause itching.
- They are usually lodged in children's nails as a result of itching and are consequently ingested again, resulting in autoinfection.

Clinical features

- Clinically, the person may remain asymptomatic.
- In children, the pathognomonic sign is intense **itching in the perianal region, especially at night**.
- The examination may reveal excoriation of perianal skin, which may also result in a secondary bacterial infection.
- Rarely, the worm may migrate to the genital areas, causing vulvovaginitis and urinary tract infections.

Laboratory diagnosis

- **Stool examination is not helpful in the diagnosis** as they are only occasionally excreted in the stool.
- *Enterobius* can be diagnosed through a cellophane tape test or using an **NIH swab** (details in Chapter 7). In this test, an adhesive tape-like material is applied to the perianal area and then examined under a microscope. Diagnosis is made by the visualisation of characteristic planoconvex ova.

Treatment

Albendazole or mebendazole or pyrantel pamoate can be used.

Strongyloidiasis

Strongyloidiasis is infection with *Strongyloides stercoralis* that can present clinically as abdominal pain, diarrhea, rash, and pulmonary symptoms such as cough and wheezing.

Agent—*Strongyloides stercoralis*

- This parasite is restricted geographically to hot and humid climates.
- The majority of cases are seen in Southeast Asia, South America, and Africa.
- *S. stercoralis* is unique as it exists in the free-living state and as a parasite.
- The female worm lives in the duodenum and the upper part of the jejunum.
- The male worms are never found in humans. For this reason, the existence of male worms is still up for debate. Some believe they do not exist while others speculate that the males may exist only in the free-living state. This former theory is supported by the fact that the female produces eggs without fertilisation with males.
- This is known as **parthenogenesis**—a natural form of asexual reproduction, in which fertilisation by males is not necessary.

Morphology (Fig. 25.32)

- While parasitic females are tiny—about 2–3 mm long, the free-living females are even smaller—1–2 mm long.
- The eggs immediately hatch into larvae (ovo-viviparous).

Life cycle (Fig. 25.32)

- The mode of infection is by the penetration of skin by the L3 larva.

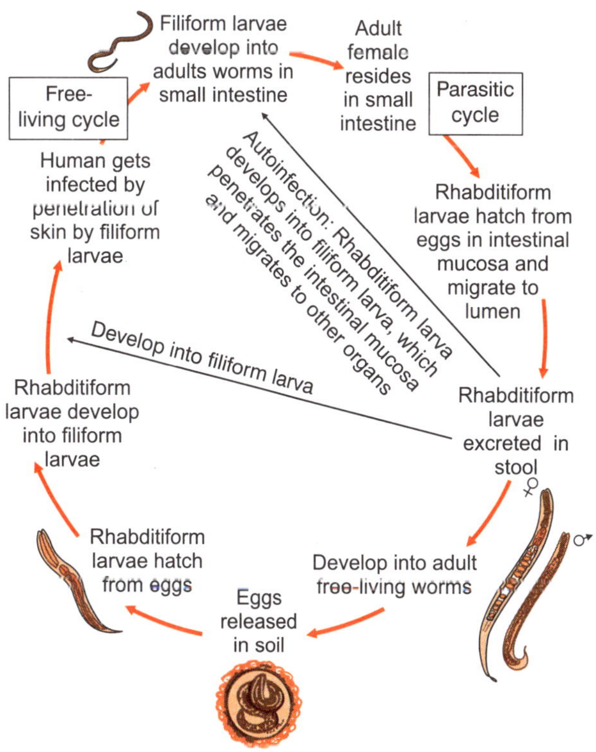

Fig. 25.32 Morphology and life cycle of *Strongyloides*

- The rhabditiform larva is passed in the feces of infected persons.
- This larva moults twice and develops in the soil to form an L3 larvae.
- Some of these may develop into adult free-living forms while others may infect humans to lead a parasitic existence.
- Humans get infected when the L3 larva **penetrates their skin.**
- The larva gains entry into the blood circulation and reaches the right side of the heart, then the lungs and finally, the alveoli. It then ascends the trachea and is swallowed.
- Subsequently, the worms reach the intestine, where they moult again to develop into adults.
- The adult female burrows into the intestinal mucosa and lays eggs that immediately hatch into larvae.
- This rhabditiform larva is passed with stools.
- Alternatively, this larva transforms into filiform larvae and penetrates the intestinal mucosa or perianal skin, resulting in autoinfection.
- Autoinfection can lead to a heavy worm load or hyperinfection.

Clinical features

- Almost 50% of infected individuals are asymptomatic.
- Patients may complain of rashes and rapidly migrating, linear, itchy lesions at the site of entry of the larva.
- These can be located on the thighs, buttocks or feet. This condition is called **larva currens or racing larva** (Chapter 28).
- **Hyper-infection syndrome and disseminated strongyloidiasis** is seen in immunosuppressed or HIV patients or in individuals who have a high steroid intake. The invasion of other organs by the parasite can be seen in the liver, kidneys, or meninges. The clinical presentation depends on the infected site. It is potentially fatal if left untreated.

Laboratory diagnosis

- **Samples:** Stool is collected to detect the larva of the parasite. A freshly prepared stool sample in saline preparation shows actively motile larvae.
- **Microscopy:** *Strongyloides* larvae are seen in stool. Eggs may be seen only in cases of hyper-infection syndrome.
- **Serology:** An antigen-capture ELISA has been developed for the detection of antigens in stool samples of suspected *Strongyloides* infection.
- **Molecular detection:** Real-time PCR is now available to detect most nematodes including *Strongyloides*. These tests are based on the detection of the cytochrome oxidase gene.
- **Other tests:** Eosinophilia and elevated IgE may be present.

Treatment

- **Drug of choice:** Ivermectin 200 mg/kg for two days. The treatment is extended up to seven days in hyper-infection.
- **Other drugs:** Thiabendazole (25 mg/kg BD) or albendazole (400 mg BD for two weeks) can be used.

Visceral Larva Migrans (VLM)

- The migration of larva in visceral organs is called visceral larva migrans.
- This is primarily caused by *Toxocara* species (*Toxocara canis* and *T. catis*) but may also be seen in infestation with hookworms of animals.
- The following organisms are implicated in VLM:
  - *Toxocara canis*
  - *T. catis*
  - *Angiostrongylus* species
  - *Gnathostoma spinigerum*
  - *Anisakis* species

Agent—*Toxocara canis*

- *Toxocara canis* is a hookworm of dogs while *T. catis* infects cats.
- In humans, *Toxocara* is **more common in children**, especially in rural areas.
- The morphology of these worms is similar to that of other nematodes.
- The adults are about 9–12 mm long, while eggs are oval with a pitted surface.
- Once ingested by humans, embryonated eggs release larvae in the small intestine which are carried by portal circulation to the liver.
- In humans, the larva is not able to develop further and get fibrosed.
- Sometimes the larva may migrate to the lungs or reach systemic circulation, causing granuloma.
- Patients present with **liver or lymph node enlargement**. Eosinophilia may be present.
- The larva can sometimes reach the eye (**ocular larva migrans**). In such cases, diffuse pan-uveitis and retinal detachment on unilateral vision loss can occur.

Laboratory diagnosis

ELISA is used for the detection of antibodies to *Toxocara*.

Treatment

The drug of choice is albendazole for five days.

Angiostrongylus cantonensis infection

- This is a **rat lungworm** that can migrate to the central nervous system or eye.
- The worm dies in the brain or eye, causing eosinophilic meningitis, encephalitis, or ocular symptoms.

- The adult worm is about 2 mm long and can sometimes be recovered from CSF.
- Antibodies can be detected by ELISA.
- Management is supportive as the administration of a larvicidal agent may result in exacerbation of symptoms.
- Abdominal angiostrongyliasis is caused by other species of *Angiostrongylus*. In this disease, the worm localises in the arteries of the intestine and lays eggs. An inflammatory reaction to eggs and worms is seen, and an eosinophilic granulomatous mass may be formed. It presents as abdominal pain and a mass in the right lower quadrant of the abdomen. Diagnosis is by biopsy, which may reveal the worm as well as eggs in HPE sections.

Infections caused by *Gnathostoma spinigerum* and *Anisakis* species

- These are parasites of fish and infect humans accidentally by the **ingestion of fish**.
- *Gnathostoma spinigerum* can cause invasive masses in the eye and brain.
- Eosinophilic meningitis and cutaneous migratory swellings may be seen.
- Anisakiasis results in abdominal infections accompanied by eosinophilia.
- Serological diagnosis can be made by detecting antibodies. The larva can be seen in biopsy tissues.

Intestinal Cestode Infections

Adult cestodes residing in the intestine are:

- *Taenia saginata*
- *Taenia solium*
- *Hymenolepis nana* and *H. diminuta*
- *Diphyllobothrium*

Taeniasis

Taeniasis is a parasitic infection in humans that is caused by the tapeworm species namely *Taenia saginata* (beef tapeworm) *and Taenia solium* (pork tapeworm).

Agent—Taenia saginata and T. solium

- *Taenia saginata*
  - *Taenia saginata* causes intestinal taeniasis.
  - This parasite is also known as the beef tapeworm as the infection is acquired by the **ingestion of beef containing encysted larvae**.
  - The disease is more prevalent in cattle-rearing countries such as Central Asia and Central and East Africa.
  - For *Taenia saginata*, humans act as definitive hosts while cattle are the intermediate hosts.
- *T. solium*
  - For *T. solium*, the intermediate hosts are pigs, while humans act as both intermediate and definitive hosts in cysticercosis.
  - *T. solium* is also called the **pork tapeworm** as it is acquired by eating undercooked pork. It can cause intestinal as well as cysticercosis in humans.
  - It is endemic in many developing countries in Southeast Asia including India and South America, Africa, and Southern Europe.

The differences between *Taenia saginata* and *T. solium* are presented in Table 25.15.

Morphology

- **Adult (Fig. 25.33 and 25.34):**
  - The adult worm is a hermaphrodite and can grow up to 5 meters in length and resides in the small intestine.
  - The scolex or head of the worm contains **four cup-like suckers** that help in attachment.
  - The segment containing the **strobila** may be immature, mature, or gravid.

Table 25.15 Differences between *T. saginata* and *T. solium*

Feature	*T. saginata*	*T. solium*
Definitive host	Humans	Intestinal taeniasis—humans Cysticercosis—humans
Intermediate host	Cattle	Intestinal taeniasis—pigs Cysticercosis—humans
Infective form	Larva in undercooked beef (cysticercus bovis)	Intestinal taeniasis—larva in undercooked pork (cysticercus cellulosae), Cysticercosis—egg
Adult worm	5–6 meters	2–4 meters
Scolex	Four suckers, no rostellum, no hooklets	Four suckers, rostellum with two rows of 25–30 hooklets
Number of proglottids per worm	1,000 2,000	<1,000
Number of eggs per gravid segment	80,000	40,000
Eggs	Acid-fast	Non-acid-fast
Clinical presentation	Intestinal taeniasis	Intestinal taeniasis and cysticercosis
Extraintestinal involvement in humans	Not seen	Cysticercosis cellulosae
Diagnosis	Eggs or proglottid in stool	Eggs or proglottid in stool
Alternative name	Beef tapeworm	Pork tapeworm, armed tapeworm

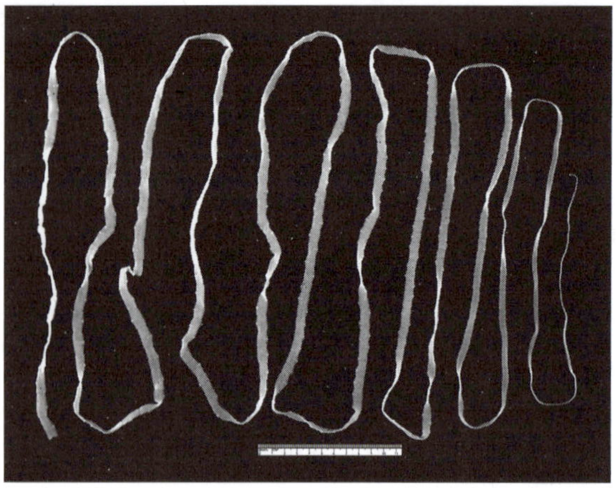

Fig. 25.33 An adult *Taenia saginata* tapeworm (*Source*: CDC, PHIL, Image ID 5260)

- The mature segments have both male and female reproductive organs and are squarish in appearance.
- On the other hand, in immature segments, the breadth is more than the length.
- The gravid segments are longer and filled with eggs.

- **Eggs (Fig. 25.34):** The eggs are round about 30–40 µm in diameter, acid-fast and double-walled, enclosing an embryo with six hooklets.

Life cycle in humans (Fig. 25.34)

- Humans get infected by ingesting the larval stages of the worm, which are found in undercooked beef.
- The larvae are released in the small intestine, where the scolices evaginate and attach to the intestinal mucosa.
- The larva grows into an adult worm and becomes sexually mature in 2–3 months.
- Fertilisation occurs in the same segment, which becomes filled with eggs (gravid segment).
- The eggs are released in stools. Sometimes, the complete gravid segment may be released in feces.

Life cycle in cattle

- The eggs are ingested by cattle and subsequently hatch into oncospheres.
- The oncosphere penetrates the intestinal wall and reaches skeletal muscles, where it changes into a bladder-like sac filled with larvae. This is known as **cysticercus bovis.**

The life cycle of *T. solium* differs from that of *T. saginata* in few aspects:

- Intermediate host—pig or cattle
- Humans can act as the definitive as well as the intermediate hosts for *T. solium*

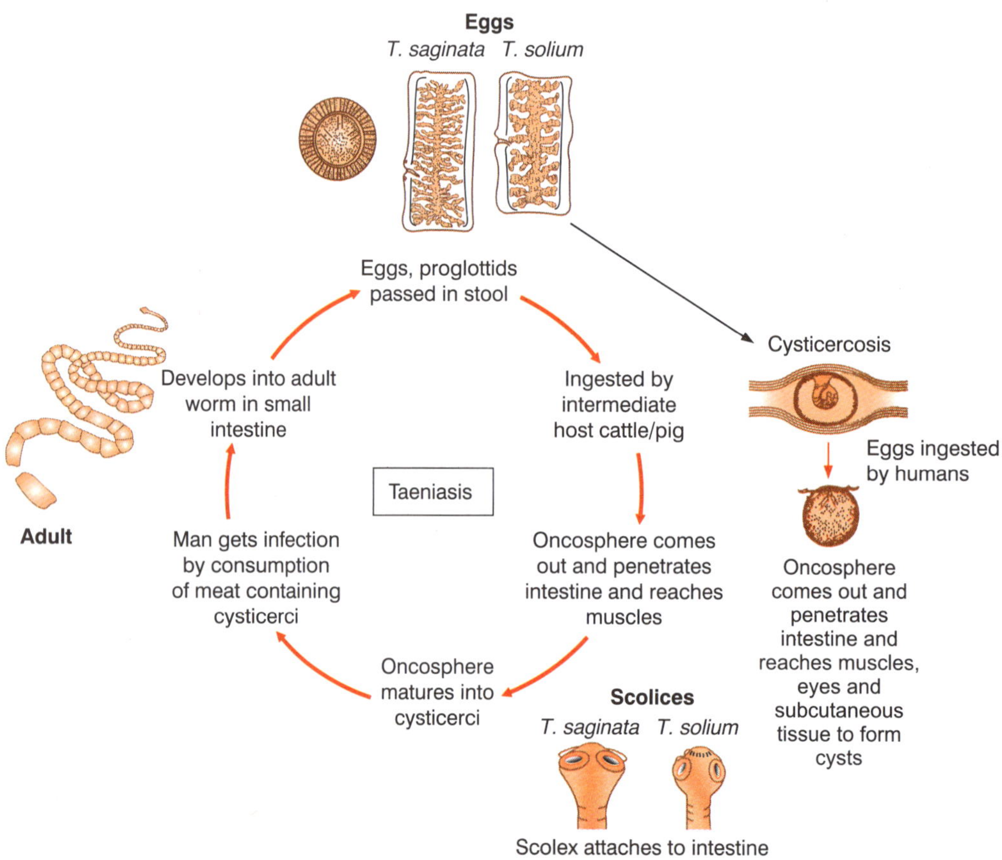

Fig. 25.34 Morphology and life cycle of *Taenia*

Mode of infection

- Ingestion of undercooked beef and pork is the comment route of infection.
- Autoinfection can occur as eggs released in stool can re-enter via **contaminated hands or by reverse peristalsis.** The gravid segments are thrown up into the stomach by reverse peristalsis, and eggs are released there.
- Cysticercosis in humans is caused by the larval forms of *Taenia solium* (see Chapter 24). The cysts show a predilection for the central nervous system (causing neurocysticercosis or NCC), subcutaneous tissue, skeletal muscles, and the eyes.

Clinical features of intestinal taeniasis

- The infection is generally asymptomatic.
- It may present with vague abdominal discomfort, weight loss, diarrhea, and pruritis in the anal region.
- Sometimes, the patient may pass proglottids in his feces, which may be the only sign of infestation.

Clinical features of cysticercosis

The presentation depends on the site of the lesion. It is discussed in Chapter 24.

Laboratory diagnosis

- **Stool examination:** The eggs of both types of *Taenia* are morphologically similar, but only *T. saginata* eggs are acid-fast. They are bile-stained and float in a saturated salt solution. Proglottids may be present in stool sometimes.
- Antigen detection for *Taenia* is available as ELISA. This technique can detect carriers and is very sensitive but fails to differentiate between species.
- Species-specific diagnosis can be made using PCR targeting mitochondrial DNA.

Hymenolepis nana infestation

- *Hymenolepis nana* is also known as the **dwarf tapeworm** as it is the smallest tapeworm to infect humans.
- It is also the most common tapeworm, with worldwide distribution.
- Another species from the same genus—*H. diminuta* (rat tapeworm)—is seen in rats and is rarely seen in humans.
- Infection with *Hymenolepis nana* is reported from South Africa, Southern Europe, Latin America, Middle East, and India.

Morphology (Fig. 25.35)

- **Adult:**
  - The adult worm is about 5–50 mm long, with a round scolex with four cup-shaped suckers and a prominent rostellum with a single circle of small hooklets.

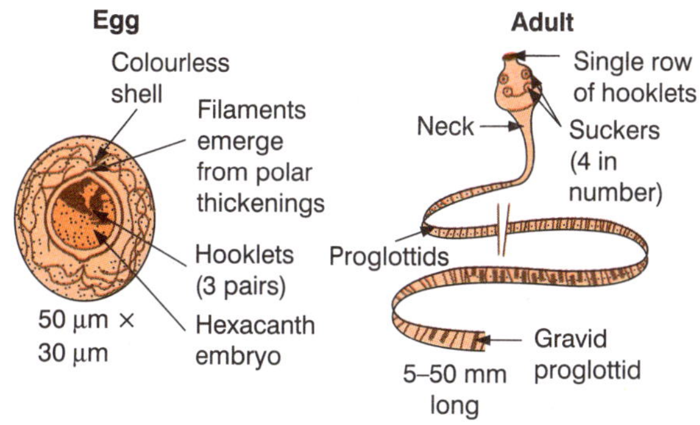

Fig. 25.35 Morphology of *Hymenolepis nana*

  - There are about 200 proglottids that overlap each other.
- **Egg:**
  - The egg is non-bile stained, double-shelled, round to oval, and about 30–50 µm in size.
  - The inner membrane has two polar thickenings, giving rise to 4–8 polar filaments.
  - Yolk granules fill up the space between the two membranes.
  - The egg is infective when laid.
- **Larva:** The larva is a solid structure with a fully developed scolex and is called a cysticercoid.

Life cycle (Fig. 25.36)

- Humans are the sole hosts. Besides humans, rodents and insects can also harbour the worm.
- Human get infected upon ingesting eggs or infected insects.
- Beetles of the genus *Tribolium* breed in cereals and grains. These are ingested by humans and rodents.
- The cysticercoids within the insect remain viable and are infective.

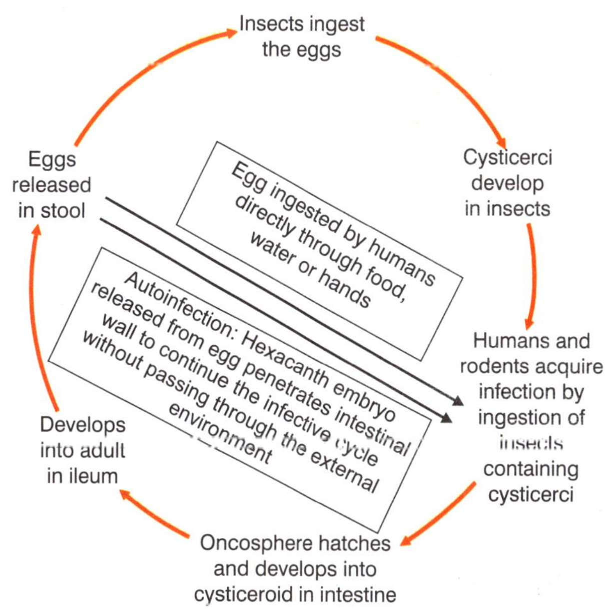

Fig. 25.36 Life cycle of *H. nana*

- Autoinfection can also occur with eggs released in the small intestine.
- Upon ingestion, the oncosphere hatches in the small intestine and penetrates the intestinal villi.
- There, it transforms into a cysticercoid larva in four days.
- The villi ruptures, and the larva develops into an adult worm in the next 1–2 months.
- Eggs laid by sexually mature adults are shed in feces or they may hatch in the small intestine to cause autoinfection.

Clinical features

- In patients with a low worm burden, the infection may remain asymptomatic.
- Non-specific symptoms may be present, such as **persistent abdominal pain, headache, dizziness, anorexia, or diarrhea.**
- Heavy infection may be seen in malnourished and immunosuppressed individuals.

Laboratory diagnosis

A wet mount of stool sample reveals the characteristic *H. nana* eggs.

Treatment

The drug of choice is praziquantel given as a single oral dose.

Hymenolepis diminuta infestation

It is a **zoonotic tapeworm of rodents**, on account of which, it is also called the '**rat tapeworm**'. The clinical presentation, diagnosis, and treatment of *H. diminuta* are similar to those of *H. nana*, but with the following differences:

- Human infection is rare and occurs due to the **ingestion of infected insects**.
- The adult is larger (10–60 cm) with more than 800 proglottids; the scolex has no hooks.
- The eggs are larger (60–85 µm) and without polar filaments; they are not infectious by the person-to-person route.
- The intermediate host of *H. diminuta* is an insect, and the definitive host is a rodent. Rodents acquire eggs by ingesting insects harbouring the 'cysticercoid stage'.
- The development in the intestine is similar to that of *H. nana*.
- The eggs are taken up by the insect host, whereby the oncosphere is released. The oncosphere then penetrates the intestinal wall to reach the hemocoel, where it develops into the cysticercoid larvae.

Diphyllobothriasis

- Diphyllobothriasis is infection with intestinal tapeworms of the family Diphyllobothriidae.
- These worms are acquired by eating raw or undercooked freshwater fish.
- The disease is more common in North America, Scandinavian countries, and Russia than in other countries.
- It is **rarely reported in India**.

Agent—Diphyllobothrium latum

- *D. latum* is the **largest known parasite to infest the human intestine**.
- It is also known as **fish tapeworm**.

Morphology (Fig. 24.37)

- **Adult:**
  - The adult worm lives in the small intestine.
  - It can grow up to 15 meters in length and may contain up to 3,000 segments.
  - The head has **two longitudinal grooves called the bothria**. It uses these grooves to attach to the intestinal mucosa.
- **Eggs:**
  - These are oval, measuring about 60 x 40 µm in size, and **operculated with a knob at one end**.
  - Freshly-passed eggs are unembryonated. Later, they develop a hexacanth oncosphere lined by a ciliated epithelium.
- **Larva**: The first stage (L1) is called the **coracidium**, the second stage is **procercoid**, and the third stage larva is known as the **plerocercoid larva.**

Life cycle (Fig. 25.37)

Definitive host: Human being

Intermediate hosts:

i. Fresh or saltwater *Cyclops* or *Diaptomus* are the first intermediate hosts
ii. Fish are the second intermediate hosts

- Mode of infection—eating undercooked fish containing plerocercoid larvae.
- The larva hatches, develops into an adult and attaches to the intestine.
- Sexual maturity is reached in four weeks, and eggs are released daily in stools.
- Eggs, upon reaching water, take about 10 days to become embryonated and release the ciliated larva (L1) in water through the operculum in the eggs.
- The coracidium has to be ingested by *Cyclops* within 12 hours, failing which, it dies.
- In *Cyclops*, the cilia are lost, and the larva burrows through the intestine to reach the body cavity. In the next 2–3 weeks, it develops into a procercoid, which is infective to fishes.
- The fish consume the *Cyclops* and release the procercoid, which migrates to the muscles and fat of the fish.

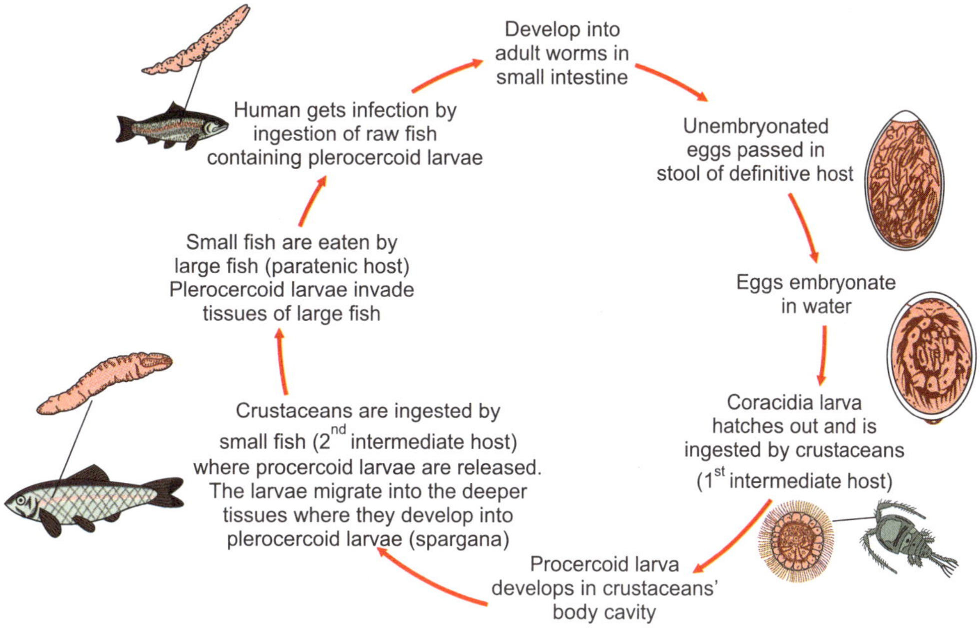

Fig. 25.37 Morphology and life cycle of *D. latum*

- The L2 larva develops into a plerocercoid larva, which is infective to humans.

Clinical features

- Although most infections are asymptomatic, the symptoms can range from minor complaints such as abdominal pain, vomiting, weakness, and weight loss to acute abdominal presentation.
- The adult worm causes disturbances in vitamin B12 metabolism, thereby reducing its absorption.

Laboratory diagnosis

- Wet mounts of stool show operculated, bile-stained eggs. In some cases, proglottids may be seen in stools. Proglottids are released in chains and not singly; they are wide rather than long.
- Eosinophilia and features of megaloblastic anemia may be seen.

Treatment

Praziquantel is the drug of choice. Alternatively, niclosamide can be given.

Intestinal Trematode Infestation

- Trematodes are hermaphrodite, leaf-like, flat and unsegmented worms varying in length from 1 mm to several cm. Females are longer and thinner than males. Adults words reside in the intestine, liver, lung, or blood vessels.
- Their life cycles are complex, often requiring **two intermediate hosts**.
- The primary host is a vertebrate, while the intermediate host, in which asexual reproduction occurs, is usually a **mollusc**.
- Most infections caused by this group are zoonotic and foodborne and have a restricted geographic distribution.
- This could be due to restricted host species or varying food habits.
- These are seen predominantly in Southeast Asia, Korea, China, Thailand, Vietnam, the Philippines, Indonesia, and India.
- The trematodes inhabiting the intestine are listed in the box below.
- All intestinal flukes reside in the small intestine in humans, except *Gastrodiscoides*, which inhabits the large intestine.
- *Heterophyes* is the smallest, whereas *Fasciolopsis buski* is the largest trematode to infect humans.
- Various types of freshwater or brackish water fish, freshwater snails, oysters, amphibians, snakes, aquatic insects, and aquatic plants act as a source of infection in humans.
- Reservoir hosts are different species of mammals or birds.

Intestinal trematodes

- *Fasciolopsis buski*
- *Echinostoma ilocanum*
- *Heterophyes heterophyes*
- *Metagonimus yokogawai*
- *Gastrodiscoides hominis*

Fasciolopsiasis

The intestinal fluke *Fasciolopsis buski*, which causes fasciolopsiasis, is the largest intestinal fluke of humans.

Agent—*Fasciolopsis buski*

- It is also known as the **giant intestinal fluke**.
- It has been reported from India (North-Eastern states, Uttar Pradesh, Bihar, Odisha and Madhya Pradesh), Pakistan, Bangladesh, China, Thailand, Malaysia, and Indonesia.

Morphology (Fig. 25.38)

- **Adult:**
  - It is dark red, fleshy, leaf-shaped, 2–8 cm long and 1–2 cm wide.
  - The oral sucker is smaller than the ventral sucker, and no cephalic cone is present.
- **Egg:**
  - The eggs are large—about 130 μm in length—elliptical and yellowish-brown with a clear thin shell.
  - They are operculated at the top and unembryonated.

Life cycle (Fig. 25.39)

- Humans are infected by consuming **uncooked aquatic plants** such as water chestnuts, lotus stem, or seeds that may be infested with encysted metacercaria.
- The metacercariae excyst in the duodenum and attach to the intestinal mucosa, where they develop into adult worms. This takes around three months.
- The adult matures, and eggs containing miracidium larvae are released in feces.
- The miracidium larva develops in the egg. After 3–6 weeks, it is released from the egg and penetrates a **snail (*Segmentina* species),** in which it multiplies asexually to develop into 2–3 generations of redia, and finally, cercariae.
- These cercariae are released from the snail after 4–7 weeks, whereafter they encyst as metacercariae on aquatic plants.

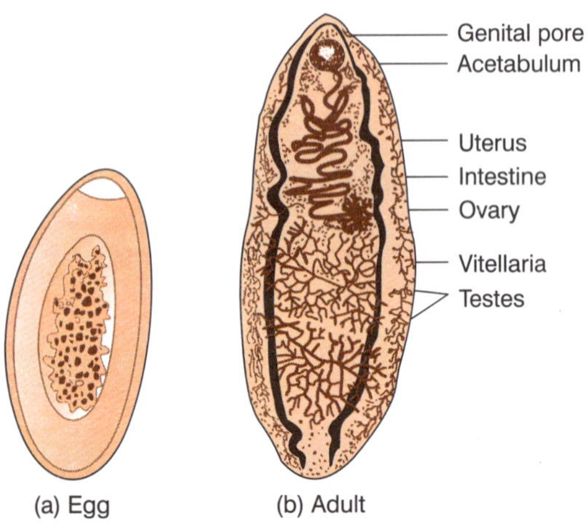

Fig. 25.38 Morphology of *Fasciolopsis buski: (a)* egg and (b) adult

Clinical features

- Infection is generally asymptomatic.
- In heavy parasitemia, abdominal pain, diarrhea, or intestinal obstruction may be present.
- Complications include ascites, anemia, and general edema. These occur due to toxic or allergic worm metabolites.
- Chronic heavy infections may cause chronic diarrhea leading to malabsorption or vitamin B12 deficiency.

Other uncommon intestinal trematode infections

- The features of some other intestinal trematodes of importance are presented in Table 25.16.
- These infections are usually light and are usually either asymptomatic or those presenting with minor problems.
- Some patients may present with abdominal pain or diarrhea due to intestinal inflammation and mild ulceration.

Intestinal schistosomiasis

- Schistosomiasis is an acute and chronic parasitic disease caused by blood flukes (trematode worms) of the genus *Schistosoma* (details in Chapter 21).
- *Schistosoma mansoni* and *S. japonicum* deposit eggs in the intestinal mucosa.
- Diarrhea and dysentery are the presenting complaints at this stage.

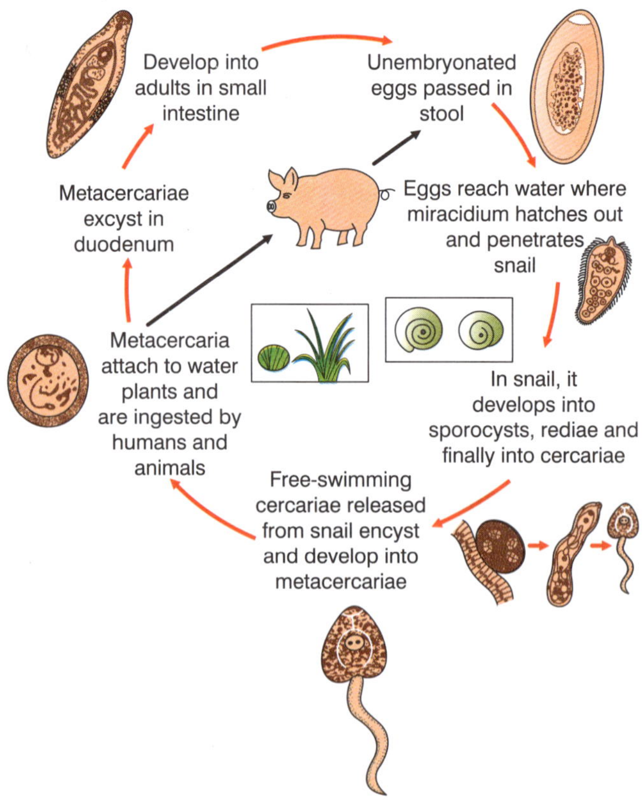

Fig. 25.39 Life cycle of *Fasciolopsis buski*

Table 25.16 Rare intestinal trematodes

Parasite	Geographical distribution	Infectious agent	Vertebrate host	Adult worm and egg	Habitat in humans
Watsonius watsoni	West Africa	Water plants	Monkeys	Adult—8 x 4 mm Eggs—operculated, 125 x 75 μm	Small intestine
Echinostoma ilocanum	Far East, including Southeast Asia,	Molluscs, fish and tadpoles	Dogs and rats	Adults—10 × 2 mm Eggs—operculum not clearly seen; 100 × 60 μm	Small intestine
Heterophyes heterophyes, Metagonimus yokogawai	China, India, Japan, Korea, Egypt, and Sudan	Freshwater fish	Fish-eating mammals	Adults—minute; 5 × 1 mm Eggs—operculum not clearly seen; 30 × 15 μm	Small intestine (jejunum and upper ileum)
Gastrodiscoides hominis	India, China, and Vietnam	Freshwater fish	Pigs, deer, mice, and rats	Adults—5 × 10 mm Eggs—pyriform; operculum not clearly seen 60 × 150 μm	Large intestine (caecum and ascending colon)

- Colorectal and liver carcinoma have been reported to result from *S. japonicum* infection.

Diagnosis of infections due to intestinal trematodes

Microscopy of stool may reveal eggs, which are difficult to differentiate except on the basis of their size.

- Eggs are operculated except those of *Schistosoma* and do not float in a saturated salt solution.
- The eggs of *Heterophyes* and *Metagonimus* are minute and may be missed if not examined carefully.
- The eggs of *S. haematobium* can be detected in urine or bladder biopsy.
- The eggs of *Schistosoma* infecting the intestines can be detected in stool samples and intestinal biopsy.
- All eggs except those of *Schistosoma haematobium* are acid-fast.

Treatment of intestinal trematode infections

- **Drug of choice:** Praziquantel is used to treat all trematode infections.
- **Alternative therapy:** Niclosamide can also be used.

Control of intestinal parasitic infections

- Control measures like health education, improved sanitation, and avoidance of the use of human feces as fertiliser are critical.
- A programme of mass treatment in highly endemic areas has been suggested, but this may not be economically feasible. Furthermore, eggs can persist in contaminated soil for three years or more.
- Certainly, improved personal hygiene among people who handle food is an important aspect of control.
- Water vegetables should be adequately washed, preferably in hot water.
- Preventing contamination of ponds and other waters with pig or human excreta is important.
- Night soil should be sterilised before it is used as a fertiliser.
- Mollusc and snail populations should be controlled by adding chlorine in appropriate dosage to the habitat.

Questions

Case review

A 16-year-old boy ate *pani-puri* from a roadside vendor. Three days later, he developed abdominal cramps with mild fever, which was followed by increased frequency of stools with blood and mucus. He visited the hospital, where a stool sample was sent for microscopic examination and culture. The sample showed plenty of pus cells. *Shigella* dysentery type 1 was isolated in culture.

i. Describe the pathogenesis of *Shigella* dysentery infection.
ii. What are the possible complications of this infection?
iii. What empiric antibiotic is suggested to treat a case of acute shigellosis in an infant?

Short answer questions

1. How do the mechanisms of toxin-mediated gastrointestinal infections by *Escherichia coli* differ?
2. Describe the epidemiology of cholera.
3. Discuss the sequence of events that occur following the entry of quadrinucleate cysts of *Entamoeba histolytica* into the gut.
4. List the common viruses causing gastroenteritis in children and mention the methods used to diagnose these infections in the laboratory.
5. Mention the genesis and type of anemia caused by hookworm.

MCQs

1. **Gastroenteritis refers to:**
   a) Inflammation of the intestines due to bacterial invasion or toxins or viral infection
   b) Inflammation of the intestines due to bacterial invasion only
   c) Toxin-mediated damage to the intestines
   d) Passage of stool with blood and mucus

2. **Food poisoning refers to:**
   a) Acute diarrhea with or without vomiting
   b) Increased frequency of bowel movement or volume of stool
   c) Diarrhea with blood and mucous in stool
   d) Toxin-mediated diarrhea or vomiting
3. **The most common non-typhoidal *Salmonella* associated with gastroenteritis is:**
   a) *Salmonella typhimurium* b) *Salmonella enteritidis*
   c) *Salmonella heidelberg* d) *Salmonella agona*
4. **Salmonellae can cause all the following clinical syndromes in humans except:**
   a) Enteric fever b) Encephalitis
   c) Gastroenteritis d) Local suppurative lesions
5. **Reiter's syndrome may be seen in infection with the following agents except:**
   a) *Salmonella* b) *Shigella*
   c) *Yersinia enterocolitica* d) *Klebsiella pneumoniae*
6. **Colonisation factor antigen (CFA) is seen in:**
   a) Enteropathogenic *E. coli*
   b) Enteroinvasive *E. coli*
   c) Enterotoxigenic *E. coli*
   d) Enteroaggregative *E. coli*
7. **The following statements about enterotoxins of *E. coli* are true except:**
   a) Three distinct types of enterotoxins are present
   b) Heat-stable toxin (ST) is also known as Shiga-like toxin (SLT)
   c) Heat-labile toxin resembles cholera toxin
   d) Heat-stable toxin acts through cGMP in the intestine
8. **'Traveller's diarrhea' is caused by:**
   a) *EPEC* b) *ETEC*
   c) *EIEC* d) *EHEC*
9. **Stool culture from a patient with dysentery grew *E. coli* that did not ferment sorbitol. The organism may be:**
   a) Enterotoxigenic *E. coli* b) Enterohemorrhagic *E. coli*
   c) Enteroinvasive *E. coli* d) Enteroaggregative *E. coli*
10. ***Shigella* are usually lactose non-fermenters. Which *Shigella* species ferments lactose?**
    a) *Shigella sonnei* b) *Shigella flexneri*
    c) *Shigella boydii* d) *Shigella dysenteriae*
11. **Mannitol fermentation is a property used for species identification of *Shigella*. Which *Shigella* is a mannitol non-fermenter?**
    a) *Shigella sonnei* b) *Shigella flexneri*
    c) *Shigella boydii* d) *Shigella dysenteriae*
12. **The infective dose of *Shigella* is:**
    a) 10–100 bacilli b) 10^3–10^5 bacilli
    c) 10^5–10^7 bacilli d) 10^7–10^9 bacilli
13. **Which of the following statement is not true regarding *Shigella*?**
    a) The mildest form of dysentery is caused by *Shigella flexneri*
    b) The most common shigellosis in developed countries is due to *Shigella sonnei*
    c) *Shigella dysenteriae* toxin has enterotoxic, neurotoxic, and cytotoxic actions
    d) Invasive property in *Shigella* is attributed to the virulence marker antigens
14. **The culture medium that is not used for the isolation of *Shigella* species is:**
    a) MacConkey agar
    b) Tellurite blood agar
    c) Deoxycholate citrate agar (DCA)
    d) Xylose lysine deoxycholate agar (XLD)
15. ***Vibrio cholerae* was first isolated by:**
    a) Filippo Pacini b) Robert Koch
    c) Louis Pasteur d) William H. Ewing
16. **The appearance of vibrios in stained smears of mucous flakes from cholera stool resembles:**
    a) Swarms of gnats b) School of fish
    c) Fish in a stream d) Thumb print
17. **Which of the following statements is not true regarding the El Tor biotypes of *Vibrio cholerae*?**
    a) Sensitive to polymyxin B
    b) Susceptible to El Tor phage 5
    c) Agglutinates chick erythrocytes
    d) Gives a positive result for the Voges–Proskauer test
18. **The false statement about vibrios is:**
    a) El Tor vibrios survive longer than classical vibrios
    b) Vibrios do not survive long in grossly contaminated waters
    c) *Vibrio cholerae* can survive at high salt concentrations
    d) Sporadic outbreaks of cholera have been caused by O-139
19. **The seventh pandemic of cholera was caused by:**
    a) Classical vibrios b) El Tor vibrios
    c) O-59 vibrios d) O-139 vibrios
20. **All the following are plating media used for *Vibrio cholera* except:**
    a) Cary Blair medium
    b) MacConkey agar
    c) Blood agar
    d) Thiosulphate citrate bile salt agar
21. **Which property of vibrios is shown by cholera red reaction?**
    a) Nitrate reduction b) Lactose fermentation
    c) Indole production d) Both (a) and (c)
22. **The first-line drug for cholera used during pregnancy is:**
    a) Ampicillin b) Doxycycline
    c) Azithromycin d) Ceftriaxone
23. **A 12-year-old girl presented with acute watery diarrhea. Hanging drop showed typical darting motility. What is the likely causative organism in this case?**
    a) Rotavirus b) *V. cholerae*
    c) *Shigella dysenteriae* d) *C. difficile*
24. **All the following are characteristic features of the classical biotype of *V. cholerae* except:**
    a) Positive hemagglutination with chick RBCs
    b) Positive Voges–Proskauer test
    c) Sensitivity to phage IV
    d) Susceptibility to polymyxin-B
25. **Cholera red reaction occurs due to:**
    a) Formation of nitroso indole
    b) Agglutination with chick RBCs
    c) Action of oxidase enzyme
    d) Sensitivity to phage V
26. **The true statement about *Vibrio parahaemolyticus* is:**
    a) It produces polar flagella only in liquid medium
    b) It does not grow in high salt concentrations
    c) It forms yellow colonies on TCBS
    d) Kanagawa phenomenon is due to the heat-stable enterotoxin
27. **Septicemia with high mortality is caused by:**
    a) *Vibrio parahaemolyticus* b) *Vibrio* vulnificus
    c) *Vibrio alginolyticus* d) *Vibrio* mimicus
28. **The food item involved in *Vibrio parahaemolyticus* food poisoning is:**
    a) Eggs b) Shellfish
    c) Raw vegetable d) Canned food
29. ***Helicobacter pylori* is:**
    a) Non-flagellated b) Amphitrichous
    c) Lophotrichous d) Peritrichous
30. **All the following are virulence factors of *H. pylori* except:**
    a) Vacuolating cytotoxin A
    b) Cytotoxin-associated gene A
    c) Urease enzyme
    d) Capsule
31. **Which of the following is true about *Campylobacter*?**
    a) Non-motile, gram-negative bacilli
    b) Pathogenic species are psychrophilic
    c) Anaerobic growth requirements
    d) Typically oxidase-positive

32. The *Campylobacter* species that is not thermophilic is:
a) *C. sputorum* b) *C. lari*
c) *C. jejuni* d) *C. coli*

33. Which *Campylobacter* causes extraintestinal infection?
a) *C. jejuni* b) *C. sputorum*
c) *C. fetus* d) *C. coli*

34. Which of the following medium is not suitable for *C. jejuni*?
a) Cary Blair medium b) Butzler's medium
c) Skirrow's medium d) Fletcher's medium

35. A 53-year-old man who was on long-term antibiotics developed diarrhea. His stool examination showed the presence of leukocytes and was positive for toxin. What is the most likely etiological agent?
a) *Campylobacter jejuni* b) *Candida albicans*
c) *Clostridium difficile* d) Enterotoxigenic *E. coli*

36. The antibiotic of choice to treat antibiotic-associated diarrhea is:
a) Ampicillin b) Vancomycin
c) Ciprofloxacin d) Cephalosporin

37. A 7-year-old girl develops vomiting and diarrhea six hours after eating sweets. Investigation reveals that the cook who made the sweet from milk had a boil on his finger. The culture from the boil grows *Staphylococcus aureus*. Which of the following virulence factor is associated with this clinical condition?
a) Staphylococcal food poisoning toxin
b) Preformed enterotoxin type A
c) Staphylococcal cytolytic toxin
d) None of the above

38. Which of the following is not true regarding the toxin responsible for the above clinical condition?
a) The toxin is antigenic and can be neutralised by an antitoxin.
b) The toxin is believed to act directly on autonomic system.
c) The toxin can be detected using ELISA.
d) It acts on gastrointestinal mucosa to cause illness.

39. All of the following are toxin-mediated *Staphylococcus aureus* diseases except:
a) Food poisoning
b) Ritter's disease
c) Intracranial thrombophlebitis
d) Toxic shock syndrome

40. Which of the following is not true of *S. aureus* food poisoning?
a) Incubation period is 2–6 hours
b) Sources are meat and milk products
c) Heat-labile enterotoxin is responsible
d) Type A toxin is responsible for most of the cases

41. The selective medium used for the isolation of *Staphylococcus aureus* from feces following food poisoning is:
a) Pike's medium b) Ludlam's medium
c) Cary-Blair medium d) Thayer-Martin medium

42. Which of the following is true of toxic shock syndrome toxin?
a) Produced by *Staphylococcus aureus* strains belonging to bacteriophage II
b) Is a superantigen
c) Causes toxic shock syndrome only in females
d) Anti-TSST antibodies are not protective

43. Cold enrichment is required for the isolation of:
a) *Klebsiella pneumoniae* b) *Proteus mirabilis*
c) *Yersinia enterocolitica* d) *S. enteritidis*

44. A group of hostel inmates develop nausea, vomiting, double vision, and difficulty in swallowing. History reveals the intake of canned tuna sandwiches, salads, and cheese one day before. Which of the following organisms is the most likely cause of this outbreak?
a) *Clostridium perfringens*
b) *Clostridium botulinum*
c) *Salmonella* species
d) *Staphylococcus aureus*

45. A 22-year-old man presents with complaints of acute-onset nausea and vomiting. His last meal was dinner ordered from a Chinese restaurant. What is the most likely etiological agent?
a) *Bacillus cereus*
b) *Clostridium botulinum*
c) *Salmonella* Typhimurium
d) *Staphylococcus aureus*

46. The mode of entry in hookworm infestation is:
a) Ingestion of larva
b) Ingestion of fertilised egg
c) Ingestion of unfertilised egg
d) Penetration of skin by larva

47. The incubation period in *C. perfringens* food poisoning is:
a) 2–3 hours b) 6–12 hours
c) 8–24 hours d) 48–72 hours

48. Which of the following bacterium produces cytotoxin-mediated gastroenteritis?
a) *Salmonella* species b) *Shigella dysenteriae*
c) *Vibrio cholerae* d) *Bacillus cereus*

49. All the following are transport media for cholera stools except:
a) Cary-Blair medium
b) Thioglycolate broth
c) Venkatraman–Ramakrishnan medium
d) Alkaline peptone water

50. An adult female develops mycotoxicosis after consuming contaminated groundnuts. What is the organism likely to be associated with this condition?
a) *Fusarium* spp. b) *Claviceps purpurea*
c) *Penicillium* spp. d) *Aspergillus* spp.

51. Mycetism refers to a condition caused by toxic effects of:
a) *Mycobacterium* infection
b) Ingestion of the food contaminated by fungus
c) Ingestion of food contaminated by fungal toxins
d) All of the above

52. Ergotism is caused by the toxic alkaloids produced by the fungus:
a) *Claviceps purpurea* b) *Aspergillus flavus*
c) *Aspergillus fumigatus* d) *Candida albicans*

53. A five-year-old child presented with nausea, vomiting, and diarrhea. Stool examination revealed viral etiology. Which one of the following is not likely to be the causative agent?
a) Rotavirus b) Calicivirus
c) Norwalk virus d) Echovirus

54. 'Cartwheel' appearance is seen in following virus:
a) Corona b) Rota
c) SARS d) Orbi

55. The virus-caused diarrhea most commonly affecting infants and children is:
a) Rota b) HIV
c) HEV d) HAV

56. Which of the following tests classifies rotaviruses based on the viral RNA migration patterns?
a) Electrophoresis b) Rocket electrophoresis
c) PAGE d) Submarine electrophoresis

57. Any of the following viruses can cause infective diarrhea except:
a) Norwalk virus b) Rotavirus
c) Parvovirus d) Calcivirus

58. A 30-year-old male presents with complaints of vague abdominal discomfort, nausea, anorexia, and weight loss. He gives a history of passing foul-smelling loose to semi-formed stools with increased frequency. Stool microscopy shows leaf-like trophozoites. What is the likely pathogen causing this disease?
a) *Entamoeba histolytica* b) *Giardia lamblia*
c) *Shigella* spp. d) *Cryptosporidium parvum*

59. The trophozoite of *Giardia* can be best described by which of the following statements?
a) It is oval with 2–4 nuclei.
b) It is a pear-shaped, non-flagellated, bilaterally symmetrical structure.
c) It is a pear-shaped, bilaterally symmetrical, flagellated structure with a sucking disc ventrally.
d) It is shapeless with pseudopodia, non-flagellated structure.

60. Which of the following shows a falling leaf motility?
a) *Entamoeba histolytica*
b) *Balantidium coli*
c) *Giardia lamblia*
d) *Toxoplasma gondii*

61. Which of the following parasite can be present in the duodenum?
a) *Entamoeba histolytica*
b) *Balantidium coli*
c) *Giardia lamblia*
d) *Toxoplasma gondii*

62. Which of the following was discovered by Antonie van Leeuwenhoek in the late 1600s when he examined his own stools?
a) *Naegleria* b) *Giardia*
c) *Pneumocystis* d) *Cryptosporidium*

63. A 37-year-old male drank unfiltered water while he was travelling. Two weeks later, he developed diarrhea without blood or pus and weight loss which continued for 5 weeks. What is the likely pathogen causing this disease?
a) *Entamoeba histolytica* b) *Giardia lamblia*
c) *Shigella* spp. d) *Cryptosporidium parvum*

64. A 46-year-old immunocompromised patient came to the ER with a history of severe, debilitating diarrhea with weight loss. Which one of the following is the most likely causative organism?
a) *Entamoeba histolytica* b) *Giardia lamblia*
c) *Shigella* spp. d) *Cryptosporidium parvum*

65. What is the most effective treatment in the above case?
a) Paromomycin b) Furamide
c) Metronidazole d) Pentamidine

66. Which of the following stains is most useful to diagnose this condition?
a) Silver stain b) H&E stain
c) Acid-Fast stain d) Stool wet mount

67. Which of the following is correct regarding *Cryptosporidium*?
a) The infective stage is sporulated thick-walled oocysts
b) The parasite resides in the large intestine
c) The life cycle takes place in two hosts
d) It commonly causes chronic diarrhea in immune-competent persons

68. A 27-year-old man who presented to the emergency has a history of two weeks of abdominal pain, diarrhea, bloody stools, weight loss, and fever. Which of the following is the most likely causative organism?
a) *Entamoeba histolytica* b) *Giardia lamblia*
c) *Salmonella* Typhimurium d) *Cryptosporidium parvum*

69. What is the most effective treatment in this case?
a) Paromomycin b) Furamide
c) Metronidazole d) Omeprazole

70. Which of the following is associated with severe dysentery?
a) *Giardia lamblia* b) *Entamoeba histolytica*
c) *Cryptosporidium parvum* d) *Cyclospora* species

71. Which of the following is the best treatment for asymptomatic amoebiasis?
a) Paromomycin b) Metronidazole
c) Diloxanide furoate d) Pentamidine

72. What is another name for *Trichuris trichiura*?
a) Pinworm b) Hookworm
c) Whipworm d) Roundworm

73. Which helminth infection can be diagnosed using cellotape?
a) Pinworm b) Hookworm
c) Whipworm d) Roundworm

74. Humans acquire roundworm infection by:
a) Penetration of skin by rhabditiform larva
b) Penetration of skin by filiform larva
c) Ingestion of egg containing rhabditiform larva
d) Ingestion of egg containing filiform larva

75. Löffler's syndrome is characterised by:
a) Irritating, non-productive cough
b) Burning substernal discomfort that is aggravated by coughing or deep inspiration
c) Mild fever
d) All of the above

76. Which of the following statements concerning*Ascaris lumbricoides* is correct?
a) *Ascaris lumbricoides* is one of the smallest nematodes.
b) *Ascaris lumbricoides* can cause pneumonia.
c) Both dogs and cats are intermediate hosts of *Ascaris lumbricoides.*
d) *Ascaris lumbricoides* is transmitted by the ingestion of freshly passed eggs.

77. Each of the following statements concerning hookworm infection is correct except:
a) Hookworm infection is transmitted by the ingestion of eggs
b) Hookworm infection is acquired by humans when filariform larvae penetrate the skin
c) Hookworm infection is caused by *Necator americanus*
d) Hookworm infection can be diagnosed by finding the egg in the stool

78. A six-year-old malnourished child presents with nausea, vomiting, and abdominal pain. He has lost his appetite, and his abdomen is distended. Plain abdominal X-rays reveal that he has dilated loops of bowel with multiple air-fluid levels. There is no air in the rectal area. Stool samples confirm a parasitic infection. Which worm is likely to cause this condition?
a) *Ascaris lumbricoides* b) *Necator americanus*
c) *Ancylostoma duodenale* d) *Enterobius vermicularis*

79. A 38-year-old woman presents with complaints of tiredness for the past six months. On examination, she looks pale, and her investigations reveal a microcytic hypochromic anemia with hemoglobin of 7.8 g/dL. Which parasitic infection is most likely still causing her anemia?
a) *Ascaris lumbricoides* b) *Taenia solium*
c) *Ancylostoma duodenale* d) *Enterobius vermicularis*

80. A 50-year-old farmer felt itching in his right foot but it resolved spontaneously in a day. After 7 days, he had difficulty breathing along with wheezing. The investigation showed hemoglobin of 8 g/dL with eosinophils 2.3%. Chest X-ray showed bilateral pulmonary infiltrates. Bronchoalveolar lavage showed pulmonary eosinophilia and filariform larvae on sputum examination. Which parasitic infection is most likely the causative agent?
a) *Ascaris lumbricoides* b) *Taenia solium*
c) *Necator americanus* d) *Enterobius vermicularis*

81. What would be the drug of choice for the treatment of this patient?
a) Single-dose albendazole b) Single-dose metronidazole
c) Clindamycin for 5 days d) Ivermectin for 3 days

82. A 11-month-old female is seen for anal pruritis that occurs especially at night. A cellophane tape test is administered and characteristic ova are seen upon microscopy. What is the most likely etiology?
a) *Ascaris* infection b) *Enterobius* infection
c) Diaper rash d) Hookworm infection

83. **Which of the following is the test of choice to confirm the suspected diagnosis for this patient?**
    a) Stool microscopy with sample collected in a container
    b) Culture of stool on MacConkey's agar
    c) Stool microscopy with sample collection using NIH swab
    d) Molecular diagnostic technique
84. **What is the most appropriate step for this patient?**
    a) Treat with a topical steroid and antifungal
    b) Treat with albendazole
    c) Treat with pyrantel pamoate
    d) No treatment is needed; it is a self-limiting illness
85. **A 10-year-old boy living in a slum presents with complaints of nausea, vomiting, and diarrhea for the last two days. On examination, he is pale and his abdomen is tender. Further examination reveals a prolapsed rectum. Which parasite infestation is likely to be associated with this condition?**
    a) *Ascaris lumbricoides* b) *Trichuris trichiura*
    c) *Cryptosporidium parvum* d) *Isospora belli*
86. **The stool examination in the above case is likely to reveal eggs which have the characteristic morphology. The main features are:**
    a) The eggs do not float in a saturated solution of common salt.
    b) The eggs are bile-stained, brown, and barrel-shaped with characteristic bipolar mucus plugs.
    c) The eggs are typically planoconvex in shape and colourless, and thin walled.
    d) They are oval and non-bile stained with a thin, clear hyaline shell containing a segmented ovum.
87. **A 14-year-old girl with leukemia presents with complaints of diarrhea and abdominal pain for the past few days. On examination, migrating, linear, itchy lesions are seen on her thighs. Stool microscopy reveals rhabditiform larvae. Which of the following organism has most likely caused the infection in this patient?**
    a) *Strongyloides stercoralis* b) *Ancylostoma duodenale*
    c) *Necator americanus* d) *Ascaris lumbricoides*
88. **Humans acquire *Strongyloides* infection by:**
    a) Ingestion of larva
    b) Ingestion of fertilised egg
    c) Penetration of skin by filiform larva
    d) Ingestion of either eggs and larva
89. **Parthenogenesis refers to:**
    a) Release of larvae from fertilised eggs in the body
    b) Production of eggs by females without fertilisation by males
    c) Production of both larva and eggs by females
    d) Migration of larva back into the body to cause autoinfection
90. **The drug of choice to treat strongyloidiasis is:**
    a) Ivermectin b) Metronidazole
    c) Pyrantel pamoate d) Praziquantel
91. **Which of the following statements is not correct regarding strongyloidiasis:**
    a) Hyper-infection syndrome and disseminated strongyloidiasis is seen in immunosuppressed or HIV patients.
    b) Patients may complain of rashes and rapidly migrating, linear, itchy lesions on the thighs, buttocks, or feet.
    c) It is potentially fatal if untreated.
    d) It can be successfully treated with metronidazole for 5 days.
92. **Which statement describes *Strongyloides stercoralis* correctly?**
    a) The adult female burrows into the intestinal mucosa and lays eggs that immediately hatch into larvae.
    b) The rhabditiform larva transforms into filiform larvae and penetrates the intestinal mucosa or perianal skin, resulting in autoinfection.
    c) It exists in the free-living state and as a parasite.
    d) All of the above.
93. **The intermediate host for *Taenia saginata* is:**
    a) Pig b) Human
    c) Cattle d) Crab
94. **Which of the following statement is not correct regarding the adult worm of *T. saginata*?**
    a) The adult worm can grow up to 5 meters in length.
    b) It resides in the small intestine.
    c) The scolex or head of the worm contains four cup-like suckers and rostellum with two rows of 25–30 hooklets.
    d) The mature segments have both male and female reproductive organs and are squarish in appearance.
95. **Which of the following differentiates the life cycle of *T. solium* from that of *T. saginata*?**
    a) The intermediate host is a pig and hence, the infections acquired by eating undercooked pork in *T. saginata*.
    b) Autoinfection can occur with *T. saginata*.
    c) Cysticercosis in humans is caused by the larval forms of *T. saginata*.
    d) Humans can act as the definitive as well as the intermediate host for *T. solium*.
96. ***Hymenolepis nana* is also known as:**
    a) Dwarf tapeworm b) Fish tapeworm
    c) Pork tapeworm d) Beef tapeworm
97. **Which statement describes the egg of *H. nana* accurately?**
    a) Non-bile stained, double shelled, round to oval with an inner membrane having two polar thickenings, giving rise to 4–8 polar filaments
    b) Round, acid-fast and double-walled, enclosing an embryo with six hooklets
    c) Bile-stained, brown, barrel-shaped with characteristic bipolar mucus plugs
    d) Oval, non-bile stained with thin a clear hyaline shell containing a segmented ovum
98. **A 20-year-old male from Sweden presents with fatigue and tiredness. He was diagnosed with anemia an year back but did not respond to iron therapy. The investigations now reveal features of megaloblastic anemia and he is now diagnosed with B12 deficiency. Which parasite is most likely to be the causative agent?**
    a) *H. nana* b) *D. latum*
    c) *T. saginata* d) *T. solium*
99. **The infection is associated with the consumption of:**
    a) Undercooked pork b) Raw vegetables
    c) Raw fish d) Raw watercress
100. **Humans acquire the infection by the ingestion of:**
    a) Plerocercoid larva in fish
    b) Procercoid larva in cyclops
    c) Miracidium larva in water
    d) Adult worm
101. **What is the drug of choice to treat *D. latum* infections?**
    a) Albendazole b) Praziquantel
    c) Metronidazole d) Any of the above
102. **A 52-year-old male presents to the with sudden onset of high fever, myalgia, nausea, abdominal cramps, and anorexia. On examination, his abdomen is tender with mild hepatomegaly. Investigations reveal high eosinophils. He gives a history of eating water chestnuts and raw fish occasionally. Which parasite is likely have caused this condition?**
    a) *Fasciolopsis buski* b) *Schistosoma mansoni*
    c) *T. saginata* d) *T. solium*
103. **The infective stage of *F. buski* in humans is:**
    a) Metacercaria larva b) Cercariae larva
    c) Miracidium larva d) Embryonated eggs
104. **A 58-year-old female presented with chronic watery diarrhea. She was seropositive for HIV, and her stool microscopy showed acid-fast oocysts measuring 5 microns in diameter. What is the likely etiological agent?**

a) *Cryptosporidium parvum*
b) *Isospora belli*
c) *Entamoeba histolytica*
d) *Balantidium coli*

105. A 28-year-old male was brought to the hospital in a state of altered mental status. He soon progressed to coma and respiratory paralysis. Family members informed the doctor that he had consumed preserved meat the previous day. A smear from the suspected meat sample showed gram-positive, sporing bacilli. What is the most likely agent responsible for this condition?
a) Enterotoxigenic *E. coli* b) *Clostridium botulinum*
c) *Campylobacter jejuni* d) *Yersinia enterocolitica*

106. A 53-year-old man who was on long-term antibiotics developed diarrhea. His stool examination showed the presence of leukocytes and was positive for toxin. What is the most likely etiological agent?
a) *Campylobacter jejuni* b) *Candida albicans*
c) *Clostridium difficile* d) Enterotoxigenic *E. coli*

107. An eight-year-old school-going boy presented with diarrhea, which he was said to have had for the past three days. His stools were greasy, bulky, and foul-smelling. Fat globules were observed on stool examination. He gave a history of consuming unfiltered water when he was out for a camp. What is the likely causative agent?
a) *Clostridium botulinum* b) *Giardia lamblia*
c) *Entamoeba histolytica* d) *Vibrio cholerae*

108. Oval terminal spores are typical of:
a) *C. perfringens* b) *C. difficile*
c) *C. tetani* d) *C. botulinum*

109. A 40-year-old traveller consumed semi-cooked egg for dinner. Early next morning, he developed severe vomiting and dysentery. His stool sample grew NLF colonies on MacConkey medium and showed acid butt/alkaline slant with plenty of H_2S and gas on TSI medium. What is the likely causative agent involved?
a) *Salmonella* Typhi b) *Shigella sonnei*
c) *Salmonella* Typhimurium d) *Salmonella newcastle*

110. A 22-year-old man develops loose motions with blood after eating at a roadside eatery. What is the main pathogenic mechanism involved in this clinical presentation?
a) Outpouring of electrolytes and fluid into the intestinal lumen causing diarrhea
b) Disruption of the intestinal epithelium, leaving raw, unprotected mucosa due to cytotoxins released by bacteria
c) Ingestion of preformed toxin causing damage to the intestinal mucosa
d) A combination of all of the above

111. A 11-year-old girl presents to the casualty with complaints of bloody diarrhea, cramping, and fever for the past two days. She is anuric for past 12 hours. On examination, she is confused and pale, and has pedal edema and small bruises on her shins and arms. She gives a history of eating a hamburger. What is the most likely agent causing this condition?
a) Enteropathogenic *E. coli* (EPEC)
b) Enterotoxigenic *E. coli* (ETEC)
c) Enterohemorrhagic *E. coli* (EHEC)
d) Enteroaggregative *E. coli* (EAEC)

Answers: 1:a, 2:d, 3:a, 4:b, 5:d, 6:c, 7:b, 8:b, 9:b, 10:a, 11:d, 12:a, 13:a, 14:b, 15:b, 16:c, 17:a, 18:c, 19:b, 20:a, 21:d, 22:c, 23:b, 24:a, 25:a, 26:a, 27:b, 28:c, 29:c, 30:d, 31:d, 32:a, 33:c, 34:d, 35:c, 36:b, 37:b, 38:d, 39:c, 40:c, 41:b, 42:b, 43:c, 44:a, 45:a, 46:d, 47:c, 48:b, 49:b, 50:d, 51:b, 52:a, 53:d, 54:b, 55:a, 56:c, 57:c, 58:b, 59:c, 60:c, 61:c, 62:b, 63:b, 64:d, 65:a, 66:c, 67:a, 68:a, 69:c, 70:b, 71:c, 72:c, 73:a, 74:c, 75:d, 76:b, 77:a, 78:a, 79:c, 80:c, 81:b, 82:b, 83:c, 84:b, 85:b, 86:b, 87:a, 88:c, 89:b, 90:a, 91:d, 92:d, 93:c, 94:c, 95:d, 96:a, 97:a, 98:b, 99:c, 100:a, 101:b, 102:a, 103:a, 104:a, 105:b, 106:c, 107:b, 108:b, 109:c, 110:b, 111:c

CHAPTER 26

Hepatobiliary Infections

(MI3.7 and 3.8)

Chapter Navigation

- Viral hepatitis
- Enteric viral hepatitis
  - HAV
  - HEV
- Bloodborne or serum hepatitis
  - HBV
  - HCV
  - HDV
  - HGV
  - Yellow fever
- Parasitic infections of the liver
  - Hydatid cyst—*Echinococcus granulosus*
  - Alveolar echinococcosis—*Echinococcus multilocularis*
  - Fascioliasis—*Fasciola hepatica*
  - Clonorchiasis—*Clonorchis sinensis*
  - Opisthorchiasis—*Opisthorchis*
- Liver abscess
- Biliary tract infections

Learning objectives

- Discuss the etiopathogenesis and viral markers in the evolution of viral hepatitis
- Discuss the modalities in the diagnosis and prevention of viral hepatitis

Many systemic infectious pathogens including hepatotropic and non-hepatotropic microorganisms have the ability to infect the liver through the systemic as well as portal circulation.

Table 26.1 lists the microorganisms that cause hepatobiliary infections.

- Many microorganisms are not primary pathogens of the liver but may affect the organ as a part of the systemic infection, with the microorganism passing through the portal circulation.
- Infections primarily affecting the liver and the biliary tract are as follows:
  - **Hepatitis (infection of the liver parenchyma)**
  - **Liver abscess (localised collection of pus)**
  - **Biliary tract infections**

HEPATITIS (INFECTION OF THE LIVER PARENCHYMA)

CLINICAL FEATURES

- The patient presents with a loss of appetite, mild-to-severe pain in the right upper quadrant, and vomiting in the early phase of the infection.
- This is followed by jaundice with yellow discolouration of the conjunctiva, skin (may not be appreciable in the darker skin tones), and urine.
- Arthralgia, fatigue, and malaise may be present.
- Invariably, low-grade fever is also present. In leptospirosis, the patient may have rashes in addition to fever.

LABORATORY DIAGNOSIS

Liver Function Tests

- Liver enzymes are almost always altered.
- Serum aminotransferases and aspartate aminotransferases are elevated.

Table 26.1 Primarily hepatotropic microorganisms

Microorganism	Examples of some clinical conditions
Bacteria	
Leptospira	Weil's disease (leptospirosis)
Klebsiella and *Enterobacter* spp.	Cholecystitis
Salmonella	
Brucella	Hepatic involvement as part of pathogenesis
Proteus and *Enterococcus*	
Q fever	
Salmonella Typhi	
Borrelia (Lyme disease)	
Viruses	
Hepatitis virus (hepatitis A, B, C, D, E and G viruses)	Viral hepatitis
Yellow fever virus	Yellow fever
Lassa fever virus	Hepatic involvement as part of pathogenesis and may present as hepatitis
Marburg virus	
Epstein Barr virus	
Cytomegalovirus	
Herpes simplex virus	
Varicella-zoster virus	
Measles virus	
Rubella virus	
Coxsackie virus	
Echovirus	
Dengue EBV and CMV	
Parasites	
Plasmodium	Hepatic involvement as part of pathogenesis
Schistosomes	
Entamoeba histolytica	Cystic mass or liver abscess
Clonorchis	Hydatid cyst of the liver
Echinococcus	
Ascaris	Biliary obstruction by parasite can cause features of obstructive jaundice
Fasciola hepatica	
Toxins	
Aflatoxin (*Aspergillus flavus*)	Aflatoxicosis—may lead to hepatocellular malignancy

- Increase in bilirubin levels coincides with the appearance of jaundice.

Microbiological Diagnosis

- Specific etiological diagnosis depends on the cause of hepatitis.
- Viral hepatitis is diagnosed by detecting viral antigens and antibodies by ELISA and immunochromatographic methods and estimating viral load by RT-PCR.
- Bacterial etiology such as *Leptospira* infections can be detected by serology or culture of blood on EMJH (Ellinghausen–McCullough–Johnson–Harris) medium.

MANAGEMENT

- Symptomatic management is the modality commonly recommended for uncomplicated infections.
- Antiviral treatment is prescribed to prevent chronicity and complications.
- Hepatitis due to bacterial causes is treated with appropriate antibiotics.

These causes have been discussed in detail in the chapters that deal with the individual organisms.

VIRAL HEPATITIS

- Many viruses infect the liver as a part of their pathogenesis; however, not all of them are hepatotropic.
- The term 'viral hepatitis' refers to a primary infection of the liver by any one of a heterogeneous group of 'hepatitis viruses' which are primarily hepatotropic—**hepatitis viruses A, B, C, D, E, and G**.
- Hepatitis viruses are **taxonomically unrelated**.
- Except for type B, which is a DNA virus, all the others are RNA viruses.
- The features common to these are their hepatotropism and ability to cause a similar icteric illness, ranging in severity from the inapparent to the fulminant, fatal forms.
- Since all types of hepatitis viruses cause a clinically indistinguishable acute illness, their differentiation is based on their serological and molecular markers.

Types of Hepatitis Viruses

- **Hepatitis A** infections occur sporadically or as epidemics, chiefly affecting children and young adults. It is transmitted through **fecal contamination of water**.
- **Hepatitis B** is a second type of viral hepatitis, **transmitted chiefly by inoculation**. It was originally observed in persons receiving **serum inoculation or blood transfusion**.
- **Hepatitis C** was identified as the cause of many **non-A non-B transfusion-associated hepatitis** cases.
- **Hepatitis D** or **delta virus is a defective** virus which depends on the helper functions of type B virus.
- **Hepatitis E** is yet another type of non-A non-B hepatitis transmitted by the fecal–oral route, prevalent in developing nations.
- **Hepatitis G** (HGV) is a rare cause of hepatic inflammation. It is a transfusion-transmitted agent but does not cause hepatitis. Although chronic infection and viremia have been documented, histologic evidence is rare, and serum aminotransferase levels are usually normal.

ENTERIC VIRAL HEPATITIS

Case 1. *A 30-year-old male presented with the chief complaint of yellowing of his skin and typically white sclera of his eyes. These symptoms were preceded by five days of progressive fatigue, nausea, and flu-like symptoms. He also complained of having produced several tan-coloured bowel movements, and dark urine. He denied any history of blood transfusion, intravenous drug abuse, unprotected sexual intercourse, diarrhea, hematuria, or rashes. Biochemical investigations revealed abnormal liver function tests with high bilirubin. A provisional diagnosis of* ***enteric hepatitis*** *was made, and the patient was further investigated to find the cause.*

i. What can be the etiological agents responsible for such conditions?
ii. How is the diagnosis established?

Enterically-transmitted hepatitis is the most common manifestation of acute hepatitis worldwide. Transmission is by the **fecal–oral route**, and infection is acquired by ingestion. Enteric hepatitis can be caused by one of the following two agents:

- Hepatitis A virus
- Hepatitis E virus

Epidemiology

- Natural infection with HAV is known to occur **only in humans**. Hepatitis A occurs sporadically or as outbreaks, which may be caused by contaminated food, water, or milk.
- Domestic or institutional spread of infection among children is common.
- Person-to-person spread is increased by **poor personal hygiene and overcrowding**.
- The epidemiology of hepatitis A resembles that of poliomyelitis.
- In developing countries, infection is acquired in childhood and by the age of 10. 90 per cent of the population possess the antibody to the virus and are immune.

- In India, type A hepatitis is the most common cause of acute hepatitis in children but is much less frequent in adults.
- In temperate regions, the disease shows an autumn–winter predilection, whereas in the tropics, no seasonal distribution is evident. In India, the disease tends to be associated with **heavy rainfall**.
- The bulk of hepatitis cases in the developing countries is transmitted enterically **through fecal pollution of drinking water** (hence the name **enterically-transmitted non-A non-B** [E-NANB]).
- It often appears as epidemics (hence also called **epidemic NANB**).
- In India, HEV is responsible for the majority of epidemics and sporadic cases of hepatitis in adults.
- Secondary attack rate among household contacts is very low in type E hepatitis—2–3 per cent as compared to 10–20 per cent in HAV infection.

Agent—Hepatitis A virus (HAV)

- Type A hepatitis (infectious hepatitis) is a subacute disease of global distribution, affecting mainly **children and young adults**.
- In 1973, **Stephen Mark Feinstone** and his co-workers demonstrated this virus in the feces of experimentally—infected human volunteers using **immunoelectron microscopy (IEM)**.
- Chimpanzees and marmosets can be infected experimentally.
- HAV can be grown in some human and simian cell cultures and is the only human hepatitis virus that can be cultivated in vitro. It has also been cloned.

Morphology

- HAV measures 27–32 nm and is a **non-enveloped RNA virus** belonging to the Picornavirus family.
- It is also known as '**enterovirus 72**'.
- Because of its unique features, HAV is now recognised as the prototype of a new genus, Hepatovirus. Only one serotype of the virus is known.

Susceptibility

HAV is inactivated by the following agents and methods:

- Heating at 60°C for one hour
- Ether and acid at pH 3
- Boiling for one minute
- 1:4,000 formaldehyde at 37°C for 72 hours
- Chlorine (bleach) 1 ppm for 30 minutes

It is not affected by anionic detergents and survives prolonged storage at a temperature of 4°C or below.

Pathogenesis

- HAV transmission is by the fecal–oral route, and infection is acquired by ingestion.
- Chronic carriers are not seen.
- Virus persistence in nature depends on the continuation of inapparent infections.
- A brief viremia occurs during the pre-icteric phase but ceases with the onset of jaundice.
- Chronic viremia does not occur.
- Parenteral or transplacental infection is not seen.
- The pathogenesis of hepatitis A is presented in Fig. 26.1.

Clinical features

- The incubation period is 15–45 days.
- Most infections are asymptomatic (Fig. 26.2).
- Overt illness is seen in only about 5 per cent of those infected.
- Two stages are seen:
  i. **Prodromal stage:** Onset may be acute or insidious, with fever (100–102°F), malaise, anorexia, nausea, and vomiting.
  ii. **Icteric stages:** Prodromal symptoms usually subside with the onset of jaundice. Dark urine and clay-coloured stool may be noted by the patient in 1–5 days. Yellow discolouration of the sclera and skin is seen when the serum bilirubin level is >2.5–3 mg/dL. Rising levels may be associated with more severe disease.
- The liver is enlarged with tenderness and pain in the right upper quadrant.
- Recovery is slow, taking 4–6 weeks.
- Very rarely, a rapidly fatal fulminant hepatitis may occur; most of the reported deaths occur in adults.
- The disease is milder in children, in whom many infections may be anicteric.
- Mortality is low (0.1–1 per cent).
- The course outcomes of hepatitis A infection are presented in Fig. 26.2.

Laboratory diagnosis

- **Specimen:** Blood, serum, or feces may be collected for the demonstration of altered liver function, antibodies, or the virus respectively.
- **Liver function tests:**
  - Serum aminotransferase and aspartate aminotransferase are elevated.
  - Aspartate transaminase (AST) and alanine aminotransferase (ALT) show variable increase initially, prior to increase in serum bilirubin.
- **Serology:** It is based on the detection of antibodies to HAV (Fig. 26.3).
  - IgM anti-HAV antibody appears during the late incubation period, reaches peak levels in 2–3 weeks and disappears after 3–4 months.

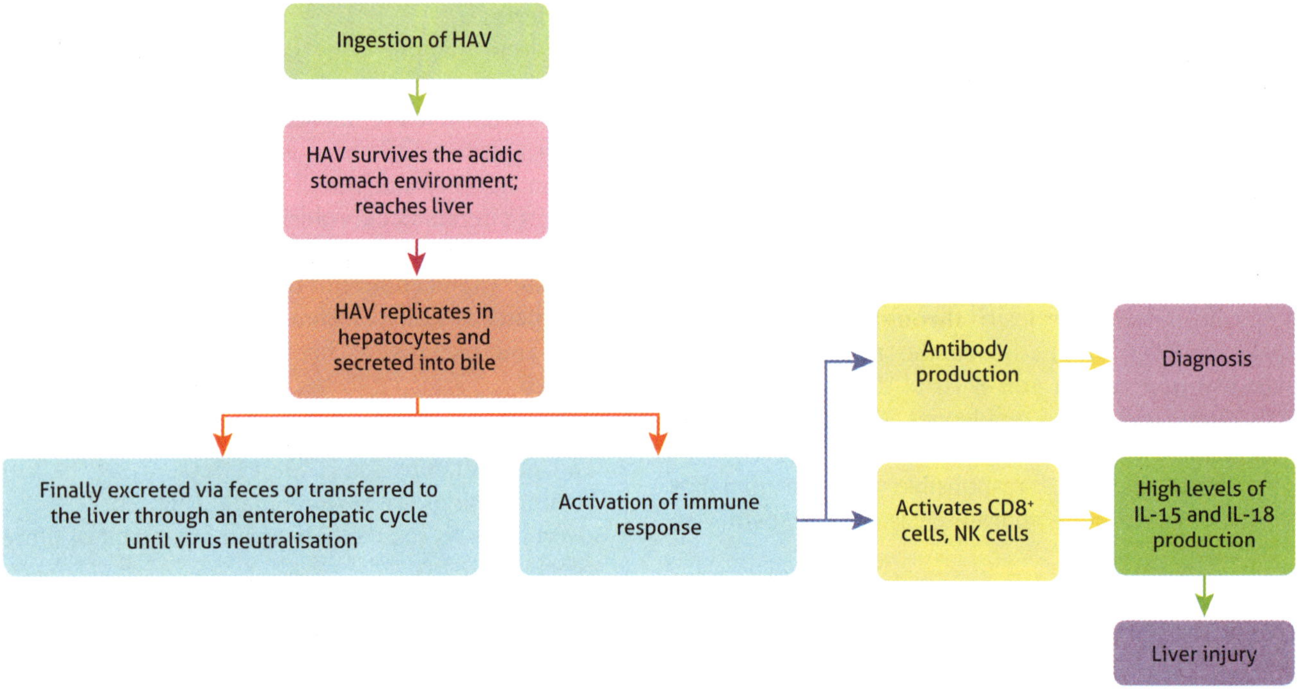

Fig. 26.1 Pathogenesis of liver injury mediated by HAV

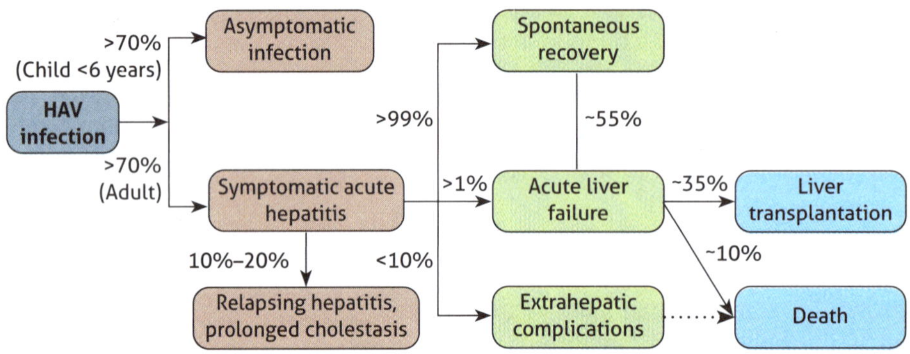

Fig. 26.2 Outcomes of hepatitis A

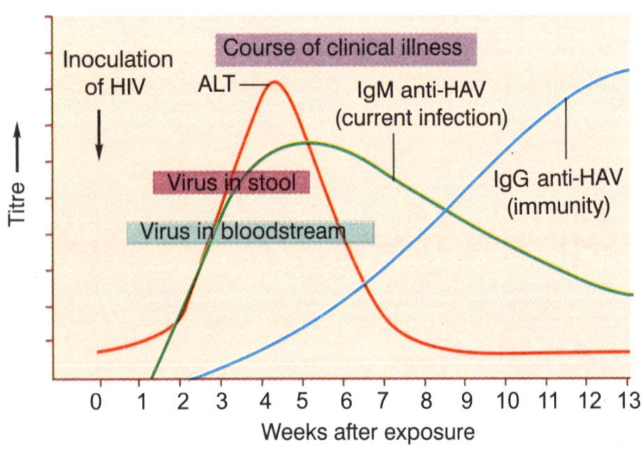

Fig. 26.3 Immunological and biological events associated with hepatitis A infection in humans

Demonstration of IgM antibody in serum indicates current or recent infection.

– The IgG antibody appears at about the same time, peaks in 3–4 months and persists much longer, sometimes, for life.

– The presence of **IgG antibody denotes recent or remote infection and immunity**.

– ELISA kits for the detection of IgM and IgG antibodies are available.

- **Direct demonstration of the virus:**
  – The virus can be visualised by IEM in fecal extracts during the late incubation period and the pre-icteric phase but seldom later.
  – This method is not commonly used for diagnosis.
- **Culture:**
  – Human cell lines and fetal monkey cell lines (FRhK-4/R) are used.
  – However, this is seldom used for laboratory diagnosis.
- **PCR:**
  – Viral RNA can be detected from stool filtrates by **real-time PCR**.
  – However, this method of diagnosis is unnecessarily expensive as the infection is usually self-limiting.

Prophylaxis

General prophylaxis

This consists of improved sanitary practices and the prevention of fecal contamination of food and water.

Specific active prophylaxis

- A safe and effective formalin-inactivated and alum-conjugated vaccine containing HAV grown in human diploid cell culture is available.
- A full course consists of two intramuscular injections of the vaccine, followed by a booster dose after 10–12 months.
- This vaccine is commonly administered to **children between 2 and 18 years**. Adults travelling to endemic areas can also be vaccinated.
- Protection begins four weeks after injection and lasts for 10–20 years.
- A combination of HAV and HBV vaccine, **Twinrix**, with the same immunogens as the individual vaccines and which immunises against both, is now available.

Specific passive prophylaxis

- Immunisation is also done with IgG-containing anti-HAV, at concentrations sufficient to protect (16% solution in a dose of 0.2–0.12 mL/kg body weight IM).
- These antibodies, when administered before exposure or in the early incubation period, can prevent or attenuate clinical illness while not necessarily preventing infection and virus excretion.
- Natural infection with HAV, clinical or subclinical, leads to lifelong immunity.
- There is no cross-immunity between HAV and any of the other hepatitis viruses.

Treatment

- Treatment is symptomatic.
- No specific antiviral drug is available.

Agent—Hepatitis E virus (HEV)

The largest HEV outbreak occurred in Delhi during the winter of 1955–56 and affected over 30,000 persons within six weeks. Several outbreaks and sporadic infections have been reported from many parts of the Indian subcontinent, Central and South Asia, North Africa, and Central America.

Morphology

- HEV is a spherical, non-enveloped virus, 32–34 nm in diameter, with a single-stranded RNA genome.
- The surface of the virion shows indentation and spikes.
- Only one serotype of the virus exists.
- It resembles Caliciviruses such as the Norwalk virus.
- It has been provisionally classified under Hepeviridae.

Clinical features

- The **incubation period** is 2–9 weeks.
- Most cases occur in **young to middle-aged adults** (15–40 years old).
- The disease is generally mild and self-limiting, with a low case fatality of about 1 per cent.
- No carrier state has yet been observed.
- High clinical severity and high case fatality rate of 20–40 per cent is seen in **pregnant women**, especially in the last trimester of pregnancy.
- The exact mechanism of liver injury in pregnant women is not clear, but there appears to be an interplay of hormonal and immunologic changes during pregnancy. This, along with a high viral load of HEV, makes the woman more vulnerable.

Laboratory diagnosis

- **Demonstration of the virus** is done by immunoelectron microscopy (IEM) in the bile and feces of patients in the incubation period or during the acute phase of the illness.
- IgG and IgM antibodies can be detected by ELISA or by rapid tests using recombinant and synthetic peptide antigens.
- RT-PCR can be performed to detect the hepatitis E virus RNA in blood and stool.

Treatment

- There is no specific antiviral therapy for hepatitis E.
- Supportive nutritional therapy and rest are advised.

Laboratory diagnosis of enteric hepatitis

- **Specimen:** Blood, serum, or feces
- **Liver function tests:** Deranged
- **Serology:**
  - *Current or recent infection*: IgM anti-HAV or anti-HEV antibody
  - **Recent or remote infection or immune:** IgG anti-HAV or anti-HEV antibody
- **Direct demonstration of the virus**: By immunoelectron microscopy (IEM) in stool
- **Culture:** Human cell lines and fetal monkey cell lines (FRhK-4/R)
- **PCR:** Viral RNA of HAV and HEV from stool filtrates by real-time PCR

Bloodborne or Serum Hepatitis

Case 2. *A 40-year-old woman who had received multiple blood transfusions over the previous six months presented with persistent fatigue, loss of appetite, nausea, vomiting, and abdominal pain for a duration of 10 days. She had a history of passing high-coloured urine. Her liver function tests showed elevated serum bilirubin and liver enzymes. A viral hepatitis panel was advised. She was diagnosed with* ***acute HBV infection*** *and was placed on supportive therapy. Her liver enzymes and viral markers were monitored every month to* *check for seroconversion from HBsAg to anti-HBsAg positivity (indicative of resolution of the infection).*

i. What are the tests included in the viral hepatitis panel in a patient of who has received repeated blood transfusions?
ii. What is the course of HBV infection?

Hepatitis may be caused by viruses that are transmitted via infected blood and blood products. These include HBV, HCV, HDV, and HEV.

Agent—Hepatitis B virus (HBV)

The discovery of HBV was serendipitous. In 1965, **Baruch Blumberg**, while studying human serum lipoprotein allotypes, observed in the serum of an Australian aborigine, a new antigen which gave a clearly defined line of precipitation with sera from two hemophiliacs who had received multiple blood transfusions. This was named the Australia antigen. By 1968, this antigen was found to be associated with serum hepatitis. It was subsequently shown to be the surface component of HBV. It was then renamed the hepatitis B surface antigen (HBsAg).

Epidemiology

- Hepatitis B is the most widespread and the most important type of viral hepatitis.
- More than a third of the world's population is estimated to be infected with HBV.
- The WHO estimates that HBV infection causes more than a million deaths a year worldwide.
- About a quarter of the infected individuals become **HBV carriers**.
- The virus is maintained in these carriers whose blood contains circulating viruses for long periods, in some, even lifelong.
- The **prevalence** of hepatitis carriers varies widely in different countries, in relation to living standards.
- India falls in the intermediate group—the carrier rate is 2–7 per cent, with higher carrier rates in the southern part of the country than the northern part.
- A quarter of infected persons develop serious liver disease, including chronic hepatitis, cirrhosis, and primary hepatic cancer.
- Since there is now an effective vaccine against HBV, hepatocellular carcinoma has become a disease that is **vaccine-preventable**.
- The infection is usually sporadic, though occasional outbreaks have occurred in hospitals, orphanages, and institutions for the mentally handicapped.

Morphology

- It is a 42 nm DNA virus of the family Hepadnaviridae with an outer envelope and an inner core, 27 nm in diameter, enclosing the viral genome and a DNA polymerase (Fig. 26.4a, b, and c).
- Electron microscopy shows three types of particles in the sera from type B hepatitis patients:
  i. **Spherical particle**, which is 22 nm in diameter, is the most abundant form
  ii. **Filamentous or tubular,** with a diameter of 22 nm and varying length
  These two particles are antigenically identical and are surface components of HBV (HBsAg) that are produced in great excess.
  iii. **Double-walled spherical structure,** measuring 42 nm in diameter; this particle is the complete hepatitis B virus; it was first described by Dane in 1970 and hence is known as the **Dane particle** (Fig. 26.5a and b)
- The envelope proteins expressed on the surface of the virion and the surplus 22 nm-diameter spherical

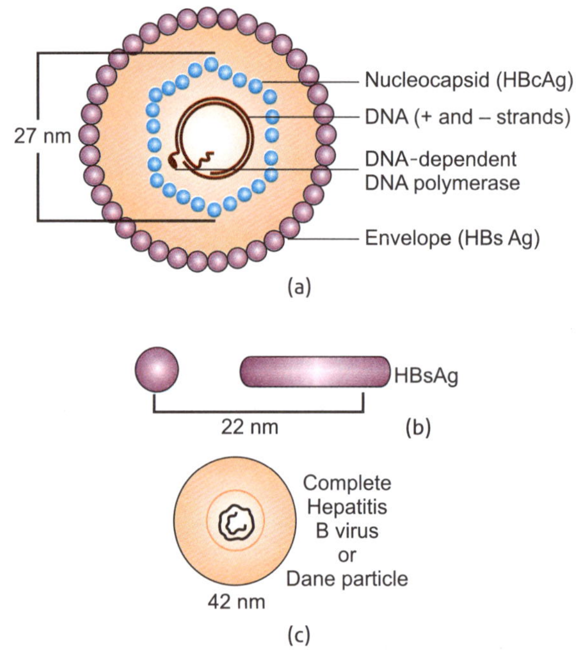

Fig. 26.4 Hepatitis B virus particles (a) 27 nm, (b) 22 nm, and (c) 42 nm

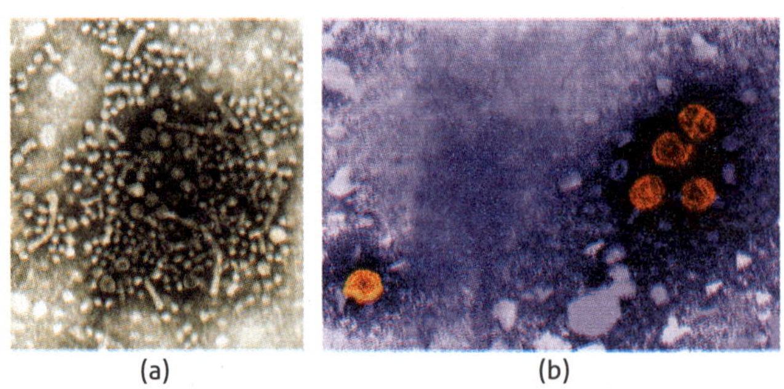

Fig. 26.5 (a) Hepatitis B virus and (b) Dane particles (*Source*: Centers for Disease Control and Prevention [CDC], Public Health Image Library [PHIL], Image IDs 10220 and 10755)

and filamentous particles constitute the hepatitis B surface antigens (HBsAg).
- HBsAg consists of two major polypeptides, one of which is glycosylated.

Susceptibility

- HBV is a relatively heat-stable virus.
- It remains viable at room temperature for long periods.
- Heat at 60°C for 10 hours reduces infectivity by 100- to 1,000-fold.
- It is susceptible to chemical agents.
- Exposure to hypochlorite (10,000 ppm available chlorine) or 2% glutaraldehyde eliminates infectivity, though HBsAg may not be destroyed by such treatment.

Antigenic diversity

1. **HBsAg (hepatitis B surface antigen):** This antigen exhibits antigenic diversity. It contains two different antigenic components:
   - The common **group-reactive antigen, a**
   - Two pairs of **type-specific antigens, d-y and w-r,** only one member of each pair is present at a time

   Four antigenic subtypes are seen, which are as follows:
   i. **ayw**: West Asia, West and North India
   ii. **adw**: Europe, America, and Australia
   iii. **adr**: South and East India
   iv. **ayr**: Very rare
2. **HBcAg (hepatitis B core antigen):**
   - Mild detergent treatment disrupts the viral envelope and exposes the core or nucleocapsid.
   - The antigen expressed on the core is called the hepatitis B core antigen (HBcAg; Fig. 26.6).
   - It is not secreted and is found only in hepatocytes.
3. **HBeAg:**
   - It is a soluble, non-particulate nucleocapsid protein secreted in serum.
   - HBcAg and HBeAg, though immunologically distinct, are coded for by the same gene.
   - Its presence indicates high levels of viral replication and infectivity.
4. **HBx:**
   - Coded by the P gene.
   - Denotes transactivating factor.
   - Enhances replication of HBV.
   - High levels seen in hepatocellular carcinoma and severe chronic hepatitis.

Viral genome

- The nucleocapsid encloses the viral genome, which comprises two linear strands of DNA held in a circular configuration.
- One of the strands (the plus strand) is incomplete and hence the DNA appears partially double-stranded and partially single-stranded.

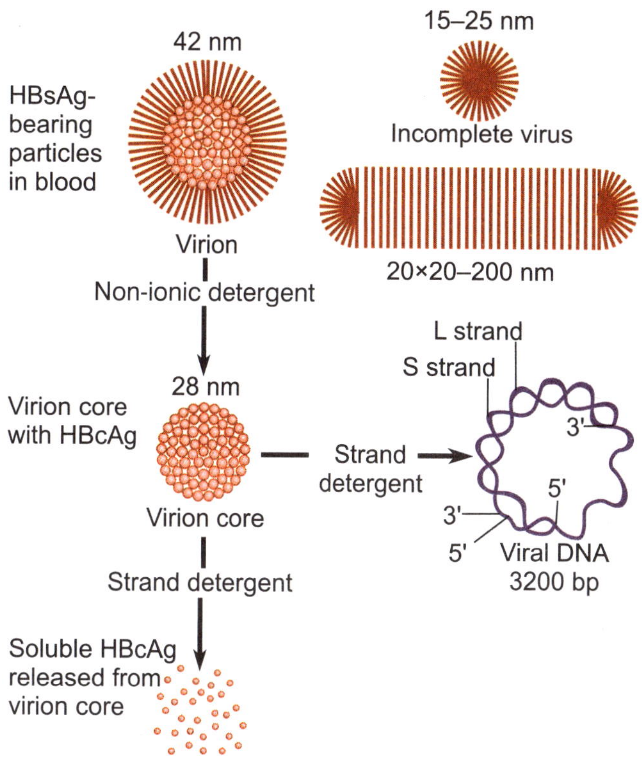

Fig. 26.6 Structural components of HBV

- Genes that code for various antigens of hepatitis B are presented in Table 26.2.

Mutations

A few cases of infection by mutant viruses have been identified. Two types of mutations have been studied:

- **Pre-core mutants:**
  – Infected patients are unable to synthesise the Hbe antigen.
  – Such patients may be positive for anti-HBe and anti-HBc.
- **Escape mutants:**
  – There is a single amino acid mutation in the HBs antigen.
  – It leads to the loss of neutralising activity by anti-HBs.

Table 26.2 HBV genes coding for different antigens

Gene	Region	Products
S	S	Major protein S (of Hbs antigen)
	S+Pre S2	Middle protein M (of Hbs antigen)
	S+Pre S1 S2	Large protein L (of Hbs antigen)
C	C	HBc antigen
	C+Pre C	HBe antigen
P	P	DNA polymerase
X	X	HBx antigen

- This mutation complicates vaccination and diagnosis.
- Escape mutatation occurs in some infants born to HBeAg-positive mothers and liver transplant recipients who had received combined immunisation with anti-HBV immunoglobulin, as a result of which the 'escape' occurs.

- **YMDD mutant:**
  - This mutant has emerged following the advent of amivudine treatment.
  - Most lamivudine-resistant strains exhibit amino acid substitutions in the YMDD (tyrosine-methionine-aspartate-aspartate) motif in the C domain of HBV polymerase.

Replication

- HBV replicates within hepatocytes.
- The viral DNA exists in the hepatocyte nucleus in the free extrachromosomal state or integrated with the cell chromosome.
- Replication resembles that seen in retroviruses in that DNA is synthesised from an RNA template by reverse transcription.
- HBV DNA and protein have also been identified in extrahepatic sites such as the bone marrow, spleen, lymph nodes, and circulating lymphocytes. However, no apparent damage is produced in these locations.
- HBV does not grow in any conventional culture system. However, limited production of the virus and its proteins has been achieved in several cell lines transfected with HBV DNA.
- HBV proteins have been cloned in bacteria and yeast.
- The chimpanzee is susceptible to experimental infection and can be used as a laboratory model.

Transmission

- The virus is present in **blood and body fluids** and excretions such as saliva, breast milk, semen, vaginal secretions, urine, bile, and feces.
- HBV is **highly infectious**, far more so than HIV.
- Any object or procedure that can transmit minute traces of infected blood or other material, as little as 0.00001 mL, can be infectious.
- There is a high occupational risk for medical and paramedical personnel, those who work at blood banks, dialysis units, medical laboratories and mental health institutions, barbers, and sex workers. HBV is is transmitted by the following routes:
- **Sexual:**
  - The infection can spread through unprotected sexual contact with infected persons (heterosexual or homosexual).
  - The risk of transmission by heterosexual and homosexual contact increases with the number of partners and the duration of such relationships.
- **Parenteral:**
  - **Blood, blood products, and organ transplantation**—The incidence of transmission is very high but is now reduced due to mandatory screening of blood, blood products, and donor organs.
  - **Contaminated needles**—In intravenous drug users who share syringes and needles and in cases of accidental needlestick injuries in healthcare workers. The use of disposable syringes, needles, and other equipment has reduced the incidence of such transmission.
- **Perinatal:**
  - Congenital or vertical transmission is quite common from carrier mothers.
  - The risk to babies is high if the mother is HBeAg-positive (60–90 per cent) and low if negative (5–15 per cent).
  - True congenital infection (in utero, transplacental) is rare.
  - Infection is usually acquired during birth by contact between maternal blood and the skin and mucosa of the fetus or in the immediate postnatal period.
  - HBV-infected neonates generally do not suffer from any clinical illness but remain carriers for life. Some of them may develop hepatocellular carcinoma after many decades.

> Transfusion of carrier blood, once the most widely known mode of infection has largely been eliminated wherever donor screening is strictly enforced. However, HBsAg screening is not a totally fail-safe method as infection has occurred even with HBsAg-negative and anti-HBc-positive blood on account of the presence of undetectable amounts of the virus.

Pathogenesis

- Natural infection occurs only in humans. There is no animal reservoir.
- The pathogenesis of hepatitis B appears to be immune-mediated.
- Hepatocytes carry viral antigens and are subject to antibody-dependent NK cell and cytotoxic T cell attack.
- In the absence of adequate immune response, HBV infection may not cause hepatitis and may lead to a carrier state instead.
- Therefore, infants and immunodeficient persons are more likely to become asymptomatic carriers following infection.

Hepatitis B virus (HBV) and HIV coinfection

- Around 70–90% of HIV-infected individuals in the United States have evidence of past or active infection with HBV.

- Both the infections are endemic in India. However, there are no country-wide data on the coinfection of HBV and HIV.
- Factors affecting the prevalence of chronic HBV include age at the time of infection and mode of acquisition, which vary geographically.
- The course of acute HBV may be modified in the presence of HIV infection with a lower incidence of icteric illness and lower rates of spontaneous clearance of HBV.
- Persons with HIV and chronic HBV coinfection have higher levels of HBV DNA and lower rates of clearance of the hepatitis B antigen.

Clinical features

- The **incubation period** of the infection is 1–6 months.
- **Clinical presentation** is similar to that of type A, but it tends to be more severe and protracted.
- The onset is insidious, and fever is not prominent.
- Extrahepatic complications like arthralgia, urticaria, and rarely, polyarteritis or glomerulonephritis may occur.
- These are ascribed to circulating immune complexes containing the viral surface antigen.

Outcomes of hepatitis B infection (Fig. 26.7)

- **Complete recovery**
  - Around 90–95 per cent of adults with acute hepatitis B infection recover within 1–2 months of onset and eliminate the virus from their body within about six months.
  - These individuals remain immune to the virus thereafter.
  - Mortality is about 0.5–2.0 per cent but may be more in post-transfusion cases.
- **Fatal fulminant hepatitis**
  - It is seen in 1 per cent of patients, particularly those who have concurrent delta virus infection.
  - They may develop hepatic encephalopathy within a short period of infection.
- **Chronic infection:** Seen in 1–10 per cent of patients in the following forms:
  - Asymptomatic carriers (2–7%)
  - Recurrent/chronic liver disease/cirrhosis (1–2%)
  - Hepatocellular carcinoma (<1%)

Laboratory diagnosis

The mainstay of diagnosis is the **detection of viral markers**. Specific diagnosis of hepatitis B rests on the serological demonstration of the viral markers. Antigen markers of hepatitis B have both diagnostic and prognostic significance. It is, therefore, necessary to understand the sequence of their appearance in blood (Fig. 26.8).

Antigen markers

i. **HBsAg:**
   - It is the **first marker to appear in blood** after infection, being detectable even before the elevation of transaminases and onset of clinical illness.
   - It remains in circulation throughout the icteric or symptomatic course of the disease.
   - In a typical case, it disappears within about two months of the onset of clinical disease but may sometimes last for six months and even beyond.

ii. **HBcAg**
   - It is not demonstrable in circulation because it is enclosed within the HBsAg coat in hepatocytes.

iii. **HBeAg**
   - It appears in blood concurrently with HBsAg or soon afterwards.
   - Circulating HBeAg is an indicator of active intrahepatic viral replication.
   - Its presence denotes high infectivity and its absence, along with the presence of anti-HBe, indicates low infectivity.

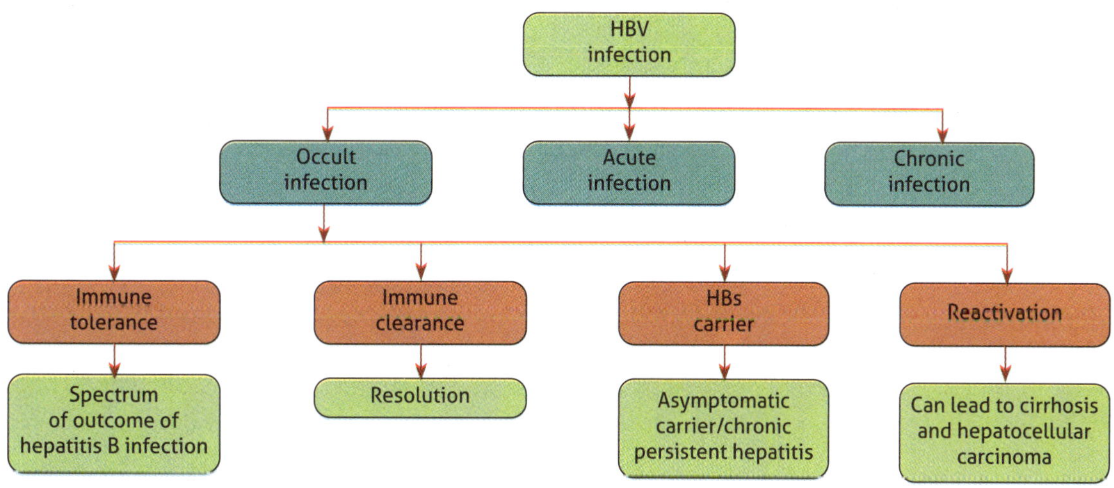

Fig. 26.7 A flowchart of possible outcomes in hepatitis B infection

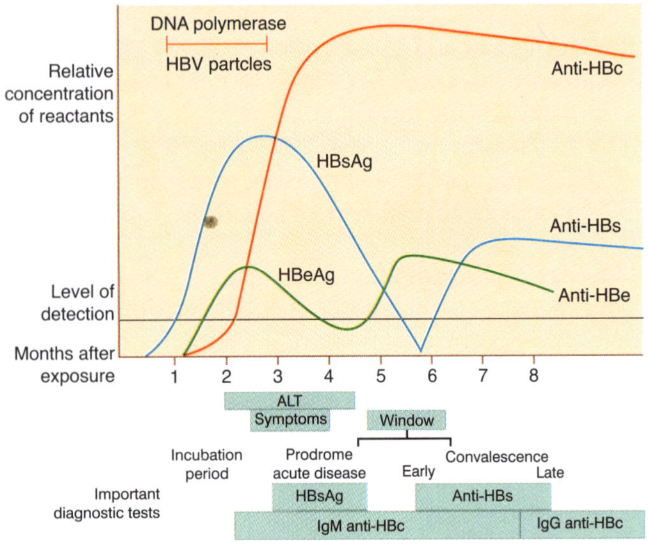

Fig. 26.8 Clinical and serological events occurring in a patient with acute hepatitis B virus infection

Antibody markers

i. **Anti-HBs**
   - When HBsAg is not detectable, the antibody anti-HBs appears and remains for very long periods.
   - Anti-HBs is the protective antibody.

ii. **Anti-HBc**
   - The antibody to HBc, anti-HBc, appears in serum a week or two after the appearance of HBsAg.
   - It is, therefore, the **earliest antibody marker** to be seen in blood, long before anti-HBe or anti-HBs.
   - As anti-HBc remains lifelong, it serves as a useful indicator of prior infection with HBV, even after all the other viral markers become undetectable.
   - Initially, anti-HBc is predominantly IgM, but after about six months, it is mainly IgG.
   - Selective tests for IgM or IgG anti-HBc, therefore, enable distinction between recent and remote infection respectively.

iii. **Anti-HBe**
   - The antibody to HBe indicates low infectivity and asymptomatic carrier stage.
   - Its appearance coincides with low HBV DNA, normal ALT, and lower hepatic inflammation.

Molecular markers

Quantitative markers: **DNA polymerase, HBV DNA and virions in blood**: The presence of these markers reflect high infectivity.

Quantitative estimation of HBV DNA

- HBV DNA level in serum reflects the degree of viral replication in the liver and hence helps to assess the progress of patients with chronic hepatitis under antiviral chemotherapy.
- Plasma HBV DNA quantification by real-time polymerase chain reaction (PCR) correlates with disease progression.
- It can be used to differentiate active HBeAg-negative disease from inactive chronic infection, and for decisions to treat and subsequent monitoring.
- It can also be used for monitoring response to antiviral therapy, as a rise may indicate the emergence of resistant variants.

> **Laboratory diagnosis of non-enteric hepatitis caused by HBV**
>
> Diagnosis is mainly serological
>
> **Antigen markers**
>
> i. **HBsAg:** First marker
> ii. **HBcAg:** It is not demonstrable in circulation
> iii. **HBeAg:**
>   - Appears in blood concurrently with HBsAg
>   - Presence denotes high infectivity, and active intrahepatic viral replication
>
> **Antibody markers**
>
> i. **Anti-HBs:** Appears with fall in HBsAg and remains high for very long periods
> ii. **Anti-HBc:** Earliest antibody marker
>   - Within a week or two after the appearance of HBsAg, remains lifelong
>   - Initially, IgM, replaced after 6 months by IgG
> iii. **Anti-HBe:**
>   - It indicates low infectivity and asymptomatic carrier stage
>   - Its appearance coincides with low HBV DNA, normal ALT, and lower hepatic inflammation
>
> **DNA polymerase, HBV DNA, and virions in blood**
>   - Reflects high infectivity

Diagnostic parameters

- The detection of HBsAg in blood is of diagnostic uncertainty.
- The disappearance of HBeAg coincides with a drop in transaminase levels in blood. This is followed by the appearance of anti-HBe.
- Generally, the presence of IgM anti-HBc indicates recent infection while, IgG anti-HBc indicates remote infection.
- Occasionally, when the level of HBsAg is too low to be detectable, diagnosis has to be made by testing for IgM anti-HBc.
- Sometimes, due to mutations in the surface antigen gene, HBsAg becomes undetectable, but the anti-hepatitis core (anti-HBc) antibody can be detected in the serum. This phenomenon is called **occult hepatitis B infection (OBI).**
- The presence of anti-HBs without any other serological virus marker indicates immunity following vaccination.

Table 26.3 presents the clinical interpretation of various serological patterns in hepatitis B.

Prophylaxis

General prophylaxis

- Infection can be prevented to a large extent by practicing standard precautions, avoiding risky practices like promiscuous sex, injectable drug abuse, and direct or indirect contact with blood, semen, or other body fluids of patients and carriers.
- Health education, the use of disposable syringes and needles, and screening of blood, semen, and organ donors have all helped to an extent, but these alone cannot eliminate the risk altogether, particularly in the developing countries.

Specific prophylaxis

- **Passive immunisation by hyperimmune hepatitis B immunoglobulin (HBIG)**
  - HBIG is prepared from the plasma of human volunteers with high-titre anti-HBs and is administered IM in a dose of 300–500 IU soon after exposure to the infection.
  - It may not prevent infection but protects against illness and the carrier state.
  - *Indications*
    - Newborns of HBsAg-positive mothers, especially if the baby is positive
    - In case of accidental percutaneous exposure
    - Unprotected sexual exposure
    - After liver transplant in case of recurrent HBV infection
- **Active immunisation by hepatitis B vaccine (HBV)**
  - *Recombinant subunit vaccine:*
    - This is a genetically engineered vaccine made by cloning the *S* gene of HBV in baker's yeast.
    - It consists of non-glycosylated HBsAg particles alone.
    - It is given with an alum adjuvant IM into the deltoid or, in infants, into the anterolateral aspect of the thigh.
    - Gluteal injection is not recommended as it may result in a poor immune response.
    - Three doses given at 0, 1, and 6 months constitute the full course.
    - Seroconversion occurs in about 90 per cent of the vaccinees. Booster doses are needed only for those at high risk.
    - The vaccine is manufactured in India and is cost-effective.
    - It has been included in the national immunisation schedule.
    - All babies receive 0.5 mL of the vaccine on the anterolateral side of the mid-thigh at birth or within 24 hours.
  - *Three-antigen vaccine*: A vaccine containing all antigenic components of HBsAg (Pre-S1, Pre-S2 and S) has been developed, which gives greater seroconversion.

Some individuals may not develop antibodies after a full course of vaccination. They are known as **non-responders.** Clinical infection confers protection which lasts much longer

Seroconversion in vaccine recipients

- This can be detected by testing for anti-HBs, which is usually detectable for about five years.
- A titre of >10 IU is said to be significant. Any titre less than that requires a booster dose.

Combined immunisation

- This type of immunisation is recommended for non-immune persons exposed to HBV.
- For babies born to carrier mothers, a single injection of 0.5 mL of HBIG given IM immediately after birth is followed by the full course of vaccine at a different anatomical site, the first dose being given within 12 hours of birth.
- When HBIG is not available, the vaccine given alone has been reported to provide protection.

Post-exposure prophylaxis guidelines

- **Unvaccinated individuals exposed to the virus through needlestick injury or exposure to blood**

Table 26.3 Interpretation of common serological patterns in HBV infection

Virus/antibody markers					
HBsAg	HBeAg	Anti-HBc	Anti-HBs	Anti-Hbe	Interpretation
+	+	IgM	–	–	Acute HBV infection; highly infectious
+	+	IgG	–	–	Late/chronic HBV infection or carrier state; highly infectious
+	–	IgG	–	+/–	Late/chronic HBV infection or carrier state; low infectivity
–	+/–	IgM	–	+/–	Seen rarely in early acute HBV infection; infectious
–	–	IgG	+/–	+/–	Remote HBV infection; infectivity nil or very low
–	–	–	+	–	Immunity following HBV vaccine

and body fluids of infected persons: A combination of HBIG and hepatitis B vaccine is advised. A single dose of 0.06 mL/kg is given immediately after exposure, followed by the full course of hepatitis B vaccination.

- **Immunocompromised individuals:** A booster dose is recommended after the regimen mentioned above.

Individuals with undetectable antibodies (<1:10) are also advised to undergo repeat vaccination.

Treatment

No specific antiviral treatment is available for acute HBV infection and the carrier state. According to the National Hepatitis Control Programme guidelines, treatment is recommended for the following:

- Patients of all ages including children with chronic hepatitis and evidence of compensated or decompensated cirrhosis should be treated, regardless of their ALT levels, HBeAg status, or HBV DNA levels
- All patients above 30 years of age with chronic hepatitis who do not have evidence of cirrhosis and have persistently abnormal ALT levels and evidence of high-level HBV replication (HBV DNA >20,000 IU/mL), regardless of HBeAg status.

Not recommended for and can be deferred in persons without evidence of cirrhosis, and with persistently normal ALT levels and low levels of HBV replication (HBV DNA <2,000 IU/mL), regardless of HBeAg status

(Ref: https://main.mohfw.gov.in/sites/default/files/National%20laboratory%20guidelines_Highress_Prepress_0.pdf)

Antivirals

- Entecavir, tenofovir disoproxil fumarate, tenofovir alafenamide, lamivudine, and famciclovir can be used.
- Interferons such as interferon alfa-2b (Intron A) and peginterferon, alone or in combination with other antiviral agents are also used.
- There is no effective treatment for the carrier states though spontaneous resolution takes place in some cases.
- Liver transplantation may be required in cases of severe damage.

Agent—Hepatitis C virus (HCV)

Epidemiology

- HCV infection is seen only in humans. Infection is transmitted by carriers, who are in abundance; it is estimated that around 200 million people worldwide are carriers of HCV.
- In general, the epidemiology of HCV resembles that of hepatitis B.
- The infection occurs across the world, with carrier rates of 1–20 per cent.
- The prevalence of the anti-hepatitis C virus (HCV) antibody in the general population is estimated to be 0.09–15%. Based on some regional level studies, it is estimated that 6–12 million people have hepatitis C in India.
- A quarter of all chronic hepatitis cases in India are believed to be due to HCV infection. In India, chronic HCV infection accounts for 12–32% of hepatocellular carcinoma and 12–20% of cirrhosis (https://nvhcp.mohfw.gov.in/common_libs/National-Action-Plan-Combating-Viral-Hepatitis-in-India.pdf).
- HCV infection usually resolves spontaneously in 15–45% patients and progresses to chronic infection in 55–85% patients.

Morphology

- HCV resembles flaviviruses in structure and organisation and has been classified as a new genus—**Hepacivirus of the family Flaviviridae.**
- It is 50–60 nm in size and contains a linear, **single-stranded RNA genome** enclosed within a core and surrounded by an envelope which bears glycoprotein spikes (Fig. 26.9).
- HCV genome consists of seven functional regions-the core, the envelope, including the E1 and E2 regions, and the non-structural region, including NS2, NS3, NS4, and NS5.
- Six distinct major genotypes and a minor genotype 7 with more than 50 subtypes within the genotypes of HCVS have been identified, indicating high mutability.
- Some genotypes are seen worldwide, while others are localised.

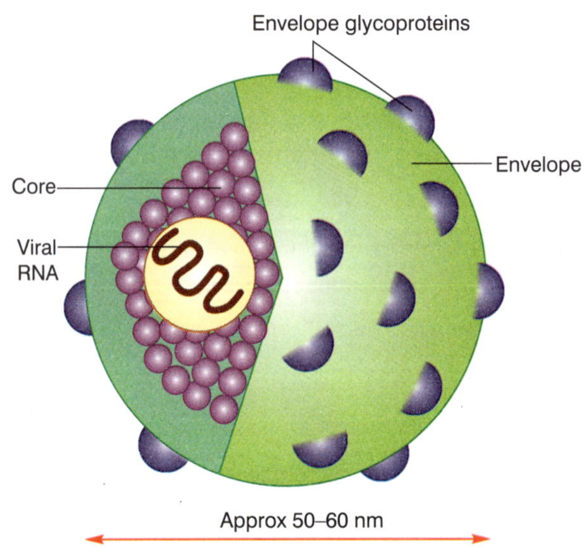

Fig. 26.9 HCV virus with its core viral RNA, envelope, and glycoprotein

- Genotype 1, especially subtype 1b, may have a greater chance of causing cirrhosis or severe liver scarring than other genotypes. Genotypes 1b and 3 may increase the risk of liver cancer.
- Genotype 3 is the second most common HCV subtype in the world, particularly in Northern Europe, South Asia, and South-East Asia.
- It usually requires longer treatment, and does not achieve strong cure rates particularly with cirrhosis.
- Because of this diversity, there is little heterologous or even homologous post-infection immunity in hepatitis C.

Transmission

- HCV is similar to HBV in that it is transmitted chiefly by **blood transfusion** and other modes of contact with infected blood or blood products.
- Sexual transmission is probably less important than it is with HBV.

Clinical features

- The incubation period is 15–160 days
- The acute illness is **mild or anicteric**.
- Overt jaundice develops in 5 per cent of patients.
- About 50–80 per cent of patients progress to chronic hepatitis, with some developing cirrhosis and hepatocellular carcinoma.

Laboratory diagnosis (Fig. 26.10 and 26.11)

Antibody detection by ELISA

- **First-generation assay:** Recombinant protein-containing NS4 (C100-3)

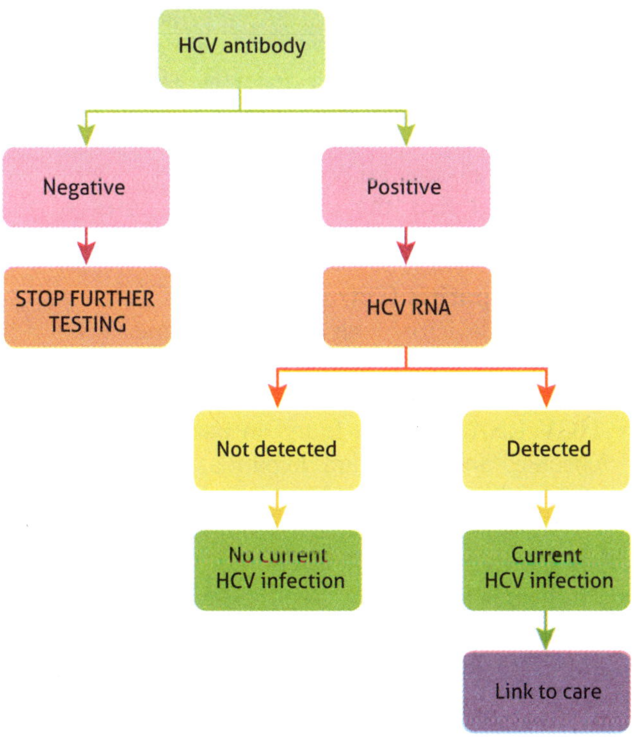

Fig. 26.10 Testing algorithm for HCV infection

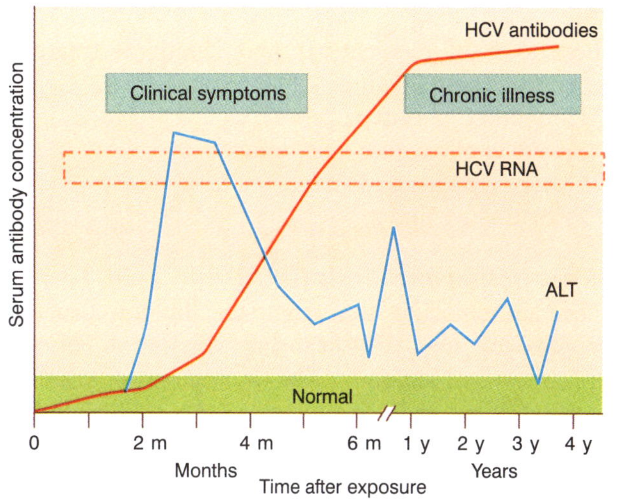

Fig. 26.11 The course of HCV infection and the appearance of HCV RNA and antibodies

- **Second-generation assay:** Recombinant proteins from core, NS3 and NS4
- **Third-generation assay:** Recombinant proteins of core, NS3, NS4, and NS5
- **Fourth-generation assay or 'antigen-antibody combo' tests**

In HCV infection, antibodies appear irregularly and late, limiting their diagnostic utility (Fig. 26.11).

Rapid assays

- Detect anti-HCV IgG

Nucleic acid testing (NAT)

- By PCR
- Has reduced window period to one week

Antigen detection

- HCV core Ag (nucleocapsid p22)
- Chemiluminescent microparticle immunoassay

Molecular testing of HCV RNA in blood

- This provides more **sensitive and specific results** within a few days of exposure to HCV.
- Molecular methods like PCR and branched DNA assay are employed for the purpose.
- HCV RNA qualitative test for molecular detection
- HCV RNA quantitative test to estimate the viral load
- HCV genotyping

No culture method is available for HCV at present.

Prophylaxis

- Only general prophylaxis such as blood screening is possible.
- No specific active or passive immunising agent is available yet.

Treatment

The choice of medications and length of treatment depend on the following:

- Hepatitis C genotype
- Presence of existing liver damage
- Other medical conditions
- Prior treatments

Direct-acting antivirals (DAA)

These are molecules that target specific non-structural proteins of the virus causing disruption of viral replication and infection. The four classes of DAA are:

i. **Non-structural proteins 3/4A (NS3/4A)**—protease inhibitors (PIs) such as telaprevir, boceprevir, asunaprevir, simeprevir, and faldaprevir
ii. **NS5B nucleoside polymerase inhibitors (NPIs)** like sofosbuvir and mericiatbine
iii. **NS5B non-nucleoside polymerase inhibitors (NNPIs)** such as deleobuvir, filibuvir, setrobuvir, and tegobuvir
iv. **NS5A inhibitors** like daclatasvir, velpatasvir, and ledipasvir

- **Sofosbuvir, velpatasvir, ledipasvir, and daclatasvir** are available in India.
- Pegylated interferon alpha-2a/2b is given to those who do not respond to antivirals.
- HCV genotypes 1, 4, and 5 require 48 weeks of therapy with a high dose of interferon.
- HCV genotypes 2 and 3 require 24 weeks of therapy while HCV genotype 1, 4, and 5 may need 48 weeks of therapy with a high dose of interferon.

Agent—Hepatitis D (delta) virus (HDV)

The hepatitis D virus (HDV), also known as delta agent, is a **defective RNA virus dependent on the helper function of HBV for its replication and expression.**

Epidemiology

- HDV is distributed worldwide but is more common in certain endemic areas.
- In the Mediterranean countries where it is endemic, infection chiefly spreads by **non-percutaneous routes, especially by close personal contact.**
- In the non-endemic areas such as northern Europe and North America, infection is more often transmitted through blood and blood products and is commonly seen in drug addicts and hemophiliacs.
- The introduction of HDV into non-endemic areas where HBV infection is common may lead to outbreaks of severe hepatitis with high mortality.

Morphology

HDV is member of genus Deltavirus. It is a spherical, 36 nm particle with an outer coat composed of the hepatitis B surface antigen surrounding a circular, single-stranded RNA genome.

Clinical features

Its mode of transmission is the same as that of HBV. Two types of infection are recognised:

- **Co-infection:** Here, delta and HBV are transmitted together at the same time. Co-infection clinically presents as acute hepatitis B, ranging from mild to fulminant disease.
- **Superinfection:** Here, delta infection occurs in a person already harbouring HBV. It usually leads to more serious and chronic illness, with deterioration of the underlying HBV infection. No association has been noted between HDV and hepatocellular carcinoma.

Laboratory diagnosis (Fig. 26.12)

- The delta antigen is primarily expressed in liver cell nuclei, where it can be demonstrated by immunofluorescence. It is only occasionally present in serum.
- Anti-delta antibodies appear in the serum and can be identified by ELISA. The IgM antibody appears 2–3 weeks after infection and is soon replaced by the IgG antibody in acute delta infection. However, in chronic infection, the IgM antibody persists for years.
- Delta RNA sequences have been cloned and DNA probes have been developed for the rapid identification of delta particles in circulation.

Prophylaxis

- No specific prophylaxis exists, but immunisation with the HBV vaccine is effective as HDV cannot infect persons immune to HBV.
- Screening of blood donors for HBsAg automatically limits bloodborne HDV infection.

Table 26.4 lists the comparative features of the various viral hepatitis types.

Agent—Hepatitis G virus (HGV)

- This is a flavivirus with an RNA genome.
- It has not been grown in culture, but its RNA genome has been cloned.
- HGV RNA has been found in patients with acute, chronic, and fulminant hepatitis, in hemophiliacs and patients who have undergone multiple transfusions or hemodialysis and in individuals who are intravenous drug abusers.
- HGV appears to be a bloodborne virus resembling HCV. Its prevalence is higher in patients infected with HIV and HCV. Its role in hepatitis is yet to be clarified.
- HGV RNA can be detected in the blood samples of infected patients using RT-PCR.

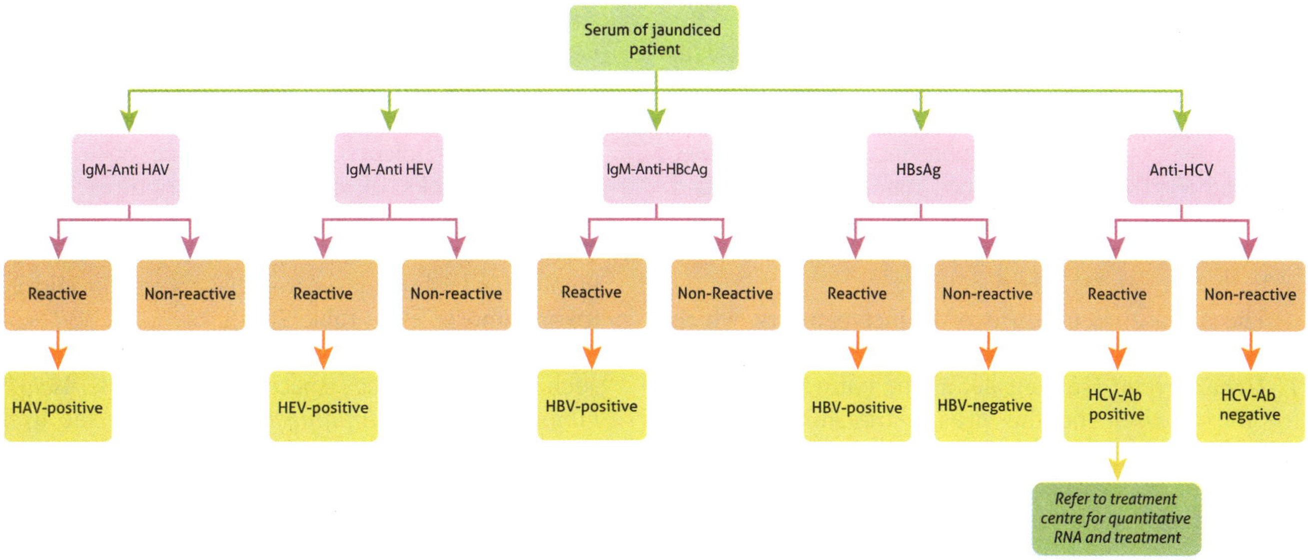

Fig. 26.12 Diagnostic workflow of a jaundiced patient suspected of hepatitis
Note:

If HBsAg is reactive and IgM anti-HBc is non-reactive: HBV-positive

If IgM anti-HBc is reactive and HBsAg is non-reactive: HBV-positive

If both reactive: HBV-positive

If both non-reactive: HBV-negative

Table 26.4 Viral hepatitis types—comparative features

Features	A	B	C	D	E
Virus	HAV, 27 nm, RNA, Picornavirus (Hepatovirus)	HBV, 42 nm, DNA (Hepadnavirus)	HCV, 50–60 nm, RNA, *Flavivirus* (Hepacivirus)	HDV, 35–37 nm, defective RNA Deltavirus	HEV, 32–34 nm, RNA, Hepevirus
Modes of infection	Fecal–oral	Percutaneous, vertical, sexual	Percutaneous	Percutaneous	Fecal–oral
Affected age group	Children	Any age	Adults	Any age	Young adults
Incubation period (days)	15–45	30–180	15–160	30–180	15–60
Onset	Acute	Insidious	Insidious	Insidious	Acute
Illness	Mild	Occasionally severe	Moderate	Occasionally severe	Mild, except in pregnancy
Carrier state	Nil	Common	Present	Nil (only with HBV)	Nil
Oncogenicity	Nil	Present specially after neonatal infection	Present	Nil	Nil
Prevalence	Worldwide	Worldwide	Probably worldwide	Endemic areas (Mediterranean, N. Europe, Central and N. America)	Only developing countries (India, Asia, Africa, Central, America)
Specific prophylaxis	Ig and vaccine	Ig and vaccine	Nil	HBV vaccine	Nil

- The presence of virus-specific antibodies in the serum of patients has been associated with viral clearance and protective immunity. However, these methods are not widely used for the diagnosis of viral hepatitis as the role of HGV in causing hepatitis is still not clear.
- Besides hepatitis G (HG), two other viruses have been detected in the blood of individuals with post-transfusion hepatitis. These are Torque **teno virus (TTV) and SENV 25–28**.

Yellow Fever

Yellow fever was recognised as a clinical entity as early as in the 17th century. However, since the early 20th century, the disease has largely been confined to certain areas of Africa and South and Central America.

Epidemiology

- Though India is a yellow fever-'receptive' area with a large population of *Aedes aegypti* and non-immune humans, the disease does not exist in India.
- In order to prevent the establishment of the disease, strict vigilance is enforced on vaccination and quarantine for travel from endemic areas. Such measures have checked the entry of the virus into India through legitimate passengers.

The disease occurs in two distinct patterns, which are as follows:

1. Urban cycle: Humans act as the natural reservoirs for the disease, and the virus is transmitted by the domestic *Aedes aegypti* mosquito.

2. Forest or sylvatic cycle: Wild monkeys act as the reservoirs and forest mosquitoes (*Haemagogus spegazzinii* in South America and *Aedes africanus* and *A. simpsoni* in Africa) act as vectors. Humans acquire the infection only when they trespass into forests or when monkeys raid villages near forests.

Clinical features

- The **incubation period** is 3–6 days.
- Yellow fever resents as a fever of acute onset with chills, headache, nausea, and vomiting.
- The pulse is usually low despite there being high temperature. Jaundice, albuminuria, and hemorrhagic manifestations develop, and the patient may die of hepatic or renal failure.
- Most cases are less severe, especially in the endemic areas, and may present as undifferentiated fever without jaundice.

Laboratory diagnosis

- Diagnosis is based on the detection of virus-specific IgM or neutralising antibody by ELISA or plaque neutralisation assay.
- Viral RNA can also be detected by real-time PCR.

Control

- **General measures:** The eradication of the vector mosquitoes would stop the transmission of the disease, but this is obviously impracticable with the concomitant existence of the sylvatic disease.
- **Specific measures:** The non-neurotropic 17D vaccine is administered by subcutaneous inoculation. Vaccination is mandatory for travel to or from endemic areas. It is valid for 10 years, beginning 10 days after vaccination. In India, the 17D vaccine is manufactured at the Central Research Institute, Kasauli.

PARASITIC INFECTIONS OF THE LIVER

- Many parasites infect the liver and the biliary tract. These infections can occur during the maturation stages of the parasite or may be caused by the adult worms.
- Parasites infiltrating the biliary tracts may result in liver abscess, space-occupying lesions, hepatitis, pancreatitis, cholecystitis, biliary tree obstruction, repeated attacks of cholangitis, strictures in the biliary tree, and cholangiocarcinoma.
- The reticuloendothelial system of the liver may be involved in leishmaniasis, malaria, cryptosporidiosis, toxoplasmosis, etc. (Table 26.5).
- In the case of parasites that are transmitted orally such as *Echinococcus* species, liver flukes, *Ascaris lumbricoides*, and *Entamoeba histolytica*, the liver is the first solid organ encountered after mucosal penetration, either directly or via portal-venous blood flow.
- Some parasites reach the liver after their larvae penetrate the skin (schistosomiasis).

Table 26.5 Types of pathology in the liver caused by parasites*

Clinical presentation	Parasite
Abscess	*Entamoeba histolytica* *Fasciola*
Hepatitis	*Toxoplasma* *Capillariasis*
Biliary obstruction	*Ascaris* *Opisthorchis* *Fasciola*
Space-occupying lesions	*Taenia solium* *Echinococcus* species
Cirrhosis	*Leishmania* *Toxocara* *Schistosoma*
Hepatomegaly	*Plasmodium* (Malaria) *Leishmania* (kala-azar) *Babesia* *Trypanosoma cruzi*
Cholangiocarcinoma	*Clonorchis sinensis* *Opisthorchis*
Biliary hyperplasia	*Ascaris* *Clonorchis sinensis*
Cholangitis	*Clonorchis sinensis* *Opisthorchiasis*
Portal hypertension	*Schistosoma*
Granulomatous hepatitis	*Toxocara* *Strongyloides*

*Details of these parasitic agents have been dealt with in the chapters dealing with their primary pathogenesis

Hydatid Cyst

> **Case 3.** *A 40-year-old male presented with abdominal pain for the past one year, with no history of vomiting, jaundice, fever, or weight loss. Abdominal ultrasound suggested* ***cystic swelling in the liver****. Routine investigations showed TLC of 8,800/μL³ and eosinophils at 6%. He was operated on to remove the cyst, which was sent to for investigations.*
> i. What is the probable diagnosis?
> ii. What is the life cycle of the parasite involved?

Hydatid cyst is caused by the **larval form of *Echinococcus granulosus*** (Fig. 26.13). It is unilocular and can range in size from a few mm to several centimetres. The cyst is most commonly found in the liver (60%), followed by the lungs (25%), muscles (5%), bones (3%), kidney (2%), spleen (1%), and other sites (1%).

Epidemiology

- *E. granulosus* is found worldwide but is more common in developing countries.
- Cases are reported from Central and South-East Asia including India and Nepal, west China, South America, the Mediterranean region, and East Africa.
- The WHO aims to eliminate the disease by 2050.
- In India, the disease has been reported from Jammu and Kashmir, Himachal Pradesh, Uttar Pradesh, Punjab, Tamil Nadu, and Andhra Pradesh.
- People at risk include animal herders, farmers, sheep and cattle rearers, zoo workers, animal handlers, and those living in the vicinity of dogs regularly feeding on sheep offal.

Agent—*Echinococcus granulosus*

Intermediate host: Human beings
Definitive host: Animals such as dogs

Human diseases are caused by four species of *Echinococcus* (Table 26.6).

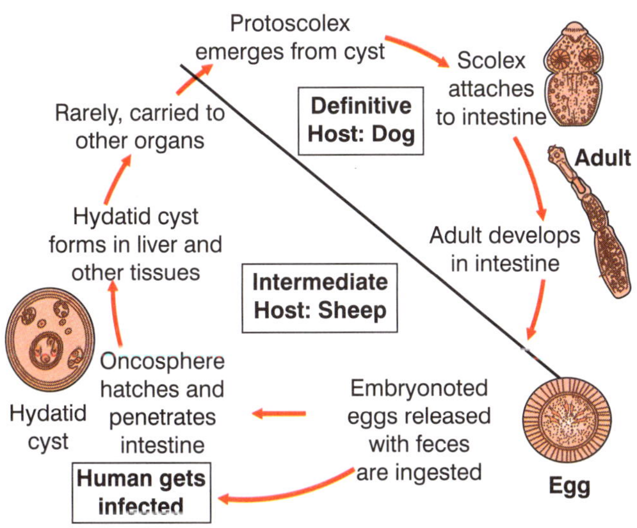

Fig. 26.13 The morphology and life cycle of *Echinococcus*

Table 26.6 *Echinococcus* species causing disease in humans and animals

Species	Definitive host	Intermediate host	Clinical presentation
E. granulosus	Dog, wolf	Sheep, horse, cattle, and camel	Single cyst
E. multilocularis	Fox	Rodents and small mammals	Polycystic, multi-locular cysts, and alveolar cysts
E. vogeli	Bush dog	Forest rodents	Polycystic
E. oligarthus	Jaguar	Rodents	Single cyst

Morphology

Three stages (Fig. 26.13) are seen—the adult, egg, and larva (cyst).

- **Adult:** It is 3–6 mm long and flat. Its body is made up of a head (**scolex**), neck, and segments (**proglottids**). The scolex has four muscular suckers and a rostellum with a double row of hooklets. The head is followed by the short neck and three segments (proglottids). The proximal segment is immature, the second segment is mature and bears the sexual organs, and the terminal segment is gravid.
- **Egg:** The fertilised egg is similar to the egg of *Taenia*. It is light brown and round with a radially striated, thick shell. It contains the oncosphere or the hexacanth larval stage with six hooklets. The eggs are highly resistant and disperse widely in the environment.
- **Larva:** The **larval form is called a hydatid cyst** (Fig. 26.14a–d). It is unilocular and can range in size from a few mm to several centimetres. The liver acts as the first filter in trapping the embryos which then develop into hydatid cysts, followed by the lungs as the second filter. Some escape from these filters and develop in other organs.

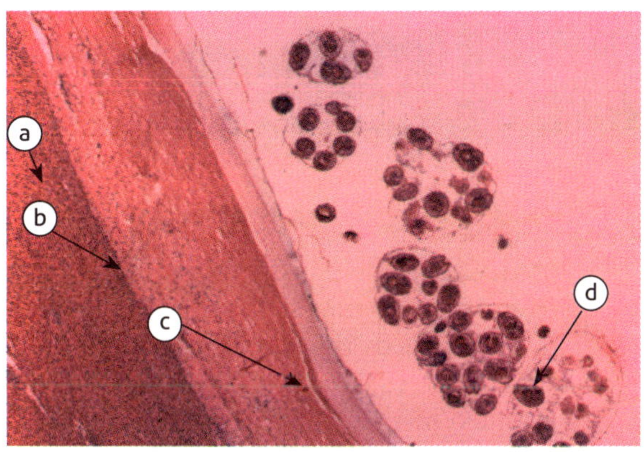

Fig. 26.14 The hydatid cyst of *Echinococcus granulosus* showing (a) the outermost adventitial layer, adjacent to the (b) inner laminated hyaline ectocyst, (c) the innermost germinal epithelium—the endocyst and (d) the brood capsules (*Source*: CDC, PHIL, Image ID 910)

Hydatid cyst

- The cyst grows at a rate of 0.3–1 cm per year and may take 5–20 years to attain sufficient size to cause symptoms.
- The cyst is surrounded by three layers, which are as follows:
  i. **Endocyst:** Inner nucleated and germinal layer. This germinal layer undergoes asexual proliferation endogenously and produces daughter cysts known as the **brood capsules**. The brood capsule gives rise to the **protoscolex**, which is capable of giving rise to an adult worm. Multiple protoscolices are found in different stages of development inside the brood capsules.
  ii. **Ectocyst:** This is the external, acellular, and laminated layer. It is a tough, elastic, and hyaline layer, which is up to 1 mm thick and resembles the white of a hard-boiled egg.
  iii. **Pericyst:** It is the host-derived fibrous adventitial layer.
- The cyst derives its nutrition by diffusion through the cyst wall.
- The cyst contains clear, colourless or pale-yellow fluid, with a specific gravity of 1.012 and pH of 6.7–7.2. The fluid is highly antigenic and anaphylactic.
- The cyst also contains **hydatid sand** which comprises scolices, brood capsules, and hooklets along with albumin, creatinine, urea, glucose, NaCl, phosphates, etc.

Life cycle (Fig. 26.13)

- Dogs and other animals acquire the infection by ingesting meat infested with hydatid cysts.
- Upon ingestion, the protoscolices are released, and an adult worm develops in the intestine.
- The worm becomes sexually mature in about 20–24 months and fertilises, releasing eggs in the dog's feces.
- These eggs are ingested by sheep, cattle, or other intermediate hosts.
- The oncospheres are released in the duodenum.
- These penetrate the intestinal wall to enter the portal circulation.
- In about 60–90% of cases, the larva is arrested in the liver and develops into a hydatid cyst. Rarely, it may be carried to other organs such as the lungs and may lodge in various body organs.
- The hydatid cyst can survive for several years until it is consumed by the definitive host to continue the life cycle.
- Humans act as accidental hosts when they replace the natural intermediate host in the life cycle.
- The eggs are ingested with food or water contaminated with dog's feces.

Clinical features

- The **incubation period** varies from months to several years.
- The patient may remain asymptomatic for years as the hydatid cyst keeps growing without causing any pathology.
- In some cases, it may regress spontaneously and calcify.
- The clinical symptoms depend on the site of the lesion and arise due to multiple reasons, which are as follows:
  - **Pressure effects:** The cyst can present clinically as a space-occupying lesion or a tumour-like mass, hepatomegaly, or abdominal distension. The cyst may grow to encroach on the biliary tree or bronchi causing cholestasis, cholangitis, or dyspnea.
  - **Acute anaphylactic reaction:** This can occur if the cyst ruptures or leaks fluid. This can present as fever, urticaria, eosinophilia, and anaphylactic shock and can be fatal.
  - **Bacterial superinfections:** Sometimes, cysts may be complicated by bacterial superinfections.
  - **Other sites:** In the case of lung cysts, the patient may complain of chest pain, cough, or hemoptysis if the cyst ruptures into the bronchi. Neurological features such as signs of raised intracranial pressure, epilepsy, and paralysis may be seen in patients with brain cysts.

Laboratory diagnosis

A combination of radiology and laboratory tests is used for diagnosis.

- **Gross examination:** The whole cyst with its fluid may be sent to the microbiology lab after surgical removal. The cyst is examined macroscopically for the characteristic **hyaline glass-like membrane filled with sand and fluid.**
- **Direct microscopy:**
  - The hydatid cyst fluid in the excised cyst can be aspirated and examined microscopically for brood capsules, protoscolices, or 'hydatid sand'.
  - In older cysts, the daughter cysts or scolices may have disintegrated and only hooklets may be present. Hydatid sand is crushed between slides and examined under the microscope.
  - Surgically removed cysts can be stained with Giemsa, H & EM, or periodic acid-Schiff (PAS) stain to demonstrate the layers of the cyst wall and the attached brood capsules.
  - Surgical aspiration of hydatid cyst is associated with a **risk of hydatid fluid leakage**, which can result in a hypersensitivity reaction or dissemination. Hence, **diagnostic aspiration is not recommended.**

- **Molecular diagnosis:** A PCR assay based on mitochondrial DNA has been developed, which can also differentiate various genotypes of *E. granulosus*.
- **Serodiagnosis:**
  - The sensitivities and specificities of serological methods range from 60–90%. These tests correlate well with radiologic investigations but cannot differentiate between recent and past infection.
  - Serology is more specific but less sensitive than imaging. Antibodies can be detected by ELISA using a crude *Echinococcus* cyst fluid antigen.
  - Recently, a recombinant antigen (B2t)-based assay has become available, which has shown better results. The test has sensitivity and specificity above 90%. Enzyme-linked immune transfer blot (EITB) assays are better than ELISA; however, these are more difficult to perform.
- **Antigen detection:** A specific echinococcal antigen can be detected in serum, urine, and CSF. Antigen detection can differentiate recent from past infection and can be used as a prognostic marker.
- **Casoni's intradermal test**: This is obsolete and only of historical importance. It is based on an immediate hypersensitivity reaction to hydatid fluid antigens.
- **X-ray and ultrasonography (USG):** Calcified cysts in the lungs and liver can be visualised on X-rays. CT scans are more accurate than X-rays (Fig. 26.15) and can detect even small cysts. USG is preferred for the diagnosis of liver cysts. It is useful in assessing efficacy to treatment also.
- **Other tests:** Eosinophilia is seen, especially if the cyst ruptures.

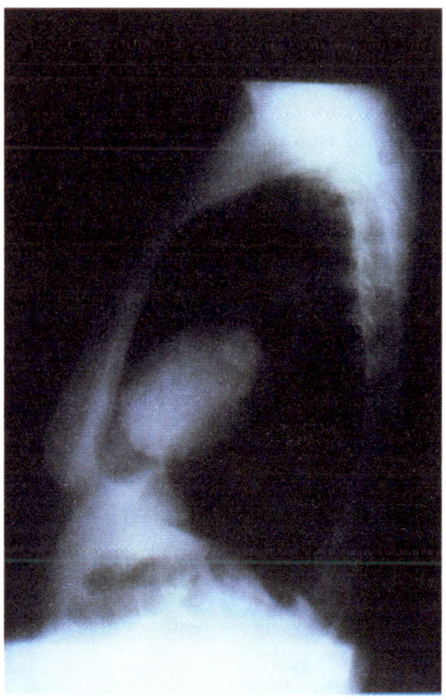

Fig. 26.15 X-ray shows the presence of a spheroid hydatid cyst located within a patient's thoracic cavity (*Source*: CDC, PHIL, Image ID 18432)

Treatment

- Effective management may require a combination of medical and surgical lines of treatment.
  - **Medical treatment:** Albendazole or mebendazole for 3–6 months are the drugs of choice. The therapy may be prolonged.
  - **Surgical treatment:** There is a high chance of recurrence with surgical removal of cysts. Surgery is recommended in advanced disease, with cysts larger than 10 cm, 6 or secondarily infected cysts. Cysts located in the brain, lung, or kidney, cysts communicating with the biliary tree, or cysts that are ruptured/exerting pressure on adjacent organs must be removed.
  - **PAIR (puncture, aspiration, injection, and re-aspiration):** The cysts are punctured percutaneously under ultrasound or CT guidance, and 10–15 mL fluid is aspirated with necessary precautions to prevent spillage of the fluid into adjoining areas. A lytic drug (either 95% ethanol or 20% hypertonic saline—protoscolicidal agents) is injected to destroy the germinal layer. This is followed by re-aspiration of fluid after a contact period of 15–20 minutes. PAIR is performed for uncomplicated, single cysts.
- No intervention is recommended for biologically inactive or degenerative cysts.
- Follow-up of patients is recommended by imaging at 3–6 monthly intervals after which they are reassessed to evaluate the need for intervention.

Alveolar Echinococcosis

- **Alveolar echinococcosis** is one of the **most fatal parasitic diseases and is caused by** *Echinococcus multilocularis*.
- Humans are infected accidentally.
- The disease has been reported from Europe and Asia and parts of USA, Canada, China, and Japan.
- It is not reported from India.
- Hunters, farmers, and workers in fur factories are at risk.

Agent—*Echinococcus multilocularis*

Morphology

- The adult is smaller than that of *E. granulosus* (1.2–3.7 mm), while the larval cyst can vary in size from 1 mm to several cm.
- It is multivesicular with little or no fluid.
- The germinal layer grows both endogenously and exogenously, creating a tube-like network enclosing cystic structures.
- The cystic vesicles may metastasise to different body sites via blood or lymph.

- The life cycle of *E. multilocularis* is quite similar to that of *E. granuloses*. Human beings acquire infection by the ingestion of the eggs of *E. multilocularis*.

Clinical features

- Primary lesions are located in the liver but metastasise to other organs.
- The cysts in the alveoli behave like malignancies and may remain asymptomatic for 10–15 years.
- Patients complain of **upper quadrant and epigastric pain**.
- Examination may show hepatomegaly and obstructive jaundice.
- If left untreated, the mortality is more than 90% within 10–15 years of diagnosis.

Laboratory diagnosis

Radioimaging, histopathology of excised cysts, and antibody detection are the mainstay of diagnosis.

Treatment

Surgical resection and albendazole therapy are used to treat *E. multilocularis* infection. PAIR is NOT used in these infections.

Liver Fluke Infestations–Trematodes Affecting the Hepatobiliary System

- Trematodes are **leaf-like, flat, and unsegmented worms** that reside in the intestine, liver, lung, or blood vessels.
- Their life cycles are complex, often requiring two intermediate hosts.
- The trematodes inhabiting the liver are known as **liver flukes**.
- Their **primary host is a vertebrate** while the **intermediate host a snail**. The **second intermediate host is a fish or crab**.

The common liver flukes are:

- *Fasciola hepatica*
- *Clonorchis sinensis*
- *Opisthorchis viverrini*
- *Opisthorchis felineus*

Fascioliasis

- Fascioliasis is a parasitic infection typically caused by *Fasciola hepatica*.
- *Fasciola* is one of the largest trematodes known to infect humans.

Agent—*Fasciola hepatica*

- Two species of *Fasciola* are pathogenic to humans—*Fasciola hepatica* or the common liver fluke and *F. gigantica*, the giant liver fluke.
- *F. hepatica* is a common zoonosis found worldwide. *F. gigantica* is endemic in Africa and Asia.
- In India, liver flukes are rarely found.
- The liver flukes are important veterinary pathogens as they infect domestic livestock such as sheep, goats, camels, and cattle in Europe, Asia, Africa, and South America.

Morphology

- **Adult** (Fig. 26.16)**:** *F. hepatica* and *F. gigantica* are similar morphologically except in the matter of size. The liver flukes are leaf-shaped with a distinctive 'shoulder' and an anterior prominence called a **cephalic cone**. An oral sucker is present on the cephalic cone and a ventral sucker is present on the anterior portion of the fluke's body, near its genital pore. *F. hepatica* measures 3 × 1.2 cm, while *F. gigantica* is 7–8 × 1.2 cm in size. Like other trematodes, adults are monoecious. In mammalian hosts, the adult worm resides in the bile ducts, gallbladder and liver tissue. The worm has a lifespan of more than 10 years.
- **Egg** (Fig. 26.16)**:** The eggs of *F. hepatica* are large, oval, yellowish-brown and ovoid, measuring 130–150 × 60–90 μm. They are operculated and unembryonated. The eggs of *F. gigantica* are larger, measuring about 160–190 × 70–90 μm.

Life cycle (Fig. 26.16)

The consumption of the metacercariae of the liver fluke with uncooked water plants causes infection in humans.

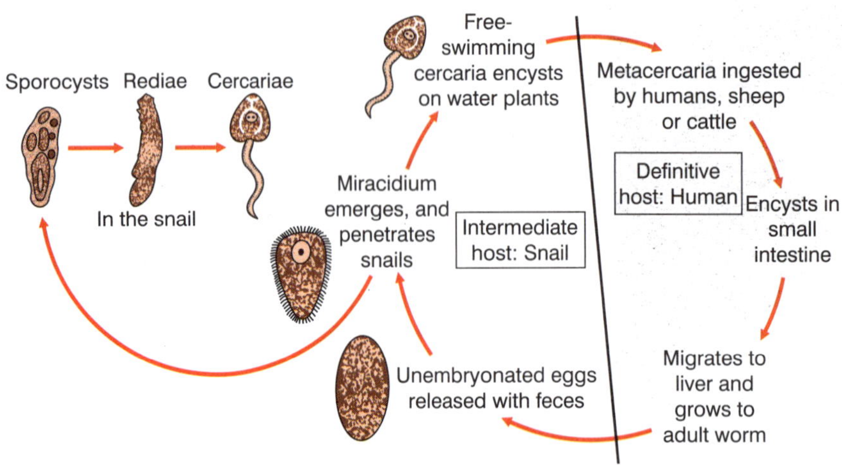

Fig. 26.16 Life cycle and stages of *Fasciola*

- **Migratory phase:** In the duodenum, the metacercariae excyst, penetrate the intestinal wall and enter the liver tissue to finally reach the bile ducts in about 6–7 weeks.
- **Sexual phase:** The worm matures and starts producing eggs. The unembryonated eggs reach the intestines with bile and are released into the environment with feces about 3–4 months after infection.
- **Asexual phase:** The eggs develop in water to release miracidia, which penetrate the snail host in 1–2 weeks. The miracidium undergoes multiplication and transforms into two generations of redia, and then cercaria. After 4–7 weeks, the cercariae are released from the snail and get encysted on aquatic plants.

Clinical features

The incubation period can vary from days to months. The presentation can be acute or chronic.

- **Acute stage:**
  - The patient presents with fever, abdominal pain, anorexia, nausea, diarrhea, and flatulence.
  - Hepatomegaly, splenomegaly, jaundice, and ascites may be present.
  - Acute symptoms can last 2–4 months.
- **Chronic stage:**
  - The worm in the bile duct elicits inflammation and granuloma formation around the eggs and worms, resulting in abscesses followed by fibrosis, bile duct hyperplasia, and deposition of collagen.
  - It can obstruct the biliary ducts, resulting in cholangitis, cholecystitis, or cholelithiasis.
  - Multiple subcapsular abscesses (known as **liver rot**), cirrhosis, and pancreatitis are known complications.

Clonorchiasis

Clonorchiasis is infection caused by the liver fluke *Clonorchis sinensis*.

Agent—*Clonorchis sinensis*

This trematode is found in eastern Russia, China, South Korea, Taiwan, and Vietnam.

Intermediate hosts: Two—a snail and a freshwater fish

Definitive hosts: Humans and domestic animals

Morphology (Fig. 26.17)

- **Adult:** The adult worm is monoecious, 1–2 cm long × 3–5 mm wide, with an oral sucker located on the anterior end of its body. The ventral sucker, which is smaller, is also situated at the anterior end.
- **Egg:** The eggs enclose a developed miracidium. They are small, yellowish-brown, ovoid, and measure 20–30 × 15 μm. They have a distinct operculum and a knob at the posterior end.

Life cycle (Fig. 26.17)

- The infection is acquired by the **consumption of raw or undercooked freshwater fish** that are infected with the metacercaria larva.
- The metacercaria excysts in the duodenum and reaches the intrahepatic ducts, where it matures in about four weeks.
- The eggs are released to reach the bile ducts and are released with feces.
- The eggs containing the mature miracidium are ingested by the first intermediate host, the snail (*Bithynia*).
- The miracidium hatches in snails, multiples, and forms **redia**. These develop into cercariae, which are released out of the snail.

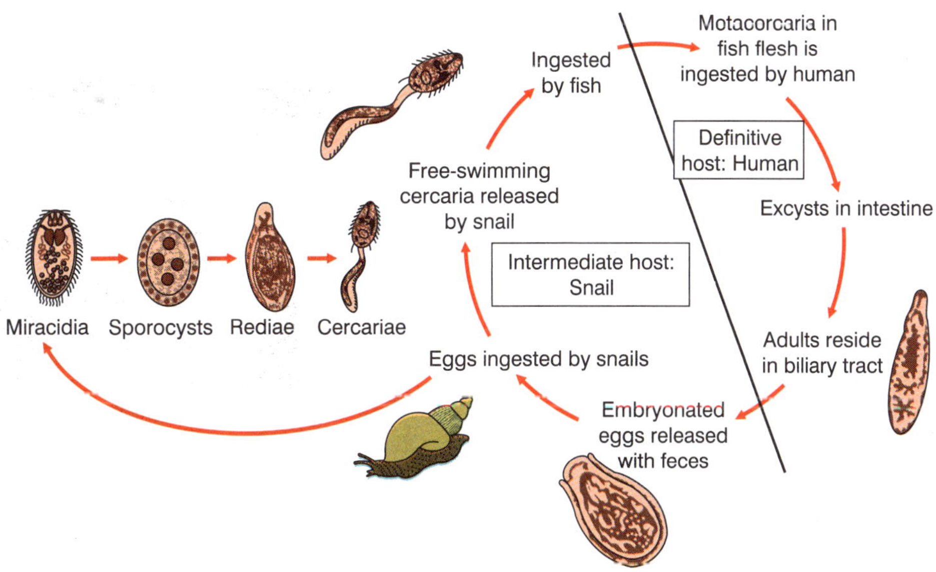

Fig. 26.17 Life cycle and stages of *Clonorchis*

- The cercaria enters the second intermediate host, a freshwater fish, in which it encysts into a metacercaria larva.

Clinical features

- **Acute stage:** This stage presents with malaise, anorexia, nausea, vomiting, abdominal pain, flatulence, and diarrhea that usually resolve within a few weeks. In heavy parasitemia, obstructive jaundice can occur, and hepatomegaly is common.
- **Chronic infection:** The patient may remain asymptomatic or may present with complications.
- **Complications:** Cholangitis, cholecystitis, obstructive jaundice, periportal fibrosis, recurrent pyogenic cholangitis, cholelithiasis, or cholangiocarcinoma may develop. Cholangiocarcinoma is the most serious complication of chronic liver fluke infection. Chronic irritation of the bile duct by the worm can lead to cancer of the bile duct.

Opisthorchiosis

Opisthorchiasis is infection with *Opisthorchis viverrini* (Southeast-Asian liver fluke) or *O. felineus* (cat liver fluke), which are acquired by eating infected raw or undercooked fish that contains infectious metacercariae (encysted stage).

Agent—Opisthorchis

Definitive hosts: Cats and dogs (*O. felineus*)
Intermediate hosts: Snails and freshwater crabs
Two species can cause disease in humans:

i. *Opisthorchis viverrini*
ii. *Opisthorchis felineus* (cat liver fluke)

- The morphology, life cycle, the acute and chronic presentation, diagnosis, and treatment for these flukes are similar to that for *Clonorchis sinensis*.
- *C. sinensis* and *O. viverrini* are found to be associated with cholangiocarcinoma with a high mortality rate.
- *Opisthorchis viverrini* is also associated with hepatocellular carcinoma.

Laboratory diagnosis of liver fluke infections

- Direct microscopy of stool, duodenal aspirate, or bile helps in visualising the characteristic eggs, ova, or larvae of the fluke. Eggs may be absent in stool during the early phase of infection. In such cases, endoscopic or percutaneous biliary aspiration may be performed to demonstrate ova or adult worms.
- Coproantigen can be detected in stool in the case of *Fasciola* infection; the circulating antigen is detected in *Clonorchis* infections.
- Serologically, the detection of IgG antibody against the crude or excretory–secretory antigen by ELISA has diagnostic value.
- Molecular detection with PCR is available at research centers to detect DNA in stool samples.
- Ultrasound, CT, or MRI are used to scan for hepatic lesions.
- Eosinophilia, anemia, and elevation of IgE and IgG can be detected by performing blood investigations.

Treatment of liver fluke infections

- *Fasciola hepatica* and *Clonorchis sinensis* infections are treated with a single dose of triclabendazole.
- Alternative drugs are nitazoxanide and bithionol.
- Praziquantel is the drug of choice for *Opisthorchis* species.

LIVER ABSCESS

Liver abscesses may be of two types, which are as follows:

i. **Pyogenic liver abscess:**

- These are suppurative infections of the liver parenchyma made up of walled-off lesions in any of the lobes.
- They contain dead polymorphs, tissue debris, and organisms—usually polymicrobial, both aerobic and anaerobic. This makes up the pus in a pyogenic liver abscess.
- The following are the causative agents of pyogenic liver abscesses:
  - *E. coli*
  - *K. pneumoniae*
  - *Staphylococcus aureus*
  - Anaerobes
  - Secondary bacterial infection of hydatid cysts

ii. **Amoebic liver abscess** (Fig. 26.18):

- The pus is usually sterile unless secondary bacterial infection occurs.
- There is an absence of inflammatory cells due to the lysis of neutrophils by the protozoan forming the typically described non-purulent **'anchovy sauce' (chocolate-coloured) pus.**

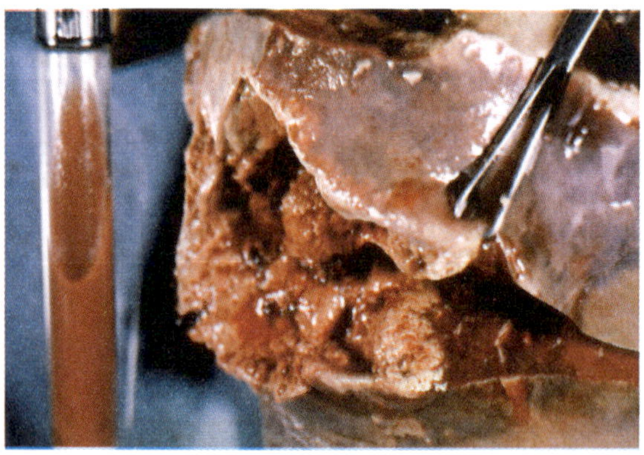

Fig. 26.18 A section of liver tissue in a case of amoebic liver abscess; note the incised live specimen, which revealed the parenchymal amoebic abscess, and on left, a test tube filled with what is termed 'anchovy sauce' (or chocolate) pus (*Source*: CDC, PHIL, Image ID 362)

- Hepatocytes undergo apoptosis or necrosis.
- The site of the abscess is usually the right upper lobe, although it can occur in the left lobe.

Clinical Features

Patients present with fever and right upper quadrant tenderness.

Laboratory Diagnosis

- Leukocytosis (predominantly neutrophils), increased alkaline phosphatase, and abnormal liver function tests are common features.
- However, these features do not differentiate pyogenic from amoebic liver abscess.

Treatment

- Pyogenic abscesses may require ultrasound-guided needle aspiration or percutaneous catheter drainage along with broad-spectrum antibiotics—either a carbapenem or a combination of cephalosporin and a beta-lactamase inhibitor to cover aerobic and anaerobic bacteria.
- Amoebic liver abscess is managed by medical treatment with anti-amoebicidal drugs.
- Drainage is required for complications of infection such as cases of secondary bacterial infection (either de novo or secondary to drainage) and patients who are at high risk of ALA rupture.

BILIARY TRACT INFECTIONS

- Biliary tract infections occur as a complication of an obstruction to the tract by gallstones. This leads to cholecystitis.
- Organisms responsible for biliary tract infections are generally the gut flora, i.e., *E. coli*, *Klebsiella*, *Enterococcus*, and anaerobic bacteria. *Salmonella* Typhi may also be isolated from carriers.
- Local infections may lead to cholangitis, which may spread from the biliary tract to produce liver abscess or invade the bloodstream, resulting in septicemia.

Laboratory Diagnosis

- Serological evaluation reveals leukocytosis, raised alanine aminotransferase (ALT) and aspartate aminotransferase (AST) elevated in cholecystitis or common bile duct obstruction. Bilirubin and alkaline phosphatase assays are used to evaluate for the presence of common bile duct obstruction.
- Imaging techniques like ultrasound may also be used for diagnosis.

Treatment

Treatment consists of surgical removal of obstruction and broad-spectrum antibiotics to cover both aerobic Enterobacteriaceae and anaerobic bacteria.

Questions

Case review

A 40-year-old woman who had received multiple blood transfusions over the previous six months presented with persistent fatigue, loss of appetite, nausea, vomiting, and abdominal pain for a duration of 10 days. She had a history of passing high-coloured urine. Her liver function tests showed elevated serum bilirubin and liver enzymes.

i) What specific tests are required to be done to know the etiology?
ii) Discuss the preventive measures that can be taken to stop transmission of the virus from person to person.

Short answer questions

1. Describe the diagnostic relevance of various markers of HBV.
2. Describe the morphology of hepatitis C virus.
3. Describe the genesis and anatomical structure of a hydatid cyst.
4. Describe the pathogensis of amoebic liver abscess.

MCQs

1. **Which of the following hepatitis viruses is a DNA virus?**
   a) Hepatitis A b) Hepatitis B
   c) Hepatitis C d) Hepatitis G
2. **Which of the following is a defective virus and depends on the helper functions of hepatitis B virus?**
   a) Hepatitis A b) Hepatitis C
   c) Hepatitis D d) Hepatitis G
3. **A 50-year-old male with a history of blood transfusion eight months ago developed jaundice. He was treated for hepatitis B virus infection. During the follow-up period, he was found to be HBsAg-negative and anti-HBsAg-positive. What does this indicate?**
   a) High infectivity of disease
   b) Progression of disease
   c) Resolution of infection
   d) Carrier status
4. **A 35-year-old commercial sex worker presents with complaints of nausea, vomiting, abdominal pain, and passing high coloured urine for four days. On investigation, her serum bilirubin and liver enzymes were high and she was HBsAg-positive and anti-HBcAg-positive. What is the likely diagnosis?**
   a) Acute HBV infection b) Acute HCV infection
   c) Chronic HBV infection d) HBV carrier
5. **The common HBV subtype present in southern and eastern parts of India is:**
   a) adw b) adr
   c) ayw d) ayr
6. **Following hepatitis B infection, the risk of becoming a carrier is highest in:**
   a) Neonates b) Children
   c) Adults d) Elders
7. **Which one of the following is false regarding hepatitis B carrier status?**
   a) Detectable HBsAg in blood for more than two months
   b) Prevalence in India is 2–7%
   c) More common among males
   d) Anti-HBc IgG positive
8. **A 68-year-old male is admitted for dialysis. He gives a history of jaundice in the previous year. As part of laboratory work-up, his serum tested for hepatitis panel showed high titres of HBsAg, HBeAg, and DNA polymerase. It indicates:**
   a) Acute HBV infection b) Chronic HCV infection
   c) Simple-carrier status d) Super-carrier status

9. **Which of the following is the infectivity marker in hepatitis B infection?**
   a) HBsAg b) Anti-HBs
   c) HBeAg d) Anti-HBeAg
10. **The presence of high HBeAg marker in the serum indicates all of the following except:**
    a) Presence of DNA polymerase in the blood
    b) Presence of HBV DNA in the blood
    c) Immunity following HBV vaccination
    d) Active viral replication
11. **Which one of the following cannot be demonstrated in blood in hepatitis B infection?**
    a) HBsAg b) HBcAg
    c) HBeAg d) Anti-HBeAg
12. **The first marker to appear and last one to disappear during the course of hepatitis infection is:**
    a) HBsAg b) HBeAg
    c) Anti-HBcAg d) Anti-HBeAg
13. **The presence of high levels of anti-HBeA antibody in the serum indicates:**
    a) Low infectivity
    b) Presence of HBV DNA in the blood
    c) Active viral replication
    d) Immunity following HBV vaccination
14. **Which marker indicates protective immunity following HBV vaccination?**
    a) HBsAg b) Anti-HBsAg
    c) HBeAg d) Anti-HBeAg
15. **Passive immunisation for hepatitis B infection is good for:**
    a) Prevention of infection
    b) Prevention of carrier status
    c) Controlling the spread of disease
    d) Controlling adverse reactions
16. **Which of the following resembles flaviviruses in structure and organisation?**
    a) Hepatitis A b) Hepatitis C
    c) Hepatitis D d) Hepatitis E
17. **Treatment on HCV hepatitis depends on:**
    a) Hepatitis C genotype
    b) Presence of existing liver damage
    c) Prior treatments
    d) All of the above
18. **A 47-year-old slum dweller presents with fever and right upper quadrant abdominal pain for the past few days. He gives a history of bloody diarrhea. A CT scan of the abdomen reveals a single cyst in the right posterior lobe of the liver. Which parasite is likely to cause this condition?**
    a) *Entamoeba histolytica* b) *Giardia lamblia*
    c) *Shigella* spp. d) *Cryptosporidium parvum*
19. **Which of the following parasites does not affect the hepatobiliary system?**
    a) *Plasmodium* species b) *Entamoeba histolytica*
    c) Schistosomes d) *Ascaris lumbricoides*
20. **Cholangiocarcinoma is a complication of infection caused by:**
    a) *Plasmodium* species b) *Entamoeba histolytica*
    c) Schistosomes d) *Clonorchis sinensis*
21. **Which of the following statement is correct regarding the adult worm of *Echinococcus granulosus*?**
    a) It is 3–6 m long and flat.
    b) The body is made up of a scolex, neck, and more than 10 proglottids.
    c) The scolex has four muscular suckers and a rostellum with a double row of hooklets.
    d) All of the above.
22. **A 45-year-old farmer with two pet dogs presents with complaints of right upper quadrant abdominal pain for the last several weeks. He is afebrile with normal vitals having mild right upper quadrant tenderness. USG abdomen shows hepatomegaly with a large multiseptated 'honeycomb' cyst in the right lobe of the liver. Changing position during the ultrasound reveals a 'sandy' matrix within the cyst. What is the causative agent?**
    a) *Plasmodium* species b) *Entamoeba histolytica*
    c) *Echinococcus granulosus* d) *Clonorchis sinensis*
23. **Which of the following statement describes the hydatid cyst caused by *E. granulosus* most accurately?**
    a) The cyst is polycystic and multilocular.
    b) It is surrounded by one thick layer.
    c) It is filled with clear, colourless fluid that is highly antigenic and anaphylactic.
    d) It contains the adult tapeworm.
24. **PAIR stands for:**
    a) Puncture, aspiration, injection, and reaspiration
    b) Puncture, aspiration, inspiration, and reaspiration
    c) Penetration, aspiration, injection, and reaspiration
    d) Puncture, aspiration, injection, and reinjection
25. **A 34-year-old female presents with abdominal pain. On examination, she is febrile with low blood pressure and tenderness in right upper quadrant of abdomen. Investigations reveal her bilirubin is 3.5 and alkaline phosphatase is 400. Which of the following parasite is most likely involved in the underlying pathology?**
    a) *Loa loa* b) *Ascariasis lumbricoides*
    c) *Enterobius vermicularis* d) *Clonorchis sinensis*
26. **The infective organism is most likely to have been transmitted to this patient by which of the following?**
    a) Consumption of water plants
    b) Consumption of unpurified drinking water
    c) Contact with contaminated soil
    d) Consumption of raw or undercooked fish
27. **Which of the following is correct regarding *Fasciola hepatica*?**
    a) It is the one of the largest cestodes to infect humans.
    b) The eggs are unoperculated and embryonated.
    c) It is leaf-shaped with two suckers.
    d) It lives in the large intestine of humans.
28. **The habitat of *Fasciola hepatica* is:**
    a) Bile ducts and gall bladder
    b) Small intestine
    c) Large intestine
    d) Duodenum
29. **Anchovy sauce pus is characteristic of liver abscess caused by:**
    a) *Entamoeba histolytica* b) *E. granulosus*
    c) *Clonorchis sinensis* d) Any pyogenic bacteria

Answers: 1:b, 2:c, 3:c, 4:a, 5:b, 6:a, 7:a, 8:d, 9:c, 10:c, 11:b, 12:c, 13:a, 14:b, 15:b, 16:b, 17:d, 18:a, 19:d, 20:d, 21:c, 22:c, 23:c, 24:a, 25:d, 26:d, 27:c, 28:a, 29:a

Part III: Microbiology as Applied to Infectious Diseases

CHAPTER 27

Genitourinary and Sexually Transmitted Infections

(MI7.1 AND 7.3)

Chapter Navigation

Learning objective

- Describe the etiopathogenesis, clinical features, and laboratory diagnosis of sexually transmitted infections

Genitourinary infections may be broadly divided into two categories:

- **Genital tract infections:** The infections that are primarily transmitted by sexual contact are known as **sexually transmitted infections (STIs)**. Non-sexually transmitted genital infections include prostatitis in men, pelvic inflammatory disease, pyogenic infections of the vulva such as Bartholin abscess, cellulitis, etc., in women.
- **Urinary tract infections (UTI):** These are infections that primarily affect the urinary tract (Chapter 28).

SEXUALLY TRANSMITTED INFECTIONS (STI)

- These infections are perhaps among the most common infections and occur across the globe. Changes in sexual behaviour, the non-availability of vaccines, and the changing socio-economic environment have made the control of these infections a challenge.
- Sexually transmitted infections are rarely caused by a single organism. For example, syphilis and gonorrhea may exist concurrently.
- Alternatively, one infection may reactivate another dormant one, e.g., gonorrhea may reactivate genital herpes. It is, therefore, prudent to investigate an individual presenting with symptoms of genital infections for multiple causative agents.
- STIs are most often caused by bacteria and viruses, followed by a few fungi and protozoa (Table 27.1).

Table 27.1 Etiology of sexually transmitted infections (STIs)

Bacteria	Viruses	Fungi	Parasites
Treponema pallidum → Syphilis	Herpes simplex virus → Genital herpes	*Candida albicans* → Vaginal thrush	*Trichomonas vaginalis* → Vaginitis, urethritis
N. gonorrhoeae → Gonorrhea	Cytomegalovirus		
Chlamydia trachomatis → L serotypes	Human papillomavirus → Warts		
Lymphogranuloma venereum			
D-K serotypes			
Non-specific urethritis			
Gardnerella vaginalis → Vaginosis	Human immunodeficiency virus → AIDS		
Haemophilus ducreyi → Chancroid	Hepatitis B virus, hepatitis C virus → hepatitis		
Klebsiella granulomatis → Granuloma inguinale	*Molluscum contagiosum* → Venereal wart		

CLINICAL FEATURES

Sexually transmitted diseases may present with the following signs and symptoms:

- Ulcerative lesions in both men and women
- Vaginal, cervical discharge, vulvovaginal itching, burning, irritation, odour in women
- Cervicitis, endometritis, salpingitis, or pelvic inflammatory disease in women
- Uretheritis, epididymitis, proctocolitis, and proctitis enteritis in men and transgender patients

Presentation in Females

1. **Vaginal discharge** is a common feature of STIs. It may give a clue about the causative agent:
   - Profuse, watery discharge → bacterial vaginosis
   - Thick, white discharge → candiasis
   - Frothy, discharge → trichomoniasis
2. **Cervical/urethral discharge** (including vaginal discharge) is a feature of *Neisseria gonorrhoeae* or *Chlamydia trachomatis* infections.
3. **Pelvic inflammatory disease** (PID) presents chiefly with lower abdominal pain, vaginal discharge, and fever. *N. gonorrhoeae* and *C. trachomatis* are the principal pathogens in this entity.

Features Common to Males and Females (Table 27.2)

Presentation in Males

- **Epididymitis** is a common presenting feature of STIs in males; it begins as urethritis in sexually active males and ascends to the epididymis. In older males, it is related to bladder infection.
- **Prostatitis** is a non-sexually transmitted infection. Bacteria ascend to the prostatic ducts by reflux from the urethra and cause infection. Pelvic pain and irritative voiding symptoms along with fever and chills are characteristic of prostatitis.

APPROACH TO DIAGNOSIS OF SUSPECTED STIs

Figure 27.1 is an algorithm of the approach to the diagnosis of suspected sexually transmitted infections.

Collection and Transport of Specimens

Specimen

- **Discharge from the infected area/lesion:** When the vaginal, cervical, or urethral discharge is profuse, it can be collected in a sterile container.
- **Swabs:** When the discharge from the infected area is scanty or very thick, it is collected with a sterile swab or endocervical curette from the base and edges of the ulcer.
- **Fluid from genital vesicles:** Sterile capillary tubes are used to collect material from incised genital vesicles.
- **Blood:** It is collected for serological tests (for *Chlamydia*) and for culture (in suspected gonococcal bacteremia).

In view of the social stigma around STIs and the association with behavioural patterns, it is imperative to respect the **patient's autonomy and maintain confidentiality of the diagnosis and reports of investigations** (details in Chapter 4).

Transport

- The causative organisms are delicate and do not remain viable for long outside the body.
- Immediate transport to the laboratory and inoculation is necessary.
- A transport medium such as the modified Stuart's medium is used when necessary.

Microscopy

A wet mount of vaginal discharge may show the following features:

- In trichomoniasis, pus cells and motile (quivering motility) trophozoites of *Trichomonas vaginalis* are noted.

Table 27.2 Ulcerative STIs seen in both males and females

Clinical entity	Features	Causative organism
Chancre (Primary syphilis)	A single ulcer that is indurated (hard), painless and not tender, with moderate lymph node swelling and no bubo	*Treponema pallidum*
Genital herpes (herpes genitalis)	Multiple, non-indurated, painful, and markedly tender ulcers with moderate-to-no lymph node swelling and no bubo	Herpes simplex virus types 1 and 2
Chancroid	Single (or multiple), non-indurated (soft), painful and markedly tender ulcer/s occur, with marked lymph node swelling and bubo formation	*Haemophilus ducreyi*
Lymphogranuloma venereum (LGV)	Transient genital ulcer(s) or papule(s), followed by the appearance of tender inguinal and/or femoral lymphadenopathy with swollen, matted lymph nodes along the course of the inguinal ligament	*Chlamydia trachomatis* serovars L1, L2 and L3
Granuloma inguinale (Donovanosis)	Painless, slowly progressive ulcerative lesions on the genitals or perineum without regional lymphadenopathy; subcutaneous granulomas (pseudobuboes) also might occur; the lesions are highly vascular and can bleed	*Klebsiella granulomatis* (formerly called *Donovania granulomatis* or *Calymmatobacterium granulomatis*)

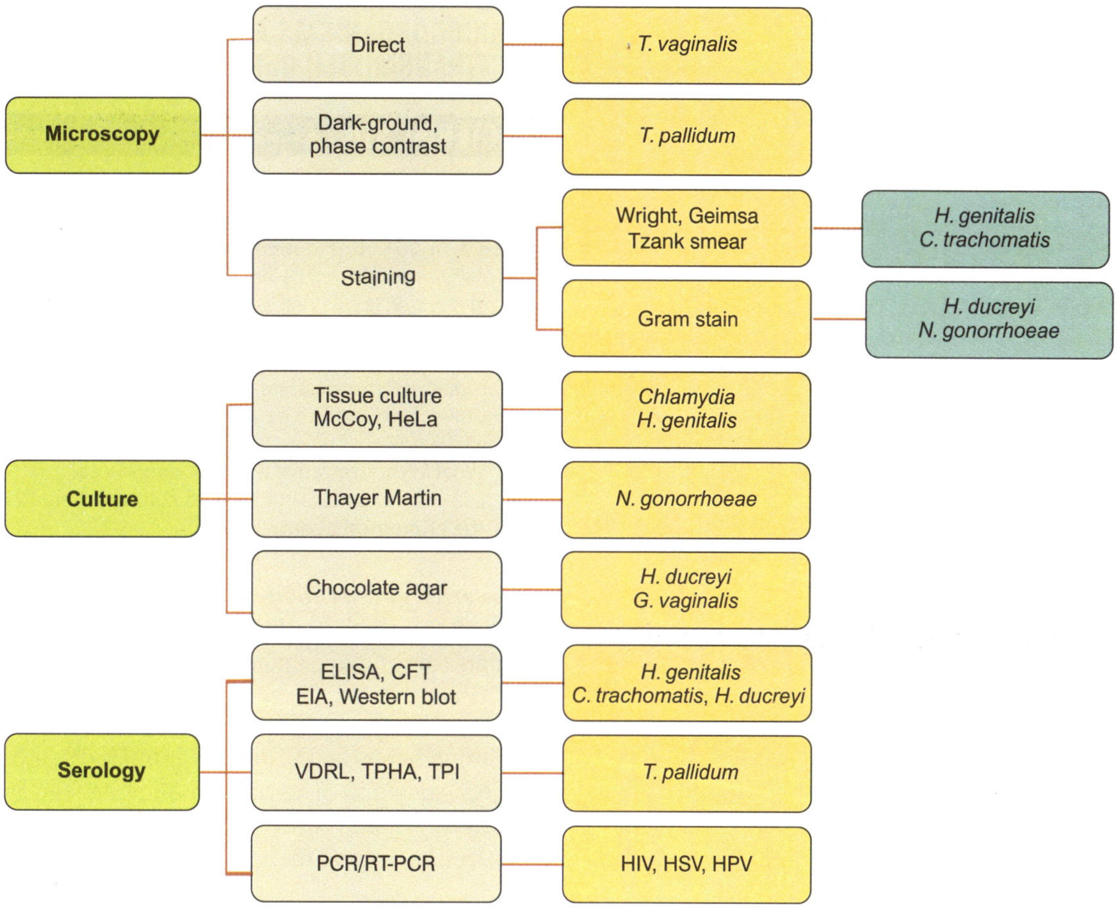

Fig. 27.1 Algorithm of the approach to the diagnosis of sexually transmitted infections (*VDRL*—Venereal Disease Research Laboratory test; *TPHA*—*T. pallidum* hemagglutination assay; *TPI*—*T. pallidum* immobilisation test)

- In candidiasis, yeast cells and pseudohyphae of *Candida albicans* are noted.

Gram staining

- Diagnosis of bacterial vaginosis can be made based on the **Nugent score**, which indicates a decreased number of lactobacilli (normal flora).
- A higher than usual number of gram-variable pleomorphic coccobacilli is suggestive of *Gardnerella vaginalis* infection.
- The presence of curved gram-negative bacilli is suggestive of *Mobiluncus* infection.

Other microscopic features

- **'Clue cells':** These are vaginal epithelial cells with adherent bacteria (*G. vaginalis*) that bear a characteristic stippled appearance; the presence of such cells and absence of pus cells suggests a diagnosis of bacterial vaginosis.
- **Gonorrhea:** Numerous polymorphs and gram-negative intracellular diplococci are seen in the cervical/urethral discharge, and sometimes in the vaginal discharge.
- **Chancroid:** Gram-negative coccobacilli are seen in pairs.
- **Candidiasis:** Gram-positive budding yeast cells can be seen in vaginal discharge.

Giemsa staining

- *Klebsiella granulomatis* is usually demonstrated in direct smears of material from lesions on the genitalia, in the groin, and on the perineum; the coccoid, rod-shaped organisms are found within histiocytes and giant cells.
- In suspected cases of primary or secondary syphilis—dark-ground microscopy is performed on preparations of material from the chancre (primary syphilis) or from mucous membranes (secondary syphilis) to demonstrate spirochetes.

Isolation

The causative organism may be isolated from an appropriate sample using the methods listed in Table 27.3.

Serology

- Serum samples are used for serological tests to establish the diagnosis of syphilis (VDRL, TPHA, TPI), chlamydial infection (CFT), and genital herpes infections (ELISA).

Table 27.3 Methods to isolate organisms responsible for sexually transmitted infections

Pathogen	Medium of isolation in the laboratory
Chlamydiae	Irradiated McCoy cells
N. gonorrhoeae	Thayer–Martin medium and NYC medium among others
H. ducreyi and *G. vaginalis*	Chocolate agar
Herpes simplex viruses types 1 and 2	Cell culture
Candida albicans	Sabouraud dextrose agar

- Molecular tests based on nucleic acid amplification tests (NAAT) have now replaced conventional culture for *Chlamydia* and *Mycoplasma* and other organisms that are difficult to cultivate such as HIV, HSV, HPV, CMV, etc.

TREATMENT

Treatment guidelines have been developed for the prevention, management, and control of reproductive tract infections including sexually transmitted infections (STI) by the Ministry of Health and Family Welfare, Government of India. These advocate following a syndromic approach to treatment (Table 27.4).

MICROORGANISMS ASSOCIATED WITH SEXUALLY TRANSMITTED GENITAL TRACT INFECTIONS

BACTERIAL STIs

Gonorrhea

Case 1. *A 20-year-old man presented with purulent urethral discharge and dysuria. He had a history of unprotected sex with a commercial sex worker seven days before. On examination of a smear of the pus, intracellular gram-negative diplococci along with abundant polymorphs were found. The culture on Thayer–Martin medium was positive, and a diagnosis of* ***gonorrhea*** *was made. The patient responded to the injection of ceftriaxone.*

i. What is the causative agent of gonorrhea?
ii. Describe the morphology and cultural characteristics of the etiological agent.

Neisseria gonorrhoeae causes the venereal disease gonorrhea, which has been known to be an infliction of human beings since ancient times.

Epidemiology

- Gonorrhea is an exclusively human disease.
- The existence of asymptomatic carriage in women makes them reservoirs, serving to perpetuate infection among their male contacts.

Table 27.4 Antibiotics for some of the common STIs (based on Operational guidelines for programme managers and service providers for strengthening STI/RTI services—available at https://naco.gov.in/sites/default/files/STI%20Operational%20Guidelines.pdf)

Color-coded kit for syndromic case management	Syndrome	Treatment
Grey	Urethral discharge, cervicitis, anorectal discharge, painful scrotal swelling	Oral azithromycin 1 g and cefixime 400 mg
Green	Vaginitis	Oral secnidazole 2 g and fluconazole 150 mg
White	Genital ulcer disease (non-herpetic)	Inj. benzathine penicillin, 2.4 million units intramuscularly and oral azithromycin 1 g (doxycycline for those who are allergic to penicillin)
Red	Herpetic genital ulcer disease, Genital herpes	Acyclovir, valacyclovir
Yellow	Lower abdominal pain/PID	Oral cefixime 400 mg and metronidazole 400 mg and doxycycline 100 mg
Black	Inguinal bubo	Oral Doxycycline and azithromycin
	Syphilis	Benzathine penicillin G
	Gonorrhea	Ceftriaxone, azithromycin
	Chlamydial infection	Ceftriaxone (single dose) + doxycycline (14 days)
	Chancroid	Azithromycin
	Lymphogranuloma venereum	Doxycycline
	Donovanosis	Doxycycline

- The recent increase in antimicrobial resistance along with increased prevalence has also affected the treatment and control of the infection.

Agent—*N. gonorrhoeae*

Morphology

- ***N. gonorrhoeae*** are gram-negative diplococci that are kidney-shaped and found predominantly within polymorphs.
- They are antigenically heterogeneous and capable of changing their surface structures in vitro.

Surface structures

- **Pili:**
  - These are hair-like structures, several micrometres long, which act as virulence factors by promoting attachment to host cells, providing motility and inhibiting phagocytosis.
  - They are composed of repeating peptide subunits (pilins) and are known to undergo antigenic and phase variation.
- The **outer membrane** contains endotoxins, which may be responsible for the toxicity in gonococcal infections.
- **Outer membrane proteins**: Gonococci possess three major outer membrane proteins named (proteins I, II, and III). These are major constituents and show antigenic diversity, which helps in typing gonococcal strains.
  - *Protein I (Porin):* **Forms pores in the surface and is used for serotyping.**
  - *Protein II (opacity associated protein or OPA):* It is related to the opacity of the gonococcal colonies outer membrane proteins. OPA helps facilitates the attachment of the organism to host cells.
  - *Protein III:* **It is complexed with PI to form porins.**
- **Lipo-oligosaccharide** attaches to sperm cells and assists in transmission.
- **Other factors** associated with pathogenicity—**IgA1 protease**, which splits and inactivates IgA.

Susceptibility

- *Neisseria gonorrhoeae* are readily killed by heat, drying, and antiseptics.
- They are strict parasites and die in 1–2 hours in exudates outside the body.
- In cultures, they die in 3–4 days but survive in slant cultures at 35°C if kept in sterile paraffin oil.
- Cultures may be preserved for years if frozen quickly and stored at –70°C.

Pathogenesis (**Fig. 27.2**)

- The infection is acquired by sexual contact.
- Initial infection involves the urethra and cervix uteri; the vaginal mucosa is not usually affected in adults because the stratified squamous epithelium is resistant to infection by the cocci and because of the acidic pH of vaginal secretions.
- The first step in infection is the adhesion of gonococci to the urethra or other mucosal surfaces with the help of pili and OPA.
- The bacteria then bind to complement receptors type 3 (CR3) on cervical/urethral epithelial cells.
- The attachment causes extensive rearrangement of the host cell actin, resulting in large projections called **ruffles**.
- Ruffles allow gonococci to enter host cells in large vacuoles and subsequently multiply within infected cells.

Clinical features

The **incubation period** is 2–8 days.

In males

- The patient presents with acute urethritis with a mucopurulent discharge, dysuria, and testicular discomfort (Case 1).
- Complication—the infections may spread to the periurethral tissues, causing abscesses and multiple discharging sinuses (**'watering can' perineum**).
- Orchitis, epididymitis, penile lymphangitis, penile edema, and post-infectious urethral strictures may develop.

In females

- **Asymptomatic carriage**—clinical disease is less severe in women, and many carry gonococci in their cervix without any symptoms.
- More than 50% of females will not manifest symptoms of their gonococcal cervix infections.

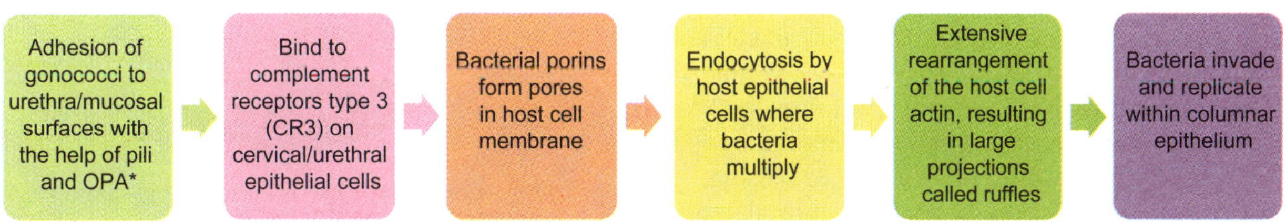

Fig. 27.2 Pathogenesis of gonococcal infections
**OPA*, opacity-associated protein

- Symptoms may be vaginal discharge, dysuria, or pelvic pain.
- Bartholin's glands infection adjacent to the vaginal introitus manifests as labial soft tissue swelling, abscess formation, and pain.
- Gonococcal infections of the urogenital tract may manifest as follows in women:
  - **Vulvovaginitis** in pre-pubertal girls
  - **Pelvic inflammatory disease (PID)**
  - **Salpingitis**, which may lead to sterility
  - **Fitz-Hugh–Curtis syndrome**—rarely, peritonitis may develop with perihepatic inflammation

Disseminated gonococcal infection (DGI)

DGI is a severe form of systemic illness and manifests as follows:

- **Proctitis** (in both sexes), may develop by direct contiguous spread in women; in men, it is usually the result of anal sex.
- **Conjunctivitis** usually occurs due to autoinoculation by the patient's fingers.
- **Blood invasion** may occur from the primary site of infection and may lead to metastatic lesions such as arthritis, ulcerative endocarditis and, very rarely, meningitis.
- Occasional cases of pyemia have been reported.
- Gonococcal bacteremia leads to skin lesions, especially hemorrhagic papules and pustules on the hands, forearm, feet, and legs and tenosynovitis and suppurative arthritis, usually of the knees, ankles, and wrists.

Ophthalmia neonatorum

- A non-venereal infection is gonococcal ophthalmia in the newborn caused by **direct infection during passage through the birth canal**.
- Once very common, this entity has been controlled by the practice of instilling 1% silver nitrate solution into the eyes of all newborn babies (**Crede's method**).

Laboratory diagnosis

Specimen

- **Females:**
  - Discharge, urethral swabs, and endocervical swabs are collected. Dacron or rayon swabs are preferred over cotton swabs since the latter may contain fatty acids that may be inhibitory to gonococci. The uretheral meatus is cleaned with a gauze soaked in saline, and a sample of the discharge is collected with a platinum loop for culture or directly onto a slide for smears.
  - Endocervical swabs are collected using a speculum. The swab is inserted and moved gently from side to side to collect secretions. Bartholin's gland secretions, if present, should be collected
  - High vaginal swabs are not satisfactory. In chronic infections, there may not be any urethral discharge.
- **Males**
  - *Acute urethritis:* Purulent material is collected directly on a slide or using a swab.
  - *Chronic urethritis:* The '**morning drop**' of secretion or exudate obtained after prostatic massage.
  - Urine is collected in cases where no urethral discharge is available. Anal or oropharyngeal samples can be collected in either sex if the history is suggestive of unusual sexual practice.

Transport

- Specimens should be inoculated immediately after collection.
- If this is not possible, specimens should be collected with charcoal-impregnated swabs and sent to the laboratory in **Stuart's or Amie's transport medium**.

Microscopy

- **Gram stain:** The presence of intracellular, gram-negative diplococci in stained smears (Fig. 27.3) provides presumptive evidence of gonorrhea in men. This test may be unreliable in women as some of the normal genital flora have similar morphology.
- **Fluorescent antibody technique:** This method has higher sensitivity and specificity of diagnosis by microscopy.

Culture (Table 27.5)

- **Media:** Chocolate agar or Thayer–Martin medium (modified Thayer–Martin medium with vancomycin, colistin and nystatin)
- **Colonies:** Small, round, translucent, convex, or slightly umbonate with a finely granular surface and lobate margins.

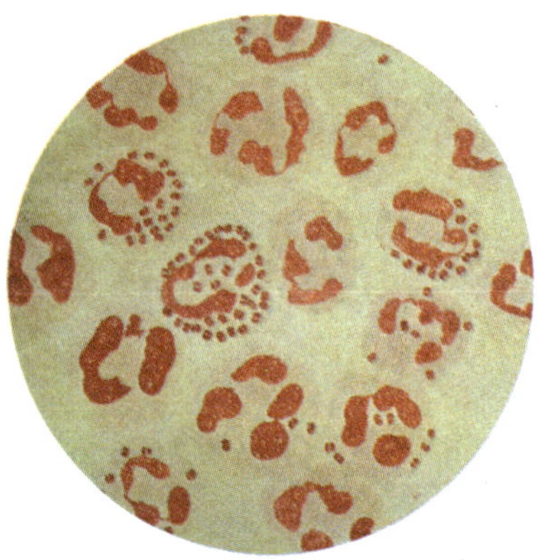

Fig. 27.3 Gonococci in urethral pus—gram-negative diplococci (*Source*: Centers for Disease Control and Prevention [CDC], Public Health Image Library [PHIL], Image ID 2108)

Table 27.5 Differential characteristics of *Neisseria* spp.

Species	Colonies	Serological classification
N. meningitidis	Round, smooth, and shiny with creamy consistency	Thirteen antigenic groups
N. gonorrhoeae	Same as above, but smaller and more opalescent	Antigenically heterogeneous
N. catarrhalis (Branhamella catarrhalis) Resembles meningococcus but is pigmented yellow	Variable, smooth, and translucent or adherent and opaque, not easily emulsifiable	Auto-agglutinates

Identification

The identification of gonococci is based on carbohydrate fermentation tests.

Serological tests

No serological test has been found to be useful for routine diagnostic purposes.

Molecular methods

PCR and molecular methods have improved the sensitivity of diagnostic assays but are not cost-effective for routine use.

Treatment

- **Empiric therapy:** A single intramuscular or intravenous dose of 500 mg ceftriaxone
- **Complicated gonococcal infections (including pelvic inflammatory disease [PID], epididymitis, and proctitis):** Single intramuscular or intravenous dose of 500 mg of ceftriaxone along with oral doxycycline 100 mg BID for seven days
- **WHO recommended treatment:** Cefixime + azithromycin or ceftriaxone + azithromycin

Due to an increase in ciprofloxacin resistance, the recommended treatment protocol is ceftriaxone 125 mg IM single dose with 1 g azithromycin single dose or doxycycline 100 mg twice a day for 7 days. The regime works well against gonococci and the frequently co-existing chlamydial infection (CDC recommendation).

Antimicrobial resistance

Antimicrobial resistance in *N. gonorrhoeae* is on the rise, with strains becoming resistant to all previous first-line therapeutic drugs as shown in Fig. 27.4.

- **Penicillinase-producing *N. gonorrhoeae* (PPNG):** Two transmissible plasmids containing genes coding for beta-lactamase, which causes resistance to penicillin, have spread widely. Lately, gonococcal isolates with ceftriaxone resistance have been described in many countries.

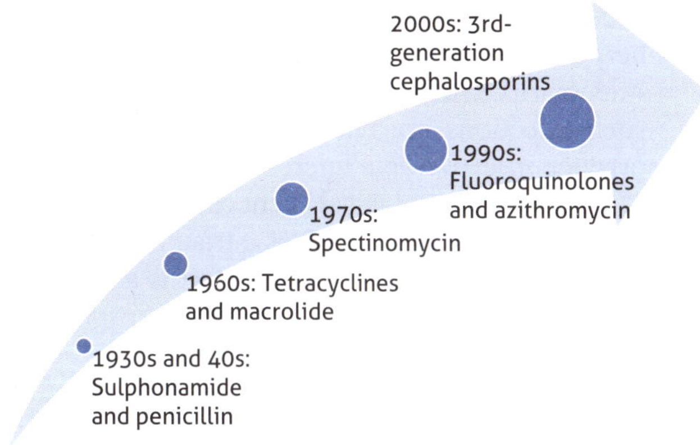

Fig. 27.4 Emergence of resistance in gonococci

- **Multidrug-resistant gonococci (MDR):** Strains resistant to multiple drugs have also emerged recently, following which, there has been an emergence of extensively drug-resistant (XDR) *Neisseria gonorrhoeae*.

Prophylaxis

- Partner management is recommended for patients regardless of whether specific etiology is identified or symptoms are present.
- Clinical disease does not confer any immunity, and vaccination has no place in prophylaxis.

Commensal neisseriae

- Several species of neisseriae inhabit the normal respiratory tract.
- The characteristic features of some of the common species are listed in Table 27.5.
- Their pathogenic significance is uncertain though some of them (e.g., *N. flavescens* and *N. catarrhalis*) have been reported occasionally as having caused meningitis.

Non-gonococcal (Non-specific) Urethritis

> Case 2. *A 20-year-old male presented to an STI clinic with complaints of urethral discharge and dysuria for two weeks. The patient was subjected to laboratory investigations. A Gram smear of the discharge showed plenty of pus cells but no bacteria; Giemsa stain showed epithelial cells with intracytoplasmic inclusion bodies. Culture was negative. The chlamydial antigen detection test was positive. A diagnosis of* ***non-gonococcal urethritis (NGU)*** *was made and treatment with tetracycline was prescribed for seven days.*
> i. What are the agents causing NGU?
> ii. Describe the spectrum of infections caused by *Chlamydia trachomatis.*

Non-gonococcal urethritis (NGU) refers to chronic urethritis in which gonococci cannot be demonstrated.

- The majority of cases are caused by *Chlamydia trachomatis* (15% to 40% of NGU cases) followed by *Mycoplasma genitalium* (15–20%). Rare causes of NGU include *Trichomonas vaginalis*, herpes simplex virus, Epstein Barr virus, and Adenovirus. Other rare agents are enteric bacteria; such infections are typically seen in males who practice insertive anal intercourse. In almost half of the cases no etiological agent can be diagnosed.

- **Reiter's syndrome** comprises a triad of **urethritis with conjunctivitis and arthritis**. It is associated with infections such as *Mycoplasma, Ureaplasma,* and *C. trachomatis*.
- The causes of NGU, and its differentiating features are listed in Table 27.6.

Agent—*Chlamydia trochomatis* serovars D-K

- Chlamydial infections have become the most common sexually transmitted disease worldwide.
- They are also the most common cause of NGU.

Table 27.6 Salient differences between gonococcal and non-gonococcal urethritis

Characteristics	Gonococcal urethritis	Non-gonococcal urethritis
Causative agent	*N. gonorrhoeae*	*C. trachomatis* serovar D-K *Ureaplasma urealyticum* *Mycoplasma genitalum* *M. hominis* Herpes simplex virus Cytomegalovirus Epstein Barr virus Adenovirus *Candida albicans* *Trichomonas vaginalis* Unknown
Transmission	Sexual	Sexual
Diagnosis	Microscopy and culture	*C. trachomatis, Ureaplasma urealyticum,* and *Mycoplasma* by immunofluorescence and antigen detection Herpes virus and cytomegalovirus by immunofluorescence, antigen detection, and PCR *Candida albicans* by microscopy and culture *Trichomonas vaginalis* by wet mount examination by microscopy
Treatment	Third-generation cephalosporin/ azithromycin	Azithromycin 1 g orally in a single dose or Doxycycline 100 mg orally twice a day for 7 days

- In India, chlamydial infection has been reported in 20–30 per cent of women with mucopurulent cervicitis and 30–60 per cent of those with salpingitis and pelvic inflammatory disease.
- *C. trachomatis* causes two types of genital infections:
  i. Urogenital syndromes caused by the oculogenital serotypes D to K, collectively referred to as genital chlamydiasis
  ii. Lymphogranuloma venereum (LGV), caused by serotypes L1, L2, and L3

Clinical features

The clinical spectrum parallels that of gonococcal infections. The two may often co-exist.

- **Males:** Urethritis (NGU), epididymitis, proctitis, conjunctivitis, and Reiter's syndrome may be seen.
- **Females:**
  – Acute urethral syndrome, bartholinitis, mucopurulent cervicitis, endometritis, salpingitis, pelvic inflammatory disease, conjunctivitis, perihepatitis (Fitz-Hugh–Curtis syndrome), and Reiter's syndrome may occur.
- **Genital chlamydiasis** may cause premature deliveries, perinatal morbidity, and postpartum fever. Untreated infection may lead to pelvic inflammatory disease (PID), infertility, and potentially fatal tubal pregnancy.

Agents—*Mycoplasma genitalium* and *M. hominis*

- *Mycoplasma* are cell wall-deficient bacteria that may be spherical or filamentous, which are associated with infections of the respiratory and urogenital tract (details in Chapter 4 and 23).
- *M. genitalium* causes symptomatic and asymptomatic urethritis among men. In women, it causes cervicitis, PID, preterm delivery, spontaneous abortion, and infertility. Men with recurrent NGU should be tested for *M. genitalium*.
- It is an extremely slow-growing organism and cultures can take up to 6 months. Molecular detection is done on urine and urethral, penile, meatal, endocervical, and vaginal swab samples using an **FDA-cleared NAAT**.
- *Mycoplasma hominis* may cause salpingitis, pelvic abscess, septic abortion, puerperal infection, and tubo-ovarian abscess.

Agent—*Ureaplasma urealyticum*

- These are strains of *Mycoplasma* frequently isolated from the urogenital tract of human beings and animals that form very tiny colonies, generally 15–50 μm in size. They are also called **T strains or T form mycoplasmas**.

- These bacteria hydrolyse urea, which is an essential growth factor in addition to cholesterol.
- They are the second most common cause of non-gonococcal urethritis after *Chlamydia trachomatis*.

Clinical features

- **Females**—acute salpingitis, pelvic inflammatory disease, cervicitis, and vaginitis. These organisms are also associated with infertility, abortion, postpartum fever, chorioamnionitis, and low birth weight of infants.
- **Males**—urethritis, proctitis, balanoposthitis, and Reiter's syndrome.

Laboratory diagnosis (Fig. 27.5)

- **Microscopy**: Neutrophils (more than four per oil immersion field in urethritis and more than 30 in cervicitis) are seen in the absence of gonococcal infection.
- **Serology**: Antigen detection by micro-IF, ELISA, or molecular techniques.

Treatment of NGU

- Azithromycin and moxifloxacin are advocated to treat NGU; they perform better than tetracycline and doxycycline.
- *Ureaplasma*, however, can be resistant even to these agents.
- *Mycoplasma* responds better to doxycycline for 7 days, followed by moxifloxacin or azithromycin.

Lymphogranuloma Venereum (LGV)

- This is an ulcerative disease of the genital area characterised by suppurative inguinal adenitis (also known as **lymphogranuloma inguinale**).
- It is caused by the *Chlamydia trachomatis* serovars L1, L2, and L3 which have a predilection for regional lymph nodes.

Clinical features

There are three stages of LGV, which are discussed below.

Primary stage

- This is characterised by the development of **painless genital ulcers or papules or papulovesicular lesions** appearing on the external genitalia (or rarely, extragenital sites).
- The primary lesion usually goes unnoticed, and the disease is usually first seen at the stage of **inguinal adenitis (bubo)**.

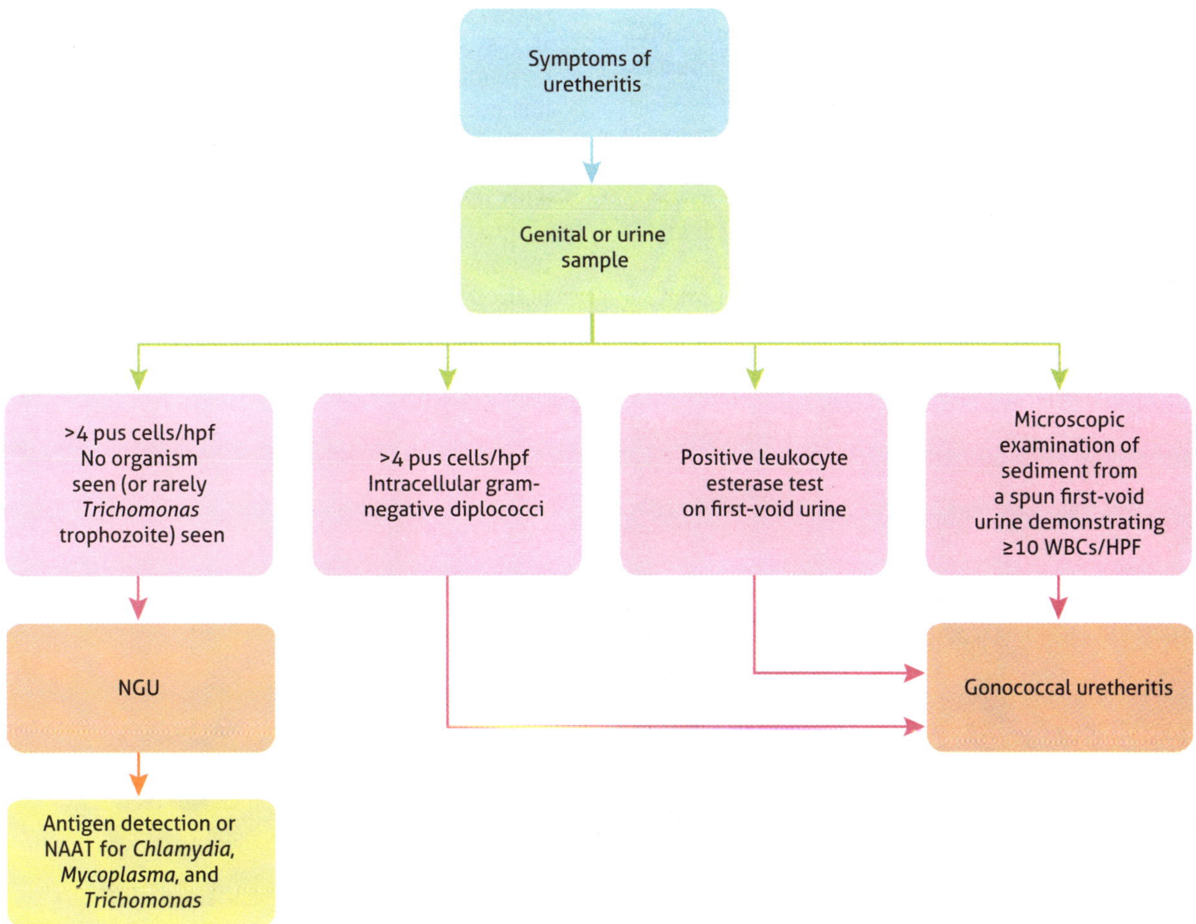

Fig. 27.5 Diagnosis of NGU

Secondary stage

- It develops two weeks later with the **spread of the disease from the lymphatics to the draining lymph nodes**. *Chlamydia trachomatis* serovars extend from the primary infection site to the regional draining lymph nodes, causing a lymphoproliferative reaction.
- There is unilateral or bilateral tender inguinal and/or femoral lymphadenopathy (also called buboes) (Fig. 27.6).
- In women, the intrapelvic and pararectal nodes are involved. Women and homosexual men may develop hemorrhagic proctitis with regional lymphadenitis. The nodes enlarge, suppurate, become adherent to the skin and break down to form sinuses discharging pus.
- Metastatic complications may sometimes occur, with the involvement of the joints, eyes, and meninges.

Tertiary stage

- The disease is **chronic, lasting for several years**, with the sequelae of scarring, fibrosis, fistulae, and lymphatic blockage of the anogenital area.
- Late sequelae are more distressing in women as they lead to rectal strictures and disfigurement of the vulva (esthiomene).

Laboratory diagnosis

- *Chlamydia trachomatis* can be visualised in genital, rectal, and lymph node specimens by culture, nucleic acid amplification test, or direct immunofluorescence.
- Microscopy has low sensitivity. Smears of material aspirated from the buboes may show the elementary bodies (**Miyagawa's granulo-corpuscles**).
- *Chlamydia* are isolated on cell cultures, which are of diagnostic value.
- LGV patients develop high titres of circulating antibodies, with titres of 1:512 or more in micro-IF test, which is diagnostic.

Treatment

LGV is treated with doxycycline 100 mg twice a day for at least three weeks.

Syphilis

> **Case 3.** *A 20-year-old male truck driver presented to a skin and venereal disease clinic with a genital ulcer of 10 days' duration. The ulcer was painless. The man had a history of unprotected sexual contact with a commercial sex worker about two weeks before this. On examination, the ulcer was found to be circumscribed, indurated, and partially healed (Fig. 27.7). The man's inguinal lymph nodes were enlarged. A provisional diagnosis of **syphilis** was made, and the patient's blood was sent for a VDRL test. This was reactive at 1:64 dilution. The patient was treated with intramuscular penicillin G (benzathine penicillin).*
>
> i. What is the clinical presentation of syphilis?
> ii. How is this condition diagnosed?

Syphilis is a systemic bacterial infection caused by the spirochete *Treponema pallidum* belonging to genus *Treponema*. The infection progresses through four stages and can affect many organ systems.

Epidemiology

- Syphilis is transmitted sexually and is worldwide in distribution.
- Humans are the only source of treponemal infection; there are no known non-human reservoirs.
- It was hoped that the spread of the disease could be contained with penicillin treatment; however, this has not been possible.

HIV and syphilis

- The AIDS pandemic had an impact on syphilis. Concurrent infection with *T. pallidum* and HIV is common and may lead to earlier evolution of neurosyphilis.
- The detection, diagnosis, and treatment of concurrent infections of HIV and syphilis and other

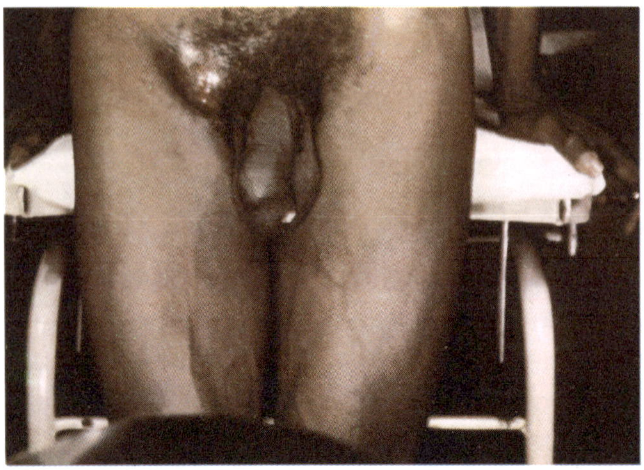

Fig. 27.6 Large, inguinal bubo on the patient's right side due to lymphogranuloma venereum (LGV) caused by the *Chlamydia trachomatis* (*Source*: CDC, PHIL, Image ID 18033)

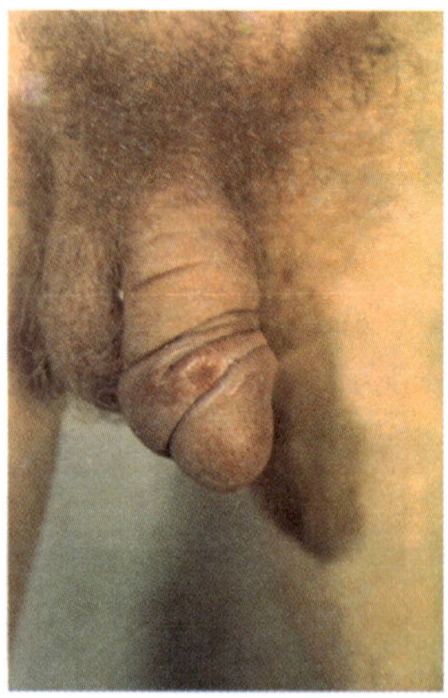

Fig. 27.7 Partially healed penile ulcer, primary syphilitic lesion (*Source*: CDC, PHIL, Image ID 17883)

sexually transmitted infections (STIs) have been discussed under control programmes.

Types of syphilis

Syphilis can be venereal, non-venereal, or congenital.

Venereal syphilis

- This is acquired by sexual contact.
- The **incubation period** is 10–90 days.
- The spirochete enters the body through minute abrasions on the mucosa or skin. *Treponema* multiplies at the site of entry.
- The clinical manifestations fall into four stages: primary, secondary, latent, and tertiary.

Primary syphilis

It is characterised by the development of chancres at the site of entry of the spirochete, usually the genitalia.

- **Chancre:**
  - It is a painless, relatively avascular, circumscribed, indurated, and superficially ulcerated lesion (Case 3).
  - It is known as a '**hard chancre**' to distinguish it from the non-indurated lesions of 'soft chancre' seen in infections caused by *H. ducreyi*.
  - It is covered with a thick, glairy exudate, which is rich in spirochetes.
  - It occurs on the genitalia and less commonly, on the mouth and nipples.
  - It invariably heals in about 10–40 days, even without treatment, leaving behind a thin scar.
- The regional lymph nodes are swollen, discrete, rubbery, and non-tender.
- Spirochetes spread from the site of entry into the lymph and bloodstream even before the chancre appears, so the patient may be infectious during the late incubation period.

Secondary syphilis

- Sets in 1–3 months after the primary lesion heals. During this interval, the patient is asymptomatic.
- The secondary lesions are due to the multiplication of the spirochetes and their dissemination through blood.
- **Condylomata** are seen at the mucocutaneous junctions (Fig. 27.8a).
- There are **roseolar or papular skin rashes, mucous patches in the oropharynx, and papulosquamous lesions on the palms and feet** (which may be confused with psoriatic lesions; Fig. 27.8b and c).
- Spirochetes are abundant in the lesions; consequently, the patient is most infectious during the secondary stage.
- There may also be ophthalmic, osseous, and meningeal involvement.
- Secondary lesions are highly variable in distribution, intensity, and duration.
- They usually undergo spontaneous healing, in some instances, over as long as four or five years.

Latent syphilis

- This is a period of quiescence seen after the secondary lesions disappear.
- Diagnosis during this period is possible only by serological tests.
- In many cases, this stage is followed by natural cure.

Tertiary syphilis

- Appears in some patients after several years.
- The disease manifests as **cardiovascular lesions** including aneurysms and chronic granulomata (gummata; Fig. 27.9) and meningovascular manifestations.
- Tertiary lesions containing few spirochetes may represent manifestations of delayed hypersensitivity.

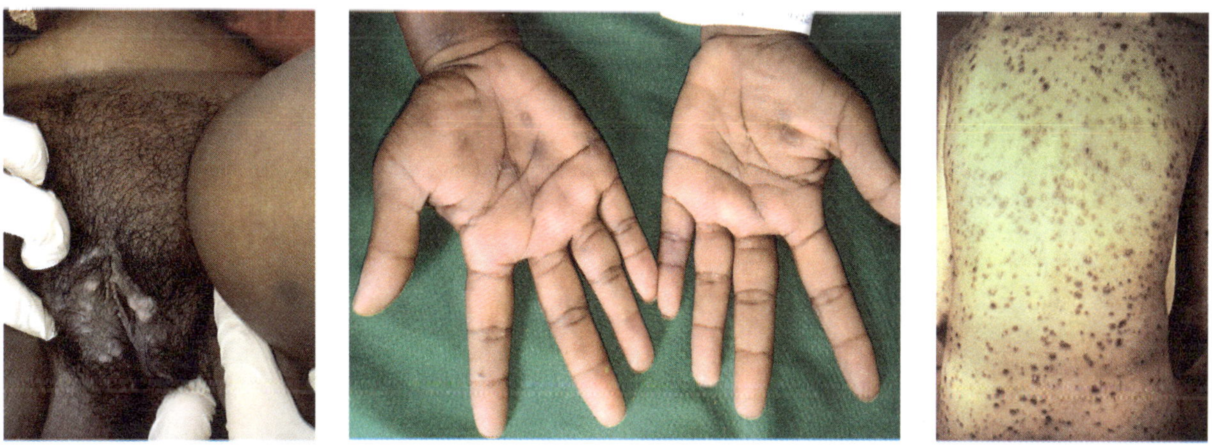

(a) (b) (c)

Fig. 27.8 Secondary syphilis (a) condylomata at the vulvovaginal mucocutaneous junction in a female patient; (b) papulosquamous rashes on the palm in secondary syphilis and (c) maculopapular rash on the back in secondary syphilis (*Source*: [a and b] Dr. Sheela Kuruvila, Department of Dermatology, Pondicherry Institute of Medical Sciences, Puducherry and [c] CDC, PHIL, Image ID 17838)

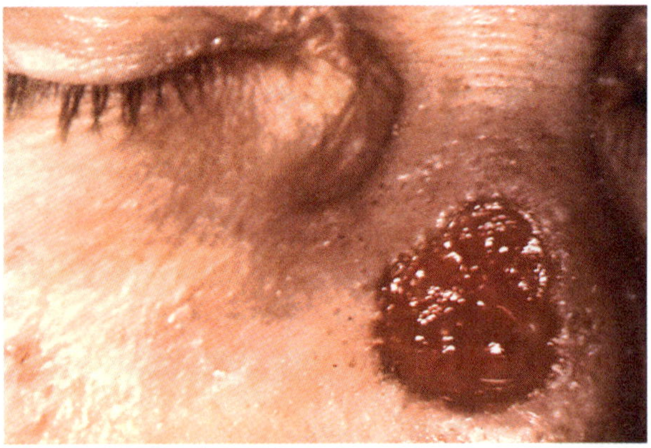

Fig. 27.9 Tertiary syphilitic lesion showing a cutaneous ulcerative lesion on the nose (*Source*: CDC, PHIL, Image ID 5330))

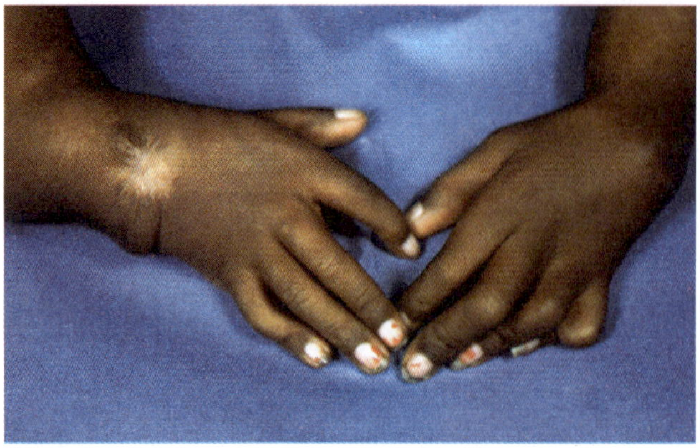

Fig. 27.10 Late congenital syphilis involving the bones—osteoperiostitis and ulnar deviation of the middle finger (*Source*: CDC, PHIL, Image ID 16750)

- The late tertiary stage may develop in a few cases several decades after the initial infection, presenting with neurological manifestations such as:
  - Tabes dorsalis
  - General paralysis of the insane (GPI)

Non-venereal syphilis

- Syphilis can also be acquired non-venereally (it is an occupational hazard for doctors or nurses).
- The natural evolution of this entity is similar to that of venereal syphilis, except that the primary chancre is extragenital, usually on the fingers.
- If syphilis is transmitted by blood transfusion, the primary chancre does not develop.

Congenital syphilis

- The infection is transmitted from an infected mother to her fetus transplacentally. This can take place at any stage of the pregnancy.
- Most neonates are asymptomatic at birth. However, in untreated infants, clinical manifestations usually appear by three months of age and may present as hepatomegaly, jaundice, rhinitis, generalised lymphadenopathy, and a maculopapular rash that progresses to desquamation and crusting.
- **Hutchinson triad** consisting of Hutchinson teeth, interstitial keratitis, and sensorineural hearing loss is relatively specific for congenital syphilis.
- Late congenital syphilis may involve the bones, resulting in osteoperiostitis (Fig. 27.10).
- Pathogenesis is the result of an immune response from the fetus. In untreated women, a history of repeated abortions is suggestive of congenital syphilis.

Agent—*Treponema pallidum*

Treponema pallidum, the causative agent of syphilis, was discovered by Fritz Schaudinn and Erich Hoffmann (1905) in the chancres and inguinal lymph nodes of syphilitic patients.

- Treponemes (*trepos*, meaning to 'turn' and *nema*, meaning 'thread') are relatively short, slender spirochetes with fine spirals and pointed or rounded ends.
- Members of the genus are almost identical in their morphology, antigenic structure, and other features, though there are differences in the clinical features and natural history of the diseases they produce.
- Some of them are pathogenic, while others are commensals in the mouth, intestines, and genitalia.
- Members of the genus *Treponema* are known to cause the following diseases in humans:
  - *Treponema pallidum* subspecies pallidum—veneral syphilis
  - *Treponema pallidum* subspecies endemicum—bejel or endemic syphilis
  - *Treponema pallidum* subspecies pertenue—yaws
  - *T. carateum*—pinta

Morphology

- *T. pallidum* is a thin, delicate spirochete with tapering ends.
- It is about 10 μm long (ranging from 4–14 μm) and 0.1–0.2 μm wide.
- It has about ten regular spirals, which are sharp and angular, at regular intervals of about 1 μm (Fig. 27.11).
- It is actively motile, exhibiting rotation around the long axis, backward and forward movements, and flexion of the whole body.
- During motion, secondary curves appear and disappear in succession, but the primary spirals are unchanged.

Ultrastructure

- The cytoplasm of *T. pallidum* is surrounded by a trilaminar cytoplasmic membrane, enclosed by a peptidoglycan cell wall, which gives the cell rigidity and shape. Outside this, it has a lipid-rich outer membrane layer (Fig. 27.12).

Fig. 27.11 *Treponema pallidum* with spirals at 1 μm intervals visualised with dark-field microscopy (*Source*: CDC, PHIL, Image ID 14969)

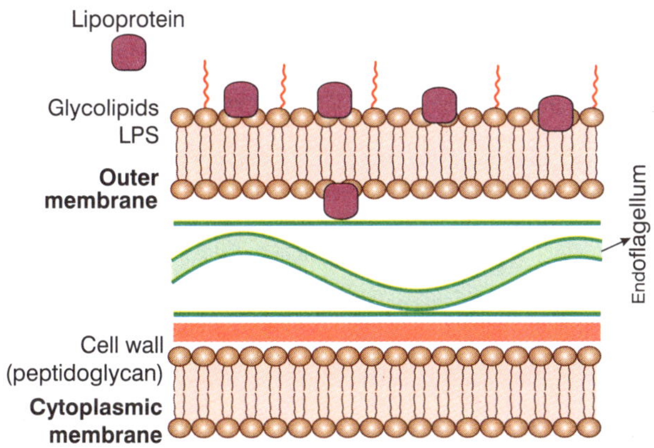

Fig. 27.12 Ultrastructure of the spirochetes showing cell wall made up of glycolipids (with LPS) and peptidoglycan; the flagellum lies in the periplasmic space, coiled around the axis of the body of the bacterium

- Originating from each end of the cell, three or four endoflagella wind around the axis of the cell in the space between the cell wall and the outer membrane layer to interdigitate at its centre.
- Endoflagella do not protrude from but remain within the outer membrane layer.

Antigenic properties

- Treponemes have a complex antigenic structure.
- They possess species-specific and group-specific antigens that evoke antibody response in an infected individual.

1. **Species-specific antigen**
   - It is probably polysaccharide in nature.
   - The antibody to this antigen is demonstrated by specific *T. pallidum* tests.
2. **Group-specific antigens**
   - These are found in *T. pallidum* as well as in non-pathogenic cultivable treponemes like the Reiter treponeme.
3. **Non-specific antigen (cardiolipin antigen)** The reagin antibody reacts with the standard or non-specific cardiolipin antigen in VDRL tests, in which a hapten, diphosphatidylglycerol, extracted from beef heart, is used as the antigen.
   - It is not known whether the reagin antibody is induced by the cardiolipin that is present in the spirochete or released from damaged host tissues.

Susceptibility

Treponemes are inactivated by distilled water, soap, mercurials, bismuth, common antiseptic agents and antibiotics.

Pathogenicity

- Natural infection with *T. pallidum* occurs only in human beings.
- Rabbits can be infected by intradermal or intratesticular inoculation, the former giving rise to chancres and the latter to granulomatous lesions 'syphilomas'.
- Hamsters are also susceptible.
- The generation time is 30–33 hours.

Immunity

- The immune mechanisms in syphilis are not adequately understood.
- Humoral immune response against the treponeme does not appear to be effective as the disease progresses even in the presence of a vigorous antibody response.
- Specifically, sensitised Th_1 cells secrete cytokines that may favour the clearance of spirochetes by activated macrophages.
- Re-infections do not appear to occur in a person with an active infection.

Laboratory diagnosis

There are two main diagnostic strategies in use:

- Demonstration of the spirochetes under the microscope
- Detection of antibodies in serum or CSF

Culture does not have a role in diagnosis.

Specimen

- Specimens from the genital ulcer should be collected with care as the lesions are highly infectious.
- Blood is collected for serology.
- CSF can be collected for investigations in the tertiary stage to confirm neurosyphilis.

Microscopy

T. pallidum cannot be seen under the light microscope in wet films. Its morphology and motility can be visualised using a dark-ground or phase-contrast microscope.

- **Dark-ground microscopy**
  - Useful in primary and secondary stages and in congenital syphilis with superficial lesions.
  - **Motility**—*T. pallidum* is identified based on its slender spiral structure, slow backward and forward movements (**cockscrew motility**), and characteristic bending at right angles.
  - Dark-ground microscopy is useful as it can be performed at the outpatient or point-of-care outlet.
  - **Sensitivity**—dark-ground microscopy has low sensitivity. Therefore, several specimens have to be screened over consecutive days.
  - A treponemal concentration of 10^4 per mL in the exudates is required to visualise the bacteria (Fig. 27.13a).
- **Direct fluorescent antibody test for *T. pallidum* (DFA-TP)**
  - It is a better and safer method of microscopic diagnosis.
  - Smears of the exudate are fixed with acetone and fluorescent-tagged monoclonal anti-*T. pallidum* antiserum is used to visualise the organisms.
  - Silver impregnation methods include **Fontana's method**, which is useful for staining films and Levaditi's method for tissue sections (Fig. 27.13b).

Cultivation

- This is not used in the laboratory confirmation of syphilis.

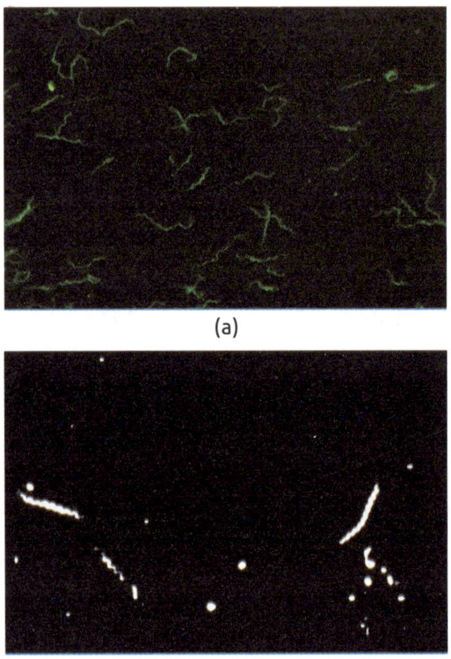

(a)

(b)

Fig. 27.13 (a) Dark-field microscopy shows closely spiralled *Treponema pallidum* and (b) Levaditi's silver impregnation method to demonstrate spirochetes in tissue (*Source*: CDC, PHIL, Image IDs 14968 and 14967)

- Pathogenic treponemes do not grow in artificial culture media.
- The cultivable strains are non-pathogenic treponemes, showing morphological and antigenic similarities with *T. pallidum*.

The **Reiter strain** (*T. phagedenis*) is widely used as the antigen in group-specific treponemal tests for the diagnosis of syphilis. It grows well in thioglycolate medium containing serum.

- Virulent *T. pallidum* strains have been maintained for many decades by serial passage in rabbit testis.
- Nichol's strain, isolated from the brain of a fatal case of general paralysis of the insane in 1912, is still being propagated and used for diagnostic and research purposes.

Serological tests

These tests form the mainstay of laboratory diagnosis. Three types of tests used to detect the antibodies:

i. Standard tests for syphilis
ii. Group-specific treponemal tests for syphilis
iii. Specific *T. pallidum* tests for syphilis

i. Standard tests for syphilis (STS)

- These utilise the cardiolipin antigen to detect the **reagin antibody**.
- The antigen is a purified lipid extract of beef heart with added lecithin and cholesterol.
- The antibody that reacts with cardiolipin is known as reagin (should not be confused with IgE antibody in atopy, which is also called reagin, though there is no connection between the two).
- The reagin antibody becomes detectable 7–10 days after the appearance of the primary chancre (or 3–5 weeks after acquiring the infection).
- Non-treponemal serological tests may be false-negative in very early and late syphilis or due to prozone phenomenon. Hence, when there is strong clinical suspicion, serum must have high dilutions to detect the prozone.
- **Sensitivity** in the primary stage is 60–75 per cent, but in the secondary stage, the sensitivity reaches 100 per cent.
- After the secondary stage, titres diminish, as a result of which, about a third of patients with late syphilis are seronegative.
- The titres may rise in patients developing cardiovascular, neurological, or gummatous lesions.
- In some cases of neurosyphilis, reagin tests may be negative with serum but positive with CSF.
- Congenital syphilis results in high titres of antibodies.
- STS becomes negative 6–18 months after effective treatment of syphilis, depending on the stage at which treatment is given. However, if treatment is started late, the tests may remain positive in low titres.

a. **VDRL** (named after the Venereal Disease Research Laboratory, USPHS, New York, where the test was developed) test is rapid and gives quantitative results.
   - **Principal**—cardiolipin, a phospholipid which exists as uniform crystals in normal serum, forms visible clumps on combining with the reagin antibody.
   - **Method**—the steps involved in VDRL are as shown in Fig. 27.14.
   - In primary and secondary syphilis, antibody titres drop following treatment. Therefore, declining antibody titres may be used to indicate therapeutic efficacy.
   - The VDRL test can be used for testing CSF in neurosyphilis also, but not in the plasma of these patients. CSF need not be heated prior to the test.

b. **Rapid plasma reagin (RPR) test**
   - It is modification of VDRL tests developed.
   - It uses a VDRL antigen containing fine carbon particles, which make the result more clear-cut and evident to the naked eye.
   - It can be performed with unheated serum or plasma but is not suitable for testing CSF. It can be conducted in field settings as no equipment is required.

c. **Automated RPR test (ART)** is available for rapid detection in a large number of samples.

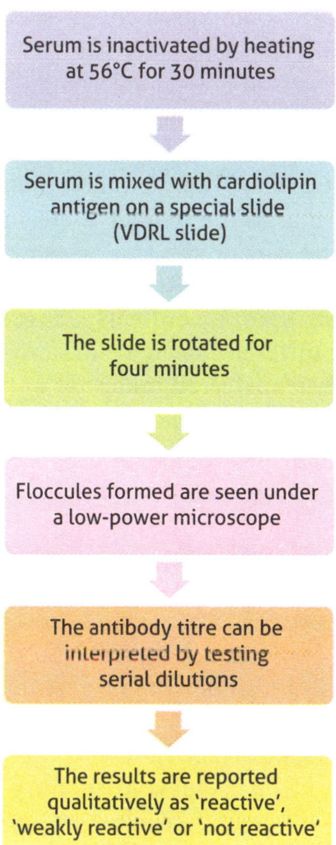

Fig. 27.14 The steps involved in the VDRL test

d. **Toluidine red unheated serum test (TRUST)**
   - It is a rapid test similar to RPR and is performed on a card.
   - The antigen is tagged with toluidine blue for easy reading.
   - CSF can also be tested, and the results are similar to those of VDRL.

e. **Unheated serum reagin test (USRT)**
   - It is a microscopic flocculation test to detect anti-cardiolipin antibodies.
   - It is also a non-treponemal screening test to diagnose and monitor treatment.

Disadvantages of STS

- **Biological false-positive (BFP) reactions** are seen due to unconnected infections because the cardiolipin antigen is a non-specific antigen
- BFP reactions may occur in about one per cent of normal sera. The BFP antibody is usually IgM, while the reagin antibody to syphilis is largely IgG.
- **Acute BFP reactions**—these last only for a few weeks or months and are usually associated with acute infections, injuries, or inflammatory conditions.
- **Chronic BFP reactions**—these persist for longer than six months and are typically seen in SLE and other collagen diseases, leprosy, malaria, relapsing fever, infectious mononucleosis, hepatitis and tropical eosinophilia

ii. Group-specific treponemal tests

These depend on the use of **Reiter's protein** as the antigen tested by complement fixation test. However, these tests are obsolete now.

iii. Specific T. pallidum tests

These tests use the virulent Nichols strain (using either whole organism or its extract) of *T. pallidum* maintained by serial inoculation in rabbit testes.

Tests using strains of T. pallidum

- **Fluorescent treponemal antibody (FTA) test:** It is an indirect immunofluorescence test using smears of Nichols strain of *T. pallidum* as the antigen.
- **FTA-absorption (FTA-ABS):**
   - It is the currently used modification of the FTA test.
   - The test serum is pre-absorbed with a sonicate of the Reiter treponemes (sorbent) to eliminate group-specific reactions.
   - FTA-ABS is as specific as the *Treponema pallidum* immobilisation test and is now accepted as a **standard reference** test. However, as it can be performed only in suitably equipped laboratories; it is not available for routine testing.
- ***Treponema pallidum* immobilisation (TPI):**
   - It uses test serum incubated with a suspension of complement and *T. pallidum* (maintained in a complex medium anaerobically).

- If antibodies are present, the treponemes are immobilised, and then can be then be viewed under dark-ground illumination.
- The TPI test has now been replaced by others such as FTA-ABS and TPHA, which are as specific and much simpler to perform.

Tests using extracts of *T. pallidum*

- ***T. pallidum* hemagglutination assay (TPHA)**
  - It uses tanned erythrocytes sensitised with a sonicated extract of *T. pallidum* as the antigen (Fig. 27.15).
  - The procedure in use is a micro-hemagglutination test (MHA-TP), which can be automated.
  - The test sera for TPHA are absorbed with a diluent containing components of the Reiter treponeme, rabbit testis, and sheep erythrocytes.
  - Sera are screened at an initial dilution of 1:80 but titres as high as 1:5,120 or more are common in the secondary stage.
  - TPHA is as specific as FTA-ABS and almost as sensitive, except in the primary stage.
  - **TPHA is used as a standard confirmatory test; TPHA kits are available commercially.**
- ***Treponema pallidum* particle agglutination (TPPA) test:**
  - It is a passive agglutination test, similar in principle to TPHA.
  - Sonicated *T. pallidum* (Nichols strain) antigens are coated onto gelatin particles.
  - These agglutinate with antibodies to *T. pallidum* in the patient's serum.

Table 27.7 shows the relative sensitivities of the serological tests commonly in use.

Enzyme immunoassays (EIA) have been developed using *T. pallidum* antigens and are available commercially (Bio-Enza Bead test; Captia Syphilis-G test).

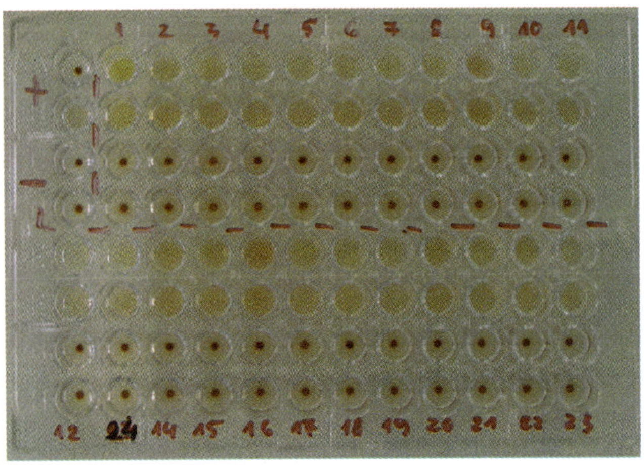

Fig. 27.15 *T. pallidum* hemagglutination assay (TPHA) on a microtitre plate; tanned erythrocytes sensitised with sonicated extracts of *T. pallidum* added to dilutions of the patient's serum showing agglutination (positive) and buttons showing negative titres

Table 27.7 Frequency of reactive serological tests in untreated syphilis (percentage)

Stage	VDRL/RPR	FTA-ABS	TPHA
Primary	70–80%	85–100%	65–85%
Secondary	100%	100%	100%
Latent/late	60–70%	95–100%	95–100%

Molecular methods

- Conventional and real-time RT-PCR have been used to detect *T. pallidum* DNA from tissues and other clinical material in syphilis.
- These methods can detect DNA at all the stages and in a short time and are used as a rapid diagnostic tool in the absence of direct immunofluorescence or dark-ground microscopy.
- *Treponema pallidum*-specific DNA detection by PCR may help to diagnose congenital neurosyphilis. However, CSF tests for VDRL and RPR are equally effective and less costly.

Laboratory diagnosis of congenital syphilis (CS)

- The diagnosis is based on a combination of laboratory criteria in both mother and infant.
- Often, there is an absence of specific signs of infection at birth; hence, serology has a crucial role in the diagnosis of congenital syphilis.
- STS are performed to detect IgM antibody titres in the infant at the time of delivery and in the mother. A serology titre of IgM in the infant that is equal to or four times that in the mother is strongly suggestive of congenital infection.
- Since **IgM does not cross the placenta, its presence in neonatal serum confirms congenital syphilis and helps differentiate it from seropositivity due to passively transferred maternal antibodies.**
- The absence of a ≥4-fold rise in titre does not rule out congenital infection.
- Selective detection of IgM antibodies by modifications of the FTA-ABS, TPHA, EIA, and VDRL tests using whole sera or separated IgM fractions have been developed.
- **Western blot**
  - An immunoblot technique which uses whole-cell antigen or recombinant polypeptide antigens of *T. pallidum* on nitrocellulose.
  - The test can detect IgM and IgG in congenital syphilis. It is a useful **confirmatory test**.
  - Besides congenital syphilis, western blot is also useful in confirming syphilis in patients with SLE and other autoimmune diseases.

According to the testing guidelines for syphilis screening (WHO, STI guidelines), **all pregnant women must be screened for syphilis at the first antenatal care visit.**

Algorithm for testing

Screening for syphilis is done first by standard tests (VDRL or RPR). The main advantage of these tests is their low cost and availability in point-of-care laboratories.

Diagnostic accuracy

- All serological tests for syphilis may be positive in non-venereal treponematoses, and some in a few other spirochetal infections as well. In Lyme disease, the VDRL test is negative, but FTA-ABS may be positive (Table 27.8).
- TPHA and FTA-ABS are helpful in excluding or confirming the diagnosis of syphilis. Both TPHA and FTA-ABS can give false-positive results, though very rarely.
- A negative TPHA virtually excludes the diagnosis of syphilis, past and present, except in the very early stages.
- In neurosyphilis, a negative CSF VDRL test may not be conclusive, but a negative TPHA test eliminates the possibility of neurosyphilis. Detection of specific IgM antibody may be helpful in some situations. A summary of the serological tests for syphilis by stage and the effects of treatment are presented in Fig. 27.16.

Prophylaxis

As transmission is by direct contact, syphilis can be prevented by taking the following precautions:

- Avoiding **sexual contact** with an infected individual
- The use of **physical barriers** (such as condoms) during sexual intercourse
- The **use of antiseptics (potassium permanganate) or antibiotics** to minimise the risk
- The use of prophylactic penicillin carries a possible danger of suppressing the primary lesion without eliminating the infection, as a result of which, recognition and treatment of the disease may become more difficult
- No vaccine is available yet

Treatment

Penicillin is uniformly effective in syphilis, but it is necessary to give an adequate dose and maintain the drug level for a sufficiently long period to establish cure.

- **Early syphilis:** A single injection of 2.4 million units of benzathine penicillin G is adequate.
- **Late syphilis:** The above dose may be repeated weekly for three weeks.
- In patients allergic to penicillin, doxycycline may be used. Ceftriaxone is effective, particularly in neurosyphilis.

> Penicillin treatment for syphilis sometimes induces the **Jarisch–Herxheimer reaction**, consisting of fever, malaisem and exacerbation of symptoms. It is frequent but harmless in primary and secondary syphilis and can be managed with bed rest and aspirin.

Monitoring after treatment

- All patients must be counselled and treated along with their partners.
- The patients must be monitored to ensure compliance with treatment.
- Patients who have HIV infection and patients who are on non-penicillin regimen should be screened and tested at 3, 6, 9, and 12 months.
- Patients with latent syphilis should be monitored with RPR or VDRL at 6-, 12-, and 24-month intervals.

Table 27.8 Reactivity of serological tests by stage of syphilis and effect of treatment

Serological tests	Antigen used	Antibodies detected	Implications
Screening tests Standard tests for syphilis (STS) VDRL RPR	Cardiolipin	IgG/IgM	Positive: Early infection (primary and secondary syphilis) Tertiary—CSF in neurosyphilis Negative: Tertiary and latent Successful treatment Non-venereal syphilis
Confirmatory tests: Specific treponemal antigen-based— FTA-ABS, TPHA, TPPA, EIA	*T. pallidum*: Reiter's or Nichols strains	IgG/IgM	Positive: All stages of syphilis Remains positive despite treatment May be positive in Lyme disease Negative: Rules out syphilis
Western blot	Whole-cell lysate of *T. pallidum* or recombinant polypeptide antigens of *T. pallidum*	IgG/IgM	Positive in infant's serum in congenital syphilis, syphilis in patients with SLE and other autoimmune diseases

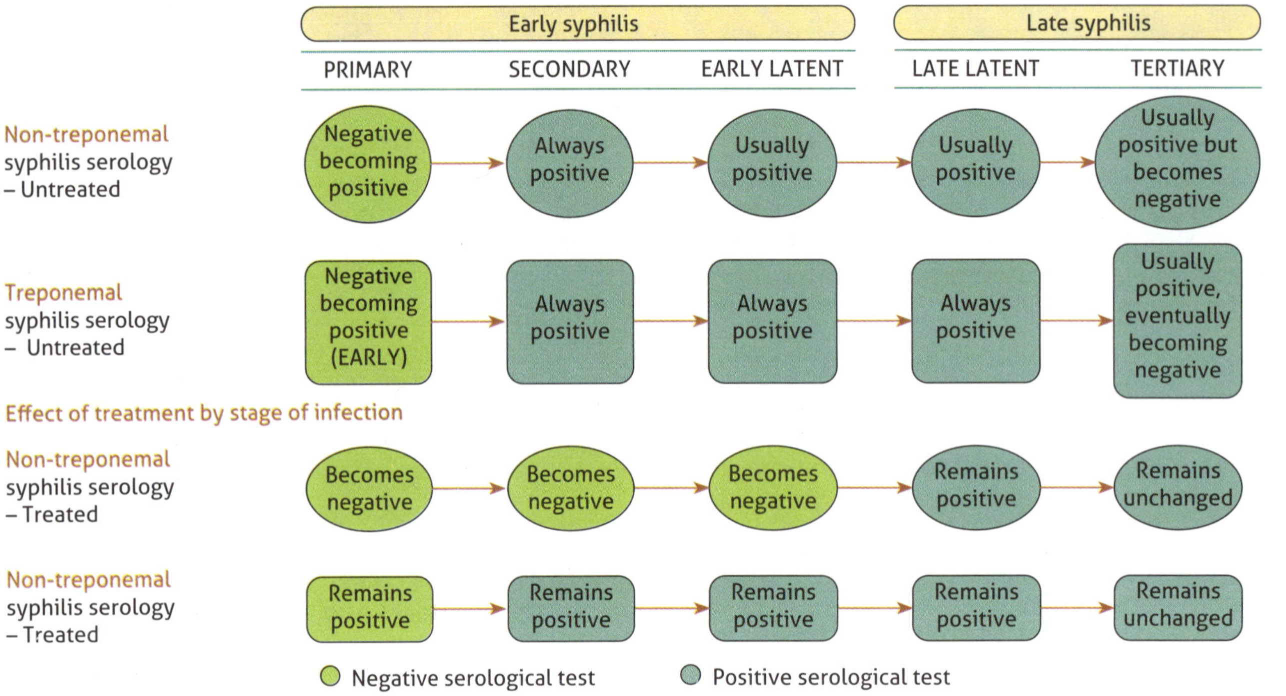

Fig. 27.16 Reactivity of serological tests by stage of syphilis and effect of treatment (*Source*: WHO Guidelines on Syphilis Screening and Treatment of Pregnant Women)

- An integrated approach to the diagnosis, treatment, and monitoring of sexually transmitted infections (STI) and HIV and TB-HIV is underway in the national programme.
- Quantitative tests are useful in monitoring the patient's response to treatment, indicating the stage of the disease and detecting re-infection.
- VDRL or RPR are preferred because they usually become negative following treatment.
- If treatment is initiated very early, the serum may not become positive at all.
- Treatment in the primary stage leads to sero-reversal in about four months; in the secondary and early latent stages, it takes 12–18 months; in later stages, it may take five years or more.
- IgG antibodies to syphilis can remain elevated despite appropriate antimicrobial treatment and a reactive result does not distinguish between recent and past infection.

Table 27.9 Comparative characteristics of venereal and non-venereal syphilis

Characteristics	Venereal syphilis	Non-venereal syphilis
Causative agent	*T. pallidum*	*T. pallidum* subsp. *endemicum* **(bejel)** *T. pallidum* subsp. *pertenue* **(yaws)** *T. carateum* **(pinta)**
Transmission	Sexual contact Congenital Non-sexual—direct contact (doctors/nurses)	Direct contact No congenital transmission reported
Laboratory diagnosis	Serological Standard tests *Treponema*-specific tests	Same as in syphilis due to antigenic similarity
Geographic location	Worldwide	Specific geographic endemicity

Non-venereal Treponematoses

Non-venereal treponemal diseases occur in endemic foci in several parts of the world, among communities with poor standards of hygiene. Three distinct forms of non-venereal treponematoses are **recognised—(i) bejel/siti or endemic syphilis, (ii) yaws and (iii) pinta** (Table 27.9). Yaws and bejel affect skin and bones while pinta affects the skin only.

Bejel/siti

- Non-venereal syphilis was endemic in several foci, It is known as bejel in Saudi Arabia and in the Balkans (nomadic tribes) or siti in Gambia. It has also been reported from India.
- The causative agent is *T. pallidum* subspecies *endemicum*.
- With the recognition of such foci and mass treatment with penicillin under the auspices of the WHO, endemic syphilis has become very rare.
- The disease is common in **young children**.
- The primary chancre is not usually seen, except sometimes on the nipples of mothers infected by their children.

- The disease is usually seen with manifestations of secondary syphilis, such as mucous patches and skin eruptions.
- It progresses to tertiary lesions, particularly gummatous lesions.
- Cardiovascular and neurological involvement is rare.
- Congenital syphilis does not occur.
- Laboratory diagnosis and treatment are as for venereal syphilis.

Yaws

- Yaws, also known as frambesia, pian, and parangi, is endemic in the tropical areas of Africa, Asia, and America.
- Yaws eradication campaigns by mass penicillin injection in endemic areas led to the virtual eradication of the disease.
- In India, cases have been identified in Andhra Pradesh, Odisha, and Madhya Pradesh. The causative agent is a *T. pallidum* subspecies, pertenue (*T. pertenue*), which is morphologically and antigenically indistinguishable from *T. pallidum*.
- Yaws is transmitted by direct skin-to-skin, non-sexual contact with the infectious lesions.
- The primary lesion (mother yaw) is an **extragenital papule which enlarges and breaks down to form an ulcerating granuloma** (Fig. 27.17).
- As in syphilis, secondary and tertiary manifestations follow, but cardiovascular or neurological involvement is rare.
- Destructive gummatous lesions of the bones are common.

Pinta

- Also known as azur, carate, and mal del pintois, pinta is endemic in Mexico, Central and South America and the neighbouring islands. Pinta is transmitted by direct skin-to-skin contact.
- The primary lesion is an extragenital papule, which does not ulcerate, but develops into a lichenoid or psoriatic patch. Secondary skin lesions are characterised by hyper- or hypopigmentation (Fig. 27.18).
- Tissues other than skin are seldom affected.
- The causative agent is *T. carateum*. It is very closely related to *T. pallidum* but is not antigenically identical, and hence cross-immunity between pinta and syphilis is only partial.

Treatment

Benzathine penicillin or azithromycin administered orally have been recommended for treatment.

Non-pathogenic Treponemes

- Several commensal treponemes (e.g., *T. denticola*) occur on the buccal and genital mucosa and may cause confusion in the diagnosis of syphilis on dark-field examination.
- They are a heterogeneous group and have not been adequately characterised.

Chancroid or soft sore

Chancroid or soft chancre, is a sexually transmitted infection caused by *Haemophilus ducreyi*.

Clinical features

- The incubation period is 4–10 days.
- **Tender, non-indurated irregular ulcers on the genitalia** occur (Fig. 27.19a). The ulcer has a friable base with yellow-gray exudate, which bleeds easily.
- Left untreated, the lesion spontaneously resolves within 1–3 months.
- The regional lymph nodes enlarge and are painful.
- Immunity does not develop following infection; instead, hypersensitivity develops, which is detectable as a positive intradermal test with killed bacilli.

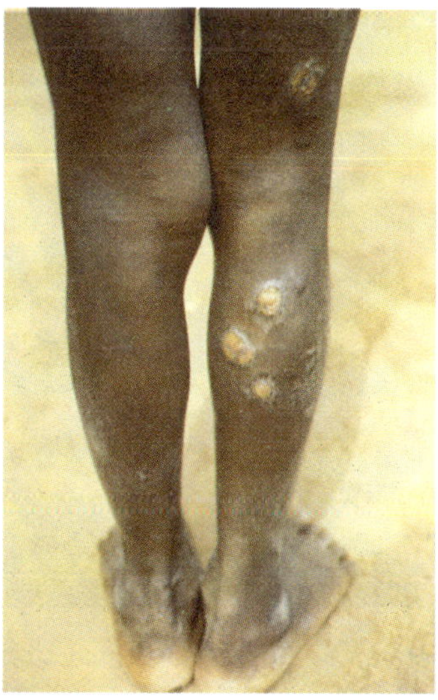

Fig. 27.17 Yaws—ulcerative granuloma (*Source*: CDC, PHIL, Image ID 3811)

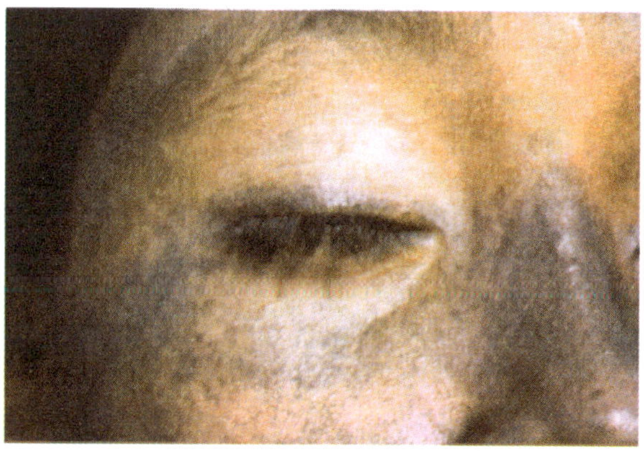

Fig. 27.18 Discoloured areas of facial skin in pinta (*Source*: CDC, PHIL, Image ID 15696)

Agent—*H. ducreyi*

- *H. ducreyi* is a short, ovoid, gram-negative bacillus (1–1.5 × 0.6 μm) with a tendency to occur in end-to-end pairs or short chains.
- It appears as groups or whorls or in parallel chains, giving a 'school of fish' or 'railroad track' appearance (Fig. 27.19b).
- It can be grown on freshly clotted rabbit blood, on the chorioallantoic membrane of a chick embryo and on chocolate agar (enriched with IsoVitaleX—which is a nutritional factor—fetal calf serum and vancomycin).
- It forms small, grey, translucent colonies after incubation at 35°C under 10% CO_2 and high humidity in 2–8 days.
- Smears made after 24–48 hours' incubation show tangled chains of bacilli.

Treatment

Erythromycin, cotrimoxazole, ciprofloxacin or ceftriaxone.

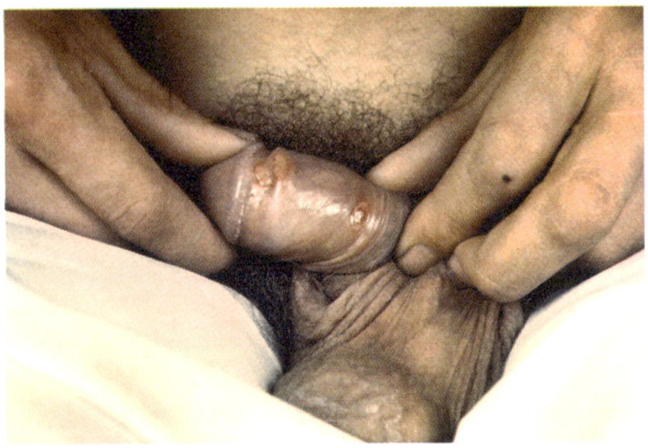

(a)

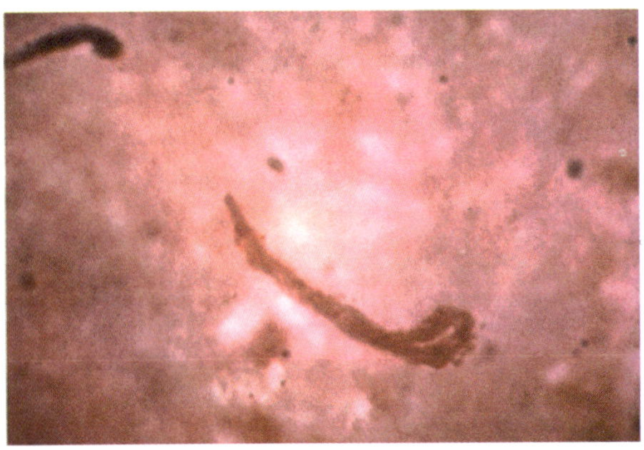

(b)

Fig. 27.19 (a) A chancroid lesion on the penis caused by *Haemophilus ducreyi* and b) a photomicrograph of sample from a chancroid showing numerous gram-negative *Haemophilus* ducreyi bacteria arranged in parallel rows—'railroad track' arrangement (*Source*: CDC, PHIL, Image IDs 15567 and 15255)

Donovanosis

- Donovanosis is caused by *Klebsiella granulomatis* (formerly known as *Calymmatobacterium granulomatis*), which is characterised by genital ulcers.
- The infection is chiefly seen in western countries but is also common in parts of Africa and South America.
- In India, it is seen in the eastern parts of the country.

Clinical features

- Donovanosis is an **ulcerating infection of the genitalia and anus**. Initially, a firm papule appears, which then ulcerates.
- Four morphological types of ulcers are described, which are as follows:
  i. **Ulcerogranulomatous lesions** are the commonest type, marked by painless, fleshy, and beefy-red ulcers
  ii. **Verrucous ulcers** are hypertrophic and less common
  iii. **Necrotic ulcers**
  iv. **Cicatricial or sclerotic forms** cause scarring without lymphadenopathy
- While genital involvement is the most common, anal involvement either alone or contiguous with the genitalia may also be seen.

Diagnosis

- The tissue smear of a biopsy reveals dark-staining **Donovan bodies** within macrophages (Fig. 27.20).
- In endemic areas, diagnosis may be made on clinical grounds alone.

Treatment

- **Drug of choice**: Doxycycline for at least three weeks or trimethoprim–sulfamethoxazole orally for three weeks or until all lesions have healed.
- **Other drugs**: Ciprofloxacin, azithromycin, or erythromycin for three weeks or until all lesions have healed.

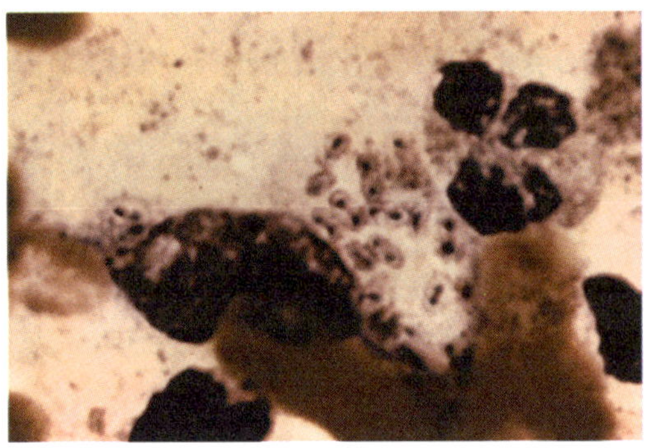

Fig. 27.20 Donovan bodies, cellular inclusions in a tissue sample from a case of granuloma inguinale or donovanosis, caused by *Klebsiella granulomatis* (*Source*: CDC, PHIL, Image ID 16202)

VIRAL SEXUALLY TRANSMITTED INFECTIONS

Genital Warts (Fig. 27.21)

Genital warts (condyloma acuminatum) are sexually transmitted infections caused by the human papillomavirus (HPV) types 6 and 11. These are spread by skin-to-skin contact, usually during sex.

Agent—Human papillomavirus (HPV)

- The papillomavirus family is a large group of DNA viruses which were formerly part of the Papovaviridae family.
- **Classification**—they comprise five genera which cause infections in humans—the alpha, beta, gamma, mupa, and nupa papillomaviruses.
- **Morphology**—papillomaviruses are icosahedral, approximately 55 nm in diameter, with double-stranded DNA.
- **Pathogenesis**—the host range is restricted to primates.

  It has a proclivity for epithelial cells of the skin and mucous membrane and stimulate cell DNA synthesis. The viral oncoproteins interact with tumour suppressor proteins of human cells. Hence the virus is associated with cervical, anogenital, and other human cancers.
- **Route of transmission**—human papillomavirus is the most common sexually transmitted infection.
  - Most sexually active men and women are exposed to the virus at some point in their lifetime.
  - The virus can be spread through oral, vaginal, or anal sex.
  - Sometimes, HPV can be transmitted to an infant during birth, causing genital or respiratory system infections.
- **Incidence**—the infection is most common among adolescents and young adults.
- **Lab diagnosis**—currently, there is no in vitro assay to detect the virus; hence neutralisation tests commonly used for other viral assays cannot be applied to the papillomavirus. HPVs are typed by molecular methods. Around 200 HPV types have been identified, several of which are specific for certain clinical lesions. HPV-16 and 18 are associated with almost 70% of cervical cancers (Table 27.10).

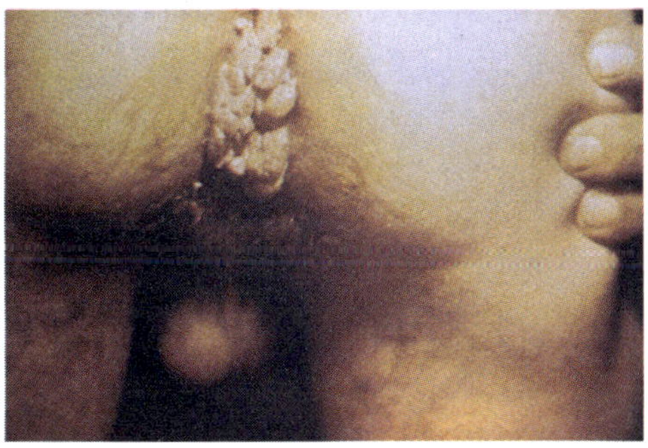

Fig. 27.21 Genital wart caused by human papillomavirus (*Source*: CDC, PHIL, Image ID 4151)

Table 27.10 Human papillomaviruses and the human infections they cause

Clinical lesions	Human papillomavirus type
Plantar warts	1
Skin warts	2, 4, 27, 57
Cutaneous lesions, anogenital condyloma, laryngeal papilloma	3, 10, 28, 49, 60, 76, 78
Genital carcinoma (including cervical carcinoma), high-grade dysplasias, laryngeal and esophageal carcinomas	16, 18, 30, 31, 35, 39, 45, 51–53, 56, 58, 59, 66, 68, 73, 82

- **Treatment**—there is no treatment available for HPV infection. Salicylates have been used to treat some warts.
- **Prevention**—two types of HPV vaccines are available:
  i. Bivalent vaccine contains HPV types 16 and 18 (Cervarix)
  ii. Quadrivalent vaccine contains HPV types 6, 11, 16, and 18 (Gardasil)
     - Adolescent children between 11–12 years of age are vaccinated against HPV.
     - Vaccination is contraindicated in pregnant women.
  iii. Gardasil 9 (which additionally protects against types 31, 33, 45, 52, and 58) is also available now.

Genital Herpes Infections (Herpes Genitalis/ Progenitalis)

Case 4. *A 25-year-old woman presented with complaints of genital blisters and fever for the last seven days and headache and flu-like symptoms for the past two days. She complained of pain in the genital and anal area, with itching and burning. On examination, there were blister-like lesions on her vulva and perianal area. Fluid was collected from the blister to confirm diagnosis. The PCR test was positive for* ***herpes virus type 2 DNA****. The patient was started on treatment with acyclovir, to which she responded. She was advised to report this episode to her doctor at any future pregnancy.*

i. What are the spectrum of diseases caused by the herpes virus?
ii. How are these diagnosed?

Herpes simplex virus type 2 (HHV type 2) is responsible for genital herpes infections and is commonly transmitted venereally. Though both types of HSV cause genital lesions, HSV-2 is more often implicated in such lesions.

Agent—Herpes viruses I and II

- The herpes virus family contains over a hundred species of enveloped DNA viruses that affect humans and animals.

- They are characterised by their ability to establish latent infections, enabling the virus to persist indefinitely within infected hosts and to undergo periodic reactivation.

Morphology

- The herpes virus capsid is icosahedral, composed of 162 capsomers, enclosing the core containing the linear double-stranded DNA genome.
- The nucleocapsid is surrounded by a lipid envelope derived from a modified host cell nuclear membrane. The envelope carries surface spikes, each about 8 nm long (Fig. 27.22a and b).
- Between the envelope and the capsid is an amorphous structure called the **tegument**, containing several proteins.
- The enveloped virion measures about 200 nm and the naked virion, about 100 nm in diameter.

Replication

Herpes viruses replicate in the host cell nucleus. They form **Cowdry type A intranuclear (Lipschutz) inclusion bodies**.

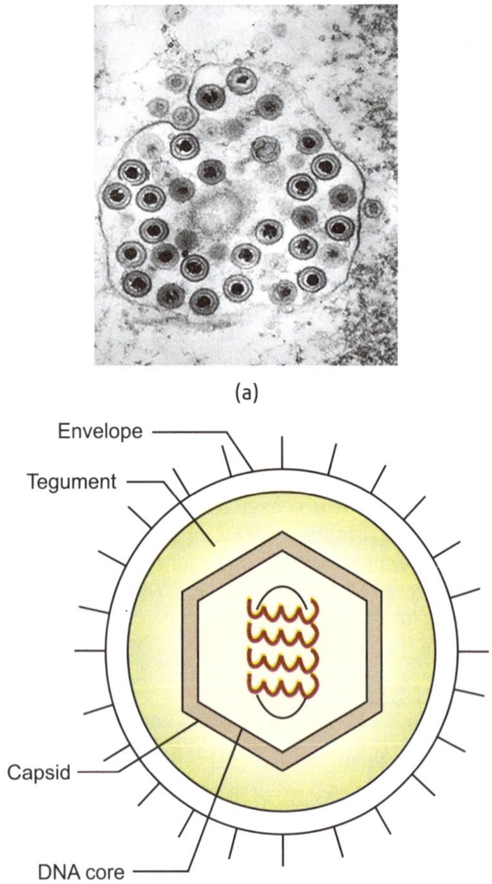

Fig. 27.22 (a) Transmission electron microscopic (TEM) image shows the presence of numerous herpes simplex virions (*Source*: CDC , PHIL, Image ID 10231) and (b) detailed structure of a herpes simplex virus

Susceptibility

- Like other enveloped viruses, herpes viruses are susceptible to fat solvents like alcohol, ether, chloroform, and bile salts.
- They are heat-labile and must be stored at –70°C.

Classification

- Eight different types of herpes viruses are known to cause human infections.
- They have been officially termed human herpesviruses (Table 27.11).

Table 27.11 Classification and biological characteristics of important herpes viruses causing infection in humans

Group	Viruses	Biological characteristics
Human herpesvirus 1	Herpes simplex virus (HSV-1)	Short replicative cycle; variable host Cytopathic Infect epithelial cells primarily
Human herpesvirus 2	Herpes simplex virus 2 (HSV-2)	Latent infection in sensory ganglia
Human herpesvirus 3	Varicella-zoster virus	
Human herpesvirus 4	Epstein–Barr virus	Narrow host range Replicate in lymphoblastoid tissue Cause latent infection
Human herpesvirus 5	Cytomegalovirus	Replicate slowly Narrow host range Cell-associated Cause cytomegaly
Human herpesvirus 6	HHV6	Attaches to mature CD4$^+$ T cell Causes roseola infantum
Human herpesvirus 7	HHV7	Acquired in infancy and childhood Can cause roseola Associated with febrile seizures and encephalitis
Human herpesvirus 8	HHV-8	Replicate slowly Inactivate tumour suppressor gene Protein synthesised by HHV-8 known as latency-associated nuclear antigen (LANA) Inactivate tumour suppressor proteins

Agent—Herpes simplex virus

The herpes simplex virus (HSV) occurs naturally only in humans. The virus is of two types, which are as follows:

1. **HSV type 1** (human herpes virus type 1 or HHV type 1) is usually isolated from **lesions in and around the mouth and is transmitted by direct contact or droplet spread from cases or carriers.**
2. **HSV type 2** (HHV type 2) is responsible for **genital herpes infections** and is commonly **transmitted venereally**.
   - The virus grows in a variety of primary and continuous cell cultures like Vero (Fig. 27.23), monkey or rabbit kidney, human amnion, and HeLa, producing cytopathic changes and well-defined foci with heaped-up cells and syncytial or giant cell formation.
   - The two types of viruses (HSV-1 and -2) cross-react serologically.

Pathogenesis

- Herpes simplex is one of the most common viral infections in humans, with about 60–90 per cent of adults showing detectable antibodies.
- **Age**—primary infection is usually acquired in early childhood, between two and five years of age.

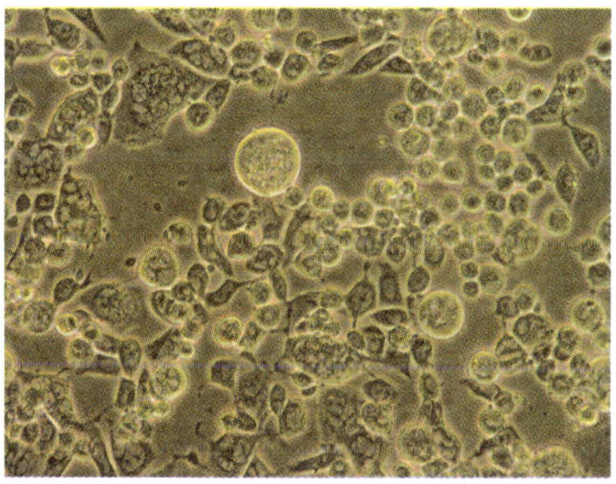

Fig. 27.23 Vero cell line infected with herpes simplex virus (note the rounded cells and giant cell formation)

- **Host**—humans are the only natural hosts.
- **Sources of infection**—the infection spreads through saliva, skin lesions, or respiratory secretions. Asymptomatic carriers form the more important sources of infection, especially in genital infection with type 2 strains.
- Herpes virus diseases are more frequent and severe in HIV-infected and other immunodeficient subjects.
- **Pathogenesis**—Fig. 27.24 outlines the pathogenesis of HSV.

Clinical features

The clinical manifestations and course depend on the site of infection, age, and immune status of the host, and the antigenic type of the virus.

- **Cutaneous infections:**
  - The most common site of lesions is the face—on the cheeks, chin, around the mouth, or on the forehead.
  - The typical lesion is the 'fever blister' or herpes febrilis, caused by viral reactivation in febrile patients (Fig. 27.25).
  - The typical herpes lesions are thin-walled, umbilicated vesicles, the roofs of which break down, leaving tiny, superficial ulcers. They heal without scarring.
  - In general, primary infections, though self-limited, are more severe and widespread and associated with systemic manifestations. Recurrent infections are more localised.
  - **Herpetic whitlow** is a localised infection generally seen on the fingertips of healthcare workers, doctors, and nurses.
  - **Eczema herpeticum** is a disseminated eruption caused by herpes infection in children with eczema. A clinically indistinguishable picture is also produced by vaccinia infection; these symptoms are designated Kaposi's varicelliform eruption for both conditions.

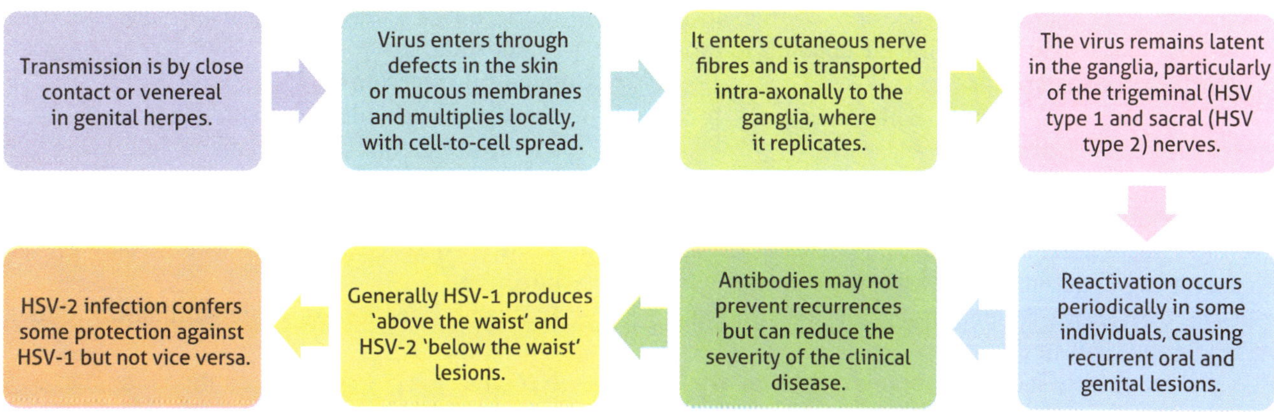

Fig. 27.24 Pathogenesis of HSV

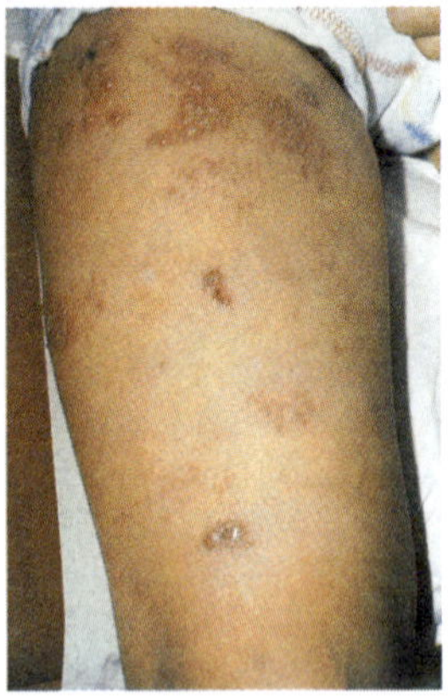

Fig. 27.25 Fever blister

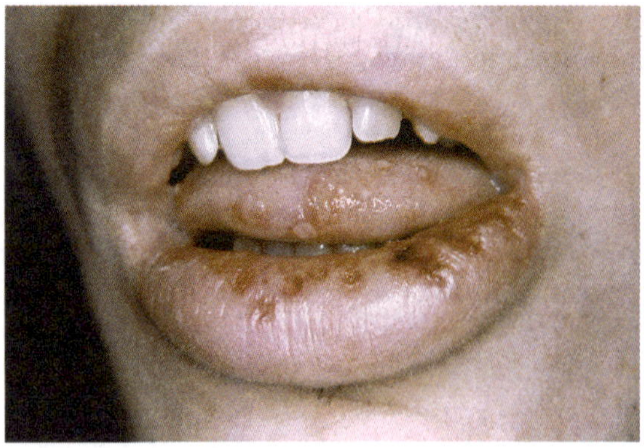

Fig. 27.26 Herpes simplex lesions on the lips (*Source*: CDC, PHIL, Image ID 12616)

- **Erythema multiforme** is a hypersensitivity reaction seen in HSV infection. It has typical target lesions on the skin.
- Mucosal infections are common, most often involving the buccal mucosa.
- **Gingivostomatitis and pharyngitis** are the most frequent conditions in primary infection and recurrent herpes labialis in recurrent infection (Fig. 27.26).
- The vesicles may ulcerate and become secondarily infected.

- **Ophthalmic infections:**
  - HSV infection is the most common cause of corneal blindness in some developed countries.
  - Acute keratoconjunctivitis may occur by itself or by extension from facial herpes.
  - Follicular conjunctivitis with vesicle formation on the lids is another manifestation.
  - The cornea may be involved, with typical branching dendritic ulcers.
  - Steroids are contraindicated as they lead to deep stromal involvement, and healing may be delayed with scarring and corneal blindness.
  - Chorioretinitis and acute necrotising retinitis are uncommon but serious manifestations.
- **Infections of the nervous system:**
  - HSV encephalitis, though rare, is the most common sporadic acute viral encephalitis in most parts of the world.
  - It has an acute onset, with fever and focal neurological symptoms.
  - The demonstration of HSV DNA in CSF by PCR is a very sensitive test in the acute stage (Case 4).
  - HSV meningitis is a self-limiting disease, usually resolving in about a week without sequelae.
  - The CSF shows lymphocytic pleocytosis and may yield the virus in culture.
  - HSV can cause sacral autonomic dysfunction and also rarely, transverse myelitis or the Guillain–Barré syndrome.
  - It has been implicated in the causation of Bell's palsy.
- **Visceral infections:**
  - HSV esophagitis may cause dysphagia, substernal pain, and weight loss.
  - It may involve the respiratory tract, causing tracheobronchitis and pneumonitis.
  - HSV is an uncommon cause of hepatitis.
  - Disseminated HSV infection may occur in patients with immunodeficiency or malnutrition or those inflicted with burns.
- **Genital infections (herpes genitalis/progenitalis)**
  - In men, the lesions occur mainly on the penis (Fig. 27.27a) or in the urethra, causing urethritis.
  - In women, the cervix, vagina, vulva (Fig. 27.27b), and perineum are affected. When only the cervix is involved, the infection may be asymptomatic. The primary infection is usually more serious, with systemic features like fever and malaise. Recurrent episodes are milder. The vesiculo-ulcerative lesions may be very painful.
  - Rectal and perineal lesions occur in homosexuals. Though both types of HSV cause genital lesions, HSV-2 is more often implicated in such lesions.
  - There have been several reports of an association between HSV-2 and carcinoma of the cervix uteri, but a causal relationship has not been established.
- **Congenital infections:**
  - Transplacental infection with HSV-1 or -2 can lead to congenital malformations, but this is rare.
  - Infection may occur during birth, particularly if the mother has genital lesions due to HSV-2. In such cases, cesarean section may prevent infection.
  - Postnatal infection is more commonly due to HSV-1.
  - Neonatal herpes may be confined to the eyes, mouth, and skin or it may affect multiple organs, with or without encephalitis.

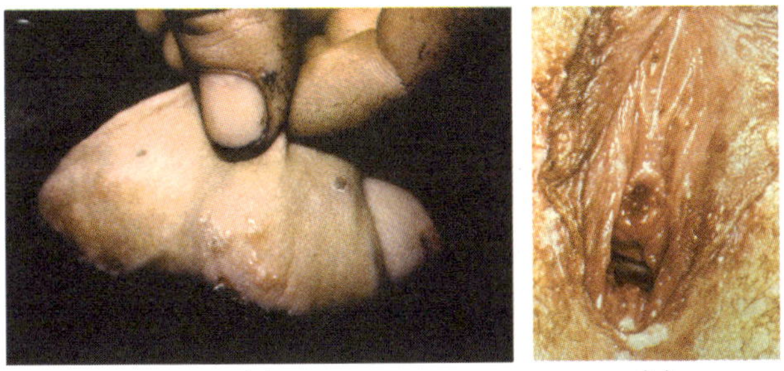

(a) (b)

Fig. 27.27 Herpes progenitalis infection: (a) penile vesicles and (b) erythematous papules on the inner surface of the labia minora (*Source*: CDC, PHIL, Image IDs 16783 and 16968)

- The mortality rate is very high, and survivors may have neurological impairment.
- Prenatal screening for antibodies in mothers using the TORCH (*Toxoplasma*, Rubella, Cytomegalovirus, and Herpes) panel can detect either IgM or IgG, denoting recent or past infection in the mother, thus predicting congenital anomalies or neonatal infection.

Laboratory diagnosis

The diagnosis of herpes virus infection may be made by microscopy, antigen or DNA detection, virus isolation, or serology.

- **Specimens:** Vesicle fluid, skin swabs, saliva, corneal scrapings, CSF, or brain biopsy on autopsy are used for diagnosis.
- **Microscopy:**
  - The Tzanck smear is a rapid, fairly sensitive and inexpensive diagnostic method.
  - Smears are prepared from the lesions, preferably from the base of vesicles and stained with 1% aqueous solution of toluidine blue 'O' for 15 seconds.
  - Multinucleated giant cells with faceted nuclei and homogeneously stained 'ground-glass' chromatin (Tzanck cells) constitute a positive smear (Fig. 27.28).
  - Giemsa-stained smears show intranuclear type A inclusion bodies.
  - Electron microscopy can demonstrate the virus particle.
  - Fluorescent antibody technique (Fig. 27.29) demonstrates the herpes virus antigen in smears or sections from lesions.
- **Virus isolation:**
  - While primary human embryonic kidney, human amnion, and several other cells can be used to cultivate the virus, human diploid fibroblasts are most preferred.
  - Syncytia formation may appear as early as 24–48 hours after infection. However, cultures should be observed for two weeks before being declared negative.
  - Intranuclear inclusion bodies are seen in the infected cells. Drug susceptibility too can be tested in cell cultures.
  - Differentiation between HSV types 1 and 2 may be made by a variety of serological techniques, by nucleic acid hybridisation or by restriction endonuclease cleavage and electrophoretic analysis of viral DNA or viral proteins.
- **Serology**:
  - It is useful in the diagnosis of primary infections.
  - Antibodies develop within a few days of infection, and a rise in titre of antibodies may be demonstrated by ELISA, neutralisation or complement fixation tests.
  - In recurrent or re-infection herpes, there may be little change in the antibody titre.
- **PCR:**
  - For the detection and genotyping of herpes simplex virus (HSV) in clinical specimens.
  - Real-time PCR is widely used as it is safer and more sensitive than isolation and immunofluorescence.

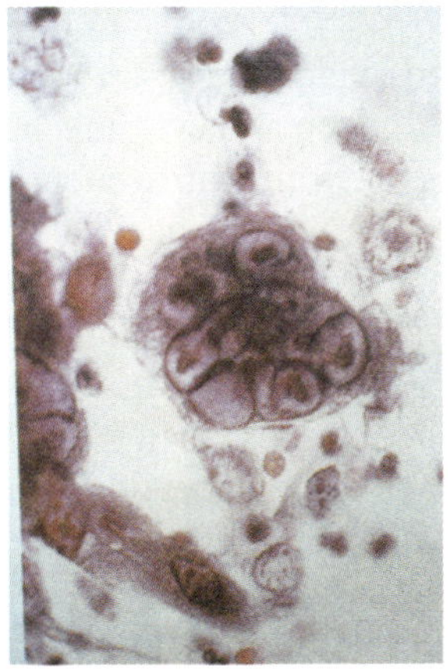

Fig. 27.28 The herpesvirus in Tzanck smear

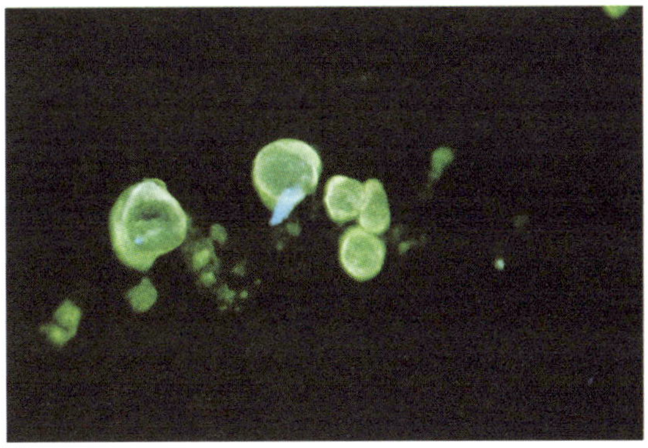

Fig. 27.29 Visualisation using fluorescein isothiocyanate (FITC)-labelled monoclonal antibody stain specific for the envelope glycoprotein of herpes simplex virus 1 and 2– HSV-1, HSV-2 (*Source*: CDC, PHIL, Image ID 16165)

Treatment

- Acyclovir and vidarabine are effective for the management of deep and systemic infections.
- Early treatment with intravenous acyclovir has improved the outcome of encephalitis.
- Oral and topical use may help in less serious conditions. Valaciclovir and famciclovir are more effective oral agents.
- When resistance to these drugs develops, drugs like foscarnet, which are independent of viral thymidine kinase action, may be useful.

Human Immunodeficiency Virus

This virus has overtaken all other infectious agents in causing sexually transmitted infections. Details of HIV and acquired immunodeficiency syndrome are discussed in detail in Chapter 21, as it primarily disseminates through lymph reticular system.

Other viral infections which may be transmitted through sexual contact are hepatitis B and C (discussed in Chapter 25).

PARASITIC SEXUALLY TRANSMITTED INFECTIONS

- A few parasites are sexually transmissible and may cause sexually transmitted infections (STIs).
- *Trichomonas vaginalis* can cause vaginitis, urethritis, and prostatitis.
- Ectoparasites such as *Sarcoptes scabiei* (scabies) and *Pthirus pubis* (pubic lice) also infest the genital tract.

Vaginitis

Agent—*Trichomonas vaginalis*

- *Trichomonas* is a flagellated protozoan that lacks a cyst stage in its life cycle.
- There are three species found in humans. Of these, only *T. vaginalis* is pathogenic to humans, in whom it causes vaginitis, urethritis, and prostatitis.
- *Trichomonas tenax* and *Pentatrichomonas hominis* are non-pathogenic and exist as normal flora in the human body.
- *T. tenax* can be found in the mouth while *P. hominis* is present in the large intestines.
- *T. vaginalis* resides in the genital tract. It has the largest genome amongst protozoa and its metabolism is similar to that of anaerobic bacteria.

Morphology (Fig. 27.30)

- The parasite exists only in the trophozoite stage; no cyst is seen.
- The trophozoite can exist in a flagellated or amoeboid form.
- The flagellated trophozoite is pear-shaped and about 7–20 μm long and 5–15 μm wide. It has four flagella at its anterior end and a single lateral flagellum.

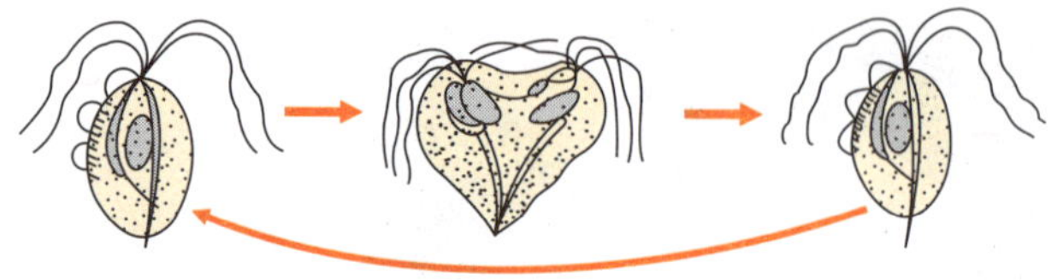

Fig. 27.30 *Trichomonas vaginalis*

- The lateral flagellum curves laterally as an undulating membrane and is known as the recurrent flagella.
- Respiration occurs anaerobically through hydrogenosomes.
- The trophozoite contains a single nucleus with a central karyosome.
- The amoeboid trophozoite is a shapeless, actively dividing form that has pseudopodia, but in which flagella are also visible. In laboratory cultures, non-flagellated forms known as pseudocysts may be seen.
- These are not seen in the clinical phase.

Life cycle

- The infective forms are the **flagellated trophozoites** which are present in the vagina and urethra of females.
- In males, the parasite resides in the seminal vesicle and prostate.
- The infection is transmitted by asymptomatic females via the sexual route.
- Upon entry, the trophozoites develop into actively dividing, tissue feeding, and amoeboid forms.
- These multiply by binary fission and evade the urogenital tract and subsequently transform into flagellated forms which can cause infection.
- The clinical presentation can be asymptomatic, vulvovaginitis, or chronic.
- The trophozoites are harboured in the genital tract without causing any symptoms. Such persons can transmit the infection silently.
- Females complain of thin, purulent, frothy, profuse, and foul-smelling discharge. Dysuria and abdominal pain may be present.
- Examination reveals small, punctate hemorrhages on the vagina and cervical mucosa, giving it a characteristic 'strawberry' appearance.
- Males present with features of non-gonococcal urethritis, prostatitis, penile ulcers and rarely, epididymitis. Chronic infections are characterised by pain during intercourse, mild genital pruritis and mucoid vaginal discharge (Fig. 27.31).

Laboratory diagnosis

- Samples such as vaginal discharge, urethral discharge, urine, and prostatic smears may be examined for trophozoites.

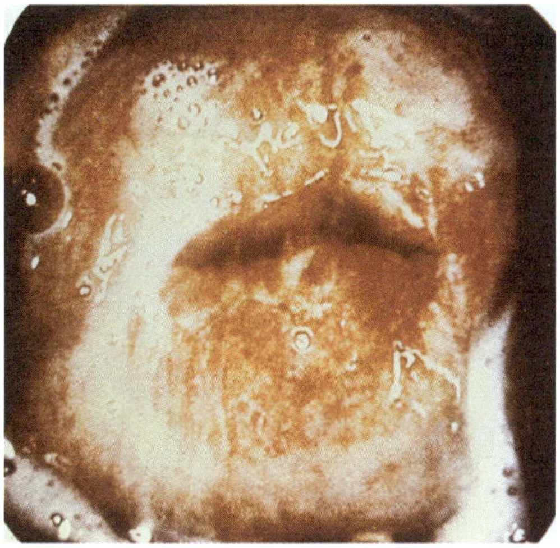

Fig. 27.31 Intravaginal, colposcopic view of a female patient's cervix showing the characteristic 'strawberry' appearance of *Trichomonas vaginalis* infection (*Source*: CDC, PHIL, Image ID 16402)

- Urine is centrifuged, and the sediment is used for diagnosis.
- Trophozoites in the urine of small children must be carefully differentiated from fecal contamination by *Pentatrichomonas hominis* as the presence of *T. vaginalis* may indicate sexual abuse.
- The vaginal pH is raised (>4.5) and the whiff test may be positive. In the whiff test, a drop of 10% KOH is added to the vaginal discharge. A fishy odour is detected in a positive case. However, these are non-specific tests and may be present in other conditions such as bacterial vaginosis.
- **Microscopy:**
  - Fresh samples (within 10–15 minutes of collection) are examined in saline under a microscope (Fig. 27.32).
  - Characteristic pear-shaped trophozoites with jerky or quivering motility along with plenty of leukocytes are seen.

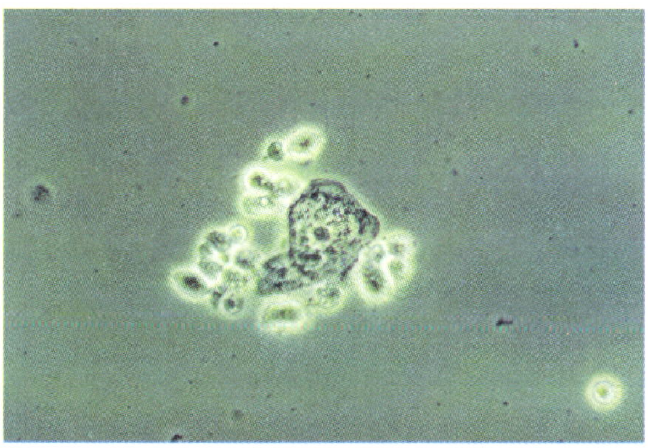

Fig. 27.32 Photomicrograph of a wet mount of a vaginal discharge specimen revealing the grouping of *Trichomonas vaginalis* (*Source*: CDC, PHIL, Image ID 17912)

  - The smears can be stained by Giemsa and Papanicolaou for visualisation of trophozoites.
  - Direct fluorescent antibody assay is also available and is more sensitive than wet mount.
  - The test uses fluorescent-labelled monoclonal antibodies to *T. vaginalis*.
- **Antigen testing:** A rapid antigen test as an immunochromatographic test is available currently. Vaginal swabs can be directly used on the test, and the result can be seen in 10 minutes.
- **Culture:** *T. vaginalis* can be cultured on the Diamond medium (which contains cysteine, ascorbic acid, maltose, yeast extract, trypticase digest, and sheep serum). The medium is incubated anaerobically for up to five days and then examined microscopically for motile trophozoites.
- **Serology:** Antibodies can be detected using ELISA in the patient's serum. The antibodies, however, persist long after infection and cannot be used to differentiate between acute and chronic infection.
- **Molecular diagnosis:** A PCR test is available for the detection of *T. vaginalis*. Multiplex PCR can also be done to diagnose syndromic cases of vaginal discharge.

Treatment

- **Drug of choice:** Metronidazole or tinidazole as a single dose of 2 g in refractory cases, treatment may be extended to five days.
- It is important to treat both partners.

Urogenital Schistosomiasis

Urogenital schistosomiasis is caused by *Schistosoma haematobium* infecting the urogenital tract. The life cycle and morphological details are given in Chapter 21.

Agent—*Schistosoma haematobium*

Pathogenesis

- The eggs are deposited in tissues of the urinary bladder (Fig. 27.33) and ureter.
- The eggs damage the tissues and induce type IV hypersensitivity reaction resulting in granuloma formations.
- These granulomas, in turn, coalesce to form nodules. Histologically, **glandular hyperplasia** is seen over the mucosa on the nodules; this is followed by fibrosis, which may lead to an obstruction in the ureter and hydronephrosis.
- **Urinary carcinoma**—constant irritation of the bladder wall results in metaplastic changes and may predispose to squamous cell carcinoma of the urinary bladder. Genital involvement may be seen in both males and females.

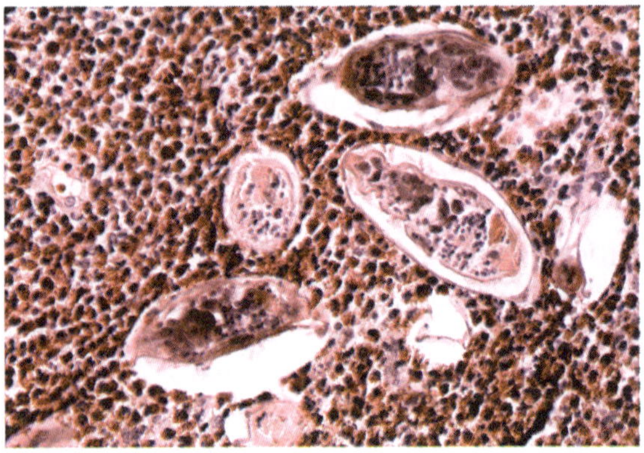

Fig. 27.33 Histopathology of bladder tissue in a case of schistosomiasis—shows the presence of clusters of *S. haematobium* ova, surrounded by intense eosinophilic infiltrates and other inflammatory cells (*Source*: CDC, PHIL, Image ID 35)

- **Intestinal schistosomiasis**—*S. mansoni* and *S. japonicum* deposit their eggs in the intestinal mucosa. As a result, granuloma, fibrosis and thickening are seen along the colon and rectum. Diarrhea and dysentery are the presenting complaints at this stage. Colorectal and liver carcinoma have been reported with *S. japonicum* and *S. mekongi* infections.

Laboratory diagnosis

- The eggs of *S. haematobium* can be detected in urine or bladder biopsy.
- The eggs of *Schistosoma* infecting the intestines can be detected in stool samples and intestinal biopsy. All eggs except those of *Schistosoma haematobium* are acid-fast.

The control of sexually transmitted diseases falls under the ambit of the **AIDS Control Programme**. This programme includes strategies to reduce the occurrence of sexually transmitted diseases through surveillance, case detection and screening (all antenatal women are screened for common STIs). Diagnosis of STIs and case management are also facilitated through this programme.

Questions

Case review

A 20-year-old male presented to an STI clinic with complaints of urethral discharge and dysuria for two weeks. The patient was subjected to laboratory investigations.

i) Mention the relevant history to be elicited from this individual with emphasis on how you will respect his autonomy and maintain the confidentiality of the test results.
ii) Present the approach to the diagnosis of STIs in the form of an algorithm.
iii) List the common etiologies of urethral discharge and their management.

Short answer questions

1. Discuss the laboratory confirmation of primary syphilis.
2. Enumerate the viral agents associated with sexually transmitted infections.
3. Discuss the role of HPV in urogenital infections and the consequences of these infections.
4. Describe the clinical features and laboratory confirmation of *Trichomonas* vaginitis.

MCQs

1. ***N. meningitidis* and *N. gonorrhoeae* are differentiated in the laboratory by:**
   a) Biochemical reactions b) Cultural characteristics
   c) Smear examination d) All of these
2. **WHO-recommended treatment for gonorrhea is:**
   a) Penicillin b) Penicillin + gentamicin
   c) Ceftriaxone d) Ceftriaxone + azithromycin
3. **Specimen not recommended for diagnosing gonorrhea is:**
   a) Urethral swab b) High vaginal swab
   c) Urine d) Endocervical swab
4. **For optimum recovery of gonococci, which of the following swabs should be used for collecting the specimen?**
   a) Cotton-tipped swabs
   b) Nylon swabs
   c) Charcoal-impregnated swabs
   d) Polyester swabs
5. **Non-specific urethritis may be caused by:**
   a) *Chlamydia trachomatis* b) *N. gonorrhoeae*
   c) *E. coli* d) *Staphylococcus aureus*
6. **All the following are features of Reiter's syndrome except:**
   a) Recurrent conjunctivitis b) Myocarditis
   c) Polyarthritis d) Urethritis
7. **Fitz-Hugh-Curtis syndrome does not include:**
   a) Conjunctivitis b) Perihepatitis
   c) Salpingitis d) Arthritis
8. **A 34-year-old man develops a small, painless, papulovesicular lesion on the external genitalia. After two weeks, his inguinal lymph nodes are swollen with pus oozing out of sinuses. The most likely diagnosis is:**
   a) Primary stage of syphilis
   b) Secondary stage of syphilis
   c) Lymphogranuloma venereum
   d) Donovanosis
9. **A 35-year-old male truck driver with a history of unprotected sex presented to the clinic with a painless, indurated ulcer on his glans that releases clear serum on pressure. Inguinal lymph nodes in both sides of the groin were enlarged and non-tender. The most probable diagnosis is:**
   a) Chancroid
   b) Syphilis
   c) Lymphogranuloma venereum
   d) Genital herpes
10. **Motility in spirochetes is by:**
    a) Flagella b) Pili
    c) Fimbriae d) Endoflagella
11. **The spirochete that does not take up any stain is:**
    a) *Treponema pallidum* b) *Treponema carateum*
    c) *Leptospira* d) *Borrelia*
12. **The best method for staining spirochetes among the following is:**
    a) Silver-impregnation b) Gram stain
    c) Acid-fast stain d) Wright stain
13. **All the following are non-venereal trepanomatoses except:**
    a) Yaws b) Syphilis
    c) Pinta d) Endemic syphilis

14. **The cardiolipin antigen of *Treponema pallidum* is:**
    a) Specific antigen
    b) Non-specific antigen
    c) An extract from non-pathogenic treponemes
    d) Used in TPHA test
15. **Chancre is a feature of:**
    a) Primary syphilis b) Secondary syphilis
    c) Tertiary syphilis d) Latent syphilis
16. **A male patient underwent treatment for penile ulcer three years back. The test used for determining his response to treatment is:**
    a) Dark-field microscopy b) TPI
    c) VDRL d) FTA-ABS
17. **The most infectious stage of syphilis is:**
    a) Primary syphilis b) Secondary syphilis
    c) Neurosyphilis d) Tertiary syphilis
18. **The highest sensitivity for reaginic antibody test is seen in:**
    a) Primary syphilis b) Secondary syphilis
    c) Tertiary syphilis d) Latent syphilis
19. **Tanned erythrocytes with treponemal antigens are used in:**
    a) TPI b) FTA-ABS
    c) TPHA d) TPPA
20. **The test by means of which neurosyphilis can be excluded is:**
    a) CSF-VDRL b) FTA-ABS
    c) TRUST d) TPHA
21. **Tabes dorsalis is a feature of:**
    a) Secondary syphilis b) Tertiary syphilis
    c) Congenital syphilis d) Late tertiary syphilis
22. **Which of the following statements is false about congenital syphilis?**
    a) Transplacental transmission occurs during the last trimester of pregnancy
    b) Lesions of congenital syphilis develop during the 4th month of gestation
    c) The mother usually gives a history of repeated abortions
    d) IgM ELISA confirms the diagnosis
23. **Virulent treponemes are maintained in:**
    a) Rabbit testes b) Foot pad of mice
    c) Thioglycolate broth d) Mouse brain
24. **Residual stigmata of congenital syphilis are all the following except:**
    a) Hutchinson's teeth b) Mulberry molars
    c) Chorioretinitis d) Saddle nose
25. **Biological false-positive reactions in VDRL are obtained in the following conditions except:**
    a) SLE b) Leprosy
    c) IMN d) Tuberculosis
26. **How many months after therapy do the standard tests for syphilis become negative?**
    a) 1–2 months b) 2–4 months
    c) 6–12 months d) 2 years
27. **Lichenoid patch without ulceration is typical of:**
    a) Yaws b) Pinta
    c) Bejel d) Parangi
28. **Donovanosis is caused by:**
    a) *Klebsiella granulomatis* b) *Lactobacillus acidovorans*
    c) *Spirillum minus* d) *Chlamydia trachomatis*
29. **A young woman presented with an ulcer on her vulva, low-grade fever, and rash on her face. Pus culture did not yield any bacterial growth. What is the recommended microscopy/staining method in this case?**
    a) Congo red b) Dark-field microscopy
    c) Giemsa stain d) India ink preparation
30. **Which is the specific test for syphilis?**
    a) VDRL b) RPR
    c) CFT d) FTA-ABS
31. **A genital pus sample showed gram-negative intracellular bean-shaped cocci. The sample was sent for culture. What is the best culture medium in this case?**
    a) Blood agar b) Chocolate agar
    c) Mueller–Hinton medium d) Thayer–Martin medium
32. **Human papilloma viruses are commonly associated with which of the following carcinomas?**
    a) Breast b) Cervix
    c) Lung d) Anogenital
33. **Which of the following HPV serotypes are associated with carcinoma of the cervix?**
    a) Types 16 and 18 b) Types 12 and14
    c) Types 20 and 24 d) Types 6 and 8
34. **Which of the following is not true regarding human papilloma virus?**
    a) It is a RNA virus.
    b) It has predisposition for epithelial cells of the skin and mucous membrane.
    c) Its host range is restricted to primates.
    d) The viral oncoproteins interact with tumour suppressor proteins of human cells.
35. **Kaposi's varicelliform eruption resembling eczema herpeticum is caused by:**
    a) Vaccinia b) HSV-2
    c) HSV-1 d) EBV
36. **Carcinoma cervix is associated with:**
    a) HSV-2 b) HPV
    c) HIV d) HSV-2 and HPV
37. **Herpetic whitlow refers to:**
    a) Hypersensitivity reaction seen in HSV infection
    b) Localised infection generally seen on the fingertips of healthcare workers
    c) Facial blisters seen in children
    d) White blisters seen in the genitals of sexually active women
38. **A patient's skin biopsy smear-stained homogenously with toluidine blue demonstrated multinucleated giant cells with faceted nuclei. This is the characteristic feature of which of the following infections?**
    a) HPV b) HSV
    c) Varicella d) Infectious mononucleosis
39. **Positive Tzanck smear shows:**
    a) Multinucleated giant cells with faceted nuclei in a homogenously stained smear (toluidine blue)
    b) Multinucleated giant cells with differentiating nuclear patterns (fluorescent stain)
    c) Uniformly-stained smear with inclusion bodies (Giemsa stain)
    d) Ground-glass appearance with abnormal cells (Gram stain)
40. **A 40-year-old dermatologist presented with multiple vesicles around the fingers. His skin smear showed multinucleated giant cells with faceted nuclei in a homogenously stained ground glass chromatin in toluidine blue stain. What is the likely organism associated?**
    a) Tinea unguium with superadded bacterial infection
    b) Herpetic infection
    c) Molluscum contagiosum
    d) Papilloma
41. **A curved anaerobic gram-positive bacillus was isolated from a case of bacterial vaginosis. The organism is:**
    a) *Mobiluncus* b) *Bifidobacterium*
    c) *Lactobacillus* d) *Veillonella*
42. **Pathognomonic cells in bacterial vaginosis are called:**
    a) Smudge cells b) Atypical cells
    c) Decoy cells d) Clue cells
43. **A 24-year-old female presents with complaints of dysuria, vulvar irritation, and painful sexual intercourse for the past few weeks. Per speculum examination reveals an inflamed cervix with profuse vaginal discharge. Microscopy of vaginal discharge shows numerous pear-shaped, motile organisms. What is the most likely causative agent?**
    a) Herpes simplex virus b) *Neisseria gonorrhoeae*
    c) *Chlamydia trachomatis* d) *Trichomonas vaginalis*

44. The drug of choice to treat this infection is:
   a) Metronidazole b) Albendazole
   c) Praziquantel d) Mebendazole

45. Which of the following is true regarding *Trichomonas*?
   a) It is a flagellated protozoan that lacks a cyst stage in its life cycle.
   b) It has four flagella at its anterior end and a single lateral flagellum.
   c) Respiration occurs anaerobically through hydrogenosomes.
   d) All of the above.

46. Which of the following is a common pathogen found in the genitourinary tract?
   a) *Entamoeba histolytica* b) *Trichomonas vaginalis*
   c) *Trypanosoma brucei* d) *Strongyloides stercoralis*

47. A 65-year-old man from Saudi Arabia presents with complaints of recurring hematuria and weight loss over past three years. Biopsy of urinary bladder reveals squamous cell carcinoma of bladder. What would be the most common parasite associated with this type of bladder cancer?
   a) *Schistosoma haematobium*
   b) *Trichomonas vaginalis*
   c) *Schistosoma mansoni*
   d) *Schistosoma japonicum*

48. Frei test is used for diagnosis of:
   a) Lymphogranuloma venereum
   b) Trachoma
   c) Inclusion conjunctivitis
   d) Infant pneumonia

49. Strawberry cervix is classically seen in:
   a) Herpes simplex virus b) *Neisseria gonorrhoeae*
   c) *Chlamydia trachomatis* d) *Trichomonas vaginalis*

Answers: 1:a, 2:d, 3:b, 4:c, 5:a, 6:b, 7:d, 8:a, 9:b, 10:d, 11:c, 12:a, 13:b, 14:b, 15:a, 16:c, 17:b, 18:b, 19:c, 20:d, 21:d, 22:a, 23:a, 24:c, 25:d, 26:c, 27:b, 28:a, 29:b, 30:c, 31:d, 32:d, 33:a, 34:a, 35:a, 36:d, 37:b, 38:b, 39:a, 40:b, 41:a, 42:d, 43:d, 44:a, 45:d, 46:b, 47:a, 48:a, 49:d

CHAPTER 28

Urinary Tract Infections

(MI7.3)

Chapter Navigation

- Types of UTIs
- Predisposing factors
- Pathogenesis
- Etiology
- Clinical features
- Laboratory diagnosis
- Treatment

Learning objectives

- Describe the etiopathogenesis and clinical features of urinary tract infections
- Discuss the laboratory diagnosis of urinary tract infections including the methods of specimen collection

The urinary tract, starting from the calyces of the kidneys to the urethra, is lined with a sheet of epithelium that is continuous with the skin. Although the flow of urine and the sloughing of these epithelial cells serve to protect the urinary tract from infection, microorganisms, particularly bacteria, may enter the tract through the epithelial surface to cause infection.

URINARY TRACT INFECTIONS (UTI)

- Urinary tract infection (UTI) is a collective term given to various bacterial infections that occur in the urinary tract.
- These are an important cause of illness, affecting all age groups.
- Unless treated early, a UTI can spread to the kidneys and progress to a secondary infection called **urosepsis**, especially in uncontrolled diabetes, and may have a fatal outcome.
- While all parts of the urinary tract may be affected, the most common UTIs are those of the bladder (cystitis) and the renal pelvis or the kidneys (pyelonephritis).

Types of UTIs

1. **Upper UTI:** Involves the kidney or ureter and includes the following conditions:
   - **Acute pyelitis**—infection of the pelvis of the kidney
   - **Acute pyelonephritis**—infection of the parenchyma of the kidney
2. **Lower UTI:** Involves infection from the urinary bladder downwards and includes the following conditions:
   - **Urethritis**—infection of the urethra
   - **Acute urethral syndrome** (in women, it is often caused by *Chlamydia*)
   - **Cystitis**—infection of the urinary bladder
   - **Prostatitis**—infection of the prostate

Based on the mode of acquisition, UTIs may be classified as follows:

- Community-acquired
- Hospital-acquired (See Chapter 12 for details on hospital-acquired UTI)
- Catheter-associated UTI (CAUTI), which refers to a UTI occurring in a person whose urinary tract is currently catheterised or has had a catheter in place within the past 48 hours

UTI can present as acute or chronic or uncomplicated or complicated (Fig. 28.1):

- **Uncomplicated UTI:** This refers to acute or recurrent lower UTI and/or upper (uncomplicated pyelonephritis) UTI, seen in non-pregnant

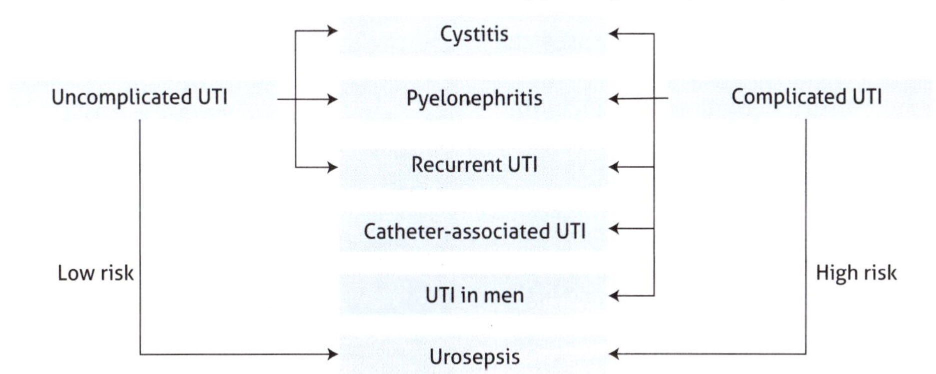

Fig. 28.1 Complicated and uncomplicated UTIs

women with no known relevant anatomical and functional abnormalities within the urinary tract or comorbidities.

- **Complicated UTI:** This term refers to UTIs that are not defined as uncomplicated. For example, UTIs in a patient with an increased chance of a complicated course—all men, pregnant women, patients with relevant anatomical or functional abnormalities of the urinary tract, indwelling urinary catheters, renal diseases, and/or with other concomitant immunocompromising diseases such as diabetes.
- **Recurrent UTI:** Defined as recurrences of uncomplicated and/or complicated UTIs with a frequency of at least three episodes/year or two UTIs in the last six months.
- **Urosepsis:** It is defined as life-threatening organ dysfunction caused by a dysregulated host response to an infection originating from the urinary tract and/or male genital organs.

PREDISPOSING FACTORS FOR UTIs

- **Age:** The incidence of UTI increases with age due to decreasing immunity, increasing incidence of comorbidities and increased colonisation of microorganisms.
- **Gender:** Male infants who are not circumcised are at a higher risk of UTI than girls. After infancy, girls are at higher risk than boys. Young and sexually active females are specifically prone to the infection due to the short urethra, the proximity of the urinary tract to the anus and urethral trauma during intercourse.
- **Pregnancy:** Dilatation of the ureters and renal pelvis, stasis, incompetence of the vesicourethral valves, and hormonal changes increase the risk of UTIs during pregnancy.
- **Structural and functional abnormalities of the urinary tract:**
  - Obstruction due to urethral stricture, calculus, prostatic hypertrophy, and tumour
  - Neurogenic bladder
  - Vesicourethral reflux
  - Genital prolapse
- **Metabolic factors:** Diabetes mellitus increases the risk of UTIs.
- **Interventions:** Instrumentation including catheterisation and any surgical procedure may result in a UTI.

Urine is an ideal medium for bacterial growth. Factors that make it less favourable for bacterial growth include a pH of less than 5, the presence of organic acids, and high urea levels. Frequent urination and high urinary volumes are also known to decrease the risk of UTIs.

PATHOGENESIS (FIG. 28.2)

- The urethral portion is colonised by commensal bacteria while the urinary bladder and upper urinary tract is completely sterile in a healthy person.
- Bacteria establish infection by ascending from the anterior urethra through the column of urine.
- They generally get eliminated by voiding. However, if there is stasis of the urine either in the bladder or ureters due to neurogenic or mechanical causes, the microorganisms colonise and establish infection.
- When bacteria invade the bladder mucosal wall, an inflammatory reaction is produced.
- Hematogenous spread to the urinary tract occurs in less than 2% of cases.

ETIOLOGY

- The etiological agents of urinary tract infections are listed in Table 28.1.
- *Escherichia coli* is the most common cause of uncomplicated UTI.

Agents

- **Bacteria:** Microorganisms associated with UTI are usually primary pathogens of some systemic infections (e.g., *Escherichia coli*—GI infections, *Staphylococcus*—skin and soft tissue infections); the details of these organisms are available in the respective chapters.
- **Fungi:** *Candida* species may cause UTI in hospitalised patients with indwelling catheters or those with diabetes.

Table 28.1 Etiological agents of urinary tract infections

Bacteria	Gram-negative bacilli • *E. coli* • *Proteus* species • *Klebsiella* • *Enterobacter* • *Pseudomonas*
	Gram-positive cocci • *S. aureus* • *Enterococcus* species • *S. epidermidis* • *S. saprophyticus*
	Gram-negative cocci • *Neisseria gonorrhoeae*
	Others • *M. tuberculosis* • *Salmonella* species • *Gardnerella vaginalis*
Viruses	Adenovirus—HSV
Fungi	*Candida albicans* and Other species of *Candida*
Parasites	*Trichomonas vaginalis*
	Schistosoma haematobium
	Filariasis *Enterobius vermicularis*

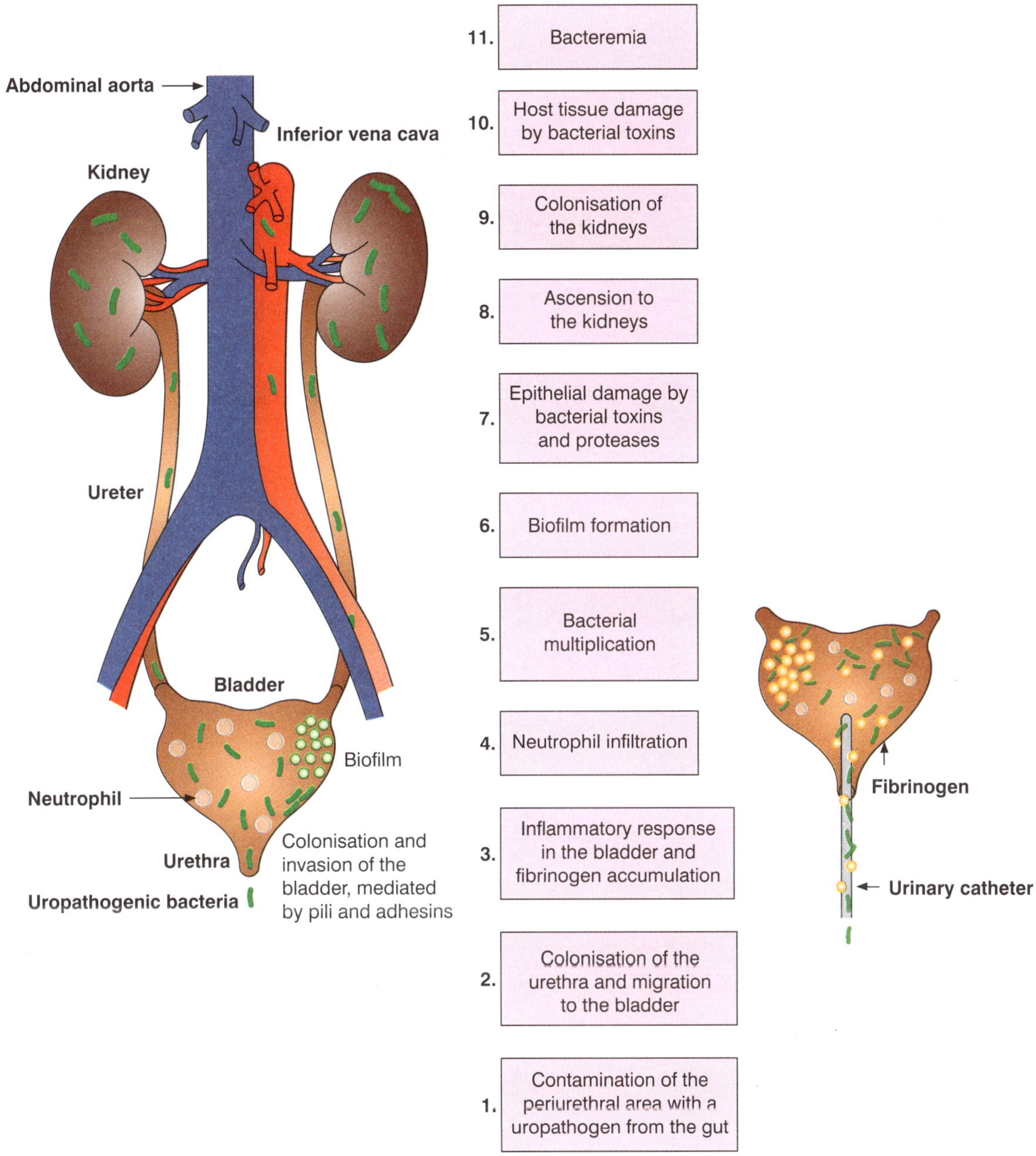

Fig. 28.2 Pathogenesis of urinary tract infections

- **Parasites:**
  - *Trichomonas vaginalis*: It colonises the female genital tract, can lead to vaginitis, and ascend into the urethra, causing UTIs.
  - *S. haematobium*
    - It causes urogenital schistosomiasis. The pathogenesis of *Schistosoma* is due to eggs deposited in the tissues of the urinary bladder and ureter, which damage the mucosa of the urinary bladder and provoke a type IV hypersensitivity reaction, resulting in granuloma formation.
    - The granulomas may coalesce to form nodules, and the mucosa over these nodules develops glandular hyperplasia. In the late stage, fibrosis can occur, causing obstruction in the ureter and leading to hydronephrosis.
    - Constant irritation of the bladder wall results in metaplastic changes and may predispose to squamous cell carcinoma of the urinary bladder.
    - Genital involvement may be seen in both males and females.
  - Filariasis can cause obstruction of lymphatic vessels, causing lymph fluid to enter the urine (**chyluria**).
  - *Enterobius vermicularis* infestation of the urinary tract can lead to recurrent UTI.

- **Virus:** Viral UTIs are extremely rare.
  - Adenovirus can cause viral UTI in disseminated adenoviral infection, when the virus gets seeded in the urinary tract.
  - Herpes simplex virus may infect the urethra, making urination painful and emptying of the bladder difficult.

CLINICAL FEATURES

Asymptomatic Bacteriuria

- Asymptomatic bacteriuria is defined as the presence of bacteria in the properly collected urine of a patient who has no signs or symptoms of a urinary tract infection. Most patients with asymptomatic bacteriuria will never develop symptomatic urinary tract infections and will have no adverse consequences from asymptomatic bacteriuria.
- However, if it goes undetected and is left untreated in in pregnancy, it may lead to symptomatic infection, pyelonephritis, and hypertension in the woman, as well as prematurity and perinatal death of the fetus.
- Because of the lack of symptoms, asymptomatic bacteriuria is diagnosed by laboratory investigation of urine specimens. Fig. 28.3 presents the strategy for the diagnosis of this condition.
- No treatment is required except in pregnant women. Treatment of all other patients (including diabetics and older patients) does not decrease the incidence of symptomatic urinary tract infections or improve survival. However, it does increase the likelihood of adverse effects from antibiotics and the development of antibiotic-resistant bacteria.

Symptomatic Uncomplicated UTI

It may present as follows:

- Pain on urination (dysuria)
- Frequent urination (frequency)
- Inability to start the urine stream (hesitancy)
- Sudden onset of the need to urinate (urgency)
- Blood in the urine (hematuria)

Pyelonephritis or pyelitis (upper UTI) characterised by:

- Flank or loin pain
- Tenderness
- High fever
- Rigor

In infants and children with UTI, fever, vomiting, diarrhea, and poor feeding may be noticed.

LABORATORY DIAGNOSIS

Significant Bacteriuria

- It is defined as the presence of at least 100,000 colony forming units (CFU) /mL of urine.
- Small numbers of bacteria are normally found in the anterior urethra and may be washed out into urine samples.
- Counts of fewer than 1,000 CFU/mL are normally considered to be urethral contaminants unless there are exceptional clinical circumstances, such as a sick, immunosuppressed patient.
- Quantitative cultures are hence necessary for the diagnosis of urinary tract infection.

Fig. 28.3 presents an algorithmic representation of the approach to the diagnosis of UTIs.

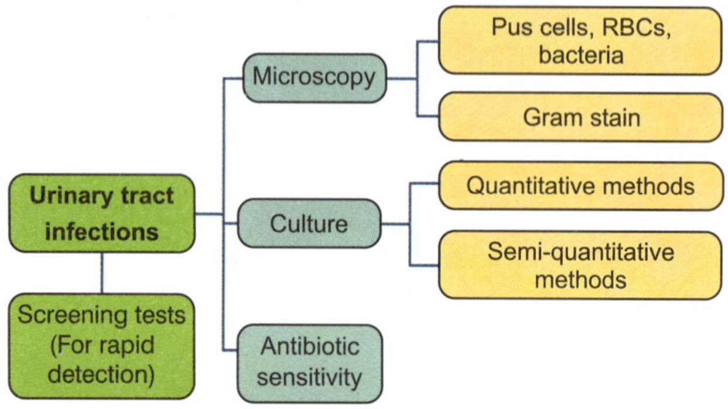

Fig. 28.3 Flowchart of the approach to the diagnosis of urinary tract infections

Specimen collection

Urine sample

As urine flows through the anterior urethra, which is in close proximity to the perineum, it is important that urine meant for culture is collected while avoiding contamination with normal urethral and skin flora as far as possible. The following types of urine samples may be collected (Fig. 28.4):

- **Midstream urine (MSU):**
  - Distal urethra contains mixed flora. These organisms are present in urine specimens unless a clean-catch, midstream specimen is obtained.
  - A midstream urine sample is appropriate for bacterial culture.
  - The **collection container** should be a wide-mouthed, screw-capped, leak-proof, and sterile universal container. It is important to provide the patient with clear instructions for proper collection.
  - The patient is advised to clean the genital area with soap and water, and to void the first part of the urine in the toilet before collecting the 'midstream' urine in the container. The last part of the urine too should be voided into the toilet.
- **Catheter sample urine (CSU):**
  - The area over the catheter is first cleaned with alcohol after donning clean gloves.
  - The catheter is clamped. With the help of a sterile syringe and needle, the urine sample is aspirated above the clamp and collected in a universal container.

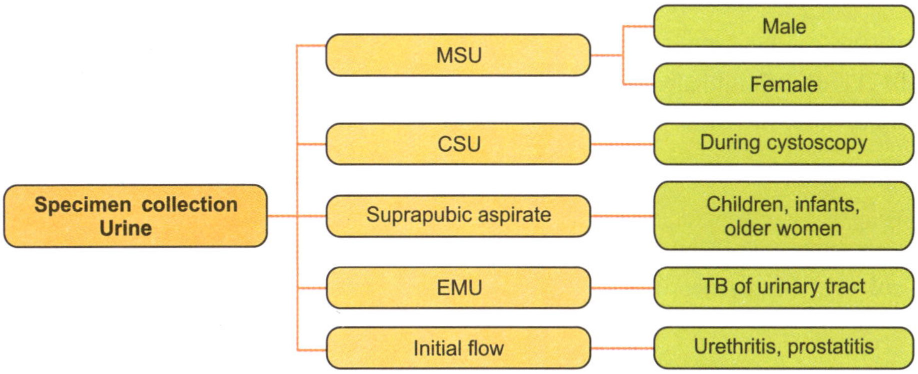

Fig. 28.4 Types of urine samples to be collected (*MSU*—midstream urine; *CSU*—catheter sample of urine; *EMU*—early morning urine)

- Urine should **never** be collected from the uro-bag or by opening the connecting draining tube.
- If the catheter is more than 14 days old, it should be replaced before collecting urine for culture.

- **Suprapubic aspirate:**
  - Urine is collected with a sterile syringe and needle from the bladder under aseptic precautions by puncturing the suprapubic area.
  - This method is useful when infants are to be tested.
- **Early morning urine (EMU):** If renal tuberculosis is suspected, three urine samples are collected on consecutive days, or the first morning urine sample is collected in entirety.
- **Blood sample:** Paired blood samples must be collected if the patient is febrile or if acute pyelonephritis or complicated UTI is suspected.

Transport

- The urine sample must be **transported at room temperature within half an hour or refrigerated at 4°C for up to six hours**. Beyond this time, such a sample should not be processed for bacterial culture.
- This because urine is a good culture medium in which contaminating bacteria may grow if it is left outside for a long time; such a sample would yield higher bacterial counts and hence, false positives.
- For the interpretation of the results, quantitative cultures need to be made.
- If the sample is from a patient who has no immediate access to a healthcare facility, and if transport to the facility would exceed four hours, a special container with 1.8% boric acid is provided, in which urine is collected as described above. Such a sample can be kept at room temperature for up to 24 hours.

Microscopy

- An uncentrifuged sample is examined by direct microscopy for **pus cells and bacteria**.
- The presence of pus cells and bacteria in large numbers reflects bacterial infection of the urinary tact.
- The presence of 10 leukocytes/mm^3 of uncentrifuged urine or 10 leukocytes/hpf of the centrifuged sample, in a clinically suspected UTI, is important for diagnosis.

Screening

i. **Dipstick test:** This is a special strip of paper that is dipped into a sample of urine and immediately produces results. It is also called a **rapid urine test**. Multiple parameters can be measured using dipstick tests such as protein, blood, leukocyte, nitrite, glucose, ketone, pH, specific gravity, bilirubin, and urobilinogen.

ii. **Griess nitrite test:** The presence of nitrite in urine indicates the presence of nitrite-reducing bacteria in urine.

iii. **Leukocyte esterase test:** It detects the presence of intact or lysed leukocytes. A positive dipstick test helps in diagnosing cystitis, whereas a negative test rules out cystitis.

iv. **Glucose test paper tests:** It is performed to look for the utilisation of minute amounts of glucose present in normal urine by infection-causing bacteria.

Other rapid tests

i. **Catalase test:** It detects the presence of catalase as evidenced by frothing on addition of hydrogen peroxide. This indicates bacteriuria, though a positive result is obtained in hematuria also.

ii. **Triphenyl tetrazolium chloride (TTC) test:** It is based on the production of a pink–red precipitate in the reagent as a result of the respiratory activity of growing bacteria.

iii. **Gram stain:** It aids the microscopic demonstration of bacteria in urine and is diagnostic.

iv. **Dip slide culture:** Agar-coated slides are immersed in urine or even exposed to the stream of urine during voiding. They are then incubated, and the growth estimated by colony counting or by colour change of the indicator.

None of the screening methods is as sensitive or reliable as culture.

Culture

This is most useful when collected from a patient with clinical features of UTI, and should always be collected before the first dose of antibiotic.

- **Media:**
  - Blood agar
  - MacConkey's agar
  - Alternatively, CLED agar can be used
- **Quantitative culture of urine:** It provides an estimate of the total number of bacteria in 1 mL of urine.
- **Pour plate method:**
  - An appropriate volume of urine is mixed with a standard volume of molten nutrient agar (between 45–48°C) and incubated.
  - The actual number of colonies is calculated using a colony counter.
  - The method is very time consuming and labour intensive and hence, not practical in the laboratory.
- **Semi-quantitative culture:**
  - **Method:** A standard loop of 4 mm diameter is used. It can hold a urine volume of 0.01 or 0.001 mL. This is cultured on media as shown in Fig. 28.5. After overnight incubation at 37°C, the colonies are counted and extrapolated to one mL.
- **Interpretation:**
  - A colony count of 10^5/mL is considered significant.
  - Counts between 10^4/mL and 10^5/mL are of uncertain significance.
  - Counts less than this are considered significant in the following cases:
    - The patient is already on antibiotics
    - There is an obstruction in the urinary tract
    - A fungal infection is present
    - Pyelonephritis is present
    - Men with pyuria
    - Symptomatic young women
    - The specimen has been collected by suprapubic aspiration
  - If ≥3 types of organisms are grown, these are considered as contaminants.
  - Identification of the isolates grown is performed using biochemical tests or automated systems.
- **Detection of antibody-coated bacteria:**
  - It differentiates upper UTI (involvement of the kidney) from lower UTI.
  - It is presumed that upper UTI results in antibody response, which allows the antibody to coat the microorganism.
  - This can be detected in the urine by immunofluorescence using antihuman immunoglobulin tagged to a fluorescent dye.

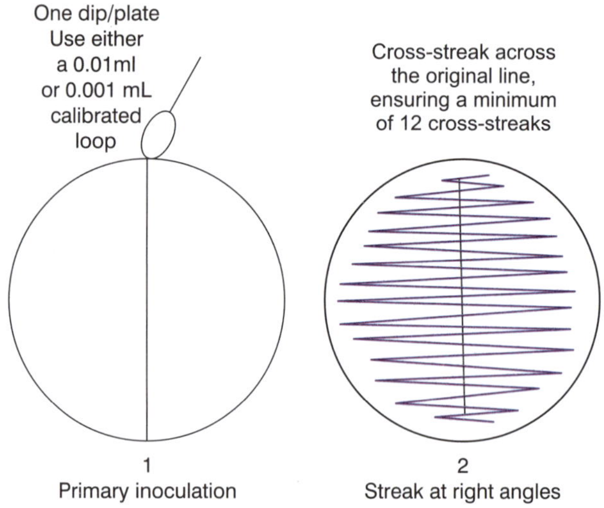

Fig. 28.5 Semi-quantitative culture of urine using a calibrated inoculation loop

Antibiotic sensitivity testing

In addition to other antibiotics, the organism must be tested against norfloxacin and nitrofurantoin which concentrate in urine and effectively kill of organisms.

Laboratory diagnosis of UTI

Significant bacteriuria: Presence of at least 100,000 colony forming units (CFU) of bacteria per mL of urine.

- **Specimen collection:**
  - **Urine sample**
    - Midstream urine (MSU)
    - Catheter sample urine (CSU)
    - Suprapubic aspirate (in infants)
  - **Early**
    - Blood sample: Paired blood samples must be collected if the patient is febrile, or acute pyelonephritis or complicated UTI is suspected
    - Transport: Transported at room temperature within half an hour or refrigerated at 4°C for up to six hours; beyond this time, such a sample should not be processed for bacterial culture
    - Microscopy: 10 leukocytes/mm^3 of uncentrifuged urine or 10 leukocytes/hpf of the centrifuged sample
- **Screening:** None of the screening methods is as sensitive or reliable as culture
- **Dipstick tests:**
  - Griess nitrite test: Nitrite in urine indicates the presence of nitrite-reducing bacteria in urine
  - Leukocyte esterase test: Intact or lysed leukocytes and helps in diagnosing cystitis
  - Glucose test paper tests
- **Other rapid tests:**
  - Catalase test: Indicates bacteriuria
  - Triphenyl tetrazolium chloride (TTC) test: Indicates growing bacteria
  - Gram stain
  - Dip slide culture

- **Culture:** Most useful when collected from a patient with clinical features of UTI, always before the first dose of antibiotic
  - **Media:** Blood agar, MacConkey's agar, or CLED agar
- **Quantitative culture of urine**
  - **Pour plate method**
  - **Semi-quantitative culture:** A standard loop that holds a urine volume of 0.01 or 0.001 mL is cultured; after overnight incubation at 37°C, the colonies are counted and extrapolated to one mL
  - **Interpretation**
    - 10^5/mL is significant
    - Counts between 10^4/mL and 10^5/mL
  - Counts less than this are considered significant in the following cases:
    - The patient on antibiotics
    - Obstruction in the urinary tract
    - Fungal infection
    - Pyelonephritis
    - Men with pyuria
    - Symptomatic young women
    - Aspirate
  - ≥3 types of organisms considered as contaminants
- **Identification:** Biochemicals or automated systems
- **Detection of antibody-coated bacteria:** Helps to differentiate upper UTI (involvement of the kidney) from lower UTI, done by immunofluorescence using antihuman immunoglobulin tagged to a fluorescent dye
- **Antibiotic sensitivity testing:** For norfloxacin and nitrofurantoin that concentrate in urine and achieve killing of organisms

TREATMENT

- The most suitable method of treatment is with antibiotics that can reach high concentrations in the renal parenchyma or urine.
- The resolution of bacteriuria has been correlated with the concentration of the antimicrobial agent in the urine rather than serum levels.
- All the antimicrobial agents approved for the treatment of urinary tract infections achieve inhibitory urinary concentrations that significantly exceed serum levels.
- Antibiotics that are intermediate sensitive can also be used as they get concentrated in urine and achieve higher concentrations that may be lethal to bacteria.
- Antibiotics that eradicate aerobic gram-negative flora but have little effect on the vaginal and fecal anaerobic flora seem to provide the best long-term cures for uncomplicated urinary tract infections, e.g., trimethoprim–sulfamethoxazole or the fluoroquinolones.
- In patients who have frequent episodes of recurrent uncomplicated cystitis, antibiotic prophylaxis should be given. Cotrimoxazole, ciprofloxacin, cephalexin and nitrofurantoin have been tried as prophylactic antibiotics. But this practice can lead to antibiotic misuse and overuse and should be used with extreme caution.

Drug of Choice (Table 28.2)

Table 28.2 Antibiotic recommended for the treatment of UTIs

Uncomplicated UTI	Cotrimoxazole or cefixime
Complicated UTI	Cefotaxime or ceftriaxone or amikacin for a minimum of 14 days
Acute cystitis	Nitrofurantoin Fosfomycin
Acute pyelonephritis	Piperacillin—tazobactam Ertapenem
Acute prostatitis	Piperacillin—tazobactam Ertapenem

Questions

Short answer questions

1. What are the techniques employed in the laboratory to culture urine? How are they interpreted?
2. Enumerate screening methods for the presumptive diagnosis of UTIs.
3. Describe the pathogenesis of urinary schistosomiasis.
4. Describe various methods of collecting urine samples for the laboratory confirmation of UTIs.

MCQs

1. **An infant was suffering from urinary tract infection. Semiquantitative culture of urine collected through suprapubic aspiration yielded 10^3 CFU/mL. Which of the following suggests the right interpretation of the test result in this case?**
   a) Suggestive of urinary tract infection
   b) Request for midstream urine sample
   c) Sample contaminated, repeat sample after proper collection
   d) Request for three days' consecutive early morning sample
2. **All of the following are presumptive methods for detection of significant bacteriuria except:**
   a) Catalase test b) Griess nitrite test
   c) Dip slide culture d) Urine microscopy
3. **All the following are urinary pathogens except:**
   a) *Enterobacter* spp.
   b) *Proteus* spp.
   c) *Schistosoma haematobium*
   d) *Entamoeba histolytica*
4. **Which one of the following is an oxidase-positive urinary bacteria?**
   a) *Staphylococcus saprophyticus*
   b) *Proteus vulgaris*
   c) *Escherichia coli*
   d) *Pseudomonas aeruginosa*

5. **One of the following is false regarding urine culture:**
   a) Quantitative urine culture is necessary
   b) Catheter urine sample is collected from urine bag
   c) Three early morning samples are collected for the diagnosis of renal tuberculosis
   d) Suprapubic aspiration of urine is suggested in infants
6. **A 28-year-old sexually active woman had frequent episodes of urinary tract infections. Which of the following common novobiocin-resistant bacteria are likely to be associated in this case?**
   a) *Escherichia coli*
   b) *Proteus vulgaris*
   c) *Pseudomonas aeruginosa*
   d) *Staphylococcus saprophyticus*
7. **Which among the following is not a screening test for UTI?**
   a) Griess nitrite test
   b) Catalase test
   c) Rapid urease test
   d) Triphenyl tetrazolium chloride test
8. **Siderophores are virulence factors present in *E. coli* strains causing:**
   a) Meningitis
   b) Urinary tract infection
   c) Septicemia
   d) Diarrhea
9. **The false statement about urine culture is:**
   a) Catheterised sample is not ideal
   b) Culture on blood agar is for obtaining bacterial count
   c) $>10^5$ bacteria/mL is considered significant from a mid-stream sample
   d) Any number of bacteria is significant in a catheterised sample

Answers: 1:a, 2:d, 3:c, 4:d, 5:b, 6:d, 7:c, 8:b, 9:d

CHAPTER 29

Skin, Soft Tissue, and Musculoskeletal Infections

(MI4.1–4.3)

Learning objectives

- Discuss the etiopathogenesis, clinical course, and the methods used for the laboratory diagnosis of skin and soft tissue infections
- Discuss the etiopathogenesis, clinical course, and the methods used for the laboratory diagnosis of bone and joint infections

INFECTIONS OF THE SKIN AND SOFT TISSUES

Skin is the largest organ of the human body, covering the body, and protecting all the structures underneath. One of its many functions is to prevent microorganisms in the surrounding environment from invading the tissues and organs underneath. However, skin itself is not immune to infections, which may be due to the following reasons:

- **Infection of intact skin:** Certain pathogens and toxins cause severe infections or reactions when they come in contact with skin (ecthyma gangrenosum).
- **Through a breach in the continuity:** Certain microorganisms breach the skin's natural defences through cuts, wounds, burns, or bites.
- **Disruption of normal microbial population of the skin:** This results in infection by new organisms in the surrounding skin and tissue (e.g., staphylococcal and streptococcal infections).

- **Pathogens colonising the skin via different routes:** Some microorganisms gain entry to the skin from other systems such as the respiratory or digestive systems and cause reactions that manifest as skin rashes or lesions (intracellular infection—herpes type I, dermal capillary plexus—varicella, cutaneous nerve route—herpes-zoster).

Skin and soft tissue infections are caused by organisms that invade the body from the external environment through breaks in skin caused by burns, injury, bites, or surgery (**primary infections**) or those that reach the site through the blood as a part of a systemic disease (**secondary infections**).

INFECTIONS OF THE DERMIS AND EPIDERMIS (TABLE 29.1)

Table 29.1 Features of different type of skin infections

Type of skin infection	Features
Cutaneous infections	
Impetigo	Involves the epidermis and usually occurs on the face; most often in children
Erysipelas	Involves the dermis as an erythematous painful lesion
Cellulitis	Affects the deep dermis; presents as a diffuse, erythematous lesion without a clear margin
Folliculitis	Involves the hair follicle; presents as a localised papule containing pus
Abscess (furuncle, carbuncle, boil)	Affects the deep dermis and presents as an inflamed nodule with a central region that is purulent with a raised point
Macule	Flat, circumscribed skin lesion usually ≤1 cm in greatest diameter, may coalesce to form a patch or rash
Papule	Elevated, circumscribed skin lesion ≤1 cm in diameter
Plaque	Papules are >1 cm in size
Nodule	A larger, firm papule with a significant vertical dimension
Pustule	Papule containing purulent fluid
Vesicle	Papule containing clear serous fluid
Bullae	Larger vesicle >1 cm
Urticaria	Wheal or hive
Subcutaneous tissue infections	
Necrotising fasciitis	Soft tissue infections involving the fascia and muscles overlying soft tissues
Progressive bacterial synergistic gangrene	Chronic, rapidly spreading necrosis of the skin occurring usually as a postoperative complication following a thoracic or abdominal surgery
Myositis	Inflammation, especially of the voluntary muscles, characterised by localised or diffuse pain, tenderness on movement or palpation, swelling and/or weakness May vary from necrosis of muscle to necrotising cutaneous myositis or anaerobic myonecrosis

Clinical Presentation

Skin infections can clinically present as:

- **Maculopapular rash, vesicles**—viral infections, zoster, shingles, and hand, foot, and mouth disease
- **Bullae**—necrotising fasciitis and gas gangrene
- **Papular and nodular lesions**—dracunculiasis, creeping eruptions, and larva migrans
- **Crusted lesions**—impetigo and ringworm

Etiology

Etiological agents of infections of the skin, subcutaneous tissue, and various types of wounds are listed in Tables 29.2 and 29.3.

INFECTION OF A BURN WOUND

Burn wounds are initially sterile due to the high temperature at which the lesions occur. However, they subsequently get colonised with the normal commensals of the skin—first the gram-positive and then the gram-negative organisms.

- A major threat to burn injuries is hospital-acquired infection by drug-resistant organisms.
- Fungal colonisation has been observed because of the widespread use of topical antibiotics.

Risk Factors

- Advanced age
- The percentage of area involved
- The involvement of lower extremities
- Scald burns
- Full-thickness burns
- Delay in treatment
- Pre-existing conditions such as diabetes

Causes of Burn Wound Infections

These are presented in Table 29.3.

Laboratory Diagnosis

- Etiological agents may be isolated from burn tissues and wound swabs.
- The antibiotic susceptibility of these organisms is important for targeted treatment.

Treatment

- Broad-spectrum antibiotics used in combination therapy are ideal to cover the wide range of bacteria that may be involved.
- Local wound toilet of the burn wound with debridement, local dressing, and systemic antibiotics/antifungal agents forms an important part of treatment.
- Infection control measures include isolation rooms, handwashing, appropriate wound care, early excision and grafting, antibiotic stewardship, and nutritional support.

Table 29.2 Etiology of infections of the epidermis and dermis

Bacteria	Viruses	Fungi	Parasites
Streptococcus pyogenes: Erysipelas Impetigo Cellulitis *Staphylococcus aureus:* Impetigo Cellulitis Furuncle *Corynebacterium minutissimum:* Erythrasma	Herpes-zoster Varicella-zoster Herpes simplex **Exanthematous lesions:** Measles Rubella Mumps (most viruses cause vesicles or rashes)	**Dermatophytes:** *Sporothrix schenckii*	Cutaneous larva migrans *A. brasiliensis*

Table 29.3 Etiology of infections of the subcutaneous tissue, postoperative wound infections, burn wound infections

Subcutaneous tissue infections			Postoperative wound infections		Burn wound infections	
Bacteria	**Fungi**	**Parasites**	**Bacteria**	**Fungi**	**Bacteria**	**Fungi**
Staphylococcus aureus Group A streptococci *Clostridium perfringens* *Clostridium novyi* *Clostridium septicum* *Bacillus* species *Bacteroides* species *Peptostreptococcus* species (Some of the above bacteria cause necrotising fasciitis and myositis)	*M. mycetomatis* *M. grisea* *Candida* species	*Trichinella* *Wuchereria bancrofti* *Dracunculus medinensis* *Taenia solium*	*S. aureus* *Streptococcus pyogenes* Coagulase-negative staphylococci *Proteus* species *Pseudomonas* species *Bacteroides* species *Prevotella* species	*Candida* species	*S. epidermidis* *Pseudomonas* species *Proteus* species Enterobacteriaceae *Bacteroides* species	*Fusarium* Zygomycetes *Candida* species *Aspergillus* species

APPROACH TO THE DIAGNOSIS OF SKIN AND SOFT TISSUE INFECTIONS (Fig. 29.1)

- **Specimen:**
  i. **Pus from the wound** is collected by the following techniques:
    - Incision and drainage
    - Needle aspiration (including vesicle fluid)
    - Sterile swab

  ii. In **dermatophyte infections**, the following specimens can be collected:
    - Scrapings from the active border of the lesion
    - Skin biopsy
    - Erysipeloid

  iii. In **subcutaneous infections**, the following specimens are collected:
    - Sample is collected from the base of the lesion
    - Surgical biopsy of deep tissues is done to obtain a positive culture
- **Microscopy:**
  - Gram stain is routinely performed on all samples.
  - KOH mount is done for suspected dermatophyte infection.
  - Tzanck smear is carried out on vesicle fluid for the detection of herpes simplex and varicella–zoster infections.
- **Culture**
  - *For aerobic bacteria*: Specimens are inoculated onto blood agar, MacConkey, and chocolate agar.
  - *For atypical mycobacteria*: LJ medium is inoculated.
  - *For anaerobic culture*: Thioglycolate broth and Robertson's cooked meat medium are inoculated and incubated in a Gaspak or anaerobic jar.
  - *For fungi*: Sabouraud dextrose agar is used with and without cycloheximide to isolate fungal pathogens.
- **Identification:** This is done on the basis of colony morphology, Gram/ZN stain, and biochemical reactions.
- **Antibiotic/antifungal sensitivity:** Antibiotic tests are performed by standard procedures like Kirby–

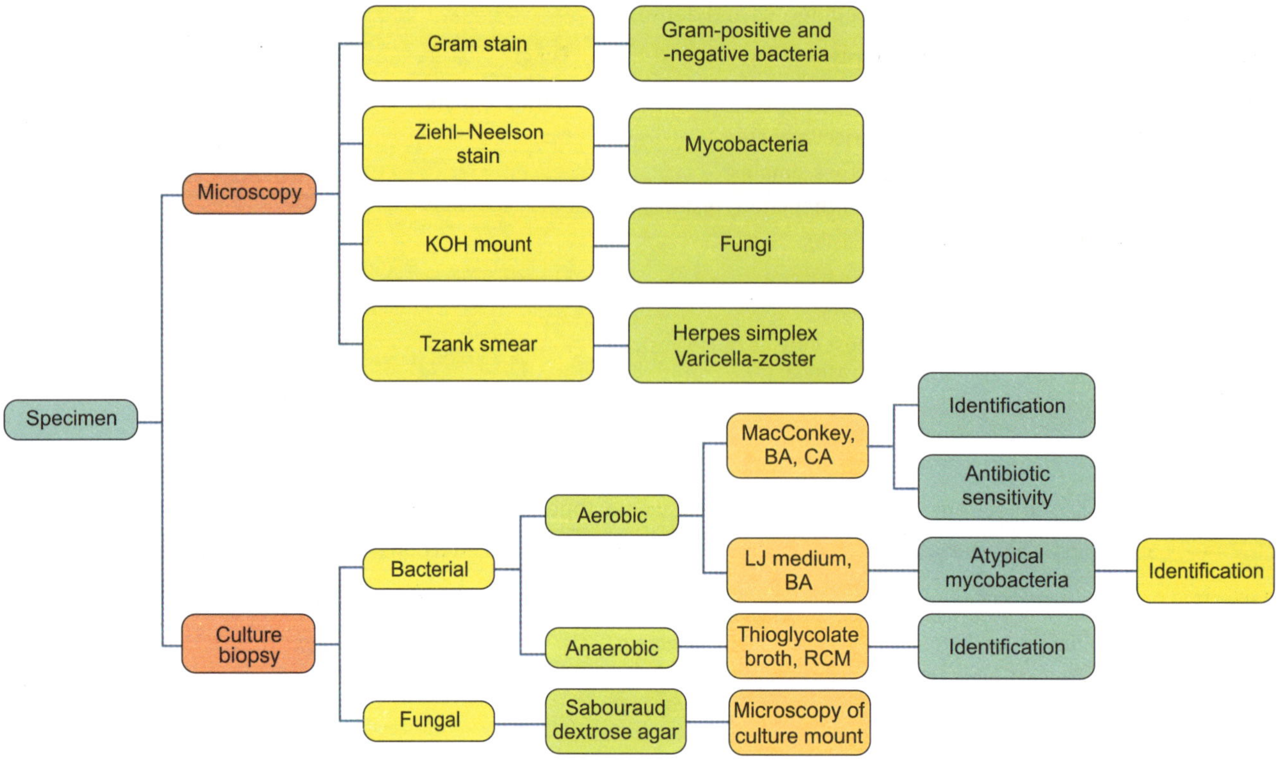

Fig. 29.1 Approach to the diagnosis of skin and soft tissue infections (*BA*—blood agar; *CA*—cetrimide agar; *RCM*—Robertson's cooked meat)

Bauer disc diffusion method for bacterial and yeast isolates (discussed in Chapters 3, 4, 5, and 6).

TREATMENT

- Most superficial skin infections are treated by **draining the localised area of pus collection**. Local dressing with topical antibiotics or oral cephalexin or dicloxacillin often suffices for small, superficial infections.
- Cellulitis, erysipelas, deep tissue infections, necrotising fasciitis (which may be a medical emergency requiring surgical debridement), myositis, etc., require systemic antibiotics (intravenous), e.g., clindamycin, piperacillin tazobactam and vancomycin (for MRSA).

MUSCULOSKELETAL INFECTIONS

These are serious infections as they may lead to the destruction of the bone and cartilage, leading to disability. The major categories of infection are as follows:

- Osteomyelitis
- Infectious arthritis (septic arthritis)
- Myositis

The most common route of entry of organisms to the bones is by **hematogenous spread**. They may also gain entry following trauma caused by a penetrating injury.

ACUTE OSTEOMYELITIS

- It is generally of **bacterial etiology**.
- In children, it often involves the ends of the long bones.
- Clinical features include abrupt-onset bone pain and localised tenderness at the site of infection.
- The most common cause of acute osteomyelitis is ***Staphylococcus***.

CHRONIC OSTEOMYELITIS

- It is of either **fungal or tubercular etiology**.
- Insidious onset—the bone is necrosed and sequestrum (avascular necrotic bone) is formed.
- In prosthetic-associated osteomyelitis, the causative organism is *S. epidermidis*.
- *Salmonella* species may cause osteomyelitis in sickle cell anemia.

SEPTIC ARTHRITIS

- Bacteria may invade the joint space and cause infectious arthritis.
- Large, weight-bearing joints like the knee or hip are often involved.
- The source of infection is often the skin, from where bacteria reach the bones through the bloodstream.
- **Clinical features:** Acute onset with fever, joint pain, redness, and swelling are the cardinal features.

- **Causative organisms:** *Staphylococcus aureus* is the most common cause of septic arthritis. *N. gonorrhoeae* may cause septic arthritis in sexually active individuals (the knee, wrist, ankle, or elbow may be involved). Other causative organisms include *S. epidermidis* and other coagulase-negative staphylococci in prosthetic joints.

MYOSITIS

- Infective myositis refers to conditions causing inflammation in muscles and may be part of a systemic infection, especially a viral infection.
- It is characterised by localised or diffuse pain, tenderness on movement or palpation, swelling, and/or weakness.

Etiology

- Bacteria such as *Staphylococcus aureus*, *Streptococcus pyogenes*, *Salmonella*, *Escherichia coli*, and *Pneumococci* can cause infective myositis. Anaerobic organism such as *C. perfringens* can cause clostridial myositis, popularly known as **gas gangrene**.
- **Lyme disease** is caused by a bite from a deer tick and is less common than osteomyelitis and septic arthritis. It is more prevalent in the Northeastern and Midwestern regions of the United States.
- **Tuberculosis** can involve the musculoskeletal system in 2—5% of cases.
- In leprosy, the skeletal muscles may be involved because of peripheral neuropathy with subsequent muscular denervation or even as a primary muscle disease eliciting an inflammatory reaction, giving rise to leprous myositis.
- Amongst **fungi**, *Candida*, *Cryptococcus neoformans*, *Histoplasma*, *Coccidioides* and *Aspergillus* may cause fungal myositis, especially in immunocompromised persons.
- The common **parasitic infections** that can cause myositis are due to *Trichinella spiralis*, *Taenia solium*, and *Toxoplasma gondii*, followed by the less common ones that include *Trypanosoma cruzi* which causes Chagas disease, *Toxocara canis*, *Schistosoma*, *Echinococcus*, *Entamoeba histolytica*, *Spirometra*, mansonoides which causes sparganosis, *Plasmodium falciparum*, and others (*Sarcocystis* spp., *Schistosoma* spp., *Echinococcus* spp., and *Microsporidia* spp.).
- **Viruses** involve the muscle and are the most common cause for infectious myositis leading to myalgia, polymyositis, or associated rhabdomyolysis. The common viruses implicated are influenza (A and B) group of viruses, parainfluenza, enteroviruses (Coxsackievirus and Echovirus), adenovirus, severe acute respiratory syndrome, coronavirus, retroviruses, herpesviruses (Varicella, Herpes simplex, Epstein–Barr, and cytomegalovirus), parvovirus B19, dengue, and hepatitis viruses (B and C).

Laboratory Diagnosis

- Bone sequestrum (in osteomyelitis) and joint aspirates (in arthritis) are stained to demonstrate bacteria and polymorphs.
- They are cultured on blood agar to yield the organism.
- Blood culture may yield the organism in around 30% of acute cases of osteomyelitis and septic arthritis.
- Molecular diagnosis is done for viruses.

Treatment

- Empiric treatment with vancomycin and cefazolin must be started early to prevent damage to the bone.
- Intravenous antibiotic treatment may be necessary in cases of acute osteomyelitis and septic arthritis.
- Removal of joint fluid by arthrocentesis may be required, in addition to antibiotic therapy.

BACTERIAL INFECTIONS OF THE SKIN AND SOFT TISSUES

STREPTOCOCCAL SKIN INFECTIONS

- Streptococci produce virulence factors such as streptolysins, DNAses, and hyaluronidase, which contribute to tissue destruction and the spread of infection. These result in various skin and systemic diseases.
- Some also release exotoxins that activate certain T cells, triggering the release of cytokines, interleukins, and other immunomodulators. These cytokines activate the complement, coagulation, and fibrinolytic systems, leading to shock, organ failure, and death.
- The most important member is *S. pyogenes* (group A streptococci or GAS). It is known to cause throat and skin infections (Fig. 29.2). Details are given in Chapter 22.
- Skin diseases caused by streptococci are listed in Table 29.4.

Laboratory Diagnosis

- Pus and tissue material are collected from the site of the infection.
- In cases of cellulitis, blood culture may often yield the organism (laboratory methods for the isolation and identification of streptococci are elaborated in Chapter 22).
- In suspected cases of necrotising fasciitis, *S. pyogenes* can be isolated from the affected site. Rising titres

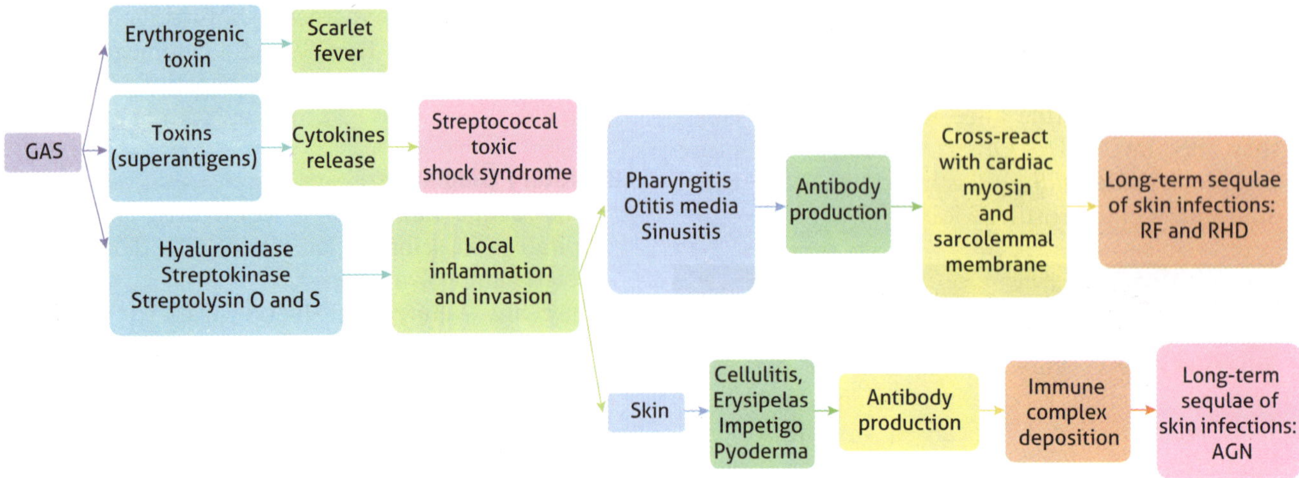

Fig 29.2 Pathogenesis of group A streptococcal infections

Table 29.4 Skin diseases caused by streptococci

Diseases	Features
Impetigo and pyoderma (Fig. 29.3)	Manifest as localised infections with induration, tenderness, and pus point Impetigo and streptococcal infection of scabies lesions are the main causes of acute glomerulonephritis in children in the tropics Antibody to DNase B and hyaluronidase are more useful in the retrospective diagnosis of pyoderma as an antecedent to acute glomerulonephritis
Erysipelas	Rapidly progressing infection with brawny edema and a rapidly advancing margin
Cellulitis (Fig. 29.4a and b)	Acute, rapidly spreading infection of the skin and subcutaneous tissue with pain, tenderness, and edema Unlike erysipelas, the margin is diffuse between involved and uninvolved tissue Systemic bacterial migration via the bloodstream can manifest as fever
Necrotising fasciitis (Fig. 29.5)	Commonly caused by a mixed aerobic and anaerobic bacterial infection, but some strains of *S. pyogenes* (particularly M types 1 and 3, which form pyrogenic exotoxin A) may cause necrotising fasciitis on their own Extensive necrosis of the subcutaneous and muscular tissues and adjacent fascia is associated with a severe systemic illness A toxic shock-like syndrome with disseminated intravascular coagulation and multiple system failure is seen
Scarlet fever	A blanching, papular rash (sandpaper rash), which is usually associated with GAS Pharyngitis in school-age and adolescent children. Strawberry tongue (swollen, bumpy and bright red) may be seen.

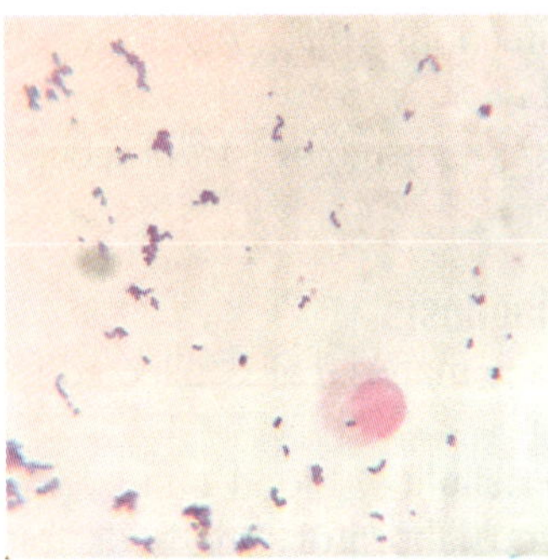

Fig. 29.3 Smear of pus from a pyoderma showing gram-positive cocci in chains with polymorphs (*Source*: Department of Microbiology, Pondicherry Institute of Medical Sciences, Puducherry)

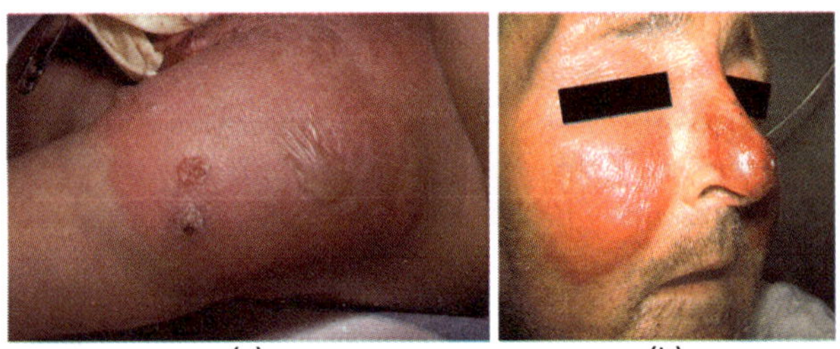

(a) (b)

Fig. 29.4 (a) Cellulitis of a child's shoulder showing spreading erythema, swelling, and accompanying areas of cutaneous purulency and (b) erysipelas affecting the face showing clear demarcation of the margin (*Source*: Centers for Disease Control and Prevention [CDC], Public Health Image Library [PHIL], Image IDs 4647 and 2874)

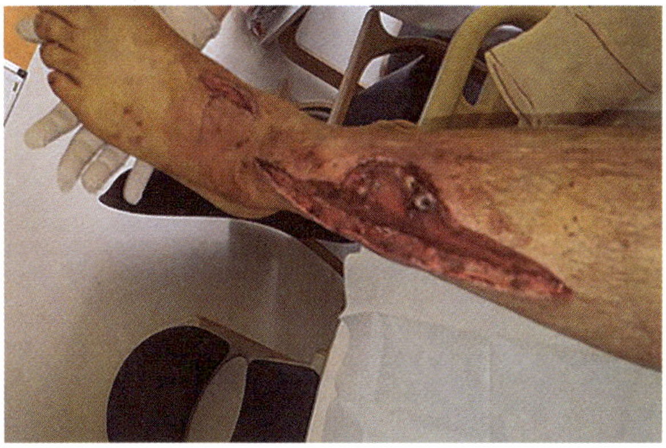

Fig. 29.5 Necrotising fasciitis after first debridement (*Source*: Wikimedia Commons)

of anti-streptolysin and anti-DNase can also be demonstrated.

Treatment

Penicillin G, clindamycin, and vancomycin are the drugs of choice in life-threatening infections.

STAPHYLOCOCCAL SCALDED SKIN SYNDROME (SSSS)

- This is an **exfoliative skin disease in which the outer layer of the epidermis gets separated from the underlying tissues**.
- It usually occurs 48 hours after birth and is rare in children older than six years. It may also occur in immunocompromised adults or those with severe renal disease.
- The severe forms of SSSS are known as **Ritter's disease** in newborns and **toxic epidermal necrolysis** in older patients.
- Milder forms are **pemphigus neonatorum** and **bullous impetigo**.

Pathogenesis (Fig. 29.6a and b)

SSSS is caused by the exfoliative (epidermolytic) toxin, also known as ET or **exfoliatin**, which:

- Cleaves desmoglein 1 complex (adhesion protein within desmosomes) in the stratum granulosum
- Disrupts keratinocyte-to-keratinocyte adhesion
- Causes separation and detachment of the superficial epidermis

Differential Diagnosis

- Toxic epidermal necrolysis
- Stevens–Johnson syndrome

Clinical Features

- The child presents with irritability, fever, and malaise. Within 24 to 48 hours, a very tender rash develops that typically starts on the face and in flexed areas of the body such as the groin, axillae, and neck with erythema and fissures.
- Soon, large, thin blisters form that may contain thin, sterile, cloudy fluid or even frank yellow pus underneath.

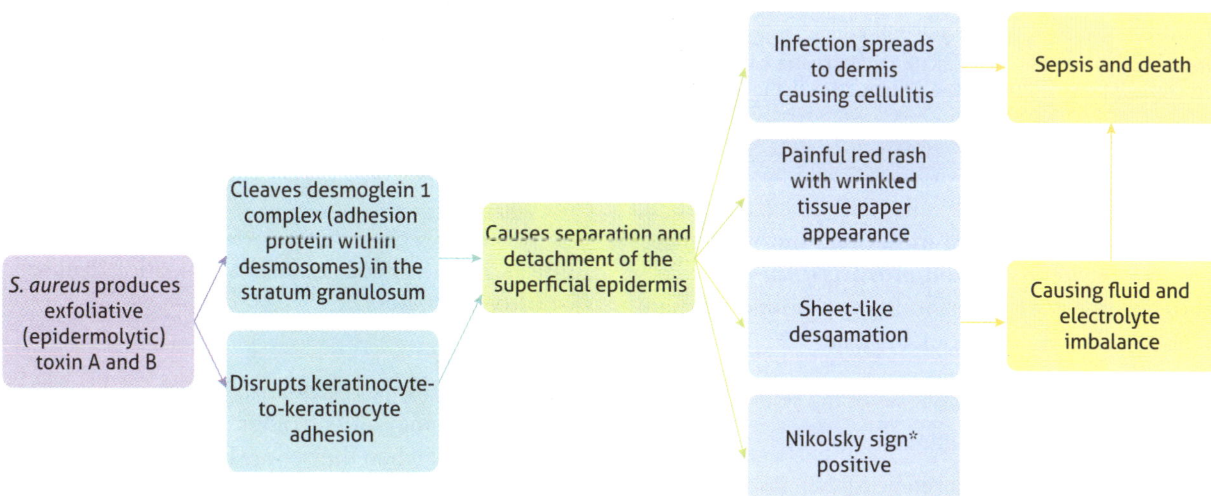

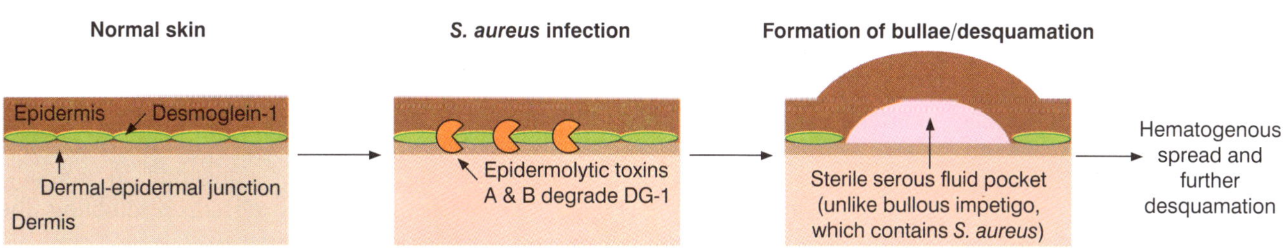

Fig 29.6a and b Pathogenesis of SSSS
**Nikolsky sign*—Slight rubbing of the skin results in shedding of outermost layer

- Unlike scarlet fever, the rash of SSSS tends to be more tender, and the skin exfoliates.

Treatment

- Topical antibiotics alone are not effective, and systemic antibiotics, depending upon sensitivity, reports must be started.
- For MRSA, vancomycin, and for MSSA, cefazolin or oxacillin may be given.

OTHER STAPHYLOCOCCAL SKIN DISEASES (TABLE 29.5)

- Staphylococcal infections characteristically cause localised pyogenic lesions, in contrast to streptococcal infections which have a spreading nature.
- Bacteria gain access to damaged skin or mucosal or tissue sites, colonise it by adhering to cells or extracellular matrix, evade host defence mechanisms, multiply, and finally, cause tissue damage.
- The dermonecrotic factor released by the organism causes necrosis of the tissue in hair follicles.
- Coagulase produced by bacteria coagulates the fibrin and localises the lesion, attracting inflammatory cells. Later, the necrotic tissue liquefies.
- If the abscess is not drained, fibrosis occurs with granulation tissue filling the cavity; it heals eventually.
- Infection might disseminate from a focus and spread via the lymphatics and bloodstream to distant organs, causing osteomyelitis, empyema, pneumonia, meningitis, endocarditis, or sepsis.

Table 29.5 lists the skin infections caused by staphylococci (details in Chapter 4 and 21).

ERYTHRASMA

Erythrasma is a common infection occurring in all age groups in people living in hot and humid climates caused by ***Corynebacterium minutissimum***.

- Predisposing factors include hot humid climate, diabetes, obesity, and poor hygiene.

Table 29.5 Staphylococcal diseases of the skin

Site of infection	Disease entities
Skin and soft tissue	Folliculitis, furuncle (boil), abscess (particularly breast abscess), wound infection, carbuncle, impetigo, paronychia, erythematous lesions (Fig. 29.7), and less often, cellulitis
Musculoskeletal system	Osteomyelitis, arthritis, bursitis, and pyomyositis
Common toxin-mediated staphylococcal disease of the skin	Toxic shock syndrome Scalded skin syndrome

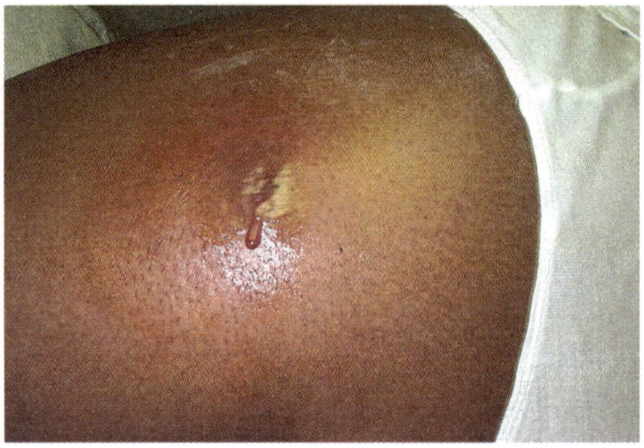

Fig. 29.7 Erythematous lesions with superficial ulceration due to *Staphylococcus aureus* (*Source*: CDC, PHIL, Image ID 7826)

- The most common sites affected are the glabrous areas of the body, skin folds, the groin, axilla, inner thighs, and the skin between the toes.
- Clinical features include red, scaly patches with a well-defined margin that later turn into a brown patch.

Treatment

Topical antibiotic lotions such as erythromycin or clindamycin or Whitefield's ointment (a mixture of salicylic acid and benzoic acid) are used.

ANTHRAX

Case 1. *A 55-year-old man working on a sheep farm developed a vesicular lesion and blister on his left hand. The lesion was painless but itchy. There was no history of fever, lethargy, or loss of appetite. Microbiological tests with swabs and culture were performed. Gram staining of the aspirate showed spore-forming, gram-positive bacilli, suggesting a diagnosis of* ***anthrax****.*

i. What is anthrax? How is it diagnosed?
ii. What are the different forms of anthrax?

Anthrax is a major zoonotic infection caused by ***Bacillus anthracis***.

Epidemiology

- **Source:** Anthrax is primarily a disease of cattle and sheep and may affect other animals also. Humans get infected when they come into contact with spores from any of these sources.
- **Route of infection:** Animals are infected by the ingestion of spores present in the soil. Infected animals shed large numbers of bacilli from the mouth, nose, and rectum. These bacilli sporulate in the soil and remain viable for many years. The disease is usually a fatal septicemia in animals, but may sometimes be localised, resembling the cutaneous disease in humans.
- **Distribution:** Anthrax is enzootic in India; there have been frequent reports of anthrax around

the Andhra Pradesh–Tamil Nadu border. There have been outbreaks in Karnataka and West Bengal.

- **Immunity:** Anthrax infection in human beings provides permanent immunity; second episodes are extremely rare.
- **Biowarfare:** *B. anthracis* is a potential tool in biological warfare and was used for this purpose in 2001, when it was sent by mail to various destinations in the USA, causing disease and many deaths.
- It can present in multiple clinical forms such as cutaneous, pulmonary, or intestinal. All types can lead to fatal septicemia or encephalopathy and meningitis.

Agent—*Bacillus anthracis*

- It is an **aerobic, gram-positive, non-motile, sporulating bacillus** that is 3–10 µm in length.
- In clinical material and culture, it appears in **chains** (end-to-end) and has the appearance of a **bamboostick or boxcar**.
- Spores are not generally seen on direct Gram staining of specimens, but they appear in culture when nutrients are depleted.
- The organism is easily recovered if samples are obtained before the administration of antibiotics.

Pathogenicity

The pathogenesis of all three forms of anthrax is presented in Fig. 29.8. Two virulence factors have been identified in the pathogenesis of *B. anthracis*.

1. **Capsule:**
   - It is a polypeptide composed of a polymer of d–glutamic acid.
   - The gene for its production is plasmid-borne and aids virulence by inhibiting phagocytosis.
   - Loss of the plasmid leads to the loss of virulence. This attenuation is used in the making of the anthrax spore vaccine (**Sterne vaccine**).
2. **Anthrax toxin:**
   - It is encoded by a separate plasmid.
   - The toxin is a complex of three fractions:
     - **Protective antigen factor (PA or factor II):** PA is the binding fraction on the target cell surface that provides attachment sites for factor I or factor III, facilitating their entry into the cell. The antibody to PA is protective.
     - **Edema factor (EF or factor I):** EF is an adenyl cyclase activated only inside the target cells, leading to intracellular accumulation of cyclic AMP. This is believed to be responsible for edema and the other biological effects of the toxin. This is the most lethal toxin.
     - **Lethal factor (LF or factor III):** Entry of LF into the target cell causes cell death, but the mechanism of this action is not known.

Clinical features

Cutaneous anthrax

- The incubation period is 1–10 days (after the inoculation of anthrax spores).
- Symptoms begin with an **itchy papular lesion** that progresses over several days into a **painless ulcer with satellite lesions (malignant pustule).**
- Cutaneous anthrax generally resolves spontaneously, but 10–20 per cent of untreated patients develop fatal septicemia or meningitis.

Malignant pustule (Fig. 29.9)

- The lesion starts as a papule 1–3 days after infection and becomes vesicular, containing fluid. This may be clear or bloodstained with several satellite lesions.
- The whole area is congested and edematous and covered with a black eschar (the **name anthrax,**

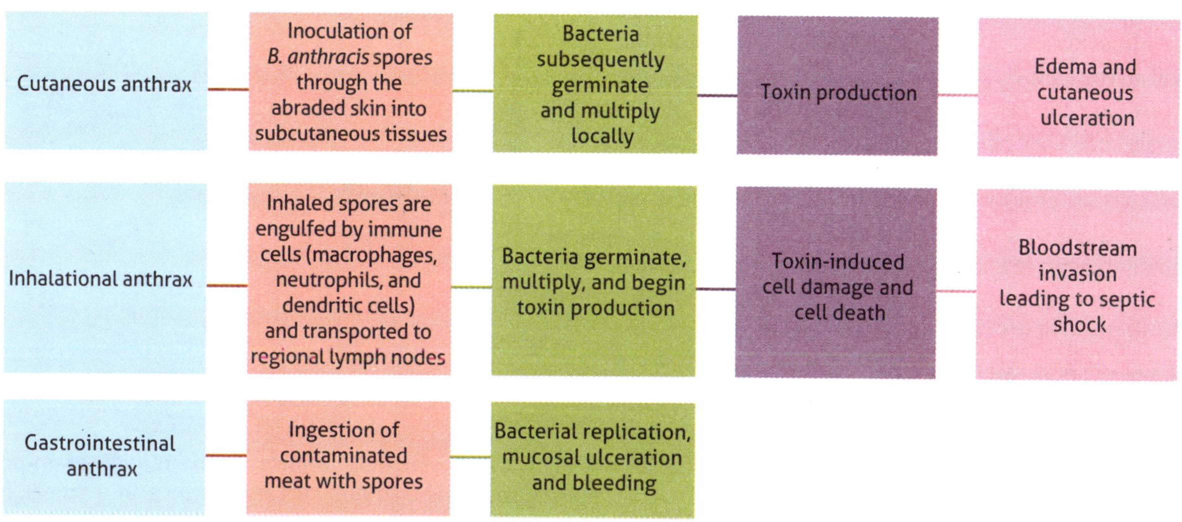

Fig. 29.8 Pathogenesis of different forms of anthrax

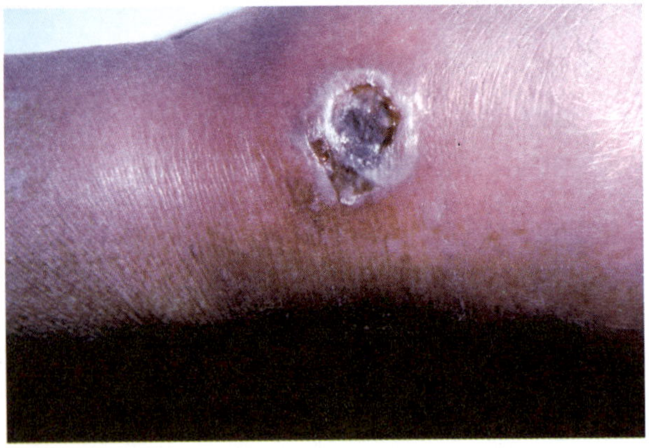

Fig. 29.9 Cutaneous lesion (malignant pustule) with black central eschar and edematous raised margin; satellite lesions are seen (*Source*: CDC, PHIL, Image ID 20334)

which means coal, comes from the black colour of the eschar).
- It is commonly seen at the site of entry of the infection on the face, neck, hands, arms, and back (exposed parts of the body).
- This disease, which is called **hide-porter's** disease, used to be common among dock workers who would carry loads of hides and skins on their bare backs.

Pulmonary anthrax
- The incubation period is 1–6 days.
- This disease is called the wool-sorter's disease because it used to be common among those who worked in wool factories where they would inhale spores from infected wool.
- **Prodromal phase**: Fever, sweats, nausea, vomiting, malaise, chest pain, and non-productive cough are the features of this phase.
- **Second stage**: It manifests as hemorrhagic lymphadenitis and mediastinitis and progresses to bacteremia. Fever, dyspnea, and stridor develop from increasing lymphadenopathy and impact the airways.
- **Complications:** Hemorrhagic pneumonia, respiratory failure, meningitis, and hemodynamic collapse are the common complications.

Intestinal anthrax
- This occurs chiefly in communities that consume **undercooked or uncooked meat**.
- It presents with fever, nausea, vomiting, and diarrhea.
- They can progress to an acute abdomen-like presentation with bloody diarrhea, hematemesis, and massive ascites.
- Severe enteritis with bloody diarrhea occur, ending fatally if left untreated.

Anthrax meningitis or meningoencephalitis
- This is a fatal condition in which the infection **disseminates to the CNS via the bloodstream**.
- The CSF is characteristically hemorrhagic, and the condition may be mistaken for cerebrovascular accident.

Laboratory diagnosis
- Since anthrax is a highly virulent organism, biosafety level class II/III cabinets (which can protect against aerosols) are mandatory to handle specimens suspected of carrying the bacilli.
- When an animal is suspected to have died of anthrax, autopsy is not permissible, as the spilt blood can contaminate the soil.
- **All clinical samples must be treated as potentially biohazardous.**

Specimen
- Pus from the cutaneous lesions, blood, and CSF in the case of bloodstream invasion and meningoencephalitis, sputum in cases of pulmonary anthrax, and feces in cases of gastroenteritis are collected for testing.
- In animals, an ear may be cut off from the carcass and sent to the laboratory. Alternatively, swabs soaked in blood, or several blood smears may be sent.

Microscopy
- The demonstration of **gram-positive bacilli with the morphology of anthrax bacilli** (Fig. 29.10) is diagnostic.
- **Spore staining** is performed to visualise spores.
- **M'Fadyean reaction with polychrome methylene blue**: When a blood film containing anthrax bacilli is stained with polychrome methylene blue for a few seconds and examined under the microscope, an amorphous purplish material is noticed around the bacilli. This represents the capsular material.
- **Sudan black B stain:** Fat globules may be made out within the bacilli.

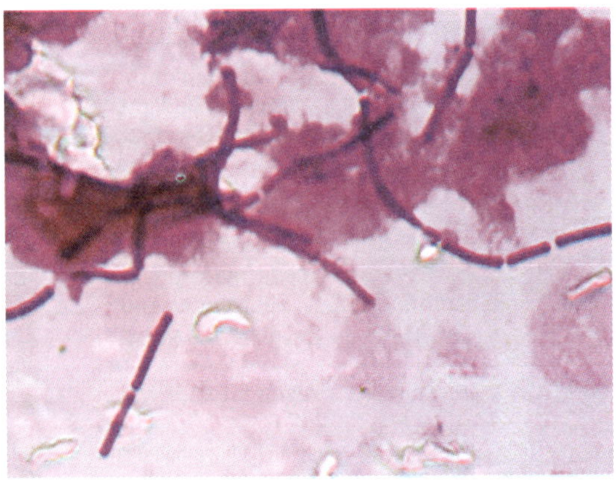

Fig. 29.10 *Bacillus anthracis* in blood—gram-positive rods, showing bamboostick appearance in a smear from a automated blood culture bottle (*Source*: CDC, PHIL, Image ID 17099)

- **Immunofluorescent microscopy:** A specific immunoglobulin tagged with a fluorescent stain can be used to confirm the identification.

Culture

- The anthrax bacillus grows on most laboratory media.
- The temperature range for its growth is **12–45°C (optimum 35–37°C)**.
- On agar plates, **irregularly round colonies** measuring 2–3 mm in diameter appear. These are raised, dull, opaque, and greyish-white with a frosted glass appearance.
- On blood agar, colonies are **non-hemolytic** though occasional strains produce a narrow zone of hemolysis.
- **Medusa-head appearance**—under a low-power microscope, the edge of the colony appears to be composed of long, interlacing chains of bacilli, resembling locks of matted hair (Fig. 29.11).
- **Virulent capsulated strains form rough cultures**, while avirulent or attenuated strains form smooth colonies.
- **String of pearls reaction—**when *B. anthracis* is grown on a solid medium containing 0.05–0.50 units of penicillin/mL, the cells become large and spherical within 3–6 hours and arrange themselves in chains on the surface of the agar, resembling a string of pearls. This reaction clearly differentiates *B. anthracis* from *B. cereus* and other aerobic spore bearers.
- **On gelatin stab culture**, a characteristic 'inverted fir tree' appearance is seen.
- In **broth**, floccular deposits with little or no turbidity are seen.
- **PLET medium**—a selective medium consisting of polymyxin, lysozyme, ethylene diamine, tetraacetic acid (EDTA), and thallous acetate added to heart infusion agar has been devised to isolate *B. anthracis* from mixtures containing other spore-bearing bacilli.
- **Animal inoculation—**experimentally, most animals are susceptible, including guinea pigs, rabbits, and mice.

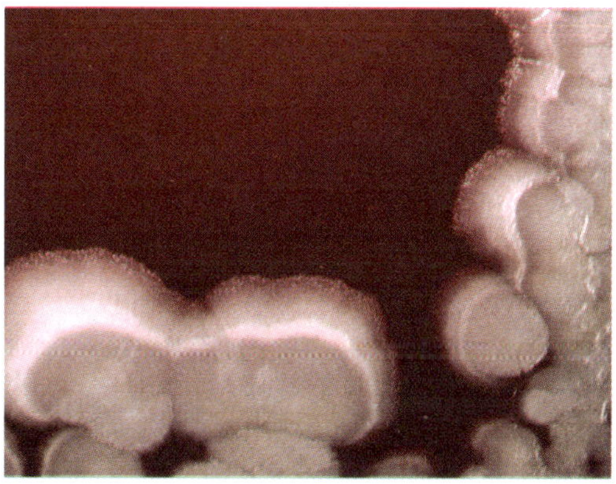

Fig. 29.11 Non-hemolytic, dry colonies of *B. anthracis* bearing the characteristic Medusa-head appearance on blood agar (*Source*: CDC, PHIL, Image ID 17099)

Serological demonstration

- The **Ascoli's thermoprecipitin test** demonstrates the anthrax antigen in tissue extracts.
- A **direct fluorescent antibody test (DFA)** is available for capsule-specific staining and polysaccharide cell wall antigen detection.
- **Serology for antibodies**—Acute and convalescent phase sera should be obtained, since antibodies to the organism can be demonstrated by gel diffusion, complement fixation, antigen-coated tanned red cell agglutination, and ELISA techniques.

Molecular typing

- For epidemiological studies and strain characterisation, MLVA (multiple-locus variable number tandem repeat analysis) or AFLP (amplified fragment length polymorphism) can be used.
- These methods help in tracking the source in case of an outbreak or to monitor the spread of the infection across the globe.

Laboratory diagnosis of anthrax

- Biosafety level class II/III cabinets (which can protect against aerosols) are mandatory to handle specimens suspected of carrying anthrax bacilli.
- When an animal is suspected to have died of anthrax, autopsy is not permissible, as the spilt blood can contaminate the soil.
- **Specimen:**
  - Pus from the cutaneous lesions, blood and CSF (bloodstream invasion and meningoencephalitis), sputum (pulmonary anthrax), and feces (gastroenteritis)
  - Animals: An ear may be cut off from the carcass, swabs soaked in blood, or blood smears
- **Microscopy:**
  - **Spore staining**
    - M'Fadyean reaction with polychrome methylene blue: For capsular material
    - Sudan black B stain: Fat globules within the bacilli
  - **Immunofluorescent microscopy** using specific immunoglobulin tagged with a fluorescent stain
- **Culture:**
  - Non-fastidious, temperature range is 12–45°C (optimum 35–37°C)
  - *Blood agar—***colonies** are **non-hemolytic, irregular, round**, raised, dull, opaque, and greyish-white with a frosted glass appearance

- *Medusa-head appearance*—under a low-power microscope
- *Virulent capsulated strains form rough cultures*, while avirulent or attenuated strains form smooth colonies
- *String of pearls reaction—B. anthracis* on a solid medium containing penicillin
- *Gelatin stab culture*: 'Inverted fir tree' appearance
- *Broth:* Floccular deposits with little or no turbidity
- *Selective medium*: **PLET** (polymyxin, lysozyme, ethylene diamine, EDTA, and thallous acetate in heart infusion agar
- *Animal inoculation*—experimentally, most animals are susceptible, including guinea pigs, rabbits, and mice

- **Antigen detection:**
  - **Ascoli's thermoprecipitin test** in tissue extracts
  - **Direct fluorescent antibody test (DFA):** For capsule-specific staining and polysaccharide cell wall antigen detection
- **Serology for antibodies**
- **Molecular typing:** MLVA (multiple-locus variable number tandem repeat analysis) or AFLP (amplified fragment length polymorphism)

Biosafety and prevention

General methods

- One of the most important preventive measures is the improvement of factory hygiene and proper sterilisation of animal products like hides and wool.
- Carcasses of animals suspected to have died of anthrax should be buried deep in quicklime or cremated to prevent soil contamination.
- The processing and isolation of the bacillus should be performed in designated reference laboratories of the veterinary facilities.

Immunisation

- **Anthrax vaccine adsorbed (AVA)** is the licensed vaccine for human use. AVA BioThrax is a vaccine for human use obtained from the culture of *B. anthracis* having the protective antigen.
- Personnel at risk including veterinarians and animal handlers are generally vaccinated to prevent occupational exposure.
- **Pre-exposure:** Only high-risk individuals are given five doses of the vaccine, followed by boosters.
- **Post-exposure:** The vaccine is given after exposure to anthrax along with antibiotics.

Vaccines for animals

- Pasteur's anthrax vaccine
- Sterne vaccine consists of spores of attenuated strains of a non-capsulated, avirulent, mutant bacillus
- Mazzucchi vaccine contains spores of a stable, attenuated Carbazoo strain in 2% saponin

Treatment (Table 29.6)

- Antibiotics have no effect on the toxin once it is formed.
- Fluoroquinolone or doxycycline can be used.
- **Human anthrax immunoglobulin (anthrasil)** is also used.
- Monoclonal antibodies to the protective antigen—obiltoxaximab or roxibacumab are now used in the treatment of anthrax.

Table 29.6 Treatment strategies of various types of anthrax

Type of condition	Treatment
Uncomplicated cutaneous anthrax	Fluoroquinolone or doxycycline orally for 7–14 days
Systemic anthrax	Human anthrax immunoglobulin (anthrasil) Monoclonal antibodies
Anthrax meningitis	At least three antimicrobial drugs including clindamycin and a full 60 days of antibiotic treatment, regardless of the vaccine strains
Pre-exposure prophylaxis for inhalation anthrax in adults	Ciprofloxacin and doxycycline for 60 days along with the vaccine (AVA BioThrax)

GAS GANGRENE

Case 2. *A 50-year-old man met with a road traffic accident in which he sustained multiple fractures with open wounds and a crush injury of the leg. He was taken to the nearest hospital two days later, at which time, he was found to be in shock. He was started on supportive therapy and antibiotics. There was edema and pain at the site of injury with increased discolouration and a serous discharge. The area around the wound exhibited crepitus on palpation. Microscopic examination of the wound discharge showed the presence of thick, brick-shaped, gram-positive bacilli along with gram-positive cocci. Based on a provisional diagnosis of gas gangrene, immediate surgical treatment with extensive excision of the local area (to prevent further spread) and intravenous penicillin with clindamycin were given. The exudate was also inoculated into Robertson's cooked meat medium.* ***Clostridium perfringens*** *and peptostreptococci grew in the culture.*

i. What is gas gangrene?
ii. Describe its pathogenesis.

Gas gangrene (Fig. 29.12) is a rapidly spreading, edematous myonecrosis, occurring characteristically in association with severe crush injuries, leading to wounds of extensive muscle masses contaminated with pathogenic clostridia, particularly ***Clostridium perfringens***. The disease has been referred to in the past as malignant edema, anaerobic (clostridial), myositis, and clostridial myonecrosis.

Polymicrobial Etiology (Table 29.7)

Table 29.7 Bacterial etiology of gas gangrene

Clostridium species	Other bacteria associated with *Clostridium* infection
C. perfringens (60% of cases) *C. novyi* (40%) *C. septicum* (20%) *C. histolyticum* (10%) *C. bifermentans* (10%) *C. fallax* (5%) *C. sordellii* *C. sporogenes* *C. tertium*	*E. coli* *Proteus* *Staphylococcus*

- The bacteriology of gas gangrene is varied.
- Rarely is this entity due to infection by a single clostridium. Generally, several species of clostridia are found in association with anaerobic streptococci and facultative anaerobes.

Pathogenesis of Gas Gangrene

The infection may be exogenous or endogenous.

- **Exogenous:** Clostridia usually enter a wound along with implanted foreign particles such as soil (particularly manured or cultivated soil), road dust, bits of clothing, or shrapnel. They may also be present on skin, especially that of the perineum and thighs.
- **Endogenous:** Infection may also develop after surgical procedures (especially amputations for vascular disease) and even injections (especially adrenaline). The source is the endogenous flora as in the gut.
- The sequence of events that lead to the development of gas gangrene are as follows (Fig. 29.13):
  - The condition favouring infection is low oxygen tension. The ionised calcium salts and silicic acid in the soil cause necrosis.
  - The crushing of tissues or tearing of the arteries produces anoxia of the muscle.
  - Extravasation of blood increases the pressure on the capillaries, further reducing blood supply.

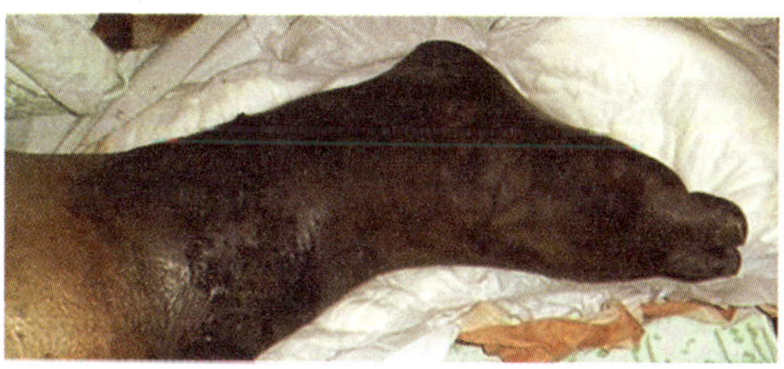

Fig. 29.12 Gas gangrene of the lower limb showing edema and discoloured skin

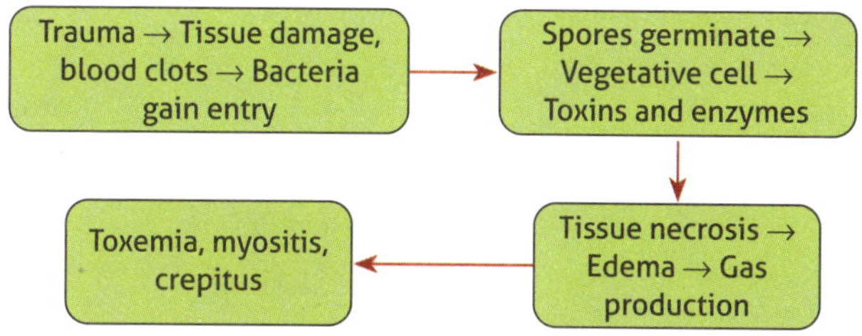

Fig. 29.13 Events leading to the formation of gas gangrene

  - The Eh (redox potential) and pH of the damaged tissues fall, and these changes, along with chemical changes including the breakdown of carbohydrates and the liberation of amino acids from proteins, provide the ideal conditions for the proliferation of anaerobes.
  - Extravasated hemoglobin and myohemoglobin are reduced and cease to act as oxygen carriers.
  - As a result, aerobic oxidation is halted and anaerobic reduction of pyruvate to lactate leads to a further fall in Eh.

Mechanism of tissue damage

Clostridia multiply and elaborate toxins which cause further tissue damage as shown in Fig. 29.14.

Clinical Features

- The incubation period can be **as short as seven hours or as long as six weeks after the wound** is established, the average being 10–48 hours with *C. perfringens*, 2–3 days with *C. septicum*, and 5–6 days with *C. novyi* infection.
- Pain, tenderness, and edema of the affected part develop, along with systemic signs of toxemia.
- A thin, watery discharge is produced from the wound, which later becomes profuse and serosanguinous (Fig. 29.15).
- The accumulation of gas makes the tissues **crepitant** (Case 2).
- In untreated cases, the disease process extends rapidly to profound toxemia and prostration, leading to death due to circulatory failure.

Agent–*Clostridium perfringens*

- *C. perfringens* (old nomenclature: *C. welchii*) is a normal inhabitant of the large intestines of humans and animals.
- It is found in feces and contaminates the skin of the perineum, buttocks, and thighs.
- Its spores are commonly found in soil, dust, and air.
- This is the most common clostridia implicated in gas gangrene. It also causes food poisoning and necrotic enteritis in humans (discussed in Chapter 24).

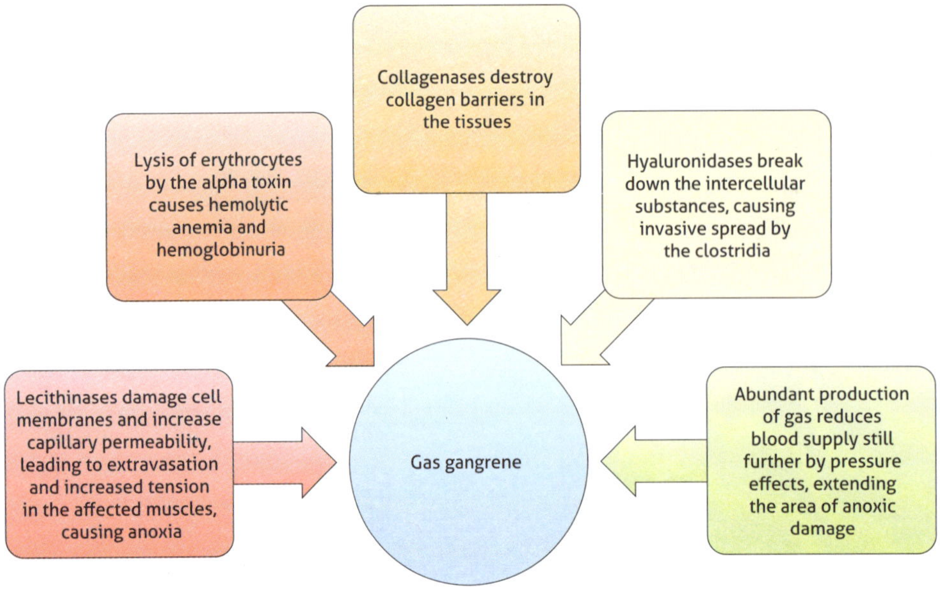

Fig. 29.14 Mechanism of tissue damage in gas gangrene

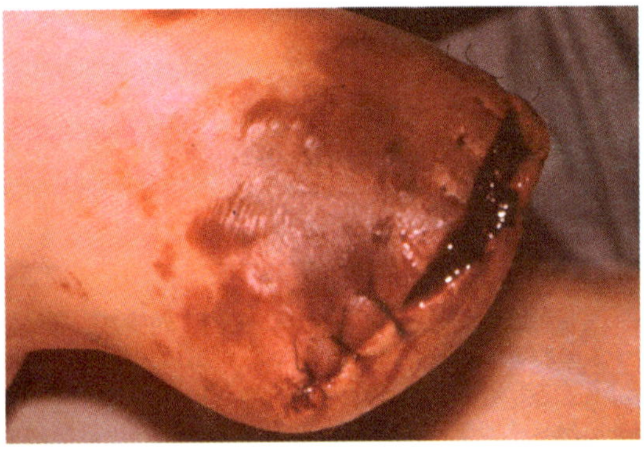

Fig. 29.15 Clostridial myonecrosis (*Source*: CDC, PHIL, Image ID 17603)

Spores

- *C. perfringens* spores are usually destroyed within five minutes by boiling.
- 'Food poisoning'-producing type A strains and certain type C strains resist boiling for 1–3 hours.
- **Autoclaving at 120°C for 15 minutes** kills them.
- Spores are resistant to commonly used antiseptics and disinfectants.

Pathogenicity

Virulence factors

- *C. perfringens* strains are classified into five types, A to E, based on the toxins they produce.
- Typing is based on neutralisation tests using intracutaneous injections of specific antitoxins in guinea pigs or intravenous injection in mice.

Toxins

C. perfringens produces at least 12 distinct toxins besides many other enzymes and biologically active soluble substances (Table 29.8).

1. **Major toxins**
   - The **alpha (α) toxin** is produced by all types of *C. perfringens* and most abundantly by type A strains.
     – It is responsible for the profound toxemia of gas gangrene. It is dermonecrotic, hemolytic, and lethal. It is a phospholipidase (lecithinase C) which, in the presence of Ca^{++} and Mg^{++} ions, splits lecithin into phosphoryl choline and diglyceride.
     – This reaction is seen as opalescence in serum or egg yolk media and is specifically neutralised by the antitoxin as described under Nagler's reaction.
   - **Beta (β), epsilon (ε), and iota (ι) toxins** have lethal and necrotising properties.
2. **Minor toxins**
   - **Gamma (γ) and eta (η) toxins have minor lethal action.**
   - **Delta (δ) toxin has a lethal effect and is hemolytic for the red cells of even-toed ungulates (sheep, goats, pigs, and cattle).**
   - **Theta (θ) toxin is an oxygen-labile hemolysin, antigenically related to streptolysin O. It is also lethal and a general cytolytic toxin.**
   - **Kappa (κ) toxin is a collagenase.**
   - **Lambda (λ) toxin is a proteinase and gelatinase.**
   - **Mu (μ) toxin is a hyaluronidase.**
   - **Nu (υ) toxin** is a deoxyribonuclease.

Enzymes

- Enzymes that destroy the blood group substance, A and H, also include neuraminidase (destroys myxovirus receptors on red blood cells),

Table 29.8 Toxins produced by *C. perfringens* types

Type	Pathogenicity	Major toxins				Minor toxins							
		α	β	ε	ι	γ	δ	η	θ	κ	λ	μ	υ
A	Gas gangrene—wound infections, septicemia	+++	–	–	–	–	–	±	±	±	–	±	±
	Food poisoning	+++	–	–	–	–	–	–	±	±	–	±	++
C	Enteritis in animals	+++	+++	–	–	–	–	–	+++	+++	–	–	±
	Enteritis necroticans in humans	+++	++	–	–	++	–	–	–	–	–	–	+++
D	Enterotoxemia of sheep	+++	–	+++	–	–	–	–	++	++	++	±	±
E	Might be a pathogen of sheep and cattle	+++	–	–	+++	–	–	–	++	++	++	±	±

hemagglutinin (active against human RBC), fibrinolysin, and hemolysin (distinct from the alpha, theta, and delta toxins).

- **Bursting factor** has a specific action on muscle tissue and is responsible for the characteristic muscle lesions in gas gangrene.
- **Circulating factor** can cause an increase in the adrenaline sensitivity of the capillary bed and inhibits phagocytosis.

Agent—*Clostridium septicum*

- It was first described by Louis Pasteur and Jules Francois Joubert (1887), who named it ***Vibrion septique.***
- It is a pleomorphic bacillus, about 3–8 × 0.6 µm in size and forms oval, central, or subterminal spores.
- It is motile due to the presence of **peritrichate flagclla**.
- Colonies are initially irregular and transparent, turning opaque on continued incubation.
- Hemolysis occurs on horse blood agar. Growth is promoted by glucose. It is saccharolytic and produces abundant gas.

Virulence factors

- *Clostridium septicum* is classified into six groups based on somatic and flagellar antigens.
- Toxins—the alpha toxin is hemolytic, dermonecrotic, and lethal; the beta toxin is a leukotoxic deoxyribonuclease; the gamma toxin a hyaluronidase; and the delta toxin an oxygen-labile hemolysin. It produces a fibrinolysin.
- *C. septicum* is found in the soil or in animal intestines and causes gas gangrene in association with other clostridia. It also causes '**braxy**' in sheep and malignant edema in cattle and sheep.

Agent—*Clostridium novyi* (*C. oedematiens*)

- This is a large, stout, pleomorphic, gram-positive bacillus with oval, subterminal spores.
- It is widely distributed in soil.
- It is a strict anaerobe, readily inactivated on exposure to air.
- Four types (A–D) are recognised, based on the production of toxins. Only type A is of medical importance, as it causes gas gangrene.

Agent—*Clostridium histolyticum*

- This is an actively proteolytic clostridium, which forms oval, subterminal, bulging spores.
- It is aerotolerant, and some growth may occur even in aerobic cultures.
- It forms at least five distinct toxins and is infrequently associated with gas gangrene in humans.

Laboratory Diagnosis of Gas Gangrene

- The diagnosis of gas gangrene must be made primarily on clinical grounds; the function of the laboratory is only to provide confirmation of the clinical diagnosis and identification and enumeration of the infecting organisms.
- The mere presence of clostridia in wounds does not constitute a diagnosis of gas gangrene.
- Bacteriological examination also helps to differentiate gas gangrene from anaerobic streptococcal myositis, which may be indistinguishable from it clinically in the early stages. (In the latter, gram-stained films show large numbers of streptococci and pus cells but not bacilli, contrasting with the scanty pus cells and diverse bacterial flora seen in films from gas gangrene.)

Specimens

- Tissue bits from the muscles at the edge of the affected area, from the necrotic area, and from the exudate in the deeper parts of the wound are collected for investigations.

- Exudates from the parts where infection appears to be most active and from the depths of the wound may be collected with a capillary pipette or a swab (which must be soaked in the exudate).
- Blood cultures may be required as they are often positive, especially in *C. perfringens* and *C. septicum* infections. However, *C. perfringens* bacteremia may occur without gas gangrene.

Microscopic examination by Gram stain

- Plump, gram-positive capsulated bacilli are seen with straight, parallel sides and rounded or truncated ends, about 4–6 × 1 μm in size.
- Spores are central or subterminal but are rarely seen in artificial culture or in material from pathological lesions.
- It can provide presumptive information about the species of clostridia.
  - *C. perfringens*: Brick-shaped, gram-positive bacilli without spores
  - *C. tetani*/*C. tetanomorphum*: Slender bacilli with round, terminal spores
  - *C. novyi*: Large bacilli with oval, subterminal spores
  - *C. septicum*: '**Citron bodies**' and boat- or leaf-shaped pleomorphic bacilli with irregular staining and no spores (Fig. 29.16)

Culture

- Clostridia are anaerobes but can also grow under microaerophilic conditions.
- **The ideal pH range** is 5.5–8.0, the optimum being 7–7.4.
- **Temperature range** is 20–50°C optimum temperature for clostridial growth is 37°C.

Cultural conditions

- The sensitivity to oxygen varies in different species. Some clostridia (e.g., *C. novyi*) are exacting anaerobes and die on exposure to oxygen, while others (e.g., *C. histolyticum*) are aerotolerant and may even grow aerobically.
- Provision of a sufficiently low redox potential (Eh) in the medium is more important than than the absence of oxygen.
- Reducing substances such as unsaturated fatty acids, ascorbic acid, glutathione, cysteine, thioglycolic acid, alkaline glucose, sulphites, or metallic iron can be added to the media to make them anaerobic.
- Growth is relatively slow on solid media and colony characteristics are variable.
- Some species are hemolytic on blood agar.
- Robertson's cooked meat broth (RCM), thioglycolate broth, and litmus milk medium are used for culture. RCM can be used as transport medium also.
- **Fresh and heated blood agar:**
  - The colonies, on overnight incubation on rabbit, sheep, or human blood agar, show target hemolysis resulting from a narrow zone of complete hemolysis due to theta toxin and a much wider zone of incomplete hemolysis due to the alpha toxin.
  - This double zone pattern of hemolysis may fade on longer incubation.

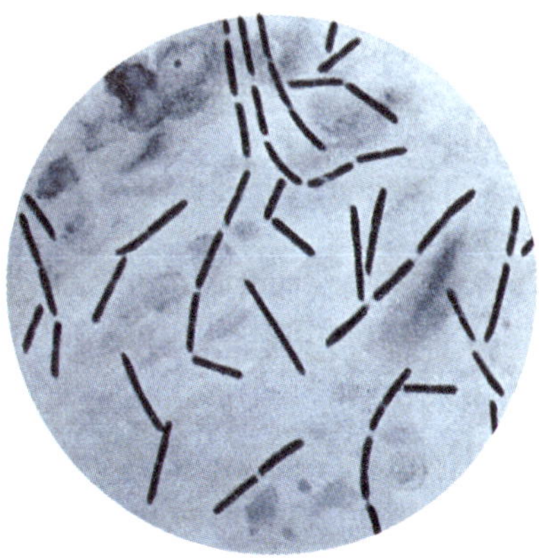

Fig. 29.16 Methylene blue-stained culture specimen revealing numerous *Clostridium septicum* bacteria—note the absence of spores (*Source*: CDC, PHIL, Image ID 217)

Nagler's reaction

- When *C. perfringens* is grown on a medium containing 6% agar, 5% Fildes' peptic digest of sheep blood, and 20% human serum, with a specific antitoxin spread on one half of the plate, colonies on the other half (without the antitoxin) are found to be surrounded by a zone of opacity (Fig. 29.17).
- Colonies on the half of the plate with the antitoxin do not produce such opacity since the antitoxin facilitates the specific neutralisation of the alpha toxin.

Reverse CAMP test

- *C. perfringens* is streaked across the centre of the plate and *Streptococcus agalactiae* is streaked at right angles to it.

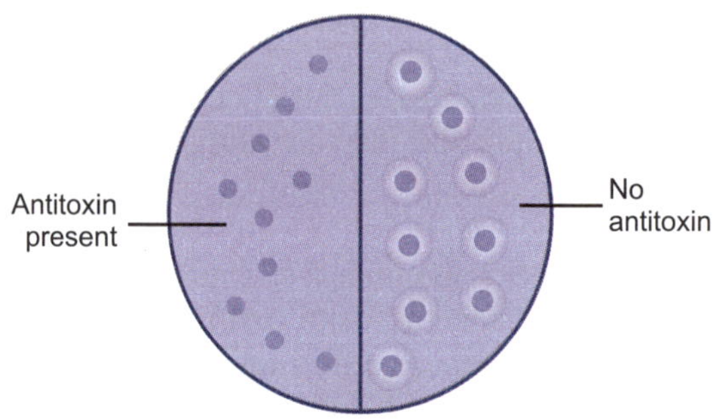

Fig. 29.17 Nagler's reaction—on the right, the presence of toxin produces opacity in the serum due to the breakdown of lecithin

- A positive reverse CAMP test is marked by the presence of an **arrow-shaped zone of enhanced hemolysis** pointing towards *C. perfringens* (Fig. 29.18).

Litmus milk test

- The fermentation of lactose leads to the formation of acid, which is indicated by a change in the colour of the litmus from blue to red.
- The acid coagulates the casein (acid clot), and the clotted milk is disrupted due to vigorous gas production. This is known as **stormy clot formation**.

Hot–cold lysis

- Alpha toxin is hemolytic on red cells due to its action on the phospholipids on the erythrocyte membranes.
- Lysis is of the hot–cold variety, best seen after incubation at 37°C followed by cooling at 4°C.

Biochemical reactions

Clostridia are identified by carbohydrate fermentation, indole, MR, VP, H_2S production, and nitrate reduction tests.

Laboratory diagnosis of gas gangrene

It is primarily clinical.

- **Specimens:**
  - Tissue bits from the muscles at the edge of the affected area, from the necrotic area, and from the exudate in the deeper parts of the wound
  - Blood cultures in *C. perfringens* and *C. septicum* infections
- **Microscopy:**
  - *Gram stain*:
    - *C. perfringens*: Brick-shaped, gram-positive bacilli without spores
    - *C. tetani/C. tetanomorphum*: Slender bacilli with round, terminal spores
    - *C. novyi*: Large bacilli with oval, subterminal spores
    - *C. septicum*: '**Citron bodies**' and boat- or leaf-shaped pleomorphic bacilli with irregular staining and no spores
- **Culture:**
  - Anaerobes but can also grow under microaerophilic conditions, **pH range**: 5.5–8.0 (optimum: 7–7.4)
  - Temperature range: 20–50°C (optimum: 37°C)
  - Fresh and heated blood agar: The colonies show target hemolysis
- **Identification:**
  - Nagler's reaction
  - Reverse CAMP test
  - Litmus milk test
  - Hot–cold lysis
- **Biochemical reactions:** Carbohydrate fermentation, indole, MR, VP, H_2S production, and nitrate reduction tests

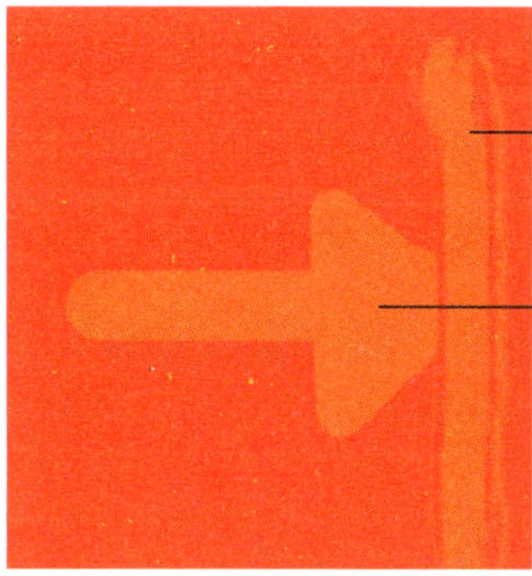

Fig. 29.18 Reverse CAMP test

Prophylaxis and Treatment

- **Early measures**—surgery is an essential prophylactic and therapeutic measure in gas gangrene. All damaged tissue should be removed promptly (debridement) and the wounds irrigated to remove blood clots, necrotic tissue, and foreign materials.
- In **established gas gangrene**, uncompromising excision of all affected parts may be life-saving.
- **Hyperbaric oxygen** may be beneficial in treatment and can be administered where facilities exist.
- **Antibiotics** are effective in prophylaxis in combination with surgical methods. The drug of choice is metronidazole given intravenously along with clindamycin and penicillin before surgery and repeated every eight hours for 24 hours.
- **Passive immunisation** with 'anti–gas gangrene serum' (equine polyvalent antitoxin in a dose of 10,000 IU *C. perfringens*, 10,000 IU *C. novyi*, and 5,000 IU *C. septicum* antitoxin given IM or in emergencies IV) is the common practice in prophylaxis.

BACTERIAL MYCETOMA

Case 3. *A 20-year-old woman presented to the surgery outpatient department with a soft, painless swelling on her left jaw, which was covered with exudates from a discharging sinus. She gave a history of trauma to the jaw. On macroscopic examination of the pus, a light-coloured granule was found in the exudate. Gram stain of the granule showed the presence of filamentous, branching, gram-positive bacilli. The condition was diagnosed as* ***cervicofacial actinomycosis****, and the woman was treated with intravenous penicillin G (8 million units per day) for two weeks followed by oral penicillin.*

i. What is actinomycosis?
ii. What are the bacterial agents that cause mycetoma?
iii. How is it diagnosed?

Mycetoma is a progressive chronic granulomatous infection of the skin and subcutaneous tissue. Mycetomas are usually caused by fungi but may be caused by bacteria as well.

Causative Agents and Types

1. **Eumycetoma (e.g., maduramycosis):** It is caused by fungi, namely *Scedosporium* (*Pseudallescheria*), *Madurella mycetomatis, M. grisea, Acremonium* spp., *Exophiala* spp., *Aspergillus nidulans,* and *Fusarium* spp.
2. **Actinomycetoma:** It is caused by actinomycetes, namely *Actinomadura*, *Streptomyces*, and *Nocardia*.
3. **Botryomycosis:** It is most commonly caused by *Staphylococcus aureus,* followed by *Pseudomonas aeruginosa, Escherichia coli,* and *Proteus*.

The differences between eumycetoma and actinomycetoma are given in Table 29.9.

Epidemiology

- Mycetoma has been referred to in the Atharva Veda as **padavalmika (foot anthill)**.
- The disease was first reported by **John Gill** (1842) from Madurai, South India. **Vandyke Carter** (1860) established its fungal origin. It came to be known as maduramycosis.
- Mycetomas are seen chiefly in the tropics, though occasional cases have been reported from the temperate countries. In India, it is common in **Tamil Nadu but rare in Kerala** and northern parts of the country.
- The disease is endemic in regions with **long dry seasons and short rainy spells** such as Central and South America, West and East Africa, India, and Sri Lanka.
- Actinomycetoma occurs more commonly than eumycetoma.
- **Persons engaged in agriculture** are especially at risk.

Actinomycosis

Actinomycosis is characterised by indurated swellings—chiefly in the connective tissue—suppuration and discharge of pale granules called **sulphur granules**. The lesions often have multiple sinuses.

- Actinomycotic mycetoma is a **localised, chronic, granulomatous involvement of the subcutaneous**

Table 29.9 Characteristics of eumycotic mycetoma and actinomycotic mycetoma

Features	Eumycetoma	Actinomycetoma
Etiology	Fungal etiology	Actinomycete (filamentous bacteria)
Granules	Mostly **hard** (tangled, tightly packed fungal hyphae and tissue) **Brown to black** *Madurella mycetomatis, Madurella grisea, Exophiala jeanselmei* **White** *Scedosporium apiospermum* *Acremonium falciforme* Can be seen in KOH as pigmented or hyaline hyphae	**Soft** (tissue elements and gram-positive filamentous bacilli) **White to yellow** *Nocardia asteroides* *Nocardia brasiliensis* *Actinomadura madurae* *Streptomyces somaliensis* **Red** *Actinomadura pelletierii*
Treatment	Surgical debridement, controlling secondary bacterial infection **Antifungals** Topical nystatin or miconazole Itraconazole, fluconazole, amphotericin B (for *Madurella* infection)	Streptomycin Trimethoprim-sulphamethoxazole Dapsone
Histopathology	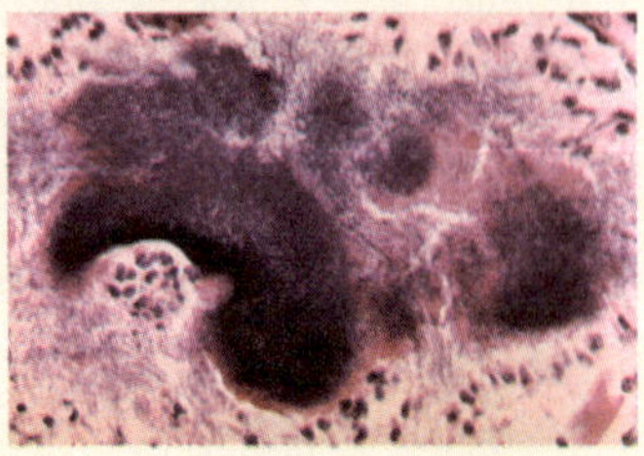 Pigmented granule tissue stained by hematoxylin–eosin reveals some of the histopathology associated with a case of black grain mycetoma caused by the fungus *Madurella mycetomatis* (*Source*: CDC, PHIL, Image ID 14739)	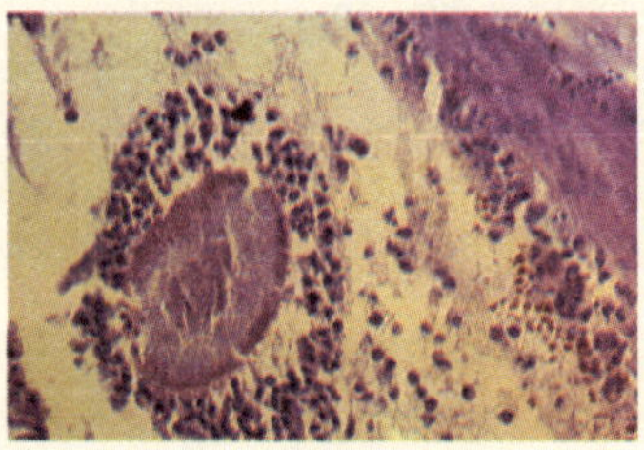 Hematoxylin–eosin-stained tissue specimen shows sulphur granule; the actinomycosis was caused by the gram-positive bacterium, *Nocardia asteroides* (*Source*: CDC, PHIL, Image ID 14745)

and deeper tissues, commonly affecting the foot and less often, the hand and other parts, and presenting as a tumour with multiple discharging sinuses with sulphur granules (Fig. 29.19).
- *Actinomyces bovis* causes 'lumpy jaw' in cattle.

Epidemiology (Actinomycotic mycetoma)

- Actinomycosis occurs throughout the world, but its incidence in developed countries has been declining, probably as a result of the widespread use of antibiotics.
- It is more common in rural areas and among agricultural workers.
- About 60 per cent of the cases of actinomycosis are cervicofacial and 20 per cent are abdominal. Pelvic actinomycosis is seen chiefly in women who use intrauterine devices.

The causative agents of actinomycosis are as follows:
- *Actinomyces* (*Actinomyces israelii*, *A. bovis*)
- *Nocardia* (*N. asteroides*, *N. brasiliensis*, *N. caviae*)
- *Actinomadura* (*A. madurae*, *A. pelletierii*)

Clinical features

Four main clinical forms of actinomycosis are seen:
1. **Cervicofacial:** This is most common form of actinomycosis. indurated lesions are seen on the cheek and submaxillary region (Case 2). Clinical presentation is in the form of hard, red relatively non-tender nodules that develops slowly, become fluctuant, and eventually drain, forming a chronic sinus tract. Lesions are often secondary to dental procedures.
2. **Thoracic:** Lesions may occur in the lung that may involve the pleura and pericardium and spread outwards through the chest wall. Lung abscess or empyema may be the presenting feature.

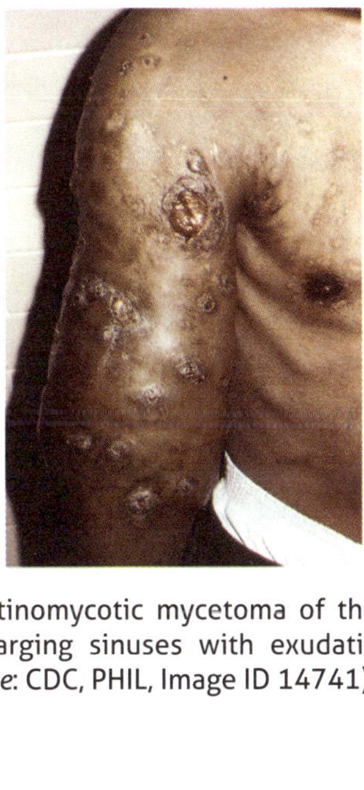

Fig. 29.19 Actinomycotic mycetoma of the upper limb—multiple discharging sinuses with exudation of pus and granules (*Source*: CDC, PHIL, Image ID 14741)

3. **Abdominal:** The abdominal lesion is usually around the caecum, involving the neighbouring tissues and the abdominal wall; sometimes, the infection spreads to the liver via the portal vein. The infection is often secondary to recent surgery, penetrating trauma or intestinal perforation.
4. **Pelvic:** These infections are associated with the intrauterine devices; generally, abscesses are produced in the bone and soft tissue, with chronic draining sinuses to the exterior.

Agent—Actinomycetes

- Actinomycetes are slender, gram-positive bacteria that break up into bacillary and coccoid forms.
- Most are free-living, particularly in the soil.
- The actinomycetes of clinical importance are *Actinomyces*, *Bifidobacterium*, *Nocardia*, and *Actinomadura* (the morphological characteristics and classification of this group have been discussed in Chapter 4).
- The name *Actinomyces bovis* was coined by Otto Harz in 1878 (*Actinomyces* in Greek means 'ray fungus').
- In 1891, M. Wolf and J. Israel isolated an anaerobic bacillus from human lesions. This was named *Actinomyces israelii*, while the *Actinomyces* causing infection in cattle was named *A. bovis*.

Actinomycetes and hypersensitivity pneumonitis

Spores of some thermophilic actinomycetes such as the *Faenia* and *Saccharomonosporia* species are present in mouldy hay, which when inhaled, may induce allergic alveolitis leading to chronic obstructive pulmonary disease (COPD also called farmer's lung).

Actinomyces have been incriminated in inflammatory diseases of the gums (gingivitis and periodontitis) and sublingual plaques leading to caries.

Laboratory diagnosis

- **Specimen:** Pus, granules, tissue in mycetoma, or sputum are collected in pulmonary disease. Gauze pads may be applied over the discharging sinuses to obtain granules.
- **Gross examination of granules:** The granules are white or yellowish and range in size from minute specks to particles measuring approximately 5 mm.
- **Microscopy:**
  - Granules are crushed between two slides, stained with Gram stain and examined.
  - A dense network of **thin, gram-positive filaments, surrounded by a peripheral zone of swollen, radiating, club-shaped structures,** presenting a **'sun's rays' appearance** are seen in the smear and tissue section (Fig. 29.20a and b).

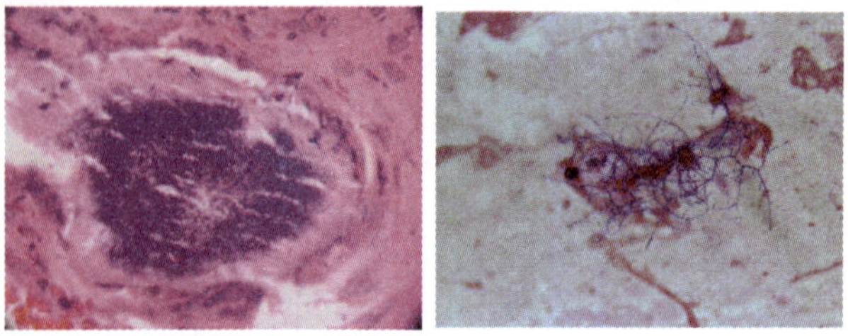

(a) (b)

Fig. 29.20 (a) Sulphur granules in section with H and E stain and (b) gram-positive filamentous bacilli in pus

- **Culture:**
  - Sulphur granules or pus-containing actinomycetes are washed and inoculated in thioglycolate liquid medium or streaked on brain–heart infusion agar and incubated anaerobically at 37°C.
  - In thioglycolate, *A. bovis* produces general turbidity, whereas *A. israelii* grows as fluffy balls at the bottom of the tube.
  - On solid media, *A. israelii* produces either small, 'spidery colonies' in 48–72 hours that become heaped up, white and irregular, or smooth, large colonies in 10 days. Other species produce different types of colonies. The colonies are dry and wrinkled on blood agar.
- **Identification of species:**
  - The isolate is identified by gram-positive filamentous appearance on microscopy, biochemical reactions (ferment arabinose, lactose, maltose, and mannitol), and fluorescent antibody methods.
  - Gel diffusion and immunofluorescence can differentiate *A. israelii* from other actinomycetes species and from other filamentous anaerobes that may produce granules in tissues.

Treatment

- Treatment includes surgical drainage or excision together with large doses of penicillin for several weeks or months.
- Actinomycetoma are treated with Dapsone (100 mg thrice daily for 6–24 months), sulphonamides, cotrimoxazole, or rifampicin.
- The disease also responds to prolonged treatment with penicillin or tetracycline.
- Treatment is continued for several months and supplemented by surgery where necessary.

NOCARDIOSIS

Case 4. *A 60-year-old, HIV-positive man developed a nodule on his hand. His CD4 count was <200/mm³). He gives a history of trauma to his hand after which a nodule developed which became swollen, painful, warm, and red. On examination, the nodules were warm and tender. Pus was drained from the nodule, which showed fine, branching filaments, which were acid-fast (to 1% sulphuric acid). Culture on blood agar showed the growth of dry, buff-coloured colonies identified as* ***N. asteroides****. The patient responded to treatment with cotrimoxazole and amikacin.*

i. What is nocardiosis? What are the clinical forms in which nocardiosis can exist?
ii. How does the causative agent differ from *Actinomyces*?

Nocardiosis is an opportunistic infection in immunocompromised hosts caused by *Nocardia.*

- *Nocardia* can affect nearly every part of the human body including skin and skin structures, the pleural, or the pulmonary system, or it can be a disseminated disease impacting other organ systems.
- Pulmonary nocardiosis and disseminated forms of the infection can occur in patients deficient in T cell-mediated immunity.

Clinical features

The infection is transmitted through contaminated soil, and not from animals or infected humans. It can occur in multiple clinical forms.

- **Cutaneous infection:** This type usually occurs following a traumatic event like a breach in skin. It may lead to local abscesses, cellulitis, or lymphocutaneous lesions. There may be purulent discharge, and later, suppurative necrosis may occur. The subcutaneous lesion is an actinomycotic mycetoma.
- **Lymphocutaneous nocardiosis:** Presentation is the same as primary cutaneous nocardiosis along with ascending regional lymphadenopathy. In severe cases, the lymph nodes may ulcerate to have weeping necrotic or purulent drainage.
- **Pulmonary nocardiosis:** It can present in acute, subacute, or chronic infection. Clinically, there can be inflammatory bronchial mass or pneumonia-like features resembling pulmonary tuberculosis. Complicated cases may develop abscess, pleural effusion, or empyema.
- **Systemic or disseminated nocardiosis:** It manifests primarily as pulmonary disease, pneumonia, lung abscess, or other lesions resembling tuberculosis. It is usually caused by *N. asteroids*. Metastatic manifestations may involve the brain, kidneys, and other organs. Systemic nocardiosis occurs more often in immunodeficient persons.

Agent—*Nocardia*

Nocardia resembles *Actinomyces* morphologically but is aerobic.

- The most common species are *Nocardia asteroids*, *Nocardia brasiliensis*, *Nocardia caviae*, *Nocardia farcinica*, and *Nocardia nova*.

- Pathogenic species are *N. asteroides*, *N. braziliensis*, and *N. caviae*.
- All species are **gram-positive**, and some such as *N. asteroides* and *N. brasiliensis* are also acid-fast to 1% sulphuric acid (Kinyoun method).
- The **natural habitat** of these organisms is soil.

The differences between *Actinomyces* and *Nocardia* are given in Table 29.10.

Morphology

- *Nocardia* are **filamentous, rod-shaped bacteria** that are non-motile and non-sporulating, gram-positive and weakly acid-fast (by Kinyoun's acid-fast staining technique).
- These organisms are obligate aerobes, and growth can often be enhanced by carbon dioxide.
- *N. asteroides* is the species most commonly involved in human disease. It is found worldwide in the soil and can infect animals.

Laboratory diagnosis

- **Direct microscopy:** The demonstration of partially acid-fast, branching, beaded filaments in smears from tissue samples is diagnostic of *Nocardia* infection.
- **Culture:** *Nocardia* are aerobic and grow readily on ordinary media, forming dry, granular and wrinkled colonies, which produce pigments ranging from yellow to red. The species can be identified by biochemical tests.

Botryomycosis

- It is a chronic inflammatory response to a bacterial infection involving the skin and sometimes internal organs.
- It is most commonly caused by *Staphylococcus aureus,* followed by *Pseudomonas aeruginosa.* Rarely, *Escherichia coli* and *Proteus* have also been implicated.
- It is a rare condition that most commonly presents as subcutaneous nodules, abscesses, ulcers, fistulae, and sinuses with purulent discharge. The lesions generally develop over several months and may drain pus. The pus may contain small yellow 'grains' similar to the sulphur granules of actinomycosis.
- The sulphur granules in botryomycosis contain bacteria (often in grape-like clusters) surrounded by an eosinophilic matrix with club-like projections. This is known as the **Splendore-Hoeppli phenomenon** and can be seen using Gram stain, giemsa stain, or silver stain.
- **Treatment** is surgical drainage, and debridement, laser vapourisation with long term antibiotic therapy.

Laboratory diagnosis of mycetoma

- **Specimen:** Pus, granules, tissue in mycetoma or sputum are collected using a gauze piece placed over the discharging sinus
- **Gross examination of granules:** White or yellowish
- **Microscopy:**
  - Granules are crushed between two slides, stained with Gram stain
  - Actinomycetes: A dense network of **thin, gram-positive filaments, surrounded by a peripheral zone of swollen, radiating, club-shaped structures,** presenting a **'sun's rays' appearance**

Table 29.10 Differences between *Actinomyces* and *Nocardia*

Features	*Actinomyces*	*Nocardia*
Morphology	Gram positive, thin filaments in V or Y branching forms that break up into coccoid or bacillary forms; non-acid fast	Partially acid-fast (1% sulphuric acid), gram-positive, branching, filamentous cells that may fragment into rod-shaped or coccoid elements or L-form
Culture	Anaerobic, grow on thioglycolate broth or brain–heart infusion agar	Aerobic, grow on simple media
Clinical spectrum	• **Cervicofacial** • **Thoracic** • **Abdominal** • **Pelvic**	• Cutaneous infections • **Lymphocutaneous nocardiosis** • **Pulmonary nocardiosis** (can mimic tuberculosis or cancer) • **Systemic or disseminated nocardiosis**
Cutaneous lesion	Hard, red, and relatively non-tender nodules that drain through sinus tracts, releasing yellow 'sulphur granules'	Local abscesses, cellulitis, mycetoma; no granules seen
Treatment	Penicillin	Sulfonamides
Habitat	Are part of the normal flora of the mouth and gastrointestinal tract	Saprophytic organism with a worldwide distribution in the soil

- *Nocardia*: Partially acid-fast (1% acid), branching, beaded filaments
- Other bacteria: Characteristic morphology seen
- **Splendore-Hoeppli phenomenon** seen (bacteria surrounded by an eosinophilic matrix with club-like projections)

- **Culture:**
  - Both aerobic and anerobic incubation; optimum temperature: 37°C
  - In thioglycolate, *A. bovis* produces general turbidity, whereas *A. israelii* grows as fluffy balls at the bottom of the tube
  - On solid media, *A. israelii* produces either small, 'spidery colonies'; other species produce dry and wrinkled on blood agar
  - *Nocardia* are aerobic and form dry, granular, wrinkled colonies with yellow to red pigments
- **Identification of species:** Morphology and acid-fastness

Treatment

- These organisms are resistant to penicillin.
- Sulphonamides, particularly sulphamethoxazole, and trimethoprim (cotrimoxazole), are used to treat nocardiosis.
- Cotrimoxazole or minocycline given for several months may be useful.
- In immunocompromised patients, amikacin and cefotaxime are advisable in addition.
- Surgery may be required to remove mycetomas.
- Therapy is most successful with skin and lung infections.

MELIOIDOSIS

- Melioidosis, a glanders-like disease, epizootic in rodents in Southeast Asia, India, and North Australia.
- The disease came into prominence after the Vietnam war and hence is called the **Vietnam time-bomb disease.**
- In India, cases are reported from the western and southern belts. The infections usually spike following the rainy season.
- The name melioidosis is derived from *melis*, a disease of asses (glanders; see next section) and *eidos*, meaning resemblance.
- It is caused by ***Burkholderia pseudomallei***.

Agent—*Burkholderia pseudomallei*

Members of the genus *Burkholderia* are small, irregularly staining, gram-negative bacilli that are abundant in nature, occupying diverse ecological niches such as soil, plants, animals, and hospital environments. Many members of the genus can cause infections in humans and animals. There are three species of *Burkholderia* that can act as human pathogens are:

- *Burkholderia pseudomallei* causes meliodosis.
- *B. mallei* causes glanders.
- *B. cepacia* causes nosocomial infections.

Pathogenesis

- *B. pseudomallei* produces **two thermolabile exotoxins**, one lethal and the other necrotising.
- Human infection is usually acquired by the **contamination of wounds, skin abrasions, or inhalation**. Soil and water containing the organism are the sources of infection.
- Underlying diseases (like diabetes mellitus and chronic liver disease) may increase the risk of infection.

Clinical features

The human disease may take one of two forms:

i. **Acute:** It may be a generalised infection presenting as acute septicemia, a subacute typhoid-like disease, or pneumonia with hemoptysis resembling tuberculosis. It has a high case fatality rate.

ii. **Chronic:** There may be multiple caseous or suppurative foci, with abscess formation in the skin and subcutaneous tissues, bones, and internal organs.

Latency:

- Long latency and reactivation may occur as the bacillus can survive intracellularly in the reticuloendothelial system. The bacillus has been isolated from water and soil in endemic areas.
- It is a soil saprophyte that causes infection in rodents and humans accidentally.

Laboratory diagnosis

The diagnosis is made when there is a high index of clinical suspicion with a history of contact with soil (as is the case with farmers and agricultural workers) followed by an ultrasound examination which may reveal multiple abscesses in the internal organs.

- **Microscopy**: Presumptive identification is made by the demonstration of the bacillus in (Fig. 29.21) exudates by microscopy (small, irregularly staining, gram-negative bacilli, showing typical bipolar **'safety-pin appearanc**e' with methylene blue stain).
- **Culture:** Isolation by culture from sputum, pus, blood, or urine shows dry, wrinkled colonies on blood and MacConkey agars. A selective medium used to isolate *B. pseudomallei* is **Ashdown's selective agar (ASA) medium**. It contains gentamicin and crystal violet to suppress other bacteria (useful in sputum culture).
- **Biochemical tests:** *B. pseudomallei* is commonly identified by its non-fermentative character, oxidase positivity, esculin hydrolysis, and reduction of nitrate. One of the tests to screen the organism is its intrinsic resistance to polymyxin B, which is used as a laboratory diagnostic marker.

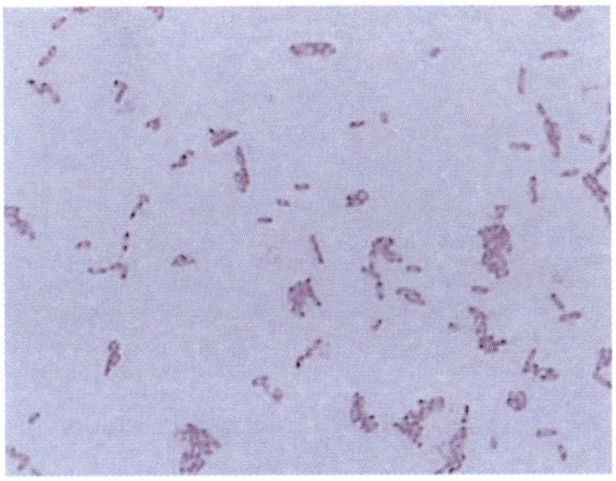

Fig. 29.21 *B. pseudomallei* with bipolar 'safety-pin appearance' on staining with methylene blue stain

- **Molecular methods:** PCR test or an automated VITEK system to identify the organism.
- **Serology:** This does not have much role in diagnosis.

Treatment

- Carbapenems (imipenem, meropenem), and ceftazidime are the drugs of choice in the acute phase of the disease.
- For maintenance therapy, cotrimoxazole, tetracycline, or amoxicillin-clavulanate are required for many months to prevent recurrence.

GLANDERS

- It is a zoonotic disease that is caused by ***Burkholderia mallei***.
- Glanders (*malleus*, in Latin), is primarily a disease of equine animals—horses, mules, and asses—and is capable of being transmitted to other animals and human beings.
- Human infection is usually occupational, found among ostlers, grooms, and veterinarians.
- It may be acute or chronic and is protean in character, with localisation in the respiratory tract, skin, or subcutaneous tissues.
- In acute glanders, there is fever, mucopurulent nasal discharge, and severe prostration.
- The acute form of the disease is usually fatal within 7–10 days of onset.
- Chronic glanders can cause death within months; while the survivors remain carriers of the disease.

While human infection is acquired only rarely from infected animals, laboratory cultures are highly infectious.

Agent–*Burkholderia mallei*

- It is a slender, non-motile, gram-negative bacillus, 2–5 × 0.5 µm in size, which stains irregularly and often has a **beaded appearance.**
- It is an aerobe and facultative anaerobe, growing on ordinary media under a wide temperature range.
- Colonies, which are small and translucent initially, become yellowish and opaque on ageing.
- This organism is quite inactive biochemically, attacking only glucose.

Agent–*Burkholderia cepacia*

- *B. cepacia* is increasingly being recognised as an opportunist pathogen, particularly for those with cystic fibrosis or chronic granulomatous disease, in whom it causes fatal necrotising pneumonia.
- It is nutritionally very versatile.
- It can grow in many common disinfectants and can even use penicillin G as its sole source of carbon. Hence, it is often associated with hospital-acquired infections.
- It is oxidase-positive and oxidatively breaks down mannitol, sorbitol, and sucrose.
- *B. cepacia* can cause urinary, respiratory, and wound infections, peritonitis, endocarditis, and septicemia. It is inherently resistant to most antibiotics.

LYME DISEASE

Lyme disease is the most common vector-borne disease in the United States. It is a zoonotic disease caused by the spirochete *Borrelia burgdorferi* and is transmitted to humans through the bite of infected blacklegged ticks (*Ixodid*).

Agent–*Borrelia burgdorferi*

- It is a spirochete with an outer membrane and inner membrane with a thin layer of peptidoglycan in between. It has a flexible cell wall, and the outer membrane lacks lipopolysaccharide.
- It is a microaerobic, motile spirochete with seven to 11 bundled perisplasmic flagella set at each end.
- It circulates between *Ixodes* ticks and a vertebrate host in an enzootic cycle. It is known to exist in ecological niches with deer and rodents, and mice can act as reserviours in some areas.

Pathogenesis

B. burgdorferi enters the skin at the site of the tick bite. After 3 to 30 days, the organisms migrate locally in the skin around the bite and spread via the lymphatics to cause regional adenopathy or disseminate in blood to organs or other skin sites. Local migration causes an inflammatory reaction known as **erythema migrans**.

Clinical features

- **Incubation period:** 3–30 days.
- The disease presents with fever, headache, fatigue, and a characteristic skin rash called **erythema migrans**. It begins as a red macule at the site of the tick bite that expands with central clearing.

Sometimes, central erythema surrounded by a pale ring may be seen. Skin lesion with central erythema surrounded by a pale ring and outer erythematous halo known as **bull's eye lesion** may be seen.
- If untreated neurologic, cardiac, or joint abnormalities can develop.

Laboratory Diagnosis

- Isolation of the organism is very difficult.
- **Serology** is the mainstay of diagnosis.
- Acute (IgM) and convalescent (IgG) antibody titers 2 weeks apart are helpful. However, seroconversion may occur after 4 weeks, or may occasionally be absent, causing a delay in diagnosis.
- Detection IgG titers alone indicate past exposure.
- PCR in CSF or synovial fluid is often positive in the later stages.

Treatment

Doxycycline or **ceftriaxone** can be used.

NON-VENEREAL TREPONEMATOSES

Non-venereal treponemal diseases occur in endemic foci in several parts of the world in three distinct forms—**(i) bejel/siti or endemic syphilis, (ii) yaws and (iii) pinta**. These are caused by treponemes and are discussed in Chapter 27.

SKIN DISEASES CAUSED BY NON-TUBERCULOUS MYCOBACTERIA (NTM)

- Mycobacterial cutaneous infections occur in **leprosy or tuberculosis**, either as localised disease or as part of a generalised infection.
- Mycobacteria other than tubercle bacilli occasionally cause human disease's especially in immunocompromised individuals and in post-organ transplant and HIV patients.
- These organisms, which were formerly known as 'atypical', 'anonymous', or 'MOTT' (mycobacteria other than tubercle bacilli), are called **non-tuberculous mycobacteria (NTM).**
- Over 80 species of NTM are found worldwide in soil and in animals.
- Saprophytic mycobacteria such as *M. phlei* are incapable of infecting human beings or animals and are distinct from NTM.
- NTM have been classified into four groups (Runyon's classification) based on pigment production and the rate of growth (discussed in Chapter 3).

Clinical Spectrum

- Medically important species are *M. fortuitum* (Fig. 29.22) and *M. chelonae*, both of which can cause chronic abscesses in human beings.

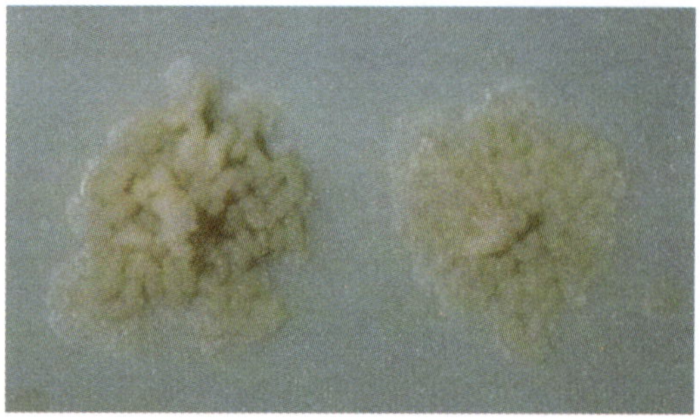

Fig. 29.22 Growth of *M. fortuitum* on LJ medium

- Outbreaks of abscesses following injection of vaccines and other preparations contaminated by these mycobacteria have been reported on a number of occasions.

> - ***M. smegmatis*** is commonly considered as a saprophyte in smegma, but is seldom seen in that location. It is frequently isolated from soft tissue lesions that follow trauma or surgery.
> - ***M. vaccae*** is reported to be an immunomodulator capable of inhibiting tissue-destroying hypersensitivity responses and stimulating protective immune processes in tuberculosis. Clinical trials of the *M. vaccae* vaccine as an adjuvant to chemotherapy in tuberculosis are underway.
> - ***M. ulcerans*** **(Group III of Runyon) and** ***M. marinum*** **(Group I)** are exclusively skin pathogens. They cause chronic ulcers and granulomatous lesions on the skin (Table 29.10). Systemic invasion does not occur, and the regional lymph glands are not involved. Cutaneous localisation occurs because these organisms multiply optimally at skin temperature.

Buruli ulcer

- Buruli ulcer is caused by *M. ulcerans* and lesions are usually seen on the legs or arms following minor injuries. After an incubation period of a few weeks, indurated nodules appear, which break down to form indolent ulcers, which slowly extend under the skin. It can destroy skin or soft tissue and also affect the bone, causing osteolylitis, scarring, and contractures.
- The disease, if untreated, can cause irreversible deformity or long-term functional disability.
- It is commonly seen in Uganda, Congo, Nigeria, Mexico, Malaysia, and New Guinea. The disease is called Barinsdale ulcer in Australia.

Agent—*M. ulcerans*

- This is a non-tuberculous *Mycobaterium* that is found in acquatic environments and is a strict pathogen of humans. It is similar to *M. marinum*. The differences between the two are presented in Table 29.11.
- A toxin—**mycolactone**—produced by *M. ulcerans* causes inflammation and necrosis.

Table 29.11 Differentiating characteristics of *M. ulcerans* and *M. marinum*

Characteristic	*M. ulcerans*	*M. marinum*
Distribution	Tropics	Temperate zones
Clinical course	Chronic, progressive ulcer	Usually, self-limited ulcer
Bacilli in ulcer	Abundant	Scanty
Rate of growth	Slower, 4–12 weeks	Faster, 1–2 weeks
Growth at 25°C	–	+
Growth at 37°C	–	+
Culture film	Bacilli in cords	No cord formation
Pigment in light	–	+
Mouse footpad lesion	Edema, rarely ulcer	Marked inflammation—purulent ulcer

Laboratory diagnosis

Diagnosis is chiefly clinical in endemic areas.

- **Samples:** FNAC from papules, nodules, plaques, edemas, or ulcers that do not have an undermined edge are collected, Swabs are most useful for ulcers that have undermined edges; rarely, punch biopsy is performed.
- **Direct smear examination for AFB:** Initially, smears from the edge of the ulcer show large clumps of bacilli which are acid-fast and alcohol-fast. Later, the immunoreactive phase sets in, and the bacilli disappear.
- **Culture:** Culture on Löwenstein–Jensen medium supplemented with 0.75% glycerol takes 4–12 weeks at 29–33°C. Colonies are yellowish, rough, and have well-demarcated edges.
- **PCR:** The most common target sequence for PCR confirmation of Buruli ulcer is the insertion sequence IS2404. It can be done on DNA extracted directly from clinical specimens or from colonies on culture media.
- **Histopathological examination:** Hallmark lesions show focally distributed extracellular clusters of AFB, massive tissue necrosis, fat cell ghosts, and remarkably low cellular infiltration in the centre of the lesions.
- Molecular tests are available for diagnosis.

Treatment

Surgical removal of lesions and skin grafting are followed by an 8-week regime of rifampicin and streptomycin.

Swimming pool granuloma

- It is also known as aquarium granuloma or fish tank granuloma.
- It is a rare skin condition caused by a non-tubercular mycobacterium known as *Mycobacterium marinum.*
- Human infection originates from contaminated swimming pools or fish tanks.
- It presents clinically as a slow-growing, inflamed, painful or sometimes painless red nodule or plaque at the site of infection that may become ulcerated or crusted.
- Hand may be affected in case the infection is acquired from an aquarium, but a swimming pool granuloma can be seen on trauma-prone body sites like the elbows, knees, feet, and the tip of the nose.

Based on the course of the disease, it can be classified into the following four types:

- **Type 1:** Crusted or ulcerated nodules or verrucous plaques developing within weeks or months after contact with the bacterium.
- **Type 2:** Distributed inflammatory nodules with a spreading pattern or with abscesses and granulomas.
- **Type 3:** Deep infections, with or without skin infection with signs of arthritis, tenosynovitis, osteomyelitis, and/or bursitis.
- **Type 4:** Lung infections and other systemic manifestations.

Agent—M. marinum

- It is a natural pathogen of cold-blooded animals and causes tuberculosis in fish and amphibians.
- Bacilli are scanty in smears from ulcers.
- Infection with *M. marinum*, but not with *M. ulcerans*, may cause a low-grade tuberculin reaction.

Laboratory diagnosis

- Diagnosis is difficult as the organisms are scanty in lesions.
- Histopathology may show sarcoid-like granuloma. The bacillus grows in 1–2 weeks at 30°C as yellow colonies.
- PCR is also available for diagnosis.
- NTB can be differentiated from tuberculous bacilli based on growth characteristics, as mentioned in Table 29.11, and certain biochemical tests.

Treatment

- Most environmental mycobacteria are resistant to the usual anti-tuberculous drugs, though pulmonary disease caused by *M. avium* complex or *M. kansasii* may respond to prolonged treatment with rifampicin, isoniazid, and ethambutol.
- Various combinations of drugs including rifabutin, clofazimine, quinolones, and newer macrolides may be used.

Leprosy

Leprosy is a chronic granulomatous disease of humans caused by *Mycobacterium leprae*. It primarily involves the skin, peripheral nerves, and nasal mucosa, but is capable of affecting any tissue or organ. It is predominantly a disease of the developing world.

Epidemiology

- Leprosy is an exclusively human disease restricted to the developing world, Asia, Africa, Latin America, and the Pacific.
- In India, leprosy is present in all states and territories. Orissa and Bihar have the highest prevalence (>5 per 1,000 population) and Haryana the least (<0.1 per 1,000).
- It is a disease of antiquity, the knowledge of which dates back to the Vedic times in India and the biblical times in the Middle East.

Mode of transmission

- The exact mode of transmission is not clear.
- Very large numbers of bacilli are shed in nasal secretions (up to 8×10^8 bacilli in one nose-blow).
- The mode of entry of the bacilli may be through the respiratory tract or through the skin.
- Asymptomatic infection appears to be quite common in endemic areas and may have a role in transmission.
- Leprosy is not highly communicable. The disease develops in only about 5 per cent of spouses living with leprosy patients.
- It is generally held that intimate and prolonged contact is necessary for infection to take place and is more likely if contact occurs during childhood.

Agent—*Mycobacterium leprae*

Mycobacterium leprae or the lepra bacillus was first observed by **Gerhard Henrik Armauer Hansen** in 1868.

Morphology

- *M. leprae* are straight or slightly curved rods, 1–8 × 0.2–0.5 µm in size, showing considerable morphological variation.
- They are acid-fast to a lesser degree than the tubercle bacillus. Hence, 5% sulphuric acid instead of 20% is used for decolourisation after staining with carbol fuchsin.
- They are bound together by a lipid-like substance, the **glia** to form masses known as '**globi**'. The globi appear in undifferentiated histiocytes called **Virchow's lepra cells** or foamy cells (Fig. 29.23).
- The morphological characterstics of *M. leprae* have been discussed in detail in Chapter 3.
- The entire genome of *M. leprae* has been mapped, and the genes for its major protein antigens have been cloned and sequenced.

Cultivation in animals

Mycobacterium leprae cannot be cultured in a cell-free medium in the laboratory. Animal models used to cultivate *M. leprae* are as follows:

- **Mice**
  - In 1960, **Charles C. Shepard** discovered that lepra bacilli could multiply in the **footpads of mice** at a low temperature (20°C). Granuloma develops at the site of inoculation in 1–6 months.
  - In thymectomised mice or following the administration of an antilymphocyte serum (to suppress CMI), generalised infection is produced, simulating lepromatous leprosy.
- **Nine-banded armadillo:**
  - It is highly susceptible to infection with the lepra bacilli.
  - Generalised infection occurs with extensive multiplication of the bacilli and the production of lesions typical of lepromatous leprosy.

Resistance

- Lepra bacilli have been found to remain viable in a warm, humid environment for 9–16 days and in moist soil for 46 days.

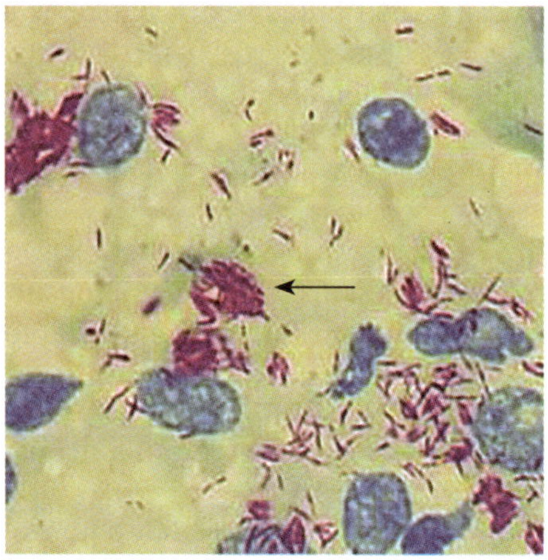

Fig. 29.23 *M. leprae*—acid-fast pink bacilli with typical 'globi' arrangement (*Source*: Department of Microbiology, Pondicherry Institute of Medical Sciences, Puducherry)

- They survive exposure to direct sunlight for two hours and ultraviolet light for 30 minutes.

Clinical features

The incubation period is 2–5 years.

Classification

Riddly and Jopling classified leprosy into five types:

1. Tuberculoid (TT)
2. Borderline tuberculoid (BT)
3. Mid-borderline (BB)
4. Borderline lepromatous (BL)
5. Lepromatous (LL)

BB leprosy is immunologically the least stable and, therefore, the rarest. This classification has been replaced now by the WHO classification.

According to the WHO, leprosy is classified on the basis of clinical manifestations and skin smear results. In the classification based on skin smears, infections showing negative smears at all sites are grouped as **paucibacillary leprosy (PB) or tuberculoid leprosy,** while those showing positive smears at any site are grouped **as multibacillary leprosy (MB) or lepromatous leprosy.** Table 29.12 lists the features of lepromatous and tuberculoid leprosy.

Borderline or dimorphous leprosy (BL)

- This refers to lesions possessing characteristics of **both tuberculoid and lepromatous types**.
- It may shift to either the lepromatous or the tuberculoid side of the spectrum depending on chemotherapy or alterations in host resistance.

Indeterminate type

- It is the early, unstable tissue reaction which is not characteristic of either the lepromatous or the tuberculoid type.
- In several persons, the indeterminate lesions undergo spontaneous healing. In others, the lesions may progress to the tuberculoid or lepromatous types.

Some diagnostic differences between these types are presented in Table 29.13.

Table 29.13 Features of different types of leprosy

Characteristics	Tuberculoid	Borderline	Lepromatous
Lepra bacilli in tissues	±	+	+++
Lepromin test	++	±	–
Mycobacterial antibodies	±	+	+++
Lymphocytic infiltration of lesions	+++	+	–
Plasma cells in lymphoid tissue	+	+	+++

Table 29.12 Features of lepromatous and tuberculoid leprosy

Type	Immune status of patient	Clinical features	Microscopy	Complications
Lepromatous leprosy (LL) (multibacillary disease)	Host resistance is low Bacilli are shed in large numbers in nasal and oral secretions Cell-mediated immunity is deficient, and the lepromin test is negative Exaggerated and broad humoral immune response Autoantibodies are common	More (>5) superficial nodular lesions (lepromata) develop, which consist of granulation tissue containing a dense collection of vacuolated cells in different stages of development ranging from mononuclear cells to lepra cells The reticuloendothelial system, eyes, testes, kidneys, and bones are also involved Erosion into the bones and cartilage of the nose gives the person's face a lion-like appearance (leonine facies; Fig. 29.24)	Bacilli are seen in large numbers or as globi inside or outside lepra cells (Fig. 29.23)	• Foot drop • Plantar ulcerations • Loss of distal digits • Loss of nasal cartilage • Lagophthalmos and corneal insensitivity (due to cranial nerve involvement) • Mild to severe testicular dysfunction in males, leading to infertility • Amyloidosis • Nerve abscesses
Tuberculoid leprosy (paucibacillary disease)	High degree of resistance Cell-mediated immunity is adequate and the lepromin test is positive	Few (1–5) and sharply demarcated, consisting of macular anesthetic patches (Fig. 29.25) Neural involvement occurs early and may be pronounced, leading to deformities, particularly in the hands and feet	Bacilli are scanty in the lesions, and infectivity is minimal	Less anti-mycobacterial and autoimmune antibodies are rare and hence the prognosis is good

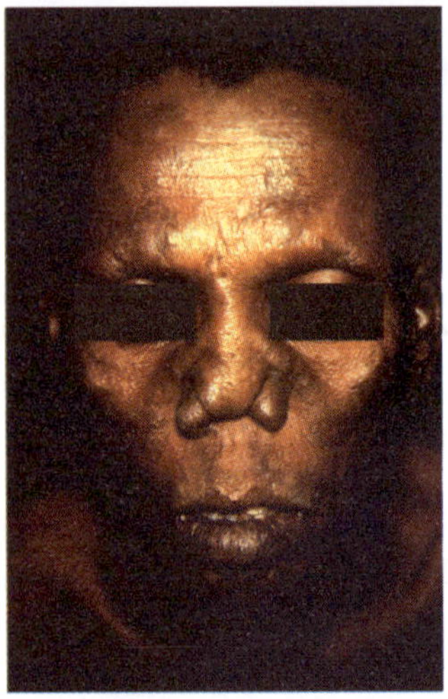

Fig. 29.24 Lepromatous leprosy with leonine facies (*Source*: CDC, PHIL, Image ID 15502)

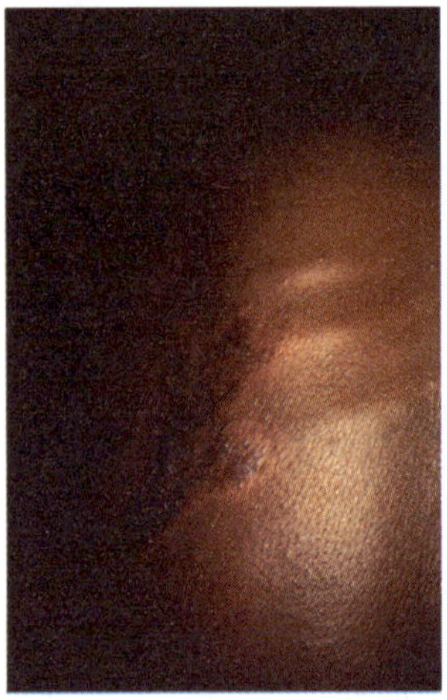

Fig. 29.25 Tuberculoid leprosy—sharply demarcated, macular patches (*Source*: CDC, PHIL, Image ID 16518)

Immunity

- **Innate immnunity**—a high degree of innate immunity against lepra bacilli seems to exist in human beings and hence, only a minority of those infected develop clinical disease.
- There is evidence of the genetic effects in the pattern of the response to the lepra bacillus infection. HLA-DR2 is seen preponderantly in persons with the tuberculoid type of reaction, while HLA-MTI and HLA-DQ1 are associated with lepromatous disease.
- Infection with lepra bacilli induces both humoral and cellular immune responses.

Humoral response

- Antibodies do not have any effect on the bacilli, while cellular immune mechanisms are capable of destroying them.
- In lepromatous leprosy, there is extensive polyclonal B cell activation. Large amounts of non-protective antibodies are produced, which include both anti-mycobacterial antibodies and autoimmune antibodies.
- The albumin:globulin ratio is reversed.

Cellular response

- The type of leprosy in an individual is determined by the status of cell-mediated immunity in that person. When cell-mediated immunity is deficient, the lepromatous type of disease develops.
- Delayed hypersensitivity to the lepra bacillus protein is absent. The deficiency in cell-mediated immunity is specific to the lepra bacillus antigens.
- Lepromatous patients have large numbers of $CD8^+$ (suppressor) lymphocytes in circulation.
- The lymphocytes present in skin granulomas are almost exclusively $CD8^+$ cells, in contrast to tuberculoid lesions, which contain predominantly $CD4^+$ cells.

Lepra reactions

Though leprosy is a chronic disease, its course may be interspersed with acute exacerbations which are of an allergic nature. Two types of such reactions occur, which are as follows:

- **Type 1, the reversal reaction or the lepra reaction**
  - Generally seen in borderline leprosy.
  - There is erythema and swelling along with pain and tenderness.
  - A similar clinical picture is seen in the downgrading reaction, which usually occurs in untreated or pregnant patients.
  - Here, the biopsy of the lesions shows a shift to the lepromatous pattern, reflecting a decrease in CMI.
- **Type 2 reaction or erythema nodosum leprosum (ENL)**
  - It occurs in the LL and BL types, usually a few months after the institution of chemotherapy.
  - Crops of tender, inflamed subcutaneous nodules appear, with fever, lymphadenopathy, and arthralgia.
  - This is an Arthus-type response to antigens released from dead lepra cells, which is characterised by neutrophil infiltration and IgG and complement deposition in the lesions.

Lepromin test

The lepromin test was first described by Kensuke Mitsuda in 1919. Standard lepromins are prepared from armadillo-derived lepra bacilli (lepromin A). Two types of reactions are seen:

- **Early reaction of Fernandez:** It consists of erythema and induration developing in 24–48 hours and usually remaining for 3–5 days. This is analogous to the tuberculin reaction.

- **Late reaction of Mitsuda:** It is the second and more specific test; it manifests in 1–2 weeks, reaches a peak in four weeks and gradually subsides in the next few weeks. The reaction consists of an indurated skin nodule, which may ulcerate. Histologically, lymphocytes, epithelioid cells and giant cells are characteristically present. The Mitsuda late reaction does not indicate pre-existing DTH but is a measure of the CMI.

Relevance of lepromin test

The lepromin test is not used to diagnose leprosy, nor does it indicate prior contact with the lepra bacillus. Lepromin test is used for the following reasons:

- **To classify the lesions in leprosy patients:** The lepromin test is positive in tuberculoid, negative in lepromatous, and variable in dimorphous and indeterminate types of disease.
- **To assess the prognosis and response to treatment:** Conversion to lepromin-positive status during treatment is evidence of improvement.
- **To assess the resistance of individuals to leprosy**
- **To identify candidate lepra bacilli**

Laboratory diagnosis

The diagnosis of leprosy is based on at least one of three cardinal signs (WHO recommendations):

i. **Hypopigmented skin patch with loss of or reduced sensation**
ii. **Enlarged nerves**
iii. **Slit-skin smear positive for leprosy bacilli**

Specimen

For routine examination, specimens are collected from the nasal mucosa, skin lesions, and ear lobules.

- **Nasal smear:** A blunt, narrow scalpel introduced into the nose is used to remove a piece of mucous membrane, which is transferred to a slide and teased out into a uniform smear. Alternately, the patient is asked to blow into a tissue paper, and the material is scraped by a scalpel and smeared on a slide.
- **Slit skin smear:** The skin over the earlobe is pinched tight to minimise bleeding, and a cut, about 5 mm long, is made with a scalpel, deep enough to get into the infiltrated layers. After wiping off blood or lymph that may have exuded, the scalpel blade is turned transversely to scrape the sides and bottom of the cut to obtain a little tissue pulp. This pulp is smeared uniformly on a slide. About 5–6 different areas of the skin are sampled, including the skin over the buttocks, forehead, chin, cheek, and ears.
- **Nerve biopsy:** A sample is collected from thickened nerves and submitted for histopathological examination.

Rarely, biopsy of the nodular lesions, thickened nerves, and lymph node puncture may be necessary in some cases.

Microscopy

- The smears are stained by the **Ziehl–Neelsen technique** using 5% instead of 20% sulphuric acid for decolourisation.
- The smears are graded based on the number of bacilli, as follows:
  - 1–10 bacilli in 100 fields = 1+
  - 1–10 bacilli in 10 fields = 2+
  - 1–10 bacilli per field = 3+
  - 10–100 bacilli per field = 4+
  - 100–1,000 bacilli per field = 5+
  - More than 1,000 bacilli, clumps and globi in every field = 6+

Bacteriological (bacterial) index (BI): It is calculated by totalling the number of pluses (+s) scored in all the smears and dividing that number by the number of smears. For calculating BI, a minimum of four skin lesions, a nasal swab, and both the ear lobes have to be examined.

Morphological index (MI): It is the percentage of solid, fragmented granular bacilli (SFGB) or uniformly stained bacilli out of the total number of bacilli counted.

Serology

It is not specific and hence is not used routinely.

Molecular method

Molecular studies are still in the research stage and have not yet been utilised as diagnostic tools.

> **Laboratory diagnosis of leprosy**
>
> WHO diagnostic strategy:
>
> i. Hypopigmented skin patch with loss of or reduced sensation
> ii. Enlarged nerves
> iii. Slit-skin smear positive for leprosy bacilli
>
> **Specimen:** Nasal mucosa, skin lesions, and ear lobules. Rare, biopsy of the nodular lesions, thickened nerves, and lymph node puncture
>
> - **Slit skin smear:** About 5–6 different areas of the skin are sampled, including the skin
>
> **Microscopy**
>
> - *Ziehl–Neelsen staining* using 5% instead of 20% sulphuric acid for decolourisation
>
> **Bacteriological (bacterial) index (BI)**
>
>   - Totalling the number of pluses (+s) scored in all the smears and dividing that number by the number of smears
>   - For calculating BI, a minimum of four skin lesions, a nasal swab and both the ear lobes have to be examined

Morphological index (MI): BI = Sum of all + sin smears/number of smears
MI = % SFGB/total number of bacilli counted
Serology: Not specific and hence is not used routinely.
Molecular methods: Not yet been utilised as diagnostic tools.

Treatment

The World Health Organization's limited duration multidrug therapy has led to the adoption of the following WHO classification for treatment purposes:

- **Single lesion paucibacillary (SLPB)**
- **Paucibacillary (PB)**, i.e., those with two to five lesions
- **Multibacillary (MB)**, i.e., those with six or more lesions

Dapsone was the first effective chemotherapeutic agent against leprosy. Its use as monotherapy for several years led to the development of resistant strains of lepra bacilli. In view of this, multiple drug therapy (MDT) is now recommended for leprosy, as it is for tuberculosis. The current recommendation is as follows:

1. **MB leprosy:** The standard adult treatment regimen is as follows:
    - Rifampicin: 600 mg once a month
    - Clofazimine: 300 mg once a month; 50 mg daily
    - Dapsone: 100 mg daily
    - Ethionamide or prothionamide may be added to this regimen or substituted for clofazimine
    - Duration: 12 months
2. **PB leprosy:** The standard adult treatment regimen is as follows:
    - Rifampicin: 600 mg once a month
    - Dapsone: 100 mg daily
    - Duration: 6 months

A minimum follow-up of four years for paucibacillary and eight years for multibacillary cases is necessary to detect any relapse.

Prophylaxis

The role of chemoprophylaxis is debatable. Instead, case finding and adequate therapy have been the methods employed for prophylaxis.

- **BCG vaccine:** No specific vaccine is available. Since there is some degree of antigenic relationship between the lepra and tubercle bacilli, field trials with different leprosy vaccines (BCG + killed lepra bacilli; ICRC bacillus) were conducted, but without conclusive evidence.
- **Leprosy vaccine:** An indigenous leprosy vaccine using the strain *Mycobacterium indicus* pranii has been approved by the Drug Controller General of India (DGCI) and Food and Drug Administration (FDA) department of the US. It is still in its pilot phase, and if found successful, will be rolled out in high-prevalence states.

National strategies to control leprosy in India

The National Leprosy Eradication Programme of the Ministry of Health and Family Welfare, Govt. of India (see Chapter 35), has been instrituted to control leprosy in India.

- It is the hub for all national programmes to detect, treat, and control leprosy.
- The programme is implemented by Indian States/ UTs. The World Health Organization and the International Federation of Anti-leprosy Associations (ILEP) along with some non-governmental organisations support the programme.

Central institutes that undertake research and programme implementation in several states are as follows:

- **Central Leprosy Teaching and Research Institute (CLTRI)** at Chengalpattu, Tamil Nadu
- **Japanese Leprosy Mission for Asia (JALMA),** which was later renamed the **National Institute for Leprosy and Other Mycobacterial Diseases**, Agra, Uttar Pradesh—conducts research on leprosy, tuberculosis and HIV; it is currently one of the ICMR institutes
- **Regional Leprosy Training and Research Institute**, Aska, Odisha—the central institute that works as a referral centre for the management of difficult-to-diagnose leprosy cases, and provides physiotherapy

Elimination of leprosy

- The Leprosy Elimination Campaign has been introduced by the **Central Leprosy Division (CLD)** of the Govt. of India's Ministry of Health and Family Welfare.
- It aims to bring about a reduction in the number of new cases of Grade II disability annually to less than 1 case per million population at the national level.

VIRAL INFECTIONS OF THE SKIN AND SOFT TISSUES

Lesions in the skin are manifestations of viral pathogenesis affecting other systems. These could be of the following etiologies:

i. The virus remains **restricted to the skin at the site of infection**, e.g., molluscum contagiosum
ii. **Intracellular infections** of the squamous epithelium (HSV1), dermal capillary plexus (varicella), cutaneous nerve roots (hepres-zoster), or viruses associated with viremia

Most of the cutaneous manifestations of systemic viral infections are in the form of vesicles or rashes due to microhemorrhages. Table 29.14 presents a list of these manifestations.

Table 29.14 Mucocutaneous lesions caused by viruses

Clinical lesions on the skin	Spread	Causative agent	Virus shed from the lesion
Vesicles			
Chickenpox	Systemic spread	Varicella-zoster	+
Shingles	Systemic spread	Varicella-zoster	+
Whitlow, cold sores	Systemic spread	Herpes simplex	+
Hand, foot, and mouth disease	Systemic spread	Enterovirus (Coxsackie A16)	+
Poxvirus lesions	Systemic spread	Various poxviruses (Table 29.15)	+
Orf	No systemic spread	Parapox	+
Papilloma (wart)	No systemic spread	Papilloma virus	+
Rashes			
Maculopapular rashes	Systemic spread	Measles, mumps, rubella (Chapter 23)	–
Cutaneous hemorrhages	Systemic spread	Dengue virus, chikungunya, and other hemorragic fever viruses (Chapter 21 and 24)	–
Erythema infectosum	Systemic spread	Parvovirus B19	–
Exanthema subitum	Systemic spread	Herpes virus 6	–

VARICELLA (CHICKENPOX)

- It is a common **childhood infection caused by the varicella–zoster virus**. The disease may, however, occur at any age. Adult chickenpox, which is more serious, is rather common in some tropical areas.
- The source of infection is a chickenpox or herpes-zoster patient.
- Infectivity is maximum during the initial stages of the disease, when the virus is present abundantly in the upper respiratory tract.
- The buccal lesions which appear in the early stage of the disease and the vesicular fluid are rich in virus content. Infectivity wanes as the disease progresses and the scabs are virtually non-infectious.
- There are no animal reservoirs of varicella.
- **One episode confers lasting immunity.**

Agent—Varicella-Zoster Virus (VZV)

- Chickenpox is seen as a primary infection in a non-immune individual.
- VZV is similar to the herpes simplex virus in its morphology.
- It does not grow in experimental animals or chick embryos.
- It can be grown in cultures of human fibroblasts, human amnion, or HeLa cells.
- The cytopathic effects are similar to but less marked than those produced by the herpes simplex virus.
- In cultures, the virus remains cell-associated and does not appear freely in the medium.
- By using highly specific antisera, it is possible to distinguish between herpes virus types 1, 2, and varicella-zoster viruses.
- Only one antigenic type of VZV is known.
- The portal of entry of the virus is the **respiratory tract or conjunctiva.**

Clinical Features

- The **incubation period** is 7–23 days.
- After about two weeks, the lesions begin to appear.
- The patient is considered to be **infectious two days before and five days after the onset of the lesions**.
- **In children**
  - Buccal lesions may not be noticed.
  - The rash usually appears on the trunk.
  - The evolution of the rash is so rapid that the various stages—macule, papule, vesicle, pustule, and scab—cannot be readily followed in individual lesions.
  - The rash is centripetal in distribution, chiefly affecting the trunk and sparing the distal parts of the limbs.
  - It is very superficial, and does not involve the deeper layers of the skin.
  - The rash typically resembles dew drops and appears in crops during the first three or four days of the disease so that lesions of varying ages can be noticed on the same patient.

- The lesions mature very quickly, beginning to crust within 48 hours.

- **In adults**
  - When varicella occurs in adults, **systemic symptoms may be severe**.
  - The rash is very profuse and overall, the disease is much more intense than in children. It may become hemorrhagic; occasionally, bullous lesions appear (Fig. 29.26).
  - **Pitted scars** on the skin may remain after recovery.

Complications

- Varicella pneumonia is more common in adults, and often fatal in the elderly.
- Other complications like myocarditis, nephritis, acute cerebellar ataxia, meningitis, and encephalitis may ensue.
- **Secondary bacterial infections**, usually due to staphylococci or streptococci, may occur.
- **Reye's syndrome** may follow varicella in some cases with a history of administration of salicylates.

Post-herpetic neuritis or shingles

- It can occur in some individuals, years after the initial infection.
- It affects a single sensory nerve ganglion, usually on the trunk and over one dermatome.
- The rashes appear as painful blisters on one side. This is thought to be the reactivation of a latent infection, but the exact mechanism of reactivation is not known.

Chickenpox in pregnancy

- This can be dangerous for both the mother and the baby.
- The disease tends to be more severe in pregnant women than in other adults.
- If maternal varicella occurs during the first half of pregnancy, the fetal infection is usually asymptomatic.

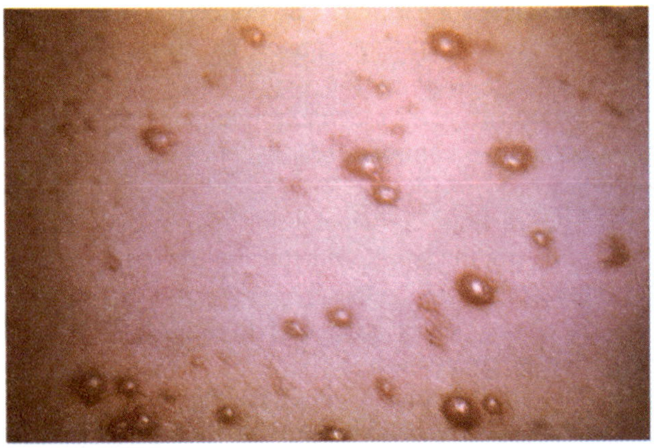

Fig. 29.26 Generalised herpes zoster outbreak due to the varicella-zoster virus (VZV) (*Source*: CDC, PHIL, Image ID 5407)

- Some infants may develop **fetal varicella syndrome,** manifesting as cicatrising skin lesions, hypoplasia of the limbs, chorioretinitis, and CNS defects. Some babies may not exhibit any defects but may carry latent VZV infection.

Neonatal varicella

When maternal varicella infection occurs near delivery, babies may develop neonatal varicella within two weeks of birth.

- If the mother's rash begins a week or more before delivery, she develops antibodies which are passed, along with the virus, to the fetus transplacentally. Such a baby, though infected, usually escapes clinical disease.
- If the mother develops chickenpox shortly before (less than a week) or within two days of delivery, the baby receives from the mother only the virus and not the antibody, and hence develops neonatal varicella. This is usually a serious, disseminated disease with a high risk of pneumonia and encephalitis. Such babies are given VZV antiserum and chemotherapy immediately after birth.

Laboratory Diagnosis

Diagnosis is usually clinical.

- **Microscopy:**
  - Multinucleated giant cells and type A intranuclear inclusion bodies may be seen in smears prepared by scraping the base of the early vesicles (Tzanck smears) and stained with toluidine blue, Giemsa or Papanicolaou stain.
  - Electron microscopy of the vesicle fluid may demonstrate the virus with typical herpes morphology.
- **Virus isolation:** This may be attempted from the buccal or cutaneous lesions in the early stages by inoculating human amnion, human fibroblast, HeLa, or Vero cells.
- **Serology:** The virus antigen can be detected in scrapings from skin lesions by immunofluorescence, and in vesicle fluid by counterimmunoelectrophoresis with zoster immune serum.
- **Molecular methods:** ELISA and PCR techniques are also in use.

Prophylaxis

Vaccine

- A lyophilised form of the vaccine is now available, which can be stored between 2°C and 8°C.
- All children should routinely receive the first dose of varicella vaccine (live-attenuated Oka strain of VZV) at **12–15 months of age**.
- The second dose of the vaccine is recommended at 4–6 years of age.

- It is safe and effective.
- Occasionally, children may develop a few vesicles, which resolve quickly.
- The vaccine is **not considered safe in pregnancy**.

Passive protection

- **Varicella-zoster immunoglobulin (VZIG)** prepared from the plasma of patients convalescing from herpes-zoster provides passive protection in immunocompromised children exposed to infection, but its availability is limited. It is not useful in treatment.
- Infection control measures—as the infection spreads by direct contact, the affected person is isolated. Barrier nursing is provided to such patients.

Treatment

- Specific treatment is indicated chiefly for immunodeficient and elderly subjects and those with complications such as varicella pneumonia, encephalitis, and disseminated zoster.
- Acyclovir and famciclovir are effective.
- Corticosteroids are contraindicated in varicella as they enhance the risk of pneumonia and disseminated disease.

HERPES-ZOSTER (SHINGLES, ZONA)

> **Case 5.** *A 50-year-old, HIV-positive man developed painful vesicular eruptions on the right side of his spine in the lumbar region. Scrapings from the base of the lesions were positive for the VZV antigen by the direct fluorescent antibody test. A culture from the fluid was inoculated on cell lines and the PCR test was positive for varicella-zoster virus. The patient was treated with acyclovir. On further enquiry, the patient revealed to the doctor that he had had chickenpox in childhood.*
> i. What is the pathogenesis of shingles?

- Herpes-zoster infection usually occurs in persons who have had chickenpox several years earlier.
- While varicella is typically a disease of childhood, herpes-zoster (from *herpein*, meaning to creep and *zoster*, meaning girdle) is one of **old age**, being common after the age of fifty years.
- The disease may, however, occur at any age; zoster has been reported, very rarely, even in the newborn.
- The virus, which remains latent in the sensory ganglia, may leak out at times, but is usually held in check by the residual immunity.
- Contact with either zoster or chickenpox may lead only to chickenpox but not zoster.
- Years after the initial infection, when the immunity has waned, the latent virus may be reactivated and triggered by some precipitating stimulus.
- Then, it travels along the sensory nerve to produce zoster lesions on the area of the skin or mucosa supplied by the nerve (Fig. 29.27).

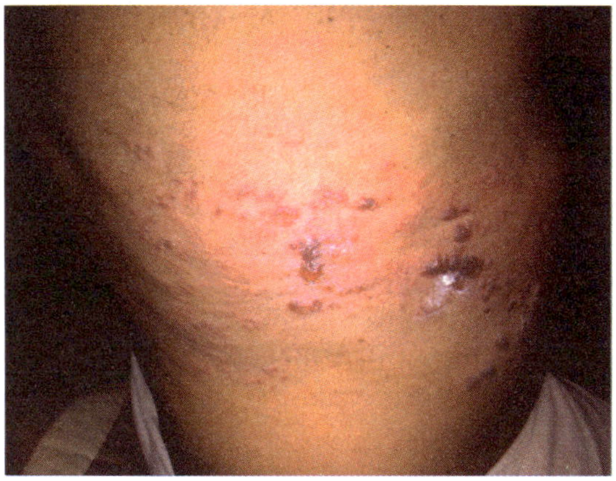

Fig. 29.27 Herpes-zoster (shingles) affecting a dermatome on the trunk of the patient (*Source*: Dr. Reba Kanungo, Department of Microbiology, Pondicherry Institute of Medical Sciences, Puducherry)

- This reactivation is associated with inflammation of the nerve, which accounts for the neuritic pain that often precedes the skin lesions.

Pathogenesis

- Herpes-zoster represents a mode of evolutionary adaptation by the VZ virus which is an obligate human parasite.
- The ability of the virus to remain latent and reappear as zoster years later confers on it a great survival advantage.
- The most commonly affected sites are the areas innervated by **spinal cord segments D3 to L2** and the **trigeminal nerve**, particularly, **its ophthalmic branch**.

Clinical Features

- The **rash is typically unilateral** and confined to the area supplied by a single sensory ganglion (Case 5).
- The lesions are identical to varicella lesions, except for their limited distribution.
- The rash heals in about two weeks. Pain and paresthesia at the affected area may persist for weeks or months.

Complications

- Lower motor neuron paralysis, meningoencephalitis, and herpes-zoster ophthalmicus may develop.
- There may be generalised zoster, where the lesions are scattered widely, perhaps due to the hematogenous dissemination of the virus.
- **Ramsay Hunt syndrome** is a rare form of zoster affecting the facial nerve, with an eruption on the tympanic membrane and the external auditory canal, and often, facial palsy. Chronic or recurrent zoster is often seen in the HIV-infected.

The laboratory diagnosis and treatment for shingles are as for chickenpox.

Hand, Foot, and Mouth Disease (HFMD)

Hand, foot, and mouth disease is a viral exanthem, and it is most commonly caused by coxsackievirus A16 and enterovirus A71.

Clinical features

- It presents as a low-grade fever, malaise, mouth or throat pain secondary to the enanthem.
- Examination reveals vesicles is surrounded by a thin halo of erythema that rupture, forming superficial ulcers with a grey-yellow base and erythematous rim.
- The exanthem can be macular, papular, or vesicular. The lesions are are non-pruritic and painless. These can occur on the dorsum of the hand, feet, buttocks, legs, and arms.
- Oral lesions commonly involve buccal and tongue ulcers but may also involve the soft palate.
- The disease is generally self-limiting, but complications such as myocarditis, aseptic meningitis, encephalitis, polio-like syndrome, and Guillain-Barre syndrome may be seen rarely.

Laboratory diagnosis

- It is mainly clinical. If required serology and PCR for coxsackievirus in skin swabs can be attempted.
- In the context of the recent increase in monkeypox cases, it has become important to differentiate HFMD cases.

Treatment is mostly symptomatic; no specific antiviral is given.

Herpetic Whitlow/Cold Sores

- Herpetic whitlow is caused by the herpes simplex virus, type 1 or type 2.
- Primary infection occurs through direct exogenous or autogenous inoculation through broken skin while recurrent infections may occur due to reactivation of the latent virus months to years following primary infection.
- After entry, the virus invades and replicates in epidermal and dermal cells. It may involve the sensory dorsal root ganglion, establishing latency. The vesicles appear around the nail, and may coalesce into large, honeycomb-like bullae in 5 to 6 days. Spread may occur proximally, involving the nail bed.
- The lesions are self-limiting and resolve in 2–4 weeks.

Clinical features

The patient presents with pain and tingling in the finger before any skin changes. This is followed by local tenderness, erythema, and edema with an initial crop of vesicles, commonly along the pulp and lateral aspect of the finger.

Diagnosis is mainly clinical. **Treatment** is mainly symptomatic as the disease is self limiting.

ERYTHEMA INFECTIOSUM

- Erythema infectiosum is also known as **fifth disease or slapped cheek syndrome**.
- It is a caused by **parvovirus B19** and is commonly seen in children between five and fifteen years of age.

Clinical Features

- A red rash appears on the cheeks after a brief period of fever, malaise, diarrhea, vomiting, and headache.
- The rash has a slapped cheek appearance with surrounding oral pallor that subsides in 4–5 days.
- It can cause a 'papular **gloves and socks syndrome**' in teenagers. The rash is pruritic, and occurs on the hands and feet.
- In pregnant women, the infection can cause severe anemia, hydrops fetalis or even stillbirth or miscarriage.

Agent–Parvovirus B19

- It is a small non-enveloped, icosahedral DNA virus that has a single-stranded linear DNA.
- **Transmission** is by respiratory droplets. In some cases bloodborne transmission may be seen.
- The virus infects progenitor red blood cells, causing anemia.
- IgM antibodies develop after two weeks when the viremia ends. Immunoglobulin G appears 1 week later, with rash and arthralgia.
- **Diagnosis** is by serology and PCR performed on blood and bone marrow.
- **Treatment** is mainly symptomatic as the disease is self-limiting.

EXANTHEMA SUBITEM OR ROSEOLA INFANTUM

Roseola infantum or exanthema subitum or sixth disease is seen globally and is caused by infection with human herpesvirus 6 (HHV-6) or, less commonly, by human herpesvirus 7 (HHV-7). It is seen mostly in children between six and 12 months of age.

Presentation includes acute onset high-grade fever that lasts for 3–5 days followed by rapid defervescence of the fever along with non-pruritic, pink papular rash that begins on the trunk.

Agent–HHV-6

- HHV-6 is a beta-herpesvirus, closely related to human cytomegalovirus and HHV-7. This group of beta-herpesviruses is known to have less cell tropism than other members of the Herpesviridae family.
- It contains a linear, double-stranded DNA genome.

- It is transmitted primarily through saliva via respiratory droplets and is known to cause fever and exanthema subitem.
- This virus was isolated from the peripheral CD4 cells of a healthy person. It shares with HIV the same CD4 receptor on T cells and could, therefore, contribute to further depletion of CD4 T cells in HIV-infected persons.

HHV-8 LESIONS

- Most individuals latently infected with HHV-8 are asymptomatic, but in children and organ transplant recipients, a primary infection syndrome characterised by fever, rash, lymphadenopathy, bone marrow failure, and occasional rapid progression to Kaposi sarcoma may be seen.
- HHV-8 is also known to cause Kaposi sarcoma, primary effusion lymphoma, Castleman's disease. It can undergo reactivation in immunocompromised hosts.
- Kaposi sarcoma manifestations vary widely, but most patients have non-tender, hyperpigmented, macular, or nodular skin lesions.
- It was initially identified by DNA sequences presumed to represent a new herpes virus identified from tissues of Kaposi sarcoma drawn from AIDS patients. It is, therefore, referred to sometimes as **Kaposi sarcoma-associated herpes virus (KSHV)**. However, a causative relationship is yet to be proved.
- The incidence of this virus is higher in men who have sex with men.
- HHV-8 has also been detected by molecular techniques in cases of basal cell carcinoma, pemphigus vulgaris, and foliaceus.
- **Diagnosis** is made by serology and PCR testing on blood, skin, lymph nodes, lungs, and the gastrointestinal tract.

POXVIRAL SKIN DISEASES

Poxviruses are the largest and most complex of viruses that infect vertebrates. They are large enough to be seen under the light microscope. Smallpox is the first disease to be totally eradicated from earth—the greatest success of medical science. There are two genera of poxviruses:

1. **Chordopoxvirinae:** These are poxviruses of vertebrates. The members causing diseases in humans are listed in Table 29.15.
2. **Entomopoxvirinae:** These are poxviruses of insects.
   - Orthopoxviruses are brick-shaped, whereas Orf and molluscum contagiosum are ovoid.
   - They are double-stranded DNA viruses.
   - They can be propagated on the chorioallantoic membrane of chick embryos, on which they produce lesions (pocks), and also in tissue culture.
   - In tissue culture, the poxviruses replicate only in the cytoplasm, forming eosinophilic cytoplasmic inclusions bodies known as **Guarnieri bodies.**
   - They prefer epidermal cells for growth.
   - The **intradermal vaccine against smallpox** was prepared from the vaccinia virus. Immunity develops 8–10 days after vaccination.
   - Cidofovir and brincidofovir have been found useful as antivirals against poxviruses.
   - In 1977, the World Health Organization (WHO) announced that smallpox was eradicated worldwide.

Monkeypox

- Monkeypox was known to be endemic in West and Central Africa but since May 2022, cases of monkeypox have been reported from countries where the disease is not endemic.

Table 29.15 Poxviruses that cause disease in humans

Genus	Virus	Primary host	Clinical presentation in humans
Orthopoxvirus	Variola	Human being	Fever with papules, vesicles, pustules causing pockmarks and scarring
	Vaccinia	Human being	Vesicular skin lesion
	Cowpox	Cattle	Vesicular lesion on hands
	Monkeypox	Monkey, squirrel	Firm/rubbery, well-circumscribed, deep-seated, and often develop umbilication
Parapoxvirus	Pseudocowpox	Cattle	Localised nodular lesions (milker's node)
Yatapox	Orf	Sheep, goat	Vesicular and granulomatous skin lesions
	Tanapox	Monkey	Vesicular skin lesions with fever
	Yabapox	Monkey	Localised skin tumors
Molluscipoxvirus	Mollascum contagiosum	Human being	Multiple skin nodules

- Most reported cases have been involved mainly, but not exclusively, men who have sex with men.
- The monkeypox virus spreads through close, personal, and often, skin-to-skin contact.
- The skin lesions are firm or rubbery, well-circumscribed, deep-seated, and often develop umbilication. In recent outbreaks, the lesions often occur in the genital and anorectal areas or in the mouth.
- The rash may be single or confined to only a few lesions. Fever and respiratory symptoms may be present. The rash is not always disseminated across many sites on the body.
- **Diagnosis** can be made by PCR on swabs obtained from skin lesions.

Orf Disease

- Orf disease, or contagious pustular dermatitis or ecthyma contagiosum, is a zoonotic and occupational infection that is acquired from **infected sheep and goats**.
- The Orf virus causes vesicular, papular, and nodular lesions on the lips and mouth of affected animals. In humans, the lesions are usually present on the dorsal aspects of the hands or fingers at areas of contact with the infected animals.
- Human beings acquire infection via direct contact with infected lesions and vesicular fluids of infected animals. The lesions progress from maculopapular rash to nodules that later dry, becoming papillomatous.
- The disease is self-limiting.

MOLLUSCUM CONTAGIOSUM

- This is a common skin infection caused by the molluscum contagiosum virus.
- The molluscum contagiosum virus is a DNA virus belonging to the **pox group** of viruses. It resembles the vaccinia virus closely.
- It causes benign viral warts that can be transmitted sexually in HIV-positive patients.
- It produces **small, pink, and pearly warts** (Fig. 29.28a) on the face back, arms, and buttocks.
- The lesions may persist for two years, and regress spontaneously.

Laboratory Diagnosis

- Molluscum contageosum can be diagnosed clinically.
- Histopathologically, the diagnosis is made based on typical changes in the stratum spinosum and granulosum (Fig. 29.28b) and the presence of inclusion bodies known as **Henderson Paterson bodies** (Fig. 29.28c).
- The virus cannot be propagated on cell lines.
- The viral DNA sequence can be detected by PCR.

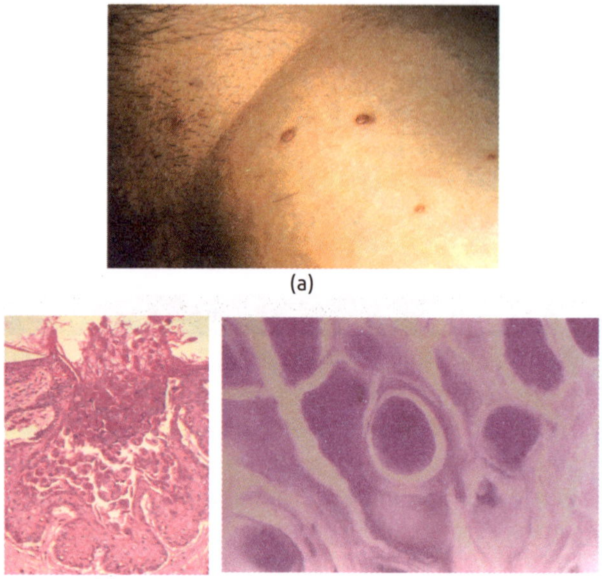

(a) (b) (c)

Fig. 29.28 (a) Pearly papules, characteristic of the viral infection, molluscum contagiosum, (b) granular, eosinophilic (pink) inclusions known as Henderson-Paterson bodies are seen within the keratinocytes of the former stratum basale, stratum spinosum, and stratum granulosum and (c) eosinophilic inclusion (*Source*: CDC, PHIL, Image ID 16687, 860 and 16689)

SKIN WARTS

- Warts or verrucae are non-cancerous skin growths that are caused by a rapid growth of cells on the outer layer of the skin.
- Skin warts are most common in childhood and typically appear and regress spontaneously over the course of weeks to months. Recurrances are common.
- All HPVs are believed to be capable of establishing long-term 'latent' infections.
- Skin warts are caused by human papilloma virus (HPV) types 2, 4, 27, and 57.

Agent–Human Papilloma Virus

The papilloma virus produces **warts** (Fig. 29.29) which are site-specific, the majority being ano-genital; a few produce plantar warts.

- These virus infections are acquired by direct contact.
- After entering through abrasions on the skin, the virus infects the basal layer of the skin and mucosa. The infection does not spread to deeper tissues.
- When the basal cells become squamified, the virus becomes infectious, and antigens appear.
- There is marked proliferation of the basal cells and prickle cells.
- These cells protrude from the surface of the skin to form a papilloma. These warts may present as flat-topped or finger-like projections.

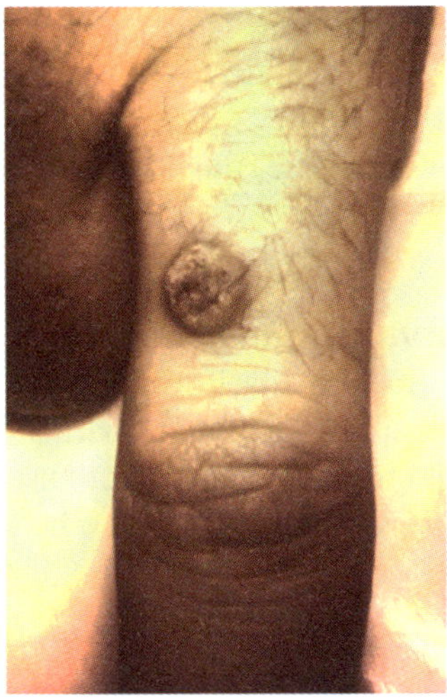

Fig. 29.29 Common wart on the finger also referred to as verruca vulgaris, caused by the human papilloma virus (HPV) (*Source*: CDC, PHIL, Image ID 19094)

Genital warts and cervical warts can appear as **cauliflower-like or flat dysplasia**. The former may progress to cervical, vulval, anal or penile cancers. Details are given in Chapter 27.

Laboratory Diagnosis

- This is chiefly based on clinical features.
- The virus cannot be cultivated, and serology does not play a decisive role.
- Viral DNA can be detected by polymerase chain reaction. There are certain genotypes which are associated with malignancy and have specific geographic distribution (details in Chapter 26).

Treatment

The treatment of skin warts involves applying karyolitic agents such as salicylic acid or cryotherapy.

PARASITIC INFECTIONS OF THE SKIN AND SUBCUTANEOUS TISSUES

A large number of parasites can affect the skin and subcutaneous tissues. They may be of the following two types:

- **Endoparasites**
- **Ectoparasites**

ENDOPARASITES

- These parasites cause clinical features while passing through the skin on their journey to their habitat.
- This association may be transient.
- Protozoa, trematodes, cestodes, and nematodes species can involve the skin or subcutaneous tissues (Table 29.16).

Table 29.16 Parasites causing skin and subcutaneous lesions

Category	Parasite	Disease caused
Protozoal parasites causing infections with skin involvement	*Trypanosoma b. rhodesiense* *T. b. gambiense*	Sleeping sickness, chancre
	T. cruzi	Chagoma, periorbital edema (Romana's sign)
	Leishmania tropica	Oriental sore
	L. braziliensis	Espundia
	L. mexicana	Chiclero ulcer
	L. donovani	Post-kala-azar dermal leishmaniasis (PKDL)
	Entamoeba histolytica	Amoebiasis cutis
Cestodes	*Schistosoma* species	Cercarial dermatitis
Trematodes	*Taenia solium*	Skin and subcutaneous cystecerci
	Ecchinococcus granulosus	Hydatid cyst in skin
Nematodes	*Ancylostoma braziliensis* *Ancylostoma caninum* *Uncinaria stenocephala* *Bunostomum phlebotomum* *Gnathostoma spinigerum* *Ancylostoma duodenale* *Necator americanus*	Cutaneous larva migrans
	Strongyloides stercoralis	Larva currens
	Trichinella spiralis	Subcutaneous nodules
	Dracunculiasis medinensis	Guinea worm disease
	Onchocerca volvulus *Loa loa* *Mansonella streptocerca*	Subcutaneous filariasis

Protozoan Infections of the Skin

Amoebiasis cutis

- This is an extremely rare clinical presentation in ***Entamoeba histolytica*** infections, in which amoebic trophozoites escape from the bowel to the skin.

- The vulnerable areas include perianal areas and the skin around a drained amoebic liver abscess.
- In some cases, pre-existing wounds may get contaminated with *Entamoeba histolytica*.
- Clinically, the lesion may present as ulceration.
- **Diagnosis** is clinical and serological.
- It can be **treated** with an amoebicide such as metronidazole.

Post-kala-azar dermal leishmaniasis

- Post-kala-azar dermal leishmaniasis (PKDL) is seen after six months to 1 or more years after a cure has been reported with kala-azar.
- It is reported in 5–10% of patients, and presents with macular, popular, or nodular rashes on the face, upper arms, or trunk.
- Such people act as a potential source of *Leishmania* infection.
- *L. donovani* can cause secondary skin lesions, especially after treatment.

Cutaneous leishmaniasis

- Caused by *Leishmania* spp., this is seen globally, with more than 600,000 to 1 million new cases occurring annually.
- The disease occurs after an incubation period of 2–8 weeks.
- It is characterised by **skin lesions that occur mainly on the exposed parts of the body**.
- The lesions begin as painless papules, become nodules, and finally ulcerate.
- They heal spontaneously, leaving lifelong scars and serious disfigurement.
- *L. tropica* lesions are singular and more swollen with a thick crust and less necrosis, and are also known as **Oriental sores, Delhi boils, Aleppo boils and Bagdad buttons**.

Nematodes Affecting the Skin

- Any nematode can present with skin lesions. This could be a part of the nematode's life cycle such as in hookworms of humans or it could be an accidental infection with zoonotic hookworms.
- Since humans are not the natural hosts of zoonotic hookworms, further development of these parasites is arrested, leaving their larvae to wander around aimlessly.
- In some parasites, the primary habitat is the skin and subcutaneous tissue.
- The skin may be the primary site of infection or may become involved secondarily.
- Many nematodes, chiefly zoonotic species infecting humans accidentally, can lead to subcutaneous lesions.
- Many nematodes that affect skin when they enter the body by piercing skin such as the human hookworm species, *Ancylostoma duodenale* and *Necator americanus,* cause a tiny, transient, itchy papular rash at the site of entry of the larva.
- Other nematodes affecting the skin include filarial worms such as *Loa loa*, *Onchocerca volvulus*, *Mansonella streptocerca*, and *Dirofilaria* spp.

Cutaneous larva migrans (creeping eruption)

- It is caused by the filariform larvae of hookworm species that are non-human parasites and which normally infect dogs, cats, and other animals.
- Cutaneous larva migrans can also be seen in human hookworm infections during the larval migration stage.
- The infection can be acquired by humans either by the ingestion of the parasite eggs or by penetration of the skin by the larvae.

Worms that cause cutaneous larva migrans are as follows:

- *Ancylostoma braziliensis*
- *Ancylostoma caninum*
- *Uncinaria stenocephala*
- *Bunostomum phlebotomum*
- *Gnathostoma spinigerum*

Human intestinal parasites (hookworms) that also manifest cutaneous lesions during the migration of the larva after penetrating the skin are as follows:

- *Ancylostoma duodenale*
- *Necator americanus*

Pathway

- Once the larvae enter the skin, the patient may develop itching, erythema, and swelling at the entry site.
- Later, the larval site may develop papular and erythematous eruptions.
- As the larva migrates to the small intestine, the lesion disappears; this usually happens in 1–2 weeks.
- The papular lesions are seen more commonly with *Necator americanus* (details in Chapter 24).

Clinical features

- The disease starts with a serpiginous, elevated rash due to the movement of the larva.
- **Ground itch**—the rash progresses upwards slowly at the rate of 1–2 cm a day and is accompanied by intense itching and urticaria.
- This rash could get secondarily infected and may present as an infected ulcer or papule.
- No visceral involvement is seen as the larva is not able to penetrate the basement membrane of the skin.

Laboratory diagnosis

- A diagnosis is made clinically by observing the characteristic rash and elevated track. However, it has to be differentiated from larva currens.

- A skin biopsy may reveal the larvae within a circular canal with secondary changes and inflammatory infiltrates. However, this method is not very sensitive and is not performed routinely.
- Eosinophilia may be present.

Treatment

- The disease is usually self-limiting.
- The topical application of thiabendazole may be prescribed. In severe cases, albendazole should be given.
- Routine deworming of pets and persons at risk is recommended.

Larva currens

- The filiform larvae of *Strongyloides stercoralis* can cause a pruritic maculopapular rash which could present as rapidly migrating linear urticaria.
- The larva produces petechial hemorrhages accompanied by intense pruritis and edema at the site of entry. This could be on the feet or in the perianal region, upper thighs, or buttocks, as in the case of autoinfection.
- In contrast to cutaneous larva migrans, these larvae move fast (2–10 cm/hour). The condition is called **larva currens or racing larva** (details in Chapter 24).

Guinea worm disease or dracunculiasis

- This disease is caused by a somatic nematode called *Dracunculus medinensis*.
- A total of 187 countries including India have been declared free from this disease. Only two countries—Angola and Chad—reported cases in 2019 (a total of 19 cases). Ethiopia, Sudan, and Mali have reported cases in dogs. This is a concern as the animals can act as natural reservoirs in nature.

Agent—*Dracunculus medinensis*

- *Dracunculus medinensis* is the largest nematode to infect humans.
- The fully developed adult female can measure up to 1–8 meters in length and 2 mm in diameter. Males are only 2 cm long.

Life cycle (*Fig. 29.30*)

Definitive host: Human being
Intermediate host: *Cyclops*
Infective form: Stage-3 larva (L3) found inside *Cyclops*

- The infection is acquired by **drinking water containing minute** *Cyclops* or by **consuming fish that have ingested** *Cyclops*.
- Once ingested, these *Cyclops* are digested in the stomach to release the L3 larvae.
- These larvae penetrate the intestinal wall and travel to the thoracic muscles, where they moult twice to mature into adult forms.
- Gravid females travel through the subcutaneous tissue to finally reach below the skin to areas that are closer to the ground such as the ankle, feet, and lower limbs.
- After a year of infection, a painful papule develops on the lower extremities, which later develops into blisters from which the worm can emerge.

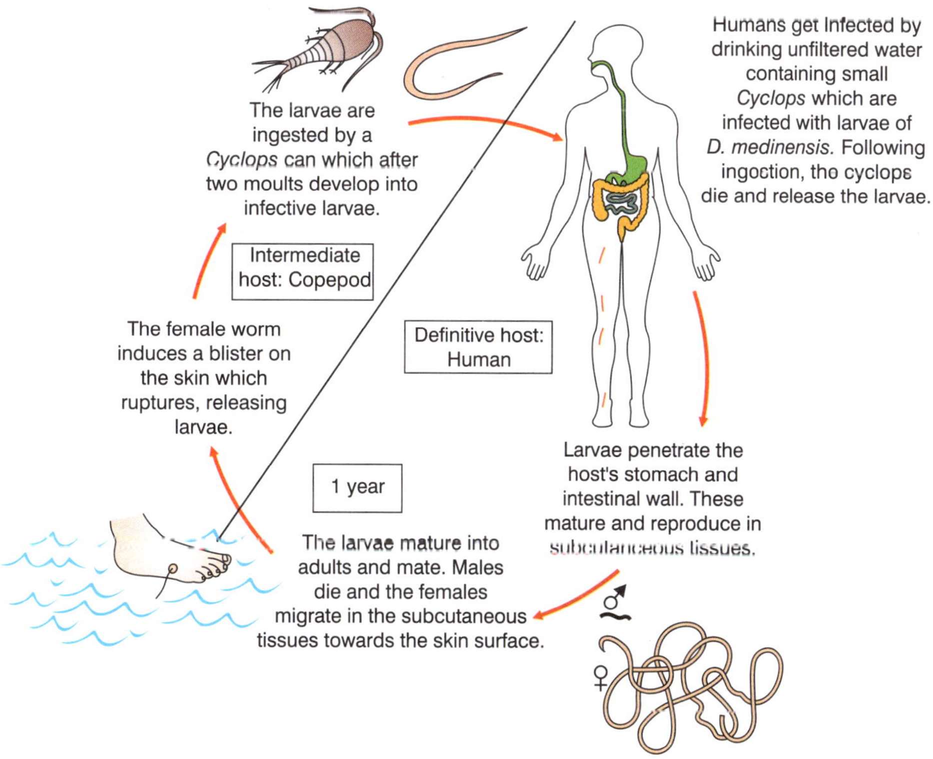

Fig. 29.30 Life cycle of *Dracunculus medinensis*

- When the blister comes in contact with water, the female prolapsed uterus contracts, releasing a large number of stage-1 larvae.
- The stage-1 larvae are motile and are ingested by *Cyclops*.
- They moult twice in the next two weeks and are once again infective to humans.

Clinical features

- The symptoms appear only after one year.
- Initially, a small papule is formed on the lower extremities associated with intense pruritis.
- The papule ruptures into a blister. Fever, nausea, and urticaria may be present.
- The blister is painful and erythematous, and the female worm may be seen emerging from it (Fig. 29.31).

Laboratory diagnosis

- This is done clinically by visualising characteristic lesions on the lower extremities in endemic areas.
- The adult worm can be demonstrated in the blister.
- The larval forms can be seen by dipping the affected area in water and then examining the water by microscopy.

Treatment

- No standard treatment is available.
- The worm can be slowly extracted over 10–15 days by wrapping it around a small stick.
- Proper wound toilet and topical antibiotics may be applied to the lesion.

Guinea worm eradication

- It was achieved in India in the year 2000.
- This was done through rigorous case finding, vector control, and advocating the use of cloth strainers for drinking water.

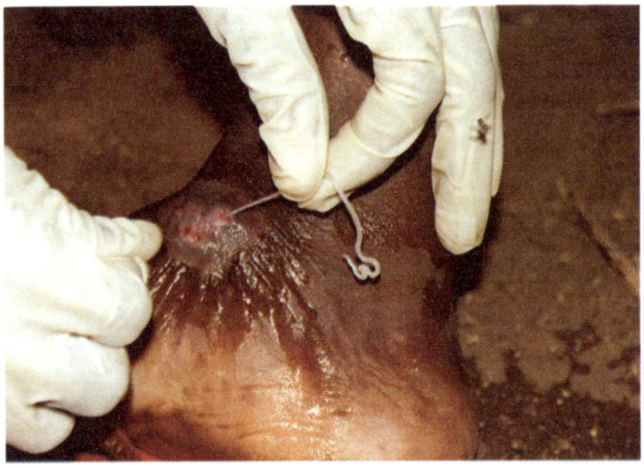

Fig. 29.31 After the rupture of the blister, the worm emerges as a whitish filament in the center of a painful ulcer which is often secondarily infected (*Source*: CDC, PHIL, Image ID 8212)

Trichinellosis

- Trichinellosis is a zoonotic infection and is primarily seen in pigs and some carnivores.
- Humans are accidental hosts.
- Many species of *Trichinella* are maintained in nature in domestic and wild animals.
- Trichinellosis in human beings is caused by *Trichinella spiralis*, which is a nematode.
- Human infections are chiefly seen in the pork-eating population of the world. The infection has been rarely reported from India.

Agent—*Trichinella spiralis*

Trichinella spiralis is the smallest nematode, measuring about 1.5–3 mm. Females are longer than males.

Life cycle (See Chapter 24)

- Humans acquire infection by eating pork that contains stage-1 larva (L1).
- The L1 larva moults four times to develop into an adult worm in the intestinal mucosa.
- After mating, the adult female releases larvae (L1) by the 5th day of infection.
- These larvae penetrate the intestinal mucosa and travel to the skeletal muscles via the lymphatics and blood over the next 4–8 weeks.
- The muscle cell develops into a cyst containing larvae.
- The fully developed muscle cysts are 400–260 μm in diameter and enclose a coiled larva that remains viable for years in animals; humans, however, are the dead end of the life cycle.
- The cyst may undergo fibrosis and calcify.

Clinical features

- Initially, only non-specific intestinal symptoms such as diarrhea, abdominal pain, nausea, or vomiting may be present.
- Later, the migration of the parasite evokes a hypersensitivity reaction with fever, eosinophilia, periorbital edema, or urticarial rash.
- Penetration and migration of the larva can elicit an immune-mediated inflammatory reaction and eosinophilia.
- In severe cases, myocarditis, encephalitis, and thromboembolic disease can also occur.
- Muscle pain and weakness may be present after the encystment stage.
- Extraocular muscles followed by the muscles in the upper trunk are the most common sites of encystment.
- The disease can be fatal in 4–8 weeks.

Laboratory diagnosis

- **Microscopy:**
  - Larvae can be seen in muscle biopsies in H & E-stained sections (Fig. 29.32).

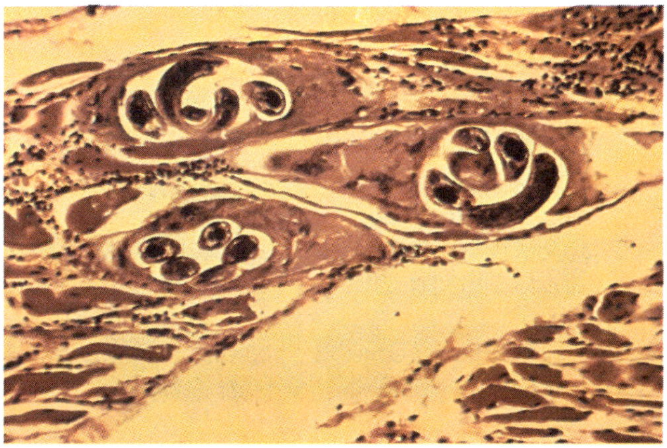

Fig. 29.32 Larvae of *Trichinella* in tissue sections (*Source*: CDC, PHIL, Image ID 5234)

  – Alternatively, the biopsy material can be pressed between slides to visualise the larva.
- **Antibody detection:**
  – *Trichenella*-specific antibodies can be detected by ELISA.
  – The excretory–secretory antigen is used for detecting IgG in the serum of infected persons, but may be negative for 3–5 weeks of infection.
  – IgG levels peak by 2–3 months post-infection and then slowly decline over several years.
- **Antigen detection:**
  – Larval somatic antigen can be detected in stool samples by ELISA.
  – The test can detect antigen from the first day of infection and for about three weeks.

Treatment

- Mebendazole or albendazole can kill the worm in the intestine.
- The drug may not be able to effectively eliminate the muscle forms.
- Trichinellosis can be prevented by eating well-cooked meat and maintaining food hygiene.

Subcutaneous filariasis

Tissue nematodes are long, slender worms with tapering ends. The adult forms are found in subcutaneous tissues and the serous cavity. Subcutaneous filariasis is caused by the following members of this group:

- *Onchocerca volvulus*
- *Loa loa*
- *Mansonella streptocerca*

Onchocerciasis/River blindness

Onchocerciasis is a skin and eye disease caused by *Oncocerca volvulus* River blindness is seen in areas of Africa, Yemen, and parts of Central and South America.

Agent—Oncocerca volvulus

- The adult worms are thin and long with tapering ends.
- Females measure around 50 cm in length, while males are smaller, about 5 cm.

Life cycle

- The vector is the black fly of the genus *Simulum*.
- These flies require the free flow of water (like in rivers) to breed and hence, the name river blindness.
- The larval forms enter via the bite of the black fly and are taken up by the dermis.
- Once there, they develop into adult worms that travel to subcutaneous tissues.
- Microfilariae are released into the blood, from where they are taken up by the blackfly in its meal.
- Microfilariae are found in the dermis or the fluids of the eye, while the adult worm resides in the subcutaneous nodules.

Clinical features

The main lesions of onchocerciasis are as follows:

- Dermatitis (onchodermatitis)
- Skin nodules
- Eye lesions

In the initial stages, the patient is asymptomatic. Later, dermatitis, subcutaneous nodules, lymphadenitis, and ocular complaints develop.

- Skin lesions such as rashes, hyper-pigmentation, atrophic changes, and micro-abscesses are seen.
- Intense itching occurs, and lesions may get secondarily infected with bacteria. Chronic dermatitis causes patches of de-pigmentation (known as **leopard skin**).
- Later, there is loss of elastic fibers and changes in skin pigmentation, resulting in a **hanging groin**.
- The adult worms in subcutaneous lesions result in granulomas.
- These subcutaneous nodules are painless, non-tender and firm, less than 2 cm in diameter, mobile, and are commonly seen on the trunk.
- In the early stages of ocular presentation, conjunctivitis with photophobia is seen. Later, blindness may ensue due to the accumulation of microfilariae (mf) in the vitreous humor or keratitis.

Laboratory diagnosis

This is based on the detection of the adult worm in the biopsy of nodules or the visualisation of microfilariae in skin or nodules (Fig. 29.33).

Treatment

Ivermectin is the drug of choice along with surgical excision of disfiguring nodules.

Loiasis

- Loiasis, also called the African eye worm, is caused by the parasitic worm *Loa loas*.

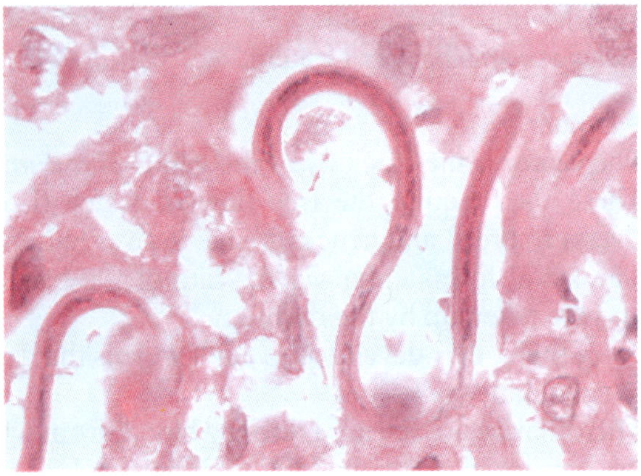

Fig. 29.33 Nodule with *Onchocerca* microfilariae

- It is transmitted to humans by bites of deerflies (also known as mango flies or mangrove flies) of the genus *Chrysops*.
- The flies that pass on the parasite breed in certain rain forests of West and Central Africa.

Loa loa

- This filarial worm is restricted to West and Central Africa. It causes subcutaneous swellings and eye lesions.
- These are small worms; the females about 50–70 mm long while males are 30–35 mm long.
- The microfilariae are sheathed and have a blunt tail tip with coarse, sparsely-placed nuclei along the length of the tail tip.

Life cycle

- The life cycle is similar to that of *W. bancrofti* except that the vector is a deerfly (*Chrysops* species).
- Larval forms injected by the bite of the fly develop into adults, which then travel to the subcutaneous tissues and eye.
- Adults live for 15–17 years, releasing microfilariae. Diurnal periodicity is seen with a peak between 12 noon and 2 pm.

Clinical features

- Clinically, lesions can develop in the eye or subcutaneous tissues.
- The ocular manifestations can be in the form of swelling of the eyelids, proptosis, and conjunctival granuloma.
- Subcutaneous lesions are commonly found on the extremities such as the knee or wrists. These are also known as **Calabar swellings or fugitive swellings**.
- These lesions occur as a reaction of the host immune response to the migrating worm in subcutaneous tissues.
- It starts within few hours of entry of the larva, and is characterised by local pain and pruritis, which typically lasts for 3–4 days.
- Rarely, neurological symptoms may develop, especially in patients treated with diethylcarbamazine (DEC) (Table 29.17).

Laboratory diagnosis

- Diagnosis is based on the demonstration of microfilariae in peripheral blood.
- The adult worm can be demonstrated in the eye or a biopsy of subcutaneous nodules.
- Eosinophilia, elevated IgE, and hypergammaglobinemia may be present.
- **Mazzoti skin test:**
  - In a mild infection, the topical application of DEC causes erythema and itching.
  - This is a reaction to a dead worm present below the skin and hence, this test must not be performed in heavy infections.

Table 29.17 Characteristics of filarial worms causing skin and subcutaneous infections

Characteristics	*Onchocerca volvulus*	*Loa loa*	*M. streptocerca*	*M. perstans*	*M. ozzardi*
Vector	*Simulum* (Black fly)	*Chrysops* (deerfly/ mango fly/red fly)	*Culicoides* (midges)	*Culicoides* (midges)	*Culicoides* (midges), *Simulum* (black fly)
Geographic distribution	Sub-saharan Africa, Yemen, parts of South and Central America	West and Central Africa	West and Central Africa	South and Central America and Africa	South and Central America
Microfilaria location	Skin	Blood, eye	Skin	Blood	Blood
Microfilaria morphology	Unsheathed	Sheathed	Unsheathed	Unsheathed	Unsheathed
Somatic cell/nuclei	Coarse, sparsely placed nuclei with no nuclei in tail tip	Coarse, sparsely placed nuclei till tail tip	Fine, diffusely scattered nuclei till tail tip	Overlapping nuclei till tail tip	Fine, diffusely scattered nuclei with no nuclei in tail tip
Tail end	Pointed tail tip	Blunt tail tip	Blunt, curved tail tip	Blunt tail tip	Hooked, pointed tail tip

Treatment

DEC or albendazole or ivermectin are available for treatment. Surgical removal of the worm or excision of the nodule is required.

Mansonellosis

Mansonellosis refers to diseases caused by the filarial nematodes *Mansonella perstans*, *M. ozzardi*, and *M. streptocerca*. These species vary in their geographic occurrence and localisation within the host. *M. ansonella ozzardi* and *M. perstans* cause **serous cavity filariasis.** The infection is transmitted by biting midges or blackflies.

Agent—*Mansonella streptocerca*

- The parasite is transmitted by biting midges (*Culicoides*).
- Adults live in the skin and subcutaneous tissues.
- Its life cycle is similar to that of other filarial worms.
- **Clinical features** include lymphadenopathy, pruritis, and hypopigmented macular patches on the trunk.
- These have to be distinguished from leprosy and granuloma multiforme.
- The majority of cases may be asymptomatic.

Agent—*M. perstans*

- The vectors are biting midges.
- The adult worms reside in body cavities and associated connective tissue.
- Microfilariae are found circulating in the blood.
- The life cycle of *M. perstans* is similar to that of other filarial worms.
- Most infections are asymptomatic.
- Complaints such as pruritus, conjunctival granulomas, joint pains, periorbital edema, angioedema, and rarely, pericarditis have been reported.
- The diagnosis can be made by detecting microfilariae in thc blood.

Agent—*M. ozzardi*

- The parasite is transmitted by midges and blackflies.
- Adult worms live in serous cavities and associated connective and subcutaneous tissues.
- The majority of cases are asymptomatic; others may present with symptoms similar to *M. perstans*.

Treatment of *Mansonella* infection

Infection with *Mansonella* species can be treated with DEC or albendazole. Ivermectin can also be used as it kills microfilariac.

Dirofilariasis

Human dirofilariasis typically presents as either subcutaneous nodules or lung parenchymal disease. It is a zoonotic disease caused by *Dirofilaria immitis* and *Dirofilaria repens*. The usual hosts of these infective nematodes are **domestic and wild carnivores**.

Agent—*Dirofilaria immitis*

- The adult females are usually 20–30 cm long by 350 µm wide; males are smaller than females (Fig. 29.34).
- Adults can live for 5–10 years.

Life cycle (Fig. 29.35)

Dirofilaria immitis or dog heartworms cannot complete their life cycle in humans, who are generally accidental hosts.

Intermediate host: Mosquito

Definitive host: Dog

- The infected mosquito introduces third-stage filarial larvae of *Dirofilaria* onto the skin of the definitive host, where they penetrate the bite wound. In the definitive host, the L3 larvae undergo two more molts into L4 and adults, the latter of which reside in subcutaneous tissues.

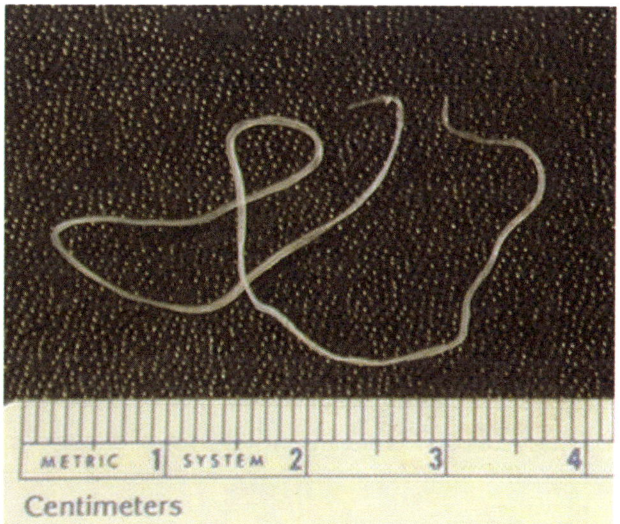

Fig. 29.34 Adult *Dirofilaria* (*Source*: https://www.cdc.gov/dpdx/dirofilariasis/images/1/Dirofilaria_eye_BAM1.jpg)

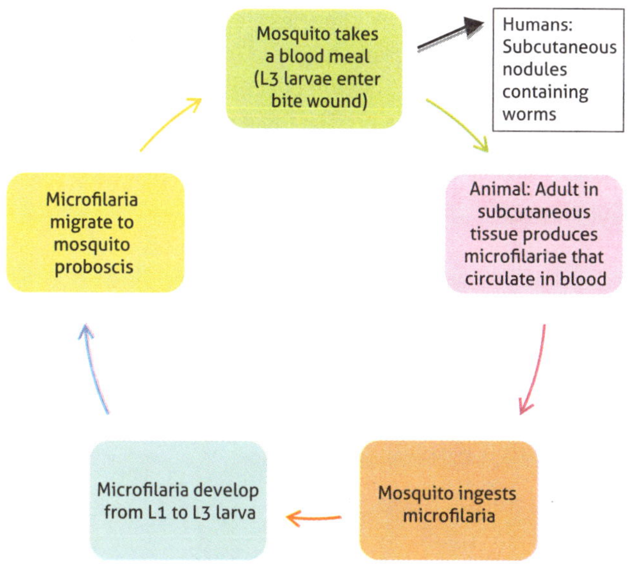

Fig. 29.35 Life cycle of *Dirofilaria*

- Females produce microfilariae that are released into the peripheral blood. A mosquito ingests the microfilariae during a blood meal.
- After ingestion, the microfilariae migrate from the mosquito's midgut, where they develop into first-stage larvae image and subsequently into third-stage infective larvae.
- The third-stage infective larvae migrate to the mosquito's proboscis and can infect another definitive host when the mosquito takes a blood meal.

Pathogenicity

- These parasites are present in granulomas or sometimes, calcified lesions.
- The lesions in the lung may be confused with tuberculosis malignancy (coin lesion).
- Allergic symptoms and eosinophilia may be present.

Cestodes Affecting the Skin

Cercarial dermatitis

- In schistosomiasis, invading cercariae can lead to a transient allergic rash that can last for 1–5 days.
- The symptoms are related to the skin in acute conditions.
- **Swimmer's itch**—the cercaria penetrates the skin, causing a maculopapular, itchy rash. This is seen in all *Schistosoma* infections. It is seen mostly with *Schistosoma japonicum* and rarely, with *S. haematobium*.
- The non-human schistosome species can also cause cercarial dermatitis.
- **Diagnosis** is clinical, based on a history of exposure to infected water (swimming or standing in water).
- The lesions are self-limiting and subside after few days.

Trematodes Affecting the Skin

- Several tapeworm species cause subcutaneous lesions in their larval stages.
- *Taenia solium*, *Echinococcus granulosus*, and *Spirometra* species are the most common tapeworms involved.

Cysticerci

- ***T. solium*** may be present in the subcutaneous tissues and are seen as small lumps under the skin.
- **Diagnosis** is by microscopy from biopsy or by visualisation of calcified cyst on X-ray, CT scan, or MRI. Serology is also useful in diagnosis.
- **Treatment** is with corticosteroids along with anthelmintic treatment.

Subcutaneous hydatid cysts

- These are relatively rare.
- They cause less discomfort and are seen as small swellings under the skin.
- Accidental rupture can result in anaphylaxis.
- Diagnosis is clinical, radiological, or serological.
- Hydatid cysts should never be aspirated or biopsied.
- Complete surgical removal is recommended with medical treatment with albendazole for 28 days.

ECTOPARASITES

- These parasites do not invade the system, but affect the skin, causing superficial infection/infestation, e.g., scabies by *Sarcoptes scabiei*.
- Ectoparasites may transmit other microbial infections, in which case, they act as vectors (Table 29.18).

Table 29.18 Ectoparasites causing human disease

Parasite	Microorganism transmitted	Disease caused
Sarcoptes scabii	Causes the disease	Scabies
Lice (*Pediculus humanus capitis/ corporis)*	*Borrelia recurrentis*	Relapsing fever
Ticks (*Ixodes*)	*Borrelia burgdorferi*	Lyme disease
Flea (*Xenopsylla cheopis*)	*Y. pestis*	Plague
Leptotrombidium	*Orientia tsutsugamushi*	Scrub typhus

FUNGAL INFECTIONS OF THE SKIN AND ITS APPENDAGES

Fungi affecting the skin and appendages may be grouped us follows:

- Superficial mycoses
- Subcutaneous mycoses

SUPERFICIAL MYCOSES

- These are a group of infections caused by fungi that live exclusively on the dead layers of the skin and its appendages.
- They have no contact with living tissue and, hence elicit no inflammatory response; the only changes produced are cosmetic effects. The following skin manifestations are produced by members of this group:
- Tinea (pityriasis) versicolor
- Tinea nigra
- Piedra
- Dermatophytosis (skin, hair, and nails)

Pityriasis Versicolor (Tinea Versicolor)

- This is a chronic, usually asymptomatic, involvement of the **stratum corneum**.
- The lipophilic, yeast-like fungus, *Malassezia furfur* (formerly known as *Pityrosporum orbiculare*), is the causative agent.

Epidemiology

The disease is worldwide in distribution but is particularly prevalent in the tropics. It occurs mainly in young adults.

Clinical features

Characteristic discrete or confluent macular areas of discolouration or depigmentation develop on the skin of the chest, abdomen, upper limbs, and back.

- The fungus may be demonstrated on normal skin.
- They have been found in human dandruff and seborrheic dermatitis.
- The skin of most, if not all, warm-blooded animals is covered with *Malassezia furfur*, and while they mostly co-exist with their hosts, they can cause serious diseases in humans and other animals; the disease may also be considered an opportunistic infection (Fig. 29.36a and b).

Laboratory diagnosis

- **Microscopy:** The examination of skin scrapings shows an abundance of yeast-like cells and short, branched filaments (spaghetti and meatball appearance; Fig. 29.36b).
- **Culture:** The fungus can be grown on Sabouraud agar, covered with a layer of olive oil at 30–35°C. The colonies appear pale yellowish-brown to cream-coloured in 5–6 days.

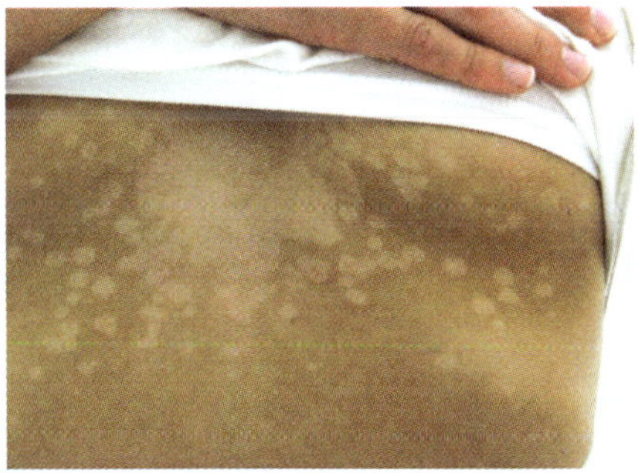

(a)

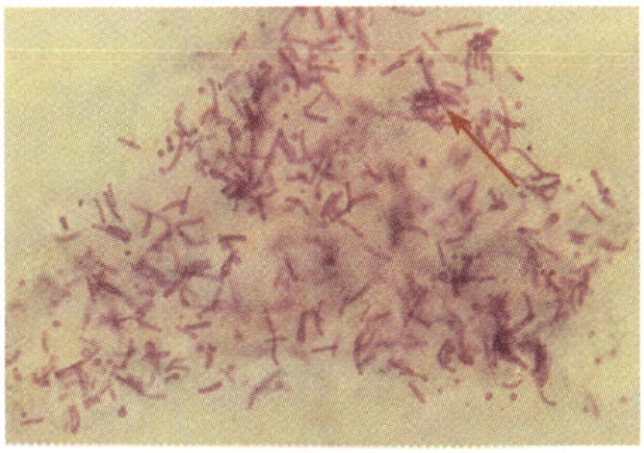

(b)

Fig. 29.36 (a) Pityriasis versicolor—macular area of discolouration on the upper chest wall and (b) photomicrograph of a skin scale tissue sample showing the presence of numerous *Malassezia furfur* yeast and short filamentous forms (spaghetti and meatball appearance) (*Source*: [a] Flickr and [b] CDC, PHIL, Image ID 2916)

Tinea Nigra

- This is a localised infection of the stratum corneum, particularly of the palms, producing black or brownish macular lesions.
- It is caused by *Exophiala werneckii* (formerly, *Cladosporium wernickii, Hortaea werneckii*) and *Exophiala castellanii*.

Epidemiology

This disease mainly occurs in the tropics.

Laboratory diagnosis

Skin scrapings show brownish, branched, septate hyphae and budding cells. Colonies on Sabouraud agar are grey or black.

Piedra

- This fungal infection of the hair is characterised by the presence of firm, irregular nodules along the hair shaft; these nodules are composed of fungal elements cemented together on the hair.
- Two varieties of piedra are recognised—black piedra (Fig. 29.37) caused by *Piedraia hortae* and white piedra caused by *Trichosporon beigelii*.

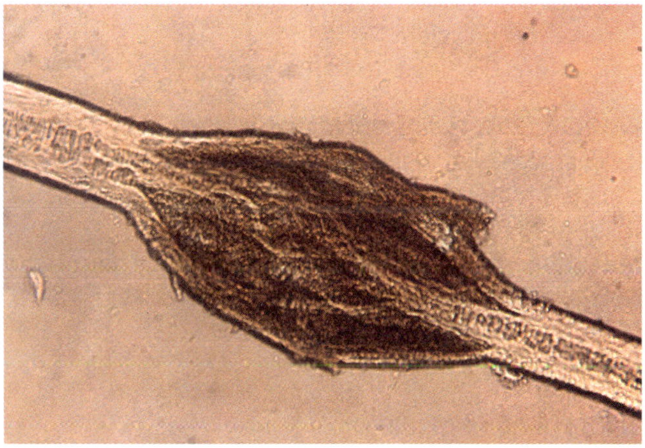

Fig. 29.37 Photomicrograph reveals pathologic morphology displayed by a hair shaft in black piedra, also known as trichosporosis, caused by *Piedraia hortae* (*Source*: CDC, PHIL, Image ID 12532)

Dermatophytosis

> **Case 6.** *A nine-year-old boy presented to the dermatology OPD complaining of a change in the colour of his toenails for about four months. He had no history of trauma or eczema of the foot. On examination, onycholysis (separation of the nail plate from the nail bed) of multiple toenails was observed. The nails were clipped, and a 20% potassium hydroxide preparation on microscopy revealed the presence of thin, septate hyphae. Culture on Sabouraud dextrose agar containing cycloheximide showed the growth of mycelial fungi, which on lactophenol cotton blue (LCB) mount, revealed the presence of slender, septate hyphae with a few pencil-shaped macroconidia and many oval microconidia. This was identified as* ***Trichophyton***

rubrum (the most common pathogen in onychomycosis). The patient was treated with oral griseofulvin and recovered.
i. What are dermatophytes?
ii. What are the pathogenic species and how are they diagnosed?

Dermatophytosis (commonly called tinea or ringworm) refers to an infection of keratinised structures (the skin, hair, and nails) caused by a group of **keratinophilic fungi called the dermatophytes.**

- The infections caused may be acute or chronic (persistent dermatophytosis that runs a chronic course with episodes of remission and exacerbation).
- The term dermatophytosis should not be confused with dermatomycosis, which refers to skin lesions produced by other fungi such as *Candida albicans* and also the cutaneous manifestations of systemic mycoses.

Characteristics of dermatophytes

- Dermatophytic fungi are hyaline filamentous fungi that digest keratin by enzymatic means but are unable to invade living tissue.
- Dermatophytes digest keratin by keratinases and are resistant to cycloheximide.
- They are classified into three genera (Table 29.19).

Clinical features of dermatophytosis

Dermatophytosis is twice as common in males as in females. Clinical classification is according to the anatomic site involved.

- Lesions on the skin tend to be circular, dry, erythematous, scaly, and itchy.
- Lesions of the hair include **kerion** (Fig. 29.38), scarring, and alopecia.

Nails (Tinea Unguium)

- Fingers and toenails infected by dermatophytes are deformed, friable, and discoloured, and there is accumulation of debris under the nails.

Table 29.19 Species of various genera of dermatophytes

Genus	Important species causing human disease
Microsporum	*M. gypseum*
	M. canis
	M. nanum
	M. equinum
Trichophyton	*T. rubrum*
	T. mentagrophytes
	T. verrucosum
	T. tonsurans
	T. violaceum
Epidermophyton	*E. floccosum*

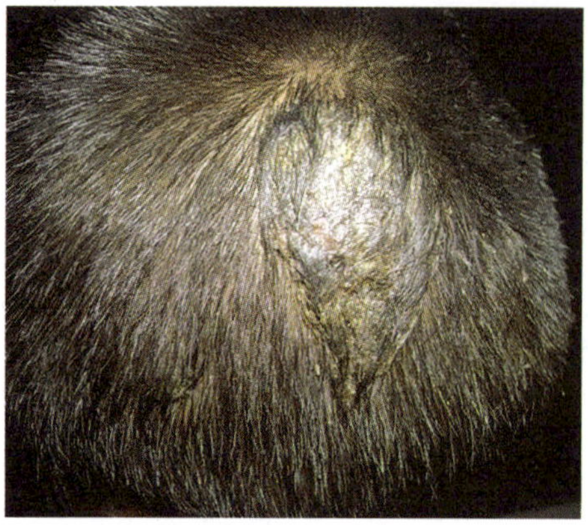

Fig. 29.38 Kerion—boggy lesions on the scalp (*Source*: Dr Sheela Kuruvila, Department of Dermatology, Pondicherry Institute of Medical Sciences, Puducherry)

- In lesions, dermatophytes appear as hyphae and arthrospores.

Table 29.20 lists the clinical types of dermatophytoses and their common causative agents.

Epidemiology

- Dermatophytoses occur throughout the world, but certain types of the disease and some species of fungi are geographically restricted.
- Social and cultural patterns also influence dermatophytoses.
- Many factors such as age, hormones, and intercurrent diseases affect the susceptibility of an individual to dermatophytoses.

Habitat

Depending on their natural habitat, dermatophytes may be classified as anthropophilic, zoophilic, and geophilic species.

Anthropophilic dermatophytes

- Human beings are the main or only hosts.
- Examples are *T. rubrum, E. floccosum,* and *M. audouinii* are examples.
- They cause mild but chronic lesions.

Zoophilic species

- These are natural parasites of animals.
- Examples incluce *T. verrucosum* in cattle and *M. canis* in dogs and cats.
- Human infections with zoophilic dermatophytes cause severe inflammation but are readily curable.

Geophilic species

- These occur naturally in soil, are relatively less pathogenic for human beings.
- Examples include *M. gypseum* and *T. ajelloi*.

Pathogenicity

- Mechanisms of pathogenesis are unclear.

Table 29.20 Clinical types of dermatophytoses and their common causative agents

Disease	Common causative agents	Involved area and main features
Tinea capitis	*Microsporum* (any species) *Trichophyton* (most species)	Scalp
Favus	*T. schoenleinii, T. violaceum, M. gypseum*	Chronic type of ringworm in which dense crusts (scutula) develop in the hair follicles, leading to alopecia and scarring
Tinea barbae	*T. rubrum, T. mentagrophytes, T. verrucosum*	Bearded areas of the face and neck
Tinea imbricata	*T. concentricum*	Extensive concentric rings of papulosquamous, scaly patches
Tinea corporis (Fig. 29.39)	*T. rubrum* and any other dermatophyte	Smooth or non-hairy skin of the body
Tinea cruris	*E. floccosum, T. rubrum*	Groin and perineum
Tinea pedis (athlete's foot)	*T. rubrum, E. floccosum*	Foot
Tinea manuum	*T. rubrum* and *T. mentagrophytes*	Hand
Tinea unguium	*T. rubrum, E. floccosum* and *T. mentagrophytes*	Nails
Ectothrix hair infection	*Microsporum* species, *T. rubrum, T. mentagrophytes*	Hair
Endothrix hair infection	*T. schoenleinii, T. tonsurans, T. violaceum*	Hair

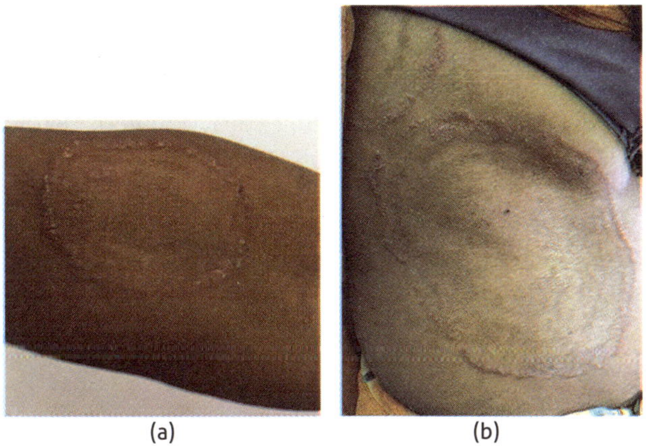

Fig. 29.39 (a) and (b) Tinea corporis--circumscribed lesions on non-hairy parts of the body (*Source*: Dr Sheela Kuruvila, Department of Dermatology, Pondicherry Institute of Medical Sciences, Puducherry)

- Fungal products may be responsible for inciting local inflammation.
- **Dermatophytids** (or the 'Id' reaction)—hypersensitivity to fungal antigens results in sterile vesicular lesions sometimes seen in sites distant from the ringworm. The reaction may follow oral antifungal therapy and can be confused with an allergic drug reaction.

Laboratory diagnosis

Diagnosis is established based on clinical features, the use of a Wood's lamp where applicable and by laboratory investigations.

Specimens

- Scrapings of the skin (from the edges of ringworm lesions) and nail and hair clippings (hair plucked from the scalp) are collected.
- In suspected tinea capitis, fungal elements are looked for in plucked hair.

Wood's lamp examination for detecting superficial fungal infections

This test is performed in a dark room by shining an ultraviolet light on the area of interest. A Wood's lamp allows only the long ultraviolet rays to pass (peak at 365 nm) through a nickel oxide filter (barium silicate with nickel oxide). Wood's lamp examination is useful for the diagnosis of certain ectothrix infections of the skin or scalp. Normally, the skin does not fluoresce or shine under ultraviolet light. However, if the region of the skin on which the Wood's lamp light is focussed is infected, that area of the skin will fluoresce. In particular, if the skin or hair is infected by the *Microsporum* species (e.g., *M. canis, M. audouinii*), the infected area will fluoresce a bright greenish-yellow. *Malassezia* infections (pityriasis versicolor) show golden-yellow fluorescence. A Wood's lamp reveals pteridine fluorescence of the infected hair.

Microscopic examination

- **KOH mount**—a wet preparation of the specimen is made by placing the scrapings in a drop of 10–20% potassium hydroxide (KOH) on a slide, which is then covered with a coverslip and left for 10–20 minutes to digest the keratin. Additional time may be required to digest nails. Digestion of keratin ('clearing') is aided by gently warming the slide.
- The presence of branching, hyaline (non-pigmented), and septate hyphae is considered positive for fungi; spores may sometimes be seen (Fig. 29.40).
- Two types of hair infection may be distinguished in wet mounts (Fig. 29.41a, b, and c)—**ectothrix**, in which arthrospores are seen as a sheath surrounding the hair, and **endothrix**, in which the spores are inside the hair shaft.

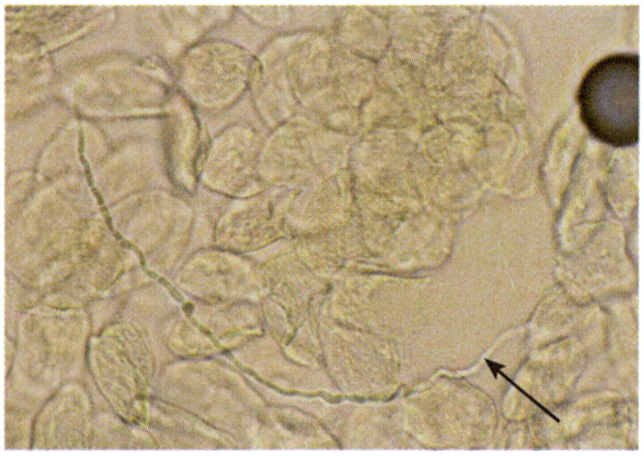

Fig. 29.40 Dermatophyte hyphae in skin scraping, KOH mount

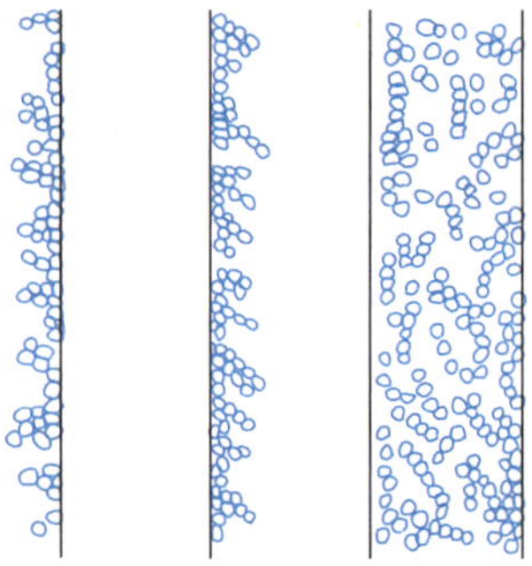

(a)

(b)

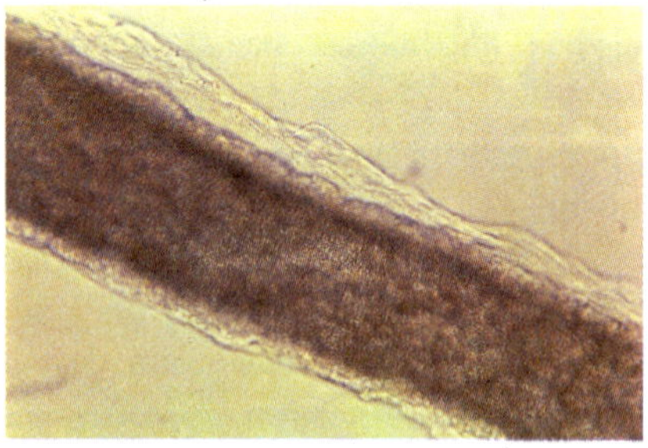

(c)

Fig. 29.41 (a) Ectothrix and endothrix types of hair infection and (b and c) photomicrographs of ectothrix and endothrix respectively (*Source*: [b and c] CDC, PHIL, Image IDs 16625 and 22026)

Hair perforation test

- This test is performed to differentiate *Trichophyton rubrum* from *Trichophyton mentagrophytes*.
- A few strands of human hair are placed in a petri dish with 20 mL of distilled water, and autoclaved.
- A few drops of sterile 10% yeast extract and a few fragments of the test fungus are added to the hair strands.
- These are incubated at 25°C for 2–3 weeks.
- The hair is then removed and examined microscopically in a lactophenol cotton blue-stained wet preparation.
- *T. mentagrophytes* causes surface erosion of the hair shaft, resulting in a wedge-shaped appearance of the shaft.

Culture

- Species identification is possible only by the examination of the culture.
- **Sabouraud dextrose ag**ar containing chloramphenicol (the antibiotic suppresses the growth of the contaminating bacteria) and cycloheximide (to suppress bacteria and non-dermatophytic filamentous fungi) and incubated aerobically at 25–30°C for up to 21 days.
- Dermatophytes are identified in the laboratory by examining the macroscopic characteristics of the fungal colonies (rate of growth, texture, and colour on the obverse and reverse; Tables 29.21 and 29.22).
- Material teased out from the fungal colony (culture mount) is stained with lactophenol cotton blue and observed under the microscope to study the hyphae and conidia. The condia (asexual spores) are of two types:
  - **Microconidia:** Small, unicellular
  - **Macroconidia:** Larger, septate

Hypersensitivity (Dermatophytid Reaction)

It can be demonstrated by skin testing with the fungal antigen, trichophyton.

Treatment

- Topical agents (ointments or gels) containing azoles (miconazole, clotrimazole, econazole) or terbinafine are the mainstay of treatment.
- Oral griseofulvin (100 mg thrice daily), azoles (ketoconazole, itraconazole), or terbinafine can be used.
  - **Mild infections:** Topical imidazole (clotrimazole or miconazole)
  - **Severe infections:** Oral griseofulvin for 4–6 weeks
  - **Hair involvement:** Treatment is for 3–6 months
  - **Nail involvement:** Treatment is for one year
- Alternative treatment—oral imidazole (ketoconazole 200 mg twice daily), triazole (itraconazole 100 mg once daily; fluconazole 150 mg once daily), or terbinafine.

Table 29.21 Culture and microscopic features of three types of dermatophytes

Characteristic	Trichophyton	Microsporum	Epidermophyton
Colonies	Powdery, velvety, or waxy with pigmentation characteristic of different species	Cotton-like, velvety, or powdery with white to brown pigmentation	Colonies are powdery and greenish-yellow
Microconidia	Abundant and arranged in clusters along the hyphae or borne on conidiophores (Fig. 29.42)	Relatively scanty and not distinctive	Absent
Macroconidia	Relatively scanty, thin, elongated, with blunt ends Distinctive shapes in different species Special hyphal characters such as spiral hyphae, racquet mycelium, and favic chandeliers in some species	Large, multicellular, spindle-shaped structures, borne singly on the ends of hyphae	Multicellular, pear-shaped, and typically arranged in clusters (Fig. 29.43)
Infection of tissues	Skin, hair, and nails	Hair and skin, but usually not the nails	Skin and nails but not the hair

Table 29.22 Common characteristics of species of dermatophytes

Species	Colony	Morphology
T. rubrum	Velvety, red pigment on reverse	Few, long, pencil-shaped macroconidia
T. mentagrophytes	White to tan, cottony or powdery Pigment variable	Clusters of microconidia Cigar-shaped macroconidia with terminal rat-tail filaments
T. tonsurans	Cream or yellow, with central furrows	Abundant microconidia Thick-walled, irregular macroconidia
T. schoenleinii	Smooth, waxy, brownish	Hyphal swellings, chlamydospores, favic chandelier
T. violaceum	Very slow-growing, waxy, violet/purple pigment	Distorted hyphae, conidia are rare
M. audouinii	Velvety, brownish, slow-growing	Thick-walled chlamydospores, conidia rare and irregular
M. canis	Cottony, orange pigment on reverse	Abundant, thick-walled, spindle-shaped macroconidia with up to 15 septae
M. gypseum	Powdery, buff-coloured	Abundant, thin-walled macroconidia with 4–6 septae
E. floccosum	Yellowish-green, powdery	Club-shaped macroconidia in clusters

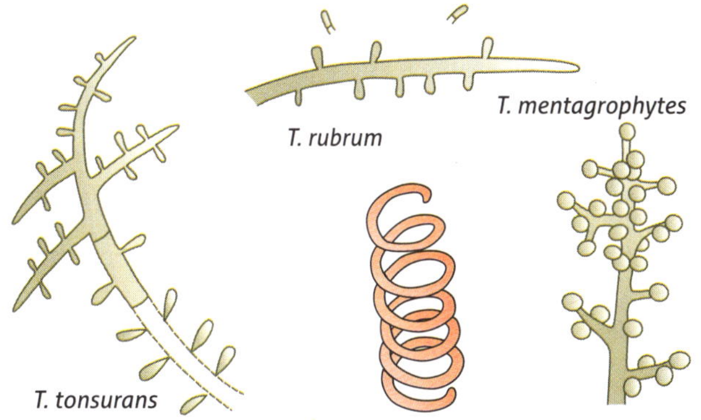

Fig. 29.42 *Trichophyton* species showing typical spiral hyphae and microconidia

Control

Personal cleanliness, avoiding contact with infectious material and effective treatment are the only control measures available.

Onychomycosis

- *Candida* infection of the nail is known as **onychomycosis**.
- This is often seen in persons who frequently come into contact with water and whose hands are often soggy. Details of the yeast-like fungus are dealt with in detail in Chapter 21.

SUBCUTANEOUS MYCOSES

- These are principally seen in tropical and subtropical areas.
- The most frequent predisposing (risk) factor is trauma.
- The disease may remain localised or may spread by contiguity.
- Fungi that cause such subcutaneous mycoses are either normally present in the soil or are pathogens of plants.

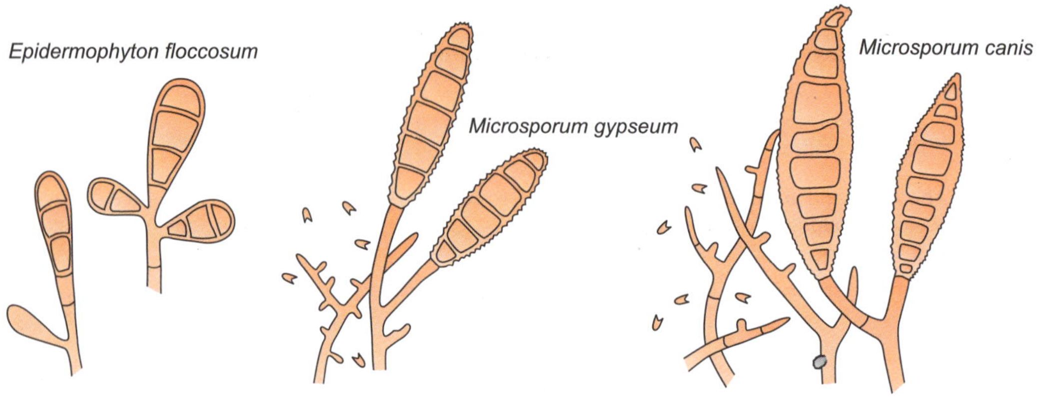

Fig. 29.43 Characteristic macroconidia

- The outcome of such infections depends on fungal virulence and host defences.

The various types of subcutaneous mycoses are as follows:

i. Mycotic mycetoma
ii. Chromomycosis (chromoblastomycosis and phaeohyphomycosis)
iii. Sporotrichosis
iv. Rhinosporidiosis
v. Subcutaneous phycomycosis (now referred to as entomophthoramycosis)

Mycotic Mycetoma

> **Case 7.** *A 45-year-old man, a farmer by occupation, presented to the surgical outpatient department with a history of multiple swellings on his foot and seropurulent pus discharge from the sinuses. The pus was subjected to Gram's stain, KOH preparation, and modified acid-fast stain (to check for a botryomycotic, eumycotic, or actinomycotic causes). KOH preparation revealed the presence of dark, pigmented, fungal hyphae;* ***Madurella mycetomatis*** *was isolated on culture. The patient was treated with oral ketoconazole; debridement of the infected tissue with split graft was also performed (Fig. 29.44).*
> i. What is this condition called?
> ii. What are the etiological agents?
> iii. How is it diagnosed?

- Mycetomas are chronic, slowly progressive, post-traumatic infections of the subcutaneous tissue caused by fungi and some bacteria found in the soil.
- They usually occur on the foot and lower extremities, and rarely infect other parts of the body. Mycetomas are usually caused by fungi but may be caused by bacteria as well. Bacterial causes have been mentioned in the earlier sections of this chapter.
- The infection is commonly known as **Madura foot**.
- Causative agents of eumycetoma: It is caused by fungi, namely *Scedosporium* (*Pseudallescheria*), *Madurella mycetomatis*, *M. grisea*, *Acremonium* spp., *Exophiala* spp., *Aspergillus nidulans*, and *Fusarium* spp.

Clinical features

- The causative agent is believed to enter through minor trauma wounds.
- The disease usually begins as a small, subcutaneous swelling of the foot, which enlarges, burrowing into the deeper tissues and tracking to the surface as multiple sinuses discharging viscid, seropurulent fluid containing granules (Fig. 29.44 and 29.45).
- The lesions are painless (Case 7).

Fig. 29.44 Mycetoma of the foot with multiple swellings and seropurulent pus discharge from the sinuses (*Source*: CDC, PHIL, Image ID 21996)

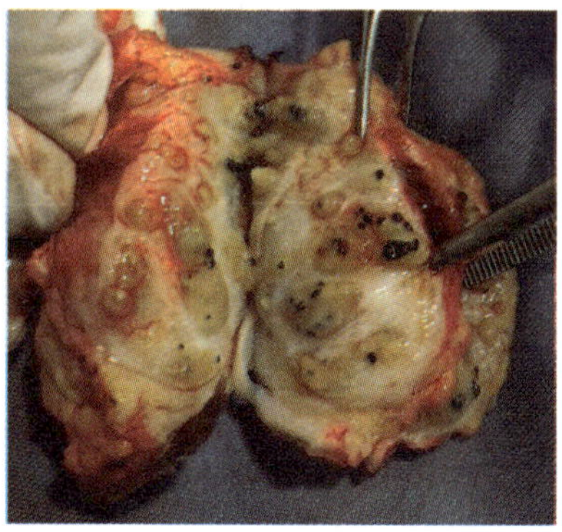

Fig. 29.45 Black granules in tissue from a case of mycetoma (*Source*: Dr Angeline Radjou, Department of Surgery, Pondicherry Institute of Medical Sciences, Puducherry)

Laboratory diagnosis

- **Specimen:** Granules from the sinus tracts on the swelling and biopsy of the lesion are collected.
- **Microscopy:**
  - The 'granules' or 'grains' are microcolonies.
  - The colour and consistency of the grains vary with the different agents causing the disease (Table 29.22).
  - Compared to actinomycetoma, eumycotic lesions are harder, broader, may be pigmented, and often show septae and chlamydospores in KOH preparation.
  - In actinomycotic mycetoma, on Gram stain, the actinomycetes appear as gram-positive, slender rods.
  - On examination of the granules, they are found to be composed of very thin (<1 μm in diameter) filaments, which are soft.
  - The growth of the organisms on SDA at 37°C or room temperature is noted—the gross appearance of the colony, the rate of growth and typical microscopic appearance of hyphae (pigments), conidiogenous cells, conidia and their arrangements help in establishing the diagnosis.

Treatment

Surgery is the mainstay of treatment of eumycetoma.

Medical therapy

- *Scedosporium apiospermum*: Miconazole and ketoconazole
- *M. mycetomatis*: Ketoconazole and griseofulvin
- *M. grisea* and *Fusarium*: Itraconazole
- Resistant lesions may require amputation
- Therapy should be continued for 18–24 months even after the disappearance of clinical or laboratory evidence of infection

Chromomycosis

It refers to a group of clinical manifestations caused by various dematiaceous (pigmented) fungi.

1. Chromoblastomycosis (verrucous dermatitis)

- It is the most common form.
- The disease is chiefly tropical and is more common among agricultural workers who work barefoot and woodcutters.

Agents

- *Fonsecaea pedrosoi, F. compacta*
- *Exophiala dermatitidis*
- *Phialophora (P. verrucosa)*
- *Cladophialophora (C. carrionii)*

Infections caused by *F. pedrosoi* and *P. verrucosa* have been reported to disseminate to other areas, especially the brain.

Clinical features

- The mode of entry is traumatic implantation.
- The disease is usually confined to the subcutaneous tissue of the feet and lower legs.
- The lesion develops slowly around the site of implantation as warty, cutaneous nodules which resemble the florets of a cauliflower (Fig. 29.46).

Laboratory diagnosis

- Histologically, the lesions show the presence of the fungus as round or irregular, dark brown, yeast-like bodies with septae called sclerotic bodies (Fig. 29.47a and b).

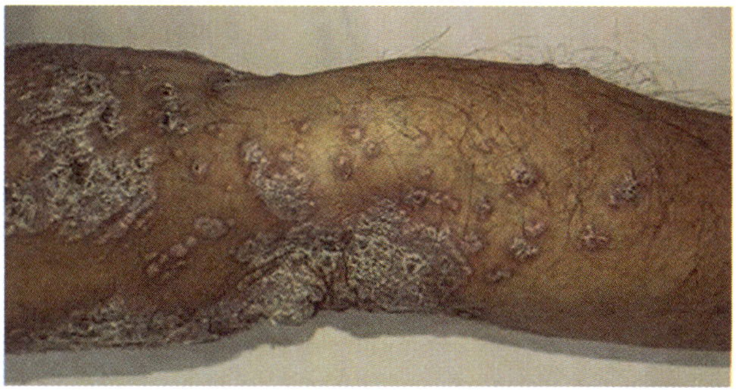

Fig. 29.46 Chromomycosis: Lesions that are warty, cutaneous nodules resembling the florets of a cauliflower (*Source*: Dr Sheela Kuruvila, Department of Dermatology, Pondicherry Institute of Medical Sciences, Puducherry)

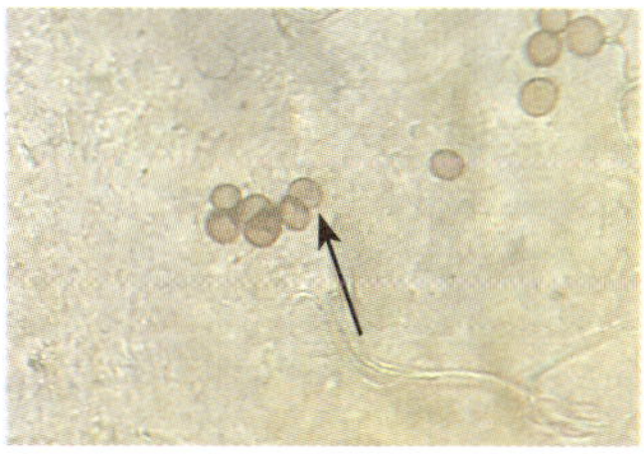

(a)

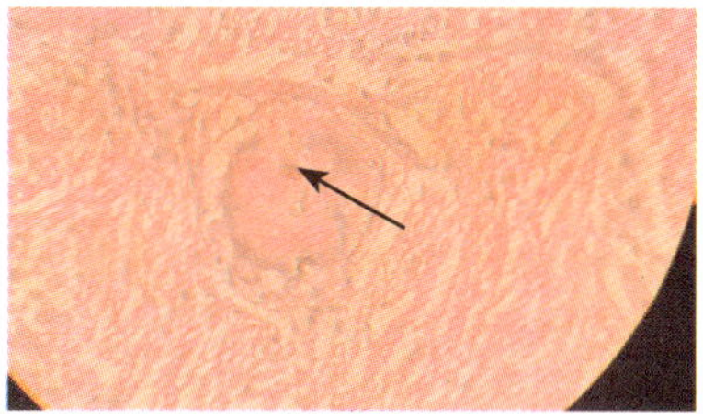

(b)

Fig. 29.47 Sclerotic bodies: (a) in KOH of tissue and (b) in the tissue section by H&E stain (*Source*: [a] Department of Microbiology, Pondicherry Institute of Medical Sciences, Puducherry and [b] Department of Pathology, Pondicherry Institute of Medical Sciences, Puducherry)

- Diagnosis can be established by the demonstration of these sclerotic bodies in KOH mounts or in tissue sections and by culture on Sabouraud agar.

Treatment

Amphotericin B, thiabendazole, 5-fluorocytosine, itraconazole, and (recently) voriconazole have been found to be useful.

2. Phaeohyphomycosis

- This group of infections caused by dematiaceous fungi includes localised or systemic infections caused by certain species such as *Phialophora*, *Cladosporium*, or other dematiaceous soil fungi, showing brown filaments in the affected tissues.
- **Sites of lesion**—cutaneous, subcutaneous, deeper tissues, or organs like the brain or lung.
- Sclerotic bodies are not found.
- The fungi appear in lesions as **distorted hyphal strands**.
- Phaeohyphomycosis is generally seen in debilitated or immunodeficient hosts.

  Clinical types of phaeohyphomycosis are as follows:
- **Brain abscess** caused by *Cladophialophora bantiana* (formerly, *Cladosporium bantianum*).
- **Subcutaneous or intramuscular lesions** develop with abscesses or cysts containing masses of brown hyphae (formerly known as phaeosporotrichosis) caused by *Exophiala* (formerly, *Phialophora*) *jeanselmei*, *E. spinifera*, *E. dermatitidis*, or *P. richardsiae*.

Sporotrichosis

- It is a chronic infection involving cutaneous, subcutaneous, and lymphatic tissues.
- It is frequently encountered in gardeners, forest workers, and manual labourers.
- It may develop in otherwise healthy individuals.
- It occurs worldwide, but especially in Central and South America and South Africa.

Agent—*Sporothrix schenckii*

Sporotrichosis is caused by *Sporothrix schenckii* (dimorphic fungus).

Pathogenesis and clinical features

- Lesions on the exposed parts of the skin follow minor trauma or thorn pricks.
- Nodules are first formed at the site of the lesion, followed by ulceration and necrosis of the nodules.
- From the lesions in the skin and subcutaneous tissue, the infection can spread via the lymphatic channels to the lymph nodes.
- This results in secondary ulcers on the lymph nodes, and the lymphatics are hardened and cord-like.
- This is called 'lymphocutaneous' sporotrichosis.
- Systemic dissemination may occur to the bones, joints, and meninges.

Laboratory diagnosis

- **Specimens:** Aspirates from the nodules and biopsy material.
- **Microscopy:**
  - Direct microscopic examination of KOH mounts of necrotic material or histopathological examination of tissue sections stained by methenamine silver stain reveals the asteroid body.
  - An **asteroid body** is a rounded or oval, basophilic, yeast-like body, 3–5 μm in diameter, with rays of an eosinophilic substance radiating from the yeast cell.
- **Culture:**
  - *S. schenckii* is a dimorphic fungus occurring in the yeast phase in tissues and in cultures at 37°C, and in the mycelia phase in nature and cultures at 25°C.
  - The septate hyphae are very thin (1–2 μm in diameter) and carry flower-like clusters of small conidia borne on delicate sterigmata (Fig. 29.48a and b).

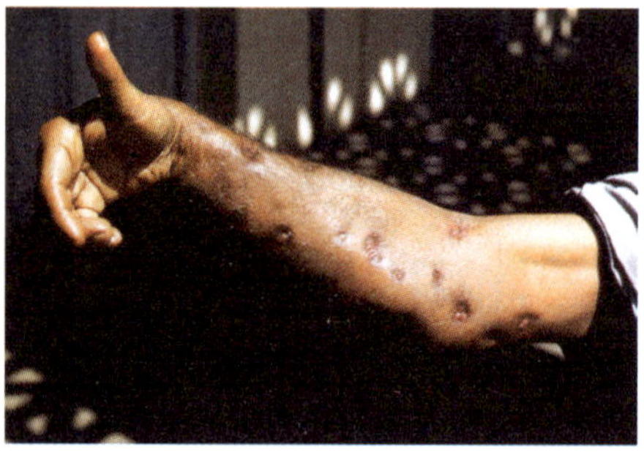

(a)

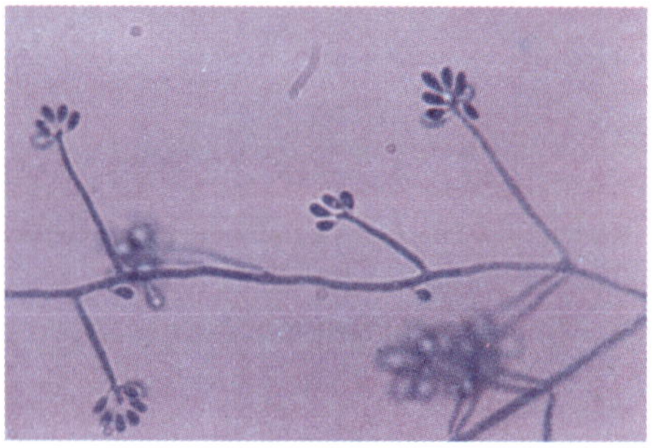

(b)

Fig. 29.48 *Sporothrix schenkii*: (a) microscopic feature of thin hyphae with flower-like arrangement of microconidia and (b) lesions in the subcutaneous tissue, spreading via the lymphatics to the draining lymph nodes (*Source*: CDC, PHIL, Image IDs 21165 and 16792)

- **Serology:**
  - Serological tests are especially helpful in the diagnosis of extracutaneous or systemic infection, where distinct clinical features are lacking. However, these tests have poor prognostic value.
  - A slide latex agglutination test, using peptido-l-rhamno-d-mannan (outer layer of the fungal cell wall) as antigen, is a reliable, sensitive and specific test. A titre of 1:4 or greater is considered as presumptive evidence of sporotrichosis.
  - In fixed cutaneous, subcutaneous, lymphocutaneous, or systemic sporotrichosis, titres of 1:4 to 1:128 occur. A rising titre or sustained high titre is seen in pulmonary sporotrichosis.

Treatment

- Spontaneous healing is possible.
- Potassium iodide, given topically or orally (250 mg thrice daily) for 2–4 months, is effective in treating cutaneous infection.
- Itraconazole is effective in over 90 per cent of patients with lymphocutaneous infection; fluconazole (400 mg/day) and terbinafine (250 mg/day) can also be administered.
- Cryotherapy with liquid nitrogen is another treatment modality.
- For disseminated infection, amphotericin B is the drug of choice.

Rhinosporidiosis

This is a chronic granulomatous disease characterised by the development of friable polyps, usually confined to the nose, mouth, or eye and, in rare cases, seen on the genitalia or other mucous membranes.

Agent—*Rhinosporidium*

- *Rhinosporidium seeberi* causes rhinosporidiosis.
- The taxonomic position of this organism is currently uncertain, with doubts being raised about whether it can be considered a fungus.

Epidemiology

Although the disease was first identified in Argentina, most cases have been reported from India and Sri Lanka.

Pathogenesis and clinical features

- The mode of infection is not known, though infection is believed to originate from stagnant water or aquatic life.
- Rhinosporidiosis is a localised condition that frequently presents as polypoidal growth in the nasal cavity; lesions can also be seen in the eye, skin, buccal cavity, or genitalia.
- While the disease is generally confined to mucous membranes, hematogenous dissemination has been recorded in very rare cases.

Laboratory diagnosis

- Histologically, the lesion is composed of large numbers of fungal spherules embedded in the stroma of connective tissue and capillaries. The spherules are 10–200 μm in diameter and contain thousands of endospores (Fig. 29.49a and b).
- *Rhinosporidium seeberi* has not been cultivated in artificial culture media.

Treatment

Excision of the polyp is the treatment of choice.

Subcutaneous Phycomycosis (Entomophthoromycoses)

Conidiobolus coronatus and *Basidiobolus ranarum* are saprophytic zygomycetes found in decaying vegetation and in the intestines of many reptiles and amphibians.

Epidemiology

It was originally reported from Indonesia and, subsequently, identified in many Asian and African countries.

Pathogenesis and clinical features

- A painless subcutaneous nodule develops, which enlarges to involve a whole limb or large areas of the body (Fig. 29.50).
- The lesions are now known to be acquired by insect bites.

Laboratory diagnosis

- The causative fungi can be isolated by culture.
- Histopathological examination shows the presence of short, broad, fragmented hyphae with beak-like projections.

Treatment

- The treatment of phycomycosis is difficult, and surgery may be required.

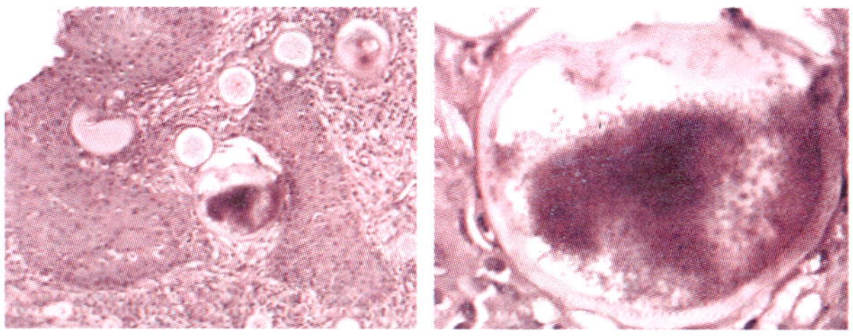

(a) (b)

Fig. 29.49 Rhinosporidiosis: (a) tissue section showing sporangia with numerous endospores (10×) and (b) thick-walled sporangium with endospores (40×) (*Source*: Department of Microbiology, Pondicherry Institute of Medical Sciences, Puducherry

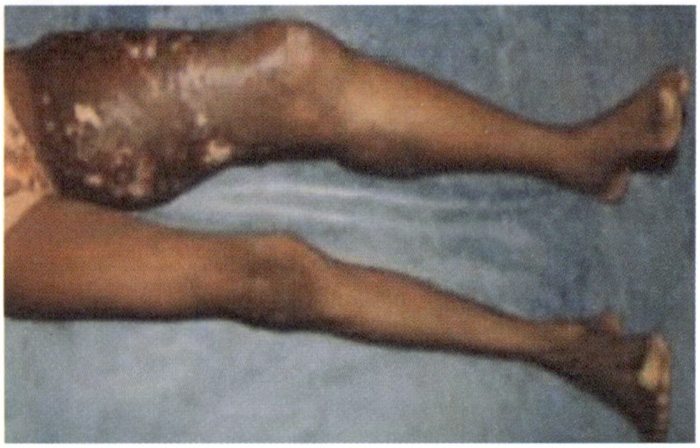

Fig. 29.50 Phycomycosis that has enlarged to involve the whole upper limb (*Source*: Dr Sheela Kuruvila, Department of Dermatology, Pondicherry Institute of Medical Sciences, Puducherry)

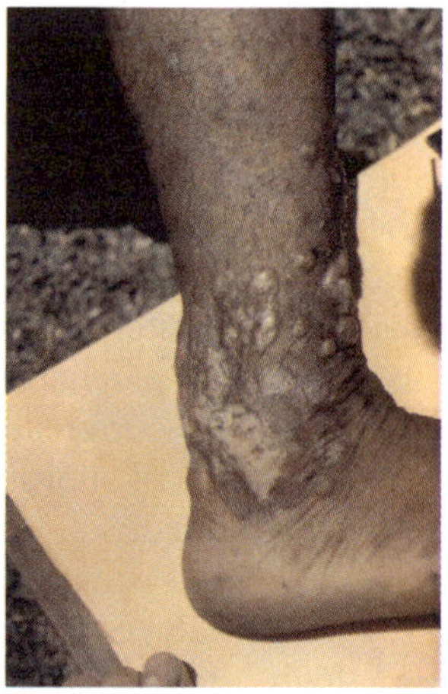

Fig. 29.51 Large, elevated ulcerative lesions in blastomycosis (*Source*: CDC, PHIL, Image ID 14873)

- Antifungal drugs like itraconazole and terbinafine hydrochloride are used for two to three months following surgery.

Blastomycosis (Details in Chapter 23)

- It is a chronic infection characterised by the formation of suppurative and granulomatous lesions with a marked predilection for the lungs and skin. It can also occur in any other part of the body.
- Blastomycosis is caused by the thermally dimorphic fungus *Blastomyces dermatitidis* (teleomorph [perfect state] is *Ajellomyces dermatitidis*).
- The cutaneous disease usually affects the skin of the face or other exposed parts of the body. The initial lesion is a papule, around which secondary nodules develop and coalesce (Fig. 29.51).

Coccidioidomycosis

- It is a primary pulmonary infection that may be inapparent, benign, severe, or even fatal. It is caused by the thermally dimorphic fungus *Coccidioides immitis*.
- A self-limited influenza-like fever (known as valley fever or desert rheumatism) may occur in some individuals.
- Less than one per cent of infected persons develop the chronic progressive disseminated disease, coccidioidal granuloma (Fig. 29.52a, b, and c).

Cutaneous Candidiasis

- Cutaneous candidiasis is commonly caused by *Candida albicans* and other species of *Candida.*
- It can involve almost any skin on the body, but most often, it occurs in warm, moist, creased areas such as the armpits and groin. *Candida* infection represents a bridge connecting superficial and deep mycoses.
- Diaper rash in infants is a common infection, seen in hot, humid climates. The fungi take advantage of the warm, moist conditions inside the diaper.
- Predisposing conditions that increase the risk of infection are diabetes, obesity, prolonged antibiotic therapy, steroid therapy, and chemotherapy.
- Diagnosis is by microscopy of skin scrapings, which show abundant budding yeast cells with pseudohyphae.
- Yeast cells can be cultured on Sabouraud dextrose agar and identified by germ tube test and carbohydrate assimilation tests.
- Treatment is by itraconazole and other azoles.

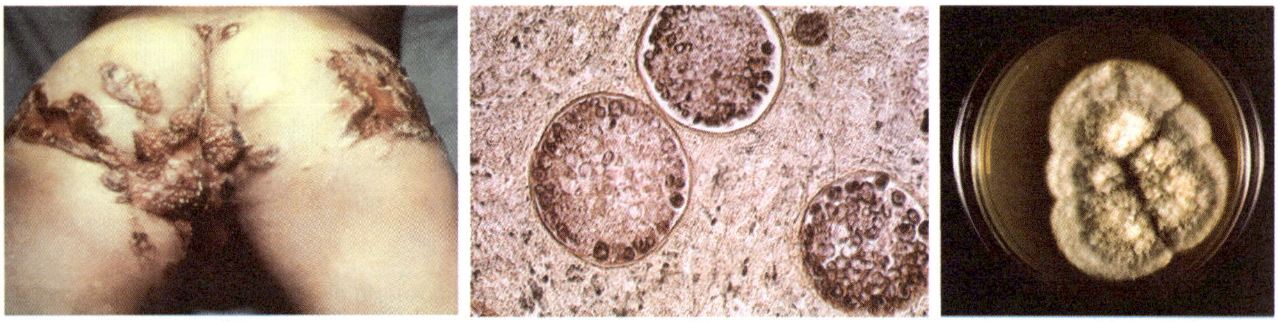

(a) (b) (c)

Fig. 29.52 (a) Coccidioidomycosis lesions on the patient's buttocks; (b) large spherule in tissue with endospores; and (c) grayish-white downy culture of *Coccidiodes immitis* (*Source*: CDC, PHIL, Image IDs 20552, 14641, and 21608)

Questions

Case review

A 50-year-old man met with a road traffic accident. He sustained multiple fractures with open wounds and a crush injury of the leg. He was taken to the nearest hospital two days later, at which time, he was found to be in shock.

i) Outline the steps required to arrive at the final diagnosis.
ii) Mention the most common organisms associated with soil-contaminated wound infection.
iii) Outline the steps of management in this case.

Short answer questions

1. Discuss the pathogenesis of shingles.
2. Describe the lesions caused by human papilloma virus.
3. Outline the laboratory diagnosis of lepromatous leprosy and mention the interpretation of slit skin smear examination.
4. Outline the laboratory methods adopted for diagnosing dermatomycoses.
5. How will you differentiate eumycotic mycetoma from actinomycotic mycetoma?

MCQs

1. **Impetigo can be caused by any of the bacteria given below except one:**
   a) *Staphylococcus aureus* b) *Streptococcus pyogenes*
   c) *Candida albicans* d) *Shigella* spp.
2. **A 30-year-old man develops an abscess on his thigh due to methicillin-resistant** *Staphylococcus aureus*. **Which antibiotic can be used to treat this patient?**
   a) Vancomycin b) Cotrimoxazole
   c) Erythromycin d) Penicillin
3. **Cellulitis in a diabetic patient caused by MRSA can be treated with all of the following antibiotics except:**
   a) Vancomycin b) Linezolid
   c) Meropenem d) Teicoplanin
4. **Which of the following is true of staphylococcal scalded skin syndrome?**
   a) The toxins have a direct effect on the stratum granulosum
   b) The mucosa is involved
   c) Its severe form is bullous impetigo
   d) It is commonly seen in adults
5. **Erythrasma is caused by:**
   a) Group A streptococci
   b) *Staphylococcus aureus*
   c) *Bacteroides* species
   d) *Corynebacterium minutissimum*
6. **Bamboo-stick appearance on Gram stain smear is the characteristic feature of:**
   a) *Bacillus anthracis*
   b) *Clostridium perfringens*
   c) *Corynebacterium diphtheriae*
   d) *Bacteroides fragilis*
7. **Medusa head type of growth on nutrient agar is produced by:**
   a) *Bacillus cereus* b) *Brucella melitensis*
   c) *Bacillus anthracis* d) *Bacillus stearothermophilus*
8. **Which of the following statements on** *Bacillus anthracis* **is not true?**
   a) First pathogenic bacterium observed under the microscope
   b) First bacillus to be isolated in pure culture
   c) First bacterium identified to produce toxin
   d) First bacterium used for the preparation of a vaccine
9. **The non-motile** *Bacillus* **species is:**
   a) *B. anthracis*
   b) *B. cereus*
   c) *B. subtilis*
   d) *B. stearothermophilus*
10. **Which one of the following does not form a component of anthrax toxin?**
    a) Protective antigen b) Bursting factor
    c) Edema factor d) Lethal factor
11. **Duckering, a method of disinfection of wool for destroying anthrax spores is by:**
    a) Use of dry heat at 140°C for 1 hour
    b) Use of 4% potassium permanganate for 15 minutes
    c) Use of 5% phenol for 10 hours
    d) Use of 2% formaldehyde at 30-40°C for 20 minutes
12. **The type of biosafety lab required to work with anthrax bacteria is:**
    a) Level 1 b) Level 2
    c) Level 3 d) Level 4
13. The selective medium used for *Bacillus anthracis* is:
    a) PLET medium b) Tellurite blood agar
    c) Cefsulodin agar d) Cetrimide agar
14. **The special stain used for spores of *Bacillus anthracis* is:**
    a) Nigrosin b) Sudan black
    c) Lugol's iodine d) Methenamine silver
15. **All the following are colony descriptions of** *Bacillus anthracis* **except:**
    a) Frosted glass b) Inverted fir tree
    c) String of pearls d) Bamboo stick
16. **The *Clostridium* species least involved in gas gangrene is:**
    a) *C. perfringens* b) *C. histolyticum*
    c) *C. novyi* d) *C. septicum*
17. **The clinical condition caused by *Clostridium perfringens* type C is:**
    a) Gangrenous appendicitis
    b) Pseudomembranous colitis
    c) Severe necrotising enteritis
    d) Neutropenic colitis
18. **Citron bodies in a Gram smear from a case of gas gangrene is suggestive of:**
    a) *Clostridium perfringens* b) *Clostridium novyi*
    c) *Clostridium septicum* d) *Clostridium sordelli*
19. **The best agent for destroying the spores of clostridia is:**
    a) 5% phenol b) 1% iodine
    c) I ppm chlorine d) 2% glutaraldehyde
20. **Dressings from a suspected anaerobic wound of a patient revealed a red fluorescence under UV light. The anaerobic bacteria that has property is:**
    a) *Prevotella* b) *Bacteroides*
    c) *Porphyromonas* d) *Fusobacterium*
21. **Lecithinase activity of alpha toxin produced by *Clostridium perfringens* can be demonstrated by:**
    a) Beta-hemolysis on blood agar
    b) Target hemolysis on blood agar
    c) Opalescence in egg yolk agar
    d) Satellitism in blood agar
22. **The lesions produced in muscle tissue in gas gangrene are due to:**
    a) Lecithinase b) Collagenase
    c) Neuraminidase d) Bursting factor
23. **Which toxin produced by *Clostridium perfringens* is responsible for the pathogenicity in gas gangrene?**
    a) Alpha b) Beta
    c) Gamma d) Delta
24. **Which of the following is correct regarding the alpha toxin produced *Clostridium perfringens*?**
    a) It is lethal, dermonecrotic and hemolytic
    b) It is a phospholipidase (lecithinase C)
    c) It can be specifically neutralised by the antitoxin
    d) All of the above
25. **Alpha toxin exhibits lysis of red blood cells at 37°C followed by cooling at 4°C. This property is known as:**
    a) Cold lysis b) Thermal lysis
    c) Hot cold lysis d) None of the above

26. Which of the following is a component of the cell wall of Actinomycetes?
a) Lipid b) Muramic acid
c) Teichoic acid d) Cysteine

27. An anaerobic actinomycete is:
a) *Actinomadura* b) *Streptomyces*
c) *Actinomyces* d) *Nocardia*

28. The most common form of actinomycosis is:
a) Thoracic b) Cervicofacial
c) Abdominal d) Pelvic

29. Spidery colonies are typically seen in:
a) *Actinomyces israelii* b) *Actinomyces bovis*
c) *Actinomyces naeslundii* d) *Actinomyces viscosus*

30. The most common *Nocardia* species to infect humans is:
a) *N. asteroids* b) *N. brasiliensis*
c) *N. caviae* d) *N. naeslundii*

31. Botryomycosis is a mycetoma-like disease caused by:
a) *Nocardia* species b) *Actinomyces* species
c) *Staphylococcus* species d) *Streptococcus* species

32. Selective isolation of *Nocardia* can be done by:
a) Culture in ordinary medium
b) Anaerobic culture methods
c) Hair baiting technique
d) Paraffin baiting

33. Pus culture from a swelling on the upper arm of a patient after 48 hours yielded growth of filamentous gram-positive bacteria on blood agar which were also acid-fast. The organism is most likely to be:
a) *Nocardia* species b) *Actinomycetes*
c) *Mycobacterium* d) *Corynebacterium*

34. The most common metastatic manifestation of nocardiosis is:
a) Arthritis b) Nephritis
c) Meningitis d) Endocarditis

35. Actinomycetes are closely related to all of the following except:
a) *Acinetobacter* b) *Corynebacterium*
c) *Mycobacterium* d) *Nocardia*

36. Which of the following is not a pathogenic mycobacterium?
a) *M. scrofulaceum* b) *M. kansasii*
c) *M. smegmatis* d) *M. genavense*

37. A 72-year-old woman has an artificial hip joint implanted. One week later, she develops fever and joint pain. Culture from the fluid aspirated from the hip shows lactose-fermenting colonies on MacConkey's agar which are acid-fast. The infecting organism may be:
a) *M. tuberculosis* b) *M. chelonae*
c) *M. ulcerans* d) *M. kansasii*

38. A 47-year-old symptomatic man has X-ray chest suggestive of tuberculosis. His sputum culture grows an acid-fast bacillus that is a photochromogen. The organism most likely is:
a) *M. tuberculosis*
b) *M. avium* intracellulare complex
c) *M. kansasii*
d) *M. gordonae*

39. The agent causing swimming pool granuloma is:
a) *M. marinum* b) *M. ulcerans*
c) *M. chelonae* d) *M. avium*

40. The non-tuberculous mycobacteria that cause pulmonary disease resembling tuberculosis is:
a) *M. scrofulaceum* b) *M. kansasii*
c) *M. gordonae* d) *M. fortuitum*

41. All these mycobacteria cause skin lesions except:
a) *M. tuberculosis* b) *M. ulcerans*
c) *M. marinum* d) *M. haemophilum*

42. The only toxin-producing mycobacteria is:
a) M. *kansasii* b) *M. gordonae*
c) *M. ulcerans* d) *M. avium* intracellulare

43. All the following are characteristics of non-tuberculous mycobacteria except:
a) Human-to-human transmission does not occur
b) Also known as environmental mycobacteria
c) They are common laboratory contaminants
d) Good response to anti-TB treatment

44. The generation time of *Mycobacterium leprae* is:
a) 20 seconds b) 20 minutes
c) 20 hours d) 20 days

45. The temperature for maintaining *Mycobacterium* leprae in the foot pad of mice is:
a) 37°C b) 55°C
c) 20°C d) 5–10°C

46. Which of the following statements is not true about lepromatous leprosy?
a) LL is seen when host resistance is low
b) Lepromatous leprosy is more infective than other types
c) Autoantibodies are common
d) Lepromin test will be positive

47. Which type of hypersensitivity reaction is erythema nodosum leprosum?
a) Type I b) Type II
c) Type III d) Type IV

48. Which of the following statements is not true about the lepromin test?
a) It does not have any diagnostic significance
b) It can be used to assess the resistance of individuals to leprosy
c) It can be used to classify lesions of leprosy
d) Mitsuda reaction is analogous to tuberculin reaction

49. What is the best clinical sample for the demonstration of varicella zoster virus infection?
a) Blood b) Saliva
c) Vesicular fluid d) Lymph node aspirate

50. GB syndrome is the complication of which of the following viral infections?
a) HPV b) HSV
c) HHV-4 d) HHV-7

51. Reye's syndrome is associated with:
a) VZV infection b) HSV infection
c) Molluscum contagiosum d) Variola

52. Ramsay–Hunt syndrome is associated with:
a) Herpes simplex virus infection
b) Herpes zoster infection
c) Epstein-Barr virus infection
d) Vaccinia

53. Whitmore disease is caused by:
a) *Burkholderia mallei*
b) *Burkholderia pseudomallei*
c) *Burkholderia cepacia*
d) *Stenotrophomonas maltophilia*

54. The selective medium used for screening soil samples for *Burkholderia pseudomallei* is:
a) Ashdown medium b) Cetrimide agar
c) Oxacillin screen agar d) King B medium

55. The drug of choice for melioidosis is:
a) Azithromycin b) Ceftazidime
c) Gentamicin d) Colistin

56. Large eosinophilic hyaline inclusion bodies called Henderson–Paterson bodies are characteristic of:
a) HSV b) Molluscum contagiosum
c) HIV d) Orf virus

57. 'HHV-8' is commonly associated with:
a) Heterosexuals
b) Men having sex with men
c) Long-term corticosteroid therapy
d) Intravenous drug users

58. Human beings acquire infection by the ingestion of contaminated water containing:
a) Embryonated eggs b) *Cyclops*
c) Adult worms d) Free-swimming larva

59. Guinea worm disease has been eradicated from India.
a) True
b) False
c) From some parts of the country
d) It can never be eradicated

60. Which of the following statement is not correct regarding trichinellosis?
a) Humans acquire infection by eating pork that contains stage-1 larva.
b) These larvae penetrate the intestinal mucosa and travel to the skeletal muscles to form cyst.
c) Extraocular muscles are the most common sites of encystment.
d) Mebendazole or albendazole can effectively eliminate the muscle forms.

61. A 22-year-old male traveller from Africa presents with complaints of itching on legs for past one month. On examination, the leg has a diffuse papular rash with plaque and atrophied skin. There is a subcutaneous nodules over his forehead. Skin biopsy shows microfilariae. Which of the following is most likely responsible for this condition?
a) Scabies b) *Ixodes*
c) *Onchocerca volvulus* d) *M. ozzardi*

62. Which statement is correct regarding *Onchocerca volvulus*?
a) *O. volvulus* is the causative agent of river blindness.
b) The infection is transmitted by biting midges.
c) Microfilariae are sheathed, with nuclei till tail tip.
d) All of the above.

63. Serous cavity filariasis is caused by:
a) *Onchocerca volvulus* b) *Loa loa*
c) *Mansonella perstans* d) *Brugia malayi*

64. Cercarial dermatitis is seen in:
a) Schistosomiasis b) Filariasis
c) Hydatid cyst d) Visceral larva migrans

65. An adult male developed brownish black, localised macular lesions on his palms. Skin scrapings showed septate hyphae and budding yeast cells. Colonies on SDA were grey-black. What is the likely causative agent?
a) *Exophiala werneckii*
b) *Trichophyton mentagrophytes*
c) *Microsporum gypseum*
d) *Rhinosporidium seeberi*

66. Which of the following is a cutaneous mycosis?
a) Favus b) Pityriasis versicolor
c) Tinea nigra d) Piedra

67. Which of the following is a subcutaneous mycosis?
a) Favus b) Pityriasis versicolor
c) Tinea capitis d) Chromomycosis

68. Which dermatophyte shows club-shaped macroconidia in clusters?
a) Trichophyton b) Microsporum
c) Epidermophyton d) All of the above

69. Which of the following is a superficial fungal infection?
a) Favus b) Pityriasis versicolor
c) Tinea capitis d) Chromomycosis

70. The causative agent of athlete's foot is:
a) *Trichophyton mentagrophytes*
b) *Trichophyton rubrum*
c) *Epidermophyton floccosum*
d) *Microsporum gypseum*

71. Black piedra is caused by:
a) *Piedraia hortae* b) *Trichosporon beigelii*
c) *Trichophyton verrucosum* d) *Microsporum gypseum*

72. White piedra is caused by:
a) *Piedraia hortae* b) *Trichosporon beigelii*
c) *Trichophyton verrucosum* d) *Microsporum gypseum*

73. Severe, boggy lesions with marked inflammatory reaction that sometimes in the scalp caused by dermatophytes are known as:
a) Favus b) Kerion
c) Unguium d) Jock itch

74. A 45-year-old farmer presented to with a history of multiple swellings on his foot and seropurulent pus discharge from the sinuses. KOH preparation from the discharge revealed the presence of dark, pigmented, fungal hyphae. Most likely agent is:
a) *Actinomadura* b) *Streptomyces*
c) *Madurella mycetomatis* d) *Nocardia*

75. Chromoblastomycosis is caused by any of the following fungi except:
a) *Sporothrix schenckii* b) *Exophiala dermatitidis*
c) *Phialophora verrucosa* d) *Fonsecaea pedrosoi*

76. All of the following are systemic fungal infections except:
a) *Sporotrichosis* b) *Rhinosporidiosis*
c) *Histoplasmosis* d) *Chromomycosis*

77. Wood's lamp examination helps in the demonstration of:
a) Fungal filaments in hair
b) Germ tubes in tissues
c) Fungal antigens in tissues
d) Favic chandelier

78. Nasal polyps are produced by:
a) *Rhinosporidium seeberi*
b) *Coxiella burnetti*
c) *Klebsiella rhinoscleromatis*
d) *Rhizopus* species

79. The presence of asteroid bodies on histopathological sections is the characteristic feature of:
a) Histoplasmosis b) Sporotrichosis
c) Rhinosporidiosis d) Zygomycoses

80. Pencil-shaped macroconidia are seen in:
a) *Trichophyton rubrum*
b) *Trichophyton mentagrophytes*
c) *Trichophyton tonsurans*
d) *Trichophyton violaceum*

81. Cigar-shaped macroconidia are seen in:
a) *Trichophyton rubrum*
b) *Trichophyton mentagrophytes*
c) *Trichophyton tonsurans*
d) *Trichophyton violaceum*

82. Rat tail filaments are seen in:
a) *Trichophyton rubrum*
b) *Trichophyton tonsurans*
c) *Trichophyton mentagrophytes*
d) *Trichophyton violaceum*

83. Favic chandelier is the characteristic feature of:
a) *T. rubrum* b) *T. schoenleinii*
c) *T. mentagrophytes* d) *T. violaceum*

84. Kerion refers to:
a) Hypersensitivity reaction to histoplasmosis
b) Marked inflammatory response in dermatophytosis
c) Hard nodule seen in sporotrichosis
d) Severe keratitis in oculomycosis

85. Ectothrix and endothrix are best demonstrated by:
a) Calcofluor white stain
b) Methenamine silver stain
c) Hair penetration test
d) Germ tube test

86. Subcutaneous phycomycosis is caused by
a) *Conidiobolus coronatus* b) *Rhinosporidium seeberi*
c) *Sporothrix schenckii* d) *Exophiala dermatitidis*

87. Madura mycosis is caused by all of the following except:
a) *Nocardia* b) *Streptomyces*
c) *Actinomadura* d) *Mucor*

88. A gardener presents with chronic progressive swelling on the leg with discharging sinuses. The most probable diagnosis is:
a) Sporotrichosis b) Histoplasmosis
c) Mucormycosis d) Mycetoma

89. The causative agent of botryomycosis is:
a) *Nocardia braziliensis* b) *Madurella mycetoma*
c) *Staphylococcus aureus* d) *Histoplasma capsulatum*

90. A 43-year-old farmer who presented with a chronic swelling on the leg with a history of sinuses discharging red granules was diagnosed with actinomycosis. What is the most probable causative agent?
a) *Actinomadura pelletieri* b) *Actinomadura madurae*
c) *Nocardia asteroides* d) *Madurella mycetomatis*

91. A gardener developed nodular lesions on his hand which gradually turned into ulcers. Histopathological examination of the necrotic tissue showed asteroid bodies and microscopic examination of the culture showed pear-shaped conidia in rosette-like clusters. What is the diagnosis?
a) Blastomycosis b) Lobomycosis
c) Sporotrichosis d) Histoplasmosis

92. Which of the following is true regarding eumycetoma?
a) Etiology is bacterial.
b) Discharging granules are soft in consistency.
c) Gram-positive filamentous bacilli are seen in direct microscopy.
d) Treatment is by surgical debridement and antimicrobials.

93. A 58-year-old agricultural worker presented with a 2-year history of right foot soft tissue mass, which had gradually increased in size over years, but it was painless. Now the lesion has started to develop discharging sinuses oozing black coloured granules. Which of the following is correct regarding actinomycotic mycetoma?
a) The lesions are harder, broader, and may be pigmented.
b) The lesions show septae and chlamydospores in KOH preparation.
c) The granules reveal gram-positive, slender rods.
d) All of the above.

94. A 68-year-old diabetic man developed high fever, pain and progressive edema of the left lower limb after an insect bite. Soon, the symptoms worsened, with involvement of the fascia and muscles of the leg. This condition could be caused by any of the following except:
a) Group A streptococci
b) *Staphylococcus aureus*
c) *Bacteroides* species
d) *Corynebacterium minutissimum*

95. A five-year-old girl developed painful vesicular lesions around the mouth. There was a history of similar lesions with two of her siblings. Examination of the vesicular fluid was positive for multinucleated giant cells. What is the likely organism associated?
a) Herpes simplex virus
b) Molluscum contagiosum virus
c) Papova virus
d) Norwalk virus

96. A 50-year-old, HIV-positive man developed painful vesicular eruptions on the right side of his spine in the lumbar region. Scrapings from the base of the lesions were positive for the VZV antigen by the direct fluorescent antibody test. Which of the following is correct regarding VZV?
a) It occurs commonly in children.
b) The rash is typically unilateral and confined to the dermatome of nerve.
c) Previous history of chicken pox is of no significance.
d) All of the above.

97. A 20-year-old female presents with ulcerative raised violaceous nodule on her shins. Her systemic examination is normal. She was treated for kala-azar three years ago. What is most likely to be seen on histopathologic examination of the lesion?
a) Amastigote forms inside macrophages
b) Amastigote forms outside macrophages
c) Promastigote forms inside macrophages
d) Promastigote forms outside macrophages

98. A 30-year-old female patient presents with complaints of a worm extruding from her left lower extremity. Initially, it was a blister before the worm popped out from it. Which of the parasite is most likely to be coming out of the blister?
a) *Dracunculus medinensis*
b) *Necator americanus*
c) *Strongyloides stercoralis*
d) *Ancylostoma duodenale*

Answers: 1:c, 2:a, 3:c, 4:a, 5:d, 6:a, 7:c, 8:c, 9:a, 10:b, 11:d, 12:c, 13:a, 14:b, 15:d, 16:b, 17:c, 18:c, 19:d, 20:c, 21:c, 22:d, 23:a, 24:d, 25:c, 26:b, 27:c, 28:b, 29:a, 30:a, 31:c, 32:a, 33:a, 34:c, 35:a, 36:c, 37:b, 38:c, 39:a, 40:b, 41:d, 42:c, 43:d, 44:d, 45:c, 46:d, 47:c, 48:d, 49:c, 50:b, 51:a, 52:b, 53:b, 54:a, 55:b, 56:b, 57:b, 58:b, 59:a, 60:d, 61:c, 62:a, 63:c, 64:a, 65:a, 66:a, 67:d, 68:c, 69:b, 70:a, 71:a, 72:b, 73:b, 74:c, 75:a, 76:b, 77:a, 78:a, 79:b, 80:a, 81:b, 82:c, 83:d, 84:b, 85:c, 86:a, 87:d, 88:d, 89:c, 90:a, 91:c, 92:d, 93:c, 94:d, 95:a, 96:b, 97:a, 98:a

CHAPTER 30

Part III: Microbiology as Applied to Infectious Diseases

Infections of the Eyes and Ears and Congenital Infections

(MI8.15)

Chapter Navigation

Learning objectives

- Enumerate common organisms associated with various ocular infections
- Describe ear infections and list the common organisms associated with them
- Enumerate common congenital infections and methods to diagnose them

INFECTIONS OF THE EYES

An eye infection is a condition in which viruses, bacteria or other microbial agents attack the external (eyelid, cornea, and conjunctiva) or internal parts of the eye. Infections of the eye commonly present as conjunctivitis, corneal ulcer, or endophthalmitis. These infections may of the following types:

- **Extraocular**—involving the external parts of the eye (ocular adnexa)
- **Intraocular**—involving the internal structures of the eye

EXTRAOCULAR INFECTIONS

The common extraocular infections are presented in Table 30.1.

INTRAOCULAR INFECTIONS

- A pathogen can gain entry into the eye through the bloodstream, the nerve route, through contaminated eye drops or contact lenses, or as a result of trauma (most common mode of entry). An important cause of ocular infection in the neonate is congenital rubella.
- **Endophthalmitis:** It is defined as an **inflammation of the inner coats of the eye**, resulting from intraocular colonisation of infectious agents with exudation within intraocular fluids. It is a potentially blinding condition that occurs following a penetrating injury to the eye (exogenous), leading to panophthalmitis or as a complication of a bloodstream (endogenous) infection. Almost any organism including bacteria and fungi can cause endophthalmitis. It can be secondary to keratitis in some cases.
- **Chorioretinitis:** It is a **type of uveitis affecting the posterior segment of the eye** (choroid and retina). It is a vision-threatening condition that is associated with many systemic conditions. The common etiological agents are tuberculosis, syphilis, bartonellosis, *Candida* infection, coccidioidomycosis, histoplasmosis, infections caused by HIV, CMV, *Toxoplasma gondii*, and *Toxocara canis*.

ORBITOL MUCORMYCOSIS

- Mucormycosis is an **aggressive opportunistic fungal infection** caused by fungi with non-septate hyphae.
- When the fungus invades the paranasal sinus mucosa, it may spread directly to the orbital apex and, from there, gain intracerebral access.
- Mucormycosis is difficult to diagnose early, as patients often present with non-specific symptoms. By the time signs of orbital apex involvement develop, it is often too late to save the patient's eye or life.
- Diagnosis is made by **imaging and histopathology and culture from biopsy specimen.**
- Treatment is surgical and supplemented with prolonged antifungals.

Molluscum contagiosum is a DNA virus that can cause eyelid nodules. It can also affect the conjunctiva through a hypersensitivity reaction leading to chronic follicular conjunctivitis, corneal changes such as a superior micropannus, and fine epithelial keratitis.

Table 30.1 Extraocular infections

Type of extraocular infection	Clinical features	Etiology				Remarks
		Bacteria	**Virus**	**Parasites**	**Fungi**	
Stye (hordeolum)	Infection of the oil glands in the eyelid or eyelashes **Symptoms**: Painful, red bump under the eyelid or at the base of the eyelashes	• *Staphylococcus aureus* • *Staphylococcus epidermidis*	–	–	–	Diagnosis is clinical and based on history; no diagnostic tests are required as they are not useful
Conjunctivitis	Infection of the conjunctiva—can be unilateral or bilateral **Symptoms:** • Eye pain • Persistent itching • Pricking or gritty sensation • Sensitivity to light • Watery eyes • Chronic redness • Discomfort • Discharge of fluids • Swelling • Dryness • Blurring vision **Examination**: Eye is congested, pink or reddish, with watering and pain (Fig. 30.1)	• *Staphylococcus* • *Streptococcus pneumoniae* *Haemophilus influenzae* • *Pseudomonas* • *Moraxella catarrhalis* • *Chlamydia trochomatis* serovar D-K • *Neisseria gonorrhoeae*	Enterovirus-70 (EV-70) Coxsackie-A24 Adenovirus	• Free-living amoeba (*Acanthamoeba*) • Larval forms of *Loa loa* • *Echinococcus granulosus* • *Toxoplasma gondii* • *Taenia solium* (which causes cysticercus cellulosae) • *Gnathostoma spinigerum* • *Onchocerca volvulus*	• *Aspergillus* • *Fusarium* • *Penicillium*	–
Keratitis	Inflammation of the cornea	• *Staphylococcus epidermidis* • *Streptococcus pneumoniae* • *Pseudomonas* • *Acinetobacter* species • *Corynebacterium diphtheriae* • *Corynebacterium xerosis* • *Neisseria* • *Haemophilus* • *Nocardia*	• Herpes simplex virus (HSV) • Herpes zoster virus (HZV) • Adenovirus	• Free-living amoeba • *Acanthamoeba* • *Naegleria* • *Microsporidia* • *Onchocerca*	• *Aspergillus Penicillium* • *Candida* • *Alternaria* • *Fusarium* • *Curvularia* • *Microsporidia*	More common in individuals who use contact lenses or occur following trauma
Uveitis	Inflammation of the middle layer of the eye, which is called the **uvea**	• Lyme disease • *Treponema* • *Pallidum* • Mycobacteria • *Leptospira*	• HSV • VZV • CMV • Rubella virus • Ebola virus • West Nile virus	• *Toxoplasma* • *Toxocara* • *Trypanosoma* • *Bartonella* spp.	–	Mostly autoimmune disorders such as rheumatoid arthritis or lupus (80%)

(Table 30.1 Contd. . .)

Table 30.1 Contd.

Type of extraocular infection	Clinical features	Etiology				Remarks
		Bacteria	Virus	Parasites	Fungi	
Blepharitis	Inflammation of the base of the eyelashes Eyelids become irritated, red, and swollen and may burn and itch	• *Staphylococcus* • *Pseudomonas*	HSV VZV	–	–	Generally bacterial; may be allergic and secondary to skin conditions (seborrheic dermatitis, rosacea, and atopic dermatitis) Mites (*Demodex* species) can also cause it
Trachoma	Chronic keratoconjunctivitis	*Chlamydia trachomatis* serovars A, B, Ba, and C	-	–	–	Scarring of the inner surface of the eyelids and erosion of the corneal surface, can leads to blindness
Ophthalmia neonatorum	Conjunctival inflammation occurring during the first month of life with severe, purulent conjunctivitis	• *Chlamydia* • *N. gonorrhoeae* • *Haemophilus* species • *Streptococcus pneumoniae* • *Staphylococcus aureus* • *Staphylococcus epidermidis* • *Streptococcus viridans* • *Escherichia coli* • *Pseudomonas aeruginosa*	Adenovirus Herpes simplex virus	–	–	Gram stain and culture can confirm bacterial etiology
Infective dacryocystitis	Swelling and blockage of the eye's tear drainage system	• *Staphylococcus* • *Streptococcus* • *Haemophilus influenzae* • *Pseudomonas aeruginosa*	–	–	–	
Corneal ulcer	Breach in the continuity of the cornea which results in the formation of ulcers	• *Staphylococcus aureus* • *Pseudomonas aeruginosa* • *Staphylococcus epidermidis*	Adenovirus HSV VZV CMV	• *Toxoplasma* microsporidia • *Acanthamoeba*	• *Aspergillus* • *Fusarium* • *Scedosporium apiospermum* • Phaeohyphomycosis • *Candida* species	

LABORATORY DIAGNOSIS OF OCULAR INFECTIONS

- The mainstay of diagnosis is smear **examination of discharge from the eye, swabs, or aspirates**.
- In case of worm infections of the eye, the diagnosis is made by **detecting microfilariae or other worms** lodged in or migrating across the eye (Fig. 30.2).
- Culture often yields the causative bacterial agent.
- As the samples are very minimal, a bedside smear and inoculation on culture media often gives better results.

TREATMENT

- **Topical antibiotics**—ciprofloxacin, azithromycin, and vancomycin drops—in conjunction with systemic therapy are used to treat endophthalmitis and other ocular infections.
- For **worm infections** of the eye, treatment involves **surgically removing the worm** and then treating the primary infection.
- In **herpes infection, acyclovir** is prescribed along with topical therapy with acyclovir eye ointment.

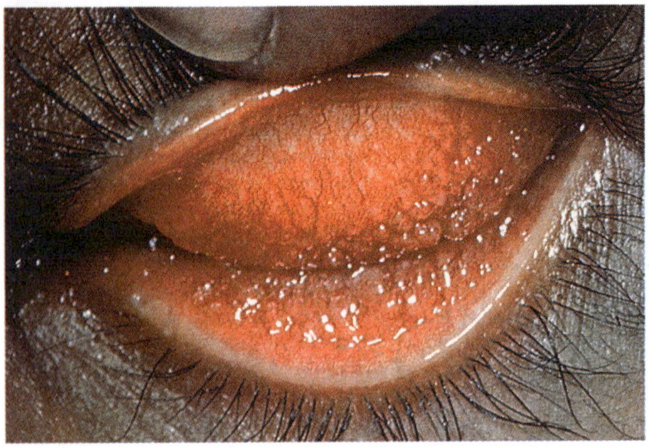

Fig. 30.1 Inclusion conjunctivitis due to *Chlamydia trachomatis* (*Source*: Centres for Disease Control and Prevention [CDC], Public Health Image Library [PHIL], Image ID 15193)

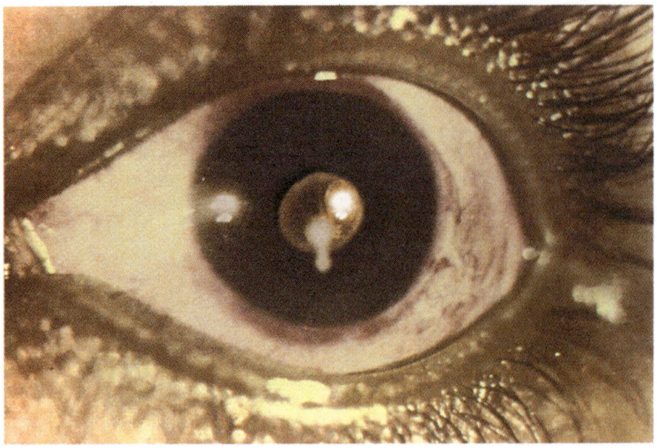

Fig. 30.2 Cysticercus, the larval form of the *Taenia solium* or pork tapeworm seen in the pupil of a patient's left eye (*Source*: CDC, PHIL, Image ID 14629)

INFECTIONS OF THE EAR

EXTERNAL EAR INFECTIONS

They may involve the skin, cartilage, mastoid cavity, periosteum, or the tympanic membrane. Some of the clinical entities of such infections are described below.

Auricular Cellulitis

- Results from **minor trauma** and involves the skin over the external ear.
- It is commonly caused by ***Staphylococcus aureus*** and **streptococci**.
- **Warm compress and first-generation cephalosporin** (cefazolin) are used to treat mild infection.
- Severe infection may require IV beta-lactamase-inhibiting antibiotics.

Perichondritis

- Infection of the perichondrium of the auricular cartilage, which is caused by **trauma to the cartilage** (e.g., ear piercing and laceration).
- The most **common organisms** involved are *S. aureus* and *Pseudomonas aeruginosa*. It is diagnosed by drainage and culture of the pus.
- **Treatment** is with quinolones, antipseudomonal penicillin (piperacillin), or cephalosporin if staphylococci are isolated.

External Otitis

- It involves the **auditory meatus** and is caused by the accumulation of moisture and macerated desquamated epithelium of the external ear.
- Acute external otitis infection is **commonly caused** by *S. aureus* or *P. aeruginosa*.
- Furuncle formation at the junction of the cartilage and the meatus with hair follicles may require **incision and drainage** followed by antipseudomonal treatment **(piperacillin) or cefazolin** for staphylococcal infection.
- Diffuse otitis externa is known as **swimmer's ear**.
- A chronic form of the infection may occur due to constant irritation caused by drainage from the middle ear.
- **Invasive otitis externa** called **malignant or necrotising otitis externa** may be seen in diabetics or immunocompromised individuals. This is a potentially life-threatening and aggressive infection. If left untreated, it may progress to the base of the skull (skull base osteomyelitis) and may involve the meninges and the brain. Organisms responsible are *Pseudomonas*, *Staphylococcus*, *Aspergillus,* and *Actinomyces.*
- **Diagnosis** is based on the microscopic examination of the ear discharge followed by culture in appropriate media.
- **Treatment** depends on the etiological agents—either organism-specific broad-spectrum antibiotics like piperacillin, tazobactam, fluoroquinolones, and aminoglycosides or antifungal amphotericin.

MIDDLE EAR INFECTIONS

Otitis Media

- This is a middle ear infection that develops following a viral throat infection or rhinosinusitis (RSV, influenza, rhinovirus infections) or bacterial infection (*S. pneumoniae*, *H. influenzae*-non-typeable, and *M. catarrhalis* and MRSA).
- **Acute otitis media** manifests as painful discharge due to fluid accumulation in the middle ear, diminished hearing, fever, vertigo, and nystagmus.
- **Chronic otitis media** manifests as persistent infection with conductive hearing loss and otorrhea. There may be perforation of the tympanic membrane.
- **Laboratory diagnosis** is based on the culture of discharge collected from the ear.

- **Treatment** is with amoxycillin–clavulanic acid or piperacillin–tazobactam or as per the sensitivity of the isolated pathogen.

Mastoiditis

- This is an infection in the mastoid air cells, which leads to the erosion of the adjoining bone and the formation of abscesses ranging from periosteal—through the mastoid tip—to deep neck abscesses.
- It generally follows untreated middle ear infection.
- **Causative agents** are *Pseudomonas*, *Staphylococcus*, *Pneumococcus*, or *Haemophilus*.
- **Symptoms** include earache accompanied by headache, fever, hearing loss, and discharge from the ear.
- **Diagnosis** is based on culture of the ear discharge.
- **Treatment** consists of empiric treatment with a broad-spectrum antibiotic, followed by drainage of pus and specific antibiotics.

CONGENITAL INFECTIONS

- Congenital infections are **infections acquired by the fetus during gestation from the mother**, who may have the active infection or a latent form of the infection. Most infections during pregnancy are asymptomatic or silent.
- Congenital infections are transmitted from mother to child during pregnancy (intrauterine) or during delivery (peripartum).
- These infections contribute to the burden of miscarriages, birth defects, stillbirths, and neonatal deaths.
- Prompt detection and treatment, vaccination if available, and other preventive strategies can reduce the number of adverse pregnancy outcomes.
- These infections result when the **infective organisms cross the placental barrier and enter the bloodstream of the fetus**.
- If the infection is acquired by the fetus in the early stage of development and organogenesis, the pregnancy generally ends in abortion. Alternatively, the baby is born with major congenital defects.
- If the infection is acquired in the later trimesters, the pregnancy may go to term and the baby may manifest the disease during the early or late neonatal period. The causes of congenital infections are listed in Table 30.2. Table 30.3 lists agents that can cause intrauterine and perinatal infections.

Table 30.2 Common microorganisms associated with congenital infections

Bacteria	Viruses	Parasites
Treponema pallidum *Streptococcus agalactiae* *Listeria monocytogenes* *Mycobacterium* *Campylobacter fetus* Gram-negative enteric pathogens *Borrelia burgdorferi*	CMV VZV Rubella HSV Parvovirus 19 Hepatitis B and C HIV Papillomavirus	*Toxoplasma* *Plasmodium*

Table 30.3 Agents that cause intrauterine and perinatal infections

Intrauterine (in utero)	Perinatal (intrapartum)
Viruses CMV, HSV, Parvovirus B19, enteroviruses, hepatitis B virus, VZV, HIV, rubella	**Viruses** CMV, HSV (type 2), hepatitis B virus, HIV
Bacteria *Treponema pallidum, Mycobacterium tuberculosis, Listeria monocytogenes, Campylobacter fetus, Borrelia burgdorferi*	**Bacteria** *Streptococcus agalactiae*, (Group B *Streptococcus*), *Enterococcus* species, *Escherichia coli, Neisseria gonorrhoeae, Listeria monocytogenes, Chlamydia trachomatis*
Fungi *Candida albicans*	**Fungi** *Candida albicans*
Parasites *Toxoplasma gondii, Plasmodium* spp., *Trypanosoma cruzi*	**Parasites** *Toxoplasma gondii, Trichomonas vaginalis*

LABORATORY DIAGNOSIS

- **Antenatal diagnosis** of congenital infections may be made by the following methods:
  - **Radiological investigations** such as ultrasound examination
  - Demonstration of **antibodies in the mother**
  - **Amniocentesis** to diagnose in-utero infections
- **Postnatal diagnosis** is made by the following methods:
  - **Specific IgM** to the infectious agent in the blood of the neonate (as it is produced in the fetus in response to the infection) by immunochromatography or ELISA
  - **Culture** of blood (in case of neonatal septicemia), CSF (meningitis), or other appropriate samples
  - Viral agents may be detected by **nucleic acid amplification** by PCR from the blood or tissues

Broadly there are three diagnostic strategies:

i. **Serology in maternal serum/blood**
   - Congenital infections due to CMV, HSV, VZV, Parvovirus B19, and Toxoplasma are usually the result of primary infection in the mother. Therefore, when IgG antibodies are detected before pregnancy, these infections are ruled out.
   - **TORCH assay** and **syphilis serology** are recommended for pregnant women. The former tests for antibodies to *Toxoplasma gondii*, rubella virus, CMV, and HSV.

- Definitive serologic diagnosis of acute infection in pregnant women requires the demonstration of seroconversion (i.e., from seronegative to seropositive). In these cases, amniotic fluid or fetal blood sampling should be performed to determine the presence of intrauterine congenital infection. **Avidity testing** may be performed to assess the duration of the infection—high avidity indicates past infection.

ii. **Microbiologic study of amniotic fluid or fetal blood sampling**
   - Cytomegalovirus, rubella, and toxoplasmosis can be diagnosed by the detection of specific IgM antibodies in fetal blood.
   - PCR performed on amniotic fluid has replaced conventional prenatal diagnostic techniques, including fetal blood sampling, in the diagnosis of these infections. B19 erythrovirus can be detected by PCR in amniotic fluid or fetal blood.
   - Definitive diagnosis of herpes simplex virus infection requires viral isolation.

iii. **Serology in children and microorganism detection by polymerase chain reaction (PCR) or culture**
   - In the newborn, congenital infections may be confirmed by measuring IgM specific antibodies.
   - Congenital varicella-zoster infection may be diagnosed on the basis of persistence of IgG antibodies after birth.

TREATMENT

- In case of **bacterial infections**, specific antibiotics are given in pediatric doses.
- **HIV** is treated with zidovudine, nevirapine, and lamivudine. Such patients have to be treated for life.
- **Toxoplasmosis** is treated with pyrimethamine and sulphadiazine.

PREVENTION

Prenatal counselling and screening help in the early detection of congenital infections. All pregnant women are screened for common infections and treated adequately to prevent congenital infections.

CONGENITAL INFECTIOUS SYNDROMES

Congenital Cytomegalovirus Infection

Epidemiology

- Primary cytomegalovirus (CMV) infection in pregnancy is associated with the highest risk of transplacental transmission.
- Congenital cytomegalovirus infection is the most common intrauterine infection in the USA. It is also the most common cause of **non-genetic sensorineural hearing loss in children**. In almost 85 to 90% cases, the disease is asymptomatic (85 to 90%).
- Primary infection of mother occurs when a previously uninfected individual acquires the disease for the first time while non-primary infections result from the reactivation of a previous infection or infection with a different strain of the virus.
- Primary infection is associated with the greatest risk of transplacental transmission (30 to 35%) in comparison to 1.1 to 1.7% for non-primary infections.

Transmission

- CMV-infected children secrete the virus in their urine and saliva for about 2 years.
- Shedding of CMV in urine and cervicovaginal secretions increases with the advancement of pregnancy. Hence the transmission of the disease is more likely at more advanced pregnancy stages, but the long-term sequelae are less likely to occur if the fetus is infected later in the pregnancy.

Clinical features

- Maternal CMV infection is frequently asymptomatic.
- Non-specific symptoms such as fever, malaise, headache, pharyngitis, lymphadenopathy, hepatosplenomegaly, arthralgias, and rashes may be present.
- Intrauterine transmission is more common, but perinatal and transmission from breast milk is also seen.
- Intrauterine infection leads to fetal death or cytomegalic inclusion disease of the newborn, which is often fatal.
- Generalised infection occurs, which is associated with hepatosplenomegaly, jaundice, thrombocytopenic purpura, and hemolytic anemia.
- The cytomegalic inclusion disease is probably the most important cause of microcephaly.
- Other manifestations include chorioretinitis and cerebral calcification resembling congenital toxoplasmosis. Survivors may show mental retardation.
- Placental inflammation and fetal death may occur.
- In neonates, the most common signs of congenital cytomegalovirus are jaundice, petechiae, hepatosplenomegaly, and microcephaly.
- Chorioretinitis with or without optic atrophy and cataracts may be present.
- Newborns could be premature or small for gestational age or both. Hearing loss may be present at birth or present later in life.

Laboratory diagnosis

The diagnostic scheme for congenital CMV infection is presented in Fig. 30.3a and b.

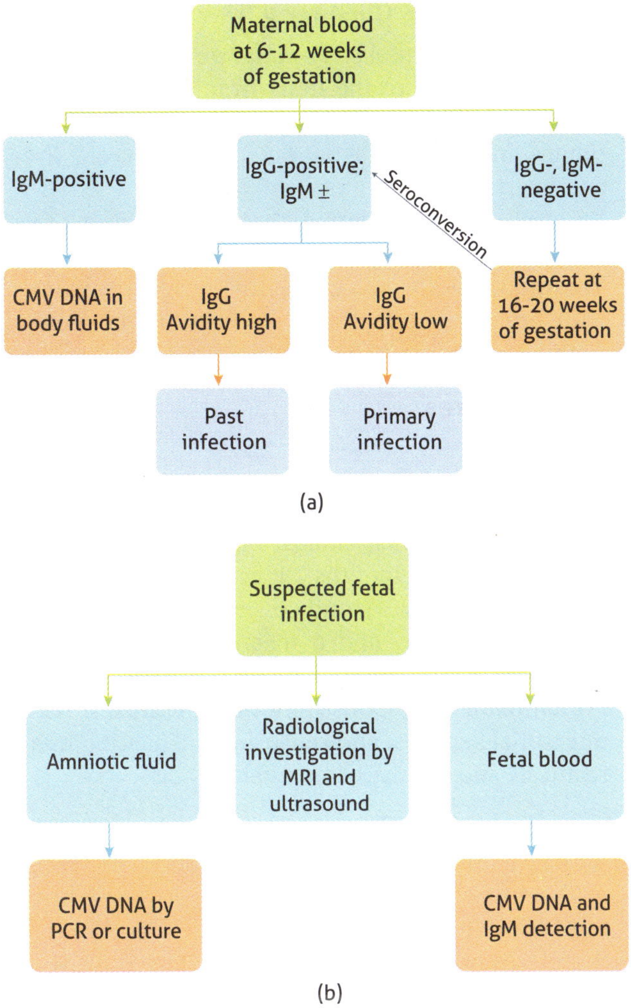

Fig. 30.3 (a) Diagnosis of maternal CMV infection and (b) diagnosis of fetal CMV infection
Note: Maternal IgM can persist up to 8 months after infection

Congenital Herpes Simplex

The infection can be acquired by fetus in the intrauterine, perinatal, or postpartum period, but is generally acquired during perinatal period.

Clinical features

- The clinical presentation can be:
  - Disease localised to the skin, eyes, and/or mouth **(SEM disease)**
  - **Encephalitis**, with or without SEM involvement
  - **Disseminated infection** involving CNS, lungs, liver, adrenal glands, skin, eyes, and/or mouth
- The infant presents with a typical triad of **cutaneous manifestations** (scarring, active lesions, hypo- and hyperpigmentation, aplasia cutis, and/or an erythematous macular exanthem), **ophthalmologic findings** (microopthalmia, retinal dysplasia, optic atrophy, and/or chorioretinitis), and **neurologic involvement** (microcephaly, encephalomalacia, hydranencephaly, and/or intracranial calcification).

Agent—herpes simplex virus

- HSVs are large, enveloped viruses with an icosahedral nucleocapsid consisting of 162 capsomeres arranged around a linear, double-stranded DNA core. The genome consists of two covalently linked components, designated L (long) and S (short).
- HSV-1 infections typically occur above waist while HSV-2 infections usually involve the genitalia and skin below the waist in sexually-active adolescents and adults. However, either type of virus can be found in either area, and both HSV-1 and HSV-2 cause herpes disease in neonates.
- Both HSV-1 and HSV-2 can establish latency following primary infection, with periodic reactivation to cause recurrent symptomatic disease or asymptomatic viral shedding.
- HSV-1 and HSV-2 are two of the eight known viruses that comprise the human herpesvirus family.

Laboratory diagnosis

- If perinatal HSV exposure is known, then **swabs from neonate's conjunctiva, oropharynx, and rectum** must be collected within 24–48 hours after delivery.
- If the neonate shows symptoms of infection in 4–6 weeks after birth, then **cultures** of vesicular, conjunctival, oropharyngeal, stool/rectal swabs, and urine, and blood must be done.
- **Microscopy:** Tzank smear on skin/mucosal lesions is an easy, quick, reproducible, and inexpensive method.
- **Antigen detection:** This test can be performed on smears from lesions and vesicular content collected from the base of vesicle using ELISA.
- **Antibody detection:** ELISA or point-of-care rapid tests can be used. Assays are now available to detect antobody specific to HSV 1 and 2.
- **Viral isolation:** The virus can be isolated in cell culture and detected using immunofluorescence.
- **Molecular diagnostics:** PCR is used to detect DNA in CSF and blood.

Fetal Varicella Syndrome

- It is a rare condition of the newborn that presents with low birth weight, limb defects, cutaneous scars, and ocular and central nervous system abnormalities.
- It is caused by **herpes zoster or varicella virus**, both of which are transmitted to the fetus following maternal varicella infection during early pregnancy. The risk of transmission is highest when the maternal infection occurs between the 13th to 20th week of gestation.

Congenital Zika Syndrome

- Zika virus (ZIKV) is a **flavivirus** (family Flaviviridae), an **RNA virus** primarily transmitted by *Aedes* mosquitoes. This infection has been detailed in Chapter 24.

- Primary Zika infection during the first and early-second trimesters of pregnancy is more commonly reported in infants with adverse outcomes that are severe. Infection in the third trimester is associated with less severe defects of the brain and eyes.

Clinical features

Congenital Zika infection can lead to a spectrum of birth defects. The unique features of Zika infection in infants are as follows:

- Severe microcephaly with a partially collapsed skull and redundant scalp with rugae (extra skin folds)
- Thin cerebral cortices with subcortical calcifications.
- Macular scarring and focal pigmentary retinal mottling
- Congenital contractures of the major joints (arthrogryposis)
- Marked early hypertonia or spasticity and symptoms of extrapyramidal involvement

Other anomalies similar to congenital CMV infection may be seen.

Laboratory diagnosis

Pregnant women living in areas with known Zika virus transmission, those likely to have been exposed through travel to an endemic area, and those who develop rashes or other symptoms consistent with possible Zika infection should be tested.

In case of clinical suspicion, specimens should be collected from mothers and liveborn infants and tested. Patients should also be tested for infection with dengue, chikungunya, and other circulating flaviviruses in endemic areas.

- **Serology:** The finding of IgM antibody to Zika should be interpreted with caution. Firstly, false-positive results may be high due to non-specific reactivity or cross-reactivity among antigens found on Zika as well as on dengue and other flaviviruses. Secondly, the IgM antibodies can persist up to two years.
- **Plaque reduction neutralisation tests (PRNTs):** These are quantitative assays that measure virus-specific neutralising antibody titres for dengue, Zika, and other flavivirus. PRNTs can resolve false-positive IgM antibody results caused by non-specific reactivity and can help in the identification of the infecting virus.
- **Molecular testing:** Viral RNA or antigen can be detected in serum, saliva, tissues, urine, or whole blood.

Congenital Parvovirus Infection

Parvovirus B19 infection transmitted during pregnancy can cause **dropsy (hydrops fetalis)** and abortion. It also causes **erythema infectiosum or slapped cheek syndrome** (details in Chapter 29).

Congenital Syphilis (Chapter 27)

- In congenital syphilis, *T. pallidum* is released into the bloodstream of the fetus from mother, causing spirochetemia and resulting in its spread to most organs including the bones, kidney, spleen, liver, and heart.
- Widespread inflammation throughout these organ systems leads to a variety of clinical manifestations.
- **Treatment** with penicillin for pregnant women with syphilis is effective for preventing congenital syphilis.

Clinical features

- In severe cases, congenital syphilis results in fetal death.
- Most neonates are asymptomatic at birth. However, in untreated infants, clinical manifestations usually appear by three months of age and may present as hepatomegaly, jaundice, rhinitis, generalised lymphadenopathy, and a maculopapular rash that progresses to desquamation and crusting.
- **Hutchinson triad** consisting of **Hutchinson teeth**, interstitial keratitis, and sensorineural hearing loss is relatively specific for congenital syphilis.

Laboratory diagnosis

- All pregnant women should be screened for syphilis in their first prenatal visit and repeat-tested in the third trimester.
- Standard tests for syphilis are often positive at birth due to maternally transferred IgG antibodies. For this reason, a single reading cannot be used for making a diagnosis. If the mother was treated appropriately during pregnancy and the newborn is clinically normal, then serial determinations of the VDRL or RPR in the newborn can be done.
- If the positive test is due to maternal transfer of antibodies, then the titer will decrease dramatically in the first two months of life.
- Another method is to determine the specific IgM against *T. pallidum*. It does not rule out congenital infection in the case of a negative result, but a positive result is indicative of active infection in the child and justifies treatment.

Congenital Listeriosis

- Congeital infection by *Listeria monocytogenes* can cause miscarriage, premature birth with chorioamnionitis (with typical brown, cloudy amniotic fluid), intrauterine death, or neonatal sepsis.
- The infection can occur transplacentally or during or after birth. It can manifest as a neonatal sepsis, pneumonia, or meningitis and has a high mortality rate.

- Severe infections can be accompanied by **granulomatous exanthema (granulomatosis infantiseptica)** or **microabscesses** all over the body, especially in the liver and spleen.
- Infection of the fetus in utero can lead to dissemination with the formation of granulomas (skin, liver, adrenal glands, lymphatic tissue, lungs, and brain).
- Neonatal infection can manifest early, in the first hours to days after birth, or late, i.e., several weeks after birth.
- Early infection is associated with low birth weight, obstetric complications, and shortly after birth, circulatory and/or respiratory insufficiency manifests itself. It usually has the appearance of sepsis or meningitis and has a high mortality rate.
- Late infection usually occurs in full-term, initially healthy newborns and occurs under the guise of purulent meningitis or sepsis.
- Details of *Listeria* are presented in Chapter 24.

Congenital Malaria

- Congenital malaria is defined as the presence of malarial parasites in the peripheral smear of the newborn within 24 hours to seven days of life.
- Clinically apparent congenital malaria is rare in areas in which malaria is endemic and levels of maternal antibody are high.
- Normally, symptoms present after 10 to 30 days postpartum. The most common clinical features are **fever, anemia,** and **splenomegaly**. Other signs and symptoms include hepatomegaly, jaundice, regurgitation, loose stools, and poor feeding. Occasionally, drowsiness, restlessness, and cyanosis may be seen.
- Typical signs and symptoms of malaria are not seen in the newborn, as these parasites can only use reticulocytes for replication and these cells are scarce in newborns due to depressed erythropoiesis after birth.
- The placenta is involved in most women who acquire malaria during pregnancy. Malaria in pregnancy can cause still birth, low birth weight, and hepatosplenomegaly.
- It is not clear whether transmission to the infant is transplacental or through direct contact with maternal blood during labour and/or parturition.
- **Diagnosis** is by smear examination.
- The drug of choice is **chloroquine**.

CONGENITAL TRYPANOSOMIASIS/LEISHMANIASIS

- Trypanosomas and leishmanias can cross the placenta. ***Trypanosoma cruzi*** may infect the placenta, with or without fetal transmission.
- Most babies born with congenital Chagas disease have mild or no signs of illness. However, infection will persist for life and these infants are at risk for developing clinical manifestations of chronic Chagas disease later in life, unless treated.
- Infants born with signs of Chagas disease may present with low birth weight, prematurity, low Apgar scores, hepatosplenomegaly, anemia, and thrombocytopenia.

Congenital rubella syndrome is elaborated in Chapter 23 and congenital toxoplasmosis is discussed in Chapter 24.

Questions

Short answer questions

1. List the types of ocular infections and the methods to identify the causative agents.
2. What are the common infections of the middle and internal ear? List the organisms associated and their treatment.
3. List the microorganisms associated with congenital infections and the consequences of these infections in relation to the trimester of pregnancy.

MCQs

1. **Which of the following is viral agent causing conjunctivitis?**
   a) Enterovirus 70 b) Enterovirus 55
   c) Orf virus d) All of the above
2. **Which of the following is not correct regrading keratitis?**
   a) It refers to the inflammation of the cornea.
   b) It can be caused by bacteria such as *Pseudomonas* and *Staphylococcus*.
   c) Viruses such as enterovirus-70 (EV-70) and coxsackie-A24 can also cause keratitis.
   d) Free living parasites in water such as free-living amoeba can be involved.
3. **Ophthalmia neonatorum is:**
   a) Conjunctival inflammation occurring in infancy.
   b) Caused by *N. gonorrhoeae*.
   c) Characterised by severe purulent conjunctivitis.
   d) Diagnosis is by PCR only.
4. **A 25-year-old villager bathed in a holy pond before visiting a temple. After 10 days, he developed photophobia, reddening and watering in his eyes. What is the likely etiological agent in this case?**
   a) *E. vermicularis* b) *D. medinensis*
   c) *A. fumigatus* d) *Acanthamoeba* spp.
5. **A newborn developed reddening of the eyes and subsequently exhibited purulent discharge 10 days after birth. All the following organisms can be the causative agents of this condition except:**
   a) *C. trachomatis* b) *N. gonorrhoeae*
   c) Cytomegalovirus d) Herpes simplex virus
6. **Dacryocystitis is clinically characterised by:**
   a) Infection of the tissues surrounding the eyeball
   b) Swelling and blockage of the eye's tear drainage system
   c) Inflammation of the base of the eyelashes
   d) Inflammation of the cornea

7. **All the following are causative organisms of otitis externa except:**
   a) *Actinomycetes* b) *P. aeruginosa*
   c) *S. aureus* d) *H. influenzae*
8. **All the following are causative organisms of otitis media except:**
   a) *Actinomycetes* b) *M. catarrhalis*
   c) *S. pneumoniae* d) MRSA
9. **A seven-day-old newborn was diagnosed with a congenital infection caused by bacteria. Which one of the following is not likely to be associated with such infections?**
   a) *T. pallidum* b) *Listeria* spp.
   c) *S. agalactiae* d) *S. saprophyticus*
10. **A 25-year-old man presented with painful diminution of vision, accompanied by redness, pain, and floaters, over a period of two weeks, in his left eye. On examination, the anterior segment revealed moderate inflammation and a fairly long live worm moving around in subconjunctival area. Which parasite is likely to be responsible?**
    a) Free-living amoeba b) Larval forms of *Loa loa*
    c) *Echinococcus granulosus* d) *Toxoplasma gondii*
11. **A pregnant woman needs to be screened for all the following organisms causing congenital infections EXCEPT:**
    a) Rubella virus b) Adenovirus
    c) Cytomegalovirus d) *Toxoplasma gondii*
12. **Slapped cheek disease is caused by:**
    a) Parvovirus B-19
    b) Human papilloma virus
    c) Human polyoma virus
    d) Herpes simplex virus
13. **A woman visited the hospital for antenatal care. Her fetal sonography showed an anomalous fetus. Which microbiological investigation should she undergo?**
    a) Widal test b) TORCH panel
    c) HIV ELISA d) HCV ELISA
14. **Which of the following is correct regarding the presentation of fetal varicella infection?**
    a) Newborn presents with low birth weight, limb defects, cutaneous scars, and central nervous system abnormalities.
    b) It is transmitted by mother during delivery or labour.
    c) Hutchinson triad may be present.
    d) Slapped cheek appearance is common
15. **Which bacteria is not associated with congenital infections?**
    a) *Treponema pallidum*
    b) *Streptococcus agalactiae*
    c) *Listeria monocytogenes*
    d) *Toxoplasma gondii*

Answers: 1:a, 2:c, 3:c, 4:d, 5:c, 6:b, 7:d, 8:a, 9:d, 10:b, 11:b, 12:a, 13:b, 14:a, 15:d

CHAPTER 31

Zoonotic and Vector-Borne Infections

(MI8.1)

Chapter Navigation

- Zoonotic infections
- Vector-borne infections
- Control of zoonotic and vector-borne infections
- Cat scratch disease—*Bartonella henselae*
- Rat bite fever—*Bartonella henselae, Streptobacillus moniliformis, Spirillum minus*
- One health

Learning objective

- Enumerate the microbial agents and vectors causing zoonotic diseases and the laboratory diagnosis of these infections

ZOONOTIC INFECTIONS

- Diseases and infections of animals that are transmitted to human hosts through direct contact, by the ingestion of a product of the animal, or through the bite of vectors are known as zoonotic infections (Table 31.1; Fig. 31.1).
- Some of these infections are distributed globally, some are endemic in certain areas, and some appear in epidemic form.
- Zoonotic infections are also closely related to the type of occupation an individual is involved in.
- Some of these infections are reportable diseases due to high transmissibility and severity.
- The epidemiology of zoonotic infection depends on the following factors:
  - **Geographic location:** Cattle- or sheep-rearing areas or areas with a high vector population, e.g., Africa, where the tsetse fly population is high
  - **Frequency of contact with animals:** Farmers, agricultural workers, and others who frequently come into close contact with animals or their products

Table 31.1 Common zoonotic infections

Bacteria	Rickettsia (vector-borne)	Viruses	Parasites	Fungi
Anthrax	Scrub typhus	Rabies	Taeniasis	Zoophilic Dermatophytoses
Plague (vector-borne)	Murine typhus	Yellow fever (vector-borne)	Echinococcosis	Cryptococcosis
Brucellosis	Tick typhus	Japanese encephalitis (vector-borne)	Leishmaniasis (vector-borne)	
Leptospirosis	Q fever	KFD (vector-borne)	Toxoplasmosis	
Salmonellosis		Chikungunya (vector-borne)	Trichinellosis	
Psittacosis (*Chlamydia*)		Cowpox	Toxocariasis	
Listeriosis				

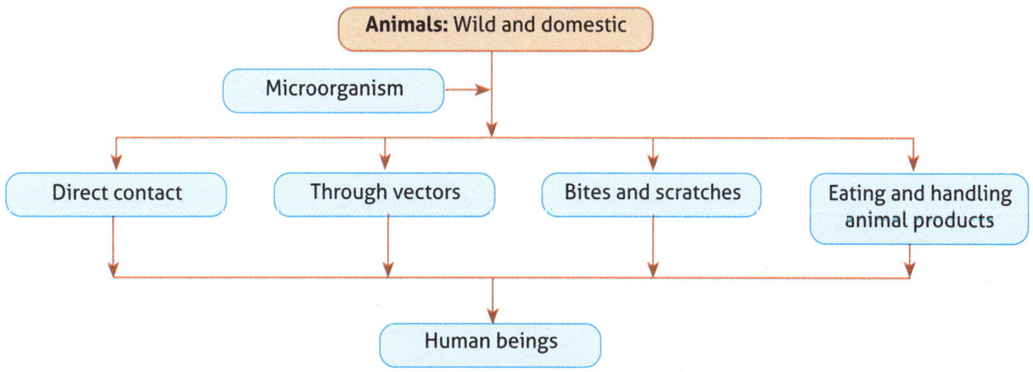

Fig. 31.1 The pathway of zoonotic infections

- **Occupation of the individual:** Tannery workers, sheep shearers, and butchers
- **Food habits of individuals:** Those who consume raw or partially cooked fish and meat
- **Domestic animals:** Hydatid disease is transmitted by dogs; cats transmit toxoplasmosis; and psittacine birds (parrots and pigeons) transmit psittacosis

Zoonotic infection may be confined to the individual who comes into contact with the infected animal or it may result in an outbreak when a group of persons are exposed at the same time.

Etiology

The most common zoonotic infections are listed in Table 31.1. The ones most commonly reported in India are as follows:

- Salmonellosis
- Leptospirosis
- Anthrax
- Brucellosis (in certain geographic areas)
- Scrub typhus
- Japanese encephalitis
- Rabies
- Chikungunya
- KFD
- Leishmaniasis
- Nipah virus infection

Factors Influencing the Emergence of Zoonotic Diseases

- **Changes in the environment** in which humans live and **agricultural practices**, as in the case of leptospirosis, plague, Rift Valley fever, and Kyasanur Forest disease
- **Increased movement or traveling** of human beings, as with amoebiasis, giardiasis, colibacillosis, salmonellosis, SAARS, and yellow fever
- **Handling of animal byproducts and waste**, e.g., anthrax, chlamydiosis, dermatophytosis, tularaemia
- Increase in the **density of animal population**, e.g., dermatophytosis and tuberculosis
- Increased **trade of animal products**, e.g., anthrax, brucellosis, salmonellosis, hantavirus infection, and bird flu
- Changing **livestock farming practices**, e.g., *E. coli* O157:H7, salmonellosis, and listeriosis
- Changing environmental conditions including **climate changes and disasters,** e.g., plague and leptospirosis
- Pathogen changes like **genetic shift and drift**, e.g., influenza and *E. coli* and *Staphylococcus* infection

Laboratory Diagnosis of Zoonotic Infections

- It is important to diagnose zoonotic infections in humans and animals, not only to treat them but also to promptly enact containment measures to prevent outbreaks.
- Besides loss of human life, economic loss may also be associated with outbreaks in animals, as has been seen with outbreaks of anthrax.
- A detailed history regarding occupation and duration of exposure to animals must be elicited from patients. History of unnatural animal deaths in the vicinity must also be enquired.

Laboratory diagnosis is based on the following:

- **Isolation** of the causative agent
- **Antigen/Antibody** detection
- **Autopsy** of the infected dead animal

Collection of specimens

Specimens are collected according to the site of the lesion. Tables 31.2, 31.3, 31.4, 31.5, and 31.6 list the types of specimens to be collected and laboratory diagnostic tests for bacterial, rickettsial, viral, parasitic, and fungal zoonoses, respectively.

Table 31.2 Laboratory procedures for the diagnosis of bacterial zoonoses

Disease	Specimen	Microscopy	Culture and appearance	Serology	Others
Cutaneous anthrax	Fluid from eschar	Gram-positive bacilli (bamboo stick appearance)	Nutrient agar—Medusa head	• Ascoli's thermoprecipitin test • CFT	Lysis by gamma phage
Pulmonary anthrax	• Sputum • Pleural fluid	Gram-positive bacilli (bamboo stick appearance)	Blood agar—string of pearls	ELISA to detect protective antigen (PA) and lethal factor (LF) IgG	Direct fluorescent antibody test
Intestinal anthrax	• Ascitic fluid • Biopsy fluid • Lesion swab (autopsy)	McFadyean's reaction	Gelatin stab—inverted fir tree	ELISA	Real-time PCR

(Table 31.2 Contd. . .)

Table 31.2 Contd.

Disease	Specimen	Microscopy	Culture and appearance	Serology	Others
Acute brucellosis (undulant fever)	Blood	Gram-negative coccobacilli	Castaneda method	Standard agglutination test, ELISA, CFT	Skin test
Bubonic plague	• Fluid from buboes • Sputum	• Gram-negative bacilli • Bipolar staining	• Nutrient agar • Blood agar	Passive hemagglutination	PCR
• Pneumonic plague • Septicemic plague	Blood	Safety pin appearance	Ghee broth—'stalactite' growth	No sensitive serological test is available	• Rapid F1 antigen detection test • PCR
Salmonellosis	• Stool • Food	Gram-negative bacilli	• MacConkey agar • Wilson and Blair medium	Widal test	
Leptospirosis	• Blood • Urine	Dark-ground microscope; spirochete	• Korthof medium • Stuart's medium • Fletcher's medium	Microscopic agglutination test	PCR
Tuberculosis (*M. bovis*)	Sputum	Acid-fast bacilli	LJ medium	No serological tests for humans, but tests are available to screen animals	PCR

Table 31.3 Laboratory procedures for the diagnosis of rickettsial zoonoses

Disease	Specimen	Culture	Serology
Scrub typhus	Blood	Yolk sac of chick embryo	Weil–Felix test
Murine typhus	Blood	Yolk sac of chick embryo	Weil–Felix test
Tick	Blood	Yolk sac of chick embryo	CFT

Table 31.4 Laboratory procedures for the diagnosis of viral zoonoses

Disease	Specimen	Microscopy	Culture	Serology/molecular detection
Rabies	• Antemortem—corneal impression smear • Skin biopsy; saliva • Postmortem—brain	• Immunofluorescence and immunoperoxidase staining • Negri bodies	Tissue culture (WI 38, BHK 21)	• Rabies virus neutralisation test to detect antibodies in CSF • Reverse transcriptase PCR
Yellow fever	Blood	Not applicable	Embryonated eggs and tissue culture	Hemagglutination inhibition
Japanese encephalitis	CSF	Not applicable	Intracerebral inoculation into suckling mice	• CFT • Neutralisation
KFD	Blood	Not applicable	Tissue culture	Immunofluorescence
Chikungunya	Blood	Not applicable	Human embryonic lung tissue and Vero cell lines	• ELISA IgM antibodies capture • ELSA (MAC-ELISA) and molecular diagnosis by nested PCR and real-time PCR

Table 31.5 Laboratory procedures for the diagnosis of parasitic zoonoses

Disease	Specimen	Microscopy	Serology
Taenia	Stool	Egg	ELISA
Echinococcus	Stool	Hydatid cysts with scolices from biopsies or after surgical removal	• ELISA • Hemagglutination • Indirect fluorescent antibody
Leishmania	Blood	Peripheral blood to visualise amastigote (LD bodies)	• Aldehyde test • Antimony test • CFT

Table 31.6 Laboratory procedures for the diagnosis of fungal zoonoses

Microscopy	Disease	Specimen	Culture
Zoophilic dermatophytes			
KOH preparation	*Microsporum canis*	Skin scraping	SDA
	Trichophyton verrucosum	Hair clipping	
	T. equinum	Nail clipping	

VECTOR-BORNE INFECTIONS

- These infections are transmitted by vectors (mosquitoes, ticks, and fleas) through bites.
- These vectors carry infective pathogens such as viruses, bacteria, and protozoa, which can be transferred from one host (carrier) to another when the insect takes its blood meal.
- Changing ecosystems and environmental degradation have resulted in the rise of vector-borne infections.
- Some common vector-borne infections and their vectors and reservoirs are presented in Table 31.7.
- The interaction between vectors, hosts and reservoirs is as presented in Fig. 31.2.

Table 31.7 Common vector-borne diseases

Diseases	Vector	Reservoir	Details in chapter
Plague	Rat flea	Rats	20
Rickettsial diseases	Ticks	Deer, rodents	20
Lyme disease	Ticks	Deer, rodents, opossums	20
Tularemia	Ticks, biting flies	Rodents	20
Japanese encephalitis	Female mosquito (*Culex*)	Water birds	23
West Nile disease	Female mosquito (*Culex*)	Wild birds	20
Dengue	Female mosquito (*Culex*)	Primates/infected persons with viremia	20
Yellow fever	Female mosquito (*Aedes aegypti*)	Primates	20
Chikungunya	Female mosquito (*Aedes aegypti/albopictus*)	Primates/birds	20
Malaria	Female mosquito (*Anopheles*)	Infected human being	20
Leishmaniasis	Sandfly	Humans for Indian kala azar Dogs are reservoirs for *L. donovani* in India, rodents in Africa and foxes in central Asia	20
Filariasis	Female mosquito (*Anopheles*)	Infected human being	20
Trypanosomiasis	Tsetse fly	Cattle, horses	20

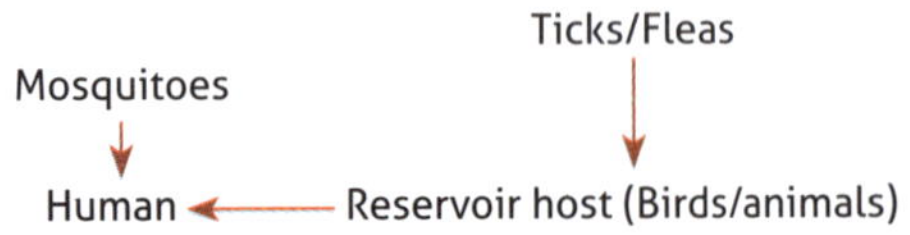

Fig. 31.2 Interaction between vectors, hosts, and reservoirs in vector-borne infections

CONTROL OF ZOONOTIC AND VECTOR-BORNE INFECTIONS (Fig. 31.3)

- Zoonotic and vector-borne infections are reportable diseases against which preventive and containment measures can be initiated.
- In India, when a disease is suspected, the health authorities are informed through the **National Vector-Borne Disease Control Programme (NVBDCP).**
- Veterinarians and public health experts visit the sites of outbreak and talk to the affected individuals and groups.
- They collect further samples for laboratory confirmation, provide technical information, and undertake preventive measures. A classic example of this activity is seen in cases of anthrax outbreaks.

Types of Preventive Measures

1. **Primary prevention:** Comprises activities directed towards preventing exposure to causal factors, e.g., quarantine and vaccination

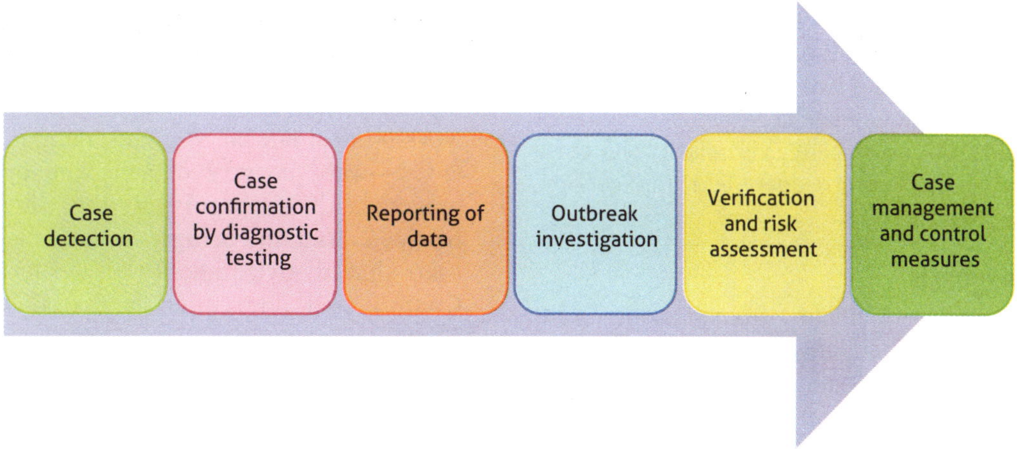

Fig. 31.3 Control of zoonotic and vector-borne infections

2. **Secondary prevention:** Consists of activities designed to detect the disease process as early as possible, before the clinical disease occurs, e.g., screening test to detect zoonotic disease, slaughter and depopulation of the animal host if required.
3. **Tertiary prevention:** Prevention by treatment

CAT SCRATCH DISEASE

- Cat scratch disease is characterised by a febrile illness **following a cat scratch associated with regional lymphadenopathy**.
- Regional lymphadenopathy is seen in the lymph nodes draining the site where a cat scratch or bite occurs. It can also lead to chronic lymphadenopathy that may spread beyond the site of inoculation and disseminate to the eye, liver, spleen, and central nervous system (CNS). In most cases, it resolves spontaneously.
- It is caused by an intracellular gram-negative rod known as *Bartonella henselae.*
- Bartonellae are Gram-negative bacilli which invade mammalian endothelial cells and blood cells. Human pathogenic strains are *B. bacilliformis* (causes **Carrion disease or Oraya fever**), *B. quintana* (causes **trench fever or five-day fever**), and *B. henselae* (causes cat scratch disease).
- *B. henselae* has also been linked with two other conditions, seen more commonly in HIV-infected and other immunodeficient persons. These are bacillary angiomatosis in which vascular nodules or tumors are seen on skin, mucosa, and other locations along with bacillary peliosis involving the liver and spleen.
- **Diagnosis** is made by careful history taking and examination. *Bartonella henselae* has been isolated from blood after prolonged incubation. It can also be demonstrated in lymph node biopsy by Warthin-Starry staining.

RAT BITE FEVER (RBF)

- Rat bite fever is characterised by relapsing lever, rash and arthralgia occurring days or weeks after a rat bite.
- Two different bacteria can cause this condition-*Streptobacillus moniliformis* and *Spirillum minus*, both of which are natural parasites of rodents.
- The infection is usually transmitted by a rat bite. Infection can occur after exposure to the urine or other bodily secretions of an infected rodent and also through food or water contaminated with feces or urine. Household pets exposed to rodents can carry the disease and subsequently infect humans.

RBF Caused by *Streptobacillus moniliformis* (Haverhill Fever or Erythema Arthriticum Epidemicum)

- It develops 2–10 days after exposure, with abrupt onset of fever, headache, and myalgia, followed by the development of a petechial rash and arthritis.
- Relapses are common in untreated cases.
- The disease can also occur as outbreaks without rat bites.
- It is believed to be caused also by the consumption of raw milk or water contaminated by rats.

Agent—*Str. moniliformis*

It is a highly pleomorphic, Gram-negative, non-motile bacillus.

RBF Caused by *Spirillum minus* (Soduku)

- It has a shorter incubation period (1–4 days). The rat bite wound, which may have healed, suppurates at the onset of fever with regional lymphadenopathy.
- The subsequent course is similar to that of the streptobacillary type.
- Moratlity rates up to 10% have been reported.

Agent—*Spirillum minus*

- It is a short, actively motile, coiled bacterium, 3–5 × 0.2–0.5 μm in size, with two or three regular spirals and 1–7 amphitrichous flagella.
- It is gram-negative but is better visualised by Giemsa or Fontana stains or by dark feld microscopy. It cannot be cultivated.

Laboratory Diagnosis

- **Specimen:** Blood, exudates from lesions, joint aspirates, and other body fluids are collected.
- **Microscopy:** Smears of the joint fluid may show pleomorphic gram-negative rods. *Sp. minus* is actively motile while *Str. moniliformis* is non-motile.
- **Culture and identification:**
  - *Str. moniliformis* grows as tangled chains of rods of various lengths, with beaded or fusiform swellings, readily developing into L forms. *Sp. minus* cannot be cultivated.
  - Growth requires the presence of blood or other body fluids.
  - *Str. moniliformis* is catalase-, oxidase-, nitrate-, urease-, and indole-negative and ferments glucose and a few other sugars, forming acid but no gas.
- **Serology:** Agglutination, CF, and fluorescent antibody tests have been used for serological diagnosis.

CONCEPT OF 'ONE HEALTH'

- 'One Health' is defined as a collaborative, multisectoral, and transdisciplinary approach—working at the local, regional, national, and global levels—with the goal of achieving optimal health outcomes by **recognising the interconnection between human beings, animals, plants, and their shared environment** (US Centers for Disease Control and Prevention and the One Health Commission).
- As discussed hitherto, the health of humans, animals, and the ecosystem is closely linked.
- In order to break the chain of transmission and control infection, it is necessary to approach the problem through a multidisciplinary, collaborative effort.
- Emerging and endemic zoonoses are responsible for the major burden of disease in the developing world.
- Additionally, antimicrobial resistance (AMR) in humans, animals, and the environment spreads from one to the other.
- A concerted approach to prevent and control zoonotic infections forms an important arm of the One Health concept of disease control.

Key Points

- ✓ Zoonoses are diseases and infections of animals that are transmitted to human hosts through direct contact, by the ingestion of a product of the animal, or by the bite of a vector.
- ✓ Zoonotic diseases are caused by bacterial (e.g., anthrax and plague), viral (e.g., dengue, chikungunya), parasitic (e.g., malaria and leishmaniasis), and fungal (e.g., zoophilic dermatomycosis and cryptococcosis) agents.
- ✓ Zoonotic infections commonly seen in India include salmonellosis, leptospirosis, anthrax, scrub typhus, and Japanese encephalitis.
- ✓ Vector-borne infections are localised to the geographic area where the vectors are found, e.g., *Anopheles*-infested areas have a high incidence of malaria, whereas *Aedes* mosquito-breeding areas report a high incidence of dengue.
- ✓ Laboratory diagnosis depends on the isolation of the causative agent, antigen/antibody detection, and autopsy of the dead infected animal.
- ✓ Rat bite fever is characterised by relapsing lever, rash, and arthralgia occurring days or weeks after a rat bite.
- ✓ Two different bacteria can cause this condition—*Streptobacillus moniliformis* and *Spirillum minus*, both of which are natural parasites of rodents.
- ✓ Zoonotic and vector-borne infections are reportable diseases which can be prevented and contained with appropriate measures.
- ✓ The control programme in India is carried out by the National Vector-Borne Disease Control Programme (NVBDCP).

Questions

Short answer questions

1. Describe the epidemiology of zoonotic infections.
2. Describe a zoonotic infection associated with cattle rearing.
3. List the important history you will elicit when you suspect an outbreak of leptospirosis.

MCQs

1. **All of the following are the examples of bacterial zoonotic diseases except:**
   a) KFD b) Anthrax
   c) Plague d) Brucellosis
2. **Zoonotic infections are caused by:**
   a) *Yersinia pestis* b) *Yersinia pseudotuberculosis*
   c) *Yersinia enterocolitica* d) All of these
3. **Which of the following is correct regrading vector-borne infections?**
   a) These infections are transmitted by vectors such as mosquitoes, ticks, and fleas through bites.
   b) These infections can be transmitted by animals.
   c) Arthropod vectors such as flies can transmit these infections as fomites.
   d) Only parasites can be transmitted as vector-borne infections.
4. ***Culex* mosquito can transmit all of the following viruses except:**
   a) Japanese encephalitis b) West Nile disease
   c) Yellow fever d) Dengue
5. **Zoophilic dermatophytes includes all except:**
   a) *Microsporum canis* b) *Trichophyton verrucosum*
   c) *T. equinum* d) *Candida albicans*

6. **Which of the following is a zoonotic infection caused by Rickettsia?**
   a) Scrub typhus b) Tick typhus
   c) Taeniasis d) All of the above
7. **Castaneda method is used for diagnosis of:**
   a) Q fever b) Brucellosis
   c) Anthrax d) Plague
8. **Acute brucellosis is also known as:**
   a) Undulant fever b) Q fever
   c) Saddleback fever d) Lassa fever
9. **Korthof medium is used for the cultivation of:**
   a) *Brucella species* b) *Yersinia pestis*
   c) *Leptospira* d) *Bacillus anthracis*

Answers: 1:a, 2:d, 3:a, 4:c, 5:d, 6:c, 7:b, 8:a, 9:c

CHAPTER 32

Part III: Microbiology as Applied to Infectious Diseases

Opportunistic Infections

(MI8.2)

Chapter Navigation

Learning objectives

- Enumerate factors leading to immunosuppression in humans
- Discuss the etiopathogenesis of opportunistic infections that affect an immunocompromised person
- Describe laboratory tests to diagnose opportunistic infections

Immunocompromised individuals lack the ability to defend themselves from microbial invasions because they lack of one or more defence mechanisms in their system. Immunodeficiency may be:

- **Congenital**
- **Acquired**

Immunodeficiency may include humoral, cell-mediated or combined immunodeficiency

INTRODUCTION

- An opportunistic infection (OI) is one that arises **due to immunosuppression**.
- A paradoxical situation also arises when certain chronic infections cause immunodeficiency and give rise to secondary infection by another organism. For example, HIV infection gives rise to several secondary infections.
- The immunosuppressed individual is prone to infection by several microbial agents.

Factors that Influence the Development of Immunosuppression

- **Malnutrition:** Leads to impaired development of all body parameters including the immune system → compromised immunity → infection
- **Ageing:** Senescence of immunocompetent cells → suboptimal immune cell function → infection
- **Leukopenia**: Cytotoxic drugs (cyclophosphamide) → impaired first line of defence → infection
- **Chronic immunodeficiency diseases:** Agammaglobulinemia, SCID (severe combined immunodeficiency) → repeated bacterial infections
- **Immunosuppressing agents:** Prescribed for autoimmune disorders, transplant patients, chemotherapy → depletion of immunocompetent cells → infection
- **Genetic predisposition:** X-linked agammaglobulinemia (XLA) → increased incidence of gram-positive infections
- **Metabolic syndromes:** Diabetes → increased incidence of infections

Physical factors that may result in deranged innate immunity are as follows:

- **Burns:** Damage to the physical barrier of the innate immune system and defects in polymorphonuclear (PMN) cellular function → leads to infections caused by *Pseudomonas*, *Staphylococcus*, *Streptococcus*, or fungi such as *Fusarium* and *Zygomycetes*
- **Traumatic or surgical wound:** There is a breach in the continuity of the skin → allows organisms to enter the body and organisms that are part of the normal flora of the skin to establish infection
- **Intravenous and prosthetic devices:** The continued presence of foreign bodies in the form of cardiac pacemakers and valves, catheters, and shunts → predispose the individual to infection by agents such as *Candida*, *Staphylococcus aureus/S. epidermidis*, and other commensals of the skin
- **Impaired clearance:** Cystic fibrosis and chronic pulmonary obstructive disease → predispose the lung to infections by *Pseudomonas*

Some infections also cause immunosuppression:

- HIV
- CMV
- EBV
- Mumps
- Measles
- Tuberculosis

Details of the types, causes, and effects of immunosuppression have been discussed in Chapter 18.

CHARACTERISTICS OF OPPURTUNISTIC PATHOGENS

- Commensals organisms (*S. epidermidis*, *S. aureus*, *S. pneumoniae*) become pathogenic when the host's defences are down; such organisms are known as **colonising opportunistic pathogens**.
- These organsisms are generally of low virulence (*Pseudomonas*, *Acinetobacter*, *Burkholderia cepacia*).

- Sometimes pathogens cause severe disease due to immunosuppression. A healthy individual recovers with a mild-to-moderate infection, whereas in an immunocompromised individual, the disease becomes fulminant and often fatal. For example, *S. aureus* (virulent organism) can cause infections in both healthy and immunocompromised individuals.
- The common organisms associated with opportunistic infections have been discussed in several chapters pertaining to their primary infections (Table 32.1).
- Besides the viruses listed in Table 32.1 and discussed in earlier chapters, two major DNA viruses that commonly cause opportunistic infections are the cytomegalovirus and Epstein–Barr virus.

OPPORTUNISTIC VIRAL INFECTIONS

Cytomegalovirus Infections

Case 1. *A 25-year-old man diagnosed with HIV and AIDS presented with a complaint of diminishing vision, which had been slowly progressing for the previous three months. His CD4 counts at the time of presentation were <100 cells/mm³. On fundus examination, retinal infiltrates which were fluffy in nature were found, along with multiple hemorrhages. Serological tests were positive for CMV IgG antibodies. PCR on blood was positive for CMV DNA. The patient was diagnosed with* ***CMV retinitis****, and treatment with ganciclovir was started. Further investigation revealed that the patient had not been taking his anti-HIV treatment regularly; he was advised to resume timely treatment. Subsequently, there was no further deterioration in the patient's condition.*

i. What are the characteristics of infections caused by CMV?
ii. How is the infection acquired?

Epidemiology

- CMV spreads slowly and probably requires close contact for transmission.
- It may spread through salivary or other secretions or by sexual contact. Another method of transmission is by blood transfusion or organ transplants.
- The virus has been detected in saliva, urine, cervical secretions, semen, blood, and milk.
- Congenitally infected infants have viruria for up to 4–5 years. They are highly infectious in early infancy.
- Up to 80 per cent of adults possess CMV antibodies, indicating the high prevalence of infection. Once infected, the person carries the virus for life.

Agent—Cytomegalovirus

- Cytomegaloviruses (CMV), formerly known as salivary gland viruses, are a group of ubiquitous herpes viruses of humans and animals like monkeys, guinea pigs, and some other species. Like other herpes viruses, they lead to prolonged latency in infected hosts.
- CMV are the **largest viruses in the herpes virus family**, with sizes ranging from 150–200 nm.
- The virus exhibits strict host specificity and infection both in vivo and in vitro, i.e., it **can be established only in homologous species**.
- Human CMV is unrelated antigenically to other herpes viruses, and even to CMV of other species, except for simian CMV with which some antigenic cross-reaction exists.
- Minor genetic and antigenic differences exist among human CMV isolates but they are of no clinical significance.

Table 32.1 Organisms associated with opportunistic infections

Bacteria	Viruses	Fungi	Parasites
Staphylococcus	CMV	*Candida*	*Cystoisospora*
Streptococcus	EBV	*Cryptococcus*	*Cyclospora*
Pneumococcus	HSV-1, -2	*Histoplasma*	*Toxoplasma*
Haemophilus	Zoster	*Pneumocystis jirovecii*	*Cryptosporidium*
Campylobacter	HPV	*Penicillium*	*Leishmania*
Flavobacterium	HHV-6, -7	*Aspergillus*	*Plasmodium falciparum*
Pseudomonas	VZV	*Encephalitozoon cuniculi*	
Atypical mycobacteria (avium intercellulare complex)	HSV-8		
	Molluscum contagiosum virus		
Salmonella			
Shigella			
Mycobacterium			
Nocardia			
Treponema			

- One of the glycoproteins on the surface of CMV acts as a receptor for the F_c portion of immunoglobulin molecules. This leads to the masking of the virus by the attachment of irrelevant immunoglobulin molecules, which in turn, prevents access to specific anti-CMV antibody.

Clinical features

- Though cytomegalovirus disease is rare, infection with the virus is extremely common.
- As with herpes simplex, the large majority of infections are inapparent, leading to prolonged latency and occasional reactivation.
- In the neonate and the immunodeficient, they cause severe disseminated disease, while in healthy/ immunocompetent children and adults, the infection is usually asymptomatic or self-limiting.
- Clinical disease may be caused by either intrauterine or postnatal infections.

Congenital infections (Chapter 30)

- **Intrauterine infection** leads to fetal death or cytomegalic inclusion disease of the newborn, which is often fatal.
- **Generalised infection** occurs, which is associated with hepatosplenomegaly, jaundice, thrombocytopenic purpura, and hemolytic anemia.
- The cytomegalic inclusion disease is probably the most important cause of microcephaly.
- Other manifestations include chorioretinitis and cerebral calcification resembling congenital toxoplasmosis. Survivors may show mental retardation.

Infection in infants

- Cytomegalic inclusion disease is seen almost exclusively in infants born to mothers who develop primary CMV infection during pregnancy.
- Infants of mothers who have CMV reactivation during pregnancy tend to have a chronic subclinical infection.
- Perinatal infection may be acquired from the infected mother through genital secretions or breast milk.

Infection in children

- Primary infections in older children and adults are usually asymptomatic.
- However, heterophile, antibody-negative, infectious mononucleosis may be seen. This is more common following transfusion of CMV-infected blood (post-transfusion mononucleosis).

In the immunocompromised host (such as an AIDS patient)

- Severe and even fatal infections develop in transplant recipients, cancer patients on chemotherapy and, most significantly, in the HIV-infected (Case 1).
- CMV is an important pathogen in AIDS patients, in whom the already weakened immune response is further damaged by the non-specific CMI-inhibiting effect of CMV.

Laboratory diagnosis

- **Specimen:** Urine, saliva, semen, and cervical secretions can be collected.
- **Microscopy**
  - Cytomegalic cells can be demonstrated in centrifuged deposits from urine or saliva in which inclusion bodies 'owl's eye' are seen (Fig. 32.1).
  - This method is simple but not very reliable.
- **Isolation**
- **Culture:** Human cytomegaloviruses can be grown in human fibroblast cultures. Epithelial cell cultures are not susceptible. However, they are affected in vivo. Cultures have to be incubated for prolonged periods—up to 50 days—as the cytopathic effects are slow to appear.
  - The virus can be isolated from urine, saliva, or other body fluids by inoculating human fibroblast cultures or by using shell vial cultures.
  - Growth in cultures can be studied microscopically by staining the cells with fluorescent-tagged antibody to early antigens of CMV.
- **Serology**
  - The demonstration of IgM antibody is useful in the diagnosis of primary infection but not in reactivation.
  - Serological techniques in use include CF, IHA, IF, and ELISA.
  - Antibody detection may be necessary for screening blood or organ donors.
  - IgM antibody detection helps in the diagnosis of congenital infections.
- **IgG avidity assay**
  - It detects binding affinity or avidity of IgG antibodies.

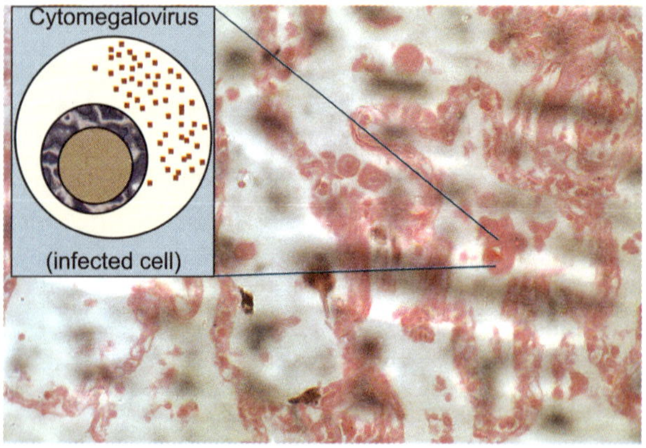

Fig. 32.1 H and E-stained lung tissue showing 'owl's eye'-like inclusion bodies of cytomegalovirus ×1,000

- It differentiates recent from recurrent infections.
- In primary infection, high levels of low avidity IgG are detected.
- Increase of IgG with high avidity occurs after recurrent infection.
- Detection of low avidity IgG along with IgM reflects risk to the fetus.

Prevention

- Prevention is indicated only in high-risk cases such as **organ transplant patients, immunodeficient persons, and premature infants**.
- Blood and organ **donors should be screened**, and CMV immunoglobulins should be administered.
- **Acyclovir** is useful in prophylaxis but not in treatment.
- No vaccine is available. Experimentally, live-attenuated vaccines (Towne 125 and AD 169 strains) and a purified CMV polypeptide vaccine have been found to be immunogenic but not effective in protecting immunodeficient subjects from CMV infection.

Treatment

Ganciclovir and foscarnet have been found to be effective and are used for the treatment of CMV disease in patients with AIDS.

Infectious Mononucleosis

> **Case 2.** *A 16-year-old girl presented with complaints of malaise and loss of appetite over the previous month. She also had a sore throat and fever of 40°C with spikes in temperature seen especially in the afternoon for the past five days. On examination, cervical lymphadenopathy and mild hepatosplenomegaly were detected. Her complete blood counts revealed absolute mononuclear lymphocytosis, with most of the lymphocytes being atypical. Her serum tested positive for the Paul–Bunnell test. Her IgM was also positive for* ***Epstein–Barr nuclear antigen (EBNA)****. The patient improved on supportive care.*
> i. What is the diagnosis?
> ii. Describe the pathogenesis of the EB virus.
> iii. What is kissing disease?

Infectious mononucleosis or 'kissing disease' is an acute, self-limiting illness, usually seen in non-immune young adults following primary infection with the **Epstein–Barr virus.**

- Epstein–Barr (EB) virus is ubiquitous in all human populations.
- EBV-infected B cells are transformed to become capable of continuous growth in vitro.
- Infection with the EB virus leads to latency, periodic reactivation, and lifelong persistence.
- EB virus antibodies are present in about 95 per cent of adults.
- EB virus infection occurs in infancy and childhood, at which stages, it is usually asymptomatic. In affluent countries, primary infection is often delayed until early childhood and adolescence, at which time, it may lead to infectious mononucleosis.

Agent—Epstein–Barr virus

- Epstein–Barr virus (EBV), also known as human herpesvirus 4, is a member of the herpes virus family.
- It is an enveloped, double-stranded DNA virus that infects B lymphocyte cells.
- It attaches to the host cells (B cells and epithelial cells) via glycoproteins present on the envelope.
- It can cause a variety of diseases and is spread mainly from saliva containing virus-infected epithelial cells.

Transmission

- Infections spreads through the saliva of infected persons, who shed the virus in oropharyngeal secretions for months following primary infection and intermittently thereafter.
- Oral contact such as kissing appears to be the predominant mode of transmission, so much so that infectious mononucleosis is also called the **'kissing disease'**.
- The EB virus is not highly contagious, and droplets and aerosols are not efficient in transmitting infection.
- Rarely, blood or marrow transfusion can transmit the disease.

Pathogenicity (Fig. 32.2)

- EB virus infection may lead to the following clinical conditions:
  - **Infectious mononucleosis**
  - **EBV-associated malignancies**
- Only human and some subhuman primate B cells have receptors (CD21 molecules, which are the same as for the C3d component of complement) for the virus.

Lymphoma formation

- Mononucleosis represents a **polyclonal transformation of infected B lymphocytes.**
- EB virus antigens are expressed on the surface of infected B cells.
- The atypical lymphocytes seen in blood smears in infectious mononucleosis are T lymphocytes undergoing blast transformation in response to such neoantigens.
- Intermittent reactivation of the latent EB virus leads to clonal proliferation of infected B cells.
- In immunocompetent subjects, clonal proliferation is kept in check by activated T cells.
- In the immunodeficient, B cell clones may replicate unchecked, resulting in lymphomas.
- Nearly half the lymphomas seen in immunodeficient subjects contain EB virus DNA sequences.
- Genetic and environmental factors are said to be important in the nasopharyngeal carcinoma seen in

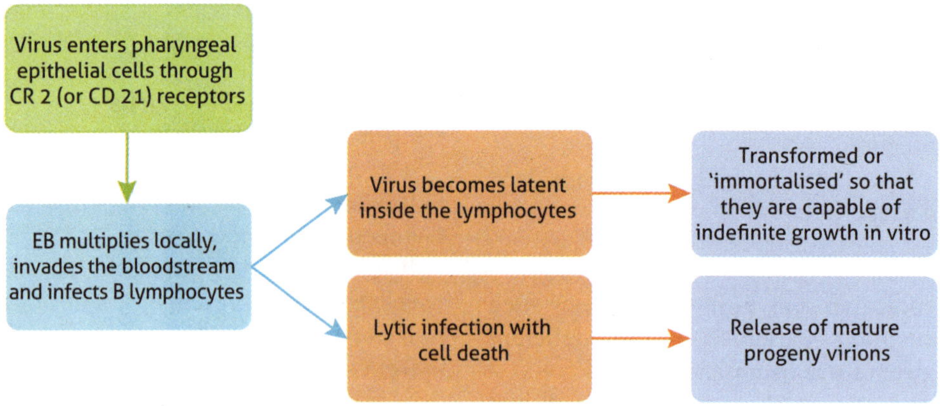

Fig. 32.2 Pathogenicity of EBV

men of Chinese origin. EB virus DNA is regularly found in the tumour cells. These patients have high levels of EB virus antibodies.

- Genetic influence is best illustrated in the X-linked lymphoproliferative (XLP or Duncan) syndrome associated with extreme susceptibility to EB virus infection.

Clinical features

Infectious mononucleosis (glandular fever)

- The **incubation period** is 4–8 weeks.
- It is an **acute, self-limiting illness**, usually seen in non-immune young adults following primary infection with the EB virus (Fig. 32.3) characterised by the following features:
  - Fever, sore throat, and lymphadenopathy
  - Abnormal lymphocytes in peripheral blood smears (Case 2)
  - A mild transient rash may be present; some patients treated with ampicillin may develop a maculopapular rash due to an immune complex reaction to the drug
  - Often associated hepatitis, which is usually subclinical
- **Complications** consist of hematological, neurological, cardiac and pulmonary conditions, and splenic rupture.
- In most cases, spontaneous resolution of the disease occurs in two to four weeks.
- In some, it may be more prolonged and lead to a state of mental and physical fatigue in convalescence.
- The infectious mononucleosis syndrome can follow infection by other agents such as cytomegalovirus and toxoplasmosis or occur as a reaction to non-infectious stimuli.

Laboratory diagnosis

- **Peripheral blood smear**
  - Initially, leukopenia is seen due to a drop in the number of polymorphs.
  - Later, there is prominent leukocytosis, with the appearance of abnormal mononuclear cells characterised by deeply basophilic, vacuolated cytoplasm, and kidney-shaped nuclei showing a lattice of fenestrated chromatin.
  - These atypical mononuclear cells are not virus-infected B lymphocytes but lymphoblasts derived from T cells reactive to the virus infection.
  - The blood picture may sometimes resemble lymphocytic leukemia.
- **Serology:**
  1. **Viral capsid antigen (VCA):** There are two capsid antigen detection tests to detect IgG and IgM respectively.
     i. ***Anti-VCA IgM***
        - Appears early in EBV infection and usually disappears within four to six weeks
        - Detected by immunofluorescence and ELISA

     ii. ***Anti-VCA IgG***
        - Indicates past or recent infection
        - Appears in the acute phase of EBV infection, peaks at two to four weeks after onset, declines slightly, then persists for life
        - Reliable indication of primary infection
  2. **Early antigen (EA)**
  - Anti-EA IgG appears in the acute phase of illness and generally falls to undetectable levels after three to six months.
  - In many people, the detection of antibodies to EA is a sign of active infection. However, 20% of healthy people may retain antibodies against EA for years.
  - Antibodies to early antigens (EA) are present in high titres in EB-associated lymphomas.
  3. **EBV nuclear antigen (EBNA)**
  - Antibody to EBNA, determined by the standard immunofluorescent test, is not seen in the acute phase of EBV infection but slowly appears two

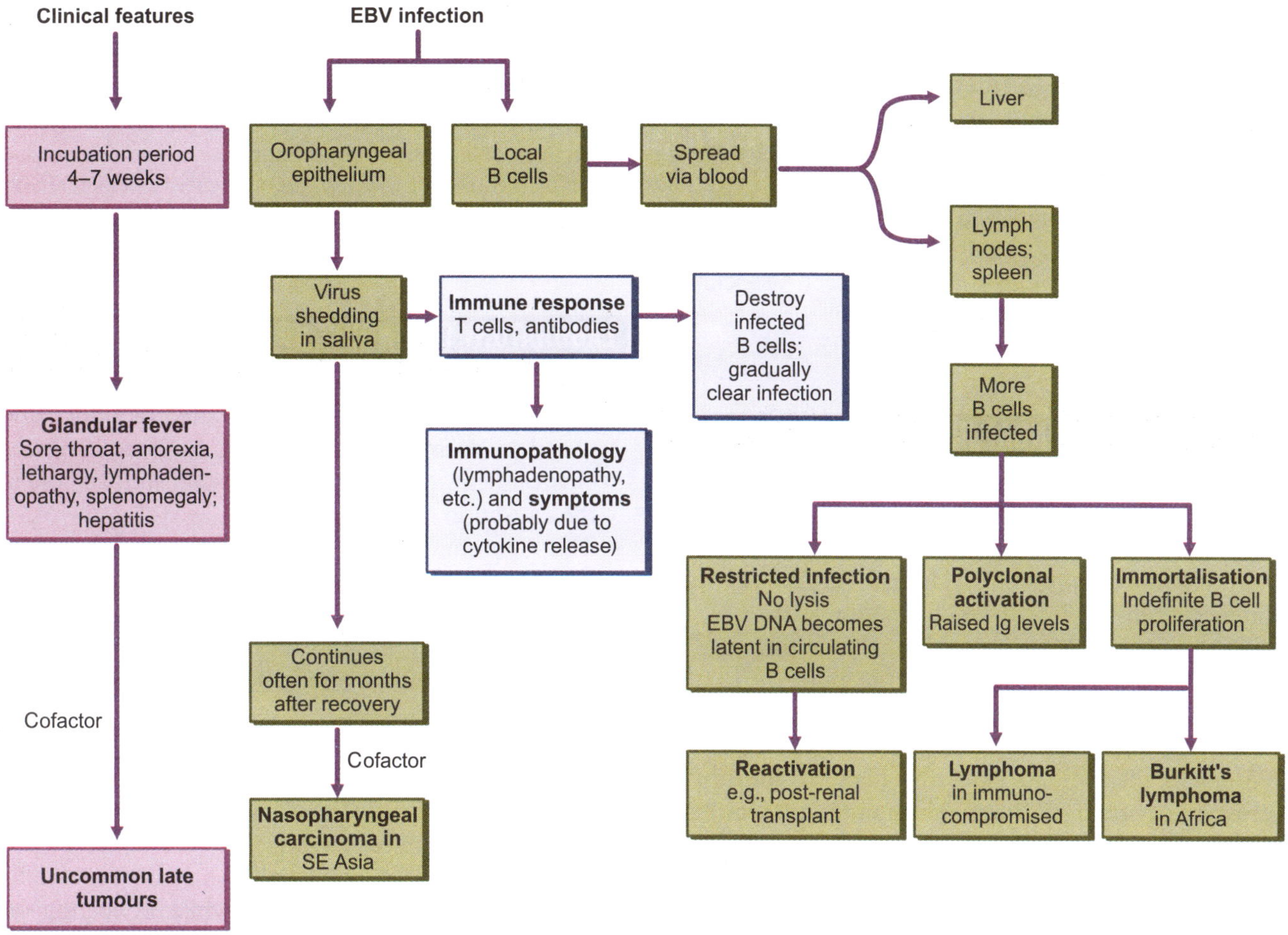

Fig. 32.3 Clinical and immunovirological events in Epstein–Barr virus infection

to four months after the onset of symptoms and persists for the rest of the person's life.
  - EBNA is a useful marker for primary infection.
- **Paul–Bunnell test**
  - Used to screen for infectious mononucleosis.
  - Heterophilic antibodies that agglutinate sheep or horse erythrocytes are detected. However, such antibodies may also occur in serum sickness and hepatitis B and sometimes, even in healthy individuals.
  - These cross-reacting Forssman antibodies can be removed by adsorbing the test serum with guinea pig kidney and ox RBCs prior to testing.
  - It appears that EBV infection alters the cell membrane of infected cells, which evoke a non-specific antibody response.
  - These antibodies disappear in six months.
  - A significant titre in the Paul-Bunnell test for the confirmation of glandular fever is >100 IU/mL.

EBV-associated malignancies

- **Burkitt's lymphoma**
- **T cell and B cell lymphomas** in immunodeficient persons such as AIDS patients and transplant recipients
- **Nasopharyngeal carcinoma** (seen commonly in persons of Chinese origin)
- **Duncan disease**, an X-linked lymphoproliferative condition which predisposes a person to the EB virus
- **Oral hairy leukoplakia (OHL)**, caused by the Epstein-Barr virus (EBV), commonly seen in HIV infection
- **Hodgkin's lymphoma**, in which EBV DNA and RNA are detected in the malignant Hodgkin Reed Sternberg cells (HRS); the viral nucleic acids present in HRS cells imply that infection of the tumour progenitor cells occurred before the cells proliferated

Polyomavirus

- This group contains a single genus. Viruses belonging to this group possess a **circular genome of double-stranded DNA within a non-enveloped capsid.**
- Human viruses are JC, BK, KI, WU, HPyV6, HPyV7, and the Merkel cell virus.
- SV40 is found in monkeys, which may pose a problem in using monkey-derived cell lines for human virus culture (it accidentally contaminates early-killed and live poliovirus vaccine grown on monkey cells).

- Human polyomaviruses are widely distributed in the population.
- JC and BK viruses may remain in the kidney and lymphoid tissue after primary infection in healthy individuals. They may get reactivated in immunosuppressed conditions, e.g., during renal transplantation, in AIDS patients, etc., causing progressive multifocal leukoencephalopathy (PML).
- KI and WU polyomaviruses have been associated with respiratory infections in children.
- Merkel cell carcinoma is a rare neuroendocrine tumour.

OPPORTUNISTIC FUNGAL INFECTIONS

- Opportunistic mycoses develop in patients who are immunosuppressed, those who have hematological malignancies or diabetes, and those who are on immunosuppressive drugs, corticosteroids, radiotherapy, or broad-spectrum antibiotics.
- These infections are caused by fungi that are of low virulence and found as contaminants in the environment (e.g., *Mucor*, *Penicillium*, *Aspergillus*) or as commensal flora of the body (e.g., *Candida*).
- Most of these fungi enter the host via the upper respiratory tract, as the spores are present in air and dust.
- The inhaled spores settle down in the nasopharyngeal mucosa, from where they may spread to the sinus and upwards to the orbit or the brain to cause rhino-orbital or rhino-cerebral mycotic infections or downwards to the lower respiratory tract to produce fungal pneumonias in susceptible immunocompromised individuals.

Parasites associated with opportunistic infections include coccidian parasites and *Strongyloides stercoralis* (Chapter 25), which causes hyperinfection.

> **Mucormycosis infection** at near-epidemic scale was reported during the COVID pandemic of 2021. The large-scale use of prednisolone for prolonged periods of time to control late-stage complications of the SARS-CoV-2 infection is thought to have predisposed individuals to sino-orbital mucormycosis. An increased incidence of *Aspergillus pneumoniae* was also observed during this period.

LABORATORY DIAGNOSIS OF OPPORTUNISTIC INFECTIONS

- As organisms associated with OI are ubiquitous and found as normal commensals of the body or in the environment, it is important to establish their role in the causation of infection.
- The immunocompromised status and type of immunodeficiency of the individual is established by hematological and immunological investigations.
- Appropriate samples as per clinical presentation are collected with sterile precaution to isolate and identify the organism.
- Direct microscopic examination is vital to demonstrate fungi or parasites in the sample.
- Isolation of the causative agents is attempted in successive samples collected over a short period of time to isolate the same organism repeatedly and rule out contamination while collecting or processing.
- Antigen detection from tissue or blood is the preferred method as it rules out contamination by commensals.
- Detection of specific microbial DNA or RNA by molecular methods has become important as they are very specific and can clinch the identification of the etiological agent.

Key Points

- ✓ An infection that arises due to immunosuppression is known as opportunistic infection (OI).
- ✓ Immunosuppression may be due to humoral, cell-mediated, or combined immunodeficiency.
- ✓ Immunosuppression depends on the following predisposing factors: ageing, leukopenia, cytotoxic drugs, and metabolic disorders like diabetes.
- ✓ Infections that cause immunosuppression are HIV, CMV, and tuberculosis.
- ✓ Organisms causing opportunistic diseases have the following characteristics:
  - They are usually commensals of the body. Such organisms are known as colonising opportunistic pathogens.
  - They are generally of low virulence (*Pseudomonas, Acinetobacter, Burkholderia cepacia*).
  - They may be found in air and soil (*Aspergillus* spores).
- ✓ Cell-mediated or combined immunodeficiency predisposes to viral opportunistic infections like CMV, EBV, VZV, etc.
- ✓ Some of the parasitic and fungal agents associated with immunocompromised individuals are coccidian parasites and *Pneumocystis jirovecii*.

Questions

Short answer questions

1. List the different types of immunodeficiency diseases that may lead to opportunistic infections.
2. Briefly outline the laboratory approach to diagnose an opportunistic infection in an individual.

MCQs

1. **A 64-year-old HIV seropositive Chinese traveller presented with an abnormal nasal growth with excessive secretions and he was diagnosed as a case of nasopharyngeal carcinoma. What is the likely causative agent?**
   a) HSV-2 b) HPV
   c) EBV d) CMV
2. **Kissing disease is associated with which one of the following infections?**
   a) HSV-2 b) EBV
   c) CMV d) HSV-1
3. **Abnormal mononuclear cells with prominent leukocytosis is the characteristic feature of:**
   a) Infectious mononucleosis
   b) Molluscum contagiosum
   c) Glandular fever
   d) Tuberculosis
4. **Significant titre by Paul-Bunnell test for confirmation of glandular fever should be above:**
   a) 20 IU/mL b) 100 IU/mL
   c) 800 IU/mL d) 1,000 IU/mL
5. **Cytomegalovirus is also known as:**
   a) Salivary gland virus
   b) Kissing disease virus
   c) Darling disease virus
   d) IM virus
6. **Which of the following is used to treat CMV infections?**
   a) Acyclovir b) Ganciclovir
   c) IFN-γ d) None
7. **Which of the following factors is not responsible for development of immunosuppression?**
   a) Malnutrition
   b) Presence of intravenous and prosthetic devices
   c) Metabolic syndromes
   d) Receiving blood transfusion
8. **Which parasite causes hyper-infection in immunosuppressed persons?**
   a) *Loa loa*
   b) *Strongyloides stercoralis*
   c) *Pneumocystis jirovecii*
   d) Roundworm

Answers: 1:c, 2:b, 3:c, 4:b, 5:a, 6:b, 7:d, 8:b

Part III: Microbiology as Applied to Infectious Diseases

CHAPTER 33

Emerging Infections

(MI8.4)

Chapter Navigation

- Emerging infections in the Indian scenario
- Re-emerging infections
- Infections leading to public health emergencies
- Antibiotic resistance
- Bioterrorism

Learning objectives

- Identify factors contributing to emerging and re-emerging infections
- Enumerate emerging and re-emerging infections
- List measures to contain emerging infections

Infectious diseases continue to be among the leading causes of death worldwide. While infections like smallpox and polio have been eradicated, the emergence of new infectious diseases and re-emergence of old infections are becoming an increasing cause for concern. Any change in the delicate balance between hosts, parasites, and ecosystems causes new infections to emerge or old diseases to re-emerge.

EMERGING INFECTIOUS DISEASES

- Diseases that have appeared for the first time in a population or that did not exist previously in an area but are rapidly increasing in incidence in that geographical range are known as emerging infectious diseases.
- The factors that contribute to the emergence of new infections are presented in Fig. 33.1.
- About two-thirds of emerging infections originate from animals (both wild and domestic), i.e., they are zoonotic.

Table 33.1 lists the new infectious diseases identified since 1997.

RE-EMERGING INFECTIONS

- These are infections that caused recognised health problems in the past and subsequently dropped in incidence to such low levels that they were no longer considered public health problems only to re-emerge with an upward trend in incidence or prevalence, either in the same geographic area or worldwide.
- Examples include cryptosporidiosis, diphtheria, malaria, pertussis, rabies, etc. Factors responsible for some of the examples of remerging infections are presented in Table 33.2.

INFECTIONS LEADING TO PUBLIC HEALTH EMERGENCIES

The World Health Organization has prioritised the following diseases for public health emergency and research:

- Crimean–Congo hemorrhagic fever (CCHF)
- Ebola virus disease and Marburg virus disease
- Lassa fever
- Middle East respiratory syndrome coronavirus (MERS-CoV)
- Severe acute respiratory syndrome (SARS)
- Nipah and henipaviral diseases
- Rift Valley fever (RVF)
- Zika
- COVID-19

EMERGING INFECTIONS IN THE INDIAN SCENARIO

The emerging bacterial and viral infections in the Indian scenario are listed in Table 33.3.

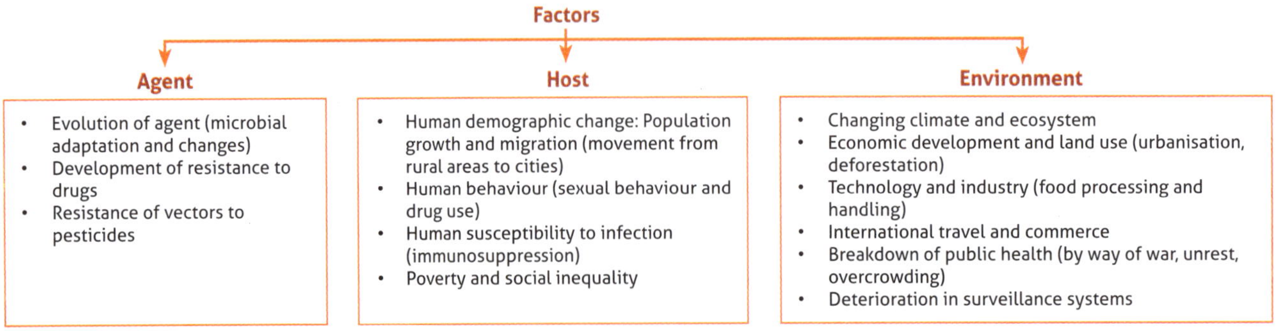

Fig. 33.1 Factors contributing to emerging infections

Table 33.1 Emerging infections recognised since 1997

Year	Agent	Disease
1997	H5N1	Avian flu (bird flu)
1999	Nipah virus (Paramyxoviridae, genus *Henipavirus*)	Encephalitis and respiratory illness in Malaysia and Singapore
2018		Outbreak in Kerala
2003	Coronavirus	SARS-CoV
2009	H1N1	Pandemic (H1N1) influenza
2011	CCHF virus	Crimean Congo hemorrhagic fever outbreak reported in Gujarat
2012	MERS-CoV	Middle East respiratory syndrome caused by a coronavirus—first identified in Saudi Arabia
2013	H7N9 Influenza virus	A novel reassortant avian-origin influenza A (H7N9) virus isolated from respiratory secretions of three critically ill patients from Shanghai, China, in March 2013
2013	Ebola virus (re-emerged)	A total of 28,616 cases with 11,310 deaths were reported in Guinea, Liberia, and Sierra Leone (2013–2016)
2016	Zika virus	First appeared in Brazil and spread to other parts of South and North America
2017	Monkeypox virus	Nigeria reported a large outbreak; since 1970, human cases of monkeypox have been reported from 10 African countries
2018	Nipah virus	Originated first in Malaysia in 1999, then reached Bangladesh and eastern parts of India; caused an outbreak in Kerala; has been detected in subsequent years and as recently as 2021
2019	COVID-19 (SARS-CoV-2) virus	Originated in the Wuhan district of China in December 2019; started spreading to other countries by January 2020 and was declared a pandemic in March 2020; it has continued in waves through 2021

Table 33.2 Examples of re-emergence and factors responsible for it

Disease	Reason for emergence
Cryptosporidiosis	Inadequate control of water supply; international travel; increased use of immunosuppressive agents (transplant, anticancer therapy)
Diphtheria, pertussis, measles	Interruption of immunisation programme due to political changes, decreased vaccine efficacy or waning immunity among vaccinated adults
Malaria	Drug resistance, climate change causing increased breeding of mosquitoes
Tuberculosis	Antibiotic-resistant pathogens; immunocompromised populations (malnourished, HIV-infected, poverty-stricken)
Rabies	Breakdown in public health measures; changes in land use (deforestation, global warming), travel

Table 33.3 Emerging and re-emerging infections in India

Bacterial infections	Viral infections
• Plague (1994, in Surat) • Leptospirosis (following rains and floods) • Brucellosis (small pockets in Karnataka) • Anthrax (outbreaks in small pockets) • Cholera (outbreaks in small pockets) • Scrub typhus (seasonal outbreaks)	• Influenza-H1N1, avian influenza • Chikungunya • Chandipura virus* • Dengue • Japanese encephalitis • SARS coronavirus, hantavirus, human enterovirus 71 • Crimean Congo hemorrhagic fever (CCHF) • Hantavirus • Zika • Nipah • COVID-19

*A few cases were reported from Maharashtra and Gujarat in 2024

ZIKA VIRUS INFECTION

- The Zika virus was first reported in three cases from the Bapunagar area of Ahmedabad, Gujarat, in January 2017.
- These cases tested positive for the virus by RT-PCR and were later confirmed by the WHO. All three cases recovered.
- The low level of transmission of the Zika virus in an area may lead to new cases among naïve populations in the future.
- Zika virus antibodies have been detected in the Indian population, indicating protective immunity.

Crimean–Congo Hemorrhagic Fever (CCHF)

- CCHF was reported for the first time in 2011 from Gujarat.

- It was reported as zoonotic in origin with ticks as the vectors.
- The outbreak was caused by person-to-person transmission in hospitals.

Nipah Virus Infection

- Nipah virus (NiV) is a zoonotic virus.
- It is a food-borne disease that is transmitted by the ingestion of dates contaminated with the urine or saliva of infected bats, following which, person-to-person transmission occurs.
- This disease originated in Malaysia in 1999 and then reached Bangladesh and the eastern parts of India including Siliguri, West Bengal, in 2001. In 2022 and 2023 Kerala reported cases of Nipah.
- Human infections range from asymptomatic infection to acute respiratory infection (mild, severe) and fatal encephalitis.

Clinical features

- The incubation period is 4–14 days (up to 45 days).
- Initial symptoms include fever, headaches, myalgia (muscle pain), vomiting, and sore throat.
- **Late symptoms** include dizziness, drowsiness, altered consciousness, and neurological signs that indicate acute encephalitis. Some people also experience atypical pneumonia and severe respiratory problems, including acute respiratory distress.
- **Complications**—encephalitis and seizures occur in severe cases, progressing to coma within 24–48 hours.
- Most people who survive acute encephalitis make a full recovery, but long-term neurologic conditions have been reported in survivors.
- Approximately 20% of patients are left with residual neurological consequences such as seizure disorder and personality changes.
- A small number of people who recover subsequently relapse or develop delayed-onset encephalitis.
- The case fatality rate is estimated to be 40–75%. This rate can vary by outbreak depending on local capabilities for epidemiological surveillance and clinical management.

Laboratory diagnosis

- Initial signs and symptoms of Nipah virus infection are non-specific, and the diagnosis is often not suspected at the time of presentation.
- In addition, the quality, quantity, type, and timing of clinical sample collection and the time needed to transfer samples to the laboratory can affect the accuracy of laboratory results.
- Diagnosis is by clinical history during the acute and convalescent phases of the disease.
- PCR, real-time polymerase chain reaction (RT-PCR) assay of body fluids, antibody detection via enzyme-linked immunosorbent assay (ELISA), and virus isolation by cell culture are also used.

Treatment

- There are currently no drugs or vaccines specific for Nipah virus infection although WHO has identified Nipah as a priority disease for the WHO Research and Development Blueprint.
- Intensive supportive care is recommended to treat severe respiratory and neurologic complications.

Human Hantavirus Infection

- Hantavirus infection was confirmed for the first time in India in 2000. Antibodies to the human hantavirus were later demonstrated in populations in South India by an ELISA test.
- In 2005, following the floods in Mumbai, some cases were found to have IgM antibodies to the virus.
- It may be that the disease exists in India, but its incidence is under-reported due to the lack of awareness.
- Early suspicion and rapid confirmation of diagnosis are vital in preventing the emergence of hantavirus infection in India.
- The availability of simple, easy-to-perform, and affordable tests is a basic necessity for proper management.
- Rapid control measures, which include contact tracing, quarantine, and active surveillance are required to contain the infection and prevent large-scale outbreaks.
- Mass vaccination, if available, must be instituted.

ANTIBIOTIC RESISTANCE

- Antimicrobial resistance in bacteria has increased in the last decade and is a cause for great concern. This is an impediment to the control of infectious diseases.
- The chief cause of the current antimicrobial resistance crisis is the uncontrolled and inappropriate use of antibiotics in both industrialised and developing countries. Of major public health concern and considered as emerging infections are the following:
  - Multidrug-resistant tuberculosis (MDR and XDR-TB)
  - Methicillin-resistant *S. aureus* (MRSA)
  - Penicillinase-producing *N. gonorrhoeae* (PPNG)
  - Vancomycin-resistant enterococci (VRE)
  - Extended-spectrum beta-lactamase (ESBL)-producing gram-negative bacteria
  - Carbapenemase-producing Enterobacteriaceae (CRE)

CONTROL OF EMERGING ANTIMICROBIAL RESISTANCE

- Rational use of antimicrobial agents
- Good infection-control practices
- Antibiotic stewardship programmes to monitor prescription and antimicrobial usage in hospitals
- Creating awareness among the public and restriction of over-the-counter sale of the anti-infective drugs
- Price and quality control of the available drugs by regulatory authorities

BIOTERRORISM

- With the increasing spectrum of the armamentarium for offence and defence across countries and regions, newer methods of combative tools such as bioterrorism are becoming a grave cause for concern.
- Microorganisms that are virulent and widely transmissible at a fast pace and microbial toxins and spores have become tools of bioterrorism.
- The category of infections thus produced would therefore fall into the category of emerging infections.
- However, as this is a matter of national concern, details of the matter have been discussed in Chapter 35 along with national programmes for disease prevention.

Key Points

- ✓ Emerging infectious diseases are infections that have appeared for the first time in a population or that did not exist previously in an area but are rapidly increasing in incidence in that geographical range.
- ✓ Factors that contribute to the emergence of such diseases are linked to behavioural, climatic, and genetic changes in humans, the environment, and infectious agents respectively.
- ✓ In India, emerging viral infections include influenza, chikungunya, Chandipura virus, dengue, Japanese encephalitis, SARS-coronavirus, and CCHF. Emerging bacterial infections include plague, leptospirosis, brucellosis, and cholera.
- ✓ Surveillance and early control measures help in the containment of these infections.
- ✓ Zika virus infection is transmitted through the bite of the *Aedes* mosquito (*A. aegypti, A. albopictus*) and is related to dengue, yellow fever and West Nile fever. It was declared as a public health emergency of international concern by the WHO in February 2016.
- ✓ Important emerging infections in the last few years are coronavirus, H1N1, CCHF, MERS-CoV, H7N9, Ebola virus, Nipah, and SARS-CoV-2.
- ✓ Some emerging infections are diseases considered public health emergencies and are prioritised for research. They are Zika, SARS-CoV-2, Nipah, etc.
- ✓ Some of the common emerging and re-emerging infections in India are dengue, Japanese encephalitis, human enterovirus 71, Crimean Congo hemorrhagic fever (CCHF), Nipah, and SARS-CoV-2. Among re-emerging bacterial agents are the ones that cause scrub typhus, plague, and cholera.

Questions

Short answer questions

1. **List the factors that influence the emergence of an infection.**
2. **Why is rising antimicrobial resistance categorised as an emerging threat?**
3. **List the emerging infections in India in the last decade and describe their epidemiology.**

MCQs

1. **Which of the following is correct regarding emerging infections?**
   a) Majority of emerging infections are zoonotic infections.
   b) Global warming is responsible for emergence of new pathogens.
   c) Changes in food production patterns cause emergence of these infections.
   d) All of the above.
2. **All of the following are considered emerging infectious agents except:**
   a) Ebola virus b) Nipah virus
   c) Zika virus d) *Salmonella* Typhi
3. **In the year 2011, Crimean Congo hemorrhagic fever was reported from which state of India?**
   a) Karnataka b) Gujarat
   c) Haryana d) Maharashtra
4. **The first case of Kyasanur forest disease was reported from:**
   a) Jammu and Kashmir b) Assam
   c) Karnataka d) Tamil Nadu
5. **Which of the following is not a zoonotic disease/infection?**
   a) Brucellosis b) Swimming pool granuloma
   c) Rabies d) Anthrax
6. **A pregnant woman returning from Brazil consulted an obstetrician for an antenatal check-up. On sonography, the fetus was found to have microcephaly. Which organism is the likely causative agent?**
   a) Cytomegalovirus b) Zika virus
   c) Herpes virus d) *Toxoplasma gondii*
7. **A person visited Malaysia for 3 months and on return, suffered from food poisoning after consuming dates and later developed pneumonia. What is the likely causative agent?**
   a) Zika virus b) *B. dermatitidis*
   c) Nipah virus d) *C. welchii*
8. **Which of the following is incorrect about Nipah virus?**
   a) It is a zoonotic virus.
   b) It is a food-borne disease that is transmitted by the ingestion of dates contaminated with the urine or saliva.
   c) There is no person to person transmission.
   d) Clinical presentation can range from asymptomatic infection to acute respiratory infection (mild, severe) and fatal encephalitis.
9. **The case fatality rate of Nipah virus is estimated to be:**
   a) <10% b) 10–20% c) 20–40% d) >40
10. **A young male was admitted in the ICU following an accident. He had to be ventilated. On the 4th day, he developed pneumonia with a multidrug-resistant organism (ESBL producer). The commonest ESBL producer is:**
    a) *Klebsiella pneumoniae* b) *Proteus mirabilis*
    c) *Yersinia enterocolitica* d) *E. coli*

Answers: 1:d, 2:d, 3:b, 4:c, 5:b, 6:b, 7:c, 8:c, 9:d, 10:a

Oncogenic Viruses

(MI8.3)

Learning objective

- Describe the mechanisms of oncogenic viruses related to malignancy

- An **oncogenic virus** is a virus that produces tumours in its natural hosts or in experimental animals or induces malignant transformation of cells on culture.
- Viral infections account for 10–20 per cent of human malignancies. Examples of viruses resulting in malignancy are as follows:
  - Hepatocellular carcinoma caused by hepatitis B and C viruses
  - Cervical cancer caused by certain types of papillomaviruses
  - Anaplastic nasopharyngeal carcinomas caused by the Epstein Barr virus
  - Adult cutaneous T cell lymphoma/leukemia by HTLV-1
- The association of viruses with malignancy was described by **Peyton Rous** in 1911 when he showed that a solid malignant tumour, fowl sarcoma, was caused by a virus—a discovery for which he was belatedly awarded the Nobel Prize in 1966.
- About a quarter of the 600 or so animal viruses possess an oncogenic potential.
- The viruses associated with cancers in human beings are presented in Tables 34.1 and 34.2.

Table 34.1 List of oncogenic viruses

RNA viruses	
I.	**Retroviruses**
	• Avian leukosis viruses • Murine leukosis viruses • Murine mammary tumour virus • Leukosis–sarcoma virus of various animals • Human T cell leukemia viruses • Hepatitis C virus
DNA viruses	
I.	**Papovaviruses**
	• Papillomaviruses of human beings, rabbits, and other animals • Polyomavirus • Simian virus 40 • BK and JC viruses
II.	**Poxviruses**
	• Molluscum contagiosum virus • Yaba virus • Shope fibroma virus
III.	**Adenoviruses**
	Many human and non-human types
IV.	**Herpes viruses**
	• Marek's disease virus • Lucke's frog tumour virus • Herpes virus pan, papio, ateles and saimiri • Epstein–Barr (EB) virus • Herpes simplex virus (HSV) types 1 and 2 • Cytomegalovirus
V.	**Hepatitis B virus**

MECHANISMS OF VIRAL ONCOGENESIS

Transformation

The transformation of a healthy cell to a malignant cell is a multistep process and may be partial or complete with the following characteristics:

- Some viral agents can **'immortalise' infected cells** (Table 34.3) so that they become capable of continuous multiplication in culture, without possessing other features of malignancy.
- Transformation is recognised primarily by a change in the morphology of cultured cells.
- Transformed cells are **altered in shape and lose the property of 'contact inhibition'** so that, instead of growing as a monolayer, they pile up, one over another, forming '**microtumours**'.
- Foci of transformation can be discerned easily and are used in the assay of oncogenic viruses (as seen in the Rous sarcoma virus).

Table 34.2 Viruses that have a definite association with human cancer

Virus family	Virus	Types of cancer
Papovaviridae	Human papillomavirus	Cervical, vulvar, penile cancers Squamous cell carcinoma
Herpesviridae	EB virus	Nasopharyngeal carcinoma African Burkitt's lymphoma
	HSV type 2	B cell lymphoma Cervical carcinoma
Hepadnaviridae	Hepatitis B virus	Hepatocellular carcinoma
Flaviviridae	Hepatitis C virus	
Retroviridae	HTL virus	Adult T cell leukemia

Table 34.4 Some oncogenes* and their chromosomal location in humans

Viral oncogene	Origin	Tumour/ malignancies	Human gene	Chromosomal location in human beings
V-src	Chicken	Sarcoma	*C-src*	20
V-ras	Rat	Sarcoma	*C-ras*	11
V-myc	Chicken	Leukemia	*C-myc*	8
V-fes	Cat	Sarcoma	*C-fes*	15
V-sis	Monkey	Sarcoma	*C-sis*	22
V-mos	Mouse	Sarcoma	*C-mos*	8

*Oncogenes are given three-letter codes from the animal or tumour from which they are derived, preceded by either V- or C-, for viral or cellular genes respectively:

• *src*, sarcoma of chicken • *ras*, rat sarcoma • *sis*, simian sarcoma • *myc*, myelomatosis of chicken • *fes*, feline sarcoma • *mos*, mouse sarcoma

Table 34.3 Properties of cells transformed by viruses

Class	Properties of transformed cells
I	Altered cell morphology: Fibroblasts become shorter, parallel orientation is lost, chromosomal aberrations appear
II	Altered cell metabolism: Increased growth rate, increased production of organic acids, and acid mucopolysaccharides
III	Altered growth characteristics: Loss of contact inhibition, formation of heaped growths (microtumours), capacity to divide indefinitely in serial culture, and grow in suspension or in semisolid agar
IV	Antigenic alterations: Appearance of new virus-specified antigens (T antigen–TSTA), loss of surface antigens, cells become agglutinable by lectins
V	Capacity to induce tumours in susceptible animals

Oncogenes

- Viral oncogenes (*V-onc*), commonly known as 'cancer genes', are genes that encode proteins triggering the transformation of normal cells into cancer cells (Table 34.4).
- Oncogenes are not essential for the replication of the virus; mutants that lack these genes replicate normally without being oncogenic.
- Genes closely resembling viral oncogenes are found in normal as well as cancer cells.
- Oncogenes isolated from cancer cells are called **cellular oncogenes (*C-onc*)**.
- Viral oncogenes are believed to have originated at some point in the distant past from the proto-oncogenes by recombination between retroviral and cellular genes. The presumed functions of many oncogenes have been identified. Some examples are as follows:
  - The oncogene *src* is related to tyrosine-specific protein kinases
  - *sis* is related to a platelet-derived growth factor
  - *myc* is related to DNA-binding proteins, all concerned with the regulation of normal cell growth
- Similar genes found in normal cells are called proto-oncogenes. They are not of viral origin. On the contrary, viral oncogenes appear to be of host cell origin.
- Cellular oncogenes contain introns characteristic of eukaryotic genes, whereas viral oncogenes do not.
- Proto-oncogenes are widespread in vertebrates and metazoa—from human beings to fruitflies. They are well conserved in their genomes, suggesting that they serve some essential functions in normal cells. They have been found to code for proteins involved in regulating cell growth and differentiation.

Transfection

- This is a useful method for the study of oncogenes.
- Certain mouse fibroblast cell lines such as NIH 3T3 can take up foreign DNA, incorporate them into their genome, and express them. This is called transfection.
- By this technique, DNA extracted from human tumour cells has been shown to transform 3T3 cells; such transforming genes have been shown to be identical to cellular oncogenes.

Anti-oncogenes

- A class of genes has been identified in the normal retinoblastoma (*Rb*) gene, the loss of which is associated with the development of retinoblastoma in children.

- The *p53* gene appears to be a tumour suppressor gene with a wide range of effects.
- Specific chromosomal deletions, recognised in association with certain types of human cancers may reflect the loss of tumour suppressor genes.
- While it is known that oncogenic viruses can transform cells in culture and induce tumours in animals under natural or experimental conditions, the exact mechanisms of viral oncogenesis are not well understood.
- Malignancy is a stable, heritable change and, as such, should be the result of a modification of the host cell genome.

ONCOGENIC DNA VIRUSES

Mechanism of Oncogenesis

- The viral DNA (or a portion of it) is integrated with the host cell genome. As the viral DNA is incomplete or 'defective', no infectious virus is produced. However, under its influence, the host cell undergoes malignant transformation.
- A virus-transformed cancer cell is in many ways analogous to a bacterium lysogenised by a defective phage. In both cases, there is no cell destruction and no virus produced.
- Acquisition of new characteristics by the transformed cell resembles lysogenic conversion in bacteria.

Human Papillomavirus (HPV)

- Human papillomavirus (HPV) infection, particularly HPV types 16 and 18, and cancer of the cervix uteri is established. A vaccine is now available to protect against these infections.
- Papillomaviruses are species-specific and infect squamous epithelial and mucous membranes, inducing different types of warts or papillomata in their hosts.
- The human papillomavirus (HPV) infects only humans and grows only in organ cultures of human skin. Over 82 types of HPV have been recognised based on genetic homology.
- There is a correlation between the virus type and the type of lesion produced.

Association of HPV with some clinical conditions (Table 34.5)

Table 34.5 Association of HPV types with clinical conditions

Clinical Condition	HPV type
Common warts (verruca vulgaris, plantar warts)	1, 2, 3, and 4
Condyloma acuminatum or genital warts	6 and 11, 42–44
Intraepithelial neoplasia	6 and 11
Cervical cancer	16 and 18
Laryngeal and esophageal carcinomas	16, 18, 30, 31, 33, 51–53

HPV vaccine

HPV vaccination prevents new HPV infection but does not treat existing infection or disease. Three recombinant vaccines are now available:

i. Quadrivalent vaccine with antigens from HPV 6, 11, 16, and 18
ii. 9-valent vaccine (HPV 6, 11, 16, 18, 31, 33, 45, 52, 58)
iii. Bivalent vaccine (HPV 16, 18)

Both vaccines are indicated in preventing persistent and precancerous genital lesions in adolescent and young adult women. They are contraindicated in pregnancy. The WHO recommends 1 or 2 doses for women aged <20 years and two doses, six months apart, for older women.

HPV vaccines available in India are:

i. Quadrivalent vaccines—Gardasil and Cervavac, which is indigenously developed in India
ii. Gardasil-9

Human Polyomaviruses

- Includes JC and BK viruses.
- Antibodies to these viruses are present widely in human sera—in about 75 per cent of adults.
- In an impaired immune system following disease or transplantation, the virus may be reactivated, leading to progressive multifocal leukoencephalopathy (PML) or renal disease.
- It has been linked to urothelial carcinoma, hemorrhagic cystitis, and urinary bladder cancer.

The JC virus

- It was isolated in 1971 from the brain of a patient with Hodgkin's disease and PML.
- It grows only in human fetal glial cell cultures.
- It is oncogenic, producing malignant gliomas following intracerebral inoculation in newborn hamsters.

The BK virus

- It was isolated in 1971 from the urine of a patient who had undergone a kidney transplant.
- It can grow in a wide range of primary and continuous cell culture.

Simian Virus 40 (SV40)

- Transformation is induced in cultured cells from several species, including human cells.
- Although no direct evidence is available, SV40 has been found to be associated with multiple tumours.

Poxvirus

- Three members of the poxvirus group induce benign tumours—rabbit fibroma, molluscum contagiosum, and Yaba virus.
- Similar tumours can be induced experimentally in many species of primates including human beings.
- The tumours regress spontaneously in a few weeks.

Adenoviruses

Though some types (12, 19, 21) may produce sarcomas in newborn rodents after experimental inoculation, these viruses do not appear to have any association with human cancer.

Herpes Viruses

Epstein–Barr virus (EBV)

- This virus is regularly found in cultured lymphocytes from Burkitt's lymphoma patients.
- In the body, the tumour cells contain no virus; however, cell lines established from them contain 5–20 per cent of cells that produce the virus.
- The virus multiplies only in human lymphoid cell lines.
- Infection is usually asymptomatic.
- In young adults without pre-existing antibodies, EB virus infection induces infectious mononucleosis.
- Lymphoma is believed to occur when the infection takes place in children whose immune systems are compromised.
- EB virus-associated lymphomas have been reported in transplant recipients.
- EBV has also been linked to nasopharyngeal carcinoma in the Chinese male population in southeast Asia and East Africa.

Herpes simplex virus

- An association has been proposed (though not proved) between herpes simplex type 2 infection and cancer of the uterine cervix.
- It has also been suggested that herpes simplex type 1 infection may be associated with cancer of the lip.
- Herpes virus type 8 has been linked to Kaposi's sarcoma.

Cytomegalovirus

It has been associated with carcinoma of the prostate and Kaposi's sarcoma.

Hepatitis B and C Viruses

Both HBV and HCV are implicated in the causation of hepatocellular carcinoma.

ONCOGENIC RNA VIRUSES

Retroviruses

- They are enveloped, spherical viruses, approximately 100 nm in size that are released by budding through the host cell membrane.
- They possess two identical, linear, single-stranded RNA molecules with RNA-dependent DNA polymerase or reverse transcriptase (hence the name retro, meaning 'reverse').
- The icosahedral nucleocapsid core encloses the helical ribonucleoprotein and is surrounded by an envelope composed of glycoprotein and lipid.
- Reverse transcriptase enzyme prepares a DNA copy of the retroviral RNA genome—initially an RNA–DNA hybrid and then its double-stranded DNA form, the provirus, which is integrated into the DNA of the infected host cell.
- It is from the provirus that all retrovirus proteins are translated.
- Infection with oncogenic retroviruses does not lead to cytolysis or death of infected cells. Instead, the provirus remains integrated with host cell DNA for the remainder of the cell's life.

Classification

While all oncogenic RNA viruses belong to the family Retroviridae, all retroviruses are not oncogenic. Retroviridae is classified into three subfamilies:

1. **Oncornvirinae**, comprising all oncogenic RNA viruses (formerly called oncornavirus)
2. **Spumavirinae**, containing the non-oncogenic 'foamy viruses' (spuma = foam) causing asymptomatic infection in several animal species
3. **Lentivirinae**, including the viruses causing 'slow infections' (lentus = slow) in animals, as well as the human and related animal immunodeficiency viruses

Types based on host range

Retroviruses are widely distributed, being found in nearly all vertebrates including animals, birds, and reptiles. Based on the host range and types of diseases caused, oncogenic retroviruses can be considered under the following groups:

- **Avian leukosis complex:** Antigenically related viruses which induce avian leukosis (lymphomatosis, myeloblastosis, and erythroblastosis viruses) or sarcoma in fowls (Rous sarcoma virus [RSV]).
- **Murine leukosis viruses:** Incude murine leukemia and sarcoma viruses.
- **Mammary tumour virus of mice:** This virus occurs in certain strains of mice that have a high natural incidence of breast cancer.
- **Leukosis–sarcoma viruses of other animals:** A number of viruses have been isolated from leukosis and sarcomas in various species of animals—cats, hamsters, rats, guinea pigs, and monkeys.
- **Human T cell leukemia (lymphotropic) viruses (HTLV):** These were isolated in 1980 from cell cultures of adult patients with cutaneous T cell lymphoma (mycosis fungoides) and leukemia (Sezary syndrome) in the USA. Similar viruses have been isolated from patients with adult T cell leukemia in Japan and the Caribbean. HTLV type I is prevalent worldwide, but the disease is limited to endemic areas. Besides adult T cell leukemia, HTLV-I has also been associated with tropical spastic paraparesis, a demyelinating disease. The

virus preferentially infects T4 (CD4) cells. Infected T cells express large quantities of interleukin-2 receptors. The closely related HTLV-II is also associated with T cell malignancy. HTLV infection is known to be spread through blood transfusions and other methods of transfer of leukocytes.

Types based on host specificity

Retroviruses usually infect only one host species, the specificity being conditioned mainly by the presence of viral receptors on the host cell surface. Depending on their ability to grow in cells from different species, retroviruses have been classified as following:

- **Ecotropic** (multiplying in cells of native host species only)
- **Amphotropic** (multiplying in cells of native and foreign species)
- **Xenotropic** (multiplying only in cells of foreign species but not of native host species)

Types based on mode of virus transmission

Two types of retrovirus transmission occur:

- Exogenous retroviruses spread horizontally, i.e., among individuals belonging to the same generation. Most oncogenic retroviruses are exogenous.
- Endogenous retroviruses are transmitted vertically, i.e., from parent to offspring—through a provirus, which is integrated with the germline cell genome. The endogenous retrovirus provirus behaves like a cellular gene and is subject to regulatory control by the host cell. Endogenous retroviruses are usually silent and do not transform cells or cause any disease. They can be detected either by 'activation' after exposure to radiation or chemicals or by nucleic acid hybridisation techniques.

Antigens

Retroviruses possess two types of antigens:

1. **Type-specific glycoprotein antigens** on the envelope
2. **Group-specific nucleoprotein antigens** in the virion core

Cross-reactions do not occur between surface antigens of retroviruses from different host species.

Genomic structure

Retroviruses have a relatively simple genomic structure (Fig. 34.1).

- The provirus of a standard retrovirus (such as a non-defective avian or murine leukemia virus) consists of three genes required for viral replication: *gag*, *pol*, and *env* in that order from the 5′ to the 3′ end.
- The *gag* gene codes for the nucleocapsid core proteins, which are group-specific antigens (hence the name).

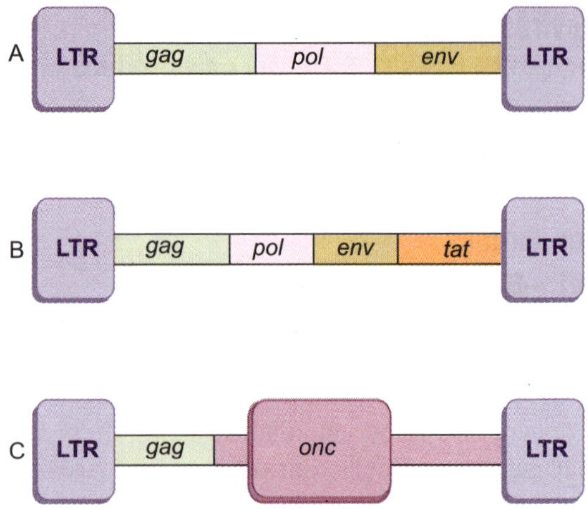

Fig. 34.1 Provirus genomic structure of different types of retroviruses; (a) basic retrovirus genome of avian leukemia viruses and slow-transforming viruses; (b) transregulating retroviruses HTLV and HIV; (c) acute transforming retrovirus, with an oncogene replacing part of the basic genome; replication defective

- The *pol* gene encodes the RNA-dependent DNA polymerase.
- The *env* gene encodes the envelope glycoproteins.
- **Long terminal repeat (LTR)** sequences are present at either end of the provirus and linked directly to the host DNA. LTRs exert regulatory control on provirus gene functions.
- Some retroviruses (transregulating viruses) such as HTLV and HIV carry a fourth gene, *tex* or *tat*, after the *env* gene. This is a transactivating gene that regulates the function of viral genes.

Resistance

Retroviruses are labile, being inactivated at 56°C in 30 minutes by mild acids, ether, and formalin but are stable at –30°C.

Mechanism of oncogenesis

Retroviruses induce tumours by any one of the following mechanisms:

- They may get introduced into the cellular genome a new transforming gene (oncogene).
- They may induce or alter the expression of a pre-existing cellular gene. Several molecular mechanisms have been suggested for the conversion of benign proto-oncogenes to cancer genes.
- The genes may get overexpressed, and the overproduced gene product may lead to abnormal growth. Recombination between retroviral and cellular genes, promoter insertion, chromosomal translocation, gene amplification, and mutation are some of the genetic processes relevant in this connection.

FUTURE APPLICATIONS

- Identification of virus association with oncogenesis has revolutionised therapeutic and preventive strategies in some malignancies, e.g., vaccine therapy for human papillomavirus.
- The study of viral oncogenesis has helped to understand the process of oncogenesis and tumour formation in several malignancies.
- It has opened up the gates for further research on anticancer drugs and vaccines.

Key Points

✓ Viruses that produce tumours in their natural hosts or in experimental animals or which induce malignant transformation of cells on culture are known as oncogenic viruses.

✓ Examples of oncogenic viruses in humans:
- Epstein–Barr virus (EBV), which causes Burkitt's lymphoma, common in central Africa
- Human papillomaviruses (HPV), associated with cervical cancer
- HTLV-1 retrovirus, linked to leukemia

✓ Chronic infection due to the hepatitis B virus may lead to hepatocellular carcinoma.

✓ Papillomaviruses are species-specific and infect squamous epithelia and mucous membranes.

✓ HPV types 1, 2, 3, and 4 cause common warts (verruca vulgaris).

✓ HPV types 6 and 11 cause condyloma acuminatum (genital warts) and are also associated with intraepithelial neoplasia.

✓ HPV-16 and 18 are causatively related to severe cervical cancer.

✓ Retroviruses are oncogenic RNA viruses. These spherical, enveloped viruses are released by budding through the host cell membrane. They have has two identical, linear, single-stranded RNA molecules with RNA-dependent DNA polymerase or reverse transcriptase (hence the name retro, meaning 'reverse').

✓ HTLV-1 causes adult T cell leukemia and tropical spastic paraparesis, a demyelinating disease.

✓ Retroviruses induce tumours either by a new transforming gene (oncogene) or by inducing or altering the expression of a pre-existing cellular gene. Several molecular mechanisms have been suggested for the conversion of benign proto-oncogenes to cancer genes.

Questions

Short answer questions

1. List and briefly discuss oncogenic viruses of humans.
2. Discuss the important papillomaviruses of medical importance and associated diseases.

MCQs

1. **Laryngeal and oesophageal carcinomas are associated with:**
   a) HPV types 16 & 18 b) HPV types 1-4
   c) HPV types 6 &11 d) HPV types 42-44
2. **Xenotropic retrovirus can multiply:**
   a) In cells of native host species only
   b) In cells of native and foreign species
   c) Only in cells of foreign species but not of native host species
   d) None of the above
3. **Which of the following statement is true regrading viral oncogenes (*V-onc*)?**
   a) These are genes that encode for proteins triggering the transformation of normal cells into cancer cells
   b) These are not essential for the replication of the virus
   c) The mutants that lack these genes cannot replicate normally without being oncogenic.
   d) All of the above.
4. **All the following are oncogenic viruses except:**
   a) Hepatitis A b) Papilloma
   c) Epstein–Barr d) HTLV-1
5. **African Burkitt's lymphoma is caused by:**
   a) Epstein–Barr virus b) Human papilloma virus
   c) Herpes simplex type-2 d) Human T cell leukemia virus
6. **Which one of the following is not an oncogenic DNA virus?**
   a) Molluscum contagiosum virus
   b) Human T cell leukemia virus
   c) Human papilloma virus
   d) Epstein–Barr virus
7. **A 63-year-old lady presented with a soft pedunculated wart on the labia. It was diagnosed as condyloma acuminatum. What is the causative agent of this condition?**
   a) Molluscum contagiosum virus
   b) Human papilloma virus
   c) Epstein–Barr virus
   d) Herpes simplex virus-2
8. **Cervical cancer is caused by:**
   a) HPV type 1–4 b) Human papilloma virus
   c) HPV type 16–18 d) Herpes simplex virus-2
9. **Which of the following causes fetal hydrops?**
   a) HPV type 1–4 b) Parvovirus B-19
   c) Epstein–Barr virus d) Human polyoma virus
10. **Parvovirus B-19 causes all of the following except:**
    a) Erythema infectiosum b) Genital tumours
    c) Aplastic crisis d) Fetal hydrops

Answers: 1:a, 2:c, 3:a, 4:a, 5:a, 6:b, 7:b, 8:c, 9:b, 10:b

Part III: Microbiology as Applied to Infectious Diseases

CHAPTER 35

National Health Programmes, Disease Detection, and Reporting

(MI8.16)

Chapter Navigation

- Communicable disease programmes under the National Health Mission
- Bioterrorism

Learning objective

- List the various national disease control programmes in India

Health is a basic human right. With this vision, national health programmes are designed to achieve the objective of ensuring accessible health for affected individuals.

In the context of infectious diseases, the objectives of these programmes are as follows:

- **Disease detection and prevention** are vital and should be accessible to all individuals in the programme. The availability of **simple, easy-to-perform, and affordable tests** is a necessity for proper management.
- **Rapid control measures** are required to contain infections and prevent large-scale outbreaks.

Control measures entail the following:

- **Contact tracing**
- **Quarantine**
- **Active surveillance**
- **Mass vaccination**, if available, for implementation

It requires a concerted effort by all stakeholders to address issues related to disease detection, treatment, control, and prevention. To this effect, the Government of India has initiated several national programmes to ensure that these measures reach the entire population.

- **Infectious diseases** are addressed under the communicable disease programmes.
- **Non-infectious diseases** like diabetes, cancer, and hypertension are grouped under non-communicable disease programmes.

COMMUNICABLE DISEASE PROGRAMMES UNDER THE NATIONAL HEALTH MISSION

National health programmes are central government-sponsored programmes that are constantly monitored and under which strategies are planned according to the disease status in the country. State health programmes are under the purview of the state governments and are implemented according to the local disease epidemiology.

1. National TB Elimination Programme (NTEP)

History

- Earlier known as the **Revised National TB Control Programme (RNTCP)**, the programme's current objective is a national strategic plan for tuberculosis elimination (2017–2025).
- It was launched in 1997 and later expanded across the country in a phased manner.
- It has now been replaced by the NTEP.

Objectives of RNTCP

- To reduce the incidence and mortality due to TB
- To prevent further emergence of drug resistance and effectively manage drug-resistant TB cases
- To improve outcomes among HIV-infected TB patients
- To involve private sector on a scale commensurate with their dominant presence in healthcare services
- To further decentralise and align basic RNTCP management units with National Rural Health Mission block level units within the general health system for effective supervision and monitoring

Goals of the NTEP

- The foremost goal is to achieve a rapid decline in the burden of TB morbidity and mortality while working towards elimination of TB in India by 2025.
- The requirements for moving towards TB elimination have been integrated into the four strategic pillars of 'Detect – Treat – Prevent – Build' (DTPB)' (Fig. 35.1).

Activities under RNTCP

RNTCP was based on the internationally recommended **Directly Observed Treatment Short-course (DOTS) strategy**.

TB–HIV coordination

- The TB–HIV collaborative activities are being undertaken in all the states since 2007.
- The NACP (National AIDS Control Programme) and RNTCP developed the 'National framework of Joint TB/HIV Collaborative activities' in 2007 and revised it in 2015. The framework articulates the policy of TB/HIV collaborative activities in the country.

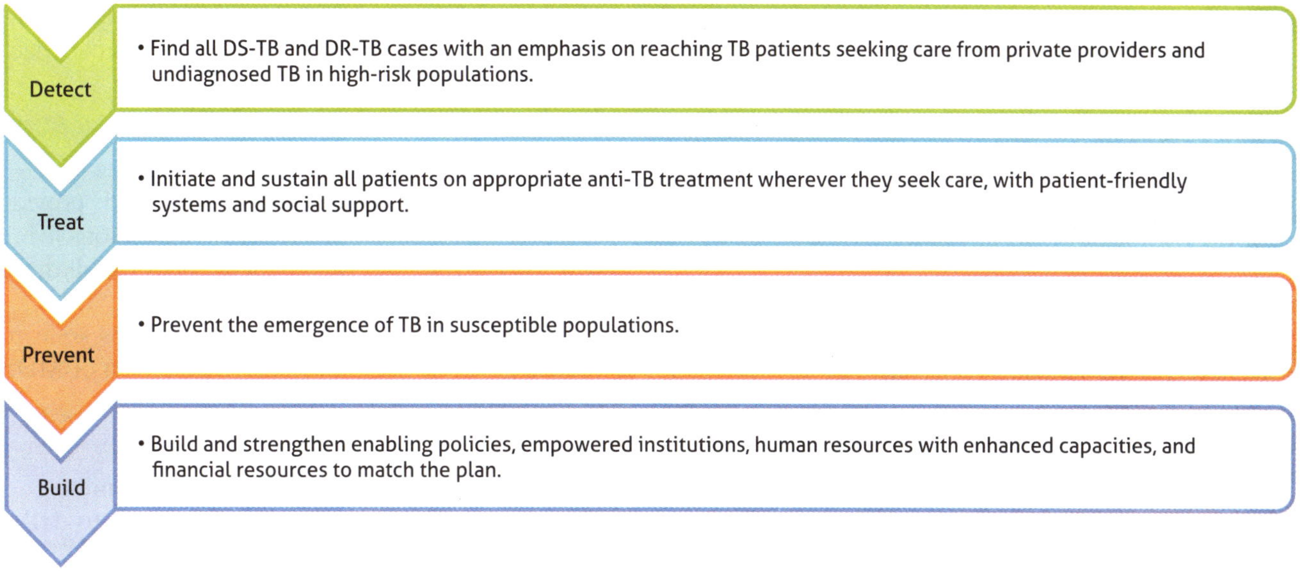

Fig. 35.1 Four strategic pillars of the NTEP

Activities under the NTEP

- Scale up free, high-sensitivity diagnostic tests and algorithms
- Scale up effective private provider engagement approaches
- Institute universal testing for drug-resistant TB
- Establish systematic screening of high-risk populations
- Prevent the loss of TB cases in the cascade of care with support systems
- Provide free TB drugs for all TB cases
- Develop a universal daily regimen for TB cases and rapid scale up of short-course regimens for drug-resistant TB and DST-guided treatment approaches
- Establish patient-friendly adherence monitoring and social support to sustain TB treatment
- Eliminate catastrophic costs by linking eligible TB patients with social welfare schemes including nutritional support
- Scale up air-borne infection control measures at healthcare facilities
- Ensure testing and treatment for latent TB infection in contacts of bacteriologically-confirmed cases and in individuals at high risk of getting TB disease
- Address social determinants of TB through intersectoral approach
- Translate high level political commitment into action through supportive policy and institutional structures

> For more information on TB control in India, visit the following websites:
> - https://tbcindia.gov.in
> - https://dghs.gov.in/content/1358_3_RevisedNationalTuberculosisControlProgramme.aspx

2. National Leprosy Eradication Programme (NLEP)

History

- The **National Leprosy Control Programme (NLCP)** was launched by the Government of India in 1954–55 with vision of a 'leprosy-free India'.
- The programme was initially taken up in endemic districts and was extended to all districts in the country from 1993–94.
- The strategy of the NLEP was based on controlling the disease through a reduction in the quantum of infection in the population and reduction in infective source, thus breaking the chain of disease transmission.

Mission

To provide quality leprosy services free of cost to all sections of the population, with easy accessibility through the integrated healthcare system including care for disability after cure of the disease.

Objectives

- To reduce the prevalence of leprosy to less than 1/10,000 population at the sub-national and district levels
- To reduce Grade II disability percentage to <1 among new cases at the national level
- To reduce Grade II disability cases to <1 case per million population at the national level
- Zero disabilities among new childhood cases
- Zero stigma and discrimination against persons affected by leprosy

Strategy

- Early detection through active surveillance by trained health workers

- Regular treatment of cases by instituting multi-drug therapy (MDT) centres in villages in moderate-to-low endemic areas/districts
- Intensified health education and public awareness campaigns to remove social stigma attached to the disease
- Appropriate medical rehabilitation and leprosy ulcer care services

For more information on the control of leprosy in India, visit https://dghs.gov.in/content/1349_3_NationalLeprosyEradicationProgramme.aspx

3. National Vector-Borne Disease Control Programme (NVBDCP)

History

This programme was launched in 2003–04 to control vector-borne diseases. Some of the existing individual programmes, which are as follows, were merged under NVBDCP:

- National Malaria Control Programme
- National Filaria Control Programme
- Kala-azar Control Programme
- Japanese B Encephalitis Control Programme
- Dengue/DHF Control Programme

Goals

- Prevention and control of six vector-borne diseases (VBDs)—**malaria, dengue, lymphatic filariasis, kala-azar, Japanese encephalitis, and chikungunya** in India.
- Out of these six diseases, three diseases, namely malaria, lymphatic filariasis, and kala-azar are targeted for elimination.

Activities

- Surveillance of vector-borne dieases
- Awareness campaigns
- Specific preventive measures like mosquito control, medication, etc.

For more information on the control of vector-borne diseases, visit the following websites:
- https://dghs.gov.in/content/1364_3_NationalVectorBorneDiseaseControlProgramme.aspx
- https://ncvbdc.mohfw.gov.in

4. National AIDS Control Programme (NACP)

The nodal agency is the National AIDS Control Organisation (NACO) in the Ministry of Health and Family Welfare.

History

- The NACP was initiated in 1992 by the Govt. of India to detect, treat, and control HIV infection (NACP I).
- It has been a successful programme which now provides testing, treatment, and prevention services throughout the country.
- It is now in the fourth phase (NACP IV).

Mission

- To reduce new infections by 50% and provide comprehensive care and support to all persons living with HIV/AIDS and treatment services for all those who require it.
- To slow down the spread of HIV infections so as to reduce the morbidity, mortality, and impact of AIDS in the country.

Objectives

- To reduce the spread of HIV infection in India
- To increase India's capacity to respond to HIV/AIDS on a long-term basis
- To build an integrated response by reaching out to diverse populations

Actvities

- **Counselling and testing services**—test and treat policy, universal viral load testing, early infant diagnosis, surveillance by testing antenatal women and populations at risk
- **Community-based screening, integrated counselling and testing services, monitoring of HIV** load and CD4 counts in patients on antiretroviral therapy
- Provision of **antiretroviral treatment**
- **Preventive services**—prevention and care, support and treatment for people living with HIV (PLHIV), awareness generation, condom promotion, prevention of parent-to-child transmission, increasing ICTC services, promotion of voluntary blood donation, and access to safe blood
- **Targeted interventions** for high-risk groups like injectable drug users, men having sex with men, female sex workers, etc.
- **Information, education, communication (IEC)**—guidelines on the management of common opportunistic infections, malignancies among adult/adolescent people living with HIV/AIDS (PLHA), and operational guidelines for ART centres to standardise ART services across the country
- **Blood and organ safety**—National Policy on Blood Banks ensures adequate supply of safe blood and blood components.
- **HIV-TB collaborative activities**

For more information on the control of HIV, visit https://naco.gov.in/nacp

5. Integrated Disease Surveillance Project (IDSP)

History

- The Integrated Disease Surveillance Project (IDSP) was launched by Hon'ble Union Minister of Health

and Family Welfare in November 2004 for a period upto March 2010.
- The project was restructured and extended up to March 2012.
- It continues in the 12th Plan with domestic budget as Integrated Disease Surveillance Programme under the National Health Mission for all States.

Mission

To strengthen the disease surveillance in the country by establishing a decentralised state-based surveillance system for epidemic-prone diseases to detect the early warning signals so that timely and effective public health actions can be initiated in response to health challenges in the country at the districts, state, and national levels.

Objectives

To strengthen/maintain decentralised laboratory-based, IT-enabled disease surveillance systems for epidemic-prone diseases to monitor disease trends and to detect and respond to outbreaks in the early rising phase through trained rapid response teams (RRTs).

Components

- **Integration and decentralisation of surveillance activities** through the establishment of surveillance units at the central, state, and district levels—under the programme, a Central Surveillance Unit (CSU) at Delhi, State Surveillance Units (SSU) at all State/UT head quarters, and District Surveillance Units (DSU) in all districts in the country have been established
- **Human resource development**—training of State Surveillance Officers, District Surveillance Officers, Rapid Response Team, and other medical and paramedical staff on principles of disease surveillance
- Use of **information communication technology** for the collection, collation, compilation, analysis, and dissemination of data—weekly disease surveillance data on epidemic-prone diseases are collected from reporting units such as subcentres, primary health centres, community health centres, hospitals including government and private sector hospitals, and medical colleges
- Strengthening of **public health laboratories**—a state-based referral laboratory network has been established by utilising the existing functional labs in the identified medical colleges and other major centers in the states and linking them with adjoining districts for providing diagnostic services for epidemic-prone diseases during outbreaks
- **Intersectoral co-ordination for zoonotic diseases** through the one-health approach
- **Surveillance** is maintained at the regional, national, and global levels with the following measures:
  - Epidemiological surveillance
  - Laboratory surveillance, e.g., influenza surveillance programme
  - Ecological surveillance
  - Anthropological surveillance
  - Investigation and early control measures
  - Monitoring and evaluation under the supervision of the global outbreak alert and response network (Global Outbreak Alert and Response Network [GOARN]) for the containment of these infections
  - Disease detection and reporting

For more information on the control of infectious diseases, visit https://idsp.mohfw.gov.in

6. National Immunisation Programmes

See Chapter 20.

7. National Viral Hepatitis Control Programme (NVHCP)

History

The National Viral Hepatitis Control Programme was launched on 28 July 2018 on the occasion of World Hepatitis Day.

Mission

- Combat hepatitis and achieve country-wide elimination of hepatitis C by 2030
- Achieve significant reduction in the infected population, morbidity, and mortality associated with hepatitis B and C viz. cirrhosis and hepatocellular carcinoma (liver cancer)
- Reduce the risk, morbidity, and mortality due to hepatitis A and E

Objectives

- Establish a **laboratory network** for the surveillance of viral hepatitis
- **Determine the prevalence** of different types of viral hepatitis
- Provide **laboratory support for outbreak investigation** of hepatitis through the established network of laboratories
- Generate **awareness** among healthcare providers and in the community about waterborne and blood-borne hepatitis by creating information and education materials.

Components

1. **Preventive component:** Increasing awareness, immunisation against hepatitis B, safety of blood and blood products, injection safety, safe socio-cultural practices, safe drinking water, hygiene, and sanitary practices
2. **Diagnosis and treatment**: Screening of pregnant women for HBsAg, free screening, diagnosis and treatment for both hepatitis B and C, provision of linkages, including with private sector and not-for-profit institutions for diagnosis and treatment, engagement with community/peer support to

enhance and ensure adherence to treatment and demand generation.

3. **Monitoring and evaluation, surveillance, and research**
4. **Training and capacity building**

For more information about the control of hepatitis, please visit https://nvhcp.mohfw.gov.in

8. National Programme on Containment of Antimicrobial Resistance (AMR)

History

- The rapid spread of multi-resistant bacteria and the lack of new antibiotics to treat infections caused by these organisms pose a rapidly increasing threat to public and animal health.
- To tackle this issue, the Government of India has launched a 'National Programme on Containment of Antimicrobial Resistance' under the 12th five-year plan (2012–2017).
- It is coordinated by Centre for Bacterial Diseases and Drug Resistance, National Center of Disease Control.

Objectives

- To establish a laboratory-based AMR surveillance system of network labs in the country and to generate quality data on antimicrobial resistance for pathogens of public health importance
- To strengthen infection control practices and promote the rational use of antibiotics
- To conduct the surveillance of antimicrobial usage in different healthcare settings and promote the rational use of antimicrobials through antimicrobial stewardship practices
- To generate awareness among healthcare providers and in the community about the rational use of antibiotics
- Activities—antimicrobial resistance, surveillance for containment of antimicrobial resistance in various geographical regions
- Development and implementation of national treatment and infection control guidelines
- Training and capacity building of professionals in relevant sectors
- IEC for dissemination of information about the rational use of antibiotics
- Development of national repository of bacterial strains/cultures (*Source*: https://www.nhp.gov.in/national-programme-oncontainment-of-anti-microbial-resistance)

For more information on the containment of antibiotic resistance, please visit: https://ncdc.gov.in

RESPONSE TO COVID-19 PANDEMIC IN INDIA

Although there is currently no single formal national programme for COVID, the Ministry of Health and Family Welfare (MoHFW), Govt. of India, the Indian Council of Medical Research (ICMR), and the National Center of Disease Control (NCDC) concertedly responded to the 2020–22 pandemic in the country by gearing up the health system at all levels to tackle the pandemic at the national and state levels through the **Integrated Disease Surveillance Programme (IDSP)**. Nationwide COVID-related preventive measures were undertaken, which included the following:

- Setting up several task forces and steering committees for COVID-19 detection
- Setting up isolation and treatment centers in several public and private sector hospitals
- Initiating the setting up of screening and testing laboratories across the country
- Instituting upgraded laboratories and capacity for molecular tests
- Following the introduction of vaccination, the Indian government initiated a programme to vaccinate all healthcare and frontline workers, who are at a greater risk of exposure to the infection due to their line of work; later, vaccination was extended to all citizens
- Constant upgradation and analysis of data helped to change policies and guidelines for prevention and control strategies

BIOTERRORISM

- Bioterrorism, as defined by the Centre for Disease Control and Prevention, is **the deliberate release, by attackers, of an agent that causes one or more of a variety of different diseases**.
- Based on the risk to national security, these agents have been categorised as A, B, and C (Table 35.1).
- These disease agents exist in nature (with the exception of smallpox, which has been eradicated in the wild), but can be manipulated to be rendered more dangerous.

Table 35.1 Categories of agents in bioterrorism

Category	Property	Examples
Category A	High-risk Pose a threat to national security Have the ability to cause social disruption	Anthrax, botulism (via botulinum), plague, smallpox, and tularemia and viruses that cause hemorrhagic fevers such as Ebola, Marburg, Lassa and Machupo
Category B	Moderately easy to disseminate and result in low mortality	Brucellosis, glanders, Q fever, ricin toxin
Category C	Agents that can be engineered for mass dissemination	Nipah virus Hantavirus

Key Points

- ✓ National health programmes are designed to achieve the objective of ensuring accessible health for affected individuals.
- ✓ The objectives of the national programmes are generally to carry out contact tracing, quarantine, active surveillance, treatment, and mass vaccination.
- ✓ Infectious diseases come under the ambit of the communicable disease programmes.
- ✓ Non-infectious diseases like diabetes are grouped under non-communicable disease programmes.
- ✓ National programmes are central government-sponsored programmes in which strategies are planned according to the disease status in the country.
- ✓ State health programmes come under the purview of the state governments and are implemented according to the local disease epidemiology.
- ✓ National programmes have been designed to prevent and control communicable diseases like TB, HIV, leprosy, malaria, filariasis, and vector-borne diseases.

Questions

Short answer questions

Mention the objectives and enumerate the role and functions of the following:

1. National AIDS Control Programme
2. National TB Elimination Programme

MCQs

1. National TB elimination programme was earlier known as:
   a) Revised National TB Control Programme
   b) Revised TB Control Programme
   c) National TB Control Programme
   d) Tuberculosis control and surveillance programme

2. Which of the following is a feature of National Leprosy Eradication Programme?
   a) Early detection through active surveillance by trained health workers
   b) Regular treatment of cases
   c) Intensified health education and public awareness campaigns
   d) All of the above

3. Many individual programmes were merged under NVDCP. Which of the following was not included in NVDCP?
   a) National Malaria Control Programme
   b) Kala-azar Control Programmes
   c) Japanese B encephalitis Control Programme
   d) National Viral Hepatitis Control Programme

4. The objective of the National AIDS Control Programme is to:
   a) Reduce new infections by 75%
   b) Provide only HIV testing facilities to general public
   c) Provide comprehensive care and support to all persons living with HIV/AIDS
   d) All of the above

5. Integrated Disease Surveillance Programme (IDSP) includes all the following EXCEPT :
   a) Epidemiological surveillance
   b) Immunological surveillance
   c) Ecological surveillance
   d) Investigation and early control

6. Which of the following activities are not a part of National Programme on Containment of Antimicrobial Resistance (AMR)?
   a) Surveillance of antimicrobial resistance in environmental isolates
   b) Development and implementation of national infection control guidelines
   c) Development of national repository of bacterial cultures
   d) Rational use of antibiotics

7. Which one of the following agents for bioterrorism does not belong to category 'A'?
   a) Q fever b) Anthrax
   c) Ebola d) Plague

8. Which of the following bioterrorism agents can be engineered for mass dissemination?
   a) *B. anthracis* b) Machupo virus
   c) Nipah virus d) *C. botulinum*

9. Which of following is an example of type B bioterrorism agent?
   a) Nipah virus b) Zika virus
   c) *Brucella* spp. d) *Yersinia pestis*

10. Biological agents that can be engineered for mass dissemination are known as bioterrorism agent of category:
   a) A b) B
   c) C d) D

11. The national Vector-Borne Disease Control Programme (NVDCP) includes all the following except:
   a) Malaria b) Filariasis
   c) Japanese encephalitis d) Brucellosis

Answers: 1:a, 2:d, 3:d, 4:c, 5:b, 6:a, 7:a, 8:c, 9:c, 10:c, 11:d

CHAPTER 36

AETCOM

(AETCOM 2.2, 2.3, 2.5)

Learning objectives

- Describe and discuss the role of non-maleficence, beneficence, and justice as guiding principles in patient care
- Describe and discuss the role of autonomy and shared responsibility as guiding principles in patient care

INTRODUCTION

The National Medical Commission has decided to implement the **Attitude, Ethics, and Communication module (AETCOM)** in all medical colleges across the country as part of CBME from 2019 batch. In accordance with this decision, the NMC has developed an AETCOM module and a facilitator's guide. The module can be accessed at https://www.nmc.org.in/wp-content/uploads/2020/01/AETCOM_book.pdf.

The new GMER 2023 guidelines have placed the foundations of bioethics and competencies 2.2, 2.3, and 2.5 under the subject domain of Microbiology.

THE FOUNDATIONS OF BIOETHICS

- **Ethics:** A scholarly discipline that examines, evaluates, and seeks to understand moral aspects (the right and wrong) of human nature and actions more deeply. Ethics permeates all that we do in medicine.
- **Bioethics:** The application of ethics to medicine and biological science.

Although the phrase "act ethically" is used every day, the ethical aspects of various actions in the medical setting are not entirely understood.

Four Principles of Bioethics

The four pillars of bioethics represent a framework for analysing the best action to take in a given situation. To use this approach, one must consider whether one's actions are in compliance with each of the four pillars (Fig. 36.1).

Autonomy	Beneficence	Non-maleficence	Justice
Respect the patient's decision to choose what's right for them	All choices are made with intent to do good.	Do no harm	Provide care to all patients in fair manner

Fig. 36.1 Four pillars of bioethics

I. Beneficence

Case 1. *A 60-year-old female presented to the casualty ward with acute abdominal pain and fever. She improved slightly with symptomatic treatment. The doctor wanted to hospitalise her, but she begged to be allowed to return home for bedrest, saying that she was extremely fearful of hospitals and procedures. The doctor insisted and finally convinced her to get admitted. She was diagnosed with acute appendicitis after full work-up and had to be operated.*

i. What ethical principle is involved here?
ii. How would you respect this principle?

The doctor here has made use of **'beneficent persuasion'**—a non-biased reframing or refocusing to encourage patients to make more advantageous treatment decisions.

- A doctor is obligated to act in the patient's best interests and to adhere to a moral code to protect and defend the patient's rights in order to prevent harm, remove conditions that will cause harm, help persons with disabilities, and rescue persons in danger.
- It requires physicians to act in a way that benefits patients.
- Because patients do not possess medical expertise and may be vulnerable because of their illness, they rely on physicians to provide sound advice and to promote their well-being. **The condition should be explained to the patient, factoring in the patient's insights, comprehension, and willingness to be treated.** This requires good communication skills.
- The physician must:
  - Work with and listen to patients
  - Note their medical problems and emotional needs
  - Recognise that **patients' non-adherence to care recommendations may be caused by their lack of comprehension, fear, pain, or other issues**
  - Not cast aside the standard of care in order to cater to unreasonable demands by the patient
  - Use simple language and avoid medical jargon
  - Consider reframing or rephrasing the information

Limits of beneficence

Patient-driven constraints

Normally, patients are motivated by their health interests. Conflicts arise when patient's views diverge from that of the doctor.

- Patients may reject treatment, but they must be made to understand the full implication of their decisions.
- The patient's understanding of what the physician tells them may be affected by their level of education,

socio-economic background, etc. These may also influence their commitment to care and their faith in the healthcare provider.

- They patient may have misgivings and misconceptions about the treatment being offered. This often gets complicated by false information available on internet and social media.
- Patients may sometimes request medical services that a doctor may consider unnecessary.

Practitioner-driven constraints and medical responsibility

- There may be external constraints like a lack of resources (waiting list for investigations, referrals, and treatments).
- Access to specialist care can take a long time, leading to ethical issues (e.g., patients dying while waiting for treatment, paying patients bypassing those covered by public healthcare plans).
- The use of non-evidence-based, experimental, or pharma industry-driven therapy may not be in the best interest of patients.

II. Non-maleficence

Case 2. *A 22-year-old female patient with complaints of recurring back pain presents to a general practitioner. She was prescribed in accordance with the standard of care but did not improve. The practitioner realised that she needs an orthopedic referral but did not refer her for the fear of losing her to someone else. He continued to treat her with a high dose of pain killers over the next eight months. Finally, the girl's friends took her to another doctor who diagnosed her with ill-managed slip disc and an addiction to pain killers.*

i. Which ethical principle is involved here?
ii. How should you respect this principle?

One of the tenets of medical practice is "*Primum non nocere*", which translates to "First, do no harm".

- It is the duty of the doctor to weigh the benefits against the risks of any intervention or treatment, to discard those that are inappropriately burdensome, and to choose the best course of action for the patient. These decisions should be based on the principles of care and not personal choices.
- It is the obligation of a physician not to harm the patient.
- It supports the doctrine exemplified by the following principles—do not kill, do not cause pain or suffering, do not incapacitate, do not cause offense, and do not deprive others of the goods of life.
- Physicians must refrain from providing ineffective treatments or acting with malice toward patients.
- Non-maleficence requires **an intention to avoid needless harm or injury that can arise through acts of commission or omission**.
- To respect this principle, physicians must do the following:
  - Be objective about the treatment of all patients.
  - Provide care but not get involved with patients.
  - Not allow money, friendship, or other personal relationships to influence adherence to the standard of care.
  - Keep patient communication focused on the evaluation of medical symptoms and the discussion of treatment options.
  - Speak frankly with the patient and document their concerns.
  - Not prescribe a controlled substance unless there is a legitimate medical requirement.
  - Maintain thorough records.
  - Re-evaluate and document the patient's level of pain and response to treatment at each visit.
- It forbids:
  - Providing ineffective therapies.
  - Acting maliciously or selfishly.

A physician must remember that if they cannot benefit the patient, they must at least, not harm them or make the situation worse.

III. Autonomy

Case 3. *A 25-year-old woman who is 12 weeks pregnant lost her husband in an accident. The girl's parents want the doctor to convince the girl to abort the child so that she can remarry easily. The girl is adamant about having the baby saying that her husband would have wanted it that way. The doctor is threatened by the woman's family and offered a bribe to conduct an MTP.*

i. Which ethical principle is involved here?
ii. How would you respect this principle?

In the case above, the patient's family holds an opinion about the patient's health that is contrary to her own wishes. Since she is an adult with full mental capacity, it is the doctor's duty to respect the patient's autonomy. Moreover, a forced abortion is not only unethical, it is also unlawful.

- Autonomy in the healthcare setting involves **respecting an individual's right to choose what's right for them**. The patient has the right to refuse or choose their treatment.
- This faith in autonomy is the central premise of the concept of **informed consent and shared decision making**.
- In practice, the principle requires respect for the decision-making capacity of competent adults. Competent adults can refuse care even if the care would be likely to save or prolong the patient's life.
- The doctor should clearly explain to the patient and their family the available treatment options and the advantages and disadvantages of options. The doctor

must make sure that the patient is able to retain this information, evaluate their options, and arrive at a decision. All such information and decisions must be documented in writing.
- The principle of autonomy does not extend to persons who lack the capacity or competence to act autonomously, such as infants and children, mentally ill or intoxicated patients, or those with developmental, mental, or physical disorders. For infants and children, the decision-making power is vested in parents or the legal guardian by default.
- Doctors may sometimes suspend or abridge autonomy for such patients. Rarely, they may do so in favour of another ethical principle—beneficence, non-maleficence, or justice.

Autonomy and paternalism

- **Favouring beneficence over patient autonomy** is known as paternalism. It makes a physician–patient relationship analogous to that of a parent and a child.
- Increasing importance of autonomy can be seen as a social reaction to '**paternalistic**' traditions within healthcare or the overriding of individual choices or intentional actions in order to provide benefit to that individual.
- Paternalism recognises the rights of individuals to self-determination.
- The differences between autonomy and paternalism are presented in Table 36.1.

Table 36.1 Differences between autonomy and paternalism

Autonomy—Capacity to think, decide, and act on the basis of such thoughts and decisions freely and independently	Paternalism—Interference with a person's liberty of action justified by reasons referring exclusively to welfare, good, happiness, needs, interests, or values of the person being coerced
Each patient is entitled to determine his own health destiny	Physician's interference with the patient's freedom of action
Agents have a right to be self-determining; individuals have a right to conduct their lives as they see fit	Its interference is justified on the ground of the patient's best interests; patient is deemed not qualified to comprehend his own medical condition due to his lack of medical expertise
Autonomy is typically taken to be a core component of a 'good life'	It is sometimes legitimate to restrict agent autonomy for their own good

Paternalism may take the form of soft or hard paternalism.

- In **soft paternalism**, the doctor acts on the grounds of beneficence or sometimes non-maleficence in situations where the patient is non-autonomous or incompetent to make the right decision. The difficulty in determining whether the patient was non-competent at the time of decision-making is a big hurdle but is ethically defensible as long as the action is in concordance with what the physician believes to be the patient's values. For example, the decision of abortion in a mentally incompetent rape victim and warnings about the dangers of cigarette smoking.
- **Hard paternalism** refers to an action by doctor that is intended to benefit a patient, but contrary to the voluntary decision of an autonomous patient who is fully informed and competent. Such a decision may become ethically indefensible. For example, preventing someone from committing suicide even if he/she has made the decision autonomously after giving it due consideration.

Shared decision-making

The correct balance between paternalism and autonomy is very important. Shared decision-making is the middle path often used by treating doctors. It involves both the doctor's medical expertise and a regard for patient autonomy. It has been shown to lead to greater patient adherence to treatment and better health outcomes.

Consumerism

An extreme end of hard paternalism is consumerism, where the doctor's role is limited to providing all the medical information and the available choices. The patient then selects from the available choices.

Delegation of autonomy

- In certain cultures or families, the decision-making power may rest with a family member other than the patient. For example, in certain households, the husband or elder son makes the decisions. The patient in such situations may be competent otherwise but may not be able to take a decision regarding their care. This may present an ethical dilemma and may result in a violation of confidentiality and privacy.
- The principle of autonomy does provide the patient the option to delegate decision-making authority to another person. The doctor, in such cases, must find a middle path tactfully by directly informing the patient and later ascertaining whether the patient would prefer to make these decisions themselves, or whether they would prefer all information to be given to her caregiver and for all decisions to be made by him.

Informed consent

- Informed consent is the **willing acceptance of a medical intervention by a patient after adequate disclosure by the physician of the nature of the intervention, its risks and benefits and also its alternatives with their risks and benefits.**

- The requirements of informed consent for a medical or surgical procedure, or for research, are that the patient or subject:
  - Is competent to understand and make decisions
  - Receives full disclosure
  - Comprehends the disclosure
  - Acts voluntarily and autonomously
  - Consents to the proposed action

Four ethical rules of autonomy (Fig. 36.2)

1. **Veracity:**
   - Honesty is a vital component of a physician–patient relationship. It can result in a loss of trust if not adhered to.
   - An autonomous patient not only has the right to know (disclosure) of his/her diagnosis and prognosis, but also has the option to forgo this disclosure. However, the physician must know which of these two options the patient prefers.
2. **Privacy:** This refers to the patient's right to privacy and confidentiality.
3. **Fidelity:** This refers to the provision of care for all, no matter who the person is and what they may have done.
4. **Confidentiality:**
   - Sharing private information on a **'need to know basis'** only.
   - Doctors are obligated not to disclose confidential information given by a patient to another party (including closest relatives) without the patient's authorisation.
   - An exception (with implied patient authorisation) is the sharing of necessary medical information for the care of the patient from the primary physician to consultants and other healthcare teams.

IV. Justice

> **Case 4.** *A 66-year-old Bangladeshi immigrant female with history of diabetes mellitus, hypertension, and vascular disease presented with c/o cough and wheezing. She could not speak Hindi or English. The physician had a very strong view that people who come to India should learn to speak Hindi or local dialect. She was turned away from care and had to be taken to a distant hospital, but died on her way there.*
> i. Which ethical principle is involved here?
> ii. How should you respect this principle?

The principle of justice is that the sick should be cared for regardless of any aspect of their identity including gender, race, religion, sexuality, socio-economic status, etc.

- Justice is generally interpreted as **fair, equitable, and appropriate treatment of all persons**.
- The most pertinent to clinical ethics is **distributive justice**. **Distributive justice** is the fair distribution of healthcare resources (which may be scarce). It is to accord each person an equal share in accordance with the following:
  - Equality
  - Requirement
  - Effort
  - Contribution
  - Merit
  - Free-market exchanges
- These factors are not mutually exclusive and are often combined in application. This creates difficulty in choosing, balancing, and refining these principles to form a coherent and workable solution to distribute medical resources.
- **Rights-based justice** refers to respect for people's rights.
- **Legal justice** refers to respect for morally acceptable laws.
- A doctor must:
  - Be sensitive and alert to cultural differences between him and his patient
  - Consider that the patient may have different beliefs about life, death, time, eye contact, personal space, relationships, and who should consent to medical treatment, etc.

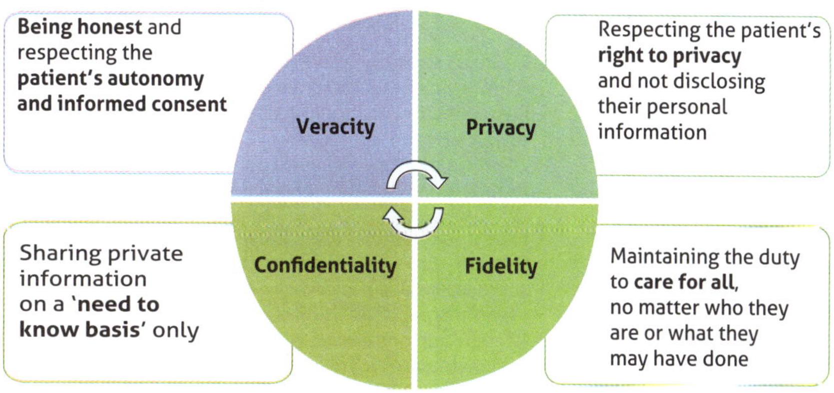

Fig. 36.2 The four rules of autonomy

- Study other cultures, especially those pertaining to a significant patient population base; this can help in communication or treatment issues that may arise
- Document objectively in patient records about language and cultural issues, including interpreter details
- Be wary of online translation tools while dealing with patients who speak different languages

Balancing the four pillars of bioethics

- Each one of the four principles of ethics is to be taken as a prima facie obligation that must be fulfilled unless it conflicts with another principle. In such a conflict, the physician has to determine the actual obligation to the patient by examining the respective weights of the competing prima facie obligations based on both content and context.
- Sometimes, the conflict may have a easy resolution. For example, consider a dehydrated patient who was treated with intravenous fluid via catheter. The intravenous catheter caused a lot of pain and discomfort to the patient. In this case, the principle of beneficence outweighs that of non-maleficence.
- Often, the doctor may face complicated situations that may not be simple to resolve. A few examples are given below:
  - A patient with a head injury who is otherwise in a sound mental state and competent to make decisions refuses a CT scan. As his doctor you are faced with conflict in principles of beneficence and autonomy. In such cases, beneficence can be taken as to include patient autonomy as 'the best interests of the patients are intimately linked with their preferences from which are derived our primary duties to them'.
  - A cancer patient refuses to undergo surgery citing a fear of surgery. As a doctor, you know that his cancer may spread, causing metastases and eventually, even death. The patient has been well-informed, and the facts have been documented. As of now, the patient is in a condition to make his decision and autonomy prevails over beneficence.

A summary of the four pillars of bioethics is provided in Fig. 36.3.

ETHICS IN CLINICAL RESEARCH AND CLINICAL MEDICAL PRACTICE

- Research has been tightly regulated because some forms of research may put a patient at risk for the benefit of others.
- Medical practice focuses on the patient's best interests, considering the benefit to risk ratio, on account of which, it is not very tightly regulated.
- There is a distinction between research and medical practice.
- The patient has to consent for any surgical or medical step of management or for accepting to be enrolled in research.

Regulations in Research

Background

In 1945, 23 German physicians were put on trial. These physicians had:

Principle	Considerations
Non-maleficence	• Type of illness (acute, chronic, reversible, terminal) • Treatment options and probability of success for each option • Adverse effects of treatment and the benefit/harm ratio • Effect of no medical/surgical treatment • Future plans for limiting or stopping treatment
Autonomy	• Competence of patient • Information explained to patient on benefits and risks of treatment • Patient understood the information and consent given • Patient preferences for treatment • Appropirate alternatives of treatment in case of refusal by patient
Justice	• Conflicts of interests between patient and doctor, organisations (clinics, hospitals), insurance, public health, and safety issues • Research or educational considerations that affect clinical decisions physician orders • Conflicts of interests based on religious beliefs/legal issues • Problems in allocation of scarce resources
Beneficience	• Expected quality of life with and without treatment • Deficits - physical, mental, social - that may develop after treatment • Assessment of a patient who cannot express himself • Identification of decision surrogate • Recognition of possible physician bias in judging quality of life • Rationale to forgo life-sustaining treatment

Fig. 36.3 Application of principles of bioethics to patient care

i. Participated in the Nazi programme to euthanise persons deemed 'unworthy of life' (the mentally ill, mentally retarded, or physically disabled) or
ii. Conducted experiments on concentration camp prisoners without their consent
  - The trial known as the **Nuremberg trial** lasted 140 days. 85 witnesses testified and almost 1,500 documents were introduced.
  - Sixteen of the doctors charged were found guilty. Seven were executed.

The Nuremberg code (1947)

- It was developed at the end of the Nuremberg trials, with the purpose of defining legitimate medical research.
- It included principles such as informed consent and absence of coercion; and properly formulation of scientific experiments.
- The code makes the voluntary consent of the human subject absolutely essential in any experiment. The duty and responsibility for ascertaining the subject's consent rests upon the individual who initiates, directs, or engages in the experiment. It is a personal duty and responsibility which may not be delegated to another with impunity.
- The experiment should be able to yield fruitful results for the good of society, and should not be random and unnecessary in nature.
- The experiment should be so designed and based on evidence that the anticipated results will justify the performance of the experiment. The degree of risk should never exceed that determined by the humanitarian importance of the problem to be solved by the experiment.
- The experiment should be conducted only by scientifically qualified persons.
- During the course of the experiment, the human subject should be at liberty to end the experiment if he deems fit.
- The protection of the experimental subject against even remote possibilities of injury disability or death must be built into the study.

Declaration of Helsinki (1964)

- It is a set of ethical principles regarding human experimentation developed for the medical community by the World Medical Association (WMA).
- It is the cornerstone document on human research ethics, experimentation, and beneficence towards experiment participants.
- It is based on the fact that medical research is subject to ethical standards that promote and ensure respect for all human subjects and protect their health and rights.
- Its basic premise is similar to that of the Nuremberg code but it also mentions the role of an ethics committees in regulating medical research.

HEALTHCARE AS A RIGHT

Case 5. *A 32-year-old intravenous drug user tests positive for HIV. A 3-year-old boy tests positive for HIV, which he has acquired from his mother congenitally. Should both the patients be given free ART?*

Case 6. *A 45-year-old patient is diagnosed with lung cancer. He gives history of smoking 10 cigarettes per day for the last 20 years. He is being treated at a free government hospital. Should smokers have the right to free treatment for lung cancer treatment?*

i. Is healthcare a right?
ii. What are the social and economic implications of healthcare as a right?
iii. What are the missing links?
iv. What are the implications for doctors?

The arguments for and against 'healthcare as a right' are plenty and varied. The real meaning of healthcare as a right requires that all of us, acting together as a community and society, take responsibility to ensure that each person can exercise this right. The right to healthcare should not be incurred at a high cost.

- Every human being has a right to health. This applies to both the right of individuals to obtain a certain standard of health and healthcare and also the government's obligation to ensure a certain standard of public health within the community generally.
- It should include access to adequate healthcare facilities and services as well as appropriate measures in relation to the socio-economic determinants of health, such as food, water and sanitation, safe and healthy working conditions, and housing.
- The right to health is closely interconnected with numerous other human rights, including the rights to food, water, housing, work, education, life, non-discrimination, privacy, access to information, and the prohibition against torture, among others.

Essential Features Included in the Right To Healthcare (https://www.escr-net.org/rights/health)

1. **Availability:** Ensure the provision of enough functioning public health and individual healthcare facilities in the country along with preventive aspects such as safe water and sanitation facilities, trained and fairly-paid medical professionals, and essential medicines.
2. **Accessibility:** Four key elements are included in access to health care:
   i. **Non-discrimination:** Health facilities and services should be accessible to everyone,

especially the most vulnerable, without discrimination or prohibition.

ii. **Physical accessibility:** The facilities and services, as well as underlying determinants of health such as water and sanitation amenities, must be within safe physical reach.

iii. **Economic accessibility:** Healthcare facilities, goods, and services must be affordable for all, in a manner that poorer households are not disproportionately burdened with health-related expenses.

iv. **Information accessibility:** It must be ensured that every person has the right to seek, receive and impart information on health, in balance with the confidentiality of medical information.

3. **Acceptability:** Health facilities should be respectful of medical ethics and the culture of individuals and communities, and also attentive to gender and life requirements.
4. **Quality:** Health facilities should be scientifically and medically appropriate and of good quality.

Social and Economic Implications of Healthcare as a Right

- Large, out-of-pocket expenditures on health pose a serious issue. In a lower, middle-income country like India, it can push vulnerable households into poverty. The catastrophic effects of healthcare spending are, however, not limited to the poor; they impact all segments of the population.
- Health insurance provides greater financial protection against health shocks by limiting the out-of-pocket expenditure.
- Providing free medical service in a vast country like India has its own financial implications. India provides healthcare regardless of the individual's ability to pay.
- Healthcare is difficult to define. It encompasses preventive care, diagnostic care, and curative care. The question is, whether cosmetic surgery, organ transplantation, infertility treatment, and the most expensive chemotherapy should be included in this.
- The rights of an individual also imply a duty on the part of others. It is the taxpayers' money that funds the healthcare delivery in most countries for its citizens. Does the duty of accessible healthcare fall on individual doctors, or hospital authorities, or governments, or only rich governments?
- To meet the total health need of an individual, it is essential that medical care include a comprehensive range of services, facilities, and personnel. However, providing this kind of service may deprive others who may have more pressing needs, leading to a colossal wastage of resources.
- Variable financing mechanisms are practised in different countries. This could be done by:
  - Compulsory purchase by employers, through private carriers, of health insurance policies
  - True national health insurance schemes, providing full coverage to all citizens and financed through taxes

Missing Links, Challenges, and Solutions

- **Missing middle:** NITI Aayog's report, 'Health Insurance for India', mentions a 'missing middle' segment. This refers to people that are not covered by a government insurance scheme because they are not poor enough to be eligible. At the same time, they are not wealthy enough to purchase private insurance schemes. It is estimated at least 30% of India's population may not get covered under any government subsidised health insurance schemes like PM-JAY, social health insurance schemes like the Employee State Insurance Scheme, or private voluntary health insurance schemes.
- **Limited and fragmented service:** This type of coverage must be avoided. For example, access to physician without access to free medicines or diagnostic services.
- **Unviable or uneconomical healthcare delivery:** The sparsely populated rural areas may have a low concentration of patients, which can make the provisions of healthcare centres uneconomical. Additionally, people may not be willing to go and serve in rural areas. Some states have an obligatory 'rural medical service' bond to make skilled human resources available in such areas.
- **Universal coverage:** Any universal health programme in India would need to include the entire population and not just be targeted at the poor.
- **Coordinated and rational service delivery:** The healthcare delivery system has to be robust so as to assure a more coordinated and more rational provision of services. This may require incentivisation to healthcare providers. For example, many states in India provide additional marks to NEET score or reserve postgraduate seats for rural service doctors.
- **Focus on primary care:** The focus of attention in healthcare delivery must shift from hospital-based care to primary care—in terms of financing, development of infrastructure, and usage. Less than 5% sick patients require hospitalisation and so the need is to have substantial investments at the primary level combined with a strong gatekeeping framework that does not allow patients to seek hospital-based care unless they have been referred by a primary-care provider.

Ayushman Bharat Pradhan Mantri Jan Arogya Yojna (PM-JAY)

One of the most important steps taken by the Government towards tackling this issue is the launch of the Pradhan Mantri–Jan Arogya Yojana (PM-JAY).

- It is the largest health assurance scheme in the world, which aims at providing a health cover of Rs 5 lakhs per family per year for secondary and tertiary care hospitalisation to over 12 crore poor and vulnerable families that form the bottom 40% of the Indian population.
- It provides cashless access to healthcare services for the beneficiary at the point of service (hospital).
- It covers up to 3 days of pre-hospitalisation and 15 days post-hospitalisation expenses such as diagnostics and medicines.
- Public hospitals are reimbursed for the healthcare services at par with the private hospitals.

Key Points

✓ **Ethics:** A scholarly discipline that examines, evaluates, and seeks to more deeply understand moral aspects (the right and wrong) of human nature and action. Ethics permeate all that we do in medicine.

✓ **Bioethics:** The application of ethics to medicine and biological science.

✓ The four pillars of bioethics are
- Autonomy: Respect patient's decision to choose what's right for them
- Beneficence: All choices are made with the intent to do good
- Non-maleficence: Do no harm
- Justice: Provide care to all patients in a fair manner

✓ Autonomy and paternalism: Favouring beneficence over patient autonomy is paternalism. It makes a physician-patient relationship analogous to that of a parent to a child.

✓ Autonomy is the capacity to think, decide, and act on the basis of such thought and decision, freely and independently, while paternalism is interference with a person's liberty of action justified by reasons referring exclusively to welfare, good, happiness, needs, interests or values of the person being coerced.

✓ **Informed consent:** It is the "willing acceptance of a medical intervention by a patient after adequate disclosure by the physician of the nature of the intervention, its risks and benefits, and also the alternatives with their risks and benefits".

✓ **Veracity**: Truth-telling is a vital component in a physician-patient relationship. It can result in loss of trust if not adhered to.

✓ An autonomous patient has not only the right to know (disclosure) of his/her diagnosis and prognosis, but also has the option to forgo this disclosure. However, the physician must know which of these two options the patient prefers.

✓ **Privacy:** The right of patient to remain private and not to disclose information.

✓ **Fidelity**: Maintaining the duty to care for all no matter who they are or what they may have done.

✓ **Confidentiality:** Doctors are obligated not to disclose confidential information given by a patient to another party (including closest relatives) without the patient's authorisation.

✓ Each one of the four principles of ethics is to be taken as a prima facie obligation that must be fulfilled, unless it conflicts with another principle. In such a conflict, the physician has to determine the actual obligation to the patient by examining the respective weights of the competing prima facie obligations based on both content and context.

✓ **Healthcare as a right**: The real meaning of healthcare as a right requires that all of us, acting together as a community and society, take responsibility to ensure that each person can exercise this right. The right to healthcare should not be at a high cost.

Questions

Case review

A 45-year-old woman is diagnosed with abdominal aortic aneurysm. The treating doctor advises surgery as the only option with a survival rate of 50%. She is scared of surgery and even after repeated counselling, is adamant to not undergo surgery. Aware of the seriousness of her condition and the need for surgery, the doctor still discharges the woman. She dies a few days later as a result of the bursting of the aneurysm. Her family sues the hospital for negligent care.

i. Which principle of bioethics is involved in this case?
ii. Is the doctor right in sending her away knowing that she might not survive? Give reasons for your answer.

Short answer questions

1. Discuss autonomy, beneficence, and non-maleficence.
2. Contrast autonomy with paternalism.
3. What are the responsibilities of patients and doctors in shared decision making?
4. What is full and reasonable disclosure?
5. Discuss the barriers to the implementation of healthcare as a universal right.

Further Reading

- AK Abbas, AH Litchman and Shiv Pillai. 2019. Basic *Immunology: Functions and disorders of the immune system*, 6th Edition. Elsevier.
- Biological and chemical terrorism: Strategic plan for preparedness and response. 2000. *MMWR*. (RR4): 1–14.
- Carol A Kauffman, Peter G Pappas, Jack D Sobel and William E Dismukes. 2011. *Essentials of Clinical Mycology*, 2nd Edition. Springer.
- CKJ Paniker. 2018. Paniker's Textbook of Parasitology, 8th Edition. Sougata Ghosh (Ed.). Jaypee Brothers.
- Dennis L Kasper and Anthony S Fauci. 2017. *Harrison's Infectious Diseases*. McGraw-Hill Education.
- Edward K Wagner, Martinez J Hewlett, David C Bloom and David Camerini. 2007. *Basic Virology*. John Wiley & Sons, Incorporated.
- Farrar J, PJ Hotez, T Junghanss, G Kang, D Lalloo and N White. 2014. *Manson's Tropical Medicine and Infectious Diseases,* 23rd edition. Saunders Ltd.
- Jenni Punt, Sharon Stranford, Patricia Jones and Judy Owen. 2019. *Kuby Immunology*, 8th Edition. W. H. Freeman, Macmillan Publishers.
- John E Bennett, Raphael Dolin, Martin J Blaser. 2016. *Mandell, Douglas, and Bennett's Infectious Disease Essentials*. Elsevier.
- Lynne Shore Garcia. 2016. *Diagnostic Medical Parasitology*, 6th Edition. ASM Press.
- Patricia M Tille. 2021. *Bailey & Scott's Diagnostic Microbiology*, 15th Edition. Elsevier Health Sciences.
- Richard Coico and Geoffrey Sunshine. 2015. Immunology: *A Short Course*, 7th Edition. Wiley–Blackwell. Saunders, Philadelphia.
- Stefan Riedel, Stephen Morse and Timothy Mietzner. 2019. *Jawetz Melnick & Adelberg's Medical Microbiology*, 28th Edition. McGraw-Hill.
- Stephen Gillespie and Kathleen Bamford. 2012. *Medical Microbiology and Infection at a Glance*, 4th Edition. John Wiley and Sons Ltd.
- Warren E Levinson, Peter Chin-Hong, Elizabeth A Joyce, Jesse Nussbaum, Brian Schwartz. 2018. *Review of Medical Microbiology And Immunology: A Guide to Clinical Infectious Diseases*, 15th Edition. McGraw Hill Education.

Web resources
- Coronavirus Disease 2019. World Health Organization. (https://www.who.int/emergencies/diseases/novel-coronavirus-2019)
- Guidelines for handling of biomedical waste for utilization-central pollution control. (https://mpcb.gov.in/sites/default/files/ biomedical-waste/Guidelines_for_Handling_of_Biomedical_Waste_for_Utilization03032020.pdf)
- Healthcare Associated Infections (HAIs). Centers for Disease Control and Prevention. (https://www.cdc.gov/hai/index.html)
- Hospital Infection Control Guidelines: Indian Council of Medical Research. (https://main.icmr.nic.in/sites/default/files/guidelines/Hospital_Infection_control_guidelines.pdf)
- Immunisation, Vaccines and Biologicals: Policy recommendations. (https://www.who.int/immunization/policy/en/)
- RNTCP National Strategic Plan for Tuberculosis Elimination 2017–25. (https://tbcindia.gov.in/WriteReadData/National%20 Strategic%20Plan%202017-25.pdf)
- Specimen processing of CSF, lymph nodes and other tissues for Xpert MTB/RIF. (https://tbcindia.gov.in/showfile.php?lid=3255)
- Treatment Guidelines for Antimicrobial Use in Common Syndromes: Indian Council of Medical Research Department of Health Research New Delhi, India 2017. (https://main.icmr.nic.in/sites/default/files/guidelines/Treatment_guidelines_2017. pdf)
- Universal Immunization Program (Govt of India). (https://main.mohfw.gov.in/sites/default/files/5628564789562315.pdf)
- Universal Immunisation Programme. (https://www.nhp.gov.in/universal-immunisation-programme_pg.)
- WHO consolidated guidelines on tuberculosis (Modules 1 to 4). (https://apps.who.int/iris/bitstream/handle/10665/331170/9789240001503-eng.pdf)
- WHO guidelines on syphilis screening and treatment of pregnant women. (https://www.who.int/reproductivehealth/publications/rtis/syphilis-ANC-screenandtreat-guidelines/en)

Websites for disease-specific guidelines
- Coronavirus Disease 2019 (COVID-19), CDC: https://www.cdc.gov/coronavirus/2019-ncov/index.html
- COVID and other epidemic and endemic infections in India: Ministry of Health and Family Welfare, Government of India: https://www.mohfw.gov.in/
- National AIDS Control Organisation (NACO), India: http://naco.gov.in/documentsreports
- National Center for Disease Control (NCDC) Guidelines, India: https://ncdc.gov.in/index1.php?lang=1 &level=0&linkid=38&lid
- National Vector Borne Disease Control Program (NVBDCP), India: https://nvbdcp.gov.in/
- World Health Organization (WHO): https://www.who.int/

Index